THE IMPACT OF AIR POLLUTION ON HEALTH, ECONOMY, ENVIRONMENT AND AGRICULTURAL SOURCES

Edited by **Mohamed K. Khallaf**

The Impact of Air Pollution on Health, Economy, Environment and Agricultural Sources
Edited by Mohamed K. Khallaf

Published by InTech

Janeza Trdine 9, 51000 Rijeka, Croatia

Publishing Process Manager Ivana Lorkovic

Technical Editor Teodora Smiljanic

Cover Designer InTech Design Team

Additional hard copies can be obtained from orders@intechopen.com

The Impact of Air Pollution on Health, Economy, Environment and Agricultural Sources,
Edited by Mohamed K. Khallaf
p. cm.
ISBN 978-953-307-528-0

For EU product safety concerns:
IN TECH d.o.o., Prolaz Marije Krucifikse Kozulić 3, 51000 Rijeka, Croatia,
info@intechopen.com or visit our website at intechopen.com.

Meet the editor

Dr. Mohamed K. Khallaf is an Associate Professor at Fayoum University, Faculty of Archaeology, Conservation Department and the head of Center of Research and Conservation of Antiquities (C.R.C.A.) at Faculty of Archaeology, Fayoum University. He obtained PhD. Degree (2004) in Restoration and Conservation of Antiquities from Cairo University, Faculty of Archaeology, Conservation Department. He is teaching many courses in restoration department from (2004) until now. He gave technical and scientific consulting in many restoration projects (2002-2005). He carried out studies and analysis for many archaeological buildings (2000-until now). He carried out the restoration work of some objects in Historic buildings (2003-2010). He participated in many international conferences such as the Sixth International Conference on Science and Technology in Archaeology and Conservation, Rome, Italy, (2008) and The 5th International Symposium of the Hellenic Society for Archaeometry, Athens, Greece, (2008). He published more than 20 papers in the field of restoration and conservation of stones and historical buildings. He evaluated some of restoration projects submitted to Science and Technological Development Fund (STDF) of the Academy of Scientific Research and Technology. He is co–supervisor of many Masters and PhDs researches. He presented many lectures in various scientific organizations.

Contents

Preface

The problem of air pollution has affects on human health, environment and many areas of life. Modern technology has led to the occurrence of air pollution and its harmful effects, but at the same time tries to push the world community to exert maximum efforts to overcome the harmful effects of it. Atmospheric pollutants have been classified according to their source, chemical composition, size and mode of release into indoor or outdoor environments. Sources of indoor pollutants are cooking and combustion, particle resuspension, building materials, air condition, consumer products, smoking, heating and biologic agents while, sources of outdoor pollutants are industrial, commercial, mobile, urban, regional, agricultural and natural. Air Pollution has an impact on the human body, such as respiratory system and heart disease. Air pollution also effects plants and agricultural products, economy and development rates in many countries all over the world. After they have been invited by the publisher, several authors from various countries, most of whom do not know each other; have collected roughly two dozen chapters which cover a wide area of Air Pollution. This book aims to strengthen the knowledge base dealing with Air Pollution. The book consists of 21 chapters dealing with Air Pollution and its effects in the field of Health, Environment, Economy and Agricultural Sources. This book is divided into four parts. The first one deals with effect of air pollution on health and human body organs, for example the impact of atmospheric pollution in the development of respiratory diseases, air pollution and cardiovascular disease, the impacts of fine particle matter on cardiovascular health, the adverse effects of air pollution on pregnancy, infant death and the association between air pollution, vitamin D status and prevalence of so-called civilization-related diseases. The second part includes the impact of air pollution on plants and agricultural sources and methods of resistance. The chapters deal with the effects of air pollutants on vegetation, current state of phytoremediation of air pollutants, such as particle matters, inorganic and organic pollutants, evaluation of existing air pollution control policies, particularly those targeted at agriculture, improving air quality, as measured by the number of exceeding levels of the CO and NO_2 standards, examining the emission gas concentration and the fundamental thermal decomposition, as well as analyzing the evolved species and the char of wood-based waste processing-materials and in the end community and occupational exposure following off-site Movement of Agricultural Fumigants. The third part includes environmental changes, geographic and climatic conditions due to air pollution. Its chapters deal with formation and reduction pollutants in combustion processes, the mobile monitoring system (MMS), as an example of a new approach for air monitoring along communication lanes, global warming or climate change, chemical pollution prediction models in function of traffic flows, emerging air pollution issues and environmental quality in the context of environmental injustice. The fourth part includes case studies concerning the impact of air pollution in the economy and development goals, such as, indoor air pollution in México, indoor air pollution and millennium development goals in Bangladesh, epidemiological and economic impact of natural gas on indoor air pollution in Colombia and economic growth and air pollution in Iran during development programs. In this book the authors explain the definition of air pollution, the most important pollutants and their different sources and effects on humans and various areas of life. The authors offer different solutions to the problems resulting from air pollution. Finally, this book is an important addition in the subject of air pollution and its impact on various fields.

Mohamed K. Khallaf
Restoration Department,
Faculty of Archaeology,
Fayoum University,
Egypt

Part 1

Effect of Air Pollution on Health and Human Body Organs

1

Impact of Air Pollution on Public Health

E. Marchwinska-Wyrwal[1,2], G. Dziubanek[1], I. Hajok[1],
M. Rusin[1], K.Oleksiuk[1] and M.Kubasiak[1]
[1]*Department of Environmental Health, Medical University of Silesia, Katowice*
[2]*University of Ecology and Management, Warsaw*
Poland

1. Introduction

The air we breathe contains emissions from many different sources: industry, motor vehicles, heating and commercial sources, household fuels as well as tobacco smoke. The effects of air pollution on health have been intensively studied in recent years. The results of these studies showed that air pollution harms human health and particularly is harmful for those who are already vulnerable because of their age as children and older people or existing health problems. The epidemiological evidence suggests that adverse health effects are dependent on both exposure concentrations and length of exposure, and that long-term exposures have larger, more persistent cumulative effects than short-term exposures [1]. Ambient air pollution has been associated with a multitude of health effects, including mortality, respiratory and cardiovascular hospitalizations, changes in lung function and asthma attacks. Current scientific evidence indicates that air pollution from the combustion of fossil fuels causes a spectrum of health effects from allergy to death. Recent assessments suggest that the public health impacts may be considerable. Air pollution is associated with a broad spectrum of acute and chronic health effects, the nature of which may vary depending on constituent of the pollutants as well as the group of the population. Current exposure to PM from anthropogenic sources leads to the loss of 8.6 months of life expectancy in Europe - from around 3 months in Finland to more than 13 months in Belgium. The most recent estimates of impacts of PM on mortality, based on PM_{10} and $PM_{2.5}$ monitoring data in 40 European countries, indicate that close to 500 000 deaths per year are accelerated due to exposure to ambient PM in those countries. According to the WHO Health Reports, air pollution at current levels in European cities is responsible for a significant burden of deaths, hospital admissions and exacerbation of symptoms, especially for cardiovascular disease. Because of the tremendous number of people affected, the impact of air pollution on cardiovascular disease represents a serious public health problem. Results from research studies have demonstrated a strong relationship between levels of airborne particles, sulfur dioxide and other fossil fuel emissions and risk of early death from heart disease. People with pre-existing conditions such as high blood pressure, previous heart disease, diabetes, respiratory disease and high cholesterol have been shown to be especially vulnerable. The results of a long-term study on influence of common air pollutants on health of US residents showed that individuals living in the more polluted cities had a higher risk of hospitalization and early death from pulmonary and heart diseases as compared to those living in the less polluted cities. The study focused on the health effects of gaseous pollutants such as sulfur dioxide, which are produced mainly by coal-burning power plants and fine particle air pollution, particles with a diameter of less than 2.5µm, that come from power plant emissions and motor vehicle exhaust. The relationship between air pollution and mortality was much stronger for the fine particle component than for the gaseous pollutants. Exposure to PM is associated with increased hospital admissions and mortality in adults. The risk increases linearly with the concentration of pollution and there is no evidence to suggest a threshold for PM below which no adverse health effects would occur. When inhaled, PM_{10} particles (with a diameter of less than 10µm) penetrate deep into the respiratory system. Finer particles (with a diameter of less than 2.5µm) then go on to penetrate the lungs and pass into the bloodstream and are carried into other body organs. Concerned that these particles cause a wide range of health impacts, WHO has developed guidelines addressing their risks. Knowledge about the links between health and air quality has significantly advanced in the last years. It was determined that short-term exposure to $PM_{2.5}$ significantly increases the risk for cardiovascular and respiratory disease among people over 65 years of age. In the U.S. the National Morbidity, Mortality and Air Pollution Study indicated a 0.41% increase in total mortality in response to a 10-$\mu g/m^3$ increase in PM_{10} in ambient air [2]. The investigators linked $PM_{2.5}$ data to hospital admissions for heart and vascular diseases, heart failure, chronic obstructive pulmonary disease and respiratory infections in an epidemiologic study of over 11.5 million Medicare participants. The study results predict that for each 100 hospital admissions for heart failure, one extra admission will occur for each 10 $\mu m/m^3$ increase in $PM_{2.5}$ [2].

New studies also indicate substantial gains in public health resulting from improvements in air quality. An improvements in air quality over the last 20 years have increased average life expectancy in the U.S. by approximately five months. Researchers at Brigham Young University and the Harvard School of Public Health tracked particulate matter air pollution in 51 major metropolitan areas from 1978 through 2001 and compared those data to death records and census data. On average, life expectancy increased by 2.72 years with about 15% of that increase due to improved air quality. Cities that had the greatest air quality improvements saw the greatest gains in life expectancy. The results shows that a reduction of 10 $\mu g/m^3$ in the ambient air concentration of particulate matter was associated with an estimated increase in average life expectancy of 0.61 years [3]. The magnitude of changes in the health state of population in the polluted part of Poland has been estimated in assuming that the ambient air pollution decrease is 10% [4]. In the case of cardiovascular diseases, a 10% reduction of lead concentration in the air will cause a decrease in the incidence by 17.6 cases per 10 000 people. A 10% reduction of cadmium concentration in the air may result in a decrease in neoplasm mortality by more than 4 cases per 10 000 inhabitants. The reduction of the concentrations of both heavy metals in the air will lower SDR; in the case of cadmium by 24.4, and in the case of lead by 31.6 people. The time scale over which the health effects develop is under investigation. Particulate air pollution is consistently related to the most serious effects, including lung cancer and other cardiopulmonary mortality. Long-term average exposure to PM is associated with both the risks of chronic effects on children's health, such as impaired development of lung function, and the frequency of acute effects, such as the aggravation of asthma or incidence of respiratory symptoms. Children who live in neighborhoods with serious air pollution problems (emissions from the automobile traffic, heavy industry) have lower IQ and score worse memory tests than children from the cleaner environments. The respiratory and cardiovascular effects of air pollution are well documented; however the possible neurodegenerative effects of air pollution have been unexplored and require further intensive research. The research in the US showed that the more heavily exposed children were to black carbon, the lower their scores on several intelligence tests. For example, the average IQ of the most heavily exposed children was 3.4 points less than children with low exposure. When the findings were adjusted for the effects of parents' education, birth weight, and exposure to tobacco smoke, the associations remained. The effects were roughly equivalent to those seen in children whose mothers smoked ten cigarettes per day while pregnant. The researchers assumed that the harmful effects may be caused by the inflammatory and oxidative effects of the black carbon particles [5]. These findings suggest additional research is needed to investigate the effects of air pollution on the development of intelligence in children and on cognitive decline for people of all ages. Globally, the prevalence of asthma and allergies has increased over the last few decades. Asthma has become the commonest chronic disease in children and is one of the major causes of hospitalization for children aged under 15 years. The increasing prevalence of allergic diseases in children throughout Europe is no longer restricted to specific seasons or environments. It has only become fully apparent in the last decade that air pollution, especially of fine particulates, plays a major role in cardiovascular disease. A half of deaths globally arises from cardiovascular disease. Even relatively small increases in the risk of cardiovascular disease will translate into huge numbers of additional people suffering more severely from the disease. There is now substantial evidence concerning the adverse effects of air pollution on pregnancy outcomes and infant death. Evidence reporting associations between maternal exposure to ambient air pollutants and adverse fetal development, in particular growth restriction, pre-term birth, and infant survival due to postnatal respiratory mortality has been growing rapidly in recent years. The association between maternal exposure to ambient air pollution and the risk of congenital anomalies, which are a significant cause of stillbirth and infant mortality has been less well studied. New evidence is also accumulating on the burden of disease due to indoor air pollution. The air pollutants such as asbestos fibers and dioxins, resulting from waste disposal, has been associated with a multitude of health effects. Asbestos fibers are dangerous to health and practically indestructible. Human exposure to asbestos fiber found in inhaled air can lead to diseases such as chronic bronchitis, asbestosis, lung cancer and mesothelioma. The World Health Organization officially recognized asbestos as a carcinogen that pollute the environment globally. It starts a process of gradual elimination of asbestos from the human environment. The building asbestos stripping operations and waste disposal, because of higher emission of asbestos fibers into environment, puts human population at enormous risk. European Union experts estimate that asbestos-related cancers will cause approximately 500 000 deaths up to the year 2030 in Western Europe alone. Dioxin and related compounds always exist in nature as complex mixtures. Dioxins are widely distributed in the environment at low concentrations, primarily as a result of air transport and deposition. Emissions of polychlorinated dibenzodioxin and dibenzofuran (PCDD/F) result from inefficiencies of combustion processes, most typically waste combustion. For uncontrolled combustion, such as open burning of household waste, chlorine content of wastes may play the most significant role in levels of dioxin emissions. Dioxin and related compounds have been shown to be developmental, reproductive, immunological, endocrinological, and cancer hazards, among others in multiple animal species. There is no reason to expect, in general, that humans would not be similarly affected at some dose, and an increasing numbers of data supports this assumption. Tetrachlorodibenzo-*p*-dioxin (TCDD) is best characterized as "carcinogenic to humans." This means that, on the basis of the weight of all of the evidence (human, animal, mode of action), TCDD meets the criteria that allow the scientific community to accept a causal relationship between TCDD exposure and cancer hazard.

2. Human exposure to air pollution: The route of air pollutants to a human organism

Air pollution is a mixture of particulate matter (PM), gases, and vapor-phase molecules [6]. The direct rout of exposure to the air pollution is a respiratory tract. In case of dust pollutants the size of particulate matter is playing an important role in the environmental health risk. PM is categorized by aerodynamic diameter. Particles below 10 μm in diameter are classified as thoracic particles PM_{10}, particles below 2.5 μm in diameter as fine particles, and particles with a diameter < 0.1 μm as ultrafine particles (UFPs) [6]. Particles larger than 10 μm are likely to land in proximal airways, but fine particles reach the lungs and are deposited in the alveoli [7]. Therefore $PM_{2.5}$ may be more harmful than larger ones [8]. Ambient fine particulate pollution was associated with increased risk of cardiovascular diseases [9]. UFPs are deposited deeply into the lungs. The study of Terzano et al. (2010), indicates that the ultrafine particles in contrast to larger-sized particles pass into the bloodstream by different transfer routes and mechanisms and then are distributed into other body organs, including the brain with potential neurotoxic effects [10]. The particulate matter is hazardous to the human health due to absorption on their surface of many harmful contaminants such as: heavy metals (lead, cadmium, mercury and the other), organic compounds (polycyclic aromatic hydrocarbons, PCBs, dioxin and furans). Gaseous pollutants, depending on their solubility in the water, are absorbed in the proximal or the distal parts of the respiratory tract. This is important from the standpoint of the health effects. Sulfur dioxide and formaldehyde are highly water-soluble gases, therefore they do not reach the lungs, and they are irritating the airway epithelium of the upper respiratory tract. For example up to 98% sulfur dioxide may be absorbed in the nasopharynx during nasal breathing [11]. NO_2 is a poorly water-soluble gas, therefore, is deposited far more peripherally in a respiratory tract compared with SO_2, but does not reach the alveoli in any significant quantities [11]. Ozone, in contrast to nitrogen dioxide, does not dissolve in water and in gaseous form reaches the lungs where it begins its malicious activity. Gaseous pollutants can be also absorbed into the body through dermal rout of exposure. However for the general population the role of this route of exposure is insignificant. The indirect rout of exposure to the air pollutants is digestive tract; it follows from the circulation of pollutants in the environment. The food chain is an important pathway of human exposure to polycyclic aromatic hydrocarbons, dioxin, PCBs and heavy metals (cadmium, lead, mercury).

3. Relation between air pollution and life expectancy (LLE)

The World Health Organization has identified ambient air pollution as a high public health priority, indicating the relationship of air pollution with increased mortality and shortened life expectancy [12]. In 2009, life expectancy at birth in twenty seven countries in the European Union [EU-27] was among the highest in the world – almost 76 years for men and 82 years for women [13,14]. In Europe there is a wide variety of life expectancy. In developed countries, located mainly in the western part of Europe and the Nordic countries, people live a few years longer than in the countries of Central and Eastern Europe and these differences are even a dozen years. According to EU experts, current exposure to PM from anthropogenic sources reduces the average life expectancy of 8.6 months in Europe - from around 3 months in Finland and Ireland to more than 13 months in Belgium [15]. It has been estimated that exposure to fine particulate matter in outdoor air leads to 725 000 years of life lost annually in Europe [16,17]. Studies in the USA have shown that people from less polluted cities live longer than those living in more polluted cities. After adjustment for other factors, an association remained between ambient annual average concentrations of fine particles (represented by $PM_{2.5}$) and age-specific risks of mortality, implying shorter life expectancy in more polluted cities [18]. It is estimated that in the European Union an increase of 1 $\mu g/m^3$ of $PM_{2.5}$ for 1 year implies an average Lost of Life Expectancy (LLE) of 0.22 days per person; the number for the United States is similar, but for Russia it is about 40% higher, and for China it is about 25% lower [19]. A study published in 2009, conducted in 51 U.S. metropolitan areas between 1970 to 2000 showed that decrease of 10 $\mu g/m^3$ in the concentration of fine particulate matter was associated with an estimated increase of average life expectancy of 0.61 year. The estimated effect of reduced exposure to pollution on life expectancy was not highly sensitive to adjustment for changes in socioeconomic, demographic, or proxy variables for the prevalence of smoking or to the restriction of observations to relatively large counties [3]. The other studies have shown that reductions in life expectancy of 1.11 years in the Netherlands, 1.37 years in Finland, and 0.80 year in Canada resulting from increases in ambient $PM_{2.5}$ concentrations of 10 $\mu g/m^3$ [20,21]. Long-term exposure to PM is particularly damaging to human health and reduces life expectancy, that is why reducing long-term PM concentrations and exposure is a priority [21,22].

4. Relation between air pollution and mortality

Air pollution is a major environmental risk for health and is estimated to cause approximately 2 million premature deaths worldwide per year. PM air pollution imparts a tremendous burden to the global public health, ranking it as the 13th leading cause of morality [23]. The estimates of health effects of PM exposure in adults are dominated by the increase in the risk of mortality due to long-time exposure to fine PM ($PM_{2.5}$) The total number of premature deaths attributed to exposure amounts to around 348 000 annually in the 25 EU countries. More than half of the burden from air pollution on human health is borne by people in developing countries [15,24]. The short-term health effects of particulate and gaseous air pollutants have been well documented, mainly through time-series studies relating short-term elevations in ambient levels of such pollutants to increases in morbidity and mortality from cardio respiratory conditions. Results of 124 studies of the largest cities in North America and Europe showed an increase in the rate of death from any cause ranging from 0.2 to 0.6% for an increase in ambient PM_{10} concentrations of 10 $\mu g/m^3$ [25]. Long-term epidemiological studies conducted in the U.S. confirm that the adverse effects of fine particulate matter ($PM_{2.5}$) on morbidity and mortality, and indicate that this effect depends on the concentration and time of exposure; long-term exposure gives higher effects than short-term exposure [26]. Long-term exposure to $PM_{2.5}$ increases the risk of no accidental mortality by 6% per a 10 $\mu g/m^3$ increase, independent of age, gender, and geographic region. Exposure to PM was also associated with an increased risk of mortality from lung cancer (range: 15% to 21% per a 10 μg/m^3 increase) and total cardiovascular mortality (range: 12% to 14% per a 10 μg/m^3 increase) [1,27]. The Medicare Cohort Air Pollution Study in the USA has estimated the relative risk of death associated with long-term exposure to $PM_{2.5}$. According to the authors Zeger et al. [28], a 10 mg/m^3 increase in six year average of $PM_{2.5}$ is associated with a 6.8–13.2% increase in mortality. Other studies [27] showed that long-term exposure to $PM_{2.5}$ increases the risk of non-accidental mortality by 6% per a 10 mg/m^3 increase, independent of age, gender, and geographic region. Exposure to $PM_{2.5}$ was also associated with an increased risk of mortality from lung cancer (range: 15–21% per a 10 mg/m^3 increase) and total cardiovascular mortality (range: 12–14% per a 10 mg/m^3 increase) [27]. People with diabetes, heart failure, chronic obstructive pulmonary disease (COPD) and inflammatory diseases such as rheumatoid arthritis are at increased risk of death when they are exposed to particulate air pollution, or soot, for one or more years. Increase of 10 μg/m^3 of PM_{10} over 2 years increased the risk of death by 32% for patients with diabetes, by 28% for patients with COPD, by 27% for patients with congestive heart failure, and by 22% for people with inflammatory diseases such as rheumatoid arthritis or lupus [29]. Significant associations were found between black smoke (BS) and SO_2 concentrations and mortality. The effects were stronger for respiratory illness than other causes of mortality for the most recent exposure periods (shorter latency times) and most recent mortality period (lower pollutant concentrations) [30]. Air pollution has long-term effects on mortality and point to continuing public health risks. They therefore have importance for policies on public health protection through regulation and control of air pollution [30].

5. The impacts of fine particulate matter on cardiovascular health

Cardiovascular disease (CVD) constitute a global problem and is the leading cause of death in the world, especially in highly developed countries. Cardiovascular disease is also a major cause of disability and of reduced quality of life [31,32]. According to forecast, almost 20 million people will die from CVDs, mainly from heart disease and stroke by 2015 [33,34]. Results from many research studies have demonstrated a strong relationship between levels of airborne particles, sulfur dioxide and other air pollutants and risk of early death from heart disease. Air pollutants have been linked with endothelial dysfunction and vasoconstriction, increased blood pressure (BP), prothrombotic and coagulant changes, systemic inflammatory and oxidative stress responses, autonomic imbalance and arrhythmias, and the progression of atherosclerosis [35].

Using data for Chicago area hospitals for years 1988 to 1993 it was found, that an increase in PM_{10} level by 10 μg/m^3 was associated with 1.27%, 1.45%, and 2% increases in hospital admissions for heart disease, chronic obstructive pulmonary disease, and pneumonia, respectively [36]. The Air Pollution and Health: a European and North American Approach (APHENA) project also examined the association between airborne particles and hospital admission for cardiac causes in eight European cities and found that the percentage increases associated with a 10 μg/m^3 elevation in PM_{10} were 0.5% for cardiac admissions in people of all ages and 0.7% for cardiac admissions in people older than 65 years [37,38]. Short-term exposure to $PM_{2.5}$ significantly increases the risk for cardiovascular and respiratory disease among people over 65 years of age. The investigators linked $PM_{2.5}$ data to hospital admissions for heart and vascular diseases, heart failure, chronic obstructive pulmonary disease, and respiratory infections in an epidemiologic study of over 11.5 million

Medicare participants. The study results predict that for each 100 hospital admissions for heart failure, one extra admission will occur for each 10 μg/m^3 increase in $PM_{2.5}$ [2]. People with pre-existing cardiovascular disease, diabetic and elderly individuals are also considered to be more susceptible to air pollution–mediated cardiovascular effects [39].

Long-term exposure to elevated concentrations of ambient $PM_{2.5}$ at levels encountered in the present-day environment (i.e. any increase by 10 μg/m^3) reduces life expectancy within a population probably by several months to a few years [40]. As $PM_{2.5}$ is most strongly associated with cardiovascular deaths in the cohort studies, the reduced life expectancy is most likely predominantly due to excess cardiovascular mortality [40]. It was found that the greater the level of the fine particulate pollution, the greater the risk of cardiovascular disease and death in post-menopausal women, who are considered to be susceptible group within the general population. The increased risk comes from the fine particulate matter typically produced by automobile exhaust. The particles damage arteries in the heart and brain. Even slight elevations in fine particulate matter concentration increased the risk significantly. The risk of dying from heart attack or stroke increased 76% for each ten microgram increase in fine particulate pollution and proved to be about three times higher than previously estimated [41]. The study also indicates that although smoking is a much larger risk factor for cardiovascular disease, exposure to fine particulate combined with smoking imposes additional effects [42]. Additional research is required to establish whether there are independent health effects of the other particulate size fractions beyond those posed by fine particles. Although the focus of the present statement is on PM, it is recognized that other air pollutants may also pose cardiovascular risk alone or in conjunction with fine-particle exposure [40]. There are some evidences that gaseous pollutants may also be a reason for hospitalizations. Hospital admissions for cardiovascular causes, particularly ischemic heart disease, were found to rise in relation to the previous-day and same-day level of SO_2, even after adjustment for PM_{10} levels [40].

Although ozone has been linked to increased cardiopulmonary mortality, strokes, and MIs in some short-term studies, long-term exposure was not associated with cardiovascular mortality after accounting for PM in a recent analysis. The recent finding that small changes in low levels of ambient carbon monoxide concentrations are related to cardiovascular hospitalizations also requires further study [40]. Several secondary aerosols (eg, nitrate and sulfate) are often associated with cardiovascular mortality; however, whether these compounds are directly harmful or are surrogate markers of toxic sources of exposure requires more investigation [40]. The results showed that the daily number of hospitalizations for cardiovascular diseases was significantly associated with daily PM_{10} and NO_2 levels, with stronger associations in the elderly (≥65 years of age) [43]. During the last 15 years air pollution induced cardiovascular toxicity has become the focus of intensive studies among cardiologists and specialists in environmental medicine. They found that long-term particulate matter exposures were most strongly associated with death due to ischemic heart disease, dysrhythmias, heart failure, and cardiac arrest. For these causes of death, a 10 μg/m^3 elevation in particulate matter was associated with 8% to 18% increases in mortality risk. Risks for smokers were comparable or larger than for non-smokers. The researchers conclude that particulate matter exposure is a risk factor for specific cardiovascular disease mortality through mechanisms that likely include pulmonary and systemic inflammation, accelerated atherosclerosis, and changes in cardiac rhythms [42]. According to more recent studies, the ultrafine particles may be translocated into the circulation and directly transported to the vasculature and heart where they can induce cardiac arrhythmias and decrease cardiac contractility and coronary flow [39]. Improving our understanding of the biological mechanisms underlying the acute cardiovascular effects of air pollution is essential to define the best prevention strategies [37]. Cardiovascular disease is very common and, as exposure to air pollution, both in the long and short term, contributes to initiation and exacerbation of disease, it is likely that even modest reductions in exposure will result in significant health gain [43].

6. The adverse effects of air pollution on pregnancy outcomes and infant death

In the last two decades there is observed an increase of the number of scientific reports about a proven influence of air pollution on an occurrence of negative health effects, connected with births. They are: preterm births, stillbirths, intrauterine fetus growth retardation, births of newborns with low birth weight and a risk of newborns' death because of respiratory system disorders [44,45]. Birth weight, gestational age, and fetal growth are important indicators of perinatal health. Low birth weight (LBW), preterm birth, or intrauterine growth retardation (IUGR) are strongly association with infant mortality and morbidity [46]. Long term study shows that low birth weight (LBW) is a risk factor for developing in adulthood coronary health diseases, hypertension and type 2 diabetes [45]. It was observed up to 20% increase in risk of LBW and preterm birth in infants born to women leaving in area with high level of air pollution, specially those exposed to higher levels of motor vehicle exhaust pollution coming from heavy-traffic roadways. Stronger effects were observed for women whose third trimester accounted for months with cold weather, when concentration of air pollutants was the highest because of an activity of local heating sources [47]. In heavily polluted environments the prematurity rate (birth before the 37th week of pregnancy) increases considerably. The study done in the 90's, in the most polluted parts of Poland (Chorzow) shows as high as 14 to 20% of the prematurity rate comparing to 8% rate for Poland this time. The average newborns' birth weight was 515g lower in Chorzow than in the Country [48]. There are more evidence each year which indicate that maternal exposures to air pollutants, including particulate matter (aerodynamic diameter 10 μm and 2.5 μm), sulfur dioxide, nitrogen dioxide and benzopyrene, are associated with adverse pregnancy outcomes [49]. Some of polycyclic aromatic hydrocarbons (PAHs), especially benzopyrene, prove carcinogenic and mutagenic effects and when penetrating through placenta, have a negative influence into fetus. The research showed the newborns, whose mothers were exposed to PAHs during pregnancy, more often born with lower birth weight and smaller head circumference [50, 51, 52]. According to Dejmek et al. (2000), the risk of delivering a growth-retarded infant increases with the level of PAHs in early gestation (first month) [53]. A bond of benzopyrene and DNA in the placenta, has an influence on intrauterine growth retardation - IUGR [54,55]. The exposure to particulate matter (PM) causes increase of risk of occurrence an intrauterine fetus growth retardation [56,57,58]. Children, whose mothers were exposed to high concentrations of PM during pregnancy, more often were born with low birth weight and 10 $\mu g/m^3$ change in PM_{10} accounted for 13.7 g less of weight [59,60,61]. Effect of mothers' exposure to high concentration (above the median 36.3 $\mu g/m^3$) of fine particles ($PM_{2,5}$) was reflected in significantly lower mean weight (128.3g) and length (0.9cm) and lower mean head circumference (0.3) of newborns

[62]. The researchers in the USA found that mothers who lived in areas with the highest levels of $PM_{2.5}$ during their pregnancy delivered slightly smaller babies than their counterparts who lived in areas with lower levels of $PM_{2.5}$ exposure. They also observed association between number of traffic-related pollutants and small for gestational birth weight as well as preterm births (before 37 weeks) [63,64]. Maternal exposure to sulfur dioxide during the first month of pregnancy increased risk of intrauterine growth retardation as well as LBW when preterm birth was associated with exposure to SO_2 during the last month of pregnancy. These results suggest an association between VLBW (below 1500g) and maternal exposures to high levels of sulfur dioxide [46,60]. Increased risk of intrauterine growth retardation was observed also in case of maternal exposure to nitrogen dioxide during the first month of pregnancy [65]. An association between exposure to levels of nitrogen dioxide above 40 $\mu g/m^3$ during the first trimester of pregnancy and a reduction in birth weight was found [66]. Study on CO influence on pregnancy [67] were the basis for estimation that one unit change in mean CO concentration during the last trimester of pregnancy increases the risk of low birth weight by 8%. Furthermore, a one unit change in mean CO concentration during the first 2 weeks after birth increases the risk of infant mortality by 2.5% relative to baseline levels [67].

7. Human's exposure to air pollution resulting from waste disposal

Asbestos is a mineral fiber that due to the unique physical and chemical properties was produced in the past and used in over 3000 products. In the 20th century, asbestos has dominated the building industry, with a maximum global production of 5 million tonnes per year. As a result, the world's asbestos (in products) is currently estimated at around 550 million tonnes. Asbestos fibers are indestructible and dangerous to health. Human exposure to mineral fiber found in inhaled air can lead to diseases such as chronic bronchitis, asbestosis, lung cancer and mesothelioma [68]. International Agency for Research on Cancer (IARC) recognized asbestos (actinolite, amosite, anthophyllite, chrysotile, crocidolite, tremolite) as Group I carcinogen [69]. This category is used when there is sufficient evidence of carcinogenicity in humans. In 1980, the US National Institute of Occupational Safety and Health (NIOSH) and the Occupational Safety and Health Administration (OSHA) working group concluded that there are no levels of exposure to asbestos below which clinical effects did not occur [70]. In the 80's of the past century, the World Health Organization officially recognized asbestos as a carcinogen that pollutes the environment globally. Environmental exposure either in the houses of asbestos workers or in the neighborhood of asbestos mines or factories has been noted in some of the cases [71]. It has been estimated that a third of the mesotheliomas occurring in the USA may be due to nonoccupational exposure [72]. The relationship between asbestos exposure and smoking indicates a synergistic effect of smoking with regard to lung cancer [69]. Further evaluations indicate that this synergistic effect is a multiplicative model [73]. Exposure to asbestos occurs through inhalation of fibers from contaminated air in the working environment, as well as from ambient air in the vicinity of point sources, or indoor air in housing and building containing asbestos materials [74]. Although, in many countries the production and utilization of asbestos-containing materials has been banned, the numerous active environmental sources still exist [70]. Exposure can also occur during installation and use of asbestos-containing products and maintenance of vehicles. Asbestos products are still in place in many buildings and continue to give rise to exposure during use, maintenance, renovation, repairs, removal and demolition [74]. The results of environmental concentrations of respirable asbestos fibers show a wide range of values. The observed discrepancy in the concentrations are dependent on the different environments, specific sampling locations and presence of more than one emission sources [70]. Unfortunately, in recent years, emissions of asbestos fibers into the urban environment has significantly intensified. This follows from the fact that the durability of asbestos-cement building products is estimated for 30 years and the possibility of exploitation of these used in the 70's and 80's of the past century is coming to an end. As is apparent from analysis the deterioration of asbestos-containing construction materials, such as asbestos-cement sheets (AC) used in residential and industrial buildings causes additional contamination of the urban environment. The study performed in a highly urbanized and densely populated town in south part of Poland, revealed that asbestos fibers identified in the air samples near buildings covered with AC panels derived from 2 groups of asbestos minerals, i.e. crocidolite and chrysotile. The observed concentrations of respirable asbestos fibres varied from 0.0010–0.0090 f/cm^3. Significantly higher values were noted in the immediate vicinity of the buildings with asbestos-containing materials, compared to sampling sites located at a distance of 100–500m from such buildings or the sites treated as an asbestos free [70]. Kovalevskiy and Tossavainen, taking measurements near a building with asbestos-containing materials in Moscow, showed that when outdoor concentrations reach the level of 0.009 f/cm^3, at the same time, indoor concentrations approach 0.049 f/cm^3 in residential premises, or even 0.57 f/cm^3 if the building was undergoing renovation [75]. The measurements of respirable fibers in the air on the playgrounds in housing estates, where path was made with admixture of asbestos-containing material showed contamination range from 0.165-0.54 f/m^3 and in apartments adjacent to the playground around 0.01 f/m^3 [76]. The significant increase in concentrations of asbestos also recorded in the immediate vicinity of buildings, at which work is ongoing disassembly of asbestos-cement facades or roofing. The work conducted by a specialized company working according to safety regulations causing dust in the workplace ranged from 1000 to 4000 f/m^3, while the same work done improperly can lead to maximum levels of respirable asbestos fibers in the amount of 80 000 f/m^3 [77]. Starting from the last decade of the 20th century, the world began a process of gradual elimination of asbestos, what in fact results the higher emission of asbestos fibers into the municipal environment. The individual disassembly of utilized asbestos panels, not obeying safe methods of removal, storage, transport and treatment of asbestos waste intensify the environmental exposure of the general population. Since 1980, the number of deaths caused by exposure to asbestos fibers increases gradually, even in countries that have banded the use of asbestos in the early 1990s [78]. Currently about 125 million people in the world are exposed to asbestos at the workplace and at least 90 000 people die each year from asbestos-related lung cancer, mesothelioma and asbestosis resulting from occupational exposures. In addition, it is believed that 7000 of deaths can be attributed to asbestos-related diseases as well as to non-occupational exposures to asbestos [74]. The Report of UN EWG shows that in the U.S. each day 30 people die, which represents 10 000 deaths per year only as a result of diseases caused by exposure to asbestos [79]. Because of long latency periods attached to the diseases, stopping the use of asbestos now will result in a decrease in the number of asbestos-related deaths after a number of decades [74]. It is estimated that in the next 40 years asbestos fibers will cause the death of about 100 000 Americans [79]. European Union experts estimate that the total number of deaths caused by asbestos-related diseases in the UK, Belgium, Germany, Switzerland, Norway, Poland and Estonia is around 15 000 annually; only in Western Europe asbestos-related diseases will caused 500 000 deaths subsequent till the end of 2030 [80]. Considering the numerous health hazards resulting from the inhalation of asbestos dust, there is no safe environmental level for his harmful factor, therefore, the exposure should be kept as low as possible [70].

8. References

[1] Pope C.A. 3rd. *Mortality effects of longer term exposure to fine particulate air pollution: Review of recent epidemiological evidence.* Inhal Toxicol. 2007, 19(Suppl. 1):33–38.

[2] Dominici F., Peng R.D., Bell M.L., Pham L., McDermott A., Zeger S.L., Samet J.M. *Fine particulate air pollution and hospital admission for cardiovascular and respiratory diseases.* JAMA. 2006 Mar 8;295(10):1127-34.

[3] Pope CA 3rd, Ezzati M., Dockery D.W. *Fine-particulate air pollution and life expectancy in the United States.* N Engl J Med. 2009 Jan 22;360(4):376-86.

[4] Marchwinska-Wyrwal E., Dziubanek G., Skrzypek M., Hajok I.. *Study of the health effects of long-term exposure to cadmium and lead in a region of Poland.* Int J Environ Health Res. 2010 Apr;20(2):81-6.

[5] Suglia S.F., Gryparis A., Wright R.O., Schwartz J., Wright R.J.. *Association of black carbon with cognition among children in a prospective birth cohort study.* Am J Epidemiol. 2008 Feb 1;167(3):280-6.

[6] Brook R.D., Franklin B., Cascio W., Hong Y., Howard G., Lipsett M., et al. *Air pollution and cardiovascular disease: a statement for healthcare professionals from the expert panel on population and prevention science of the American Heart Association.* Circulation 2004; 109:2655–71.

[7] Kim, C.S., Fishe.r, D.M., Lutz., D.J., Gerrity, T.R., 1994. *Particle deposition in bifurcating airway models with varying airway geometry.* J. Aerosol. 25, 567–581

[8] Farina F., Sancini G., Mantecca P, Gallinotti D., Camatini M., Palestini P. *The acute toxic effects of particulate matter in mouse lung are related to size and season of collection.* Toxicol Lett. 2011 Mar 1. [Epub ahead of print]

[9] Pope C.A. 3rd,Muhlestein J.B., May H.T., Renlund DG, Anderson J.L., Horne B.D. *Ischemic heart disease events triggered by short-term exposure to fine particulate air pollution.* Circulation. 2006 Dec 5;114(23):2443-8. Epub 2006 Nov 13.

[10] Terzano C., Di Stefano F., Conti V., Graziani E., Petroianni A. *Air pollution ultrafine particles: toxicity beyond the lung.* Eur Rev Med Pharmacol Sci. 2010 Oct;14(10):809-21.

[11] Sandström T. *Respiratory effects of air pollutants: experimental studies in humans.* Eur Respir J, 1995, 8, 976–995

[12] Samet J, Krewski D. *Health effects associated with exposure to ambient air pollution.* J Toxicol Environ Health A. 2007; 1;70(3-4): 227-42.

[13] http://epp.eurostat.ec.europa.eu

[14] *Systematic Review of Health Aspects of Air Pollution in Europe.* Copenhagen, Denmark, WHO Regional Office for Europe 2004

[15] *Exposure of children to air pollution (particulate matter) in outdoor air.* Copenhagen, WHO Regional Office for Europe 2009 (ENHIS fact sheet 3.3)

[16] Cohen A, Anderson H. *The Global Burden of Disease Due to Outdoor Air Pollution.* Journal of Toxicology and Environmental Health. 2005; 68:1301–1307

[17] EEA 2010. The European Environment - State and outlook 2010: Synthesis. http://www.eea.europa.eu/soer/synthesis/synthesis
[18] Dockery, D, Pope C. A, Xu, X. *An association between air pollution and mortality in six US cities.* New England Journal for Medicine 1993; 329: 1753-1759
[19] Leksell I, Rabl A. *Air Pollution and Mortality: Quantification and Valuation of Years of Life lost.* Risk Anal. 2001; 21 (5):843-57
[20] Coyle D, Stieb D, Burnett RT, et al. *Impact of particulate air pollution on quality-adjusted life expectancy in Canada.* J Toxicol Environ Health A 2003; 66:1847-1863
[21] Krewski D. *Evaluating the Effects of Ambient Air Pollution on Life Expectancy.* N Engl J Med 2009; 360:413-415
[22] Craig L, Brook JR, Chiotti Q, et al. *Air Pollution and Public Health: A Guidance Document for Risk Managers.* J Toxicol Environ Health A 2008;71:588-698
[23] Brook RD. *Cardiovascular effects of air pollution.* Clin Sci (Lond). 2008; 115(6): 175-87
[24] http://www.who.int/mediacentre/factsheets/fs313/en/index.html
[25] Samoli E, Peng R, Ramsay T, et al. *Acute effects of ambient particulate matter on mortality in Europe and North America: results from the APHENA study.* Environ Health Perspect 2008;116:1480-6
[26] Puett R, Hart J, Yanosky J et al. *Chronic Fine and Coarse Particulate Exposure, Mortality and Coronary Heart Disease in the Nurses' Health Study.* Environ Health Perspect. 2009;117(11): 1702–1706
[27] Chen H, Goldberg MS, Villeneuve PJ. *A systematic review of the relation between long-term exposure to ambient air pollution and chronic diseases.* Rev Environ Health. 2008;23(4):243-97, 2008
[28] Zeger SL, Dominici F, McDermott A, Samet JM. *Mortality in the Medicare population and chronic exposure to fine particulate air pollution in urban centers (2000-2005).* Environ Health Perspect. 2008;116(12):1614-9
[29] ATS 2006 International Conference: *Air Pollution Increases Mortality Risk.*
[30] Elliott P, Shaddick G, Wakefield JC, et al. *Long-term associations of outdoor air pollution with mortality in Great Britain.* Thorax 2007; 62(12):1088-94
[31] World Health Organization Regional Office for Europe. European Heart Health Charter. 2007 (http://www.heartcharter.org/)
[32] The European Heart Network. Annual Report 2009 fighting heart disease and stroke (http://www.ehnheart.org/publications/annual-reports.html)
[33] http://www.who.int/mediacentre/factsheets/fs317/en/index.html
[34] http://www.allcountries.org/health/cardiovascular_diseases.html
[35] Brook RD, Rajagopalan S, Pope A, et al. *Particulate Matter air Pollution and Cardiovascular Disease: An Update to the Scientific Statement From the American Heart Association.* Circulation 2010; 121: 2331-2378
[36] Schwartz J. *Is there harvesting association of airborne particles with daily deaths and hospital admissions?* Epidemiology 2001; 12: 55– 61
[37] Franchini M, Mannucci PM. *Short-term effects of air pollution on cardiovascular diseases: outcomes and mechanisms.* J ThrombHaemost 2007;5:2169-2174
[38] Le Tertre A, Medina S, Samoli E, Forsberg B, Michelozzi P, Boumghar A, Vonk JM, Bellini A, Atkinson R, Ayres JG, SunyerJ,Schwartz J, Katsouyanni K. *Short-term effects of particulate air pollution on cardiovascular diseases in eight European cities.* J Epidemiol Community Health 2002; 56: 773–9
[39] Simkhovich B, Kleinman M, Kloner R. *Air Pollution and Cardiovascular Injury.* Journal of the American College of Cardiology 2008; 52 (9): 719-726

[40] Miller KA, Siscovick DS, Sheppard L et al. *Long-term exposure to air pollution and incidence of cardiovascular events in women.* N EnglJ Med 2007 Feb 1;356(5):447-58
[41] Pope CA 3rd, Burnett RT, Thurston GD et al. *Cardiovascular mortality and long-term exposure to particulate air pollution: epidemiological evidence of general pathophysiological pathways of disease.* Circulation 2004 Jan 6;109(1): 71-7. Epub 2003 Dec 15
[42] Larrieu S, Jusot J-F, Blanchard M et al. *Short term effects of air pollution on hospitalizations for cardiovascular diseases in eight French cities: The PSAS program.* Science of the Total Environment 2007, 387: 105-112
[43] EU Department of Health. *Cardiovascular Disease and Air Pollution.* A Report by the Committee on the Medical Effects of Air Pollutants 2006
[44] Rankin J, Chadwick T, Natarajan M, Howel D, Pearce MS, Pless-Mulloli T. *Maternal exposure to ambient air pollutants and risk of congenital anomalies.* Environ Res 2009; 109: 181-187.
[45] Ricciardi C, Guastadisegni C. *Environmental inequities and low birth weight.* Ann Ist Super Sanita 2003; 39(2): 229-234.
[46] Liu S, Krewski D, Shi Y, Chen Y, Burnett RT. *Association between gaseous ambient air pollutants and adverse pregnancy outcomes in Vancouver, Canada.* Environ Health Perspect 2003; 111(14): 1773-1778.
[47] Wilhelm M, Ritz B. *Residential proximity to traffic and adverse birth outcomes in Los Angeles county, California, 1994-1996.* Environ Health Perspect 2003; 111(2): 207-216.
[48] Kasznia-Kocot J, Buszman Z. *The influence of pollution on children's health in Chorzów.* Pol J Environ Stud 1995; 4(2): 29-32.
[49] Salam MT, Millstein J, Li Y, Lurmann FW, Margolis HG, Gilliland FD. *Birth outcomes and prenatal exposure to ozone, carbon monoxide and particulate matter: results from the children's health study.* Environ Health Perspect 2005; 113(11): 1638-1644.
[50] Barakat AO. *PAHs and petroleum biomarkers in the atmospheric environment of Alexandria City, Egypt.* Water Air Soil Pollut 2002; 139: 289- 310.
[51] Perera F.P., Whyatt R.M., Jędrychowski W., et al. *A study of the effects of environmental polycyclic aromatic hydrocarbons on birth outcomes in Poland.* Am J Epidemiol 1998; 147: 309-314.
[52] Perera F.P., Rauh V., Tsai W.Y., et al. *Effects of transplacental exposure to environmental pollutants on birth outcomes in a multi-ethnic population.* Environ Health Perspect 2003; 111: 201–205.
[53] Dejmek J., Solansky I., Benes I., et al. *The impact of polycyclic aromatic hydrocarbons and fine particles on pregnancy outcome.* Environ Health Perspect 2000; 108: 1159–1164.
[54] Sram R.J., Binkova B., Rossner P., et al. *Adverse reproductive outcomes from exposure to environmental mutagens.* Mutat Res 1999; 428: 203-215.
[55] Topinka J., Binkova B., Stavkova Z., et al. *DNA adducts in human placenta as related to air pollution and to GSTM1 genotype.* Mutat Res 1997; 390: 59-68.
[56] Boy E, Bruce N, Delgado H. *Birth weight and exposure to kitchen wood smoke during pregnancy in rural Guatemala.* Environ Health Perspect 2002; 110(1): 109-114.
[57] Dejmek J, Selevan SG, Benes I, Solansky I, Sram RJ. *Fetal growth and maternal exposure to particulate matter during pregnancy.* Environ Health Perspect 1999; 107(6): 475-480.
[58] Chen L, Yang W, Jennison BL, Goodrich A, Omaye ST. *Air pollution and birth weight in Northern Nevada, 1991-1999.* Inhal Toxicol 2002; 14: 141-157.
[59] Wang X, Ding H, Ryan L, Xu X. *Association between air pollution and low birth weight: a community-based study.* Environ Health Perspect 1997; 105(5): 514-520.

[60] Rogers JF, Thompson SJ, Addy CL, McKeown RE, Cowen DJ, Decoufle P. *Association of very low birth weight with exposure to environmental sulphur dioxide and total suspended particulates.* Am J Epidemiol 1999; 151(6): 602-613.
[61] Gouveia N, Bremner SA, Novaes HDM. *Association between ambient air pollution and birth weight in Sao Paulo, Brazil.* J Epidemiol Community Health 2004; 58: 11-17.
[62] Jędrychowski W, Bendkowska I, Flak E, et al. *Estimated risk for altered fetal growth resulting from exposure to fine particles during pregnancy: an epidemiologic prospective cohort study in Poland.* Environ Health Perspect 2004; 112(14): 1398-1402.
[63] Parker JD, Woodruff TJ, Basu R, Schoendorf KC. *Air pollution and birth weight among trem infants in California.* Pediatrics 2005; 115(1): 121-128.
[64] Brauer M, Lencar C, Tamburic L, Koehoorn M, Demers P, Karr C. *A cohort study of traffic-related air pollution impacts on birth outcomes.* Environ Health Perspect 2008; 116(5): 680-686.
[65] Ballester F, Estarlich M, Iniguez C, et al. *Air pollution exposure during pregnancy and reduced birth size: a prospective birth cohort study in Valencia, Spain.* Environ Health 2010; 9(6).
[66] Maroziene L, Grazuleviciene R. *Maternal exposure to low-level air pollution and pregnancy outcomes: a population-based study.* Environ Health 2002; 1(6).
[67] Currie J, Neidell M, Schmieder JF. *Air pollution and infant health: Lessons from New Jersey.* J Health Econ 2009; 28: 688-703.
[68] Szeszenia-Dąbrowska N., Sobala W. *Environmental pollution by asbestos. Health effects.* The Nofer Institute of Occupational Medicine. ISBN 978-83-923517-5-7, Lodz, 2010.
[69] IARC Monographs, Vol. 14, 1977.
[70] Krakowiak E, Górny R, Cembrzyńska J, Sąkol G, Boissier-Draghi M, Anczyk E. *Environmental Exposure to Airborne Asbestos Fibres in a Highly Urbanized City.* Ann Agric Environ Med, 16, 121-128, 2009.
[71] Bignon J., Sebastien P., di Menza L., Nebut M., Payan H. *French registry of mesotheliomas 1965-1978.* Rev Fr Mal Respir, 7, 223-242, 1979.
[72] Enterline P.E. *Cancer produced by nonoccupational asbestos exposure in the United States.* J Air Pollution Control Assoc., 33, 318-322, 1983.
[73] Saracci R. *Asbestos and lung cancer: an analysis of the epidemiological evidence on the asbestos-smoking interaction.* Int J Cancer, 20, 323-331, 1977.
[74] WHO, Special Report: *Elimination of asbestos – related diseases*, 2006.
[75] Currie G, Watt S, Maskell N. *Reconstruction of Moscow: allergy, dust, asbestos and other medical aspects.* BMJ, 339, 2009.
[76] Więcek E. *Asbestos-exposure and health effects.* Work Safety, 2, 2004.
[77] Jawecki B. *Programming asbestos removal at the local level – a proposal guidlines. Infrastructure and Ecology of Rural Areas.* Polish Academy of Sciences, Krakow, 9, 73-83, 2008.
[78] Łuniewski A, Łuniewski S. *Asbestos-historical burden of the 20th century. Publishing house of Environmental and Resources Economists.* ISBN 978-83-61643-28-9, Białystok, 2008.
[79] Kazan-Allen L. *Asbestos and mesothelioma: Worldwide trends.* Lung Cancer, 49, 3-8, 2005.
[80] Vogel L. *Special Report - Asbestos in the world.* HESA, 2005.

2

Impact of Air Pollution on Vitamin D Status and Related Health Consequences

Alina Kurylowicz
Department of Human Epigenetics, Mossakowski Medical Research Centre, Polish Academy of Sciences
Poland

1. Introduction

During the last decade a growing interest in vitamin D-related problems can be observed and a few thousands of scientific publications are dedicated to this task every year.

The alarming statistical data regarding propagation of vitamin D deficiency in all continents and our developing knowledge about novel, non-calcemic actions of vitamin D seem to be two main causes of this phenomenon. Indeed, recent epidemiological studies suggest that worldwide prevalence of inadequate vitamin D status is higher than expected, however it may result from different factors, including geographical location and cultural background.

Since vitamin D skin synthesis, which takes place under the influence of the sun's ultraviolet B (UVB) radiation, covers as much as 90% of total body needs, inadequate radiation or insufficient cutaneous absorption of UVB are regarded as major causes of vitamin D deficiency.

Air pollution is a chief factor determining the extent of solar UVB that reaches earth surface and several epidemiological data, which represents different populations living in different geographic latitudes, indicate that atmospheric pollution (especially high tropospheric ozone content) may play a significant independent role in the development of vitamin D deficiency. As a result prevalence of D hypovitaminosis among urban residents can be more than twice higher than that of rural inhabitants.

The problem of vitamin D insufficiency has been underestimated for many years and it has been predominantly associated only with bone diseases. Discovery of vitamin D receptor and its identification in a wide number of tissues other than bones led to the designation of novel, so-called "non-calcemic" (e.g. antiproliferative and immunomodulatory), actions of cholecalciferol. Subsequently, there is a growing number of diseases and conditions, development and/or progression of which can be associated with inadequate vitamin D status, including different types of cancers, autoimmune disorders as well as cardiovascular events. Interestingly, prevalence of many of these diseases also positively correlates with the intensity of air pollution. Therefore one can conclude that with the increasing atmospheric pollution grows a number of people in the world who are vitamin D deficient or insufficient and indirectly exposed to several poor health outcomes.

This chapter gives an overview of the literature on this topic and assesses the available data about the association between air pollution, vitamin D status and prevalence of so-called civilization-related diseases. A general information regarding vitamin D metabolism is given first. Next, current definitions of vitamin D status are presented, followed by the alarming statistical data about vitamin D deficiency pandemic and its correlation with the intensity of air pollution and prevalence of common human diseases. To understand possible mechanism linking vitamin D deficiency with these poor health outcomes, first a brief presentation of traditional (calcemic) and novel ("non-calcemic") actions of vitamin D is given. Subsequently, epidemiological and experimental data regarding the association between vitamin D status and prevalence of bone diseases, different types of cancers, autoimmune disorders and cardiovascular diseases is presented. Finally, therapeutic perspectives and general guidelines about vitamin D supplementation and requirements are included. The chapter ends with a short summary which facts about vitamin D can be generally accepted and which still require more solid scientific background.

2. Vitamin D metabolism

In mammals, vitamin D_3 (cholecalciferol) is either produced in the skin by the non-enzymatic conversion (photochemical cleavage) of provitamin D (7-dehydrocholesterol) to previtamin D_3 under the influence of ultraviolet radiation (290-315nm), or delivered from food sources. In the skin previtamin D_3 can also undergo isomerization that results in creation of biologically inactive compounds (lumisterol and tachysterol) and this mechanism prevents from the vitamin D_3 overproduction and intoxication after prolonged exposure to sunlight. In the liver vitamin D_3 undergoes enzymatic hydroxylation to 25-hydroxyvitamin D_3 ($25(OH)D_3$) in the reaction catalyzed by 25-hydroxylase (CYP2R1).
In the tissues that have 1α-hydroxylaze activity (predominantly in kidneys, but also in activated macrophages, colon, prostate, breast, brain as well as in other tissues), $25(OH)D_3$ is converted to the active vitamin D metabolite - $1,25(OH)_2 D_3$ (calcitriol), whereas hydroxylation in position 24 (by 24-hydroxylase - CYP24) initiates degradation of vitamin D metabolites (Horst & Reinhardt, 1997) [Figure 1]. Efficiency of renal vitamin D_3 hydroxylation is regulated by the level of calcium and phosphate ingestion, parathyroid hormone and, in a negative feedback, by circulating levels of $1,25(OH)D_3$. The extra-renal hydroxylation is determined by local factors e.g cytokines, growth factors as well as by $25(OH)D_3$ concentration, making it particularly sensitive to vitamin D deficiency (Marques et al., 2010).
The serum $25(OH)D_3$ concentration is the parameter of choice for the assessment of vitamin D status for several reasons. First of all, it reflects total vitamin D derived from dietary intake and sunlight exposure as well as the conversion of vitamin D from adipose stores in the liver. Furthermore, it is relatively stable and it has a long (2-3 weeks) half-life in circulation. Finally, it was shown in several epidemiological studies that $25(OH)D_3$ levels correlated best with several clinical conditions.

3. Definition and prevalence of vitamin D deficiency, insufficiency and adequacy

Definition of vitamin D status has evolved during the last decade. Whereas $25(OH)D_3$ level below 10 ng/μl (25nmol/l) is still, indisputably defined as vitamin D deficiency, the ranges of vitamin D sufficiency has been recently modified. For years, $25(OH)D_3$ levels between 10 ng/μl and 20 ng/μl have been identified as insufficiency (WHO report 2003), however with the recent changes in laboratory reference rages, nowadays vitamin D adequacy is defined as 30 to 76 ng/μl (75nmol/l) and $25(OH)D_3$ concentrations between 10 and 30 ng/μl are described as insufficiency. Vitamin D intoxication which is extremely rare, occurs when $25(OH)D_3$ concentrations exceed 150 ng/μl.

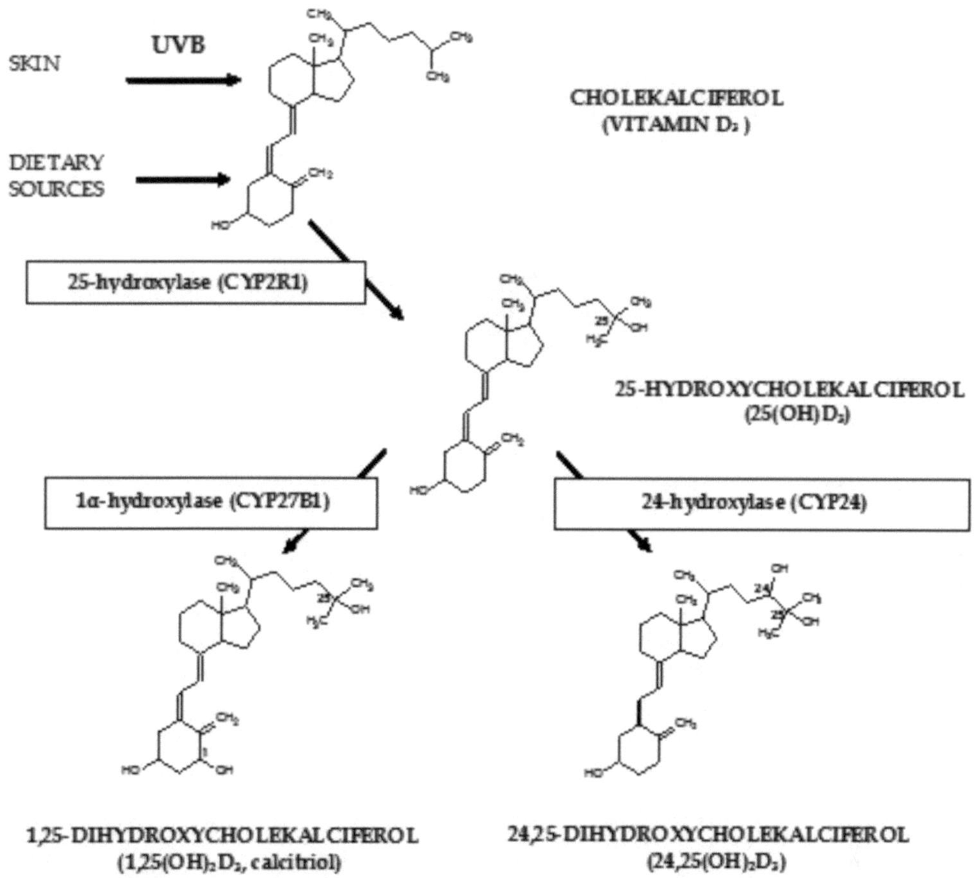

Fig. 1. The simplified scheme of the vitamin D synthesis
Cholekalciferol (synthesized in the skin upon the UV radiation or delivered with food undergoes several hydroxylations, firstly in the position 25 in the liver. The next hydroxylation catalyzed by the 1α-hydroxylase (CYP27B1) leads to the synthesis of the active metabolite – 1,25-dihydroxycholekalciferol. The alternative metabolic pathway of 25-hydroxycholekalciferol leads via hydroxylation in position 24 (the reaction is catalyzed by the vitamin D 24-hydroxylase – CYP24).

There are at least two reasons for setting the low end of the normal range of vitamin D levels at 30 ng/μl. First of all, it is suggested that serum concentration of vitamin D over 30 ng/μl guarantees optimal calcium absorption and below this concentration levels of parathyroid hormone (PTH) rise (Rosen, 2011). Secondly, this concentration provides an adequate amount of substrate for the nonrenal conversion of $25(OH)D_3$ to $1,25(OH)_2D_3$. It is assumed that children have the same requirements as adults, however no comparable studies have been carried out on intestinal calcium absorption and parthyroid hormone levels in children.

During last years vitamin D deficiency and insufficiency has been recognized as a 21st century pandemic and after changing the ranges of vitamin D concentrations defined as normal, it is estimated that this problem concerns as much as 30-50% (according to some sources up to 80%) of the general population (Hollick & Chen, 2008, Ovesen et al., 2003). In a Euronut-Seneca study that compared vitamin D status in elderly citizens in Europe living in latitudes from 35^0 to more than 60^0 the mean $25(OH)D_3$ concentrations ranged from 20 to 60 nmol/l and surprisingly, were higher in the northern countries than in the Mediterranean area (Wielen et al., 1995). This data was confirmed by the results of MORE study conducted in women with osteoporosis (Lips, 2010). In Middle East and in India, vitamin D status correlates with clothing style and is believed to be the lowest among all continents (Arya et al., 2004, Sachan et al., 2005). In South East Asia vitamin D status is generally better (Fraser, 2004). In studies performed in North America several ethnic and life-style related differences in vitamin D status were observed. However, in general, vitamin D deficiency was higher in non-hispanic whites compared to non-hispanic blacks and was in between in Mexican-American (Looker et al., 2008). In Africa, it was reported that the vitamin D status improves from north to south (Prentice et al., 2009) whereas in Australia and Oceania, the trend is opposite (Rockell et al., 2006).

Although studies performed in different continents cannot be exactly compared due to e.g. (i) different methods used to asses $25(OH)D_3$ concentrations (precision and accuracy may vary depending on the kind of the assay) (ii) seasonal fluctuations in the vitamin D dietary intake (iii) seasonal variations in efficacy of cholekalciferol skin synthesis (it is estimated that in northern latitudes serum $25(OH)D_3$ levels decline from late summer to midwinter by 20%), some general patterns suggested by these studies can be recognized. In adults, groups at the highest risk of vitamin D deficiency are elderly (especially those in nursing homes) and obese subjects, and those who due to religious habits are wearing clothes covering most of the body (Lips, 2010).

Based on epidemiological studies it is estimated that an average dietary intake of cholecalciferol is as small as only 200 IU per day, thus the main source of vitamin D remains its skin-derived synthesis that covers as much as 90% of body's total requirements. However it should be emphasized that its efficiency is highly variable and anything that hampers transmission of solar UVB radiation to the earth surface and anything that diminishes the penetration of UVB radiation into the skin will affect efficacy of vitamin D skin synthesis.

Since melanin is extremely efficient in absorbing UVB radiation, individuals with increased skin pigmentation have reduced abilities of vitamin D skin synthesis and are at higher risk of vitamin D deficiency. Appropriate application of sunscreens results in the similar effect. The angle at which the sun reaches the earth has a significant effect on the number of UVB photons that reach the earth surface therefore geographic latitude and season (autumn and winter) as well as agegeographic latitude and season (autumn and winter) as well as age (elderly people have 75% lower concentrations of vitamin D precursor 7-dehydrocholesterol), sex (vitamin D status is usually lower in women) and cultural conditionings (especially practice of purdah that totally prevents exposition of the body to sunlight) are another factors potentially reducing efficacy of vitamin D skin synthesis (Webb, 2006).

Apart from UVB-related vitamin D deficiency, there are several medical or physical conditions which may impair vitamin D status, that include, among others: fat malabsorption, use of anticonvulsant drugs that induce catabolism of vitamin D active metabolites, chronic kidney disease or obesity (fat tissue is known to entrap vitamin D) (Zhang & Naughton, 2010).

Recently the scientists focused their attention on the problem of air pollution as an important factor reducing the amount of UVB radiation reaching the earth surface and therefore correlating with prevalence of vitamin D deficiency.

4. Correlation of air pollution and prevalence of vitamin D deficiency, insufficiency and adequacy

Air pollution is regarded as a dominant factor influencing the extent of solar UVB reaching earth surface. It was proved by both: observational and prospective studies, performed in different populations living in different geographic latitudes that atmospheric pollution (especially high tropospheric ozone content) may play a significant independent role in the development of vitamin D deficiency. Tropospheric ozone can efficiently absorb UVB radiation and decrease the amount of photons reaching ground level. It was proved that the level of air pollution is inversely related to the extent of solar UVB that reaches earth surface. Since industrial areas are those, of the highest intensity of air pollution, prevalence of D hypovitaminosis among urban residents can be more than twice higher compared to rural inhabitants.

To date, there have been published only 3 studies that attempted to correlate the intensity of air pollution with the prevalence of vitamin D deficiency, however their results are unequivocal.

In the tropics, children who live in regions with higher levels of ambient air pollution have been shown to be at increased risk of developing vitamin D-dependent rickets, compared to those living in less polluted areas (Agarwal et al., 2002). Children living in the highly polluted part of Delhi, despite similar types of housing, had significantly lower mean serum concentrations of 25(OH)D_3 compared to those living in the less polluted areas of the city. The prevalence of vitamin D deficiency was correlated with the intensity of air pollution that in turn correlated inversely with the amount of UVB light reaching the ground level. None of the children included into the study used any vitamin D supplementation or consumed vitamin D fortified food, so the differences in the vitamin D status of the children cannot be explained by different dietary habits. However the authors did not collect data regarding time which each of individuals spent on exposition to sunlight that seems to be a chief limitation of this work.

In another study performed in adult European citizens of urban areas above 35^0 N (where cutaneous vitamin D synthesis in winter is nearly negligible), prevalence of vitamin D insufficiency was significantly higher compared to rural residents (38% *vs* 18%) (Manicourt & Devogealer, 2008). The compared groups did not differ in the mean age, body mass index and amount of time spent outdoor. Therefore the authors concluded that the observed difference in vitamin D status must be due to the efficacy of vitamin D cutaneous synthesis. The main reason of this phenomenon was 3 times higher tropospheric ozone concentration in urban compared to rural areas. These results were confirmed by a subsequent study performed in urban and rural areas of Iran (Hosseinpanah et al., 2010), where air pollution (determined as the high tropospheric ozone content) was found to be an independent, significant risk factor of vitamin D deficiency.

For many years the problem of vitamin D insufficiency has been underestimated, associated only with calcium/phosphorus metabolism and regarded as an area of interest reserved only for pediatricians or orthopedists. However, the last years brought a number of epidemiological studies revealing the unexpected connection between the vitamin D hypovitaminosis and prevalence of several diseases that have forced the scientist to revise their opinions on the mechanism of vitamin D action and its role in the maintenance of body homeostasis (Walters, 1992).

5. Vitamin D mechanisms of action

1,25$(OH)_2D_3$ may act in target tissues via both genomic and non-genomic mechanisms. The non-genomic mechanism which is still not fully understood, is associated with stimulation of the enzymatic activity of a nonreceptor protein tyrosine kinase Src that results in activation of the mitogen activated protein kinase (MAP) signaling pathway (Gniadecki, 1998). Much more is known about the interaction of 1,25$(OH)_2D_3$ with its nuclear receptor VDR (vitamin D receptor).

VDR together with thyroid hormone receptor (TR) and retinoid-X receptor (RXR) belongs to the class II of nuclear receptors family that act as a transcription factors modulating expression of vitamin D-directed genes. Upon binding with 1,25$(OH)_2D_3$, the VDR forms a heterodimer with the retinoid-X receptor and translocates from cytoplasm to nucleus where it interacts with vitamin D responsive elements (VDRE) in promoter regions of target genes and regulates their expression (Dusso & Brown, 1998) [Figure 2].

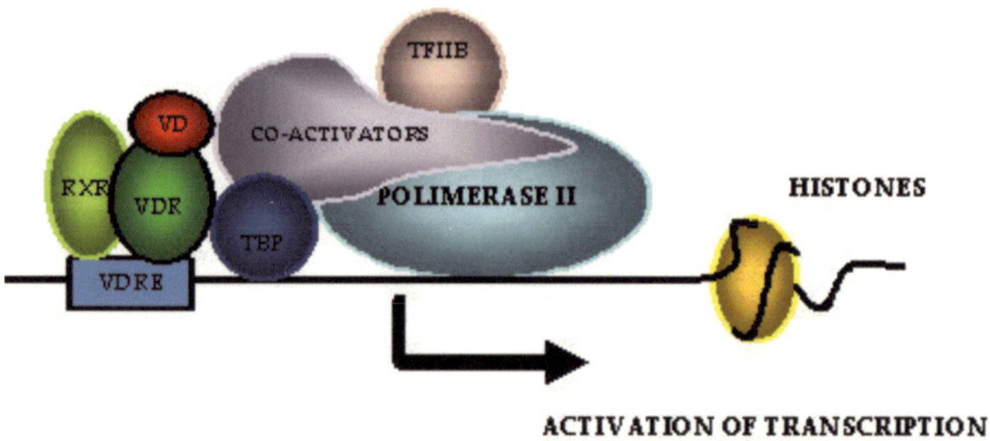

Fig. 2. The simplified scheme of activation of the target gene by the vitamin D receptor. The vitamin D receptor (VDR) with vitamin D (VD) binds as a heterodimer with the retinoid X receptor (RXR) specific sequence in the promoter region of the target gene – the vitamin D responsive element (VDRE). *Via* TATA binding protein (TBP) transcription factor II B (TFIIB), VDR gets in touch with the RNA II polymerase and other transcription factors (not present on these scheme). The transcription level is regulated by the complex of co-activators which bind VDR.

Identification of VDR in the intestine, kidney, bones and parathyroid glands (organs traditionally associated with mineral homeostasis) was not surprising. It is an undisputable fact that vitamin D is essential for maintenance of calcium and phosphorus homeostasis. In response to hypocalcemia-induced secretion of parathyroid hormone, and subsequent increase of *CYP27B1* expression and conversion of 25$(OH)D_3$ to an active 1,25$(OH)_2D_3$, calcitriol, secreted to circulation, reaches the target cells and, after binding with VDR interacts with VDR-responsive elements in target genes that results in increased calcium and phosphate absorption in gut and in release of calcium and phosphate from the mineral phase of bones. Other actions of vitamin D related to bone metabolism include: inhibition of type 1 collagen synthesis, induction of osteocalcine production, stimulation of monocytes-macrophages differentiation into osteoclasts and production of RANK ligand which mediates in maturation of osteoclast precursors into osteoclasts, responsible for calcium mobilization from bones (Clarke & Kohsla, 2010).

The identification of VDR expression as well as the demonstration of the vitamin D 1α-hydroxylaze (CYP27B1) activity in a wide number of tissues other than bones (including, e.g.: skin, blood cells, prostate, breast, brain and skeletal muscles) started a new era in the understanding of vitamin D action and led to the designation of novel, so-called "non-calcemic" actions of calcitriol.

Nowadays, $1,25(OH)_2D_3$ can be defined as both a hormone (when it is synthesized in kidneys and secreted to the circulation) and as a cytokine. As a cytokine calcitriol participates in the regulation of innate immunity. It is synthesized locally by monocytes-macrophages and in an intracrine manner, *via* interaction with VDR, modulates immune response towards microbial agents. *In vitro* studies suggest also crucial role of $1,25(OH)_2D_3$ in regulation of differentiation, maturation and function of other antigen presenting cells - dendritic cells. Other functions of vitamin D in the immune system include: regulation of the differentiation and activation of CD4 lymphocytes, increase in the number and function of regulatory T cells (Treg) , reduction in the production Th1-derived cytokines, stimulation of the Th2 helper and natural killer (NK) T cells and probably many others (Marques et al., 2010].

$1,25(OH)_2D_3$ has been also shown to have antiproliferative and antiapoptotic properties. Although the exact mechanism by which $1,25(OH)_2D_3$ regulates cellular proliferation is not fully understood and may differ between tissues and cell lines, several pathways have been proposed. In *in vitro* studies, $1,25(OH)_2D_3$ *via* interaction with VDR, increases expression of cyclin-dependant kinase (CDK) inhibitors - e.g.: proteins p21 and p27. It results in keeping the cell in G1/S phase and in prevention of DNA synthesis (see Figure 3) as it was shown in experiments performed on human prostate adenocarcinoma (lymph node, carcinoma, prostate - LNCaP) and on human leukemia U937 cell lines (Zhuang & Burnstein, 1998; Liu et al., 1996). The inhibition of mitogenic signals transmitted via e.g. epithelial growth factor (EGF) receptor, inhibition of prostaglandins, activation of transforming growth factor β (TGF-β) and proteins binding insulin-like growth factor (IGF-BP3) are examples of other (and not only) mechanisms by which vitamin D is able to regulate cell cycle progression (Desprez et al., 1991, Moreno et al., 2006).

Apoptosis is an example of another cellular process which can be regulated by $1,25(OH)_2D_3$, however the exact mechanisms of this phenomenon are still being investigated. Experiments conducted on human breast cancer and on chronic lymphatic leukemia cell lines revealed that $1,25(OH)_2D_3$ is able to inhibit expression of the protooncogen *bcl-2* and increase expression of the pro-apoptotic protein Bax (Elstner et al., 1995, Xu et al., 1993). In addition, in breast and prostate cancers cell lines, calcitriol was shown to induce release of cytochrome c in the mechanism that does not depend on kaspases (Spina et al., 2006).

Apart from its engagement in cell proliferation and apoptosis, vitamin D has been also found to be involved in the regulation of cell adhesion and angiogenesis, two other processes important for cancer development and progression. "Anti-invasive" properties of

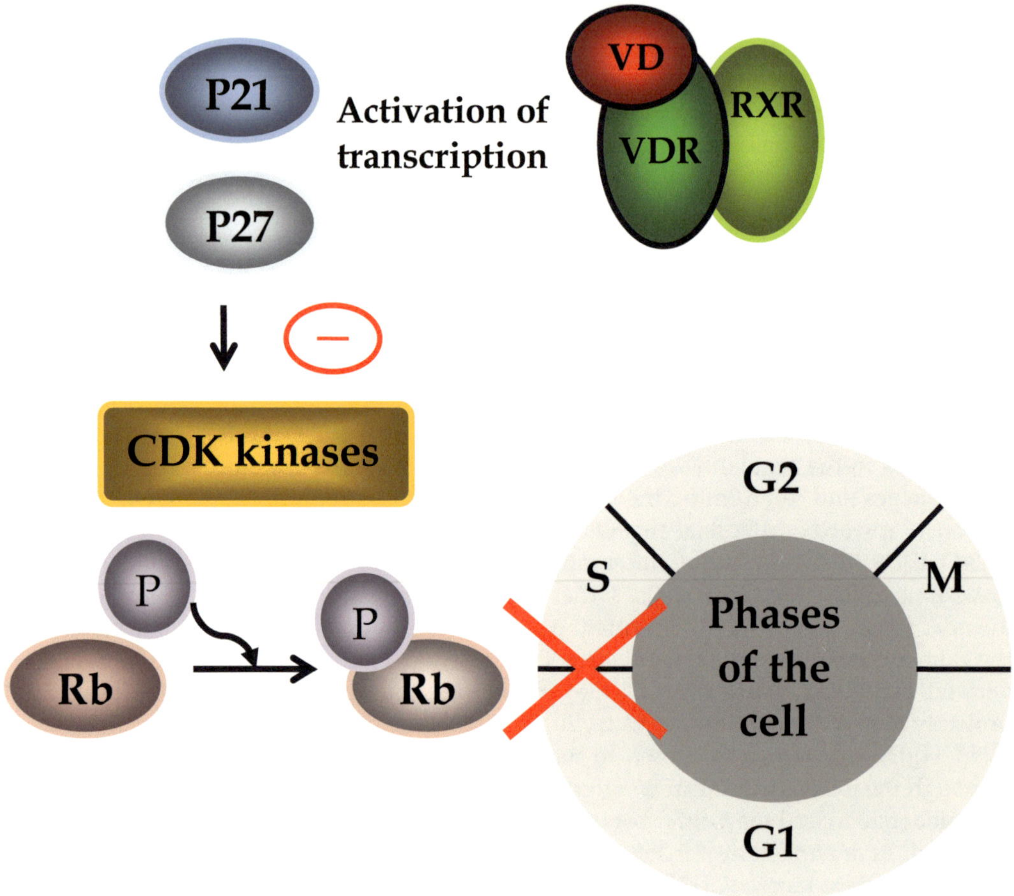

Fig. 3. Inhibition of cell cycle progresion by vitamin D
One of the factors determining the switch between G1 and S phases of the cell cycle is retinoblastoma (Rb) protein phosphorylation that leads to the release of several transcription factors activating genes involved in the cell cycle progression. Rb phosphorylation is catalyzed by the G1 cyclines and cyclin dependent kinases - CDKs. The CDKs activity can be inhibited by the p21 and p27 proteins. Vitamin D with the vitamin D receptor (VDR) binds to the regulatory sequences in the promoters of *p21* and *p27* genes activating their transcription that leads to the inhibition of the CDKs, lack of Rb phosphorylation and inhibition of the cell cycle progression.

$1,25(OH)_2D_3$ has been shown both in *in vitro* experiments (performed on human breast and lung cancer cell lines) as well as on animal models of prostate and bladder cancers, and include:

i. inhibition of metalloproteinases and serine proteases,
ii. up-regulation of cadherin E expression,
iii. down-regulation of integrin α6 and β4 expression (Bao, et al., 2006a, Hansen et al., 1994, Konety et al., 2001).

Inhibition of interleukin 8 gene expression (*via* interaction of $1,25(OH)_2D_3$ with p65 subunit of nuclear factor κB) is proposed as one of potential mechanisms by which calcitriol may interfere with the process of angiogenesis (Bao, et al., 2006b).
Knowledge of novel, non-calcemic actions of vitamin D that were proved in studies *in vitro* and on animal models, helps to understand the connection between the vitamin D and several human diseases that was reported by epidemiological studies.

6. Expected and unexpected consequences of vitamin D deficiency

Knowledge of traditional and novel, non-calcemic actions of vitamin D that were proved in studies *in vitro* and on animal models, helps to understand the connection between the vitamin D and several human diseases which was reported by epidemiological studies.

6.1 Musculoskeletal symptoms

For years vitamin D deficiency deficiency has been predominantly associated with impaired bone mineralization and development of rickets in children as well as osteopenia, osteoporosis and fractures in adults.
The consequences of vitamin D deficiency in mothers can be observed in the fetal skeleton just at the beginning of the 19th week of human gestation, resulting in rachitic phenotype, the severity of which is directly associated with a decreasing $25(OH)D_3$ level in the maternal circulation (Mahon et al., 2009). Vitamin D levels in infants correlate with vitamin D levels in their mothers during the first two months of life. Later, vitamin D status in babies, like in adults, depends on their diet and exposure to sunlight.
In rickets caused by vitamin D deficiency (a leading cause of rickets), both intestinal calcium absorption and renal phosphate reabsorption are significantly reduced. The decreased levels of serum calcium and phosphorus levels result in decreased bone mineralization. In addition, phosphorus deficiency contributes to the failure of the expected apoptosis of hypertrophic chondrocytes that results in deorganization of the growth plate. Clinical presentation of nutritional rickets depends on the duration and severity of vitamin D deficiency. At the beginning hypocalcemic symptoms are predominant, whereas skeletal deformities become obvious in later, more advanced stages. Classical symptoms of so-called "blooming" rickets include: craniotabes in infants older than 2-3 months, delayed fontanel closure, wrists enlargement, rachitic rosary, delayed teething, carious teeth, enamel hypoplasia, "O"- or "X"-type leg deforminty, kyphosis and narrow pelvis, chest deformities, costal or lower extremity fractures, caput quadratum, frontal bossing, fractures, brown tumor and extremity pain. Extra-skeletal symptoms include: hypotonia, constipation, proximal myopathy, cardiomyopathy and heart failure, myelofibrosis and pancytopenia, growth retardation, hypocalcemic convultions and benign intracranial hypertension (Levine, 2009).
Although full-symptomatic rickets seems to be nowadays a curiosity, epidemiological data indicate that rickets is not a disease of the past, nor it is limited to developing countries. It is estimated that nowadays 5 per 1,000,000 children aged between 6 months and 5 years have rickets, and the main risk factors for its development are breastfeeding and dark carnation. The peak prevalence of vitamin D-deficient rickets is characteristically between 6 and 18 months of age, with a further smaller peak occurring during adolescence (Nield et al., 2006).
In adults vitamin D deficiency can also cause a skeletal mineralization defect, resulting in osteomalacia demonstrating with isolated or global bone discomfort accompanied by joints

and muscle pain. However some observational studies conducted in adults and concerning the association between skeletal health and 25(OH)D_3 serum levels had conflicting results, suggesting both a strong correlation with the incidence of fractures and falls or only a fair relation (Chung et al., 2009). In general, it is assumed that the decrease of serum 25(OH)D_3 concentration that results in the persistent secondary parathyroidism enhances osteoclastogenesis and subsequent bone resorption.

The critical role of vitamin D in bone mineralization is well established, but our growing knowledge about non-calcemic actions of calcitriol led to the elongation of the list of diseases and conditions which development and/or progression can be associated with inadequate vitamin D status, including: different types of cancers, autoimmune disorders, cardiovascular events and probably many others. Interestingly, prevalence of this diseases frequently correlates with intensity of environmental pollution and many of them are diagnosed more frequently in inhabitants of industrial districts compared to rural areas (Pope, 2002, Pope, 2003, Grant, 2006, Ritz, 2010).

6.2 Vitamin D deficiency and cancer risk

The observational study conducted in 1941 reported that living at higher latitudes (where vitamin D skin synthesis is impaired) is associated with an increased risk of the development and dying of many common cancers including colon, prostate, ovarian and breast cancers (Apperly, 1941).

This observation was followed by the epidemiological studies performed in Europe and in the North America, assessing influence of many potential risk factors for incidence of cancers (e.g. smoking, alcohol consumption, economic burdens and environmental pollution) revealing that morbidity of several cancers (including colon, gastric, lung and breast carcinomas) is inversely associated with exposition to UVB radiation (Grant, 2003, Grant, 2005, Grant & Garland, 2006, Grant, 2009).

Next, on the one hand, another epidemiological prospective, study proved that vitamin D insufficiency (25(OH)D_3 concentrations below 20ng/ml) is associated with 30-50% higher risk of colorectal cancers (Garland et al., 2009). On the other hand, it was shown in a meta-analysis that appropriate (>400 IU) or increased (>1000 IU) vitamin D intake may be associated with a decreased risk of colon, breast, pancreas, esophagus and non-Hodgkin lymphomas (Garland et al., 2006, Giovannucci et al., 2006, Holick & Chen, 2008). It is estimated that the daily intake of 2000 IU of vitamin D would lead to 25% reduction in incidence of breast cancer and 27% reduction in incidence of colorectal cancer (Garland, 2009). Association of vitamin D deficiency with prevalence of breast, skin and prostate carcinomas has been also reported, however the concentrations at which the increased risk has been observed varied. Moreover, there are studies suggesting a U-shaped association between vitamin D concentration and cancers risk, pointing that some individuals may be adversely affected by elevated 25(OH)D_3 concentrations with respect to risk of prostate, breast, pancreas and esophagus cancers (Toner et al., 2010). This inconsistency may be caused by inadequate consideration of modifiers of 25(OH)D_3 concentrations. Therefore, to date, there is not yet sufficient evidence to recommend high-dose vitamin D supplementation for the prevention of cancer.

The hypothesis, about the connection between the vitamin D deficiency and cancer development, based on the observational data, has been confirmed by the *in vitro* studies described in the previous section and by experiments on animal models. For example: in vitamin D deficient Balb/c mice, injected with MC-26 colon cancer cells, the tumor growth

was accelerated, compared to vitamin D sufficient animals. In addition, vitamin D sufficient animals presented higher intra-tumor expression of VDR and CYP27B1, which suggests possible autocrine/paracrine cell growth regulation by vitamin D (Tangpricha et al., 2005). In turn mice with *vdr* knock-out (mutation that eliminates genomic action of $1,25(OH)_2D_3$) have been found to be more susceptible to develop leukemias, breast and skin cancers under the influence of common carcinogens compared to wild-type animals (Welsh et al., 2004).
Discovery of anti-cancer properties of vitamin D (that has no antioxidant properties) required a modification of a classic two-hit model of cancerogenesis, that provides that development of cancer depends on both: activation of proto-oncogenes and deactivation of tumor suppressing genes. Therefore, in context of vitamin D, a novel model of cancer pathogenesis has been proposed - a so-called: the Disjunction-Initiation-Natural Selection-Metastasis-Involution-Transition (DINO-MIT) model. This model is based on the classical concepts of carcinogenesis like initiation and promotion, however it includes also the life cycle of malignancies and provides an explanation of the ability of vitamin D to prevent or arrest the development of cancer (described in detail by Garland et al., 2006).

6.3 Vitamin D deficiency and autoimmune disorders

Similarly to carcinomas, prevalence of some autoimmune disorders such as multiple sclerosis and type 1 diabetes mellitus also correlates with the geographical latitudes. It has been proposed that vitamin D deficiency can disturb the immunological equilibrium and therefore contribute to the development of autoimmunity (e.g. by exacerbation of Th1 immune response) (Cantorna, 2004). In order to understand the complexity of these mechanism, again, insights from animal models of common human autoimmune diseases occurred to be particularly helpful.
Vitamin D deprivation leads to the acceleration of the development of experimental allergic encephalomyelitis (EAE - an animal model of multiple sclerosis) in mice immunized with myelin antigens (e.g. MOG35-55) (Cantorna et al., 1996). In turn, supplementation with active vitamin D metabolite and its analogs may favorably influence the course of autoimmune diseases or even prevent their occurrence. In mice immunization with type 2 collagen leads to the development of collagen induced arthritis (CIA - an animal model of rheumatoid arthritis), however, if the animals are given $1,25(OH)_2D_3$ till the 14th day after immunization, they do not present any symptoms of arthritis at all. If $1,25(OH)_2D_3$ is given to mice with symptomatic artritis, it may alleviate the disease course (Cantorna et al., 1998). Similar evidence comes from the studies performed in nonobese diabetic (NOD) mice that due to the autoimmune inflammation and destruction of pancreatic islets develop insulin-dependent type 1 diabetes. In young subjects, administration of $1,25(OH)_2D_3$ till the third week after birth may completely prevent development of diabetes whereas in mature mice leads to the 50% reduction of inflammatory infiltrates in the pancreatic islets, compared with placebo-treated animals (Gregori et al., 2002). Reduction of T cells infiltration and decrease in the number of demyelinization sites has been also found in spinal cords of mice with experimental allergic encephalomyelitis treated with vitamin D analogs, compared to wild type animals (Mattner et al., 2000).
In humans, epidemiological data revealed that multiple sclerosis morbidity in Caucasians is significantly higher in less insolated areas, and the course of disease correlates with seasonal fluctuation of $25(OH)D_3$ serum concentrations (exacerbations in spring when $25(OH)D_3$ concentrations are the lowest) (Cantorna, 2008). Similarly, in patients with systemic lupus

erythematosus (SLE) severe vitamin D deficiency has been reported by many authors, and it correlates with disease activity (Kamen, 2010).

In contrary, like in epidemiological studies performed in different cancers, adequate consumption of cholecalciferol (400 IU per day) was shown, in a prospective study, to decrease a risk of multiple sclerosis development (relative risk – RR 0.59) (Munger et al., 2004). Vitamin D deficiency during the first year of life was found also to be a severe risk factor of type 1 diabetes whereas its proper administration in early childhood may reduce the risk of the development of this disease by 78% (RR 0.12) (Hypponen et al., 2001). A correlation between the adequate vitamin D intake and lower risk of the development of rheumatoid arthrits (RR=0.67) was reported (Merlino et al., 2004), however it was not confirmed by further studies [Costenbader, 2008]. Inconsistent are also results of the studies regarding the association between vitamin D intake during pregnancy and risk of type 1 diabetes in the offspring as well as with prevalence of other autoimmune disorders.

These discrepancies may be owed to the fact that, like in carcinogenesis, vitamin D deficiency is probably not the chief trigger in the development of autoimmunity. This hypothesis is supported by the *in vivo* experiments, where animals with vitamin D deficiency or with the knock-out of *vdr* gene do not present autoimmune disorders spontaneously and require other stimuli to develop autoimmunity. However these animals may present more severe disease phenotype of autoimmune disorders, as it was shown in mice depleted with IL-10 gene (animal model of inflammatory bowel diseases – IBD) (Froicu & Cantorna, 2007).

6.4 Vitamin D deficiency and cardiovascular health

Cardiovascular diseases are the most common cause of mortality and morbidity worldwide. There is also a growing number of evidence that vitamin D deficiency can be associated with development of several cardiovascular diseases including hypertension, ischemic heart disease and cardiac hypertrophy.

In vitro studies revealed that addition of $1,25(OH)_2D_3$ to the cardiomyocyte cells culture resulted in: inhibition of cell proliferation, enhanced cardiomyocyte formation, decrease of apoptosis and cell-cycle associated genes' expression. In turn animals with *vdr* gene knock-out were shown to develop cardiac hypertrophy, display hypertension as well as increased serum angiotensin and tissue renin levels. These studies suggest that vitamin D *via* its influence on cardiac and vascular structure and function may modulate cardiovascular risk (Artaza et al., 2009).

Observational studies reported a strong association between vitamin D hypovitaminosis and other traditional cardiovascular risk factors. Moreover, according to other reports, vitamin D deficiency is a predictor of all-cause and cardiovascular mortality, whereas vitamin D adequacy is associated with 43% reduction in cardiometabolic disorders (Parker et al., 2010).

The importance of vitamin D status in context of coronary artery disease is well established based on observational studies that reported:

i. inverse correlation of mortality from ischemic heart disease and the exposure to sunlight (Fleck, 1989),
ii. high prevalence of vitamin D deficiency in patients with ischemic heart disease (Kendrick et al., 2009),

iii. increased risk of sudden cardiac death, heart failure and fatal stroke in patients with ischemic heart disease who were vitamin D deficient (Pilz et al., 2008).

A correlation between the time of exposition to UVB radiation and decrease in blood pressure was also observed (Krause et al., 1998) and vitamin D replacement in deficient subjects led to a significant improvement of flow-mediated dilatation of the brachial artery that suggested an influence of vitamin D on vascular smooth cells (Tarcin et al., 2009). Indeed, it was shown on animal models that active vitamin D metabolite inhibits rennin expression in the juxtaglomerular apparatus and may prevent proliferation of vascular smooth muscle cells (Li et al., 2002; Carthy et al., 1989). Both, small retrospective observational studies and a large, cross-sectional study, confirmed that the mean blood pressure vary inversely with serum 25(OH)D_3 concentrations and the association remained significant after adjustment for age, gender, race, ethnicity and physical activity (Scragg et al., 2007, Reddy Vanga et al., 2010).

Additionaly, individuals with vitamin D insufficiency have been found to higher prevalence of peripheral artery disease (Melamed et al., 2008) and worse outcomes in end-stage heart failure (Zittermann et al., 2008).

7. Vitamin D in treatment

Encouraging epidemiological data coming from observational studies, linking vitamin D deficiency with common human diseases, rose a hope that calcitriol and its derivates may be useful in everyday clinical practice (Holick, 2004, Holick, 2007). However, to date bone diseases are the only examples where administration of vitamin D is generally accepted for treatment.

7.1 Rickets and osteoporosis

In order to restore vitamin D reserve in rickets two methods of treatment are proposed. In a low dosage and long-term therapy model, vitamin D is administered 1000-10000 IU/day (dose depends on age) for 2-3 months. After that period 400 IU/day therapy is recommended to maintain the serum vitamin D level. Another regimen, reserved mainly for patients suspected of poor compliance, includes 100 000 - 600 000 IU of vitamin D in a single dose (Wharton & Bishop, 2003).

Prospective, randomized, placebo-controlled trials provided evidence supporting the benefit of vitamin D supplementation in patients with osteoporosis (LaCroix et al., 2009, Meier et al., 2004, Jackson et al., 2006). Data regarding a beneficial effect of calcium plus vitamin D on bone density in postmenopausal women and older men is consistent, however reports on fracture risk are more variable. Reduction in frequency of fractures was reported in some studies (Larsen et al., 2004) whereas large randomized trials have not shown any association between vitamin D supplementation and reduction in fracture risk. Additionally, in many of these trials, it is difficult to differentiate the effect of calcium from that of vitamin D. Nevertheless, in the Women's Health Initiative (the largest of the available trials) subgroup analysis revealed that calcium and vitamin D supplementation was associated with reduced fracture incidence in those subjects who were most compliant (Jackson et al., 2006).

Although the optimal serum concentration of 25(OH)D_3 in patients with osteoporosis is not clear, based upon meta-analyses, one approach to vitamin D supplementation would be to give 400-800 IU daily with a target serum 25(OH)D_3 concentration >20 ng/ml. Older individuals with greater risk may require higher amounts that allow to maintain serum levels of 25(OH)D_3 from 30 to 40 ng/ml (Dawson-Hughes, 2005). Yearly high-dose of vitamin D (e.g. 500,000 IU) for the osteoporosis treatment is not recommended.

7.2 Cancers

In observational studies, sufficient exposition to UVB radiation and vitamin D adequacy have been found to improve prognosis in cancer patients. A study assessing the survival rates for several cancers (including breast, colon, prostate, lung carcinomas and Hodgkin's lymphoma) revealed that patients diagnosed in autumn had approximately 30% higher 18-month survival rate than those diagnosed in winter or spring (Robsahm et al., 2004, Porojnicu et al, 2007).

Results of recent observational studies, reporting that maintenance of 25(OH)D_3 levels about 30ng/ml may be crucial in the prevention of breast, colon and prostate carcinomas (Garland, 2009), together with newly discovered, proapoptotic and anti-proliferative properties of vitamin D stimulate hope that 1,25$(OH)_2D_3$ and its analogues will have practical application in treatment of cancers. Severe hypercalcemia has occurred to be the chief limitation for the common use of vitamin D derivates in oncology and therefore vitamin D analogues are used mainly in combination with other anti-cancer drugs in reduced doses. Both *in vitro* and *in vivo* studies suggest that addition of vitamin D may potentializate action of many drugs commonly used in cancer treatment, including e.g: dexamethasone, docetaxel, paclitaxel, tamoxifen, retinoids, platinium derivatives and others (Gewirtz et.al., 2002). Administration of 1,25$(OH)_2D_3$ may also sensitize cancer cells to radiotherapy as it was shown in prostate cancer line LNCaP or breast cancer line MCF-7 (Dunlap et al., 2003, Polar et al., 2003).

Numerous vitamin D analogs with minimal calcemic activity have been designed with intent to be applied in cancer therapy and some of them have been found to be effective in inhibiting tumor growth (e.g. colon) in animal models. Some of them are tested for treatment (e.g. calcipotriol administered locally in skin metastases of breast cancer) however to date, no active analogs of vitamin D have been proved to be efficacious for the treatment of any human cancer by themselves.

No prospective, randomized, double-blind study has been conducted in order to asses the anticancer potential of vitamin D analogues in humans. However, in a small pilot study performed in 7 patients with prostate cancer, daily administration of 0.5-2.5 µg/24h of 1,25$(OH)_2D_3$ per 6 to 15 months, resulted in significant reduction of prostate specific antigen (PSA) concentration in 6 individuals (Gross et al., 1998). In another study, 37 patients with prostate cancer insensitive to androgens, were given sequentially 0.5 µg of 1,25$(OH)_2D_3$ (day 1) and 36 mg/m^3 of docetaxel (day 2) during 6 following weeks and put on diet containing 400-500 mg of calcium per day. In 30 of them (81%) a significant reduction of PSA concentration was observed after 8 weeks from the beginning of the treatment (Beer et al., 2003).

7.3 Autoimmune diseases

Despite solid clinical and experimental evidence regarding the positive influence of vitamin D derivates on the development and progression of autoimmune disorders, to date vitamin D is not routinely administered in their treatment.

The only disease with partially autoimmune pathogenesis, where vitamin D and its analogs are generally accepted for the treatment is psoriasis. Both systemic and topical administration of vitamin D and its analogs leads to the significant improvement in 70-80% of the treated patients, measured by the Total Severity Score (TSS) or Psoriasis Area Severity Index (PASI) (Ashcroft et al., 2000). Administration of vitamin D for treatment of other autoimmune diseases is still experimental since in humans only small and non-controlled studies have been conducted. For instance, in a an open study performed in 19 patients with rheumatoid arthritis, addition of alphacalcidol ($1\alpha(OH)D_3$) to the traditional disease-modifying anti-rheumatic drugs for three months, resulted in a significant reduction of symptoms in 89% of patients (45% achieved complete remission and 44% had satisfactory results) without side effects (Andjelkovic et al., 1999). There is also one report on a positive influence of $1,25(OH)_2D_3$ on thyroid hormones levels in patients with autoimmune hyperthyroidism, who acquired the treatment in order to improve bone mineral density (Kelman & Lane, 2005).

7.4 Cardiovascular health

To date, studies evaluating influence of vitamin D supplementation on cardiovascular health are few and have inconsistent results. Observational studies reported that individuals who take oral vitamin D supplementation have lower blood pressure and some interventional trials found a correlation between oral vitamin D administration or increased exposure to UVB and decrease of blood pressure. However, this phenomenon was not confirmed by a large prospective studies and a meta-analysis on prevalence of hypertension and vitamin D intake (Forman et al., 2005, Witham et al., 2009). Vitamin D supplementation seemed also to have no effect on the risk of cardiovascular mortality (La Croix et al., 2009).

8. Guidelines regarding vitamin D supplementation

The recognition of vitamin D deficiency as a worldwide problem rises a task of means of its supplementation. Historically, fortification of milk in the 1930s with vitamin D was effective in eradicating rickets in Europe and US. However, after an unexpected outbreak of hypercalcemia in British children in the 1950s caused by the excessive vitamin D consumption from fortified food, fortification of dairy products with vitamin D had been forbidden (Holick, 2010).This situation was probably caused by the imperfection of methods applied to assess the vitamin D concentration in food at that time.

Naturally the main source of vitamin D are wild-caught, not farm-raised oily fish, e.g. salmon, mackerel, herring, and oils from fish. However, since no quality control over vitamin D content in natural products is available, they are not recommended as reliable sources of vitamin D supplementation. In many countries dairy products, juices, bread and other products are fortified with vitamin D.

The recommendations of American Institute of Medicine from 1999 state that all children and adults up to the age of 50 should consume 200 IU of vitamin D daily, whereas individuals aged 51-70 and those who are over 71 years old, 400 IU and 600 IU, respectively. However, these amounts have been insufficient in prevention of osteomalacia and osteoporosis, and therefore many experts now suggest that, in the absence of the adequate sun exposure, as much as 800-1000 IU of vitamin D per day is needed for healthy individuals, regardless of age (Holick and Chen, 2008). In case of diseases leading to vitamin D mal-absorption or sequestration in fat tissue (obesity), higher doses may be required. In aspect of cancer prevention, the National Academy of Sciences-Institute of Medicine (USA) recommends even 2000-4000 IU/day (Garland, 2009). It is estimated that raising the mean population level of 25(OH)D_3 up to 42 ng/ml would result in 18% reduction in all-cause mortality, 25% reduction in prevalence of cancers and cardiovascular diseases, 15% in morbidity of diabetes and a prevalence of multiple sclerosis would be reduced by half (Grant & Schuitemaker, 2010), however approaching this vitamin D status would require supplementation of 2500-4000 IU of cholecalciferol per day.

According to the nutritional guidelines established during the 14th vitamin D workshop in October 2009, appropriate vitamin D status should be achieved rather through supplementation than diet fortification and rather through the use of vitamin D_3 than vitamin D_2. It is suggested that vitamin D_2 is in half less effective than vitamin D_3 in maintaining the 25(OH)D_3 serum levels and therefore it should be administered in higher doses. Moreover, all evidence reported to date on the efficacy of vitamin D in e.g. cancer prevention has been based on vitamin D_3. To verify this theory, a number of prospective vitamin D replacement trials with vitamin D_2 and vitamin D_3 has been conducted (Adams & Hewison, 2010).

In the treatment of evident vitamin D deficiency 50 000 IU of vitamin D is recommended weekly for 8 weeks, resulting usually in 25(OH)D_3 concentration of 30 ng/ml, and after 50 000 IU of vitamin D every 2 weeks to maintain its level. It has been estimated that for every 100 IU of vitamin D ingested, there is an increase in the blood level of 25(OH)D_3 of 1 ng/ml and sensible exposure to sunlight is more effective in raising blood levels of 25(OH)D_3 than 1000 IU vitamin D_3 taken daily for adults of most skin types. However, with the growing intensity of atmospheric pollution, which as it was shown above, is a chief factor limiting the amount of UVB radiation reaching the earth surface, obtaining appropriate amount of vitamin D from skin synthesis is getting more and more challenging.

9. Conclusions

Based on the available observational and experimental data regarding the connection between the air pollution, vitamin D status and related health consequences, following statements can be formulated:

i. the worldwide problem of vitamin D deficiency can be partially explained by the growing intensity of atmospheric pollution, however this observation is based only on the results of relatively small studies and should be replicated in larger scale,
ii. both: vitamin D deficiency and intensity of air pollution correlate with prevalence of common human diseases including different types of cancers, cardiovascular and autoimmune diseases that was confirmed by large, prospective studies and meta-analyses,
iii. association between vitamin D status and extra-skeletal health outcomes can be explained by the molecular mechanism of vitamin D action that exceed bone metabolism,
iv. despite promising results of *in vitro* and *in vivo* studies regarding the favourable influence of vitamin D supplementation on prevalence and course of different diseases, its therapeutic application is still limited only to the treatment of bone diseases,
v. novel definition of vitamin D insufficiency together with the increasing intensity of the atmospheric pollution that impairs vitamin D skin synthesis and with epidemiological data regarding the benefits of vitamin D intake should result soon in the novel guidelines regarding vitamin D supplementation.

10. References

Adams JS & Hewison M. (2010) Update in vitamin D. *Journal of Clinical Endocrinology & Metabolism*, Vol. 95, No. 2, (Febuary 2010), pp. 471-478. PMID: 20133466

Agarwal KS, Mughal MZ, Upadhyay P, Berry JL, Mawer EB & Puliyel JM. (2002) The impact of atmospheric pollution on vitamin D status of infants and toddlers in Delhi, India. *Archives of Disease in Childhood*, Vol. 87, No. 2, (August 2002), pp. 111-113, PMID: 12138058

Andjelkovic Z, Vojinovic J, Pejnovic N, Popovic M, Dujic A, Mitrovic D, Pavlica L & Stefanovic D. (1999) Disease modifying and immunomodulatory effects of high dose 1 alpha (OH) D3 in rheumatoid arthritis patients. *Clinical and Experimental Rheumatology*. Vol. 17, No. 4, (July,August 1999), pp. 453-456, PMID: 10464556

Apperly FL. (1941) The relation of solar radiation to cancer mortality in North America. *Cancer Research*, Vol 1, No. 1, (1941), pp. 191–195.

Artaza JN, Mehrotra R & Norris KC. (2009) Vitamin D and the cardiovascular system. *Clinical Journal of the American Society of Nephrology*, Vol. 4, No. 9, (September 2009), pp. 1515-1522, PMID: 19696220

Arya V, Bhambri R, Godbole MM &Mithal A.(2004) Vitamin D status and its relationship with bone mineral density in healthy Asian Indians. *Osteoporos International*, Vol. 15, No. 1, (January 2004), pp. 56-61. PMID: 13680103

Ashcroft DM, Po AL, Williams HC & Griffiths CE. (2000) Systematic review of comparative efficacy and tolerability of calcipotriol in treating chronic plaque psoriasis. *British Medical Journal*, Vol. 320, No. 7240, (April 2000), pp. 963-967, PMID: 10753146

Bao BY, Yeh SD & Lee YF. (2006a) 1alpha,25-dihydroxyvitamin D3 inhibits prostate cancer cell invasion via modulation of selective proteases. *Carcinogenesis*, Vol. 27, No. 1,(January 2006), pp. 32-42, PMID: 15987715

Bao BY, Yao J & Lee YF. (2006b) 1alpha, 25-dihydroxyvitamin D3 suppresses interleukin-8-mediated prostate cancer cell angiogenesis. *Carcinogenesis*. Vol. 27, No. 9, (September 2006), pp. 1883-1893, PMID: 1662482

Beer TM, Eilers KM, Garzotto M, Egorin MJ, Lowe BA & Henner WD. (2003) Weekly high-dose calcitriol and docetaxel in metastatic androgen-independent prostate cancer. *Journal of Clinical Oncology*, Vol. 21, No. 1, (January 2003), pp. 123-128, PMID: 12506180

Cantorna MT, Hayes CE & DeLuca HF. (1996) 1,25-Dihydroxyvitamin D3 reversibly blocks the progression of relapsing encephalomyelitis, a model of multiple sclerosis. *Proceedings of the National Academy of Sciences of the United States of America*, Vol. 93, No. 15, (July 1996), pp. 7861-7864, PMID: 8755567

Cantorna MT, Hayes CE & DeLuca HF. (1998) 1,25-Dihydroxycholecalciferol inhibits the progression of arthritis in murine models of human arthritis. *The Journal of Nutrition*, Vol. 128, No. 1, (January 1998), pp. 68-72, PMID: 9430604

Cantorna MT & Mahon BD. (2004) Mounting evidence for vitamin D as an environmental factor affecting autoimmune disease prevalence. *Experimental Biology and Medicine (Maywood, N.J.)*, Vol. 229, No. 11, (December 2004), pp. 1136-1142. PMID:15564440

Cantorna MT. (2008) Vitamin D and multiple sclerosis: an update. *Nutrition Reviews*, Vol 66, No. 10, Suppl. 2, (October 2008), pp. S135-138, PMID: 18844840

Carthy EP, Yamashita W, Hsu A & Ooi BS. (1989) 1,25-Dihydroxyvitamin D3 and rat vascular smooth muscle cell growth. *Hypertension,* Vol.13, No. 6, Pt. 2, (June 1989), pp.954-959, PMID: 2786849

Chung M, Balk EM, Brendel M, Ip S, Lau J, Lee J, Lichtenstein A, Patel K, Raman G, Tatsioni A, Terasawa T & Trikalinos TA. (2009) Vitamin D and calcium: a systematic review of health outcomes. *Evidence Report - Technology Assessment,* Vol. 183, pp.1-420, PMID: 20629479

Clarke BL & Khosla S.(2010) Physiology of bone loss. Radiologic Clinic of North America. Vol. 48, No. 3, (May, 2010), pp. 483-495, PMID: 20609887

Costenbader KH, Feskanich D, Holmes M, Karlson EW & Benito-Garcia E. (2008) Vitamin D intake and risks of systemic lupus erythematosus and rheumatoid arthritis in women. *Annals of Rheumatic Diseases,* Vol. 67, No. 4, (April 2008), pp. 530-535, PMID: 17666449

Dawson-Hughes B, Heaney RP, Holick MF, Lips P, Meunier PJ & Vieth R.(2005) Estimates of optimal vitamin D status. *Osteoporos International,* Vol. 16, No. 7, (July 2005), pp. 713-716, PMID: 15776217

Desprez PY, Poujol D, Falette N, Lefebvre MF & Saez S.(1991) 1,25-Dihydroxyvitamin D3 increases epidermal growth factor receptor gene expression in BT-20 breast carcinoma cells. *Biochemical and Biophysical Research Communications,* Vol. 176, No. 1, (April 1991), pp. 1-6, PMID: 167333

Dunlap N, Schwartz GG, Eads D, Cramer SD, Sherk AB, John V & Koumenis C. (2003) 1alpha,25-dihydroxyvitamin D(3) (calcitriol) and its analogue, 19-nor-1alpha,25(OH)(2)D(2), potentiate the effects of ionising radiation on human prostate cancer cells. *British Journal of Cancer,* Vol. 89, No. 4, (August 2003), pp. 746-753, PMID 12915889

Dusso AS & Brown AJ.(1998) Mechanism of vitamin D action and its regulation. *American Journal of Kidney Diseases,* Vol.32, No. 2, Supplement 2, (October 1998), pp. 13-24, PMID: 9808140

Elstner E, Linker-Israeli M, Said J, Umiel T, de Vos S, Shintaku IP, Heber D, Binderup L, Uskokovic M & Koeffler HP.(1995) 20-epi-vitamin D3 analogues: a novel class of potent inhibitors of proliferation and inducers of differentiation of human breast cancer cell lines. *Cancer Research,* Vol. 55, No. 13, (July 1995),pp. 2822-2830, PMID: 7796409

Fleck A.(1989) Latitude and ischaemic heart disease.*Lancet.* Vol.1, No. 8638, (March 1989), pp. 613, PMID: 2564129

Forman JP, Bischoff-Ferrari HA, Willett WC, Stampfer MJ & Curhan GC. (2005) Vitamin D intake and risk of incident hypertension: results from three large prospective cohort studies. *Hypertension,* Vol. 46, No. 4, (October 2005), pp: 676-682, PMID: 16144983

Fraser DR.(2004) Vitamin D-deficiency in Asia. *The Journal of* Steroid Biochemistry *and Molecular* Biology, Vol. 89-90, No..1-5, (May 2004), pp. 491-495, PMID: 15225826

Froicu M & Cantorna MT. (2007) Vitamin D and the vitamin D receptor are critical for control of the innate immune response to colonic injury. *BMC Immunology,* Vol. 8, No. 5, (March 2007), pp. 1-11, PMID: 17397543

Garland CF, Garland FC, Gorham ED, Lipkin M, Newmark H, Mohr SB & Holick MF.(2006) The role of vitamin D in cancer prevention. *American Journal of Public Health,* Vol. 96, No. 2, (February 2006), pp. 252-261, PMID: 16380576

Garland CF, Gorham ED, Mohr SB & Garland FC. (2009) Vitamin D for cancer prevention: global perspective. *Annals of epidemiology*, Vol. 19, No. 7, (July 2009), pp. 468-83, PMID 19523595

Gewirtz DA, Gupta MS & Sundaram S. (2002) Vitamin D3 and vitamin D3 analogues as an adjunct to cancer chemo-therapy and radiotherapy. *Current Medical Chemistry: Anticancer Agents*, Vol. 2, No. 6, (November 2002), pp.683-690, PMID: 12678720

Giovannucci E, Liu Y, Rimm EB, Hollis BW, Fuchs CS, Stampfer MJ & Willett WC. (2006) Prospective study of predictors of vitamin D status and cancer incidence and mortality in men. Journal of the National Cancer Institute, Vol. 98, No. 7, (April 2006), pp. 451-459, PMID: 1659578

Gniadecki R. (1998) Nongenomic signaling by vitamin D: a new face of Src. *Biochemical Pharmacology*, Vol. 56, No. 10 (November 1998), pp. 1273-1277, PMID: 9825725

Grant WB. (2003) Ecologic studies of solar UV-B radiation and cancer mortality rates. *Recent Results in Cancer Research*, Vol. 164, (2003), pp. 371-377. PMID: 12899536.

Grant WB, Garland CF& Holick MF.(2005) Comparisons of estimated economic burdens due to insufficient solar ultraviolet irradiance and vitamin D and excess solar UV irradiance for the United States. *Photochemistry and Photobiology*, Vol. 81, No. 6, (November - December 2005), pp. 1276-1286, PMID: 16159309

Grant WB & Garland CF. (2006) The association of solar ultraviolet B (UVB) with reducing risk of cancer: multifactorial ecologic analysis of geographic variation in age-adjusted cancer mortality rates. *Anticancer Research* Vol. 26, No. 4A, (July - August 2006), pp. 2687-2699, PMID: 16886679

Grant WB. (2006) Epidemiology of disease risks in relation to vitamin D insufficiency. *Progress in Biophysics and Molecular Biology*, Vol. 92, No. 1, (September 2006), pp. 65-79. PMID: 16546242

Grant WB & Mohr SB. (2009) Ecological studies of ultraviolet B, vitamin D and cancer since 2000. *Annals of Epidemiology*, Vol. 19, No. 7, (July 2009), pp. 446-454, PMID: 19269856

Grant WB & Schuitemaker GE. (2010) Health benefits of higher serum 25-hydroxyvitamin D levels in The Netherlands. *The Journal of Steroid Biochemistry and Molecular Biology*, Vol.121, No. 1-2, (July 2010), pp. 456-458, PMID: 20398763

Gregori S, Giarratana N, Smiroldo S, Uskokovic M & Adorini L. (2002) A 1alpha,25-dihydroxyvitamin D(3) analog enhances regulatory T-cells and arrests autoimmune diabetes in NOD mice. *Diabetes*, Vol. 51, No. 5, (May 2002), pp.1367-1374, PMID: 11978632

Gross C, Stamey T, Hancock S & Feldman D. (1998) Treatment of early recurrent prostate cancer with 1,25-dihydroxyvitamin D3 (calcitriol). *The Journal of Urology*, Vol. 159, No. 6, (June 1998), pp. 2035-2039, PMID: 9598513

Hansen CM, Frandsen TL, Brünner N & Binderup L.(1994) 1 alpha,25-Dihydroxyvitamin D3 inhibits the invasive potential of human breast cancer cells in vitro. Clinical & Experimental Metastasis, Vol. 12, No. 3, (May 1994), pp. 195-202, PMID: 819419

Holick MF. (2004) Sunlight and vitamin D for bone health and prevention of autoimmune diseases, cancers, and cardiovascular disease. *The American Journal of Clinical Nutrition*, Vol. 80, No. 6, Supplement, (December 2004), pp. 1678-8168, PMID: 15585788

Holick MF. (2007) Vitamin D deficiency. *The New England Journal of Medicine*, Vol. 357, No. 3, (July 2007), pp. 266-281, PMID:17634462

Holick MF & Chen TC. (2008) Vitamin D deficiency: a worldwide problem with health consequences. *The American Journal of Clinical Nutrition*, Vol. 87, No. 4, (April 2008), pp. 1080S-1086S, PMID: 18400738

Holick MF. (2010) Vitamin D: Evolutionary, Physiological and Health Perspectives. *Current Drugs Targets*, Vol. 12, No. 1, (January 2010), pp. 4-18, PMID: 20795941

Horst RL & Reinhardt TA. (1997) Vitamin D metabolism in *Vitamin D* Feldman D, pp. 13-31, Academic Press, ISBN 0-12-252685-6, San Diego

Hosseinpanah F, Pour SH, Heibatollahi M, Moghbel N, Asefzade S & Azizi F. (2010) The effects of air pollution on vitamin D status in healthy women: a cross sectional study. *BMC Public Health*, Vol. 10, (August 2010), pp. 519-525. PMID: 20799984

Hyppönen E, Läärä E, Reunanen A, Järvelin MR & Virtanen SM. (2001) Intake of vitamin D and risk of type 1 diabetes: a birth-cohort study. *Lancet*, Vol. 358, No. 9292, (November 2001), pp. 1500-1503, PMID: 11705562

Jackson RD, LaCroix AZ, Gass M, Wallace RB, Robbins J, Lewis CE, Bassford T, Beresford SA, Black HR, Blanchette P, Bonds DE, Brunner RL, Brzyski RG, Caan B, Cauley JA, Chlebowski RT, Cummings SR, Granek I, Hays J, Heiss G, Hendrix SL, Howard BV, Hsia J, Hubbell FA, Johnson KC, Judd H, Kotchen JM, Kuller LH, Langer RD, Lasser NL, Limacher MC, Ludlam S, Manson JE, Margolis KL, McGowan J, Ockene JK, O'Sullivan MJ, Phillips L, Prentice RL, Sarto GE, Stefanick ML, Van Horn L, Wactawski-Wende J, Whitlock E, Anderson GL, Assaf AR & Barad D; Women's Health Initiative Investigators. (2006) Calcium plus vitamin D supplementation and the risk of fractures. *The New England Journal of Medicine*, Vol. 354, No. 7, (February 2006), pp. 669-683, PMID: 16481635

Kamen DL. (2010) Vitamin D in Lupus. *Bulletin of the NYU Hospital for Joint Diseases*, Vol. 68, No. 3, (2010), pp. 218-22, PMID: 20969555

Kelman A & Lane NE. (2005) The management of secondary osteoporosis. *Best Practice and Research: Clinical Rheumatology*. Vol. 19, No. 6, (December 2005), pp.1021-1037, PMID: 16301195

Kendrick J, Targher G, Smits G & Chonchol M. 25-Hydroxyvitamin D deficiency is independently associated with cardiovascular disease in the Third National Health and Nutrition Examination Survey. (2009) *Atherosclerosis*, Vol. 205, No. 1, (July 2009), pp. 255-260, PMID: 19091317

Konety BR, Lavelle JP, Pirtskalaishvili G, Dhir R, Meyers SA, Nguyen TS, Hershberger P, Shurin MR, Johnson CS, Trump DL, Zeidel ML & Getzenberg RH.(2001) Effects of vitamin D (calcitriol) on transitional cell carcinoma of the bladder in vitro and in vivo. *The Journal of Urology*, Vol 165, No. 1, (January 2001), pp. 253-258, PMID: 11125420

Krause R, Bühring M, Hopfenmüller W, Holick MF& Sharma AM.(1998) Ultraviolet B and blood pressure. *Lancet*, Vol. 352, No. 9129, (August 1998), pp. 709-710, PMID: 9728997

LaCroix AZ, Kotchen J, Anderson G, Brzyski R, Cauley JA, Cummings SR, Gass M, Johnson KC, Ko M, Larson J, Manson JE, Stefanick ML & Wactawski-Wende J.(2009) Calcium plus vitamin D supplementation and mortality in postmenopausal women: the Women's Health Initiative calcium-vitamin D randomized controlled trial. *Journal of Gerontology: Biological Sciences*, Vol. 64, No. 5, (May 2009), pp. 559-567, PMID: 19221190

Larsen ER, Mosekilde L & Foldspang A. (2004) Vitamin D and calcium supplementation prevents osteoporotic fractures in elderly community dwelling residents: a pragmatic population-based 3-year intervention study. *Journal of Bone and Mineral Research,* Vol. 19, No. 3, (March 2004), pp. 370-378, PMID: 15040824

Levine MA (2009) Common bone and mineral disorders of childchood, in: *Manual of Endocrinology and metabolism,* N Lavin (Fourth Edition), 381-413, Lippincott Williams & Wilkins, a Wolters Kluwer business, ISBN-13: 978-0-7817-6886-3, Baltimore USA

Li YC, Kong J, Wei M, Chen ZF, Liu SQ & Cao LP. (2002) 1,25-Dihydroxyvitamin D(3) is a negative endocrine regulator of the renin-angiotensin system. *The Journal of Clinical Investigation,* Vol. 110, No. 2, (July 2002), pp. 229-38, PMID: 12122115

Lips P. (2010) Worldwide status of vitamin D nutrition. *The Journal of Steroid Biochemistry and Molecular Biology,* Vol. 121, No. 1-2, (July 2010), pp. 297-300, PMID: 20197091

Liu M, Lee MH, Cohen M, Bommakanti M & Freedman LP. (1996) Transcriptional activation of the Cdk inhibitor p21 by vitamin D3 leads to the induced differentiation of the myelomonocytic cell line U937. *Genes & Development,* Vol. 10, No. 2, (January 1996), pp. 142-153, PMID: 856674

Looker AC, Pfeiffer CM, Lacher DA, Schleicher RL, Picciano MF & Yetley EA.(2009) Serum 25-hydroxyvitamin D status of the US population: 1988-1994 compared with 2000-2004. *The American Journal of Clinical Nutrition,* Vol. 88, No. 6, (December 2009), pp. 1519-1527, PMID: 19064511

Mahon P, Harvey N, Crozier S, Inskip H, Robinson S, Arden N, Swaminathan R, Cooper C & Godfrey K; SWS Study Group. (2010) Low maternal vitamin D status and fetal bone development: cohort study. *Journal of Bone and Mineral Research,* Vol. 25, No. 1, (January 2010), pp. 14-19, PMID: 19580464

Manicourt DH & Devogelaer JP. (2008) Urban tropospheric ozone increases the prevalence of vitamin D deficiency among Belgian postmenopausal women with outdoor activities during summer. *The Journal of Clinical Endocrinology & Metabolism,* Vol. 93, No. 10, (October 2008), pp. 3893-3899, PMID: 18628525

Marques CD, Dantas AT, Fragoso TS & Duarte AL. (2010) The importance of vitamin D levels in autoimmune diseases. *Revista Brasileira de Reumatologia,* Vol 50, No. 1, (February 2010), pp.67-80, PMID: 21125142

Mattner F, Smiroldo S, Galbiati F, Muller M, Di Lucia P, Poliani PL, Martino G, Panina-Bordignon P & Adorini L. (2000) Inhibition of Th1 development and treatment of chronic-relapsing experimental allergic encephalomyelitis by a non-hypercalcemic analogue of 1,25-dihydroxyvitamin D(3). *European Journal of Immunology,* Vol. 30, No. 2, (February 2000), pp. 498-508, PMID: 10671205

Meier C, Woitge HW, Witte K, Lemmer B & Seibel MJ.(2004) Supplementation with oral vitamin D3 and calcium during winter prevents seasonal bone loss: a randomized controlled open-label prospective trial. *Journal of Bone and Mineral Research,* Vol. 19, No. 8, (August 2004), pp.1221-1230, PMID: 15231008

Melamed ML, Muntner P, Michos ED, Uribarri J, Weber C, Sharma J & Raggi P. (2008) Serum 25-hydroxyvitamin D levels and the prevalence of peripheral arterial disease: results from NHANES 2001 to 2004. *Arteriosclerosis Thrombosis and Vascular Biology,* Vol. 28, No. 6, (June 2008), pp. 1179-1185, PMID: 18417640

Merewood A, Mehta SD, Grossman X, Chen TC, Mathieu JS, Holick MF & Bauchner H. (2010) Widespread vitamin D deficiency in urban Massachusetts newborns and their mothers. *Pediatrics*, Vol. 125, No. 4, (April 2010), pp. 640-647, PMID: 20308219

Merlino LA, Curtis J, Mikuls TR, Cerhan JR, Criswell LA & Saag KG; Iowa Women's Health Study. (2004) Vitamin D intake is inversely associated with rheumatoid arthritis: results from the Iowa Women's Health Study. *Arthritis and Rheumatism*, Vol. 50, No. 1, (January 2004), pp. 72-77, PMID: 14730601

Moreno J, Krishnan AV, Peehl DM & Feldman D. (2006) Mechanisms of vitamin D-mediated growth inhibition in prostate cancer cells: inhibition of the prostaglandin pathway.*Anticancer Research*, Vol. 26, No. 4A, (July-August 2006), pp. 2525-2530, PMID: 16886660

Munger KL, Zhang SM, O'Reilly E, Hernán MA, Olek MJ, Willett WC & Ascherio A.(2004) Vitamin D intake and incidence of multiple sclerosis. *Neurology*, Vol. 62, No. 1, (January 2004), pp.60-65, PMID: 14718698

Nield LS, Mahajan P, Joshi A & Kamat D. (2006) Rickets: not a disease of the past. *American Family Physician*, Vol. 74, No. 4, (August 2006), pp. 619-626, PMID: 16939184

Ovesen L, Andersen R & Jakobsen J. (2003) Geographical differences in vitamin D status, with particular reference to European countries. *The Proceedings of Nutrition Society*, Vol. 62, No. 4, (November 2003), pp. 813-821, PMID: 15018480

Parker J, Hashmi O, Dutton D, Mavrodaris A, Stranges S, Kandala NB, Clarke A & Franco OH. (2010) Levels of vitamin D and cardiometabolic disorders: systematic review and meta-analysis, *Maturitas*, Vol. 65, No. 3, (March 2010), pp. 225-236, PMID: 20031348

Pilz S, März W, Wellnitz B, Seelhorst U, Fahrleitner-Pammer A, Dimai HP, Boehm BO & Dobnig H (2008) Association of vitamin D deficiency with heart failure and sudden cardiac death in a large cross-sectional study of patients referred for coronary angiography. *Journal of Clinical Endocrinology and Metabolism*, Vol. 93, No. 10, (October 2008), pp. 3927-39-35, PMID: 18682515

Polar MK, Gennings C, Park M, Gupta MS & Gewirtz DA. (2003)Effect of the vitamin D3 analog ILX 23-7553 on apoptosis and sensitivity to fractionated radiation in breast tumor cells and normal human fibroblasts. *Cancer Chemotherapy and Pharmacology*, Vol. 51, No. 5, (May 2003), pp. 415-421, PMID: 12690516

Pope CA 3rd, Burnett RT, Thun MJ, Calle EE, Krewski D, Ito K & Thurston GD. (2002) Lung cancer, cardiopulmonary mortality, and long-term exposure to fine particulate air pollution. *JAMA: the journal of American medical Association*, Vol. 287, No. 9, (March 2002), pp. 1132-1141, PMID: 11879110

Pope CA 3rd, Burnett RT, Thurston GD, Thun MJ, Calle EE, Krewski D & Godleski JJ.(2004) Cardiovascular mortality and long-term exposure to particulate air pollution: epidemiological evidence of general pathophysiological pathways of disease. *Circulation*, Vol. 109, No. 1, (January 2004), pp. 71-77, PMID: 1467614

Porojnicu A, Robsahm TE, Berg JP & Moan J. (2007) Season of diagnosis is a predictor of cancer survival. Sun-induced vitamin D may be involved: a possible role of sun-induced Vitamin D. *The Journal of Steroid Biochemistry and Molecular Biology*, Vol. 103, No. 3-5, (March 2007), pp. 675-678, PMID: 17229569

Prentice A, Jarjou LM, Goldberg GR, Bennett J, Cole TJ & Schoenmakers I. (2009) Maternal plasma 25-hydroxyvitamin D concentration and birthweight, growth and bone mineral accretion of Gambian infants. *Acta Paediatrica*, Vol. 98, No. 8, (August 2009), pp. 1360-1362, PMID: 1959447

Reddy Vanga S, Good M, Howard PA & Vacek JL. (2010) Role of vitamin D in cardiovascular health. *American Journal of Cardiology*, Vol. 106, No. 6, (September 2010), pp. 798-805, PMID: 2081612

Ritz SA. (2010) Air pollution as a potential contributor to the 'epidemic' of autoimmune disease. Medical Hypotheses, Vol. 74, No. 1, (January 2010), pp.110-117, PMID: 19665849

Robsahm TE, Tretli S, Dahlback A & Moan J. (2004) Vitamin D3 from sunlight may improve the prognosis of breast-, colon- and prostate cancer (Norway). Cancer Causes & Control, Vol. 15, No. 2 (March 2004), pp. 149-158. PMID: 15017127

Rockell JE, Skeaff CM, Williams SM & Green TJ. (2006) Serum 25-hydroxyvitamin D concentrations of New Zealanders aged 15 years and older. *Osteoporosis International*, Vol. 17, No. 9, (July 2006), pp. 1382-1389, PMID:16832715

Rosen CJ. (2011) Clinical practice. Vitamin D insufficiency. *The New England Journal of Medicine*, Vol. 364, No. 3, (January 2011), pp.248-254, PMID: 21247315

Sachan A, Gupta R, Das V, Agarwal A, Awasthi PK & Bhatia V. (2005) High prevalence of vitamin D deficiency among pregnant women and their newborns in northern India. *The American Journal of Clinical Nutrition*, Vol. 81, No. 5, (May 2005), pp. 1060-1064, PMID:15883429

Scragg R, Sowers M & Bell C.(2007) Serum 25-hydroxyvitamin D, ethnicity, and blood pressure in the Third National Health and Nutrition Examination Survey. *American Journal of Hypertension*, Vol. 20, No. 7, (July 2007), pp. 713-719, PMID: 17586404

Spina CS, Tangpricha V, Uskokovic M, Adorinic L, Maehr H & Holick MF. (2006) Vitamin D and cancer. *Anticancer Research*, Vol. 26, No. 4A, (July-August 2006), pp. 2515-2524, PMID: 16886659

Tangpricha V, Spina C, Yao M, Chen TC, Wolfe MM & Holick MF. (2005) Vitamin D deficiency enhances the growth of MC-26 colon cancer xenografts in Balb/c mice. *The Journal of Nutrition*, Vol. 135, No. 10, (October 2005), pp. 2350-2354, PMID: 1617719

Tarcin O, Yavuz DG, Ozben B, Telli A, Ogunc AV, Yuksel M, Toprak A, Yazici D, Sancak S, Deyneli O & Akalin S. (2009) Effect of vitamin D deficiency and replacement on endothelial function in asymptomatic subjects. *The Journal of Clinical Endocrinology and Metabolism*, Vol. 94, No. 10, (October 2009), pp. 4023-4030, PMID: 19584181

Toner CD, Davis CD & Milner JA. (2010) The vitamin D and cancer conundrum: aiming at a moving target. *Journal of American Dietetic Association*, Vol. 110, No. 10, (October 2010), pp. 1492-1500. PMID: 20869488

Walters MR. (1992) Newly identified actions of the vitamin D endocrine system. *Endocrine Reviews*, Vol. 13, No. 4, (November 1992), pp. 719-764. PMID: 1333949

Webb AR. (2006) Who, what, where and when-influences on cutaneous vitamin D synthesis. *Progress in Biophysics and Molecular Biology*, Vol. 92, No. 1, (September 2006), pp. 17-25. PMID: 16766240

Welsh J. (2004) Vitamin D and breast cancer: insights from animal models. *The American Journal of Clinical Nutrition*, Vol. 80, No. 6 (Supplement), pp. 1721S-1724S, PMID: 15585794

WHO Scientific Group on the Prevention and Management of Osteoporosis.(2003) Prevention and management of osteoporosis: report of WHO scientific group. *World Health Organ Technical Report Series,* Vol. 921, (2003), pp. 1-164, PMID: 15293701

van der Wielen RP, Löwik MR, van den Berg H, de Groot LC, Haller J, Moreiras O & van Staveren WA. (1995) Serum vitamin D concentrations among elderly people in Europe. *Lancet.,* Vol. 346, No. 8969, (July 1995), pp. 207-210, PMID: 7616799

Willheim M, Thien R, Schrattbauer K, Bajna E, Holub M, Gruber R, Baier K, Pietschmann P, Reinisch W, Scheiner O & Peterlik M. (1999) Regulatory effects of 1alpha,25-dihydroxyvitamin D3 on the cytokine production of human peripheral blood lymphocytes. *The Journal of Clinical Endocrinology and Metabolism,* Vol. 84, No. 10, (October 1999), pp. 3739-3744, PMID: 10523023

Witham MD, Nadir MA &Struthers AD. (2009) Effect of vitamin D on blood pressure: a systematic review and meta-analysis.Journal of Hypertension, Vo. 27, No. 10, (October 2009), pp. 1948-1954, PMID: 19587609

Wharton B & Bishop N. (2003) Rickets. *Lancet,* Vol. 362, No. 9393, (October 2003), pp.1389-1400, PMID: 14585642

Xu HM, Tepper CG, Jones JB, Fernandez CE & Studzinski GP.(1993) 1,25-Dihydroxyvitamin D3 protects HL60 cells against apoptosis but down-regulates the expression of the bcl-2 gene. *Experimental Cell Research,* Vol. 209, No. 2, (December 1993), pp. 367-374, PMID: 8262155

Zhang R & Naughton DP. (2010) Vitamin D in health and disease: current perspectives. *Nutrition Journal,* Vol. 9, No. 65, (December 2010), pp. 1-13, PMID: 21143872

Zhuang SH & Burnstein KL (1998) Antiproliferative effect of 1alpha,25-dihydroxyvitamin D3 in human prostate cancer cell line LNCaP involves reduction of cyclin-dependent kinase 2 activity and persistent G1 accumulation. *Endocrinology,* Vol. 139, No. 3, (March 1998), pp. 1197-1207, PMID: 949205

Zittermann A, Schleithoff SS, Götting C, Dronow O, Fuchs U, Kuhn J, Kleesiek K, Tenderich G, & Koerfer R (2008) .Poor outcome in end-stage heart failure patients with low circulating calcitriol levels. *European Journal of Heart Failure,* Vol. 10, No. 3, (March 2008), pp. 321-327, PMID: 18304873

3

Air Pollution and Its Effects in the Respiratory System

Fortoul, T.I. et al.*
Departamento de Biologia Celular y Tisular, Facultad de Medicina
Universidad Nacional Autonoma de Mexico (UNAM)
Mexico

1. Introduction

1.1 Respiratory system structure and normal function

The respiratory system is the gateway for 15,000 liters of air that enter through the nostrils, and after its appropriate conditioning, arrives to the delicate net structured by the lung parenchyma. At this site gas diffusion takes place, and oxygen diffuses through the Pneumocyte I cytoplasm to reach the erythrocyte's hemoglobin, passing by the endothelial cell.

To make this possible, a well-structured conducting system leads the air (trachea, bronchi, bronchioles, alveoli) from the nostrils to the alveoli. A specialized epithelium blankets the tubes, with some local modifications (Fortoul et al., 2010). This epithelium is constituted by a variety of cells with different functions (Figure 1).

In the submucose under the bronchi, mucous and serous glands are located, and liberate its secretion to the surface of the epithelium (Figure 2). These glands, as well as the goblet cells, produce mucus that in normal situations is Alcian blue/ PAS+ (Davis & Dicker, 2008). The biochemistry of the mucus, produced by the goblet or the submucosal glands is important because they determinate their viscosity, pH, and charge, as well as its stain affinity (Figure 3) (Rose & Voynow, 2006).

With the help of intercellular junctions, this tubes system separates spaces between the epithelium, which is known as interstitial space and it houses connective tissue, fibroblast, lymphocytes, macrophages, and other cells that may migrate from the capillaries.

Along these tubes the air that enters through the nostrils becomes moistened, filtered and temperated, to arrive into the alveoli for gas diffusion. A change in the composition and function of the epithelial components will be associated with a variety of diseases which are associated to air pollutants exposure (Figure 4) (Mussali-Galante & Fortoul, 2008).

---

* Rojas-Lemus, M.[1], Rodriguez-Lara V.[1], Cano-Gutierrez, G.[1], Gonzalez-Villalva, A.[1], Ustarroz-Cano, M.[1], Garcia-Pelaez, I.[1], Lopez-Valdez, N.[1], Falcon-Rodriguez C.I.[1], Silva-Martinez, J.[1], Gonzalez-Rendon, E.S.[1], Montaño, L.F.[1], Cano-Gutierrez, B.[1], Bizarro-Nevares P.[1], Colin Barenque L.[2]
[1]*Departamento de Biologia Celular y Tisular, Facultad de Medicina, Universidad Nacional Autonoma de Mexico (UNAM), Mexico*
[2]*Departamento de Neurociencias, Iztacala, UNAM, Mexico*

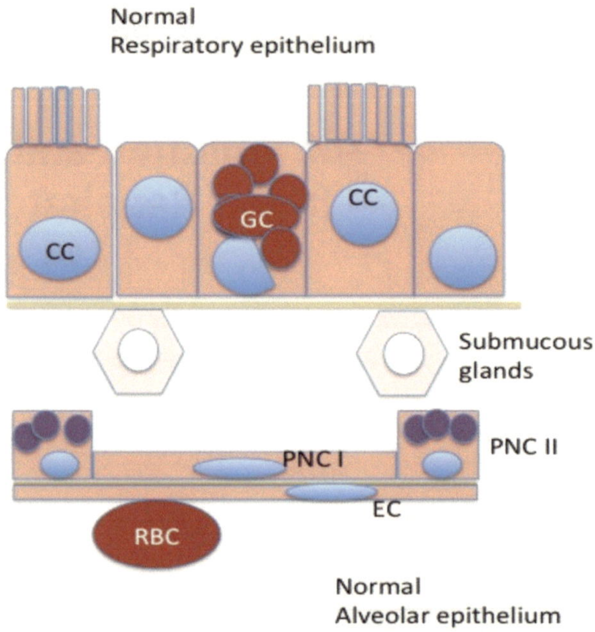

Fig. 1. Respiratory system epithelial cells. Ciliated Cell (CC), Goblet Cell (GC), Pneumocyte I (PNCI), Pneumocyte II (PNCII), Red Blood Cell (RBC), Endothelial Cell (EC).

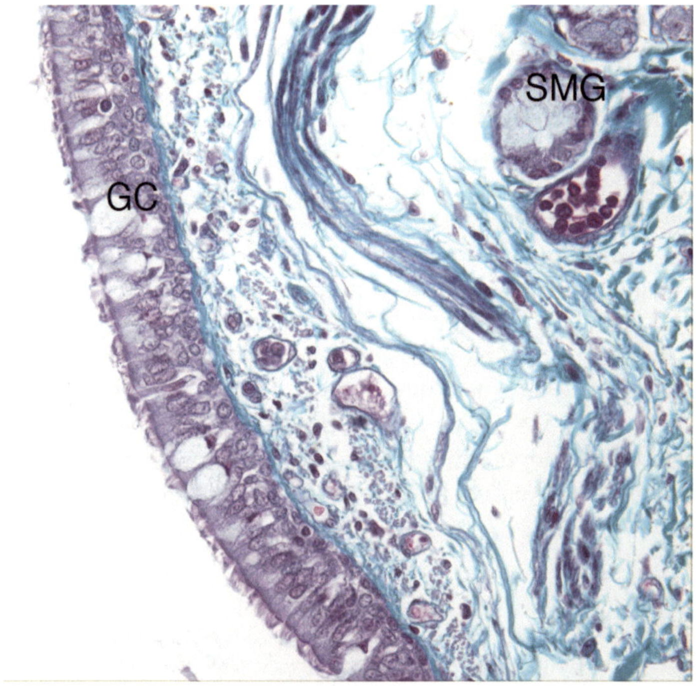

Fig. 2. Bronchial epithelium with Ciliated and Goblet Cells (GC). Submucose glands (SMC) with serous demilunes are also observed.

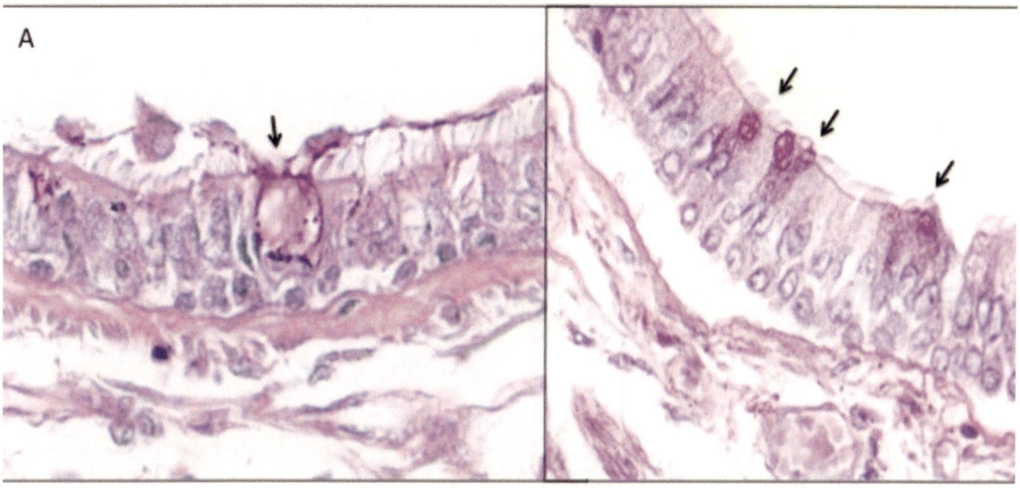

Fig. 3. Respiratory epithelium stained with PAS Schiff. (A) In normal situations, few PAS+ cells are observed (arrow). (B) In disease an increase in these cells is observed (arrows).

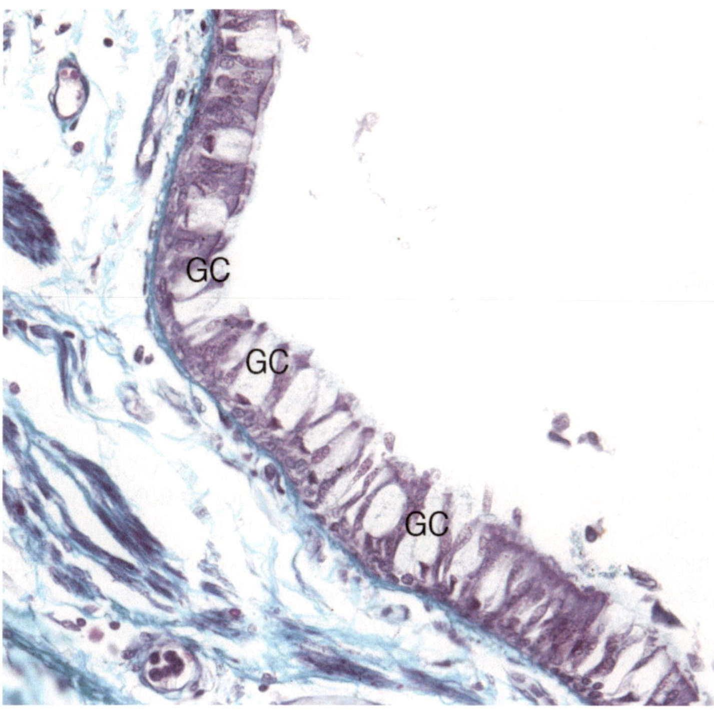

Fig. 4. The increase in the number of Goblet cells is a response of the respiratory epithelium to atmospheric aggressions.

1.2 Diseases associated with air pollutants (inflammation, fibrosis, COPD, asthma, cancer, immunologic modifications)

Air pollutants, described latter in this chapter, induce different reactions and diseases. Inflammation as a consequence of the exposure to irritants such as Ozone, NOx, or particulate matter has been described (Yang & Omaye, 2009). Also, fibrosis as a consequence of the exposure to gases is explored with more detail in a next section (Figure 5).

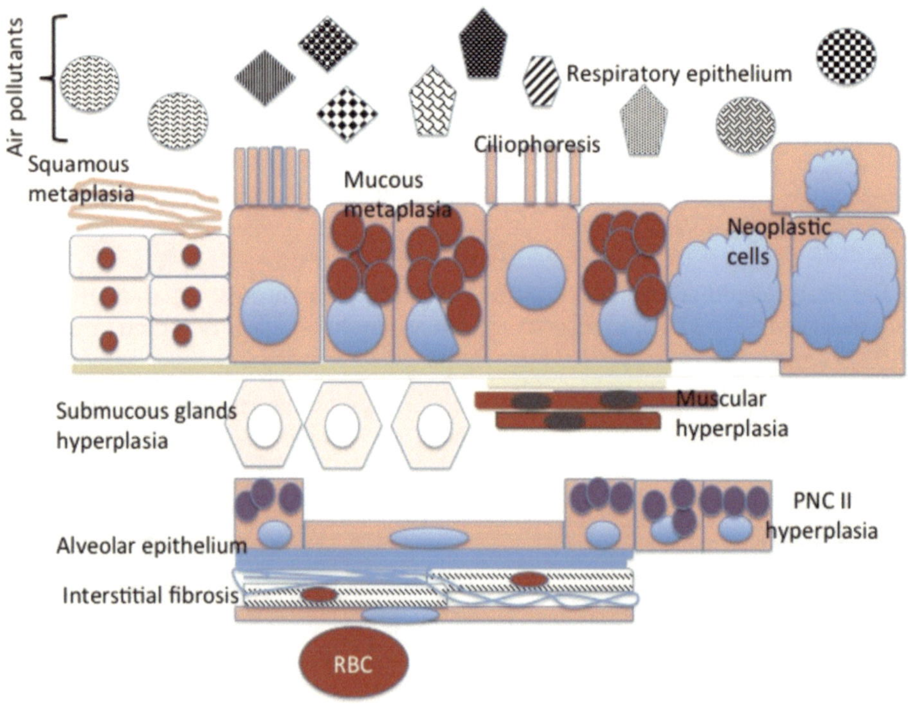

Fig. 5. Schematic representation of the changes that may be found in the respiratory epithelium, after its exposure to air pollutants.

Chronic Obstructive Pulmonary Diseases (COPD) included asthma, increases its frequency in cities with high atmospheric pollution (Olivieri & Scoditti, 2005). The epithelium in these cases may increase the release of IL-6, TNFα, PGE_2, PDGF, TGFβ, VEGF, and a variety of chemokines, and other mediators capable of inducing proliferation of fibroblasts and mucus production (Holgate, 2008).

In addition other components of air pollutants, such as metals, Volatile Organic Compounds (VOCs) are carcinogenic; other pathologies are associated with the exposure (Yang & Omaye, 2009).

In the next sections the different responses of the lung to air pollutants will be explored.

2. Air pollutants

Air Pollutants are natural constituents of the air. Animals produce carbon dioxide as the end result of respiration, volcanic action produces sulfur oxides, and wind movement ensures the presence of suspended particulates. Pollutants are part of our everyday life, and it is difficult to remove them from the respirable air. However, man has caused a severe imbalance, in the natural mechanisms for atmosphere clearance, increasing the discharges of pollutants in the atmosphere, resulting in severe effects on human health (Atash, 2007).

2.1 Classification of air pollutants

Atmospheric pollutants have been classified according to their source; chemical composition, size and release form into indoor or outdoor environments (Bernstein et al., 2004):

A. Primary - secondary pollutants
   1. Primary: pollutants emitted directly into the atmosphere
   2. Secondary: pollutants that form in the air as a result of chemical reactions with other pollutants and gases.

B. Indoor - outdoor pollutants
   1. Indoor pollutants
      1.1 Sources: cooking and combustion, particle resuspension, building materials, air condition, consumer products, smoking, heating, biologic agents
      1.2 Products: combustion products, CO, CO_2, Specific volatile organic compounds, microbial agents and organic dusts, radon, manmade vitreous fibers
   2. Outdoor pollutants
      2.1 Sources: industrial, commercial, mobile, urban, regional, agricultural, natural.
      2.2 Products: SO_2, ozone, NO_x, CO, PM, Specific volatile organic compounds

C. Gaseous - particle pollutants
   1. Gaseous: SO_2, NO_x, ozone, CO, Specific volatile organic compounds
   2. Particle: coarse PM (2.5 - 10 µm; regulatory standard = PM_{10}), fine PM (0.1 - 2.5 µm regulatory standard = $PM_{2.5}$); ultrafine PM (<0.1 µm; no regulated).

2.2 Gases

Problems in air pollution were associated with high concentrations of sulfur dioxide (SO_2) in 20th Century. Through controlled exposure, human studies to SO_2 (0.25 ppm) for only 5 minutes a rapid bronchoconstriction, in both healthy and asthmatic subjects was described. In patients exposed to inhalation of SO_2 a relationship with TNF-α promoter polymorphism was identified, which is know to be associated also with asthma (Winterton et al., 2001; Bernstein et al., 2004).

Ozone is formed in the troposphere through a complex series of reactions involving the action of sunlight on nitrogen dioxide and hydrocarbons. The global concentration of O_3 has increased due to an increase in nitric oxide emissions associated with the switch to fossil fuels during the industrial period. Nitric oxide is rapidly transformed into nitrogen dioxide by atmospheric oxidants such as ozone (Finlayson & Pitts, 1997). Exposure to ozone causes a decreased in forced vital capacity and FEV_1 associated with chest discomfort on inspiration and increased nonspecific airway hyper hyperresponsiveness (Bernstein et al., 2004).

NO_2 is emitted directly into the atmosphere by combustion processes; however, the main source is the oxidation of NO by reactive species. Once NO is converted to NO_2, a variety of reactions can generate nitrate radical (NO_3), dinitrogen pentoxide (N_2O_5) (Finlayson & Pitts, 1997). NO_2 exposure (2 - 6 ppm) induces an inflammatory response in the airways characterized by neutrophil influx and reduced lymphocyte subpopulations, also it might play a more prominent role as sensitizing agent to inhaled allergens (Strand et al., 1997).

Carbon monoxide (CO) is produced by the incomplete combustion of carbon or carbon compounds. This gas can be bound to hemoglobin, forming carboxy hemoglobin and immobilizing hemoglobin function. Another compound of carbon, carbon dioxide is needed in plants' life cycle, in the reaction of photosynthesis. But its rise in the atmosphere in the presence of other gases such as methane and chlorofluorocarbons, is the cause of the greenhouse effect, (Raub et al., 2000).

2.3 Suspended particles

Particle air pollution is a mixture of solid, liquid or solid and liquid particles suspended in the air. The size of suspended particles varies, from a few nm to tens of µm. The PM 10 (thoracic) are particles smaller than 10 µm in diameter that can penetrate into lower respiratory tract; PM 2.5 (respirable) particles smaller than 2.5 µm that can penetrate into gas-exchange region of the lung, and ultrafine particles smaller than 100 nm which have a limited contribution to particle mass, but which in terms of numbers are more abundant than the other sizes, and offer a very large surface area, with increased degrees of lung penetration.

Major natural sources of particles include organic material terrestrial dust caused by winds, sea spray, biogenic emissions, volcanic eruptions and wild fires; the contamination through anthropogenic or technogenic is produced by combustion, industrial waste, nuclear energy, anthropogenic fire, and burning of household waste (Finlayson-Pitts & Pits, 2000). Metal particles (mercury, cadmium, nickel and lead) are part of PM2.5.

3. Lung development modifications by air pollutants

The respiratory system begins its development at the 4th week of pregnancy, in the fetal larynx as a respiratory *primordium*. This structure is covered by endoderm that will differentiate into the respiratory epithelium, and the respiratory glands.

Other structures will be developed from the splanchnic mesoderm that surrounds the endoderm.

The regulation of the respiratory system morphogenesis is coordinated by different structures, in different time periods. It has been reported the expression of molecules such as Fibroblast Growth Factor (FGF), sonic hedgehog, Bone Morphogenic Protein (BPM), retinoic acid and Wnt signaling pathways, as well as various transcription factors as part of this regulatory network (Cardoso & Lü, 2006).

At the end of the 7-week, the lungs are already developed, but its maturation is extended during the pre and postnatal periods. 80% of the alveolar tissue ends its development until the end of adolescence. This large maturation time exposes the lung to suffer damage and modify its development; this also increases the risk for developing pulmonary diseases early in adulthood such as restriction or a decrease in lung function (Wang & Pinkerton, 2008; Rojas-Martinez, et al., 2007).

Prenatal exposure to Environmental Tobacco Smoke (ETS) has been associated with an increased risk of asthma, and during the postnatal period, the exposure increases asthma exacerbations, as well as the risk for respiratory infections (Wang & Pinkerton, 2008). Studies in mice indicate that prenatal exposure to ETS increases allergic responses in postnatal period (Penn et al., 2007; Gern, 2010).

The exposure to outdoor pollutants such as Particulate Matter (PM), carbon monoxide (CO), sulphur (SO_2), nitrogen (NOx), and ozone, decreases lung function during childhood

(Mortimer et al., 2008). Also O_3 exposures increase hyper-reactivity along with an increase in TNF-**a**, IL-1**b**, KC, IL-6, and MCP-1 with non-visible structural lung modifications (Auten et al., 2009).

4. Xenobiotic metabolism

4.1 Lung metabolic active cells

The lung is one of the main sites for xenobiotics metabolism and in some cases, biotransformation (Castell et al., 2005). Biotransformation is the process by which cells modify xenobiotics, with the ultimate goal of facilitating the elimination of lipophilic substances. These reactions are classified in three phases. Phase I: Enzymes encoded by cytochrome P450 (hemoproteins actively involved in the biotransformation of xenobiotics). Phase II: Reactions tend to render more water-soluble products and less active metabolites, and finally, Phase III: Elimination (Choudhary et al., 2005).

More than 40 different types of cells have been described in the lung (Pavek & Dvorak, 2008), and some are metabolically active. Immunohistochemical analysis has evidenced the presence of cytochrome P450 (Cytochrome Proteins) (CYP) in lung cells such as: macrophages (Pavek & Dvorak, 2008), endothelium, alveolar cells types I and II, ciliated cells (Castell et al., 2005) and Non-Ciliated Bronchiolar Cell or Clara cell. Clara cells are the leading cells for xenobiotic metabolism in the lung because its profuse cytochrome P450 mono oxygenase activity (Katavolos et al., 2009).

4.2 CYPs and lung metabolism (CYPs and cell metabolism)

CYP families and subfamilies are responsible of the oxidative metabolism of the majority of xenobiotics such as: drugs, environmental pollutants and carcinogens. The main families are: CYP1, CYP2 and CYP3 and comprise about half of the total CYPs. The subfamilies are classifieds according to the degree of nucleonic and amino acid sequence homology (Castell et al., 2005). All CYPs are localized in the cell's smooth endoplasmic reticulum. CYP 450 enzymes act as mono oxygenases, and use one atom of molecular oxygen to oxidase xenobiotics, requiring the aid of NADPH-cytochrome P450 reductase, to provide the electrons required for the reduction of the second oxygen atom to H_2O_2 (Ioannides, 2008). The lung contains a variety of subfamilies such as: CYP1A1, CYP1B1, CYP2A6, CYP2B6, CYP2E1, CYP3A5 (Castell et al., 2005), CYP2B1, CYP3A1 (Pons et al., 2000), CYP2F1, CYP2F2 (Carlson, 2008) and CYP2S1 (Deb & Bandiera, 2010).

4.3 Air pollution modification of CYPs

Air pollution increases or decreases the CYPs amount in the lung. Exposure to environmental factors such as: dioxins as 2,3,7,8-Tetrachlorodibenzo-p-dioxin (TCDD), polycyclic aromatic hydrocarbons (PAH) as benzo [a] pyrene (BaP) and tobacco smoke (Chang et al., 2006), enhances CYP1A1 activity and has been reported to be a prognostic factor for lung cancer development. (Ioannides, 2008). CYP1B1 and CYP2S1 also increase its activity after the exposure to PAH and BaP (Deb & Bandiera, 2010), also CYP1A1 and CYP1B1 are localized in the Clara cell (Chang et al., 2006). PAH and BaP also bind to arylhidrocarbonyl receptor (AhR) that translocates to the nucleus and acts as a transcription factor that binds to a specific DNA recognition sequence, termed the xenobiotic responsive element (XRE) (Anwar-Mohamed et al., 2009) and all CYPs can generate ROS during NADPH-dependent CYP catalysis (Ioannides, 2008) (Figure 6).

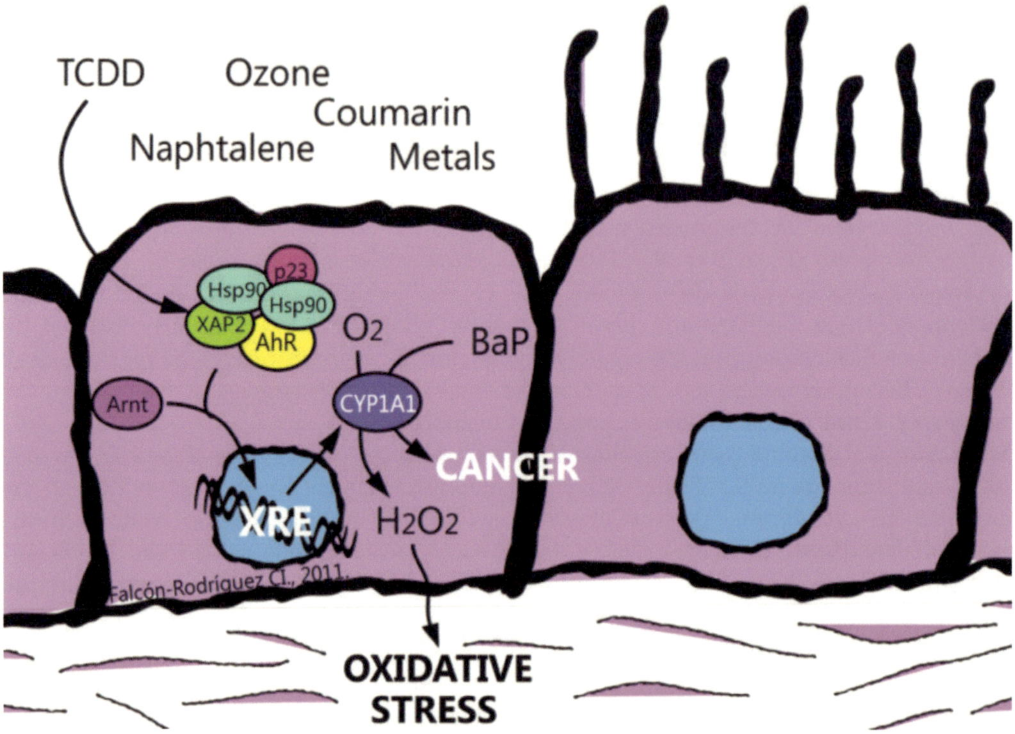

Fig. 6. Molecular mechanism of CYP1A1 activation by AhR signal transduction pathway in Clara cell (Bronchiolar epithelium. A ligand enters to the cell (TCDD) and binds to the cytosolic complex of AhR, chaperones hsp90, co-chaperone p23. Ligand-AhR complex translocates into the nucleus. AhR-ARNT heterodimer then binds to xenobiotic response element (XRE).

Other hydrocarbons as toluene, occurs naturally in crude oil and in the tolu tree. It is also produced in gasoline process, fuels from crude oil, making coke from coal and it is used in paints and paint thinners. Toluene exposure enhances CYP2B1 activity in rat lung (Pons et al., 2000). Naphthalene, derived from petroleum, coal, and tobacco smoke (Morris & Buckpitt, 2009), increases CYP2F2 activity. In mouse lung CYP2F2 the metabolism rate was 107 nmol/min/nmol P450, whereas for human lung CYP2F1 was 0.045 nmol/min/nmol P450 (Carlson, 2008). Other products such as coumarin, a natural product used widely as a fragrance ingredient, and also been used clinically at high dosages in humans for the treatment of high-protein lymphedemas, and as an antineoplastic agent in the treatment of renal cell carcinoma and malignant melanoma. The Coumarin-Epoxidation increases the expression of CYP2F2, and shows a specific metabolism in mouse Clara cell (Born et al., 2002). Furthermore, Clara cell does not contain CYP2E1, but, when rats are exposed to ozone, the CYP2E1 is induced in the bronchiolar epithelium (Ioannides, 2008). In general Particulate Matter induces gene expression of CYP1A1 and CYP2E1 (Abbas et al., 2009). Finally, metals are components of Total Suspended Particles (PST). Scarce information is available on the metal effects on lung CYPs, although, demonstrated recently that V^{5+} was able to decrease the TCDD-mediated induction of CYP1A1 mRNA, protein and catalytic activity (Anwar-Mohamed et al., 2009).

5. Air pollutant effects on respiratory system

There are changes associated with atmospheric pollutants that are evidenced with different patterns and diseases. Changes in the local lymphoid tissue, carcinogenic and genotoxic changes, fibrosis, COPD, are some of the outcomes observed after air pollution. Inflammation is core event in all the changes observed after the contact of the epithelial cells with the air pollutants, so it will be the first described in this section.

5.1 Inflammation and air pollution

The first suggestion of the consequences of air pollution date back to the serious events that occurred in Belgium in 1948 in the so called "killer fog" incident, and in the so called "London fog" that happened in the first week of December 1952 and provoked 4000 more deaths than those expected. After those episodes air pollution control policies were introduced since it became clear that air pollution was associated with an increase in morbidity and mortality in individuals with cardiovascular disease or with chronic obstructive pulmonary disease (Mills et al., 2009). The root of this increased mortality lay in the lung chronic inflammation that affects the pulmonary vascular endothelium, the thrombotic potential and the fibrinolitic balance in exposed individuals. These processes favor atherosclerotic plaque rupture, thrombosis and translocation of particulate matter to the blood, through the pulmonary capillaries; the latter affects directly the body vascular endothelium leading to the loss of endothelium integrity thus initiating a pulmonary and a systemic inflammatory reaction (Mills et al., 2006). A fraction of the translocated particulate matters accumulates in the liver, the spleen, the thymus and others (Nemmar et al., 2002).

Ambient particulate matter (PM) is a mixture of inhalable particles that are considered as serious contaminants. These particles come from the combustion of biomass fuels used for cooking and heating homes, emissions from internal combustion motors and industrial machinery, and forest fires (Torres-Duque et al., 2008). This matter has been grouped in coarse (2.5-10μm) and fine (2.5 μm or less) depending on their diameter.. Both types induce serious health consequences in exposed individuals. Both types of particles penetrate into the lung, however the coarse particles are more dangerous because of their mass (Hetland et al., 2005). Obviously there are many inherent conditions in the exposed individual that increase the risk to develop cardiovascular or pulmonary disease: alfa-1 antitrypsin deficiency, family history of chronic obstructive pulmonary disease or atherosclerosis, personal history of frequent upper respiratory tract infections, hypersensitivity to inhaled irritants, tobacco, asthma, and being female (Ekici et al., 2005). Also, we must consider the enhanced air pollution in overcrowded cities with serious traffic problems. These cities usually have a higher concentration of PMs in the air. An increase of 7 μg/m3 in the PM2.5-10 concentration is associated with a 5% decrease in FEV1 and an odds ratio of 1.33 for chronic obstructive pulmonary disease in women compared with men (Schikowski et al., 2005).

Both types of particles recruit and activate neutrophils, but PM2.5-10 induces a higher proinflammatory activity (Wegesser & Last, 2009). There are important differences in the relative concentration and type of components (metals, organic compounds, ultrafine particles adhered to larger particles) (Donaldson et al., 2005) between coarse and fine particles that depend on the season and geographic site where the sample is collected (Seagrave et al., 2006).

Other constituents of ambient particulate matter are biological materials, especially b-glucans, fungi spores and endotoxins that derive from gram-negative bacteria (Schwarze et al., 2006). It has been shown that PM constituent responsible for the pro-inflammatory activity induced by PM2.5-10 are endotoxins and particularly a soluble fraction known as lipopolysaccharides (LPS) (Schins et al., 2004). The concentration of endotoxin in PM depends on the site and season, as we have already mentioned, but environmental humidity is associated with endotoxin concentration in particulate matter (Spaan et al., 2008). One example of the impact of biological components in PM is the heat shock protein HSP60 derived from *Chlamydia pneumoniae,* detected in PM2.5-10 that promotes lung inflammation and pulmonary dendritic cells activation through the innate immune response receptor TLR4 and the MyD88 pathway. The inflammation induced by HSP is secondary to an increase in the number of immune-related cells in the broncho alveolar lavage (BAL), enhanced recruitment of neutrophils, increased synthesis of IL-6 and over-expression of CD80 and CD86 in BAL dendritic cells (Bulut et al, 2009). HSP60 also activate pulmonary macrophages and endothelial cells through TLR4 in a MyD88-dependent pathway (Bulut, 2002). Interestingly, and independently from its endotoxin content, PM2.5 per se activate macrophages through TLR-2 y TLR-4 (Shoenfelt et al., 2009). Similarly, the respiratory epithelium, the first point of contact for inhaled foreign organism, also express TLR4 and secrete, upon activation, pro-inflammatory cytokines and chemokines that recruit neutrophils and T lymphocytes to the infection site (Parker & Prince, 2011).

Pulmonary inflammation induced by exposition to or inhalation of PM is closely related to particle processing by alveolar macrophages. Once these macrophages become activated a cascade of pro-inflammatory cytokines is initiated leading to endothelial damage. A recent study shows that in the presence of PM2.5-10 the amount of alveolar macrophages is duplicated and the amount of activated macrophages triplicates leading to persistent lung and systemic inflammation that were both associated with vascular endothelial dysfunction (Tamagawa et al., 2008). The increase in IL-6 serum concentration in PM2.5-10 exposed individuals was important during the first two weeks of exposure and was directly related to the amount of activated macrophages; afterwards, the concentration of IL-6 was similar in exposed individuals and controls. Alveolar macrophages activated with PMs secrete TNF-a, granulocyte-monocyte colony stimulating factor, and IL-1b (van Eeden et al., 2001). It is highly probable that these cytokines act not only locally but systemically thus generating a more organized inflammatory response that include the bone marrow. The acute exposure to PMs induces a rapid bone marrow response liberating leukocytes and platelets into the systemic circulation. As far as inflammatory cytokines in the pulmonary inflammation process, it has been shown that IL-6 inhibits directly the expression of eNOS thus diminishing nitric oxide production by the endothelium (Saura et al., 2006). Chemokine secretion is also altered in individuals exposed to particulate matter. Interleukin 8, a chemokine responsible for the recruitment and activation of pulmonary neutrophils in inflammation sites, binds to PMs (Seagrave, 2008). Activated neutrophils and macrophages secrete IL-8 and IL-1b.The excessive recruitment of neutrophils is clearly detected in BAL samples of individuals exposed to particulate matter.

A recent study analyzed the effect of daily changes in particulate matter air pollution upon the inflammatory cells response. The results showed that as the concentration of PM2.5-10 increases in the air the serum concentration of fibrinogen and the expression of E-selectin in exposed individuals increases whereas the concentration of prothrombin and von Willebrand antigen diminishes. The serum concentration of C reactive protein, coagulation factor VII, amyloid A and the soluble ICAM-1 fraction was not modified (Hildebrandt et al., 2009). Fibrinogen is an acute phase protein and a coagulation factor synthetized by the liver in response to elevated IL-6 serum concentration (Gabay & Kushner, 1999). E-selectin reflects the activation of the vascular endothelium and it is well known that the increased expression of this adhesion molecule is associated with the serum concentration of inflammatory cytokines such as TNF-a, IFN-y and IL-6 (Rice & Bevilacqua, 1989) as well as the recruitment of leukocytes and T lymphocytes in the inflammation zone. The presence of neutrophils in an IL-6 enriched environment enhances the expression of E-selectin in endothelial cells but also its apoptotic death thus perpetuating the inflammatory damage (Barnes et al., 2011). It has also been shown that prolonged exposure to PM2.5 is associated with an important plasmatic increase in endothelin-1 concentration and in an increased pulmonary artery pressure (Calderon-Garcidueñas et al., 2007). The contribution of all these processes in the lung keeps the inflammatory process alive.

Heat inactivation of PM2.5-10 diminishes the expression of CD14, CD11b/CR3 y HLA-DR and the phagocytic activity of alveolar macrophages (Alexis et al., 2006) but its influence in neutrophil recruitment into the airways is still controversial. Exposures to viruses, increase the serum concentration of INF-g, previous to the inhalation of particulate matter induce a stronger pulmonary inflammation response that include oxidative damage to the lung. The result of these changes is a loss in the antibacterial function of neutrophils and alveolar macrophages and consequently, an increase in the local content of bacterial endotoxins (Sigaud et al., 2007) that perpetuates the pulmonary inflammation. Sigaud and coworkers (2007) also demonstrated that macrophages exposed to PMs have an enhanced expression of multiple inflammation-related genes: MCP-5, IL-9, IL-17B, IL-1b, MIP-1b, MIP-3b, IL-8R, C10, CCR-1, CCR-2 and MDC.

In summary, exposure to particulate matter induces excessive production of pro-inflammatory cytokines and chemokines by alveolar macrophages and lung dendritic cells, both of which are activated through TLR2 and TLR4 and the MyD88 signaling pathway. The excessive amount of these inflammatory molecules directly affect the pulmonary and systemic vascular endothelium by diminishing its capacity to regulate properly its vascular tone and permeability, triggering abnormal coagulation and fibrinolysis mechanisms, and increasing the adhesion of inflammatory cells to the vascular endothelium.

5.2 Bronchial associated lymphoid tissue (BALT) modifications

Bronchus-associated lymphoid tissue (BALT) is a constitutive mucosal lymphoid tissue adjacent to major airways. BALT is composed by B cells surrounded by a parafollicular region of T cells, dendritic (DCs) and macrophages. As the result of air pollution BALT can acquire antigens, allergens or contaminants from the airways, then complex interactions occur increasing its efficiency. For example, BALT can initiate local immune responses and the amount of BALT increases (Randall, 2010).

The presence of BALT in adult mammals depends on species, antigen stimulation and age. BALT is found in normal lungs of most healthy adult rabbits, rats, guinea pigs and old adult mice. In contrast, the presence and frequency of BALT in normal lungs of healthy adult humans is controversial (Kawamata et al., 2009).

In humans BALT is neither found at birth nor in healthy adults but transiently arises during childhood and adolescence. In both humans and mice, air pollution can induce BALT, data derives from splenectomized lymphotoxin α-deficient mice, which lack all secondary lymphoid organs but do develop BALT. This suggests that BALT can serve as induction sites for adaptive immune responses to contaminants. However, mechanisms that control the development and maintenance of BALT are largely unknown (Halle et al., 2009).

5.3 Carcinogenic and mutagenic effects

Lung cancer is the leading cause of cancer death worldwide. Smoking is the major risk factor for lung cancer. Although some subjects who have never smoked get lung cancer, smoking causes 9 out of 10 cases of this pathology. So there are other factors that promote the carcinogenic process. It has been reported that air pollution exposure may cause lung cancer. In a six US cities study, Dockery and coworkers (1993) found that the greatest effects were for lung cancer and cardiopulmonary disease, between the least and most polluted cities. Pope et al. (1995) found increased long-term effects on cardiopulmonary mortality and lung cancer in a 17-year follow-up (Naess et al., 2007). Also, Cao and coworkers (2011) analysis provides the first prospective evidence in China that air pollution (e.g., SO2) may contribute to the increased risk for lung cancer mortality. In Europe, the proportion of lung cancers attributable to urban air pollution is estimated to be 11% (Molina et al., 2008). Likewise, it have been reported that the type of lung cancer is related to air pollution, for example, in a Spain study, individuals living near industries displayed an excess risk of lung cancer (OR=1.49; 95%CI=0.93-2.39), which attained statistical significance for small cell carcinomas (OR=2.23; 95%CI=1.01-4.92), residents in urban areas showed a statistically significant increased risk for adenocarcinoma (OR=1.92; 95%CI=1.09-3.38). In Aviles a health area, no differences in risk was found (López-Cima et al., 2011).

Urban air, particularly in densely populated urban environments, contains inorganic particulates (arsenic, asbestos, chromium, cadmium, lead and nickel), radionuclides (210Pb, 212Pb and 222Rn), gaseous and particulate organic species (benzene, benzo[α]pyrene, 1,3-butadiene and benzene-soluble organics), oxidants such as ozone and sulfur and nitrogen oxides in particle form. These substances are present as components of complex mixtures proceeding basically from industries emissions, combustion of fossil fuels for power generation or transportation, and all are related with carcinogenesis process (Naess et al., 2007).

Particulate matter (PM), especially fine particles of less than 2.5 μm in diameter (PM2.5) is related with lung cancer (Perez-Padilla et al., 20010). Pope and coworkers (2002) reported that each 10-μg/m3 elevation in long-term average PM2.5 ambient concentrations was associated with approximately a 8% increased risk of lung cancer mortality, although the magnitude of the effect somewhat depended on the time frame of pollution monitoring. Particulate matter contains a lot of compounds that are considered human carcinogens like heavy metals such as cadmium, cobalt, chromium and nickel. Several epidemiological studies have clearly demonstrated that exposure to metals has toxic and carcinogenic affects in animals and humans; some of them have been demonstrated to be lung carcinogenesis promoters (Salnikow & Zhitkovich, 2008). Human cadmium exposure is associated with lung cancer. Also in animal models, cadmium induces lung carcinomas after inhalation (Beyersmann & Hartwig, 2008). At the cellular level, this metal affects cell proliferation, differentiation, apoptosis, and other cellular activities and may cause numerous molecular lesions that would be relevant to carcinogenesis (Mates et al., 2010). Inorganic cobalt compounds, both soluble and particulate forms, caused lung tumors in animal experiments (Beyersmann & Hartwig, 2008). The IARC recently classified the mixture cobalt/tungsten carbide (Co/WC) as carcinogenic to humans (Mates et al., 2010).

Chromium is other carcinogenic metal. Epidemiological studies have consistently shown that the lower respiratory tract is the target organ of Cr(VI) compound exposure, and occupational exposure to these compounds is strongly associated with a higher incidence of lung cancer. Chromium exists in the environment in two major valence states, Cr(VI) and chromium (III) [Cr(III)], and Cr(VI) is actively transported into cells by the anionic transport system. The reduction of Cr(VI) to Cr(III) can lead to the formation of DNA-chromium adducts, DNA-DNA and DNA-protein cross-links, DNA-Cr(III)-amino acid ternary complexes and radical-mediated DNA strand breaks. In addition, it has been reported that lung cancer from workers exposed to Cr(VI) has a high percentage of G to T transversion mutations in the non-transcribed strand of the p53 gene (Feng et al., 2003). Chromium compounds also are capable to induce oxidative stress and the deregulation of cell proliferation (Beyersmann & Hartwig, 2008). On the other hand, several epidemiological studies demonstrated a strong correlation between nickel exposure and risk of lung and nasal cancer, especially in the case of workers at nickel refineries. Evidence from experimental animals has demonstrated the carcinogenicity of metallic nickel, which is also classified as possibly carcinogenic to humans (Group 2B). Nickel carcinogenesis involves epigenetic alterations, disruption of cellular iron homeostasis (by interfering with iron-dependent enzymes), generation of ROS, and activation of the hypoxia-signalling pathway (Salnikow & Zhitkovich, 2008).

Sulfur oxide pollution (as measured by sulfate particles and/or sulfur dioxide) is significantly associated lung cancer mortality. Elevated mortality risks have been associated primarily with measures of fine particulate and sulfur oxide pollution (Naess et al., 2007; Pope et al., 2002). Inhalation exposure to air pollutants, e.g., SO2, has been associated with the DNA damage of multiple organs including the lung, providing a possible biological pathway through which air pollution may affect lung cancer incidence (Cao et al., 2011).

Exposure to nitrogen oxide (NOx and NO2) also is related to lung cancer (Naess et al., 2007; Raaschou-Nielson et al. 2010). Studies have been shown that women had particularly large effects for lung cancer in the young age group, somewhat less so for the old exposed to nitrogen oxide (Naess et al., 2007). Moreover NOx exposure has significant correlations with adenocarcinoma (AC) type cancer incidence rates for both genders (Chen et al., 2009), although women seem to be more susceptible (Liaw et al., 2010). NOx may potentially trigger mutagenic and carcinogenic activity and play significant roles in the metabolism and behavior of AC type lung cancer (Fujimoto et al., 1998). It appears that the higher the NOx concentration, the higher the AC incidence rate. Recent studies on the role of NO in tumor progression suggest that NO is an important bioregulatory and signaling molecule and may play a role in the process of carcinogenesis (Tamir & Tannenbaum, 1996). NO is an endothelial growth factor that specifically mediates tumor vascularization (Jenkins et al., 1995) and tumor blood flow (Tozer et al., 1997). Exposure of cells to high NO concentrations cause DNA damage and apoptosis. Moreover, recent results have shown that NO stimulates p53 accumulation (Forrester et al., 1996). In summary, according to previous literature, NO and its derivatives can cause DNA damages (Wink et al., 1993) and play important roles in human lung AC (Fujimoto et al.,1998) (Liaw et al., 2010).

Airborne polycyclic aromatic hydrocarbons (PAHs) are emitted when organic matter is burned. It has long been known that several PAHs can produce cancers in experimental animals, and epidemiologic studies of exposed workers, especially in coke ovens and aluminum smelters, have shown clear excesses of lung cancer (Bostrom et al., 2002; Ben et al., 2004). PAHs have been suggested as being responsible for the initiation and development of lung cancer. PAHs and their metabolites are involved in mechanisms of carcinogenesis; produce early chromosomal changes, transformation of cells in culture, cytotoxicity and mutagenicity. The benzo(a)pyrene (BAP) for example, directly damages p53. In smoking-related lung cancer, 40% of the p53 gene mutations are G to T transversions, and 90% of this type of mutation can be attributed to the non-transcribed (coding) strand. G to T transversion has been regarded as a hallmark of PAH-induced mutations in smoking-related lung cancer. Activated metabolites of PAHs in cigarette smoke, including benzo[a]pyrene diol epoxide (BPDE), preferentially form DNA adducts at methylated CpG sites along the p53 gene corresponding to the afore mentioned major mutational hotspots in smoking-related lung cancer (Feng et al., 2003).

On the other hand, radon gas is a naturally occurring radioactive gas that can seep out of the soil into buildings, also emanates from uranium-bearing soil and porous rock. Radon is the second biggest cause of lung cancer after smoking. The cancer risk from radon increases the risk from smoking. Radon induces damage to a checkpoint tumor suppressor gene such as Tp53 (which codes for p53) since alpha particle radiation is a key mechanism for radon-related lung cancer (Harley et al., 2008; Bissett & McLaughlin 2010). Some studies suggest a relationship between the AGGARG-ATGMET transversion in codon 249 of P53 from people exposed to high radon concentrations (Ruano-Ravina et al., 2009). The damage done to epithelial cells of the lung occurs when radiation interacts either directly with DNA in the cell nucleus or indirectly through the affect of free radicals (UNSCEAR, 2000). Recently, *in vitro* studies of cells exposed to alpha-particle radiation gave evidence that more cells showed damage than those that were traversed by alpha-particles (Sawant et al., 2001; Alavanja 2002). Radon produces oxidative stress, although some similarities in the increased frequency of p53 mutations at a later stage in this process, DNA damage in the form of sister-chromatid exchange and mutations, have been observed for both smokers and those exposed to radon gas (Alavanja 2002). Other studies provided evidence of a GSTM1 and radon interaction in the increasing risk for lung cancer. Glutathione-S-transferase M1 (GSTM1) conjugates known carcinogens such as epoxides of polycyclic aromatic hydrocarbons (Risch & Plass, 2008). Other studies show that radon exposure in miners induces gene mutations and chromosomal aberrations. Numerous *in vitro* cytogenetic studies demonstrated that radon induces different types of genetic and cytogenetic damage that is likely to play a role in radon lung carcinogenesis (Al-Zoughool & Krewski, 2009).

5.4 Genotoxic effects

Generation of DNA damage is considered an important initial event in carcinogenesis. Multiples assays exist for the detection of different genotoxic effects of compounds in experimental systems, or for exposure investigation for genotoxic agents in environmental or occupational settings (Moller 2005). Cells with DNA damage are more susceptible to develop mutations after exposure to xenobiotics (Olive et al. 2001). For this reason, genotoxic evaluation of environmental pollutants is necessary in the respiratory system, because it is the first contact for inhaled xenobiotics.

Studies on environmental pollution (and their components), genotoxicity and respiratory tract, have been carried out *in vivo* and *in vitro*. Table 1 summarizes the results obtained in the last years.

A number of studies have considered DNA damage as an endpoint for the effects of air pollutants (Vineis & Husgafvel- Pursiainen 2005). In this report, we evidence the results obtained by genotoxic profiles, indicating that air pollutants cause alterations in the genetic material of the tested cells (strand breaks [illustrate in Figure 7], oxidative damage, adducts and micronucleus). DNA damage could provoke mutations in any cell from the respiratory tract and, may facilitate the development of neoplastic events.

	Exposure	System/Tissue	End Point	Results	Reference
In vivo					
	DEP Inhalation	Big Blue rats Male Lungs	Mutant frequency Mutation spectra DNA adducts 8-OHdG levels	+ + + +	Sato et al., 2000
	DEP Inhalation	Mice Female Lungs Bronchioalveolar lavage cells	8-OHdG levels DNA strand breaks (comet assay)	+ +	Risom et al., 2003
	DEP Inhalation	Mice	DNA adducts DNA strand breaks (comet assay) DNA oxidative damage Mutation frequency	+ + + -	Dybdahl et al., 2004
	Environmental Pollution	Human Females/males Nasal epithelial cells	DNA strand breaks (comet assay)	+	Fortoul et al., 2010
In vitro	DEP	A549 cells	DNA strand breaks (comet assay)	+	Dybdahl et al., 2004
	Particulate Matter (PM2.5 and PM 10)	A549 cells	DNA strand breaks (comet assay)	+	Gutierrez-Castillo et al., 2006
	Particulate Matter (PM10)	A549 cells	Micronucleus	+	Roubicek et al., 2007
	Fine Particles	A549 cells	DNA strand breaks (comet assay)	+	Sharma et al., 2007
	Nitrogen dioxide (NO_2)	Nasal epithelial cells	DNA strand breaks (comet assay) Micronucleus	+ -	Koehler et al., 2010

DEP: Diesel Exhaust Particles. A549 cells: human lung epithelial cell line. +Positive results. - Negative results. $PM_{2.5}$: Particulate Matter with aerodynamic diameter ≤2.5 μm. PM_{10}: Particulate Matter with aerodynamic diameter ≤10 μm.

Table 1. Genotoxic damage studies associated with air pollutants.

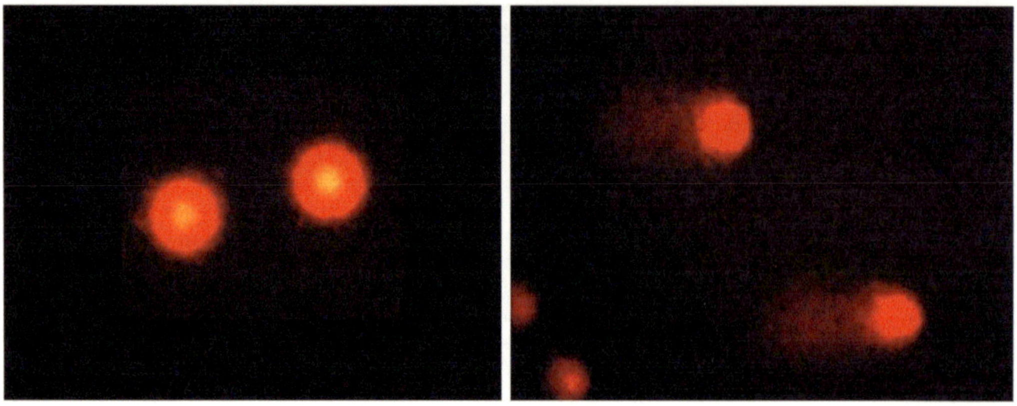

Fig. 7. Comet assay in nasal epithelial cells. Left: Undamaged Cells. Right: Damaged Cells (comets).

5.5 Pulmonary fibrosis and atmospheric pollution

Pulmonary fibrosis is the end result of a diverse group of lung disorders. Although there are multiple initiating agents for pulmonary fibrosis, including toxins, fibres/particles, autoimmune reactions, drugs and radiation, the etiology of the majority of cases of pulmonary fibrosis is unknown and these cases are referred to as Idiopathic Pulmonary Fibrosis (IPF). The harmful effects of environmental pollution on the respiratory system are undeniable. The combustion of fuels and its derivatives are the main cause for pollutant emission through engines and industrial plants (Zuurbier, et al., 2011). Chronic exposure to particulate matter, ozone and cigarette smoke can produce long-term effects on the lungs. This source of pollutants sets the lungs in a situation of constant aggression that result in a state of chronic inflammation, which could lead to pulmonary fibrosis (Churg & Wright, 2002).

5.5.1 Fibrotic lung reactions due to air pollution

The etiology for lung's fibrotic reactions is still unknown and it has been proposed a variety of environmental stimuli, such as metals, cigarette smoke, drugs and infectious agents (Araya & Nishimura, 2010). The degree of fibrosis will depend on the response of each individual, even with similar exposures, indicating that host genetic factors influence the fibrotic response of the patient (Westergren-Thorsson, et al., 2010).

Diesel exhaust particulate matter in polluted environments derived from internal combustion engines, increases the transcription of inflammatory cytokines and antimicrobial peptides, contributing to increased inflammatory response of airways in patients with chronic obstructive pulmonary diseases (COPD) (Nam, et al., 2006). This induces an oxidative state caused by lung macrophages that increase pulmonary responses and could result in irreversible lung fibrosis (Figure 8).

It has now been implicated the Transforming Growth Factor β1 (TGF-β1) as a key factor in the lung fibrotic response (Datta et al., 2011; Koli et al., 2008). The changes caused by TGF-β1 could be observed in repeated injuries of the airways, as occurs in asthma, chronic obstructive pulmonary disease and pulmonary fibrosis (Araya & Nishimura, 2010).

Exposure to cigarette smoke produces a large amount of reactive oxygen species and activates latent TGF-β1. The inhalation of this smoke promotes the recruitment of macrophages and neutrophils that are important sources of reactive oxygen species and also

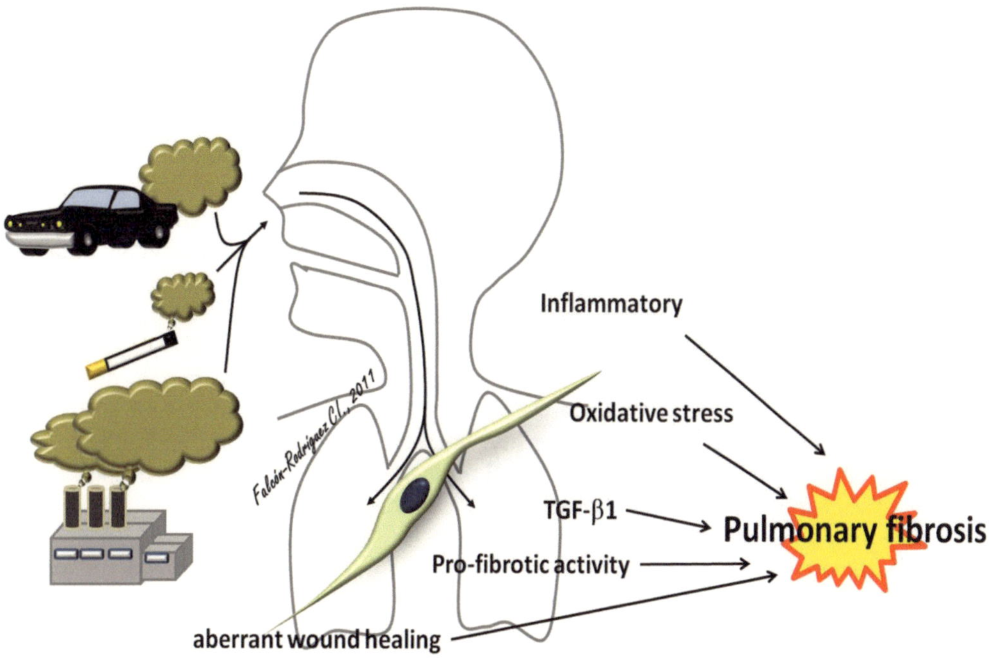

Fig. 8. Environmental pollution from various sources, such as transition metals, particle exposure, noxious drugs, cigarette smoke, induces repetitive cellular injury and inflammation. This leaves the tissue susceptible to increased TGF-β1, inflammation, profibrotic signals and aberrant wound healing, all of which may contribute to the progression of pulmonary fibrosis.

contribute to activate TGF-β1 (Westergren-Thorsson, et al., 2010). On the other hand, TGF-β1 induces transdifferentiation of fibroblasts to myofibroblasts and perpetuates the fibrogenic process, also providing a protective effect against myofibroblasts apoptosis, which are not removed once the lung lesions healed (Song et al., 2011).

Asbestos can induce lung fibrosis in occupational and experimental exposures, (Ross and Murray.2004; Dai & Churg, 2001). Mineral dusts can directly induce fibrosis in the airway wall, and this has been studied *in vivo*. Coexposures to cigarette smoke or ozone increase the fibrogenic effect of mineral dusts (Churg & Wright, 2002; Churg et al., 1996). Asbestosis and mineral dusts exposure increase the gene expression of profibrotic factors: TGF-β 1 and platelet derived growth factor (PDGF) that also increases procollagen. These changes may explain fibrosis progression (Churg and Wright, 2002; Churg et al., 1999). There is also human evidence of small airway remodeling in chronic exposure to high levels of particulate air pollution (Churg et al., 2003).

Oxidative stress, as a consequence of an inflammatory stimulus such as air pollution, plays a critical role in the pathogenesis of IPF (Park et al., 2009). Fibrotic stimuli of unknown origin are thought to create an imbalance between oxidant production and antioxidant protection, resulting in the accumulation of reactive oxygen species (ROS) (Rahman et al., 1999). The precise pathways leading from injury to fibrosis are not well established, but oxidants may contribute to the production of profibrotic factors such as TGF β, and oxidized proteins have been reported in human subjects with IPF. In addition, some studies have reported that various antioxidant enzyme systems protect against lung fibrosis (Gao et al., 2008; Khang et al., 2003).

5.6 Chronic obstructive pulmonary disease (COPD)

Chronic Obstructive Pulmonary Disease (COPD) is a slowly progressive condition characterized by airflow limitation which is usually reversible (MacNee, 2007). Although smoking habit is the main factor associated with COPD, air pollution has been implicated in its pathogenesis, and there is enough evidence to support the association between air pollution and COPD exacerbations and worsening of those with pre-existing COPD. Particulate Matter (PM), Nitrogen Oxides (NOx) are associated with its development (Salvi & Barnes, 2009).

Some of the observed changes are the increase in the presence of goblet cells (mucous metaplasia) as well as submucose glands hyperplasia. This increase in mucus production will result in reduced mucocliliary clearance. Emphysema, the enlargement and destruction of the alveolar spaces, is also part of this entity, because the decrease in the elastic recoil of the lung the air is trapped in the alveoli. Small airways are very important components in this entity. The changes, such as inflammation and fibrosis modify the prognosis: the greater the damage the poorer the prognosis.

The main participant in the progression of the bronchial damage is the inflammatory response. The epithelial cells generate this response when they are exposed to air pollutants or other irritants such as cigarette smoke, then a variety of pro-inflammatory, pro-fibrotic and mitotic factors are liberated. These factors increase endothelial permeability; activate macrophages, CD8+ Lymphocytes, and neutrophils. These cells produce more inflammatory factors, which, in chronic exposure inflammation leads to airway remodeling. (Figure 9) (Roth, 2008).

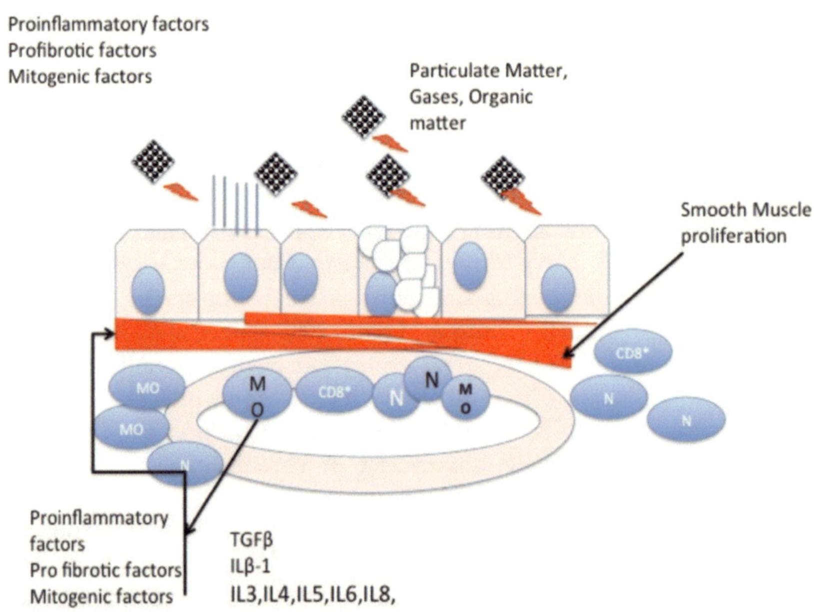

Fig. 9. The respiratory epithelium produces proinflammatory factors that will activate macrophages (MO), will increase neutrophil (N) and CD8* lymphocytes migration (CD8+). Also, smooth muscle proliferation is observed, as well as an alteration of the extracellular matrix.

Remodeling includes smooth muscle and fibroblast proliferation, synthesis of extracellular matrix (MMPs), decrease in proteases synthesis and increase in gelatinases. The sum of all these events will maintain the obstruction observed in these patients.

6. Conclusion

Further studies are needed to assess the impact of atmospheric pollution in the development of respiratory diseases, and to explore new therapeutic approaches to reverse the progression of the chronic changes that currently are observed in asthma, COPD, fibrosis and cancer.

7. Perspectives

More information about the differences in the response to air pollutants is needs, as well as the possible treatments, if any of the biomarkers is found. This drives us to the need for more specific biomarkers to identify the severity of the inflammation, or the type of inflammation. Because air Pollution is a complex mixture of organic and inorganic elements, more detailed information about specific characteristics of each pollutant response would be very helpful.

The lung has its own responses and support for more research in this field of interest, must be encouraged.

8. Acknowledgements

Authors thank to Veronica Rodriguez-Mata for her technical work in the technical work for the histological preparations.

9. References

Abbas, I.; Saint-Georges; F.; Billet, S.; Verdin, A.; Mulliez, P.; Shirali, P. & Garçon, G. (2009). Air Pollution Particulate Matter (PM 2.5)-Induced Gene Expression of Volatile Organic Compound and/or Polycyclic Aromatic Hydrocarbon-metabolizing Enzymes in an In Vitro Coculture Lung Model. *Toxicology in Vitro,* Vol.23, No.1, (February 2009), pp. 37–46, ISSN 0887-2333

Alavanja, M.C.R. (2002). Biologic Damage Resulting from Exposure to Tobacco Smoke and from Radon: Implication for Preventive Interventions. *Oncogene,* Vol.21, No.48, (October 2002), pp.7365 - 7375, ISSN 0950-9232

Alexis, N.E.; Lay J.C.; Zeman, K.; Bennet, W.E; Peden, D.B.; Soukup, J.M.; Devlin, R.B. & Becker, S. (2006). Biological material on inhaled coarse fraction particulate matter activates airway phagocytes in vivo on healthy volunteers. *Journal of Allergy Clinical Immunology,* Vol.117, No. pp. 1367-1373 ISSN 00916749

Al-Zoughool, M.& Krewski, D. (2009). Health Effects of Radon: A Review of the Literature. *International Journal Radiation Biology,* Vol. 85, No. 1, (2009),pp. 57-69 ISSN 0955-3002

Anwar-Mohamed, A.; Elbekai, R.H. & El-Kadi, A.O.S. (2009) Regulation of CYP1A1 by heavy metals and consequences for drug metabolism. *Expert Opinion on Drug Metabolism & Toxicology,* Vol.5, No.5, (May 2009), pp. 501-521, ISSN 1742-5255

Araya, J. & Nishimura, S. L. (2010). Fibrogenic reactions in lung disease. *Annual Review of Pathology,* No.5, pp. 77-98, *ISSN*: 1553-4014,

Atash, F. (2007). The deterioration of urban environments in developing countries: Mitigating the air pollution crisis in Theran, Iran. *Cities,* Vol.24, No.6, (February 2007), pp. 339 – 409, *ISSN* 0264-2751

Auten, R.L.; Potts, E.N.; Mason, S.N.; Fischer, B.; Huang, Y. & Foster, W.M. (2009). Maternal Exposure to Particulate Matter Increases Postnatal Ozone-induced Airway Hyper reactivity in Juvenile Mice. *American Journal Respiratory Critical Care Medicine,* Vol.180, No.12, (December 2009), pp. 1218-1226, ISSN 1073-449X

Barnes, T.C., Spiller, D.G., Anderson, M.E; Edwards, S..W., Moots, R.J.(2011). Endothelial activation and apoptosis mediated by neutrophil-dependent interleukin 6 trans-signaling: a novel target for systemic sclerosis. *Annals of the Rheumatic Diseases* (February, 2011) Vol. 70, No. 2, pp. 366-372. ISNN 1468-2060

Ben A.; Hutchinson E.; Unwin, J.; & Fletcher, T. Lung Cancer Risk after Exposure to Polycyclic Aromatic Hydrocarbons: A Review and Meta-Analysis. *Environmental Health Perspectives,* Vol.112, No. 9, (January 2004), pp. 970-78, ISSN: 0091-6765

Bernstein, J.; Neil, A.; Barnes, Ch.; Bernstein, L.; Bernstein, J.; Nel, A.; Peden, D.; Diaz S.D.; Tarlo, S. & Williams, B. (2004). Health effects of air pollution. *Environmental and Occupational Respiratory Disorders,* Vol.114, No. 5 (August 2004), pp. 1116 -1123

Beyersmann, D. & Hartwig, A. (2008). Carcinogenic Metal Compounds: Recent Insight into Molecular and Cellular Mechanisms. *Archives of Toxicology,* Vol. 82, (May 2008), pp. 493–512, ISSN 0340-5761

Bissett, R.J. & McLaughlin J.R. Radon. *Chronic Disease Canada,* Vol.29, Suppl 1, (2010), pp.38-50, ISSN 0228-8699

Born, S.L.; Caudill D.; Fliter K.L. & Purdon M.P. (2002). Identification of the Cytochromes P450 that Catalyze Coumarin 3,4-epoxidation and 3-hydroxylation. *Drug Metabolism and Disposition,* Vol.30, No.5, (May 2002), pp. 483–487, ISSN 0090-9556

Bostrom, CE.; Gerde, P.; Hanberg, A.; Jernstrom, B.; Johansson, C.; Kyrklund, T.; Rannung, A.; Tornqvist, M.; Victorin, K. & Westerholm R. (2002). Cancer Risk Assessment, Indicators, and Guidelines for Polycyclic Aromatic Hydrocarbons in the Ambient Air. *Environmental Health Perspectives,* Vol.110, (Jun 2002), pp.451–488, ISSN 0091-6765

Bulut, Y.; Shimada, K.; Wong, M.H; Chen, S; Gray, P; Alsabeh, R.; Doherty, T.M.; Crother, T.R; & Arditi, M.(2009). Chlamydial heat shock protein 60 induces acute pulmonary inflammation in mice via the Toll-like receptor 4- and MyD88-dependent pathway. *Infection and Immunity,* Vol. 77, No. 7, (April, 2009) pp. 2683-2690. ISSN:0019-9567

Calderon-Garcidueñas, L.; Vincent, R; Mora-Tiscareño, A; Franco-Lira, M; Henriquez-Roldan, C; Barragan-Mejia, G; Garrido-Garcia, L; Camacho-Reyes, L; Valencia-Salazar, G; Paredes, R; Romero, L.; Osnaya, H.; Villareal-Calderon, R; Torres-Jardon, R; Hazucha, M.J; & Reed, W. (2007). Elevated plasma endothelin-1 and pulmonary arterial pressure in children exposed to air pollution. *Environmental Health Perspectives,* (August, 2007) Vol. 115, No. 8, pp. 1248-1253. ISSN:1552-9924

Cao, J.; Chunxue Yang,; Jianxin Li,; Renjie Chen,; Bingheng Chen,; Dongfeng Gu & Haidong Kan J. (2011). Association Between Long-Term Exposure to Outdoor Air Pollution and Mortality in China: A Cohort Study. *Journal of Hazardous Materials* . Vol.186, No. 2-3, (January 2011), pp.1594–1600. ISSN: 0304-3894

Cardoso, W.V. & Lü, J. (2006). Regulation of Early Lung Morphogenesis: Questions, Facts and Controversies. *Development,* Vol.133, No.9 (May 2006),pp. 1611-1624, ISSN 0950-1991

Carlson, G.R. (2008). Critical Appraisal of the Expression of Cytochrome P450 Enzymes in Human Lung and Evaluation of the Possibility that such Expression Provides Evidence of Potential Styrene Tumorigenicity in Humans. *Toxicology*, Vol.254, No.1-2, (December 2008), pp. 1–10, ISSN 030-483X

Castell, J.V.; Donato, M.T. & Gomez-Lechon, J.M. (2005). Metabolism and Bioactivation of Toxicants in the Lung. The In Vitro Cellular Approach. *Experimental and Toxicologic Pathology*, Vol.57, (July 2005), pp.189–204, ISSN 0940-2993

Chang, H.; Chang, L.W.; Cheng, Y.H.; Tsai, W.T.; Tsai, M.X. & Lin, P.P. (2006). Preferential Induction of CYP1A1 and CYP1B1 in CCSP-Positive Cells. *Toxicological Sciences*, Vol.89, No.1, (January 2006), pp. 205–213, ISSN 1096-6080.

Chen, F.; Jackson, H. & Bina, WF. (2009). Lung Adenocarcinoma Incidence Rates and Their Relation to Motor Vehicle Density. *Cancer Epidemiology Biomarkers Prevention*. Vol.18, No.3, (March 2009), 760–764, ISSN 1055-9965

Choudhary, D.; Jansson, I.; Stoilov, I.; Sarfarazi, M. & Schenkman J.B. (2005). Expression Patterns of Mouse and Human CYP Orthologs (Families 1–4) During Development and in Different Adult Tissues. *Archives of Biochemistry and Biophysics*, Vol.436, No.1, (April 2005), pp. 50–61, ISSN 0003-9861

Churg, A. & Wright, J.L. (2002). Airway wall remodeling induced by occupational mineral dusts and air pollutant particles. *Chest* ; Vol.122, No.6 suppl (December 2002), pp. 306S-309S, ISSN 0012-3692

Churg, A.; Brauer M. & Keeling, B. (1996). Ozone enhances the uptake of mineral particles by tracheobronchial epithelial cells in organ culture. *American Journal of Respiratory and Critical Care of Medicine, No.*153, No.4, pp.1230–1234, *ISSN* 1073-449X

Churg, A.; Brauer, M.; Avila-Casado, M.C.; Fortoul, T.I. & Wright, J.L. (2003). Chronic exposure to high levels of particulate air pollution and small airway eemodeling. *EnvironmentalHealth Perspectives*, Vol.111, No.5, (May 2003) pp. 714–718, *ISSN: 0091-6765.*

Churg, A.; Gilks, B. & Dai, J. (1999). Induction of fibrogenic mediators by fine and ultrafine titanium dioxide in rat tracheal explants. *American Journal of Physiology*, Vol.277, No. 5 pt 1, pp. L975–L982, *ISSN*: 0363-6143;

Dai, J. & Churg, A. (2001). Relationship of fiber surface iron and active oxygen species to expression of procollagen, PDGF-A, and TGF 1 in tracheal explants exposed to amosite asbestos. *American Journal of Respiratory Cell and Molecular Biology*, Vol.24, No.4, (April 2001)pp. 427–435, *ISSN*: 1535-4989

Datta, A.; Scotton, C. J. & Chambers, R. C. (2011). Novel Therapeutic Approaches for Pulmonary Fibrosis. *British Journal of Pharmacology*. Epub ahead of print, ISSN *0007-1188.*

Davis, C.W. & Dicker, B.E. (2008). Regulated Airway Goblet Cell Mucin Secretion. *Annual Review of Physiology*, Vol.70, (March 2008), pp.487-512, ISSN 0066-4278

Deb, S. & Bandiera, S.M. (2010). Characterization of a New Cytochrome P450 Enzyme, CYP2S1, in Rats: Its Regulation by Aryl Hydrocarbon Receptor Agonists. *Toxicology*, Vol.267, No.1-3, (January 2010), pp. 91–98, ISSN 0300-483X

Dockery, DW.; Pope, CA; Xu, X.; Spengler, JD.; Ware, JH,; Fay, ME,; Ferris, BG. Jr. & Speizer, FE. (1993). An Association Between Air Pollution and Mortality in Six U.S. Cities. *New England Journal Medicine*, Vol.329, (December 1993), pp.1753–9, ISSN 0028-4793

Donaldson, K; Mills, N; MacNee, W.; Robinson, S; & Newby, D. (2005). Role of inflammation in cardiopulmonary health effects of PM. *Toxicology and Applied Pharmacology*, Vol. 207, No. 2 (Sept, 1, 2005) pp. 483-488. ISSN:0041-008X

Dybdahl, M.; Risom, L.; Bornholdt, J.; Autrup, H.; Loft, S. & Wallin, H. (2004). Inflammatory and Genotoxic Effects of Diesel Particles In Vitro and In Vivo. *Mutation Research*, Vol.562, No.1-2, (August 2004), pp.119-131, ISSN 1383-5718

Ekici, A.; Ekici, M., Kurtipek, E., Akin, A., Arsian, M., Kara, T., Apaydin, Z., Demir, S. (2005) Obstructive airway diseases in women exposed to biomass smoke. *Environmental Research*, Vol. 99, No.1, (September, 2005) pp. 93-98. ISSN:0013-935

Feng, Z.; Hu, WW.; Rom, WN.; Costa, M. & Tang, MS. (2003). Chromium(VI) exposure enhances polycyclic aromatic hydrocarbon DNA binding at the p53 gene in human lung cells. *Carcinogenesis*, Vol.24, No.4, (April 2003), pp.771-778, ISSN 0143-3334.

Finlayson-Pitts, B. J. & Pitts, J. N. (1997). Tropospheric air pollution: ozone, airborne toxics, polycyclic aromatic hydrocarbons and particles. *Science*, Vol.276, No. 5315, (May 1997), pp.1045-1052, *ISSN 0036-8075*

Finlayson-Pitts, B.J. & Pitts, J.N. (2000). Chemistry of the Upper and Lower Atmosphere, Academic Press, San Diego, California, xxii+969 pp., ISBN 0-12-257060xMills, N.L. Donaldson, D., Hadoke, P.W., Boon, N.A., MacNee, W., Cassee, F.R., Sandström, T., Blomberg, A. & Newby D.E. (2009). Adverse cardiovascular effects of air pollution. Nature Clinical Practice Cardiovascular Medicine, Vol. 6, No.1, (November, 2009), pp. 36-44, ISSN: 1759-5002

Forrester, K.; Ambs, S.; Lupold, SE.; Kapust, RB.; Spillare. EP.; Weinberg, WC.; Felley-Bosco, E.; Wang, XW.; Geller, DA.;Tzeng, E. & Billar, TR. (1996). Harris CC. Nitric Oxide-Induced p53 Accumulation and Regulation of Inducible Nitric Oxide Synthase Expression by Wild-type p53. *Proceedings of the National Academy Science USA*. Vol. 93, (March1996), pp. 2442–2447, ISSN 0027-8424

Fortoul, T.I.; Rodriguez-Lara, V.; Lopez-Valdez, N. & Falcon-Rodriguez, C.I. (2010). Sistema Respiratorio, In: *Histologia y Biologia Celular*. T.I. Fortoul van der Goes, (Ed.), 187-195, McGraw-Hill, ISBN: 978-607-15-0340-4, China

Fortoul, T.I.; Rojas-Lemus, M.; Avila-Casado M.C.; Rodriguez-Lara V.; Montaño, L.F.; Muñoz-Comonfort, A. & Lopez-Zepeda, S. (2010), Endogenous Antioxidants and Nasal Human Epithelium Response to Air Pollutants: Genotoxic and Inmmunocytochemical Evaluation. *Journal of Applied Toxicology*, Vol.30, No.7, (October, 2010), pp.661-665, ISSN 0260-437X

Fujimoto, H.; Sasaki, J.; Matsumoto, M.; Suga, M.; Ando, Y.; Iggo, R.; Tuda, M.; Saya, H. & Anas, M. (1998). Significant Correlation of Nitric Oxide Synthase Activity and p53 Gene Mutation in Stage I Lung Adenocarcinoma. *Japanese Journal of Cancer Research*.Vol. 89, (July 1998), pp. 696–702, ISSN 0910-5050

Gabay, C., & Kushner, I. (1999) Acute-phase proteins and other systemic responses to inflammation. *New England Journal of Medicine*, (February, 11, 1999) Vol. 340, No.6, pp.448-454. ISSN:0028-4793

Gao, F.; Kinnula, V.L.; Myllarniemi, M. & Oury, T.D. (2008). Extracellular superoxide dismutase in pulmonary fibrosis. *Antioxidants &. Redox Signal*, Vol.10, No. 2, (February 2008), pp.343–54, *ISSN*: 1523-0864

Gern, J.E. (2010). The Urban Environment and Childhood Asthma Study. *Journal of Allergy and Clinical Immunology*, Vol.125, No.3, (March 2010), pp. 545-549, ISSN 0091-6749

Gutierrez-Castillo, M.E.; Roubicek, D.A.; Cebrian-García M.E.; Vizcaya-Ruiz, A.D.; Sordo-Cedeno, M. & Ostrosky-Wegman, P. (2006). Effect of Chemical Composition on the Induction of DNA Damage by Urban Airbone Particulate Matter. *Environmental and Molecular Mutagenesis,* Vol.47, No.3, (April 2006), pp. 199-211, ISSN 0893-6692

Halle, S.; Dujardin, H.; Bakocevic, N.; Fleige, H.; Danzer, H.; Willenzon, S.; Suezer, Y.; Hammerling, G.; Garbi, N.; Sutter, G.; Worbs, T. & Forster, R. (2009). Induced bronchus-associated lymphoid tissue serves as a general priming site for T cells and is maintained by dendritic cells. *Journal of Experimental Medicine,* Vol.206, No.12, (November 2009), pp. 2593-2601, *ISSN: 0022-1007*

Harley, NH.; Chittaporn, P.; Heikkinen, MSA.; Meyers, OA. & Robbins, ES. (2008). Radon Carcinogenesis: Risk Data and Cellular Hits. *Radiation Protection Dosimetry.* Vol.130, No.1, (June 2008), pp.107–109, ISSN 0144-8420

Hetland, R.B; Cassee, F.R; Lag, M., Refsnes, M; Dybing, E; & Schwaze, P.E. (2005) Cytokine release from alveolar macrophages exposed to ambient particulate matter: heterogeneity in relation to size, city and season. *Particle and Fibre Toxicology,* Vol. 17, No.2, (August, 2005) pp. 1-15, SSN:1743-8977

Hildebrandt, K; Ruckerl, R.; Koeening, W.; Schneider, A.; Pitz, M.; Heinrich, J; Marder, V.; Frampton, M.; Oberdorster, G; Wichmann, G; & Peters, A. (2009). Short-term effects of air pollution: a panel study of blood markers in patients with chronic pulmonary disease. *Particle and Fibre Toxicology* (September, 2009). Vol. 26, No.25, pp. 1-13. ISSN:1743-8977

Holgate, S.T. (2008). The Airway Epithelium is Central to the Pathogenesis of Asthma. *Allergology International,* Vol.57, No.1, (March 2008), pp. 1-10, ISSN 1323-8930

Ioannides, C. (2008). *Cytochromes P450. Role in the metabolism and toxicity of drugs and others xenobiotics,* RSC Publishing, ISBN: 978-0-85404-274-6, Cambrigde, UK

Jenkins, DC,; Charles, IG.; Thomsen, LL.; Moss, DW.; Holmes, LS.; Baylis, SA.; Rhodes, P.; Westmore, K.; Emson, PC. & Moncada, S. (1995). Roles of Nitric Oxide in Tumor Growth. *Proceedings of the National Academy Science USA,* Vol.92, (May 1995), pp.4392–6, ISSN 0027-8424

Kang, S.K.; Rabbani, Z.N.; Folz, R.J.; Golson, M.L.; Huang, H.; Samulski; T.S.; Dewhirst, M.W.; Anscher, M.S. & Vujaskovic, Z. (2003). Overexpression of extracellular superoxide dismutase protects mice from radiation-induced lung injury. *International Journal of Radiation Oncology Biology Physics,* Vol.15, No. 57, (November 2003), pp. 1056–66, *ISSN*: 0360-3016.

Katavolos, P.; Ackerley, C.A.; Viel, L.; Clark, M.E.; Wen, X. & Bienzle, D. (2009). Clara Cell Secretory Protein Is Reduced in Equine Recurrent Airway Obstruction. *Veterinary Pathology,* Vol.46, No.4, (July 2009), pp. 604–613, ISSN 0300-9858

Kawamata, N.; Xu, B.; Nishijima, H.; Aoyama, K.; Kusumoto, M.; Takeuchi, T.; Tei, Ch.; Michie, A. & Matsuyama, T. (2009). Expression of endothelia and lymphocyte adhesion molecules in bronchus-associated lymphoid tissue (BALT) in adult human lung. *Respiratory Research,* Vol.10, No.97, (October 2009), pp. 1-11, *ISSN*: 1465-9921

Koehler, C.; Ginzkey, C.; Friehs, G.; Hackenberg, S.; Froelich, K.; Scherzed, A.; Burghartz, M.; Kessler, M. & Kleinsasser, N. (2010). Aspects of Nitrogen Dioxide Toxicity in Environmental Urban Concentrations in Human Nasal Epithelium. *Toxicology and Applied Pharmacology,* Vol.245, No.2, (June 2010), pp. 219-225, ISSN 0041-008X

Koli, K.; Myllarniemi, M.; Keski-Oja, J. & Kinnula, V.L. (2008). Transforming growth factor-ß activation in the lung: focus on fibrosis and reactive oxygen species. *Antioxidants & Redox Signal*, Vol.10, No.2, (February 2008), pp. 333–42, *ISSN*: 1523-0864

Liaw, YP.; Ting, TF.; Ho, CC. & Chiou, ZY. (2010). Cell Type Specificity of Lung Cancer Associated with Nitric Oxide. *Science of the Total Environment*, Vol.408, No.21, (October 2010), pp.4931-4, ISSN 0048-9697

López-Cima, MF.; García-Pérez, J.; Pérez-Gómez, B.; Aragonés, N.; López-Abente G.; Tardón, A. & Pollán, M. (2011). Lung Cancer Risk and Pollution in an Industrial Region of Northern Spain: a Hospital-based Case-Control Study. *International Journal Health Geographics*, Vol.25, No.10, (January 2011), pp.10-23, ISSN 1572-9893

MacNee, W. (2007) Pathogenesis of Chronic Obstructive Pulmonary disease. *Clinics in Chest Medicine, Vol.2, No.4, (November 2005)*, pp. 479-513, *ISSN*: 0272-5231

Mates, JM. ; Segura, JA.; Alonso, FJ. & Marquez, J. (2010). Roles of Dioxins and Heavy Metals in Cancer and Neurological Diseases Using ROS-Mediated Mechanisms. *Free Radical Biology & Medicine,* Vol. 49, (November 2010), pp. 1328–1341. ISSN: 0891-5849

Mills, N.L; Amin, N; Robinson, S.D; Anand, A; Davies, J; Patel, D; de la Fuente, J.M; Cassee, F.R; Boon, R.A; MacNee, W; Millar, A.M; Donaldson, K. & Newby D.E. (2006). Do Inhaled Carbon Nanoparticles Translocate Directly into the Circulation in Humans? *American Journal of Respiratory and Critical Care Medicine* Vol. 173, No. 4, (February, 2006) pp. 426–431, ISSN 1535-4970

Molina, JR.; Yang, P.; Cassivi, SD.; Schild, SE. & Adjei, AA. (2008). Non-Small Cell Lung Cancer: Epidemiology, Risk Factors, Treatment, and Survivorship. *Mayo Clinic Proceedings*, Vol.83, No.5, (May 2008), pp.584-594, ISSN 0025-6196

Moller, P. (2005). Genotoxicity of Environmental Agents Assessed by the Alkaline Comet Assay. *Basic & Clinical Pharmacology and Toxicology*, Vol.96, Suppl1, (January 2005), pp.1-42, ISSN I742-7835

Morris, J.B. & Buckpitt, A.R. (2009). Upper Respiratory Tract Uptake of Naphthalene. *Toxicological Sciences,* Vol.111, No.2, (October 2009), pp. 383–391, ISSN 1096-6080

Mortimer, K.; Neugebauer, R.; Lurmann, F.; Alcorn, S.; Balmes, J. & Tager, I. (2008). Air Pollution and Pulmonary Function in Asthmatic Children: Effects of Prenatal and Lifetime Exposures. *Epidemiology,* Vol.19, No.4, (July 2008), pp. 550-562, ISSN 1044-3983

Mussali-Galante, P. & Fortoul, T.I. (2008). Atmospheric pollution, In: *Environmental Research Progress,* F.P. Maes, (Ed.), 147-160, Nova Science Publishers, ISBN: 978-1-60456-080-0, New York, USA

Næss Øyvind; Per Nafstad; Geir Aamodt; Bjørgulf Claussen & Pa Rosland. (2007). Relation Between Concentration of Air Pollution and Cause-Specific Mortality:Four-Year Exposures to Nitrogen Dioxide and Particulate Matter Pollutants in 470 Neighborhoods in Oslo, Norway. *American Journal of Epidemiology*. Vol.165, No.4, (February 2007), pp. 435–443, ISSN 0002-9262

Nam, H. Y., Ahn, E. K., Kim, H. J., Lim, Y., Lee, C. B., Lee, K. Y. & Vallyathan,V. (2006). Diesel exhaust particles increase IL-1beta-induced human beta-defensin expression via NF-kappaB-mediated pathway in human lung epithelial cells. *Particle & Fibre Toxicology,* Vol.3, No.9 (May 2006) (available on line), *ISSN*: 17438977

Nemmar, A; Vanbilloen, H; Hoylaerts, M.F; Hoet, P.H; Verbruggen, A.; & Nemery, B. (2002). Passage of intratracheally instilled ultrafine particles from the lung into the systemic circulation in hamster. *American Journal of Respiratory and Critical Care Medicine,* Vol. 165, No.12, (June, 2002) pp. 1671–1672, ISSN 1535-4970

Olive, P.L.; Durand, R.E.; Banath, J.P. & Johnston, P.J. (2001). Analysis of DNA Damage in Individual Cells. *Methods in Cell Biology,* Vol.64, (2001), pp.235- 249, ISSN 0091-679X

Olivieri, D. & Scoditti E. (2005). Impact of Environmental Factors on Lung Defences. *European Respiratory Review,* Vol.14, No.95, (December 2005), pp. 51-56, ISSN 0905-9180

Park, H.S.; Kim, S.R. & Lee, Y.L. (2009). Impact of oxidative stress on lung diseases. *Respirology, Vol.*14, No.1, (January 2009), pp. 27- 38, *ISSN*: 13237799

Parker,D., & Prince, A.(2011) Innate Immunity in the Respiratory Epithelium. *American Journal of Respiratory and Critical Care Medicine,* (Februarry, 2011)(Epublish ahead of print) ISSN:1044-1549

Pavek, P. & Dvorak, Z. (2008). Xenobiotic-Induced Transcriptional Regulation of Xenobiotic Metabolizing Enzymes of the Cytochrome P450 Superfamily in Human Extrahepatic Tissues. *Current Drug Metabolism,* Vol.9, No.2, (February 2008), pp. 129-143, ISSN 1389-2002

Penn, A.L.; Rouse, R.L.; Horohov, D.W.; Kearney, M.T.; Paulsen, D.B. & Lomax, L. (2007). In Utero Exposure to Environmental Tobacco Smoke Potentiates Adult Responses to Allergen in BALB/c Mice. *Environmental Health Perspectives,* Vol.115, No.4, (April 2007), pp. 548-555, ISSN 0091-6765

Perez-Padilla, R.; Schilmann, A. & Riojas-Rodriguez, H. (2010). Respiratory Health Effects of Indoor Air Pollution. *International Journal Tuberculosis Lung Disease,* Vol. 14, No. 9, (September 2010), pp. 1079–1086, ISSN 1081-1206

Pons, F.; Haag, M.; Corcos, L.; Bonnet, P.; Guillouzo, A.; Lugnier, A. & Frossard, N. (2000). Inhalation of Toluene Diisocyanate Affects Cytochrome P4502B1 Expression in Rat Lung. *Archives of Toxicology,* Vol.74, No.7, (September 2000), pp. 397-403, ISSN 0340-5761

Pope, CA.; Burnett, RT.; Thun, MJ.; Calle, EE.; Krewski, D.; Ito, K. & Thurston, GD. (2002). Lung Cancer, Cardiopulmonary Mortality, and Long-term Exposure to Fine Particulate Air Pollution. *Journal of the American Medical Association,* Vol. 287, No. 9, (March 2002), pp. 1132-1141, ISSN 0098-7484

Pope, CA.; Thun, MJ.; Namboodiri, MM.; Dockery, DW.; Evans, JS.; Speizer, FE. & Heath, CW. (1995). Particulate Air Pollution as a Predictor of Mortality in a Prospective Study of U.S. Adults. *American Journal Respiratory Critcal Care Medicine,* 151, (March 1995), pp.669–74, ISSN 1073-449X

Raaschou-Nielson, O.; Bak, H.; Sorensen, M.; Jensen, SS.; Ketzel, M.; Huiberg, M.; Schnohr, P.; Tjonneland, A. Overvad, K. & Loft, S. (2010). Air Pollution from Traffic and Risk for Lung Cancer in Three Danish Cohorts. *Cancer Epidemiology Biomarkers and Prevention.* Vol. 19, No.5, (May 2010), pp. 1284-91, ISSN1055-9965

Rahman, I.; Skwarska, E.; Henry, M.; Davis, M.; O'Connor, C.M.; FitzGerald, M.X.; Greening, A. & MacNee, W. (1999). Systemic and pulmonary oxidative stress in idiopathic pulmonary fibrosis. *Free Radical Biology & Medicine,* Vol.27, No.1-2, (July 1999), pp. 60–8, *ISSN: 0891-5849.*

Randall, T.D. (2010). Bronchus-associated lymphoid tissue (BALT) structure and function. *Advances in Immunology,* No. 107, (November 2010), pp. 187-241, *ISSN* 0065-2776

Raub, J.; MathieuNolf, M.; Hampson N. & Thom S. (2000). Carbon monoxide poisoning - a public health perspective. *Toxicology*, Vol.145, No.1, (April 2000), pp 1- 14, ISSN 0300-483X

Rice, G.E., & Bevilacqua, M.P. (1989) An inducible endothelial cell surface glycoprotein mediates melanoma adhesion. *Science*, (December,8, 1989) Vol .246, No. 4935, pp.1303-1306. ISSN:0193-4511

Risch, A. & Plass, C. (2008). Lung Cancer Epigenetics and Genetics. *International Journal of Cancer*, Vol. 123, No. 1, (July 2008), pp. 1–7, ISSN 0020-7136

Risom, L.; Dybdahl, M.; Bornholdt, J.; Vogel, U.; Wallin, H.; Moller, P. & Loft, S. (2003). Oxidative DNA Damage and Defence Gene Expression in the Mouse Lung after Short-Term Exposure to Diesel Exhaust Particles by Inhalation. *Carcinogenesis*, Vol.24, No.11, (November 2003), pp.1847-1852, ISSN 0143-3334

Rojas-Martinez, R.; Perez-Padilla, R.; Olaiz-Fernandez G.; Mendoza-Alvarado, L.; Moreno-Macias, H.; Fortoul, T.; McDonell, W.; Loomis, D. & Romieu, I. (2007). Lung Function Growth in Children With Long-term Exposure to Air Pollutants in Mexico City. *American Journal of Respiratory and Critical Care of Medicine, No.*176, pp. 377-384, *ISSN* 1073-449X

Rose, M.C. & Voynow, J.A. (2006). Respiratory Tract Mucin Genes and Mucin Glycoproteins in Health and Disease. *Physiological Reviews*, Vol.86, No.1, (January 2006), pp. 245-278, ISSN 0031-9333

Ross, M.H. & Murray, J. (2004). Occupational Respiratory disease in mining. *Occupational Medicine*, Vol.54, No.5, pp. 304–310, *ISSN: 0962-7480.*

Roth, M. (2008) Pathogenesis of COPD. Part II. Inflammation in COPD. The *International Journal of Tuberculosis and Lung Disease*, Vol.12, No.4 (April 2008), pp. 375-380, *ISSN* 1027-3719

Roubicek, D.A.; Gutierrez-Castillo, M.; Sordo, M.; Cebrian-García M.E. & Ostrosky-Wegman, P. (2007). Micronuclei Induced by Airbone Particulate Matter from Mexico City. *Mutation Research*, Vol.631, No 1, (July 2007), pp.9-15, ISSN 1383-5718

Ruano-Ravina, A.; Faraldo-Valles, MJ. & Barros-Dios, JM. (2009). Is there a Specific Mutation of p53 Gene Due to Radon Exposure? A Systematic Review. *International Journal Radiation Biology.* Vol.85, No. 7, pp.614–621, ISSN 0955-3002

Salnikow, K. & Zhitkovich, A. (2008). Genetic and Epigenetic Mechanisms in Metal Carcinogenesis and Cocarcinogenesis: Nickel, Arsenic, and Chromium, *Chemical Research Toxicology*, Vol. 21, (January 2008), pp. 28–44, ISSN 0893-228X

Salvi, S.S., & Barnes, P. (2009) Chronic Obstructive Pulmonary Disease in Non-smokers. *The Lancet*, Vol.374, No. 9691 (August 2009), pp. 733-743, *ISSN 0140-6736*

Sato, H.; Sone, H.; Sagai, M.; Suzuki, KT. & Aoki, Y. (2000). Increase in Mutation Frequency in Lung of Big Blue(R) Rats by Exposure to Diesel Exhaust. *Carcinogenesis*, Vol.21, No.4, (April 2000), pp. 653-661, ISSN 0143-3334

Saura, M; Zaragoza, C; Bao, C.; Rodriguez-Puyol, M; & Lowenstein, C.J. (2006). Stat3 mediates interleukin-6 [correction of interelukin-6] inhibition of human endothelial nitric-oxide synthase expression. *Journal of Biological Chemistry*, (October,1, 2006). Vol.281, No.40, pp. 30057-30062. ISSN: 1083-351X

Sawant, S.G.; Randers-Pherson, G.; Geard, CR.; Brenner, D.J. & Hall EJ. (2001). The Bystander Effect in Radiation Oncogenesis: I. Transformation in C3H 10T1/2 Cells in vitro can be initiated in the Unirradiated Neighbors of Irradiated Cells. *Radiation Research*, Vol.155, No.3, (March 2001), pp. 397-401, ISSN 0033-7587

Schikowski, T; Sugiri, D; Ranft, U; Gehring, U; Heinrich, J; Wichmann, H.E.; & Kramer, U. (2005) Long-term air pollution exposure and living close to busy roads are associated with COPD in women. *Respiratory Research*, Vol. 22, No. 6 (December, 2005) pp. 152-161, ISSN:1465-9921

Schins, R.P; Lightbody, J.H; Borm, P.J; Shi, T; Donaldson, K.; & Stone, V. (2004) Inflammatory effects of coarse and fine particulate matter in relation to chemical and biological constituents. *Toxicology and Applied Pharmacology*, Vol. 195, No. 1 (Feb, 2004) pp. 1-11. ISSN:0041-008X

Schwarze, P.E; Oyreyik, J; Lag, M; Refsnes, M.; Nafstad, P.; Hetland, R.B.; & Dybing, E. (2006) Particulate matter properties and health effects: consistency of epidemiological and toxicological studies. *Human and Experimental Toxicology*, Vol. 25, No. 10 (October, 2006) pp.559-579. ISSN:0960-3271

Seagrave, J. (2008). Mechanisms and implications of air pollution particle associations with chemokines. *Toxicology and Applied Pharmacology*, (November, 1, 2008).Vol. 232, No. 3, pp. 469-477. ISSN: 1096-0333

Seagrave, J; McDonald, J.D; Bedrick, E; Edgerton, E.S; Gigliotti, A.P; Jansen, J.J; Ke, L.; Naeher, L.P; Seilkop, S.K; Zheng, M; & Mauderly, J.L. (2006) Lung toxicity of ambient particulate matter from southeastern U.S. sites with different contributing sources: relationships between composition and effects. *Environmental Health Perspectives*, Vol. 114, No. 9, (September, 2006) pp. 1387-1393. ISSN:0091-6765

Sharma, A.K; Jensen, K.A; Rank, J; White, P.A; Lundstedt, S; Gagne, R; Jacobsen, N.R; Kristiansen, J; Vogel, U. & Wallin, H. (2007). Genotoxicity, Inflammation and Physic-Chemical Properties of Fine Particle Samples from an Incineration Energy Plant and Urban Air. *Mutation Research-Genetic Toxicology and Environmental Mutagenesis*, Vol.633, No.2, (October 2007), pp. 95-111, ISSN 1385-5718

Shoenfelt, J.; Mitkus, R. J.; Zeisler, R.; Spatz, R. O.; Powell, J.; Fenton, M. J.; Squibb, K. A.; Medvedev, A. E. (2009). Involvement of TLR2 and TLR4 in inflammatory immune responses induced by fine and coarse ambient air particulate matter. *Journal of Leukocyte Biology*, Vol. 86 No.2, (August, 2009) pp. 303-312. ISSN:0741-5400

Sigaud, S.; Goldsmith, C.A.; Zhou, H.; Yang, Z.; Fedulov, A.; Imrich, A.; & Kobzik, L. (2007). Air pollution particles diminish bacterial clearance in the primed lungs of mice. *Toxicology and Applied Pharmacology*, (August, 1, 2007).Vol. 223, No. 1, pp. 1-9. ISSN: 1096-0333

Song, J. W, Hong, S. B, Lim, C.M, Koh, Y. & Kim, D. S. (2011). Acute exacerbation of idiopathic pulmonary fibrosis: incidence, risk factors and outcome. *European Respiratory Journal*, Vol. 37, No.2, (February 2011) pp. 356-363, *ISSN*: 0903-1936

Spaan, S.; Schinkel, J.; Wouters, I.M.; Preller, L.; Tielemans, E.; Nij, E.T.; & Heederik, D. (2008). Variability in endotoxin exposure levels and consequences for exposure assessment. *Annals of Occupational Hygiene*, Vol.52, No.5 (July 2008) pp. 303-316. ISSN:0003-4878

Strand, V, Rak, S, Svartengren, M. & Bylin, G. (1997). Nitrogen dioxide exposure enhances asthmatic reaction to inhaled allergen in subjects with asthma. *American Journal of Respiratory and Critical Care Medicine*, Vol.155, No. 3, (March 1997), pp 881 -887, ISSN 1073-449XWinterton, DL.; Kaufman, J, Keener CV, Quigley, S, Farin, FM.; Williams PV. & Koenid, JQ. (2001). Genetic Polymorphism as a Biomarkers of Sensitivity to Inhaled Sulfur Dioxide in Subjects with Asthma. (2001). *Annals of Allergy Asthma & Immunology*, Vol.86, No. 2, (February 2001), pp 232-238, ISSN 1073-449X

Tamagawa, E.; Bai, N.; Morimot, K.; Gray, C.; Mui, T.; Yatera, K.; Zhang, X.; Xing, L.; Laher, I.; Sin, D.D.; Man, S.F.; & van Eeden, S.F. (2008). Particulate matter exposure induces persistent lung inflammation and endothelial dysfunction. *American Journal of Physiology-Lung Cellular and Molecular Physiology*. (July, 2008) Vol. 295, No. 1, pp. L79-L85. ISSN:1040-0605

Tamir, S. & Tannenbaum, S.R. (1996). The Role of Nitric Oxide (NO.) In the Carcinogenic Process. *Biochimica et Biophysica Acta*. Vol.1288, No.2, (October 1996), pp. F31–F36, ISSN 0304-419X

Torres-Duque, C.; Maldonado, D.; Perez-Padilla, R.; Ezzati, M.; & Viegi, G. (2008). Biomass fuels and respiratory diseases: a review of the evidence. *Proceedings of the American Thoracic Society*, Vol. 5, No. 5 (August, 2008) pp. 577-590, ISSN:1546-3222

Tozer, G.M.; Prise, V.E.; & Chaplin, D.J. (1997). Inhibition of Nitric Oxide Synthase Induces a Selective Reduction in Tumor Blood Flow That is Reversible with L-arginine. *Cancer Research*, Vol. 57, (March 1997), pp.948–955, ISSN: 0008-5472

United Nations Scientific Committee on the Effects of Atomic Radiation (UNSCEAR). (2000). Sources and Effects of Ionizing Radiation. New York: United Nations. Available from http://www.unscear.org/unscear/publications/2000_1.html

van Eeden, S.F.; Tan, W.C.; Suwa, T.; Mukae, H.; Terashima, T.; Fujii, T.; Oui, D.; Vincent, R.; Hogg, J.C. (2001). Cytokines involved in the systemic inflammatory response induced by exposure to particulate matter air pollutants (PM(10). *American Journal of Respiratory and Critical Care Medicine* (September, 1, 2001) Vol. 164, No. 5, pp.826-830. ISSN: 1535-4970

Vineis, P. & Husgafvel-Pursiainen, K. (2005). Air Pollution and Cancer: Biomarker Studies in Human Populations. *Carcinogenesis*, Vol.26, No.11, (November 2005), pp.1846-1855, ISSN 0143-3334

Wang, L. & Pinkerton KE. (2008). Detrimental Effects of Tobacco Smoke Exposure During Development on Postnatal Lung Function and Asthma. *Birth Defects Research Part C: Embryo Today*, Vol.84, No.1, (March 2008), pp. 54-60, ISSN 1542-975X

Wegesser, T.C. & Last, J.A. (2009) Mouse lung inflammation after instillation of particulate matter collected from a working dairy barn. *Toxicology and Applied Pharmacology*, Vol. 236, No. 3 (March, 2009) pp. 348-357. ISSN:0041-008X

Westergren-Thorsson, G.; Larsen, K.; Nihlberg, K.; Andersson-Sjoland, A.; Hallgren, O.; Marko-Varga, G. & Bjermer, L. (2010). Pathological airway remodeling in inflammation. *The Clinical Respiratory Journal*, Vol.4, No.1 suppl, (May 2010), pp. 1-8, *ISSN*: 1752-6981

Wink, DA.; Darbyshire, JF.; Nims, RW.; Saavedra, JE. & Ford, PC. (1993). Reactions of the Bioregulatory Agent Nitric Oxide in Oxygenated Aqueous Media: Determination of the Kinetics for Oxidation and Nitrosation by Intermediates Generated in the NO/O2 Reaction. *Chemical Research Toxicology*, Vol. 6, (Jan-Feb 1993), pp.23–7, ISSN 0893-228X

Yang, W. & Omaye, S.T. (2009). Air Pollutants, Oxidative Stress and Human Health. *Mutation Research/ Genetic Toxicology and Environm Mutagenesis*, Vol.674, No.1-2, (March 2009), pp. 45-54, ISSN 1383-5718

Zuurbier, M.; Hoek, G.; Oldenwening, M.; Meliefste, K.; van den Hazel, P., & Brunekreef, B. (2011). Respiratory Effects of Commuters' Exposure to Air Pollution in Traffic. *Epidemiology*. Vol. 22, No.2, (March 2011), pp. 219-27, *ISSN*: 1044-3983

4

Air Pollution and Cardiovascular Disease

Jan Emmerechts[1], Lotte Jacobs[2] and Marc F. Hoylaerts[1]
[1]*Center for Molecular and Vascular Biology*
[2]*Occupational & Environmental Medicine, Unit of lung toxicology*
University of Leuven
Belgium

1. Introduction

Numerous epidemiological studies report consistent associations between exposure to urban air pollution and cardio-respiratory morbidity and mortality. One of the important discoveries of these epidemiological studies during the last decade was that the increased mortality associated with enhanced air pollution exposure was not due only to pulmonary diseases, but mainly to *cardiovascular* diseases. (Zanobetti et al. 2003, Samet et al. 2000, Dockery et al. 1993, Jerrett et al. 2005, Pope et al. 2004a, Pope et al. 2002, Simkhovich, Kleinman and Kloner 2008, Nawrot, Nemmar and Nemery 2006, Hoek et al. 2002, Katsouyanni et al. 2001, Dominici et al. 2003).

The focus in the initial epidemiological research was directed towards the association between both short-term and long-term exposure to air pollution and arterial cardiovascular effects, such as myocardial infarction. These landmark studies, in the beginning of the 90's, were quickly followed by experimental studies in humans and in rodents, to unravel the underlying pathophysiological mechanisms. The number of publications in this field increased exponentially, so that by the beginning of 2011, a search through PubMed using the MeSH terms 'air pollution' and 'cardiovascular disease' retrieved almost 1300 hits.

Ambient environmental air pollutants include gaseous (carbon monoxide, nitrogen oxides, sulfur dioxide, ozone) and particulate components. The particulate component, particulate matter (PM), is subdivided based on size ranges into 'thoracic particles' (PM_{10}, with a mean aerodynamic diameter <10 µm), 'coarse particles' (>2.5 µm and <10 µm), 'fine particles' ($PM_{2.5}$, <2.5 µm), and ultra-fine particles (UFP, <0.1 µm). Although exposure to some gaseous components has been linked to cardiovascular events, the larger body of evidence points towards the deleterious effects of the particulates in air pollution. Therefore, this chapter will focus mainly on the cardiovascular morbidity induced by PM exposure.

Active cigarette smoking has been established as a major independent cause of cardiovascular disease (HHS 2004). The inhaled dose of fine particles from ambient air pollution, as from secondhand cigarette smoke, is extremely small compared with that from active cigarette smoking. Accordingly, the estimated relative risks from active smoking, even at relatively light smoking levels, are substantially larger than the risks from ambient air pollution or secondhand smoke. However, the risks induced by these latter 2 types of exposure are higher than would be expected from a simple linear extrapolation based on the amount of inhaled PM from active smoking (Pope et al. 2009), and have important public health implications (Nawrot et al. 2011).

Arterial and venous thrombosis share common risk factors (Lowe 2008). The role of air pollution exposure as a risk factor for *arterial* events now being beyond discussion, a few years ago, epidemiologists started investigating a possible association with *venous* thrombotic events. Thus, in 2008, Baccarelli et al. demonstrated a link between chronic exposure to elevated levels of air pollution and deep vein thrombosis (DVT) for the first time. To understand the pathophysiological mechanisms underlying the observed link between air pollution and cardiovascular morbidity, one should take into account the complex interplay of prohemostatic and antihemostatic mechanisms, with different protagonists for the arterial and the venous vasculature. The human cardiovascular system consists of a functional vascular network for blood distribution, subdivided in a systemic and pulmonary circulatory system. The systemic circulation transports oxygenated blood through the arteries from the left heart to the organs and returns oxygen-depleted blood through the veins to the lungs. The pulmonary circulation subsequently transports the oxygen-depleted blood from the heart to the lungs, where it is oxygenated and returned to the heart.

Vascular integrity throughout the vascular tree is maintained by the vessel wall itself, as well as by a complex hemostatic mechanism involving blood platelets and coagulation factors.

The critical need to rapidly form a stable, localized clot in response to vascular injury (='hemostasis') must be balanced with the need to maintain blood flow within the vessels. Different antihemostatic mechanisms prevent clot formation under resting physiological conditions, and limit clot growth to the site of vascular injury. When prohemostatic tendencies proceed beyond the physiological need to maintain vascular integrity, a pathological thrombus may form, obstructing the normal blood flow (='thrombosis'). In the arterial system, thrombus formation induces oxygen-deprivation (ischemia) of the downstream tissues, such as myocardial infarction and cerebral ischemia. The formation of an arterial thrombus largely depends on the activation of blood platelets, and is most often triggered by the rupture of an atherosclerotic plaque. Indeed, the chronic localized deposition of lipids into the arterial vessel wall (atherosclerosis) leads to the formation of plaques that can rupture when unstable, hereby exposing their procoagulant contents to the circulation (Ross 1999). Hence, while often being asymptomatic in itself over many years, atherosclerosis formation may cumulate into an acute burst of symptomatic arterial thrombus formation.

In the venous system, thrombus formation results from a decrease in blood flow, in conjunction with a hypercoagulable state and endothelial dysfunction (Virchow's triad), and most often affects the deep veins of the legs (deep vein thrombosis, DVT). The most serious complication of DVT is the embolisation of clot dislodgements to the lungs (pulmonary embolism, PE).

The following paragraphs will describe how air pollutants affect arterial and venous functionality and lead to pathophysiological manifestations.

2. Particle triggered pathophysiological mechanisms

Inhaled particles deposit in various segments of the human respiratory tract. While the larger PM_{10} particles impact to a large extent in the nasopharyngeal and tracheal region, the smaller $PM_{2.5}$ particles penetrate deeper into the bronchi and bronchioli, whereas the UFP reach the alveolar regions. Inhaled particles are believed to affect the cardiovascular system through 3 different pathways: interference with the autonomic nervous system, direct translocation of UFP into the systemic circulation and pulmonary inflammation.

PM inhalation deranges the autonomic nervous control of the heart (Brook et al. 2004). Numerous studies (e.g. (Park et al. 2010, Pope et al. 1999)) have shown that, by reducing the heart rate variability, PM may increase the risk for cardiac arrhytmias and sudden death . In addition, elevations in air pollution have been associated with ST-segment depression (Pekkanen et al. 2002, Mills et al. 2007), an impaired cardiac deceleration capacity (Schneider et al. 2010), hypertension (Ibald-Mulli et al. 2001) and increased diastolic blood pressure (Urch et al. 2005). The exact underlying mechanisms remain to be elucidated, but stimulation of irritant receptors in the airways and subsequent reflex activation of the nervous system as well as direct effects of pollutants on cardiac ion channels have been suggested (Brook et al. 2004, Pope et al. 2004b).

A second mechanism of action comprises the translocation of inhaled particles into the systemic circulation. Direct effects may occur via UFP that readily cross the pulmonary epithelial barrier, along with soluble constituents released from the larger particles (e.g. transition metals). Systemic translocation of particles was demonstrated in experimental animal models (Nemmar et al. 2001) (Oberdorster et al. 2002). Although evidence of systemic translocation from human studies is less clear, with both positive (Nemmar et al. 2002, Pery et al. 2009) and negative (Mills et al. 2006) findings, it is likely that this pathway also exists in humans, given the deep penetration of UFP into the alveoli and the close apposition of the alveolar wall and the capillary network. Radioactivity in the systemic circulation was already detected 1 minute after the inhalation of radioactively labelled carbon particles in humans, with peak radioactivity levels between 10 and 20 minutes (Nemmar et al. 2002). When measured in rats under resting conditions, only a small fraction (1.6-2.5%) of intratracheally instilled UFP translocated into the circulation. However, this fraction increased to 4.7% following pretreatment of the lungs with lipopolysaccharides, suggesting a role for pulmonary inflammation in enhancing the extrapulmonary translocation of particles (Chen et al. 2006). Different translocation mechanisms, ranging from endocytosis by alveolar type I and endothelial cells, over phagocytosis by macrophages to passage through widened tight junctions are recognized and depend on the particle surface chemistry (Oberdorster, Oberdorster and Oberdorster 2005). However, a detailed description of the different pathways is beyond the scope of this article. Once UFP have translocated to the blood circulation, they can be distributed throughout the body, and interact with the vascular endothelium or circulating cells, such as blood platelets and leukocytes.

Inhaled PM executes its deleterious effects also via a third, more chronic mechanism, namely pulmonary inflammation and oxidative stress. Exposure to PM induces a proinflammatory response in human lungs (Ghio, Kim and Devlin 2000), consistent with observations in *in vivo* animal models (Nemmar et al. 2003c, Emmerechts et al. 2010) and in vitro cellular models (Mitschik et al. 2008, Alfaro-Moreno et al. 2008). The presence of soluble transition metals in PM enhances the inflammatory responses via increased oxidative stress (Jiang et al. 2000). The PM-induced pulmonary inflammation is followed by the release of inflammatory cytokines, such as interleukin (IL)-1β, IL-6 and granulocyte macrophage colony-stimulating factor (van Eeden et al. 2001) in the circulation, resulting in the release of bone-marrow derived neutrophils and monocytes (Tan et al. 2000).

The generation of a systemic inflammatory response, mostly demonstrated by increases in C-reactive protein (CRP) (Peters et al. 2001b, Hertel et al. 2010), is of major importance in the pathogenesis of cardiovascular events. Upon PM exposure, IL-6 translocates from the lung into the systemic circulation (Kido et al. 2011) and is directly involved in the regulation of the synthesis of CRP in the liver. Elevated concentrations of IL-6 are associated with an increased risk of cardiovascular events (Ridker et al. 2000, Lindmark et al. 2001) and mortality (Volpato et al. 2001). Knock-out mice that lacked IL-6 were protected against the prothrombotic effects of PM exposure (Mutlu et al. 2007). Increasing evidence points to an extensive cross-talk between inflammation and hemostasis, whereby inflammation leads to activation of blood platelets and of coagulation, and activated blood platelets and coagulation factors also considerably contribute to the inflammatory action (Levi and van der Poll 2010).

In the following paragraphs, the deleterious effects of PM exposure on arterial and venous parameters will be discussed. By virtue of their respective protagonist roles, blood platelet activation will mainly be discussed in the paragraph on arterial events, while coagulation activation will mainly be discussed in the paragraph on venous events. Formally, arterial thrombosis, the basis for myocardial infarction, is the result of vessel wall injury and formation of a platelet-rich thrombus. Venous thrombosis, the basis for VTE (venous thromboembolism) results from coagulation activation and formation of a fibrin-rich thrombus. It should be noted, however, that both blood platelet and coagulation activation intervene in arterial and venous thrombosis, and that both systems highly interact with each other (Prandoni 2009).

3. Air pollution and arterial events

Over the last 2 decades, a vast number of epidemiological studies (reviewed in (Maitre et al. 2006)) have provided convincing evidence to conclude that chronic exposure to PM enhances atherosclerosis and that acute exposure increases the risk of atherosclerotic plaque rupture, triggering arterial thrombosis, myocardial infarction and cardiovascular mortality. Relative risk levels for cardiovascular disease may differ between different studies, due to differences in study design. Short-term effects have been most often studied in time-series and case-crossover studies, while long-term effects have been studied in case-control and cohort studies. Relative risk levels are generally lower in time series studies than in other epidemiological designs. Nevertheless, the associations between cardiovascular disease and PM exposure are consistent, whatever the type of method used (Maitre et al. 2006).

An initial landmark report was that of the Harvard Six Cities study (Dockery et al. 1993), in which a cohort of 8111 adults were followed up for 14 to 16 years. The adjusted overall mortality rate for the most polluted city vs. the least polluted was 1.26 (95%CI 1.08-1.47). Cardiovascular deaths accounted for the largest single category of increased mortality. Each 10 $\mu g/m^3$ increase in long-term levels of $PM_{2.5}$ has been associated with a 8 to 18% increase in cardiovascular mortality (Pope et al. 2004a). An association with mortality was also found for traffic-related air pollution and several traffic exposure variables, although relative risks were small (Beelen et al. 2008). The effects of long-term PM exposure on cardiovascular mortality have been shown elegantly by the demonstration of a parallelism between air quality improvement and reduction in cardiovascular events on a population-based level (Laden et al. 2006, Boldo et al. 2011). A potential benefit in general mortality can be expected within 2 years after the reduction of PM exposure (Schwartz et al. 2008).

The magnitude of these associations appeared to be more pronounced for the smaller $PM_{2.5}$ fraction than for the larger PM_{10} fraction (Puett et al. 2009). Considering a large body of evidence, a recent updated version of the American Heart Association scientific statement on 'Air Pollution and Cardiovascular Disease' (Brook et al. 2010) concluded that per 10 $\mu g/m^3$ increase in long-term levels of $PM_{2.5}$, all-cause mortality increased by an approximate 10%. The mortality risk specifically related to cardiovascular disease appears to be elevated to a similar, or perhaps even greater extent, ranging from 3 to 76% over different studies.

3.1 Chronic PM exposure and atherosclerosis

What etiological agent can explain the link between chronic air pollution exposure and cardiovascular mortality? Künzli et al. provided the first epidemiological evidence for an association with atherosclerosis: living in the areas of Los Angeles with highest annual mean concentrations of ambient $PM_{2.5}$ was associated with an increased intima-media thickness of the carotid artery (Kunzli et al. 2005).

Distance from the residence to a major road correlated with the degree of coronary artery calcification, a measure for atherosclerosis (Hoffmann et al. 2007).

Another study in 5172 adults investigated 20-year PM exposure and found an association, although weaker than in the previous studies, with carotid intima media thickness, but not with other measures of atherosclerosis i.e. coronary calcium and ankle brachial index (Diez Roux et al. 2008).

A recent study demonstrates that long-term PM exposure is not only related to the degree, but also to a faster progression rate of atherosclerosis (Kunzli et al. 2010).

Along with this epidemiological evidence, experimental research also established a link between exposure to PM and the development of atherosclerosis. Repeated exposure to PM_{10} in rabbits was associated with both systemic inflammation and the progression of the atherosclerotic process, the extent of which correlated with the extent of PM_{10} phagocytosed by alveolar macrophages (Suwa et al. 2002).

Exposing genetically susceptible apolipoprotein E-null mice for 6 months to an equivalent concentration of 15.2 $\mu g/m^3$ $PM_{2.5}$ over a lifetime, enhanced abdominal aortic plaque formation as compared to mice exposed to filtered air (Sun et al. 2005). Interestingly, ultrafine (<0.18 μm) particle-exposed mice exhibited significantly larger atheroslerotic lesions than mice exposed to fine (<2.5 μm) particles or filtered air (Araujo et al. 2008).

Atherosclerosis is now considered an inflammatory disease with low density lipoprotein (LDL) cholesterol accumulation in the arteries as the primary risk factor (Ross 1999). However, up to 50% of the patients who develop atherosclerosis do not have high cholesterol (Braunwald 1997). Therefore, it is the relationship between the accumulated lipids and other harmful components of inflammation in the arterial vessel wall that is of concern. LDL infiltration of the arterial vessel wall is followed by oxidative modification to oxidized LDL (ox-LDL) in the subendothelial space and chemotaxis of monocytes. These monocytes differentiate into macrophages and the subsequent phagocytosis of ox-LDL leads to the formation of foam cells and the release of inflammatory mediators, inducing a vicious cycle of inflammation. Further stages include smooth muscle cell proliferation, formation of a fibrous cap with necrotic core and calcification (Ross 1999). Thickening of the vessel wall and obliteration of the vascular lumen induces downstream ischemia of the tissues.

PM exposure can induce atherosclerosis via different pathways: systemically translocated UFP or their chemical constituents induce activation of proatherogenic molecular pathways in endothelial cells, by oxidative stress. Inflammatory mediators released from the lungs may promote chemotaxis of monocytes into the vessel wall. PM induces high-density lipoprotein (HDL) dysfunction with loss of its anti-inflammatory properties (Araujo and Nel 2009).

Oxidative transformation of LDL into ox-LDL is a key step in the initiation and progression of atherosclerosis (Stocker and Keaney 2004), and circulating levels of ox-LDL are therefore an early marker, and a risk factor for the disease (Wallenfeldt et al. 2004). The correlation between PM exposure and circulating levels of ox-LDL on an individual level was shown by Jacobs et al., demonstrating a dose-dependent association between this parameter and the carbon load of airway macrophages, a personal marker for chronic exposure to fossil fuel derived ultrafine particles (Jacobs et al. 2011).

It has been previously shown that particles can induce oxidative stress both in vitro (Jimenez et al. 2000, Carter et al. 1997) and in exposed animals (Costa and Dreher 1997, Kadiiska et al. 1997, Tao, Gonzalez-Flecha and Kobzik 2003, Araujo et al. 2008).

In agreement with epidemiological findings (Puett et al. 2009), experimental studies suggest that the smaller particles are more pathogenic, as a result of their greater propensity to induce systemic prooxidant and proinflammatory effects (Araujo et al. 2008). Indeed, ambient UFP trigger the induction of the antioxidant gene heme oxygenase 1 (HO-1) to a higher degree than ambient $PM_{2.5}$ or coarse particles, both *in vitro* (Li et al. 2004) and *in vivo* (Araujo et al. 2008, Araujo and Nel 2009). Several mechanisms contribute to the greater proatherogenic potential of UFP: because of their small size, particles < 0.1-0.2 μm contribute very little to overall $PM_{2.5}$ mass. However, they represent >85-90% of the total $PM_{2.5}$ particle number (Sioutas, Delfino and Singh 2005). The high number of UFP, in conjunction with a large surface-to-mass ratio increases the bioavailability of the pro-oxidant chemicals (polycyclic aromatic hydrocarbons, transition metals etc.) present on the UFP's surface. The number of chemicals that are displayed on the surface of particles increases exponentially as the size shrinks below 100 nm (Oberdorster et al. 2005). Deep penetration in the lung and subsequent translocation of UFP into the circulation make these pro-oxidant chemicals more bioavailable at the contact sites of the particles with cells and tissues.

3.2 Acute PM exposure and arterial thrombosis

Not only chronic, but also short-term PM exposure has been linked to cardiovascular mortality: Both the American NMMAPS (National Morbidity, Mortality, and Air Pollution Study (Dominici et al. 2003)) and the European APHEA2 (Air Pollution and Health: A European Approach (Katsouyanni et al. 2001, Zanobetti et al. 2003)) studies (approximately 50 million and 43 million persons included respectively) demonstrated small increases in cardiovascular mortality with increasing PM exposure. In an attempt to evaluate the coherence of studies across continents, the APHENA (A Combined European and North American Approach) analyzed data of these 2 aforementioned studies and Canadian studies (Samoli et al. 2008). The combined effect on all-cause mortality ranged from 0.2% to 0.6% for a 10 μg/m^3 increase in daily levels of ambient PM_{10}, with greater effects for the elderly (>75 years) and the unemployed. An extensive review of studies investigating a link between short-term PM exposure and cardiovascular mortality is provided in (Brook et al. 2010).

Peters et al. (Peters et al. 2001a) demonstrated an increased risk of myocardial infarction in association with elevated concentrations of fine $PM_{2.5}$, both in the previous 2-hours period and the day before the onset. Likewise, the onset of myocardial infarction was linked to participation in traffic, as soon as 1 h afterward (odds ratio 2.92, 95%CI 2.22-3.83) (Peters et al. 2004).

Exposure to ambient $PM_{2.5}$ is associated with short-term increases in hospital admission rates for cerebrovascular, peripheral and cardiac ischemic disease, heart rhythm and heart failure, with the strongest association for heart failure (1.28 % 95%CI 0.78-1.78% increase in risk per 10 $\mu g/m^3$ increase in same-day $PM_{2.5}$) (Dominici et al. 2006).

The risk of mortality from coronary heart disease related to PM exposure appears to be higher in women (RR 1.42, 95%CI 1.06-1.90) than in men (RR 0.90, 95%CI 0.76-1.05 per 10 $\mu g/m^3$ increase in $PM_{2.5}$)(Chen et al. 2005). In a study of 65893 postmenopausal women with a median follow-up of 6 years, each increase in long-term levels of $PM_{2.5}$ of 10 $\mu g/m^3$, measured at the women's residence, was associated with a 24% (95%CI 09-41%) increase in the risk of a cardiovascular event, and a 76% (95%CI 25-147%) increase in the risk of death of cardiovascular disease (Miller et al. 2007).

Although the magnitude of the risk on myocardial infarction induced by short-term PM exposure is rather small on a personal level, it is of major importance on a population level, by virtue of the large number of people exposed. Taking into account both risk magnitude and risk prevalence by measurement of the population attributable fraction (PAF), Nawrot et al. showed that a short-term increase in air pollution exposure is an important trigger for myocardial infarction, of similar magnitude (PAF 5-7%) as other well accepted triggers such as physical exertion, alcohol and coffee (Nawrot et al. 2011).

Epidemiological studies suggest an association between short-term increases in PM exposure and atherosclerotic plaque rupture, causing arterial thrombosis and myocardial infarction. In contrast to the growing number of mechanistic studies investigating the role of chronic PM exposure on atherogenesis, the precise mechanisms explaining the role of short-term PM exposure in acute plaque rupture largely remain to be elucidated. However, several epidemiological and mechanistic studies demonstrated that, in parallel to atherosclerotic plaque rupture, direct or indirect activation of circulating blood platelets by PM contributes to the arterial thrombosis risk. Indeed, the extent to which a growing thrombus occludes the vascular lumen may in part depend on platelet hyperactivity.

Under physiological circumstances, the high blood pressure generated on the arterial side of the circulation requires a powerful, almost instantaneous prohemostatic response in order to minimize blood loss from sites of vascular injury. Blood platelets play a critical role in this response. Upon damage of the endothelial cell layer covering the luminal side of blood vessels, circulating blood platelets adhere to the exposed subendothelial matrix through the binding of the glycoprotein (GP) Ib-IX-V receptor to exposed von Willebrand factor (vWF). Blood platelet adhesion is further enhanced by the binding of different GP receptors to other subendothelial matrix proteins, such as collagen and fibrin(ogen). Upon adhesion and activation of the blood platelets by various agonists, vWF and fibrinogen molecules cross-link different platelets, resulting in blood platelet aggregation and the formation of an initial platelet plug which covers the site of endothelial lesion. The simultaneous activation of the coagulation cascade leads to the formation of a network of insoluble fibrin strands that further stabilize the initial platelet plug.

Air pollution exposure can induce an inappropriate activation of blood platelets beyond the physiological need to restore vessel damage, resulting in arterial thrombosis (Fig. 1).

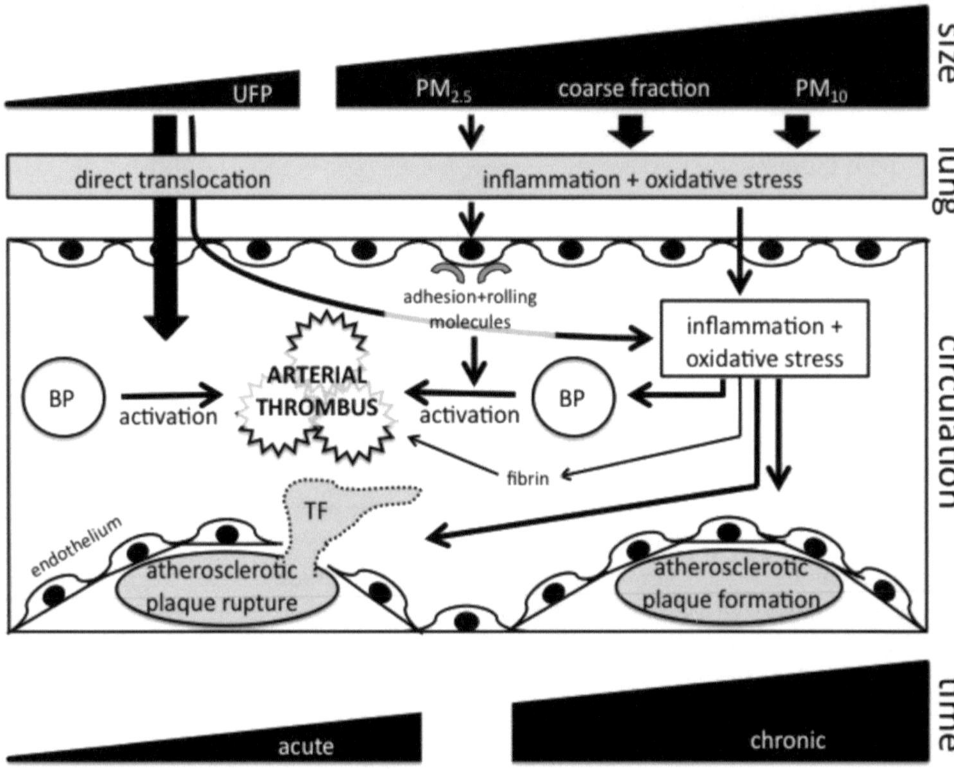

BP: blood platelet, PM: particulate matter, TF: tissue factor, UFP: ultra-fine particles

Fig. 1. Biological pathways linking PM exposure and arterial thrombosis

By exposing healthy volunteers to diluted diesel exhaust, Lucking et al. showed an association with enhanced platelet activity and thrombus formation in an *ex vivo* perfusion chamber, 2 hours and 6 hours after exposure, in conjunction with increased numbers of platelet-neutrophil (+52%) and platelet-monocyte (+30%) conjugates (Lucking et al. 2008).

Short-term, but not long-term PM exposure was found to enhance platelet function, as measured *ex vivo* by a shortening of the closure time of the Platelet Function Analyzer (PFA-100, Siemens Healthcare Diagnostics), in patients with diabetes (Jacobs et al. 2009). In this study, an interquartile range (39.2 μg/m³) increase in the PM_{10} concentration, measured 2 hours before the clinical investigation at the entrance of the hospital, was associated with a decrease of 21.1 sec (95%CI -35.3 to -6.8) in the PFA-100 closure time. Platelet function was not correlated with leukocyte counts, suggesting that short-term PM exposure may have effects on platelet function independently of systemic inflammation, as was also shown in experimental animal models (Nemmar et al. 2003c).

Ambient PM_{10} levels have also been associated with augmented platelet aggregation 24 to 96 hours after exposure in healthy adults, in the absence of increased CRP or fibrinogen (Rudez et al. 2009). In patients with coronary heart disease, mean concentrations over 24 hours of ambient UFP, but not $PM_{2.5}$ or PM_{10} were positively associated with the levels of soluble CD40 ligand, a marker for platelet activation. No assocations were found with longer time frames, up to 5 days (Ruckerl et al. 2007b).

In experimental conditions using DEP, Nemmar et al. demonstrated a prothrombotic tendency and activation of circulating blood platelets (confirmed by PFA-100), as well as lung inflammation, which persisted up to 24 hr after intratracheal instillation of DEP in hamsters (Nemmar et al. 2003a, Nemmar et al. 2003c).

However, different pathophysiological mechanisms seem to be responsible for the observed prothrombotic risk at different time points. Pretreatment of hamsters with a histamine H1-receptor antagonist, an anti-inflammatory drug, abolished pulmonary inflammation at all time points and reduced DEP-induced thrombosis at 6 and 24 hours post-instillation, indicating a crucial role for inflammation in thrombogenicity at these time points. Likewise, the administration of other anti-inflammatory drugs, such as dexamethasone and selective inhibitors of basophils, macrophages and neutrophils, also significantly reduced the PM-induced prothrombogenicity at 24 hours (Nemmar et al. 2004, Nemmar et al. 2005).

In contrast, pretreatment with the histamine H1-receptor antagonist did not reduce thrombosis as soon as 1 hour after DEP exposure (Nemmar et al. 2003c). Therefore, the early prothrombotic tendency appears not to result from pulmonary inflammation, but possibly from direct effects of systemically translocated particles on the blood platelets and/or the (pulmonary) vessel wall (Nemmar et al. 2003c). The direct activating effect of PM on blood platelets was shown by the addition of as little as 0.5 μg/mL DEPs to untreated hamster blood, significantly shortenening the PAF-100 closure time (Nemmar et al. 2003a), as well as by a dose-dependent (0.1-1 μg/mL) effect of PM on in vitro platelet aggregation in rat blood (Nemmar et al. 2010), although similar experiments in human blood were negative (Rudez et al. 2009).

In agreement with these results, 1 hour after intratracheal instillation, well-defined positively charged ultrafine (60 nm) polystyrene particles significantly enhanced platelet-rich thrombus formation, while 400 nm particles, incapable of systemic translocation, did not affect thrombus formation, despite similar increases in neutrophils, lactate dehydrogenase and histamine levels in the bronchoalveolar lavage fluid (Nemmar et al. 2003b).

Pulmonary instillation of carbon nanotubes elevated platelet-leukocyte conjugates at 6 hours and increased the peripheral thrombotic potential at 24 hours after exposure. Inhibition of P-selectin abrogated these responses (Nemmar et al. 2007). P-selectin is found in storage Weibel-Palade bodies of endothelial cells and in α-granules of platelets, from where it can be expressed on the outer membrane upon activation. Surface expression of P-selectin initiates capture and rolling of circulating leukocytes over stimulated endothelium (Theilmeier et al. 2002) and the formation of platelet-leukocyte conjugates (Yokoyama et al. 2005). Increased levels of platelet-leukocyte conjugates have been demonstrated in Indian women who used biomass as cooking fuel, producing higher levels of PM, as compared to women cooking with a cleaner fuel (liquefied petroleum gas) (Ray et al. 2006). In a panel study of 60 elderly subjects with coronary artery disease, Delfino et al. demonstrated associations between soluble P-selectin levels and the mean 1 to 5-day concentrations of ambient finer particles ($PM_{0.25}$ and $PM_{2.5}$), but not the bigger PM_{10} (Delfino et al. 2009). Taken together, these studies suggest that the release of pulmonary cell-derived mediators (eg. histamine) and the expression of endothelial and platelet surface proteins (eg. P-selectin) after several hours, along with the more rapid activation of circulating platelets by direct contact with UFP may mediate peripheral prothrombotic effects.

4. Air pollution and venous thromboembolism

4.1 Epidemiology

In addition to the well-recognized PM-related adverse effects on the arterial vascular system, recent epidemiological evidence also suggests an association between exposure to PM and venous thromboembolism (VTE). Baccarelli et al. reported a 70% increase in the risk of deep vein thrombosis (DVT) for each 10 $\mu g/m^3$ increase of the annual mean level of PM_{10} in the areas of residence of the study subjects (OR 1.70, 95%CI 1.30-2.23) (Baccarelli et al. 2008). The observed exposure-response relationship was approximately linear over the observed PM_{10} range, so that PM_{10} at the higher concentrations within the international limits can still increase the risk of DVT, as compared to the lowest concentration measured. These authors found, in the same study subjects, that living near major traffic roads was also associated with an increased risk of DVT, even after controlling for the community-level PM pollution (Baccarelli et al. 2009). Very recently, exposure to PM has also been associated with hospital admission for VTE in Chile. Both for DVT and for PE, pooled estimates of relative risk of hospitalization were 1.05 (95%CI 1.03-1.06) for a 20.02 $\mu g/m^3$ increase in $PM_{2.5}$ (Dales, Cakmak and Vidal 2010).

However, these initial epidemiological reports on the association between air pollution exposure and venous thrombosis were followed by a number of prospective cohort studies that failed to demonstrate an association: 26,450 post-menopausal women, enrolled in the Women's Health Initiative (WHI) Hormone Therapy trials, were randomized to treatment with either hormone therapy or placebo. Regardless of the treatment category, no evidence was found of an association between short- or long-term (up to 1 year) PM exposure and VTE (Shih et al. 2010). Of note, the aforementioned study of Baccarelli et al. also observed lower PM-induced VTE risk among women compared to men (Baccarelli et al. 2008). A prospective study in 13,134 middle-aged persons, including men and women, also provided evidence against an association between VTE and long-term air pollution exposure, as assessed by residential distance to a major road (Kan et al. 2011).

Hence, in contrast to the well-accepted and documented deleterious effects of air pollution exposure on arterial events, data are scarce and the link with venous thrombosis is less straightforward, prompting further epidemiological investigation.

4.2 Pathophysiology

At lower rates of shear found in the venous circulation, the contribution of blood platelets to clot formation is of lesser importance than in the arterial circulation, leaving a protagonist role for the coagulation cascade in venous hemostasis. Activation of the coagulation cascade is initiated by activation of coagulation factor VII (FVII) by binding to tissue factor (TF), expressed on subendothelial cells such as fibroblasts and vascular smooth muscle cells. The complex of TF and activated FVII (FVIIa) initiates a cascade of subsequent coagulation factor activations, resulting in the generation of thrombin. Thrombin (FII) is a key enzyme, converting fibrinogen monomers to fibrin polymers that clot into a fibrin plug, and amplifying the coagulation cascade through activation of FV, FVIII and FXI.

The mechanisms underlying the observed increase in venous thrombosis in association with exposure to air pollution remain largely unknown, and published results with regard to markers of secondary hemostasis activation are conflicting. Although some epidemiological and controlled exposure studies demonstrated an association between PM exposure and shortening of the prothrombin time (PT) or increased levels of fibrinogen and vWF, others failed to demonstrate positive associations with these or other markers of coagulation, in humans (Table 1). In fact, disappointingly few studies reported on PM-induced coagulatory changes that could form the basis for the observed link between air pollution and DVT. How can this conundrum of PM-induced DVT in the absence of a procoagulant phenotype be explained?

One explanation for the lack of positive associations between PM exposure and measurement of parameters of coagulation might be found in the short observation time frame that was used in most studies. While short-term PM exposure enhances blood platelet activation, a more chronically sustained exposure appears to be necessary to induce significant changes in the coagulatory cascade.

This hypothesis is corroborated by epidemiological findings in which the risk for DVT was only associated with the mean PM concentration over a one year period, and not with any shorter time-point (Baccarelli et al. 2008). This was confirmed by animal studies in which short- term exposure of healthy mice to intratracheally instilled DEP or UPM enhanced arterial, but not venous thrombosis (Emmerechts et al. 2010). In this study, even very high doses of PM (up to 200 μg/mouse), given as a single dose, induced only mild increases in the levels of FVII, FVIII and fibrinogen. Likewise, exposure of rats to concentrated PM from New York City air did not alter levels of fibrinogen, FVII or thrombin-antithrombin complexes (TAT) (Nadziejko et al. 2002).

Significant increases in the level of fibrinogen, or decreases in the levels of the anticoagulant proteins activated protein C or tissue factor pathway inhibitor (TFPI) upon short-term PM exposure have been observed in rodents, but at doses of 100 μg or higher per mouse (Cozzi et al. 2007, Inoue et al. 2006). One study stands out among other studies on procoagulant changes and PM exposure: Mutlu et al. observed a pronounced prothrombotic phenotype in mice upon a single intratracheal instillation of as few as 10 μg of PM_{10}, characterized by shortenings in bleeding time, PT and aPTT, and relatively high increases in the levels of circulating blood platelets, FVII, FVIII, FX and fibrinogen (Mutlu et al. 2007). The reason for the discrepancy between this and other studies being unclear, this study is of value since it demonstrated the absence of a PM-induced prothrombotic phenotype in interleukin-6 (IL-6) knock-out mice, recognizing a major role of inflammatory factors in the induction of procoagulant changes following PM exposure.

Indeed, although some studies suggest a short-term effect of directly translocated UFP through the activation of the coagulation cascade via contact activation, as demonstrated *in* vitro (Kilinc et al. 2010), evidence seems to favor a more prominent role for inflammatory changes related to chronic PM exposure. In this context, it is of interest that the only coagulation factor for which the associations with air pollution were consistent over different studies in humans is fibrinogen (Table 1), an acute phase protein that is upregulated during inflammatory processes.

However, although considered to be a (minor) risk factor, elevated levels of fibrinogen seem unlikely to be solely responsible for the PM-induced increased risk of DVT.

Through the expression of procoagulant proteins and lipids on their surface, microvesicles (also called microparticles, a term we prefer to avoid in the context of pollution by particles) could offer an alternative explanation. Microvesicles are circulating vesicles released from stimulated or apoptotic cells in the vasculature, or during thrombogenesis in the bone marrow, with a mean diameter smaller than 1 μm. Through their surface expression of

reference	subjects					exposure		coagulatory changes	
author + year	n	type	mean age (SD or range)	gender (% male)	controlled exposure	type of air pollution	exposure time	significant changes	no significant changes
(Seaton et al. 1999)	112	NA	70 (7)	?	no	ambient PM_{10}	3 days	FVII (-), fbg (-)	
(Ghio et al. 2000)	38	healthy subjects	26 (0.7)	95	yes	concentrated $PM_{2.5}$	2h	fbg	
(Pekkanen et al. 2000)	7205	healthy subjects	NA	69	no	ambient PM_{10}	1-3 days	fbg	
(Ghio et al. 2003)	20	healthy subjects	25 (0.8)	70	yes	concentrated $PM_{2.5}$	2h	fbg	D-dim, PC, vWF
(Riediker et al. 2004)	9	healthy subjects	27 (23-30)	100	no	in-vehicle $PM_{2.5}$	9h	vWF	
(Beckett et al. 2005)	12	healthy subjects	35 (23-52)	50	yes	zinc oxide particles	2h		fbg, FVII, vWF
(Blomberg et al. 2005)	15	COPD patients	66 (56-72)	NA	yes	diesel exhaust	1h		fbg, D-dim, vWF
(Barregard et al. 2006)	13	healthy subjects	34 (20-56)	46	yes	wood smoke	4h	FVIII	fbg, FVII, D-dim, vWF
(Ruckerl et al. 2006)	57	CHD patients	66 (6)	100	no	ambient $PM_{2.5}$ and PM_{10}	1-5 days	FVII (-), vWF	fbg, D-dim
(Baccarelli et al. 2007)	1218	healthy subjects	44 (11-84)	40	no	ambient PM_{10}	t0 - 30days	PT	aPTT, fbg, AT, PC, PS
(Carlsten et al. 2007)	13	healthy subjects	25 (20-42)	85	yes	diesel exhaust	2h		D-dim, vWF
(Chuang et al. 2007)	76	healthy students	21 (18-25)	65	no	ambient $PM_{2.5}$ and PM_{10}	1-3 days	fbg	
(Ruckerl et al. 2007a)	1003	MI survivors	65 (45-78)	69	no	ambient $PM_{2.5}$ and PM_{10}	1-4 days	fbg	
(Scharrer et al. 2007)	20	healthy subjects	29 (8)	60	yes	welding fume	1h		FVIII, vWF, AT
(Brauner et al. 2008)	41	healthy subjects	67 (60-75)	51	yes	indoor $PM_{2.5}$ and PM_{10}	2 days		fbg, FII+VII+X
(Lucking et al. 2008)	20	healthy subjects	26 (21-44)	NA	yes	diesel exhaust	1-2h		PT, aPTT
(Rudez et al. 2009)	40	healthy subjects	41 (15)	35	no	ambient PM_{10}	6h-4days		fbg,TG
(Samet et al. 2009)	19	healthy subjects	18-35	53	yes	concentrated ambient UFP	2h	D-dim	fbg, FIX, FXII, vWF
(Bonzini et al. 2010)	37	steel plant workers	42 (7)	100	no	occupational PM_{10}	1-3 days	PT,TG	aPTT, D-dim
(Stewart et al. 2010)	19	T2DM patients	48 (9)	47	yes	carbon UFP	2h		FVII, FIX, D-dim, TF
(Thompson et al. 2010)	45	healthy subjects	27 (19-48)	49	no	ambient $PM_{2.5}$	t0 - 7 days		fbg
(Jacobs et al. 2011)	70	DM patients	57 (14)	53	no	carbon load in alveolar Mφ	NA		vWF

COPD: chronic obstructive pulmonary disease, MI: myocardial infarction, CHD: coronary heart disease, DM: diabetes mellitus, T2DM: type 2 diabetes mellitus, PM: particulate matter, UFP: ultra-fine particles, Mφ: macrophages, PT: prothrombin time, aPTT: activated partial prothrombin time, AT: antithrombin, PC: protein C, PS: protein S, F: factor, fbg: fibrinogen, D-dim: D-dimers, vWF: von Willebrand factor, TG: thrombin

Table 1. Associations between PM exposure and coagulatory changes according to different studies.

negatively charged phospholipids and of tissue factor (TF), they create a procoagulant surface on which coagulation factors can bind and be activated (Morel et al. 2006). Indeed, the initial concept that TF presence is limited to a hemostatic envelope surrounding blood vessels has been challenged by the identification of 'blood borne' TF, either on circulating white blood cells or microvesicles, as a soluble protein, or possibly on stimulated endothelial cells (Pawlinski et al. 2010).

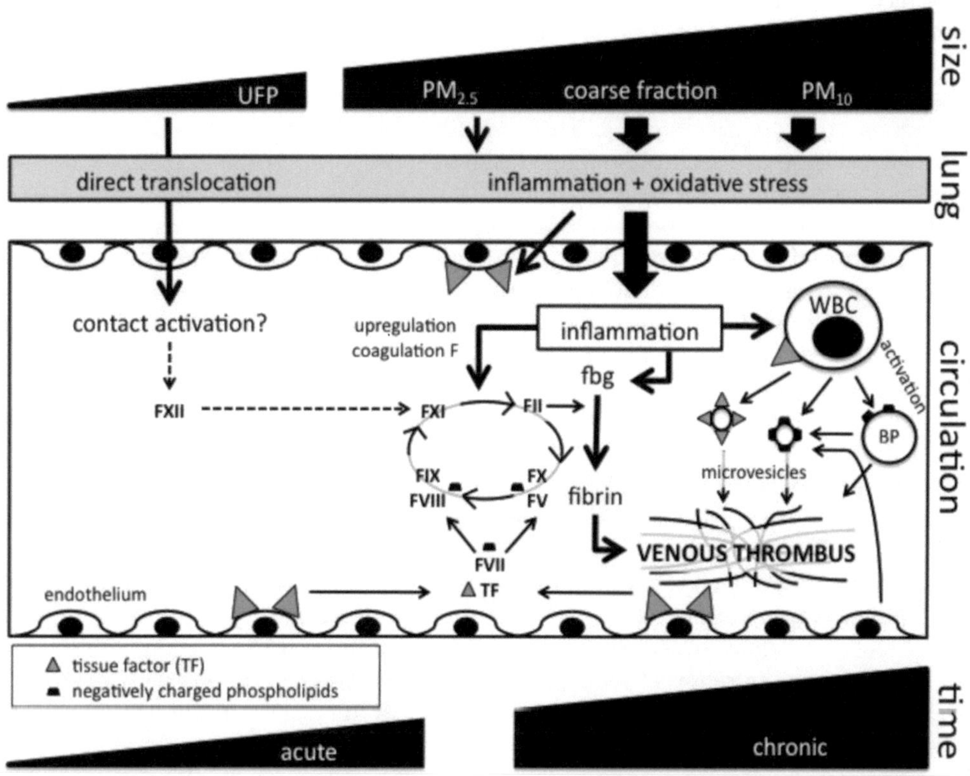

BP: blood platelet, F: factor, fbg: fibrinogen, PM: particulate matter, TF: tissue factor, UFP: ultra-fine particles, WBC: white blood cell

Fig. 2. Biological pathways linking PM exposure and venous thrombosis

A role for microvesicles has been suggested by the work of Bonzini et al., investigating blood samples collected in steel-production plant workers. Besides shortening the PT, elevated PM exposure also enhanced thrombin generation, but only when measured in an assay without the exogenous addition of a coagulation trigger or negatively charged phospholipids (Bonzini et al. 2010). These findings suggest that PM exposure may induce the release of small amounts of endogenous TF and/or negatively charged phospholipids that may function as triggers of thrombin generation in the assay system. Circulating microvesicles might well be the source of these triggers. This hypothesis is corroborated by animal studies demonstrating elevated numbers of procoagulant microvesicles, 24 hours after intratracheal instillation of carbon nanotubes in mice (Nemmar et al. 2007). Likewise, when stimulated *ex vivo*, blood platelets from mice exposed to concentrated ambient PM for 2 weeks released more microvesicles relative to platelets from ambient air-exposed control animals (Wilson et al. 2010). However, observational or controlled exposure studies in humans are needed for further confirmation. Figure 2 summarizes the possible pathophysiological pathways linking PM exposure and venous thrombogenicity.

5. Endothelial function and fibrinolysis

The effects of air pollution on the endothelial function and the fibrinolytic system have mainly been investigated in controlled exposure studies by 2 research groups who joined forces. The groups of Newby and Blomberg used exposure chambers to expose healthy and compromised volunteers to the diluted exhaust of an iddling diesel engine for several hours in randomized cross-over studies. They demonstrated an impaired bradykinin-induced endothelial release of tissue plasminogen activator (t-PA) upon diesel exhaust inhalation (estimated reduction of net t-PA release of 34%) (Mills et al. 2005, Mills et al. 2007), in addition to an attenuated agonist-induced increase in blood flow at 6 hours post-inhalation, in the absence of inflammatory changes (Mills et al. 2005). At 24 hours post-inhalation, endothelium-dependent vasodilatation (induced by acetylcholine and bradykinin) remained impaired, while endothelium-independent vasodilatation (using sodium nitroprusside and verapamil) and t-PA release were unaffected, in the presence of mild sytemic inflammation (Tornqvist et al. 2007).

These and other (Bonzini et al. 2010, Chuang et al. 2007, Ghio et al. 2003, Samet et al. 2009) studies did not demonstrate an assocation between PM exposure and baseline levels (not bradykinin-induced) of t-PA.

While studies, based on controlled exposure to diluted diesel exhaust (Mills et al. 2007, Tornqvist et al. 2007, Carlsten et al. 2007) or concentrated ambient particles (Ghio et al. 2003), did not observe increases in the levels of plasminogen activator inhibitor-1 (PAI-1), some epidemiological or animal studies, focussing on urban PM, did: a study in 76 young healthy students demonstrated elevated PAI-1 concentrations in association with the mean $PM_{2.5}$ or PM_{10} concentration at their university's campus over 1 to 3 days (Chuang et al. 2007). Likewise, urban PM upregulated PAI-1 levels, 24 hours after intratracheal instillation in mice (Cozzi et al. 2007).

PM exposure could also impair the endothelial repair mechanisms by reducing the number of endothelial progenitor cells, as demonstrated by a recent report (O'Toole et al. 2010).

Taken together, these studies indicate a potential deleterious effect of PM inhalation on the endothelial and fibrinolytic function that may aggravate the prothrombotic phenotype induced by blood platelet and coagulation activation.

6. Conclusions

A wide array of epidemiological and experimental studies have provided persuasive evidence that air pollutants, the PM fraction in particular, contribute to cardiovascular morbidity and mortality. By virtue of the heterogeneity in both study design and the composition of the PM considered by these studies, it is not surprising that not all findings have been consistent. However, considering the overall weight of scientific evidence, some general conclusions can be drawn: through the induction of inflammation and oxidative stress, the inhalation of particulates, especially the finest fractions ($PM_{2.5}$ and UFP), over longer time periods contributes to atherosclerotic plaque formation. At shorter time points (<24 h), these particles may induce plaque rupture and activate blood platelets, leading to acute peripheral arterial events such as myocardial infarction. Blood platelet activation within the first few hours is inflammation-independent, most probably resulting from direct contact with systemically translocated particles and/or activated endothelium. Thereafter, inflammatory changes are responsible for further platelet activation.

Although evidence linking PM exposure with venous thromboembolic events is less established than with arterial events and warrants further investigation, recent findings suggest that chronic air pollution exposure is also a risk factor for venous thrombosis. Inflammatory changes, along with the generation of circulating procoagulant microvesicles might be of larger importance than coagulation factor upregulation, favoring a role for the larger particles (PM_{10}) with higher pro-inflammatory endotoxin content on their surface.

Air pollution exposure may not be the highest risk factor for arterial or venous thrombosis on an individual level. However, because of the huge number of persons exposed, on a global scale it is a major, and more importantly, a modifiable risk factor for cardiovascular disease and mortality.

7. References

Alfaro-Moreno, E., T. S. Nawrot, B. M. Vanaudenaerde, M. F. Hoylaerts, J. A. Vanoirbeek, B. Nemery & P. H. Hoet (2008) Co-cultures of multiple cell types mimic pulmonary cell communication in response to urban PM10. *Eur Respir J,* 32, 1184-94.

Araujo, J. A., B. Barajas, M. Kleinman, X. Wang, B. J. Bennett, K. W. Gong, M. Navab, J. Harkema, C. Sioutas, A. J. Lusis & A. E. Nel (2008) Ambient particulate pollutants in the ultrafine range promote early atherosclerosis and systemic oxidative stress. *Circ Res,* 102, 589-96.

Araujo, J. A. & A. E. Nel (2009) Particulate matter and atherosclerosis: role of particle size, composition and oxidative stress. *Part Fibre Toxicol,* 6, 24.

Baccarelli, A., I. Martinelli, V. Pegoraro, S. Melly, P. Grillo, A. Zanobetti, L. Hou, P. A. Bertazzi, P. M. Mannucci & J. Schwartz (2009) Living near major traffic roads and risk of deep vein thrombosis. *Circulation,* 119, 3118-24.

Baccarelli, A., I. Martinelli, A. Zanobetti, P. Grillo, L. F. Hou, P. A. Bertazzi, P. M. Mannucci & J. Schwartz (2008) Exposure to particulate air pollution and risk of deep vein thrombosis. *Arch Intern Med,* 168, 920-7.

Baccarelli, A., A. Zanobetti, I. Martinelli, P. Grillo, L. Hou, S. Giacomini, M. Bonzini, G. Lanzani, P. M. Mannucci, P. A. Bertazzi & J. Schwartz (2007) Effects of exposure to air pollution on blood coagulation. *J Thromb Haemost,* 5, 252-60.

Barregard, L., G. Sallsten, P. Gustafson, L. Andersson, L. Johansson, S. Basu & L. Stigendal (2006) Experimental exposure to wood-smoke particles in healthy humans: effects on markers of inflammation, coagulation, and lipid peroxidation. *Inhal Toxicol,* 18, 845-53.

Beckett, W. S., D. F. Chalupa, A. Pauly-Brown, D. M. Speers, J. C. Stewart, M. W. Frampton, M. J. Utell, L. S. Huang, C. Cox, W. Zareba & G. Oberdorster (2005) Comparing inhaled ultrafine versus fine zinc oxide particles in healthy adults: a human inhalation study. *Am J Respir Crit Care Med,* 171, 1129-35.

Beelen, R., G. Hoek, P. A. van den Brandt, R. A. Goldbohm, P. Fischer, L. J. Schouten, M. Jerrett, E. Hughes, B. Armstrong & B. Brunekreef (2008) Long-term effects of traffic-related air pollution on mortality in a Dutch cohort (NLCS-AIR study). *Environ Health Perspect,* 116, 196-202.

Blomberg, A., H. Tornqvist, L. Desmyter, V. Deneys & C. Hermans (2005) Exposure to diesel exhaust nanoparticles does not induce blood hypercoagulability in an at-risk population. *J Thromb Haemost*, 3, 2103-5.

Boldo, E., C. Linares, J. Lumbreras, R. Borge, A. Narros, J. Garcia-Perez, P. Fernandez-Navarro, B. Perez-Gomez, N. Aragones, R. Ramis, M. Pollan, T. Moreno, A. Karanasiou & G. Lopez-Abente (2011) Health impact assessment of a reduction in ambient PM(2.5) levels in Spain. *Environ Int*, 37, 342-8.

Bonzini, M., A. Tripodi, A. Artoni, L. Tarantini, B. Marinelli, P. A. Bertazzi, P. Apostoli & A. Baccarelli (2010) Effects of inhalable particulate matter on blood coagulation. *J Thromb Haemost*, 8, 662-8.

Brauner, E. V., L. Forchhammer, P. Moller, L. Barregard, L. Gunnarsen, A. Afshari, P. Wahlin, M. Glasius, L. O. Dragsted, S. Basu, O. Raaschou-Nielsen & S. Loft (2008) Indoor particles affect vascular function in the aged: an air filtration-based intervention study. *Am J Respir Crit Care Med*, 177, 419-25.

Braunwald, E. (1997) Shattuck lecture--cardiovascular medicine at the turn of the millennium: triumphs, concerns, and opportunities. *N Engl J Med*, 337, 1360-9.

Brook, R. D., B. Franklin, W. Cascio, Y. Hong, G. Howard, M. Lipsett, R. Luepker, M. Mittleman, J. Samet, S. C. Smith, Jr. & I. Tager (2004) Air pollution and cardiovascular disease: a statement for healthcare professionals from the Expert Panel on Population and Prevention Science of the American Heart Association. *Circulation*, 109, 2655-71.

Brook, R. D., S. Rajagopalan, C. A. Pope, 3rd, J. R. Brook, A. Bhatnagar, A. V. Diez-Roux, F. Holguin, Y. Hong, R. V. Luepker, M. A. Mittleman, A. Peters, D. Siscovick, S. C. Smith, Jr., L. Whitsel & J. D. Kaufman (2010) Particulate matter air pollution and cardiovascular disease: An update to the scientific statement from the American Heart Association. *Circulation*, 121, 2331-78.

Carlsten, C., J. D. Kaufman, A. Peretz, C. A. Trenga, L. Sheppard & J. H. Sullivan (2007) Coagulation markers in healthy human subjects exposed to diesel exhaust. *Thromb Res*, 120, 849-55.

Carter, J. D., A. J. Ghio, J. M. Samet & R. B. Devlin (1997) Cytokine production by human airway epithelial cells after exposure to an air pollution particle is metal-dependent. *Toxicol Appl Pharmacol*, 146, 180-8.

Chen, J., M. Tan, A. Nemmar, W. Song, M. Dong, G. Zhang & Y. Li (2006) Quantification of extrapulmonary translocation of intratracheal-instilled particles in vivo in rats: effect of lipopolysaccharide. *Toxicology*, 222, 195-201.

Chen, L. H., S. F. Knutsen, D. Shavlik, W. L. Beeson, F. Petersen, M. Ghamsary & D. Abbey (2005) The association between fatal coronary heart disease and ambient particulate air pollution: Are females at greater risk? *Environ Health Perspect*, 113, 1723-9.

Chuang, K. J., C. C. Chan, T. C. Su, C. T. Lee & C. S. Tang (2007) The effect of urban air pollution on inflammation, oxidative stress, coagulation, and autonomic dysfunction in young adults. *Am J Respir Crit Care Med*, 176, 370-6.

Costa, D. L. & K. L. Dreher (1997) Bioavailable transition metals in particulate matter mediate cardiopulmonary injury in healthy and compromised animal models. *Environ Health Perspect*, 105 Suppl 5, 1053-60.

Cozzi, E., C. J. Wingard, W. E. Cascio, R. B. Devlin, J. J. Miles, A. R. Bofferding, R. M. Lust, M. R. Van Scott & R. A. Henriksen (2007) Effect of ambient particulate matter exposure on hemostasis. *Transl Res,* 149, 324-32.

Dales, R. E., S. Cakmak & C. B. Vidal (2010) Air Pollution and hospitalization for venous thromboembolic disease in Chile. *J Thromb Haemost.*

Delfino, R. J., N. Staimer, T. Tjoa, D. L. Gillen, A. Polidori, M. Arhami, M. T. Kleinman, N. D. Vaziri, J. Longhurst & C. Sioutas (2009) Air pollution exposures and circulating biomarkers of effect in a susceptible population: clues to potential causal component mixtures and mechanisms. *Environ Health Perspect,* 117, 1232-8.

Diez Roux, A. V., A. H. Auchincloss, T. G. Franklin, T. Raghunathan, R. G. Barr, J. Kaufman, B. Astor & J. Keeler (2008) Long-term exposure to ambient particulate matter and prevalence of subclinical atherosclerosis in the Multi-Ethnic Study of Atherosclerosis. *Am J Epidemiol,* 167, 667-75.

Dockery, D. W., C. A. Pope, 3rd, X. Xu, J. D. Spengler, J. H. Ware, M. E. Fay, B. G. Ferris, Jr. & F. E. Speizer (1993) An association between air pollution and mortality in six U.S. cities. *N Engl J Med,* 329, 1753-9.

Dominici, F., A. McDermott, M. Daniels, S. L. Zeger & J. M. Samet (2003) Mortality among residents of 90 cities. In Revised Analyses of Time-Series Studies of Air Pollution and Health. *Boston, MA: Health Effects Institute,* 9-24.

Dominici, F., R. D. Peng, M. L. Bell, L. Pham, A. McDermott, S. L. Zeger & J. M. Samet (2006) Fine particulate air pollution and hospital admission for cardiovascular and respiratory diseases. *JAMA,* 295, 1127-34.

Emmerechts, J., E. Alfaro-Moreno, B. M. Vanaudenaerde, B. Nemery & M. F. Hoylaerts (2010) Short-term exposure to particulate matter induces arterial but not venous thrombosis in healthy mice. *J Thromb Haemost,* 8, 2651-61.

Ghio, A. J., A. Hall, M. A. Bassett, W. E. Cascio & R. B. Devlin (2003) Exposure to concentrated ambient air particles alters hematologic indices in humans. *Inhal Toxicol,* 15, 1465-78.

Ghio, A. J., C. Kim & R. B. Devlin (2000) Concentrated ambient air particles induce mild pulmonary inflammation in healthy human volunteers. *Am J Respir Crit Care Med,* 162, 981-8.

Hertel, S., A. Viehmann, S. Moebus, K. Mann, M. Brocker-Preuss, S. Mohlenkamp, M. Nonnemacher, R. Erbel, H. Jakobs, M. Memmesheimer, K. H. Jockel & B. Hoffmann (2010) Influence of short-term exposure to ultrafine and fine particles on systemic inflammation. *Eur J Epidemiol,* 25, 581-92.

HHS (2004) The Health Consequences of Smoking: A Report of the Surgeon General. *US Department of Health and Human Services, Centers for Disease Control and Prevention, National Center for Chronic Disease Prevention and Health Promotion, Office of Smoking and Health.*

Hoek, G., B. Brunekreef, S. Goldbohm, P. Fischer & P. A. van den Brandt (2002) Association between mortality and indicators of traffic-related air pollution in the Netherlands: a cohort study. *Lancet,* 360, 1203-9.

Hoffmann, B., S. Moebus, S. Mohlenkamp, A. Stang, N. Lehmann, N. Dragano, A. Schmermund, M. Memmesheimer, K. Mann, R. Erbel & K. H. Jockel (2007) Residential exposure to traffic is associated with coronary atherosclerosis. *Circulation,* 116, 489-96.

Ibald-Mulli, A., J. Stieber, H. E. Wichmann, W. Koenig & A. Peters (2001) Effects of air pollution on blood pressure: a population-based approach. *Am J Public Health,* 91, 571-7.

Inoue, K., H. Takano, M. Sakurai, T. Oda, H. Tamura, R. Yanagisawa, A. Shimada & T. Yoshikawa (2006) Pulmonary exposure to diesel exhaust particles enhances coagulatory disturbance with endothelial damage and systemic inflammation related to lung inflammation. *Exp Biol Med (Maywood),* 231, 1626-32.

Jacobs, L., J. Emmerechts, M. F. Hoylaerts, C. Mathieu, P. H. Hoet, B. Nemery & T. S. Nawrot (2011) Traffic Air Pollution and Oxidized LDL. *PLoS One,* 6.

Jacobs, L., J. Emmerechts, C. Mathieu, M. F. Hoylaerts, F. Fierens, P. H. Hoet, B. Nemery & T. S. Nawrot (2009) Air pollution related prothrombotic changes in persons with diabetes. *Environ Health Perspect,* 118, 191-6.

Jerrett, M., R. T. Burnett, R. Ma, C. A. Pope, 3rd, D. Krewski, K. B. Newbold, G. Thurston, Y. Shi, N. Finkelstein, E. E. Calle & M. J. Thun (2005) Spatial analysis of air pollution and mortality in Los Angeles. *Epidemiology,* 16, 727-36.

Jiang, N., K. L. Dreher, J. A. Dye, Y. Li, J. H. Richards, L. D. Martin & K. B. Adler (2000) Residual oil fly ash induces cytotoxicity and mucin secretion by guinea pig tracheal epithelial cells via an oxidant-mediated mechanism. *Toxicol Appl Pharmacol,* 163, 221-30.

Jimenez, L. A., J. Thompson, D. A. Brown, I. Rahman, F. Antonicelli, R. Duffin, E. M. Drost, R. T. Hay, K. Donaldson & W. MacNee (2000) Activation of NF-kappaB by PM(10) occurs via an iron-mediated mechanism in the absence of IkappaB degradation. *Toxicol Appl Pharmacol,* 166, 101-10.

Kadiiska, M. B., R. P. Mason, K. L. Dreher, D. L. Costa & A. J. Ghio (1997) In vivo evidence of free radical formation in the rat lung after exposure to an emission source air pollution particle. *Chem Res Toxicol,* 10, 1104-8.

Kan, H., A. R. Folsom, M. Cushman, K. M. Rose, W. D. Rosamond, D. Liao, F. Lurmann & S. J. London (2011) Traffic exposure and incident venous thromboembolism in the atherosclerosis risk in communities (ARIC) study. *J Thromb Haemost.*

Katsouyanni, K., G. Touloumi, E. Samoli, A. Gryparis, A. Le Tertre, Y. Monopolis, G. Rossi, D. Zmirou, F. Ballester, A. Boumghar, H. R. Anderson, B. Wojtyniak, A. Paldy, R. Braunstein, J. Pekkanen, C. Schindler & J. Schwartz (2001) Confounding and effect modification in the short-term effects of ambient particles on total mortality: results from 29 European cities within the APHEA2 project. *Epidemiology,* 12, 521-31.

Kido, T., E. Tamagawa, N. Bai, K. Suda, H. H. Yang, Y. Li, G. Chiang, K. Yatera, H. Mukae, D. D. Sin & S. F. Van Eeden (2011) Particulate matter induces translocation of IL-6 from the lung to the systemic circulation. *Am J Respir Cell Mol Biol,* 44, 197-204.

Kilinc E, van Oerle R, Borissoff JI, Oschatz C, Gerlofs-Nijland ME, Janssen NA, Cassee FR, Sandstrom T, Renne T, Ten Cate H, Spronk HM. Factor XII Activation is Essential to Sustain the Procoagulant Effects of Particulate Matter. *J Thromb Haemost.* 2011.

Kunzli, N., M. Jerrett, R. Garcia-Esteban, X. Basagana, B. Beckermann, F. Gilliland, M. Medina, J. Peters, H. N. Hodis & W. J. Mack (2010) Ambient air pollution and the progression of atherosclerosis in adults. *PLoS One,* 5, e9096.

Kunzli, N., M. Jerrett, W. J. Mack, B. Beckerman, L. LaBree, F. Gilliland, D. Thomas, J. Peters & H. N. Hodis (2005) Ambient air pollution and atherosclerosis in Los Angeles. *Environ Health Perspect,* 113, 201-6.

Laden, F., J. Schwartz, F. E. Speizer & D. W. Dockery (2006) Reduction in fine particulate air pollution and mortality: Extended follow-up of the Harvard Six Cities study. *Am J Respir Crit Care Med,* 173, 667-72.

Levi, M. & T. van der Poll (2010) Inflammation and coagulation. *Crit Care Med,* 38, S26-34.

Li, N., J. Alam, M. I. Venkatesan, A. Eiguren-Fernandez, D. Schmitz, E. Di Stefano, N. Slaughter, E. Killeen, X. Wang, A. Huang, M. Wang, A. H. Miguel, A. Cho, C. Sioutas & A. E. Nel (2004) Nrf2 is a key transcription factor that regulates antioxidant defense in macrophages and epithelial cells: protecting against the proinflammatory and oxidizing effects of diesel exhaust chemicals. *J Immunol,* 173, 3467-81.

Lindmark, E., E. Diderholm, L. Wallentin & A. Siegbahn (2001) Relationship between interleukin 6 and mortality in patients with unstable coronary artery disease: effects of an early invasive or noninvasive strategy. *JAMA,* 286, 2107-13.

Lowe, G. D. (2008) Common risk factors for both arterial and venous thrombosis. *Br J Haematol,* 140, 488-95.

Lucking, A. J., M. Lundback, N. L. Mills, D. Faratian, S. L. Barath, J. Pourazar, F. R. Cassee, K. Donaldson, N. A. Boon, J. J. Badimon, T. Sandstrom, A. Blomberg & D. E. Newby (2008) Diesel exhaust inhalation increases thrombus formation in man. *Eur Heart J,* 29, 3043-51.

Maitre, A., V. Bonneterre, L. Huillard, P. Sabatier & R. de Gaudemaris (2006) Impact of urban atmospheric pollution on coronary disease. *Eur Heart J,* 27, 2275-84.

Miller, K. A., D. S. Siscovick, L. Sheppard, K. Shepherd, J. H. Sullivan, G. L. Anderson & J. D. Kaufman (2007) Long-term exposure to air pollution and incidence of cardiovascular events in women. *N Engl J Med,* 356, 447-58.

Mills, N. L., N. Amin, S. D. Robinson, A. Anand, J. Davies, D. Patel, J. M. de la Fuente, F. R. Cassee, N. A. Boon, W. Macnee, A. M. Millar, K. Donaldson & D. E. Newby (2006) Do inhaled carbon nanoparticles translocate directly into the circulation in humans? *Am J Respir Crit Care Med,* 173, 426-31.

Mills, N. L., H. Tornqvist, M. C. Gonzalez, E. Vink, S. D. Robinson, S. Soderberg, N. A. Boon, K. Donaldson, T. Sandstrom, A. Blomberg & D. E. Newby (2007) Ischemic and thrombotic effects of dilute diesel-exhaust inhalation in men with coronary heart disease. *N Engl J Med,* 357, 1075-82.

Mills, N. L., H. Tornqvist, S. D. Robinson, M. Gonzalez, K. Darnley, W. MacNee, N. A. Boon, K. Donaldson, A. Blomberg, T. Sandstrom & D. E. Newby (2005) Diesel exhaust inhalation causes vascular dysfunction and impaired endogenous fibrinolysis. *Circulation,* 112, 3930-6.

Mitschik, S., R. Schierl, D. Nowak & R. A. Jorres (2008) Effects of particulate matter on cytokine production in vitro: a comparative analysis of published studies. *Inhal Toxicol,* 20, 399-414.

Morel, O., F. Toti, B. Hugel, B. Bakouboula, L. Camoin-Jau, F. Dignat-George & J. M. Freyssinet (2006) Procoagulant microparticles: disrupting the vascular homeostasis equation? *Arterioscler Thromb Vasc Biol,* 26, 2594-604.

Mutlu, G. M., D. Green, A. Bellmeyer, C. M. Baker, Z. Burgess, N. Rajamannan, J. W. Christman, N. Foiles, D. W. Kamp, A. J. Ghio, N. S. Chandel, D. A. Dean, J. I. Sznajder & G. R. Budinger (2007) Ambient particulate matter accelerates coagulation via an IL-6-dependent pathway. *J Clin Invest,* 117, 2952-61.

Nadziejko, C., K. Fang, L. C. Chen, B. Cohen, M. Karpatkin & A. Nadas (2002) Effect of concentrated ambient particulate matter on blood coagulation parameters in rats. *Res Rep Health Eff Inst*, 7-29; discussion 31-8.

Nawrot, T. S., A. Nemmar & B. Nemery (2006) Air pollution: To the heart of the matter. *Eur Heart J*, 27, 2269-71.

Nawrot, T. S., L. Perez, N. Kunzli, E. Munters & B. Nemery (2011) Public health importance of triggers of myocardial infarction: a comparative risk assessment. *Lancet*, 377, 732-40.

Nemmar, A., S. Al-Salam, S. Zia, S. Dhanasekaran, M. Shudadevi & B. H. Ali (2010) Time-course effects of systemically administered diesel exhaust particles in rats. *Toxicol Lett*, 194, 58-65.

Nemmar, A., P. H. Hoet, D. Dinsdale, J. Vermylen, M. F. Hoylaerts & B. Nemery (2003a) Diesel exhaust particles in lung acutely enhance experimental peripheral thrombosis. *Circulation*, 107, 1202-8.

Nemmar, A., P. H. Hoet, P. Vandervoort, D. Dinsdale, B. Nemery & M. F. Hoylaerts (2007) Enhanced peripheral thrombogenicity after lung inflammation is mediated by platelet-leukocyte activation: role of P-selectin. *J Thromb Haemost*, 5, 1217-26.

Nemmar, A., P. H. Hoet, B. Vanquickenborne, D. Dinsdale, M. Thomeer, M. F. Hoylaerts, H. Vanbilloen, L. Mortelmans & B. Nemery (2002) Passage of inhaled particles into the blood circulation in humans. *Circulation*, 105, 411-4.

Nemmar, A., P. H. Hoet, J. Vermylen, B. Nemery & M. F. Hoylaerts (2004) Pharmacological stabilization of mast cells abrogates late thrombotic events induced by diesel exhaust particles in hamsters. *Circulation*, 110, 1670-7.

Nemmar, A., M. F. Hoylaerts, P. H. Hoet, J. Vermylen & B. Nemery (2003b) Size effect of intratracheally instilled particles on pulmonary inflammation and vascular thrombosis. *Toxicol Appl Pharmacol*, 186, 38-45.

Nemmar, A., B. Nemery, P. H. Hoet, N. Van Rooijen & M. F. Hoylaerts (2005) Silica particles enhance peripheral thrombosis: key role of lung macrophage-neutrophil cross-talk. *Am J Respir Crit Care Med*, 171, 872-9.

Nemmar, A., B. Nemery, P. H. Hoet, J. Vermylen & M. F. Hoylaerts (2003c) Pulmonary inflammation and thrombogenicity caused by diesel particles in hamsters: role of histamine. *Am J Respir Crit Care Med*, 168, 1366-72.

Nemmar, A., H. Vanbilloen, M. F. Hoylaerts, P. H. Hoet, A. Verbruggen & B. Nemery (2001) Passage of intratracheally instilled ultrafine particles from the lung into the systemic circulation in hamster. *Am J Respir Crit Care Med*, 164, 1665-8.

O'Toole, T. E., J. Hellmann, L. Wheat, P. Haberzettl, J. Lee, D. J. Conklin, A. Bhatnagar & C. A. Pope, 3rd (2010) Episodic exposure to fine particulate air pollution decreases circulating levels of endothelial progenitor cells. *Circ Res*, 107, 200-3.

Oberdorster, G., E. Oberdorster & J. Oberdorster (2005) Nanotoxicology: an emerging discipline evolving from studies of ultrafine particles. *Environ Health Perspect*, 113, 823-39.

Oberdorster, G., Z. Sharp, V. Atudorei, A. Elder, R. Gelein, A. Lunts, W. Kreyling & C. Cox (2002) Extrapulmonary translocation of ultrafine carbon particles following whole-body inhalation exposure of rats. *J Toxicol Environ Health A*, 65, 1531-43.

Park, S. K., A. H. Auchincloss, M. S. O'Neill, R. Prineas, J. C. Correa, J. Keeler, R. G. Barr, J. D. Kaufman & A. V. Diez Roux (2010) Particulate air pollution, metabolic syndrome, and heart rate variability: the multi-ethnic study of atherosclerosis (MESA). *Environ Health Perspect,* 118, 1406-11.

Pawlinski, R., J. G. Wang, A. P. Owens, 3rd, J. Williams, S. Antoniak, M. Tencati, T. Luther, J. W. Rowley, E. N. Low, A. S. Weyrich & N. Mackman (2010) Hematopoietic and nonhematopoietic cell tissue factor activates the coagulation cascade in endotoxemic mice. *Blood,* 116, 806-14.

Pekkanen, J., E. J. Brunner, H. R. Anderson, P. Tiittanen & R. W. Atkinson (2000) Daily concentrations of air pollution and plasma fibrinogen in London. *Occup Environ Med,* 57, 818-22.

Pekkanen, J., A. Peters, G. Hoek, P. Tiittanen, B. Brunekreef, J. de Hartog, J. Heinrich, A. Ibald-Mulli, W. G. Kreyling, T. Lanki, K. L. Timonen & E. Vanninen (2002) Particulate air pollution and risk of ST-segment depression during repeated submaximal exercise tests among subjects with coronary heart disease: the Exposure and Risk Assessment for Fine and Ultrafine Particles in Ambient Air (ULTRA) study. *Circulation,* 106, 933-8.

Pery, A. R., C. Brochot, P. H. Hoet, A. Nemmar & F. Y. Bois (2009) Development of a physiologically based kinetic model for 99m-technetium-labelled carbon nanoparticles inhaled by humans. *Inhal Toxicol,* 21, 1099-107.

Peters, A., D. W. Dockery, J. E. Muller & M. A. Mittleman (2001a) Increased particulate air pollution and the triggering of myocardial infarction. *Circulation,* 103, 2810-5.

Peters, A., M. Frohlich, A. Doring, T. Immervoll, H. E. Wichmann, W. L. Hutchinson, M. B. Pepys & W. Koenig (2001b) Particulate air pollution is associated with an acute phase response in men; results from the MONICA-Augsburg Study. *Eur Heart J,* 22, 1198-204.

Peters, A., S. von Klot, M. Heier, I. Trentinaglia, A. Hormann, H. E. Wichmann & H. Lowel (2004) Exposure to traffic and the onset of myocardial infarction. *N Engl J Med,* 351, 1721-30.

Pope, C. A., 3rd, R. T. Burnett, D. Krewski, M. Jerrett, Y. Shi, E. E. Calle & M. J. Thun (2009) Cardiovascular mortality and exposure to airborne fine particulate matter and cigarette smoke: shape of the exposure-response relationship. *Circulation,* 120, 941-8.

Pope, C. A., 3rd, R. T. Burnett, M. J. Thun, E. E. Calle, D. Krewski, K. Ito & G. D. Thurston (2002) Lung cancer, cardiopulmonary mortality, and long-term exposure to fine particulate air pollution. *JAMA,* 287, 1132-41.

Pope, C. A., 3rd, R. T. Burnett, G. D. Thurston, M. J. Thun, E. E. Calle, D. Krewski & J. J. Godleski (2004a) Cardiovascular mortality and long-term exposure to particulate air pollution: epidemiological evidence of general pathophysiological pathways of disease. *Circulation,* 109, 71-7.

Pope, C. A., 3rd, M. L. Hansen, R. W. Long, K. R. Nielsen, N. L. Eatough, W. E. Wilson & D. J. Eatough (2004b) Ambient particulate air pollution, heart rate variability, and blood markers of inflammation in a panel of elderly subjects. *Environ Health Perspect,* 112, 339-45.

Pope, C. A., 3rd, R. L. Verrier, E. G. Lovett, A. C. Larson, M. E. Raizenne, R. E. Kanner, J. Schwartz, G. M. Villegas, D. R. Gold & D. W. Dockery (1999) Heart rate variability associated with particulate air pollution. *Am Heart J,* 138, 890-9.

Prandoni, P. (2009) Venous and arterial thrombosis: Two aspects of the same disease? *Clin Epidemiol,* 1, 1-6.

Puett, R. C., J. E. Hart, J. D. Yanosky, C. Paciorek, J. Schwartz, H. Suh, F. E. Speizer & F. Laden (2009) Chronic fine and coarse particulate exposure, mortality, and coronary heart disease in the Nurses' Health Study. *Environ Health Perspect,* 117, 1697-701.

Ray, M. R., S. Mukherjee, S. Roychoudhury, P. Bhattacharya, M. Banerjee, S. Siddique, S. Chakraborty & T. Lahiri (2006) Platelet activation, upregulation of CD11b/ CD18 expression on leukocytes and increase in circulating leukocyte-platelet aggregates in Indian women chronically exposed to biomass smoke. *Hum Exp Toxicol,* 25, 627-35.

Ridker, P. M., N. Rifai, M. J. Stampfer & C. H. Hennekens (2000) Plasma concentration of interleukin-6 and the risk of future myocardial infarction among apparently healthy men. *Circulation,* 101, 1767-72.

Riediker, M., W. E. Cascio, T. R. Griggs, M. C. Herbst, P. A. Bromberg, L. Neas, R. W. Williams & R. B. Devlin (2004) Particulate matter exposure in cars is associated with cardiovascular effects in healthy young men. *Am J Respir Crit Care Med,* 169, 934-40.

Ross, R. (1999) Atherosclerosis--an inflammatory disease. *N Engl J Med,* 340, 115-26.

Ruckerl, R., S. Greven, P. Ljungman, P. Aalto, C. Antoniades, T. Bellander, N. Berglind, C. Chrysohoou, F. Forastiere, B. Jacquemin, S. von Klot, W. Koenig, H. Kuchenhoff, T. Lanki, J. Pekkanen, C. A. Perucci, A. Schneider, J. Sunyer & A. Peters (2007a) Air pollution and inflammation (interleukin-6, C-reactive protein, fibrinogen) in myocardial infarction survivors. *Environ Health Perspect,* 115, 1072-80.

Ruckerl, R., A. Ibald-Mulli, W. Koenig, A. Schneider, G. Woelke, J. Cyrys, J. Heinrich, V. Marder, M. Frampton, H. E. Wichmann & A. Peters (2006) Air pollution and markers of inflammation and coagulation in patients with coronary heart disease. *Am J Respir Crit Care Med,* 173, 432-41.

Ruckerl, R., R. P. Phipps, A. Schneider, M. Frampton, J. Cyrys, G. Oberdorster, H. E. Wichmann & A. Peters (2007b) Ultrafine particles and platelet activation in patients with coronary heart disease--results from a prospective panel study. *Part Fibre Toxicol,* 4, 1.

Rudez, G., N. A. Janssen, E. Kilinc, F. W. Leebeek, M. E. Gerlofs-Nijland, H. M. Spronk, H. ten Cate, F. R. Cassee & M. P. de Maat (2009) Effects of ambient air pollution on hemostasis and inflammation. *Environ Health Perspect,* 117, 995-1001.

Samet, J. M., F. Dominici, F. C. Curriero, I. Coursac & S. L. Zeger (2000) Fine particulate air pollution and mortality in 20 U.S. cities, 1987-1994. *N Engl J Med,* 343, 1742-9.

Samet, J. M., A. Rappold, D. Graff, W. E. Cascio, J. H. Berntsen, Y. C. Huang, M. Herbst, M. Bassett, T. Montilla, M. J. Hazucha, P. A. Bromberg & R. B. Devlin (2009) Concentrated ambient ultrafine particle exposure induces cardiac changes in young healthy volunteers. *Am J Respir Crit Care Med,* 179, 1034-42.

Samoli, E., R. Peng, T. Ramsay, M. Pipikou, G. Touloumi, F. Dominici, R. Burnett, A. Cohen, D. Krewski, J. Samet & K. Katsouyanni (2008) Acute effects of ambient particulate matter on mortality in Europe and North America: results from the APHENA study. *Environ Health Perspect,* 116, 1480-6.

Scharrer, E., H. Hessel, A. Kronseder, W. Guth, B. Rolinski, R. A. Jorres, K. Radon, R. Schierl, P. Angerer & D. Nowak (2007) Heart rate variability, hemostatic and acute inflammatory blood parameters in healthy adults after short-term exposure to welding fume. *Int Arch Occup Environ Health,* 80, 265-72.

Schneider, A., R. Hampel, A. Ibald-Mulli, W. Zareba, G. Schmidt, R. Schneider, R. Ruckerl, J. P. Couderc, B. Mykins, G. Oberdorster, G. Wolke, M. Pitz, H. E. Wichmann & A. Peters (2010) Changes in deceleration capacity of heart rate and heart rate variability induced by ambient air pollution in individuals with coronary artery disease. *Part Fibre Toxicol,* 7, 29.

Schwartz, J., B. Coull, F. Laden & L. Ryan (2008) The effect of dose and timing of dose on the association between airborne particles and survival. *Environ Health Perspect,* 116, 64-9.

Seaton, A., A. Soutar, V. Crawford, R. Elton, S. McNerlan, J. Cherrie, M. Watt, R. Agius & R. Stout (1999) Particulate air pollution and the blood. *Thorax,* 54, 1027-32.

Shih, R. A., B. A. Griffin, N. Salkowski, A. Jewell, C. Eibner, C. E. Bird, D. Liao, M. Cushman, H. G. Margolis, C. B. Eaton & E. A. Whitsel (2010) Ambient Particulate Matter Air Pollution and Venous Thromboembolism in the Women's Health Initiative Hormone Therapy Trials. *Environ Health Perspect.*

Simkhovich, B. Z., M. T. Kleinman & R. A. Kloner (2008) Air pollution and cardiovascular injury epidemiology, toxicology, and mechanisms. *J Am Coll Cardiol,* 52, 719-26.

Sioutas, C., R. J. Delfino & M. Singh (2005) Exposure assessment for atmospheric ultrafine particles (UFPs) and implications in epidemiologic research. *Environ Health Perspect,* 113, 947-55.

Stewart, J. C., D. C. Chalupa, R. B. Devlin, L. M. Frasier, L. S. Huang, E. L. Little, S. M. Lee, R. P. Phipps, A. P. Pietropaoli, M. B. Taubman, M. J. Utell & M. W. Frampton (2010) Vascular effects of ultrafine particles in persons with type 2 diabetes. *Environ Health Perspect,* 118, 1692-8.

Stocker, R. & J. F. Keaney, Jr. (2004) Role of oxidative modifications in atherosclerosis. *Physiol Rev,* 84, 1381-478.

Sun, Q., A. Wang, X. Jin, A. Natanzon, D. Duquaine, R. D. Brook, J. G. Aguinaldo, Z. A. Fayad, V. Fuster, M. Lippmann, L. C. Chen & S. Rajagopalan (2005) Long-term air pollution exposure and acceleration of atherosclerosis and vascular inflammation in an animal model. *JAMA,* 294, 3003-10.

Suwa, T., J. C. Hogg, K. B. Quinlan, A. Ohgami, R. Vincent & S. F. van Eeden (2002) Particulate air pollution induces progression of atherosclerosis. *J Am Coll Cardiol,* 39, 935-42.

Tan, W. C., D. Qiu, B. L. Liam, T. P. Ng, S. H. Lee, S. F. van Eeden, Y. D'Yachkova & J. C. Hogg (2000) The human bone marrow response to acute air pollution caused by forest fires. *Am J Respir Crit Care Med,* 161, 1213-7.

Tao, F., B. Gonzalez-Flecha & L. Kobzik (2003) Reactive oxygen species in pulmonary inflammation by ambient particulates. *Free Radic Biol Med,* 35, 327-40.

Theilmeier, G., C. Michiels, E. Spaepen, I. Vreys, D. Collen, J. Vermylen & M. F. Hoylaerts (2002) Endothelial von Willebrand factor recruits platelets to atherosclerosis-prone sites in response to hypercholesterolemia. *Blood,* 99, 4486-93.

Thompson, A. M., A. Zanobetti, F. Silverman, J. Schwartz, B. Coull, B. Urch, M. Speck, J. R. Brook, M. Manno & D. R. Gold (2010) Baseline repeated measures from controlled human exposure studies: associations between ambient air pollution exposure and the systemic inflammatory biomarkers IL-6 and fibrinogen. *Environ Health Perspect,* 118, 120-4.

Tornqvist, H., N. L. Mills, M. Gonzalez, M. R. Miller, S. D. Robinson, I. L. Megson, W. Macnee, K. Donaldson, S. Soderberg, D. E. Newby, T. Sandstrom & A. Blomberg

(2007) Persistent endothelial dysfunction in humans after diesel exhaust inhalation. *Am J Respir Crit Care Med,* 176, 395-400.

Urch, B., F. Silverman, P. Corey, J. R. Brook, K. Z. Lukic, S. Rajagopalan & R. D. Brook (2005) Acute blood pressure responses in healthy adults during controlled air pollution exposures. *Environ Health Perspect,* 113, 1052-5.

van Eeden, S. F., W. C. Tan, T. Suwa, H. Mukae, T. Terashima, T. Fujii, D. Qui, R. Vincent & J. C. Hogg (2001) Cytokines involved in the systemic inflammatory response induced by exposure to particulate matter air pollutants (PM(10)). *Am J Respir Crit Care Med,* 164, 826-30.

Volpato, S., J. M. Guralnik, L. Ferrucci, J. Balfour, P. Chaves, L. P. Fried & T. B. Harris (2001) Cardiovascular disease, interleukin-6, and risk of mortality in older women: the women's health and aging study. *Circulation,* 103, 947-53.

Wallenfeldt, K., B. Fagerberg, J. Wikstrand & J. Hulthe (2004) Oxidized low-density lipoprotein in plasma is a prognostic marker of subclinical atherosclerosis development in clinically healthy men. *J Intern Med,* 256, 413-20.

Wilson, D. W., H. H. Aung, M. W. Lame, L. Plummer, K. E. Pinkerton, W. Ham, M. Kleeman, J. W. Norris & F. Tablin (2010) Exposure of mice to concentrated ambient particulate matter results in platelet and systemic cytokine activation. *Inhal Toxicol,* 22, 267-76.

Yokoyama, S., H. Ikeda, N. Haramaki, H. Yasukawa, T. Murohara & T. Imaizumi (2005) Platelet P-selectin plays an important role in arterial thrombogenesis by forming large stable platelet-leukocyte aggregates. *J Am Coll Cardiol,* 45, 1280-6.

Zanobetti, A., J. Schwartz, E. Samoli, A. Gryparis, G. Touloumi, J. Peacock, R. H. Anderson, A. Le Tertre, J. Bobros, M. Celko, A. Goren, B. Forsberg, P. Michelozzi, D. Rabczenko, S. P. Hoyos, H. E. Wichmann & K. Katsouyanni (2003) The temporal pattern of respiratory and heart disease mortality in response to air pollution. *Environ Health Perspect,* 111, 1188-93.

Ambient Air Pollution and Reproductive Health

Xiaohui Xu[1], Haidong Kan[2] and Sandie Ha[1]
[1]*Department of Epidemiology, College of Public Health and Health Professions, University of Florida,*
[2]*Department of Environmental Health, School of Public Health, Fudan University,*
[1]*USA*
[2]*China*

1. Introduction

Adverse health effects of ambient air pollution on mortality and morbidity in adults and children have been extensively studied across the world (Barnett et al. 2006; Bell et al. 2004; Dominici et al. 2003; Gold et al. 1999; Jerrett et al. 2005; Middleton et al. 2008; Pope 1999; Pope et al. 1991; Pope and Kanner 1993; Samoli et al. 2007; Wietlisbach et al. 1996). Considerable consistency across studies has been observed for many health endpoints including total mortality, cardiopulmonary mortality and morbidity. Moreover, air pollution studies suggest that the opposite ends of the age spectrum are more susceptible than the general population (Dockery and Pope 1994; Saldiva et al. 1995; Schwartz et al. 1994). Therefore, fetuses are thought to be a vulnerable subgroup of the population who could be most endangered by the effects of air pollution (Pope 2000). Early studies had shown that maternal active and passive smoking could impair reproductive outcomes. Thus, there is a strong belief that prenatal exposure to air contaminants, which is similar to the effects of maternal smoking, can also lead to some adverse pregnancy outcomes. The number of studies linking air pollution with adverse pregnancy outcomes has recently grown steadily since the late 1990s. The adverse effects of air pollutants including particulate matter (PM), nitrogen oxide (NOx), sulfur dioxide (SO_2), carbon monoxide (CO), and ozone (O_3) on measures of fetal size, gestational duration and other reproductive outcomes have been studied.

Adverse reproductive outcomes including low birth weight (LBW: birth weight <2,500g) or preterm delivery (PTD: birth at <37 weeks of gestation) have arisen fairly consistently in recent years. During 1990-2006, the rates have risen 21% for PTD and 19% for LBW in the United States, respectively (Martin et al. 2008). Studies have suggested that LBW or PTD has been associated with not only childhood mortality and morbidity but also the risk of diseases in adulthood such as heart diseases and diabetes (Clapp Iii and Lopez 2007; Osmond and Barker 2000; Rinaudo and Lamb 2008; Thompson 2007). The prevention of adverse pregnancy outcomes is a renewed national and international priority in maternal and child health (Damus 2008). There is an emerging need to identify the etiological factors of adverse birth outcomes such as environmental exposures, which could be modifiable in order to help reverse the increasing rates.

In this chapter, we review the health effects of ambient air pollution on pregnancy outcomes such as LBW, very low birth weight (VLBW: <1,500g), small for gestational age or intrauterine growth retardation (SGA or IUGR, i.e. birth weight below the 10th percentile for that gestational age), PTD, and birth defects. Further, we discuss the potential biological mechanisms underlying the associations between air pollution and adverse reproductive outcomes. Finally, we briefly introduce the methodology including study designs, exposure measurements, windows of exposure, confounders, and future directions for study of air pollution and pregnancy outcomes.

2. Health effects of air pollution on birth outcomes

Birth outcomes, including outcomes of fetal size such as LBW, VLBW, SGA, and IUGR, outcomes of gestational age such as PTD, and birth defects, have frequently been studied in epidemiological research on health effects of air pollution. The associations between air pollution and these birth outcomes may suggest different etiologic and pathogenic mechanisms. Moreover, air pollution is a mixture of pollutants, which also vary in nature and possible health effects. Therefore, existing evidence of health effects of air pollution on birth outcomes are briefly examined by types of birth outcome and/or criteria air pollutants including PM, NO_x, SO_2, and CO as below. Although the associations between O_3 exposure and birth outcomes have also been investigated by several studies, available evidence seems only to support an association with birth defects (Shah and Balkhair 2011). Therefore, the health effects of O3 are merely discussed for birth defects.

2.1 Health effects on fetal size - LBW, VLBW, SGA and IUGR

The effects of ambient air pollutants on fetal size, specifically LBW have been well studied during the past decades. Overall, there were some positive findings, which suggest that air pollution exposure, through all periods during pregnancy, is associated with adverse health effects on fetal size including LBW, VLBW, SGA and IUGR. However, air pollution consists of many components, making it difficult to pinpoint which pollutant can actually affect fetal size. Therefore, we present below the results from some selected studies by criteria air pollutants individually.

2.1.1. Particulate matter (PM)

PM generally attracts much more attention in epidemiological research of air pollution because of its physical and chemical characteristics of consisting of organic and inorganic components. The effects of PM on birth outcomes have been reported to be more pronounced compared to other pollutants. Several exposure indices of PM such as total suspend particles (TSP), particulate matter smaller than 10μm aerodynamic diameter (PM_{10}) and smaller than 2.5μm aerodynamic diameter ($PM_{2.5}$) have been used in studies. Wang et al firstly reported that exposure to TSP in the third trimester of pregnancy was significantly associated with LBW (Wang et al. 1997). After that, many studies have been conducted to report associations between particulate air pollution and LBW, SGA, as well as IUGR. Bobak et al found that annual concentration of TSP during the year of birth was associated with an increased risk of LBW (Bobak and Leon 1999). Rogers et al also found that annual TSP in the year of birth is associated with the risk of LBW (Rogers et al. 2000). Consistent evidence on adverse health effects of PM on birth weight has been reported when PM_{10} and $PM_{2.5}$ have been used as measures of exposure. Mannes et al found that each 1 $\mu g/m^3$

increase in PM_{10} is associated with on average a birth weight reduction of 4 grams (95% confidence interval (CI): 3g to 6g) after adjusting for other important covariates (Mannes et al. 2005). Dugandzic et al also found that PM_{10} exposure in the highest quartile during the first trimester increased the risk of LBW by 33% (OR= 1.33, 95% CI: 1.02 to 1.74) compared to exposure in the lowest quartile (Dugandzic et al. 2006). Similar results of health effects of PM_{10} on LBW were also observed in Xu et al's study (Xu et al. 2010). In addition, $PM_{2.5}$ is found to be associated with a reduction in birth weight by Parker et al in California in 2005. They found that exposure to $PM_{2.5}$ in the highest quartile during pregnancy is associated with a reduction in infant birth weight of -36.1 g (95% CI: -16.5 g to -55.8 g) compared to exposure in the lowest quartile of <11.9 µg/m³ after adjusting for other covariates (Parker et al. 2005). Another study reported by Rich et al also found that every 4 µg/m³ increase in $PM_{2.5}$ exposure during the first and the third trimester increases the risk for SGA by 4.5% (95% CI: 0.5-8.7) and 4.1% (95% CI: 0.3-8.0), respectively (Rich et al. 2009). In addition, Dejmek et al 1999 also found that exposure to a high tertile of $PM_{2.5}$ during the first gestational month is associated with an increased risk of IUGR (OR=2.11, 95% CI: 1.20-3.70) among northern Bohemian women (Dejmek et al. 1999). Liu et al (2007) reported that every 10 µg/m³ increase in $PM_{2.5}$ is associated with roughly six to seven percent increase (95% CI: 3-10%) in the risk of IUGR after adjusting for other risk factors (Liu et al. 2007). However, other studies have not found a significant association between particulate air pollution and fetal size (Maisonet et al. 2001; Salam et al. 2005).

2.1.2 NO_x

Health effects of nitric oxide (NO) and nitrogen dioxide (NO_2) on birth weight have also been studied. Studies on health effects of NO on fetal size is limited (Bobak 2000; Bobak and Leon 1999; Landgren 1996) and no existing evidence shows adverse health effects of NO on fetal size. Several studies show the associations between NO_2 exposure and LBW (Ballester et al. 2010; Bell et al. 2007; Ha et al. 2001; Lee et al. 2003). Some other studies also show that exposure to NO_2 is associated with increased risks of SGA. For example, an earlier report also found that each 10 parts per billion (ppb) increase in NO_2 exposure during the first month of pregnancy increases the risk of IUGR by 5% after adjusting for important maternal, infant and other environmental factors (Liu et al. 2003). Similar results were found by Mannes et al in 2005. This study indicated that each 1 ppb increase in NO_2 exposure during pregnancy is associated with 1 to 34 grams reductions in birth weight among infants in Sydney, Australia (Mannes et al. 2005). Liu et al (2007) also reported that every 20ppb increase in NO_2 exposure during the first, second, and third trimesters is associated with increased risks of IUGR by 16%, 14%, and 16%, respectively (Liu et al. 2007). However, several recent studies did not find a significant association between NO_2 exposure and fetal size (Gehring et al. 2010; Madsen et al. 2010).

2.1.3 SO_2

SO_2 is another important component of air pollution, and has been found to be associated with fetal size. Liu et al 2003 found that every 5.0 ppb increase in SO_2 exposure during the first month of pregnancy increases the risk of IUGR by approximately 7% (OR = 1.07, 95% CI: 1.01–1.13) among Canadian women (Liu et al. 2003). Several other studies have also linked SO_2 exposure with LBW. Wang et al have found that each 100 µg/m³ increase in SO_2 exposure during pregnancy is associated with 11% increase (OR= 1.11, 95% CI: 1.06-1.16) in the risk of LBW and a 7.3 gram reduction in birth weight among Chinese women (Wang et al. 1997). Bobak and Leon also found that every 50 μg/m^3 increase in SO_2 exposure during the whole pregnancy period is associated with a 10 % increase (OR=1.10, 95% CI: 1.02 –1.17) in the risk of having a LBW infant (Bobak and Leon 1999). Dugandzic et al 2006, in their cohort study in Canada, found that SO_2 exposure in the highest quartile during the first trimester is associated with a 36% increase in the risk of having a LBW infant (OR=1.36, 95% CI:1.04 to 1.78) (Dugandzic et al. 2006).

2.1.4 CO

Carbon monoxide is found to be associated with decreased birth weight. The earliest study reported by Alderman suggests a potential link between maternal CO exposure and risk of LBW (Alderman et al. 1987). Later, Ritz and Yu reported a significant association between CO exposure and risk of LBW in 1999 (Ritz and Yu 1999). Liu et al (2003) also found that each 1 ppb increase in CO exposure during the first month of pregnancy is associated with a 6% increase in the risk of IUGR (OR = 1.06, 95% CI:1.01–1.10) among Canadian women (Liu et al. 2003). Another later report by Liu et al in 2007 found that after adjusting for important maternal, infant and environmental factors, every 1ppb increase in CO exposure during the first, second, and third trimester of pregnancy is associated with a 18%, 15%, and 19% increase in the risk of IUGR, respectively (Liu et al. 2007). Significant findings were also reported in other studies (Lee et al. 2003; Maisonet et al. 2001). However, the evidence of health effects of CO exposure on fetal size is less consistent. Some studies have not shown a significant association (Huynh et al. 2006; Salam et al. 2005). Moreover, one other study found a protective effect of CO exposure on LBW (Lin et al. 2004).

2.2 Health effects on gestational age - PTD

PTD has also been extensively studied in the past decades. Studies have reported the effects of various air pollutants such as PM, NO_x, SO_2, and CO on PTD. The effect size and time of exposure of a specific air pollutant are inconsistent across studies because of the differences in study design, exposure assessment, and confounders being controlled for.

2.2.1 PM: TSP, PM_{10} and $PM_{2.5}$

Two studies had used TSP as a measure of exposure. One study reported that a decrease in the gestation age of 0.042 weeks was associated with each 100 μg/m^3 increase in TSP with a 7-day lag (Xu et al. 1995) while the other showed no significant association between TSP and PTD (Bobak 2000). Meanwhile, several studies have reported a significant association between PM_{10} exposure and PTD. A cohort study showed 16% and 20% increased risks of preterm birth for a 50 μg/m^3 increase in TSP exposure during the first month of pregnancy and during the 6 weeks before birth, respectively (Ritz et al. 2000). A study conducted in Australia suggested that exposure to PM_{10} during the first trimester was associated with a 15% increase in the risk of PTD (Hansen et al. 2006). A time-series analysis conducted in Pennsylvania also suggested a significant association between PTD and PM_{10} exposure with 2-day as well as 5-day lag (Sagiv et al. 2005). Another time-series study conducted in Shanghai also shows that a 4.4% increase in the risk of PTD was PTD was observed for each 10 μg/m^3 increase in PM_{10} exposure during the 8 weeks before pregnancy (Jiang et al. 2007). However, some other studies did not observe a significant effect of PM_{10} on PTD (Brauer et al. 2008; Kim et al. 2007; Lee et al. 2008). The effect of exposure to $PM_{2.5}$ was also assessed by several studies. Huynh et al (2006) found that exposure to to the highest quartile of $PM_{2.5}$ (> 22.1μg/m^3) during pregnancy is associated with a 15% increase in the risk of PTD compared to the lowest quartile of exposure. In the same study, similar results were found for exposure during the first month (OR=1.21, 95%CI: 1.12, 1.30) and last two weeks of pregnancy (OR=1.17 , 95%CI: 1.09, 1.27) (Huynh et al. 2006). A two-stage design study being conducted in Los Angeles reported that exposure to $PM_{2.5}$ > 21.4 µg/m³ during the first trimester increased the risk of PTD by about 10% (OR=1.10, 95%CI: 1.01, 1.20).

2.2.2 NO_x

The association of NO_2 exposure and PTD was explored in several studies. Bobak et al reported that 10% and 11% increased risks of PTD were observed for each 50 µg/m³ increase of NO_x during the first and third trimesters, respectively (Bobak 2000). NO_2 is also found to be associated with PTD among residents in Lithuania (Maroziene and Grazuleviciene 2002). Each 10 µg/m³ increase in NO_2 exposure during pregnancy period and first trimester is associated with 25% (OR = 1.25, 95%CI: 1.07-1.46) and 67% (OR = 1.67, 95% CI: 1.28-2.18) increase in the risk of PTD, respectively. A time series analysis indicated that daily PTD rate is associated with average exposure to NO_2 during the 6 weeks preceding delivery (Darrow et al. 2009). A recent cohort study conducted in Spain reported that the risk of PTD was shown to be significant when women were exposed to NO_2 levels >46.2 µg/m³ during the second (OR=1.11, 95% CI: 1.03-1.25) and third trimesters (OR=1.10, 95% CI: 1.00-1.21) as well as throughout the entire pregnancy (OR=1.29, 95% CI: 1.13-1.46) (Llop et al. 2010). However, evidence on the effect of NO_2 on PTD is inconsistent. Several other studies did not show a significant association (Gehring et al. 2010; Liu et al. 2003; Ritz et al. 2007).

2.2.3 SO_2

The earliest study conducted in China reported a significant effect of SO_2 on PTD. It shows that each 100 µg/m³ increase in SO_2 during the 7 days preceding delivery is associated with .075 wk (12.6h) decrease in duration of gestation (Xu et al. 1995). Bobak et al reported that exposure to SO_2 during all trimesters was significantly associated with the risk of PTD (Bobak 2000). Liu et al in 2003 also reported the significant effects of SO_2 on PTD among women living in Vancouver, Canada. This study indicated that each 5.0 ppb increase in SO_2 exposure during the last month of pregnancy is associated with a 9% increase in risk of PTD (Liu et al. 2003). A time series analysis also revealed that a 15% increased risk for PTD was observed for each 15 ppb increase in average SO_2 in the 6 weeks before birth in Pennsylvania (Sagiv et al. 2005). Similar findings from another time-series analysis conducted in Shanghai were observed (Jiang et al. 2007). Leem et al found that exposure to the highest quartile of SO_2 during first trimester increased the risk of PTD by 21% compared to the lowest quartile of exposure (OR= 1.21, 95% CI: 1.04-1.42) in Korea (Leem et al. 2006). A study in Australia also revealed that SO_2 levels in the first and third trimester of pregnancy were a significant predictor of preterm birth (ORs between 2.30 to 3.15) (Jalaludin et al. 2007). However, no significant effect of SO_2 was observed in other studies (Brauer et al. 2008; Darrow et al. 2009; Landgren 1996).

2.2.4 CO

CO is among one of the air pollutants that have been demonstrated to have negative effects on PTD. Ritz et al reported that exposure to CO during the first month of pregnancy and 6 weeks before birth was significantly associated with an increased risk of PTD in the inland regions of Southern California (Ritz et al. 2000). A study reported by Liu et al showed that an 8% increase (95% CI: 1%-15%) in the risk of PTD was observed for each 1.0 parts per million (ppm) increase in exposure to CO during the last month of pregnancy (Liu et al. 2003). Leem et al found that exposure to the highest quartiles of CO during first trimester increased the risk of PTD by 26% (OR= 1.26, 95% CI: 1.11-1.44) in Korea (Leem et al. 2006). Similar results reported by Wilhelm and Ritz in 2005 suggest that exposure to the highest quartile of CO during the first trimester increases the risk of PTD by 27% (95% CI: 7%-50%) (Wilhelm and Ritz 2005). Another cohort study found that exposure to high level of CO during the first trimester of pregnancy (> 1.25 ppb) is associated with 1.25 (95% CI: 1.12, 1.38) times of the risk of PTD compared to low level of CO exposure (< 0.59ppb) after adjusting for important maternal, infant, and environmental characteristics (Ritz et al. 2007). However, several other studies including a recent study did not observe significant effects of CO on PTD (Darrow et al. 2009; Huynh et al. 2006; Rudra et al. 2011).

2.3 Health effects on birth defects

Birth defects are another group of adverse birth outcomes that have recently been found to be associated with air pollution in several studies. Types of birth defects that have been reported to be associated with air pollution include heart defects and cleft palate. Ritz et al examined the effects of air pollutants including CO, NO_2, O_3, and PM_{10} on cardiac and orofacial defects in southern California. They found that second-month CO and O_3 exposures were significantly associated with cardiac defects after adjusting for maternal age, ethnicity, education, access to prenatal care, infant gender, decade of infant birth, parity, time since last pregnancy, birth type, and other air pollutants (Ritz et al. 2002).

Gilboa et al in 2005 investigated the effects of air pollutants such as CO, PM_{10} and SO_2 on different types of heart defects (e.g tetralogy of Fallot, and atrial and ventricular septal defects) among Texas women (Gilboa et al. 2005). Specifically, this study found that exposure to CO in the highest quartile (>0.7ppm) during pregnancy was associated with 2.04 times of the risk of tetralogy of Fallot compared to the lowest quartile after adjusting for infant sex, plurality, maternal education, maternal race, and season of conception (Gilboa et al. 2005). In the same study, PM_{10} and SO_2 were also found to be significantly associated with isolated atrial septal defects (OR = 2.27, 95% CI: 1.43-3.60) and isolated ventricular septal defects (OR =2.16, 95% CI: 1.51-3.09), respectively.

Marshall et al in 2010 examined the effects of criteria air pollutants including CO, O_3, NO_2, SO_2, PM_{10} and $PM_{2.5}$ on oral cleft defects in New Jersey. This study suggested that exposure to the highest quartile (>0.033ppm) of O_3 during the 3-8th month of pregnancy is associated with 2.2 times (95% CI: 1.0-4.9) greater risk of cleft palate compared to exposure to the lowest quartile (<0.015ppm). However, this study found no significant effects of other air pollutants on cleft defect. On the contrary, CO was surprisingly found to be a protective factor of cleft palate defect (OR=0.4, 95% CI: 0.2-0.7) (Marshall et al. 2010).

Another study conducted in the northeast of England had examined the effects of black smoke and SO_2 on congenital heart defects. This analysis found a weak association between maternal exposure to black smoke and heart defects but not for SO_2 exposure (Dadvand et al. 2011).

In summary, criteria air pollutants have been linked with adverse birth outcomes of fetal size, gestational age and birth defects. Several recent publications had also systematically reviewed epidemiologic evidence on the associations between maternal air pollution exposure and birth outcomes (Bosetti et al. 2010; Glinianaia et al. 2004; Maisonet et al. 2004;

Shah and Balkhair 2011). Although the heterogeneity and/or absence of associations between individual air pollutant and birth outcomes exists in published studies, current epidemiological evidence suggests that maternal exposure to PM and SO_2 is associated with several adverse birth outcomes. Reviewing the biologically plausible mechanistic pathways underlying the impacts of air pollution on birth outcomes would greatly contribute to our understanding of these associations.

3. Biologically plausible mechanisms

Although the specific mechanisms through which air pollution affects birth outcomes remain to be fully understood, with emerging bio-molecular technologies, an increasing number of studies show evidence towards potential mechanisms linking air pollution with birth outcomes. Air pollution has been frequently shown to affect the respiratory, cardiovascular, circulatory, and nervous system through multiple potential pathways, some of which might be appropriate for linking air pollution with birth outcomes (Block and Calderon-Garciduenas 2009; Brook 2008; Kunzli and Tager 2005). Several publications have described the biologically plausible mechanistic pathways through which air pollution impacts on birth outcomes (Kannan et al. 2006; Slama et al. 2008). We also used a figure from Slama et al's publication to show the possible biological mechanisms (see figure 1). These mechanisms are described in the following section.

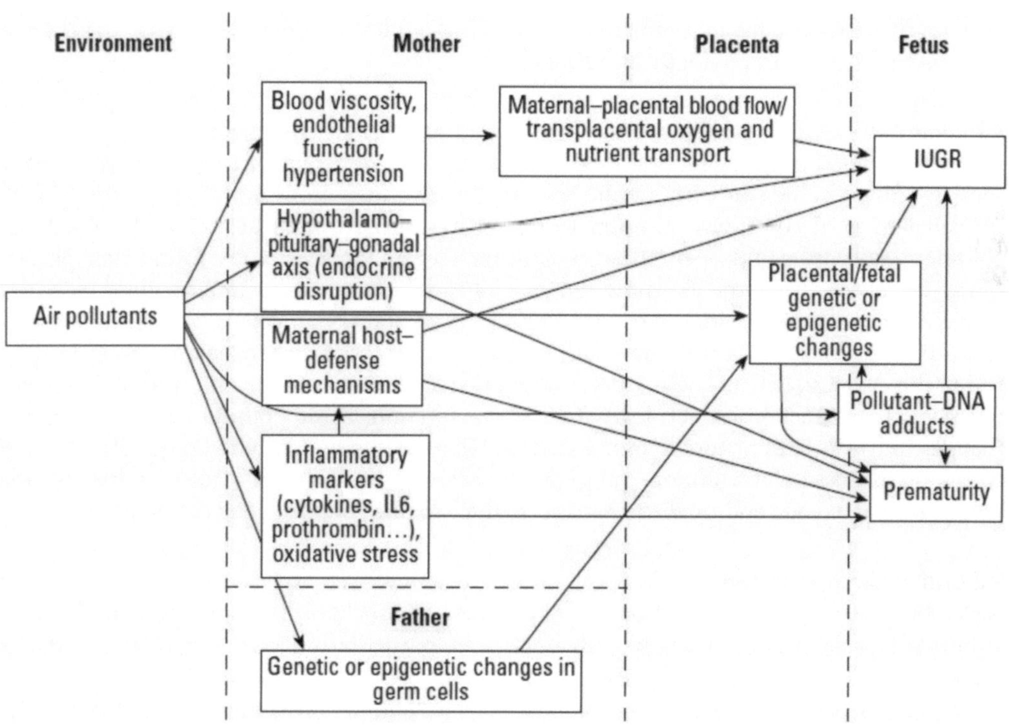

Source: Slama et al. Envrionmental Health Perspective, 2008, 116(6):795

Fig. 1. Possible biological mechanisms by which air pollutants may affect birth outcomes

3.1 Oxidative stress and inflammation

It has been believed that oxidative stress and inflammation play central roles in adverse health effects of air pollution (Donaldson et al. 2001; Tao et al. 2003). Oxidative stress is a condition in which the body encounters more reactive oxygen species than its ability to properly remove them, resulting in excess peroxides and free radicals that can ultimately damage cell components and biological processes (Leeuwenburgh and Heinecke 2001). For example, several studies reported that transition-metal constituents in PM can induce oxidative stress (Kadiiska et al. 1997; Prahalad et al. 2001). PM can also activate inflammatory cells to generate of oxidative stressors such as reactive oxygen species and reactive nitrogen species (Tao et al. 2003). Consequently, oxidative stress can directly lead to DNA damage and may affect the embryo at its earlier stage of growth (Mohorovic 2004; Risom et al. 2005). Moreover, oxidative stress may also influence sperm motility and concentration, which is relevant for male reproductive health (Agarwal et al. 2006).

Air pollution-induced inflammation is considered as another potential biological mechanism through which air pollution cause adverse effects in organs such as the lung (Kunzli and Tager 2005). Inflammation could be a direct result of oxidation or up-regulation of pro-inflammatory mediators induced by air pollutants (Risom et al. 2005). Several studies have shown that air pollutants can enter the blood stream from lung and get deposited onto various body organs via active transport or passive diffusion (Chen et al. 2008; Peters et al. 2006). If air pollutants reach the placenta, they will also induce acute placental inflammation, which subsequently results in impaired transplacetal nutrient exchanges (Bobak 2000). In addition, inflammation could change maternal immunity, reduce host defense, thus increase the maternal risk of infections, which may in turn increase the risks of adverse birth outcomes (Wilhelm and Ritz 2005).

3.2 Changes in rheological factors of blood and endothelial function

Changes of blood coagulability and viscosity as a result of exposure to air pollutants have been suggested in the studies of cardiovascular health effects of air pollution (Coppola et al. 1989; Peters et al. 1997). In addition to the changes in rheological factors of blood, air pollution exposure could also influence endothelial functions. A study found that plasma concentrations of asymmetric dimethyl arginine, which suggests an impaired vascular function, were increased after exposure to $PM_{2.5}$ (Valkonen et al. 2001). Another study also found that exposure to air pollution could cause conduit arterial vasoconstriction in healthy adults (Brook et al. 2002). These studies suggest that exposure to air pollution may trigger endothelial dysfunction, which leads to vasoconstriction. If air pollution could cause the changes in rheological factors of blood such as blood viscosity and blood coagulability, and artery vasoconstriction among pregnancy women, it in turn could influence the transplacental oxygen and nutrient transport which further impacts fetal development.

3.3 Endocrine disruption

Air pollution has also been linked to endocrine dysfunctions that can potentially have negative impacts of birth outcomes. Air pollutants, particularly PM, may interfere with the endocrine system and affect progesterone production (Furuta et al. 2004; Takeda et al. 2004). Although this pathway has been less extensively studied, the possibility of the effects of air pollution on the endocrine system was suggested in recent investigations. Growing evidence suggests that inhalation of environmental tobacco smoke (ETS) contributes to disruptions in thyroid function (Carrillo et al. 2009). Air pollution, similar to ETS, could also interfere with thyroid function. Disruption of thyroid function is associated with fetal development (Blazer et al. 2003; Patel et al. 2011). In addition, polycyclic aromatic hydrocarbon (PAH) was also found to have anti-estrogenic activity and can disrupt endocrine functions (Tran et al. 1996). Studies also found that endocrine disruption among pregnant women might cause IUGR (Kanaka-Gantenbein et al. 2003). Therefore, the link between air pollution and IUGR might involve this pathway. Moreover, a study found that air organic compounds such as PAH can also directly affect epidermal growth factor (EGF) and insulin-like growth factor type I and II (IGF-1 and IGF-2) receptors, leading to inhibition of placental cell growth and proliferation (Dejmek et al. 2000). This could cause a decreased fetal-placental exchange of oxygen and nutrients, which are the critical factors regulating fetal growth.

3.4 Hemodymanic changes

Air pollution exposure has been linked to hemodynamic responses such as changes in blood pressure, heart rate and rhythm, and cardiac autonomic tone (Ibald-Mulli et al. 2004). Specifically, PM exposure is associated with increase in heart rate, heart rate variability, and the frequency of ectopic beats. It is also associated with increase in systolic and diastolic blood pressure (Linn et al. 1999; van den Hooven et al. 2011). The effects of air pollution on hemodymanic changes can suggest a potential mechanism through which it can affect birth outcomes. The hemodymanic changes discussed are all risk factors for hypertension and other cardiovascular diseases. Therefore, exposure to air pollution can increase risks of health conditions (e.g. hypertension) that can ultimately increase the risk of adverse birth outcomes. For example, exposure to PM_{10} has been found to increase the risk of pregnancy-induced hypertension (van den Hooven et al 2011), which is found to be associated with elevated risk of preterm birth (OR=3.30), low birth weight (OR=4.68), fetal growth restriction (OR=2.94), and low Apgar scores 1 minute (OR=2.99) and 5 minutes (OR=2.08) (Olusanya and Solanke 2011).

3.5 Germ-line mutations

Although most of the mechanisms suggested are maternally related, recent studies suggest that paternal factors, especially germ-line mutation resulted from pollution exposure can also greatly influence birth outcomes. In animal studies, it was found that, germline mutation frequencies in wild herring gulls nesting At a polluted site were 2.8 to 8.5 times higher than birds that nested in non-polluted sites. This is consistent with what was discussed earlier—air pollution can adversely impact DNA replication process (Somers and Cooper 2009). In mice, similar results are found where there is an increased frequency of expanded simple tandem repeats among mice that are exposed to air pollution (Somers and Cooper 2009). Moreover, Somers et al 2002 also found evidence that air pollution is capable of inducing heritable DNA mutations (Somers et al. 2002). Since the sentinel animals that were used to study the association of air pollution and DNA mutations were from areas populated with humans, it is very likely that air pollution can also lead to germline mutations among humans, which ultimately can affect birth outcomes. Moreover, several studies have been conducted to support the hypotheses that air pollution is associated with elevated levels of DNA fragmentation in human sperm (Jafarabadi 2007; Rubes et al. 2005).

In conclusion, research on potentially biological mechanisms through which air pollution can affect birth outcomes is still limited. We review several potential mechanisms above which have been suggested in previous studies. Meanwhile, it is important to recognize the need of further research to increase our understanding about the biological mechanisms underlying the impact of various air pollutants.

4. Methodology

As we reviewed above, evidence on the associations between air pollution and birth outcomes are not consistent in all cases. The heterogeneity and/or absence of association could be due to the differences in study design, exposure measurement, identification of windows of exposure, and selection of confounders. We discussed some common methods in these areas below.

4.1 Study designs

To assess whether air pollution exposure has an effect on pregnancy outcomes, the investigators need to compare the pregnancy outcome occurrence of individuals who have been subject to different exposure levels and, in particular, by comparing the occurrence or prevalence rate of pregnancy outcomes of exposed persons with those of unexposed or less exposed persons (Strickland et al. 2009). Specifically, descriptive studies, ecological studies, case-control studies, and cohort studies can be used to examine the relationship between air pollution and pregnancy outcomes.

The first step in investigating the pregnancy outcomes associated with air pollution exposure could be a **descriptive study**. Descriptive studies examine the distribution of pregnancy outcomes in a defined population. They are helpful in assessing the possibility that an association exists, and identifying hypotheses to be evaluated in analytical studies. Any type of pregnancy outcomes can be used for conducting descriptive studies. Descriptive data are commonly applied to examine patterns of pregnancy outcomes by place, time and person. *Geographic comparisons* based on standardized morbidity rates can be made among different geographic regions. The variations between these regions concerning occurrence of pregnancy outcomes contribute to the basis of causal hypotheses. *Temporal trends* in pregnancy outcomes rates can also be valuable to indicate the possible effects of air pollution.

Ecological studies are studies in which the investigators analyze hypothesized associations between air pollution and pregnancy outcomes using groups of people, rather than individuals, as the unit of analysis. It compares aggregate measures of exposure, such as average exposure or proportion of population exposed, with aggregate measures of pregnancy outcomes rates, for the same population. A traditional approach is to use *geographical areas* (Bobak and Leon 1999) as the basis for defining the study groups, and then correlations between aggregate measures of exposure and pregnancy outcomes at the same geographical location are analyzed. The *time series* design is a special and emerging type of ecological study with widespread application in epidemiological studies of short-term exposure to air pollution and pregnant outcomes (Darrow et al. 2009; Jiang et al. 2007; Lee et al. 2008; Sagiv et al. 2005; Zhao et al.). It investigates the relationship between air pollution levels and pregnancy outcomes - each measured and aggregated over the same time units (e.g., days, weeks) during a specified time period. Modeling time-series data is challenging because the relatively small effect of air pollutants is hard to be identified and quantified in the presence of strong confounding. Typically, one of the most widely-used statistical approaches for time series analysis of air pollution and health is the generalized additive model (GAM). A strength of time-series analyses is the inherent control of individual-level risk factors that do not vary temporally. However, in studies of adverse pregnancy outcomes, risk factors considered time-invariant at the individual level may vary seasonally when aggregated into a pregnancy risk set. Therefore, time-series investigations of seasonally-varying exposures and adverse pregnancy outcomes should consider the potential for bias due to seasonal heterogeneity in the risk set. Moreover, this study design must be used with caution due to the important but hardly predictable ecological bias for which group-level associations do not accurately reflect individual-level associations.

A **case-control study** examines associations between air pollution exposures and adverse pregnancy outcomes by comparing cases, or individuals who developed the outcome with controls who are a sample of the source population from which the cases were identified. Controls are usually individuals who are similar to the cases in terms of risk characteristics, but who have not developed the pregnancy outcomes. Having selected cases and controls, the investigators then determine the prior air pollution exposure of the cases and controls. Hospital-based cases are usually selected for their high accessibility and cooperative attitude but this method is subject to bias. A population-based case control study (Rogers et al. 2000) is the principal alternative, for which all incident cases of the outcome in a defined geographic area are included as cases. The selection of controls can be more challenging. The most commonly used control groups are a random sample of the source population from which the cases are selected or persons seeking medical care at the same institutions as the cases for conditions believed to be unrelated to the health outcome of interest. Individual or frequency matching is another efficient strategy to select controls so that the distributions of some a priori selected risk factors are identical or nearly the same for the controls as for the cases (Dadvand et al. 2011; Hansen et al. 2009).

A cohort study selects subjects who are at the risk of developing a particular pregnancy outcome, and then are divided into groups according to their air pollution exposure status. The study groups are then followed over time to determine the subsequent incidence of the pregnancy outcome within each group. This study design enables investigators to measure incidence rates and to estimate all effect measures, such as rate ratio (RR) and rate difference. Cohort studies have been an effective method to assess the long-term health effects of acute or chronic exposure to air pollution. A fixed cohort consists of a group of individuals who are identified at a point or interval of tine and then followed over time. A dynamic cohort allows the inclusion of members over time as they fit the selection criteria. A historical or retrospective cohort study is conceptually identical, but less expensive and easier to conduct, to a prospective cohort study except that the study takes place after the causal events have unfolded. Most studies of air pollution and birth outcomes have used a population-based cohort study to examine their associations by linking information from birth certificate records with data from air monitoring (Aguilera et al. 2008; Brauer et al. 2008; Kim et al. 2007; Ritz et al. 2007). This approach allows conducting a study with a large study sample size at very low cost because of the utilization of existing datasets. However, it also easily suffers from exposure misclassification and inadequate controlling of confounders.

Two-phase (stage) designs nesting a sample within a cohort, for which both outcome and some exposure information are available, have a long history in epidemiology. In this approach, some additional information at the individual level is obtained for a sample nested within a cohort. Furthermore, the additional information can be used to assess potential confounding effects on the estimated relationships between exposures of interest and outcomes within the context of the larger cohort. Therefore, this is a special case control design and can better control confounding by use of survey data and minimization of any selection or response bias. The two-phase study design has been recently applied to investigate the associations between air pollution and birth outcomes (Ritz et al. 2007). This approach holds promising because it provides a better capability of controlling for confounding than a classical population-based cohort study of air pollution and birth outcomes.

4.2 Methods of exposure assessment

Air pollution exposure assessment in most studies of air pollution and birth outcomes relies on existing networks of ambient monitoring stations. This approach of exposure assessment is mostly to assign exposure to maternal residential location based on the nearest monitor site or the monitor sites within the same administrative unit of maternal address (e.g. county) (Alderman et al. 1987; Dejmek et al. 1999; Xu et al. 1995; Zhao et al. 2011). However, the method is more likely to result in exposure misclassification because the monitoring networks are relatively sparse and insufficient for capturing all spatial variations in exposure and individual time-activity pattern is also not considered. Another approach for maternal exposure assessment is to develop surrogate measures, i.e. proximity methods , for example, the distance to sources of air pollution for estimating the level of air pollution exposure (Kashima et al. 2011; Yorifuji et al. 2011). Although proximity methods are straightforward, they have largely been thought as a form of exploratory analysis in traffic-related air pollution studies because they have considerable limitations such as invalid assumption of isotropic dispersion of pollution and inadequate consideration of other factors of meteorological condition and topography. Recently, interpolation methods have been applied to estimate maternal air pollution exposure, such as the inverse distance weighting (Xu et al. 2010) and Kriging method (Leem et al. 2006; Seo et al. 2010). Interpolation models produce a continuous surface of pollution concentration, and then extract individual exposure from the surface based on residential locations. This approach is limited by issues regarding exaggerated variation of predictions, spatial coverage and representatives of the existing monitoring sites, and incapability of taking other possible predictors into accounts. Furthermore, dispersion models have been developed to address a number of limitations for air pollution exposure assessment (Batterman et al. 2010; Hoffmann et al. 2009a; Hoffmann et al. 2009b). These models have been regarded as a more realistic representation of the problem by making use of data on emissions, meteorological conditions, and topography in conjunction with information from empirical monitoring systems in predicting spatial exposure estimates of air pollution concentration. Some studies have applied this method for maternal air pollution exposure assessments (Ihrig et al. 1998; Madsen et al. 2010; Wu et al. 2009). However, the features of dispersion models such as assumption of Gaussian dispersion, relatively costly data input, and complicated models, validation, impede their wide applications in estimating air pollution exposure. Land Use Regression (LUR) Models are the methods which have been developed to predict air pollutant concentrations at a given site based on the surrounding land use and traffic characteristics (Briggs et al. 1997; Briggs et al. 2000; Clougherty et al. 2008). This approach establishes a multivariate regression model, where monitored air pollution data serves as a dependent variable and traffic information, land use/cover, and other geographic information are considered as independent variables (Ryan and LeMasters 2007). The model can be expanded using GIS approaches, and thus is capable of predicting the level of air pollution at any location of an area. The advance in GIS technology and availability of detailed geo-reference data further increase the feasibility of implementing LUR models and reduce the costs of modeling. Although many studies have demonstrated their success in predicting air pollution levels, a number of limitations remain. First, the LUR models have been primarily utilized for urban areas, but few of them have been expanded to suburban or rural areas because of lack of information. Second, LUR models are limited to explaining spatial variation of air pollutants, but few of them have addressed temporal variations. This is because the model predictors are temporally stable, such as land use, elevation, and population density (Molter et al. 2010; Rose et al. 2011). A recent review has highlighted the potential for improving the LUR approach by including a spatial and a temporal component in the models (Hoek et al. 2008). Several studies have used this method for maternal exposure assessments (Gehring et al. 2011). The main characteristics of the exposure methods described above are summarized in Table 1.

Although several exposure assessment models have been developed to improve maternal air pollution exposure, personal air pollution exposure remains challenge in the field because limited studies have considered microenvironment in personal exposure estimate. Some methods such as personal dosimetry and biomarkers may need to be considered in future studies.

Methods	Characteristics	Strengths	Limitations
Air monitoring networks	Using existing air data for individual exposure estimation	Easy to access the data; Low cost;	Lack of space coverage; increase likelihoods of misclassification
Proximity methods	Developing surrogate measures such as distance to the road as exposure index	Straightforward; Requires basic GIS techniques;	Invalid assumption of isotropic dispersion of pollution; inadequate consideration
Geo-statistical methods	Using geo-statistical methods to created predicted surfaces of air pollutants	Continuous space coverage; Require few other spatial data;	Exaggerated variation of predictions; required extensive monitoring network; incapability of controlling for other covariates
Dispersion models	Making use of data on emissions, meteorological conditions, topography and information from empirical monitoring systems for spatial prediction	Continuous space coverage; More realistic representation of the problem;	Invalid assumption of Gaussian dispersion; relatively costly data input; Complicated model validation
Land Use Regression (LUR) Models	Predicting air pollutant concentrations at a given site based on the surrounding land use and traffic characteristics	Continuous space coverage; Simple regression model;	Requires temporal variation of predictor; Unmatched of existing spatial data;

Table 1. Characteristics of common methods of exposure assessment in studies of air pollution on birth outcomes

4.3 Windows of exposure

It is believed that the fetus at some particular period of pregnancy is more sensitive to adverse health effects of air pollution. While some studies attempted to determine the effects of air pollution exposure during specific phases of pregnancy, evidence currently available does not support any firm conclusions. Most studies of air pollution and birth outcomes have examined windows of exposure such as the whole pregnancy, each trimester and/or each month of pregnancy. This inconsistency in defining windows of exposure across studies makes it extremely difficult to determine a specific critical period of exposure for each birth outcome. However, according to previous research, negative effects of various air pollutants can be seen as early as the first month of pregnancy. For example, the effects of SO_2, $PM_{2.5}$, and PM_{10} exposure are seen as soon as during the first gestation month for low birth weight, preterm delivery and intrauterine growth restriction (Dejmek et al. 1999; Huynh et al. 2006; Liu et al. 2003; Ritz et al. 2000). For studies which considered all three pregnancy trimesters, exposures during the first and third trimesters appeared to be significantly associated with some birth outcomes. For example, Bukowski et al 2007 demonstrated in their study that variation in birth weight might be determined by fetal size during the first 12 weeks after conception (Bukowski et al. 2007).

Most of the studies reviewed had only examined exposure during pregnancy. However, pre-pregnancy exposure to air pollutants which may affect genetic or epigenetic component in parents see **Germ-line mutations)**, might in turn impact birth outcomes. Therefore, future studies might also consider examining pre-pregnancy windows of exposure.

4.4 Confounders

Confounding threats the validity of observational research. This concern is particularly pertinent in studies of air pollution and birth outcomes as the effects of air pollution on birth outcomes are small. Strickland et al had thoroughly discussed the issue of confounding in epidemiological studies of air pollution and birth outcomes (Strickland et al. 2009). Under the counterfactual disease model, the effect of a factor on a disease is determined by comparing the risk of disease in a population exposure to the factor with one in the same population without exposure to the factor (i.e. under a counterfactual condition). As Strickland et al suggested, risk in the population under a counterfactual condition cannot be observed. A substitute population requires representing the risk in the population under a counterfactual condition. Confounding occurs if the risk in the substitute population is different from the risk in the population under a counterfactual condition.

Birth certificates have been used as the primary data source for information of confounders in most of studies of air pollution and birth outcomes. The birth records usually have information on maternal and infantile characteristics such as maternal age, race, educational attainment, pregnancy complications, and infant gender, etc. Confounding in the studies may arise from imperfect classification of existing information, i.e. residual confounding and unmeasured other variables such as socioeconomic status and maternal nutrition. Residual confounding refers to the confounding due to an imperfect classification of the groups of variables, e.g. maternal age groups and educational groups. Within the group, the confounding of the variable may still exist. For confounding effects of unmeasured factors in birth records, the two-phase study, as we described above, may provide a way to deal with the issue. However, selecting potential confounders requires caution. Some factors which are the intermediate factors in the pathway of air pollution and birth outcomes should not be selected as confounders. For example, some pregnancy complications such as preeclampsia might be caused by air pollution and also risk factors of adverse birth outcomes (Wu et al. 2009).

5. Recommendations of future studies

Several limitations exist in previous air pollution studies on birth outcomes. A few papers have already discussed these issues and provide directions for future studies (Slama et al. 2008; Woodruff et al. 2009). Specifically, the following—but not limited to—issues are of major concern: 1) Accurate individual air pollution exposure including identification of windows of exposure remains a major challenge in the field; 2) The effects of specific components of PM have rarely been investigated; 3) Research of air pollution on fetal health rarely considers the concurrent effects of multiple air pollutants; 4) Little has been done to examine the potential effect modifiers of the relationship between air pollution and birth outcomes; 5) Research including experimental and epidemiological studies on biologically plausible mechanisms is limited. These issues might need to be adequately addressed in future.

1. Development of accurate air pollution exposure assessment remains an important work. The exposure models are in need of improvement in terms of spatial and temporal resolution by incorporating some new technologies such as remote sensing. Individual time-activity patterns in combination with the information from the exposure models might be used together for personal exposure assessment. Some methods of personal monitoring and biomarkers might be considered in prospective cohort studies.
2. Few studies have considered PM composition in relation to birth outcomes (Bell et al. 2010). Different sources of PM may have different chemical constituents and thus have different toxicity. The studies on PM composition might provide clues to explain the heterogeneity of health effects of PM across studies. Further, it would provide information on target chemicals or sources for prevention efforts.
3. Individual is simultaneously exposed to multiple air pollutants. As we reviewed, several criteria air pollutants have adverse effects on birth outcomes. It may be more reasonable to assume that there is a mixture of pollutants that is considered harmful to birth outcomes. However, few studies have investigated the adverse health effects of an air pollution mixture on birth outcomes. It is an important area of ongoing research of air pollution on fetal health. Some sophisticated models need to be developed to examine the effect of a mixture of pollutants because of high collinearity between air pollutants.
4. Air pollution studies on mortality and morbidity have suggested that people with different personal characteristics might respond to air pollution in different ways (Dubowsky et al. 2006; Zanobetti et al. 2000; Zeka et al. 2006). The National Research Council has identified the potential effect modifiers as a key data gap and has emphasized the need for continued study of the most susceptible populations related to air pollution (NRC 2004). However, little has been done to examine the potential modifiers of the health effects of air pollution on birth outcomes. Limited evidence suggests that the effect sizes of air pollution on adverse pregnancy outcomes may be different by fetal gender (Ghosh et al. 2007; Jedrychowski et al. 2009), pregnancy complication (Rich et al. 2009), and maternal nutrition (Jedrychowski et al. 2007; Kannan et al. 2006). Much remains to be learned about these relationships and what else could cause fetuses to be especially sensitive to the adverse health effects of air pollution.

5. Research to identify relevant biological mechanisms is needed to augment our understanding about the relationships. Epidemiological studies could consider other perinatal end points such as preeclampsia and placental size in future studies. The findings from these studies would provide important supporting evidence on the potential biological mechanisms as discussed above.

6. Acknowledgements

Publication of this chapter was funded in part by the University of Florida Open-Access Publishing Fund.

7. References

Agarwal A, Sharma RK, Nallella KP, Thomas AJ, Jr., Alvarez JG, and Sikka SC. 2006. Reactive oxygen species as an independent marker of male factor infertility. Fertil Steril 86, 878-885.

Aguilera I, Sunyer J, Fernandez-Patier R, Hoek G, Aguirre-Alfaro A, Meliefste K, et al. 2008. Estimation of outdoor NO(x), NO(2), and BTEX exposure in a cohort of pregnant women using land use regression modeling. Environ Sci Technol 42, 815-821.

Alderman BW, Baron AE, and Savitz DA. 1987. Maternal exposure to neighborhood carbon monoxide and risk of low infant birth weight. Public Health Rep 102, 410-414.

Ballester F, Estarlich M, Iniguez C, Llop S, Ramon R, Esplugues A, et al. 2010. Air pollution exposure during pregnancy and reduced birth size: a prospective birth cohort study in Valencia, Spain. Environ Health 9, 6.

Barnett AG, Williams GM, Schwartz J, Best TL, Neller AH, Petroeschevsky AL, et al. 2006. The effects of air pollution on hospitalizations for cardiovascular disease in elderly people in Australian and New Zealand cities. Environ Health Perspect 114, 1018-1023.

Batterman SA, Zhang K, and Kononowech R. 2010. Prediction and analysis of near-road concentrations using a reduced-form emission/dispersion model. Environ Health 9, 29.

Bell ML, Belanger K, Ebisu K, Gent JF, Lee HJ, Koutrakis P, et al. 2010. Prenatal exposure to fine particulate matter and birth weight: variations by particulate constituents and sources. Epidemiology 21, 884-891.

Bell ML, Ebisu K, and Belanger K. 2007. Ambient air pollution and low birth weight in Connecticut and Massachusetts. Environ Health Perspect 115, 1118-1124.

Bell ML, McDermott A, Zeger SL, Samet JM, and Dominici F. 2004. Ozone and short-term mortality in 95 US urban communities, 1987-2000. Jama 292, 2372-2378.

Blazer S, Moreh-Waterman Y, Miller-Lotan R, Tamir A, and Hochberg Z. 2003. Maternal hypothyroidism may affect fetal growth and neonatal thyroid function. Obstet Gynecol 102, 232-241.

Block ML, and Calderon-Garciduenas L. 2009. Air pollution: mechanisms of neuroinflammation and CNS disease. Trends Neurosci 32, 506-516.

Bobak M. 2000. Outdoor air pollution, low birth weight, and prematurity. Environ Health Perspect 108, 173-176.

Bobak M, and Leon DA. 1999. Pregnancy outcomes and outdoor air pollution: an ecological study in districts of the Czech Republic 1986-8. Occup Environ Med 56, 539-543.

Bosetti C, Nieuwenhuijsen MJ, Gallus S, Cipriani S, La Vecchia C, and Parazzini F. 2010. Ambient particulate matter and preterm birth or birth weight: a review of the literature. Arch Toxicol 84, 447-460.

Brauer M, Lencar C, Tamburic L, Koehoorn M, Demers P, and Karr C. 2008. A cohort study of traffic-related air pollution impacts on birth outcomes. Environ Health Perspect 116, 680-686.

Briggs DJ, Collins S, Elliott P, Fischer P, Kingham S, Lebret E, et al. 1997. Mapping urban air pollution using GIS: a regression-based approach. International Journal of Geographical Information Science 11, 699-718.

Briggs DJ, de Hoogh C, Gulliver J, Wills J, Elliott P, Kingham S, et al. 2000. A regression-based method for mapping traffic-related air pollution: application and testing in four contrasting urban environments. Sci Total Environ 253, 151-167.

Brook RD. 2008. Cardiovascular effects of air pollution. Clin Sci (Lond) 115, 175-187.

Brook RD, Brook JR, Urch B, Vincent R, Rajagopalan S, and Silverman F. 2002. Inhalation of fine particulate air pollution and ozone causes acute arterial vasoconstriction in healthy adults. Circulation 105, 1534-1536.

Bukowski R, Smith GC, Malone FD, Ball RH, Nyberg DA, Comstock CH, et al. 2007. Fetal growth in early pregnancy and risk of delivering low birth weight infant: prospective cohort study. Bmj 334, 836.

Carrillo AE, Metsios GS, and Flouris AD. 2009. Effects of secondhand smoke on thyroid function. Inflamm Allergy Drug Targets 8, 359-363.

Chen L, Yokel RA, Hennig B, and Toborek M. 2008. Manufactured aluminum oxide nanoparticles decrease expression of tight junction proteins in brain vasculature. J Neuroimmune Pharmacol 3, 286-295.

Clapp Iii JF, and Lopez B. 2007. Size at Birth, Obesity and Blood Pressure at Age Five. Metab Syndr Relat Disord 5, 116-126.

Clougherty JE, Wright RJ, Baxter LK, and Levy JI. 2008. Land use regression modeling of intra-urban residential variability in multiple traffic-related air pollutants. Environ Health 7, 17.

Coppola L, Giunta R, Grassia A, Misso L, Verrazzo G, Violano PF, et al. 1989. Air pollution by gasoline exhaust fumes: effect on platelet function and blood viscosity. Med Lav 80, 187-191.

Dadvand P, Rankin J, Rushton S, and Pless-Mulloli T. 2011. Association between maternal exposure to ambient air pollution and congenital heart disease: A register-based spatiotemporal analysis. Am J Epidemiol 173, 171-182.

Damus K. 2008. Prevention of preterm birth: a renewed national priority. Curr Opin Obstet Gynecol 20, 590-596.

Darrow LA, Klein M, Flanders WD, Waller LA, Correa A, Marcus M, et al. 2009. Ambient Air Pollution and Preterm Birth: A Time-series Analysis. Epidemiology.

Dejmek J, Jelinek R, Solansky I, Benes I, and Sram RJ. 2000. Fecundability and parental exposure to ambient sulfur dioxide. Environ Health Perspect 108, 647-654.

Dejmek J, Selevan SG, Benes I, Solansky I, and Sram RJ. 1999. Fetal growth and maternal exposure to particulate matter during pregnancy. Environ Health Perspect 107, 475-480.

Dockery DW, and Pope CA, 3rd. 1994. Acute respiratory effects of particulate air pollution. Annu Rev Public Health 15, 107-132.

Dominici F, McDermott A, Zeger SL, and Samet JM. 2003. Airborne particulate matter and mortality: timescale effects in four US cities. Am J Epidemiol 157, 1055-1065.

Donaldson K, Stone V, Seaton A, and MacNee W. 2001. Ambient particle inhalation and the cardiovascular system: potential mechanisms. Environ Health Perspect 109 Suppl 4, 523-527.

Dubowsky SD, Suh H, Schwartz J, Coull BA, and Gold DR. 2006. Diabetes, obesity, and hypertension may enhance associations between air pollution and markers of systemic inflammation. Environ Health Perspect 114, 992-998.

Dugandzic R, Dodds L, Stieb D, and Smith-Doiron M. 2006. The association between low level exposures to ambient air pollution and term low birth weight: a retrospective cohort study. Environ Health 5, 3.

Furuta C, Suzuki AK, Taneda S, Kamata K, Hayashi H, Mori Y, et al. 2004. Estrogenic activities of nitrophenols in diesel exhaust particles. Biol Reprod 70, 1527-1533.

Gehring U, Wijga AH, Fischer P, de Jongste JC, Kerkhof M, Koppelman GH, et al. 2010. Traffic-related air pollution, preterm birth and term birth weight in the PIAMA birth cohort study. Environ Res 111, 125-135.

Gehring U, Wijga AH, Fischer P, de Jongste JC, Kerkhof M, Koppelman GH, et al. 2011. Traffic-related air pollution, preterm birth and term birth weight in the PIAMA birth cohort study. Environ Res 111, 125-135.

Ghosh R, Rankin J, Pless-Mulloli T, and Glinianaia S. 2007. Does the effect of air pollution on pregnancy outcomes differ by gender? A systematic review. Environ Res 105, 400-408.

Gilboa SM, Mendola P, Olshan AF, Langlois PH, Savitz DA, Loomis D, et al. 2005. Relation between ambient air quality and selected birth defects, seven county study, Texas, 1997-2000. Am J Epidemiol 162, 238-252.

Glinianaia SV, Rankin J, Bell R, Pless-Mulloli T, and Howel D. 2004. Particulate air pollution and fetal health: a systematic review of the epidemiologic evidence. Epidemiology 15, 36-45.

Gold DR, Damokosh AI, Pope CA, 3rd, Dockery DW, McDonnell WF, Serrano P, et al. 1999. Particulate and ozone pollutant effects on the respiratory function of children in southwest Mexico City. Epidemiology 10, 8-16.

Ha EH, Hong YC, Lee BE, Woo BH, Schwartz J, and Christiani DC. 2001. Is air pollution a risk factor for low birth weight in Seoul? Epidemiology 12, 643-648.

Hansen C, Neller A, Williams G, and Simpson R. 2006. Maternal exposure to low levels of ambient air pollution and preterm birth in Brisbane, Australia. Bjog 113, 935-941.

Hansen CA, Barnett AG, Jalaludin BB, and Morgan GG. 2009. Ambient air pollution and birth defects in brisbane, australia. PLoS One 4, e5408.

Hoek G, Beelen R, de Hoogh K, Vienneau D, Gulliver J, Fischer P, et al. 2008. A review of land-use regression models to assess spatial variation of outdoor air pollution. Atmospheric Environment 42, 7561-7578.

Hoffmann B, Moebus S, Dragano N, Stang A, Mohlenkamp S, Schmermund A, et al. 2009a. Chronic residential exposure to particulate matter air pollution and systemic inflammatory markers. Environ Health Perspect 117, 1302-1308.

Hoffmann B, Moebus S, Kroger K, Stang A, Mohlenkamp S, Dragano N, et al. 2009b. Residential exposure to urban air pollution, ankle-brachial index, and peripheral arterial disease. Epidemiology 20, 280-288.

Huynh M, Woodruff TJ, Parker JD, and Schoendorf KC. 2006. Relationships between air pollution and preterm birth in California. Paediatr Perinat Epidemiol 20, 454-461.

Ibald-Mulli A, Timonen KL, Peters A, Heinrich J, Wolke G, Lanki T, et al. 2004. Effects of particulate air pollution on blood pressure and heart rate in subjects with cardiovascular disease: a multicenter approach. Environ Health Perspect 112, 369-377.

Ihrig MM, Shalat SL, and Baynes C. 1998. A hospital-based case-control study of stillbirths and environmental exposure to arsenic using an atmospheric dispersion model linked to a geographical information system. Epidemiology 9, 290-294.

Jafarabadi M. 2007. Episodic air pollution is associated with increased DNA fragmentation in human sperm without other changes in semen quality. Hum Reprod 22, 3263; author reply 3264.

Jalaludin B, Mannes T, Morgan G, Lincoln D, Sheppeard V, and Corbett S. 2007. Impact of ambient air pollution on gestational age is modified by season in Sydney, Australia. Environ Health 6, 16.

Jedrychowski W, Masters E, Choi H, Sochacka E, Flak E, Mroz E, et al. 2007. Pre-pregnancy dietary vitamin A intake may alleviate the adverse birth outcomes associated with prenatal pollutant exposure: epidemiologic cohort study in Poland. Int J Occup Environ Health 13, 175-180.

Jedrychowski W, Perera F, Mrozek-Budzyn D, Mroz E, Flak E, Spengler JD, et al. 2009. Gender differences in fetal growth of newborns exposed prenatally to airborne fine particulate matter. Environ Res 109, 447-456.

Jerrett M, Burnett RT, Ma R, Pope CA, 3rd, Krewski D, Newbold KB, et al. 2005. Spatial analysis of air pollution and mortality in Los Angeles. Epidemiology 16, 727-736.

Jiang LL, Zhang YH, Song GX, Chen GH, Chen BH, Zhao NQ, et al. 2007. A time series analysis of outdoor air pollution and preterm birth in Shanghai, China. Biomed Environ Sci 20, 426-431.

Kadiiska MB, Mason RP, Dreher KL, Costa DL, and Ghio AJ. 1997. In vivo evidence of free radical formation in the rat lung after exposure to an emission source air pollution particle. Chem Res Toxicol 10, 1104-1108.

Kanaka-Gantenbein C, Mastorakos G, and Chrousos GP. 2003. Endocrine-related causes and consequences of intrauterine growth retardation. Ann N Y Acad Sci 997, 150-157.

Kannan S, Misra DP, Dvonch JT, and Krishnakumar A. 2006. Exposures to airborne particulate matter and adverse perinatal outcomes: a biologically plausible mechanistic framework for exploring potential effect modification by nutrition. Environ Health Perspect 114, 1636-1642.

Kashima S, Naruse H, Yorifuji T, Ohki S, Murakoshi T, Takao S, et al. 2011. Residential proximity to heavy traffic and birth weight in Shizuoka, Japan. Environ Res.

Kim OJ, Ha EH, Kim BM, Seo JH, Park HS, Jung WJ, et al. 2007. PM10 and pregnancy outcomes: a hospital-based cohort study of pregnant women in Seoul. J Occup Environ Med 49, 1394-1402.

Kunzli N, and Tager IB. 2005. Air pollution: from lung to heart. Swiss Med Wkly 135, 697-702.

Landgren O. 1996. Environmental pollution and delivery outcome in southern Sweden: a study with central registries. Acta Paediatr 85, 1361-1364.

Lee BE, Ha EH, Park HS, Kim YJ, Hong YC, Kim H, et al. 2003. Exposure to air pollution during different gestational phases contributes to risks of low birth weight. Hum Reprod 18, 638-643.

Lee SJ, Hajat S, Steer PJ, and Filippi V. 2008. A time-series analysis of any short-term effects of meteorological and air pollution factors on preterm births in London, UK. Environ Res 106, 185-194.

Leem JH, Kaplan BM, Shim YK, Pohl HR, Gotway CA, Bullard SM, et al. 2006. Exposures to air pollutants during pregnancy and preterm delivery. Environ Health Perspect 114, 905-910.

Leeuwenburgh C, and Heinecke JW. 2001. Oxidative stress and antioxidants in exercise. Curr Med Chem 8, 829-838.

Lin CM, Li CY, and Mao IF. 2004. Increased risks of term low-birth-weight infants in a petrochemical industrial city with high air pollution levels. Arch Environ Health 59, 663-668.

Linn WS, Gong H, Jr., Clark KW, and Anderson KR. 1999. Day-to-day particulate exposures and health changes in Los Angeles area residents with severe lung disease. J Air Waste Manag Assoc 49, 108-115.

Liu S, Krewski D, Shi Y, Chen Y, and Burnett RT. 2003. Association between gaseous ambient air pollutants and adverse pregnancy outcomes in Vancouver, Canada. Environ Health Perspect 111, 1773-1778.

Liu S, Krewski D, Shi Y, Chen Y, and Burnett RT. 2007. Association between maternal exposure to ambient air pollutants during pregnancy and fetal growth restriction. J Expo Sci Environ Epidemiol 17, 426-432.

Llop S, Ballester F, Estarlich M, Esplugues A, Rebagliato M, and Iniguez C. 2010. Preterm birth and exposure to air pollutants during pregnancy. Environ Res 110, 778-785.

Madsen C, Gehring U, Walker SE, Brunekreef B, Stigum H, Naess O, et al. 2010. Ambient air pollution exposure, residential mobility and term birth weight in Oslo, Norway. Environ Res 110, 363-371.

Maisonet M, Bush TJ, Correa A, and Jaakkola JJ. 2001. Relation between ambient air pollution and low birth weight in the Northeastern United States. Environ Health Perspect 109 Suppl 3, 351-356.

Maisonet M, Correa A, Misra D, and Jaakkola JJ. 2004. A review of the literature on the effects of ambient air pollution on fetal growth. Environ Res 95, 106-115.

Mannes T, Jalaludin B, Morgan G, Lincoln D, Sheppeard V, and Corbett S. 2005. Impact of ambient air pollution on birth weight in Sydney, Australia. Occup Environ Med 62, 524-530.

Maroziene L, and Grazuleviciene R. 2002. Maternal exposure to low-level air pollution and pregnancy outcomes: a population-based study. Environ Health 1, 6.

Marshall EG, Harris G, and Wartenberg D. 2010. Oral cleft defects and maternal exposure to ambient air pollutants in New Jersey. Birth Defects Res A Clin Mol Teratol 88, 205-215.

Martin JA, Kung HC, Mathews TJ, Hoyert DL, Strobino DM, Guyer B, et al. 2008. Annual summary of vital statistics: 2006. Pediatrics 121, 788-801.

Middleton N, Yiallouros P, Kleanthous S, Kolokotroni O, Schwartz J, Dockery DW, et al. 2008. A 10-year time-series analysis of respiratory and cardiovascular morbidity in

Nicosia, Cyprus: the effect of short-term changes in air pollution and dust storms. Environ Health 7, 39.

Mohorovic L. 2004. First two months of pregnancy--critical time for preterm delivery and low birthweight caused by adverse effects of coal combustion toxics. Early Hum Dev 80, 115-123.

Molter A, Lindley S, de Vocht F, Simpson A, and Agius R. 2010. Modelling air pollution for epidemiologic research--part II: predicting temporal variation through land use regression. Sci Total Environ 409, 211-217.

NRC. 2004. National Research Council. Research priorities for airborne particulate matter. IV. Continuing research progress. Washington, DC: The National Academies Press, 2004.

Olusanya BO, and Solanke OA. 2011. Perinatal Outcomes Associated with Maternal Hypertensive Disorders of Pregnancy in a Developing Country. Hypertens Pregnancy.

Osmond C, and Barker DJ. 2000. Fetal, infant, and childhood growth are predictors of coronary heart disease, diabetes, and hypertension in adult men and women. Environ Health Perspect 108 Suppl 3, 545-553.

Parker JD, Woodruff TJ, Basu R, and Schoendorf KC. 2005. Air pollution and birth weight among term infants in California. Pediatrics 115, 121-128.

Patel J, Landers K, Li H, Mortimer RH, and Richard K. 2011. Thyroid hormones and fetal neurological development. J Endocrinol 2011.

Peters A, Doring A, Wichmann HE, and Koenig W. 1997. Increased plasma viscosity during an air pollution episode: a link to mortality? Lancet 349, 1582-1587.

Peters A, Veronesi B, Calderon-Garciduenas L, Gehr P, Chen LC, Geiser M, et al. 2006. Translocation and potential neurological effects of fine and ultrafine particles a critical update. Part Fibre Toxicol 3, 13.

Pope CA, 3rd. 1999. Mortality and air pollution: associations persist with continued advances in research methodology. Environ Health Perspect 107, 613-614.

Pope CA, 3rd. 2000. Epidemiology of fine particulate air pollution and human health: biologic mechanisms and who's at risk? Environ Health Perspect 108 Suppl 4, 713-723.

Pope CA, 3rd, Dockery DW, Spengler JD, and Raizenne ME. 1991. Respiratory health and PM10 pollution. A daily time series analysis. Am Rev Respir Dis 144, 668-674.

Pope CA, 3rd, and Kanner RE. 1993. Acute effects of PM10 pollution on pulmonary function of smokers with mild to moderate chronic obstructive pulmonary disease. Am Rev Respir Dis 147, 1336-1340.

Prahalad AK, Inmon J, Dailey LA, Madden MC, Ghio AJ, and Gallagher JE. 2001. Air pollution particles mediated oxidative DNA base damage in a cell free system and in human airway epithelial cells in relation to particulate metal content and bioreactivity. Chem Res Toxicol 14, 879-887.

Rich DQ, Demissie K, Lu SE, Kamat L, Wartenberg D, and Rhoads GG. 2009. Ambient air pollutant concentrations during pregnancy and the risk of fetal growth restriction. J Epidemiol Community Health 63, 488-496.

Rinaudo PF, and Lamb J. 2008. Fetal origins of perinatal morbidity and/or adult disease. Semin Reprod Med 26, 436-445.

Risom L, Moller P, and Loft S. 2005. Oxidative stress-induced DNA damage by particulate air pollution. Mutat Res 592, 119-137.

Ritz B, Wilhelm M, Hoggatt KJ, and Ghosh JK. 2007. Ambient air pollution and preterm birth in the environment and pregnancy outcomes study at the University of California, Los Angeles. Am J Epidemiol 166, 1045-1052.

Ritz B, and Yu F. 1999. The effect of ambient carbon monoxide on low birth weight among children born in southern California between 1989 and 1993. Environ Health Perspect 107, 17-25.

Ritz B, Yu F, Chapa G, and Fruin S. 2000. Effect of air pollution on preterm birth among children born in Southern California between 1989 and 1993. Epidemiology 11, 502-511.

Ritz B, Yu F, Fruin S, Chapa G, Shaw GM, and Harris JA. 2002. Ambient air pollution and risk of birth defects in Southern California. Am J Epidemiol 155, 17-25.

Rogers JF, Thompson SJ, Addy CL, McKeown RE, Cowen DJ, and Decoufle P. 2000. Association of very low birth weight with exposures to environmental sulfur dioxide and total suspended particulates. Am J Epidemiol 151, 602-613.

Rose N, Cowie C, Gillett R, and Marks GB. 2011. Validation of a spatiotemporal land use regression model incorporating fixed site monitors. Environ Sci Technol 45, 294-299.

Rubes J, Selevan SG, Evenson DP, Zudova D, Vozdova M, Zudova Z, et al. 2005. Episodic air pollution is associated with increased DNA fragmentation in human sperm without other changes in semen quality. Hum Reprod 20, 2776-2783.

Rudra CB, Williams MA, Sheppard L, Koenig JQ, and Schiff MA. 2011. Ambient Carbon Monoxide and Fine Particulate Matter in Relation to Preeclampsia and Preterm Delivery in Western Washington State. Environ Health Perspect.

Ryan PH, and LeMasters GK. 2007. A review of land-use regression models for characterizing intraurban air pollution exposure. Inhal Toxicol 19 Suppl 1, 127-133.

Sagiv SK, Mendola P, Loomis D, Herring AH, Neas LM, Savitz DA, et al. 2005. A time-series analysis of air pollution and preterm birth in Pennsylvania, 1997-2001. Environ Health Perspect 113, 602-606.

Salam MT, Millstein J, Li YF, Lurmann FW, Margolis HG, and Gilliland FD. 2005. Birth outcomes and prenatal exposure to ozone, carbon monoxide, and particulate matter: results from the Children's Health Study. Environ Health Perspect 113, 1638-1644.

Saldiva PH, Pope CA, 3rd, Schwartz J, Dockery DW, Lichtenfels AJ, Salge JM, et al. 1995. Air pollution and mortality in elderly people: a time-series study in Sao Paulo, Brazil. Arch Environ Health 50, 159-163.

Samoli E, Touloumi G, Schwartz J, Anderson HR, Schindler C, Forsberg B, et al. 2007. Short-term effects of carbon monoxide on mortality: an analysis within the APHEA project. Environ Health Perspect 115, 1578-1583.

Schwartz J, Dockery DW, Neas LM, Wypij D, Ware JH, Spengler JD, et al. 1994. Acute effects of summer air pollution on respiratory symptom reporting in children. Am J Respir Crit Care Med 150, 1234-1242.

Seo JH, Leem JH, Ha EH, Kim OJ, Kim BM, Lee JY, et al. 2010. Population-attributable risk of low birthweight related to PM10 pollution in seven Korean cities. Paediatr Perinat Epidemiol 24, 140-148.

Shah PS, and Balkhair T. 2011. Air pollution and birth outcomes: a systematic review. Environ Int 37, 498-516.

Slama R, Darrow L, Parker J, Woodruff TJ, Strickland M, Nieuwenhuijsen M, et al. 2008. Meeting report: atmospheric pollution and human reproduction. Environ Health Perspect 116, 791-798.

Somers CM, and Cooper DN. 2009. Air pollution and mutations in the germline: are humans at risk? Hum Genet 125, 119-130.

Somers CM, Yauk CL, White PA, Parfett CL, and Quinn JS. 2002. Air pollution induces heritable DNA mutations. Proc Natl Acad Sci U S A 99, 15904-15907.

Strickland MJ, Klein M, Darrow LA, Flanders WD, Correa A, Marcus M, et al. 2009. The issue of confounding in epidemiological studies of ambient air pollution and pregnancy outcomes. J Epidemiol Community Health 63, 500-504.

Takeda K, Tsukue N, and Yoshida S. 2004. Endocrine-disrupting activity of chemicals in diesel exhaust and diesel exhaust particles. Environ Sci 11, 33-45.

Tao F, Gonzalez-Flecha B, and Kobzik L. 2003. Reactive oxygen species in pulmonary inflammation by ambient particulates. Free Radic Biol Med 35, 327-340.

Thompson JN. 2007. Fetal nutrition and adult hypertension, diabetes, obesity, and coronary artery disease. Neonatal Netw 26, 235-240.

Tran DQ, Ide CF, McLachlan JA, and Arnold SF. 1996. The anti-estrogenic activity of selected polynuclear aromatic hydrocarbons in yeast expressing human estrogen receptor. Biochem Biophys Res Commun 229, 101-108.

Valkonen VP, Paiva H, Salonen JT, Lakka TA, Lehtimaki T, Laakso J, et al. 2001. Risk of acute coronary events and serum concentration of asymmetrical dimethylarginine. Lancet 358, 2127-2128.

van den Hooven EH, de Kluizenaar Y, Pierik FH, Hofman A, van Ratingen SW, Zandveld PY, et al. 2011. Air pollution, blood pressure, and the risk of hypertensive complications during pregnancy: the generation R study. Hypertension 57, 406-412.

Wang X, Ding H, Ryan L, and Xu X. 1997. Association between air pollution and low birth weight: a community-based study. Environ Health Perspect 105, 514-520.

Wietlisbach V, Pope CA, 3rd, and Ackermann-Liebrich U. 1996. Air pollution and daily mortality in three Swiss urban areas. Soz Praventivmed 41, 107-115.

Wilhelm M, and Ritz B. 2005. Local variations in CO and particulate air pollution and adverse birth outcomes in Los Angeles County, California, USA. Environ Health Perspect 113, 1212-1221.

Woodruff TJ, Parker JD, Darrow LA, Slama R, Bell ML, Choi H, et al. 2009. Methodological issues in studies of air pollution and reproductive health. Environ Res 109, 311-320.

Wu J, Ren C, Delfino RJ, Chung J, Wilhelm M, and Ritz B. 2009. Association between local traffic-generated air pollution and preeclampsia and preterm delivery in the south coast air basin of California. Environ Health Perspect 117, 1773-1779.

Xu X, Ding H, and Wang X. 1995. Acute effects of total suspended particles and sulfur dioxides on preterm delivery: a community-based cohort study. Arch Environ Health 50, 407-415.

Xu X, Sharma RK, Talbott EO, Zborowski JV, Rager J, Arena VC, et al. 2010. PM10 air pollution exposure during pregnancy and term low birth weight in Allegheny County, PA, 1994-2000. Int Arch Occup Environ Health 84, 251-257.

Yorifuji T, Naruse H, Kashima S, Ohki S, Murakoshi T, Takao S, et al. 2011. Residential proximity to major roads and preterm births. Epidemiology 22, 74-80.

Zanobetti A, Schwartz J, and Gold D. 2000. Are there sensitive subgroups for the effects of airborne particles? Environ Health Perspect 108, 841-845.

Zeka A, Zanobetti A, and Schwartz J. 2006. Individual-level modifiers of the effects of particulate matter on daily mortality. Am J Epidemiol 163, 849-859.

Zhao Q, Liang Z, Tao S, Zhu J, and Du Y. 2011. Effects of air pollution on neonatal prematurity in Guangzhou of China: a time-series study. Environ Health 10, 2.

Part 2

Environmental Changes, Geographic and Climatic Conditions due to Air Pollution

6

Reduction of Air Pollution by Combustion Processes

Aurel Gaba and Stefania Felicia Iordache
Valahia University of Targoviste
Romania

1. Introduction

Among human activities, those in which fuel combustion processes intervene are those who contaminate the atmosphere greatly.

The combustion process is a process of rapid oxidation, followed by light phenomena and the release of large amounts of energy, able to maintain it at high temperatures. Compared with slow oxidation processes, it is characteristic to the combustion process sudden acceleration of the reaction rate to achieve theoretically infinite values. This applies, for example, to the stoichiometric mixture of methane - oxygen heated to a temperature of 560°C in a sealed container. Heating the same mixture to a temperature of only 200°C, result in a slow oxidation process, which produces methanol, formic acid, formaldehyde, carbon monoxide and carbon dioxide gas, with an overall response rate with an evolution with measured values up to a maximum, after which rate value decreases with the depletion of reagents. In everyday life we encounter slow oxidation processes at every step. Thus, minerals are subject to slow oxidation process which occurs at ambient temperature by consumption of oxygen from atmospheric air, with production of oxides in a state of maximum stability. Such a process is carbon steel corrosion under the action of atmospheric oxygen at ambient temperatures, which is transformed first into ferrous oxide (FeO) and then in a more stable substance, ferric oxide (Fe_2O_3). Also, living organisms consume oxygen in the atmosphere, at room temperature to oxidize nutrients over a slow but very complex process. In both examples above, as in any oxidation process, there are necessary two substances: the oxidant, which has the ability to quickly combine with the substance subject to oxidation, respectively, the substance that is oxidized, called fuel.

The transformation of chemical energy of fossil fuels in forms of energy directly useable, primarily mechanical energy, electrical energy and heat energy, is practically done only by means of combustion. In the production of electrical and heat energy, are consumed by burning, at present, 87% fossil fuels, the remainder being nuclear energy and regenerative energy (hydraulic, wind energy, solar, geothermal and marine) 6%, respectively 7%. At this consumption of the fossil fuels the consumption to produce mechanical energy in transports and the technological consumption, e.g. consumption of coal to produce metallurgical coke and for injection in blast furnaces, is added.

In 2008, according to the World Energy Outlook (2010), world consumption of fuels was 12 300 million toe (tons oil equivalent), of which 30 660 million barrels of oil and 3100 billion cubic meters of natural gas. For the year 2035, according to the script "new scenario" of the

World Energy Outlook (2010), total consumption will be increased up to16 700 million toe, of which 36 135 million barrels of oil and 4500 billion cubic meters of natural gas. These significant increases in world consumption of fuels results in conditions that anticipate an increase in the share of nuclear energy to 8% and regenerative energy to 14% in electrical and heat energy production. As a result, World Energy Outlook (2010), consider that the regulations of Copenhagen Accord (The United Nations Climate Change Conference in Copenhagen, 7-19 December 2009) to reduce CO2 emissions are not sufficient to halt the temperature increase with up to 2°C, worldwide.

In the fuel combustion processes the maximum release of energy is obtained with the complete combustion with, minimum excess of air, closer to stoichiometric combustion, which generally leads to the exclusive formation of CO_2, H_2O, SO_2 and N_2 gases, respectively O_2 and N_2 of the air excess. In practice, combustion processes slide away more or less from this ideal, both due to the complex structure of used fuels, such as coals and inferior oils, also due to imperfection of the combustion installations. Thus are formed unburned substances as solid particles, rich in carbon, which in most cases are accompanied by unburned gases. Unburned substances discharged into the atmosphere with the combustion gases, lead to heat energy loss, which leads to a decrease of the thermal efficiency of the process but at the same time, they have polluting action (Ghia & Gaba, 2000).

Sources of air pollution from burning fuels are classified as follows: stationary sources, mainly consisting of boilers, furnaces and gas turbines and mobile sources consisting of transport means, whose internal combustion engines are diesel (diesel-consuming, also called compression-ignition engines) and spark ignition engines, consuming gasoline. Lately there are used as stationary sources in heat energy and electricity production in cogeneration and internal combustion engines. Classification of stationary sources is done by industry type:

- Energy Industry (heat energy and electricity production, petroleum refining, manufacture of solid fuel and other industries);
- Manufacturing Industry and construction (iron and steel, chemicals, pulp, paper and printing, food processing, beverage and tobacco), called the EEA Industry (energy) sector. Also in EEA (European Environment Agency) publications mobile sources are classified in road transport and other transport.

Wastes scattered in the atmosphere by fuel combustion processes are called emissions (some are returning back on the ground forming immissions) and form two major categories: gases and solid particles.

Gases released into the atmosphere by fuel combustion process are: CO_2, SO_2, H_2O, N_2, O_2, CO, NO_x, PAHs (polycyclic aromatic hydrocarbons), VOCs (volatile organic compounds), dioxins, furans. Lately it was started the separate monitoring of CH_4 methane from volatile organic compounds, these remaining without CH_4 becoming NMVOCs (non-methane volatile organic compounds). With the exception of H_2O, N_2 and O_2, all other elements pollute the atmosphere directly and some by induced forms (usually synergistic, their effect being amplified, more than their cumulative effect) such as the greenhouse effect, smog and acid rain (acid deposition more accurately).

The greenhouse effect is due to more than 50% to CO_2 and in very small proportion to nitrous oxide N_2O, mainly as gases from combustion processes. 65% of CO_2 in the atmosphere comes from burning fuels. Carbon dioxide recently considered as pollutant, prevents the heat energy discharge from the surface of the Earth which is received by radiation from the sun. With the passage of time, even a few degrees increase in the Earth temperature could lead to reduction of the ice cap, the spread of deserts, rising the levels of oceans and sea, as important changes of the climate. In consequence, inside United Nations it was negotiated a convention on climate change which was signed at the Rio Conference in 1992, with the objective of stabilizing greenhouse effect concentrations.

Smog is a synergy, which may be:

- Reducing, also called sulfur dioxide, wet or London type;
- Photochemical, also called an oxidizer or Californian type.

Reducing smog is formed due to the presence, in high concentrations, of SO_2 and soft particles in suspention (smoke) in a calm atmosphere with high humidity, at relatively low temperatures. Component substances that contribute to forming reducing smog are solid particles of soot, cenospheres, sulfur oxides and carbon oxide, derived from incomplete combustion of coals or oils. Cold water vapors condense around the solid particles of smoke, which have a large surface and can act as catalysts to form soft particles of fog. Sometimes reducing smog also contains soft particles of silicon, aluminum as metal oxides.

Photochemical smog is formed due to the presence, in high concentrations, of NO_x and VOCs, in a calm atmosphere, with temperatures over 18°C, in the absence of rain and wind. By the action of solar radiation on primary pollutants, NO_x and VOCs, mainly derived from heavy road traffic, secondary pollutants are formed as NO_2, which decompose under the action of solar radiation in highly toxic compounds, such as peroxiacetil nitrate (PAN).

Acid rain is also a form of pollution induced by pollutants emitted by combustion processes of fuels, SO_x and NO_x. Arrived in the atmosphere, sulphur and nitrogen oxides react with oxygen, water vapors and other substances forming acid compounds, such as dilute solutions of H_2SO_4, HNO_3 and their salts, sulfates and nitrates. These acid compounds reaching the Earth by rain, snow and fog, forming wet acidic depositions, respectively by particles, wind and stormwater forming dry acidic depositions. Acidic depositions have a negative impact on population health status and adverse effects on aquatic ecosystems, on forests and crops. Also, acidic depositions attack buildings and works of art.

Another form of pollution induced by pollutants emitted by fuel combustion processes, is the formation of tropospheric (high altitude) ozone (O_3). Precursors of tropospheric ozone O_3 formation are: nitrogen oxides (NO_x), non-methane volatile organic compounds (NMVOC), carbon monoxide (CO) and methane (CH_4). Because of the ozone in the atmosphere vegetation growth is reduced and aging occurs on the crops and forests. Solid particles in the air are classified by their size, their capacity of sedimentation and the possibility to penetrate the organisms, as following:

- Particles larger than 10 μm are deposited after the law of gravity with an uniform accelerated motion, also known as dust;
- Particles between 10 μm and 0.1 μm that are deposited after Stokes law, consisting mainly of ash and black smoke from burning fuel results;
- Particles smaller than 0.1 μm that are deposited and move after Brownian movement law.

Especially in large steam boilers, through the burning of coals and oils, solid particles driven by exhaust gases may contain non-ferrous metals (Pb, Zn, Ni, Cd, Sn etc.) and metalloids (F and As).

In diesel engines, through incomplete combustion the fumes can appear white, blue respectively, which consist of a suspension in gases of liquid particles of unburned fuel or

partially oxidized. Particles with diameter over 1 μm are found in the white smoke and particles with a diameter of about 0.5 μm appear in the blue smoke.
Black smoke from diesel engines contains solid carbonaceous particles, with average diameters of 1 μm.
Some experts consider that solid primary particles ranging in size from 0.01 to 0.05 μm, during the expansion and exhaustion of burning gases, find suited conditions to aggregate in soot particles with dimensions up to 0.6 μm occupying a specific surface of about 200 m^2/g. Experimentally it was found that these particles are formed under certain specific conditions, of: 20-90% amorphous carbon, 2-5% ash and 10-80% carcinogenic hydrocarbons.
In order to diminish air pollution by combustion processes, the following solutions are recommended (Baukal, 2001, Ghia & Gaba,2000):

- reducing the pollutant emission at origin, at the producing sources (primary procedures) by changing-replacing the combustion processes (use of renewable energy, production reducing, increase of production/technological lines efficiency, replace of production technologies), by using pretreatment procedures of fuels and oxidizer substances (changing fuels and oxidizers, improving the fuels characteristics by processing, preparation and use of additives) and by using improvement - changing procedures of combustion processes, both in case of combustion installations - especially the burners and burning chambers with reference to their construction (mixed stages, flue gases recirculation, auto-carburizing, steam or water spraying) and in case of combustion adjustment (control of excess air, burners out of service, adjustment of combustion air temperature);
- collecting and /or reducing the pollutants resulted from flue gasses before their suppression into the air by using the so-called secondary pollution reducing techniques or flue gases treatment (catalytic and non catalytic reactors, active carbon processes, SNOX and DESONOX processes, electrostatic precipitators, bag filters, scrubbers etc).

2. Formation and reduction pollutants in combustion processes

2.1 Carbon monoxide

The emission of carbon monoxide (CO) pollutant in combustion processes is produced when some conditions occur which lead to incomplete burning of fuels that contain carbon. Among heavy users of the most polluting fossil fuels, through emission of CO, are the means of transportation driven by spark ignition engines and compression ignition engines, and the lesser CO polluting are power plants.
If oxygen is missing during the combustion, compared to stoichiometric requirements, in an area of a fuel-air mixture, it will result CO even at a sufficiently high temperature. The possibility of CO subsequent oxidation, when the contribution of necessary oxygen to a lower temperature appears, is slower than forming CO.
In boilers and in furnaces, the presence of CO in the exhausted flue gases highlight essentially a disorder of the combustion equipment: burner, sprayer, combustion air blow installation, etc. Carbon oxide appears usually in the flue gases discharged from old, rudimentary combustion plants and is almost absent in the exhaust gases from modern industrial combustion plants of high-capacity. Also, at these facilities is taken into account that reducing the excess of air, which reduces the emissions of NO_x, can increase CO emission. As a result an optimal value for the air excess coefficient is established (Gaba, 2005).

Ensuring complete combustion is a measure considered to be BAT (best available techniques), which ensure the maintenance of the CO emission levels below 100 mg/Nm^3, for 15% O_2 (for internal combustion stationary engines and gas turbines) and in boilers and furnaces, for 3% O_2, to run on liquid and gaseous fuels and 6% O_2 to run on solid fuels (Directive 2008/1/EC, 2008).

For mobile sources, which are the biggest pollutants with CO, as reduction techniques were developed, among others, multi-point injection engines that work at high pressures and oxidation catalysts (Duca et al., 2002).

The amount of CO emitted by transportation in European Union countries (EU-27) represents 89% of the total CO emissions at their level. Due to equipping vehicles with better catalysts and application in EU-27 of new regulations that produced engines up to Euro-6, CO emissions diminished, in 2008, compared to 1990, with 71% (EEA-TERM 003, 2011).

2.2 Sulfur oxides

Beside some gaseous fuels, the other industrial fossil fuels generally contain sulfur S. For engines, diesel has a low sulfur content, which is null to petrol diesel prompts. Sulfur content in diesel determines SO_2 polluting emissions, which on diesel engines represent about 10% of the polluting emissions of these engines, but are being endangered as a result of production, lately, of sulfur-free diesel fuels.

Maximum sulfur content from main fossil fuels is regulated, e.g. France, $S \leq 4\%$ for industrial liquid fuels and $S \leq 1\%$ for solid fuels.

Primarily coals and oils, which are used in industrial plants to produce electricity and heat, are responsible for environmental pollution by sulfur oxides, strong acidity of air and acid rains. Sulfur dioxide tendency is to reduce the "greenhouse effect" by "albedo" effect.

Crude oils contain variable quantities of sulfur depending on their origin. During refining operations, sulfur tends to gather in waste fractions which can be an important element to the industrial liquid fuels. The difficulty to reduce the sulfur content in crude oils and petroleum residues consist especially in their content of asfaltene and metals which lead to deactivation of catalysts. Researches are made in this respect. At a large scale of solid fuels the sulfur content is close to the values for heavy liquid fuels. Also, desulphurization of solid fuels with high content of sulfur is a major activity, that develops both through practical achievements and through researches.

Discharged in the atmosphere, SO_2 react in proportion of 1-2 ‰/h with oxygen under the action of solar ultraviolet radiation resulting SO_3 which combines with water vapor in the atmosphere to form sulfuric acid. In very wet days, with fog, the degree of conversion to sulfuric acid is up to 15.7%.

Also, starting from SO_2 gas, sulfate aerosols forms but they can be washed by the rain which becomes acid. To ensure a satisfactory dilution and dispersion of the SO_2 gas in the atmosphere, the combustion gases must be evacuated from the funnel with an adequate speed and to a convenient height from the ground. Thus the large thermoelectric oil plants funnels exceed 200m height frequently. Because of this the pollution due to pollutants produced by sulfur in the fuel may occur at great distances from where they were generated.

Generally, the growth of sulfur dioxide quantity in the atmosphere has the effect of increasing the overall morbidity rate of the population. At concentrations of sulfur dioxide in the cities air over 0.046 ppm (annual average) lead to an increasing frequency of respiratory diseases, at concentrations of 0.52 ppm in the presence of other solid particles, overall mortality is increased. Sulfur dioxide and sulfuric acid aggravate the respiratory system of animals, but this is much more toxic than SO_2, its action being dependent on the size of aerosols. In a series of countries, particularly those located in northern area, the specific flora is affected even by the traces of sulfur dioxide, moss and lichens around large cities from Europe and America, disappear. Sulfur dioxide absorbed by plants leads to acute and chronic effects. Acute effects, that follow high concentrations and the relatively short exposures, manifest by changing the color (to yellow-ivory or red-brown) of damaged tissues. Chronic effects, results of prolonged exposure to low concentrations, manifest by perpetual yellowing of the plant foliage following the alteration of the chlorophyll production mechanism. It seems that, to the acute effects, the plant defend itself transforming sulfur dioxide into sulfuric acid and then in sulfates that are deposited in certain portions of tissues, but soon the balance of sulfates is disturbed and sulfuric acid appears in the plant cellular system which attacks its cells. Chronic effects are due to sulfate aggregations. Sulfur dioxide in small concentration (on the order of the size of 100 - 500 $\mu g/m^3$) react synergistically with both ozone and nitrogen dioxide, which affects the plants tissue, especially as these are more sensitive.

Sulfur dioxide and sulfuric acid formed have very corrosive action on metals, constructions, leather, paper and textiles, especially if relative humidity of air exceeds 70%. Attack of construction materials by SO_2 is due to next mechanism: SO_2 in the atmosphere prompts a condensation of water vapors as fog even when their partial pressure is lower than the saturation pressure at that temperature, due to the fact that SO_2 form with water a solution which a lower vapor pressure than pure water, slip by easily as fog state; the phenomenon can be facilitated by existing aerosols as particles in the smoke, which forms condensation nuclei. In soft drops of fog with SO_2 is obtained sulfurous acid and its oxidation to sulfuric acid. Reducing character of SO_2 manifests also by changing the color of some paint pigments. As such it is affected lead white. Calcium carbonate, a constituent of many construction materials, some of which serve to works of arts, is transformed into calcium sulfate. Destructive effect is accelerated by the acid rains.

In 1983 the first symptoms of forests etching have appeared in Western Europe, due to acid rains which gather both acid pollution and photooxidant pollution. It has been established that acidity in the atmosphere, most often originates in a ratio of 2/3 due to emissions of sulfur oxides, less than 1/3 because of NO_x, and the rest due to pollutants as fluorine or chlorine. Photooxidant pollution is due transforming NO_x in the presence of hydrocarbons and solar radiation.

Beside the negative effects outlined above, the presence of sulfur oxides in the air space reduces the visibility when the fog photochemical processes occurring and determines the appearance of an unpleasant smell.

In the atmosphere, oxidation of SO_2 to SO_3 takes place mainly through a photochemical process, although there are some opinions that it could take place catalytic processes also. During these reactions the presence of other pollutants such as nitrogen oxides or metal oxides may play an important role to accelerate the reactions. But there are occuring other factors such as concentration, residence time in the atmosphere, temperature, humidity, intensity and spectral distribution of radiations. In boilers furnaces sulfur trioxide, or sulfuric anhydride SO_3, derives from the oxidation sulfur dioxide. Actually, in the combustion gases is found 3-5% of the initial sulfur contained in the fuel, as SO_3.

Under certain circumstances SO_2 combines with water vapors, giving sulfurous acid gas H_2SO_3 and SO_3 combines still with water vapors, giving gaseous sulfuric acid H_2SO_4. Acid

vapors condensation on cold walls attack metals. Beside corrosion, due to SO_4H_2 vapors condensation, especially in thermal uninsulated funnels, it is favored the agglomeration of solid particles of soot and flying coke, who create acid rain too.
The fight against acid corrosion and pollution is done in three ways:

- Preventing the forming of anhydrides, especially SO_3;
- Neutralizing the formation of acids;
- Preventing the condensation of formed acid vapors.

Avoiding actual forming of anhydrides, means burning fuels with a low sulfur content. Heavy petroleum desulphurization is not generalized to an industrial scale today because of the high prices required. Fuels with low sulfur content can be subjected to industrial operations only when sulfur compounds are extremely harmful. Sometimes it is possible to have them burned out at low excesses of air, thus forming reduced SO_2 and SO_3 quantities. Limiting the excess of air at maximum 3% is difficult to do because:

- the burner must ensure the right mix between combustion air and fuel over the whole tuning;
- industrial boilers being equipped with numerous burners, each burner must have its own system of adjustment and control, which raise the cost of the related equipment. So the solution is applied only to very large boilers from thermoelectric power plants, which have numerous personnel for maintenance and exploitation.

Spraying silica allows to stop the catalytic action of deposits that form on the heating and overheating pipes, which reduces the amount of formed SO_3. Frequent cleaning of depositions of solid particles completes the action mentioned before, but does not reduce the formation of SO_3 from the flame. SO_3 action may be partially neutralized by injection into the combustion products of calcium carbonate, zinc oxide, dolomite and of ammonia, which allows some boilers to reduce significantly the corrosion of air preheaters. Injection technique is sensible and it must be assured a good distribution of the substances mentioned above, avoiding obstruction of the flue gas flow and solid particle entrainment in atmosphere. When using ammonia, neutralization does not stop at the stage of acid sulfate of ammonia, but it reaches the stage of neutral sulfate.
Primary procedures for reduction of SO_2 pollutant emissions make the desulphurization, reducing the content of SO_2 in the flue gas by injecting absorbent substances sprayed into the furnace, such as limestone, lime or dolomite. Desulphurization combustion can be seen as a primary technique for major reduction of SO_2, applied to certain types of boilers. For the first achieved version, when the flame is obtained from a conventional burner, the U.S. has developed an extensive research program LIMB (Limestone Injection in Multistage Burners) with several forms of application.
Desulphurization efficiency depends particularly on the magnitude of the temperatures in the furnace, but meritorious performances are obtained using high consumption of absorbent.
For the second achieved version, the combustion is developing in dense or circulating fluidized bed. Such techniques can achieve efficiencies of combustion and desulphurization with high values having a consumption of absorbent lower than the first version. It should be noted that circulating bed technique is superior as efficiency than the dense bed technique.
Dense fluidized bed is characterized by the flow speed, generally reduced, where the solid particles move relatively slow and require a moderate recirculation. Increasing the speed of flow, the volume occupied by the layer will increase, as the movement speed of solid

particles, which will require an intense recirculation of particles realizing the circulating fluidized bed. These particles are of two types:

- a kind of incompletely burned fuel, because the coal particles introduced into the furnace having a high temperature generate gaseous volatile mater and carbon tailing which require a much longer combustion time than the combustion gases;
- absorbent material, which is mostly limestone, which has three forms: $CaCO_3$, CaO and $CaSO_4$.

Because the system operates in stationary state, the continuously exhausted gases will have to contain very small quantities of combustible substances to obtain the highest possible energy efficiency. Also, the exhaust gases will contain $CaCO_3$ and CaO as well as a very high content of $CaSO_4$ for a good desulphurization. Conventional circulating fluidized bed boilers realize the burning and desulphurization in the same furnace known as reactor that is fed simultaneously with fuel and absorbent material. The differences between these boilers refer to the ways of extracting produced energy and the means of control operation of variation with load. IFP, Lardet-Babcock and Cecar made a boiler called self desulphurization to which combustion and desulphurization are produced in different rooms. This boiler has a circulating fluidized bed furnace which is fed with combustion gases, having the temperature of 500-850°C, instead of air, as conventional boilers are operating. Hot combustion gases, rich in SO_2, from the furnace, are routed through the loop of circulation of solid particles in suspension, where is added limestone. This loop does not have exchange surfaces for useful transfer of heat, so it is protected from the effects of the sulfuric corrosion. The combustion gases thus sweetened are directed to the heat exchange surfaces of a exhaust-heat boiler. Whatever the system, in principle, all types of boilers with circulating layer are operating with relatively soft particles of limestone, of a few hundred microns, rather than a few millimeters bed as the dense bed boilers user. We mention that circulating bed systems allow a better use of calcium than dense systems.

Although there are few studies of the kinetics for desulphurization in circulating bed, there are global and fragmentary results often achieved on industrial boilers where the degree of desulphurization was measured in given operational conditions, knowing the nature of absorbent and the fuel burned and the Ca/C ratio, respectively.

Secondary procedures for reducing pollutant emissions of SO_2 use desulphurization plants of exhaust gases. The most used are dry FGD (flue gas desulphurization) plants or wet FGD plants and to a lesser extent plants combined, for desulphurization and NOx reduction of exhaust gases, as DESONOX, which are regarded as techniques associated with BAT. FGD plants with lime or limestone wet scrubber have a rate of SO_2 reduction by 92-98%, those who dry scrubbers, with spray, by 85-92%.

Depending on the type of installation of boilers, old or (new), as its power, by applying the BAT techniques, the emission levels of SO_2, in mg/Nm^3, are limited, e.g. for liquid fuel (calculated at 3% O_2) to 350 (350) for powers under 100MWt, to 250 (200) for powers under 300MWt and to 200 (150) for powers over 300MWt (Directive 2008/1/EC, 2008).

The amount of SO_2 emitted in producing electricity and heat sector in European Union countries (EU-27) represents 69% of the total SO_2 emissions at their level (energy using industry sector is 10.2% and transportation 3.8%). Due to replacing solid and liquid fuels with high sulfur content with natural gases, the use of technologies of flue gas desulphurization and application in EU-27 of the new regulations which limited the sulfur content of some liquid fuels, the emissions of SO_2 have decreased from approx. 26000 kt, in 1990, by 71%, to approx. 7500 kt in 2007 (EEA -APE 001, 2010).

2.3 Nitrogen oxides

Among heavy users of the most polluting fossil fuels, through the NO_x emissions, are the means of transportation driven by spark ignition engines and compression ignition engines, followed by thermoelectric power plants and then by boilers and furnaces in the sector of energy use in Industry (EEA -APE 002, 2010).

NOx emission depends on the residence time of the molecules in the flame, most of NOx is formed in the second part of flame development, where the temperature is sufficiently high. Usage of the excess air raise the NOx equilibrium value, and the time needed to achieve balance is net superior to the stationary time of the molecules in the furnace, which explains the experimental values of NOx lower than the theoretical front.

There are three possible mechanisms of formation the NOx (De Soete,1982, Kramer, 1986, Skunca, 1981):

- thermal NOx from using the nitrogen from the combustion air;
- fuel NOx, by conversion of chemically bound nitrogen in the fuel;
- prompt NOx.

NOx forming synthesis is detailed schematically as follows:

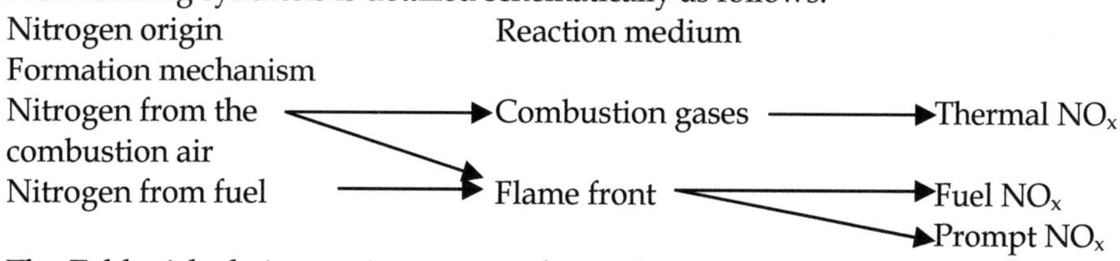

The Zeldovich chain reaction pattern shows the exponential dependence for temperature of the emission of NO_x, until the temperature threshold is reached. The emission is directly proportional with the residence time in the furnace of the molecules and with the oxygen square concentration (Zeldovich, 1980).

The temperature is the strongest factor to influence the NO_x emission, exceeding significantly the influence of the oxygen concentration and the stationary time of the molecules. Stepped combustion is mainly intended to avoid forming high temperatures for reactants before reaching equilibrium, thus nitrogen oxides overall emission to be lower.

The wrong placement of the burners may determine through the jets interaction phenomenon local increases in temperature, which can lead to increased emissions of nitrogen oxides. Fuel nitrogen is present as organic compounds in the amines (NH and NC) and cyanides (CN) family.

These compounds react in two directions by:

- Reaction with substances that contain oxygen, forming NO and
- Reaction with substances containing nitrogen, forming molecular nitrogen.

Therefore not all the fuel nitrogen from combustion passes in NO_x.

Thermal NO formation occurs in the flame in the post-reaction area, according to distinct mechanisms corresponding to mixture of fuel with poor or rich air.

The main factors that influence the development of thermal NO production reactions are: the atomic oxygen concentration, the residence time of molecules and the temperature of the furnace, with higher values at 1300°C, exert the strongest influence.

The reaction of molecular nitrogen with atomic oxygen is the slowest. Dissociation of molecular oxygen at normal combustion temperatures is insignificant. Thus, for example at 2000 K, atomic oxygen is formed having a concentration below 10 ppm.

NO formation from fuel nitrogen occurs in the flame after a complex mechanism, partially unknown. It is known that first CN radicals are formed, whose evolution in the presence of oxygen leads to the formation of NO. The main factors that influence the forming of NO in this case are the nitrogen content of the fuel, oxygen concentration in the fuel, residence time and the flame temperature.

NO formation from the nitrogen contained in the fuel flows a little faster than the formation of thermal NO, but is considerably slower than the prompt NO formation.

NO_2 formation occurs in the exhaust ducts of combustion gases, funnel and free atmosphere at temperatures below 650°C, the main factors being: the temperature value, molecular oxygen O_2 concentration, residence time, air pollution and solar radiation.

Primary procedures for reducing pollutant emissions of NO_x have known three generations in development (ECE NOx Task Force, 1992)

- First generation consisting in reducing the temperature of the combustion air preheating (RAP), flue gas recirculation application in the furnace (FGR), use of low excess of combustion air (LEA) and burners-out-of-service (BOOS);
- The second generation consisting in production of low NO_x burners 1 (LNB) or air staging at burner, flue gas recirculation at burner (FGR) and over fire air (OFA);
- Third generation consisting in production of low NO_x burners 2 (LNB) or air and fuel - staging at burner and in furnace NOx reduction (IFNR) as well as burners improvement for NO_x decreased by increasing the internal flue gas recirculation.

In addition to these techniques there can be mentioned:

- Emulsification of oil with water;
- Injection of aerosols;
- Water-steam injection;
- Addition of fuel, combustion air, or spraying additives in flame (Ghia & Gaba, 2000).

NO_x emission depends on the size of volume of the furnace in which take place the development of flames produced by burners. Large furnaces allow the existence of higher temperatures in and after the area occupied by the flame, so the NO_x emission will increase, and the phenomenon is amplified by increasing the residence time. As a result, thermal charges of the volume and the cross section of the furnace represent important criteria for assessing the emissions of NO_x.

In large furnaces, which burn gaseous fuels, the NO_x emission is reduced by lowering the preheating temperature of the combustion air. Using this technique to oil burning is not recommended because it increases the percentage of unburned, especially soot production, as a result of lowering the burning temperature.

The amount of preheating temperature of the combustion air must have an optimum determined by the combustion to be more completely, NO_x emission below the normalized limit and high thermal efficiency for the heat production plant (Ghia & Gaba, 2000).

For industrial furnaces, where high temperatures in the furnace are required by the technological process, the amount of preheated temperature of the combustion air is dependent on this need taking into account, firstly, the type of the fuel used. Primary reduction of NO_x emission is done especially by removing the temperature peaks that are achieved mainly by the combustion gases recirculation.

Flue gases recirculation in the combustion air represent an efficient method of reduction of NO_x emissions for fuels and furnaces that allow development of high temperatures in the flame. The researches have shown that the most effective method of decreasing the levels of

NO_x is the recirculation gases mixture with the primary air or gaseous fuel, because in this case it acts on the maximum of temperature of the flame core. Thus the burning rate is reduced, which increase the length of the flame and reduces the NO_x content. As result there are necessary constructive changes to the burners and limiting the quantity of recirculation gases, so the complete combustion to be assured.

But it is noticed that by increasing the recirculation degree, the flue gas CO content is increasing, while the NO_x content is decreasing. Reducing the temperature of the flame decrease the useful transfer of the heat by convection, because of the increase of the gases flow from the combustion of fuel with recirculation flue gas flow.

To the Hrenox process, recirculation gases mixture with some cold combustion air is introduced into the burner separately from the rest of the combustion air that is preheated. For a high degree of flue gas recirculation of 15% and a proportion of cold air up to maximum 10%, the NO_x emission was below 200 mg/m^{3}N for the combustion of gaseous fuels and below 250 mg/m^{3}N for the combustion of high quality liquid industrial fuels.

Another way to reduce NO_x emissions is in using tertiary air injected through special slits, located above or below them.

The burners will work with decreased excess of air, the combustion being done by tertiary air which represent 15-30% of total combustion air. Often the application of this combustion technique is related to the existence of large furnaces, equipped with several rows of burners located at different heights, where only to the upper burners it may be introduced by blowing tertiary air.

Also, to the individual burners of natural gas, or oil, the burning in steps is a safe technique reduce the NO_x emissions. Combustion by-steps can be obtained by introducing fuel gradually by FS (air- fuel staged) process. The fuel added outside the primary burner forms a more under-stoichiometric flame after which, in the second stage NO_x is reduced by NH_3, HCN and CO radicals in N_2. Optimal proportion of the secondary fuel from total fuel consumption is 20-30%. The third stage follows when the process completes when the final combustion with reduced NO_x is done.

Insertion by-steps of the combustion air or fuel, reduces the NO_x emissions, but this process must be implemented by optimal constructive solutions, whereas the NO_x reduction processes are contrary to those of decreasing the CO and partially oxidized hydrocarbons.

The time required for combustion and reduction being relatively high, the FS process is recommended to apply to large furnaces. At the oil combustion, the FS process application may produce significant soot quantities in the primary flame, so it must be applied judiciously in case of combined oil-gas burners.

INFR combustion by-steps process implies the fuel injection in the furnace above the main combustion zone thereby causing a secondary under-stoichiometric area, after which it is added downstream the secondary air which completes the combustion. As a result, the combustion is divided into three zones. Hydro-carbonate radicals, formed in the second zone, occur to a temperature of over 1200°C in the reduction atmosphere, reacting with nitrogen oxides produced in the main combustion zone so the N_2 and other components to be formed. In practice, reducing NO_x emissions by spraying water is rarely used, in large furnaces and only with superior fuels that form relatively high combustion temperatures and who have virtually no sulfur in the composition.

In the case of heat engines, primary measurements to reduce NO_x emissions refer to optimizing the combustion chamber, the injection or carburetion systems, engine operation with an appropriate adjustment, using recirculation of exhaust gases, to which it is added the use of fuels treated by different methods (desulphurization, emulsifying with water, ultrasonic treatment).

Most of the researches conducted at the boilers to reduce the emissions of NO_x produced by burners referred to the oil - natural gas mixed burners to which the constructive solutions adopted are independent of fuel that they use. Primarily it is used the combustion by-steps and the combustion gases internal recirculation, as the ASR (Axial Stage Return Flow) burner is done. Part of the combustion air flow passes through the central tubing of the burner as primary air. The secondary air is repressed from the burner box through pipes equidistant distributed on a peripheric circumference, each pipe having two outputs for secondary air I and secondary air II, the supply being performed by a separate cold air fan. To delay the combustion with primary air, due to the ejection caused by axial repression with higher speed of secondary air, flue gases are absorbed from the furnace through a annular transversal section, forming a separating layer between the primary step and the secondary step. In this burner there were integrated essential components of techniques to reduce NO_x emissions:

- Supply with combustion air in three stages;
- Aspiration of the flue gas from the furnace ensuring internal recirculation;
- Separation of primary air from secondary air through a stream of combustion inert gases that reduces the burning rate;
- Liquid and gaseous fuels flows are adjustable allowing the control of the combustion processes to conditions of reduced NO_x.

Very low NO_x emissions were obtained by using the ASR burner alone and by applying additional techniques such as the addition of tertiary air, external flue gas recirculation and special spray nozzles for liquid fuel. To the combustion of oils with high sulfur content, according to the quality of these fuels, for entry into the solid particles and sulfur oxides emissions standards it is required the application of post-combustion control techniques that reside in the use of desulphurization systems, filters and cyclones.

For the furnaces of boilers burning solid fuel with low volatile content and a lot of bituminous ash having the liquid discharge of ash, NOx emissions level is relatively high. Since it is necessary to maintain a minimum value of the combustion gases temperature required for the liquids discharge of ash, NO_x emission control techniques are restricted in operation. Cyclone furnaces with second chamber downstream offer very good conditions for combustion by-steps. When inserted into the cyclone furnace the required air for combustion there are obtained high temperatures for oxygen excess, so the NOx emissions have high values to 1500-1800 mg/Nm^3. To the combustion by-steps the procedure is as follows:

- in the cyclone furnace is inserted about 80% of stoichiometric air;
- it is injected in the secondary chamber, as secondary and tertiary air, the remaining air required for fuel combustion. The optimal conditions for injection with tertiary air are met when the injection nozzles are designed and located so to provide the intimate mixture of the flue gas from the cyclone furnace and air. Respecting the above conditions, it was obtained a reduction in NO_x emissions of 30-40% depending on the operating mode. Further reduction in NO_x emissions was achieved by applying the external flue gas recirculation.

For mixed gas-oil burners the reduction of NO_x emissions is done by the swirling the combustion air to a turbulence degree of $n<3$ to prevent high temperature peaks as a result of excessive shortening of the flame.

Intensifying the internal recirculation of combustion gases is done with repressed secondary air with great speed through nozzles located at the end of the entry into splay, creating an important vacuum that absorb primary combustion products. Combustion air admission is done best in three steps to avoid local temperature increase to high values. It is recommended that the primary air to represent 55-60% of the total combustion air. Oil spraying is done especially with slightly overheated steam using Y-type injectors.

To reduce NO_x emissions and pollutants in general, it can be used the self-carburizing process for both boilers and furnaces (Gaba, 2010).

A low burner of original conception, with solid carbon particles formation in flame, is fitted on three condensation boilers for steam and hot water. The main burner elements provide the formation of annular jet of combustion air mixed with natural gas and of central fuel jet, which, by thermal cracking, forms subsequently particles of solid carbon. These particles lead to increase of flame emissivity. Having in view the NOx emission, the characteristics of this burner are the following:

- use of natural gas as fuel;
- the solid carbon particles formation in flame;
- three stages mixing of the fuel in the combustion air;
- the swirling motion of the ventilated combustion air.

The low NOx burner for boilers demonstrated excellent NOx and CO performance, producing emission levels below 60 ppm for CO(1.05 air ratio at nominal power and 1.09 air ratio at minimum power) and below 50 ppm for NOx (1.03 air ratio at nominal power and 1.09 air ratio at minimum power), operating with natural gas (Gaba, 2010).

Another low NOx burner of original conception, with solid carbon particles formation in flame, is fitted on forge and treatment furnaces (Gaba, 2009). The main constructive elements of this low NOx recuperative burner are: a ceramic quarl, a zone for the natural gas-ventilated air mixture, a zone for fuel atomization by collision with the compressed air, an air preheater and an ejector for exhaust gases. The ceramic quarl burner has a cylindrical combustion chamber and in the peripherical zone a few cylindrical channels for exhaust gases entrances. Between combustion chamber and cylindrical channels are another channels for exhaust gases recirculation. The zone for the natural gas-ventilated combustion air includes three concentric pipes: the central pipe with different orifices and nozzles for natural gas and two pipes for divided combustion air in primary, secondary and tertiary air. The air preheater ensures the heat exchange from exhaust gases to combustion air. Having in view the NO_x emission, the characteristics of this burner are the following:

- use of natural gas or heavy fuel;
- three stages mixing of the fuel with the combustion air;
- the solid carbon particles formation in flame;
- the exhaust gases recirculation.

The low NO_x recuperative burner demonstrated excellent NO_x and CO performance, producing emission levels below 30[50] ppm for CO(1.03 [1.08]air ratio, and 700^{0}C furnace temperature) and below 120 [160] ppm for NO_x (1.30 [1.35] air ratio, and 1200^{0}C furnace temperature), operating with natural gas[heavy fuel] (Gaba, 2009).

In conclusion, the solutions for upgrading the combustion plants for fossil fuels for boilers and furnaces can be grouped into:

- the completion of the measurement, control and automation equipment and establishing optimal operating parameters;

- the development of new types of burners;
- the development of new combustion installations with high complexity.

The following can be a part of the first type of solutions:

- operation with optimal values of the coefficient of combustion excess;
- operation with optimal values of the preheating temperature of the combustion air and the gaseous fuel, like BFG (blast furnace gas) that is used;
- operation with an optimal number of burners with a given unitary capacity.

The second group refers to making of:

- burners with the insertion by-steps of combustion air or fuel;
- burners with internal recirculation of flue gas;
- burners having gradual combustion.

In the last group of solutions there are:

- making of combustion plants with external recirculation;
- making of combustion plants with water-fuel liquid emulsions or additived aerosols for fluid ionization in the furnace;
- replacing of the solid fuels with liquid or gaseous fuels that are less pollutant. By applying primary techniques, the rate of reduction of NO_x emissions is 10-44% for operating with reduced air excess, 10-70% at the insertion by-steps of the fuel or oxidizer, 20-50% at flue gas recirculation, 20-30% at low preheating of the combustion air and 25-60% to use low NO_x burners.

Secondary procedures for reducing NO_x pollutant emissions use flue gases treatment plants. The most commonly used installations are SNCR (selective non-catalytic reduction) and SCR (selective catalytic reduction). In a less extent combined plants are used for desulphurization and NO_x reduction of exhaust gases, DESONOX and active carbon plants (ECE NO_x Task Force, 1992). Selective Catalytic Reduction occurs at temperatures of 300-400°C in the presence of catalysts, by injecting ammonia (NH_3). Catalysts used: TiO_2, WO_3, V_2O_5.

The rate of reduction of nitrogen oxides is 80-95% for SCR installations. Selective non-catalytic reduction occurs at temperatures of 900-1000°C in the absence of catalysts, by ammonia injection. The rate of reduction of nitrogen oxides is 30-50% (ECE NO_x Task Force, 1992).

The reactions presented in the literature are selective, indicating oxidation of ammonia and sulfur dioxide (SO_2) may not occur, the presence of oxygen being essential in the development of some reactions.

Treatment of exhaust gases by passing them in the activated carbon reactor is done at temperatures of 100-150°C.

Depending on the type of installation, old or (new), by applying BAT (best available techniques), the emission levels of NO_x, calculated as NO_2, in mg/Nm^3, are limited to 100 (100) for boilers, operating on gaseous fuels (calculated at 3% O_2), to 90 (50) for gas turbines (calculated at 15% O_2) and to 100 (75) for internal combustion engines, stationary, for gaseous fuel, calculated at 15% O_2. In case of using liquid fuels on boiler installations, depending on the type of installation, old or (new), and its power by applying the BAT techniques, emission levels of SO_2, in mg/Nm^3, calculated at 3% O_2, are limited to 350 (350) for power below 100MWt, 250 (200) for power below 300MWt and 200 (150) for power over 300MWt (Directive 2008/1/EC, 2008).

The amount of NO_x emitted in the means of transport sector in European Union countries (EU-27) represent 46.3% of the total of NO_x emissions at their level (electricity and heat generation sector represent 20.5% and energy using industry represents 13.9%). Due mainly to use low NO_x combustion technology, to the replacement of solid fuels to natural gases, to the use flue gases treatment (selective non-catalytic and selective catalytic) emissions of SO_2 decreased from approx. 16900 kt, in 1990, with 31% to about 10300 kt in 2007 (EEA - APE 002, 2010).

2.4 Solid particles

Particles released into the atmosphere come both from combustion processes and the entrainment by the combustion gases from the raw material, e.g. rotary drum furnaces for clinker, limestone, dolomite, etc.

Result of the combustion processes, the black smoke is a heterogeneous mixture and variable in structure, between soft solid particulates, water vapors and gases that result from the incomplete combustion of fuels. As a result of the reduced size of suspended solid particles, the smoke refract light and color depending on the concentration and color of substances component in it. That is why the smoke comes in large range of dark colors from the weak gray to black, depending especially by the degree of perfection of the combustion and the quality of burned fuels. The determination of the index that characterizes smoke color is used for technical control of combustion quality, in combating environmental pollution, resulted in particular from burning of solid fuels and lower oils (Bacharach no.). Generally, the formation of unburned solid materials in fossil fuels flames is pointing out especially the wrong mixing between fuel with air that is insufficient to achieve combustion even before carbon oxide is obtained. To liquid fuels without mineral content these solid particles can virtually disappear by burning that if the oxygen is in adequate proportions, in warm enough areas of the furnaces and for a sufficient residence time. In black smoke it is noticed the emergence of two categories of unburned solid: very soft particles of soot and particles much larger than the first, sometimes called cenospheres from intermediate and heavy fuel combustion, or especially flying coke by combustion of solid fuels.

The mechanism of the formation process of soot particles in gaseous fuels flames is explained by theories that belong to three main groups. Thus, the first group includes theories that explain formation by thermal decomposition of hydrocarbons in fuel in carbon and hydrogen, with subsequent polymerization of the carbon. The second group of theories explains soot formation by oxidation of hydrocarbons until peroxides are formed and then by decomposition of peroxides free radicals are divided, which favors higher hydrocarbon formation, which under the high temperatures in the flame, are decomposing to form solid carbon particles. The third group of theories involves the formation of soot particles in flame by polymerization of the C_2 radical.

All these theories have a qualitative character, experiments verifying only certain assumptions that are made. The main factors that prompt the formation, quantity, quality and size of soot particles are as following (De Soete & Feugier, 1974):

- the degree of fuel enrichment (dosage), or lack of required air for stoichiometric combustion;
- physico-chemical properties of fuels, determined by the gravimetrically report C/H of carbon and hydrogen from fuels and by the molecular mass;
- the conditions of mixing fuel with combustion air which may be sufficient but unevenly distributed in relation to fuel gas;
- the temperature level in the combustion chamber, as the temperature rises the polymerization and hydrocarbon cracking processes intensify;

- the pressure in the combustion chamber which, by increasing, it intensifies the formation of soot process in flame.

The burner of industrial liquid fuels create the fuels spraying as drops, firstly, to increase their surface in contact with the oxidizer. Spraying of liquid fuels in the form of drops jets that ignite in a gaseous environment that lacks the oxygen required for complete combustion, can generate soft soot particles and relatively large solid particles with average diameter up to 100 μm.These last ones having a spongy aspect and spherical form, thus called cenospheres, are more numerous as the liquid fuel is more heavy, meaning that it contains molecules with more carbon atoms.

The cenospheres result from cracking hydrocarbons in liquid phase as soft drops, through decomposition reactions that lead to release of gaseous products and building of solid residue rich in carbon, but which also contains minerals from the original fuel. Similarly, the flying coke is made from soft coal particles, sprayed with air blast at coal dust operating burners. The heavy liquid fuels, in particular, build solid tailings from sediments and organ-metallic compounds that associate with carbon cracking residues. Overall mass of solid particles thus formed for heavy fuels may be in the order of a few decigrams per kilogram of fuel.

Cenospheres are in smaller numbers than particles of soot, but their overall mass is generally much bigger. Essentially, the presence of these relatively large solid particles, sometimes make the regulations compliance on pollution from the combustion of heavy liquid fuels difficult (Ghia & Gaba, 2000).

Given the size of cenospheres, their combustion is slower than the combustion of soot particles. This combustion is essentially controlled by the diffusion phenomena. Thus, its influence is decisive on: diffusion rate of oxygen to particle, diffusion rate of gases resulted from combustion by the ambient gaseous environment, the particle temperature and the activation energy of substances at work. It should not be missed the fact that the hydrocarbon skeleton of cenospheres contains mineral elements originally present in the fuel. These elements may be particularly harmful in contact with the heating furnaces charge, with the elaboration furnaces bath, refraction bricks from the furnace's walls, boilers or chemical industry ovens pipes. A negative example is the harmful action of vanadium salts on steel at temperatures above 600°C. In flame, the vanadium is present mainly as V_2O_5 because V_2O_4, V_2O_3 and V_3O_5 oxides can be considered as intermediate steps to V_2O_5. When using a good quality combustion plant and properly exploited, the solid particles mostly disappear by burning, leaving behind ashes of the initial mineral mass of fuel. To the heat generators operating on industrial liquid fuels is imposed that solid suspensions should not exceed 100-200 mg/kg of fuel. This prevents the excessive pollution of the environment and the excessive increase of maintenance costs associated with deposition cleaning on the heat change pipes. Oil use with high sulfur content, requires operation at low excess of air close to the stoichiometric value, to reduce corrosion at low temperature. It is favored this way the increase of losses by unburned particles related to the soot and cenospheres and flying coke respectively. Reducing these polluting emissions is directly influenced by the ratio between carbon C and hydrogen H of the initial fuel and the presence of asfaltenes in oil.

Gaseous fuels with a high ratio C/H produce soot easier than those with the lower C/H ratio. Soot production in the flame makes this to be brighter and to transfer more useable heat by radiation. As a result of this advantage, there was developed a process (called self - carburizing) by which carbon particles produced in the initial zone of the flame are completely oxidized to the top of the flame. This procedure has been used on some burners of boilers and furnaces, proving at the same time with the increasing efficiency very effective to reduce the emissions of CO, NO_x and partially oxidized hydrocarbons (Gaba, 2010).

Primary procedures for reducing particles emissions, are to replace the solid and heavy liquid fuels (oil) with light liquid fuel combustions and to use complete combustion methods.

Primary processes that reduce particle emissions resulting from incomplete combustion of liquid fuel are:

- prior physical treatment of liquid fuel consisting in preheating, filtration, centrifugation to remove the largest part of sediments, water, soft particles of minerals;
- high quality spraying of which are essential choosing of optimal spraying system, well adjusted preheat temperature of the fuel. After spraying by mechanical injection under pressure of heavy liquid fuels, a variation of the viscosity of dozen centistokes modify essentially the spraying features, with great repercussions on the process of combustion;
- fuel combustion with preheated air with a moderated air excess (below 30%), using especially overheated steam spraying;
- Improvement, by addition, of the combustion characteristics of industrial fuels used;
- Use of modern combustion techniques such as aerosols combustion, or water - liquid fuels emulsions combustion (with 3% - 5% water).

Secondary procedures for reducing particles emissions, most used by boilers and furnaces are:

- dry mechanical refinement, by which the particles settling is done under the action of gravity and/or centrifugal force;
- Wet refinement by passing the gases under a liquid mist;
- Refinement by filtration, by passing gases through porous materials;
- Electrical refinement by gas ionization and dispersed particles sedimentation in a high-voltage electrical field. There are considered BAT techniques the use of electrical filters (ESP), textile fabric filters (FF), the wet scrubbers and cyclones, when the latter are used together with ESP or FF. The retention efficiency of ESP for particles larger than 1μm, is 96.5%, the FF is 99.95%, the wet scrubber is 98.5%, and the cyclone is over 85% for particles larger than 5μm (Directive 2008/1/EC, 2008).

Depending on the type of installation of boilers, old or (new), also on its power, by applying the BAT techniques, the levels of particles emission, in mg/Nm^3, are limited to 30 (20) for powers below 100MWt, 25 (20) for powers below 300MWt and 20 (10) for powers over 300MWt, for liquid fuel (calculated at 3% O_2) and for solid fuels (coal and lignite), calculated at 6% O2. For internal combustion engines, stationary, liquid fuel, BAT associated filtering techniques can lead to a maximum value of 30 mg/ Nm^3, gas oil and 50 mg / Nm^3, heavy liquid fuel (Directive 2008/1/EC, 2008).

The amount of PM 10 particles, with diameters less than 10μm, in European Union countries (EU-27), as primary products (directly discharged into the atmosphere) and secondary (derived from the precursors NO_x, SO_2 and NH_3, transformed by photochemical reactions into particles) were reduced by 53% between 1990-2005. The direct amount of PM10 were reduced from 4779 kt in 1990 to 2491 kt in 2005, and the secondary from 33588 kt to 19629 kt. The sector of combustion processes represents 70% of the overall direct emissions of particulates and 90% of the total secondary products. The sector of energy using industry represented 39% of the overall direct emissions of PM10 particulates in 1990 and reached at 13% in 2005, the industry of power generating represented 14% of the overall emissions of PM10 particles in 1990 and reached 10% in 2005, and the means of transport reached from 13% to 21%. The sector of energy using industry represented 39% of the total secondary emission of PM10 particles in 1990 and reached 26% in 2005, the industry of power generation represented 15% of the total secondary emissions of PM 10 particles in 1990 and reached 11% in 2005, and the means of transport reached from 26% to 29%. These substantial reductions were done, due to the dusting techniques of the exhaust gases, the replacement of solid and liquid fuels with high content in sulfur with natural gases, the introduction of the primary techniques for achieving combustion with reduced NO_x emissions, the use of flue gases treatment and the application in EU-27 of new regulations which limited the content of sulfur from some liquid fuels and which used advanced catalysts in internal combustion engines for transportation (EEA– EN 07,2010).

2.5 Volatile organic compounds (VOCs) and persistent organic pollutants (POPs)

Volatile organic compounds are aliphatic and aromatic hydrocarbons, alcohols, ketones, esters, aldehydes, benzene, toluene, acetones, methanol and formaldehydes.
POPs are mainly composed of polycyclic aromatic hydrocarbons (PAHs), resulted in combustion processes of fuels in a smaller extent, but with particularly harmful effects, from dioxins and furans, resulted from the incineration of waste, pesticides and products of chemical industry.
Partially oxidized hydrocarbons are generated by the lack of oxygen (combustion air) in combustion process, or from slow oxidation.
Lack of oxygen in the combustion process can be explained locally or temporarily, because combustion installation from industry are supplied with excess of combustion air. Due to poor mixing of fuel-combustion air, in the flame can occur rich zones in fuels which are formed unsaturated hydrocarbons, aldehydes and acids and, in some cases, the process can continue until the formation of soot. These unburned elements may evolve, behind the flame front, depending on thermodynamic parameters and the interaction of oxidants.
To achieve sufficiently high oxidation rates the following are necessary:

- to set the mixture of fuel gas - combustion air between flammability limits;
- achieving a level of temperature high enough ahead of the flame front;
- creating a bigger turbulence;
- ensuring that the various intermediate elements stay a sufficient period of time in the flame (Gaba, 2005).

In combustion processes, the objective of reducing VOCs is their volatilization and oxidation to CO_2 and H_2O, by thermal oxidation, catalytic or bio-oxidation. In the case of incinerators, especially, is practiced burning in secondary outbreaks, in which the catalysts can be introduced, after the primary focus.
Among the sectors of fuel combustion processes that produce VOCs, transport is the most significant with 18,6% of total EU emissions, the rest being insignificant. Between 1990 and 2007, road transport has reduced VOCs by 60%, as a result of using the three-way catalytic converter and the improvement of diesel quality (EEA - APE 004, 2010).
Among the sectors of fuel combustion processes that produce PAHs, residential heating sector is the most significant with 41% of total EU emissions, and transport represents 15%. Between 1990 and 2007, residential heating sector reduced the production of PAHs by 14% and road transport by 47%, mainly due to the advanced catalyst (EEA - APE 006, 2010).

2.6 Atmospheric greenhouse gas concentrations

Among the gases that produce greenhouse effect, the carbon dioxide has the most important contribution, its percentage being 65%. It appears in an overwhelmingly proportion after fuel combustion processes. In the year 2009 the concentration of CO_2 in the atmosphere was 387ppm. Carbon dioxide recently considered as pollutant, prevents the discharge of heat energy from the surface of the Earth which is acquired by radiation from the sun. As the time passes, even the increase with a few degrees of the temperature could lead to reduction of the land ice cap, the spread of deserts, raising sea and oceans levels, as well as important climate changes. The concentration of CO_2 in the atmosphere has increased by 38% compared to the pre-industrial age, when it was 280 ppm (EEA - CSI 013, 2010).

In 1990, a report of a United Nations committee has confirmed the tendency of heating of the globe and considered that, if the world will not take measures to stabilize the gas flow released in the atmosphere (carbon dioxide, methanol, gases from chemical fertilizers, various hydrocarbons), the temperature will rise by three degrees until 2000. A United Nations Framework Convention on Climate Changes was negotiated between 1991 - 1992 and signed at Rio Conference in 1992. The Convention entered into force in 1995. The main objective of the Convention was to stabilize the greenhouse gas concentrations from atmosphere, especially carbon dioxide, at a level that will not endanger climate system.

Kyoto Protocol, signed in 1997, with the exception of USA, is an extension of the UNFCCC (United Nations Framework Convention on Climate Change) from 1993 and is the first step in an effort to reduce the long-term changes in emissions to prevent climate changes. Both at the global (UNFCCC, 2009) and the EU level (October 2008 Environment Council conclusions) this "dangerous anthropogenic interference" has been recognized by formulating the objective of keeping the long-term global average temperature rise below 2°C compared to pre-industrial times.

There is a consensus in EU-27 countries for collective action to reduce greenhouse effect gas emanations by: improving techniques for burning fossil fuels, the use of renewable sources for energy production (water, sun, wind, geothermal energy), improving industrial technologies, decreasing the consumption of fuels for road transport, increasing the share of public transport, reduce emanation of animal manure, reduce the fertilizers based on nitrogen, stopping the clearance of forest and accelerated action of forestation, reducing waste and gases from municipal waste dumps, etc. The commitment taken by the EU-27 for 2008-2012 compared to 1997, is a reduction of 8% versus 5%, as was stipulated in the Kyoto Protocol (EEA- EN 07, 2010).

3. Determination of pollutant emissions in combustion processes

Quantitative determination of pollutant emissions refers to:

- The determination of pollutant concentration, c_i [mg/Nm^3], representing the concentration of pollutant i in the gases released in atmosphere, required to screening verification in the emission limits stipulated by the regulations;
- Determination of pollutant quantity emission, which represents the amount of pollutants released into the atmosphere over a specified period, usually one year, E_i [t], required to establish the activities contribution from combustion processes at general air pollution.

Determination of pollutant concentration, c_i [mg/Nm^3] and the amount of pollutants E_i [t] produced by a combustion plants requires knowledge of the effluent gaseous composition in point of measurement. In addition, at determination of pollutant quantity emission, it is necessary to know the amount of fuel consumed during the period, in m^3N, for gaseous fuels, or kg for liquid or solid fuels [1].

The measurements are based on national norms, which are prescribed the measuring and calculation methodologies, type of measurement devices that are accepted, as well as the execution range for measurements. For large combustion plants requires continuous monitoring. The automatic devices that are used:

- gas analyzer to determine the volumetric concentrations, from the flue gas mixture, O_2 [% volumetric], CO_2 [% volumetric], CO [ppm], NO_x [ppm], NO_2 [ppm], and SO_2 [ppm];
- automatic device to determine the volatile organic compounds;
- automatic instrument, with isokinetic probe for determining particulate matter (Gaba, 2005).

Gaseous effluent concentrations measured values are required, on the one hand to determine the quantity of emissions of pollutants and on the other hand directly to determine the concentration (in mg/Nm^3) of the emission of pollutants. Calculation of concentrations of different pollutants emissions i, c_i [mg/Nm^3], related to reference conditions, it is required to be done to compare it with the limit values set by national regulations, for periods of time, depending on the type, age and power plant and fuel type consumed.

E.g. if the analyzer indicates a concentration of 60 ppm SO_2, 80 ppm NO_x and 10.5% O_2, to measurement of gaseous effluents produced by burning a liquid fuel, it can be calculated the concentration of pollutant emission of SO_2, CSO_2, NO_x, C_{Nox}, respectively, related to reference conditions, as [1.10]:

$$c_{SO_x}\left(mg/Nm^3\right) = c_{SO_x mas}\left(ppm\right)\cdot k1\cdot\frac{21-c_{O_{2ref}}}{21-c_{O_{2mas}}} \qquad (1)$$

$$c_{NO_x}\left(mg/Nm^3\right) = c_{NO_x mas}\left(ppm\right)\cdot k2\cdot\frac{21-c_{O_{2ref}}}{21-c_{O_{2mas}}}$$

where:

$c_{SO2\ mas}$ - SO_2 emission concentration measured in ppm;

$c_{Nox\ mas}$ - NO_x emission concentration measured in ppm;

k1 - 2.85 – ppm conversion factor → mg/Nm^3 for SO_2;

k2 - 2.05 – ppm conversion factor →mg/Nm^3 for NO_2;

$C_{O2\ ref}$ – 3% volumetric - reference oxygen concentration for a gaseous fuel according to EU norms;

$C_{O2\ mas}$ – measured oxygen concentration in the gaseous effluent, in volumetric %.

Replacing in (1) with numeric values it is obtained:

$$C_{SO_x} = 60\cdot 2{,}85\cdot\frac{21-3}{21-10{,}5} = 293{,}1\, mg/Nm^3 \qquad (2)$$

$$C_{NO_x} = 80\cdot 2{,}05\cdot\frac{21-3}{21-10{,}5} = 281{,}1\, mg/Nm^3$$

The amount thus calculated is compared with the emission limit for SO_2 and NO_x respectively, according to national norms, which are usually set for some period of time, depending on the type, age and power of the plant and fuel type consumed.
In the case of large combustion plants, which consume liquid fuels, in Table 1 are presented associated BAT limit values, for NO_x emissions calculated as NO_2 and SO_2, in mg/Nm³, for 3% O_2, depending on the type and power of the plant (Directive 2001/80/EC, 2001).

Crt. No.	Plant type	Pollutant	Measure unit	Thermal power, MWt		
				<100	100-300	>300
1	Old	SO_2	mg/Nm³	350	250	200
2	Old	NO_x	mg/Nm³	450	200	150
3	New	SO_2	mg/Nm³	350	200	150
4	New	NO_x	mg/Nm³	300	150	100

Table 1. Limit values of SO_2 and NO_x emissions for electricity and heating generation.

It is found that the values calculated for the exemplified installation would not exceed the limits unless it would have a thermal power less than 100 MW.
To determine the exact amount of a pollutant i, Ei, in t for a certain period, the general relation can be used:

$$E_i = ki \cdot V_{gu} \cdot B \cdot c_i \cdot 10^9 [t] \tag{3}$$

where:
ki - conversion factor for the i pollutant, ppm → mg/Nm³;
V_{gu} - volume of dry burned gases produced by burning the fuel,in m³N/m³N or m³N/kg;
B - fuel consumption, in Nm³ of consumed gas in the respective period, or kg of liquid or solid fuel;
c_i - measured concentration of i pollutant, in ppm
The specialty papers shows calculating relations for the volume of dry flue gases resulting from combustion of various fuels [10]. For example, for combustion of gaseous fuels to find the volume of dry flue gases produced from burning a Nm³ of fuel, use the following formula:

$$V_{gu} = \frac{CO_2^{(c)} + CO^{(c)} + \Sigma m C_m H_n^{(c)}}{CO_2^{(g.a)} + CO^{(g.a)} + \Sigma m C_m H_n^{(g.a)}} \tag{4}$$

Values in the numerator results from the initial analysis of fuel gas composition and the denominator values are measured at the exhaust of plant flue gases, for example with a flue gas analyzer.
For less accurate calculations, but faster, it is used the relationship:

$$E_i = fi \cdot B \cdot PCI \cdot 10^6 [t] \tag{5}$$

where:
fi - emission factor, in g/GJ;
B - fuel consumption, in Nm^3 of consumed gas in the in the respective period, or kg of liquid or solid fuel;
PCI - low calorific power of the fuel, in GJ/Nm^3, GJ/kg respectively.
The factors *fi* (in g/GJ, respectively lb/10^6 Btu), are used in the EU and U.S. respectively, to calculate the amount of pollutants emitted into the atmosphere over a period of time and are tabled according to the many types of fuels, installation, etc. (US EPA- AP 42, 1995). In Table 2 are summarized the approximate amounts of various pollutants to be emitted into the atmosphere worldwide, calculated with the emission factors *fi* (taking the values for coals, oils and natural gas) for fuel consumption in 2030 provided by the World Energy Outlook -2008.

crt. no.	Fuel type		Pollutant				
			SO_2	NO_x	CO	PM10	VOCs
1	Coals	Emission factor [g/GJ]	10746.5	4083.67	1308.49	3094.99	21.49
		Pollutant quantity [kt]	2207331	838786	268766	635713	4414
2	Oils	Emission factor [g/GJ]	7307.62	1590.48	143.57	343.89	38.68
		Pollutant quantity [kt]	1562457	340064	30697	73528	8270
3	Natural gas	Emission factor [g/GJ]	0.0	601.80	150.88	12.89	25.79
		Pollutant quantity [kt]	0.0	92430	23173	1979	3961

Table 2. Approximate amounts of various pollutants to be emitted into the atmosphere worldwide, calculated with emission factors fi, for fuel consumption in 2030.

To these annual amounts of pollutants emission the amount of CO_2 is added, approximate for the year 2030, to approx. 30 million kt, which would lead to a planetary temperature exceeding by more than 2°C.To reduce the quantity of air pollutants emissions it is necessary that reduction measures and techniques to be generalized on a planetary scale Also there should be mentioned the EU-27 achievements in the period 1990-2007, that have significantly reduced emissions of pollutants into the atmosphere: CO by 63%, NO_x by 31%, SO_2 by 69%, VOCs by 41%, PAHs by 63% and PM10 by 53%. For the future, the key to success in reducing the air pollution is increasing energy efficiency and using regenerative resources. In 2009, the EU agrees a Climate and Energy Package to:

- reduce greenhouse gas emission by 20% by 2020;
- increase the share of renewable energy by 20% by 2020;
- improve energy efficiency by 20% by 2020 (EEA - EN 07, 2010).

4. References

Baukal, C. (2001). *The John Zink Combustion Handbook*, CRC Press, New York.

De Soete, G. & Feugier, A. (1974). *Connaissance actualles sur les mecanismes de formation des principaux polluants dus à la combustion*. Technip, Paris.

De Soete, G. (1982). *Physikalisch-chemische Mechanismen bei der Stickstoffoxidbildung in industriellen Flammen*. Gas Wärme International, vol. 31, no.1, (Jan. 1982), pp.15-23,

Directive 2008/1/EC - *Integrated Pollution Prevention and Control, Reference Document on Best Available Techniques for LargeCombustionPlants,*July2006. www.eippcb.jrc.es/reference

Duca, Gh., Crăciun, A., Sajin, T., Gaba, A., Păunescu, L. (2002). *Modern combustion and air pollution reduction techniques*. USM, ISBN 9975-70-077-2, Chisinau.

ECE (Economic Commission for Europe) NOx Task Force. (1992). *Operating Experience with NOx Abatement at Stationary Sources*. IIP University of Karlsruhe.

EEA (European Environment Agency). 2010. *Sulfur dioxide SO2 emissions* (APE 001), Feb. 2010. www.eea.europa.eu/APE.

EEA. 2010. *Emissions of nitrogen oxides NOx* (APE 002), Feb. 2010. www.eea.europa.eu/APE

EEA. 2010. *Non-metane volatile organic compounds (NMVOC) emissions* (APE-004), Feb. 2010. www.eea.europa.eu/APE

EEA. 2010. *Persistent organic pollutant (POP) emissions* (APE-006), Feb. 2010. www.eea.europa.eu/APE

EEA. 2010. *Atmospheric greenhouse gas concentration* (CSI-013), Nov. 2010. www.eea.europa.eu/CSI

EEA. 2010. *Energy related particulate matter emissions* (EN-07), Jan.2010. www.eea.europa.eu/EN

EEA. 2011. *Transport emissions of air pollutants* (TERM 003), Jan. 2011. www.eea.europa.eu/TERM

IEA (International Energy Agency). (2008). *World Energy Outlook-2008*. www.iea.org/text base/nppdf (www. worldenergyoutlook.org.)

IEA. (2010). *World Energy Outlook-2010*. www.iea.org/text (www. worldenergyoutlook.org.)

Gaba, A. (2005). *Combustion of fuels and environment*. Bibliotheca, ISBN 973-712-061-2, Targoviste.

Gaba, A. (2009). *Low NOx recuperative burner for forge and treatment furnaces*. Metalurgia International, vol. 14, no.8, (August, 2009), pp.23-30, ISSN 1582-2214.

Gaba, A. (2010) *Air pollution reduction by using low NOx burners for furnaces and boilers*. Environmental Engineering and Management Journal, vol. 9, no.1, (January, 2010), pp.165-170, ISSN 1582-9596.

Ghia, V. & Gaba, A. (2000). *Primary measures for air pollution reduction in combustion processes*. AGIR, ISBN 973-8130-06-09, Bucharest.

Kramer, H. (1986). *Grundlangen de NOx Entstehung und Minderung*. Gas Wärme International, vol. 35, no. 5, (Mai-June, 1986), pp.239-247, ISSN 0020-9384.

Skunca, I. (1981). *Vereinfachten Verfahren zur Berechnung der Gleichgewichts- konzentrationen von Stickstoffmonoxid und Kohlenmonoxid in Flammen*. Gas Wärme International, vol. 30, no. 1, (Jan., 1981), pp.48-52, ISSN 0020-9384.

U.S. EPA, AP-42. (1995).*Compilation of Air Pollutant Emission Factors*, Jan. 1995. http://www.epa.gov/ttn/chief/ap42/index.html
Zeldovici, Ia. B. (1980). *Matematicescaia teoria gorenia i vrzva*, Nauka, Moskva.

7

Monitoring of Gaseous Air Pollution

Ryszard J. Katulski[1], Jacek Namieśnik[2], Jarosław Sadowski[1], Jacek Stefański[1] and Waldemar Wardencki[2]
[1]Gdansk University of Technology,
[1]Department of Radiocommunication Systems and Networks
[2]Gdansk University of Technology, Department of Analytical Chemistry
Poland

1. Introduction

Air pollution can be defined as the presence in the atmosphere of one or more contaminants in such quality and for such duration as is injurious, or tends to be injurious, to human health or welfare, animal or plant life. It is the contamination of air by the discharge of harmful substances. Industrialization have led to air getting more and more polluted over the years. Industries, vehicles, increase in the population, and urbanization are some of the major factors responsible for air pollution. The following industries are among those that emit a great deal of pollutants into the air: thermal power plants, cement, steel, refineries, petro chemicals, and mines.

Air pollution can cause health problems and it can also damage the environment and property. It has caused thinning of the protective ozone layer of the atmosphere, which is leading to climate change. The source of pollution may be in one country but the impact of pollution may be felt elsewhere. The discovery of pesticides in Antarctica, where they have never been used, suggests the extent to which aerial transport can carry pollutants from one place to another.

The most important tools in environmental protection is monitoring and modeling. Both the monitoring and modeling of air pollution is essential to provide a picture of the damage humans are doing to the environment, and to enable pollution problems to be discovered and dealt with. An environmental monitoring could be defined as a system of detection, measurements, evaluations and forecasts of environmental states, and the collecting, processing and spreading of information on the environment.

Air pollution and its control is a global issue demanding international cooperation. Monitoring of air pollution is a very important source of data. However, measurement of the air pollutant concentrations, in comparison to monitoring of other elements in the environment, is the most difficult. The difficulties arise from the large dynamics of the atmosphere, causing that it constitutes the main route of pollution spreading and their transport between remaining environmental compartments and universal exposure for a large population without a chance for isolation, which is possible in the case of polluted waters and soil. Another problem is low concentration of air pollutants and their interaction with other gases.

This chapter is divided into two parts. In the first, a general information on gaseous air pollution, parameters used to characterize the pollution level and equipment applied for air monitoring and analysis is presented. Finally, the mobile monitoring system (MMS), as an example of a new approach for air monitoring along communication lanes, proposed by the authors is described.

2. History of air pollution legislation

Concern about urban air quality is not new. First complaints were recorded in the 13th century when coal was first used in London. Today the emphasis has shifted from the pollution problems caused by industry to the pollution in urban areas. A growing concern over the influence of different air pollutants on human health were the main driving force to develop and implement of air quality criteria and standards.

In 1967, the US Congress enacted the Air Quality Acts, the first modern environmental law.

The Clean Air Act (Environmental Protection Agency, 2008), which was last amended in 1990, requires the United States Environmental Protection (US EPA) to set National Ambient Air Quality Standards (NAAQS) for pollutants considered harmful to public health and the environment. The Clean Air Act established two types of national air quality standards. *Primary standards* set limits to protect public health, including the health of "sensitive" populations such as asthmatics, children, and the elderly. *Secondary standards* set limits to protect public welfare, including protection against decreased visibility, damage to animals, crops, vegetation, and buildings. The EPA has established NAAQS for six principal pollutants, which are called criteria pollutants: sulphur dioxide, particulate matter, nitrogen oxide, carbon monoxide, ozone and lead. These standards are threshold concentrations based on a detailed review of scientific information related to effects.

In Europe, the first international air quality standards were introduced by the European Commission in 1980 for SO_2 and suspended particulates, mainly aimed at protecting human health. A few years later, the World Health Organization, recognizing ecological damage as being relevant to human health, introduced air quality guidelines for Europe which included the former as well as the latter, later revised in 2000 (World Health Organisation, 2000). The newest directive on ambient air quality and cleaner air of the European Union entered into force in June 2008 (European Union, 2008). It merges four earlier directives and one Council decision into a single directive on air quality. The new directive of the European Union on air quality takes into account concerns from the latest WHO air quality guidelines (WHO, 2005) on fine particles. Reflecting the latest WHO air quality guidelines that identify fine particles (PM2.5) as one of the most dangerous pollutants for human health, the new EU directive sets objectives and target dates for reducing population exposure to PM2.5. It also maintains limits for concentrations of coarser particles known as PM10 and other main pollutants already subject to legislation.

Pollutant levels, especially gaseous, which negatively affect life on earth are nowadays well defined. Table 1 presents examples of air quality standards issued by the Environmental Protection Agency (EPA), the Word Health Organization (WHO) and some states (ARMAAG).

3. Regulated air pollutants

The contaminants in ambient air that are of concern are basically categorized as criteria and non criteria pollutants (Grifin, 2007).

Criteria air pollutants are those air contaminants for which numerical concentration limits

Pollutant	Averaging time	WHO	EPA	UE	UK	France	Germany	Poland
SO_2	10-15 min	500	-	-	266 (not more than 35 times)	-	-	-
	30 min	-	-	-	-		-	-
	1h	-	-	350 (not more than 24 times)	350 (not more than 24 times)	350 (not more than 24 times)	350	350 (not more than 24 times)
	3h	-	1300	-	-	-	-	-
	24h	125	365	125 (not more than 3 times)	125 (not more than 3 times)	125 (not more than 3 times)	125	125 (not more than 3 times)
	year	50	80	20	20	20	20	20
NO_2	30 min	-	-	-	-	-	200	-
	1h	200	-	200 (not more than 18 times)	200 (not more than 18 times)	230 (not longer than 0.2% of time)	-	200 (not more than 18 times)
	24h	-	-	-	-	-	100	-
	year	40	100	40	40	46	-	40
PM10	30 min	-	-	-	-	-	-	-
	24h	20	150	50 (not more than 35 times)	50 (not more than 35 times)	50 (not more than 35 times)	50	50 (not more than 35 times)
	year	50	-	40	40	40	40	40
CO	10-15 min	100000	-	-	-	-	-	-
	30 min	60000	-	-	-	-	-	-
	1h	30000	4000	-	-	-	-	-
	8h	10000	1000	1000	1000	1000	1000	1000
	24h	-	-	-	-	-	-	-
	year	-	-	-	-	-	-	-
O_3	30 min	-	-	-	-	-	-	-
	1h	-	235	-	-	-	-	-
	8h	100	157	120	100 (not more than 10 times)	120	120	120 (not more than 25 days)
Pb	24h	-	-	-	-	-	5	-
	3 months	-	1.5	-	0.5	-	-	-
	year	0.5	-	0.5	-	0.5	0.5	0.5
Benzene	year	-	-	5	16.25	8	-	5

Table 1. Comparison of limit values [μm/m^3] for a given averaging time, number of exceedances per year issued by different countries and organizations (ARMAAG)

have been set as the dividing line between acceptable air quality and poor or unhealthy air quality. Criteria pollutants include five gases/vapours and two solids: nitrogen oxides (NOx), sulphur dioxide (SO_2), carbon monoxide (CO), ozone (O_3), benzene, particulate matter (PM10) and lead (Pb). Non criteria pollutants are those contaminants designated as toxic or hazardous by legislation or regulation. They fall in two further subcategories, depending on the legislation that defines them. In general, hazardous air pollutants may pose a variety of health effects, whereas the toxic focus on one physiological response.

4. Characteristics of criteria air pollutants

Air pollutants arise from a wide variety of sources although they are mainly a result of the combustion process (Friedrich & Reis, 2004). The largest sources include power generation, motor vehicle and industry. The emissions of pollutants into the atmosphere badly influence vegetation, human and animal life, agriculture and climate. Emissions of carbon monoxide (CO), nitrogen oxides (NOx) and hydrocarbons (HC) are controlled by catalytic converters on new gasoline driven cars. Emissions of sulphur oxides are being reduced through lower sulphur content in gasoline. However, emissions of particulate matter are not decreasing. Any successful strategies for controlling or countering these problems must be based on reliable air quality monitoring data for management, to make informed decisions on air pollution control.

Volatile organic compounds (VOCs) is a collective name for a very large number of different chemical species, including hydrocarbons, halocarbons and oxygenates that have different physico-chemical properties and are directly emitted from both anthropogenic and natural sources, and which can contribute to the formation of secondary pollutants with different efficiencies. For vehicular emissions, the list of compounds is long and variable depending on fuel, type of engine, and operating conditions. Hydrocarbons such as ethane, ethyne, higher aliphatic hydrocarbons, benzene, toluene, and xylenes (BTX) are typical emissions in most cases. Each of these compounds can be released unreacted or can undergo oxidation reactions. One of them, benzene, is found in the highest concentrations. Ambient concentrations are typically between 1-50 ppb but close to major emissions can be as high as several hundred ppb. In the unreacted state, it has undesirable ecotoxicological properties. Besides causing annoying physiological reactions such as dizziness and membrane irritation, it is known to be a human carcinogen.

The two nitrogen oxides, NO and NO_2 (together called NO_x) from anthropogenic sources are present as a consequence of various combustion processes from both stationary sources, i.e. power generation (21%) and mobile sources, i.e. transport (44%). These species have very short atmospheric lifetimes, around five days, being ultimately converted to nitric acid and removed in rainfalls. However, nitrogen oxide is important because it is a precursor to tropospheric ozone. Whereas NO does not affect climate, ozone does. A typical sea-level mixing ratio of NO is 5 ppt (parts per trillion, $1:10^{12}$) but in urban regions, NO mixing ratios reach 0.1 ppm in the early morning, but decrease to zero by midmorning due to their reaction with ozone. A major source of NO_2 is oxidation of NO, being intermediary between NO emission and O_3 formation. Nitrogen dioxide is one of the six criteria air pollutants for which ambient standards are set by the US EPA under CAAA70 (Clean Air Act Amendments of 1970). In urban regions, the mixing ratio of NO_2 ranges from 0.1 to 0.25 ppm. It is more prevalent during midmorning than during midday or afternoon because sunlight breaks down most NO_2 past midmorning. Exposure to high concentrations of NO_2

harms the lungs and increases respiratory infections. It may trigger asthma by damaging or irritating and sensitizing the lungs, making people more susceptible to allergens. At higher concentrations, it can result in acute bronchitis or death.

Sulfur dioxide (SO2) is a strong smelling, colorless gas that is formed by the combustion of fossil fuels, smelting, manufacture of sulfuric acid, conversion of wood pulp to paper, incineration of refuse and production of elemental sulphur. Power plants, which may use coal or oil high in sulfur content, can be major sources of SO_2 accounting for about 50 % of annual global emissions. SO_2 and other sulfur oxides contribute to the problem of acid deposition and can be major contributors to smog. Natural background levels of SO_2 are about 2 ppb. Hourly peak values can reach 750 ppb on infrequent occasions. Sulphur dioxide can lead to lung diseases. SO_2 is a criteria air pollutant.

Ozone (O_3) is not directly emitted from both anthropogenic and natural sources. Its only source into air is chemical reaction. In urban air, ozone mixing ratios range from less than 0.01 ppm at night to 0.5 ppm (during afternoons in the most polluted cities world wide), with typical values of 0.15 ppm during moderately polluted afternoons. Ozone causes headaches at concentrations greater than 0.15 ppmv, chest pains at mixing ratios greater than 0.25 ppm, and a sore throat and coughing at mixing ratios greater than 0.30 ppm. Exceeding a level of 0.30 ppm decreases lung functions. Respiratory symptoms include coughing and breathing discomfort. Ozone can also accelerate the aging of lung tissue. It also interferes with the growth of plants and deteriorates organic materials, such as rubber, textiles and some paints and coatings. Furthermore, ozone increases plant and tree stress and their susceptibility to disease, infestation and death.

Particulate matter (PM) in the atmosphere, frequently described simply as particle pollution, arises from natural sources, such as windborne dust, sea spray, and volcanoes and from anthropogenic activities, such as combustion of fuels. Particle pollution in the air includes a mixture of solids and liquid droplets and come in a wide range of sizes. Those less than 10 micrometers in diameter (PM10) are so small that they can enter the lungs, potentially causing serious health problems. Particles less than 2.5 micrometers in diameter are called fine particles. These particles are so small they can only be detected with an electron microscope. Sources of fine particles include all types of combustion, including motor vehicles, power plants, residential wood burning, forest fires, agricultural burning, and some industrial processes. Particles between 2.5 and 10 micrometers in diameter are referred to as coarse. Sources of coarse particles include crushing or grinding operations, and dust stirred up by vehicles traveling on roads. After being released into the air, particles can change their size and composition by condensation of vapour species or by evaporation, by coagulating with other particles or by chemical reaction. Particles smaller than 1 μm in diameter generally have atmospheric concentrations in a range from around ten to several thousand per cm^3; those exceeding 1 μm diameter are usually found at concentrations less than 1 cm^{-3}.

Carbon monoxide (CO) is a colourless, odourless gas that is produced by the incomplete burning of carbon-based fuels including petrol, diesel and wood. It is also produced from the combustion of natural and synthetic products such as cigarettes. Natural background levels of CO fall in a range of 10-200 ppb. Levels in urban areas are highly variable, depending upon weather conditions and traffic density. 8-hour mean values are generally less than 10 ppm but sometimes can be as high as 500 ppm. Carbon monoxide lowers the amount of oxygen that enters the blood. It can slow human reflexes and may make people confused and sleepy.

5. Air pollution monitoring networks

For the purpose of monitoring and reporting of air pollution, most industrialised countries have been divided into regions or zones and urban areas or agglomerations, e.g. in Europe in accordance with EC Directive 96/62/EC (European Union, 1996). This Directive sets a framework for ways to monitor and report ambient levels of air pollutants. Other directives set ambient air limit values for particular pollutants:

- Directive 99/30/EC for nitrogen dioxide and oxides of nitrogen, sulphur dioxide, lead and particulate matter,
- Directive 2000/69/EC for benzene and carbon monoxide,
- Directive 2002/3/EC for ozone.

The monitoring sites are organized into automatic and non-automatic networks (national, regional and local) that gather a specific kind of information using a particular method. For example, across the UK there are over 1500 monitoring sites which monitor air quality and these are organized into networks (automatic and non-automatic) that gather a specific kind of information, using a particular method. The pollutants measured and the method used by each network depend on the reason for setting up the network, and what the data is to be used for. In Poland, the monitoring of air quality has been performed systematically since 1992, mainly by using automatic air monitoring stations. National network comprises 81 specialized monitoring stations, where basic pollutants are measured everyday (CO, CO_2, SO_x, NO_x, HCl and particulate matter PM10 and PM2,5). Regional and local networks comprises 390 stations localized in cities over 20 000 citizens and additionally 60 stations in spas.

The obtained information concerns different types of concentrations of investigated pollutants depending on applied sampling techniques and a measuring period. The results of measurements may be referred to in real time (instantaneous concentrations) or to a selected period of time (e.g., 1-hr, 8 hrs, 24-hrs, month, year). Final measurements represent averaged concentrations. The obtained data from air monitoring is used in air quality inventories and bulletins.

Considering the frequency of discrete sampling, periodic and instantaneous measurements are distinguished. Taking into account space, parameter measurements are divided to a point, averaged along a defined part of space and averaged on the selected area. Point monitoring is inadequate to measure poorly mixed gases such as fugitive emissions over large areas. If the point instrument is wrongly placed, measurement results are not representative. Final measurements enable the determination of weighted average concentrations over the sampling period.

6. General characteristic of instruments for air monitoring

The earliest air quality measurements used relatively simple, manual techniques. Often a sample was collected at a site and then returned to a laboratory for analysis. This is, in fact, still a common practice today, due to the relative low cost and high reliability of these types of measurements. Since seventieth of the XX century the automatic analyzers were introduced to the monitoring networks. These had the benefit of low labour costs and could provide highly resolved, continuous measurements. These continuous measurements became a requirement for regulatory purposes and so, in many cities urban monitoring network was established to monitor compliance with the emerging EC Directive limit values on air quality. The networks were subsequently expanded, following commitments by different governments to expand urban monitoring in particular countries and improve public availability of air quality information.

The basic requirement of the analyzers for air monitoring is high measurement sensitivity, i.e. a low limit of detection (LOD) and low limit of quantitation (LOQ). It gives a chance for the detection of pollutants at required levels. Instruments which should acquire analytical data in real time or only with a small time delay, have to be equipped with the following additional capabilities:

- provide high resolution data (characterized by a low response time),
- provide automatic calibration and zeroing,
- extended operation without service or maintenance.

The last demand means that they should be equipped with an independent power supply and be able to automatically regenerate or exchange worn out filters and, depending on the type of detector (sensor) used, fulfill special demands, e.g., for electrochemical sensors, exchange or supplement the working solution and reagents, and in devices with FID or FPD detection, protect against flame extinguishing.

Analytical instruments currently used for the detection and determination of atmospheric pollutants can be classified according to various criteria (Wardencki et al., 2008).

Recently, measuring techniques based on a physical (or physicochemical) principle, have been used more frequently in the assessment of air quality. Such methods involve the direct determination of a physical property of a pollutant, sometimes after its interaction with another compound. In this approach, better stability, sensitivity and reliability may be easily achieved. Furthermore, the practical application requires less maintenance. Instruments based on this principle can be easily automated, which enables their use in providing continuous measurements needed for up-to-date assessments of air quality. It is especially relevant to environmental monitoring because many existing standards refer to a specified period of time, i.e. 1 hour, 24 hours or a year.

According to the location where measurements are taken, instruments can be stationary or *on-site*. In the first case, analysis is performed in the laboratory and sophisticated instruments are used, such as mass, electron mobility or X-ray fluorescence spectrometers. *On-site* systems enable measuring pollution levels in the field. Since access to a sophisticated laboratory is not required, the devices (usually uncomplicated, relatively cheap and portable) hold great promise for use in remote locations. The main advantage of on-site analysis is the potential for rapid assessment and response to a particular problem.

All monitoring systems can be classified as mobile or stationary. Most existing systems which monitor gaseous pollutants of atmospheric air and ambient aerosols, both automatic and manual, usually perform stationary measurements, i.e. they are directly linked to a specific point or space in the vicinity of that point. Based on the data obtained from a single monitoring site, it is not possible to assess spatial and temporal variations of air pollutants.

Mobile refers to a continuous-monitoring instrument that is portable or transportable. They are usually designed to perform analytical measurements without preliminary operations. Portable refers to a self-contained, battery-operated device, which may be worn or carried by the person using it, or may require the use of special vehicles for placement in a specific area to be monitored. Transportable gas monitors can be mounted on a vehicle such as a car, plane, balloon, ship or space shuttle, but not to a mining machine or an industrial truck.

For mobile systems, registered values of pollutants have to be correlated with information about the geographical site and actual meteorological conditions (temperature and humidity).

Several contributions published during the last decade have proven the advantages of mobile systems in getting information concerning the spatial and temporal distribution of atmospheric trace gases, without the need of a dense network of stationary stations. Most of the proposed systems are based on the application of mobile laboratories (Bukowiecki et al., 2002), (Gouriou et al., 2004), (Pirjola et al., 2004), equipped with appropriate monitors. There are also systems which allow the measurement of pollutants in a stream of vehicles, but a measuring unit is installed on any vehicle (Katulski et al., 2006, 2007) rather than attached to a dedicated van (Seakins et al., 2002), (Bogo et al., 1999).

The general trend in the field of creating instruments for air quality assessment is combining several instruments into one system and forming so called hybrid multisensor systems, controlled by a microprocessor capable of transferring the obtained data to a central station, frequently using a wireless system. In the central station, the data is collected both from single objects (houses, plants) or from larger areas. Many systems are equipped with devices for testing the sensors and for providing a diagnosis of the whole instrument. They frequently have alarms which warn the user of any dangerous situation due to the breeching of some value limit. Such systems are battery-powered and able to work continuously for several days or months.

The environments in which analyzers are used differ from the relative calm of the laboratory. Analyzers have to withstand wide ambient temperatures, fluctuations and vibrations. Due to this, many systems are completely sealed so as to operate independent of outside conditions and to be able to withstand the onslaught of monkey-wrench mechanics.

7. Description of mobile system for on-road measurements of air pollutants

The developed prototype of monitoring system consists of an original equipment measuring in motion the gaseous pollution of atmospheric air, further called ARPOL (the acronym stems from ARMAAG - Agency of Regional Monitoring of Atmosphere of Gdańsk Agglomeration and Gdańsk POLytechnic), which is designed to test the pollution of atmospheric air with the following chemical compounds: C_6H_6, NO_2, NO_x, CO and CO_2. The equipment can be installed on any mobile vehicle like a bus or taxicab, relay the results of concentration measurements in air of the substances outlined above to a server which plays the role of a database, along with information on vehicle location and speed as well as the temperature and humidity of air at the measurement spot. Data transmission between the equipment and the server is effected with the use of a radio modem operating in GSM/GPRS systems.

An ARPOL monitoring station has been developed and built in the laboratory of the Department of Radiocommunication Systems and Networks at Gdańsk University of Technology. The set of commercially available gas concentration sensors in air has been selected by researchers from the Chemical Faculty of Gdańsk University of Technology and the ARMAAG Foundation. The construction of the system falls into line with requirements of the sensors and the anticipated application. In the result of decisions taken it has been established to use semiconductor (thick-film) sensors of the newest generation, made of nanostructures with grain-size distribution from 30 to 50 nm. After preliminary investigations the following sensor (all from Figaro Engineering Inc from Osaka, Japan) were chosen: TGS2442 for CO, TGS4161 for CO_2, TGS823 for benzene, TGS2106 for NO_2 and TGS2201 for NO_x. Some of these sensors were designed for measurement of gases in ppm (parts per million, 10^{-6}) concentration, but the concentration of NO_x, CO or C_6H_6 in typical air condition is of the order of ppb (parts per billion, 10^{-9}). However, comparison of results from several months of measurements proved that after proper calibration and compensation of air temperature and humidity, they were enough stable and precise for determining the sources and regions of highest pollution caused by road traffic. All of the gas concentration sensors are heated using internal heaters as a substrate for sensing structure. The temperature of substrate cannot be controlled because no temperature sensor is built in the substrate of gas-sensing element. Correction of substrate temperature change caused by change in temperature of air flowing around the sensor is made by measuring the temperature of air using additional semiconductor sensor and proper calibration equations. Proper heating of gas concentration sensors takes up to several tens of minutes. During this period results of measurements cannot be considered as a reliable results. Also any break in powering the sensors makes results unstable for some unpredictable time (up to several minutes) so monitoring device with such sensors need to provide uninterrupted heating of sensors during whole measurement campaign. Special electronic circuits ensure high sensitivity and high accuracy of measurements.

In the design of ARPOL a microcontroller of the AtMega series (supplier: Atmel Corporation from San Jose, USA) with built-in analog-to-digital converters has been used; the microcontroller circuit has been connected to a ready-made GSM/GPRS modem board (supplier: Enfora from Richardson, USA) and a GPS signal receiver (supplier: Trimble Navigation Ltd. from Sunnyvale, USA), as well as a series RS232 interface (RS232 signal converter based on Maxim chip, supplier: Maxim Integrated Products Inc. from Sunnyvale, USA) and the required additional supply circuits. The main board has been also provided with a connector for the attachment of a board with possible additional sensors not foreseen in the original design, and transoptor circuits and actuators which allow to apply control signals.

To ensure highest immunity of the equipment against shocks it was decided that the whole electronic circuitry of the main part of ARPOL will be assembled on one printed board. As some devices of the equipment (power supply circuits in particular) may heat up during normal operation, it has been decided to place all chemical sensors and the external temperature and air humidity sensors at one edge of the main board, so that the microprocessor part with power supplies could be separated in the housing from the sensors by means of an appropriate dividing partition.

The equipment has been enclosed in a housing which provides protection of the system from the influence of precipitation, while simultaneously allowing access of atmospheric air to the gas-concentration, temperature and humidity sensors. The housing permits the unit to be mounted both in a car and stationary at selected points in town. Part of the electronic board with sensors has been enclosed in a plastic channel terminated with an air in- and outlet, provided with additional "roofs" for protection from rain. In order to reduce the effect of exhaust gas of the engine the unit has forced sampling system opening towards the driving direction. To ensure an uniform air exchange, in the channel with sensors a blower has been placed, provided with revolution control – a blower failure or reduction of its efficiency resulting in a decrease of air flow in the measurement channel is registered by a microcontroller as an alarm state.

A schematic description of the operation of the mobile monitoring system (MMS) for testing and analysis of atmospheric air pollution along road traffic arteries is presented in Fig.1. This system consists of an ARPOL unit mounted on a vehicle and a WWW server in which a database for collection and processing of measurement data has been established. The values of concentration of chemical compounds polluting the atmosphere are transmitted through radio, through the mobile GSM/GPRS network and the Internet network to a WWW server located anywhere, e.g. at Gdańsk University of Technology (Gdańsk Polytechnic). There, these data are processed and verified as to their credibility. Subsequently, they are sorted according to the measurement date and, if needed, displayed on the monitor of an authorized user by a common WWW page browser.

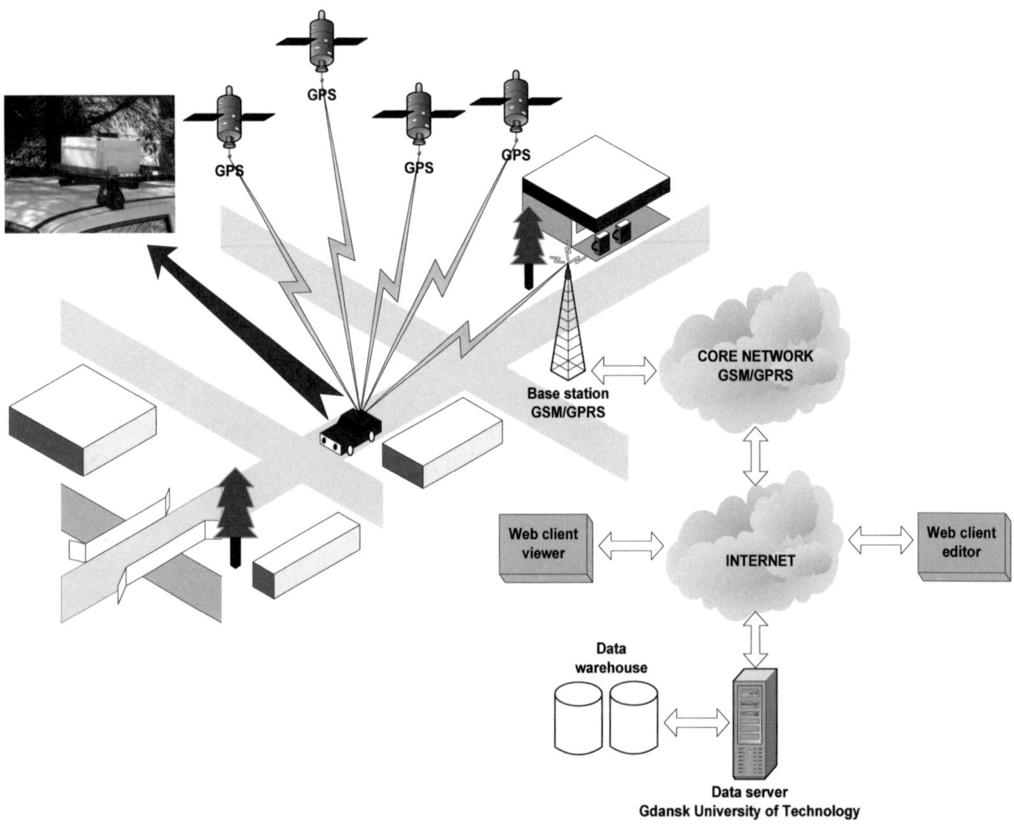

Fig. 1. The concept of the mobile monitoring system for chemical agents control in the air.

To achieve simplicity of operation of the monitoring system, the configuration process of ARPOL unit and the WWW server is kept to a minimum. This means that this system can be implemented on a wide range of servers.

The proposed monitoring unit can be described as maintenance-free. Properly configured and set-up monitoring station can operate for several months without intervention. Information about the state of a device and possible errors in the functioning of electronic components or problems with data collection are transmitted to respective person via short messages (SMS).

8. Sensors description

Road traffic is responsible for emission of several air pollutants, the most important being nitrogen oxides (NO and NO_2, together called NO_x) which may harm the lungs and trigger asthma, sulphur dioxide (SO_2), particulate matter (PM), carbon monoxide (CO), carbon dioxide (CO_2) and volatile organic compounds (VOCs) which can contribute to the formation of secondary pollutants with different efficiencies. The measurement of particulate matter in air requires some air filtering and mass measuring during specified time which cannot be made using semiconductor sensors and is not applicable to mobile monitoring units. Other air pollutants can be measured using thick film semiconductor sensors. Prototypes of monitoring units for proposed systems use selected sensors made by Figaro:

- TGS2442 - CO sensor, response: resistance of SnO_2 layer on electrical insulation layer. Response time: tens of seconds;
- TGS4161 - CO_2 sensor, response: electromotive force which is the result of electrochemical reaction of CO_2 and electrodes made of gold and lithum carbonate with cation (Na+) solid electrolyte. Response time: up to 1 minute, recovery time: up to 2.5minutes;
- TGS823 - VOC sensor. This sensor reacts to the presence of several volatile organic compounds, such as: Benzene (C_6H_6), Ethanol, n-Hexane, Acetone, Isobutane and others, but is used mostly for detection of Benzene which is found in highest concentrations. Response: resistance of SnO_2 layer, response time: tens of seconds;
- TGS2106 - NO_2 sensor, response: resistance of metal-oxide semiconductor layer. Response time: several seconds;
- TGS2201 - NO_x and CO sensor (double) used only for NO_x measurement, response: resistance of metal-oxide semiconductor layer. Response time: tens of seconds.

Some of these sensors were designed for measurement of gases in ppmv (parts per million,10^{-6}) concentration, but the concentration of NO_x, CO or C_6H_6 in typical air condition is of the order of ppbv (parts per billion, 10^{-9}). However, comparison of results from several months of measurements proved that after proper calibration and compensation of air temperature and humidity, they are enough stable and precise for determining the sources and regions of highest pollution caused by road traffic.

9. Air monitoring unit

Block diagram of mobile monitoring unit is presented in Fig.2. This version of measurement device is made of sensor unit, A/D converters and microcontroller to convert sensor's electric response to gas concentrations. Microcontroller is also responsible for linking measurement results with geographic coordinates from GPS and sending these data to server using GSM/GPRS link. The interior of one of monitoring units built in Gdansk University of Technology laboratory is shown in Fig.3.

The gas concentration sensors together with temperature and humidity sensors are put inside manifold (at the top of Fig.3). Air flow inside manifold is forced by fan, which rotate speed is controlled by microprocessor because change in flow rate will change gas sensor's substrate temperatures as air is cooling the casings of sensors.

Electrical response of gas sensors is converted into voltage signals using very high impedance operational amplifier (for EMF output) or by resistive voltage divider (for

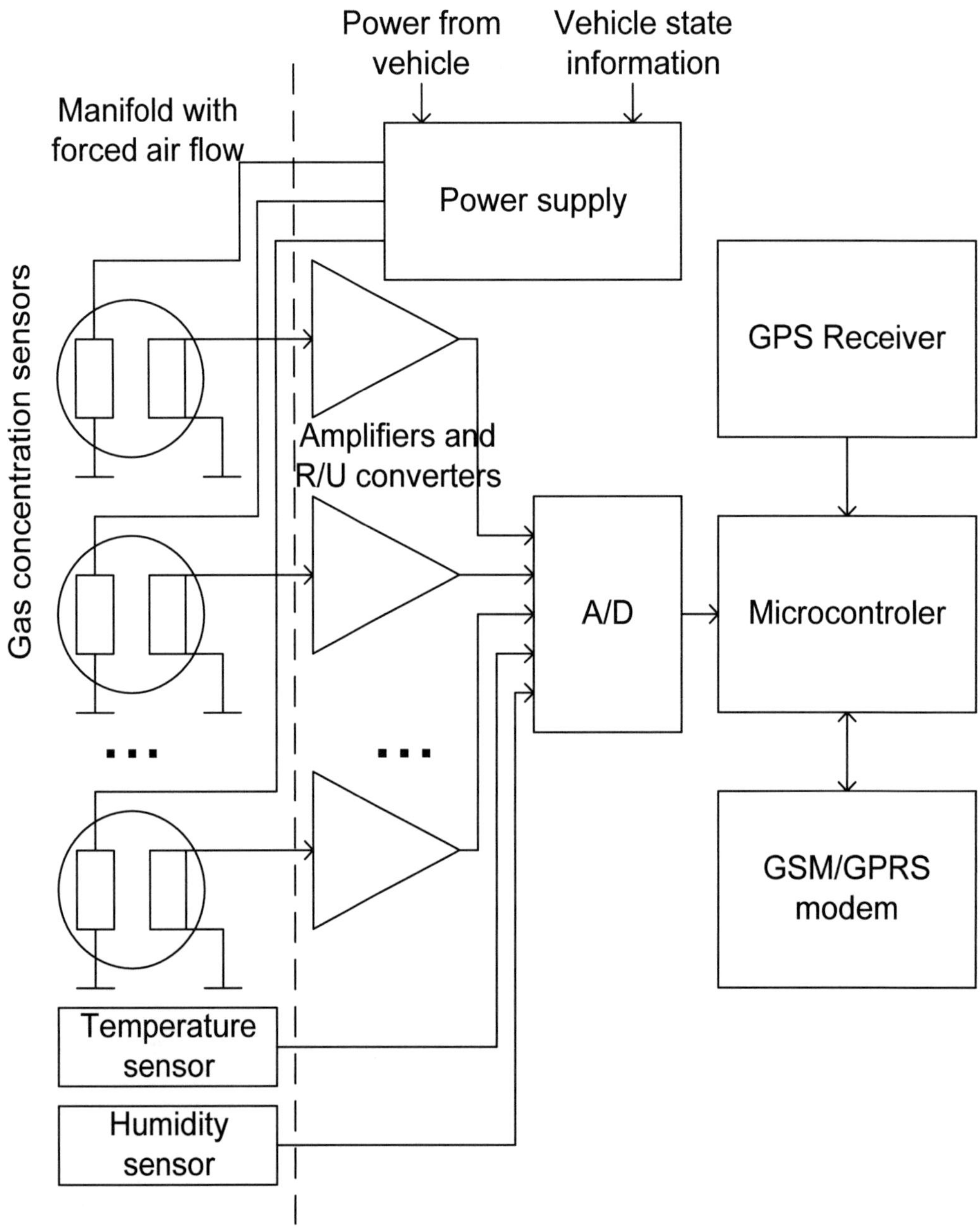

Fig. 2. Block diagram of mobile monitoring unit.

Fig. 3. Mobile monitoring unit prototype.

resistive output). Then the signals are passing low-pass filters with cutoff frequency about 10Hz for all sensors except TGS2442 which requires pulsed resistance measurement and therefore filter cutoff frequency has to be greater than 100Hz. Additional noise-filtering is made by microprocessor which averages digital values after A/D conversion during 10 seconds period. Averaged digital values are being transmitted to server without any additional conversion. Calculation of gas concentration is made by server using nonlinear equations derived during calibration procedure. Other data sent to server in every packet are:

- monitoring unit identification number;
- geographical coordinates of measurement point (from GPS receiver);
- speed and direction of movement (from GPS receiver);
- date/time of measurement (from GPS receiver);
- temperature and humidity of air in manifold;
- temperature of electronic circuits outside manifold;
- external or internal power supply voltage;
- state of vehicle engine in case of mounting the device on car/bus;
- error flags which indicate condition of monitoring unit.

Data transmission uses GPRS/EDGE radio link and TCP/IP protocol stack. To avoid writing special software at server side, data records are encoded as special HTTP request. Such requests are easily handled by PHP script and after decoding raw data records are written to MySQL database.

10. Conclusions

Polluted air negatively influences health and in some cases may even lead to death. Therefore, the issue of air quality is now a major concern for many countries which have been working to improve air quality by controlling emissions of harmful substances into the atmosphere, improving fuel quality, and by integrating environmental protection requirements into the transport and energy sectors. Despite these improvements in air quality over recent years, the problem of air pollution still remains. Therefore, more needs to be done at local, national, European and international levels.

Monitoring of air pollution is a prerequisite of air quality control and is carried out by a wide variety of analytical methods employing different measurement instruments which have different sensitivities and specificities. Monitoring plays a critical role in protecting the environment and is a key element of all actions related with the management and protection of ambient air.

According to existing law, member countries of the European Union are obliged to carry out measurements of atmospheric air cleanness and to introduce preventive measures against excessive pollution exceeding allowable threshold limits. To meet these obligations, the member countries are required to prepare inventories of zones and agglomerates where transgression of limiting norms occurs, and to prepare and introduce plans and programs how to adhere to the parameters required by law. Thus the mobile monitoring system described here, called ARPOL, unique on the national scale, can upgrade the existing system of measuring atmospheric air cleanness and facilitate taking appropriate decisions through analysis of various emission scenarios.

The developed device forms part of an expert system which undoubtedly contributes to the enhancement of ecological safety in a given area. It may be a cheap and useful tool (considering low investment and exploitation costs) for assessing ecological effects of new efforts for limiting pollutants emissions rather than for assessment of air quality. It contributes to more careful environmental planning by extending the measurement basis with information from non-stationary sources.

11. References

Environmental Protection Agency (2008). Clean Air Act, in: *U.S. Code*, 03.2011, Available from http://www.epa.gov/air/caa/

World Health Organisation (2000). Air Quality Guidelines for Europe, 2nd edition, WHO Regional Publications European Series No.91, Copenhagen, 2000.

European Union (2008). Directive 2008/50/EC of the European Parliament and of the Council of 21 May 2008 on ambient air quality and cleaner air for Europe, Official Journal of the European Union, 11.6.2008, L. 152/1-152/44

WHO (2005). WHO Air quality guidelines for particulate matter, ozone, nitrogen dioxide and sulfur dioxide - Global update 2005 - Summary of risk assessment

ARMAAG. Emission limits values (in Polish), 03.2011, Available from http://armaag.gda.pl/normy.htm

Griffin, R.D. (2007). *Principles of air quality management*, Taylor & Francis Group, Boca Raton, 2007

Friedrich, R. & Reis, S. (2004). *Emission of Air Pollutants.* Springer_Verlag Berlin Heidelberg, 2004

European Union (1996). Council Directive 96/62/EC of 27 September 1996 on ambient air quality assessment and management . Official Journal L 296 , 21/11/1996 P. 0055 - 0063

Wardencki, W., Katulski, R. Stefański J. & Namieśnik, J. (2008). *The state of the art in the field of non-stationary instruments for the determination and monitoring of atmospheric pollutants.* Critical Reviews in Analytical Chemistry, 2008, 38, 1-10.

Bukowiecki, N., Dommen, J., Prevot, A.S.H, Richter, R., Weingartner, E. & Baltensperger U. (2002). *A mobile pollutant measurement laboratory-measuring gas phase and aerosol ambient concentrations with high spatial and temporal resolution.* Atmospheric Environment, 2002, 36, 5569-5579.

Gouriou, F., Morin, J-P. & Weill, M-E. (2004). *On-road measurements of particle number concentrations and size distributions in urban and tunnel environments Atmospheric Environment*, 2004, 38 (18), 2831-2840.

Pirjola, L., Parviainen, H., Hussein,T., Valli, A., Hameri, K., Aaalto, P., Virtanen, A., Keskinen, J., Pakkanen,T.A., Makela,T. & Hillamo, R.E. (2004). *''Sniffer''-a novel tool for chasing vehicles and measuring traffic pollutants.* Atmospheric Environment, 2004, 38, 3625-3635.

Katulski, R., Stefański, J., Wardencki, W. & Żurek J. (2006). Concept of the mobile monitoring system for chemical agents control in the air, Proceedings of the *IEEE Conference on Technologies for Homeland Security – Enhancing Transportation Security and Efficiency*, June 7-8, 2006, Boston, USA, 181-184.

Katulski, R., Stefański, J., Wardencki, W. & Żurek J. (2007). The Mobile Monitoring System (MMS) – a Useful Tool for Assessing Air Pollution in Cities, Proceedings of the *Pittsburgh Conference on Analytical Chemistry and Applied Spectroscopy PITTCON'2007*, February 25 – March 2, 2007, Chicago, USA.

Seakins, P.W, Lansley, D.L, Hodgson, A., Huntley A. & Pope, F. (2002), *New Directions: Mobile laboratory reveals new issues in urban air quality,* Atmospheric Environment 2002, 36, 1247-1248.

Bogo H., Negri R.M. & San Roman, E. (1999). *Continuous measurement of gaseous pollutants in Buenos Aires city*. Atmospheric Environment, 1999, 33, 2587 - 2598.

8

Influence of the Air Engine on Global Warming Issues - 21st Century Fuel Technology

Bharat Raj Singh[1] and Onkar Singh[2]
[1]*SMS Institute of Technology, Lucknow*
[2]*Harcourt Butler Technological Institute, Kanpur*
India

1. Introduction

Global warming or climate change is undoubtedly one of the most important challenges for our future generation, and quite possibly any generation in history. The worldwide scientific community is unanimous in its agreement that global warming is happening, that is our fault, and that the opportunity to stop it is slipping away. If we let it get out of our control, the consequences - which are already evident in most of our lifetimes - will be catastrophic. For example some of the consequences that can be reasonably expected are rising sea levels, frequent and severe natural disasters, large-scale food and water shortages, plagues, massive species extinctions, unprecedented numbers of refugees, intensified ethnic and political tensions, and a global economic depression the likes of which no one has ever seen.

1.1 Current situation

The situation is still within our grip, but we must act now, we must act strongly, and we must act together. Individuals, companies, and governments across the globe must each do what they can to reverse climate change. We will never get a second chance.

1.2 What can we do?

There are plenty of things that one can reduce the alarming carbon footprint shown in **Fig.1**. The key word is *reduce*. Anyone can greatly lessen the impact on climate change by using the planet's resources more responsibly. There are many things which can be reduced, and many ways that can reduce them, but three of the major ones are: reduce the amount of animal products being consumed (e.g., meat, dairy, eggs, leather, etc.), reduce the amount of fuel being used in cars, air travels, etc., and reduce the amount of electricity being used for domestic as well as industrial purposes utilizing some alternatives. There are plenty of good resources as detailed below encourages to-do the research.

Since the use of transport has become a part of life in the current civilization of developed and developing countries, the population of transport vehicles are increasing 2 to 3 times in every 5-7 years. Approximately 80% transport population in the developing countries such as; India, China, Taiwan, Romania, Bulgaria etc., are two wheelers, that is also adding about 77.8% of total emission in the atmosphere.Hence the transport sector alone, is responsible to the faster consumption of hydrocarbon fuel and releases heavy tail pipe emissions, thereby adding billion tonnes of excessive carbon dioxide, carbon monoxide and unburnt hydrocarbon etc. (see **Table-1**), in the atmospheric air every day. This is causing a serious threat to the global warming. Also the fuel reserves are depleting very fast.

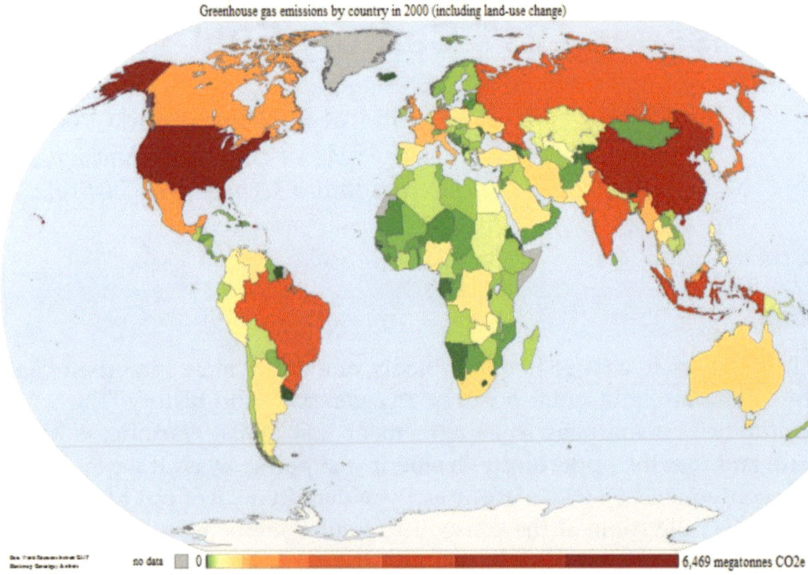

Fig. 1. Contrywise Green House Emission (Carbon Foot Print)

S.No.	Name of Major Countries	Population In Millions	Total CO2 Yearly Release (In Million Tonnes)	Remarks
1.	China	1334.8	6,126.7	
2.	USA	311.3	5,983.0	
3.	**Russia**	**138.9**	**1,572.4**	
4.	**India**	**1183.3**	**1,526.5**	
5.	**Japan**	**126.6**	**1,316.6**	
6.	**Germany**	**82.1**	**878.5**	
7.	Canada	33.8	561.0	
8.	Iran	77.5	519.3	
9.	Italy	57.9	486.4	
10.	South Korea	48.7	475.8	
11.	Mexico	113.5	440.0	
12.	South Africa	49.2	415.3	
13.	France	64.6	409.6	
14.	Australia	21.6	392.0	
15.	Saudi Arabia	26.0	385.6	

(Source: www.Breathingearth.net Feb'2011)

Table 1. List of 15-Countries Contributing Highest CO_2

In 1956, the Marion King Hubbert a noted geophysicist predicted that US Fuel reserves may peak by 1975 and fuel crisis will be noticed by 1995. He illustrated the projection with a bell shaped *Hubert Curve* based on the availability and consumptions of the fossil fuel. Large fields are discovered first, small ones later. After exploration and initial growth in output, production plateaus and eventually declines to zero (Hubbert, 1995). Thereafter in 2003, Aleklett and Campbell expressed their views that most of the countries will pass through peak oil days by 2010-12 and fuel consumption will reach to 80% by 2020-30 with the current rate of consumption (Alekett & Campbell, 2003).
In India, vehicular pollution is estimated to have increased eight times over the last two decades. This source alone is estimated to contribute about 70 per cent to the total air pollution. With 243.3 million tons of carbon released from the consumption and combustion of fossil fuels in 1999, India is ranked fifth in the world behind the U.S., China, Russia and Japan. India's contribution to world carbon emissions is expected to increase in the coming years due to the rapid pace of urbanization, shift from non-commercial to commercial fuels, increased vehicular usage and continued use of older and more inefficient coal-fired and fuel power-plants. Thus, peak oil year may be the turning point for mankind which in turn lead to the end of 100 year of easy growth, if self-sufficiency and sustainability of energy is not maintained on priority. It may end up a better world as per proceedings of conference held in Architectural Institute, 2004, Concept and Technology- First International Workshop on Sustainable Habitat Systems, at Japan. Presently major thrust is being given to explore wind energy, hydro-power, tidal and nuclear power generation etc. Efforts are also given to energy storage system for the clean energy by conversion system and its better utilization to run prime-moves for light vehicles.
This paper describes about the energy conversion and energy storage system. One of the dominant option is storage of compressed air from easily available atmospheric air which may be compressed by electricity or by alternative energy sources like wind, solar energy etc. or disaster energy sources and stored in the air receiver or storage tanks of suitable size. Such energy could be reutilized as clean energy source for running prime-movers of domestic appliances and light vehicles as a nearly zero pollution fuel sources. The design of air turbine / engine and its different parametric and performance aspect are also elaborated in this article. To maintain sustainability in 21st Century and check the global warming issues to the extent of 50-60%, the storage energy systems and its utilization to run the lighter vehicles or motorbikes by novel and efficient air engine / turbine could be one of the dominant technology. This may also leads to environmentally and ecologically better future.

2. Types of energy storage system

The Power Conversion System (PCS) is a vital part of all energy storage systems. It interfaces the energy storage, the energy storage device and the load (the end-user). PCS cost is significant and it can be greater than 25% of the overall energy storage system. PCS cost range from Rs. 4,700/kW for UPS markets to Rs.56, 000/kW for Standalone markets have been seen. Some of the major PCS markets include:

- Motor drives
- Power supplies
- UPS (uninterruptible power supply)
- Electric vehicles
- Inverters/Converters for solar-hybrid systems

- Micro-turbines
- Fuel cells
- Wind turbines

However, power conversion system technology has been evolving slowly due to the limited distributed energy resource (DER) market. As a result, Energy Storage System cost has been high with low profit margins and the manufacturing volume has been low impacting reliability and quality of the Power Conversion System designs. What is needed is a significant reduction in overall cost with improved reliability, development of state-of-the-art Power Conversion System with multiple uses, which increases production volumes for DER applications, improve controls and adaptability, and improve manufacturing.

2.1 Battery energy storage system

Sandia National Laboratory has a broad technical base of battery expertise focusing on integrated storage systems. These storage systems operate in varying environments and electrical conditions. In these storage systems there are many different types of battery technologies. With the different designs are having advantages under specific operational conditions. It is important to understand the capabilities and limitations of each storage technology.

- Lead-Acid Battery- short cycle life
- Li-Ion -Lithium Ion Battery-High energy and 100% efficiency
- NaS -Sodium Sulfur Battery- can run at high temperature of 300 deg centigrade.
- PSB - Polysulfide Bromide Flow Battery- 75% efficiency
- VRB -Vanadium Redox Flow Battery- 85% efficiency.

2.2 Super capacitor energy storage

Electrochemical capacitors (EC) store electrical energy in the two series capacitors of the electric double layer (EDL), which is formed between each of the electrodes and the electrolyte ions. The distance over which the charge separation occurs is just a few angstroms. The capacitance and energy density of these devices is thousands of times larger than electrolytic capacitors. The electrodes are often made with porous carbon material. The electrolyte is either aqueous or organic. The aqueous capacitors have a lower energy density due to a lower cell voltage but are less expensive and work in a wider temperature range. The asymmetrical capacitors that use metal for one of the electrodes have a significantly larger energy density than the symmetric ones and have lower leakage current.

2.3 Fly wheel energy storage

Most modern flywheel energy storage systems consist of a massive rotating cylinder (comprised of a rim attached to a shaft) supported on a stator by magnetically levitated bearings that eliminate bearing wear and increase system life. To maintain efficiency, the flywheel system is operated in a low vacuum environment to reduce drag. The flywheel is connected to a motor/generator mounted onto the stator that interact with the utility grid. Some of the key features of flywheels are little maintenance, long life (20 years or 10s of thousands of deep cycles) and environmentally inert material. Flywheels can bridge the gap between short term ride-through and long term storage with excellent cyclic and load following characteristics. The choice of using solid steel versus composite rims is based on the system cost, weight, size, and performance trades of using dense steel (200 to 375 m/s

tip speed) vs. a much lighter but stronger composite that can achieve much higher rim velocities (600 to 1000 m/s tip speed). Actual delivered energy depends on the speed range of the flywheel as it cannot deliver its rated power at very low speeds. For example, over 3:1 speed range, a flywheel will deliver ~90% of its stored energy to the electric load.
While high-power flywheels are developed and deployed for aerospace and UPS applications, there is an effort, pioneered by Beacon Power, to optimize low cost commercial flywheel designs for long duration operation (up to several hours). 2kW / 6kWh systems are in telecom service today. Megawatts for minutes or hours can be stored using a flywheel farm approach. Forty 25kW / 25 kWh wheels can store 1MW for 1 hour efficiently in a small footprint.
The stored energy can be approximated by:

$$E = (Iw^2)/2 = (mr^2w^2)/2 = (mv^2)/2$$

where w is the rotational velocity (rad/sec), I the moment of inertia for the thin rim cylinder, m is the cylinder mass and v is linear rim velocity.

2.4 Compressed air energy storage

The Technology of air engine is not new.The Sterling air engine was developed in 1790-1810, but due to its limitation no much work was carried out. In view of fire problems in Coalmines and other volatile places, where high flammable fuel like fossil fuel vehicles are not adviseable, compressed air operated vehicles are normally being put in use. Now from last two decades major thrust is being given by the researchers for development of compressed air engine. Some technical developments, which may be considered to work on 21st Century Energy Storage system and can work on compressed air or hybrid system as an alternative to fossil fuel for running light vehicles, are listed below:-

2.4.1 Reciprocating compressed air engine

Guy Negre, a French Scientist, in 1998 developed compressed air- 4- cylinders engine run on air and gasoline, claims zero pollution cars and got 52- patents registered since 1998 to 2004.The car was demonstrated in Oct.'2004 publically(Negre, 2004).

2.4.2 Rotary hot air motor (Quasi-turbine)

Saint Hilaire G., an inventor developed zero pollution cars using Quasiturbine with a set of 14- engines parameters and disclosed on Sept'2005 using gasoline (Saint-Hilaire, 2005). In the basic single rotor Quasiturbine engine, an oval housing surrounds a four-sided articulated rotor, which turns and moves within the housing. The sides of the rotor seal against the sides of the housing, and the corners of rotor seal against the inner periphery, dividing it into four chambers.

3. Compressed air as alternative and zero emission source to fossil fuel

3.1 History of air engine

The compressed air as an energy and locomotion vector is precisely not a recent technology. In fact at the end of the 19th century the first approximations to what could one day become a compressed air driven vehicle already existed, through the arrival of the first pneumatic locomotives. Yet even two centuries before that Dennis Papin apparently came up with the idea of using compressed air (Royal Society London, 1687).

The first recorded compressed-air vehicle in France was built by the Frenchmen Andraud and Tessie of Motay in 1838. A car ran on a test track at Chaillot on the 9th July 1840, and worked well, but the idea was not pursued further. In 1872 the Mekarski air engine was used for street transit, consisting of a single-stage engine. It represented an extremely important advance in terms of pneumatic engines, due to its forward thinking use of thermodynamics, ensuring that the air was heated, by passing it through tanks of boiling water. Numerous locomotives were manufactured and the first in Nantes in 1879. The H. K. Porter Company in Pittsburgh sold hundreds of these locomotives to coal-mining companies in the eastern U.S. With the hopeful days of air powered street transit over, the compressed air locomotive became a standard fixture in coal mines around the world because it created no heat or spark and was therefore invaluable in gassy mines where explosions were always a danger with electric or gas engines.

Also in 1896, Porter supplied ten compressed air motor cars for the Eckington System in Washington, D.C. There was a tank on the front of the engine and it was recharged at the station. Between 1890 and 1902 ten compressed air trams circulated in Bern, Switzerland. In 1892, Robert Hardie introduced a new method of heating that at the same time served to increase the range of the engine. However, the first urban transport locomotive was not introduced until 1898, by Hoadley and Knight, and they introduced a two stage engine. Later on, in 1912 the American's method was improved by Europeans, adding a further expansion stage to the engine (3 stages).

In 1926, Lee Barton Williams of Pittsburg USA presented his invention: an automobile which he claimed run on air. The motor starts on gasoline, but after it has reached a speed of ten miles an hour the gasoline supply is shut off and the air starts to work. At the first test his invention attained a speed of 62 miles an hour.

In January 1932 what appears to be the first journalistic article ever written about a car driven by compressed air was published. In 1934, a 21-year-old researcher Johannes Wardenier announced the development of world's first fuel-less automobile. For weeks Dutch newspapers reported of an incredible invention that would change the world for ever.

After the Second World War the term "air engine" was never used again in textbooks in reference to compressed air or pneumatic locomotives and, whenever they mentioned it the article would go on to state that these engines are of little use or efficiency.

In 1970's, Willard Truitt presented his invention in McKees Rocks, USA. But because he did not have the financial means to develop his compressed air car further, he gave the rights of his invention to NASA and the US Army in 1982.

In 1979, Terry Miller decided that compressed air was the perfect medium for storing energy. He developed Air Car One, which he built for $ 1,500. Terry's engines showed that it was feasible to manufacture a car that could run on compressed air. He patented his method in 1983 (US4370857).

In the 1980's Carl Leissler developed a motor that was able to function on air. The retired horticulturalist had been working from his garage in Hollywood for over 15 years. He says that to use his motor in a car you might have to use a small electric or gas energy source to help drive the air compressor. 'We might be able to get 2000 miles per gallon; air is a power in itself' Leissler comments.

Until 1987 the German company Arnold Jung Lokomotivenfabrik GmbH produced locomotives functioning on compressed air to be used in mines. In the 1980's they were still selling and renovating locomotives.

Currently the tram association in Bern Switzerland (BTG) is developing a locomotive according to the original plans.
At present (2008) various researchers and industries are developing compressed air engines/motors applicable to transportation, apart from the many industries that produce and commercialize compressed air motors for industrial purposes.

3.2 Advantages / disadvantages of air engine

There are advantages and disadvantages of air engines in comparison to the electric motors.

3.2.1 Advantages

- Air turbines do not need any electric power to run and hence, they can easily be used in volatile atmospheres also.
- Air turbines of the same size as that of electric motor can deliver more power as power density of the air engine is higher and thus, air engine the same size as that of electric motor can deliver more power.
- The air turbines can operate without any type of auxiliary speed reducers unlike electric motors.
- Overloads cause no harm to the air turbines whereas in the case of electric motor, it can easily trip the circuits, which have to be restored.
- Air engines can easily be controlled by simple flow control by means of valves whereas electric motor need electronic speed controls. By regulating the pressure, the torque produced by the air turbines can easily be varied.
- Air turbines do not need any magnetic starters like the electric motors.
- Air turbines generate very less heat as compared to the electric motors.

3.2.2 Disadvantages

- There are some drawbacks of air engines as the flow of air or the pressure is disturbed; the torque being produced by the air engines suffers in a great way.

3.3 Working principles of air engine / air turbine

3.3.1 Thermodynamic principle

The high pressure jet of air when injected into the rotor of vaned air turbine causes its rotation. Rotor movement is caused due to impact of air jet and the expansion of high pressure air admitted in turbine. Thus the high pressure air entering through the inlet passage pushes the vane for producing rotational movement through this vane and thereafter air so collected between two consecutive vanes of the rotor is gradually expanded till the rotation brings vanes in front of exit passage. This can be considered like isobaric admission and adiabatic expansion of high pressure air both contributing in producing the shaft work from air turbine. Compressed air leaving the air turbine after expansion is sent out from the exit passage.
The proposed air turbine is considered to work on the reverse working principle of vane type compressor and transforms the energy of the compressed air into shaft work. The multi-vane type air turbine has casing diameter 100 mm and rotor diameter 75 mm as shown in **Fig. 2**. In this arrangement total shaft work is cumulative effect of isobaric admission of compressed air jet on vanes and the adiabatic expansion of high pressure air.

In an earlier study conducted by authors a prototype of air turbine was developed (Singh & Singh, 2008c).
A cylinder for the storage of compressed air with a capacity of storing air for the requirement of 30 minutes running at initial stage and maximum pressure of 20 bar is used as a source of compressed air. The compressed air storage cylinder is designed to produce constant pressure for the minimum variation of torque at low volumes of compressed air and attached with filter, regulator and lubricator. The clean air then admits into air turbine through inlet nozzle. Vanes of novel air turbine are placed under spring loading to maintain their regular contact with the casing wall to minimize leakage which is proposed as improvement over the currently available vane turbine. A study on high efficiency energy conversion system for liquid nitrogen (Honton,2004) design and verification of airfoil and its tests, influence of tip speed ratios for small wind turbine and parabolic heat transfer and structural analysis were also carried out for conceptualizing the energy conversion system and design of the air turbine (Faglsang, 2004; Gorla, 2004; Knowlen, 1998; Schreck, 2004; Selig, 2004). Studies have shown feasibility of vane type novel air turbine (Singh & Singh 2006; 2007; 2008 a, b; 2009 a-f).

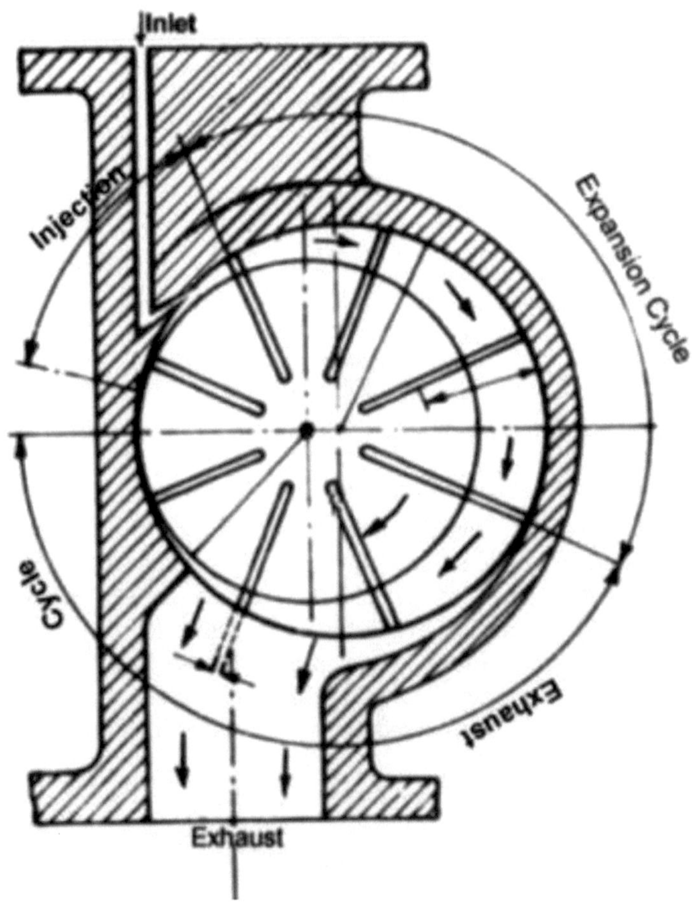

Fig. 2. Air Turbine- Model

The present objective is to investigate the performance of an air turbine with the variation of injection angle, i.e. angle at which air should be admitted into the turbine between first two consecutive vanes. The air turbine considered here has capability to yield output of 5.25 to 6.50 HP at 4-6 bar air pressure for speed of 2000–2500 rpm, which is suitable for a motorcycle.
It is assumed that the scavenging of the rotor is perfect and the work involved in recompression of the residual air is absent. The assumptions made for thermodynamic analysis are as under.

- Admission of air takes place isobarically.
- High pressure air admitted in turbine rotor expands adiabatically.
- There is no residual air in rotor after the expanded air is discharged from rotor. Thus no recompression of air takes place.
- Loss of work due to rubbing friction between vanes and casing walls is neglected.
- Leakage of air across the vane tip and rotor casing contact interface is negligible.
- Mechanical losses due to shaft friction are negligible.
- Trace of the lubricant added to the air being admitted is negligible compared to total mass flow of air.
- Loss due to throttling of air at admission and exit is negligible.
- Exit of expanded air occurs isochorically
- Flow across the turbine is of steady flow type.

3.3.2 Mathematical modeling

The mathematical model shown here is already presented in the author's publications (Singh & Singh, 2010a, b, c, d, e, f, g and h). But for the benefits of readers it is again reproduced here in brief. The high pressure jet of air at ambient temperature drives the rotor in novel air turbine due to both isobaric admission and adiabatic expansion. Such high pressure air when enters through the inlet passage, pushes the vane for producing rotational movement through this vane and thereafter air so collected between two consecutive vanes of the rotor is gradually expanded up to exit passage. This isobaric admission and adiabatic expansion of high pressure air both contribute in producing the shaft work from air turbine. Compressed air leaving the air turbine after expansion is sent out from the exit passage. It is assumed that the scavenging of the rotor is perfect and the work involved in recompression of the residual air is absent.
From **Fig. 3**, it is seen that work output is due to isobaric admission (E to 1), adiabatic expansion (1 to 4) and steady exit flow work (4 to 5). Thus, total work done due to thermodynamic process may be written as:
[Area under (E145CE)] = [Area under (E1BOE) +Area under (14AB1) - Area under (4AOD4) + Steady Flow (45CD4)]
Total Work output = [Thermodynamic expansion work (w_1)] + [Exit steady flow work (w_2)]

$$w = [(w_1) + (w_2)\} \quad (1)$$

Now thermodynamic expansion work (w_1), can be written as:

$$w_1 = p_1.v_1 + \left(\frac{p_1.v_1 - p_4.v_4}{\gamma - 1}\right) - p_4.v_4 \text{ or } w_1 = \left(\frac{\gamma}{\gamma - 1}\right)(p_1.v_1 - p_4.v_4)$$

For adiabatic process, $p.v^{\gamma} = p_1.v_1^{\gamma} = p_4.v_4^{\gamma}$ = constant

$$\text{or } v_4 = \left(\frac{p_1}{p_4}\right)^{\frac{1}{\gamma}}.v_1$$

Thus thermodynamic expansion work output would be:

$$w_1 = \left(\frac{\gamma}{\gamma - 1}\right).p_1.v_1.\left\{1 - \left(\frac{p_4}{p_1}\right)^{\frac{\gamma-1}{\gamma}}\right\} \quad (2)$$

The exhaust of expanded air to atmosphere (4-5) takes place after the expansion process (E-4) as shown in **Fig.3**. In this process; till no over expansion takes place pressure p_4 can't fall below atmospheric pressure p_5. Thus at constant volume when pressure p_4 drops to exit pressure p_5, no physical work is seen. Since turbine is functioning as positive displacement machine, hence under steady fluid flow at the exit of the turbine, the potential work is absorbed by the rotor and flow work (w_2), can be written as:

$$w_2 = \int_4^5 v.dp = v_4(p_4 - p_5) \quad (3)$$

Substituting equations (2), (3) into equation (1), the net work output will be:

$$w = (w_1 + w_2) = \left(\frac{\gamma}{\gamma - 1}\right).p_1.v_1.\left\{1 - \left(\frac{p_4}{p_1}\right)^{\frac{\gamma-1}{\gamma}}\right\} + (p_4 - p_5).v_4 \quad (4)$$

When air turbine is having n number of vanes, then shaft output can be written as:

$$w_n = n.\left(\frac{\gamma}{\gamma - 1}\right).p_1.v_1\left\{1 - \left(\frac{p_4}{p_1}\right)^{\frac{\gamma-1}{\gamma}}\right\} + n.(p_4 - p_5).v_4 \quad (5)$$

where w_n is work output (in Nm), for complete one cycle.

Therefore, the total power output (work done per unit time) (W_{total}), for speed of rotation N rpm will be mentioned as (Book; Singh O., 2009):

$$W_{total} = n.(N/60).\left(\frac{\gamma}{\gamma - 1}\right).p_1.v_1.\left\{1 - \left(\frac{p_4}{p_1}\right)^{\frac{\gamma-1}{\gamma}}\right\} + n.(N/60).(p_4 - p_5).v_4 \quad (6)$$

where $$W_{\exp} = n.(N/60).\left(\frac{\gamma}{\gamma - 1}\right).p_1.v_1.\left\{1 - \left(\frac{p_4}{p_1}\right)^{\frac{\gamma-1}{\gamma}}\right\}$$

and $$W_{flow} = n.(N/60).(p_4 - p_5)v_4$$

Fig. 2. shows that if vanes are at angular spacing of θ degree, then total number of vanes will be n = $(360/\theta)$. The variation in volume during expansion from inlet to exit (i.e. v_1 to v_4) can be derived by the variable extended length of vane as shown in Figure 3 at every point of movement between two consecutive vanes.

From **Fig. 4**, it is seen that when two consecutive vanes at OK and OL moves to position OH and OB, the extended vane lengths varies from SK to IH and LM to BG, thus the variable length BG at variable α_i is assumed as $X_{at'variable'\alpha}$ can be written from the geometry:

$$BG = X_{at,variable"\alpha"} = (1/2).D.cos\left[\sin^{-1}\left\{\left(\frac{D-d}{D}\right).\sin\alpha\right\}\right] + (1/2).(D-d).\cos\alpha - d/2 \quad (7)$$

where D is diameter of casing and d is diameter of rotor, α is angle $\angle$BOF, β is angle $\angle$BAF and θ is angle $\angle$HOB or $\angle$KOL, between two consecutive vanes and ϕ is angle $\angle$KOJ at which injection pressure enters the air turbine.

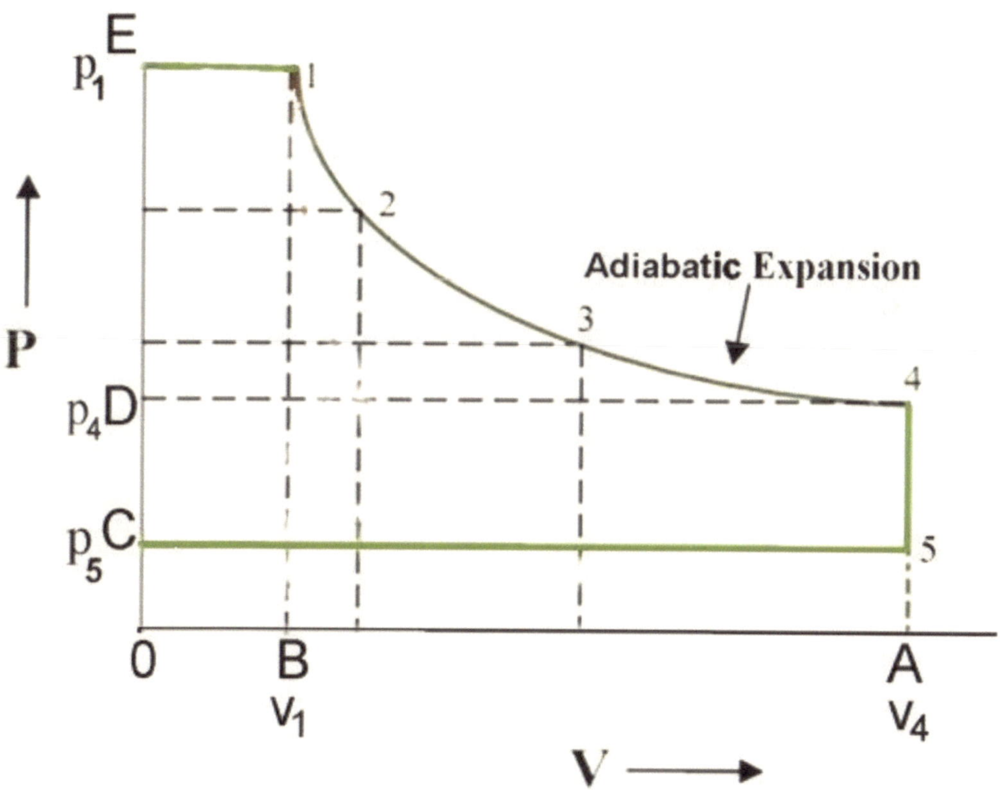

Fig. 3. Thermodynamic Processes (Isobaric, adiabatic and Isochoric Expansion)

Variable volume of cuboids B-G-I-H-B can be written as:

$$v_{cuboids} = L.\left\{\frac{(X_{1i} + X_{2i})(d + X_{1i})}{4}\right\}.\sin\theta \tag{8}$$

Where BG= X_{1i} and IH= X_{2i} are variable projected lengths of vanes when rotor rotates in the turbine as shown in **Fig. 4**.

The lengths (IG, HB and LK, SM.), are considered linear whereas all are chords of circles. This approximation is done in mathematical model which has very least impact on the overall values.

The volume at inlet v_1 or v_{min} will fall between angle $\angle$ LOF= $\alpha_{1min} = (180 - \theta - \phi)$ and angle $\angle$ KOF= $\alpha_{2min} = (\alpha_{1min} + \theta) = (180 - \phi)$ as seen in **Fig. 3**, when air is admits into turbine at angle ϕ .

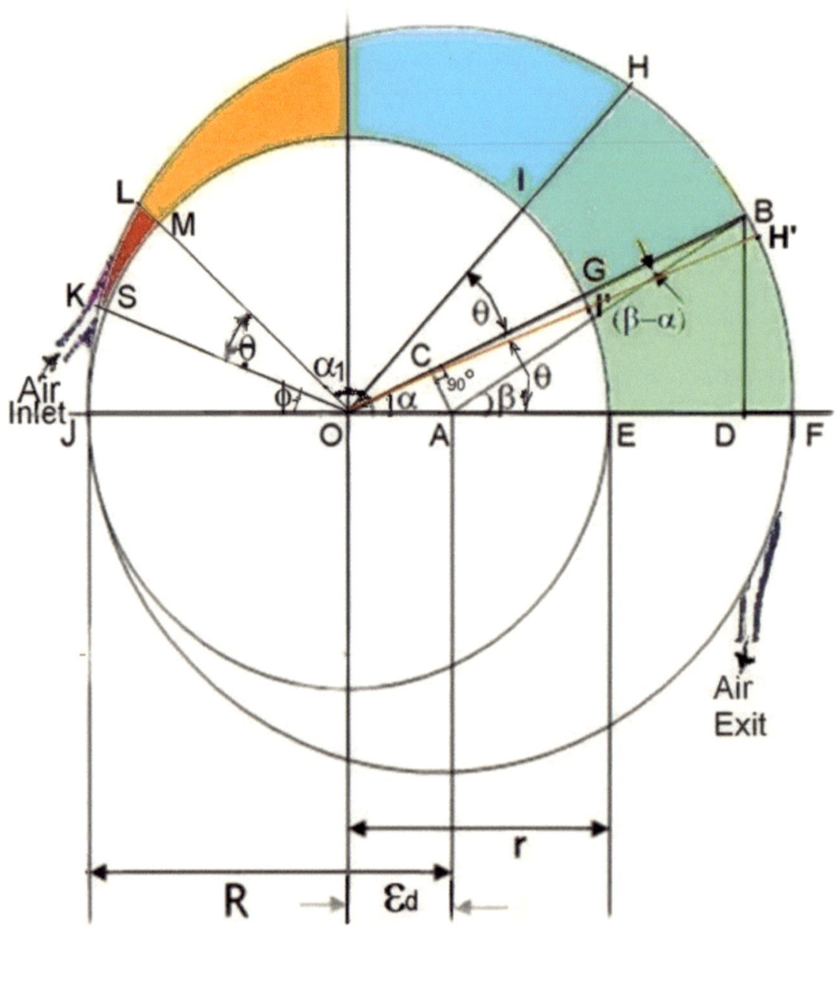

Fig. 4. Variable length BG and IH and injection angle ϕ

Applying above conditions into equations (7), then LM=X_{1min} and SK=X_{2min} can be written as:

$$X_{1\min} = R.cos\left[\sin^{-1}\left\{\left(\frac{R-r}{R}\right).\sin(180-\theta-\phi)\right\}\right] + \left[(R-r).\cos(180-\theta-\phi)-r\right] \quad (9)$$

$$X_{2\min} = R.cos\left[\sin^{-1}\left\{\left(\frac{R-r}{R}\right).\sin(180-\phi)\right\}\right] + \left[(R-r).\cos(180-\phi)-r\right] \quad (10)$$

Applying values of X_{1min} and X_{2min} to equation (8),

$$v_1 = v_{\min} = L.\left\{\frac{(X_{1\min}+X_{2\min})(2r+X_{1\min})}{4}\right\}.\sin\theta \quad (11)$$

The Volume at exit v_4 or $v_{\max}$ will fall between angle $\angle$ BOF $\alpha_{1\max} = \alpha = 0$ and angle $\angle$ HOF $\alpha_{2\max} = (\alpha_{1\max} + \theta) = \theta$

Applying above conditions into equations (7), then FE=X_{1max}=Corresponding to BG at α =0 degree and I'H'=X_{2max}=Corresponding IH at $(\alpha+\theta)=\theta$ degree can be written as:

$$X_{1\max} = (D-d) = 2(R-r) \quad (12)$$

$$X_{2\max} = R.cos\left[\sin^{-1}\left\{\left(\frac{R-r}{R}\right).\sin\theta\right\}\right] + \{(R-r).\cos\theta\} - r \quad (13)$$

Applying values of X_{1max} and $X_{2\,max}$ to equation (8),

$$v_4 = v_{\max} = L.\left\{\frac{(X_{1\max}+X_{2\max})(2r+X_{1\max})}{4}\right\}.\sin\theta \quad (14)$$

Substituting values of v_1 and v_4 from equations (11) and (14) to equation (6), the total power output available W_{total} can be written as:

$$W_{total} = n.(N/60).\left(\frac{\gamma}{\gamma-1}\right).\left\{1-\left(\frac{p_4}{p_1}\right)^{\frac{\gamma-1}{\gamma}}\right\}p_1.\left[L.\left\{\frac{(X_{1\min}+X_{2\min}).(2r+X_{1\min})}{4}\right\}.\sin\theta\right]$$
$$+n.(N/60).(p_4-p_5).\left[L.\left\{\frac{(X_{1\max}+X_{2\max}).(2r+X_{1\max})}{4}\right\}.\sin\theta\right] \quad (15)$$

4. Input parameters and assumptions

In this study the vane angle (θ) of air turbine is considered 30°, 45°, 51.4°, 60°, 72° and 90° (i.e. 12, 8, 7, 6, 5 and 4 vanes respectively) and injection angle (ϕ) is kept 30°, 45° and 60°. The outer diameter of Casing is considered 200 mm, 150 mm and 100 mm and correspondingly Rotor diameter is chosen 150 mm, 100 mm and 75 mm correspondingly.

Other various input parameters are listed in **Table 2, 3, 4** for investigation of optimum shaft power output at different vane angles, and injection angles.

4.1 To Optimize the power by selecting various options of vane angles (θ)

4.1.1 Input parameters for investigation of effect of vane angle

Symbols	Parameters
$D=2R$	150 mm (outer)
$d=2r$	100 mm (inner) corresponding
p_1	2bar (=30psi), 3bar (=45 psi), 4bar (=60 psi), 5bar (=75 psi), 6bar (=90 psi)
p_4	$=(v_1/v_4)^{\gamma}.p_1$ $> p_5$ assuming adiabatic expansion
p_5	1 atm = 1.0132 bar
θ	30 °, 36 °, 45 °, 51.4 °, 60 °, 72 °, 90 ° angles between 2- consecutive vanes (i.e. rotor contains correspondingly 12, 10, 8, 7, 6, 5, 4 vanes)
N	2500 rpm
L	35 mm length of rotor
γ	1.4 for air
n	Number of vanes (360 / θ)
Ø	Injection angles at which compressed air enters through nozzle into rotor *Case-1:* 30°, *Case-2*: 45° and *Case-3:* 60°

Table 2. Input Parameters for Investigation of Vane angles

4.1.2 Results and discussion

From **Fig. 5** to **Fig. 7**, it is noted that there is an optimum power output value for each set of injection pressure for different vane angles. Thus the optimum total power output is obtained as:

Case-1. For injection angle 30°

Total work power from **Fig. 5,** is seen to increase with increasing vane angles from 36° to 60° or with the decrease in number of vanes from 12 to 6 on rotor and thereafter it declines from vane angles (θ) = 72° to 90°. With increase in injection pressure the work output increases gradually as shown in graphical patterns. Thus the optimal total shaft output is found at vane angle (θ) = 60° (i.e. 6- vanes).

Case-2. For Injection angle 45°

Similarly, total power output from **Fig. 6,** is seen to increase with increasing vane angle up to 36° to 51.4° or with the decrease in number of vanes from 12 to 7 on rotor and thereafter it declines from vane angles (θ) = 60° to 90°. With increase in injection pressure the work output increases gradually as shown in graphical patterns. Thus the optimal total shaft output is found at vane angle (θ) = 45° and 51.4° (i.e. 8-7 vanes).

Case-3. For Injection angle 60°
Here also the total power output from **Fig. 7**, is seen to increase with increasing vane angle up to 36° or with the decrease in number of vanes from 12 to 10 on rotor and thereafter it declines for vane angles (θ) = 45° to 90°. With increase in injection pressure the work output increases gradually as shown in graphical patterns. Thus the optimal total shaft output is found at vane angle (θ) = 36° (i.e. 10 vanes).

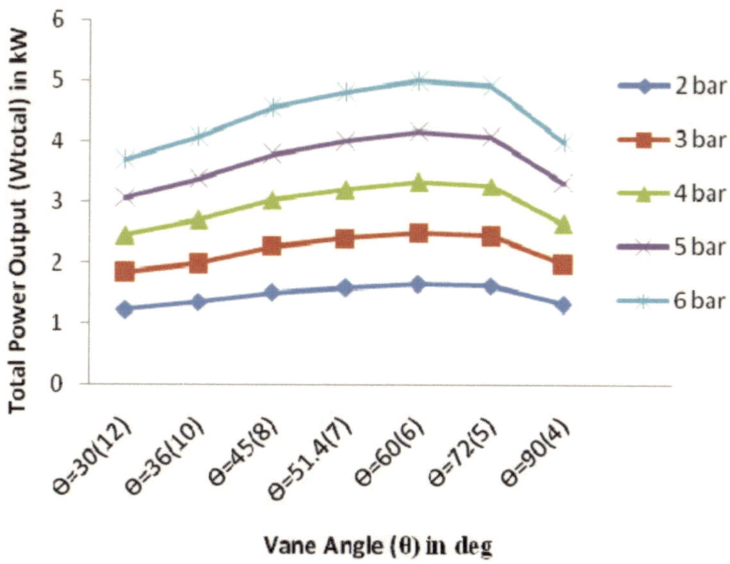

Fig. 5. Total power output versus different vane angles, at injection air pressure 2-6 bar, injection angle of 30° and speed of rotation 2500 rpm

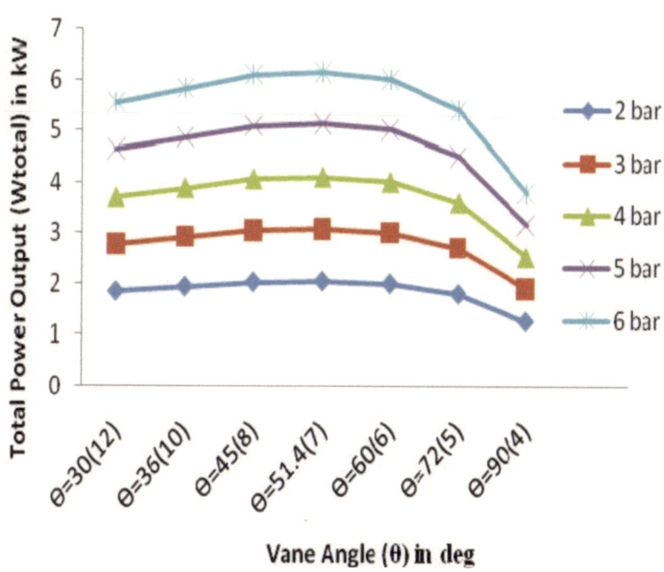

Fig. 6. Total power output versus different vane angles, at injection air pressure 2-6 bar, injection angle of 45° and speed of rotation 2500 rpm

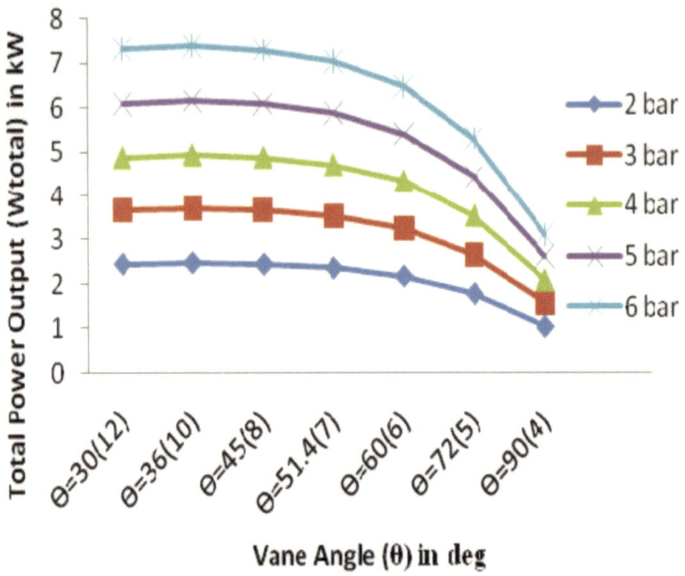

Fig. 7. Total power output versus different vane angles, at injection air pressure 2-6 bar, injection angle of 60° and speed of rotation 2500 rpm

Thus, total optimum power is observed as:
Case-1: 5.00 kW at vane angle (θ) = 60° (i.e. 6 vanes) when injection angle is kept 30°,
Case-2: 6.1-6.15 kW at vane angle (θ) = 45°, 51.4° (i.e. 8-7 vanes) when injection angle is 45° and
Case-3: 7.39 kW at vane angle (θ) = 36° (i.e. 10 vanes) when injection angle is 60°.

4.2 To optimize the power by selection of injection angle (Ø)
4.2.1 Input parameters for investigation of injection angle

Symbols	Parameters
D=2R	200 mm (outer)
d=2r	160 mm (inner) corresponding
p_1	2 bar (=30 psi), 3 bar (=45 psi), 4 bar (=60 psi), 5 bar (=75 psi), 6 bar (=90 psi)
p_5	1.0132 bar (atmospheric pressure)
p_4	$(v_1 / v_4)^{\gamma} .p_1 > p_5$ assuming adiabatic expansion
θ	45° (i.e. rotor contains correspondingly 8 number vanes)
N	2500 rpm (as total power is directly proportion to rpm)
L	35 mm length of rotor
γ	1.4 for air
n	Number of vanes = (360 / θ)
ϕ	10°, 15°, 20°, 25°, 30°, 45°, 60°, 75°, 90° angles at which compressed air enters through nozzle into rotor

Table 3. Input Parameters for Investigation of Injection angle

4.2.2 Results and discussion

From the **Fig. 8** it is observed that total power output is seen to increase with small pace at increasing injection angles from 10° to 30° at 2 bar admission pressure, it gradually increases with increase in injection angles from 15° to 60° and thereafter it suddenly decreases with increase in injection angles from 60° to 90°. Also total power output is seen to be large for higher injection pressure 3 - 6 bar and follow the same trend as shown in graphical patterns, which attributes to the large power capacity.

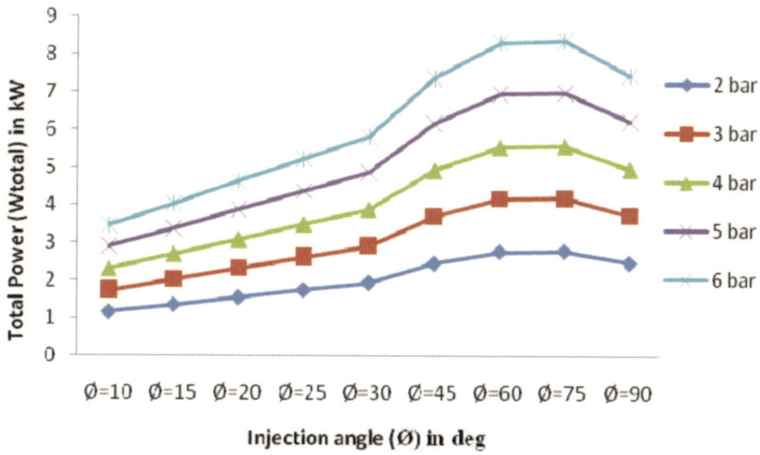

Fig. 8. Total turbine power output versus different injection angles, at injection air pressure 2-6 bar and speed of rotation 2500 rpm

It is also observed that the multi-vane turbine develops maximum shaft power output when injection angle is kept 60° or 75° for all injection pressures in the range of 2-6 bar.

4.3 To optimize the power by critically selecting vane and injection angles

4.3.1 Input parameters for investigation of vane and injection angle

Symbols	**Parameters**
D=2R	150 mm (outer)
d =2r	100 mm (inner) corresponding
p_1	2bar (=30psi), 3bar (=45 psi), 4bar (=60 psi), 5bar (=75 psi), 6bar (=90 psi)
p_4	$=(v_1 / v_4)^{\gamma} . p_1$ assuming adiabatic expansion
p_5	1 atm = 1.0132 bar
θ	30 °, 36 °, 45°, 51.4°, 60 °, 72 ° and 90 ° angles between 2- consecutive vanes (i.e. rotor contains correspondingly 12, 10, 8, 7, 6, 5 and 4 vanes)
N	3000 rpm
L	45 mm length of rotor
γ	1.4 for air
n	(360 / θ) Number of vanes
Ø	30 °, 45 ° and 60 ° Injection angles at which compressed air enters through nozzle into rotor

Table 4. Input Parameters for Investigation of Vane and Injection Angles Both

4.3.2 Results and discussion

From **Fig, 9**, it is noted that optimum total power output is obtained as:

i. 5.0 kW when vane angle (θ) = 60° (i.e. 6 vanes) and injection angle (Ø) = 30°,

ii. 6.1-6.15 kW when vane angle (θ) = 45°-51.4° (i.e. 8-7 vanes) and injection angle (Ø) = 45 ° and

iii. 7.39 kW at vane angle (θ) is 30°-45° (i.e. 12-8 vanes) and injection angle (Ø) = 60°.

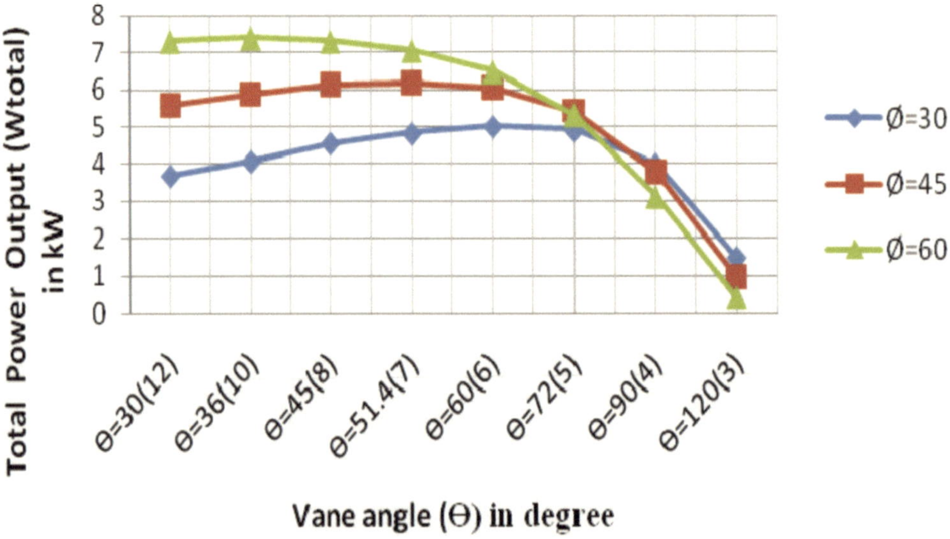

Fig. 9. Total power output versus different vane angles at air injection pressure 6 bar, at injection angle of 30°, 45°, 60°, and speed of rotation 3000 rpm

Thus the maximum shaft power output 6.1 kW to 7.39 kW is achievable at 60° injection angle when number of vanes of rotor is kept between 12-8 (vane angle is 30°-45°). Such type turbine can be used as replacement to internal combustion engines and it can easily be used to run light vehicle or motorbikes with zero pollution.

5. Conclusions

From the above study, it is noticed that the heavy consumptions of fossil fuel in the transport sector is causing serious threat to mankind causing global warming and damage to the environment and ecology. Following conclusions are observed from the study:

- Fossil fuel reserves are depleting due to growing energy requirements.
- Vehicular pollution is estimated to increase many fold in near future.
- Energy storage system could be utilized for the clean energy future.
- The shaft power output of small air engine is achievable from 6.1 kW to 7.39 kW, which is sufficient to run motorbikes or any light vehicle.
- Compressed air technology is going to be the most economical and dominating energy source for running light vehicles / motorbike as tail pipe emission is zero.

The air engine technology, if implimented widely in developing countries to run the light vehicles / motobikes, it might reduce 50-60% global warming issues which is a big challenge of 21st century.

6. Acknowledgement

The authors would like to express their sincere thanks to the SMS Institute of Technology, Lucknow and Harcourt Butler Technological Institute, Kanpur for extending the support of laboratory setup and other facilities.

7. Nomenclature

d	: diameter of rotor (2r) in meter
D	: diameter of outer (2R) cylinder in meter
L	: length of rotor having vanes in meter
n	: no. of vanes = $(360/\theta)$
N	: no. of revolution per minute
p	: pressure in bar
$p_1 v_1$	: pressure and volume respectively at which air strike the Turbine
p_4, v_4	: pressure and volume respectively at which maximum expansion of air takes place
p_5	: pressure at which turbine releases the air to atmosphere.
v	: volume in cum
w	: theoretical work output in (J) Joules
W	: theoretical power output (W) Watts
X_{1i}	: variable extended lengths of vane at point 1
X_{2i}	: variable extended lengths of vane at point 2

Subscripts

1, 2...4, 5	: subscripts – indicates the positions of vanes in casing
exp	: expansion
min	: minimum
max	: maximum

Greek symbols

α	: angle BOF (see Fig.3)
α_1	: angle LOF (=180 - ϕ) (see Fig.3)
α_2	: angle KOF (=180 - θ - ϕ) (see Fig.3)
β	: angle BAF (see Fig.3)
γ	: 1.4 for air
θ	: angle between 2-vanes (BOH) (see Fig.3)
ϕ	: angle at which compressed air enters into rotor through nozzle
ξ_d	: eccentricity *(R-r)*

8. References

Aleklett K. and Campbell C.J., 2003, The Peak and Decline of World Oil and Gas Production- Minerals and Energy, Raw Materials Report, Volume 18, Number 1, 2003, pp. 5-20(16).

Architectural Institute, 2004, Concept and Technology- First International Workshop on Sustainable Habitat Systems – Kyushu Branch, Japan, December, 10, 2004 (10:00-18:00), Venue: ACROS Fukuoka.

Fuglsang P., Bak C. and Gunna M., 2004, Design and verification of the Ris0-B1 Airfoil-family for Wind Turbines, *Journal of Solar Energy Engg.*, ASME- Nov'2004, Vol.126 pp 1002-1008.

Gorla R., and Reddy, S., 2005, Probabilistic Heat Transfer and Structural Analysis of Turbine Blade, *IJTJE*, Vol. 22, pp 1- 11.

Honton E. J., 2004, Hydrogen Fuel Cell Car, presented at *15th Annual US Conference and Hydrogen Expo*, April'2004, USA.

Hubbert, M. K., 1956, Nuclear energy and the fossil fuels, Amer., Petrol Inst. Drilling and Production Practice, *Proc. Spring Meeting*, San Antonio, Texas, pp 7-25.

Knowlen C., Bruckner A. P., Mattick A.T. and Hertzberg A., 1998, High Efficiency Energy Conversion Systems for Liquid Nitrogen Automobiles, *Society of Automotive Engineers, Inc., AIAA* 98-1898.

Negre Guy and Negre Cyril, 2004, Compressed Air-The Most Sustainable Energy Carrier for Community Vehicles, Speech in front of assembly at Kultur gathered for *Fuel Cells World*, Tuesday 29th June '2004.

Royal Society London, (1687), The History of the Compressed Air Vehicles, *Website: http://www.aircarfactories.com/air-cars/compressed-air-history.html.*

Saint Hilaire G., Saint Hilaire R. and Saint Hilaire, Y., 2005, Quasiturbine zero pollution car using gasoline. *Festival at Le Lundi, Montreal Gazette*, 26 September 2005.

Schreck S. and Robinson M., 2004, Tip Speed Ratio Influences on Rationally Augmented Boundary Layer Topology and Aerodynamic Force Generation, *Journal of Solar Energy Engg.*, ASME-Nov' 2004-Vol.126 pp1025-1033.

Selig Michel S. and Bryan D. McGranahan, 2004, Wind Tunnel Aerodynamics Tests of Six Airfoils for use on Small Wind Turbines, *Journal of Solar Energy Engg.*, ASME-Nov'2004, Vol.126, pp986-1000.

Singh B.R. and Singh Onkar, 2006, Necessity and Potential for Bio-Diesel Use in India, *International Conference on Bio-Fuel Vision-2015*, October'13th -15th, 2006 at Bikaner, India- Proceedings pp 71-89.

Singh B.R. and Singh Onkar, 2007, Uses of Wind Power as a Non-Conventional / Renewable Energy for Sustainability, *National Conference on State of Art Technology in Mechanical Engineering, STEM-2007*, held on October 29-31, 2007 at College of Technology, G.B. Pant University, Pant Nagar, UP-Proceedings pp 503-515.

Singh B.R. and Singh Onkar, 2008a, A concept for Development of a Vaned Type Novel Air Turbine, *12th International Symposium on Transport Phenomena and Dynamics of Rotating Machinery* - held on February 17-22, 2008 at Pacific Center of Thermal-Fluids Engineering, Sheraton Mohana Surfrider Hotel Honolulu, Hawaii - Paper No. ISROMAC-12-20046.

Singh B.R. and Singh Onkar, 2008b, A Study to Optimize the Output of Vaned Type Novel Air Turbine, *4th International Conference on Energy Research and Development*, held on 17-19 November, 2008 at State of Kuwait, Kuwait- Paper No. ICERD - 4 -1353.

Singh B.R. and Singh Onkar, 2008c, Development of a vaned type novel Air Turbine, International Journal of Mechanical Engineering Science (The manuscript was received on 21st December 2007 and was accepted after revision for publication on 03rd June 2008), International Journal of IMechE, Part C, Vol.222, 2008, pp 2419-2426.

Singh B.R. and Singh Onkar, 2008d, Energy Storage System to meet Challenges of 21st Century- an Overview, *All India Seminar on Energy Management in Perceptive of Indian Scenario*-held on October 17-19, 2008 at Institution of Engineer (India), State Centre, Engineer's Bhawan, Lucknow-Proceedings Chapter15, pp 157-167.

Singh B.R. and Singh Onkar, 2008e, Parametric Evaluation of Vane Angle on performance of Novel Air Turbine, *Journal of Science, Engineering and Management*, SITM , December, 2008,Vol. 2, pp 7-18.

Singh B.R. and Singh Onkar, 2009a, Analytical Study on a Vaned Type Novel Air Turbine for Different Conditions of Casing and Rotor Diameters, *2009 ASME International Conference on Energy Sustainability* - held on July 17-23, 2009, at San Francisco, California, USA, Paper No. ES2009 -90207.

Singh B.R. and Singh Onkar, 2009b, Applications of Compressed Air as an Alternative Energy to Meet Challenges of 21st Century- Global Warming, *International Conference on Engineering Congress on Alternatives Energy Applications: Option or Necessity?*, held on 3-5 November, 2009 at State of Kuwait, Kuwait- Paper No. EC2009 Kuwait-1082.

Singh B.R. and Singh Onkar, 2009c, Numerical Analysis of Pressure Admission Angle to Vane Angle Ratios on Performance of a Vaned Type Novel Air Turbine, *International Journal of Natural Science and Engineering, IJNSE*, France, Vol. 1-1-2009, pp 20-27.

Singh B.R. and Singh Onkar, 2009d, Optimization of Power Output of a Vaned Type Novel Air Turbine With Respect to Different Injection Angles-Under Ideal Adiabatic Expansion, *International Journal of Mechanical Engineering, Serials Publications, New Delhi, India*, Vol. 2 Number 2, July-December'2009, pp 205-211.

Singh B.R. and Singh Onkar, 2009e, Parametric Evaluations of Injection Angles and Vane Angles on Performance of a Vaned Type Novel Air Turbine, *International Journal of Mathematical, Physical and Engineering Science, IJMPES*, France, Vol. 3-4-2009, pp 226-233.

Singh B.R. and Singh Onkar, 2009f, Theoretical Investigations on Different Casing and Rotor Diameters Ratio to Optimize Shaft Output of a Vaned Type Air Turbine , *International Journal of Natural Science and Engineering, IJNSE*, France, Vol. 1-1-2009, pp 28-35.

Singh B.R. and Singh Onkar, 2010a, Effect of Different Vane Angle on Rotor - Casing Diameter Ratios to Optimize the Shaft Output of a Vaned Type Novel Air Turbine, *International Journal of Engineering Science and Technology, Chennai, India, IJEST*-ISSN-0975-5472,Vol. 2, Number 3 (2), pp 114-121.

Singh B.R. and Singh Onkar, 2010b, Study of Effect of Injection Angle to Rotor-Casing Diameter Ratios on Performance of a Vaned Type Novel Air Turbine, *International Journal of Engineering Science and Technology, Chennai, India, IJEST*-ISSN-0975-5472,Vol. 2, Number 4 (10), pp 409-417.

Singh B.R. and Singh Onkar, 2010c, Critical Effect of Rotor Vanes with Different Injection Angles on Performance of a Vaned Type Novel Air Turbine"- *International Journal of Engineering and Technology, Chennai, India,* IJET-ISSN: 0975-4024, Vol. 2 Number 2(28), 2010, pp. 118-123.

Singh B.R. and Singh Onkar, 2010d, Study of Effect of Rotor Vanes to Rotor-Casing Dimensions on Performance of a Zero Pollution Vane Type Novel Air Turbine,

International Journal of the Physical Sciences, 5170-00200, Nairobi-73023 Victoria Island, Lagos, Vol.5(5), 2010, pp. 547-556..

Singh B.R. and Singh Onkar, 2010e, Analytical Investigations on Different Air Injection Angles to Optimize Power Output of a Vaned Type Air Turbine, *International Journal of Power and Energy* (The manuscript was received on 11th June' 2009 and was accepted after revision for publication on 07th October' 2009), *Proceedings of* IMechE, Part A, Vol.224, 2010, pp 305-311.

Singh B.R. and Singh Onkar, 2010f, Compressed Air Energy Storage Power Plant and its Feasibility in India- An Overview, *Proceedings of National Conference on Advances in Management of Energy Efficiency and Clean Environment,* April 3-4, 2010, pp 55-61.

Singh B.R. and Singh Onkar, 2010g, Study of Influence of Vane Angle on Shaft Output of a Multi Vane Air Turbine, *International Journal of Renewable and Sustainable Energy,* AIP, New York, USA, Vol.2 Number 3, pp. 033101-16.

Singh B.R. and Singh Onkar, 2010h, Performance Investigations for Power Output of a Vaned Type Novel Air Turbine, *MIT- International Journal of Mechanical Engineering,* Moradabad, India, ISSN : 2230-7699; (Accepted: Dec. 17, 2010; Published: Jan 11. 2011), Vol.1(1),2011, pp 9-16 .

Singh Onkar, 2009, Reciprocating and Rotary Compressor, Applied Thermodynamics, *New Age International (P) Ltd., Publishers, New Delhi, India,* ISBN: 978-81-224-2583-3, Feb2009, pp 727-728.

9

Air Pollution in the Niger Delta Area: Scope, Challenges and Remedies

Godson Rowland Ana
Department of Environmental Health Sciences, Faculty of Public Health, University of Ibadan, Nigeria

1. Introduction

Air pollution is one of the major environmental problems confronting the Niger Delta Area (NDA) yet information regarding this is very scanty. Aside from data collected by a few individuals and corporate organizations at scattered locations, there is no comprehensive and empirical database on the magnitude of the hazard and its deleterious effects on the ecosystems and people in the region. Also the existing network of meteorological stations is too coarse to provide data covering the whole of the region. The NDA is Nigeria's most endowed region in terms of oil mineral reserves and one of the most industrialized after Lagos, Nigeria's former administrative capital. The operations of these industries especially the upstream and downstream petroleum sectors as well as a variety of other anthropogenically related activities including biomass combustion, refuse burning and traffic emissions releases a barrage of substances like volatile organics, oxides of carbon, nitrogen, sulphur, particulate matter, heavy metals and other toxics at levels that most times exceed both the national and international guidelines. Apart from compromising the quality of the atmosphere most of the air pollution in the region is observed to have local and regional effects such as the formation of acid rain, water pollution, soil pollution, impacts on plants and wild life, effects on materials and artifacts and recently contributing to the global warming effects. While most of these claims may be true a great majority has not been substantiated scientifically and is rather based on assumptions or being speculative. There are numerous challenges facing air quality studies in NDA. Like in most developing nations the issues are multifaceted and the most acute have been lack of equipment, inadequate expertise, lack of infrastructure and weak policy frame work. This and many other factors has been the bane to achieving the much desired goals towards preserving the quality of the ecosystems and safeguarding the health of the public in the region. In this chapter a conscious effort has been made to give a brief description of the Niger delta area, the possible sources of air pollution in the region with highlights of a few case scenarios of air pollution from community automobile traffic and industries as well as common related health problems. Again as a way forward to most of the problems and drawbacks hitherto mentioned several models and schemes have been advocated.

2. The Niger Delta area

The Niger Delta area in Nigeria (Fig 1) is situated in the Gulf of Guinea between longitude 50E to 80E and latitudes 40N to 60N.It is the largest wetland in Africa and the third largest in the world consisting of flat low lying swampy terrain that is criss-crossed by meandering and anastomosing streams, rivers and creeks. It covers 20,000 km^2 within wetlands of 70,000 km^2 formed primarily by sediment deposition. It has an equatorial monsoon climate influenced by the south west monsoonal winds (maritime tropical) MT airmass coming from the South Atlantic Ocean. It is home to 20 million people drawn from nine states namely Abia, Akwa Ibom, Bayelsa, Cross River, Delta, Edo, Imo, Ondo and Rivers states with 40 different ethnic groups. This floodplain makes up 7.5% of Nigeria's total land mass. The Delta's environment can be broken down into four ecological zones: coastal barrier islands, mangrove swamp forests, freshwater swamps, and lowland rainforests. This incredibly well-endowed ecosystem, contains one of the highest concentrations of biodiversity on the planet, in addition to supporting abundant flora and fauna, arable terrain that can sustain a wide variety of crops, lumber or agricultural trees, and more species of freshwater fish than any ecosystem in West Africa. The vegetation of the Niger Delta consists mainly of forest swamps. The forests are of two types, nearest the sea is a belt of saline/brackish Mangrove swamp separated from the sea by sand beach ridges. Numerous sandy islands occur with fresh water vegetation. Fresh water swamps gradually supersede the mangrove on the landward side. More than 70% of Nigeria's crude oil and gas production is from the area. The region produces over 90% of Nigeria's foreign earnings through oil exploration activities. It plays host to most of the upstream and downstream oil related industries and non oil related industries that release tons of pollutants into the ecosystems. The pollution from the Niger Delta on a scale could be regarded as one of the worst among similar delta areas in the world.

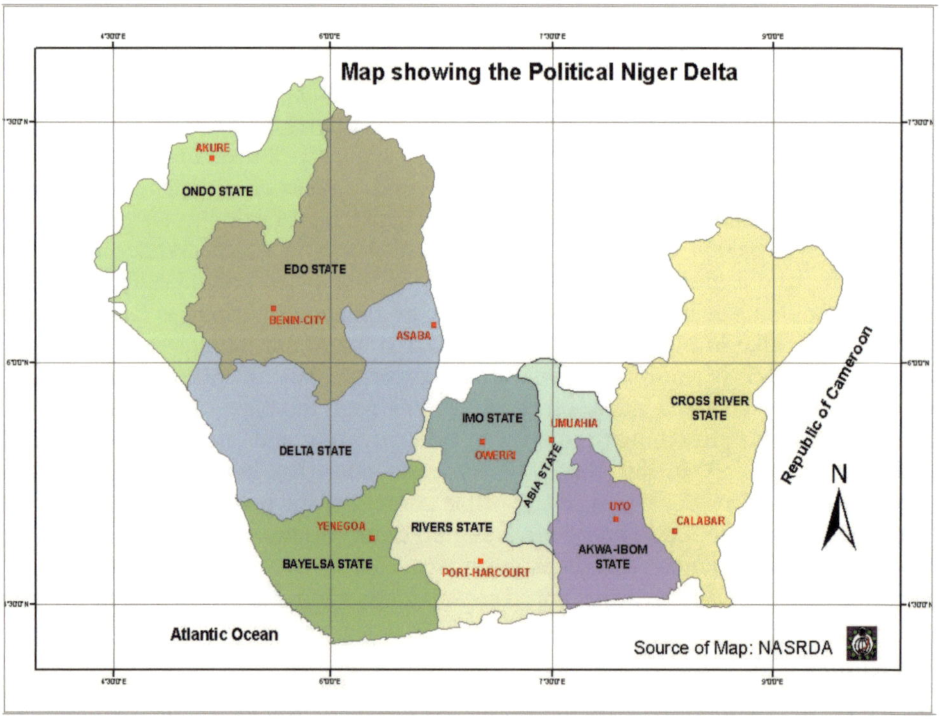

Fig. 1. Political Map of the Niger Delta Area

3. Air pollution sources

3.1 Biomass combustion

Biomass in form of firewood, coal, bamboo trunks and dead leaves are commonly used sources of cooking fuel in the region. Of all these the most frequently used is firewood. This fuel woods are usually logged from nearby bushes and forests or are collected as dead branches within the residential vicinities by the women and sometimes children and are used to generate energy for cooking(Pate 1). The combustion of firewood releases gaseous pollutants and particulate matter. From literature the gaseous pollutants from cooking emissions are carbon monoxide CO, carbon dioxide CO_2, sulphur dioxide SO_2, nitrogen dioxide NO_2, volatile organic compounds VOCs and particulate matter. The particulate matter generated is in the form of carbon black, sooth and fly ash which are major components of smoke and are most often within the 10μm size range.

Plate 1. Cooking Fuel emissions from biomass combustion

3.2 Bush burning

Bush burning is a common phenomenon in the Niger delta communities. Most times the burning of bush is done prior to land cultivation as one of the initial steps of land preparation. This constitutes part of the pre planting preparation. The process of bush burning leads to the release of various types of gaseous pollutants and particulate matter. Very often the gas stream is inundated with volatile organics and oxides of carbon (COx), sulphur(SOx) and nitrogen(NOx) depending on the fuel composition and intensity of the flame. Particulate matter usually within the 10μm size range is also produced in the course of the combustion process.

3.3 Refuse burning

Refuse disposal is a major environmental problem in the Niger delta communities particularly in the urban areas (Plate 2). The refuse is usually from multiple sources including domestic, municipal, agricultural and industrial sources. One of the environmentally unfriendly methods of managing the waste is by open burning either on nearby lands or open dumps within the residential vicinities. The composition of the refuse, age of the dump and intensity of the flame usually determines the nature of the air pollutants. Often times the air within refuse burning sites is inundated with VOCs, COx, SOx, NOx, total hydrocarbons(THCs), as well as various classes of toxic and hazardous compounds viz polycyclic aromatic hydrocarbons(PAHs), dioxins, PCBs(Polychloro Biphenyls) and heavy metals such as lead, nickel and mercury.

Plate 2. Emissions from open burning of solid waste

3.4 Traffic emissions

Over 600 million people globally are exposed to hazardous level of traffic - generated pollutants UN, (1998). Human exposure to these air pollutants is believed to have posed severe health problems especially in urban areas where pollution levels are on the increase. Pollution due to traffic constitute up to 90 - 95% of the ambient CO levels, 80 - 90% of NOx, hydrocarbon and particulate matter in the world, posing a serious threat to human health Savile, (1993). Research has shown that transportation sources in the USA were responsible for 77% of CO levels, 80 -90% of NOx, 36% of volatile organic compounds and 22% of particulate matter USEPA, (1993). Similarly, in UK the average concentration of NO_2 was found to increase by 35% from 1986 to 1991 due to increase in vehicular emission CEC, (1992). On the global scene, Seneca and Tausig, (1994) concluded that transportation is the major culprit of air pollution accounting for over 80% of total air pollutants.
In Nigeria much attention is focused on general industrial pollution and pollution from the oil industries, with little attention on the effects of air pollution from mobile transportation sources (Faboye, 1997; Iyoha, 2000 and Magbabeola, 2001).Increased pollution from mobile

Plate 3. Traffic emissions from busy urban roads

sources is on the increase with per capita increase in vehicle ownership(Plate 3). The consequence of this is the congestion of most Nigeria city roads and a corresponding increase in the burden of air pollutants and their associated effects. Studies conducted by Akpan and Ndoke, (1999) in Northern Nigeria show higher values of CO_2 concentration (1780ppm- 1840ppm) in heavily congested areas in Kaduna and (1160ppm-1530ppm) in Abuja. A study of the impacts of urban road transportation on the ambient air was conducted by Koku and Osuntogun, (1999) in three cities in south western Nigeria. Air quality indicators namely CO, SO_2, NO_2, and total suspended particulates (TSP) were determined. The highest levels obtained for the air pollution indicators were CO- 233ppm , SO_2-2.9ppm , NO_2-1.5ppm and total particulates 852ppm in Lagos . In Ibadan the highest levels obtained were CO-271ppm, SO_2 -1.44 ppm and NO_2- 1.0ppm. In Ado-Ekiti the highest levels obtained were CO-317ppm, NO_2 -0.6ppm and SO_2-0.8ppm. These results were found to be higher than FEPA limits for CO-10ppm, SO_2-0.01ppm, NO_2-0.04-0.06ppm.

A comparative study of emission levels in Lagos and the Niger Delta area was reported by Jerome, (2000). Two major cities, Port-Harcourt and Warri were considered for the NDA. The results obtained showed that the concentrations of TSP, NOx, SO_2, and CO in Lagos and the Niger Delta communities were above the FEPA recommended limits. The CO levels for Lagos 10 - 250ppm were higher than the levels 5.0 - 61.0ppm and 1.0 - 52ppm recorded in the two selected communities in the Niger Delta. The TSP concentrations were also high for both locations when compared to WHO standard. A similar traffic related emission study was carried out in Calabar, a major city in NDA and the results presented in the box indicate that the overall average value(OAV) of each pollutant for the 72 hours experimental duration was found to be in the following range: CO :5.0 - 6.1ppm, NO_2 :0 -0.05ppm, PM10: 202 - 230$\mu g/m^3$.

Lagos Area Pollutant	Non-Traffic Urban zone	Traffic Zone	Niger Delta Area Oil Communities	Cities	FEPA Standards
TSP μ/m^3	31.4 -746.5	72 – 950	92.2 – 348.5	396.8 – 583.3	250
NOX(ppm)	81 – 81.5	34 – 131.6	22.0 – 295.0	35 – 370	40 – 60
SO_2(ppm)	0.5 – 43	20 – 250	7.0 – 97.0	16 – 300	100
CO(ppm)	0.5 – 3.9	10 – 250	5.0 – 61.0	1.0 – 52	10
CO/NO_X(ppm)	0.0 – 6.0	50 – 200	20	15 – 130	-

Source: Jerome 2000

Table 1. Ambient air pollutants in Lagos and Niger Delta Area

At the National level, available data on the total number of vehicles registered in Nigeria shows an increase from 38,000 to 1.6 million between 1950 and 1992(Enemari, 2001). Data from the Federal Road Safety Commission (FRSC) of Nigeria however, indicates that between 1999 and 2004 about six million vehicles (6,000,000) were registered in Nigeria (Fig 2) of which 70% of the registered vehicles were cars and 30% busses and trucks. The reason for the large increase of registered vehicles in 1999 was due to the deadline given for the registration of all vehicles in the country by FRSC. The subsequent years indicate vehicles coming in for the first time into the country. The increase recorded in 2002 was the climax due to favourable government policies on importation of fairly used vehicles at the inception of democratic rule in Nigeria. Although there was a decline in 2004, it is obvious that the number of vehicles plying Nigerian roads had continued to increase. In Lagos alone about 223,764 vehicles were registered in 2008 compared to 160,134 vehicles registered in 2007. Also in the first quarter of 2009 more than 58,000 vehicles were registered and a total of 27,587 motorcycles were also registered between July 2006 and July 2009 (http//alafrica.com/stories). From the above data it is apparent that the average emission concentration from motor vehicles and motorcycles in the country is in the increase.

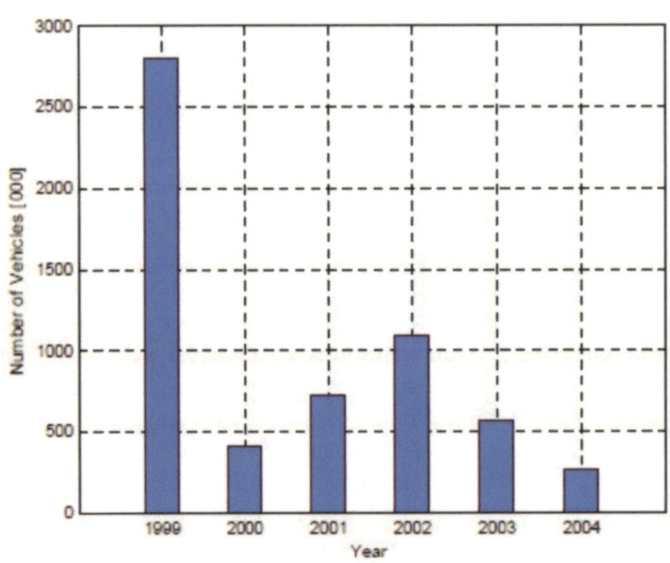

Fig. 2. Total number of vehicles Registered in Nigeria (1999-2004)

Box 1: Traffic Emission Scenario in Calabar

The overall average values (OAV) of each pollutant for the 72 hours experimental duration was found to be in the following range: CO: 5.0-6.1ppm, NO2 - 0 -0.05ppm, PM10 - 202 - 230 $\mu g/m^3$ and noise level was 60.5 - 63dB. Comparing these values with AQI CO, range from moderate to very poor; PM10 and SO_2 were very poor. The OAV value of NO_2 is constant for the three experimental days comparing with AQI rating the value is still poor. The reason for this constant OAV value of NO_2 could be the recent government policy of replacing the intercity transport system with new efficient cars with catalytic reduction system. The overall AQI rating for ambient air quality data indicates that the air quality standard in Calabar is below the USEPA standard for ambient air quality rating. .

3.5 Industrial emissions

Apart from Lagos which was the former political capital but still regarded as Nigeria's economic capital, the next most industrialized part of the country is the Niger Delta Area. The region boasts of oil and non oil related industries including refinery, petrochemical, liquefied natural gas, chemical fertilizer, aluminum smelter, paper, cement, flour, wood, battery and textile industries etc which emit various kinds of air pollutants (Box 2). The pollution from these industries adds to the burden of gaseous and particulate pollutants in the air. In this chapter information regarding air pollution from three of these industries namely the chemical fertilizer plant, petroleum refinery plant and the petrochemical complex would be reported.

Box 2: Major Air pollutants in NDA

Gases: CO_2, CH_4, SO_2, N_2O, NO_2, NH_3
VOCs

Particles: PM10, PM2.5, PAHs, Dioxins
PAHs, Heavy metals

3.5.1 Chemical fertilizer industry

The chemical fertilizer industry in Nigeria is still in its infant stage. The National Fertilizer Company of Nigeria (NAFCON) described here is located at Onne (4.49^0 and 4.50 N and 6.59^0 and 7.00 East of Greenwich Meridian), 30 km away from Port Harcourt in the Niger delta area. It utilizes a variety of raw materials such as natural or synthetic gas (methane), atmospheric nitrogen, steam and sand as filler material in the production of various formulations. The major products (per day) are ammonia (1,000 tonnes), urea (1,500 tonnes), and NPK formulations (1,000 tonnes).The main sections where gaseous and particulate emissions are reported are the ammonia plant, urea plant, NPK plant, bulk blending plant, bulk storage plant, and the bagging area. Potential sources for leaks in the ammonia plant are the refrigeration loop, storage areas, flanges, valve packing, and the pump and compressor seals. The gaseous emissions from the ammonia plant include reformer and boiler flue gases, excess carbon dioxide, condenser stripper vapour and ammonia discharge. The gases coming from the reformer are vented into the atmosphere with a mixture of CO_2 (20%) and O_2 (3-4%). Most of the CO_2 is recovered and recycled in the process. The condensate of the process water contained some CO_2, methanol and other dissolved gases, which are usually stripped. In the urea plant there are two possible sources of emissions, viz. the high-pressure scrubber and the granular stack. Some amounts of methane, hydrogen, ammonia and CO_2 are released. The wet scrubbers also release some particulate urea through the vent. In the NPK plant the tail gas scrubber is the only source of emission of ammonia, particulates and small quantity of fluoride. In various operations, e.g. granulation, drying, mixing and cooling, some amounts of dust and fumes are generated. In a study carried out by Ana et al,2005 the highest mean levels of ammonia (459.1ppm), was recorded at the Urea synthesis plant and the lowest levels (7.29ppm) recorded at a control point about 5km away from the point. For particulate matter, the highest concentration of 260875.7 $\mu g/m^3$ and the least concentration of 7.29ug/m^3 were recorded.

3.5.2 Refinery and petrochemical complex

The quality of air in Nigeria's largest oil refinery (PHRC) and petrochemical complex (EPCL) was assessed based on key priority pollutants in a study carried out by Ana et al(2009)(Plate 4).Three points each were observed per industrial location At the refinery, the highest PM10 level (130.3$\mu g/m^3$) was recorded while at the petrochemical complex the highest PM10 level (81.3 $\mu g/m^3$) was recorded. The study indicated that the average PM10 levels at PHRC were higher than that recorded at EPCL .In terms of the level of heavy metals, the highest Pb and Ni levels of 0.20 mg/m^3 and 0.86 mg/m^3 respectively was recorded at the PHRC. At the petrochemical complex, the highest Pb and Ni levels of 0.16 mg/m^3 and 0.05 mg/m^3 respectively was recorded at EPCL. Overall, the average heavy metal concentrations were higher at PHRC.The highest concentration of benzo (a) pyrene (1.63x102ng/m^3) was recorded at PHRC compared to (1.61x102 ng/m^3) that was recorded at EPCL. However, the total PAH concentration taken as the sum of benzo(a)pyrene and indeno(123)-cd pyrene was found to be higher at EPCL when compared with PHRC(Table 2).

3.6 Gas flaring

According to Cedigaz (2000), Nigeria holds the highest record (19.79%) of natural gas flaring globally and is responsible for about 46 % of Africa's total gas flared per tonne of oil produced(Plate 5). Until present there are not less than 123 flaring sites in the region making Nigeria one of the highest emitter of green house gases in Africa (Uyigue and Agho, 2007). Similarly, analytical assessment of the statistical bulletin of the Central Bank of Nigeria, (2004) showed that the average rate of gas flaring in Nigeria during the period 1970-1979 stood at 97%, while 97% and 95% were flared between 1980 and 1989, and 1990 to 1999 respectively. Between 2000 and 2004 51% was flared. In the same vein, between 1970 and 2004 Nigeria has flared an average of 76% of the total gas produced. Much of the natural gas extracted in oil wells in the Niger Delta is immediately flared into the environment at a rate that approximates 70 million /m^3 per day. This is equivalent to 40% of African natural gas consumption and forms the single largest source of greenhouse gas emissions on the planet (Wikipedia 2007, Moffat and Linden, 1995). Orubu, (2002b), who undertook a comparison of concentrations of ambient air pollutants in the region and Lagos State concludes that

Plate 4. Emissions form a Petrochemical plant

Parameter	PHRC1 (Mean ± SD)	PHRC2 (Mean ± SD)	PHRC3 (Mean ± SD)	Average/ Location	Guideline Limits
PM_{10} ($\mu g/m^3$)	130 ± 3.32	36.5 ± 1.27	27.9 ± 2.33	64.8	100(FMENV)
Pb (mg/m^3)	0.16 ± 0.12	0.20 ± 0.03	0.11 ± 0.10	0.16	-
Cd (mg/m^3)	0.008 ± 0.002	0.003 ± 0.001	0.009 ± 0.006	0.007	-
Ni (mg/m^3)	0.004 ± 0.002	0.009 ± 0.004	0.86 ± 0.34	0.291	-
Benzo (a) pyrene ng/m^3	-	-	1.63×10^2	54.3	<0.1–100ng/m^3 (WHO, 1997)
Indeno (1,2,3-cd) pyrene ng/m^3	1.53×10^3	2.53×10^3	-	0.0014	<0.1–100ng/m^3 (WHO, 1997)
^{T}PAH* (ng/m^3)	1.53×10^3	2.53×10^3	1.63×10^2	54.3	<0.1–100ng/m^3 (WHO, 1997)

* Total PAH (sum of benzo(a)pyrene + indeno(123-cd)pyrene)

Table 2. Levels of Priority Air pollutants in Nigeria's Refinery and Petrochemicals pollutant concentrations are highest in the Niger Delta and argues that some of the green house gases (such as methane and carbon dioxide) emitted at flare sites contribute to global warming. The largest proportion of these flare sites are located in the Niger Delta.

Plate 5. Gas flaring from an oil exploration process

3.7 Pipeline explosion

The explosion of pipelines occurs either accidentally or by sabotage(Plate 6). In the Niger Delta area much of the pipeline explosions are a product of the later and they are usually accompanied most times with fire outbreak. The burning flame and smoke from the oil pipelines releases large concentrations of gaseous substances and particulate matter. The substances in most cases include COx, NOx VOCs, THCs, carbon black, soot and some heavy metal residues.

3.8 Multiple air pollution sources

Within the Niger delta communities there are diverse sources of air pollution. The common air pollution sources are biomass combustion, bush burning, automobile emissions, generator emissions, pipeline explosions, industrial emissions and gas flaring. The measurement of air quality for some specific priority pollutants viz ammonia, particulate matter and Polycyclic aromatic hydrocarbons at specific locations within the communities has been carried out using standard procedures (Ana et al, 2005, 2009). The mean values of the parameters are presented in Table 3.

3.9 Meteorological conditions in the Niger Delta

The Niger Delta has a humid, semi-hot equatorial climate. Temperature ranges are small and constant throughout the year. It's hottest month records 28 °C (82.4 °F) and its coolest month 26 °C (78.8 °F) with the temperature range of not more than 2 °C (5 °F). Precipitation

Plate 6. Emissions from fire outbreak following pipeline explosion

Air quality at Eleme and Ahoada East

Sample Parameter	Eleme Mean ± SD	Range	Ahoada East Mean ± SD	Range	National Guideline Limits	*t*-statistic	*p*-value
TSP(mg/m^3)	123.4 ± 14.9	37.2–260.6	53.5 ± 26.3	8.84–161.4	100 (mg/m3)	–6.118	0.0001**
Fe (mg/m^3)	0.09 ± 0.06	0.09–0.10	0.027 ± 0.006	0.01–0.05	None	–2.764	0.0171*
Zn (mg/m^3)	0.05 ± 0.03	0.02–0.10	0.05 ± 0.02	0.03–0.42	-do-	0.000	1.000
Pb (mg/m^3)	0.06 ± 0.03	0.03–0.10	0.14 ± 0.03	0.03–0.13	-do-	4.939	0.0003**
Cd (mg/m^3)	0.009 ± 0.01	0.002–0.02	0.006 ± 0.003	0.004–0.01	-do-	–0.760	0.4618
Ni (mg/m^3)	0.03 ± 0.02	0.02–0.05	0.03 ± 0.02	0.01–0.05	-do-	0.000	1.000
†PAH(ng/m^3)	7.99×10^3	3.48×10^{-1}–	1.21×10^{-3}	2.50×10^{-5}–	<0.1 –	7.62×10^6	0.000**
	$\pm 1.31 \times 10^4$	1.88×10^4	$\pm 2.77 \times 10^{-3}$	7.49×10^{-2}	100 ng/m^3		
	(1.91×10^2)		(6.43×10^{-4})				

* Significant ** Highly Significant

† Values in brackets represent median values for PAH in the various locations

Table 3. Levels of Air Quality Parameters in Two Niger Delta Communities is heavy, between 1824 millimeters to over 4000 millimeters along the coast. Rain falls throughout the year with a short break in August and a longer break from December to January. The heavy rainfall experienced in the area leads to floods in urban areas and communities along the rivers. Also due to the heavy and frequent rainfalls, soils consisting mainly of silt and clay become saturated, reducing infiltration to the barest minimum and encouraging run-off. Rainfall induced run-off is directed mainly to the back swamps in the upper and middle parts of the delta. In the coastal zone, there is an even smaller topographic gradient that makes run-off difficult.

4. Environmental and health problems associated with air pollution in the delta area

4.1 Environmental problems

4.1.1 Formation of acid rain

From the release of large concentration of oxides of nitrogen and sulphur in the Niger delta environment it is glaring that some cases of acidified rain occur in the region. Although literature is deplete in this area it is obvious that such phenomenon is experienced. Olobaniyi and Efe (2007) also show elevated levels of lead (0.56 mg/l) and low pH values ranging from 5.10 - 6.35 in rain water collected in Warri and environs. Akpoborie, *et al*, (2000) also report low pH values from water obtained from shallow hand dug wells in Ughelli, Warri, and Okurekpo all in Delta State.

4.1.2 Visibility Impairment

Most combustion processes lead to the release of oxides of nitrogen and lower fractions of hydrocarbons and in the presence of uv radiation the resultant smog could lead to decrease in visibility. Although not readily reported this is a common phenomenon in a region like the Niger delta that has large hydrocarbon deposits and various episodes of combustion processes.

4.1.3 Ambient thermal conditions

Some 45.8 billion kilowatts of heat are discharged into the atmosphere of the Niger-Delta from 1.8 billion cubic feet of gas everyday (Aaron, 2006). Ogbuigwe (1998), reports that temperatures produced at flare sites could be as high as 1,600^{o}C. Temperature as high as 400 ^{o}C at an average distance of 43.8metres from flare sites in Isoko, Delta State have been recorded.

4.1.4 Effects on vegetation and animal life

Studies by Alakpodia (1989, 1995) showed that flares have negative effects on vegetation growth, animal life and ecological equilibrium in the Niger Delta area. Heat Production kills vegetation around the heat area. It suppresses the growth of flowering plants and reduces agricultural productivity and wild life biodiversity.

4.1.5 Corrosion effects on materials and other artifacts

The presence of acidified rain water in the environment increases the corrosion rate of roofing sheets, monuments and other economic structures. In the Niger delta area there is glaring evidence of the impacts of corrosion on several building structures and arts work and these cases have been observed to deteriorate at rather alarming rates.

4.1.6 Coastal erosion and flooding

Awosika and Foluronsho, (2006) documented projected impacts of sea level rise SLR on the Nigerian coastal area and resources to include: large scale inundation, increased coastal erosion, salt water intrusion into coastal aquifers, habitat modification with direct effects on wild life distribution, increased frequency of high intensity rainfall events and associated increased run off, elevated erosion rates, flash floods and increased frequency of ocean storm surges. Other models projected a loss of up to 15,000 km^2 by the year 2100 as a result of a 1m (SLR) while Magbagbeola, (2002) projects flooding of over 18,000km^2 land area within the delta in an SLR of 1m in the next 100 years.

4.1.7 Air quality impairment

Oluwole et al.(1996) in a typical air quality assessment of the Niger Delta showed that the levels of volatile oxides of carbon, nitrogen and sulphur and total particulates exceed existing Federal Environmental Protection Agency's (FEPA,1991) standards. Also Olobaniyi and Efe (2007) showed elevated levels of lead at concentrations of 0.56 mg/l in the atmosphere.

4.2 Health effects

4.2.1 Common health problems in the general population

A survey was carried out on the health conditions of populations' resident and working in the Niger Delta communities (Ana et al, 2009).A summary of the prevalent air pollution-related morbidities is shown in Box 3. Two communities, one with high industrial presence, Eleme and the other, Ahoada East, with low industrial presence were selected. Questionnaire and hospital records were employed for this survey and morbidities that have direct bearing with air pollution were focused on. The questionnaire survey indicated that at Eleme (39, 60.9%) as compared to Ahoada East (10, 4.5%) reported contaminated air as the major reason for ill health among the residents. Also there were more reported cases of skin outgrowths among Eleme residents as compared to Ahoada East. The morbidity conditions from hospital records for respiratory disorders showed 3.85% in males and 4.39% in females (Fig 3a). At Ahoada East respiratory disorder was(3.68% male; 4.18% female) (Fig 3b). In terms of respiratory disease, there was a high significant difference ($p < 0.0001$) between the male and female population at Ahoada East only. Also between Eleme and Ahoada East there was a significant difference among the male population ($p = 0.04$) and the female population ($p = 0.05$) (Figures 4 and 5).In addition the results indicated that for the combined male and female population, there was a significant difference between the two communities for skin disorders ($p = 0.023$) and disease of the respiratory tract ($p = 0.045$).

Box 3: Prevalent Air Pollution related Morbidities
Breathing Difficulties
Bronchitis
Aggravation of Asthma
Cardio-respiratory disorders
Pulmonary edema
Eye disorders
Skin disorders

4.2.2 Common health problems in the industrial community

In a study carried out by Ana et al (2005) in which 384 plant workers were interviewed in a chemical fertilizer industry at Onne in the Niger Delta, 70.5% spend 8 hours per day at work. 66.1% reported respiratory disorders, 24.4% reported skin disorders and 22.6% reported eye disorders. There were strong associations ($p < 0.05$) between eye and respiratory disorders and the industrial activities carried out by the workers. Also Ana et al (2009) in another survey carried out among refinery (PHRC) and petrochemicals (EPCL) plant workers reported that 70.8% workers at PHRC as compared to 67.2% at EPCL reported various health problems related to poor air quality conditions. In both cases, the major cause was attributed to gas flaring. Symptoms associated with exposure to fumes and sprays such as respiratory disorder were widely reported among 40.8% PHRC and 27.6% EPCL workers. There were reports of respiratory symptoms associated with exposure to dust and smoke, leading to irritation among 65.7% PHRC and 57.1% EPCL workers. Both the eyes and the skin were among other parts of the body also affected by the irritants. Further statistical analysis indicated that the duration of stay of PHRC workers in their residential communities was significantly associated with respiratory health problem ($p=0.000$), with Cancers ($p=0.000$). At EPCL, the duration of residence in the community was significantly associated with miscarriages ($p=0.000$), with deformed children ($p=0.000$) and with symptoms related to health effects from air contaminants ($p= 0.000$).

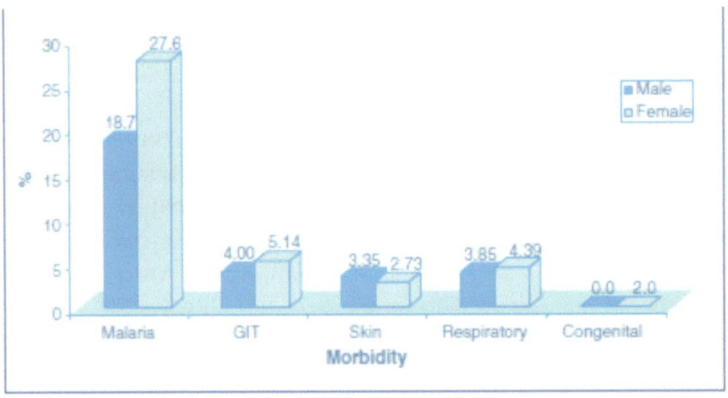

Fig. 3a. Typical morbidity pattern in a more industrialized NDA community

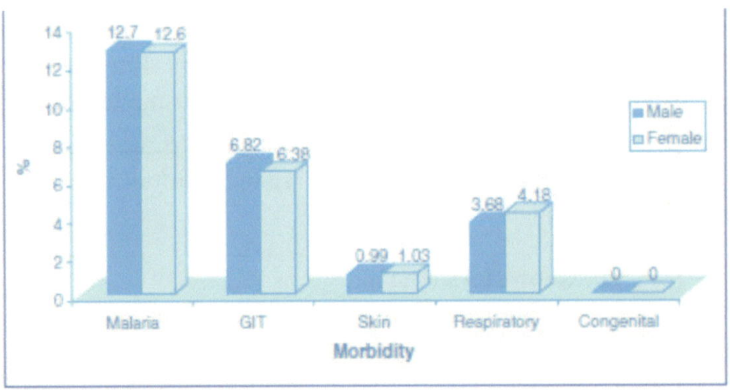

Fig. 3b. Typical morbidity pattern in a less industrialized NDA community

4.2.3 Emerging health problems

The exposure of populations to a barrage of pollutants and pollution episodes in the Niger delta area is believed to cause several health problems, of which etiologies of most cannot be ascertained. Cancer prevalence is believed to be on the increase. Studies by Ana et al(2010) revealed that of the two cancer reference centres Ibadan in the South west and Port Harcourt in the Niger delta the ratio of reporting was 1:4 for UPTH (904) and UCH (3521) respectively. The results indicate that apart from prostate and breast cancers that were higher in Ibadan (79.1%) than in Port Harcourt (75.4%) both the lung and skin cancers were more prevalent in Port Harcourt than in Ibadan(Table 4). This observation is consistent with studies by Ana et al.(2009) which indicated increased lung cancers in the Port Harcourt environment due likely to exposure to atmospheric insults. Similarly, the higher percentage of skin cancer in Port Harcourt above that recorded in Ibadan could be explained though with some degree of uncertainty by the increased environmental risk factors in the more industrialized Port Harcourt area.

Cancer types	Port Harcourt (N=904)	Ibadan(N=3521)	X^2	p- value
Lung	33(3.7)	113(3.2)	0.44	0.508
Skin	172(19.0)	365(10.4)	50.59	<0.001
Eye	15(1.7)	258(7.3)	39.92	<0.001
Others	684(75.4)	2787(79.1)	5.18	0.023

Table 4. Comparison of Cancer cases between two prominent cities in Nigeria

5. Challenges of air quality studies in the Delta area

Air quality studies in Nigeria and particularly the Niger Delta area is still in its infant stage and encumbered with several challenges. According to Taiwo (2005) air pollution studies in Nigeria are few and independently carried out. The same author also indicated that government is not involved in systematic and consistent air quality assessment programmes as is being done in other parts of the world such as that carried out by the Environmental Protection Agency (EPA) in the United States. There is the problem of insecurity and difficulty in terrain that militates against most community based air sampling initiatives and then the lack of requisite and adequate technical manpower to carry out the multifaceted and complex air quality studies in the region. Other obvious and prominent drawbacks are as follows

- Lack of emissions inventory/database due to lack of consistent and systematic measurements
- Unavailability of air pollution and GHG monitoring stations in the Niger Delta. This is based on information from World Data Centre for Greenhouse Gases
- Few independent and research-based measurement data are not readily available for general public use
- Lack of collaboration between key regulatory authorities
- Laxity in the enforcement of emission regulations
- Air quality assessment and air pollution studies have focused mainly on urban centres (Ajao& Anurigwo, 2002)

6. Conclusions

Air pollution has been identified as one of the most critical environmental problems confronting the Niger delta Area. Traffic, industry and gas flaring are the major air pollution sources in the region. Impact on socioeconomic, ecosystems health, properties and climate have been linked with the pollution episodes in the area but which requires substantial scientific and empirical evidence. Multiple factors especially lack of equipment, inadequate skilled personnel and poor policy frame work has militated against effective and qualitative air quality studies in the area. To achieve long term goals for the region in this regard that would lead to the overall benefit of the people in the area, there is need therefore to employ a holistic and integrated approach to air pollution management that will involve all the stakeholders.

7. Future directions

In order to improve on the current air quality monitoring and assessment programmes in the Niger Delta area there is need to embark on the following:

- Develop monitoring mechanisms, regulations and enforcement measures
- Institute planning policies to minimize pollution that may be caused by future development.
- Government agencies such as the Niger Delta Development Commission(NDDC) should collaborate with other multinationals and stakeholders in air pollution management to come up with a comprehensive AQM scheme for the region.
- There should be a focus on the reduction of pollution levels from vehicles, industry, gas flaring and domestic burning of timber, to permissible levels as defined in national and international standards.
- The impact of air pollution from industrial and vehicular sources on the health of the communities in the region and its biodiversity needs to be researched in-depth.
- Existing air quality monitoring programmes should be re-examined and new ones introduced to determine the most effective means of mainstreaming national programmes with regional projects to improve air quality.
- Research on air quality should focus on source apportionment of the pollutants in the region
- Indepth epidemiological and toxicological studies using risk and exposure assessment tools need to be carried out to establish causalities between the air pollution exposure factors and the associated health problems.
- Motor vehicles annual testing and other regulations must be created or re-introduced and strictly enforced.
- Focus should be on air pollution models, real-time monitoring of pollutants and speciation of pollutants from the particulate matter.
- There is a need to engage in renewable energy, clean energy and cleaner air initiatives.
- Usage of emissions abatement control mechanisms by polluters should be enforced.

8. References

Aaron, K. K. (2006). Human Rights Violation and Environmental Degradation in the Niger-Delta in Elizabeth Porter and Baden Offord(eds), Activating Human Rights, Oxford, Barne, New York.

Ajao EA. and Anurigwo S(2002). Land - based sources of pollution in the Niger Delta, Nigeria. Ambio Journal of the human environment. 31 (5):442 - 445.

Akpan, U.G., Ndoke, P. N. (1999). Contribution of vehicular traffic emission to CO2 emission in Kaduna and Abuja.Federal University of Technology Minna Nigeria.

Akpoborie, I.A.; Ekakite, A. O.; Adaikpoh, E.O.(2000), The Quality of Groundwater from Dug wells in Parts of the Western Niger Delta, Knowledge Review 2: 72-79.

Alakpodia, I J (1989), The Effects of Gas Flaring on the Micro-climate and Adjacent Vegetation in Isoko Area of Bendel State" (Unpublished M.Sc Thesis, University of Ibadan).

Alakpodia, I. J. (1995), "The Oil Industry and the Economic Environment of the Niger Delta"Paper presented at the 30th Annual Conference.Nigeria Geographical Association, at the University of Benin. Benin City. Nigeria.

Ana, Godson R.E.E, Mynepalli K.C.Sridhar and Elijah A. Bamgboye (2009) Environmental risk factors and health outcomes in selected communities of the Niger Delta Area, Nigeria. Perspectives in Public Health. Vol 129(4); 183-191

Ana, G.R.E.E and Sridhar, M.K.C(2009) Industrial Emissions and Health Hazards among selected factory workers at Eleme, Nigeria. Journal of Environmental Health Research, Vol9(1):43-51

Ana, Godson R. E. E, Sridhar, Mynepalli K.C and Asuzu, Michael C(2010) Environmental risk factors and hospital based cancers in two Nigerian cities Journal of Public Health and Epidemiology Vol. 2(8), pp. 216-223,

Awosika, L.; Foluronsho, R. (2006), Climate Change and Impact on the coastal environment of Nigeria, In, Sustainable Environmental Management in Nigeria, Eds. Ivbijaro *et al.*Mattivi Productions, Ibadan, pp 85-103.

CEC, 1992. The state of the environment in the European Community. Overview, Vol. 3. Commission of the Communities, Brussels and Belgium.

Cedigaz, (2000) www.cedigaz.org.Central Bank of Nigeria. (2004), Statistical Bulletin 15

Enemari, E., 2001. Vehicular emissions: Environmental and health implications. National conference on the phase-out leaded gasoline in Nigeria.

Faboya, O.O., 1997. "Industrial pollution and waste management" pp 26-35 in Akinjide Osuntokun(ed),Dimensions of Environmental problems in Nigeria, Ibadan Davidson press.

Federal Environmental Protection Agency, (1991), National Policy on the Environment, The Presidency, Abuja, Nigeria

Gill, L. S.; Nyawuame, H. G. K.; Ehikhametalor, A. O. (1992), Effect of Crude oil on growth and anatomical features of *Chromolaena odorata* L. Newsletter 5:46-50 http//alafrica.com/stories.

Iyoha, M. A., 2009. The Environmental effects of oil industry activities on the Nigerian Economy:A theoretical Analysis: Paper presented at National Conference on the management of Nigeria's petroleum Resources, organised by the Department of Economics, Delta State University.

Jerome, A. 2000. ''Use of Economic instruments for Environmental Management in Nigeria'' Paper presented at workshop on Environmental Management in Nigeria and Administration (NCEMA).

Koku, C.A., Osuntogun, B.A. 2007. Enviromental-Impacts of Road Transportation in Southwestern States of Nigeria. Journal of Applied Sciences 7 (16): 2536-2360, 2007.

Magbagbeola, N. O. 2001. The use of Economic Instruments for Industrial pollution Abatement in Nigeria: Application to the Lagos Lagoon. Selected papers, Annual Conferences of the Nigerian Economic Society held in Port-Harcourt.

Magbagbeola, I (2002), Environmental Underdevelopment of the Niger Delta: An Eclectic view, In: The Petroleum Industry, Economy and the Niger-Delta Environment. (Eds), Orubu. C, Ogisi, D.O. and Okoh R.N. 32-40.

Moffat, D; Linden O (1995), "Perception and Reality Assessing Priorities for Sustainable Development in the Niger Delta, Journal of the Human Environment, 24 No 7-8 327-538.

Ogbuigwe, A (1998), "Judicial Activism in the Enforcement of Environmental Regulations in the Petroleum Industry: Past, Present and the Future." Proceedings of 1998 International Conference on the Petroleum and the Nigerian Environment 83-123.

Olobaniyi, S.B.; Efe, S.I. (2007), Comparative assessment of rainwater and groundwater quality in an oil producing area of Nigeria: environmental and health implications, Journal of Environmental Health and Res. (6)2: 111-118.

Oluwole, A.F.; Olaniyi, H.B.; Akeredolu, F.A.; Ogunsola, O.J.; Obioh, I.B. (1996), Impact of the Petroleum Industry on Air Quality in Nigeria, Presented at the 8th Biennial International Seminar on the Petroleum Industry and the Nigerian Environment, Port Harcourt, 17-21 November.

Orubu, C O (2002b), Oil Industry activities, Environmental Quality, and the Paradox of Poverty in Niger Delta. In: The Petroleum Industry, the Economy and the Niger-Delta Environment. (Eds), Orubu, C.O., Ogisi, D.O. and Okoh, R.N. 17-31.

Saville, S. B., 1993. Automotive options and quality Management in developing Countries Industrial Environment. 16(1-2); 20, 32.

Seneca, J.J. and M.K. Tausig. 1994. *Environmental Economics*, Engle Wood Cliffs, Prentice Hall.

Taiwo, Olukayode(2005). "The case of Lagos –Air quality improvement project." LAMATA.

USEPA. 1993. Guide to Environmental Issues, Doc. No 520/B-94-01. United States Environmental Protection Agency, Washington, DC, USA.

United Nation. Prospect of World Urbanization 1989 (Population Study No. 112) New York.

Uyigue, E. and Agho, M. (2007). Coping with Climate Change and Environmental Degradation in the Niger-Delta of South Nigeria. Benin. Community Research and Development Centre (CREDC).

Wikipedia (2007), http://en.Wikipedia.org/Wiki/Environmental Issues in Nigeria/Niger Delta.

World Bank (1995), Defining an Environmental Strategy for the Niger Delta. Washington D.C.Industry and Energy Operations Division (West/Central Africa Department

10

Emerging Air Pollution Issues in Changing Pearl River Delta of South China

Mang Lin[1], Iat-Neng Chan[2], Chuen-Yu Chan[*1], Xue-Mei Wang[1] and Han-Ying Dong[1]
[1]*School of Environmental Science and Engineering, Sun Yat-sen University*
[2]*Faculty of Science and Technology, University of Macao*
P.R. China

1. Introduction

1.1 Background

The Pearl River Delta (PRD) in Southern China is one of the most developed regions in China. Administratively, the delta region is constituted by nine cities in Guangdong Province, and the Special Administrative Regions of Hong Kong and Macao. The PRD has developed into a manufacturing center of the world since the onset of China's economic reform in the late 1970's. The important role of this region in China is highlighted by the high percentage contribution of gross domestic product (GDP) of Guangdong Province, which is mostly contributed by the PRD cities in the mainland, to that of the nation (>10%, National Bureau of Statistics China, 2009).

As an important manufacturing center and city cluster, the PRD region is inevitably affected by severe air pollution problems, which attract much public attention. In response to this situation, a regional air quality monitoring network compositing of 16 automatic stations has been established by the joint effort from Guangdong and Hong Kong governments since late 2005 with an aim to examine the air pollution problem of the region (GDEPMC & HKEPD, 2005). A series of large scale research programs including comprehensive aircraft and ground-base measurements and modeling exercises have been conducted by the scientific community to better characterize and analyse the problem and to find scientific evidence for formulating possible control strategies (e.g. the Program of Regional Integrated Experiments on Air Quality over the PRD of China-PRIDE-PRD in 2004 and 2006) (Zhang et al., 2008; Hua et al., 2008).

1.2 Changing characteristics of the PRD region

Fig. 1 depicts the land use change of the PRD from 1992 to 2004. The major urban areas were mainly found in the capital Guangzhou and the former British colony Hong Kong in the early 1990's. However, in the 21st century, almost the whole PRD region was characterized by urban areas. This dramatic urban expansion was found to be one of the most key factors affecting the air quality and climate of the region (Wang et al., 2007; Kaufmann et al., 2007). Furthermore, the steady increasing trends of GDP and electricity consumption also

---

*corresponding author

highlight the tremendous progress of urbanization and industrialization of this region (Fig. 2). Concurrent atmospheric composition changes and air quality degradation have been observed by space-borne satellite (e.g., Richter et al., 2005; Van der A et al., 2006; Ghude et al., 2009) and limited long-term ground-based measurements. Chan et al. (2004) first deduced that the regional background ozone (O_3) concentration in south China including the PRD region had been increased at a rate of 1.5% per year from 1984 to 1999 based on the urban measurements in Hong Kong. This finding was recently proven by Wang et al. (2009) using the data recorded from a background station in Hong Kong during the period 1994 to 2007. Besides, the metropolitan Guangzhou was also inevitably suffered from the air quality and visibility degradation during recent half century (Deng et al., 2008). The China's "12th five-year's plan" intends the PRD to transform its labor intensive and resource consuming light industry to heavy industry, telecom, equipment manufacture, auto and petrochemical industries for examples. With such a strategy, the traditional industry and enterprises are forced to move out from this region. Moreover, "cleaner" fuels such as liquefied petroleum gas (LPG) and low sulfur gasoline have been promoted by the government. Such massive changes of industry structure and prevalent fuel are anticipated to have strong effects on air quality. For instance, Tang et al. (2008) reported that there have been around 1.5 times increases in propane concentration in Guangzhou due to the introduction of LPG to the city in 2003. This pilot study highlights the scientific importance of clear understanding of the emerging air pollution issues along with these economic and industrial changes.

1.3 Aim and objectives

The objective of this chapter is to unveil the long-term air pollution changes in three typical commercial and industrial cities of the PRD region, Hong Kong, Macao and Foshan. The possible impacts of these changes on the atmospheric environment and their implications on the regional and urban air quality control strategy will be discussed. These cities are selected simply because they are the cities that we have accessible data. Hong Kong and Macao are located at the southern part of the PRD (Fig. 1). With the second-largest stock market in Asia, Hong Kong is well-known as an international financial center, while Macao is famous as a gambling city with flourishing tourism. On the other hand, Foshan is located at the central-western of the region, where urbanization process is most rapid (Fig. 1). This city is highly industrialized as reflected by its high industrial production value compared with those of Hong Kong and Macao (Table 1). The city is especially famous for ceramic industry, which produced approximately one fourth of the world's ceramic products in 2007 (Peng 2007; Yin 2008). A statistical summary on the general information of these three cities and data available periods are shown in Table 1.

2. Air pollution patterns and occurrence of episodes

Fig. 3 and Fig. 4 summarize the mean concentrations of major air pollutants and the occurrence of episode days in Hong Kong, Macao and Foshan. As expected, the highest SO_2 and PM_{10} concentrations were found in the industrial city Foshan due to the high SO_2 and PM_{10} emissions from industrial activities. Note that the average SO_2 concentration (60 μg m^{-3}) reached the National Ambient Air Quality Standard (Level II) (GB3095-1996) (NAAQS II) of 60 μg m^{-3} (monthly mean). In fact, more than 50% of the days during the period 2001-2007 in Foshan had daily SO_2 concentrations exceeding the NAAQS I of 50 μg m^{-3} (Table 2 and Fig. 4). Besides, the high percentages of PM_{10} episode days in Foshan highlight the severe

	Hong Kong	Macao	Foshan
Population[a]	7.0 million	0.5 million	5.7 million
Area[a]	1100 km^2	21 km^2	3800 km^2
Vehicle number[a]	0.70 million	0.18 million	0.60 million
Value of industrial production[b]	160 billion (HKD)	16 billion (MOP)	940 billion (CNY)
Visibility data	1968.01-2008.12	1990.01-2007.12	1998.01-2008.12
SO_2 data	1984.01-2008.12	1998.04-2007.12	2001.01-2008.12
NO_2 data	1984.01-2008.12	1998.04-2007.12	2001.01-2008.12
O_3 data	1984.01-2008.12	1998.04-2007.12	n.a.[c]
PM_{10} data	1993.01-2008.12	1998.04-2007.12	2001.01-2008.12

a. statistical data at 2008; b. statistical data at 2007; c. n.a. stands for data not applicable

Table 1. Summary of the general statistic information and the periods with data available for analysis

particle pollution at this industrial region. Hong Kong and Macao are both commercial cities with limited manufacturing industries. The SO_2, PM_{10} and O_3 concentrations of Macao were slightly higher, which could be attributed to the fact that Macao is located at the southeast of Hong Kong (Fig. 1). The prevailing northeasterly wind from September to April will carry aged and polluted air from Hong Kong and other PRD cities like Shenzhen and Dongguan to Macao to deteriorate air quality.

The average NO_2 concentration was highest in Hong Kong and lowest in Foshan. This result is reasonable as Hong Kong is a commercial city with dense traffic (Table 1), which results in the elevated NO_x emission and ambient NO_2 concentration. However, it is noted particularly that the average NO_2 concentration of Foshan during the period of 2006-2008 (60 μg m^{-3}, not shown) was higher than that in Hong Kong and Macao, which suggested that the remarkable increases of vehicle number and vehicular emission in Foshan was becoming more important (Wan et al., 2011). Moreover, although the number of episode day with daily NO_2 concentration larger than 80 μg m^{-3} (NAAQS I & II) in Foshan was close to that in Macao (~7.3%), the frequency of daily NO_2 concentration exceeding 120 μg m^{-3} (NAAQS III) in Foshan was significantly higher than that in Macao (Fig. 3). This fact emphasizes the needs of implementation of stricter NO_2 control measures in Foshan.

	Level I	Level II	Level III
SO_2 (Daily Mean)	50	150	250
NO_2 (Daily Mean)	80	80	120
O_3 (Hourly Mean)	120	160	200
PM_{10} (Daily Mean)	50	150	250

Table 2. Summary of the National Ambient Air Quality Standard (GB3095-1996) in China (unit: μg m^{-3})

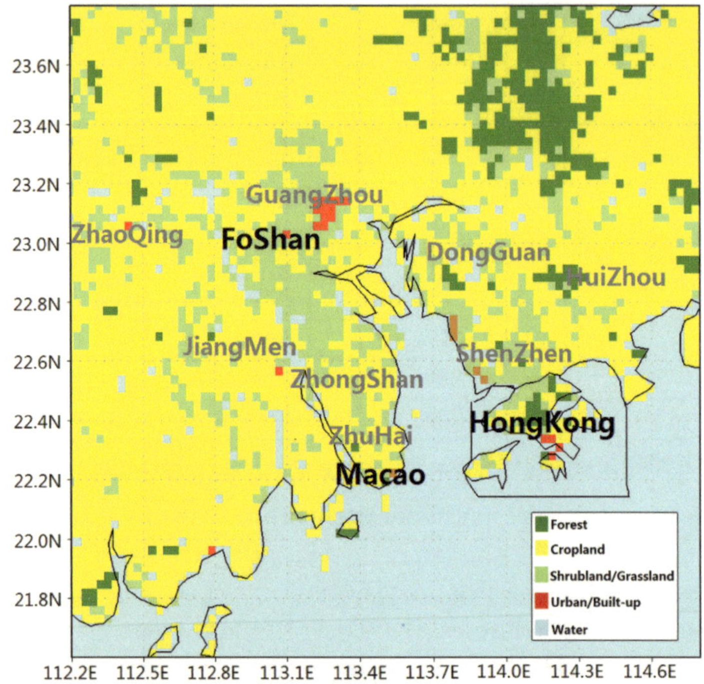

(a)

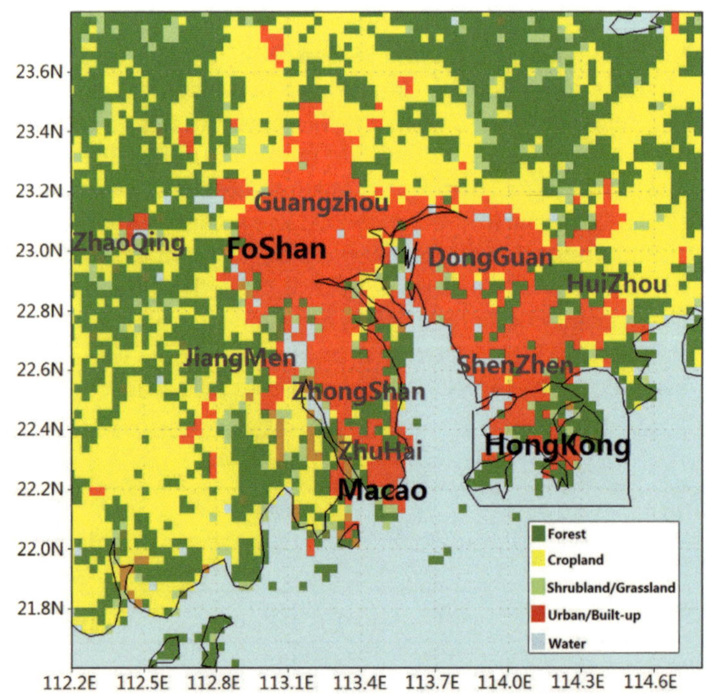

(b)

Fig. 1. Urban expansion in the PRD region from (a) 1992 to (b) 2004

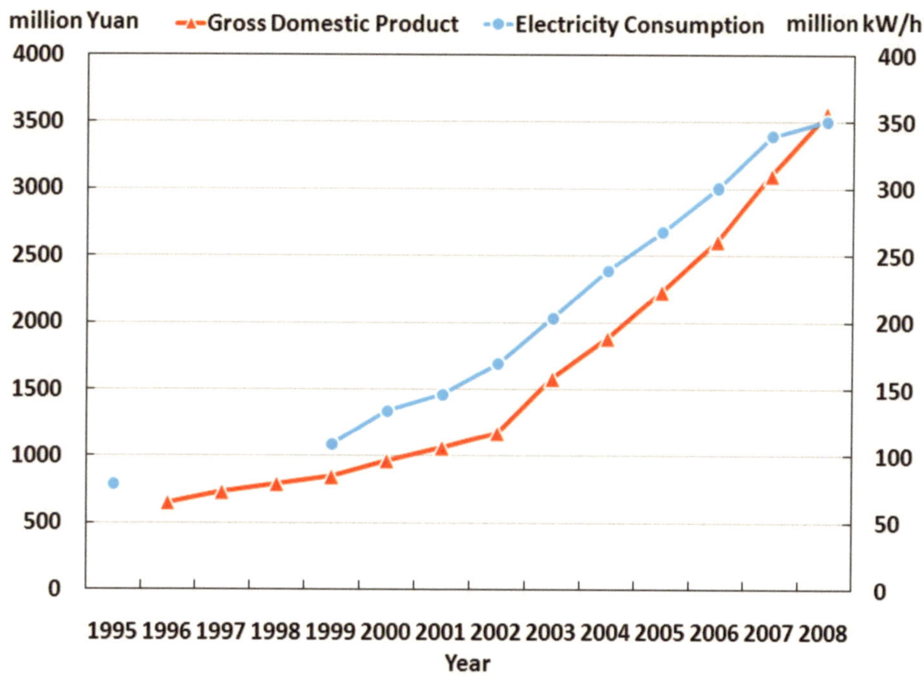

Fig. 2. Trends of annual gross domestic product and electricity consumption in Guangdong Province from 1995 to 2008.

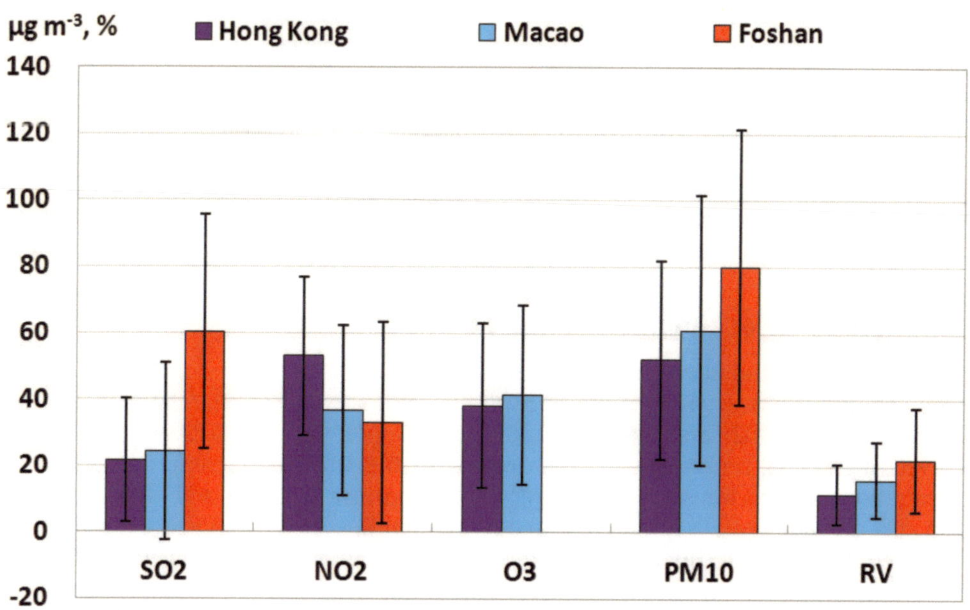

Fig. 3. Comparison of average air pollutants concentrations (2001-2007) and reduced visibility percentage (1998-2007) in Hong Kong, Macao and Foshan (error bars stand for one standard deviation)

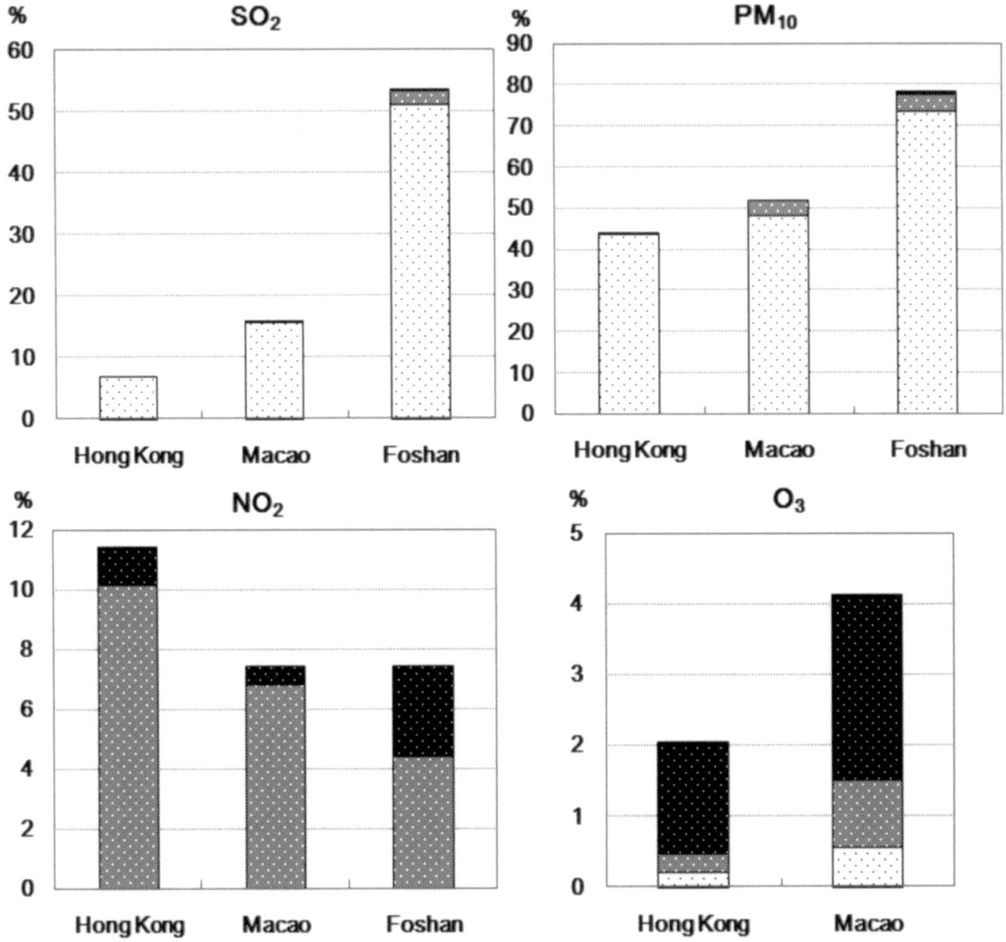

Fig. 4. Comparison of percentages of SO_2, PM_{10} and NO_2 exceeding days as well as O_3 exceeding hours in Hong Kong, Macao and Foshan during the period of 2001-2007 (White, gray and black bars stand for the percentages exceeding NAAQS I, II and III respectively)

3. Changes of major atmospheric compositions

3.1 Surface ozone and nitrogen dioxide

Ozone concentrations in Hong Kong and Macao both showed substantial increases with rates of 1.00 (R^2=0.752) and 1.06 (R^2=0.324) μg m^{-3} per year respectively (Fig. 5). The observed increasing trends are mostly caused by the regional increases in surface O_3 concentrations due to transport of its precursors from south China and East Asia as the urban area of Hong Kong is shown to be a net sink of inflowing O_3 due to the titration effect of nitrogen monoxide freshly emitted from traffic (Chan et al. 2004; Wang et al. 2009). NO_2 concentration in Macao showed a smoothly increasing trend with a rate of 1.26 μg m^{-3} per year (R^2=0.712). We believe that this change is probably accounted by the simultaneous increasing trend of vehicle number in Macao, which increased from 0.11 million in 2000 to 0.18 million in 2007 (Macao Statistics and Census Service, www.dsec.gov.mo). In contrast, NO_2 in Hong Kong kept a rather stable level after the 1990s.

Although long-term O_3 data were not available in Foshan, NO_2 concentration showed sharp increases from 18 μg m^{-3} in mid 2005 to 80 μg m^{-3} in late 2006 contradicting to the general decreases before 2005 (Fig. 5). This increasing rate reached 440% although NO_2 decreased to lower levels from 2007 to 2008. The overall increasing rate during 2001-2008 was 6.34 μg m^{-3} per year (R^2=0.496). The continuous increases in coal burning and road traffic, as well as lack of an efficient NO_2 control strategy, led to the remarkable increases of NO_2 level. More detailed discussions can be found in Wan et al (2011).

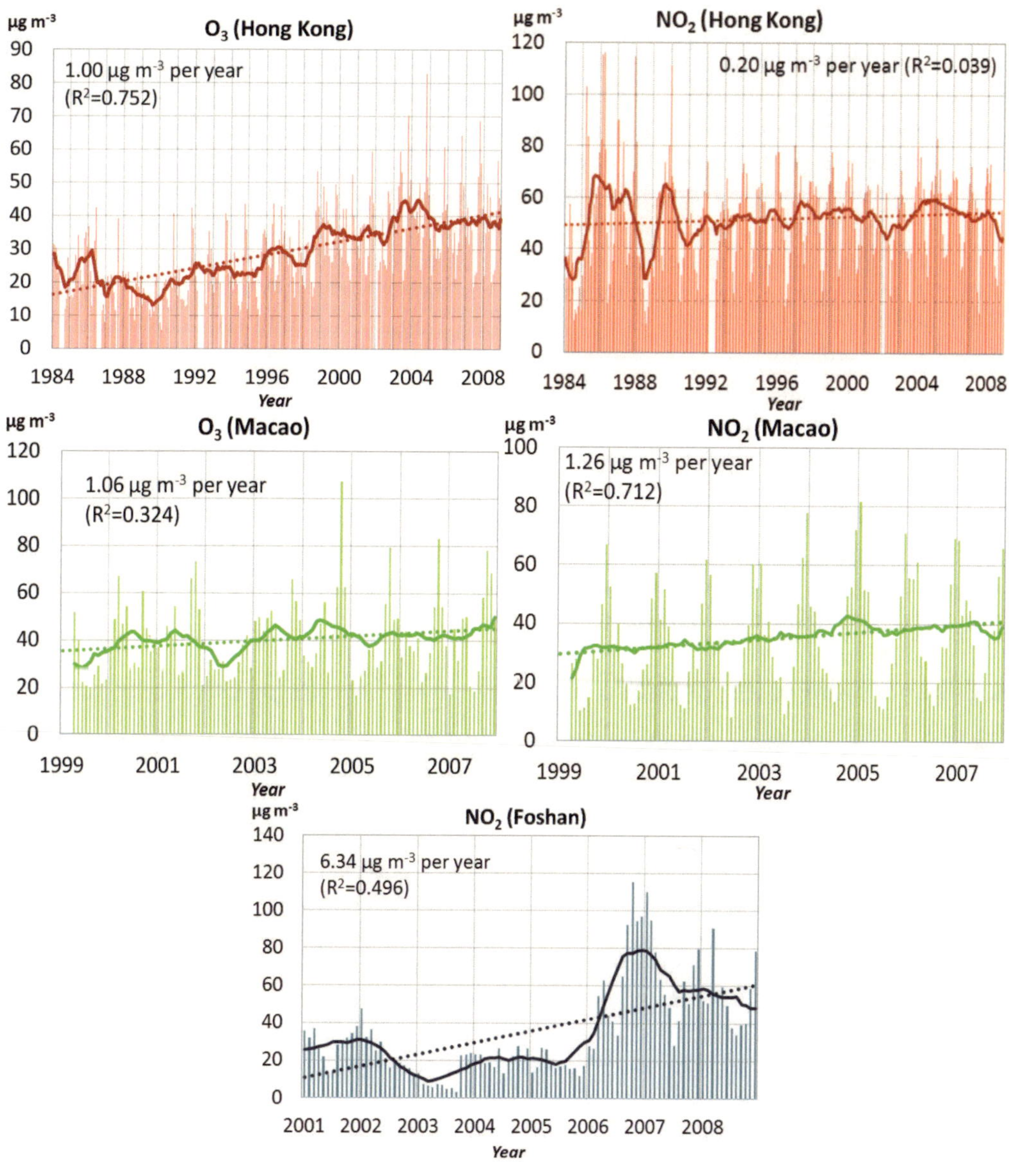

Fig. 5. Trends of O_3 and NO_2 concentrations in Hong Kong, Macao and Foshan (bars indicate the monthly means; solid lines indicate 12 months running averages; dot lines indicate linear regression lines)

3.2 Particulate matter and sulfur dioxide

In Macau, PM_{10} concentrations showed a steady increasing trend with a rate of 2.39 μg m^{-3} per year (R^2=0.716, Fig. 6). As high portion of organic carbon, which is one of the important species in particle primarily emitted by automobile in urban area, was found in PM_{10} in Macao (Wu et al., 2003), the increases of PM_{10} concentration were probably attributed to the increases of vehicle emission as mentioned above. Apart from the local emission, the airborne particulate matter in Macao was also influenced by transported secondary aerosols of regional scale (Wu et al., 2002, 2003). On the contrary, PM_{10} concentrations in Foshan displayed a steady decreasing trend with an overall rate of 3.47 μg m^{-3} per year in 2001-2008, which was believed to be due to continuous implementation of stricter emission control measures on various industrial sources. For instance, many high emitting factories including cement plants and small-scale ceramic plants were forced to close down by the city government in the past decades (Wan et al., 2011).

The SO_2 concentrations in Hong Kong showed a growth rate of 0.28 μg m^{-3} per year in the period of 1984-2008 (Fig. 6). Specifically, the decreases from 1993 were due to the effective banns of high-sulfur fuel usage and reduction of power plant emissions in Hong Kong (HKEPD, www.epd.gov.hk), while the increases in recent years were owing to the increases in SO_2 emissions in both Shenzhen city and Hong Kong (Chan and Yao 2008). Apart from the local SO_2 emission, the increasing trend of regional SO_2 emission also needed to be considered (Lu et al., 2010). Although the SO_2 concentration in Foshan was high when compared to other two commercial cities, a slightly decreasing trend (0.28 μg m^{-3} per year, R^2=0.364) was observed in the 21st century, which was due to the wide application of flue-gas desulfurization (FGD) devices in power plants in response to a new policy of China's government (Lu et al., 2010). In fact, the overall SO_2 removal efficiency of exhaust stream passing the treatment system in various major industrial sectors in Foshan was up to 64% (Fig. 7). However, the continuous growth in coal consumption triggered the substantial increases in 2006 (Wan et al., 2011).

4. Enhancement of atmospheric oxidizing power and its possible impacts

Given that O_3 is a vital atmospheric oxidant and NO_2 is one of the most important O_3 precursors, the remarkable increases of O_3 in Hong Kong and Macao as well as the tremendous augment of NO_2 in Foshan highlight the enhancement of atmospheric oxidizing power in the PRD region. It is reasonable to hypothesize that such changes would exert impacts on atmospheric environmental quality and chemistry. In this section, we used visibility as a proxy of atmospheric environmental quality changes to investigate the impacts of air pollution changes on the atmospheric chemical processes. Following the definition of the Hong Kong Observatory, the observed visibility data were transformed to Reduced Visibility (RV) which is defined as the percentage of hours with visibility below 8 km, excluding the data of rainy days or those affected by fog or high relative humidity (≥95%) (Leung et al., 2004). With this definition, the resulted RV would better indicate the frequency of visibility degradation caused by elevated pollutant level in the atmosphere. As expected, the highest RV was found in Foshan while the lowest RV occurred in Hong Kong (Fig. 3).

Three multiple regression models and the partial correlations between RV and each air pollutant were calculated. The model results were subjected to the F-test and student t-test with a significant level $p<0.05$. After that, the air pollutants with significant partial correlation were kept at the model equations while others were removed from the equations

(Table 3). Generally, visibility in Hong Kong is primarily influenced by concentration levels of O_3, SO_2 and NO_2 in recent 16 years. Particularly, RV would increase 0.28%, 0.27% and 0.15% if concentrations of SO_2, NO_2 and O_3 increase 1 μg m^{-3} respectively. In Macao, the RV was significantly affected by PM_{10} concentration with a value of increasing 0.33% of RV per 1 μg m^{-3} of PM_{10}. Similar relationship exists in Foshan with a value of increasing 0.50% of RV per 1 μg m^{-3} of PM_{10}. Besides, the RV in Foshan correlated with NO_2 concentration with a value of increasing 0.14% of RV per 1 μg m^{-3} of NO_2.

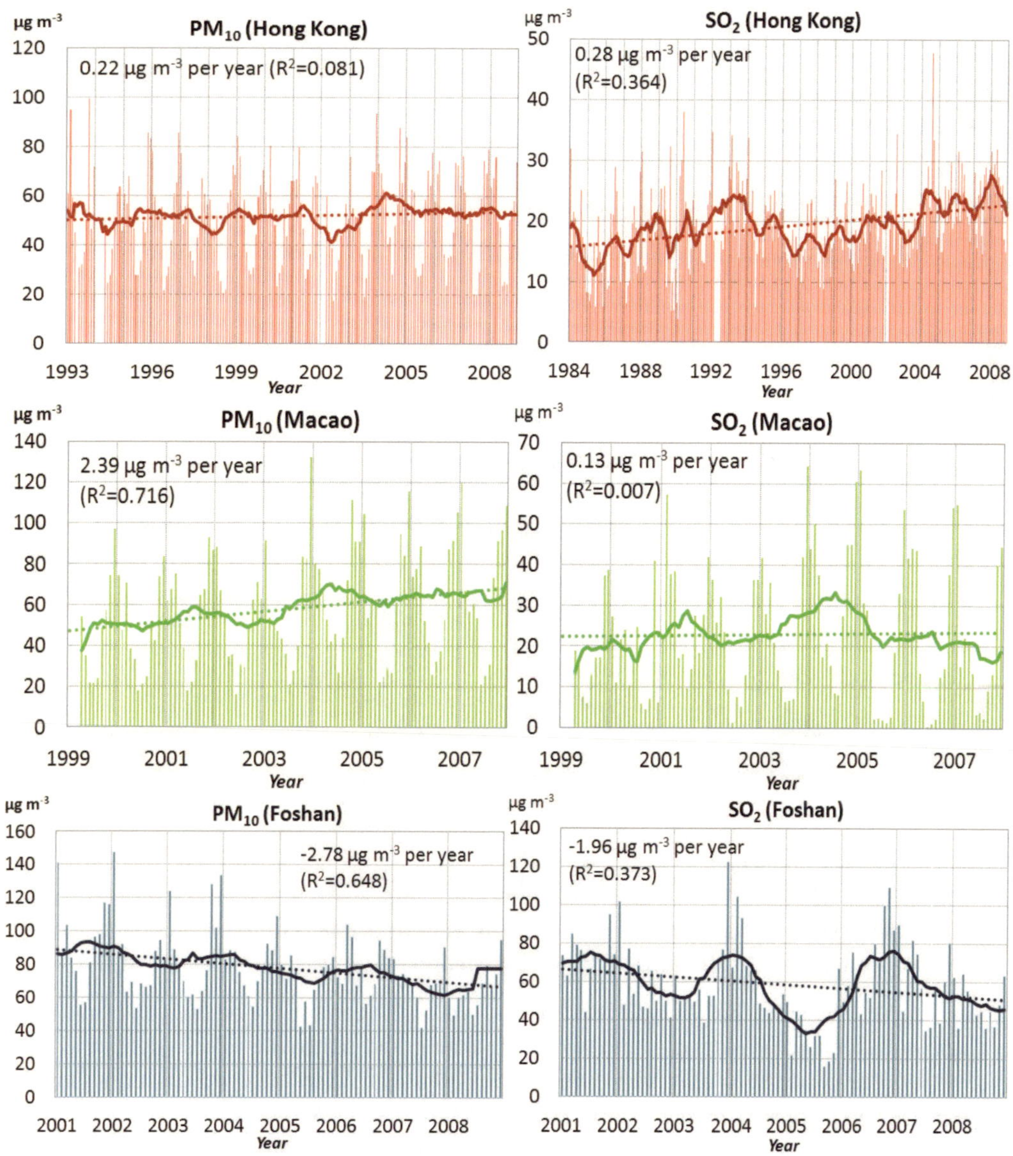

Fig. 6. Trends of PM_{10} and SO_2 concentrations in Hong Kong, Macao and Foshan (bars indicate the monthly means; solid lines indicate 12 months running averages; dot lines indicate linear regression lines)

It is well-known that visibility impairment is always proportion to the loading of fine particles owing to the scattering capacity of aerosol. Besides, NO_2 is another important species in the atmosphere which can degrade visibility due to its strong blue-light-absorbing capacity. Therefore, it is easy to explain the correlation between RV and PM_{10} as well as NO_2.

Regression Model	Independent Variables	Equation[a]
1 (Hong Kong 1993 - 2008)	O_3, SO_2, NO_2, PM_{10}	[RV] = 0.28 [SO_2] + 0.27 [NO_2] + 0.15 [O_3] - 15.80
2 (Macao 1999 - 2007)	O_3, SO_2, NO_2, PM_{10}	[RV] = 0.33 [PM_{10}] - 5.70
3 (Foshan 2001 - 2008)	SO_2, NO_2, PM_{10}	[RV] = 0.14 [NO_2] + 0.50 [PM_{10}] - 21.24

a. Units of RV and air pollutant are % and $\mu g\ m^{-3}$, respectively.

Table 3. Summary of equations calculated by multiple linear regression method

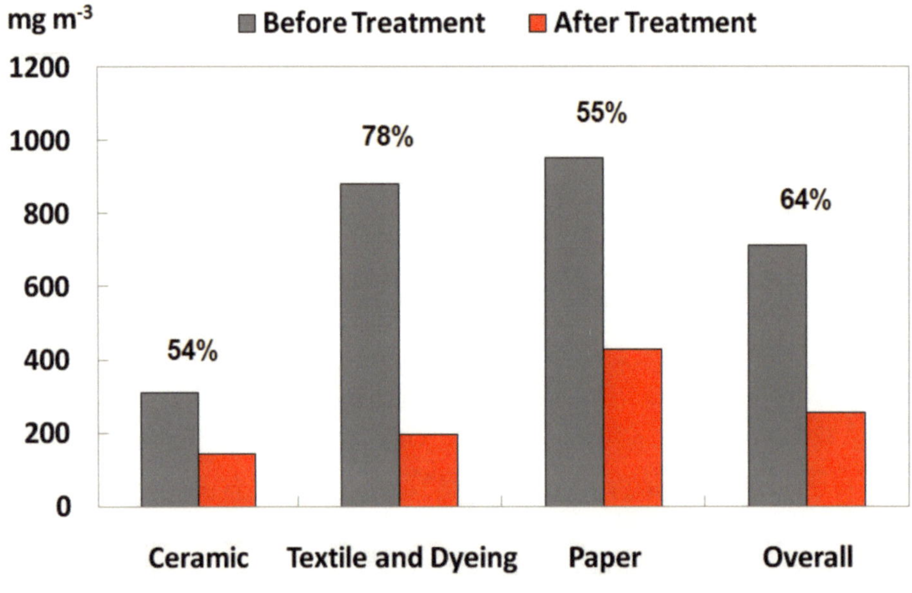

Fig. 7. Emission data of major industrial sectors in 2007 and 2008: comparison of SO_2 concentrations before and after emission control treatment (Source: Wan et al., 2011)

However, the correlation between RV and SO_2 as well as O_3 can hardly be directly interpreted because both O_3 and SO_2 are rather weak in absorbing and scattering visible light, and they thus will not produce strong influence on visibility directly. In fact, this result highlights the changes of atmospheric chemistry due to the tremendous increases of O_3 concentration in Hong Kong and Macao. In the enhanced O_3 atmosphere, the increasing oxidation rate would transform SO_2 to secondary fine sulfate particles more rapidly, which could effectively degrade visibility (Fig. 8, Meng et al. 1997, Shao et al. 2006). The chemical mechanism of S(IV) to S(VI) transformation has been extensively studied in literatures (e.g. Penkett et al, 1979;

Stockwell and Calvert, 1983; Seinfeld and Pandis, 1998; Liang and Jacobson, 1999). The enhanced formation of secondary sulfate in the O_3-rich atmosphere mentioned above actually agrees with the findings from a recent study, in which SO_4^{2-} concentrations in East Asia were found to increase at a greater rate than SO_2 emissions and SO_2 concentrations because "East Asia is a less oxidant-limited area than other parts of the world" (Lu et al. 2010).

In addition, although long-term O_3 data in Foshan were not available, the elevated level of NO_2 concentration more or less highlights the enhancement of oxidizing power of atmosphere in this region as well. It is well known that O_3 is generated from the photolysis of NO_2 and reacts mainly with NO to regenerate NO_2 quickly. If CO or VOCs are present, the simply photostationary state mentioned above is modified in a way that NO_2 can be regenerated through the reaction between NO and HO_2 (or RO_2). Net formation of O_3 occurs because the conversion of NO to NO_2 is accomplished by the HO_2 (or RO_2) radical rather than by O_3 itself. Therefore, the enhanced NO_2 concentration in the atmosphere also reflects the presences of other secondary pollutants like fine particle which significantly contributes to visibility impairment. The significant correlations between RV and NO_2 concentration in Hong Kong and Foshan thus underline the important role of NO_2, or even secondary pollutants, play in visibility reduction. In Macao, RV did not correlate well with O_3 and NO_2 concentrations. However, the high portion of secondary aerosols in PM_{10} and steady increasing trends of O_3 and NO_2 also emphasize the vital role of high O_3 levels play in photochemical formation of secondary pollutants such as sulfate, which would exert notable impacts on visibility, public health and climate.

5. Summary and discussion

Changes of major atmospheric compositions and air pollution levels due to the rapid economic and industrial development have been observed in the PRD region. The data such as those presented in this paper suggest that these long-term air pollution trends have vital implications on the formulation of regional air quality control strategies. Generally, the elevated concentration levels of NO_2 and O_3 in the three commercial and industrial cities highlight the enhancement of atmospheric oxidizing power in the PRD region, which may exert profound effects on atmospheric chemistry, public health as well as regional climate.

Specifically, both of SO_2 and O_3 concentrations in Hong Kong show significant increasing trends. Given that SO_2 would be oxidized to secondary sulfate more easily in the O_3 enhanced atmosphere, the simultaneously increasing O_3 and SO_2 are reasonably expected to trigger the formation of secondary sulfate and resulted in the pronounced visibility degradation. Wang et al. (2009) have suggested that short-term O_3 control strategies should be aimed at O_3 precursor sources in Hong Kong and the adjacent PRD while the long-term ones need to consider distant sources. Given that the increase of SO_2 concentration was due to the enhanced local SO_2 emissions of Hong Kong and Shenzhen (Chan and Yao, 2008) as well as elevated regional SO_2 emissions in East Asia (Lu et al., 2010), similar strategy should be adopted for SO_2 control.

In Macao, the increasing trends of secondary pollutants like O_3 and NO_2 highlight the influences of increasing O_3 precursor emission in its upwind regions such as eastern PRD and the east Chinese coast. Therefore, the air pollution controls of Hong Kong, as well as other upwind regions are vital. Furthermore, the augment of NO_2 concentration also implies that the local vehicular emission control of Macao should not be neglected. As PM_{10} in Macao was mainly affected by local vehicular emission and regional transport, its increasing trend also emphasized the importance of both local and regional controls.

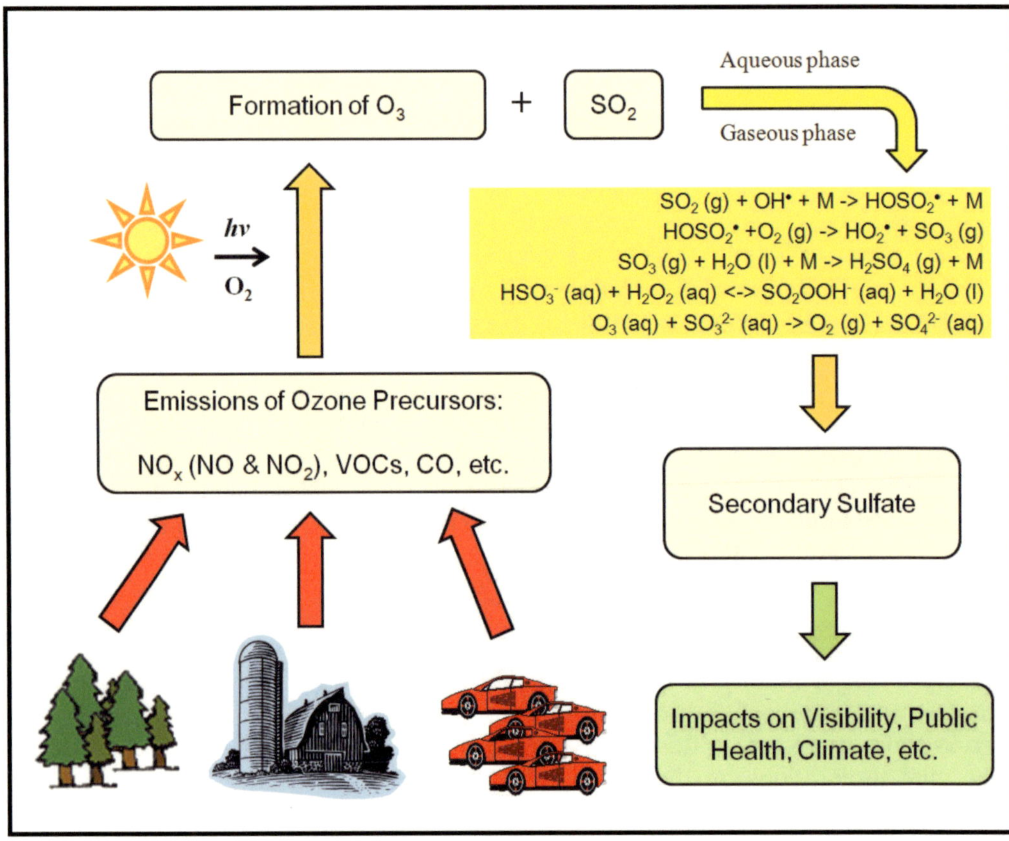

Fig. 8. Schematic diagram showing photochemical ozone formation from its precursors emitted from anthropogenic and natural sources and its impact on secondary sulfate formation and visibility degradation through aqueous and gaseous phase chemical processes in the ozone-enhanced atmosphere.

On the other hand, decreasing trends of PM_{10} and SO_2 were observed in Foshan. These are due to the efforts of more efficient control in industrial sectors. In a recent paper, Wan et al. (2011) emphasized that the implementation of stricter PM_{10} and SO_2 control measures in industrial exhaust had resulted in reduced ambient air pollution level. However, this industrial city is still severely polluted compared to the commercial cities like Hong Kong and Macao due to its high coal burning and industrial emissions. Such fact suggested that the SO_2 as well as particles emission control measures must be strengthened. In contrast, the surprising increases of NO_2 concentration due to the continuous increasing in road traffic and lack of efficient NO_2 control in industrial sectors should be noticed. Such increasing trend is an evident of atmospheric oxidizing power enhancement in this region. If the air quality control strategies are not changed, the NO_2 concentration will keep increasing especially with the fact that the vehicle number keeps on increasing sharply in recent years.

Overall, due to the rapid economical and industrial development, the nature of air pollution in the PRD region has been changing sharply. Essentially, the findings reported in this chapter indicated that the atmospheric oxidizing power in the PRD region has been continuously increasing. Even in the industrial city, which was more affected by primary pollutants like SO_2, the vehicle emission and resulted photochemical pollution should be noticed. The enhancing atmospheric oxidizing power could change the atmospheric chemistry by triggering the secondary pollutant formation via increasing oxidation rate. The increasing secondary pollutants may exert a number of significant impacts on public health and regional climate. As most of secondary pollutants were related to transport of aged air mass, such findings emphasize the critical impacts of trans-boundary transportation and regional air pollutions. With an aim to control the secondary pollutants, cooperation with other provinces and even countries become more and more important. Besides, due to the enhanced atmospheric oxidation capacities, primary pollutants like SO_2 and some of volatile organic compounds will be transformed to secondary fine particles more efficiently, and therefore, stricter implementation of local primary pollutant control was also necessary. In summary, in order to efficiently improve the regional air quality in the PRD region, local and regional efforts on air pollution control are both required.

6. Acknowledgment

We thank Hong Kong Observatory, Hong Kong Environmental Protection Department, Meteorological and Geophysical Bureau of Macao, Environmental Protection Bureau of Foshan and Meteorological Bureau of Foshan for providing the data. This study was supported by a key project of the Natural Science Foundation of Guangdong Province, China (No. 825102501000002), the National Natural Science Foundation of China (No. 40875075), a joint fund of the National Natural Science Foundation of China and Natural Science Foundation of Guangdong Province, China (No. U0833001), and Rosa Luxemburg Foundation.

7. References

Chan, C.Y. & Chan, L.Y. (2000). Effect of meteorology and air pollutant transport on ozone episodes at a subtropical coastal Asian city, Hong Kong. *Journal of Geophysical Research*, 105(D16), 20707-20724.

Chan, C.Y., Chan, L.Y. & Harris, J.M. (2004). Urban and background ozone trend in 1984-1999 at subtropical Hong Kong, South China. *Journal of Ozone: Science and Engineering* 25, 513-522.

Chan, C.K. & Yao, X.H. (2008). Air pollution in mega cities in China. *Atmospheric Environment* 42, 1-42.

Chan, L.Y., Chan, C.Y. & Qin, Y. (1998). Surface ozone pattern in Hong Kong. *Journal of Applied Meteorology*, 37, 1153-1165.

Che, H.Z., Zhang, X.Y., Li, Y., Zou, Z.J. & Qu, J.J. (2007). Horizontal visibility trends in China 1981-2005. *Geophysical Research Letters* 34, doi:10.1029/2007GL031450.

Che, H.Z., Zhang, X.Y., Li, Y., Zou, Z.J., Qu, J.J. & Hao, X.J. (2009). Haze trends over the capital cities of 31 provinces in China, 1981-2005. *Theoretical and Applied Climatology* 97(3-4), 235-242.

Deng, X.J., Tie, X.X., Wu, D., Zhou, X.J., Bi, X.Y., Tan, H.B., Li, F. & Jiang, C.L. (2008). Long-term trend of visibility and its characterizations in the Pearl River Delta (PRD) region, China. *Atmospheric Environment* 42(7), 1424-1435.

Guangdong Provincial Environmental Protection Monitoring Center & Hong Kong Environmental Protection Department (GDEPMC & HKEPD) (2005). Pearl River Delta Regional Air Quality Monitoring Network Operational Report for the Second Half of 2005. Report number : PRDAIR-2005-1.

GDEPMC & HKEPD (2008). Pearl River Delta Regional Air Quality Monitoring Network Report. Report number : PRDAIR-2008-2.

Ghude, S.D., Van der A, R.J., Beig, G., Fadnavis, S. & Polade, S.D. (2009). Satellite derived trends in NO_2 over the major global hotspot regions during the past decade and their inter-comparison. *Environmental Pollution* 157, (1873-1878).

Hua, W., Chen, Z.M., Jie, C.Y., Kondo, Y., Hofzumahaus, A., Takegawa, N., Chang, C.C., Lu, K.D., Miyazaki, Y., Kita, K.,Wang, H.L., Zhang, Y.H., Hu, M., 2008. Atmospheric hydrogen peroxide and organic hydroperoxides during PRIDE-PRD'06, China: their concentration, formation mechanism and contribution to secondary aerosols. Atmospheric Chemistry and Physics 8, 6755e6773.

Kaufmann, R.K., Seto, K.C., Schneider, A., Liu, Z., Zhou, L. & Wang, W. (2007). Climate response to rapid urban growth: evidence of a human-induced precipitation deficit. *Journal of Climate*, 20: (2299-2306).

Leung, Y.K., Wu, M.C. & Yeung, K.K., 2007. A preliminary study on the relationship between visibility and atmospheric suspended particulate concentration in Hong Kong. Hong Kong Observatory Reprint No. 692. (In Chinese with English abstract)

Liang, J.Y. & Jacobson, M.Z. (2000). Effects of subgrid segregation on ozone production efficieny in a chemical model. *Atmospheric Environment* 34(18), (2975-2982).

Lu, Z., Streets, D.G., Zhang, Q., Wang, S., Carmichael, G.R., Cheng, Y.F., Wei, C., Chin, M., Diehl, T., & Tan, Q. (2010). Sulfur dioxide emissions in China and sulfur trends in East Asia since 2000. *Atmospheric Chemistry and Physics*, 10, 6311-6331.

Meng, Z., Dabdub, D. & Seinfeld, J.H. (1997).Chemical coupling between atmospheric ozone and particulate matter. *Science*, 277, 5322, 116-119.

National Bureau of Statistics China (2009). China Statistical Yearbook 2008. Beijing: China Statistics Press.

Peng, J.N. (2007). The ceramic production transfer in Foshan. Available at URL: http://finance.sina.com.cn/roll/20071206/15211840962.shtml. (In Chinese)

Penkett, S.A., Jones, B.M.R. & Eggleton, A.E.J. (1979). A study of SO_2 oxidation in stored rainwater samples. *Atmospheric Environment* 13(1), 139-147.

Qiu, J.H. & Yang, L.Q. (2000). Variation characteristics of atmospheric aerosol optical depths and visibility in North China during 1980–1994. *Atmospheric Environment* 34(4), 603-609.

Richter, A., Burrows, J.P., Nüß, H., Granier & C., Niemeier, U. (2005). Increase in tropospheric nitrogen dioxide over China observed from space. *Nature* 437, 129-132.

Seinfeld, J.H. & Pandis, S.N. (1997). Atmospheric chemistry and physics: From air pollution to climate change. New York: Wiley-Interscience.

Shao, M., Tang, X.Y., Zhang, Y.H. & Li, W.J. (2006). City clusters in China: air and surface water pollution. *Front Ecol Environ* 4(7): 353-361.

Sisler, J.F. & Malm, W.C. (2000). Interpretation of trends of $PM_{2.5}$ and reconstructed visibility from the IMPROVE network. *Journal of the Air & Waste Management Association* 50, 775-789.

Stockwell, W.R. & Calvert, J.G. (1983). The mechanism of the HO-SO_2 reaction. *Atmospheric Environment* 17(11), 2231-2235.

Tang, J.H., Chan, L.Y., Chan, C.Y., Li, Y.S., Chang, C.C., Wang, X.M., Zou S.C., Barletta, B., Blake, D.R. & Wu, D. (2008). Implications of changing urban and rural emissions on non-methane hydrocarbons in the Pearl River Delta region of China. *Atmospheric Environment* 42, 3780-3794.

Van der A, R.J., Peters, D.H.M.U., Eskes, H., Boersma, K.F., Van Roozendael, M., De Smedt, I. & Kelder, H.M. (2006). Detection of the trend and seasonal variation in tropospheric NO2 over China. *Journal of Geophysical Research* 111, doi:10.1029/2005JD006594.

Wan, J.M., Lin, M., Chan, C.Y., Zhang, Z.S., Engling, G., Wang, X.M., Chan, I.N. & Li, S.Y., (2011). Change of air quality and its impact on atmospheric visibility in central-western Pearl River Delta. *Environmental Monitoring and Assessment*, 172, 339-351.

Wang, X.M., Lin, W.S., Yang, L.M., Deng, R.R. & Lin, H. (2007). A numerical study of influences of urban land-use change on ozone distribution over the Pearl River Delta region, China. *Tellus B*, 59,3 633-641.

Wang, T., Wei, X.L., Ding, A.J., Poon, C.N., Lam, K.S., Li, Y.S., Chan, L.Y. & Anson, M. (2009). Increasing surface ozone concentrations in the background atmosphere of southern China, 1994-2007. *Atmospheric Chemistry and Physics Discussions* 9, 10429-10455.

Wu, Y., Hao, J.M., Fu, L.X., Wang, Z.S., Tang, U.W. (2002). Vertical and horizontal profiles of airborne particulate matter near major roads in Macao, China. *Atmospheric Environment* 36, 4907-4918.

Wu, Y., Hao, J.M., Fu, L.X., Hu, J.N., Wang, Z.S., Tang, U.W. (2003). Chemical characteristics of airborne particulate matter near major roads and at background locations in Macao, China. *The Science of the Total Environment* 317, 159-172.

Yin, H. (2008). Analysis and prospect of the tile industry in 2008. *China Ceramics*, 44(5), 3-4. (In Chinese with English abstract)

Zhang, Y.H., Hu, M., Zhong, L.J., Wiedensohler, A., Liu, S.C., Andreae, M.O., Wang, W. & Fan, S.J. (2008). Regional Integrated Experiments on Air Quality over Pearl River Delta 2004 (PRIDE-PRD2004): Overview. *Atmospheric Environment* 42, 6157-6173.

Zheng, J., Zhong, L., Wang, T., Louie, P.K.K. & Li, Z. (2010). Ground-level ozone in the Pearl River Delta region : Analysis of data from a recently established regional air quality monitoring network. *Atmospheric Environment* 44, 814-823.

11

Causes and Consequences of Air Pollution and Environmental Injustice as Critical Issues for Science and Environmental Education

Anastasia Dimitriou[1] and Vasilia Christidou[2]
[1]Democritus University of Thrace
[2]University of Thessaly
Greece

1. Introduction

1.1 Humanity facing complex environmental issues

Modern societies suffer from a diversity of problems involving crucial environmental aspects. These problems concern biodiversity conservation, clean water and clean unpolluted air availability, healthy food, personal and public health and well-being. The causes of these problems lay in the way individuals -as well as entire societies- interact with the natural and form the man-made environment by adopting specific practices and developmental choices. They are also connected with the way people perceive the relationship between human beings, nature and society, in other words the way in which we perceive nature, ourselves and others. For decades -even for centuries- the kind of this relationship has been determined by the conviction of man's dominance over nature and of the more powerful over the weak parts of the population. Thus, in order to face environmental problems and protect the environment it is essential that the relationship between human beings, nature and society is revised. As Einstein aptly put it *"we can't solve problems by using the same kind of thinking we used when we created them"*.

Confronting environmental problems is not an unequivocal process, since they involve multiple dimensions and are determined by a variety of factors. They entail physical, chemical, biological, ecological, social, economic, and political dimensions, intrinsically interrelated and interacting, and often conflicting. Also, environmental problems are usually ideologically 'charged', incorporate different social groups' views and interests, and are connected with decisions taken at economic, political and social level.

Therefore, facing environmental problems is a complex procedure requiring –among others-determining their causes and consequences, their management, the development of alternative solutions and decision-making. It also requires disentangling their components and characteristics and critically determining their inherent interrelations, mechanisms, and conflicts. For this procedure to be effective, an understanding of the environment, its functions and constituent systems –both natural and social- as a whole is imperative. Hence, addressing environmental issues calls for interdisciplinary, systemic and holistic approaches. It also calls for promoting citizens' critical and systemic thinking, in order for them to identify the causes –and not simply the symptoms- of these issues, to understand the conflicts embedded in their consideration and treatment by different social, political, or interest groups (Dimitriou, 2009; Flogaitis, 2006; Orr, 1992; Scott & Oulton 1998; Sterling, 2004; Tilbury & Ross, 2006).

In this context, the role of science and environmental education is important. These fields of study concentrate on mechanisms and phenomena of the natural and man-made environment and provide students –as future citizens- the knowledge, skills and methodological tools necessary for studying the world around them. Both science and environmental education can contribute to the development of citizens capable of analytical and synthetic thinking, of investigating their position in the world they live in, of understanding the ways in which people change their environment and are affected by it and its changes, and conscious of the relationships between the individual, society, and the environment. Citizens who will be equipped with knowledge and skills essential for appreciating contemporary environmental problems, making relevant decisions and taking actions for their solution in the perspective of a sustainable future. Moreover, students, as future citizens could be supported in perceiving themselves as equivalent in their relationships with other people and natural entities, with a view to counterbalancing the dominating view of man's dominance over nature.

1.2 Sustainability: A key idea to cope with environmental degradation and social injustice

It is widely acknowledged that modern lifestyles and developmental models exceed earth's ability to support human beings and other organisms in terms of providing clean and adequate natural resources. An important concept related to modern lifestyles and the resulting environmental degradation is that of the 'ecological footprint'. Ecological footprint is a measure of people's demand on natural ecosystems. It is a measure of the area of biologically productive[1] land and water an individual, population or activity needs to produce the resources it consumes and absorb the waste it generates using current technology and resource management practices (Global Footprint Network, 2011). In other words, ecological footprint represents the land area an individual, a city or a nation take to support their lifestyles. It is tightly connected with the amounts of waste and pollutants produced by human activities as well as their energy demands. Probably the personal ecological footprint differs from one person to another as well as from one city (or country) to another. More particularly, the ecological footprint of developed countries is much larger than that of developing ones (Global Footprint Network, 2005). This indicates that ecological footprint is related with the notion of environmental injustice.

This has lead societies to recognize the necessity of changing the way of thinking about human activities, focusing on practices that sustain and protect the environment. Adopting sustainable everyday practices and developmental interventions is a crucial challenge for modern societies. Sustainability means the capacity to endure and refers to both physical and man-made systems, living and non-living things. Thus the term "sustainability" is used as determinative for that kind of development that ensures healthy ecosystems and the potential of long-term preservation of humanity.

---

[1] The land and water area that supports significant photysynthetic activity and biomass accumulation used by humans.

The terms of sustainability and sustainable development have been defined in various ways. The most common definition is given in the report of the World Commission on Environment and Development entitled "Our Common Future", also known as the Brundtland Report, according to which *"Sustainable development is development that meets the needs of the present without compromising the ability of future generations to meet their own needs"* (WCED, 1987, paragraph 27). In particular, as it is quoted (WCED, 1987, paragraph 27):
"The concept of sustainable development does imply limits -not absolute limits but limitations imposed by the present state of technology and social organization on environmental resources and by the ability of the biosphere to absorb the effects of human activities. But technology and social organization can be both managed and improved to make way for a new era of economic growth. The Commission believes that widespread poverty is no longer inevitable. Poverty is not only an evil in itself, but sustainable development requires meeting the basic needs of all and extending to all the opportunity to fulfil their aspirations for a better life. A world in which poverty is endemic will always be prone to ecological and other catastrophes".
Sustainability contains within it, at least, three crucial concepts:

- the fulfilment of human needs -and in particular the essential needs of the world's poor- that support improvement of people's lives;
- the fulfilment of the needs of present and future generations that support the value of intergeneration equity;
- the ability of the biosphere to absorb the effects of human activities.

So, sustainability is an evolving concept that contains the key ideas of improving everyone's quality of life (including present and future generations), in the frame of the limits that are posed by the carrying capacity of the ecosystems that support lives. The carrying capacity of an ecosystem is the population size of the species that the ecosystem can support indefinitely, providing the necessities (food, habitat, water, air, and ground) without being damaged (Hui, 2006).
A crucial factor for accomplishing sustainability is the appreciation of social injustice resulting from environmental degradation. Both researchers and international organisations agree that the degree to which the quality of life of individuals and societies is influenced by environmental degradation is different between members of a society as well as between societies (Bard et al 2007; Cifuentes, & Frumkin, 2007; UNICEF, 2008; WHO, 2003a, 2006, 2008a).
These differences create or amplify existing inequalities both between and within countries, related to the quality of basic sanitary infrastructures, food sufficiency, or people's access to clean air and potable water. These commodities constitute fundamental rights for life and health. Nowadays -despite technological and cultural achievements- people suffer from lack of clean, potable water, food insufficiency, polluted air, and deficient sanitary infrastructure. Conditions intensifying environmental injustice threaten fundamental human rights. Klaus Toepfer, Executive Director of the United Nations Environment Programme appositely described this reality at the 57th Human Rights Commission meeting in Geneva in 2001 (Shelton, 2002):
"Environmental conditions clearly help to determine the extent to which people enjoy their basic rights to life, health, adequate food and housing, and traditional livelihood and culture. It is time to recognize that those who pollute or destroy the natural environment are not just committing a crime against nature, but are violating human rights as well. Human rights cannot be secured in a degraded or polluted environment. The fundamental right to life is threatened by soil degradation and deforestation and by exposures to toxic chemicals, hazardous wastes, and contaminated drinking water".

The study of environmental quality in relation to social inequalities produced by environmental degradation is supported and promoted by various researchers. Some underline the role of environmental science in this direction and focus on the development of scientific methods to estimate pollution of environmental recipients, i.e. air, ground, and water (Stephens, 2007). Others emphasise the relation of environmental quality with quality of life and particularly with human health (Donohoe, 2003; Heynen, 2003; Mitchell & Dorling, 2003).

In this chapter we aim at elaborating on the idea of environmental justice as central in organising and implementing relevant educational interventions in the context of science and environmental education at school, as a means to cope with environmental and social injustice and promote sustainability. In the following sections of this chapter we discuss the problem of air pollution and further develop the issues of environmental justice and injustice. More particularly, air pollution, its causes and consequences are discussed, with an emphasis on human health and quality of life. Data demonstrating environmental injustice caused by air pollution are presented and the idea of environmental justice is introduced as a means to counterbalance it. The discussion then focuses on the contribution of science and environmental education to facilitate students' and (future) citizens' understanding of the causes and consequences of air pollution as critical dimensions of environmental injustice. Suggestions and implications for developing relevant educational programs at school are also provided, aiming at enhancing students' scientific knowledge and skills necessary to understand air pollution and determine actions for resolving it in view of sustainability and environmental justice.

2. Air pollution and environmental injustice

Air pollution is one of the most pressing environmental problems and is among the critical challenges facing modern societies. Air pollution is responsible for major harmful effects on human health, animal lives, natural ecosystems and the man-made environment. It is also responsible for climate change due to the enhanced greenhouse effect, acid rain, and the depletion of the ozone layer that constitute important global environmental problems. Air pollution occurs both outdoors and indoors and is caused by human activities and natural mechanisms. Although air quality degradation because of air pollution is a global problem that threatens humanity's wellbeing overall, it is largely associated with environmental injustice.

2.1 Outdoor air pollution

Outdoor air pollution is defined as the release of several substances -called air pollutants- into the atmosphere, in concentrations that threaten the wellbeing of living organisms or disrupt the function of the environment as a system leading to human health damages in various ways. Air pollutants can be gaseous, solid particles, or liquid droplets. They can be products of either natural processes or human activities. Natural sources of air pollutants include volcanic activity, forest fires, organic decay or soil dispersion into the air by the wind. The main anthropogenic sources of pollutants released in the atmosphere are human activities such as transportation (motor vehicles, aircrafts), burning coal or other fossil fuels for energy demands, industrial processes, or use of chemicals in agriculture, and facilities like power plants, incinerators, landfills for waste deposition (EEA, 2006a; 2006b; EPA, 2007; Valent et al., 2004).

The major outdoor air pollutants produced by human activities include, among others (EEA, 2006a; 2006b; EPA, 2007):

- Carbon oxides, especially carbon monoxide (CO) and carbon dioxide (CO_2) produced by the transportation sector (motor vehicle exhaust) and combustion of fossil fuels;
- Nitrogen oxides (NO_X), especially nitrogen dioxide (NO_2) emitted from high temperature fossil fuel combustion and electricity production;
- Sulphur oxides (SO_X), produced in various industrial processes such the smelting of sulphur-bearing ores for extracting metals and electricity production;
- Volatile organic compounds (hydrocarbons, VOCs) that include a variety of substances released from power plants and from industries producing numerous products such as painting colours, cleaning products, pesticides, building materials, and furniture;
- Particulate matter (PM), that is solid or liquid air pollutants mainly emitted by power plants and the transportation sector (aircrafts, motor vehicles), mining, and incinerators;
- Ground level ozone (O_3), an air pollutant that results from photochemical reactions between nitrogen oxides (NOx) and volatile organic compounds (VOCs) in the presence of sunlight;
- Toxic metals particularly lead (Pb), cadmium (Cd) and cooper (Cu) that are emitted from the transportation sector (motor vehicle exhaust), as well as from industrial procedures (production of painting colours, mining processes).

Outdoor air pollution is a major problem mainly in urban areas because of overpopulation. Increasing population[2] in urban areas results in increasing demands in transportation, industrial production and energy, which constitute the main sources of outdoor air pollution (UN, 2004). Moreover, this problem is intensified due to inadequate green open spaces in towns and cities and their restricted possibility to improve air quality and reduce air pollution (Givoni, 1991).

2.2 Indoor air pollution

Indoor air pollution refers to the amount of chemical, biological and physical contaminants in the air inside a building. Indoor air pollution can often be worse than outdoor air pollution. There is a wide range of indoor air pollutant sources in houses including building materials, pressed wood products and furniture, central heating and cooling systems, several personal care or household cleaning products, painting colours, solvents, heating or cooking appliances (stoves, wood and gas burning fireplaces, gas heaters), tobacco smoke, office machines and a variety of other products used in daily activities (EPA, 2007, 2008, 2009). In addition, outdoor pollutants are also traced in indoor spaces, such as radon (a natural trace component of soil and ground). The major source of indoor air pollution is household combustion of coal or biomass for cooking and heating. It is estimated that more than half of the world's population relies on animal dung, wood, crop waste, or coal to meet their most basic energy needs (WHO, 2005).

The major indoor pollutants that can practically be traced in any home are -among others- formaldehyde, asbestos, carbon monoxide, nitrogen oxides, sulphur oxides, benzene, polyaromatic compounds, particulate matter, toxic metals (lead, mercury, cadmium, and chromium), and volatile organic compounds (EPA, 2007, 2008, 2009; WHO, 2005).

---

[2] It is estimated that in 2010, 50.5 per cent or 3.5 billion of the people on Earth were living in cities (UN, 2004).

Several factors affect indoor air quality and are related to the type and maintenance of their sources, the level of hazard of the emitted pollutants, the quality of house ventilation and the quality of the building as well. For example, an old air pollutant source (a stove, a gas heating system, etc.) that is not properly maintained can emit significantly larger amounts of air pollutants than a source that is properly maintained (EPA, 2009). Indoors air temperature and humidity, biological contaminants (bacteria, mold, pollen, and viruses) as well as overcrowding are also important factors determining the quality of indoors air. In particular, high air temperature and humidity can affect the rate of indoors pollutant emissions. Moreover, these factors combined with inadequate building ventilation can further degrade air quality since indoors pollutants cannot be removed. Indoor air degradation in buildings may result in several health problems such as headaches, tiredness, hypersensitivity, coughing, eye, nose, and throat irritation, skin and mucosal dryness, dizziness, nausea, lethargy, inability to concentrate and mental fatigue. The occurrence of these symptoms is defined as the "sick building syndrome"(Burge, 2004; Stolwijk, 1991).

The "sick building syndrome", is also related to workplaces. In order to ensure healthy conditions at workplaces international organizations have defined upper limits of both environmental and chemical factors that could cause serious problems or disturbances to human health. Among these factors thermal conditions (air temperature and humidity) as well as the levels of dust and chemicals concentration in the air are recognized as important factors affecting health and safety conditions at workplace (ILO, 1993; OSHA, 2009).

2.3 Air pollution, health and environmental injustice

Numerous outdoor and indoor air pollutants affect human health and pose significant threats to individuals worldwide, such as cardiovascular or respiratory disorders, asthma and lung cancer, which can be fatal (WHO, 2002, 2003a, 2006, 2009). Given that an individual inhales more than 14.000 litres of air per day, it becomes obvious that air pollutants pose significant dangers for human health. According to the World Health Organization (WHO, 2006, 2009) more than two million premature deaths each year can be attributed to the effects of urban outdoor and indoor air pollution. In particular, every year indoor air pollution is responsible for the death of 1.6 million people while 800.000 deaths from lung cancer, cardiovascular and respiratory diseases worldwide are attributed to outdoor air pollution (Valent et al., 2004; WHO, 2005). It is estimated that currently air pollution reduces average life expectancy of Europeans by 9 to 24 months (CEC, 2006).

Do urban populations equally enjoy environmental quality? Intense social injustice is often located within cities and is related to the way in which the man-made environment is formulated, functions and develops in time and space. This evolution is a result of various social, cultural, economic, and political processes, along with environmental components and their interrelations, which combine to shape space consumption in an urban environment determining its residential quality from an environmental perspective. These processes are typically regulated by individuals belonging in higher socio-economic strata, having access to decision-making and economic activity centres, which allows them to act in favour of financial profit and against lower socio-economic status populations (Bard et al., 2007; FOE, 2001; Harner, Warner, Pierce & Huber, 2002).

Some dimensions of environmental quality of an urban area entailing social injustice include the spatial distribution of functions and activities, the distribution of outdoor and green spaces and the quality of residential and natural environments. Distribution of functions and activities in urban spaces frequently involves environmental injustice, creating environmentally degraded areas. Noisy facilities (industries, landfills, biological treatment facilities, highways) serving large urban populations are usually located in degraded residential areas. These facilities are responsible for pollutant emissions, noise, and also entail other dangers (e.g. explosions, fires) threatening nearby residents' physical and mental health. Air quality in these regions is typically low due to toxic gaseous substances emitted (Bard et al., 2007). According to international epidemiological studies, the populations living in the vicinity of large-scale urban facilities like industrialized regions exhibit higher frequencies of respiratory and cardiovascular incidents compared to inhabitants of other regions of the same cities. These incidents are related with pollutants traced in the air of these regions (Bard et al., 2007; Harner, Warner, Pierce & Huber, 2002; WHO, 2003a, 2003b).

As far as green and free open spaces are concerned, apart from being generally inadequate, they are not evenly distributed in all city districts; lack of public open spaces is more intense in densely inhabited and degraded quarters. Green space development requires free open spaces that are not always available, mainly because of the density of an urban area. Nevertheless, density is differentiated from one district to another within the same city, which results in further degradation of unprivileged neighbourhoods compared to privileged ones. As Heynen (2003, p. 980) points out: *"Social production of urban environments explicitly leads to uneven urban environments and environmental injustice. Environmental inequalities clearly exist within cities".*

Furthermore, air pollution occurs in both rural and urban areas because of transportation, industrial emissions and energy use. What is common between highly polluted rural and urban areas is that they are both correlated with low income residential quarters, which indicates another facet of social and environmental injustice (FOE, 2001; Mitchell & Dorling, 2003). In other words, higher socio-economic status social groups' actions influence life quality of lower socio-economic status groups within countries. In a study implemented in London respiratory problems have been found to concentrate in the poorest areas and to correlate with high traffic levels even though the responsibility for the causes of these problems is unequal -car ownership was lower in areas with worse traffic levels (Stevenson et al., 1999).

Environmental differences and inequalities related to air pollution also appear between countries. In developing countries the issue of air quality is mainly related with indoor air pollution. In most of these countries organic fuel, such as wood and coal are the most common energy resources available for domestic use –e.g. for heating and cooking. Their combustion releases harmful pollutants inside houses, especially solid particles (WHO, 2005). Chronic exposure to solid particles is considered to contribute to the development of cardiovascular and respiratory diseases, as well as of lung cancer, while it is related with mortality mostly among children (WHO, 2006). Actually, 20%[3] of child mortality for children under 5 years is attributed to indoor pollution (WHO, 2008a). Also, it is estimated that indoor air pollution associated with using biomass fuels (such as wood and grasses) causes an average of 871,500 deaths annually from childhood pneumonia (Valent et al., 2004).

As already indicated, dangers of air pollution are particularly increased in the case of children, who –in proportion to their body weight and surface- inhale larger quantities of air than adults, thus taking in larger amounts of the dangerous substances they are being exposed to. Moreover, their body systems are not mature, but are still evolving during

---

[3] This percentage corresponds to two million children under five years of age annually.

childhood. This implies that children are more vulnerable to incoming toxic substances than adults. Their tissues directly absorb these substances, and -due to their immaturity- they are not able to restore the damages caused. So, if their brain, immune, or endocrine system are affected by toxic substances, the damage caused can be permanent and irreversible (Landrigan et al., 1998).

Data relate the development of leukemia in children living in areas with high concentrations of suspended particles. Also, children suffering from asthma living in areas with increased air pollution-especially pollution from nitrogen oxides, acid vapours and particulate matter- exhibit a higher probability of deterioration (e.g. development of bronchitis symptoms). These data are further supported by research indicating that when children suffering from respiratory disorders leave areas with high particulate air pollution, their lung function improves, while this is not the case for those who remain in polluted areas (WHO, 2002, 2003a).

In addition, epidemiological studies in European countries indicate that morbidity and mortality are higher in low socio-economic status populations[4] suffering unhealthy environmental conditions. This especially concerns populations living in industrialised regions and nearby power plants (Bard et al., 2007; Cifuentes & Frumkin, 2007; UNICEF, 2008; WHO, 2003a, 2006, 2008a, 2009). Usually these regions are inhabited by low socio-economic status populations intensively experiencing the consequences of environmental degradation without being responsible for its emergence. On the contrary, they suffer from choices made by individuals of higher socio-economic status, pursuing economic growth. These choices concern investments entailing excessive use or degradation of natural resources. As a result, those who produce environmental degradation enjoy profits, while others suffer from its consequences. In this frame air pollution poses a threat to the fundamental right of life in a healthy environment and creates environmental injustice.

Another issue related with air quality degradation which entails social injustice involves air quality in working environments. These constitute local environments frequently containing harmful indoor air components, which may directly influence employees' health. They include various physical, chemical and biological components[5] released during the production process and degrade the quality of the air inhaled causing severe problems (ILO, 1993). It is estimated that in the US approximately 135 employees die every day from diseases caused by long-term exposure to toxic[6] substances at work (Faber & McCarthy, 2002). According to data available by European and World organisations high rates of morbidity and mortality are recorded among employees due to their exposure to environmental dangers in their workplaces. Obviously, low income employees -typically in lower ranking occupations- are those who suffer indoor air pollution in their workplaces the

[4] In Western Europe danger of disease caused by degraded environmental quality is 1.5-2.5 times higher in low socio-economic status populations.

[5] Physical environmental components include light, noise, radiation (ionizing or non-ionizing), dust, temperature, and humidity of indoor air. Chemical components are substances belonging in the production process and may be dangerous gaseous, toxic, erosive, or explosive substances. Biological components involve pathogens, bacteria, or viruses that may be traced in the air of a working environment.

[6] Toxicity is the degree to which a substance is able to externally or internally damage an exposed organism. Toxicity danger is the probability of harmful effects caused by the use of a chemical compound.

most compared to those in highly ranked occupations –i.e. close to management (FACTS 29, 2002; WHO, 2003b, 2004).

2.4 Environmental justice

The issues raised in the previous paragraphs suggest that air pollution is directly related with social injustice. People are exposed to polluted air and their health and wellbeing are affected in socially unjust ways. Mitigating and overcoming this injustice does not mean equal exposure to environmental danger; it means elimination of every kind of pollution endangering the environment and living organisms. The challenge here lies in ensuring air quality, health and wellbeing for everyone. Health is not simplistically meant as absence of illness, but as a broader concept involving prevention as well as promoting physical, mental, and social wellbeing. It is a basic human value and a fundamental right of every individual regardless of race, religion, political beliefs, social, or economic status. It is the foundation of modern citizens', and accordingly our societies' and civilization's development and progress (WHO, 1992).

Environmental justice can thus bridge the gap between environmental quality and social injustice. Environmental justice is the fair treatment and meaningful involvement of all people with respect to the development, implementation, and enforcement of environmental laws, regulations, and policies. It will be achieved when everyone enjoys the same degree of protection from environmental and health hazards and equal access to decision-making processes to have a healthy environment in which to live, learn, and work (EPA, 2001).

The demand for environmental justice is expressed by organised social groups in the frame of the 'environmental justice movement'[7] which has been increasingly established and reinforced in people's consciousness. It originates from hunan rights movements and is organised on the basis of three central principles: equality in environmental danger distribution, recognition of otherness and diversity between individuals, and active participation in political processes designing and applying environmental policies (Faber, & McCarthy, 2002; Harner, Warner, Pierce & Tuber, 2002; Scholosberg, 2004).

Environmental justice thus encompasses a variety of environmental, social and political dimensions and components. It involves issues such as natural resources management and distribution, social justice, development for all citizens, the right to participation in co-constructing an environment of quality. It is therefore apparent that environmental justice and the relevant issues are considered as directly connected to human rights. The UN Declaration of Human Rights and the Environment recognized the strong relationship between human rights and environmental quality that is reflected in the following principles (UN, 1994, pp. 3-4):

1. Human rights, an ecologically sound environment, sustainable development and peace are interdependent and indivisible.
2. All persons have the right to a secure, healthy and ecologically sound environment. This right and other human rights, including civil, cultural, economic, political, and social rights, are universal, interdependent and indivisible.
3. All persons shall be free from any form of discrimination in regard to actions and decisions that affect the environment.

---

[7] The Environmental Justice Movement started in the United States in the early '80s and has since spread in European and other countries.

4. All persons have the right to an environment adequate to meet equitably the needs of present generations and that does not impair the rights of future generations to equitably meet their needs.
5. All persons have the right to freedom from pollution, environmental degradation and activities that adversely affect the environment, threaten life, health, livelihood, well-being or sustainable development within, across or outside national boundaries.

3. Environmental quality in the context of science and environmental education for the promotion of environmental justice

3.1 Environmental education and environmental quality

Environmental education has long been established as a premise for the development and promotion of education about the environment. In the beginning -during the '70s- it focused on improving environmental quality in view of the urgent and unprecedented environmental problems. During that period the environment was identified with 'nature' and its protection was a moral imperative. In this context, environmental education concentrated on the development of knowledge about nature and its functions; on the development of the skills necessary to face environmental problems on the basis of identifying their causes; and on encouraging appropriate behaviours for resolving them. In other words, environmental education focused on causal relationships and suggested that a change in behaviour (the cause) could amend the effect for the benefit of the environment (UN, 1973; UNESCO, 1976, 1978).

During the '80's and '90's environmental education was associated with developmental issues, since environmental problems were considered as instigated by developmental choices made by modern societies. Environmental education was enriched with concepts related with development, economy, and society, and aimed at supporting learners' capacity to investigate further from the physicochemical components of environmental problems, the socio-economic dimensions inherent in their creation in the framework of the relationships between individuals, society and the environment. The role of human activities and choices in environmental quality was acknowledged as critical in the formation and resolution of environmental problems (UNEP, 1992; UNESCO, 1988; UNESCO-EPD, 1997). At the same time the notion of sustainability -involving social processes, individual and collective values- was dynamically adopted as central in organising and achieving environmental education objectives. As indicated in the Declaration of Thessaloniki (UNESCO-EPD, 1997): *"The concept of sustainability encompasses not only environment but also poverty, population, health, food security, democracy, human rights and peace. Sustainability is, in the final analysis, a moral and ethical imperative in which cultural diversity and traditional knowledge need to be respected"* (article 10).

Since the early '00's the idea of sustainability has been further promoted as a crucial organising principle redefining the scope of environmental education. In the context of sustainability, development is not merely conceptualised as economic growth, or economic profit, but also takes into account environmental quality and social cohesion. Therefore, sustainable development encompasses populations, animal and plant species, ecosystems, natural resources (water, air, energy, agriculture, biodiversity), climate change, rural development, urbanization, and integrates concerns such as poverty reduction, human rights, health, human security, culture diversity and intercultural understanding (UNESCO, 2005, 2007).

In order to accomplish sustainable development environmental education aims at preparing individuals for being capable of co-constructing their future in terms of environmental and social sustainability by integrating values, activities and principles inherently linked to sustainable development and encourage a change in attitudes, behaviours and values to ensure a more sustainable future in social, environmental and economic terms (UNESCO, 2007). For this general aim to be accomplished particular objectives are considered as crucial for each of the following domains (UNESCO, 2005, p. 6):

- Society: an understanding of social institutions and their role in change and development, as well as the democratic and participatory systems which give opportunity for the expression of opinion, the selection of governments, the forging of consensus and the resolution of differences.
- Environment: awareness of the resources and fragility of the physical environment and the effects on it of human activity and decisions, with a commitment to factoring environmental concerns into social and economic policy development.
- Economy: a sensitivity to the limits and potential of economic growth and their impact on society and on the environment, with a commitment to assess personal and societal levels of consumption out of concern for the environment and for social justice.

Finally, sustainability is organised with reference to the notions of ecological limits, social justice, environmental justice, and individual participation in environmental actions. The formation of developmental processes based on environmental quality, balance of ecological processes, and quality of socio-political structures is imperative for ensuring a sustainable future aiming at the improvement of life quality of citizens (Huckle, 2006; Jacobs, 2004; Orr, 1992; Scott, & Gough, 2004). In this framework, both individuals and society should respect the boundaries set by ecological processes within and between natural systems. This requires an appropriate understanding of the concepts of carrying capacity and ecological footprint (see subsection 1.2) that determine the sustainable function of ecosystems and the degree to which human societies exert pressure on them in order to fulfill their diverse needs.

Therefore, environmental education can play an important role in supporting individuals to critically consider environmental decisions of contemporary societies and enabling them to investigate and challenge the social structures and political decisions that preserve or reinforce social and environmental injustice. At the same time, it can strengthen students' roles as future citizens by encouraging participation in the design and implementation of actions to resolve environmental problems, training them in democratic processes, and involving them in issues concerning their local communities. Such orientations build up citizens' confidence, mutual help and solidarity for the improvement of life quality. These values are the foundations of a democratic society with social and environmental justice.

3.2 Science education and environmental quality

Science has developed tools and knowledge for understanding the complex physicochemical mechanisms underlying environmental problems, the ways in which human interventions and everyday habits affect natural ecosystems and the ways in which ecosystems react to these interventions. Ecological footprint, ecosystems' carrying capacity, atmospheric and hydrosphere mechanisms are among these important tools for studying environmental problems, and estimating the degree to which human activities, practices and developmental interventions create or enhance them (AAAS, 1989; Hodson, 2003). Science teaching can address these issues to support students in constructing the environmental knowledge necessary for understanding environmental problems and undertaking appropriate actions for their resolution. Scientific knowledge and methodological tools can therefore enhance students' ecological literacy, which involves understanding natural phenomena and their interrelations, the interdependence between society, economy, the natural environment and the ways in which human activities affect the environment and are affected by it (Orr, 1992). In this framework natural systems are considered as constituents of the environment (also including the social, man-made, and economic environment). Therefore, science teaching can enhance students' ability to connect scientific knowledge with everyday issues, such as environmental problems which nowadays -more than ever before- are part of individuals' and society's daily life.

Besides, such an integration of scientific knowledge with the multiple components of environmental problems is expected to stimulate students' involvement in science, since these problems require an interdisciplinary connection of science teaching with education for the environment; on the one hand integrated scientific knowledge provides the means for understanding environmental issues by means of a holistic and systemic approach; on the other it provides insight to the human-environment interaction that enables environmental sustainability. From this perspective science teaching may become more meaningful for students, since it also incorporates a strong affective dimension and provides possibilities to develop evaluative judgments and take up environmentally appropriate actions in real life situations (AAAS, 1989; Brody, 1991; Bybee, 1993; Gough, 2002; Hodson, 2003; Iozzi, 1989; Schneider et al., 2002).

In this context science teaching enhances students' critical thinking, essential for understanding the complexity of environmental problems. Critical thinking requires that an individual is both mentally and emotionally engaged when facing a problem in order to end up with rational conclusions, judgments, and choices. It is a process of data collection, processing, analysis and synthesis with existing knowledge to explain concepts and phenomena, to understand personal conceptions and evaluate them against the ideas of others, and to question prejudices (Jones, Merritt & Palmer, 1999; Tilbury & Ross, 2006). Furthermore, data analysis and interpretation to extract conclusions are basic components of critical thinking, and are also considered as central in science education (Facione, 2007).

As already mentioned, systemic thinking is crucial for understanding environmental problems. Understanding the environment -as a 'system' interacting with other systems – involves conceptualising its organisation in time, its influxes, the decisions and measures taken by individuals and organisations about its management, and the way in which it reacts to the actions it is imposed to (Ragou & Souchon, 2000). Thus, a systemic approach of environmental problems requires systemic thinking, which is substantiated by means of the complementary processes of analytical and synthetic thinking (Bartlett, 2001; Espejo, 1994; Sterman 1994; Richardson, 1994; Richmond, 1993, 1994). The first process is essential in understanding the parts of a problem; the second enables comprehension of the way in which these parts interact within a system, as well as with surrounding systems. According to Richmond (1994) systemic thinking is a dynamic way of reasoning and learning, the 'art' and science of creating reliable conclusions about a system's behaviour by deeply understanding the structures regulating it. It is therefore evident that science education can provide essential knowledge and tools of thinking for studying natural systems and their interrelations with other -natural or man-made- systems.

Moreover, the study of environmental problems through the prism of environmental justice can emphasise the social role of science education in improving life quality and ensuring individual and public health. More particularly, an environmental perspective in science education allows for

- Application of scientific tools and knowledge in estimating environmental degradation and its consequences on health, and in revealing pertinent social injustice;
- Development of instruments to consider alternative ways of environmental management that ensure everyone's right to security and health;
- Application of scientific knowledge in considering the environment as a system comprising other natural and social systems and in understanding their interrelations as well as the relations producing social injustice;
- Development of scientific method skills to support critical and systemic thinking and promote the idea of the social role of science and technology in dealing with environmental injustice.

3.3 Connecting environmental and science education to promote environmental justice

The preceding analysis indicates that the connection of science with environmental education can elucidate environmental and social injustice inherent in environmental problems. This connection can serve basic objectives regarded as preconditions for students' engagement in understanding environmental problems and dealing with environmental injustice. These objectives include (Brody, 1991; Bybee, 1993; Orr, 1992; Peloso, 2007; Scott, 2007):

1. Supporting students in developing thinking skills (e.g. systematic observation and investigation, problem solving, critical synthesis of different parameters in decision making) to allow them satisfy basic personal needs and use natural resources rationally aiming at achieving environmental justice;
2. Supporting students in developing communication and collaboration skills so that they are capable of formulating and comparing different views, of constructing new, synthetic and consensual ideas ensuring environmental quality for all citizens;
3. Enhancing students' interest in science by revealing its relevance to real and important issues concerning life quality;
4. Developing learning processes that support students in constructing more adequate explanatory models about the natural world and the environment;
5. Revealing interrelations between science, technology, the environment, and society, and promoting positive attitudes towards the environment, aiming at improving its quality to the benefit of all people;
6. Presenting science and technology as useful means for the adaptation of humanity to the environment, since they provide rationally structured explanations of reality -as interpreted in every period and developmental phase- to ensure environmental quality for all;
7. Making clear that science and technology entail -apart from the necessary information and knowledge- value and ethical issues.
8. Supporting students' understanding of the importance of concepts like interdependence, sustainable development, conservation of natural resources, population control. Awareness of the sensitive environmental equilibrium could support the development of a 'universal ethos' to counteract social injustice and adopt sustainable actions for protecting, conserving and improving environmental quality.

4. Air pollution in the framework of science and environmental education to promote environmental justice

As already mentioned air pollution is related with environmental injustice inherent in its causes as well as its consequences. On the one hand, the causes of air pollution lie in people's lifestyles, which are differentiated according to their socio-economic possibilities. On the other, they are related to developmental choices and financial investments made by higher socio-economic strata, and to political decisions forming their institutional context. At the same time air pollution is associated with the way in which cities are organised, with their infrastructures and green spaces available determining the quality of urban environment and producing inequalities between residents. Thus, air quality is recognised as one of the priorities for healthy and sustainable cities.

Apart from knowledge about scientific concepts and the complexity of environmental phenomena, the development of educational programs related to environmental issues - such as air pollution- also requires that students' ideas and ways of thinking are taken into account. An analysis of learners' thinking and conceptions can provide valuable insight into the way in which individuals or social groups come to organise their reasoning and concern about the environment and the relevant problems. For instance, if one attributes environmental problems to the development of industry and believes that they can be faced through technological applications, s/he adopts a technocratic approach. On the other hand, those who attribute environmental problems to human activities embrace a more politicised stance (Wals, 1992). In the first case an individual may not develop further interest or an intention to act to protect the environment, given that technology can solve any problem. Therefore, students' thinking about environmental problems can guide the development of teaching material and activities that depart from their conceptions.

Students' ideas about air pollution have been recorded by a number of studies. Their main findings can be summarised as follows:

- Pupils conceptualise air pollution as presence of harmful substances into the atmosphere (Ali, 1991). From 10-11 years on they appreciate that air is polluted with 'extra gas(es)', 'fumes', or 'smoke', some of them identifying specific pollutants such as CFCs, carbon monoxide and dioxide, sulphur dioxide, and nitrogen oxides (Batterham et al., 1996; Boyes and Stanisstreet, 1997; Thornber et al., 1999).
- Students describe cars and transportation, industry, waste, fires and sprays as the primary anthropogenic pollution sources (Boyes and Stanisstreet, 1997; Thornber et al., 1999). They regard air pollution as exclusively man-made (Ali, 1991; Boyes and Stanisstreet, 1994; Brody, 1991; Brody, 1994; Dove, 1996; Thornber et al., 1999), adopt a restricted and local view of the issue, being unable to appreciate its global dimension (Brody, 1994), and perceive air pollutants as inert, i.e. not participating in physical or chemical processes in the atmosphere (Marinopoulos and Stavridou, 2002).
- Air pollution is considered to cause plant and animal diseases and/or death, as well as respiratory problems to humans (Ali, 1991; Batterham et al., 1996; Brody, 1994; Thornber et al., 1999). Some students attribute other global environmental problems to air pollution (Batterham et al., 1996). In this case, they often hold a generalised idea of global environmental issues, confusing them at the levels of causes, mechanisms and consequences (Boyes and Stanisstreet, 1997; Christidou & Koulaidis, 1996; Francis et al., 1993; Hillman et al., 1996; Koulaidis & Christidou, 1999; Plunkett and Skamp, 1994; Rye

et al., 1994). They often consider air pollutants in general as enhancing the greenhouse effect and ozone depletion.

- Research outcomes from a Greek study (Dimitriou & Crhistidou, 2007) exploring students' understanding of air pollution indicate that they have a generalised understanding of the problem. They are familiar with a variety of its components - such as sources of pollution, pollutant types, consequences on living things and on the environment- but they do not seem to comprehend the relationships between these components. At the same time they do not have an adequate understanding of the degree to which individual and social decisions and practices influence the intensity of air pollution. On the other hand, they seem to have a more appropriate understanding of the effects of air pollution on human health and explicitly express their concern about them.

The preceding analysis reveals that students are not aware of all the dimensions of air pollution and therefore cannot fully understand the interrelations and interactions between its components. Understanding the framework of relations between natural and human dimensions composing an environmental phenomenon such as air pollution is -in our view- a prerequisite for developing critical reasoning about it and should constitute a key educational objective for supporting future citizens in thinking about the world. Therefore, students should become competent in recognising and interpreting the framework of physical, ecological, social, economic and political relations underlying environmental issues, the contradictions inherent in those components, as well as possible alternative explanations and solutions to these issues. They should also become capable of sincerely examining their personal views, practices, prejudices and stereotypes, make appropriate decisions and implement changes at individual and collective level. In other words, students should be equipped with knowledge, skills and thinking tools necessary for interpreting the world around them in order to change it, since the world cannot change unless it is adequately explained (Huckle, 2004).

In this context, educational programs on air pollution should ultimately aim at determining all relevant components, indicating their interrelations and the ways in which these interrelations enhance or reduce the problem and its consequences. In other words, educational programs need to be designed in the framework of a systemic approach for air pollution, considering all the systems with which it interacts. Such a systemic approach would also contribute to understanding and resolving environmental injustice inherent in air pollution and related with people's rights, lifestyles, financial activities, along with state and governmental decisions. These systems reflect the social, political, and financial dimensions of air pollution and are tightly connected with the way individuals and organised societies (institutions, legislation, etc.) decide and act. The functions of these systems correspond to influxes to the air pollution 'system' and in turn influence the way it functions (at the level of physicochemical and biological mechanisms), the way it reacts to these influxes and its effects on the natural and man-made environment.

This approach provides the possibility of a holistic study of air pollution in order to understand natural and human components causing it and their interrelations. At the same time it provides an opportunity for understanding the temporal evolution of this issue due to human interventions -be it in terms of managerial practices or protection practices for its resolution.

The previous discussion allows for developing educational programs about air quality degradation and the relevant social injustice by studying indoor and outdoor local environments. Such programs would enhance students' critical understanding of environmental quality in the context of individual and social activities, practises and political actions. Within this context, environmental justice in the perspective of sustainability is a crucial component for maintaining a healthy environment for all, and for creating sustainable places to live and work, now and in the future. By integrating scientific knowledge with life experiences students would be encouraged to make personal decisions and become active members of society to ensure better air quality and environmental justice. More particularly, educational programs about air quality degradation and the relevant social injustice could aim at:

- **Investigating different lifestyles and the ways these affect air quality and enhance – or reduce-air pollution**
- Lifestyle describes the way an individual lives. It includes, among others, individual consumption behaviours and everyday life practises. Almost all of our daily activities and habits eventually produce indoors and outdoors air pollutants. Investigating their different lifestyles would give students the opportunity to understand how themselves and others affect air quality in their daily lives. Moreover, students could be encouraged to investigate alternative habits that would contribute to air pollution reduction, such as using alternative energy sources (e.g. solar, or wind energy), saving energy, or using eco-friendly products.
- **Assessing the personal ecological footprint on air and comparing ecological footprints of different socio-economic status individuals**
- This approach would provide students the opportunity to explore the way individuals of different economic status consume natural resources consequently affecting air quality. Since socio-economic status is plays a crucial role in determining consumption behaviours and everyday practices, such an assessment would aid students develop a deeper understanding of social and environmental injustice related to air pollution.
- **Estimating the factors influencing urban space quality (urban design, transportation networks, green spaces) and evaluating its quality in privileged and underprivileged areas within the same city to reveal ensuing injustice**
- Involving students in investigating urban space gives them the opportunity to realise that the way a city is organised affects the quality of citizens' lives. They are also assisted in recognising the relationship between environmental quality and human wellbeing and to locate environmental injustices in their local environments to concretise that urban quality is not the same within the same city. Students living in underprivileged areas, in areas of injustice, should be empowered to develop and participate in communities of resistance and planning (Peloso, 2007). As Bullard, one of the major researchers and organizers in the environmental justice movement, suggested: *"We have to educate young people that it is their right to have access to open space, green space, parks, outdoors, as opposed to people thinking that their supposed to be living in an area where the only park is a basketball court with no net. We have to give people this idea that it's their right to have access to open space and green space and we have to provide funds to make sure that we get them early on and take them on field trips, take them to a wilderness area, a refuge, a reserve, to a park-a real park and to integrate this information into our curriculum"* (Schweizer, 1999).

- **Estimating the effects of air pollution on human health and the distribution of relevant diseases in respect to environmental quality in privileged and underprivileged areas within the same city**
  Estimating the distribution of diseases that are attributed to urban air pollution gives students the opportunity to understand that environmental justice is the most significant tool for a healthy environment for all people regardless of socio-economic status.
- **Exploring pollution sources and their distribution in the local environment in relation to the socio-economic status of residents**
  This approach gives students the opportunity to realize that people of low income usually live in degraded residential areas where the main air pollution sources, such as industries, landfills, biological treatment facilities, highways, are usually located.
- **Determining the factors influencing air quality in different workplaces**
  Students could be encouraged to investigate air quality at different workplaces to locate related environmental injustice experienced by specific groups of employees. In this framework, the "sick building syndrome" could also be explored based on real case studies.
- **Measuring air pollutant concentrations in the school, local and broader environment**
  This activity would give students the opportunity to locate air pollutant sources in their immediate environment and suggest alternative interventions to their school in order to render it sustainable. Such interventions could include -among others- using building material, heating sources, furniture, and equipment with a low impact on air quality. Such actions would also enable all students participate in school life and in decision-making processes, to cooperate with local citizens and broaden their activities to improve the quality of the local environment and thus improve wellbeing of the local community.
- **Studying national and international legislation referring to air quality and engaging students in proposing legislative directions for the formation of urban environments, improvement of urban air quality and elimination of environmental injustice**
  Student's involvement in studying environmental legislation provides the opportunity of deeper understanding of the political dimension of air pollution and environmental injustice.
- **Familiarising students with national and local institutions in order to understand their role in the formation and conservation of the local environment**
  Student's familiarisation with institutions enables greater interaction with the community. It is also expected to increase their interest in local issues and their willingness to work together on reducing air pollution and promoting sustainability.

To determine environmental injustice at an international level, educational programs to assist students in understanding injustice between developed and developing countries could be developed. Indicative objectives of such programs would include:

- **Recording differences in energy sources used for satisfying basic daily necessities of developed and developing countries' inhabitants and identifying indoor air pollution in each case;**
- **Studying indoor air pollution effects on human health and especially on children;**
- **Studying geographical distribution of diseases related with air pollution in developed and developing countries;**
- **Comparing personal ecological footprints on air between inhabitants of developed and developing countries;**

- **Understanding poverty experienced by people in developing or unprivileged countries as determinative factor for the environmental injustices they experienced.**

Probably, one could object that studying local environmental problems and pertinent injustices solutions is much more meaningful to students than grasping -for instance- environmental injustices between development and developing countries. Are students able to cope with these injustices and promote air quality for the people in developing countries? We strongly believe that education has to promote global citizenship and allow students to understand that environmental problems are linked to the social, economic and political situations in countries all over the world. Therefore, science and environmental programs about air pollution could start from the local environment and deal with real, everyday and concrete problems facing students and their community and subsequently be enriched with activities that enable students to develop a deeper and global understanding of poverty as a critical dimension of environmental injustice (UNCSD, 2003). After all, "education is one of the most effective forces to bring about the changes in knowledge, values, behaviour and lifestyles required to achieve sustainability and stability within and among countries, and to guarantee democracy, human security and peace" (UNESCO, 2005b, p. 10).

5. Conclusions

In this chapter we attempted to present environmental quality in the context of environmental injustice. The discussion was based on data provided by international institutions mapping environmental degradation, its consequences on human health, wellbeing and life quality. These data reveal that environmental injustice within countries is particularly experienced by low socio-economic level groups of the population living in underprivileged areas. Moreover, environmental injustice between countries is more intense and degrades health quality of their population, indeed contributing to increase of mortality.

The chapter particularly focused on the environmental problem of air pollution aiming at highlighting environmental injustice caused by the types, sources, and distribution of air pollutants along with their effects on human health and wellbeing. These issues were discussed in the scope of sustainability, which in our view is the only means to ensure environmental quality and justice for all. This is a requirement and a fundamental challenge for modern societies.

Sustainability and environmental justice can only be accomplished by means of active involvement of citizens in environmental actions. In other words engaging people in making crucial decisions about the quality of their environment is a prerequisite for environmental sustainability and justice to occur. Therefore, individuals should be supported in demanding life quality based on respect and justice for all human beings; in ensuring equal access to natural resources and health for all; in deciding upon a development of their cities in harmony with nature and with respect to the cultures of different communities.

In this direction we propose the design and implementation of relevant educational programs deriving knowledge, methods, and modes of thinking from science and environmental education. These two fields provide the necessary tools for understanding environmental degradation and pursuing sustainability and environmental justice. More particularly, science education supplies tools and methods for estimating environmental quality and protection, and for creating a healthy environment ensuring life quality and health for human beings. Environmental education can complement this endeavour by critically supporting individuals in understanding environmental challenges and their causes, developing actions, and adopting practices to resolve them. Injustice, inherent in environmental degradation, can constitute an organising principle in designing and implementing educational programs for the environment, and more particularly in relation to air pollution.

The conjunction of science and environmental education organised along the concepts of sustainability and environmental justice also opens new grounds for research. Application of the general principles and stances outlined in this chapter in real school environments in the context of specific educational activities would allow a systematic and reliable estimation of their value. More specifically, evaluation of educational applications of the ideas presented in this chapter could investigate students' understanding of air pollution; their relevant speculations; the alternative solutions they propose for reducing air pollution along with pertinent environmental injustice; their views of the role of science in this enterprise. Systematic research on these topics is expected to yield documented, valuable, and productive outcomes to further promote education for sustainable development.

6. References

Ali, I. (1991). How do English pupils understand pollution? *Environmental Education and Information*, Vol. 10, No. 4, pp. 203-220, ISSN 0144-9281.

American Association for the Advancement of Science (AAAS) (1989). *Project 2061: Science for all Americans*, 03.01.2010, Available from http://www.project2061.org

Bard, D., Laurent, O., Filleul, L., Havard, S., Deguen, S., Segala, C. et al. (2007). Exploring the joint Effect of atmospheric pollution and socioeconomic status on selected health outcomes: An overview of the PAISARC project. *Environmental Research Letters*, Vol. 2, No. 4, 28.09.2008, Available from http://iopscience.iop.org/1748-9326/2/4/045003/pdf/1748-9326_2_4_045003.pdf

Bartlett, G. (2001). *Systemic Thinking, a Simple Thinking Technique for Gaining Systemic Focus*, 10.03.2007, Available from http://probsolv.com/systemic-thinking.doc

Batterham, D., Stanisstreet, M., & Boyes, E., (1996). Kids, cars and conservation: children's ideas about the environmental impact of motor vehicles, *International Journal of Science Education*, Vol. 18, No. 3, pp. 347-354, ISSN 0950-0693

Boyes, E., & Stanisstreet, M. (1994). The ideas of secondary school students concerning ozone layer damage. *Global Environmental Change*, Vol. 4, No. 4, pp. 311-324, ISSN 0959-3780

Brody, M. (1991). Understanding of pollution among 4th, 8th and 11th grade students. *International Journal of Science Education*, Vol. 22, No. 2, pp. 24-33, ISSN 0950-0693

Brody, M. (1994). Student science knowledge related to ecological crisis. *International Journal of Science Education*, Vol. 16, No. 4, pp. 421-435, ISSN 0950-0693

Brody, M., Chipman, E., & Marion, S. (1989). Student knowledge of scientific and natural resource concepts concerning acidic deposition. *Journal of Environmental Education*, Vol. 20, No. 2, pp. 32-42, ISSN 0095-8964

Bybee, R. W. (1993). *Reforming Science Education - Social Perspectives and Personal Reflections.* Teachers College Press, New York

Christidou, V., & Koulaidis, V. (1996). Children's Models of the Ozone Layer and Ozone Depletion. *Research in Science Education*, Vol. 26, No. 4, pp. 421-436, ISSN 157-244X

Cifuentes, E., & Frumkin, H. (2007). Environmental injustice: case studies from the South. *Environmental Research Letters,* Vol. 2, 26.09.2008, Available from http://iopscience.iop.org/1748-9326/2/4/045034/pdf/1748-9326_2_4_045034.pdf

Commission of the European Communities (CEC) (2006). *Cohesion Policy and cities: the urban contribution to growth and jobs in the regions.* COM (2005)718

Dimitriou, A. (2009). *Environmental education: Environment, Sustainability. Theoretical and Pedagogical Approaches.* Epikentro, ISBN 978-960-458-214-3, Thessaloniki (in Greek)

Dimitriou, A., & Christidou, V., (2007). Pupils' understanding of air pollution. *Journal of Biological Education,* Vol. 42, No. 1, pp. 24-29, ISSN 0021-9266

Donohoe, M. (2003). Causes and health consequences of environmental degradation and social injustice. *Social Science & Medicine,* Vol. 56, pp. 573–587, ISSN 0277-9536

Dove, J. (1996). Student teacher understanding of the greenhouse effect, ozone layer depletion and acid rain. *Environmental Education Research,* Vol. 2, No. 1, pp. 89-100, ISSN 1350-4622

EEA (European Environment Agency) (2006a). *Air pollution at street level in European cities* Copenhagen, EEA

EEA (European Environment Agency) (2006b). *Application of the emmisions trading directive by EU member states. Technical report No 6.*, Copenhagen, EEA

Environmental Protection Agency (EPA) (2001). *Environmental Justice,* 20.03.2007, Available from http://epa.gov/swerosps/ej/index.html

Environmental Protection Agency (EPA) (2007). *Terms of environment.* Office of Communications, Education, and Public Affairs, Washington DC

Environmental Protection Agency (EPA) (2008). Care for Your Air: A Guide to Indoor Air Quality, 20.02.2011, Available from http://www.epa.gov/iaq/pdfs/careforyourair.pdf

Environmental Protection Agency (EPA) (2009). Residential Air Cleaners. Indoor Air Quality, 10.01.2010, Available from http://www.epa.gov/iaq/pdfs/residential_air_cleaners.pdf

Espejo, R. (1994). What is systemic thinking? *System Dynamics Review,* Vol. 10, No. 2-3, pp. 199-212, ISSN 1099-1727

Faber, D., & McCarthy, D. (2002). The evolving structure of the environmental justice movement in the United States: New models for democratic decision-making. *Social Justice Research,* Vol. 14, No. 4, pp. 405-421, ISSN 0885-7466

Facione, P. A. (2007). *Critical thinking: what it is and why it counts,* 15.07.2008, Available from http://www.insightassessment.com/t.html

FACTS 29, (2002). *Health Care Sectors.* European Agency for Safety and Health of Work, Geneva

Flogaitis, E. (2006). *Education for the environmenta and sustainability.* Ellinika Grammata, ISBN 960-442-238-3, Athens (in Greek)

Francis, C., Boyes, E., Qualter, A., & Stanisstreet, M. (1993). Ideas of elementary pupils about reducing the 'Greenhouse Effect'. *Science Education,* Vol. 77, No. 4, pp. 375-392, ISSN 1098-237X

Friends of the Earth (FOE) (2001). *Pollution and Poverty - Breaking the Link.* Friends of the Earth, London

Givoni, B. (1991). Impact of planted areas on urban environmental quality: A review. *Atmospheric Environment, Part B, Urban Atmosphere*, Vol. 25, Issue 3, pp. 289-299, ISSN 1352-2310

Global Footprint Network (2005). EUROPE 2005. *The Ecological Footprint*. Belgium, WWF

Global Footprint Network (2011). *The Ecological Footprint*, 04.03.2011, Available from http://www.footprintnetwork.org/en/index.php

Gough, A. (2002). Mutualism: a different agenda for environmental and science education. *International Journal of Science Education*, Vol. 24, No. 11, pp. 1201-1215, ISSN 0950-0693

Harner, J., Warner, K., Pierce, J., & Huber, T. (2002). Urban environmental justice indices. *The Professional Geographer*, Vol. 54, No. 3, pp. 318-331, ISSN 0033-0124

Heynen, N. (2003). The scalar production of injustice within the urban forest. *Antipode*. Vol. 35, No. 5, pp. 980-998, ISSN 0066-4812

Hodson, D. (2003). Time for action: Science education for an alternative future. *International Journal of Science Education*, Vol. 25, No. 6, pp. 645-670, ISSN 0950-0693

Huckle, J. 2006. *Education for Sustainable Development*. A briefing paper for the Training and Development Agency for Schools, 15.06.2007, Available from http://john.huckle.org.uk

Hui, C. (2006). Carrying capacity, population equilibrium, and environment's maximal load. *Ecological Modelling*, Vol. 192, No. 1-2, pp. 317-320, ISSN 0304-3800

International Labour Organization (ILO), (1993). *Encyclopedia of occupational health and Safety*. ILO, Geneva

Jacobs, M. (2004). Sustainable development: A contested concept. In: *Fairness and futurity. Essays on Environmental sustainability and social justice* A. Dobson (Ed.), 21-54, Oxford University Press, London

Jones, P. C., Merritt, Q., & Palmer, C. (1999). Critical thinking and interdisciplinarity in envrionmental higher education: The case for epistemological and values awareness, *Journal of Geography in Higher Education*, Vol. 23, No. 3, pp. 349-357, ISSN 0309-8265

Landrigan, P. J., Carlson, J. E., Bearer, C. F., Spyker Cranmer, J., Bullard, R. D., Etzel, R. A., et al. (1998). Children's health and the environment: A new agenda for preventive research, *Environmental Health Perspectives*, Vol. 106, No. S3, ISSN 0091-6765

Iozzi, A. L. (1989). What research says to the educator. Part One: Environmental education and the affective domain, *Journal of Environmental Education*, Vol. 20, No. 3, pp. 3-9, ISSN 0095-8964

Koulaidis V., & Christidou V. (1999). Models of Students' Thinking Concerning the Greenhouse Effect and Teaching Implications. *Science Education*, Vol. 83, pp. 559-576, ISSN 0036-8326

Marinopoulos, D., & Stavridou, H., (2002). The influence of a collaborative learning environment on primary students' conceptions about acid rain. *Journal of Biological Education*, vol. 37, No. 1, pp. 18-25, ISSN 0021-9266

Mitchell, G., & Dorling, D. (2003). An environmental justice analysis of British air quality. *Environment and Planning A*, Vol. 35, No. 5, pp. 909 -929, ISSN 0308-518X

Orr, D. (1992). *Ecological Literacy: Education and the transition to a postmodern world*. SUNY Press, ISBN 0-7914-874-4

Occupational Safety and Health Administration (OSHA) (2009). *Assigned Protection Factors for the Revised Respiratory Protection Standard,* 15.01.2010, Available from http://www.osha.gov/Publications/3352-APF-respirators.pdf

Peloso, J. (2007). Environmental justice education: Empowering students to become environmental citizens. *Perspectives on Urban Education,* Vol. 5, No. 1, 02.02.2010, Available from http://www.urbanedjournal.org

Ragou P. et Souchon C. (2000). L'approche systémique: Une méthode d' apprentissage et un outil d' évaluation en education relative à l' environnement. *Education Relative à l'Environnement: Regards, Recherches, Reflexions,* Vol. 2, pp. 195-200, ISSN 1373-9689

Richardson, G. (1994). Introduction: System thinkers, system thinking. *System Dynamics Review,* Vol. 10, No. 2-3, pp. 95-99, ISSN 1099-1727

Richmond, B. (1993). Systems Thinking: Critical thinking skills for the 1990s and beyond. *System Dynamics Review,* Vol. 9, No. 2, pp. 113-133, ISSN 1099-1727

Richmond, B. (1994). Systems thinking/System dynamics: Let's just get on it. *System Dynamics Review,* Vol. 10, No. 2-3, pp. 135-157, ISSN 1099-1727

Scholosberg, D. (2004). Preconceiving environmental justice: Global movements and political theories. *Environmental Politics,* Vol. 13, No. 3, pp. 517-540, ISSN 0964-4016

Scott, W. (2007). *Raising Standards: Making Sense Of The Sustainable Schools Agenda,* Specialist Schools and Academies Trust, UK,

Scott, W., & Gough, S. (2004) (Eds.), *Key Issues in Sustainable Development and Learning. A Critical Review.* Routledge Falmer, ISBN 0-415-27649-7, London and N. York

Schneider, R. M., Krajcik, J., Marx, R. W., & Soloway, E. (2002). Performance of students in project-based science classrooms on a national measure of science achievement, *Journal of Research in Science Teaching,* Vol. 39, No. 5, pp. 410-422, ISSN 0022-4308

Schweizer, E. (1999). *Environmental justice: An interview with Robert Bullard.* 04.03.2007, Available from http://www.ejnet.org/ej/bullard.html

Shelton, D. (2002). *Human Rights, Health & Environmental Protection: Linkages in Law and Practice.* Notre Dame London Law Centre, London

Stephens, C., Bullock, S., & Scott, A. (2001). Environmental justice: Rights and means to a healthy environment for all, *Special Briefing Paper Economic and Social Research Council (ESRC) Global Environmental Change Programme* (Brighton: ESRC Global Environmental Change Programme, University of Sussex), 04.06.2008, Available from www.foe.co.uk/resource/reports/environmental_justice.pdf

Stephens, C. (2007). Environmental justice: A critical issue for all environmental scientists everywhere. *Environmental Research Letters,* Vol. 2, No. 1-2, pp. 1-3, ISSN 1748-9326

Sterman, J. (1994). Learning in and about complex systems. *System Dynamics Review,* Vol. 10, No. 2-3, pp. 291-330, ISSN 1099-1727

Stevenson, S., Stephens, C., Landon, M., Fletcher, T., Wilkinson, P., & Grundy, C. (1999). Examining the inequality of car ownership and the effects of pollution and health outcomes, presented at the "Healthy Planet Forum", June, Environmental Epidemiology Unit, School of Hygiene and Tropical Medicine, London

Stolwijk, J. A.(1991). Sick-Building Syndrome. *Environmental Health Perspectives,* Vol. 95, pp. 99-100, ISSN 0091-6765

Thornber J., Stanisstreet, M. & Boyes, E. (1999). School students' ideas about air pollution: Hindrance or help for learning? *Journal of Science Education and Technology,* Vol. 8, No. 1, pp. 67-73, ISSN 1059-0145

Tilbury, D., & Ross, K. (2006). *Living Change: Documenting Good Practice in Education for Sustainability in NSW.* Macquarie University, Sydney, and Nature Conservation Council, NSW

United Nations UN, (Ed.) (1973). *Report of the United Nations Conference on the Human Environment* (Stockholm 5-16 June 1972). UN, New York

United Nations (UN) (1994). Draft Principles on Human Rights and the Environment *E/CN.4/Sub.2/1994/9, Annex I*

United Nations. (2004). World population to 2300 United Nations Department of Economic and Social Affairs, Population Division

United Nations Education Scientific and Cultural Organisation (UNESCO), (1976). *The international workshop on environmental education.* Belgrade, Yougoslavia, 13-22 Oct 1975. Final report. UNESCO, Paris

United Nations Education Scientific and Cultural Organisation (UNESCO), (1978). *Intergovernmental Conference on Environmental Education (Tiblisi USSR, 14-26 October 1977),* UNESCO, Paris

United Nations Education Scientific and Cultural Organisation (UNESCO), (1982). *Environmental Education: Selected Activities of Unesco-UNEP International Environmental Education Programme 1975-1982,* UNESCO, Paris

United Nations Education Scientific and Cultural Organisation (UNESCO), (1988). *International strategy for action in the field of environmental education and training for the 1990s.* UNESCO, Paris

United Nations Education Scientific and Cultural Organisation (UNESCO), (1990). *Environmental Education: Selected Activities of Unesco-UNEP International Environmental Education Programme 1975-1990.* UNESCO, Paris

United Nations Education Scientific and Cultural Organisation (UNESCO), (1992). *United Nations Conference on Environmental and Development: Agenda 21.* UNESCO, Switzerland

United Nations Education Scientific and Cultural Organisation (UNESCO- EPD), (1997). *Declaration of Thessaloniki. UNESCO-EPD97/CONF.401/CLD.2,* UNESCO, Paris

United Nations Education Scientific and Cultural Organisation (UNESCO), (2002). *Teaching and learning for a sustainable future: A multimedia teacher education programme,* 23.04.2006, Available from http://www.unesco.org/education/tlsf/.

United Nations Education Scientific and Cultural Organisation (UNESCO), (2005). UN Decade of Education for Sustainable Development 2005-2014 International Implementation Scheme. DRAFT. UNESCO, Paris

United Nations Education Scientific and Cultural Organisation (UNESCO), (2005b). *UNESCO and sustainable development,* UNESCO, Paris

United Nations Commission on Sustainable Development (UNCSD), (2003). *Plan of Implementation of the World Summit on Sustainable Development,* New York: United Nations.

United Nations Education Scientific and Cultural Organisation (UNESCO), (2007). The UN Decade of Education for Sustainable Development (DESD 2005-2014). The First Two Years. UNESCO, Paris

UNICEF, (2008). *Childhood poverty in industrial world,* 10.10.2008, Available from http://www.unicef.gr.

Valent, F., Little, D., Tamburlini, G., & Barbone, F. (2004). *Burden of disease attributable to selected environmental factors and injuries among Europe's children and adolescents.*

World Health Organization, (WHO Environmental Burden of Disease Series, No. 8), Geneva

World Commission on Environment and Development (WCED) (1987). *Our Common Future.* Oxford, Oxford University Press.

Wiedmann, T., & Barrett, J. (2010). A review of the ecological footprint indicator—perceptions and methods, *Sustainability,* Vol. 2, 1645-1693, ISSN 2071-1050

World Commission on Environment and Development (WCED), (1987). *Our Common Future,* Oxford University Press, Oxford

World Health Organization, (WHO), (1992). *Ottawa Charter for health promotion.* World Health Organization, Geneva,

World Health Organization (WHO), (2002). *World health report 2002. Reducing risks, promoting healthy life.* World Health Organization, Geneva,

World Health Organization (WHO), (2003a). *Health Aspects of Air Pollution with Particulate Matter, Ozone and Nitrogen Dioxide.* World Health Organization, Geneva

World Health Organization, (WHO), (2003b). *Technical meeting on exposure - response relationships on health.* World Health Organization, Regional Office for Europe

World Health Organization, (WHO), (2004). *Development of Environment and Health Indicators for European Union Countries: Results of a Pilot Study.* Report on a WHO Working Group Meeting. Bonn, Germany

World Health Organization, (WHO), (2005*). Indoor air pollution and health. Fact sheet N°292, 10.03.2011*, 12.11.07, Available from http://www.who.int/mediacentre/factsheets/fs29

World Health Organization (WHO), (2006). Air quality guidelines for particulate matter, ozone, nitrogen dioxide and sulfur dioxide. Global update 2005. Summary of risk assessment. World Health Organization, Geneva

World Health Organization (WHO), (2008a). *The world health report 2008: primary health care now more than ever*. World Health Organization, Geneva

World Health Organization (WHO), (2008b). Water, sanitation and hygiene links to health, 10.10.2008, Available from http://www.who.org

World Health Organization (WHO), (2009). Global health risks: mortality and burden of disease attributable to selected major risks. World Health Organization, Geneva

Part 3

The Impact of Air Pollution on Plants, Agricultural Sources and Methods of Resistance

12

The Effects of Air Pollutants on Vegetation and the Role of Vegetation in Reducing Atmospheric Pollution

Iuliana Florentina Gheorghe[1] and Barbu Ion[2]
[1]*Ecological University of Bucharest, Faculty of Ecology and Environmental Protection*
[2]*Forest Research and Management Institute*
Romania

1. Introduction

The main air pollutants are represented by gases forms, particles in suspension, different ionizing radiation and noise.

The gases forms are: oxidized and reduced forms of carbon (CO_2, CO, CH_4), of nitrogen (NO_2, NO, N_2O_4, NH_3, NH^{4+}), SO_2, O_3, C_6H_6 vapours, Hg, volatile phenols, Cl_2, etc.

The particulate forms are: PM10 and PM2.5 particulate matter, heavy metals with toxic effect (Pb, Ni, Cd, As), polycyclic aromatic hydrocarbons PAHs, etc.

Atmospheric pollutants have a negative effect on the plants; they can have direct toxic effects, or indirectly by changing soil pH followed by solubilization of toxic salts of metals like aluminum. The particulate matters have a negative mechanical effect. They cover the leaf blade reducing light penetration and blocking the opening of stomata. These impediments influence strongly the process of photosynthesis which rate declines sharply.

Also the leaves of the trees have an important role in retention of the particulate matters; they are mostly affected when the wet and dry atmospheric deposition increase.

The vegetation plays an important positive role in atmospheric purification and air pollutants reduction.

The primary producers represented by plants are an important component in biogeochemical cycles. The vegetation made exchanges with a part of the atmospheric gases by photosynthesis, respiration processes, and the final stage of litter decomposition which mineralization.

The plants play an important role in reducing atmospheric CO_2 content, by photosynthesis. This reduction of atmospheric CO_2 content has an important role in reducing of greenhouse gases, participating in reducing greenhouse effect and its consequences on climatic changes. The carbon stored in plants is the result of balance between carbon fixed by photosynthesis and carbon released in the atmosphere by respiration.

As the structure of vegetation is more complex, the carbon stock in plants biomass is higher and the period of storage is longer. The most efficient type of vegetation in storing carbon in terms of carbon stored in plants alive is the temperate-continental forest; and in terms of carbon stored in dead organic matter are peat lands.

Trees have also been planted to reduce the intensity of ionizing radiation and noise in different urban and industrial areas. The existence of vegetation in an area creates a microclimate where the temperature differentials between day and night are buffered.
This prevents the occurrence of warmer temperatures which stimulate the production of volatile pollutants into the atmosphere.

2. General information about air pollution

Environmental pollution is any discharge of material or energy into water, land, or air that causes or may cause acute (short-term) or chronic (long-term) detriment to the Earth's ecological balance or that lowers the quality of life. Pollutants may cause primary damage, with direct identifiable impact on the environment, or secondary damage in the form of minor perturbations in the delicate balance of the biological food web that are detectable only over long time periods.
Air pollution is the process which the substances and the energy forms are not present in normal atmospheric composition reach the atmosphere, or are present but in much lower concentrations.Air pollution is the introduction of chemicals, particulate matter, or biological materials that cause harm or discomfort to humans or other living organisms, or cause damage to the natural environment or built environment, into the atmosphere.
More than 3,000 substances that are not part of the atmospheric composition, falling in the atmosphere can be considered air pollutants.
Some substances that are normally present in the atmosphere in a certain concentration can be considerate pollutants because their concentration is much higher than usual concentration.
Also certain substances that are normally present in certain layers of the atmosphere (e.g. ozone in the stratosphere), once arrived in the troposphere is pollutant.
Some gases, such as oxides of nitrogen may have beneficial effect on vegetation, after hydration may affect the leaf fertilizer.
The air pollutants factors can be chemical (chemicals), mechanics (particles in suspension) physical (ionizing radiation) and acoustic (noise).

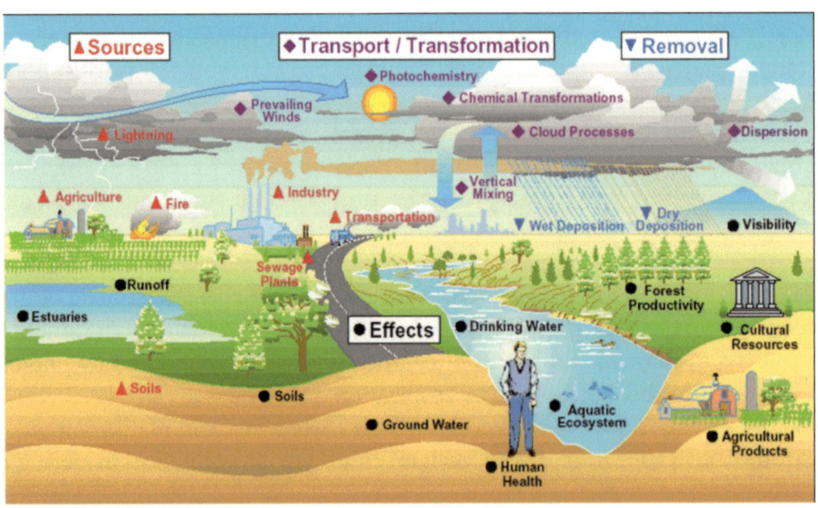

Fig. 1. Global polluants circuit (www.cleartheair.nsw.gov.au, 2011)

Pollutants describe a global circuit; they are produced by different sources, are transported and transformed into atmosphere, some of them being removed, another part is reaching the earth having different effects on different biocoenosis of ecosystems (fig. 1).

An analysis done at the global level revealed a diversification of pollutants agents and sources of air pollution. This diversification and increasing concentrations are in strict correlation with industrialization and the increasing of amounts use as fossil energy (non-renewable sources).

At the beginning, the pollution has been felt in urban areas and the forms of relief that favored the accumulation of pollutants and long stay (depressions, closed valleys, etc.). Currently, air pollution has become a larger area, sometimes to disperse across multiple continents.

Air pollution can be analyzed on three spatial scales: global pollution, regional pollution and local pollutants.

The global pollution is the result of cumulative effects of various sources, located on the entire surface of the globe, manifested by global effects: the stratospheric ozone depletion; greenhouse effect - emission of greenhouse gases (CO_2, methane, CFCs, etc.); formation of aerosols (pollutant clouds which suspended particles and chemical compounds).

The regional pollution is in part the result of local air pollution--including that produced by individual sources, such as automobiles - that has spread out to encompass areas of many thousands of square kilometers. Meteorological conditions and landforms can greatly influence air-pollution concentrations at any given place, especially locally and regionally. For example, cities located in bowls or valleys over which atmospheric inversions form and act as imperfect lids are especially likely to suffer from incidences of severe smog. Oxides of sulfur and nitrogen carried long distances by the atmosphere and then precipitated in solution as acid rain, can cause serious damage to vegetation, waterways, and buildings.

The local pollutants (smog) can be loosely defined as a multi-source, widespread air pollution that occurs in the air of cities. Smog, a contraction of the words smoke and fog, has been caused throughout recorded history by water condensing on smoke particles, usually from burning coal.

In terms of the effects of pollutants can be acidifying agents - sulphur dioxide (SO_2), nitrogen oxides (NOx) ammonia (NH_3) fluoride and Cl_2, hydrogen chloride (HCl) - and oxidizing agents - carbon monoxide (CO), PAN (peroxyacetylnitrate-$CH_3CO.O_2.NO_2$), ozone (O_3).

3. Sources of pollutants

Air pollution comes from natural and anthropic sources; these sources generate pollutants with different effects at global level or on individuals of plants and animals (tab. 1).

Natural processes that affect air quality include volcanoes, which produce sulfur, chlorine, and ash particulates. Wildfires produce smoke and carbon monoxide. Cattle and other animals emit methane as part of their digestive process. Even pine trees emit volatile organic compounds (VOCs).

Many forms of air pollution are **human-made**. Industrial plants, power plants and vehicles with internal combustion engines produce nitrogen oxides, VOCs, carbon monoxide, carbon dioxide, sulfur dioxide and particulates. In most mega-cities, cars are the main source of these pollutants. Stoves, incinerators, and farmers burning their crop waste produce carbon monoxide, carbon dioxide, as well as particulates. Other human-made sources include aerosol sprays and leaky refrigerators, as well as fumes from paint, varnish, and other solvents. One important thing to remember about air pollution is that it doesn't say in one place. Winds and weather play an important part in transport of pollution locally,

Name of pollutants	Origin	Effects
Natural sources		
sulfur, chlorine, and ash particulates, smoke and carbon monoxide methane volatile organic compounds (VOCs) Aerosol from deforestation and burning: CO, CO_2, NO, NO_2, N_2O, NH_4	Volcanoes, wildfires, cattle and other animals, pine trees	- acid rain, smog, - respiratory irritant - increased respiratory - diseases - damage cell membranes of plants The effects are high only for volcanoes.
Anthropic sources		
Carbon monoxide, carbon dioxide, sulphur dioxide, nitrogen oxides, fluorides and substances with fluorine, chlorine (Cl_2), bromine (Br_2) and iodine (I_2), small dust particles, VOC, methane, ammonia and radioactive radiation.	Industry: the mining industry, oil and natural gas extraction, the energy industry based on fossil fuels - coal, oil, natural gas, the production of brick, tile, enamel frit, ceramics, and glass; the manufacture of aluminium and steel; and the production of hydrofluoric acid, phosphate chemicals and fertilizers. central heating, chemical and metallurgical industry, engineering internal combustion machinery industry, industrial waste, noises	- respiratory irritant, - acid rain, - smog, - increased respiratory - formation of secondary pollutants (PAN, O_3) - effect on soil fertilizer - Respiratory diseases - toxic effects on living cells - greenhouse gas effect - toxic effects - carcinogenic proprieties - accumulation in tissues - blocking of different processes - stratospheric ozone depletion
CO, CO_2, NO, NO_2, NH_3, CH_4, SO_2, oxides of heavy metals, H_2SO_4, SPM, HC, VOC, background aerosols: sea salt oxidation of sulphur containing gases, same organics, nitrous oxide (N_2O) pesticides	Agriculture: the vegetation fire, the denitrification process, in soils excessively fertilized and excessive use the pesticides , paddy field, intensive husbandry, deforestation	- formation of secondary pollutants (PAN, O_3) - effect on soil fertilizer - respiratory diseases - greenhouse gas effect - toxic effects - acid rain, - stratospheric ozone depletion
Aerosols from transport and constructions NO_x, CO, HCl, Lead and other heavy metals , SPM	The motor vehicle pollution, noises	- smog - increased respiratory diseases - damage cell membranes of plants - carcinogenic proprieties - accumulation in tissues - blocking of different processes - stratospheric ozone depletion
Domestic aerosols CFC, HC, FC, H_2S, CH_4 CO_2	sewage plans, authorized landfill site	- carcinogenic proprieties - accumulation in tissues - blocking of different processes - stratospheric ozone depletion

Table 1. Type of pollutants, origin and effect at global level or on plants end animals individuals

regionally, and even around the world, where it affects everything it comes in contact with. The major anthropic sources of air pollution are:

- industry and conventional energies (the mining industry, the energy industry based on fossil fuels - coal, oil, natural gas, central heating, chemical and metallurgical industry, engineering internal combustion machinery industry, industrial waste, noises, etc);

- agriculture (the vegetation fire, denitrification in soils excessively fertilized, paddy field, intensive husbandry, deforestation, etc)
- transportation (motor vehicle pollution, noises, etc)
- and urbanization (sewage plans, authorized landfill site, etc) (fig. 2)

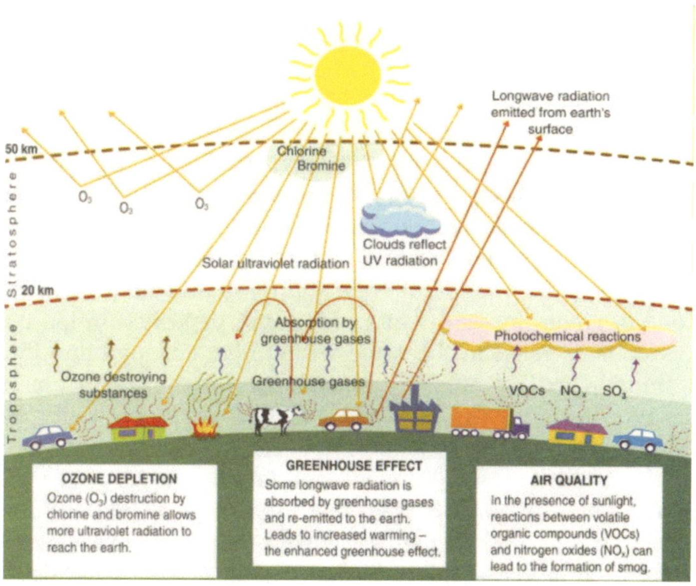

Fig. 2. The major anthropic sources of air pollution (www.cleartheair.nsw.gov.au, 2011)

4. The most important atmospheric pollutants

The air pollutants are represented by gases forms, particles in suspension, different ionizing radiation and noise.

5. Major gaseous pollutants

Sulphur dioxide (SO_2) is the most important and common air pollutant produced in huge amounts in combustion of coal and other fuels in industrial and domestic use. It is also produced during smelting of sulphide ores. Sulphur dioxide concentrations in air have decreased in the past two decades, mainly because we use more non-sulphur-containing fuels for the generation of energy. Sulphur dioxide is a stinging gas and as a result it can cause breathing problems with humans. In moist environments, sulphur dioxide may be transferred to sulphuric acid. This acid causes acidification and winter smog.

Nitrogen oxides (NO_x) and nitrous oxide (N_2O) Using catalysers in car exhausts can prevent emissions of nitrogen oxides. Nitrogen oxides are gasses that react with other air pollutants when they are present in air. For example, nitrogen oxides play an important rolein the formation of ozone in the lower atmosphere, and in acidification and eutrophication processes. They can deeply penetrate the lungs and damage human lung functions.

Fluorides Common gaseous fluoride pollutants are HF, SiF_6, CF_4 and F_2. Particulate fluoride pollutants include Ca_3AlF_6 (Cryolite), CaF_2, NH_3F, AlF_6, CaSiF, NaF and Na_2SiF_6. Aerosols are often formed from NaF, $NaAlF_6$ and AlF_6. Chief sources of fluoride pollutants are brickworks, aluminium factories, glassworks, steelworks, ceramic factories, phosphate fertilizer plants and uranium smelters. Some fluorine pollution also occurs during combustion of coal. Most injurious fluoride pollutant is gaseous hydrogen fluoride (HF).

Fluoride is released into the air in large quantities by aluminum reduction plants, phosphate processors, steel mills, coal burning operations, brick and tile manufacturers, and various less significant sources[1]. It can cause adverse effects when ingested by domestic animals or absorbed by plants. There are also reports that fluoride air pollution can adversely affect human health, though these are less well documented than those concerning sensitive animals and plants. Fluorides are released into the air in both a gaseous state (as hydrogen fluoride and silicon tetra-fluoride) and in solid particles. The particles fall on, and the gases are absorbed by, vegetation near the polluting industry. If this vegetation includes forage crops which are fed to cattle, sheep, horses, or pigs, serious problems may ensue, since these animals, particularly the cattle are vulnerable to fluoride [2]. In fact, according to the U.S. Ninety-six percent of the ingested fluoride that accumulates in the bodies of animals is incorporated into the crystal structure of bone and tooth mineral [3], [4]. When fluoride is ingested with food or water, most of that which is not deposited in the bones, teeth, and other calcified tissue is excreted in the urine within hours of ingestion [5]. Thus it is not surprising that fluoride mainly affects the bones and teeth. Teeth are more markedly affected by ingested fluoride than are bones, but their high sensitivity is limited to the period of their formation. Thus a cow that has not been exposed to excessive fluoride before the age of two and one-half to three years will not develop the severe dental lesions which would occur in the same animal exposed at a younger age [6]. The developing tooth exposed to small amounts of fluoride may experience color variations ("mottling") that have little or no effect on the animal's ability to eat. Higher levels of fluoride result in more serious dental abnormalities, ranging from small, brittle, chalky areas on the tooth surface to pitting of enamel and easily eroded teeth [2]. Even more serious effects, including severe pain and the wearing down of the tooth right to the gum, can prevent the cattle from drinking cold water or eating. Localized or generalized enlargement of certain bones in the legs (metacarpals and metatarsals) and the lower jaw (mandible) of cattle are common symptoms of excessive fluoride ingestion [7]. As highly abnormal bone tissue replaces normal bone, [8] overall enlargement occurs, and the normally smooth bone surfaces take on a chalky, white, irregular appearance [2]. Hard ground can cause fluorotic hoof (pedal) bones to fracture, resulting in severe lameness [6]. Cattle with advanced fluorosis may also be crippled by mineralization of ligaments, tendons, and the structures surrounding the joints [9]. Enlargement of the joints themselves may also contribute to lameness. Fluoride-induced tooth destruction, lameness, and stiff joints affect the animal's ability to stand, eat, and graze, and all tend to lower the milk yield of dairy cattle or the weight of beef cattle.

Chlorine (Cl_2) Although chlorine concentrations change very rapidly in the atmosphere due to atmospheric chemistry and light rain can remove all the chlorine from the air in a very short time, chlorine injury can occur to plants near the source of pollution.

The impact of chlorine pollution increases in bright sunlight and decreases in drought and low temperature.

Many particulate and gaseous fluorides are produced when ores containing fluorine are processed and used in industries. Common gaseous fluoride pollutants are HF, SiF_6, CF_4 and F_2. Particulate fluoride pollutants include Ca_3AlF_6 (Cryolite), CaF_2, NH_3F, AlF_6, CaSiF, NaF and Na_2SiF_6. Aerosols are often formed from NaF, $NaAlF_6$ and AlF_6. Chief sources of fluoride pollutants are brickworks, aluminium factories, glassworks, steelworks, ceramic factories, phosphate fertilizer plants and uranium smelters. Some fluorine pollution also occurs during combustion of coal. Most injurious fluoride pollutant is gaseous hydrogen fluoride (HF).

Hydrogen chloride (HCl) HCl gas is released in large quantities in combustion of PVC and all chlorinated hydrocarbon material in large fires or incinerators. The HCl gas is very hygroscopic and quickly changes to hydrochloric acid by reacting with atmospheric moisture and forms aerosol droplets.

Ammonia (NH_3) Continuous releases of ammonia from the sources are rarely high enough to cause acute injury but occasional high release or spillage may cause ammonia pollution. High concentrations of ammonia are sometimes found around intensive farm units e.g. chicken batteries. Extent of injury reduces rapidly with increase in distance form the source. Under certain conditions the ammonia may remain as a cloud above ground level causing more injury to trees than to the ground flora. Injury symptoms may take up to 9 days to develop. In most plant species, recovery may occur in about 2 weeks after exposure is stopped.

Ammonia forms during agricultural activities. EEA-32 emissions of NH_3 have declined by 24% between the years 1990 and 2008. Agriculture was responsible for 94% of NH_3 emissions in 2008. The reduction in emissions within the agricultural sector is primarily due to a reduction in livestock numbers (especially cattle) since 1990, changes in the handling and management of organic manures and from the decreased use of nitrogenous fertilizers. The reductions achieved in the agricultural sector have been marginally offset by the increased emissions which have occurred during this period in sectors such as transport and to a lesser extent the energy industry and other (non-energy) sectors.

Environmental context: NH_3 contributes to acid deposition (plays an important role in acidification) and eutrophication. The subsequent impacts of acid deposition can be significant, including adverse effects on aquatic ecosystems in rivers and lakes and damage to forests, crops and other vegetation. Eutrophication can lead to severe reductions in water quality with subsequent impacts including decreased biodiversity, changes in species composition and dominance, and toxicity effects. NH_3 also contributes to the formation of secondary particulate aerosols, an important air pollutant due to its adverse impacts on human health.

VOC (Volatile Organic Compounds) VOC can be a range of different contaminants, such as carbohydrates, organic compounds and solvents. These compounds usually derive from petrol and gasoline reservoirs, industrial processes and fuel combustion, paint and cleanser use, or agricultural activities. VOC play an important role in ozone shaping in the lower atmospheric layer, the main cause of smog. VOC can cause various health effects, depending on the kind of compounds that are present and their concentrations. Effects can vary from smell nuisance to decreases in lung capacity, and even cancer.

Organic gases (Ethylene) and Methane (CH_4) Among organic gaseous pollutants, ethylene is most common. Other organic gases are propylene, butylenes and acetylene. Ethylene is continuously emitted from many sources involving combustion or processing of petroleum or its products or burning of organic materials e.g. straw burning. Other organic gases are also produced in various chemical industrial processes.

Ethylene is a natural plant growth substance so the injury effects produced by it on plants are very similar to growth abnormality symptoms. Other organic gases also produce symptoms similar to those of ethylene pollution. However, the sensitivities of species to different gases are variable.

Ethylene is a byproduct of automobile exhaust and can be a noticeable problem in urban environments.

Chlorofluorocarbons (or CFCs) is an organic compound that contains carbon, chlorine, and fluorine, produced as a volatile derivative of methane and ethane. A common subclass is the **hydro-chlorofluorocarbons** (HCFCs), which contain hydrogen, as well. They are also commonly known by the DuPont trade name **Freon**. The most common representative is dichlorodifluoromethane (R-12 or Freon-12). Many CFCs have been widely used as refrigerants, propellants (in aerosol applications), and solvents. The manufacture of such compounds is being phased out by the Montreal Protocol because they contribute to ozone depletion.

Minor gaseous pollutants Many other air pollutants which are highly injurious to animals and human beings also cause damage to plants. However, plants are affected by these gases at quite higher concentrations than the animals. Common such gaseous pollutants are CO, H_2S, Br_2, I_2 and Hg - vapors.

Hydrogen sulphide (H_2S) It is a colorless, very poisonous, flammable gas with the characteristic foul odor of rotten eggs at concentrations up to 100 parts per million. It often results from the bacterial breakdown of organic matter in the absence of oxygen, such as in swamps and sewers (anaerobic digestion). It also occurs in volcanic gases, natural gas, and some well waters. The human body produces small amounts of H_2S and uses it as a signaling molecule.

Carbon monoxide (CO) **carbon dioxide** in excess (CO_2), this gas consists during incomplete combustion of fuels. When we let a car engine run in a closed room, carbon monoxide concentrations in the air will rise extensively. Carbon monoxide contributes to the greenhouse effect, smog and acidification. The gas can bind to hemoglobin in blood, preventing oxygen transport through the body. This results in oxygen depletion of the heart, brains and blood vessels, eventually causing death.

It is highly toxic to humans and animals in higher quantities, although it is also produced in normal animal metabolism in low quantities, and is thought to have some normal biological functions.

Carbon monoxide consists of one carbon atom and one oxygen atom, connected by a triple bond which consists of two covalent bonds as well as one dative covalent bond. It is the simplest ox-carbon. In coordination complexes the carbon monoxide ligand is called carbonyl.

Carbon monoxide is produced from the partial oxidation of carbon-containing compounds; it forms when there is not enough oxygen to produce carbon dioxide (CO_2), such as when operating a stove or an internal combustion engine in an enclosed space. In the presence of oxygen, carbon monoxide burns with a blue flame, producing carbon dioxide [10]. Coal gas, which was widely used before the 1960s for domestic lighting, cooking and heating, had carbon monoxide as a significant constituent. Some processes in modern technology, such as iron smelting, still produce carbon monoxide as a byproduct [11]. Worldwide, the largest source of carbon monoxide is natural in origin; due to photochemical reactions in the troposphere which generate about 5×10^{12} kilograms per year [12], other natural sources of CO include volcanoes, forest fires, and other forms of combustion.

In biology, carbon monoxide is naturally produced by the action of heme oxygenase 1 and 2 on the heme from hemoglobin breakdown. This process produces a certain amount of carboxyhemoglobin in normal persons, even if they do not breathe any carbon monoxide. Following the first report that carbon monoxide is a normal neurotransmitter in 1993, as well as one of three gases that naturally modulate inflammatory responses in the body (the other two being nitric oxide and hydrogen sulfide), carbon monoxide has received a great deal of clinical attention as a biological regulator . In many tissues, all three gases are known to act as anti- inflammatory, vasodilators and promoters of neo-vascular growth [13]. Clinical trials of small amounts of carbon monoxide as a drug are on-going.

Bromine (Br_2) and Iodine (I_2)

At high temperatures, organo-bromine compounds are easily converted to free bromine atoms, a process which acts to terminate free radical chemical chain reactions. This makes such compounds useful fire retardants and this is bromine's primary industrial use, consuming more than half of world production of the element. The same property allows volatile organo-bromine compounds, under the action of sunlight, to form free bromine atoms in the atmosphere which are highly effective in ozone depletion. This unwanted side-effect has caused many common volatile brominated organics like methyl bromide, a pesticide that was formerly a large industrial bromine consumer, to be abandoned. Remaining uses of bromine compounds are in well-drilling fluids, as an intermediate in manufacture of organic chemicals, and in film photography.

Iodine and its compounds are primarily used in nutrition, the production of acetic acid and polymers. Iodine's relatively high atomic number, low toxicity, and ease of attachment to organic compounds have made it a part of many X-ray contrast materials in modern medicine. Like the other halogens, iodine occurs mainly as a diatomic molecule I_2, not the atom. In nature, iodine is a relatively rare element, ranking 47th in abundance. It is the heaviest essential element utilized in biological functions. Its rarity in many soils has led to many deficiency problems in land animals and inland human populations, with iodine deficiency affecting about two billion people and being the leading preventable cause of mental retardation [14]. As a component of thyroid hormones, iodine is required by higher animals. Radioisotopes of iodine are concentrated in the thyroid gland. This property of thyroid-concentration, along with its mode of beta decay, makes iodine-131 one of the most carcinogenic nuclear fission products.

Hg and Mercury vapors Pre-industrial deposition rates of mercury from the atmosphere may be about 4 ng/ (1 l of ice deposit). Although that can be considered a natural level of exposure, regional or global sources have significant effects. Volcanic eruptions can increase the atmospheric source by 4–6 times [15]. Natural sources, such as volcanoes, are responsible for approximately half of atmospheric mercury emissions. The human-generated half can be divided into the following estimated percentages:

- 65% from stationary combustion, of which coal-fired power plants are the largest aggregate source (40% of U.S. mercury emissions in 1999). This includes power plants fueled with gas where the mercury has not been removed. Emissions from coal combustion are between one and two orders of magnitude higher than emissions from oil combustion, depending on the country [15].
- 11% from gold production. The three largest point sources for mercury emissions in the U.S. are the three largest gold mines. Hydro-geochemical release of mercury from gold-mine tailings has been accounted as a significant source of atmospheric mercury in eastern Canada [16]
- 6.8% from non-ferrous metal production, typically smelters.
- 6.4% from cement production.
- 3.0% from waste disposal, including municipal and hazardous waste, crematoria, and sewage sludge incineration. This is a significant underestimate due to limited information, and is likely to be off by a factor of two to five.

- 3.0% from caustic soda production.
- 1.4% from pig iron and steel production.
- 1.1% from mercury production, mainly for batteries.
- 2.0% from other sources [15], (EPA report, 2007).

The above percentages are estimates of the global human-caused mercury emissions in 2000, excluding biomass burning, an important source in some regions [15]. Current atmospheric mercury contamination in outdoor urban air is (0.01–0.02 $\mu g/m^3$) indoor concentrations are significantly elevated over outdoor concentrations, in the range 0.0065–0.523 $\mu g/m^3$ (average 0.069 $\mu g/m^3$) [17]. Mercury also enters into the environment through the improper disposal (e.g., land filling, incineration) of certain products.

6. Particulate pollutants

Particles in suspension

Dust particles. Dust particles form a complex of organic compounds and minerals. These can derive from natural sources, such as volcanoes, or human activities, such as industrial combustion processes or traffic. Particles are categorized according to particle size. The smallest particles have the ability to transport toxic compounds into the respiratory tract. Some of these compounds are carcinogenic. The upper respiratory tract stops the larger dust particles. When they are released into the environment, dust particles can cause acidification and winter smog.

Cement-kiln dust Cement factories are the chief source of cement dust pollution. The composition of such dust varies with the source. Main component of cement dust is CaO and varying amounts of K_2O, Na_2O and KCl and traces of Al, Fe, Mn, Mg, S and silica. Dust with more than 24% CaO is more injurious to plants. Fine particles cause more damage than larger particles. Cement-kiln dust is alkaline in nature and dissolves in atmospheric moisture forming a solution of pH 10-12.

Lime and gypsum Lime and gypsum processing industries and mining deposits are chief sources from where fine particles of these substances are blown away to great distances.

Soot Burning of fossil fuels, organic matter or natural forest fires produce huge quantities of fine carbon particles which form the soot pollution. Soot can be dispersed over a quite wide area and transported to great distances by blowing winds.

Magnesium oxide Magnesium roasters are the chief sources of such pollution. The magnesium oxide dust may be carried by winds and deposited even at a distance of 5 km from the source. In the atmosphere, magnesium sulphate ($MgSO_4$) combines with carbon dioxide and water to form $Mg(CO_3)_2$. Both these compounds are alkaline and slightly soluble in water.

Boron Boric acid and borax are common raw materials in many industries. Oven and refrigerator manufacturing industries are chief sources of boron pollution.

Chlorides of sodium, potassium and calcium Sodium and calcium chlorides are commonly used in cold countries on the roads during winters to melt ice and snow. Potash industry produces aerial emission of KCl and NaCl in ratio of 3:1. All such chlorides are carried away by winds and deposited on the soil and plants.

Sodium sulphate, with an annual production of 6 million tones, it is a major commodity chemical and one of the most damaging salts in structure conservation: when it grows in the pores of stones it can achieve high levels of pressure, causing structures to crack.

Sodium sulfate is mainly used for the manufacture of detergents and in the Kraft process of paper pulping. About two-thirds of the world's production is from mirabilite, the natural mineral form of the decahydrate, and the remainder from by-products of chemical processes such as hydrochloric acid production.

Pesticides, insecticides and herbicides Pesticide use in the agricultural industry began in earnest in the early 1940s. Although pesticide use had been quite popular for more than twenty years, government officials first became aware of the potential danger of pesticide runoff to humans in the early 1960s when Rachel Carson's famous and influential *Silent Spring* was published. Though this book warned mainly of the detrimental effects of DDT (a popular insecticide developed in the early 1940s) for birds and other non-human victims, Carson's work inspired health officials to speculate about the effects of pesticide runoff on humans. Recently, exposure to DDT was linked to Parkinson's disease. Because of concern over DDT's adverse effects on the environment and on people, this pesticide was banned in 1972. Despite the ban of DDT, pesticide use continues, and the effects of some modern insecticides and herbicides can be just as debilitating. Even through careful use, runoff from pesticides continues to makes its way into drinking water sources.

7. Secondary pollutants

Photo-oxidants In presence of strong sunlight and in hot weather a series of complex chemical reactions involving nitrogen oxides and hydrocarbons may produce certain photo-oxidant chemicals. These chemicals do not have any specific anthropogenic source but are formed over wide areas in which suitable environmental conditions are prevailing. Two such photo-oxidants that can reach ambient concentrations toxic to plants are PAN (Peroxyacetylnitrate) and ozone.

PAN (Peroxyacetylnitrate-$CH_3CO.O_2.NO_2$) Impact of this secondary pollutant is not affected by humidity. However, the impact decreases with lowering of temperature and increasing drought conditions. The impact also increases in the morning and in bright sunlight. Young plants and young rapidly expanding leaves are more sensitive to this pollutant. PAN interacts with SO_2 and O_3 in complex manner producing variable impact conditions.

Ozone (O_3) is the main pollutant in the oxidant smog complex.

Ozone is formed in the troposphere when sunlight causes complex photochemical reactions involving oxides of nitrogen (NOx), volatile organic hydrocarbons (VOC) and carbon monoxide that originate chiefly from gasoline engines and burning of other fossil fuels. Woody vegetation is another major source of VOCs. NOx and VOCs can be transported long distances by regional weather patterns before they react to create ozone in the atmosphere, where it can persist for several weeks. Seasonal exposures at low elevations consist of days when ozone concentrations are relatively low or average, punctuated by days when concentrations are high. Concentrations of ozone are highest during calm, sunny, spring and summer days when primary pollutants from urban areas are present. Ozone concentrations in rural areas can be higher than in urban areas while ozone levels at high elevations can be relatively constant throughout the day and night.

Middle aged leaves and young plants are more sensitive to ozone. This pollutant interacts with SO_2, NO_2, PAN and heavy metals in complex manner.

Ozone is created through photochemical transfer of oxygen. This process takes place under the influence of ultra violet sunlight (UV), aided by pollutants in the outside air (fig. 2). Ozone causes smog and contributes to acidification and climate change. Ozone is an aggressive gas. This can easily penetrate the respiratory tract, deeply. When humans are exposed to ozone, the consequences may be irritation of the eyes and the respiratory tract.

Acid deposition Various acid gases, aerosols and other acidic substances released into the atmosphere from the industrial or domestic sources of combustion of fossil fuels eventually come down to the ground. These substances are deposited directly on the water bodies. In addition, these substances also reach the water bodies along with run-off rainwater from the polluted soil. Deposition of acidic substances causes acidification of water by lowering its pH below 6.0. The sulphates, nitrates and chlorides have been reported to make water bodies like lakes, rivers and ponds acidic in many countries.

Acid deposition is not merely characterized as acid rain; it can also be snow and fog or gas and dust. Acid deposition mainly forms during fossil fuel combustion. When emissions of sulphur dioxide and nitrogen oxides come in contact with water, they will become sulphuric acid and nitric acid.

When acidifying agents, such as sulphur dioxide, nitrogen oxides and ammonia, end up in plants, surface water and soils, this has a number of consequences:

- availability of nutrients and metal spores is likely to decrease
- when acidity is high more metals will dissolve in water. This can cause surface water to become polluted, which has serious health effects on aquatic plants and animals. For example, high aluminum (Al) concentrations can complicate nutrients uptake by plants. This makes aluminum one of the prior causes of forest decay. Mercury can be dispersed by transport through surface water, causing it to accumulate in fish. Mercury can bio magnify up the food chain, to be taken up by humans eventually
- Buildings and monuments may be damaged through erosion. Sulphur dioxide breaks down limestone by reacting with calcium carbonate, causing limestone to absorb water during rainfall. Limestone will than fragment

Noise pollution has a relatively recent origin. It is a composite of sounds generated by human activities ranging from blasting stereo systems to the roar of supersonic transport jets. Although the frequency (pitch) of noise may be of major importance, most noise sources are measured in terms of intensity, or strength of the sound field. The standard unit, one decibel (dB), is the amount of sound that is just audible to the average human. The decibel scale is somewhat misleading because it is logarithmic rather than linear; for example, a noise source measuring 70 dB is 10 times as loud as a source measuring 60 dB and 100 times as loud as a source reading 50 dB. Noise may be generally associated with industrial society, where heavy machinery, motor vehicles, and aircraft have become everyday items. Noise pollution is more intense in the work environment than in the general environment, although ambient noise increased an average of one dB per year during the 1980s. The average background noise in a typical home today is between 40 and 50 decibels. Some examples of high-level sources in the environment are heavy trucks (90 dB at 15 m/50 ft), freight trains (75 dB at 15 m/50 ft), and air conditioning (60 dB at 6 m/20 ft).

Radiation pollution is any form of ionizing or no ionizing radiation that results from human activities. The most well-known radiation results from the detonation of nuclear devices and the controlled release of energy by nuclear-power generating plants (see nuclear energy). Other sources of radiation include spent-fuel reprocessing plants, by-products of mining operations, and experimental research laboratories. Increased exposure to medical x rays and to radiation emissions from microwave ovens and other household appliances, although of considerably less magnitude, all constitute sources of environmental radiation.

Radioactive radiation. Radioactive radiation and radioactive particles are naturally present in the environment. During power plant incidents or treatments of nuclear waste from a war where nuclear weapons are used, radioactive radiation can enter the air on account of humans. When humans are exposed to high levels of radioactive radiation, the chances of serious health effects are very high. Radioactive radiation can cause DNA alteration and cancer.

8. Flow of atmospheric pollutants at global level

The air pollutants are produced by different sectors of the economy like: industry, agriculture, transports and urbanization. The burning of hydrocarbons in motor vehicle engines gives rise to CO_2, CO, SO_2 (sulfur dioxide), NO_x (NO [nitrogen monoxide]) and NO_2-- in varying proportions-and C_2H_4 (ethylene), as well as a variety of other hydrocarbons. Additional SO_2 originates from domestic and industrial burning of fossil fuels. Industrial plants, such as chemical works and metal-smelting plants, release SO_2, H_2S, NO_2, and HF (hydrogen fluoride) into the atmosphere. Tall chimney stacks may be used to carry gases and particles to a high altitude and thus avoid local pollution, but the pollutants return to Earth, sometimes hundreds of kilometers from the original source.

Photochemical smog is the product of chemical reactions driven by sunlight and involving NO_x of urban and industrial origin and volatile organic compounds from either vegetation (*biogenic* hydrocarbons) or human activities (*anthropogenic* hydrocarbons). Ozone (O_3) and peroxyacetylnitrate (PAN) produced in these complex reactions can become injurious to plants and other life forms, depending on concentration and duration of exposure. Hydrogen peroxide, another potentially injurious molecule, can form by the reaction between O_3 and naturally released volatiles (terpenes) from forest trees.

The concentrations of polluting gases, or their solutions, to which plants are exposed are thus highly variable, depending on location, wind direction, rainfall, and sunlight. Experiments aimed at determining the impact of chronic exposure to low concentrations of gases should allow plants to grow under near-natural conditions. One method is to grow the plants in open-top chambers into which gases are carefully metered, or where plants receiving ambient, polluted air are compared with controls receiving air that has been scrubbed of pollutants.

These pollutants emitted into the atmosphere can react with components of the atmosphere and transform into more or less aggressive or toxic compounds. The air pollutants can accumulate and manifest directly effects in the atmosphere (greenhouse effect, ozone layer depletion, etc) or can to transform in other pollutants and manifest indirectly effects on ecosystem biocoenosis, plants, animals and human health (fig. 3)

9. Effect of pollutants on vegetation, direct effects, and indirect effect, gas toxicity, wet and dry deposition, and deposition mixtures

Dust pollution is of localized importance near roads, quarries, cement works, and other industrial areas. Apart from screening out sunlight, dust on leaves blocks stomata and lowers their conductance to CO_2, simultaneously interfering with photosystem II. Polluting gases such as SO_2 and NO_x enter leaves through stomata, following the same diffusion pathway as CO_2. NO_x dissolves in cells and gives rise to nitrite ions (NO_2^-, which are toxic at high concentrations) and nitrate ions (NO_3^-) that enter into nitrogen metabolism as if they had been absorbed through the roots. In some cases, exposure to pollutant gases,

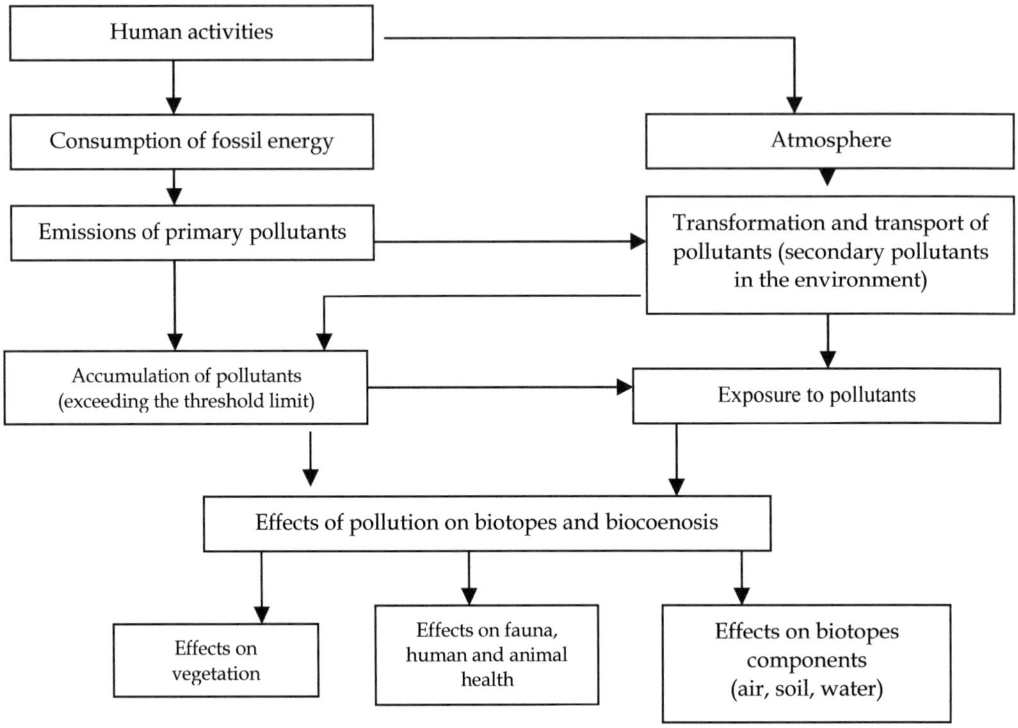

Fig. 3. The air pollutants flow diagram

particularly SO_2, causes stomatal closure, which protects the leaf against further entry of the pollutant but also curtails photosynthesis. In the cells, SO_2 dissolves to give bisulfite and sulfite ions; sulfite is toxic, but at low concentrations it is metabolized by chloroplasts to sulfate, which is not toxic. At sufficiently low concentrations, bisulfite and sulfite are effectively detoxified by plants, and SO_2 air pollution then provides a sulfur source for the plant. In urban areas these polluting gases may be present in such high concentrations that they cannot be detoxified rapidly enough to avoid injury. Ozone is presently considered to be the most damaging phytotoxic air pollutant in North America [18], [19]. It has been estimated that wherever the mean daily O_3 concentration reaches 40, 50, or 60 ppb (parts per billion or per 10^9), the combined yields of soybean, maize, winter wheat, and cotton would be decreased by 5, 10, and 16%, respectively. Ozone is highly reactive: It binds to plasma membranes and it alters metabolism. As a result, stomatal apertures are poorly regulated, chloroplast thylakoid membranes are damaged, rubisco is degraded, and photosynthesis is inhibited. Ozone reacts with O_2 and produces reactive oxygen species, including hydrogen peroxide (H_2O_2), superoxide (O_2^-), singlet oxygen ($^1O_2^*$), and the hydroxyl radical (^-OH). These denature proteins and damage nucleic acids (thereby giving rise to mutations), and cause lipid peroxidation, which breaks down lipids in membranes. Reactive oxygen species form also in the absence of O_3, particularly in electron transport in the mitochondria and chloroplasts, when electrons can be donated to O_2. Cells are protected, at least in part, from reactive oxygen species by enzymatic and nonenzymatic defense mechanisms [20], [21]. Defense against reactive oxygen species is provided by the scavenging properties of molecules, such as ascorbic acid, α-tocopherol, phenolic compounds, and glutathione. Superoxide dismutases (SODs) catalyze the reduction of superoxide to hydrogen peroxide.

Hydrogen peroxide is then converted to H_2O by the action of catalases and peroxidases. Of particular importance is the ascorbate-specific peroxidase localized in the chloroplast. Acting in concert, ascorbate peroxidase, dehydroascorbate reductase, and glutathione reductase remove H_2O_2 in a series of reactions called the *Halliwell–Asada pathway*, named after its discoverers. Glutathione is a sulfur-containing tripeptide that, in its reduced form, reacts rapidly with dehydroascorbate and becomes oxidized in the process. Glutathione reductase catalyzes the regeneration of reduced glutathione (GSH) from its oxidized form (GSSG) in the following reaction:

$$GSSG + NADPH + H^+ \rightarrow 2\,GSH + NADP^+$$

Exposure of plants to reactive oxygen species stimulates the transcription and translation of genes that encode enzymes involved in protection mechanisms. In Arabidopsis, exposure for 6 hours per day to low levels of O_3 induces the expression of several genes that encode enzymes associated with protection from reactive oxygen species, including SOD, glutathione S-transferase (which catalyzes detoxification reactions involving glutathione), and phenylalanine ammonia lyase (an important enzyme at the start of the phenylpropanoid pathway that leads to the synthesis of flavonoids and other phenolics) [22].

In transgenic tobacco transformed with a gene from *Escherichia coli* to give additional glutathione reductase activity in the chloroplast, short-term exposure to high levels of SO_2 is much less damaging than for wild-type tobacco [23]. Environmental extremes may either accelerate the production of reactive oxygen species or impair the normal defense mechanisms that protect cells from reactive oxygen species. In water-deficient leaves, for example, greater oxygen photoreduction by photosystems I and II increases superoxide production, and the pool of glutathione, as well as the activity of glutathione reductase, increase-presumably as part of the cell defense mechanism. In contrast, levels of ascorbate, another antioxidant, generally decline with mild water stress. Transgenic plants overexpressing mitochondrion superoxide dismutase (Mn-SOD), the isozyme localized in the mitochondrial matrix, show less water-deficit damage and, significantly, improved survival and yield under field conditions [24]. In other experiments, transgenic alfalfa overexpressing Mn-SOD was found to be more tolerant of freezing. Conversely, winter rye, wheat, and barley acclimated at 2 °C for several weeks, were found to have developed resistance to the herbicides, paraquat and acifluorfen, which generate reactive oxygen species. Such investigations support the hypothesis that tolerance of oxidative stress is an important factor in tolerance to a wide range of environmental extremes. Many deleterious changes in metabolism caused by air pollution precede external symptoms of injury, which appear only at much higher concentrations. For example, when plants are exposed to air containing NO_x, lesions on leaves appear at a NO_x concentration of 5 ml/l, but photosynthesis starts to be inhibited at a concentration of only 0.1 ml/l. These low, threshold concentrations refer to the effects of a single pollutant. However, two or more pollutants acting together can have a synergistic effect, producing damage at lower concentrations than if they were acting separately. In addition, vegetation weakened by air pollution can become more susceptible to invasion by pathogens and pests. Unpolluted rain is slightly acidic, with a pH close to 5.6, because the CO_2 dissolved in it produces the weak acid, H_2CO_3. Dissolution of NO_x and SO_2 in water droplets in the atmosphere causes the pH of rain to decrease to 3 to 4, and in southern California polluted droplets in fog can be as acidic as pH 1.7. Dilute acidic solution can remove mineral nutrients from leaves, depending on the age of the leaf and the integrity of the cuticle and surface waxes. The total annual contributions to the soil of acid from acid rain (*wet deposition*) and from particulate matter falling on the soil plus direct absorption from the atmosphere (*dry deposition*) may reach 1.0 to 3.0 kg H^+ per hectare in parts of Europe and the northeastern United States [25]. In soils that lack free calcium carbonate, and therefore are not strongly buffered, such additions of acid can be harmful to plants. Furthermore, the added acid can result in the release of aluminum ions from soil minerals, causing aluminum toxicity. Air pollution is considered to be a major factor in the decline of forests in heavily polluted areas of Europe and North America. There are indications that fast-growing pioneer species are better able to tolerate an acidifying atmosphere than are climax forest trees, possibly because they have a greater potential for assimilation of dissolved NO_x, and more effective acid buffering of the leaf tissue cell sap. Air pollution injury to plants can be evident in several ways. Injury to foliage may be visible in a short time and appear as necrotic lesions (dead tissue), or it can develop slowly as a yellowing or chlorosis of the leaf. There may be a reduction in growth of various portions of a plant. Plants may be killed outright, but they usually do not succumb until they have suffered recurrent injury.

Major primary air pollutants gases are sulphur dioxide, oxides of nitrogen particularly NO_2, HF, HCl, chlorine, ammonia, ethylene and other organic substances. Particulate air pollutants are soot, dust, fine particles of cement and various other substances. Various fertilizers, pesticides and insecticides used in aerial spray are also important air pollutants. The common sources of the pollutants, factors affecting the effect of pollutant and the injury symptoms produced in plants are discussed below.

10. Major gaseous pollutants

Sulphur dioxide (SO_2)

Sulfur dioxide is a major component in acid rain. One of the byproducts of sulfur dioxide is sulfuric acid, and both can be extremely damaging to plants that are exposed to these chemicals. Exposed leaves can begin to lose their color in irregular, blotchy white spots. Some leaves can develop red, brown or black spots. When the pigments in enough tissue are damaged or killed, plants can begin to lose their leaves. Crop output is greatly reduced and growth can be stunted. This is especially noticeable in young plants.

It is the most important and common air pollutant produced in huge amounts in combustion of coal and other fuels in industrial and domestic use. It is also produced during smelting of sulphide ores. Major sources of sulfur dioxide are coal-burning operations, especially those providing electric power and space heating. Sulfur dioxide emissions can also result from the burning of petroleum and the smelting of sulfur containing ores.

SO_2 effects increase in high humidity, windy conditions, in the early morning, in the deficiency of K and Cl_2 and excess of sulphur in the soil. It interacts with ozone, NO_2 and HF. The nature of interaction depends on the relative proportion of gases. The impact of SO_2 decreases in low soil moisture, low temperature, deficiency of nitrogen, sulphur and phosphorus and sometimes in excess of nitrogen also.

In angiosperms, young leaves and in conifers, needles are most sensitive to SO_2 pollution. In general, seedlings are more sensitive than older plants. The effect of the gas usually decreases with age of the plant and lesser morphological and physiological symptoms appear in older plants.

Injury symptoms: The gas is a strong reducing agent. In low concentration, it is oxidized and used in protein synthesis of the plant. However, in high concentration, it causes

swelling of thylakoids and interferes with electron transport chain. In SO_2 pollution, plants show initial reduction of photosynthesis and increased respiration. The gas reduces stomatal opening and thus causes general water stress in plants. SO_2 replaces oxygen in cellular materials and changes their nature. It affects structural proteins in the cell membrane and thus changes the membrane permeability. High concentration of the gas causes accumulation of sulphydril and decrease of sulphides in plants. SO_2 interferes with amino acid metabolism and reduces the synthesis of proteins and enzymes. It stimulates the oxidation of PGA and increases the pentose phosphate cycle activity. It reduces the level of keto acids, ATP, sucrose and glutamate in plants and increases the level of glucose, fructose and glycolate. It inactivates many enzymes either by breaking their S-S bonds or by changing their stereo structure. In lichens, the gas induces photooxidation in the phycobiont part. Most common visible symptom of SO_2 injury is water-soaked appearance of leaves which later become necrotic changing into brown spots. Color and shape of necrotic spots is characteristic in different species and NO_2 concentrations. In some species, characteristic intraveinal chlorosis is caused. In general, SO_2 pollution results in abscission of older leaves and tip necrosis in flower and sepals.
Sulfur dioxide enters the leaves mainly through the stomata (microscopic openings) and the resultant injury is classified as either acute or chronic. Acute injury (fig. 4) is caused by absorption of high concentrations of sulfur dioxide in a relatively short time. The symptoms appear as 2-sided (bifacial) lesions that usually occur between the veins and occasionally along the margins of the leaves. The colors of the necrotic area can vary from a light tan or near white to an orange-red or brown depending on the time of year, the plant species affected and weather conditions. Recently expanded leaves usually are the most sensitive to acute sulfur dioxide injury, the very youngest and oldest being somewhat more resistant.

Fig. 4. Acute sulfur dioxide injuries to raspberry [26].

Chronic injury is caused by long-term absorption of sulfur dioxide at sub-lethal concentrations. The symptoms appear as a yellowing or chlorosis of the leaf, and occasionally as a bronzing on the under surface of the leaves. Different plant species and varieties and even individuals of the same species may vary considerably in their sensitivity to sulfur dioxide. These variations occur because of the differences in geographical location, climate, stage of growth and maturation. The following crop plants are generally considered susceptible to sulfur dioxide: alfalfa, barley, buckwheat, clover, oats, pumpkin, radish, rhubarb, spinach, squash, Swiss chard and tobacco. Resistant crop plants include asparagus, cabbage, celery, corn, onion and potato. Plants damaged by sulfur dioxide can be as far as 30 miles from its source, but the most severe damage, defoliation and discoloring is typically found within five miles. For some plants, it can take exposure of only four hours to suffer damage. A wide variety of plants are vulnerable, from alfalfa and carrots to crab apple and fir trees.

Nitrogen dioxide (NO_2)

NO_2 mostly affects the leaves and seedlings. Its effects decrease with increasing age of the plant and tissue. Conifers are found to be more sensitive to this gas during spring and summer than in winters. Older needles are more sensitive to the gas than young ones.

Injury symptoms: The gas causes formation of crystalloid structures in the stroma of chloroplasts and swelling of thylakoid membrane. As a result the photosynthetic activity of the plant is reduced.

Most common visible injury symptoms are chlorosis in angiospermic leaves and tip burn in conifer needles. In angiosperms, most of the species produce water-soaked intraveinal areas that later become necrotic. Tip burn is common symptom in bracts, sepals and awns.

Fluorides

Fluirides in general, are accumulated in the plant tissues over long times. They are first accumulated in the leaves and then are translocated towards tips and margins of the leaves. The injury symptoms are produced only after a critical level of fluoride is attained. Due to such accumulation over long times, flurides generally and HF particularly can induce injury at very low atmospheric concentrations. Critical concentration for fluoride injury is 0.1 ppm for several days. Toxicity of particulate fluorides depends upon the particle size, their solubility and humidity of the atmosphere.

HF gas is much lighter than air and so can cause damage in plants even at a distance of 30 km from the source. It is a hygroscopic gas and forms acidic cloud near the source. Generally the impact of HF pollution increases with humidity and excess of P in soil while decreases in low temperature, drought and deficiency of N and Ca in the soil. In some species, impact of HF has been reported to decrease with excess of N and Ca in the soil.

In most of the species, recovery from moderate fluoride injury can occur within few days if exposure to pollutant stops. However, some highly sensitive species e.g. pine and spruce can never recover fully. HF generally affects immature leaves in angiosperms and needles in conifers.

Injury symptoms: Fluorides combine with metal components of proteins or inhibit them otherwise and thus interfere with the activity of many enzymes. As a result the cell wall composition, photosynthesis, respiration, carbohydrate synthesis, synthesis of nucleic acids and nucleotides and energy balance of the cell are affected. In the leaves subjected to HF exposure, endoplasmic reticulum is reduced, ribosomes are detached from ER, number of ribosomes is reduced and mitochondria become swollen. Chlorophyll synthesis and cellulose synthesis are inhibited. Activities of UDP-glucose-fructose transglucosylase, phosphoglucomutase, enolase and polyphenol oxidase are reduced. On the other hand activities of catalase, peroxidase, pyruvate kinase, PEP-carboxylase, glucose-6-phosphate dehydrogenase, cytochrome oxidase and pentose phosphate pathway are stimulated. In conifer needles common visible injury symptoms are chlorosis later turning into red/brown discolouration, tip burn later turning into necrosis of whole needle, formation of sharply defined red/purple bands between healthy and injured tissue. Similar symptoms are common in angiospermic leaves also. In addition, the angiospermic leaves in many species also show zonation of necrotic areas, leaf cupping, curling of leaf edges and ragged leaf margins. In sepals, petals, bracts and awns, water-soaked margins and later tip and marginal necrosis are observed. Fluorides absorbed by leaves are conducted towards the margins of broad leaves (grapes) and to the tips of monocotyledonous leaves (gladiolus).

Little injury takes place at the site of absorption, whereas the margins or the tips of the leaves build up injurious concentrations. The injury (fig. 5) starts as a gray or light-green water-soaked lesion, which turns tan to reddish-brown. With continued exposure the necrotic areas increase in size, spreading inward to the midrib on broad leaves and downward on monocotyledonous leaves.

Fig. 5. Fluoride injuries to plum foliage [26].

The fluoride enters the leaf through the stomata and is moved to the margins where it accumulates and causes tissue injury. Note, the characteristic dark band separating the healthy (green) and injured (brown) tissues of affected leaves. Studies of susceptibility of plant species to fluorides show that apricot, barley (young), blueberry, peach (fruit), gladiolus, grape, plum, prune, sweet corn and tulip are most sensitive. Resistant plants include alfalfa, asparagus, bean (snap), cabbage, carrot, cauliflower, celery, cucumber, eggplant, pea, pear, pepper, potato, squash, tobacco and wheat.

Chlorine (Cl_2)

Older plants are more sensitive to chlorine than seedlings. The age of tissue has little effect on the sensitivity and older as well as young tissues are almost equally afected by chlorine pollution.

Injury symptoms: Chlorine injury symptoms can appear from 18 hours to 8 days after exposure. In most plant species, recovery from chlorine injury can occur 3 to 4 days after exposure is stopped. Chlorine injury symptoms are quite variable in different species. Most common visible symptoms in conifers are chlorosis, tip burn and necrosis is needles. In angiosperm leaves, marginal or intraveinal necrosis, water-soaked appearance, leaf cupping and abscission are common.

Hydrogen chloride (HCl)

The HCl injury can be caused to plants even at a distance of 800 meter from the source. Like fluorides, the chloride from HCl is accumulated in the leaves and translocated towards their margins and tips. Symptoms of HCl injury appear after a critical concentration is reached, usually between 24 and 72 hours after the exposure.

Impact of HCl pollution decreases with increase in humidity, deficiency of Mg and excess of Ca. Mature plants are more sensitive to HCl than seedlings. Similarly, young fully expanded leaves are more sensitive than immature unexpanded leaves.

Injury symptoms: Most common visible injury symptoms in conifer needles are red or brown discolouration and tip burn. In angiosperm leaves, common symptoms are intraveinal water-soaked streaks, yellow or brown necrosis, tip necrosis, bleached areas around the necrosis and shot-holing. Tip burn, necrotic stipple and discolouration in sepals and petals are also observed.

Ammonia (NH_3)

Impact of ammonia on plants generally increases with humidity and decreases with drought. Effect of darkness on ammonia sensitivity is highly variable among species. Some species are more sensitive to low concentrations of ammonia than to its high concentration. Age of tissue has little effect on sensitivity and both young and old tissues are equally sensitive to ammonia.

Injury symptoms: Most common visible symptoms in conifers are black discolouration, usually sharply bordered tip burn and abscission of needles. In angiosperm leaves, common symptoms are water-soaked appearance later turning black, intercostal necrosis, slight marginal and upper surface injury, glazong/bronzing of upper surface, desiccation and abscission. Ammonia injury to vegetation has been observed frequently in Ontario in recent years following accidents involving the storage, transportation or application of anhydrous and aqua ammonia fertilizers. These episodes usually release large quantities of ammonia into the atmosphere for brief periods of time and cause severe injury to vegetation in the immediate vicinity. Complete system expression on affected vegetation usually takes several days to develop, and appears as irregular, bleached, bifacial, necrotic lesions. Grasses often show reddish, interveinal necrotic streaking or dark upper surface discolouration. Flowers, fruit and woody tissues usually are not affected, and in the case of severe injury to fruit trees, recovery through the production of new leaves can occur (fig. 6). Sensitive species include apple, barley, beans, clover, radish, raspberry and soybean. Resistant species include alfalfa, beet, carrot, corn, cucumber, eggplant, onion, peach, rhubarb and tomato.

Fig. 6. Severe ammonia injuries to apple foliage and subsequent recovery through the production of new leaves following the fumigation [26].

Organic gases (Ethylene)

Ethylene injury symptoms develop in plants only in exposure to high concentrations and take several days to develop. After exposure to the gas is stopped, level of recovery is variable in different species. Generally, younger plant parts recover but older parts do not. Much 'acute' damage to plants is caused on the fringes of polluted area or by a steady leakage of gas in low concentration.

Injury symptoms: In injuriously high concentrations of ethylene, growth of plants is stopped. In low concentrations, growth abnormalities appear. In conifers, yellow tips in needles and abscission of branches and cones are common. In angiosperms, common symptoms are epinasty or hyponasty, loss of bark, abscission of leaves and flowers, premature flower opening and fruit ripening. Ethylene affects the growth hormones and regulatory process that takes place in the plant and results in a number of outward manifestations of infection. Leaves can begin to curl and die; ethylene causes the leaves of plants to curl down and fold under as they shrivel and are stuck with necrosis. On flowering plants, buds can stop opening or flowers can begin to show signs of discoloration or die and drop sooner than expected. Even in more resistant plants like evergreen conifers, growth of the plant will be stunted, needles will be small and few pine cones will be produced. Plants such as peach trees, marigolds, blackberries and tomatoes are extremely vulnerable to damage from exposure to ethylene.

11. Minor gaseous pollutants

Hydrogen sulphide (H_2S)
Plants show wilting on exposure to this gas but the symptoms develop after about 48 hours. No injury occurs below the exposure of 40 ppm for 4 hours.

Carbon monoxide (CO)
Like ethylene this gas produces epinasty, chlorosis and abscission. However, concentration of over 1000 times that of ethylene is needed to produce same degree of damage. No injury to plants occurs below exposure of 100 ppm for 1 week.

Bromine (Br_2) and Iodine (I_2)
Studies show these gases are highly toxic to plants. HI and I_2 are readily absorbed and accumulated by plants producing visible injury symptoms similar to those of SO_2. Injury occurs at exposure of 0.1 ppm for 18 hours.

Common injury symptoms of bromine in angiosperms are necrosis of leaf margins, leaf tips and tendrils; brown discoloration and black spots later spreading to entire leaf. In conifers, yellow/white needle tips or red/brown discoloration later becoming grey/brown are common symptoms.

Mercury vapors (Hg)
Unlike other pollutants, flowers are more sensitive to Hg than leaves. Injury symptoms usually appear within 24 hours of Hg exposure but often go on increasing up to 5 days.

Common injury symptoms due to Hg-vapors pollution are abscission of oldest leaves, interveinal necrosis, chlorosis around veins, flower abscission, loss of petal colors, buds remaining closed and later becoming necrotic, blackening of stamens, pistils and peduncles.

Particulate pollutants
Different types of solid particulate materials are also important air pollutants. Each of these affects the plants in characteristic manner. Some common particulate air pollutants have been discussed below.

Cement-kiln dust
In generals, plants having hairy surface of leaves trap more dust and are, therefore, damaged more than the plants with shiny leaf surface. The cement dust forms crusts on the surface of leaves, twigs and flowers. This inhibits gaseous exchange from the surfaces of plant parts. Such crust on the leaves also inhibits light penetration and consequently reduces photosynthesis. Such crusts are especially thicker in conditions of dew, mist or light rains. In dry conditions, dust blowing with wind is highly abrasive and damages the cuticle of leaves. Cuticle is also damaged due to alkalinity of cement dust. Due to damaged cuticle plants become more susceptible to infection by pathogens.

Lime and gypsum
Lime and gypsum deposited on the soil from the air, these change the pH of the soil and thus affect the nutrient availability to plants. Such deposition usually causes appearance of various nutrient deficiency symptoms in the plants. Lime and gypsum are less adhering as compared to cement-kiln dust. However, these are also trapped and deposited on the

surface of plant parts particularly the leaves with hairy surfaces and produce injury symptoms similar to cement dust. Lime and gypsum particles blowing with wind are also highly abrasive for cuticle.

Soot

Soot deposited on the surface of leaves may be washed away by rains so its damage may be reduced. However, in bright sunlight and high temperature, the damage is increased.

Soot deposited on the surface of leaves inhibits light penetration, increased surface temperature due to absorption of heat and clogging of stomata. The result of these is reduced gaseous exchange, reduced photosynthesis and general weakening of the plant growth. Necrotic spots also develop in many species due to soot deposition.

Magnesium oxide

Deposited on the soil these compounds can soon increase the soil pH to levels injurious to plants. Deposition of these substances on the soil prevents germination of seedlings. The seedlings that are able to emerge usually have yellow/brown tips of leaves and their roots are stunted. In areas of heavy pollution, composition of the vegetation changes completely.

Boron

Severe injury to plants is observed even at a distance of 200 meters from the source and mild injury may be observed up to 500 meters in all the directions from the source.

Impact of boron pollution is more severe on older leaves than on younger leaves. Boron is also accumulated in the leaves and produces injury symptoms quite similar to fluoride pollution.

Chlorides of sodium, potassium and calcium Injury symptoms produced by these chlorides in plants are very similar to those produced by SO_2 and fluoride pollution.

Sodium sulphate dust can cause necrosis of leaves of the plants. The damage increases in moist condition.

Pesticides, insecticides and herbicides

A large variety of such chemicals are sprayed on the crops these days. These substances may drift with wind to nearby areas. Generally, these chemicals are deposited on the soil and form important soil pollutants. However, in frosty conditions when crops and other plants damaged by early frost are quite susceptible to foliar spray of these chemicals, these may also be injurious air pollutants. Injury symptoms vary with the plant species and the type of chemical. Generally, the symptoms are produced on foliage and are quite similar to those produced when these substances act as soil pollutants.

12. Secondary pollutants and plants

Many of the primary pollutants under specific environmental conditions may interact with each other and produce secondary environmental pollutants or certain complex environmental conditions that are injurious to plants. Such secondary pollutants and pollution conditions are discussed below.

Photo-oxidants

PAN (Peroxyacetylnitrate-$CH_3CO.O_2.NO_2$)

The common visible symptoms of exposure to PAN are chlorosis and necrosis in leaves. It also interferes with photosynthesis, respiration and absorption and synthesis of carbohydrates and proteins. It inhibits photorespiration, NADP reduction, carbon dioxide fixation, cellulose synthesis and the enzymes associated with photosynthesis and respiration.

Ozone (O_3) is released into the atmosphere from the burning of fossil fuels and is one of the most harmful pollutants to plants. It can be carried for long distances and is readily absorbed as a part of the photosynthetic process. Plants exposed to large amounts of ozone can develop spots on their leaves. These spots are irregular and often tan, brown or black. Some leaves can take on a bronze or red appearance, usually as a precursor to necrosis. Depending on the concentration of ozone in the environment, plants can show different amounts of discoloration before the leaves begin to die.
Studies by the National Crop Loss Assessment Network show that ozone in the environment also has a detrimental effect on crop production. While crops such as cotton, soybeans and other dicots are more sensitive than monocot crops, all crops sampled over the decades-long studies show significant loss of productivity when exposed to ozone. Cotton crops show significantly less yield when exposed to levels of ozone in the atmosphere. Middle aged leaves and young plants are more sensitive to ozone. This pollutant interacts with SO_2, NO_2, PAN and heavy metals in complex manner.
Common symptoms of ozone pollution are yellowing, flecking and blotching in leaves, premature senescence and early maturity. It interferes with pollen formation, pollination, pollen germination and growth of pollen tubes. Increase in the level of RNA, starch, polysaccharides and number of polysomes is observed in ozone pollution. Ozone stimulates respiration, inhibits oxidative phosphorylation and changes membrane permeability. In some species, it inhibits the synthesis of glucon and cellulose and reduces the level of reducing sugars, ascorbic acid and ATP while in other species the effect is opposite to it. The impact of ozone on plants increases with humidity and decreases with drought, darkness, low temperature, high soil salinity, deficiency of soil phosphorus and excess of soil sulphur. Throughout the growing season, particularly July and August, ozone levels vary significantly. Periods of high ozone are associated with regional southerly air flows that are carried across the lower. Localized, domestic ozone levels also contribute to the already high background levels. Injury levels vary annually and white bean, which are particularly sensitive, are often used as an indicator of damage. Other sensitive species include cucumber, grape, green bean, lettuce, onion, potato, radish, rutabagas, spinach, sweet corn, tobacco and tomato. Resistant species include endive, pear and apricot.

Fig. 7. Ozone injuries to soybean foliage [26].

Ozone symptoms (fig. 7) characteristically occur on the upper surface of affected leaves and appear as a flecking, bronzing or bleaching of the leaf tissues. Although yield reductions are usually with visible foliar injury, crop loss can also occur without any sign of pollutant stress.

Conversely, some crops can sustain visible foliar injury without any adverse effect on yield. Susceptibility to ozone injury is influenced by many environmental and plant growth factors. High relative humidity, optimum soil-nitrogen levels and water availability increase susceptibility. Injury development on broad leaves also is influenced by the stage of maturity. The youngest leaves are resistant. With expansion, they become successively susceptible at middle and basal portions. The leaves become resistant again at complete maturation

Ground-level ozone causes more damage to plants than all other air pollutants combined. This web page describes the ozone pollution situation, shows classical symptoms of ozone injury and shows how ozone affects yield of several major crops.

Ozone enters leaves through stomata during normal gas exchange. As a strong oxidant, ozone (or secondary products resulting from oxidation by ozone such as reactive oxygen species) causes several types of symptoms including chlorosis and necrosis. It is almost impossible to tell whether foliar chlorosis or necrosis in the field is caused by ozone or normal senescence. Several additional symptom types are commonly associated with ozone exposure, however. These include flecks (tiny light-tan irregular spots less than 1 mm diameter), stipples (small darkly pigmented areas approximately 2-4 mm diameter), bronzing, and reddening.

Ozone symptoms usually occur between the veins on the upper leaf surface of older and middle-aged leaves, but may also involve both leaf surfaces (bifacial) for some species. The type and severity of injury is dependent on several factors including duration and concentration of ozone exposure, weather conditions and plant genetics. One or all of these symptoms can occur on some species under some conditions, and specific symptoms on one species can differ from symptoms on another. With continuing daily ozone exposure, classical symptoms (stippling, flecking, bronzing, and reddening) are gradually obscured by chlorosis and necrosis.

Studies in open-top field chambers have repeatedly verified that flecking, stippling, bronzing and reddening on plant leaves are classical responses to ambient levels of ozone. Plants grown in chambers receiving air filtered with activated charcoal (CF) to reduce ozone concentrations do not develop symptoms that occur on plants grown in non-filtered air (NF) at ambient ozone concentrations. Foliar symptoms shown on this web site mainly occurred on plants exposed to ambient concentrations of ozone (either in NF chambers or in ambient air).

Yield Loss Caused by Ozone

Field research to measure effects of seasonal exposure to ozone on crop yield has been in progress for more than 40 years. Most of this research utilized open-top field chambers in which growth conditions are similar to outside conditions. The most extensive research on crop loss was performed from 1980 to 1987 at five locations in the USA as part of the National Crop Loss Assessment Network (NCLAN). At each location, numerous chambers were used to expose plants to ozone treatments spanning the range of concentrations that occur in different areas of the world. The NCLAN focused on the most important agronomic crops nationally

The strongest evidence for significant effects of ozone on crop yield comes from NCLAN studies [18] (fig. 8). The results show that dicotyledonous species (soybean, cotton and peanut) are more sensitive to yield loss caused by ozone than monocot species (sorghum, field corn and winter wheat).

Particulate Matter

Particulate matter such as cement dust, magnesium-lime dust and carbon soot deposited on vegetation can inhibit the normal respiration and photosynthesis mechanisms within the leaf. Cement dust may cause chlorosis and death of leaf tissue by the combination of a thick crust and alkaline toxicity produced in wet weather. The dust coating (fig. 9) also may affect the normal action of pesticides and other agricultural chemicals applied as sprays to foliage. In addition, accumulation of alkaline dusts in the soil can increase soil pH to levels adverse to crop growth.

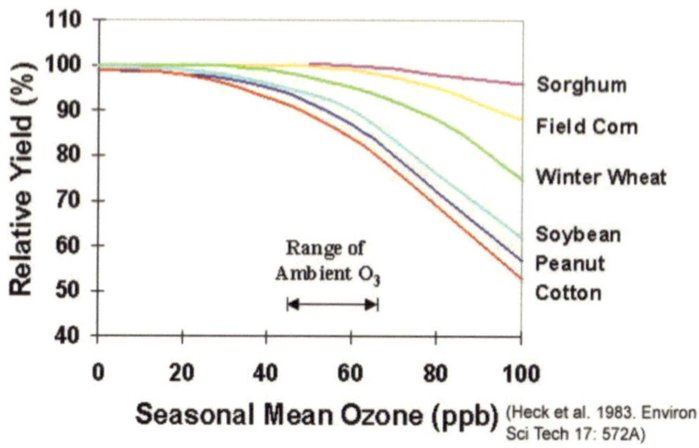

Fig. 8. Effect of ozone on yield of crops [18]. cotton image by arklite06 from

Fig. 9. Cement-dust coating on apple leaves and fruit. The dust had no injurious effect on the foliage, but inhibited the action of a pre-harvest crop spray [26].

13. Effect of atmospheric pollutants on vegetation monitoring system, why forestry monitoring system?

Because the crop plants are mostly annual plants they can not show the long-term effects produced by air pollutants. Therefore to monitor the effects of air pollution are recommended the trees, the changes in forest structure highlight the harmful effects of different air pollutants.

The evident decline of the health state of the forest in Europe since the beginning of the 1980 due to the negative impact of air pollution were illustrated by numerous publication from this period (see litt.). In the efforts to obtain objective and comparable data concerning the health of the European forests were developed a common methodology for the assessment of the forest state under the influence of air pollution. This network is known under the short term ICP- Forest (International Cooperative Programme for the investigation of the trans-boundary Pollution Influence on the Forests).

The poor health status of the forests in Central Europe concerns all the Europe. The pictures of the forests on large area were dominated by tree with defoliated crowns and an increasing rate of the death trees (fig 10). The assessment of the causes of the "new damages"- neuartiges waldschaden, in germ- is not easy because the symptoms of the decline were different from the symptoms of the damages caused by natural (biotic and abiotic) and anthropic causes.

Under the umbrella of ICP Forest Programme, were developed and implemented an European network of plots for the assessment of the parameters of the trees crowns condition known as Level I plots. The grid of the European Level I plots were established at 16 * 16 km, arranged on a transnational unified grid over Europe. In comparison with the national grids used by each country the obtained data were relevant for the evaluation of the forest health state at European level.

After 1996 were put in function the Level II monitoring plots used for the intensive monitoring and collection of comparable data related to the changes in forest ecosystems which are directly connected to specific environment at factors such as atmosphere pollution and acid deposition. Such data can help in a better understanding at the relation causes and effects in the forests decline.

Fig. 10. General aspects of silver fir crowns affected by decline in the border of the northern Carpathians (Forest District Solca)

The monitoring results contribute to the scientific basis of air pollution control policies of UN/ECE and the European Commission (EC). Fifteen years of monitoring forest condition and two decades of forest damage research have shown, however, that the discussion of recent forest damage must not be confined to the effects of air pollution alone. The comprehensive monitoring programme corresponds to the complex interrelations between natural and anthropogenic factors in forest ecosystems. Infrastructure and data of the programme are thought to be relevant for other processes of international forest policies, e.g.

those on biodiversity, climate change and sustainable forest management. In this respect the monitoring pursues the objectives of Resolution SI of the Strasbourg, Resolution HI of the Helsinki and Resolution, L2 of the Lisbon Ministerial Conference on the Protection of Forests in Europe, and contributes to global forest policies such as the United Nations Forum on Forests (ICP- Report 2007).

The monitoring results obtained each year are summarized in annual Executive Reports. The methodological background and detailed results of the individual surveys are described in Technical Reports (www.icp-forests.org).

Methodology for the crown health condition assessment of forests

The state of health of forest trees can be determined by assessing the foliage loss. With a little practice, this can be accurately estimated by the foresters or other trained personnel. The development of forest damage can be traced through repeated assessments of the same trees.

Loss of needles or leaves should be assessed after sprouting in spring or early summer and before broadleaves and larch display autumn coloration, at best in July and August. Evergreen conifers (fig. 11-18) may also be assessed in their winter state as long as they are free of snow. Assessments should be made under good light conditions in good weather: rain and fog render assessments inaccurate. Leaf or needle loss is estimated for the entire crown. The crown is considered to reach from the peak of the tree to, the lowest strong green branch forming part of the crown as such; epicormic shoots on the stem are not considered, while those in the crown are.

Fig. 9.-12. Defoliation in % of Oak crowns [27].

Fig. 13.-16. Defoliation in % of Beech crowns [27]

A forest tree can spread its crown to a greater or lesser extent depending on the room available within the stand. Consequently, spatial conditions must be considered in crown assessment; that is, the maximum foliage that each tree could possibly produce must be taken as a basis. The photo series (fig. 19 - 26) depicts trees of the upper storey with well developed crowns enjoying optimum light conditions. It is therefore applicable to trees of the middle and lower strata only to a limited extent. Foliage loss may be determined by comparing the tree under consideration with the corresponding photo series. The appearance of the crown is matched with one of the photos and the foliage loss estimated to a degree of 5 percent accuracy. Assessments should be made with field-glasses from a distance of at least one tree-length. Field-glasses permit precise identification of bare branches and twigs and discoloration. In subsequent surveys it is important that the tree always be observed from the same side; this should either be marked on the tree itself or noted in terms of compass direction. Leaf or needle loss due to known causes, e.g., hail, lightning, whipping, insect attack, etc., should not be included but separately inventoried [27].

14. Political background and objectives of ICP forests

The objectives and the strategy of ICP Forests are based on the draft long-term strategy and the work plan for the effects-oriented activities of the Working Group on Effects (WGE) of

Fig. 19.-22. Defoliation in % of Oak crowns [27].

Fig. 23.-26. Defoliation in % of Beech crowns [27].

CLRTAP (Convention on Long-range Transboundary Air Pollution). The draft long-term strategy of WGE specifies the following long-term aims to which all ICP are expected to contribute:

Assessment of knowledge on

- The present status, long-term trends and dynamics, and the degree and geographical extent of the impact of air pollution, particularly, but not exclusively, its long range trans-boundary impact
- Exposure-response relationships for agreed air pollutants;
- Critical loads, levels and limits for agreed air pollutants;
- Interactive effects of air pollution and climate change on forest ecosystems
- Moreover, the long-term strategy of WGE specifies the following long-term priorities of special relevance to ICP Forests:
- Derivation of exposure-response functions for chemical and biological effects of air pollutants including investigation of nutrient nitrogen, acidifying compounds and ozone effects on ecosystem functions and on biodiversity, including combinations with other stresses (e.g. climate change and land use practices);
- Further development of models and mapping procedures, particularly for effects of nitrogen and ozone on the environment and for the description of dynamic processes of damage and recovery (acidification, eutrophication, heavy metal accumulation) by including to a larger extent biological effects;
- Evaluation of environmental benefits of air pollution control policies.
- In order to meet the information needs of the Working Group on Effects of Atmospheric Pollution (WGE), ICP Forests pursues the following two main objectives:
- Objective 1: A periodic overview on the spatial and temporal variation of forest condition in relation to anthropogenic and natural stress factors (in particular air pollution) by means of European-wide and national large-scale representative monitoring on a systematic network.
- Objective 2: A better understanding of the cause-effect relationships between the condition of forest ecosystems and anthropogenic as well as natural stress factors (in particular air pollution) by means of intensive monitoring on a number of selected permanent observation plots spread over Europe and to study the development of important forest ecosystems in Europe.

These objectives imply in accordance with the long-term priorities of WGE contributions to calculations of critical loads and levels and the assessment of their exceedances. They imply also dynamic modeling of the response of forest ecosystems to deposition scenarios expected for the future. Additional insight is gained by compiling available studies from the National Focal Centers (NFCs) and from related programmes inside and outside of Convention on Long-range Trans-boundary Air Pollution.

15. Strategy of ICP forests

Monitoring activities

In order to meet its data generation and reporting obligations, ICP Forests employs data collection at two levels.

- Large-scale monitoring (Level I) provides a periodic overview of the spatial and temporal variation in a range of attributes related to forest condition. Level I plots, national forest

inventory (NFI) plots, and other related inventory plots may be combined when appropriate, feasible and necessary, according to defined and agreed procedures.
- Intensive monitoring (Level II) is carried out on plots installed in important forest ecosystems.

These plots are dedicated to in-depth investigation of the interactive effects of anthropogenic and natural stress factors on the condition of forest ecosystems.

Quality assurance and control

All monitoring activities are harmonized by ICP Forests among the participating countries and are laid down in this Manual. This ensures a standard approach for data collection and evaluation and can form the nucleus for a future common European forest monitoring programme. A consistent quality assurance approach is applied within the programme covering the set up of methods, data collection, submission and investigation as well as reporting. Quality assurance and control is supervised by the Programme Coordinating Group through its Quality Assurance Committee. A set of Expert Panels cares for data quality assurance within the specific surveys and for the further development of monitoring methods and standards. This includes field checks, inter-calibration courses, laboratory ring tests, and data validation.

Data evaluation and reporting

A range of monitoring variables is required to meet the information requirements of Convention on Long-range Trans-boundary Air Pollution and other international institutions. The Programme Coordinating Group and the Expert Panels are responsible for a data evaluation and reporting approach which takes the medium term work-plan of Working Group on Effects of Atmospheric Pollution into account. International and national data from other programmes and institutions should be included in combined analysis. The main topics for data analysis are:

- Forest condition
- Effects on forest ecosystems from
- Acidity and nitrogen
- Ozone

Contributions in the fields of

- Climate change
- Biodiversity

Trends in deposition and their interactive effects on the adaptation and vulnerability of forest ecosystems are evaluated. This includes spatial and temporal changes and cause-effect relationships with special emphasis on critical loads and their exceedances. Dynamic models and transfer functions derived from suitably selected intensive monitoring plots are used to investigate the effects of climatic factors and greenhouse gases on forest ecosystems and applied to the large scale monitoring plots. These models are validated against measured data collected at the plots. Furthermore, data gathered at the plots are used in an integrated manner to investigate the carbon sequestration potential of forests, ozone fluxes to forests and contribute to assess status and trends of forest biodiversity at the pan-European level.

Results

The integrative monitoring approach of ICP Forests using the Level I and Level II networks provides robust data on the health and stability of forests. This facilitates an understanding of the effects of deposition on the role and functioning of forest ecosystems in protecting soils and water. Furthermore the programme surveys can contribute to the understanding and forecast of climate change effects on forests and can be used to supply information on the sequestration of carbon and are going to provide information on forest biodiversity as an integral part of forest ecosystems. Results are published via reports and a website (www.icp-forests.org).

ICP Forests aims to provide periodic overviews on the spatial and temporal variation of forest condition in relation to man-made and natural stress factors (particularly air pollution); to contribute to a better understanding of the cause-effect relationships between the condition of forest ecosystems and man-made and natural stress factors (particularly air pollution); and to study the development of important forest ecosystems in Europe.

More specifically, to support harmonized forest monitoring by linking existing and new monitoring mechanisms at the national, regional and EU level (tab. 2); to collect quantitative and qualitative forest data related to climate change, air pollution, biodiversity, and forest condition; and to contribute information on sustainable forest management to the Ministerial Conference on the Protection of Forests in Europe.

Survey	Number of plots		Assessment frequency
	installed	Data submitted for 2007	
Crown condition	836	462	Annually
Foliar chemistry	904	200	Every two years
Soil condition	615	0	Every ten years
Soil solution chemistry	302	169	Continuously
Tree growth	811	70	Every five years
Deposition	657	353	Continuously
Ambient air quality (active)	84	27	Continuously
Ambient air quality (passive)	254	167	Continuously
Ozone induced injury	114	43	Annually
Meteorology	265	191	Continuously
Phenology	186	58	Several times per year
Ground vegetation	777	67	Every five years
Litterfall	262	105	Continuously
Remote sensing	National data		Preferably at plot installation

Table 2. Surveys and number of plots for Level II monitoring. The variation in assessment frequency results in different numbers of plots with data submission for the different surveys (after www.icp-forests.org)

Conclusions after 25 years of forest monitoring at European level

For 25 years, forest condition has been monitored by ICP Forests in close cooperation with the European Commission. The system combines an inventory approach with intensive monitoring. It provides reliable and representative data on forest ecosystem health and vitality and helps to detect responses of forest ecosystems to the changing environment. The data collected so far provide a major input for several international programmes and initiatives, such as the Convention on Long-range Trans-boundary Air Pollution and the Ministerial Conference for the Protection of Forests in Europe.

Participating countries	Forest area (x 1000 ha)	% of forest area	Grid size (km X Uin)	No. of sample plots	No. of sample trees	Defoliation of all species by class (aggregates), national surveys		
						0	1	2 - 4
Albania	1063	37	no survey in 2009					
Andorra	18		16 X 16	3	73	60.3	32.9	6.8
Austria	3878	46.2	no survey in					
Belarus	7921	38.2	16 X 16	409	9620	27,7	63.9	8.4
Belgium	700	23.1	42/82	122	2858	30,7	49.1	20.2
Bulgaria	3699	33.3	42/82/162	159	5560	29,6	49.3	21.1
Croatia	2061	36.5	16 X 16	83	1991	37.2	36.5	26.3
Cyprus	298	32.2	16 X 16	15	362	3,0	60.8	36.2
Czech Republic	2647	33.6	82/162	133	5284	11.7	31.5	56.8
Denmark	527	12.2	72/162	16	384	69,0	25.5	5.5
Estonia	2213	49.1	16 X 16	92	2202	44.3	48.5	7.2
Finland	20150	66.3	162* 24 X 32	886	7182	58,2	32.7	9.1
France	15840	28.9	16 X 16	500	9949	28.7	37.8	33.5
Germany	11076	31	162*42	424	10376	36,4	37.1	26.5
Greece	2034	19.5		89	2098	42.2	33.5	24.3
Hu ngary	1904	22.5	16 X 16	78	1872	54,8	26.8	18.4
Ireland	680	6.3	16x 16	30	599	69.9	17.5	12.5
Italy	8675	28,8	16 X 16	257	6966	24.5	39.7	35.8
Latvia	3162	49	8x8	340	8036	17	69.2	13.8
Liechtenstein	8	50	no survey in 2009					
Lithuania	2150	32.9	8X8 / 16X16	983	5961	18.6	63.7	17.7
Luxembourg	89	34.4			no survey in	2009		
FYR of Macedonia					no survey in	2009		
Rep. of Moldova	318	9.4	2x2	622	13676	43.1	31.7	25.2
The Netherlands	334	9.6			no survey in	2009		
Norway	12000	37.1	32/ 92	1622	9332	43.1	35.8	21
Poland	9200	29.4	16 X 16 ,	1923	38460	24.1	58.2	17.7
Portugal	3234	36.4	no survey in 2009					
Romania	6233	26.1	16 X 16	227	5448	44.1	37	18.9
Russian Fed.	809090	73.2		365	11016	80	13.8	6.2
Serbia	2360		16 X 16/4x4	130	2765	68.1	21.6	10.3
Slovak Republic	1961	40	16x 16"	. 108	4049	9.3	58.6	32.1
Slovenia	1099	54.2	16 X 16	44	1056	18.2	46.4	35.5
Spain	11588	30.9	16 X 16	620	14880	17.8	64,4	17,7
Sweden	28300	69	Varying	3217	7097	59,9	25.1	15,0
Switzerland	1186	28.7	16 X 16	48	1040	32.3	49.4	18.3
Turkey	21389	27.5	16 X 16	563	12290	25.1	56,2	18.7
Ukraine	9400	15.4	16 X 16	1483	34498	66.4	26.8	6.8
United Kingdom	2837	11.7	no .survey in 2009					
Total	**1011322**		**Varying**	**15591**	**236980**			

Table 3. Forest surveys and defoliation classes for all tree species in European countries (2009). Results of national surveys as submitted by National Focal Centres (after www.icp-forests.org)

In the early 1980s, a dramatic deterioration in forest condition was observed in Europe and this initiated the implementation of forest condition monitoring under Convention on Long-range Trans-boundary Air Pollution. Today, the monitoring results indicate that, at the large scale, forest condition has deteriorated far less severely than was feared at that time. Stress factors like insects, fungi and weather effects have been shown to affect tree health. The drought in the Mediterranean region in the mid-1990s and the extremely warm and dry summer across large parts of Europe in 2003 led to increased levels of defoliation as a natural reaction of trees to this type of stress. The programme has also reported on acidifying deposition which is regionally correlated with defoliation and on atmospheric inputs that are accentuating other stress factors. In the past three years there has been little change in the mean levels of defoliation for the main European tree species. However, long-term trends show more deterioration than improvement (tab. 3).

The health status of forest trees in Europe is monitored over large areas by surveys of tree crown condition. Trees that are fully foliated are regarded as healthy. The Ministerial Conference on the Protection of Forests in Europe uses defoliation as one of four indicators for forest health and vitality.

- In 2009, crown condition data were submitted for 7193 plots in 30 countries. In total, 136 778 trees were assessed. This constitutes the programmer's largest number of plots for which annual data were submitted.
- In 2009, 20.2 % of all trees assessed had a needle or leaf loss of more than 25 % and were thus classified as either damaged or dead (fig. 27). This represents no change relative to 2008.
- Of the main tree species, European and sessile oak had the highest levels of damaged and dead trees, at 31.8 %.
- There were no significant changes in crown condition over the past ten years on two-thirds of the plots, but deterioration prevailed on the remaining third.
- In 2009, a fifth of the 136 778 trees studied were considered damaged or dead
- Trends vary between species, with European and sessile oak the most frequently damaged species. However, both have shown some recovery over the past five years. The health of Norway spruce and Scots pine has improved over the past 18 years. Defoliation in common beech, Holm oak and maritime pine has increased.
- There has been no significant change in tree health on most plots monitored over the past ten years. Defoliation increased on 24.4 % of plots monitored and decreased, indicating an improvement in crown condition, on only 14.9 % (fig. 28).
- Over the past 18 years there has been a clear improvement in crown condition for Scots pine and a slight improvement for Norway spruce. European and sessile oak have shown the highest mean defoliation over the past decade.
- Defoliation peaked after the extremely dry and warm summer in 2003 and has been slowly recovering since 2007. Defoliation of common beech peaked in 2004, while Holm oak showed a sharp deterioration in crown condition in the mid-1990s and again in 2005. Unfavorable weather conditions are thought to be responsible for these trends. There was a reasonably consistent increase in defoliation of maritime pine up to 2005, followed by a short period of recovery after which crown condition again deteriorated in 2009 [28], [29], [30].

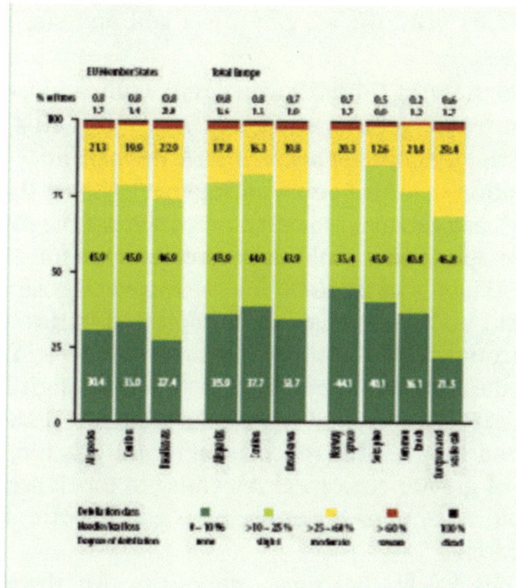

Fig. 27. Extent of defoliation for the main European tree species. Total Europe and EU, 2009. (after www.icpforests.org)

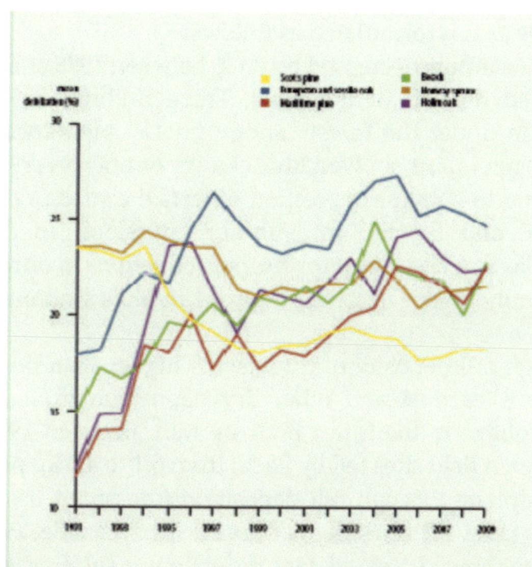

Fig. 28. Mean percentage defoliation for the most frequent tree species in European forests (after www.icpforests.org)

- Defoliation is an indicator of tree health and vitality that can be easily monitored over large areas and which reacts to many different factors, including climatic conditions and weather extremes as well as insect and fungal infestations.
- Defoliation represents a valuable early warning system for the response of the forest ecosystems to change - this is particularly relevant as climatic extremes are predicted to occur more frequently in the relatively near future.

- Deposition of pollutants from the air can affect soil and site conditions and thus the condition of forest trees.
- The status and trends in forest condition vary regionally and for different species. Local conditions may differ from the European average.

Conclusions concerning the dynamic of atmospheric deposition

ICP Forests began deposition measurements on intensive (Level II) monitoring plots in the latter half of the 1990s. Measurements are carried out within the forest stands (through fall deposition) and in nearby open fields (bulk deposition). In the forest canopy, some elements can be leached from the foliage and increase the measured deposition load, whereas others are taken up by leaves and needles and are so not detected in through fall. Bulk deposition is mostly lower than through fall deposition because of the additional deposition loads filtered from the air by the forest canopy. Thus, neither through fall deposition nor bulk deposition is equal to the total deposition received by the forest stands. However, through fall deposition is presented here as this reflects the inputs reaching the forest floor and so these measurements are of greater ecological relevance to forest ecosystems than open field measurements. On the plots, samples are collected weekly, fortnightly or monthly and are analyzed by national experts.

After intensive quality checks, annual mean deposition for the years 1998 to 2007 was calculated for plots with complete data sets. Slopes of plot wise linear regressions of deposition over time were tested for significance. Plot-specific means were calculated for the period 2005 to 2007.

The most relevant trends cam is formulated as follows:

- Mean annual sulphur inputs decreased by 30 % between 1998 and 2007, with significant reductions measured on half of the plots. These findings are based on deposition measurements made under the forest canopy on 157 plots located mostly in central Europe. Mean nitrogen inputs showed little change or only a very small decrease.
- The downward trend in sulphur deposition reflects the success of the clean air policies under the UNECE and the EU for sulphur emissions. In contrast, the nitrogen deposition data indicate a clear need for further reductions in nitrogen emissions.
- Deposition is generally higher on central European plots than on plots in northern and southern Europe.
- On average, through fall deposition in forests is higher than deposition on open field sites because trees filter dust and other dry deposition from the air which is then washed from the foliage to the forest floor by rain. Between 1998 and 2007, sulphate deposition on the open field sites fell by 26 %; from 6.1 to 4.5 kg per hectare per year.
- The decrease in sulphate through fall deposition (measured below the forest canopy) was higher at 34 %; from 10.0 to 6.6 kg per hectare per year (fig. 29).
- About half the plots showed a significant reduction in sulphur inputs over the 10-year study period. The data are mean values from around 150 measurement stations located mainly in central Europe.
- Mean nitrogen deposition within the forest stands fluctuated (for nitrogen measured as nitrate and ammonium) and few plots showed significant changes in through fall deposition.
- Slight decreases in mean nitrogen deposition at the open field plots were observed (fig. 30). The deposition data show the success of the clean air policies in Europe for sulphur emissions, and show the need for further reductions in nitrogen emissions [31], [32], [33].

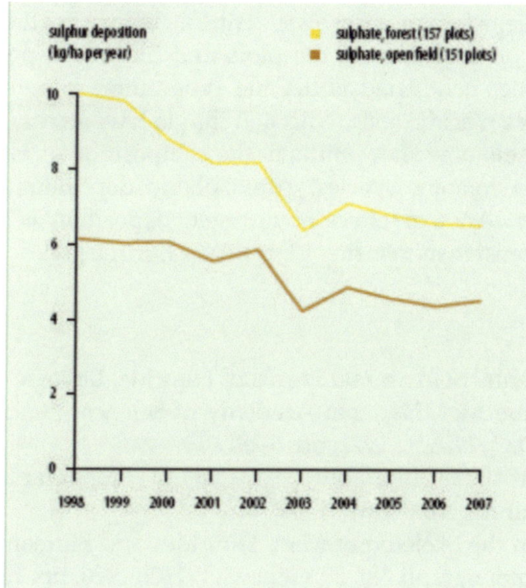

Fig. 29. Development of mean deposition of sulphate from 1998 to 2007. The forest canopy alters pollutants from the air. Inputs within the forest stands are higher than in the open field. In 2003 there was less precipitation and thus less deposition. (after ww.icpforests.org)

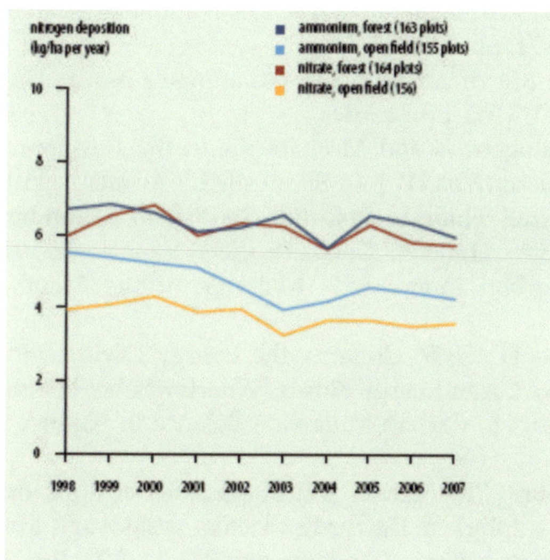

Fig. 30. Development of mean plot deposition of nitrogen compounds plots) from 1998 to 2007. Some reduction are visible in open field measurements. There was little change in deposition for the forest stands over the10 years of observation. (after www.icpforests.org)

Atmospheric deposition has been the specific focus of the programme since its inception. Current evaluations show decreasing sulphur inputs on about 50% of around 150 intensive monitoring plots since 1998, which is a result of clean air policies under the LRTAP

Convention and EU legislation. However, critical limits in the soil water are still substantially exceeded on a quarter of the plots and indicate a potential threat to forest vegetation. Earlier studies conducted under the programme have shown that the risk of storm damage is higher on acidic soils. Nitrogen inputs have hardly changed over the past ten years and the data sets now show shifts in the composition of forest ground vegetation towards more nitrogen tolerant species. Atmospheric deposition is a driver for these changes in biodiversity. Another effect of nitrogen deposition is increased tree growth which was found on intensive monitoring plots across Europe [34].

16. References

[1] Stein, L., Environmental Sources and Forms of Fluoride, Biologic Effects of Atmospheric Pollutants - Fluorides, National Academy of Sciences-National Research Council, Washington, D.C., N.A.S., 1971, pp. 5-28.
[2] Shupe, J. L., Fluorosis of Livestock, Air Quality Monograph No. 69-4, American Petroleum Institute, New York, 1969.
[3] Zipkin, I., Effects on the Skeleton of Man, Fluorides and Human Health, World Health Organization monograph No. 59, Geneva, WHO, 1970, pp. 185-201.
[4] Zipkin, I., Air Pollutants Affecting the Performance of Domestic Animals, Agricultural Handbook No. 380, U.S. Dept. of Agriculture, revised 1972, 109 pp.
[5] Zipkin, I., W. A. Lee, and N. C. Leone, Amer. J., Air pollutants and fluorides effects, Pub. Hlth. , 1957, 47:848.
[6] Suttle, J. W., Effects of Fluoride on Animals, Biological Effects of Atmospheric Pollutants - Fluorides, National Academy of Sciences-National Research Council, Washington, D.C., N.A.S., 1971, pp. 133-162.
[7] Hobbes, C. S., G. M. Merriman, Fluoride and animal's health, Univ. of Tenn. Agri. Exp. Sta. Bull. no. 351, 1963, pp 24:964.
[8] Johnson, L. C., Histogenesis and Mechanisms in the Development of Osteofluorosis, Fluorine Chemistry, Vol. IV, J. H. Simons (ed.), Academic Press, N.Y., 1965.
[9] Greenall, L., Industrial Fluoride Pollution in British Columbia, Canadian Scientific Pollution and Environmental Control Society, Vancouver, 1971.
[10] Thompson M., Carbon Monoxide - Molecule of the Month, Winchester College, UK, 2009.
[11] Robert U., Edward H., 2009, Crossing the Energy Divide: Moving from Fossil Fuel Dependence to a Clean-Energy Future, Wharton School Publishing, pp.36.
[12] Weinstock, B.; Niki, H., Carbon Monoxide Balance in Nature, Science 176 (32), 1972, pp. 290.
[13] Li L., Hsu, A. Moore, PK, Actions and interactions of nitric oxide, carbon monoxide and hydrogen sulphide in the cardiovascular system and in inflammation--a tale of three gases!, Pharmacology & therapeutics 123 (3), 2009, pp. 386–400.
[14] McNeil, Donald G. Jr., In Raising the World's I.Q., the Secret's in the Salt, New York Times, 2006.
[15] Pacyna E. G, Pacyna J. M, Steenhuisen F, Wilson S., Global anthropogenic mercury emission inventory for 2000". Atmos Environ 40 (22), 2006, pp. 4048.
[16] Solnit R., Winged Mercury and the Golden Calf, Orion Magazine, Retrieved 2007-12-03, 2006.

[17] Maprani A. C., Al T. A., MacQuarrie K. T., Dalziel, J. A., Shaw, S. A.; Yeats, Ph. A., Determination of Mercury Evasion in a Contaminated Headwater Stream, Environmental Science & Technology 39, 2005, pp. 1679.
[18] Heagle, A. S., Ozone and crop yield. Annu. Rev. Phytopathol. 27, 1989, pp. 397–423.
[19] Krupa, S. V., Gruenhage, L., Jaeger, H.-J., Nosal, M., Manning, W. J., Legge, A. H., and Hanewald, K., Ambient ozone (O_3) and adverse crop response: A unified view of cause and effect. Environ. Pollut. 87, 1995, pp. 119–126.
[20] Bowler, C., Van Montagu, M., and Inze, D., Superoxide dismutase and stress tolerance. Annu. Rev. Plant Physiol. Plant Mol. Biol. 43, 1992, pp. 83–116.
[21] Elstner, E. F., and Osswald, W. Mechanisms of oxygen activation during plant stress. Proc. R. Soc. B Biol. 102B, 1994, pp. 131–154.
[22] Taiz L., and Zeiger, E., Plant physiology, SINUAER, 2006,
[23] Aono, M., Kubo, A., Saji, H., Tanaka, K., and Kondo, N., Enhanced tolerance to photooxidative stress of transgenic Nicotiana tabacum with high chloroplastic glutathione reductase activity. Plant Cell Physiol. 34, 1993, pp. 129–135.
[24] McKersie, B., Bowley, S. R., Harjanto, E., and Leprince, O., Water-deficit tolerance and field performance of transgenic alfalfa overexpressing superoxide dismutase. Plant Physiol. 111, 1996, pp. 1177–1181.
[25] Schwartz, S. E., Acid deposition: Unraveling a regional phenomenon. Science 243, 1989, pp. 753.
[26] Heather G., Air Pollution on Agricultural Crops, Factsheet, No. 85-002, 2003,
[27] Muler, E., Stierlin, H., Water-deficit tolerance and field performance of transgenic alfalfa over expressing super-oxide dismutase. Plant Physiol. 111, 1990, pp. 1177–1181.
[28] Barbu, I., Cercetările de ecofiziologie forestieră şi fenomenul de uscare a răşinoaselor din Bucovina. Referat prezentat la Conferinţa naţională de fiziologia plantelor, ASAS Bucureşti, 1985.
[29] Barbu, I., Tipuri de uscare la bradul din România. Lucrare prezentată la Conferinţa a III-a de Ecologie Arad. Ziridava. Vol. XVI, 1986.
[30] Barbu, I., Moartea bradului - simptom al degradării mediului. Ed. Ceres, Bucureşti, 1991.
[31] Geambaşu, N., Seceta şi fenomenul de uscare a bradului din unele păduri din Bucovina. Revista Pădurilor nr. 2, 1988.
[32] Hildebrandt, G., Inventur und Beobachtung aktueler Waldschaden auf Landesebene. In: Schriftenreihe der Forstlichen Facultat Freiburg. AFZ nr. 12, Forstliche Hochschulwoche, 1983, pp. 148
[33] Hutte, P., Vom Waldsterben bedrohte Walder Polens. AFZ nr. 33, 1986, pp. 819.
[34] Spiecker, H. et al., Growth trends in European Forests. EFI Research Report nr. 5. Berlin, Springer, 1996.
1983 : Belastung durch Luftschadstoffe in der BRD., AFZ nr. 51/52, p. 1 385 - 1 386
1988 : Forest decline in Europa in 1987 attributed to an pollutants,
UNITED NATIONS, Economic and Social Council Economic Commission for Europe, Sesiunea 46, 11-14 oct. 1988
2007, 2008, 2009, United States Environmental Protection Agency (EPA) Report.
www.cleartheair.nsw.gov.au
(www.icp-forests.org).

www.icp-forests.org/Reports.htm
www.cleartheair.nsw.gov.au/science_and_research/cause_and_effect/what_effects_air_quality_.aspx
www.windows2universe.org/milagro/effects/wildlife_forests.html
http://5e.plantphys.net/article.php?ch=e&id=262

13

Phytoremediation on Air Pollution

Hua Yang and Yanju Liu
Beijing Center for Physical and Chemical Analysis
China

1. Introduction

Air pollution has been becoming a necessary evil with rapid industrialization and urbanization around the world, after it results in kinds of human health problems, such as ophthalmic, respiratory and cardiovascular diseases (Brunekreef & Holgate, 2002; Giles et al., 2011; Gudmundsson, 2011; Jamrozik & Musk, 2011; Miller et al., 2007; Nandasena, 2010). The direst threat posted by air pollution may be its hard controlling caused by its strong flowability. Either could it be spread from one source location to a larger region, even the whole planet, and the sweeping radiation pollution in air originated from Fukushima in Japan is unfortunately in this case, or conversely diluted with changes of climatic conditions (Sample, 2007).

Among all types of treatments of contaminants, including microbial bioremediation, phytoremediation stands out for its benefits yielded from self-maintaining, soil stabilization and other advantages to meet greater public approval (Doty et al., 2007). And different phytoremediation process is responsible for specific pollutant. Air pollutants can be divided into anthropogenic and natural pollutants according to their sources, or primary and secondary pollutants, which stem from reactions of primary pollutants, when taking production process into account (UNEP, 2004). However, physical, chemical and biologic pollutants belong to three general categories as usually discussed in air pollution treatment, and particulate matters, organic and inorganic chemicals, and microorganism were referred to the above categories respectively. This chapter will start from conception of phytoremediation, current state of phytoremediation of air pollutants, such as particulate matters, inorganic and organic pollutants, and a case study from recent authors' research on phytoremediation of benzene and toluene of indoor air will be finally presented.

2. Phytoremediation

Phytoremediation is a way to mitigate environment pollutions, such as in air, water and soil pollution in virtue of plants, more often than not, combined with their associated microorganisms. This concept has been widely applied to treat pollutants in soil and water (Cunningham & Ow, 1996; Schröder et al, 2002).

2.1 Principles of phytoremediation

Briefly, for phytoremediation, roles from roots to shoots are played in phytostabilization, rhyzodegradation, phytoextraction, phytodegradation, phytovolatilization and phytofiltration

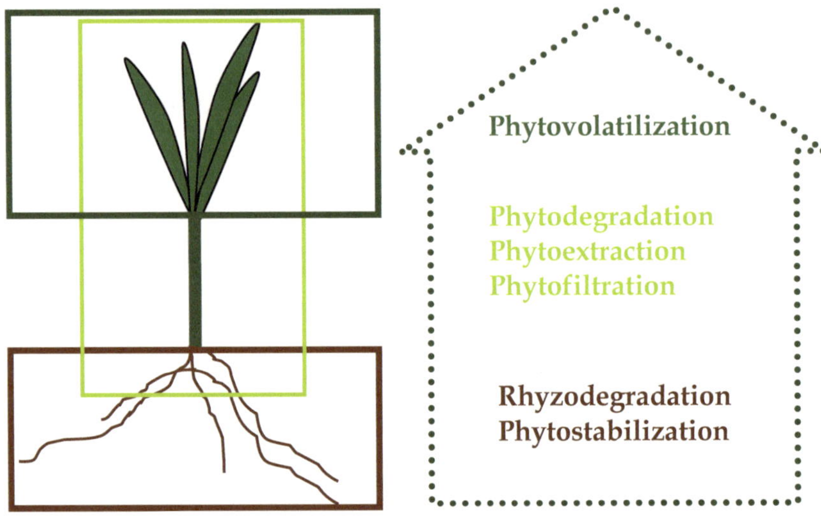

Fig. 1. Different parts of the plants are responsible for different processes of phytoremediation.

(Morikawa & Erkin, 2003) (see Fig. 1.). It is the process for phytostabilization that contaminants were sequestrated, but not degraded, in the root zones. The narrow region of 1-3 mm from root surface is rhyzosphere, where there are much more profuse microorganisms than in bulk soil. Root secretions and rhyzospheric microorganisms are joint-agents to degrade the pollutants in soil in the process of rhyzodegradation. Phytoextraction is with regard to extracting pollutants from soil and translocating them from the roots to shoots by plants. Phytodegradation is to transform and/or degrade pollutants, which enter the plants from soil, water or air. Phytovolatilization means the pollutants are volatile out of the stoma of the plants and degraded by hydroxyl radicals, otherwise they will become air pollutants. It is phytofiltration that particulate matters in air are reduced through the surface of plants (Singh & Tripathi, 2007).

2.2 Characteristics of phytoremediation

2.2.1 Main advantages

Phytoremediation is proved to be helpful to reduce pollutants in soil, water and air; it is energy-economizing and cost-effective compared with those physical or chemical methods since it is solar-energy driven cleanup technology; it is a preference for people as its aesthetic value and less environment disruption, which letting nature take its course.

2.2.2 Primary disadvantages

Plants are also living things, and there exists tolerance limits for the toxicity of pollutants; it is a kind of "leave it to chance" thing since it is too natural to disrupt the environment: optimum concentrations of pollutants for remediation varying with species of plants, intermediates possibly leading to second pollutants, time-consuming when considering growth period of plants.

2.3 Examples of phytoremediation of air pollutants

2.3.1 Particulate matters

Vegetation has been used to shield dust in many countries, and the process is called phytofiltration. A 8-m wide green belt may reduce 2 to 3 times of dust fall (Novoderzhikina et al., 1966 as cited in Singh & Tripathi, 2007). For phytofiltration, morphological characteristics, such as orientation of leaf on the main axis, size and shape, surface nature, the presence or absence of trichomes and wax deposition, are factors to trap or capture dust from ambient air.

2.3.2 Inorganic pollutants

NO_X

NO_X are air pollutants, and they are one of precursors of photochemical reaction. They can reach plant system through wet/dry deposition to foliar or roots. Plant species, plant age, NO_X concentration and other environmental conditions are factors which influence the leaf penetration of NO_X. After NO_2 enter into the plant, most of them are metabolized to organic compounds, e.g. amino acid, through nitrate assimilation pathway. The enzymes in plants, such as nitrate reductase, nitrite reductase or glutamine synthetase may play an important role in this process. Genetic engineering could turn plants into magic sink for NOx by making the enzymes overexpressed.

SO_2

Nearly 70% SO_2 in the atmosphere originates from fossil fuel combustion. SO_2 enters into plants mainly through stoma, and can be utilized in a reductive sulfur cycle in plants. They are changed into SO_4^{2-} or SO_3^{2-} in cell walls. Adenyl-5-phosphosulphate, carrier protein, carrier protein with bound sulphite and carrier protein with bound sulphide are the intermediates in the reduction pathway. The final products are cysteine or other organic compounds.

2.3.3 Organic pollutants

Formaldehyde

Formaldehyde, a ubiquitous air pollutant, is so harmful that it is classified as a mutagen and carcinogen. In the 1980's, an NASA's research in the USA revealed that low level of formaldehyde in air could be removed by plant leaves alone, while higher concentrations of the toxic chemical can be filtered by activated carbon firstly and the plant roots and associated microorganisms degrade and assimilate remained chemicals (Wolverton,1988). In cell-culture experiments (Giese,1994), when the concentration of formaldehyde is low enough ($8.5mg \cdot m^{-3}$), spider plants (*Chlorophytum comosum* L.) shoot can metabolize it to organic acids, amino acids, free sugars, lipids and cell-wall components. In soybean (*Glycine max* L.) cell cultures, serine and phosphatidylcholine are the major metabolic products for formaldehyde. NAS^+-dependent formaldehyde dehydrogenase and glutathione-dependent formaldehyde dehydrogenases have been isolated from cell-suspension cultures of some plants. It appears that formaldehyde can be oxidized and then degraded in C1 metabolism in phytodegradation.

Benzene and toluene

As members of VOCs (Volatile Organic Compounds, VOCs), benzene is genotoxicity and carcinogenicity while toluene is a neurotoxic chemical (Pariselli et al., 2009). Benzene and toluene have been reported to be removed from air and be assimilated by plants (Porter, 1994; Ugrekhelidze, 1997). Porter (1994) suggested at the lower toluene exposure, a fairly modest amount light change was credited with significant increase or decrease of removal rate and the plant, the only photoresponsive organisms in the system, may be responsible for the phenomenon. Furthurmore, hydroxylation is considered to be the first step of the aromatic ring cleavage of benzene and toluene in higher plants (Ugrekhelidze, 1997). Besides, a substantial role of microorganisms in the growth medium in removing benzene and toluene was also reported by previous researches (Orwell et al., 2004; Orwell et al., 2006; Wood et al., 2002; Wood et al., 2006).

Two compartments Model

Foliage-air exchange is believed to be the primary route of plant uptake particularly for semi-volatile and volatile organics. Two-compartment model has been used to explain atmosphere-foliage bioaccumulation phenomenon, namely a relatively fast initial uptake followed by a period of slower uptake (Keymeulen, 1995; Mackay et al, 2006). In the model, the leaf was divided into two compartments according to its structure. The first compartment can be referred to the non-living plant cuticle, where physico-chemical sorption of airborne lipophilic compounds can occur. The second compartment is within the leaf interior, where further sorption and metabolism of organic compounds may exist.

3. Removal of benzene and toluene from indoor air by ornamental houseplants—a case study (Yang, 2010)

3.1 Indoor air pollution

Indoor air pollution in urban environments has become a significant health concern, as city dwellers often spend over 90% of their time indoors (USEPA, 1987; Abbritti & Muzi, 1995; American Lung Assoc., 2001). "Sick building syndrome" (SBS) or "building-related illness" is partly attributed to the indoor chemicals, particularly in air-conditioned buildings (Burge et al., 1987; Mendell and Smith, 1990; Carpenter, 1998; Brasche et al., 1999; Carrer et al., 1999; Jones, 1999; Sundell, 2004). As indoor pollutants, volatile organic compounds (VOCs) with 300 species having been detected, should not be negligible (Orwell et al., 2004), after the cocktail of VOCs might lead to additive effects on human health (Wolkoff, 1995; Weschler and Shields, 1997; Pariselli *et al.*, 2009). Thus, removing these chemical mixtures from indoor air merits serious consideration. Many experimental studies have investigated the effects of plants on VOCs (Cape, 2003). This part presents a study on applicability of phytoremediation of benzene and toluene from indoor air.

3.2 Experiments design

Indoor ornamental plants were purchased from flower markets for experiments. Plants were well watered and allowed to drain for 24 hours before fumigation.

A series of experiments were designed as follows (see Fig. 2.):

1. **Plants-Screening experiments**

This experiment aimed to acquire plants which are capable of effective-removing binary mixture of benzene (0.26 mg · m^{-3}) and toluene (0.3 mg · m^{-3}). For this purpose, initial screening experiment and replication experiment for validation were carried out in sequence. Allowing for vast amounts of species of plants, only one individual for every species of plants was fumigated in the initial screening experiment. After some ineligible ones could inevitably have slipped into candidates in the initial screening experiment, further three replicates for those species which showed relative high removal rates were

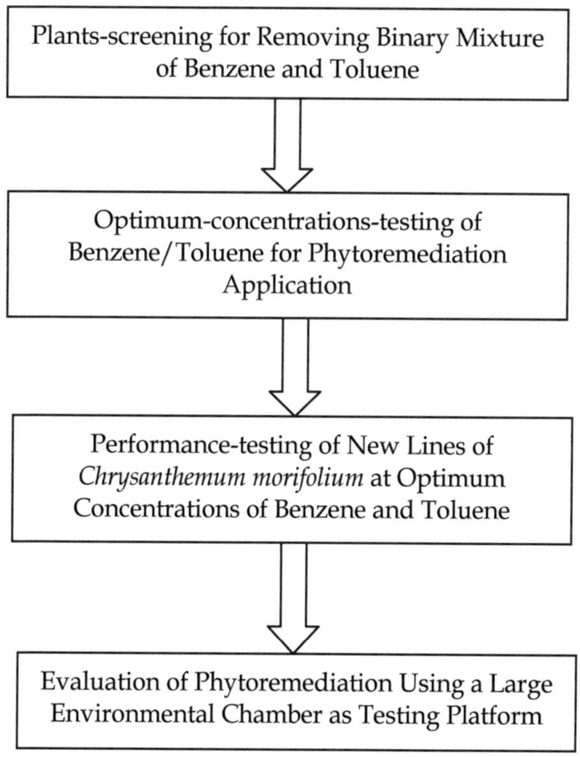

Fig. 2. Processes of the case study on Removal of Benzene and Toluene from Indoor Air by Ornamental Houseplants

established in replication experiment to confirm the removal effects and to quantitatively estimate the rate and efficiency of phytoremediation.

2. **Fumigating-experiments at gradient concentrations**

The gradient concentrations (Table 1.) were set up to investigate optimum concentrations of benzene and toluene for phytoremediation application. The plants were chosen from those plants with relatively high removal efficiency from plants-screening experiment.

3. **Evaluating-experiments for new lines of *Chrysanthemum morifolium***

It is one object in the experiments to evaluate removal performance of new lines of *Chrysanthemum morifolium* at the optimum concentrations of benzene and toluene acquired in the experiments mentioned above. *Chrysanthemum morifolium*, a Chinese traditional flower, has been so popular they are available everywhere in the country. The phytoremediation on indoor air pollution is believed to be easy to be promoted, if *Chrysanthemum morifolium* demonstrates ideal removal effects.

The other object is to establish a large environmental chamber as a testing-platform to evaluate phytoremediation on indoor air pollution.

Gradient concentrations ($mg \cdot m^{-3}$)	
Benzene	Toluene
0.26	0.31
0.64	0.76
1.3	1.5

Table 1. Gradient concentrations of benzene and toluene set up in the fumigating experiments

3.3 Fumigation system

A set of eight cylindrical, Plexiglas chambers (41 cm in diameter, 70 cm in height, ca. 0.09 m^3 in volume) were lined with Teflon film to reduce possible absorption by Plexiglas material (see Fig. 3.). Stainless steel or Teflon Pipes were used to carry gasses in order to minimize the loss of benzene/toluene. Simultaneously, compressed purified air and standard mixture of benzene and toluene entered a cylindrical, stainless steel mixing chamber. The flow rate of air and benzene/toluene was regulated by Float Flow Meter and Mass Flow Meter respectively, which were calibrated by Bubble Flow Meter before use. Thus, air containing expected concentrations of benzene and toluene was continuously and evenly distributed to fumigation chambers through 8 Float Flow Meters. It took about 2 hours to achieve expected concentrations of benzene and toluene of outlet air based on the continuous measuring the benzene/toluene concentrations at the inlets and outlets of the eight empty chambers. Fig. 4. shows the fumigation system in the study.

Fig. 3. Photograph of a Plexiglas chamber housing one potted plant

Among the eight chambers, two chambers were designated randomly as controls, one holding only pots and soil similar to those used in the experimental chambers; another empty one used to monitor stability of gasses flow. And the other six chambers housed six different plant species per trial. After three hours (from 9 a.m. to 12 noon) and six hours

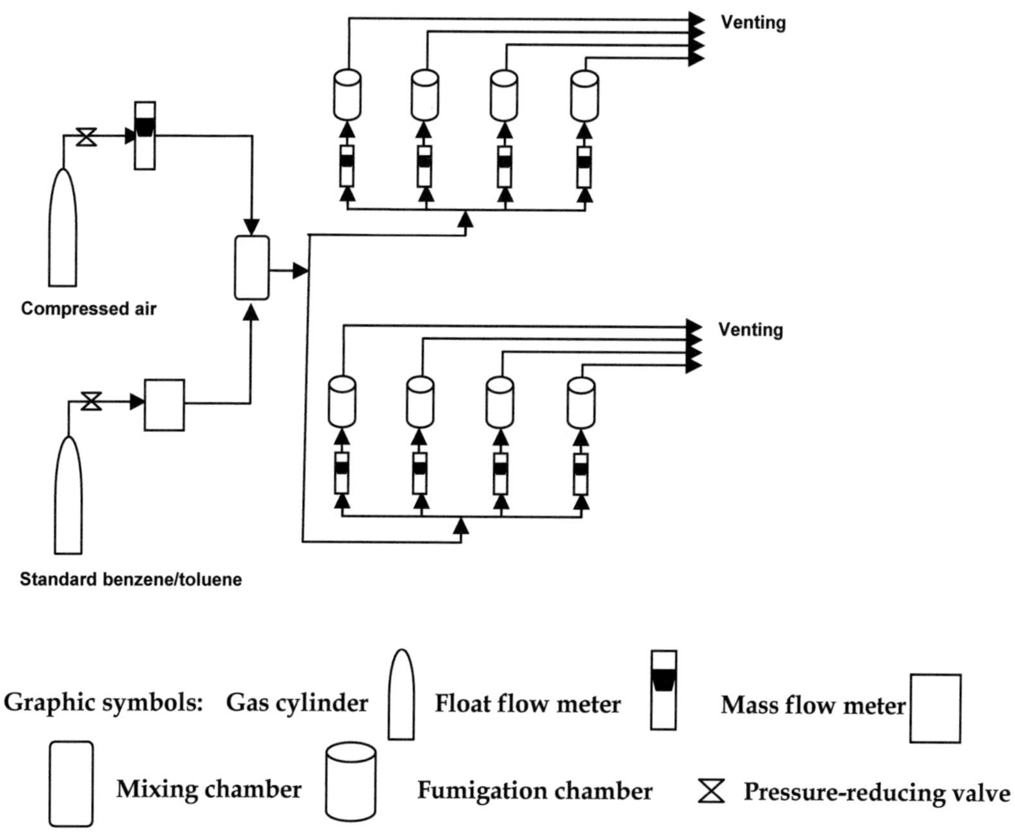

Fig. 4. Schematic diagram for fumigation system

(continue to extend from 12 noon to 3 p.m.) of exposure, gasses were sampled using a 100 mL gastight syringe for immediate analysis. The benzene/toluene removal rate was calculated as follows:

$$(Ccontrol - Csample)/ Ccontrol \times 100\% \quad (1)$$

Where C represents benzene or toluene concentration ($mg \cdot mL^{-1}$).

After sample collection, chambers were evacuated and prepared to test the next plants. The same sampling procedure with 8 empty chambers was inserted every 5 trials to test leak to guarantee uniform concentrations of benzene/toluene in all eight chambers.

14 m^3 stainless steel of environmental chamber was made to meet ASTM 1333-96 (reaproved 2002) (see Fig. 5.). Parameters inside the chamber, such as temperature (25°C), relative humidity (45%) and air exchange rate (once every four hours), were maintained during the whole testing period. In the large chamber, concentrations in three steps, namely, background of empty chamber, only chemical evaporation and plants-fumigating were analyzed. 1 mL of benzene and toluene liquid were dropped in a 90 cm-high Petri dish to evaporate. The first sampling was taken after 10 minutes of evaporation, followed by next 8

Fig. 5. The photograph of inside of the environmental chamber

samplings every one hour. Gas samples were collected from 6 air sampling ports, each of them with 15 L of sample. When plants-fumigating, four replicates of the same species of *Chrysanthemum morifolium* were placed on the four corners of the chamber floor. Ventilation and scrubbing should be conducted between fumigation experiments.

3.4 Results and discussion

1. Plants-Screening experiments

In the initial screening experiment, 94 species and cultivars were tested, among which 15 species removed more than 30% of benzene or toluene at either 3-hour or 6-hour sampling. Eight species of plants, *Rhododendron hybrids, Hosta plantaginea, Ctenanthe oppenheimiana* cv. Tricolor, *Ficus elastica, Calathea rotundifolia* cv. Fasciata, *Codiaeum variegatum, Chrysanthemum morifolium, Hemerocallis fulva,* were chosen for the following replication experiment to confirm the removal effects, since they are more easily accessible. On the other hand, fixed time dependence on removal rates was not found for all species in this phase, just as Ugrekhelidze (1997) pointed out the ability of different plants to absorb and utilize exogenous hydrocarbons varied.

In the initial screening experiment, *Crassula portulacea, Cymbidium Golden Elf.* and *Dieffenbachia amoena* cv. Tropic Snow did not display well as in previous experiment (Liu et al., 2007), less than 10% removal rate of benzene. Apart from toluene interference, fumigation fashion (kinetic/static and flow rate at inlet of each fumigation chamber), chamber size, concentration of benzene may all contribute to different removal effects

(Table 2.). In reference to studies on the effects of plants on VOCs, two main types of fumigation chambers were applied – static chambers and kinetic chambers. For a static chamber, more accessories have to be equipped with, such as thermostat and stirring devices needed to accelerate atmospheric equilibration, so that difference of concentrations at the same spot in the chamber at different time period can be evaluated effects of plants on VOCs, whereas effects of plants on VOCs can be determined by the difference of concentrations at inlet and outlet of a kinetic chamber with continuous gas-flow through (e.g. Liu et al., 2007).

For replication experiment, removal rates of 8 species of plants of benzene or toluene at either sampling (3-hour or 6-hour) were showed to be more than 30% in the replication experiment. Furthermore, for the convenience of comparison, the removal rates of benzene and toluene was normalized for leaf area (m^{-2}). For most of 8 species of plants, the removal rates of benzene after 6-hour fumigation increased compared with those after 3-hour fumigation. The case of toluene reversed. A competitive reduction might exist when both benzene and toluene penetrating into leaves were under metabolic control, but combined removal remained essentially unchanged (Porter, 1994). This could be used to elaborate the reversal trend of removal rate with time showed by benzene and toluene.

Chrysanthemum morifolium displayed an excellent performance in removing both benzene and toluene at twice samplings in replication experiment. So it was chosen for further Evaluating-experiment.

2. Fumigating-experiments at gradient concentrations

From those 15 species which removed more than 30% of benzene or toluene at 3-hour or 6-hour sampling in the initial screening experiment, six species of plants such as *Rhododendron hybrids, Ficus elastica, Codiaeum variegatum, Hemerocallis fulva, Euphorbia pulcherrima and Cymbidium sinense* were chosen to test removal capacities at gradient concentrations of benzene and toluene (Yang et al., 2011) (see Table 1). The removal rates at different condition varied. At 0.64 mg m^{-3} of benzene and/or 0.76 mg $m^{-3}$of toluene, better removal effects were showed, and considering the household application, it is better to arrange some species of plants to achieve the best removal effects taking account of concentrations, fumigation time and pollutants.

3. Evaluating-experiment for new lines of *Chrysanthemum morifolium*

Not only did *Chrysanthemum morifolium* show an excellent performance in removing both benzene and toluene according to the results showed in plants-screening experiments, but also it is a kind of symbolic Chinese traditional flower. So it was chosen for evaluating-experiment. The new lines of *Chrysanthemum morifolium* were provided by Institute of Vegetable and Flower, the Chinese Academy of Agricultural Sciences. They are bred via anther culture and radiation mutation from breeding parents of *Chrysanthemum* with spray forms. It was the first performance-testing in China to remove gaseous benzene-toluene using new lines of *Chrysanthemum morifolium* (Yang et al., 2010).

While selective removal of benzene or toluene was not significantly showed in nine lines of *Chrysanthemum*, most of them removed more toluene than benzene. Removing rates varied significantly with lines. Samples from 6-hour of fumigation suggested that above 30% of removal rates of both benzene and toluene were reached by lines of the number 2, 4, 5, 9, in the meanwhile removal rate of toluene from line of the number 7 dropped to 0 from 13% at 3-hour of fumigation (Yang et al., 2010). From the results of removal amount, lines of the number 2, 4, 5, 8, 9 should be taken for application performance testing in real world considering better performance-testing results.

Chrysanthemum morifolium and *Calathea rotundifolia* cv. Fasciata were chosen for performance evaluation in the large environmental chamber. After 10 minutes of evaporation, 10 mg m^{-3} of benzene and 8.5 mg m^{-3} of toluene were detected in the chamber. At first three hours, benzene and toluene were reduced significantly when fumigating *Chrysanthemum morifolium* compared with only chemical fumigation. It suggested that phytoremediation on indoor air pollution should be an effective way to remove organic pollutants. Besides, the environmental chamber is stable in performance, low in adsorption and background value of pollutants, convenient and safe in operation so that it could provide a testing platform for evaluating phytoremediation of gaseous pollutants.

Plant species	Fumigation fashion	Chamber size	VOCs member	Concentration level	Removal rate
Crassula portulacea Cymbidium Golden Elf. and *Dieffenbachia amoena* cv. Tropic Snow	Kinetic fumigation (0.5 L/min at inlet of each chamber)	Cylindrical; 40 cm in diameter, 60 cm in height,	Benzene	150 ppbv	More than 20% after 2-h fumigation
Epipremnum aureum	Static fumigation	Cubic; 60 × 60 × 60 cm	Benzene	25 ppmv	Less than 10% (Average removal rate in initial dose) after 3-h fumigation

Table 2. Fumigation methods and results in previous studies (Liu et al., 2007; Orwell et al., 2004)

4. Future perspective

As a technology, phytoremediation on air pollution is still in its infancy, and study on the applicability of phytoremediation is scarce. First, it is necessary that appropriate testing and evaluating platform should be established for different application. Static and kinetic fumigation fashion, fumigation box and large environmental chamber should be adopted according to goals of experiments. Second, pollutants-removing mechanism needs further studying from morphology to metabolism, from plant alone to associated microorganisms. Third, based on mechanism studies, prediction models can be established and accuracy be improved. In the meanwhile, the study of optimum concentrations of pollutants depending on species of plants should not be neglected for the technology promotion.

5. Conclusion

Phytoremediation is an eco-friendly, cost-effective way to remove pollutants from environment. Phytostabilization, rhyzodegradation, phytoextraction, phytodegradation, phytovolatilization and phytofiltration are main principles of phytoremediation. As physical and chemical air pollutants, phytoremediation of particulate matters, NO_X, SO_2, formaldehyde, benzene and toluene have been reported. Moreover, the authors studied applicability of phytoremediation from aspects of plants-screening, optimum concentrations and evaluation system. For all this, the study on the applicability of phytoremediation still needs work for future technology promotion.

6. Acknowledgements

Authors thank National Natural Science Funds (Grant No. 40875082) for supporting this project.

7. References

Abbritti, M.C. & Muzi, G. (1995). Indoor Air Quality and Health Effects in Office Buildings, In: *Proceedings of Healthy Buildings `95, An International Conference on Healthy buildings in a Mild Climate*. Maroni, M. (Ed). Vol. 1, pp. 185-195, University of Milano and International Centre for Pesticide Safety, Milan, Italy

American Lung Association. (2001). When you can't breathe, nothing else matters. In: *Air Quality*. Available from www.lungusa.org/air/

ASTM 1333-96. (Reapproved 2002). Standard Test Method for Determining Formaldehyde Concentrations in Air and Emission Rates from Wood Products Using a Large Chamber, pp. 1-12

Brasche, S.; Bullinger, M.; Gebhardt, H.; Herzog, V.; Hornung, P.; Kruppa, B.; Meyer, E.; Moreld, M.; Schwab, R.; Mackensen, S.; Winkens, A. & Bischof ,W. (1999). Factors Determining Different Symptom Patterns of Sick Building Syndrome Results From a Multivariate Analysis. In: *Proceedings of Indoor Air `99, The 8th International Conference on Indoor Air Quality and Climate*. Vol.5, pp. 402—407, Edinburgh, UK

Brunekreef, B & Holgate, S. T. (2002). Air Pollution and Health. *Lancet*, 2002; Vol.360, No.9341, (October 2002), pp. 1233–1342, ISSN 1470 2045

Burge, S.; Hedge, A.; Wilson, S.; Harris, B.J. & Robertson, A. (1987). Sick Building Syndrome: a Study of 4373 Office Workers. *Annals of Occupational Hygiene*, Vol.31, pp. 493-504, ISSN 0003-4878

Cape, J.N. (2003). Effects of Airborne Volatile Organic Compounds on Plants. *Environmental Pollution*, Vol.122, No.1, (March 2003), pp. 145—157, ISSN 0269-7491

Carpenter, D.O. (1998). Human Health Effects of Environmental Pollutants: New Insights. *Environmental Monitoring and Assessment*, Vol.53, No. 1, (October 1998), pp. 245—58, ISSN 0167-6369

Carrer, P.; Alcini, D.; Cavallo, D.; Visigalli, F.; Bollini, D. & Maroni, M. (1999). Home and Workplace Complaints and Symptoms in Office Workers and Correlation with Indoor Air Pollution. In: *Proceedings of the 8th International Conference on Indoor Air Quality and Climate*, Vol.1, pp. 129-134, Edinburgh, UK

Cunningham, S.D. & Ow, D.W. (1996) Promises and Prospects of Phytoremediation, *Plant Physiol*. Vol.110, (March 1996), pp. 715-719, ISSN 0032-0889

Doty, S. L.; Andrew, J. C.; Moore, A. L.; Vajzovic, A.; Singleton, G. L.; Ma, C.P.; Khan, Z; Xin, G.; Kang, J. W.; Park, J. Y.; Meilan, R.; Strauss, S. H.; Wilkerson, J.; Farin, F. & Strand, S. E. (2007). Enhanced Phytoremediation of Volatile Environmental

Pollutants with Transgenic Trees. *Proceedings of the National Academy of Sciences*, Vol.104, No.43, (October 2007), pp. 16816–16821, ISSN 0027-8424

Giese, M.; Bauer-Doranth, U.; Langebartels, C. & Sandermann H. (1994). Detoxification of Formaldehyde by the Spider Plant *(Chlorophytum comosum* L.) and by Soybean *(Clycine max* L.) Cell-Suspension Cultures. *Plant Physiol.* Vol.104, No.4, (April 1994), pp. 1301-1309, ISSN 0032-0889

Giles, L.V.; Barn, P.K.N.; Romieu, I.; Mittleman, M.A.; Eeden, S.; Allen, R.; Carlsten, C.; Stieb, D.; Noonan, C.; Smargiassi, A.; Kaufman, J.D.; Hajat, S.K.T.; & Brauer, M. (2011). From Good Intentions to Proven Interventions: Effectiveness of Actions to Reduce the Health Impacts of Air Pollution. Environmental Health Perspectives, Vol.119, No.1, (January 2011), pp. 29-36, ISSN 0091-6765

Gudmundsson, G. (2011). Respiratory Health Effects of Volcanic Ash with Special Reference to Iceland. A review. *The Clinical Respiratory Journa*l, Vol.5, pp. 2–9, (January 2011), ISSN 1752-6981

Jamrozik, E. & Musk, A.W. (2011). Respiratory Health Issues in the Asia–Pacific Region: An Overview, *Respirology*, Vol.16, No.1, (January 2011),pp. 3–12, ISSN 1440-1843

Jones, A.P. (1999). Indoor Air Quality and Health. *Atmospheric Environment*, Vol.33, No. 28, (December 1999), pp. 4535-4564, ISSN 1352-2310

Keymeulen, R. ; Schamp, N. & Langenhove, H.V. (1995). Uptake of Gaseous Toluene in Plant Leaves: A Two Compartment Model, *Chemosphere*, Vol. 31, No.8, (October 1995),pp. 3961-3975, ISSN 0045-6535

Liu, Y.J.; Mu, Y.J.; Zhu, Y.G.; Ding, H. & Arens, N.C. (2007). Which Ornamental Plant Species Effectively Remove Benzene From Indoor Air? *Atmospheric Environment*, Vol.41, pp. 650-654, ISSN 1352-2310

Mackay, D.; Foster, K.L.; Patwa, Z. & Webster, E. (2006). Chemical Partitioning to Foliage: The Contribution and Legacy of Davide Calamari, *Environ Sci & Pollut Res*, 2006, Vol.13, No.1, (January 2006), pp. 2-8, Vol. 30, pp. 786-791, ISSN 0944-1344

Mendell, M.J. & Smith, A.H. (1990). Consistent Pattern of Elevated Symptoms in Air-conditioned Office Buildings: A Re-analysis of Epidemiological Studies. *American Journal of Public Health*, Vol.80, No.10, pp. 1193-1199, ISSN 0090-0036

Miller, K.A.; Siscovick, D.S.; Sheppard, L.; Shepherd, K.; Sullivan, J.H.; Anderson, G.L.; & Kaufman, J.D. (2007). Long-Term Exposure to Air Pollution and Incidence of Cardiovascular Events in Women. *The new england journal of medicine,* Vol. 356, No. 5, (February 2007), pp. 447-458, ISSN: 0954-7762

Morikawa, H. & Erkin Ö. C. (2003). Basic Processes in Phytoremediation and Some Applications to Air Pollution Control. *Chemosphere* , Vol.52, No.9, (September 2003), pp. 1553–1558, ISSN: 0045-6535

Nandasena, Y. L. S.; Wickremasinghe A. R. & Sathiakumar N. (2010). Aira rPollution and Health in Sri Lanka: a Review of Epidemiologic Studies. *BMC Public Health*, Vol.10, (June 2010), Available from http://www.biomedcentral.com/1471-2458/10/300)

Orwell, R.L.; Wood, R.A.; Tarran, J.; Torpy, F. & Burchett, M.D. (2004). Removal of Benzene by the Indoor Plant/Substrate Microcosm and Implications for Air Quality. *Water, Air, and Soil Pollution*, Vol.157, (September 2004), pp. 193-207, ISSN: 0049-6979

Orwell, R.L.; Wood, R.A.; Burchett, M.D.; Tarran, J. & Torpy, F. (2006). The Potted-plant Microcosm Substantially Reduces Indoor Air VOC Pollution: II. Laboratory Study. *Water, Air, and Soil Pollution*, Vol.177, (November 2006), pp. 59-80, ISSN 0049-6979

Pariselli, F.; Sacco, M.G.; Ponti, J.; & Rembges, D. (2009) Effects of Toluene and Benzene Air Mixtures on Human Lung Cells (A549). *Experimental and Toxicologic Pathology*, Vol.61, No. 4, (July 2009), pp. 381-386, ISSN: 0940-2993

Porter, J.R. (1994). Touene Removal From Air by Dieffenbachia in a Closed Environment. *Adv Space Res*, Vol.14, No.11, (November 1994), pp. 99–103, ISSN 0273-1177

Sample, I. (April 2007). Urban Air Pollution More Dangerous than Chernobyl, In: *The Guardian*, Available from http://www.guardian.co.uk/science/2007/apr/03/uknews.pollution

Schröder, P.; Harvey, P.J.; & Schwitzgébel, J.P. (2002). Prospects for the Phytoremediation of Organic Pollutants in Europe, *Environ Soi & Pollut Res*, Vol.9, No.1, pp. 1-3, ISSN 0944-1344

Singh, S.N & Tripathi, R D. (2007). *Environmental Bioremediation Technologies*, Springer, ISBN 9783540347934, Berlin, Germany

Sundell, J. (2004). On the History of Indoor Air Qulity and Health. *Indoor Air*, Vol.14, No. 7, pp. 51-58, ISSN 0905-6947

Ugrekhelidze, D.; Korte, F., & Kvesitadze, G. (1997). Uptake and Transformation of Benzene and Toluene by Plant Leaves. *Ecotoxicology and Environmental Safety*, Vol.37, No. 1, (June 1997)pp. 24-29, ISSN: 0147-6513

UNEP. Malé Declaration on Control and Prevention of Air Pollution and its likely Transboundary Effects for South Asia,(March 2004), pp. 9-16, Available from http://www.rrcap.unep.org/male/manual/national/02chapter2.pdf

USEPA. (1987). The Total Exposure Assessment Methodology (TEAM) Study. In: *EPA600/S6-87/002, USA*. Available from http://www.epa.gov/ncepihomS

Weschler, C.J. & Shields, H.C. (1997). Potential Reactions Among Indoor Air Pollutants. *Atmospheric Environment*, Vol.31, No.21, pp. 3487–3495, ISSN 1352-2310

Wolkoff, P. (1995). Volatile Organic Compounds Sources, Measurements, Emissions and the Impact on Indoor Air Quality. *Indoor Air*. Vol. Suppl. 3, pp. 1-73, ISSN: 0905-6947

Wolverton, B.C. (July 1988). Foliage Plants For Improving Indoor Air Quality. In: *National Foliage Foundation Interiorscape Seminar*, pp. 29-45, National Aeronautics and Space Administration, John C. Stennis Space Center, Hollywood, FL, United States

Wood, R.A.; Orwell, R.L.; Tarran, J.; Torpy, F. & Burchett, M. (2002). Potted-plant/Growth Media Interactions and Capacities for Removal of Volatiles from Indoor Air. *Journal of Horticultural Science & Biotechnology*, Vol.77, No.1, (January 2002), pp. 120-129, ISSN 1462-0316

Wood, R.A.; Burchett, M.D.; Alquezar, R.; Tarran, J.; Orwell, R.L., & Torpy, F. (2006). The Potted-plant Microcosm Substantially Reduces Indoor Air VOC Pollution: I. Office Field-study. *Water, Air, and Soil Pollution*, Vol.175, No.1-4, (September 2006), pp. 163-180, ISSN: 0049-6979

Yang, H. (2010). *Removal of Benzene/Toluene from Indoor Air by Ornamental Plants and its Application and the Mechanism*, Postdoctoral Research Report, Zhejiang University and Beijing Academy of Science and Technology

Yang, H.; Liu, Q.Y.; & Liu, Y.J. (2011). Removal effects of benzene/toluene in indoor air using potted-plants and its application analysis, *Environmental Chemistry*, ISSN 0254-6108

Yang, H.; Liu, Z.Y.; Ge, H.; Yang, S.H.; Ge, W.Y. & Liu, Y.J. (2010). Performance-Testing in Removing Benzene-Toluene Binary Gas Using New Lines of *Chrysanthemum*. *Northern Horticulture*, No. 226, (October 2010), pp.5-8, ISSN 1001-0009

14

Combustion Emissions Analysis of Wood-Based Waste Processing-Materials

Han Chien Lin
Department of Forest Products Science and Furniture Engineering, College of Agriculture, National Chiayi University Taiwan

1. Introduction

The pursuit of economic development by the human has come at a great cost to the animals and plants of this planet, and has damaged the quality of our air and water. In the last 100 years more problems have come to threaten the environment, such as vanishing forests, acid rain, the burning of fossil-fuels and global warming. Currently, researchers are focusing on so-called "eco-materials" that make effective and environmentally friendly use of natural resources, such as the recycling of industrial wastes, the development of new environmentally friendly materials, etc. The evaluated items are included the frontier, amenity and adjustment of materials (Okuma, 1998). Besides, Life Cycle Analysis (LCA), started in the early 1970s, is a method that considers energy and raw material consumption, different types of emissions and other important factors related to a specific product's entire life cycle. The Society of Environmental Toxicology and Chemistry (SETAC) published the widely accepted series of guidelines and definitions of LCA, and the ISO 14040-14043 is considered the LCA standard (Bras & Román, 2006). Recycling of wood products, as one item of the LCAs, generally includes two kinds of methods. The first type is recycling of discarded forest products or resource-recovered (renewable) materials, which means to develop new wood-based materials in accordance with the principle of cascade utilization (Akiyama, 1998). The second type is the waste disposal options for spent various wood materials, including demolition, combustion and/or landfilling (Jang, 1997; Kobayashi et al., 2000). However, discarded the wood-based waste processing-materials, especially wood or/and wood-based materials that have been processed with different types of adhesives, coatings and preservatives, are extremely difficult to be recycled as a raw material, because these materials for wood products used cannot presently be used as a raw material again (Lin et al., 2001), and the waste stage has a very significant impact on the life cycle assessment results (Esser et al., 2000).

As we know, carbon dioxide (CO_2) is one of elemental compounds in the atmosphere (air) and a factor in photosynthesis and absorption. CO_2 is also the best-known greenhouse gas because it is capable of increasing the average global temperature depending on the sensitivity of the atmosphere (such as the concentration of CO_2) to greenhouse forcing (Allaby, 2000; Aplin et al., 2000). Carbon monoxide (CO), one of tasteless, colorless and non-odorous gases, is produced primarily by car exhaust and incomplete combustion of organic materials. Hence, CO can easily be inhaled and combined with the hemoglobin in blood, which prevents oxygen (O_2) transportation in the blood and causes intoxication effects such as headache, giddiness, tiredness and buzzing in the ears etc (Lin, 1992). Sulfur dioxide (SO_2) is one of the precursors for acid rain because not only a sulfurous acid (H_2SO_3) reaction is formed when SO_2 is compounded as mist in the atmosphere, but also SO_2 has a chemical reaction with O_2 and forms sulfur trioxide (SO_3). SO_3 then becomes liquid SO_3 because its' dew point is 22 °C. At the dew point a sulfuric acid (H_2SO_4) reaction is formed. When inhaling SO_2 at about 30 to 40 ppm, the body will feel giddy and experience breathing difficultly (Lin, 1992; Aplin et al., 2000). Nitrogen oxide (NOx) is so-called "thermal induced compound" because it has two reactions during combustion. The first reaction is from oxidization in which a nitric (N) substance after combustion is formed with the numerator of O_2. The second reaction involves the reaction of O_2 and N_2 from the air during the combustion process (Li et al., 1993; Jang, 2000; Takabashi et al., 1995).

On the other hand, the thermal behavior (pyrolysis, thermal decomposition or combustion) of every combustible material or product produces a toxic smoke atmosphere. This toxic smoke atmosphere, in sufficiently high concentrations, presents a hazard to exposed animals and humans. Discarded woods treated with any of these above mentioned wood-based waste processing-materials from the combustion are therefore need to be disposed of in an environmentally acceptable manner (Ibach, 1999; Lin et al., 2001; Cooper, 2001; Lin et al., 2006). The thermal composition of these wood-based waste processing-materials during combustion, including such things as, decomposition temperature, compounds emitted, total amount of volatiles evolved (Kercher & Nagle, 2001) and the amount of char formed (residual elements) (Lin et al., 2007) is need to be investigated. This is due to the fact that some of the materials are not organic compounds, and their precise molecular composition may vary. In addition, the thermal decomposition of all combustible materials produces a toxic smoke. Combustion toxicology indicates adverse effects produced from exposure to fire-generated toxic species (Hartzell, 1996), but they are not included in the wood recycling industry, basically because of technical problems and regulatory concerns, and the same applies to waste wood management (Falk, 1997). In many cases the final treatment for used wood products is destruction by the combustion (Jang, 1997). The emitted gases from burning the wood-based waste processing-materials, such as CO_2, CO, SO_2 and NOx, and so on are discharged and contribute to environmental pollution.

To provide an understanding of the emitted gases and the fundamental thermal decomposition, as well as to analyze the evolved species and the char of wood-based waste processing-materials, including wood adhesives, wood coatings and wood preservatives, the focus for this chapter is divided into three categories. The first category involves using a Flue-gas Analyzer from the exit of a 45° flammability testing cabinet as specified in the Continuous Emission Monitoring (CEM) technique; the emission gas concentrations during the combustion are analyzed. The second category involves the study of thermal decomposition, and examines the effectiveness of the thermal properties of these materials, using Thermogravimetric Analysis (TGA). In the third category two kinds of spectrometric techniques are applied. One applies Thermogravimetric Analysis with Infrared Spectrometry (TGA-IR) to monitor the effectiveness, on an analytical scale, of these processing-materials and also to examine the emissions of certain compounds (evolved species) in each particular analysis. The others apply, Fourier Transform Infrared Spectroscopy (FTIR), Energy Dispersive X-ray Spectrometry (EDX), and Elemental Analysis (EA) etc, to examine certain residual elements (organic and inorganic materials) of the char, before/after thermal decomposition. The main goal of this chapter is to realize fundamental knowledge of the thermal decomposition of the wood-based waste processing-materials, while at the same time understanding an air pollution reference for gas emissions, evolved species and a disposal-end practice (of the residual elements) reference for the char.

2. Analysis methods of emission gas, thermal decomposition and evolved species

2.1 Continuous emission monitoring

Detecting flue-gas emissions is required by strict environmental legislation. Combustion better detects the flue-gas emission value for analyzing waste products (Jang, 2000). Continuous Emission Monitoring (CEM) techniques are utilized widely for detecting gaseous emissions such as CO, NOx, SO_2 and O_2 and CO_2 concentration at the chimney exit of furnaces to determine the concentrations of various pollutants (Jang, 2000). This portion is to introduce the emission gas concentrations during the combustion of wood-based waste processing-materials by using the CEM technique. A Flue-gas Analyzer (MSI, 2000 W. Germany) was used to examine the gas emitted of wood adhesives, wood coatings and wood preservatives from the exit of a 45° flammability testing cabinet (with reference to CNS 7614 A3125, NCYU handmade). The fuel used was liquid propane. The gas flow direction during the combustion was determined using a gas-measuring probe (MSI 2000 type B). The design and methods are shown in Fig. 1.

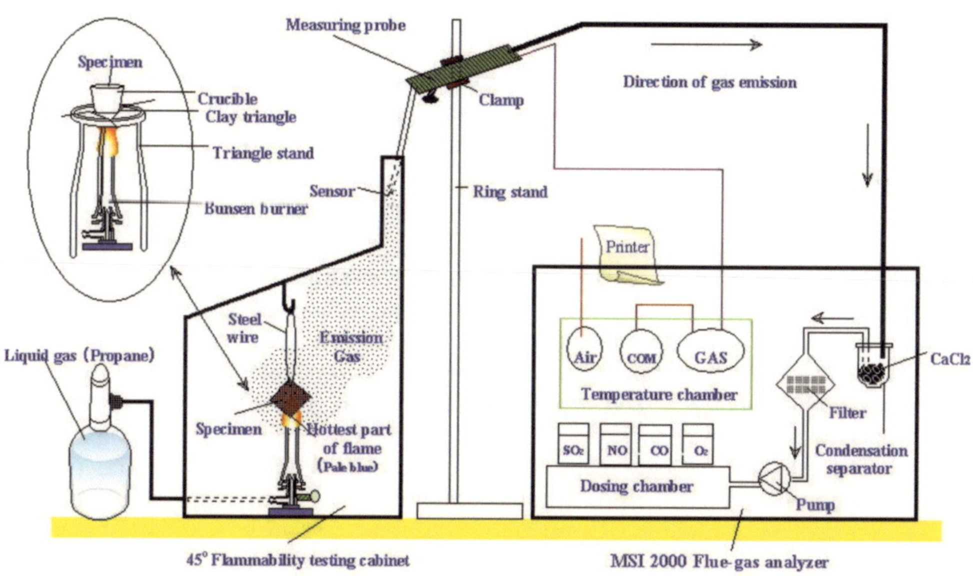

Fig. 1. Diagrams of emissions gas measuring apparatus.

The method involved two combustion treatments in accordance with each type of material. The first used for particleboards (PBs) with 4 types of wood adhesives, coatedwoods with three types of wood coatings and the coatings on a stainless steel, as well as preservative-treated woods with 2 types of wood preservatives. The specimen is hung inside the test cabinet with a steel wire and burned with a Bunsen burner adjusted to (pale blue flame)

about 850 to 980 °C. For the second treatment, the adhesives, the preservatives or wood particles was placed into a crucible on a clay triangle with a triangle stand, respectively. This method is displayed in the oval figure at the upper left in Fig. 1. The flame temperature was verified using a temperature-element apparatus (SUNTEX, ST-54) for both methods.
The combustion measureed method is described as follow. The first cycle run in the Flue-gas Analyzer involved no combustion with a specimen in the test cabinet to be certain that the measured value of the emission gas was zero and the gas temperature was room temperature. The material was ignited and then examined for 10 cycles continuously. A measured cycle for the emission gas concentrations and the emission gas temperature during specimen combustion was about 40 sec. The maximum experimental value was obtained and compared with the value from each cycle. The CO, SO_2 and NOx gases were determined according to parts per million (ppm). The O_2 and CO_2 concentrations were defined as the volume percent (VOL %) of the air in the emission exit. The gas temperature was measured simultaneously.

2.2 Thermogravimetric analysis

Thermogravimetric analysis (TGA) is a common technique for measuring the weight change of a material as a function of temperature. Throughout this chapter, TGA was carried out using a Perkin-Elmer TGA 1, fully supported by computer-controlled software options, from Perkin-Elmer Thermal Analysis Systems, for control and data handling. The wood-based waste processing-materials, including wood adhesives and wood preservatives, approximately 5.0-10.0 mg in weight, were placed into an open experimental sample pan, and suspended from a sensitive microbalance. A furnace surrounded each material sample to provide accurate heating from 50 to 850 °C, while the measurement was taking place. In order to isolate the pyrolysis step, considered the most important step in the burning process for materials or products, a nitrogen atmosphere was selected for TGA analysis (Hsu et al., 2000). All runs were purged in nitrogen atmosphere, at a flow rate of 50 mL/min, with several heating rates of 5, 20 and 40 °C/min. The thermal behavior for the wood-based waste processing-materials was then analyzed.

2.3 Thermogravimetric analysis with infrared spectrometry

TGA, described in the previous section, is a quantitative technique, which does not identify evolved compounds. To understand the evolved species produced by the pyrolysis of the wood-based waste processing-materials, analysis was carried out using a TGA (Perkin-Elmer Pyris 7 TGA model), linked to a secondary method of analysis, Fourier-Transform Infrared Spectroscopy (using a Perkin-Elmer FTIR), which was able to identify the evolved species during the TGA measurements (Statheropoulos& Kyriakou, 2000). The TGA and FTIR (TGA-IR) were connected using a transfer line with a continuous scanner in a gaseous cell. The Infrared spectrometric (IR) response data was recorded by a personal computer. The time-based software obtained and analyzed the evolved species as they occurred. The samples, weighing about 5.0 mg, were placed in an open pan and suspended from a sensitive microbalance in the Pyris 7 TGA model. The conditions were set the same as for the Perkin-Elmer TGA 1, described above, for heating and nitrogen flow rates. The evolved species analysis was performed on the TGA-IR. However, the IR response obtained for various evolved species using the TGA-IR appeared about 10 sec later, when compared to the TGA traces. In order to discuss the criteria and limitations of both TGA-IR and TGA for each particular analysis, and to ensure accurate results, it was necessary to carefully study and compare the results obtained between the pyrolysis temperature (obtained from TGA) and the decomposition time of the evolved species from the IR response (obtained from TGA-IR).

3. Emission gas, thermal decomposition and evolved species

3.1 Wood adhesives

3.1.1 Emission gas of various adhesive types of particleboard

For particleboard (PB) with various adhesives, the particles used were shaving-type wood particles (involved import mixer wood particles 85 % and dismantling material 15 %) from a PB plant in Japan, the Dantani Corporation. Urea formaldehyde resin (UF, Oshika Resin-180), melamine-urea formaldehyde copolymer resin (MUF, PB-1230), phenol formaldehyde resin (PF, PB-1310) and diphenylmethane diisocyanate adhesives (MDI, PB-1605), from the Oshika Shinko CO. LTD, Japan, were used to project as the manufacture the PBs. The preparation and manufacture of PB please see the reference (Lin et al., 2001).

The results of the relationships between emission gas temperatures, the concentration of O_2 and CO_2 and the combustion time for PB manufactured with 4 types of adhesives are shown in Fig. 2.

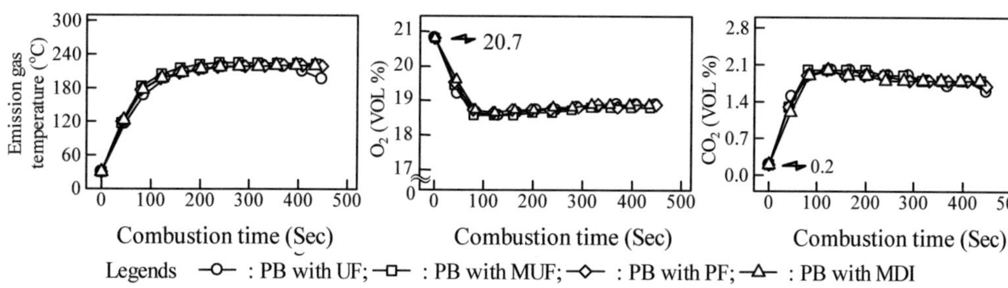

Fig. 2. Relationship between combusion time and emission gas temperature, O_2, CO_2 concentration in particleboard manufacturing with various adhesives.

The increase in emission gas temperature corresponded to the increase in combustion time for each PB. The emission gas temperature increased rapidly up to about 175 °C, then slowed down in the temperature range of 180 to 240 °C, shown as a plateau curve. From the Flue-gas Analyzer used, the standard value of the O_2 and CO_2 concentrations (VOL%) is 20.7 % and 0.2 %. The concentration of O_2 decreased from 20.7 % to about 18 % linearly, and on the contrary the concentration of CO_2 increased from 0.2 % to about 2.0 %. Both O_2 and CO_2 approached the shape of a plateau curve and had a close relationship during combustion. It was indicated that the influence of adhesive types for manufacturing the PB on the gas temperature and either of the concentration of O_2 or CO_2 during the combustion was insignificantly.

It is well known that PB is basically manufactured from particles and adhesives. In order to understand the above-mentioned PB which produced serious pollution gases, the combined results for emission gases for these PB, 4 types of adhesives, and wood particles are shown in Fig. 3. The peak for CO gas was found at about 100 sec during combustion and then the decreasing tendency slowed down to the end of combustion treatment for all of the materials. The CO had the highest point at about 100 sec for about 20 ppm and the decreasing tendency was reached quickly after 100 sec. In other words, the influence of different type adhesives on CO was insignificantly during the combustion. For the adhesives, MDI was more significant on the emission of CO for about 70 ppm (at about 100 sec) during the combustion than the others because it was considered that the reason of incomplete combustion in which MDI had the higher resin solids, 96.3 % (Lin et al., 2001). The result showed that CO was also produced from the wood particles during the combustion at about 100 sec for about 16 ppm. This is attributed to the particles from the wood and a carbon-hydrogen chemical compound.

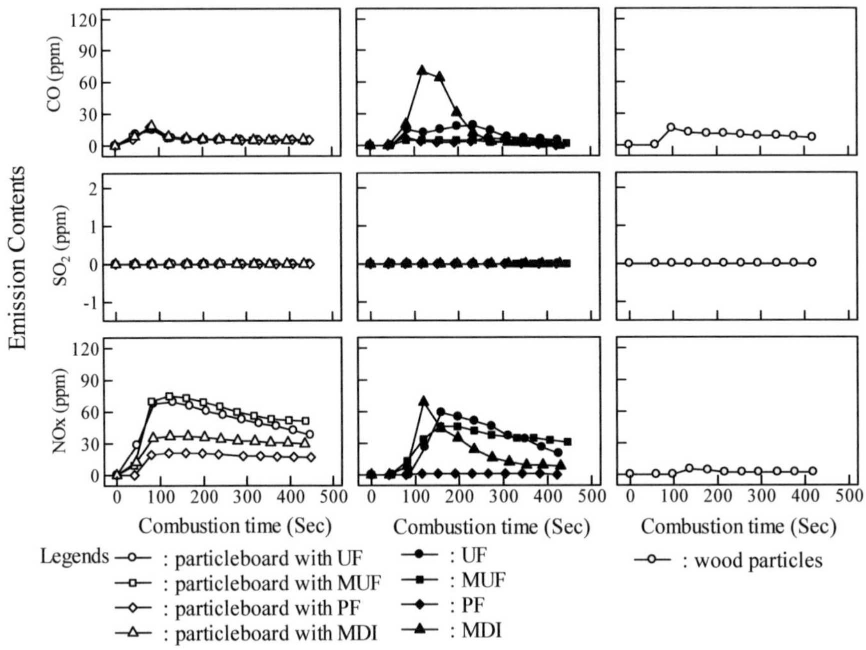

Fig. 3. Relationship between combustion time and emmission contents (CO. SO_2, Nox) in particleboard manufacturing with various adhesives, various adhesives and wood particles.

The emission SO_2 did not produce in the emission gas during the combustion for the PB with various adhesives and the adhesives because the sulphur (S) content was not included in 4 types of adhesive from the typical molecular structure (Chen, 1996; Lin et al., 2004). It is suggested that PB and wood adhesives was not the origin of the SO_2 in environment when it was concerned with final disposal (combustion).

The results showed that the PB and the adhesives were all produced the emission gas, NOx, but it for the wood particles was insignificantly during the combustion. It is considered that NOx is a main pollution origin. The reason for NOx occurring is because N is included in both wood particles and adhesives. Either the O_2 and N_2 reaction from the air and particles or the fuel, propane, produced the N residues. The typical molecular structure UF, MUF and MDI, in addition to PF, was involved a nitric-base that was easily to produce the oxidization with the numerator of O_2 during the combustion (Lin et al., 2004). Moreover, the NOx emissions for PB had the same tendency in which the peak point was about 100 sec from the start of combustion with a decreasing tendency with the combustion time from 100 sec to the end of the experiment. The quantities of emission gases, NOx, for the PB after about 100 sec were followed by MUF (70 ppm), UF (67 ppm), MDI (35 ppm), and PF (20 ppm) in that order. MDI produced the highest quantities of NOx at about 100 sec. It is inferred that the ignition point or/and heat decomposition of the adhesive might influence the combustion time and the quantity of emission gas (Lin et al., 2001).

As the results, the emission contents of CO, SO_2 and NOx from the combustion of PB and adhesives were found that the PB manufactured with PF adhesive had the least pollution gas during the combustion than the others. It is suggested that PF is the best adhesives with the least environmental impact to use for manufacturing PB, when considering the final treatment by the combustion.

3.1.2 Thermal decomposition of various adhesives

To comprehend the thermal behavior (pyrolysis, thermal decomposition) of the four types of adhesives, the specimens were measured by TGA, after being air-dried at ambient temperatures. Decomposition profiles were obtained while heating at a rate of 20 °C/min in nitrogen, between 50 °C (initial temperature) and 850 °C (end temperature). The relationships of TGA and DTG curves between temperature and weight loss, and the derivative weight for each adhesive specimen, UF, MUF, PF and MDI, are shown in Fig. 4.

As expected, the four types of adhesives typically showed gradual weight loss. The thermogram, resulting from the analysis of UF (solid line) and MUF (dotted line), showed that there were two subtle changes (shoulders) at temperatures of 50 to 390 °C, suggesting that both had almost identical initial (the first sharp step) and secondary decomposition temperatures, because UF and MUF are copolymers. The adhesive specimen weight for PF (dash line) was significantly less at temperatures between 50 °C and 500 °C in the TGA tests; the weight loss for MDI (dash-dot-dash line) was significant at temperatures from

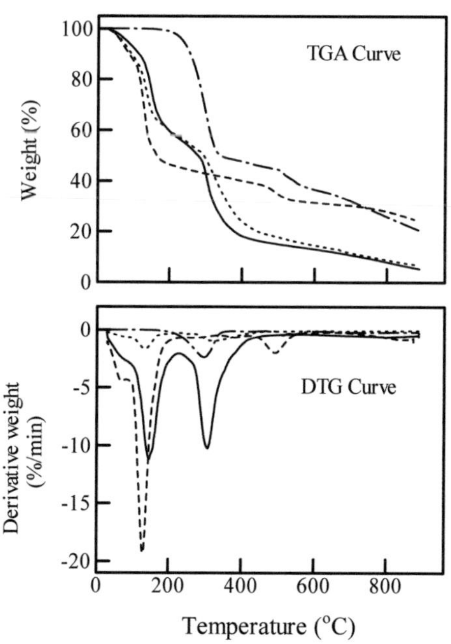

Fig. 4. TGA and DTG curves for 4 types of adhesives (UF, MUF and PF and MDI) in nitrogen at 20°C/min.

200 to 400°C. This indicates that the chemical structure, in the molecular chain, or chemical composition, of PF and MDI is different (Chen, 1996; Lin et al., 2004). The results also found that MDI exhibited the highest initial decomposition at a temperature of 297°C and better overall thermo-stability than the other types of adhesives. PF and MDI provided a better yield of char at about 24.88 and 20.61 wt%, respectively, at 850°C. This showed that theirthermal behavior was better than that of UF (5.40 wt%) or MUF (6.99 wt%), as the proposed action mechanism is based on the charred layer acting as a physical barrier, which slows down heat, and results in mass transfers between the gas and condensation phases (Camino et al., 1993, Hornsby, 1996 and Zhu & Shi, 2001).

3.1.3 Evolved species of various adhesives

Figure 5 shows the Infrared spectrometric (IR) response of the four types of adhesives in the TGA-IR measurements. The change in absorbance (concentration of evolved species %) corresponded to the increase in decomposition time for the four adhesives. The IR response of all adhesives in the TGA-IR tests showed that the absorbance of UF, MUF and MDI was higher than that of PF. This indicated that the concentration of evolved species of PF was less than that of MDI, and less than that of the urea copolymer series adhesives, UF and MUF. Of them all, PF was considered to be the best adhesive with the least environmental impact. A similar conclusion was reached in Fig. 3 and the previous research (Lin et al., 2001), which indicated that PF was the most environmentally friendly adhesive for manufacturing particleboard, because the emission contents (such as CO, SO_2 and NOx gases) of PF were less than those of UF, MUF and MDI, when the used PB was destroyed by combustion. Therefore, it has been suggested that PF is the most environmentally friendly adhesive, because it has the least environmental impact when used in many industrial and domestic applications, such as for structural members in furniture or architecture.

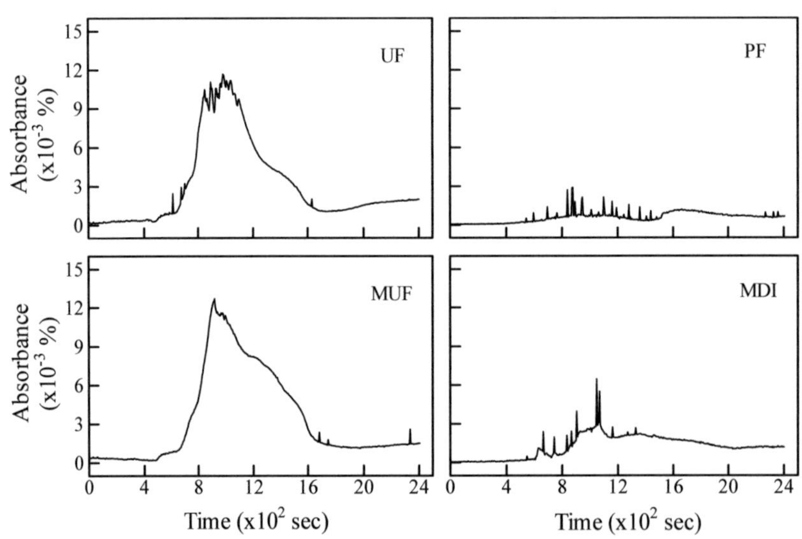

Fig. 5. IR response of 4 types of adhesives (UF, MUF and PF and MDI) from TGA - IR analysis.

In recent years, there has been much attention paid to the use of adhesives (formaldehyde copolymer compounds), because, during combustion, they emit toxic gases and corrosive smoke; exposure to these toxic and acidic fumes can cause great harm to people and can also damage costly equipment (Ladomersky, 2000). With the use of TGA-IR, it is possible to simultaneously monitor evolved species, pyrolysis temperature and concentrations formed (absorbance), as well as the evolution profile of certain compounds. Moreover, IR spectrometry is a common means of evaluating a functional base when a material undergoes degradation. The evolution of decomposition time (or pyrolysis temperature) on absorbance and wavenumbers (basically, IR spectra) for the various types of adhesive specimens (UF, MUF, PF and MDI) is shown in Fig. 6.

For TGA-IR spectra of UF, the CO_2 peak, located at 2300 and 2400 cm^{-1} was shown beginning from about 240 sec (at a temperature of about 127 °C in Fig. 4) to the end. CO_2 increased linearly to the top point of absorbance at about 960 sec (367 °C), then decreased at about 1680 sec (607 °C) and kept the same concentration of CO_2 till the end of the tests. 3545 and 3509 cm^{-1} was assigned to the N-H stretching of NH_2 moieties (Zhu & Shi, 2001) from about 360 sec (167 °C) to the end. A higher concentration of N-H stretching was seen between 840 and 1440 sec (327 to 527 °C). The range from 480 to 1180 sec (207 to 440 °C) showed a weak peak at 2921 and 2851 cm^{-1}, respectively, which was attributed to the C-H stretching of CH_2 moieties. Another weak peak, at 1447 cm^{-1}, was attributed to C-H stretching of CH_2 moieties, from 480 sec. The weak peaks at 1701 and 1625 cm^{-1} were assigned to the C=O (carbonyl group), stretching from 480 sec to the end. Moreover, several weak peaks at 972 or 962 cm^{-1} also emerged due to C=O stretching, during the period from 480 sec to the end. The top point of absorbance for C=O stretching at 972 or 962 cm^{-1} was found at 960 sec, which was the same as the top point of absorbance for CO_2. The results of MUF indicated that UF and MUF present the same tendency in the left side of Fig. 6, as both belong to the same copolymer urea formaldehyde series. The results for the absorbance peak and the time location (decomposition temperature) of CO_2, N-H stretching, C-H stretching and C=O stretching for MUF were the same as for the TGA-IR spectra of UF; the N-H deformation vibration of NH_2 moieties was found at 1605 and 1649 cm^{-1}, from about 360 sec (167 °C) to the end. The weak peak at 1449 cm^{-1} was attributed to C-H deformation vibration of CH_2 moieties, while at 1414 cm^{-1}, it was assigned to C-N stretching. Both of these began close to 480 sec (207 °C) to the end. A weak peak of C=O stretching, located at 1736 cm^{-1}, was also found from 600 sec to the end (247 to 850 °C). In addition, two weak peaks of triazine, located at 1365 and 1565 cm^{-1}, began from about 480 sec to the end.

For the TGA-IR spectra of PF, the weak peaks of CO_2 were also located at 2300 and 2400 cm^{-1} and began from about 600 to 1440 sec (247 to 527 °C), almost disappearing towards the end of the tests. Some of weak board absorbance at 3300 to 3600 cm^{-1} was assigned to OH groups from about 720 sec (287 °C) to 2160 sec (767 °C). A weak peak at 2980 cm^{-1} was assigned to out-of-plane bending of C-H on the benzene (aromatic) ring, from 1440 sec (527 °C) to the end. The spectrum results in this study are similar to those reported by Reghunadhan Nair et al (2001). Another weak peak at 1100 cm^{-1} was attributed to the stretching vibration of benzene C-C, from 1560 sec (567 °C) to the end. Besides, the results of MDI indicated that the two weak peaks of CO_2 were the same for UF, MUF and MDI, located at 2300 and 2400 cm^{-1} and which began from about 480 to 1560 sec (207 to 567 °C), almost disappearing towards the end of the tests. Some weak board absorbance located at 3700 cm^{-1}, was assigned to N-H stretching from about 600 sec (247 °C) to 1440 sec (527 °C). In addition, some weak board absorbance, found at 3100 cm^{-1}, was assigned to OH group,

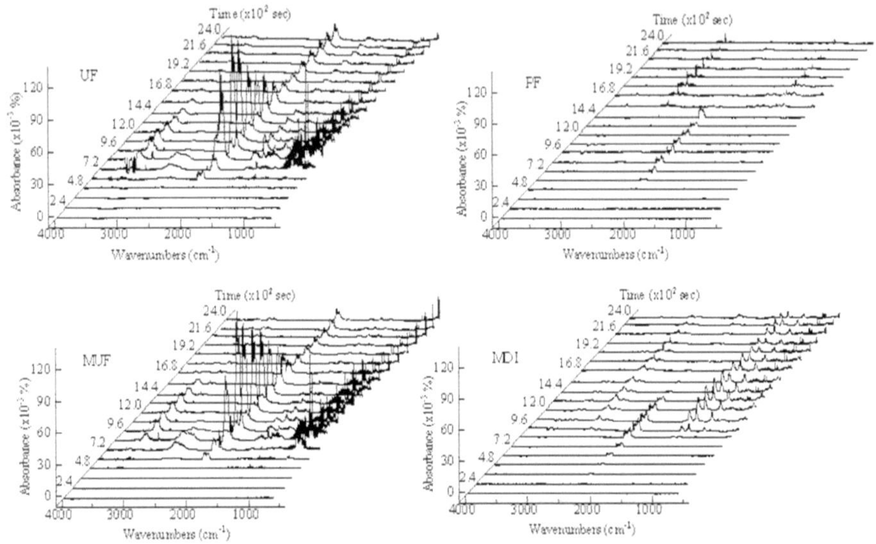

Fig. 6. TGA - IR spectra of 4 types of adhesives.

from about 960 sec (367 °C) to the end. Other weak peaks at 1624 cm^{-1} were assigned to C=O (carbonyl group) stretching from 840 sec (327 °C) to the end. Two weak peaks at 972 or 927 cm^{-1} also emerged as C=O stretching, from 1200 sec (447 °C). Results obtained also showed several weak peaks at 1114 and 1206 cm^{-1}, which were assigned to the C-NH group and N=C group (Levchik et al., 1995) from 840 sec (327 °C) to the end, respectively.

To clarify the effectiveness of the analytical scale towards the four adhesives, and to identify the evolved compounds of each particular analysis, the most probable structures for the different polymers found in these adhesives (Chen, 1996; Lin et al., 2004). UF analysis of the TGA-IR response showed that the thermal decomposition of the NH groups at a temperature of about 167 °C was initially determined as a possible degradation species. The same results were previously reported in literature (Zhu & Shi, 2001). The CH_2 and C=O groups were then degraded at about 207 °C; the final contents of the char, therefore, were practically negligible (5.40 wt%). Some of the MUF evolved species are the same as the UF evolved species, because they belong to the same series of copolymers. The TGA-IR results showed that the N-H groups were degraded at a temperature of 167 °C, while the CH_2 and triazine groups were degraded at 207 °C. In the case of PF, the thermal decomposition of the OH groups was found to be significant at 287 °C, although it could be degraded earlier, at temperatures of 65 and 129 °C; this information was obtained from the results of the TGA tests. In addition, the CH_2 groups were degraded at 527 °C and the benzene C-C at 567 °C. For MDI, the C=O and N=C groups were degraded at about 327 °C, which was analogous to the thermal decomposition of the OCN group during MDI pyrolysis, because of the resultant chemical reactions. Also, the char left at 850 °C in the TGA tests for PF and MDI was 24.88 and 20.61 wt%, respectively. The benzene (aromatic groups) structures are for PF and MDI; that is, the thermal performance was enabled when the chemical compounds included aromatic groups.

As the results, when compared to the molecular structures of the adhesives, enabled us to point out the kinds of evolved species (functional bases or the change in molecular structures) for referencing the thermo-degradation processes of the adhesives.

3.2 Wood coatings

3.2.1 Emission gas of wood coatings specimens

In the style-conscious wooden furniture industry, good finishing is perhaps the most critical factor. The sale of furniture is highly influenced by its decorative value and attractive appearance. However, the expectation of life for wooden furniture is about 17 years (Akiyama, 1998). Discarded wooden furniture are extremely difficult to recycle because the coatings on the furniture surface cannot presently be used as a raw material (particles or chips) again when it is considered with the bond quality of reuse materials, such as various PBs. In many cases the final treatment for used wooden furniture is destruction and then treats with combustion. Moreover, wood coatings are generally produced with various chemical compounds. The follow portion is to introduce the combustion emissions from wood coatings, and also to explore the modification of chemical structure using a potential application of Fourier Transform Infrared Spectroscopy (FTIR) from wood coated with various wood coatings before/after an accelerated weathering test, which is the QUV artificial accelerated-degradation (QUV degradation). Three types of commercial coatings, nitrocellulose lacquer coatings (NC lacquer, Kuo Rong Corporation in Taiwan), polyurethane coatings (PU coatings, Kuo Rong Corporation in Taiwan), and ultraviolet curing coatings (UV curing coatings, Chia Tai Corporation in Taiwan) were coated on the tangential sections of *Cryptomeria japonica* D. DON., and on a stainless steel, respectively. Before combustion, the nondegraded and degraded coating woods were analyzed using the FTIR. The data produced are to comprehend and compare the environmental pollutions from the coated materials with wood coatings before/after QUV degradation when they are discarded by combustion, and to as a basic reference to develop the environmentally friendly wood coatings hopefully. The basic properties of coatings, the finishing process and the accelerated weathering tests please see the references (Lin & Huang, 2004; Lin, 2005).

The emission gas temperature of all coated materials was rapidly up to about 160-170 °C, increased slowly up to 180-200 °C and then kept in the temperature range of 200-210 °C, shown as a plateau curve. The concentration of O_2 decreased from 20.7 % to about 18-19 % linearly, and on the contrary the concentration of CO_2 increased from 0.2 % to about 1.2-2.0 %. Both O_2 and CO_2 then approached the shape of a plateau curve as well as had a close relationship during combustion.

It is well known that wood coatings are generally combined with various chemical compounds. To understand the coated specimen produced serious pollution gases after QUV degradation, the combined results of emission contents are shown in Fig. 7. The highest emission quantity of CO and NOx for the coated specimens of NC lacquer coatings was about 198 and 102 ppm, 110 and 47 ppm for PU coatings and 90 and 9 ppm for UV curing coatings. The emission SO_2 in this experiment as expected did not produce in the emission gas during the combustion for the coated specimens before/after QUV degradation because the sulphur content was not included in 3 types of wood coatings from the typical molecular structure (Lay, 1985). It is suggested that the coatings are not the origin of the SO_2 in environment when it was concerned with final disposal (combustion). Three types of coatings still produced those pollutants (CO, SO_2 and NOx) after QUV degradation, especially for NOx gas. This is because the chemical structure of these coatings produced a series of photochemical chain reaction and the coating film separated (peeled) from the coated substrate (Chang & Chou, 1999). However, the pollutants were relatively a decreasing tendency for the coatings after QUV degradation. The above-mentioned results

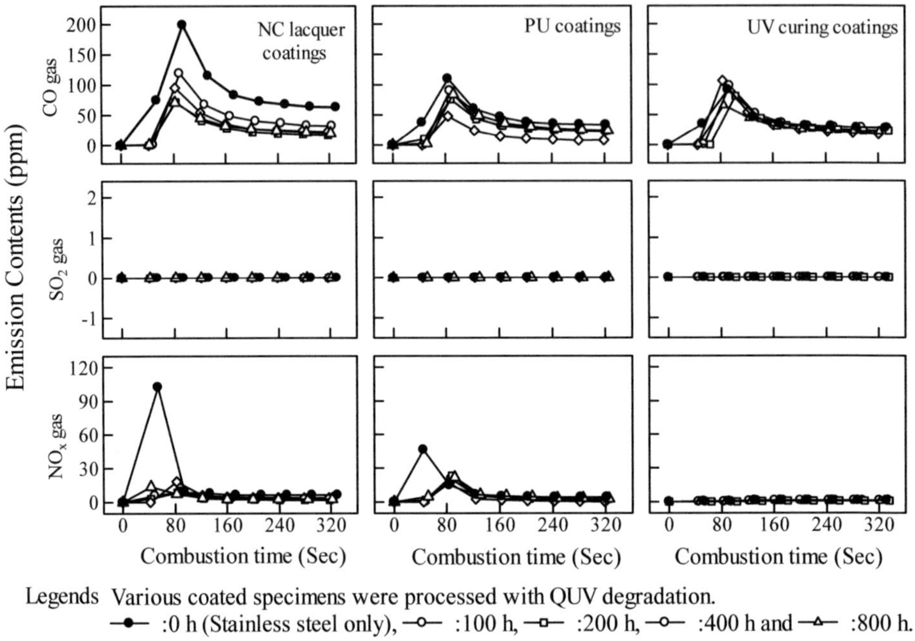

Fig. 7. Relationship between combustion time and emission contents (CO2, SO 2, NOx) for stainless steel and various coated specimens.

suggested that UV curing coatings is the most environmentally friendly coatings while the coatings are considered to discard process with combustion.

3.2.2 FTIR analysis of wood coatings

Chang (1990) reported that FTIR, possessing many unique advantages, is a powerful and potential technique for the analysis and characterization of micro-specimens of various wooden materials, coatings and coatedwood, especially with particular attention to the chemical modifications of materials after weathering and photodegradation. Moreover, It is well known that wood coatings are generally composed with various resin and chemical compounds. Basically, the manufactured principle of NC lacquer is to substitute NO_2+ for OH of CH_2OH or vascular cellulose and then became NC lacquer. In other wards, NC composed with alkyd resin and then added some of additives to become NC lacquer. To comprehend and compare the production of combustion emissions from three types of wood coatings, the coatedwoods before/after degraded at 0, 100, 200, 400 and 800 h with the QUV accelerated weathermeter were analyzed using the FTIR, shown in Fig. 8. The FTIR spectra of NC lacquer for QUV 0 h showed that the peaks at 1643 and 1273 cm^{-1} were assigned to the asymmetric NO_2 stretching vibration and symmetric NO_2 vibration of CH_2ONO_2 or vascular cellulose, and another peak at 830 cm^{-1} was assigned to N-O (nitrogen-oxygen) stretching vibration (Pouchert, 1981). Results showed that these peaks disappeared on the FTIR spectra when the NC lacquer coatedwoods after QUV degradation of 100, 200, 400 and 800h. This is because CH_2ONO_2 of NC lacquer is easily degraded by photodegradation (Chang, 1994). Therefore, the results in Fig. 7 obtained that the emission of NOx for the NC lacquer after QUV degradation decreased significantly more than control ones.

In general, the main compound of commercial NC lacquer is alkyd resin. To improve the properties of NC lacquer, it added some of additives, such as plasticizer etc. This indicated that from the FTIR spectra of NC lacquer (Fig. 8), the control coatedwood was showed a location of peak at 1720 cm^{-1}, assigned to C=O (carboxylic group) stretching vibration that was probably from alkyd resin and plasticizer, or NC degradation to become -COOH base. After QUV degradation, this peak frequency was changed its intensity and location. This is because the change in the chemical structure of NC lacquer occurred, as well as C=O produced some degraded derivatives (Chang, 1990). The results in Fig. 7 indicated that the emission of CO for the NC lacquer coatedwoods after QUV degradation decreased significantly more than control ones. The results in Fig. 8 also showed that the FTIR spectra of NC lacquer coatedwoods after QUV degradation at 100 h appeared a peak at 1015 cm^{-1}, assigned to -COOH stretching. It is indicated that this commercial NC lacquer due to QUV degradation produced a degraded derivative, -COOH base, but this degraded derivative disappeared at QUV degradation of 200, 400 and 800 h. Hence, the results in Fig. 7 showed that for the NC lacquer coatedwoods after QUV degradation at 100 h, the emission of CO decreased after 120 sec during combustion and the decreasing tendency then slowed down to the end of combustion. It is inferred that because -COOH base decomposed easily during the combustion.

FTIR spectra in the middle of Fig. 8 showed that PU coatings degraded by QUV degradation at 0, 100, 200, 400 and 800 h. The FTIR spectrum for the coatings control ones showed that the location of peak at 1716 cm^{-1} was assigned to C=O (carboxylic group) stretching vibration of urethane and peak at 1217 cm^{-1} was assigned to -N-C-O-R (Pouchert, 1981). After QUV degradation, FTIR spectra showed that both C=O produced the change in the intensity and the shift of peak frequency, and then increased its width. Chang (1990) reported that PU coatings due to photodegradation occurred yellowing, generated quinine-imides structure, on the coating film and then produced the degraded derivatives of ester or/and carboxyl. Hence, the results in Fig. 7 showed that for the PU coated specimens after various QUV degradations the emission of CO decreased significantly. Moreover, the peak at 1531 cm^{-1} was assigned to -C-NH of R-N-C-O-R' for QUV degradation at 0 h. Results also showed that this peak disappeared on the FTIR spectra when the PU coatings coatedwoods after QUV degradation at 100, 200, 400 and 800 h. Therefore, the results in Fig. 7 obtained that the emissions of CO and NOx for the PU coatings coatedwoods decreased more than control ones.

FTIR spectra in the bottom of Fig. 8 showed that UV curing coatings coatedwoods degraded by various time of QUV degradations. Because the main compounds of UV curing coatings used are epoxy oligomer and acrylate monomer, the FTIR spectrum of the UV curing coatings coatedwoods after QUV degradation at 100 h showed that the location of peak at 1724 cm^{-1} was assigned to C=O stretching vibration and at 1604, 1505, 824 cm^{-1} was assigned to benzene ring. Moreover, the peaks at 1178 and 1175 cm^{-1} were assigned to C-H stretching vibration (Chang & Chou 1999). From the results of FTIR spectra, the UV curing coatings coatedwoods after QUV degradation at 100, 200, 400 and 800 h showed that the above-mention functional bases (group) do not change significantly, but from the emission of CO decreased (Fig. 7) the study is inferred that the benzene rings in UV curing coatings are extremely stable during the heat decomposition (combustion), but the UV curing coatings coatedwoods after QUV degradation are possible to produce the oxidation. It is inferred that the emission of CO, therefore, might decrease relatively during the combustion.

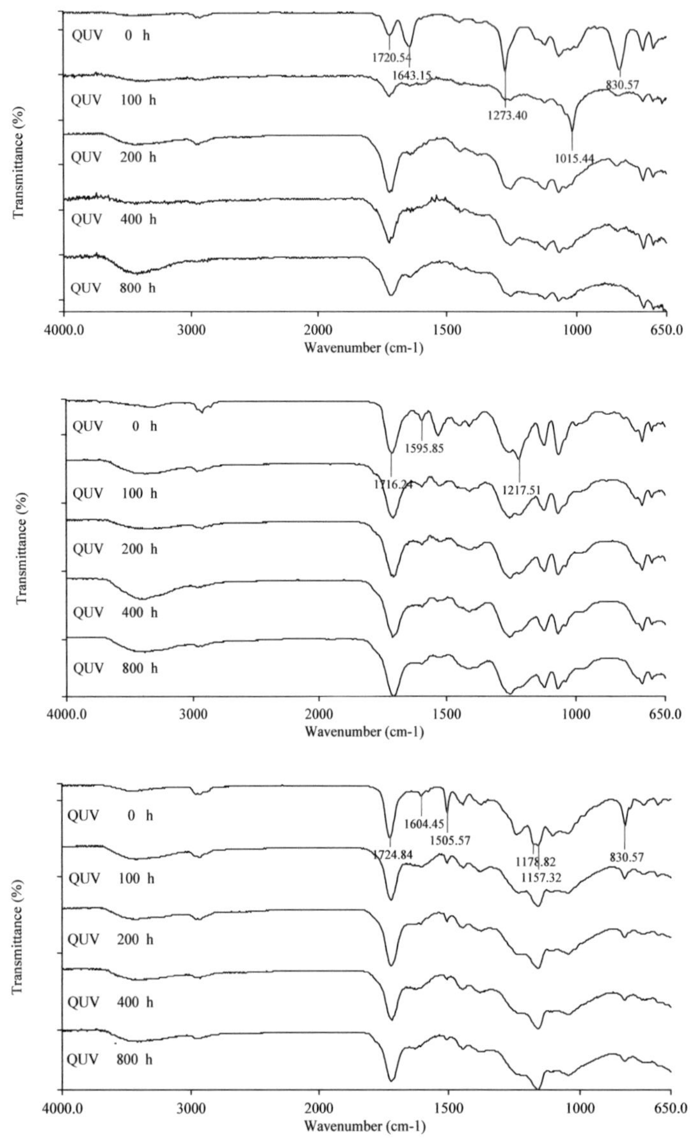

Fig. 8. FTIR spectra of NC lacquer (top of Fig), PU coatings (middle of Fig) and UV curing coatings (bottom of Fig) coatedwoods degraded by various time of QUV degradations.

To analyze the QUV degradation of coatedwoods, using FTIR showing the change in the functional bases or the molecular structures in combination with CEM analysis has led to useful results, involving the identification of emission products. Three types of coatings still produced those pollutants (CO and NOx) after QUV degradation, but the pollutants are relatively a decreasing tendency while comparing to the control ones. When analyzed to the QUV degradation of the coatings, the results enabled us to point out the kinds of functional base or the change in molecular structures for referencing the emissions of the wood coatings.

3.3 Wood preservatives

3.3.1 Emission gas of preservative-treated woods

Extra protection is needed when wood is exposed to adverse environments. This is because wood becomes subject to degradation by a variety of natural causes when left untreated in many outdoor applications. For the past few decades, wood can be protected from the attack of decay fungi, harmful insects, or marine borers by applying chemical wood preservatives with long-lasting results (Connell, 1991). Wood preservatives can be divided into two general classes: (1) oilborne (oil-based) preservatives, such as coal-tar creosote (creosote oil) and petroleum solutions of pentachlorophenol and (2) waterborne (water-based) preservatives that are applied as water solutions, such as chromated copper arsenate (CCA) and ammoniacal copper Quats (ACQ) (Ibach, 1999). However, environmental concerns have drastically changed the active ingredients of wood preservatives, resulting in restricted use (Sanders & Windom, 1980; Sanders & Riedel, 1987; Weis & Weis, 1992a, 1992b, 1996, 1999; Cooper, 2001), especially since substantial amounts of CCA or creosote oil remain in the wood for many years and the disposal of scrap wood has become an ever increasing problem (Cox, 1991; Cooper et al., 2003; Solo-Gabriele et al., 2003). Some years ago, alternative copper-based preservatives, ACQ, comprised of a combination of copper and organic biocides were developed relatively, and are being used increasingly for wood preservation in many countries.

Because wood preservatives are generally produced with various chemical compounds, it is necessary to understand the composition of the gases that are emitted when burning these preservative-treated woods, and the amount of char formed (residual metal) during combustion (Jang, 1997; Kercher & Nagle, 2001; Humphrey, 2002). The recognition of the need for abatement of air pollution has led to further interest in, and investigation of combustion as a major issue when disposing of cellulose wastes and cellulose residual materials (Helsen & Bulck, 1998). Regardless of waste minimization efforts, improved disposal-end management practices will play a key role in minimizing the impacts of disposed of CCA-treated wood in the next 25 to 40 years (Cooper et al., 2003), and of the ACQ-treated woods as well (Lin et al. 2006). To provide an understanding of the disposal of CCA- and ACQ-treated woods and the basic emitted gases generated by burning these materials, our goal in this portion is to acquire fundamental knowledge of the disposal-end practice (residual elements) reference for the char. The emissions from the discharge gas during the preservative-treated specimen combustion and the residual elements of the char after combustion are interpreted, respectively.

One of the studies used *Taiwania cryptomerioides* Hay (Taiwania) as a specimen to be treated with chromated copper arsenate (CCA) or ammoniacal copper quats (ACQ) preservatives. CCA and ACQ-treated woods, the preservatives themselves as well as their individual ingredients during combustion were investigated (Lin et al. 2007). Basically, both woods treated with CCA or ACQ are considered to be a type of cellulose material (Humphrey, 2002). The recognition of the need to reduce air pollution has led to further interest in, and the investigation of the feasibility of combustion to dispose of cellulose wastes and residual materials (Helsen & Bulck, 1998). The results of the emission contents for Taiwania and the two types of treated woods are shown in Fig. 9.

The peak for CO gas for the samples of Taiwania (T) was found to be 40 sec with 116.0 ppm after the beginning of combustion, after which it showed a decreasing tendency until it became a plateau curve until the end of the combustion time,. However for the CCA specimens a maximum value of CO was obtained at about 115 sec with 108.0 ppm, and for the ACQ specimens the measurement of the CO gas increased linearly to about 80 sec with 159.3 ppm from the start of the combustion. After that, both became a plateau curve and remained so until the end of the combustion time. This indicated that the emission of CO gas not only was concerned with the ignition time (Ladomersky, 2000; Lin et al., 2001), but also because it resulted in complete combustion at a different combustion time due to the different types of preservatives.

In middle of Fig. 9, The results also found no SO_2 being emitted gas by any of the specimens during combustion, but the specimens whether or not they were impregnated with preservatives all produced NOx gas (the bottom of Fig. 9). Although it was insignificant for the amount of wood and CCA specimens during the combustion, the peak of NOx gas for the ACQ specimens was about 26-26.5 ppm at about 75 to 115 sec of the combustion time. NOx is considered the main source of pollution from burning wood treated with ACQ. The reason for NOx occurring from the burning of preservative-treated wood is because N is included in both the wood (Shirokae & Samezima, 1996; Kercher & Nagal, 2001; Humphrey, 2002) and in the preservatives (Lin et al., 2006). Either the O_2 or the N_2 reaction from the air and the samples or the fuel, propane, produced the N residues (Lin, 1992; Takabashi et al., 1995). It is inferred that the ignition point and/or the heat decomposition of the preservatives may influence the combustion time and the quantity of emission gases, NOx.

To understand which preservative-treated wood produced the more serious pollution, the combined results of the emission contents from CCA and ACQ preservatives and their main ingredients of chromium trioxide (CrO_3), cupric oxide (CuO), arsenic pentoxide (As_2O_5), and N-alkyl benzyl dimethyl ammonium chloride (BKC) are shown in Fig. 10.

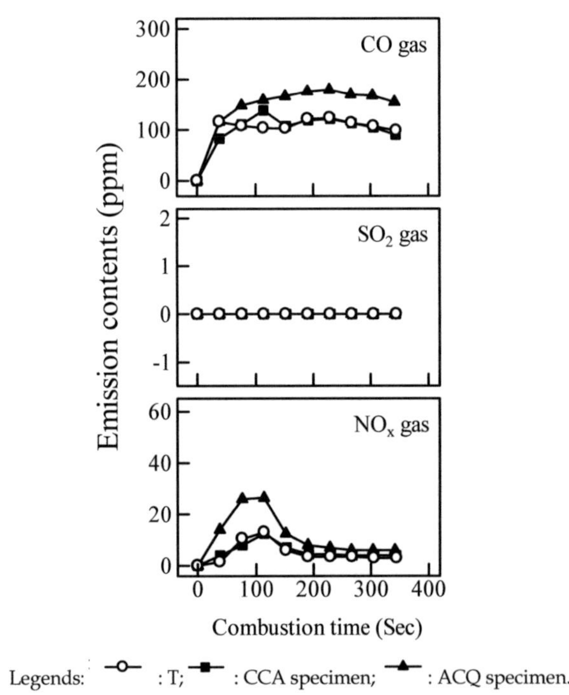

Fig. 9. Relationship between combustion time and emission contents for various experimental specimens.

The above-mentioned situation of incomplete combustion for wood particles (Taiwania) produced CO during combustion. This is attributed to the fact that the particles are wood and are a carbon-hydrogen compound. (Shirokae & Samezima, 1996; Helsen & Bulck, 1998). The same tendency was evident for the CCA and ACQ specimens (Fig. 10). In other words, the influence of different types of preservatives on the production of CO was insignificant during combustion. However, among the ingredients of the preservatives, AS_2O_5 (157.0 ppm) and BKC (85.5 ppm) were more significant on the emission of CO during combustion than the others. As expected there was no SO_2 in the emission gases in this experiment during the combustion of the wood, the preservatives or their ingredients because there was no sulphur content in either of the 2 types of preservatives (Nicholas & Schultz, 1994). It is suggested that CCA and ACQ are not the source of the SO_2 in the environment. The results also showed that the ingredients of the preservatives all produced NOx in the emission gases. However, during the combustion of the wood particles, CCA-3, AS_2O_5, CrO_3, and CuO, produced less NOx than the ACQ specimens and ACQ. Of these materials, ACQ and BKC were more significant for the emission of NOx during combustion than the others chemicals (ingredients). The maximum NO_x of the ACQ and ACQ specimens was about 23.5-26.5 ppm, and was about 32.0 ppm for BKC. This is because the main combinations of ACQ are BKC, and in addition nitrogen (N) is also included in the BKC (Nicholas & Schultz, 1994). This indicates that NOx is a main source of pollution in wood with ACQ preservatives.

Moreover, wood is considered an ecological material (Takahashi et al., 1995), but it is necessary to determine the thermal composition of these preservatives during combustion,

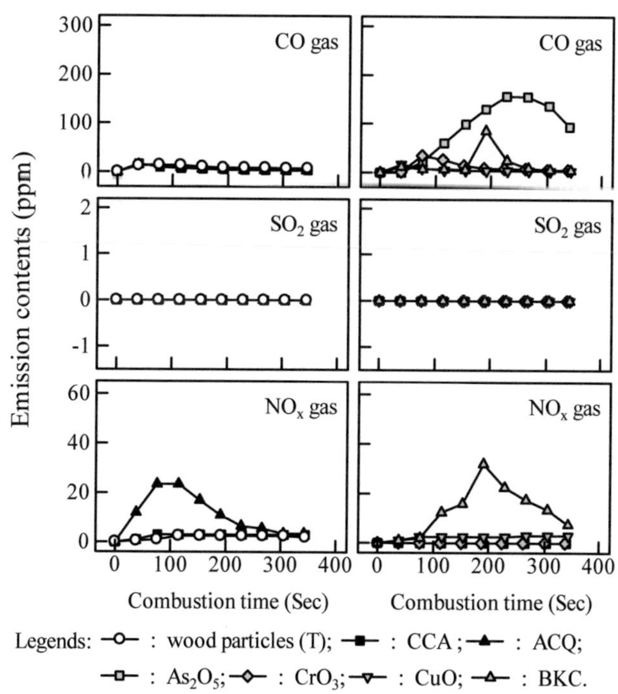

Fig. 10. Relationship between combustion time and emission contents for various wood preservatives and the main components of CCA and ACQ.

including such things as, decomposition temperature, compounds emitted, total amount of volatiles evolved (Kercher & Nagle, 2001) and the amount of char formed (residual elements) (Lin et al., 2007). This is due to the fact that both CCA and ACQ are not organic compounds, and their precise molecular composition may vary. To clarify the residual elements (inorganic and organic materials) of discarded woods treated by either CCA or ACQ, the main ingredients, the wood preservatives, their make-up and the char before/after combustion are analyzed with an Energy Dispersive X-ray Spectrometer (EDX), and are analyzed after combustion using an Elemental Vario CHNS/O analyzer (EA, Germany). The result for the carbon (C), nitrogen (N), sulfur (S), and hydrogen (H) elements was then determined. The EDX results are shown in Table 1.

The main elements of Taiwania (untreated specimens), both preservative woods and preservatives were C, 56.35 % for the untreated specimens, 47.96 % for the CCA specimens, and 50.24 % for the ACQ specimens. For the preservatives, CCA was 4.30 %, and ACQ was 3.74 %. For both of their ingredients, AS_2O_5 was 3.25 %, CrO_3 was 1.50 %, CuO was 3.76 %, but BKC was 89.35 %. After combustion, the relative proportion of C was higher, 89.42 % for the untreated specimens, 60.42 % for the CCA specimens, 77.89 % for the ACQ specimens, and 2.70 % for CCA, and 22.13 % for ACQ. In addition AS_2O_5 was 1.47 %, CrO_3 was 1.33 %, CuO was 14.02 %, and BKC was unmeasured (trace amount). The results of element O showed that it had relatively decreased for all samples after being burned and the amount of chromium (Cr) for the CCA specimens was 7.38 %, for CCA it was 34.99 %, and the CrO_3 was 64.15 % after being burned. Three of them were obviously more than the relative amount of As (Arsenic, 4.08 % for the CCA specimen, 11.89 % for CCA, and 36.92 % for AS_2O_5), and Cu (copper, 5.18 % for the CCA specimen, 16.69 % for CCA-3, and 58.39 % for CuO). This is because the Cr from the char of the CCA preservatives is a stable element (Hirata et al., 1993; Kercher & Nagle, 2001; Cooper et al, 2003; Helsen & Bulck, 2005; Lee et al., 2005). The char of the ACQ specimens left an amount of inorganic metal elements, Cu (2.67 %). As the results (Table 1) indicated, the main char of the ACQ and CuO was Cu, 50.14 and 58.39 % which was almost unable to be combusted toward the end of the combustion (Kercher & Nagle, 2001; Lee et al., 2005; Lin et al., 2006).

Element	*Taiwania cryptomerioides* Hay (wt %)		CCA Specimen (wt %)		CCA (wt %)		As_2O_5 (wt %)		CrO_3 (wt %)		CuO (wt %)		ACQ Specimen (wt %)		ACQ (wt %)		BKC (wt %)	
	before	after	before	after	before	after	before	after	before	after	before	after	before	after	before	after	before	after
C	56.35	89.42	47.96	60.42	4.30	2.70	3.25	1.47	1.50	1.33	3.76	14.02	50.24	77.89	3.74	22.13	89.35	–[1)]
O	43.45	8.09	49.43	20.55	59.78	33.50	28.59	49.23	28.14	34.52	24.69	21.89	49.30	7.17	29.94	26.38	–	–
Cr	–	–	0.08	7.38	7.41	34.99	–	–	70.36	64.15	–	–	–	–	–	–	–	–
Cu	–	–	1.75	5.18	16.29	16.69	–	–	–	–	71.55	58.39	0.45	2.67	65.76	50.14	–	–
As	–	–	0.72	4.08	10.21	11.89	68.16	36.92	–	–	–	–	–	–	–	–	–	–

[1)] Race amount (can not be measured).

Table 1. EDX analysis of various experimental samples before and after the combustion

Furthermore, the results of EA are shown on Table 2. In addition to the high level of carbon (C), the amount of N (1.42 %) was higher for ACQ than that for the other specimens, and the BKC was almost burned out (could not be measured). This inferred that the ACQ specimen could greatly produce NOx gases (Fig. 10) during combustion. The result of the EA analysis inferred that the ACQ specimen could produce NOx gases during combustion, because the amount of N (1.42 %) was higher for ACQ than for the other specimens, and the BKC was almost burned out. The above results suggest that by using EDX and EA to analyze the residual elements in char, the results can provide experimental data for referencing the char of preservative-treated woods in CEM tests.

Element	*Taiwania cryptomerioides* Hay	CCA specimen)	CCA	As_2O_5	CrO_3	CuO	ACQ specimen	ACQ	BKC
C	79.58 (0.03)	75.38 (0.54)	0.03 (0.01)	– 1)	87.04 (0.17)	63.73 (0.08)	80.93 (0.20)	4.51 (0.11)	–
N	0.28 (0.04)	0.15 (0.01)	0.05 (0.01)	–	0.97 (0.05)	0.03 (0.02)	0.33 (0.01)	1.42 (0.01)	–
S	1.17 (0.62)	0.60 (0.04)	0.36 (0.02)	–	3.22 (0.02)	0.53 (0.17)	0.54 (0.08)	0.41 (0.07)	–
H	3.27 (0.03)	2.90 (0.04)	0.19 (0.01)	–	2.68 (0.20)	0.36 (0.59)	3.13 (0.03)	0.55 (0.03)	–

1) Race amount (can not be measured).

Table 2. EA analysis of various experimental samples after the combustion

3.3.2 Thermal decomposition of wood preservatives

To comprehend the thermal properties (thermal decomposition, decomposition step, and char) of the three types of preservatives, the preservatives samples were measured by TGA, after having been oven-dried. All runs were purged in an air atmosphere, at a flow rate of 50 mL/min, with 2 kinds of heating rate, at 5 °C/min (as a slow-heating regime) or at 40 °C/min (fast-heating regime), in order to isolate the decomposition step. That is to say, the slow-heating and fast-heating regimes were applied to compare the differences in their thermal properties and their char. The relationships of the TGA and the DTG curves between temperature and weight loss, and the derivative weight for each preservatives sample, creosote, CCA and ACQ, are shown in Fig.11.

The results obtained that regardless of the heating rate, the three types of wood preservatives typically showed gradual weight loss and the decomposition temperature of each of the preservatives for the heating rate at 5 °C/min was lower than that at 40 °C/min. Regardless of different heating rate, the char (wt %) of the CCA was more than that of the ACQ, and the amount of char for the creosote was lower than that for either of the water-based preservatives. Hirata et al. (1993) reported that Cr and Cu from the CCA preservatives were largely present in the char and that arsenic (As) was easily volatilized with the increase in temperature during combustion. Some researches (Kercher & Nagle, 2001; Cooper et al., 2003; Helsen & Bulck, 2005) stated that in the TGA tests, when CCA preservatives were analyzed with air, AS_2O_5 was formed in a gaseous state and that it disappeared at a temperature of over 600 °C. In addition, CuO produced Cu_2O when the temperature reached about 800 °C both in air atmosphere, CO was then further volatilized and only Cu remained. Also, Cr (III) is a stable element with an unproductive thermal decomposition (Lin et al., 2006). In other words, it is evident that after combustion the Cr and Cu from CCA are remained. The results of ACQ show that the weight loss drastically decreases in air at either heating rate. The composition temperature for the heating rate of 5 °C/min was lower than that of 40 °C/min. These results are the same as reported previously (Lin et al., 2006). ACQ preservatives, a combination of copper and organic biocides, indicated that the main residual element was Cu (Lin et al., 2007), that is, the other elements were volatilized with the increase in temperature during combustion, such as BKC, one of the main compounds used in this study for ACQ. Therefore, the char of CCA was higher than that of ACQ (the top of Fig. 11), based on the different heating rates (5 or 40 °C/min) set.

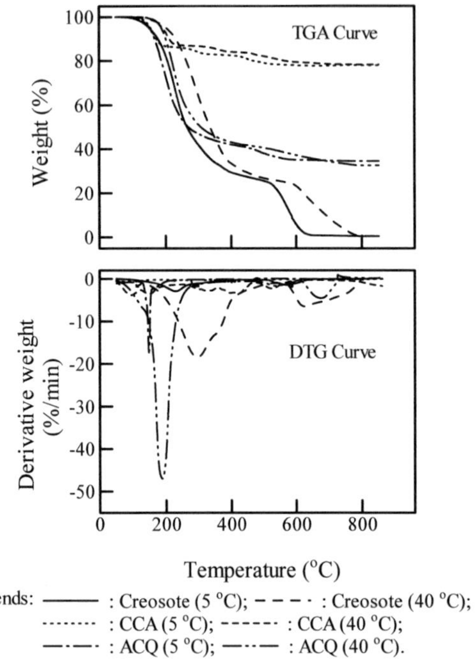

Fig. 11. TGA and DTG curves for 3 types of wood preservatives (Creosote, CCA and ACQ) in air at 5 or 40°C/min.

3.3.3 Evolved species and char of wood preservatives

Because the thermal decomposition of all combustible materials produces a toxic smoke, and combustion toxicology indicates adverse effects produced from exposure to fire-generated toxic species (Hartzell, 1996), the understanding of the evolved species and the char of wood preservatives is analyzed. Figure 12 shows the infrared spectrometric (IR) response of the three types of preservatives for both heating rates, at 5 °C/min and 40 °C/min in the TGA-IR measurements.

The change in absorbance (concentration of evolved species, %) corresponds to the increase in decomposition time for the three preservatives. Comparing the IR response of all preservatives at a heating rate of 5 °C/min, the absorbance levels for creosote were higher than for CCA and ACQ. This indicated that the concentration of evolved species of water-based preservatives was less than those of oil-based ones. Of them all, CCA was inferred to be the best preservatives with the least environmental impact, if the disposal process of the preservatives was combustion and considering the results of air pollution only. It is indicated that CCA is the better environmentally friendly preservatives for wood protection, because the emission contents of CCA are less than those of ACQ, considering their environmental impact only during combustion.

Furthermore, the right side of Fig. 12 also shows the IR response of the three types of wood preservatives with a heating rate of 40 °C/min in the TGA-IR measurements. The change in concentration of evolved species corresponds to the increase in decomposition time for the three preservatives. The IR response of ACQ showed a greater peak at the end than that of creosote and CCA. This suggests that at 40 °C/min of the heating rate, the concentration of evolved species for ACQ is higher than that for either creosote or CCA. Comparing the IR

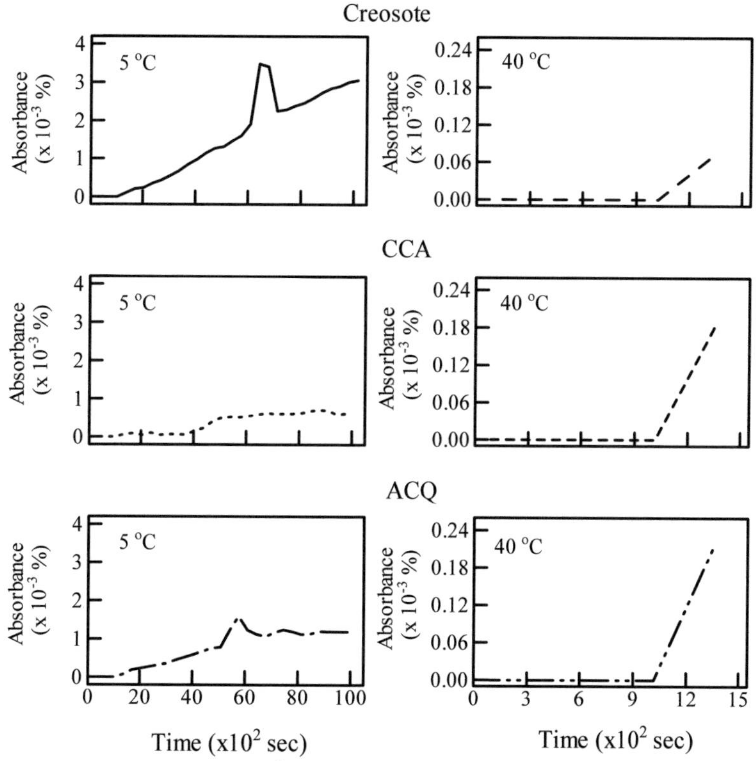

Fig. 12. IR response of 3 types of wood preservatives in air at 5 or 40°C/min from TGA-IR analysis.

response of all preservatives at both heating rates, the absorbance levels at 5 °C/min for oil-based preservatives were higher than for water-based one, but at 40 °C/min the results were reversed. This indicates that for wood preservatives the concentration of evolved species is related to the heating rate, because of the variation in decomposition steps.

At present much attention is being given to the use of wood preservatives, because of the fact that during combustion they emit toxic gases and produce a corrosive smoke. Exposure to these toxic emissions is harmful to humans and in addition can cause damage to expensive equipment (Ladomersky, 2000; Lin et al., 2007). With the use of TGA-IR, it is possible to simultaneously monitor evolved species, decomposition temperature and the concentrations formed (absorbance), as well as the evolution profile of certain compounds. Moreover, IR spectrometry is a common means of evaluating a functional base when a material undergoes degradation. The evolution of decomposition time (or temperature) on absorbance and wavenumbers (basically, IR spectra) for three types of preservatives is shown in Fig. 13. These results indicate that from the TGA-IR response the evolved species for the slow-heating regime (5 °C/min) were less than those for the fast-heating regime (40 °C/min), even for the different preservatives.

For the creosote, the CO_2 peak, located at 2296 and 2356 cm^{-1}, started showing from about 3600 sec (at a temperature of about 349 °C in the TGA tests) to the end for a heating rate of 5 °C/min and was steady from beginning to the end for a heating rate of 40 °C/min. The other CO_2 peak was located at 617 cm^{-1} for the heating rate of 5 °C/min, and was from about 2400 sec (249 °C) to the end, but at 666 cm^{-1} for a heating rate of 40 °C/min the other CO_2 peak was from about 400 sec (82.5 °C) to the end. In addition, the peaks at 1585 and 3780 cm^{-1} were due to the N-H stretching of NH_2 moieties (Zhu and Shi, 2001) from about 1200 sec (149 °C) to the end for the heating rate of 5 °C/min. A higher concentration of N-H stretching was seen between 6000 and 6600 sec (549 to 599 °C). For the heating rate of 5 °C/min the range from 5400 to 9000 sec (499 to 799 °C) showed a slight peak at 718 cm^{-1}, which was attributed to the groups of - CH_2 - rocking and another slight peak, at 3670 cm^{-1}, was attributed to the OH group, from about 1800 sec (199 °C) to the end. However, for the heating rate of 40 °C/min, some of the weak peaks, at 3741, 1794, 1316 and 814 cm^{-1}, were due to the OH group, C=O (carbonyl group) stretching, the group of C-NO_2 stretching and N-H deformation vibration of NH_2 moieties, all ranging from the beginning of the combustion to the end.

For CCA, these results indicate that the fast-heating regime presented many more evolved species. The absorbance peaks and the time location (including the decomposition temperature) of CO_2 at 2296 and 2356 cm^{-1} were from about 2400 sec (249 °C) to the end for the heating rate of 5 °C/min, but from the beginning to the end for the heating rate of 40 °C/min. C=O stretching at 1725 cm^{-1} was from about 1200 sec (149 °C) to the end for the heating rate at 5 °C/min and from the beginning to the end for the heating rate of 40 °C/min. One of the lower peaks was from the beginning to the end for the heating rate of 5 °C/min and from about 200 sec (176 °C) to the end for the heating rate of 40 °C/min.

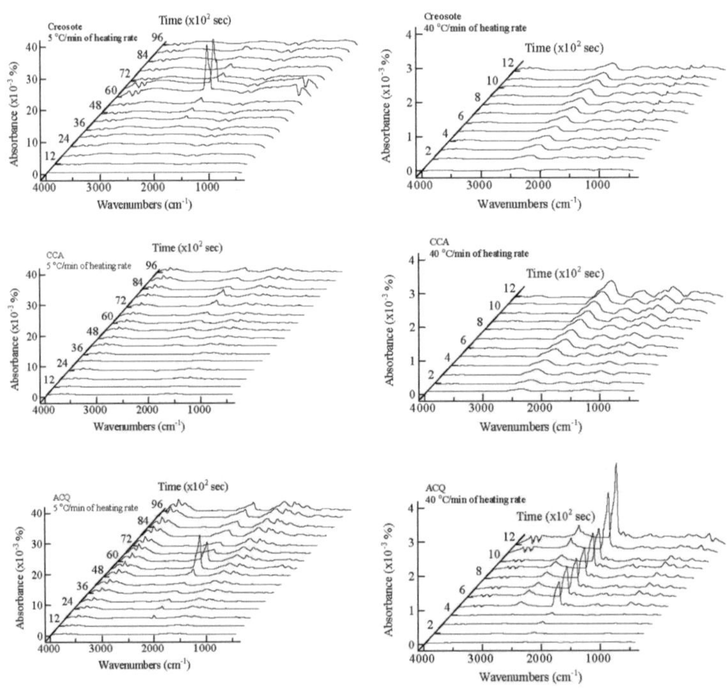

Fig. 13. TGA-IR spectra of 3 types of wood preservatives in air with 5 or 40°C/min.

This is attributed to N-H stretching at 3780 cm^{-1}. Other low peaks for the groups of - CH_2 - rocking at 710 to 718 cm^{-1} from the beginning to the end showed up in the slow and faster-heating regimes. Except those evolved species, several weak peaks, for the heating at a rate of 40 °C/min only, at 1805, 1437, 1300, 932 and 817 cm^{-1} emerged due to $=CH_2$, OH groups, C-NO_2 stretching, -OH out-of plane wagging, -NH_2 moieties etc.

For ACQ at the heating rate of 5 °C/min (bottom left in Fig. 13), the peaks of CO_2 were located at 2296 and 2356 cm^{-1} and began from about 1800 sec (199 °C) to the end of the tests, and the highest absorbance showed at 5400 sec (499 °C). The other CO_2 weak peak, located at 666 cm^{-1}, was from about 2400 sec (249 °C) to the end and a higher concentration of CO_2 was also seen at that point (5400 sec.). Another wavenumber at 3747 cm^{-1} was assigned to the N-H stretching of NH_2 moieties (Zhu and Shi, 2001) from about the beginning to the end and the highest point was located at the end. For the heating rate of 40 °C/min, the CO_2 peak was located at 2296 and 2356 cm^{-1} from the beginning to about 1210 sec (850 °C). The range from about 400 sec (310 °C) to the end showed two weak peaks of absorbance at 3609 cm^{-1} and 3747 cm^{-1}, which were attributed to the N-H stretching of NH_2 moieties and the OH groups. In addition, several weak peaks, for the heating at a rate of 40 °C/min only, at 1731, 1385, 1273, 1075, 930, and 716 cm^{-1} were due to C=O stretching, -CH_3, N-NO_2, C-O and -OH out-of plane wagging, and groups of - CH_2 - rocking etc. from about 400 sec (310 °C) to the end, during the period of the TGA-IR tests.

From the above results, it is suggested that using TGA-IR, to analyze the thermal behavior of wood preservatives, in combination with TGA analysis, has led to useful results, involving the identification of pyrolysis products.

4. Conclusion

To realize fundamental knowledge of the thermal decomposition of the wood-based waste processing-materials, including wood adhesives, wood coatings and wood preservatives, while at the same time understanding an air pollution reference for emission gases, evolved species and a disposal-end practice (of the residual elements) reference for the char. This chapter is mainly focused on examining the emission gas concentration and the fundamental thermal decomposition, as well as analyzing the evolved species and the char of wood-based waste processing-materials. The following results were obtained.

By using the CEM, the larger or greater these materials are, the greater the emission contents and the longer the combustion time. The concentration of O_2 was negatively and CO_2 was positively related to the emission gas temperature. Both O_2 and CO_2 had a close relationship during combustion. The main pollutants during combustion were from the ingredients of these materials, and the incomplete combustion was one of main variable factors related to the quantities of emission content (CO) and change O_2 and CO_2 concentration in the air. SO_2 was not produced during combustion for these materials because their ingredients did not have a sulfur-base. The emission contents of CO, SO_2 and NOx were found that for the adhesive, the PB manufactured with PF adhesive had the least pollution gas during the combustion than the others. This suggested that PF is the best adhesive with the least environmental impact to use for manufacturing PB. For wood coatings and wood preservatives, the emission contents were still produced after having been accelerated weathering degradation, but they were relatively a decreasing tendency. Furthermore, the results of FTIR, when analyzed to the degradation of the coatedwoods for referencing the emissions of the coatings, indicated that the change of various functional bases or/and molecular structures were occurred. The result of the EA indicated that prior to combustion, the amount of nitrogen (N) is higher for the ACQ-treated woods than that for the CCA ones; therefore, the NOx gases during combustion were higher for the ACQ-treated woods.

By using the TGA, the results of wood adhesives found that MDI exhibited the highest initial decomposition and better overall thermo-stability than the other types of adhesives. PF and MDI provided a better yield of char and their thermal behavior was better than that of UF or MUF. The thermal properties of CCA- or ACQ-treated wood were similar to that of untreated wood in either nitrogen or air atmosphere, but the pyrolysis temperature for the slow-heating regime was lower than that of fast-ones. Regardless of different atmosphere or heating rate, the char of the CCA-treated wood and the CCA was more than that of the ACQ ones and the ACQ. The EDX analysis of both-treated woods indicated that the relative proportion of carbon was higher after thermal decomposition. Chromium (Cr) from the char of CCA is a stable element, and the amount of Cr is obviously more than that of Arsenic (As) and copper (Cu), and the char of ACQ left a high amount of Cu.

By using the TGA-IR, the IR response of the adhesives showed that the absorbance of UF, MUF and MDI was higher than that of PF; this indicated that the concentration of evolved species of PF was less than that of MDI, and less than that of UF and MUF. These results, when compared to the molecular structures of the adhesives, enabled us to point out the kinds of evolved species (functional bases or the change in molecular structures) for referencing the thermo-degradation processes of the adhesives. The preservatives showed that considering the decomposition step (thermal time or temperature) at different heating rates, the results obtained showed that at the slow-heating regime, the concentrations of CO_2 produced by creosote and ACQ were higher than that of CCA, but at the fast-heating regime, the concentration of CO_2 for ACQ was higher than that of creosote and CCA. Moreover, these results, when compared to the chemical ingredients of the preservatives and EDX analysis, enabled us to point out the kinds of evolved species for referencing the thermo-decomposition processes of the preservatives before/after time degradation.

The use of CEM for investigating the emission gases, TGA for analyzing the thermal properties of the wood-based waste processing-materials, in combination with TGA-IR, FTIR, EDX and EA analysis, has led to useful results, regarding the identification of thermal decomposition products (evolved species and char).

5. Acknowledgment

The author is grateful to Dr. Yasuhide Murase, Professor Emeritus of Kyushu University, Japan for his research assistance, and offers my sincere appreciation to the National Science Council, Taiwan through Grant No. NSC 94-2313-B-415-008 for experimental financial support.

6. References

Akiyama, T. (1998). *Wood Recycles*, Sunchoh Publication, ISBN 4-88282-186-9 C3052, Japan, pp. 7-29

Allaby, M. (2000). *Basics of Environmental Science 2ND*, Routledge, ISBN 9780203147085, Australia, pp. 44-50

Aplin, G.; Beggs, P.; Brierley, G.; Cleugh, H.; Curson, P.; Mitchell, P.; Pitman, A. & Rich, D. (2000). *Global Environmental Crises Second Edition*, Oxford, ISBN 0-19-550827-0, Australia, pp. 30-60

Bras, B. & Román, F. (2006). An Introduction to Life Cycle Analysis/Assessment (LCA), In: *LCA intro*, Available from http: www.srl.gatech.edu/education/ME4171/LCA_intro .ppt

Camino, G.; Costa, L. & Luda, M. P. (1993). Mechanistic aspects of intumescent fire retardant system. Makromol. Chem., *Macromol. Symp.*, Vol.74, pp. 71-83, ISSN1022-1360

Chang, S. T. (1990). Chemical Characterization of Woody Materials and Coating Using Fourier Transform Infrared Spectroscopy, *Q. Jour. Chin. For.*, Vol.23, No.1, pp. 87-103, ISSN 0578-1345

Chang, S. T. (1994). Study on the Lightfastness and the Weatherability of Finishes on Wood Surface, *Forest Productions Industries, Taiwan ROC.*, Vol.13, No.2, pp. 275-294, ISSN 0254-6523

Chang, S. T. & Chou, P. L. (1999). Photo-Discoloration of UV-Curable Acrylic Coatings and the Underlying Wood, Polymer Degradation and Stability, Vol.63, pp. 435-439, ISSN 0141-3910

Chen, C. M. (1996). *Wood adhesives*, National Institute for Compilation and Translation, Taiwan, ROC, pp. 280-286

Connell, M. (1991). Industrial Wood Preservatives - The History, Development, Use, Advantages and Future Trends. In: *Thompson R, ed. The chemistry of wood preservation*, pp. 16-34, Cambridge, Royal Society of Chemistry

Cooper, P. A. (2001). Analysis of Consumer Lumber Waste Management Options, *Final report prepared for Wood Preservation Strategic Options Process*, Manufacturers / Treaters Steering Committee and Consumer Waste Lumber Working Group

Cooper, P.; Ung, T.; Kazi, F. & Qi, D. (2003). Two Approaches of CCA Treated Wood: Extraction for Recycling and Wood Cement Composites, *Presented at the AWPA Annual Meeting*, Boston, MA

Cox, C. (1991). Chromated Copper Arsenate, *J. of Pesticide Reform*, Vol.11, No.1, pp. 2-6

Esser, P.; Eggels, P. & Voss, A. (2000). Waste Management of Wood Products in Life Cycle Assessment, *Presented at the 31st Annual IRG Meeting*, Kona, Hawaii, US, IRG/WP 00-50154

Falk, B. (1997). Opportunities for the Woodwaste Resources, *Forest Products J.*, Vol.47, No.6, pp. 17-22, ISSN 0015-7473

Hartzell, G. E. (1996). Overview of combustion toxicology, *Toxicology*, Vol.155, pp. 7-23, ISSN0300-483x

Helsen, L. & Bulck, E. V. D. (1998). The Microdistribution of Copper, Chromium and Arsenic in CCA Treated Wood and Its Pyrolysis Residue Using Energy Dispersive X-ray Analysis in Scanning Electron Microscopy, *Holzforschung*, Vol.52, No.6, pp. 607-614, ISSN 0018-3830

Hirata, T.; Inoue, M. & Fukui, Y. (1993). Pyrolysis and Combustion Toxicity of Wood Treated with CCA, *Wood Sci Technol*, Vol.27, pp. 35-47, ISSN 1435-0211

Hornsby, P. R. (1996). The application of hydrated mineral fillers as fire retardand and somke suppressing additives for polymers, *Macromol Symp.*, Vol.108, pp. 203-219, ISSN 1022-1360

Humphrey, D.G. (2002). The Chemistry of Chromated Copper Arsenate Wood Preservatives, *Reviews in Inorganic Chemistry*, Vol.2, No.1, pp. 1-40, ISSN 0193-4929

Hsu, C. K.; Cheng, K. B. & Lee, J. S. (2000). The applications of the thermal analysis techniques on polymer materials, *J. Spinning and Weaving Engineering, Taiwan ROC*, Vol.8, pp. 43-58

Ibach, R. E. (1999). *Wood Handbook - Wood as an Engineer Material. Chapter 14 Wood preservatives*, Forest Products Society, ISBN 1-892529-02-05, USA, pp. 14-2 - 14-9

Jang, Y. M. (1997). *Waste Processing*. WEN-JING Book Corp, ISBN 957-512-726-9, Taipei, pp. 49-85

Jang, Y M. (2000). *Technique of Burning Process*, WEN-JING Book Corp, ISBN 957-512-264-x, Taipei, pp. 213-221, 421-430

Kercher, A. K. & Nagle, D. C. (2001). TGA modeling of the thermal decomposition of CCA treated lumber waste, *Wood Science and Technology*, Vol.35, pp. 325-341, ISSN 0043-7719

Kobayashi, N.; Thikae, M. & Tominaga, Y. (2000). The properties and manufacture of renewable board from particleboard. In: *Abstracts of the 50th Annual Meeting of the Japan wood Research Society*, Japan Wood Research Society, Kyoto, Japan, April 3-5, 2000, pp. 285

Ladomersky, J. (2000). Emission analysis and minimization from the wood waste combustion, *Drevarsky Vyskum*, Vol.45, No.4, pp. 33-44, ISSN 0012-6136

Lee, H. L.; Shiah, T. C.; Hsu, F. L. & S. Lin, J. (2005). Effects of New Waterborne Wood Preservatives on the Durability and Thermal Properties of Wood, *Taiwan J. For Sci.*, Vol.20, No.2, pp. 139-156, ISSN 1026-4469

Levchik, S. V.; Camino, G.; Luda, M. P.; Costa, L.; Costes, B.; Henry, Y.; Morel, E. & Muller, G. (1995). Mechanistic study of thermal behavior and combustion performance of epoxy resin: I Homopolymerized TGDDM, *Polymer for Advanced Technologies*, Vol.6, No.2, pp. 53-62, ISSN 1099-1581

Lay, K. Y. (1985). *Coatings and Finishing in Practice*, Fu Han Publication, Taiwan, pp. 84-107

Li, Y. D.; Yan, W. Y. & Juang, T. S. (1993). *Property of Polymer Substance*, KOU-LIH Book Corp Taipei, pp. 325-328

Lin, H. C.; Huang, J. C.; Fujimoto, Y. & Murase, Y. (2001). Analysis of gases emitted from particleboard combustion, *Forest Products Industries, Taiwan ROC.*, Vol.20, No.2, pp. 165-174, ISSN 0254-6523

Lin, H. C. & Huang, J. C., (2004). Analysis of Combustion Emissions from Wood Coatings after QUV Degradation, *J. Agric. For., NCYU*, Vol.1, No.1, pp. 50 – 68, ISSN 1029-0206

Lin, H. C.; Ohuchi, T. & Murase, Y. (2004). Estimating Thermal Behavior and Analyzing Evolved Species of Adhesives through Thermogravimetric Analysis combined with Spectrometric Techniques, *J. of Fac. Agr., Kyushu Univ.*, Vol.49, No.2, pp. 449 – 459, ISBN 0023-6152

Lin, H. C. (2005). Potential Application of FTIR to Combustion Emissions Analysis of Degradation Coatedwood. *J. NCYU*, Vol.75, pp. 15-40, ISSN 1029-0206

Lin, H. C.; Ohuchi, T.; Murase, Y.; Shiah, T.-C.; Gu, L. T.; Lee, M. J. & Wu, Y. D. (2006). Application of TGA and EDX Analysis to Evaluate the Process of Preservative-Treated Woods, *J. of Fac. Agr., Kyushu Univ.*, Vol.51, No.2, pp. 337-344, ISBN0023-6152

Lin, H. C., Ohuchi, T. & Murase, Y. (2007). Analysis of Combustion Emissions and Char from CCA- and ACQ-Treated Woods, *J. of Fac. Agr., Kyushu Univ.*, Vol.52, No.1, pp. 91-98, ISBN 0023-6152

Lin, J. F. (1992). *Precaution of Special Chemical Substance*, Labors Department in Taiwan, pp. 105-107

Nicholas, D. D. & Schultz, T. P. (1994). Biocides That Have Potential as Wood Preservatives - An Overview. *Proceeding from the Conference sponsored by the Forest Products Society*, Georgia, USA, pp. 169-173

Okuma, T. (1998). *Wood-Based Resource Materials*. Tokyo University Publication, Japan, pp. 17-20

Pouchert, C. J. (1981). *The Aldrich Library of Infrared Spectra Edition III*, Aldrich Chemical Company INC., ISBN10-9999780024, Wisconsin, USA. pp. 233, 435-436, 1065

Reghunadhan Nair, C. P.; Bindu R. L. & Ninan K. N. (2001). Thermal characteristics of addition-cure phenolic resins, *Polymer Degradation and Stability*, Vol.73, pp. 251-257, ISSN 0141-3910

Sanders, J.G. & Windom H. L. (1980). The Uptake and Reduction of Arsenic Species by Marine Algae, *J. Environmental science*, Vol.10, pp. 555-567, ISSN 1001-0742

Sanders, J. G. & Riedel G. F. (1987). Control of Trace Element Toxicity by Phytoplankton, *Recent Adv Phytochem*, Vol.21, pp. 131-149, ISSN 9780080451251

Shirokae, S, & Samezima I. (1996). *Wood Science Series 4 Chemistry*, Kaiseisha, ISBN 978-4-906165-43-8, Japan pp. 15-19

Solo-Gabriele, H. M.; Townsend T. G. & Schert J. (2003). Environmental Impacts of CCA Treated Wood: A Summary from Seven Years of Study Focusing on the US Florida Environment, *Presented at the 34th Annual IRG Meeting*, Brisbane, Australia, IRG/WP 03-50205

Statheropoulos, M. & Kyriakou S. A. (2000). Quantitative thermogravimetric-mass spectrometric analysis for monitoring the effects of fire retardants on cellulose pyrolysis, *Analytica Chimica Acta.*, Vol.409, pp. 203-214, ISSN 0003-2670

Takabashi, T.; Suzuki, M. & Nakano, T. (1995). *Wood Science Series 5 Environment*, Kaiseisha, ISBN 978-4-906165-89-6, Japan pp. 7-15.

Weis, J. S. & Weis P. (1922a). Transfer of Contaminants from CCA-Treated Lumber to Aquatic Biota, *J Exp Mar Biol Ecol*, Vol.161, pp. 189-99, ISSN 0022-0981

Weis, J. S. & Weis P. (1922b). Construction Materials in Estuaries Reduction in the Epibiotic Community on Chromated Copper Arsenate (CCA)-treated wood, Mar *Ecol Prog Series*, Vol.83, pp. 45-53, ISSBN 0171-8630

Weis, J. S. & Weis P. (1996). Reduction on Toxicity of Chromated Copper Arsenate (CCA)-Treated Wood as Assessed by Community Study, *Mar Environment Res*, Vol.41, pp. 15-25, ISSN 01411136

Weis, J. S. & Weis P. (1999). Accumulation of Metals in Consumers Associated with Chromated Copper Arsenate-Treated Wood Panels, *Mar Environment Res*, Vol.48, pp. 73, ISSN 01411136

Zhu, S. & Shi W. (2001). Hyperbranched polyurethane acrylates containing phosphate as radiation curable flame-retardant coating material, *Conference Proceedings, China*, pp. 108-114

15

An Assessment of the Effectiveness of California's Local Air Pollution Controls on Agricultural Sources

C.-Y. Cynthia Lin
University of California at Davis
USA

1. Introduction

Air pollution has been recognized as a significant environmental problem in California since the early 20th century. Between 1905 and 1912, regulations were enacted by the city council of Los Angeles to regulate emissions [1]. As motor vehicle traffic increased, a new type of smog was observed and described: the "Los Angeles" or photochemical smog, as distinct from the "London" smog that resulted from coal combustion. Arie Haagen-Smit from Caltech characterized the chemistry of this smog and identified ozone as the principal oxidant in the early 1950's. Meanwhile, the first air district in the U.S. was created in Los Angeles in 1947, and was later merged with other local districts in 1977 to form the South Coast Air Quality Management District [1].

Negative effects of air pollution have been extensively documented, and include impairment of human lung function, degradation of materials, and injury to plants. In addition to affecting human health, the high ambient ozone levels found in Southern California and the San Joaquin Valley also cause yield reductions up to 30% for some crops [2].

In addition to having some of the nation's most polluted air basins, California also has the nation's most stringent set of state and local air quality standards. Although regulation has led to improvements in air quality [3], exceedances of air quality standards still take place. For example, between 1990 and 1998, the San Joaquin Valley Air Basin experienced an average of 97 days per year above the eight-hour ozone standard, while the Sacramento Valley Air Basin experienced an average of 30 days per year above the standard during the same time period [4].

Farming and livestock operations are significant sources of emissions in California, and bear the negative effects of specific air pollutants as well. Agriculture-related air pollution results from primary emissions from machinery and vehicles employed in production, chemical compounds used in the course of production, e.g. pesticides, as well as emissions from the agricultural systems themselves. For example, agricultural livestock emit nitrogen compounds such as oxides of nitrogen (NO_x) and ammonia. Vehicles used in agricultural production emit volatile organic compounds (VOCs), NO_x and carbon monoxide (CO) [5]. These emissions may lead to the formation of secondary air pollutants, such as ozone, that are deleterious to workers as well as crops [6].

This chapter evaluates whether existing air pollution control policies, particularly those targeted at agriculture, have succeeded in improving air quality, as measured by the number of exceedances of the CO and NO_2 standards. The following air pollution control policies are examined: policies and regulations for agricultural burning, visible emissions, fugitive dust, emission of particulate matter (PM) and PM precursors, emissions of nitrogen compounds, orchard and citrus heaters that release black carbon, and penalty fees.

This chapter builds upon the existing environmental economics literature on air quality, most notably the econometric analysis of the impact of federal particulate matter regulation on infant health conducted by Chay and Greenstone [7] and the study of the impact of air pollution on infant death in California by Currie and Neidell [8], in several ways. First, this paper focuses on the effects of regulation rather than on the effects of air quality. The results therefore have direct implications for policy. Second, the econometric methodology used in this paper exploits the natural variation in policy among the different air districts in California to identify the effects of these policies. Third, this paper examines multiple policies, not just one.

Results from the multivariable regressions point to mixed effects of the air pollution control policies on air quality. Agricultural burning policies and penalty fees reduce the pollution from CO. Other policies such as the prohibition on visible emission, fugitive dust, particulate matter, nitrogen and the reduction of animal matter are correlated with higher levels of CO. Regulations on orchard and citrus heaters have no significant effect on the number of exceedances of the CO and NO_2 standards.

Results of this research will lead to a better understanding of the regulations affecting air quality, and will provide insight into the appropriate development of management practice to mitigate air pollution problems.

The remainder of the chapter proceeds as follows. The next section provides background information on the state of California with respect to air quality management. Section 3 describes the data. Section 4 presents the methods and results from the econometric analysis. Section 5 concludes.

2. California's air pollution policy

The state of California is divided into fifty-eight different counties and is overseen by the Governor of California. California is divided into thirty-five air districts, which are called either Air Pollution Control Districts or Air Quality Management Districts. These air districts are responsible for controlling air pollution from stationary sources. Several air districts span the areas of many counties, while some counties belong to different air districts. Some air districts, such as the Great Basin Unified air district, trace the division lines marking the regional air basin [9].

The different air districts in California have their own set of laws and regulations regarding stationary sources. These laws and regulations are written by the authorities within the air district and are applicable to the entire air district. These laws and regulations must be at least as stringent as the standards set by the federal government [10].

The fact that each air district has its own set of laws and regulations is crucial to the analysis of the policies presented in this paper. The variation between the policies of the different air districts within the state of California provides a setting for a natural experiment. Since the different air districts all fall in the same state, they are subject to the same federal and state laws. The districts also share many other similar characteristics, for example climate, geographical location, etc. The variation in air quality over the time the different policies take place, then, must be due mostly to the policies themselves and not due to geographical location, climate or different state laws. Thus, the variation in similar policies across air districts enables one to better single out the effect of policies on air quality and health.

Among the many different laws and regulations governing each of the thirty-five air districts in California, this paper focuses on the following eight types of policy:

Agricultural burning: This policy regulates open outdoor fires used in agricultural operations in the growing of crops, the raising of animals, the disposal of agribusiness waste, or for purposes such as forest management, range improvement, irrigation system management, etc. The policy requires burning permits and imposes no-burn days.

1. *Visible emissions*: This policy provides limits for visible emissions. In many districts, emissions from agricultural operations are exempt.
2. *Fugitive dust*: The purpose of this policy is to reduce the amount of particulate matter entrained in the air as a result of anthropogenic fugitive dust sources by requiring actions to prevent, reduce or mitigate fugitive dust emissions.
3. *Particulate matter*: This policy imposes limits on particulate matter emissions.
4. *Nitrogen*: This policy imposes limits on NO_x emissions.
5. *Reduction of animal matter*: This policy requires that the gases, vapors and gas-entrained effluents from any article, machine equipment, or other contrivance used for the reduction of animal matter to be incinerated or processed.
6. *Orchard and citrus heaters*: This policy regulates orchard and citrus heaters.
7. *Penalty fee*: Stationary sources with the potential to emit regulated pollutants (including nitrogen oxides, VOCs, CO and PM10) above a certain amount need to obtain permits to operate consistent with the requirements of Title V of the federal Clean Air Act as amended in 1990. This policy requires operators of units requiring Title V permits to pay a penalty if they fail to pay for their permit within a certain number of days after it is issued.

These eight policies were chosen because they were similar in theme across the different counties in the different air districts but they either vary in whether they were implemented at all, or differ in the date of implementation. For example, the prohibition on fugitive dust applies in Amador and Imperial Counties but not in El Dorado and Monterey Counties. Of Amador and Imperial Counties, the dates of implementation of the prohibition policy differ. In Amador County the policy did not take effect until the year 2000, while the policy was implemented in Imperial County prior to 1994 [11].

The similarity of theme in the different policies chosen and the difference in the details of each of the policy add to the quality of the analysis. Since the policies are similar in theme but are different in details, comparisons can be made and the effect of each policy can be examined. For example, once controls are taken into account, differences between air quality in Amador County and El Dorado County can be said to be attributable to the prohibition on fugitive dust which is effective in Amador County but not in El Dorado County. Likewise, the differences in air quality in Amador County and Imperial County can be attributed to the differences in the date of implementation of the prohibition on fugitive dust.

3. Data description

We use annual county-level data from 1980-2000.

The policy variables used in this paper are constructed from the California Air Resources Board's online database of state and county laws and regulations concerning air quality in the different air districts [11]. For each of the eight policies chosen, dummy variables for whether or not the policy is in place for each county for each year since 1980 to 2000 were constructed. Summary statistics for the policy variables are presented in Table 1.

Policy variable	**# obs**	**mean**	**s.d.**
Visible Emissions	1178	0.698	0.459
Fugitive Dust	1178	0.338	0.480
Particulate Matter	1178	0.690	0.463
Nitrogen	1178	0.397	0.490
Reduction of Animal matter	1178	0.518	0.500
Orchard and citrus heaters	1178	0.296	0.457
Agricultural Burning	1537	0.856	0.351
Penalty Fee	510	0.720	0.450

Table 1. Summary statistics for policy variables

To measure air quality, this paper focuses on two agriculture-related air pollutants: CO and nitrogen dioxide (NO_2). For each pollutant, data on the number of days per year exceeding the state standards as compiled by the California Air Resources Board are used. In particular, for CO, the number of days exceeding the state's 1-hour standard for CO of 20 parts per million (ppm) and the number of days exceeding the state's 8-hour standard for CO of 9.0 ppm are used. For NO_2, the number of days exceeding state's 1-hour standard for NO_2 of 0.18 ppm is used. Summary statistics for the air quality variables are in Table 2. On average 0.696 days exceeded the state standard for 1-hour CO, 4.452 days exceeded the state standard for 8-hour CO, and 0.542 days exceeded the state standard for 1-hour NO_2. The number of days exceeding the state standard for all three measures of air pollution has decreased significantly over time.

Air quality variable	**# obs**	**mean**	**s.d.**	**min**	**max**	**trend**
# days exceeding the state standard for:						
CO, 1-hour	1187	0.696	3.283	0	43	-0.999* (0.014)
CO, 8-hour	1187	4.452	14.631	0	133	-0.524* (0.063)
NO_2, 1-hour	1069	0.542	3.273	0	42	-0.089* (0.015)

*Note: The "trend" column gives the coefficients on year of the regression of air quality on time and a constant. The standard errors are given in parentheses under the coefficients. * indicates the coefficient is significant to the 5% significant level.*

Table 2. Summary statistics for air quality variables

The socio-economic data used in this paper are obtained from the U.S. Bureau of Health Professions' Area Resource File. Socio-economic data used in this paper include population, and the percentage of county area occupied by farms, and per capita income. Summary statistics for these variables are presented in Table 3. Both population and per capita income have a significant upward trend over time, while there is no significant trend over time for the percentage of county area occupied by farms.

Variable	Obs	Mean	Std. Dev.	Min	Max	Trend
Population (thousand people)	1102	291.97	971.64	0.011	9519.34	45.78* (5.25)
Percentage of area that are farms	1218	378.3276	259.4809	0	922	0.011 (1.115)
Per capita income	1160	17164.32	6748.864	6807	57982	804.273* (21.071)

Table 3. Summary statistic for socio-economic variables

4. Econometric analysis

In order to identify the effects of the chosen air quality policies on air quality, multivariable regressions that exploit the natural variation in policy among the different air districts in California are used.

It is possible that the correlation between the regulations and air quality reflect some omitted characteristics such as socio-economic status that are correlated with both regulations and air quality. To address the possibility of omitted variables bias, we use a method similar to that used by Currie and Neidell [8]: we include a rich set of covariates and employ county fixed effects to capture any unobserved characteristics of counties that are constant over time. The control variables used are year, population, population density, per capita income, and acres of farmland. The Hausman test was used to determine whether controlling for fixed effects was more appropriate than controlling for random effects.

The natural variation of policy among air districts mitigates the potential endogeneity of regulation, since two neighboring counties that may share similar characteristics and that may have similar levels of pollution prior to the implementation of a particular policy, all else equal, may still belong to different air districts.

The econometric model is:

$$pollution_{it} = \beta_0 + \beta_1 policy_{it} + x_{it}'\beta_2 + u_i + \varepsilon_{it} \qquad (1)$$

where $pollution_{it}$ is the value of the pollution variable (number of days exceeding the state's 1-hour standard for CO, the number of days exceeding the state's 8-hour standard for CO, or the number of days exceeding state's 1-hour standard for NO_2) in county i in year t; $policy_{it}$ is a dummy variable denoting whether or not the particular policy under consideration was in place in county i in year t; x_{it} is a vector of controls (year, population, population density, per capita income, and percentage of land area that are farms); and u_i is a county fixed effect.

The results for the effects of policy on air pollution are presented in Table 4. Because the results of the Hausman test favored the fixed effects model for most of the regressions, only the fixed effect results are reported.

Policy variable	*Dependent variable is # days exceeding the state standard for:*		
	CO, 1-hour	CO, 8-hour	NO_2, 1-hour
agricultural burning	-0.777*	-3.013*	-0.788
	(0.226)	(1.123)	(0.306)
p-value (Pr >F)	0.00*	0.00*	0.00*
R^2	0.017	0.014	0.030
# obs	626	626	553
visible emission	0.632*	3.013*	0.138
	(0.238)	(1.178)	(0.351)
p-value (Pr >F)	0.00*	0.00*	0.00*
R^2	0.000	0.003	0.012
# obs	660	660	561
fugitive dust	0.643*	3.725*	0.102
	(0.258)	(1.272)	(0.364)
p-value (Pr >F)	0.00 *	0.00*	0.00*
R^2	0.004	0.004	0.015
# obs	660	660	561
particulate matter	0.632*	3.013*	0.138
	(0.238)	(1.178)	(0.351)
p-value (Pr >F)	0.00*	0.00*	0.00*
R^2	0.001	0.004	0.012
# obs	660	660	561
nitrogen	0.580*	2.424*	0.392
	(0.171)	(0.846)	(0.236)
p-value (Pr >F)	0.00*	0.00*	0.00*
R^2	0.003	0.000	0.015
# obs	660	660	561
reduction of animal matter	0.545*	2.626*	0.020
	(0.234)	(1.158)	(0.334)
p-value (Pr >F)	0.00*	0.00*	0.00*
R^2	0.001	0.000	0.014
# obs	660	660	561
orchard and citrus heaters	0.058	-0.114	0.045
	(0.225)	(1.111)	(0.337)
p-value (Pr >F)	0.00*	0.00*	0.00*
R^2	0.001	0.000	0.013
# obs	660	660	561
penalty fee	-0.221*	-4.495*	0.014
	(0.091)	(0.756)	(0.017)
p-value (Pr >F)	0.00*	0.00*	0.75
R^2	0.043	0.146	0.001
# obs	376	376	378

*Notes: Standard errors are in parentheses. The controls are year, population, population-density, per capita income, percentage of land-area that are farms, and county fixed effects. * indicates the coefficient is significant to the 5% significant level.*

Table 4. The effects of policy on air quality

The results point to mixed effects of the chosen air pollution control policies on air pollutants. An agricultural burning policy significantly reduces the number of days exceeding the state standard for both 1-hour and 8-hour CO. Agricultural burning policies also reduce the number of days exceeding the state standard for NO_2, but the effect is not significant at a 5% level.

The results also suggest that having prohibitions on visible emission, fugitive dust, particulate matter, nitrogen and the reduction of animal matter seem to be significantly correlated with increasing pollution from both 1-hour and 8-hour CO. These policies also increase pollution from NO_2, but the effect is not significant at a 5% level.

The regulations on orchard and citrus heaters has no significant effect on the number of days exceeding the state standard for CO or NO_2.

A penalty fee has a significant negative effect on the number of days exceeding the state standard for both 1-hour and 8-hour CO, but no significant effect on the number of days exceeding the state standard for NO_2.

None of the policies examined had a significant effect on the number of days exceeding the state standard for NO_2.

According to the results, the regulations that were most effective in improving air quality were the regulations on agricultural burning and the penalty fees for noncompliance with the standards.

5. Conclusion

This chapter evaluates whether existing air pollution control policies, particularly those targeted at agriculture, have succeeded in improving air quality, as measured by the number of exceedances of the CO and NO_2 standards. The following air pollution control policies are examined: policies and regulations for agricultural burning, visible emissions, fugitive dust, emission of particulate matter (PM) and PM precursors, emissions of nitrogen compounds, orchard and citrus heaters that release black carbon, and penalty fees.

Results from the multivariable regressions point to mixed effects of air pollution control policies on air quality, as measured by the number of exceedances of the CO and NO_2 standards. Agricultural burning policies and penalty fees reduce the pollution from CO. Other policies such as the prohibition on visible emission, fugitive dust, particulate matter, nitrogen and the reduction of animal matter are correlated with higher levels of CO. Regulations on orchard and citrus heaters have no significant effect on the number of exceedances of the CO and NO_2 standards.

The regulations that were most effective in improving air quality were the regulations on agricultural burning and the penalty fees for noncompliance with the standards.

Results of this research will lead to a better understanding of the regulations affecting air quality, and will provide insight into the appropriate development of management practice to mitigate air pollution problems.

6. Acknowledgment

Kanittha Tambunlertchai provided excellent research assistance. I thank Gary Chamberlain and John Karlik for discussions. I received financial support from the Giannini Foundation for Agricultural Economics, an EPA Science to Achieve Results graduate fellowship, a National Science Foundation graduate research fellowship, and a Repsol YPF - Harvard Kennedy School Pre-Doctoral Fellowship in energy policy. All errors are my own.

7. References

[1] Jacobson, M.Z. (2002). Atmospheric Pollution. Cambridge University Press.

[2] Hall, J.V.; Winer, A.M.; Kleinman, M.T.; Lurmann, F.W.; Brajer, V. & Colome, S.D. (1992). Valuing the health benefits of clean air. *Science*, Vol.255, pp. 812-817.

[3] Lin, C.-Y.C.; Jacob, D.J. & Fiore, A.M. (2001). Trends in exceedances of the ozone air quality standard in the continental United States, 1980-1998. *Atmospheric Environment*, Vol. 35, pp. 3217-3228.

[4] California Air Resources Board [ARB]. (2006). Top 4 summary: Top 4 measurements and days above the standard, September 2006, Available from: http://www.arb.ca.gov/adam/cgi-bin/db2www/adamtop4b.d2w/start)

[5] Canadian Environmental Protection Agency [EPA]. (2004). Environment Canada, July 2004, Available from: http://www.ec.gc.ca/environment_e.html

[6] Winer, A.M.; Olszyk, D.M. & Howitt, R.E. (1990). Air quality impacts on California agriculture, 1990-2010, In: *Agriculture in California: On the Brink of a New Millennium, 1990-2010*, 89-112, University of California Agricultural Issues Center.

[7] Chay, K. & Greenstone, M. (2003). Air quality, infant mortality, and the Clean Air Act of 1970. NBER Working Paper Series (Working Paper No. 10053), Cambridge, MA.

[8] Currie, J. & Neidell, M. (2005). Air pollution and infant health: What can we learn from California's recent experience? *Quarterly Journal of Economics*, Vol.120, No.3, pp. 1003-1030.

[9] California Air Resources Board [ARB]. (2004). Organizations in the California Air Resources Board, July 2004, Available from: http://www.arb.ca.gov/html/org/org.htm

[10] U.S. Environmental Protection Agency [EPA]. (1993). The plain English guide to the Clean Air Act, July 2004, Available from: http://www.epa.gov/oar/oaqps/peg_caa/pegcaain.html

[11] California Air Resources Board [ARB]. (2004). Laws and regulations database, July 2004, Available from: http://www.arb.ca.gov/html/lawsregs.htm

16

Air Pollution Control in Municipal Solid Waste Incinerators

Margarida J. Quina[1], João C.M. Bordado[2] and Rosa M. Quinta-Ferreira[1]
[1]*Research Centre on Chemical Processes Engineering and Forest Products*
Department of Chemical Engineering, University of Coimbra
[2]*Department of Chemical and Biological Engineering, IBB, Instituto Superior Técnico*
Portugal

1. Introduction

Municipal solid waste (MSW) remains a major problem in modern societies, even though the significant efforts to prevent, reduce, reuse and recycle. At present, municipal solid waste incineration (MSWI) in waste-to-energy (WtE) plants is one of the main management options in most of the developed countries. The technology for recovering energy from MSW has evolved over the years and now sophisticated air pollution control (APC) equipment insures that emissions comply with the stringent limits established in developed countries. This chapter shows the role of incineration in WtE processes in the ambit of MSW management, giving an overview of the MSWI technologies and APC devices used for cleaning the gaseous emissions. The main focus is on the key air pollutants, such as dioxins and furans. At the end, the impact of emission on health risks is also briefly considered.

2. Contribution of MSWI in modern solid waste management systems

The waste hierarchy in force in European Union, Directive 2008/98/EC, and in other developed countries sets out the following options for waste management: prevention, re-use, recycling, other recovery (e.g. energy recovery) and disposal. Indeed, nowadays modern systems embrace in general different methodologies aiming as much as possible to achieve sustainable global solutions. Life Cycle Assessment (LCA) tools have been used to assess the potential environmental burdens of different waste management strategies, from environmental, energetic and economic point of view. These calculations have shown that landfilling, even if gas is recovered and leachate is collected and treated, should be avoided, due to the fact that resources in the waste are inefficiently utilised (Sundqvist, 2005). Environmental sound alternatives include incineration, material recycling, anaerobic digestion or composting.
Incineration is a combustion process at high temperature that allows rather complete oxidation of solid wastes, liquids or gases. Combustion systems may be very complex involving simultaneous coupled heat and mass transfer, chemical reaction and fluid flows. A global equation for representing combustion of wastes in air, may take the following form (Jenkins et al., 1998):

$$C_{x1}H_{x2}O_{x3}N_{x4}S_{x5}Cl_{x6}Si_{x7}K_{x8}Ca_{x9}Mg_{x10}Na_{x11}P_{x12}Fe_{x13}Al_{x14}Ti_{x15} + n_1\, H_2O + n_2\,(1+e)(O_2+3.76N_2) \rightarrow n_3\, CO_2 + n_4\, H_2O + n_5\, O_2 + n_6\, N_2 + n_7\, CO + n_8\, CH_4 + n_9\, NO + n_{10}\, NO_2 + n_{11}\, SO_2 + n_{12}\, HCl + n_{13}\, KCl + n_{14}\, K_2SO_4 + n_{15}\, C + \ldots \quad (1)$$

It is important to note that the empirical formula represented in Eq. (1) is incomplete since it includes only 15 elements and a real waste may contain a lot more, some of them found in traces; the molar indices x1 to x15 can vary widely; n_1 corresponds to the moisture in waste; n_2 is related with the amount of air (considered as a binary mixture of O_2 and N_2) used in the combustion; (1+e) is the excess of air in relation to the stoichiometric amount, usually ranges from 1.2 to 2.5 (depending on whether the fuel is gas, liquid or solid) (BREF, 2006); n_3 to n_{15} correspond to the stoichiometric coefficients of the different species that can be found as reaction products, among many others that can be released in the emissions. If the incinerated material is represented by a simpler formula, like $C_uH_vO_wN_xS_y$, then the combustion equation may be simplified and represented by Eq. (2)

$$C_uH_vO_wN_xS_y + (u+v/4-w/2+y)\, O_2 \rightarrow u\, CO_2 + v/2\, H_2O + x/2\, N_2 + y\, SO_2 \quad (2)$$

In the scope of thermal treatments of solid wastes, Fig. 1 shows the difference in terms of pyrolysis, gasification and incineration by taking into account the amount of air present.

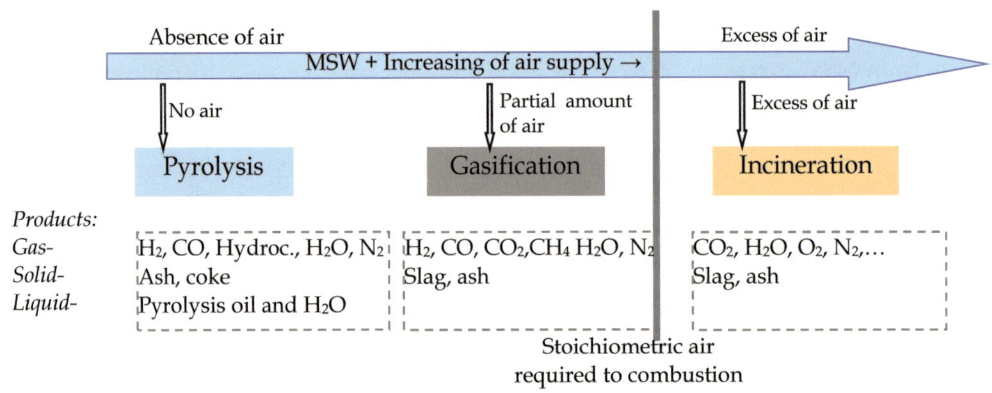

Fig. 1. Classification of the thermal technologies for treating MSW (based on DEFRA, 2007).

These thermal processes correspond to very different technologies in the way of treating waste and energy recovery. In incineration, the energy is released through the oxidation reactions, and its recovery occurs directly from the gases formed.

At present, municipal solid waste incineration (MSWI) in waste-to-energy (WtE) has confirmed to be an environmentally friendly solution and a common alternative to landfilling, while allowing recovery of a large part of the energy contained in MSW. In practice, MSWI has several advantages and disadvantages as reported in Table 1. Nevertheless, the main problems associated to these processes are probably the large volume of gaseous emissions which may pose environmental health risks (Moy et al., 2008) and hazardous solid wastes that remain after incineration as fly ash or air pollution control (APC) residues (Quina et al., 2008a,b).

MSW is generated by households and other similar wastes in nature and composition, which in general is collected and managed by or on behalf of municipal authorities, and includes materials such as paper, plastics, food, glass and household appliances. Fig. 2 shows typical composition of MSW usually associated to these waste streams, based on Gentil et al. (2009), and information reported by environmental agencies from Portugal (APA) and from USA (EPA) for the reference year 2009.

Advantages	Disadvantages
- handle waste without pre-treatment - reduce landfilling demand for MSW - reduce waste volume by 90% - reduce waste weight by 70% - possibility of recovering energy (electricity or heat) - if well managed, low air pollution is released - destroy potential pathogens and toxic organic contaminants - can be located close to the centre of gravity of MSW generation - reduce cost of waste transportation - require minimum land - stack emissions are odour-free - reduce organic materials mainly to CO_2 instead CH_4 and other VOC	- originate hazardous waste (APC residues), that requires safe disposal - originate slags (bottom ashes) - originate huge volume of flue gases - high investment and operating costs - high maintenance costs - require skilled staff - require suitable composition for autocombustion - negative public perception (so far)

Table 1. Advantages and disadvantages of municipal solid waste incineration.

According to Eurostat data for EU-27 State Members, MSW produced in 2008 was on average about 524 kg per capita, but it is possible to find values between 800 kg in Denmark to 300 kg in the Czech Republic (Eurostat 2010). Globally, in 2008, the EU-27 countries produced the huge amount of 259 Mt of MSW, whereas 221 Mt was accounted for in the EU-15. Figs. 3-4 depict the way that MSW stream has been treated in various countries, and in particular Fig. 3 shows the evolution in the EU-27 from 1995 to 2009 taken into account landfill, incineration, composting and recycling. It is important to note that, in 2009, about 20% of waste was incinerated, which correspond to 50.9 Mt. Considering that the average lower calorific value (LCV) should not be less than 7 MJ/kg of waste, in order to occur a chain of reactions able to self-supporting combustion, and assuming that in Europe the LCV

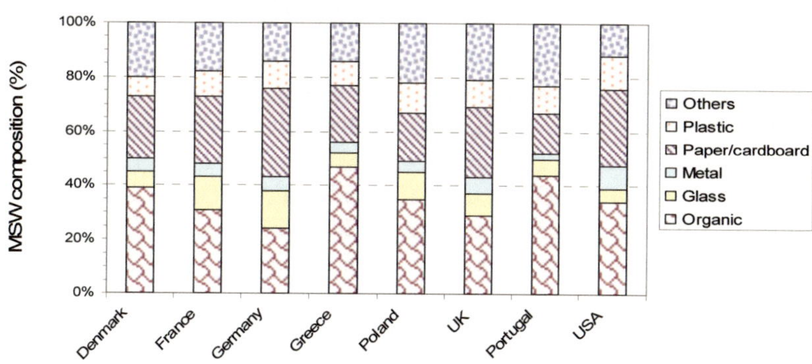

Fig. 2. Composition of MSW (based on Gentil et al., 2009, Portugal APA, US EPA).

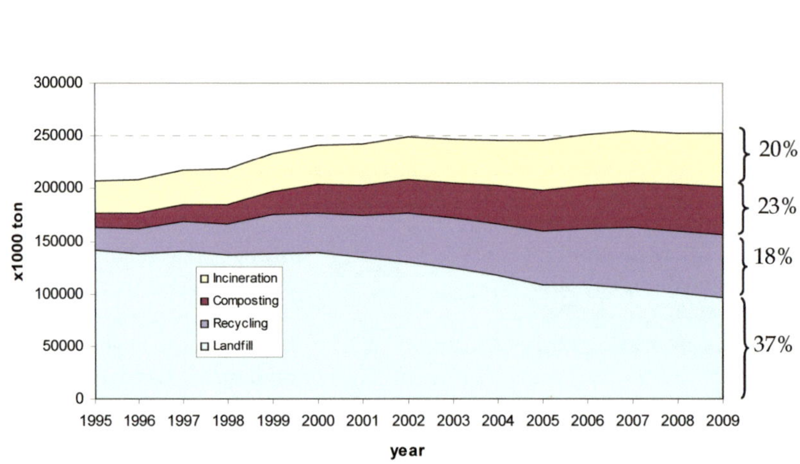

Fig. 3. Contribution of landfill, incineration, composting and recycling in EU-27 State Members (Source: based on Eurostat databases).

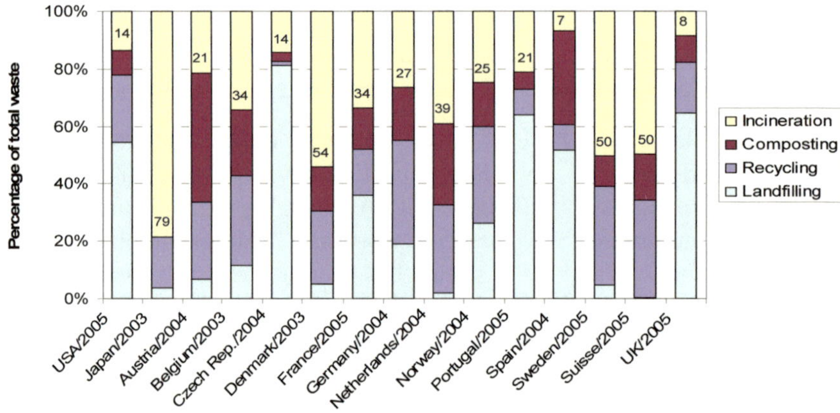

Fig. 4. Contrasting MSW management practice in selected countries, in landfills, incineration, composting and recycling (Source: based on OECD statistic databases).

is in the range of 9-13 MJ/kg (Worl Bank Report, 1999), the combustion of 50.9 Mt led to an enormous amount of energy available for recovery. Fig. 3 points out that landfilling has been gradually decreasing since 1995, and in 2009 its contribution accounts for 37%. According to Fig. 4, Japan is the country where incineration has higher contribution (79%) and in Europe, countries such as Denmark (54%) and Sweden (50%) have the highest rates.

By taking into account the information from BREF (2006) for waste incineration, Table 2 summarizes the number and total capacity of the existing incinerators in 17 European countries.

It is important to note that these numbers may vary according to the source of information used, and the year of reference. According to DEFRA (2007), in 2000, about 291 incineration sites with energy recovery located in 18 Western European countries, processed about 50

Country	Number of MSWI	Capacity Mt/year	Country	Number of MSWI	Capacity Mt/year
Austria	5	0.5	Luxembourg	1	0.15
Belgium	17	2.4	Portugal	3	1.2
Denmark	32	2.7	Spain	9	1.1
Finland	1	0.07	Sweden	30	2.5
France	210	11.7	Netherlands	11	5.3
Germany	59	13.4	UK	17	3.0
Greece	0	0	Norway	11	0.65
Ireland	0	0	Switzerland	29	3.3
Italy	32	1.7			

Table 2. Number and total capacity of the existing incinerators in 17 European countries.

million ton of waste and 50 TWh of energy recovered (40 million ton of oil equivalents). According to Directive 2008/98/CE, a formula is indicated, Eq. (3), to clarify when the incineration of MSW is energy-efficient and may be considered a recovery operation. Indeed, the energy efficiency must be equal or above 0.6 or 0.65 depending on the installation permitted before or after 31 December 2008, respectively.

$$\text{Energy efficiency} = (E_p-(E_f+E_i))/(0.97x(E_w+E_f)) \qquad (3)$$

where E_p is the annual energy produced as heat (multiply by 1.1) or electricity (multiply by 2,6), GJ/year, E_f the annual energy input to the system from fuels contributing to the production of steam (GJ/year); E_w the annual energy contained in the treated waste calculated using the net calorific value of the waste (GJ/year); and E_i the annual energy imported excluding E_w and E_f (GJ/year). A corrective factor of 0.97 is introduced to accounting for energy losses due to radiation and bottom ash. It is worthwhile to refer that high efficiency is not easy to reach only through production of electricity. Hot water usage should be considered also, whenever feasible at the location.

3. Municipal solid waste incinerators and air pollution control technologies

Different technologies can be applied to MSW including mass burning with travelling grate, rotary kilns, modular-two stage combustion and fluidised bed (BREF, 2006). In Europe, grate incinerators are used in more than 90% of the installations and in the specific case of fluidised bed, MSW has to be pre-treated. The incineration technology used for MSW has been changing over the last 10 to 15 years, mainly driven by legislation requirements, which has forced low emission limits to air. According to Directive 2000/76/EC, a 'incineration plants' *correspond to any stationary or mobile technical unit dedicated to the thermal treatment of wastes with or without recovery of the combustion heat generated. This includes the incineration by oxidation of waste as well as other thermal treatment processes such as pyrolysis or gasification in so far as the substances resulting from the treatment are subsequently incinerated.* This description comprises the site and the entire incineration plant including:

- waste reception and handling (storage, on site pre-treatment facilities),
- combustion chamber (waste-fuel and air-supply systems),
- energy recovery (boiler, economizer, etc.),
- facilities for clean-up gaseous emissions,

- on-site facilities for treatment or storage of residues and waste water, stack,
- devices and systems for controlling incineration operations, recording and monitoring incineration conditions.

These areas may be distributed as indicated in Fig. 5, which represents a scheme of a typical mass burning MSW incinerator (IAWG, 1997).

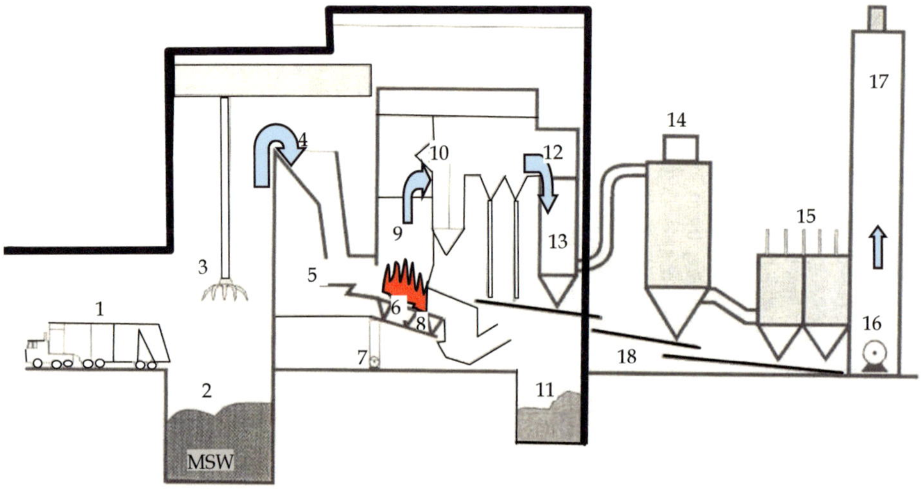

1- waste collection vehicle
2- waste storage pit
3- waste handle crane
4- feed hopper
5- feeder
6- grate
7- forced-draft fan
8- undergrate air zone
9- furnace
10- boiler
11- bottom ash bunker
12- superheater
13- economiser
14- dry scrubber
15- fabric filter baghouse
16- induced-draft fan
17- stack
18- APC residues conveyor

Fig. 5. Simplified scheme of a MSW incinerator (adapted from IAWG, 1997).

Considering the diagram of Fig. 5, a brief description of the mass flow into the incinerator is given below. MSW is in general delivered in trucks (1) and discharged into the storage pit "as-received" (2), in enough amounts for providing a continuous feeding material to the WtE plant. Then, waste is randomly picked up through a handle crane (3), and dropped into the feed hopper (4). The waste flows through the feeder (5) onto the moving grate (6) where combustion takes place. The plant should be controlled in order to optimize the combustion conditions, to ensure, as much as possible, complete carbon burn-out, and for this the residence time on the grate is usually no more than 60 min. The forced-draft fan (7) forces primary air through undergrate air zone (8) into the furnace (9), in order to supply oxygen to promote oxidation reactions, e.g. Eq (1). The primary air is in general taken from the storage pit (2) to lower the air pressure and eliminate most odour emissions from the storage area. Although it is not represented in Fig. 5, a secondary air supply system is common in the furnace, to guarantee turbulence of flue gases (secondary-air) and to ensure complete combustion. About 10-20% (v/v) of flue-gas is recirculated as secondary air. The reactions involved in this process are exothermic and release a high amount of energy that is carried over by the flue gases as heat. Indeed, for example, the upper calorific values of MSW in Germany are usually in the range of 7-15 MJ/kg (BREF, 2006). Energy recovery occurs mostly in boiler (10), superheater (12) and economizer (13). The burned-out bottom ashes are normally quenched and transported to a storage bunker (11). In most of the incinerators, the bottom ashes are transported on conveyors and ferrous metals sorted, and thus at the same time metals recycling and improvement of the slag properties take place. Slag is partly vitrified and can be handled as non-hazardous or special waste in many countries. The huge amount of gases produced during combustion contains air pollutants harmful for the environment that must comply with the stringent regulatory limits. Thus, depending on the desired cleaning degree, different air pollution control (APC) systems may be used. As an example, in Fig. 5, a dry scrubber (14) and fabric filters (15) are used. In these units, APC residues are produced and further transported through a conveyor (18) for a silo (not represented). Most of the modern incinerators treat APC residues before disposed of in monofills. Finally, by using induced-draft fan (16), the cleaned flue gas is released via the stack. Concerning air pollution, it is extremely important to note that combustion includes very fast reactions (fractions of seconds) that take place in gas phase, and self-supporting combustion is possible if heat value of the waste and oxygen concentration is sufficient. Thus, grate length should ensure the phases indicated in Fig. 6.

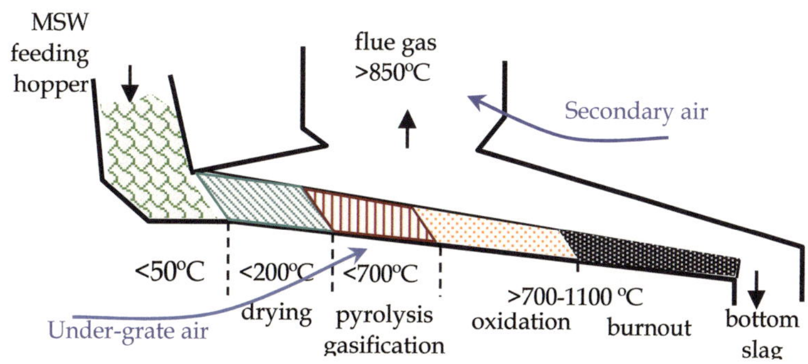

Fig. 6. Phases and temperatures on the grate of a MSWI (adapted from IAWG, 1997).

In practice, water and volatile content of MSW are evolved in the initial phase of drying at temperatures until the range of 200ºC, where no oxygen is required. The next phase corresponds to pyrolysis and gasification of the organic materials, in which organic compounds are transferred to the gaseous phase. After that, in the oxidation, the combustible gases react with oxygen releasing heat and other lower weight molecules. In fully oxidative conditions, reactions are almost complete and the major gases are water, nitrogen, carbon dioxide and oxygen. It is very important to note that those phases overlap in spatial and temporal sense. Nonetheless, some in-furnace technical measures (e.g. air distribution and furnace design) can be taken to influence those stages in order to reduce pollutants in gaseous emissions (BREF, 2006). European legislation imposes a minimum gas phase combustion temperature of 850ºC and at least 2 s for the residence time.

In general, MSW incineration plants operate 24 h/day and close to 365 days/year. Availability of the plant is very often over 98% and this imposes several redundancies of equipment and on-operation maintenance procedures. Fig.7 summarizes the main inputs and outputs of MSW incineration process, where it is indicated that 1 t of waste originates nearly 300 kg of bottom ashes, 30 kg of APC residues, and the rest is emitted as flue gas. The air flow rate is a control variable that is adjusted depending on the characteristics of the stack gases. However, in general about 4000 to 4500 m^3/t of air is required to guarantee fully oxidizing atmosphere (IAWG, 1997). The flue gas volume originated in MSWI is dependent on technology, in particular concerning the existence of flue gas recirculation. However, in literature there are indications that these values may be in the range of 4500 to 6000 Nm^3/ton of waste (Achternbosch and Richers, 2002; BREF, 2006). Even though large local variations can be observed, typically 1 ton of MSW allows energy recovery in the range of 400-700 kWh of electricity and additional 1205 kWh as heat (BREF, 2006). In general, liquid effluents can occur from: APC devices mainly if wet systems are used, the collection and storage of bottom ash, the water/steam cycle, sanitary areas, rainwater, and cooling water. However, if re-circulation is maximized reduced amounts of wastewater are produced. Chemical reactants used in incinerator plants may be ammonium hydroxide (NH_4OH) or ammonia as gas (NH_3) for reducing NO_x in furnace, neutralizing agents (e.g. $Ca(OH)_2$) and adsorption materials (e.g. activated carbon). Typical consumptions rates per ton are 0.8 kg of NH_4OH, 8 kg of $Ca(OH)_2$ and 0.5 kg of activated carbon.

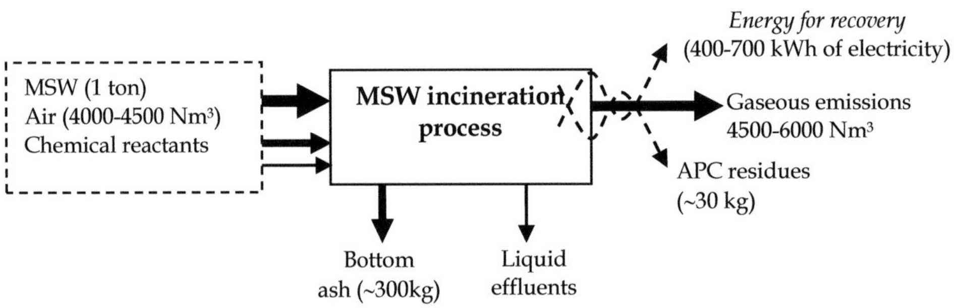

Fig. 7. Main inputs and outputs of MSW incinerators.

3.1 Gas cleaning systems for waste incinerators

The gas mixture leaving the incineration furnace has three main types of components that have to be removed to the possible extent before the exit at the stack:

- Fly ash which is composed by particles pneumatically transported by the gaseous flow;
- Acids and acids precursors, such as sulphur dioxide, nitrogen oxides, hydrochloric acid;
- Dioxins and analogues that are compounds formed by radical recombination with structures such as polychloro dibenzodioxins and the respective furan analogues.

The hot gas mixture leaving the furnace, exchanges heat at the surface of the heat exchanger vertical tubes, inside which the high or medium pressure steam is generated and before entering the cleaning systems, part of this gas is diverted through a booster to be injected in the cameras below the moving grate of the furnace. This gas recycling is essential in terms of overall energy recovery and also very important to promote one easier control of the stoichiometric excess of oxygen in the furnace. The remaining effluent gas mixture has to be cleaned by several unit operations in the gas cleaning systems.

3.2 The main targets

During the heat exchange at the surface of the boiler tubes, the effluent gas is cooled and some additional solid formation occurs, increasing the particulate fraction. This raw gas effluent of the boiler system has to be cleaned up to the environmental standards applied at the location where the incinerator is installed. The main ranges for the concentrations in the raw gas and the mandatory final emission limits most commonly used are included in

Table 3 together with the calculated efficiency required for the cleaning of each type of polluting agent. This high removal efficiency required, indicated in the Table 3, imposes the use of multiple systems for the gas cleaning, usually in sequence.

Pollutant	Concentration in raw gas from boiler (mg/Nm^3)	Max admissible at exhaust (mg/Nm^3)	Removal efficiency required (%)
Fly Ash	1500 - 2000	10	99,9
HCl	300 - 2000	10	> 99
SO_2	200 - 1000	5	99.5
NO_x	200 - 500	70	86%
HF	2 - 25	1	96%
Hg	0.2 – 0.8	0.01	99%
Cd + other metals	2 -15	0.05	>99.5
Dioxins (ng I-TEQ/Nm^3)	0.5 – 5	0.1	98%

Table 3. Required efficiency for flue gas cleaning systems.

3.3 Unit operations for gas cleaning

A large number of unit operations based on primary separation processes can be used for the gas cleaning of the flue gas generated in waste incineration systems. In Table 4 for each type of flue gas pollutant, a combination of unit operations is indicated with the respective typical range of reduction. The well designed sequence of gas cleaning methods allows for a drastic reduction of pollutants as stated by the waste incineration BREF Table 5 (adopted with comments from Table 5.2 of the BREF, 2006).

Pollutant	Process Steps	Reduction (%)
SO_x	Wet scrubber or dry multicyclone	50 - 90
HCl	Wet scrubber or semi-dry	75 - 95
NO_x	Selective catalytic reduction	10 - 60
Heavy metals	Dry scrubber + electrostatic precipitator	70 - 95
Fly ash *	Electrostatic precipitator + fabric hose filter	95 - 99.9
Dioxins & Furans	Activated carbon + fabric hose filter	50 - 99.9

*Very often the fly ash surface has adsorbed other pollutants such as dioxins and heavy metals

Table 4. Gas cleaning processes and typical range of specific pollutant reduction by combination of unit operations.

3.4 Separation of fly ash and activated carbon

Fly ash generated at power plants where the composition of the fuel is reasonably constant, is very often collected and used as raw material for the production of Portland cement. Fly ash generated at waste incinerators is usually contaminated with heavy metals and other dangerous substances and have to be treated as a hazardous residue, requiring inertization before the disposal is controlled landfill.

Substance	Non-Continuous samples	Half-hour mean value	Daily mean value	Notes
Dust		1-20	1-5	Lower levels achieved with fabric filters as hose bags filters
HCl		1-50	1-8	Use of wet processes preferred
SO_2		1-150	1-40	Use of wet processes preferred
NOx with SCR		40-300	40-100	Additional energy demand and costs
NOx with SCR		30-250	120-180	At high raw gas NOx levels, NH_3 slip to be taken into account, preferred method in conjunction with wet processes
TOC		1-20	1-10	Optimum combustion conditions are required
CO		5-100	5-30	Optimum combustion conditions
Hg	<0.05	0.001-0.03	0.001-0.02	Input control, carbon-based adsorption processes
PCDD/PCDF (ng ITEQ/Nm^3)	0.01-0.1			Optimum combustion conditions, temperatures controls to reduce synthesis, carbon-based adsorption processes.

Table 5. Attainable emission levels for waste incineration facilities (excerpt from Table 5.2 of BREF (2006) "Operational emission levels associated with the use of BAT" for air pollutants expressed in mg/Nm^3).

The activated carbon, in powder form, is very often used to adsorb organic pollutants such as dioxins and furans, as it will be detailed later on, this powder is collected together with the fly ash (small systems) in a specific stage of cleaning. The main type of equipments used for the removal of solid particles of the fly ash and activated carbon are: cyclones, electrostatic precipitators, and fabric hose filters, Table 6. Construction details, as well as operating modes have been described extensively (Niessen, 2002) and therefore it was considered more relevant to focus the detail here on the range of applicability for the fly ash separation.

3.4.1 Cyclones

Cyclones are rather efficient for the removal of solid particles with average diameter over 100 µm from gaseous flows. The design of cyclones was reviewed elsewhere (Morcos, 1989; Lee and Huffman, 1996; Amutha Rani et al., 2008) and since for their construction steel or stainless steel can be used the range of permissible operating temperatures is rather wide. In gas cleaning of flue gas from incinerators cyclones are very often used as primary separators followed by other separating units designed for the retention of particles of the lower size present in the fly ash.

3.4.2 Electrostatic precipitators (ESP)

Nearly one century ago, Dr. Frederick Cottrell introduced the concept of particle separation by the application of an electric field. Overall efficiency of ESP depends mainly on the plates and rapper design, and collection zones have to be carefully designed to assure the adequate thickness of laminar boundary layer, in order to prevent reentrainment into the gas stream the aggregate formed with the collected particles. Typical gas velocities inside the precipitation zone are always below 1 m/s and very often below 0.5 m/s. Electrostatic precipitation, as many other process equipment items in an incineration plant have to be adjusted to steady state conditions but with the control system prepared to modulate under unbalance to peak conditions. Dynamic control is often required and the fine control of efficiency is of utmost importance (Bordado and Gomes, 1999).

Equipment	Typical efficiency for fly ash	Typical efficiency for AC	Typical pressure drop range	Maximum operating temperature	Range of particles sizes
Cyclones	up to 80%	up to 50%	10 to 1000 Pa	1300 °C	≥ 20 μm
ESP	up to 99%	up to 80%	50 to 300 Pa	450 °C	0.08 to 20 μm
Fabric hose filters	up to 99%	up to 99%	500 to 2000 Pa usually with a booster fan	240 °C	0.04 to 50 μm

Table 6. Fly ash and activated carbon (AC) separation.

3.5 Separation of acids

The acids present in the flue gas, such as HCl and HF and the precursor of acid SO_2 can be separated by different processes: dry process (with the use of a solid adsorbent); semi dry (with the use of a spray absorber); wet process (with aqueous solutions).

Criteria	Wet FGT (W)	Semi-Wet FGT (SW)	Dry line FGT (DL)	Dry Sodium bicarbonate FGT (DS)
Air emissions performance	+	0	-	0
Residue production	+	0	-	0
Water consumption	-	0	+	+
Effluent production	-	+	+	+
Reagent consumption	+	0	-	0
Ability to cope with inlet variations of pollutant	+	0	-	0
Plume visibility	-	0	+	+
Process complexity	-	0	+	+
Costs - capital	generally higher	medium	generally lower	generally lower
Costs operational	medium	generally lower	medium	generally lower

Note

+ : means that the use of the technique generally offers an advantage in respect of the assessment criteria considered

0 : the technique offers no significant advantage or disadvantage in respect of the assessment criteria considered

- : means that the use of the technique generally offers a disadvantage in respect of the assessment criteria considered

Table 7. An example assessment of some IPPC relevant criteria that may be taken into account when selecting between wet/semi-wet/dry flue gas treatment (FGT) options.

As solid adsorbents, the dry lime (calcium hydroxide) is more commonly used but dry sodium bicarbonate can also be an option. A comparison of the specific processes for acids separation is qualitatively presented in Table 7. Each one of these alternatives has nowadays several variants and refinements but it was considered out of the scope of this review the fine detail of the multiple available alternatives.

3.6 Nitrogen oxides reduction

Nitrogen oxides content in flue gas are usually reduced by two reactive processes: SNCR (selective non-catalytic reduction) and SCR (selective catalytic reduction). Other chemical process focused on the reduction of NO_x formation during the combustion have been extensively studied at laboratory and pilot plant level, and may are based on radical quenchers that minimize N_2 oxidation by radical reactions. However most, if not all those processes, have a large negative impact on combustion efficiency and consequently they cause an increase content of VOC and namely PAH in the flue gas.

The SNCR process uses ammonia (NH_3) as reactant or for smaller systems urea ($CO(NH_2)_2$) as reducing agents, with direct injection into the furnace. At high temperature, urea decomposes with ammonia formation, Eq.(4), and the main overall reduction reactions take place between 850 °C and 1050 °C according with the schemes Eq.(5)-(6):

$$CO(NH_2)_2 + H_2O \rightarrow 2\ NH_3 + CO_2 \tag{4}$$

$$4\ NO + 4\ NH_3 + O_2 \rightarrow 4\ N_2 + 6\ H_2O \tag{5}$$

$$6\ NO_2 + 8\ NH_3 \rightarrow 7\ N_2 + 12\ H_2O \tag{6}$$

The SCR process (Jungtten et al., 1988) uses a solid catalyst usually in a fixed bed and operates between 200 as 400 °C, in presence of ammonia. This range of operating temperature and the fact that catalysts deactivate in presence of strong acid media, imposes the SCR module to be installed after particulate material, as well as acidic components are already removed from flue gas. The more representative overall reactions taking place at the surface of the catalyst are indicated by Eq. (7)-(9). A comparison of the main features SNCR and SCR processes is presented in Table 8.

$$4\ NO + 4\ NH_3 + O_2 \rightarrow 4\ N_2 + 6\ H_2O \tag{7}$$

$$6\ NO + 4\ NH_3 \rightarrow 5\ N_2 + 6\ H_2O \tag{8}$$

$$2\ NO_2 + 8\ NH_3 + O_2 \rightarrow 3\ N_2 + 6\ H_2O \tag{9}$$

3.7 Carbon monoxide and volatile organic compounds

The increase of carbon monoxide (CO) and/or volatile organic compounds (VOC) content in the flue gas is a strong indication of inappropriate burning conditions in the furnace. Several adjustments by the control systems can be adopted, but the more common are:

i. Increase of raw air inlet to the furnace;
ii. Reduction of flue gas recycling to the furnace;
iii. Slight increase of pressure below the grid.

Both i) and ii) will increase the stoichiometric excess of air in the furnace, allowing for a more complete oxidation of the wastes, and of the respective volatiles, such as thermal

	SNCR	SCR
Advantage	Lower investment cost Lower corrosion problems	More efficient
Disadvantage	Limited efficiency	Higher investment cost Higher pressure drop Requires higher O_2 excess
Efficiency for NO_x reduction	Up to 70% Typical 30 to 60%	Up to 85% Typical 50 to 80%

Table 8. SNCR versus SCR processes for NO_x reduction.

degradation products. For wastes with a rather high content of halogens or phosphorus (for example in presence of increased content of flame retardants and/or PVC), the adoption of i) and specially ii) will be very effective, but increase of stoichiometric oxygen over 10% usually results in an overall energy recovery efficiency reduction due to energy consumption to heat-up the unavoidable N_2 inlet. The adoption of iii) is especially effective for wastes with high moisture content, resulting in a more effective lifting of the waste material from the moving grate. It is important to refer that measure iii) can only be adopted with the so called high pressure drop grates in which the grate perforations and gaps are rather small (for example VON ROLL type grates).

Separation of carbon monoxide and volatile organic compounds present in the flue gas is not an easy task. One of the very few specific processes, worth to refer, is the catalytic oxidation in heterogeneous converter. In general terms, additional O_2 is injected in the flue gas stream and this mixture is further oxidized over a fixed bed of a strong oxidation catalyst such as platinum non woven felts.

For rather small waste incinerators, and when the waste has one high content of PVC disposables (e.g. hospital waste) this oxidation step is included in the gas cleaning system, usually close to the boiler since the oxidation is more effective at high flue gas temperatures. For larger systems such as ones used for mass burning of municipal wastes this systems is very seldom included in the design due to the very high specific investment cost and also due to the additional loss of energy due to boosting requirements to compensate the pressure drop.

3.8 Moisture condensation in gas cleaning systems

Moisture condensation is one of the major problems that can occur in gas cleaning systems. The occurrence of condensation causes drastic corrosion problems as well as fly ash aggregation as a mud or "paste" and induces the malfunction of several gas cleaning steps mainly the hose fabric filters. Both in the design as well as in the operation of incineration systems, to avoid moisture condensation is mandatory and in case of continuous occurrence the life span of the overall plant will be reduced as well as the average availability due to more frequent cleaning and maintenance requirements.

Incidence on operating costs is therefore also significant, and although several mitigating measures can be adopted from the control point of view, the best may be of course to prevent its occurrence by a careful design of the plant. Not only the overall water balance has to be considered for steady and dynamic conditions but careful estimation of the dew point of the flue gas in different critical points must be performed.

The main contributions for the total water present in the flue gas come from:

i. Moisture in the inlet air;

ii. Moisture vaporized from the wastes;
iii. Water formed in the combustion reactions;
iv. Water vaporized in the flue gas stream in the wet or semi-wet cleaning steps;

Rather high values of i) and ii) are expected to occur simultaneously in rainy days or if snow accumulates on the waste at the collection points, and therefore active prevention of condensation is of utmost importance at the design stage.

The flue gas temperature, from the outlet of the boiler until the stack, is becoming lower, partly due to energy recovering systems, and therefore the probability of the occurrence of condensation becomes higher toward the last stages of the gas cleaning system.

Re-heating of the flue gas, to avoid condensation can be achieved in practice by different processes. The more common ones are:

i. Injection of limited flow of hot flue gas boosted stream in the main stream: this is very effective system of re-heating, but it has the disadvantage that hot flue gas from the boiler is more contaminated, and the overall cleaning efficiency is slightly reduced;
ii. By heat exchanging with hot flue gas: in this case the two streams are kept separated and the only drawback is the increased pressure drop thought the heat exchanger and the higher investment cost.

In both cases the heat-up to, say 2 to 5 degree centigrade over the higher dew point temperature that can occur is usually effective to prevent condensation since the pressure profile is, as a rule, very stable. To prevent condensation, the re-heating is very often considered in the design at least in two points of the cleaning system:

i. Immediately before the exhaust stack to avoid the formation of "plume" caused by condensation during cooling by the cold air;
ii. Before the hose fabric filters to avoid clogging by mud formation in the inside wall of the filter.

The use of corrosion resistant materials and special coatings where condensation is more prone to occur, is indeed a good design practice, but to prevent the occurrence is certainly also of major importance.

3.9 Emerging technologies for gas cleaning

3.9.1 Oil and emulsion scrubbing

Dioxins and polyaromatic hydrocarbons (PAH) have a very limited solubility in aqueous solutions and therefore their removal by aqueous wet scrubbers is rather limited. A small depletion is observed due to condensation in the rather cold water solution, as well as in the sludge of solid particles at which surface they remain adsorbed, but due to their marked lipophilic character they tend to float and eventually be removed from the scrubbing solution by the gas flow.

High boiling oils, partly unsaturated proved to be an efficient scrubbing media, as well as their respective oil-in-water emulsions stabilized by non-ionic surfactants with HLB between 7 and 10 (Encyclopedia of emulsion technology, 1988). The oil emulsion retains mostly dioxins, furans and PAH and it is considered a good practice to dispose the emulsion as soon as total content of those pollutants reaches 0.1 mg/L. The design of the scrubbing systems considers a buffer volume of emulsion that allows for the emulsion exchange up to 4 times per year in the worst operating conditions. Emulsion preparation vessel and transfer pump is therefore to be included in the auxiliary equipment.

Disposal of the contaminated emulsion includes an emulsion breaker, liquid-liquid separation and absorption of the oil phase in wood particles or other lipophilic absorber, such as open cell polyurethane (Clark et al., 2008) or phenolic foam, and the feeding of the solid material for the furnace for further energy recovery.
A separate burner for the oil into the furnace can also be installed, but operators of waste incinerators do not like to feed liquids or oils to the furnace for a large number of operational problems that can occur.

3.9.2 Dedusting and particle aggregation

Although very efficient for over 99% of the particles the electrostatic precipitators and the fabric hose filters are not effective for nanoparticles as well as for the smaller sub-micron particles. Flue gas exhaust at the stack, although very clean has however a small content of nanoparticles. Recent concerns on the physiological effect of respirable nanoparticles induced research on their study and on methods of avoiding their presence in the exhaust flue gas. Improving particle aggregation and deducting seems the more promising alternative and intense research is being conducted in that direction. So far it was proved that very fine dust and nanoparticles can be effectively aggregated by water dispersible polymers with controlled water compatibility (Nunes et al., 2011).

4. Main pollutants released during MSWI and the European legislation

The main direct impacts caused by incineration are emissions to air, residues production (bottom ash and APC residues), emissions to water, energy production/consumption, material consumption, noise and vibration, fugitive emissions (including odour mainly from waste storage), storage/handling/processing risks of wastes. Among these, emissions to air are really significant, since the incineration of 1 ton of MSW produces a huge volume of flue-gas. Some measurements revealed that 4500-6000 Nm^3/ton (at 11% O_2) are usually emitted. Depending on the technology, operating conditions and the composition of waste incinerated, diverse pollutants are formed and emitted in flue gas:

- particulate matter – with a broad distribution size;
- acids and other gases– HCl, HF, HBr, HI, SO_2, NO_X, NH_3, ...;
- heavy metals – Hg, Cd, Tl, As, Ni, Pb, Sb, Se, Sn, Zn, ...;
- carbon compounds – CO, VOC, PCDD/PCDF, PCB, PAH,...;

In this scope, measures for reducing pollutant precursors in wastes, such as chlorine and volatile heavy metals, may have positive effects. In practical terms, redesign products and packages could be an important aspect for reducing emissions. With respect to heavy metals, it is important to avoid batteries, leather, pigments, cans, and for reducing chlorine, wastes such as plastics (PVC) must be avoided.
Since the early 1970s Europe has been strongly committed to the environment protection of air. In this ambit, it should be mentioned the Directive 2000/76/EC of the European Parliament and of the Council, on the incineration of waste that aims "*to prevent or to limit as far as practicable negative effects on the environment, in particular pollution by emissions into air, soil, surface water and groundwater, and the resulting risks to human health, from the incineration and co-incineration of waste.*" This Directive, known as Waste Incineration Directive (WID), states that continuous measurements of NO_x, CO, total dust, TOC, HCl, HF and SO_2 should be carried out, and at least twice a year for heavy metals (Cd, Tl, Hg, Sb, As, Pb, Cr, Co, Cu, Mn, Ni, V) and dioxins and furans. In fact, WID imposes very restrictive limit values for the

main air emission pollutants. For example, the emission limit value allowed to the total concentration of dioxins and furans calculated using the concept of toxic equivalence is 0.1 ng I-TEQ/Nm^3. Table 9 shows the air legal limits values indicated in WID, as well as the usual ranges that can be observed before treatment of flue-gas (crude flue-gas).

	Before treatment (BREF, 2006)	Ref.1	After treatment (BREF, 2006)	Legal limits*
Dust (mg/Nm^3)	1000-5000	2000-10000	0.1-4	10
CO (mg/Nm^3)	5-50		2-45	50
TOC (mg/Nm^3)	1-10		0.1-5	10
HCl (mg/Nm^3)	500-2000	400-1500	0.1-6	10
HF (mg/Nm^3)	5-20	2-20	0.01-0.1	1
SO_2 (mg/Nm^3)	200-1000	200-800	0.2-20	50
NO+NO_2 (mg/Nm^3)	250-500	200-400	20-180	200 #
NO (mg/Nm^3)	<40			not defined
Hg (mg/Nm^3)	0.05-0.5	0.3-0.8	0.0002-0.05	0.05
Cd,Tl (mg/Nm^3)	<3	3-12	0.0002-0.03	0.05
Other heavy metals (Sb, As, Pb, Cr, Co, Cu, Mn, Ni, V)	<50		0.0002-0.05	0.5
Dioxins and Furans (ng I-TEQ/Nm^3)	0.05-10	<1-5	0.0002-0.08	0.1
CO_2	5-10%			not defined
H_2O	10-20%			not defined

Ref.1 - Achternbosch and Richers (2002), *set on Directive 2000/76/EC, #for nominal capacity ≥ 6 ton/h or new incineration plants.

Table 9. Concentrations of gaseous emissions before and after treatment and legal limits.

To prove compliance with the emission values set in WID, measured concentrations, C_M, shall be standardised (indexed with S), C_S, at the following conditions: 273 K, 101.3 kPa, 11 % O_2, and dry gas (M_S=0) by using the following equation:

$$C_S = C_M \frac{21 - O_{2,S}}{21 - O_{2,M}} \frac{100 - M_S}{100 - M_M} \frac{T_M}{T_S} \frac{P_S}{P_M} \qquad (10)$$

where oxygen concentration ($O_{s,M}$), moisture (M_M), temperature (T_M) and pressure (P_M) represent the measured conditions.

The emissions of pollutants such as HCl, HF, SO_2, NO_x and heavy metals depend mostly on the waste characteristics and air pollution control devices. Furnace technology determine to a great extent CO, VOC and NO_x. The total dust emitted is determined upon APC devices. Fig. 8 shows a mass balance in respect to some pollutants per 1 ton of MSW incinerated.

4.1 Dust

During the incineration process, part of the waste is incombustible and removed from the incinerator as solid wastes, usually referred as bottom ash or slag, APC residues and very low quantity of residual particles in the flue gas. The matrix compounds of these particles are mainly aluminium and silicon oxides. Concerning air pollution from MSWI, the most import fraction is the finest inorganic particles that remain after the combustion of the

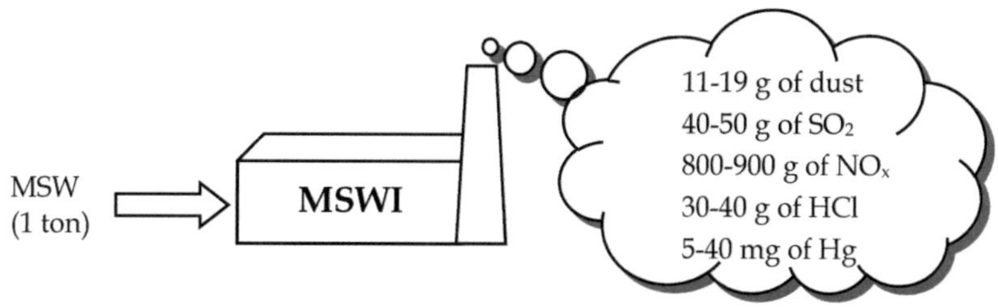

Fig. 8. Mass balance in respect to some pollutants in stack gases.

waste and pass with the flue gas out of the furnace. Some of these particles settle as boiler ash and are removed in the hoppers. A very important aspect concerning air pollution is the fact that as flue gas cools diverse gaseous compounds (some pollutants) condense on the particle surfaces. Since high concentration of HCl is present in the boiler, the condensate phases may include diverse metals chlorides such as $ZnCl_2$, $PbCl_2$ and $CdCl_2$ (World Bank Report, 1999).

Particulate matter is often grouped into categories based on their diameters (e.g. $PM_{2.5}$ with a diameter less than 2.5 μm). For example, $PM_{2.5}$ may persist in the atmosphere for long periods, travelling long distances and they are capable to penetrate deep into the respiratory tract. Their toxicity is normally correlated with the respective chemical composition.

Air pollution control (APC) residues produced in incineration plants are classified as hazardous solid waste mainly due to its high alkaline pH, high content in volatile heavy metals and soluble salts, and trace amounts of very toxic compounds such as dioxins and furans (Quina et al 2008b). The current APC disposal methods in use are very limited, and in the near future they are likely to become unsustainable, due to stringent environmental regulations. Development of new and safer inertization processes will be a must in the future. When APC residues are formed in devices with $Ca(OH)_2$ injection, reaction products such as CaOHCl, CaF_2, $CaSO_3$ or $CaSO_4$ may be present in considerable concentration, as result of HCl, HF and SO_2 gases neutralization. Often activated carbon is also added which also ends up in APC residues.

4.2 Carbon monoxide

Carbon monoxide (CO) is a gas that can be found in the flue-gas as a product of the incomplete combustion of organic compounds that is measured on-line and usually used to check the incineration efficiency. In MSWI it is usual to assume that when CO concentration is low in flue-gas then the gas burn out has high quality, and the TOC emissions are low in those conditions. CO arises in combustion chambers when there is not enough O_2 for full oxidation or if temperature is not high enough to guarantee full reaction to CO_2. Carbon monoxide oxidation to carbon dioxide occurs after some time of its release to the atmosphere (BREF, 2006).

4.3 Volatile organic compounds

Volatile organic compounds (VOC) are products of incomplete combustion and may cover a wide range of compounds, since they can include carbon chains or rings that have high vapour pressure (e.g. 0.27 kPa at 250 °C). When combustion operates at high efficiency, very

low concentrations of organic compounds are expected into air emissions. Indeed, in general, flue-gas from MSWI only contains trace quantities of volatile organic compounds (VOC) that are quantified on-line in the flue-gas as total organic carbon (TOC). Commonly speciation of VOC may include compounds such as methane, ethane, propane, butane, pentane, hexane, heptane, ethylene, benzene, toluene and xylene (BTX), ethylbenzene, acetylene, formaldehyde, acetone, etc. Methane (CH_4) is not formed as long as oxidative conditions are observed in WtE process. In fact, CH_4 may only arise in the waste bunker, due to anaerobic processes during long periods of storage.
Beyond the organic compounds referred above, traces of polycyclic aromatic hydrocarbons (PAH), polychlorinates biphenyls (PCB) and dioxins (PCDD/PCDF - discussed in section 4.10) may occur in flue-gas, mostly adsorbed into particles.

4.4 Acid chloride and fluoride

Chlorine and fluorine may be present in MSW in significant quantities. The main sources of chlorine are plastics such as PVC, salty food waste and other inorganic chlorides. Fluorine can arise into plastics such as PTFE, fluorinated textiles and other inorganic fluorides. Under the furnace conditions chlorine and fluorine are mostly converted into acid hydrogen halides, HCl and HF, and part of these may react to form metal chlorides.
In general, HCl and HF are removed by the same technology, by adding neutralization agents, for instance lime ($Ca(OH)_2$), sodium hydroxide (NaOH), or calcium carbonate ($CaCO_3$). The following reactions show the separation of acid pollutants with $Ca(OH)_2$

$$2\,HCl + Ca(OH)_2 \rightarrow CaCl_2 + 2\,H_2O \tag{11}$$

$$2\,HF + Ca(OH)_2 \rightarrow CaF_2 + 2\,H_2O \tag{12}$$

These reactions may entail complex mechanisms due to various gas/solid phases involved, and some studies refer that instead $CaCl_2$ the main phase formed is CaOH (Quina et. al. 2008b). Gaseous emissions concentrations are in general in the range of 0.1 -6 mg/Nm^3 for HCl and 0.01 -0.1 mg/Nm^3 for HF, and thus below the regulatory limits.

4.5 Sulphur dioxide

Sulphur dioxide is a reaction product of S-compounds contained in the waste with oxygen, and its concentration in the flue-gas is proportional to the amount present in the waste:

$$C_xH_yS + w\,O_2 \rightarrow CO_2 + H_2O + SO_2 \tag{13}$$

Regardless of whether in organic or organic forms, most of the sulphur oxides end up in stack gases as SO_2. According to BREF (2006), in spite of the regulatory threshold of 50 mg/Nm^3, most of the installations reveal emissions lower than 20 mg/Nm^3. This pollutant can be removed by the reactions:

$$SO_2 + Ca(OH)_2 \rightarrow CaSO_3 + 1/2\,H_2O \tag{14}$$

It is important to note that SO_2 is a highly reactive gas with short half life indoors. Although it is a known respiratory irritant and bronchoconstrictor, its effects seem limited to patients with asthma and bronchitis, while sensitivity to exposure varies widely (Katsouyanni et al., 1997). In some cases, SO_2 is highly correlated with fine particles, and thus these pollutants may be surrogates for each other. Sulphur dioxide is also known as highly toxic for Koch's bacillus and thus, the incidence of tuberculosis is recognized as lower in areas of SO_2 release.

4.6 Nitrous dioxides

Nitrogen oxides, NO_x, are well known acidic and global warming gases, and during incineration process three main different mechanisms can lead to their formation. Indeed, part of the air nitrogen and part of the nitrogen contained in the waste (or fuel) may be oxidized to NO_x. The nitrous oxides produced from air are known as *thermal* NO_x, and the ones originated from waste as *fuel* NO_x. These reactions can significantly occur for temperatures above 1300 °C and whenever oxygen is not a limiting reagent. In WtE plants, thermal NO_x is often much greater than fuel NO_x. The formation of nitrogen oxides through radical reactions with organic compounds is known as *prompt* NO_x, and has little contribution in WtE incineration plants.

Legislation set a regulatory limit of 200 mg/Nm^3, and to comply with this limit, primary and secondary measures may be taken. The main primary techniques can involve both over supply air and excessive high furnace temperatures. Secondary measures may demand injection of reduction agents such as ammonia (as 25% aqueous solution) or urea, for promoting SNCR indicated in Eq. (5)-(6) (section 3.6).

The reduction of NO_x can attain 85% at temperature of 1000°C, and thus some NH_3 can also be released in the flue-gas (BREF, 2006). When urea is used as reducing agent in SNCR, a little N_2O may arise. Nitrous oxide (N_2O) is usually correlated with CO, since it is formed under low oxygen concentrations, but it is not included in NO_x emissions. Alternatively, SCR by using specific catalysts can be integrated in the APC area, after acid neutralization and dedusting. Depending on the catalyst, SCR reactions can take place in the range of 180 to 450°C. Stack emissions typically exhibit 95% of NO and 5% of NO_2 (BREF, 2006). Specific emission of NO_x may be in the range of 800-900 g/ton of MSW.

4.7 Mercury

Mercury is by far the most thermally mobile metal, highly toxic, and at 357 °C it is all volatilized into the flue gas. Although restriction in uses of Hg, some peak loads of MSW with batteries, electrical devices, thermometers and specific wastes may lead to incompliance with the regulatory threshold of 0.05 mg/Nm^3.

The oxidation state of Hg depends mostly on the characteristics of the flue-gas, but in general, the most significant forms in MSWI are Hg^0 and $HgCl_2$. When the HCl concentration in flue-gas is higher than SO_2 (a reduction agent), mercury is mainly in the form $HgCl_2$, which is easier to remove from the gas phase than Hg^0.

For this element, the primary measures are those which avoid as much as possible to introduce in the furnace contaminated wastes that contain Hg. Nevertheless, secondary techniques are usually implemented in order to comply with the regulatory limit at stack emissions. The lowest emissions of Hg are observed when sulphur doped activated carbon is used. Some analysis showed that wet acidic scrubbers can serve as a sink for Hg if it occurs in chloride form (BREF, 2006). The amount of Hg in MSW is in the range of 0.5- 5 g Hg/ton MSW, and from this less than 2% goes to bottom ashes, 70-90% is released into APC residues, and less than 10% is emitted to air emissions, which correspond to 0.0002-0.05 mg/Nm^3 (IAWG, 1997).

4.8 Cadmiun and tallium

Cadmium is an environmental problematic toxic element that may arise in air emissions of MSWI due to its thermal mobility. Although some countries restrict applications of Cd, frequent sources in MSWI are electronic devices (including accumulators), paints, Ni-Cd batteries, and cadmium-stabilised plastics. Under the conditions usually observed in the furnace, Cd is mainly converted into $CdCl_2$. The APC devices normally lead to a concentration of Cd in APC residues, and less than 1% is released to the atmosphere. The reported range is usually 0.0002-0.03 mg/Nm^3 (11% of O_2) (BREF, 2006).

The thallium amount in municipal solid waste is virtually non-existent and very often below any detection limit.

4.9 Other heavy metals

During the main stages of incineration, combustion temperatures attained in the furnace determine the evaporation degree of heavy metals and most of inorganic salts. Indeed, besides Hg, Cd and Tl also Sb, As, Pb, Cr, Co, Cu, Mn, Ni and V must be under control in stack emissions. These are very toxic metals since they are carcinogenic or may provoke respiratory damages. Most of these metals are in flue-gas as oxides or chlorides, which condense onto dust particles as the gases cool down downstream of the boiler. Therefore, it is expected that the part of these metals that vaporize in furnace, probably ends as condensed phases in APC residues. Thus, according to measurements in diverse European incinerators, the limit of 0.5 mg/Nm^3 is usually complied (BREF, 2006). Lead and chromium are currently the most problematic heavy metals found in APC residues that lead to a classification as hazardous waste (Quina et al, 2008b).

4.10 Dioxins

Polychlorinated dibenzo-p-dioxins (PCDD) and polychlorinated dibenzofurans (PCDF) are commonly referred as *dioxins* and include a set of tricyclic aromatic persistent organic pollutants (POPs), that are ubiquitous in the environment and may provoke significant risk to human health (WHO, 1989). Recently, Hites (2011) presented a short but very interesting overview and history about dioxins, where it is referred that PCDD/PCDF is now a well-known environmental contaminant family. In fact, depending on the position of chlorine in the benzene ring, 210 congeners can be formed. Fig. 9 (a)-(b) shows the generic structure of dioxins and furans, where chlorine may be attached at positions m=1- 4 and n=6-9, and thus, it is possible to form 75 dioxin and 135 furan congeners. The most toxic dioxin is 2,3,7,8-tetrachlorodibenzo-*p*-dioxin (2,3,7,8-TCDD) and it has been considered as molecule characterized with the lowest lethal dose to 50% of a population (LD_{50}). For example, for guinea pigs, LD_{50} is only 0.6 µg/kg. PCDD/PCDF are in general very lipophilic, solid at room temperature, with low volatility and insoluble in water. Thus, their dispersion in atmosphere is likely to occur mainly in particulate matter.

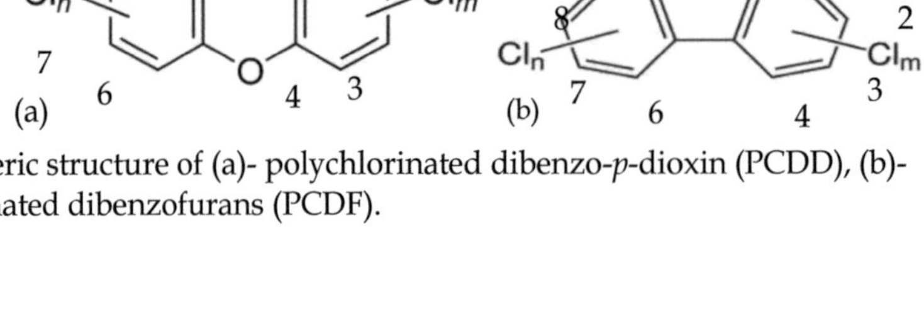

Fig. 9. Generic structure of (a)- polychlorinated dibenzo-*p*-dioxin (PCDD), (b)- polychlorinated dibenzofurans (PCDF).

PCDD/PCDF are considered unintentionally produced persistent organic pollutants (POPs), in respect to which the Stockholm Convention obliges Parties to decrease the total release. According to UNEP Chemical (2005) these by-products are formed in various processes and activities (e.g. waste incineration, metal and mineral production, heat and power generation, transport, open burning, production and use of goods) or they can be introduced into processes as contaminants in raw materials (e.g. disposal of wastes). In 1977, Olie and co-workers showed that PCDD/PCDF are in trace quantities in fly ashes and flue gases of MSW incinerators (Olie et al., 1977). Since then, it is well known that dioxins are formed in trace quantities in incineration processes (UNEP Chemical, 2005). McKay (2002) reviewed methods to minimise the formation of these very toxic compounds in MSWI. Indeed, in addition to PCDD/PCDF originally present in the furnace feedstock, three formation theories should be considered:

- precursor compounds such as chlorinated hydrocarbons already in the MSW feed or formed in furnace by reacting rapidly with other groups (e.g. chlorobenzene) ;
- *de-novo* synthesis of smaller and relatively innocuous chemical molecules combining together in the low temperature range;
- radical recombination with prevalence of the structures with longer life-span (more symmetrical and more stable).

The maximum formation rate of dioxins is in the range 300–400 °C, and the knowledge of this fact associated with the information about the reaction mechanisms have been the basis of designing combustion systems (McKay, 2002). Therefore, two strategies have been adopted for complete destruction in combustion and for minimize their formation. In this scope, it is very important that the combustion temperature should be above 1000 °C, the residence time should exceed 2 s, and high turbulence in the combustion chamber should be promoted (Reynolds number higher than 50,000). As an end-of-pipe treatment, it should be assured that a very rapid gas cooling from 450 to 250 °C is able to occur. Currently, semi-dry lime scrubbing and bag filters coupled with activated carbon injection have been playing an important role in prevention or minimisation of dioxins in stack gas emission to the environment (McKay, 2002). In addition, after the regulatory limit of 0.1 ng I-TEQ/Nm^3 entered into force, MSWI is no longer a significant sources of dioxins. In fact, some nineties studies refer that MSWI seem to be the most important source of PCDD/PCDF to air emissions (Olie et al., 1998). However, the emissions of MSWI of units that comply with the limit established in Directive 2000/76/EC are no longer so important (Abad et al., 2003; Lee et al., 2007). Indeed, measurements for MSWI plants, equipped with up-to-date APC systems showed that average concentration is around 0.048 ng I-TEQ/Nm^3, and the monitored values are all in the range 0.01-0.08 ng I-TEQ/Nm^3 (Abad et al., 2006). Table 10 shows the concentration of the seventy congeners that should be taken into account to calculate the equivalent toxicity (I-TEQ), from four different references (Ref1-Ref4).

At this point it is important to note that there is evidence that some additional treatment systems for flue gas may result in a dramatic decrease in dioxins concentrations emitted in stack flue gas, having been reported reductions of 99.98% (Lee et al., 2007). On the other hand, while the concentrations are very low, these chemicals are highly toxic. In 1997, the International Agency for Research on Cancer declared the 2,3,7,8-TCDD to be carcinogenic to humans (Abad et al., 2006). PCDD/PCDF are among the most widely studied organic chemical and there are a large number of publications on the toxicological effects (WHO 1989; Van Den Berg et al., 1994; Mukerjee, 1998). The effect of dioxin on humans is associated with increased risk of chloracne and hyperpigmentation, changed liver function

Congener	Ref1 (pg/Nm³)*	Ref2 (ng/Nm³)	Ref3 (ng/Nm³)	Ref4 (ng/Nm³) 1996	2000
2,3,7,8-TeCDD	0.34 ± 0.23	0.00259	0.0003	1.29	0.007
1,2,3,7,8-PeCDD	3.53 ± 2.23	0.00886	0.0004	15.41	0.026
1,2,3,4,7,8-HxCDD	5.92 ± 3.13	0.0109	0.0010	21.10	0.032
1,2,3,6,7,8-HxCDD	27.0 ± 16.6	0.0299	0.0010	48.88	0.085
1,2,3,7,8,9-HxCDD	11.2 ± 6.83	0.0155	0.0000	55.28	0.074
1,2,3,4,6,7,8-HpCDD	204.7 ± 104.4	0.213	0.0000	227.66	0.533
OCDD	403.6± 198.4	0.340	0.0000	251.24	0.665
2,3,7,8-TeCDF	9.25 ± 14.2	0.0190	0.0000	94.10	0.029
1,2,3,7,8-PeCDF	12.6 ± 8.1	0.0222	0.0000	10.81	0.010
2,3,4,7,8-PeCDF	30.0 ± 15.3	0.0388	0.0010	81.04	0.028
1,2,3,4,7,8-HxCDF	25.4 ± 13.1	0.0306	0.0000	121.76	0.052
1,2,3,6,7,8-HxCDF	46.3± 24.1	0.0350	0.0010	69.21	0.050
2,3,4,6,7,8-HxCDF	86.4± 46.9	0.0516	0.0010	10.47	0.003
1,2,3,7,8,9-HxCDF	21.4 ± 13.9	0.00446	0.0010	135.83	0.102
1,2,3,4,6,7,8-HpCDF	145.0± 93.5	0.110	0.0000	227.06	0.231
1,2,3,4,7,8,9-HpCDF	52.7 ± 34.9	0.0240	0.0000	68.01	0.040
OCDF	219.1 ± 210.0	0.0906	0.0000	198.91	0.345
ΣPCDD/PCDF	1305	1.05	0.0067	1638.1	2.312
I-TEQ (PCDD/DF)	46.9 ± 22.5	0.0511	0.0016	111.39	0.086

Ref.1- Abad et al. (2006); Ref.2 -Chen et al. (2008); Ref.3- Lee et al. (2007); Ref.4- Meneses et al (2004);

*-average value ± standard deviation.

Table 10. Concentration of PCDD/PCDF in air emissions from MSWI.

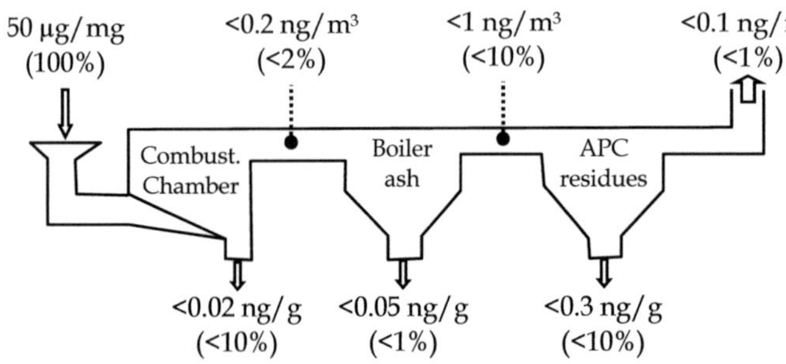

Fig. 10. PCDD/PCDF mass balance in modern MSW incinerator plant (adapted from McKay, 2002).

and lipid metabolism, changes in activities of various liver enzymes, depression of the immune system, and endocrine-and nervous-system abnormalities. It is a potent teratogenic and fetotoxic chemical in animals (Mukerjee, 1998). Indeed, 2,3,7,8-TCDD seems to be a very potent promoter of carcinogenesis liver of rat, and populations occupationally exposed to dioxins have increased incidences of soft-tissue sarcoma and non-Hodgkin's lymphoma.

Fig. 10 shows a representative mass balance in a modern MSWI unit, where it is implicit that overall the incineration process is now a sink of dioxins. In general, if high technology combustion is combined with sophisticated APC devices an emission factor of 0.5 µg I-TEQ/ton of MSW can be assumed to air, 15 µg I-TEQ/ton to fly ashes and 1.5 µg I-TEQ/ton to bottom ashes (UNEP, 2005).
At the end, it is very important to note that the 2,3,7,8-TCDD concentration in human tissue is decreasing by about a factor of 7 over a 25-year period (Hites, 2011).

4.11 Green house gases

Green house gases enclose carbon dioxide (CO_2), methane (CH_4), nitrous oxide (N_2O), hydrofluorocarbons (HFC) perfluorocarbons (PFC) and sulphur hexafluoride (SF_6). Specific emissions of CO_2 usually observed in MSWI are in the range of 0.1 to 1.7 ton/ton of MSW. The organic compounds of MSW are biomass and fossil wastes, and so the fraction of CO_2 from fossil must be taken into account for the climate changes. Concerning methane, since the furnace is kept at high oxidative conditions, no emissions of CH_4 are expected in stack gases. Some methane can be formed during long periods of waste storage and under anaerobic conditions. In the cases where primary air is supplied from the storage bunker, the CH_4 is oxidized to CO_2 and H_2O.
In Table 11 a summary of the most important techniques/technologies is indicated for the reduction of the main incineration pollutant described above (BREF, 2006).

5. The impact on health of emissions to air from MSWI

The activity of MSW incineration may pose environmental problems at various levels, concerning either ambient contamination provoked mainly by gaseous emissions or coming from the landfilling of APC residues after being submitted to solidification/stabilization treatment. Highly toxic PCDD/PCDF or harmful heavy metals such as Cd, Pb and Cr can be spread in the surroundings of incinerators plants and landfill disposal facilities. Ultimately, population health living nearby may be adversely affected by carcinogenic and non-carcinogenic compounds so that assessing their risk on the potential increase of human diseases is truly advisable. In this ambit some literature studies have addressed this theme.
Cangialosi et al. (2008) described the health impact associated to the gaseous emissions of a municipal solid waste incinerator in Italy, based on an atmospheric dispersion model involving pollutants content and human exposure level. Food intake revealed to be the major threat when compared with air inhalation and dermal contact, being, however, the carcinogenic (dioxins, furans, Cd) and non-carcinogenic (Pb and Hg) contamination values lower than those accepted. Therefore, their final conclusions supported that no relevant health problems may be endorsed to modern MSWI systems. Volatile organic compounds and bioaerosols possibly attributed to a MSWI plant in Spain (Vilavert et al., 2011) were checked and potential risks for human health were not identified once pollutants concentrations were similar to those observed in other regions. In a comparative study, Moy et al. (2008) reported that the landfill impact in human health through inhalation is five times higher than the one associated to the incineration operation, with cancer risk for both alternatives considered as acceptable when analysing data from New York City, whereas non-cancer health problems showed to be unacceptable. Concerning landfill sites for disposal of air pollution control (APC) residues from municipal solid wastes in UK, health risks for direct and indirect human exposure were also assessed with dispersion models by

Macleod et al. (2006) revealing that irrelevant danger was associated to inhalation while tolerable daily soil intake (TDSI) could be possibly surpassed due to the high deposition levels of those residues. Investigations about the impact of harmful PCDD/PCDF released

Pollutant	Techniques
Particles	Electrostatic precipitators Wet electrostatic precipitators Condensation electrostatic precipitators Ionization wet scrubbers Fabric filters Cyclones and multi-cyclones
Acid gases (HCl, HF, SO_x, …)	Wet-scrubber Semi-dray scrubber (e.g. suspension of lime) + bag filter Dry-scrubber (e.g. lime or sodium bicarbonate)
Direct desulphurisation	Injection of adsorbents (e.g. calcium compounds) directly into the incineration chamber
Oxides of nitrogen (NOx)	Primary techniques: air and temperature control, flue-gas recirculation, Secondary techniques: Selective Non- Catalytic Reduction (SNCR) and Selective Catalytic Reduction (SCR).
Hg	Primary techniques: separate collection, restrictions of receipt contaminated wastes Secondary techniques: scrubber by adding oxidants, activated carbon, furnace coke or zeolites
Other heavy metals	Converted into no-volatile oxides and deposited into fly ash, all techniques referred to remove particles can be applied. Activated carbon injection into scrubbing units.
Organic carbon compounds	Adsorption on activated carbon. SCR used for NOx. Catalytic bag filters Static bed filters Rapid quenching of flue-gas
Greenhouse gases (CO_2, N_2O)	All techniques used for NOx. Increase energy recovery efficiency
APC residues	Treated (e.g. solidification/stabilization and disposed of) Thermal treatment (vitrification, melting, sintering) Extraction and separation Chemical stabilisation
Bottom ash	Separation of metals Screening and crushing Treatment using ageing conditions High temperature slagging rotary kilns

Table 11. Most important techniques for the reduction of the main incinerator pollutants.

from a MSWI in Korea (Oh et al., 2006) showed that those emissions affected more directly atmosphere quality, with furan levels in air closer to the emitted values than those found in soil. In order to deepen consequences of environment contamination of MSWI, Cordier et al. (2004) analysed birth malformations in the population surrounding 70 incinerators in France. Though congenital problems were not significantly more relevant in those groups, higher incidence of some anomalies associated to renal, urology, cardiology and skin pathologies could be ascertained to persistent pollutants accumulated from the old incineration methodologies and road traffic pollution as well. The effect resulting from the application of more restrictive environment EU legislation regarding MSWI emissions was considered by Meneses et al. (2004), by determining PCDD/Fs concentrations in air, soil and vegetation, and their danger for human health, having concluded that both pollutants emissions and cancer risk was reduced after those measures implementation. Other contamination sources were, nevertheless, pointed out since pollutants emission levels diminished more significantly than health problems after the installation of new cleaning gas systems. Also, the replacement of old systems (electrostatic precipitator) by new technologies (semi-dry scrubber, fabric filters and charcoal injection) in Spain (Abad et al., 2003) revealed a significant decrease on dioxins emissions spread in stack gas, bottom ash and air pollution control residues, where the higher dioxin content was found. Heavy metal deposition around a MSWI area in USA showed quite different pollution concentrations during fall and winter (Feng et al., 2000). Methodologies envisaging minimisation of dioxins released from MSWI processes were reviewed by McKay (2002) accounting for three possible sources, furnace feedstock, precursor compounds or innocuous molecules forming dioxins by heterogeneous catalytic reactions or homogeneous thermal reactions. Finally, it is worthy to refer that in September 2009, the Health Protection Agency from UK analysed the on-going research about health impact of municipal solid waste incinerators and conclusions about the low influence in humans diseases of modern technologies were justified by the fact that emissions are nowadays well controlled, imparting, therefore, low contribution to global air pollution.

6. Conclusion

At present, municipal solid waste incineration (MSWI) in waste-to-energy (WtE) is considered to be an environmentally friendly technology, while allowing recovery of a large part of the energy enclosed in MSW. Indeed, MSWI is one of the main management options in most of the developed countries. Nonetheless, the main problems associated to these processes are the large volume of gaseous emissions which may pose environmental health risks (if not well cleaned) and the hazardous fly ash or air pollution control (APC) residues that remain after incineration. The main air pollution formed in MSWI that are nowadays under regulatory observation are: dust, CO, TOC, HCl, HF, SO_2, NO_x, Hg, Cd, Tl, Sb, As, Pb, Cr, Co, Cu, Mn, Ni, V and PCDD/PCDF.

However, it is very important to note that air pollution induced by MSWI has been strongly reduced by some up-to-date sophisticated air pollution control equipments. In fact, process engineering of the gas cleaning systems can still be extended to lower emission limits if the environmental legislation so requires. It is only a question of overall cost of operation per ton of incoming waste to be incinerated.

Some studies that analysed in detail the health impact associated to the gaseous emissions of MSWI have revealed that no relevant health problems may be associated to modern MSWI systems.

7. References

Abad, E. Martinez, K. Caixach, J. Rivera, J., (2006). Polychlorinated dibenzo-p-dioxins, dibenzofurans and 'dioxin-like' PCBs in flue gas emissions from municipal waste management plants, *Chemosphere*, 63, pp. 570–580.

Abad, E., Caixach, J., Rivera, J. (2003). Improvements in dioxin abatement strategies at a municipal waste management plant in Barcelona, *Chemosphere*, 50, pp. 1175–1182.

Achternbosch, M., Richers, U. (2002). *Materials flows and investment costs of flue gas cleaning systems of municipal solid waste incinerators*, Forschungzentrum Karlsruhe GmbH, ISSN 0947-8620, Karlsruhe, Germany.

Amutha Rani, D., Boccaccini, A.R., Deegan, D., Cheeseman, C.R. (2008). Air pollution control residues from waste incineration: Current UK situation and assessment of alternative technologies, *Waste Management*, 28, 2279-2292.

Bordado, J.C.M., Gomes, J.F.P. (1999). Practical determination of the efficiency of electrostatic precipitators. *Powder and Bulk Engineering*, 2 , 37-42.

BREF (2006). *Integrated Pollution Prevention and Control – Reference Document on the Best Available Techniques for Waste Incineration*, European Commission.

Cangialosi, F., Intini, G. , Liberti, L., Notarnicola , M., Stellacci, P., (2008). Health risk assessment of air emissions from a municipal solid waste incineration plant - A case study, *Waste Management* , 28, pp. 885–895.

Chen, C.-K., Lin, C., Lin, Y.-C., Wang, L.-C., Chang-Chien, G.-P. (2008). Polychlorinated dibenzo-*p*-dioxins/dibenzofuran mass distribution in both start-up and normal condition in the whole municipal solid waste incinerator, *Journal of Hazardous Materials*, 160, pp. 37–44.

Clark, J.W., Bordado, J.C.M., Correia, C.S.R., Correia, N.D.S., Oil absorption foam, WO2008043545, Appl. 17April 2008.

Cordier, S., Chevrier, C., Robert-Gnansia, E., Lorente, C., Brula, P., Hours, M. (2004). Risk of congenital anomalies in the vicinity of municipal solid waste incinerators, *Occupational and Environmental Med*icine, 61, pp. 8–15.

DEFRA (2007). *Incineration of Municipal Solid Waste*, Report of Department for Environment, Food & Rural Affairs, UK.

Directive 2000/76/EC of The European Parliament and the Council, of 4 December 2000, on the incineration of waste.

Directive 2008/98/EC of The European Parliament and the Council of 19 November 2008 on waste and repealing certain Directives.

Encyclopedia of emulsion technology, Ed. Paul Becher, Marcel Dekker, New York (1988).

Eurostat (2010). Environmental statistics and accounts in Europe, Eurostat Statistical books, European Commission. <http://www.epp.eurostat.ec.europa.eu>

Feng, X., Melander, A.P., Klaue, B., (2000). Contribution of municipal waste incineration to trace metal deposition on the vicinity, *Water, Air, and Soil Pollution*, 119, pp. 295–316.

Gentil, E., Clavreul, J., Christensen, T.H. (2009). Global warming factor of municipal solid waste management in Europe, *Waste Management and Research*, 27, pp. 850-860.

Hites, R.A. (2011). Dioxins: an overview and history, *Environment Science and Technology*, 45, pp. 16-20.

IAWG (1997). *Municipal Solid Waste Incinerator Residues*, The international ash working group, Studies in environmental Science 67, Elsevier, The Netherlands.

Jenkins, B.M., Baxter L.L., Miles Jr., T.R., Miles T.R. (1998). Combustion properties of biomass, *Fuel Processing Technology*, 54, pp. 17–46.

Jungten, H. Richter, E. Knoblauch K., Hoang-Phou T., (1988) Catalytic NOx reduction by ammonia and carbon catalysts, *Chemical Engineering Science*. 43, 419-428.

Katsouyanni, K. , Touloumi, G., Spix, C., Schwartz, J., Balducci, F., Medina, S., Rossi, G., Wojtyniak, B., Sunyer, J., Bacharova, L., Schouten, J. P., Ponka, A., Anderson, H. R. (1997) . Short term effects of ambient sulphur dioxide and particulate matter on mortality in 12 European cities: results from time series data from the APHEA project, *British Medical Journal* , 314, pp. 1658-1663.

Lee, C.C., Huffman, G.L. (1996). Medical waste management/incineration, *Journal of Hazardous Materials*, 48, 1-30.

Lee, S.-J., Choi, S.-D., Jin, G.-Z., Oh, J.-E., Chang, Y.-S., Shin, S. K., (2007). Assessment of PCDD/F risk after implementation of emission reduction at a MSWI, *Chemosphere*, 68, pp. 856–863.

Macleod, C., Duarte-Davidson, R., Fisher, B., Ng, B., Willey, D., Shi, J.P., Martin, I., Drew, G., Pollard, S. (2006). Modeling human exposures to air pollution control (APC) residues released from landfills in England and Wales, *Environment International*, 32, pp. 500–509.

McKay, G. (2002). Dioxin characterisation, formation and minimisation during municipal solid waste (MSW) incineration: review, *Chemical Engineering Journal*, 86, pp. 343-368

Meneses, M., Schuhmacher, M., Domingo, J.L. (2004). Health risk assessment of emissions of dioxins and furans from a municipal waste incinerator: comparison with other emission sources, *Environment International*, 30, pp. 481–489.

Morcos, V.H. (1989). Energy recovery from municipal solid waste incineration - a review. *Heat Recovery Systems and CHP*, 9, 115-126.

Moy, P. Krishnan, N. Ulloa, P. Cohen, S. Brandt-Rauf, P.W. (2008). Options for management of municipal solid waste in New York City: A preliminary comparison of health risks and policy implications, *Journal of Environmental Management*, 87, pp. 73–79.

Mukerjee, D. (1998). Health impact of polychlorinated dibenzo-p-dioxins: A critical review, *Journal of the Air & Waste Management Association*, 48, pp. 157-165.

Niessen, W.R. (2002). *Combustion and incineration processes*. 3rd Ed., Marcel Dekker Inc., New York, USA.

Nunes, A. Duarte, A. Bordado, J., PCT Patent, 2011.

Oh, J.-E., Choi, S.-D., Lee, S.-J., Chang, Y.-S., (2006). Influence of a municipal solid waste incinerator on ambient air and soil PCDD/Fs levels, *Chemosphere*, 64, pp. 579–587.

Olie, K., Addink, R., Schoonenboom, M., (1998). Metals as catalysts during the formation and decomposition of clhorinated dioxins and furans in incineration processes, *Journal of the Air & Waste Management Association*, 48, pp. 101-105.

Olie, K., Vermeulen, P., Hutzinger, O., (1977). Chlorodibenzo-p-dioxins and chlorodibenzofurans are trace components of fly ash and flue gas of some municipal incinerators in The Netherlands, *Chemosphere*, 6, pp. 455-459.

Quina, M.J., Bordado, J.C.M., Quinta Ferreira, R.M., (2008a) Treatment and Use of Air Pollution Control Residues from MSW Incineration: an Overview, *Waste Management*, 28, pp. 2097-2121.

Quina, M.J., Santos, R.C., Bordado, J.C.M., Quinta Ferreira, R.M. (2008b). Characterization of air pollution control residues produced in a municipal solid waste incinerator in Portugal, *Journal of Hazardous Materials,* 152, pp. 853-869.

Sundqvist, J.O. (2005). *How should municipal solid waste be treated – a system study of incineration, material recycling, anaerobic digestion and composting,* Report of Swedish Environmental Research Institute.

UNEP Chemicals (2005). *Standardized Toolkit for identification and quantification of dioxin and furan releases*. Genebra, Switzerland.

Van Den Berg, M., Dejongh, J., Poiger, H., Olson, J.R. (1994). Toxicokinetics and metabolism of polychlorinated dibenzo-*p*-dioxins (PCDDs) and dibenzofurans (PCDFs) and their relevance for toxicity. *Critical review in toxicology,* 24, pp. 1–74.

Vilavert, L., Nadal, M., Figueras, M. J., Kumar, K. , Domingo, J.L. (2011). Levels of chemical and microbiological pollutants in the vicinity of a waste incineration plant and human health risks: temporal trends, *accept for publication in Chemosphere.*

WHO (1989). *Polychlorinated dibenzo-para-dioxins and dibenzofurans.* Geneva, World Health Organization, (Environmental Health Criteria, No. 88).

World Bank Report (1999). *Municipal Solid Waste Incineration,* Technical Guidance Report of World Bank, Washington D.C.

Part 4

The Impact of Air Pollution in the Economy and Development Goals: Case Studies

17

Indoor Air Pollution in Mexico

Iván Nelinho Pérez Maldonado, Lucia Guadalupe Pruneda Álvarez, Fernando Díaz-Barriga, Lilia Elisa Batres Esquivel, Francisco Javier Pérez Vázquez and Rebeca Isabel Martínez Salinas
Laboratorio de Toxicología, Facultad de Medicina, Universidad Autónoma de San Luis México

1. Introduction

General health conditions in Mexico have improved considerably over the last 50 years. This is clearly seen in life expectancy at birth. Between 1950 and 2000, the country added 25 years to this indicator. However, the respective differences among the states show the inequality prevailing in the country (SSA, 2005).

The differences found per region are an important fact to consider in understanding the Mexican situation. In the modern and industrialized northern region, the population is concentrated mostly in urban zones. The southern region is clearly traditional, unindustrialized and with a high indigenous population living in small, dispersed rural communities. This explains why the health indicators for the northern region are similar to those of developed countries, while the same indicators for the southern region are similar to those of developing countries with social and economic problems.

In 2002, there were 43,719,756 persons under 19 years of age in Mexico, representing just over 40 percent of the total population. The birth rate per 1,000 inhabitants has decreased from 45 in 1960 to 17 in 2000, while the child mortality rate in the first year of life decreased from 19.1 per 1,000 inhabitants in 1998 to 16.78 per 1,000 inhabitants in 2002. The leading causes of death also changed radically over the past 50 years. Transmissible diseases and congenital illnesses were displaced as the primary causes of death by non-transmissible diseases and injuries. In the same period, the percentage of deaths due to intestinal infections decreased by a factor of 14 (from 14.3 percent to 1 percent), while deaths from heart disease quadrupled (from 4 percent to 16 percent; SSA, 2005).

Another fact pointing to a clear epidemiological transition in the country is the changing pattern of causes of mortality among children under one year of age. Previously, infant deaths were concentrated in the postnatal period (between one month and one year of age), mostly caused by acute respiratory infections and diarrhea. Presently, infant deaths are concentrated in the first 28 days of life, due to perinatal causes and congenital anomalies requiring high-technology intervention (SSA, 2005).

Common infections and congenital events continue to cause major harm to health in highly marginalized groups. Over the last 10 years, deaths from diarrhea in children under five years of age have decreased by 85 percent. However, there are southern states with mortality rates from diarrhea above 40 per 100,000 inhabitants under five years of age, i.e., five times higher than the rates found in the more developed northern states. The same is found with deaths from acute respiratory infections, another clear example of the persistently lower quality of life (SSA, 2005).

Despite the major progress in health systems, the problems of poverty, social inequality, marginalization, the lack of services, and environmental air, water and soil pollution constitute important factors associated with a poor quality of life for a high percentage of the child population, primarily in the country's rural areas.

Air pollution is a generalized problem in Mexico's major metropolitan areas. However, current demographic growth, industrial concentration, greater numbers of vehicles, increased fuel consumption and inadequate urban mobility patterns have caused the problem to increase in other areas, such as medium-size cities.

Indoor air pollution caused by the burning of wood or charcoal used for cooking or heating constitutes a public health problem with repercussions for the population under five years of age and women of reproductive age, especially in the country's marginalized areas. In 1990, one of every three Mexicans (91 percent of rural inhabitants and 11 percent of urban inhabitants) used wood for cooking. In 1993, 25.6 million persons were estimated to use wood as household fuel, decreasing to 17.2 million inhabitants in 2000.

2. Biomass as energy source

The use of solid fuels for cooking and heating is likely to be the largest source of indoor air pollution on a global scale. Nearly half the world continues to cook with solid fuels such as wood, crop residues, agricultural wastes, and animal dung, with wood being the most commonly used (Bruce et al. 2000). When used in simple cooking stoves, these fuels emit substantial amounts of toxic pollutants. In households with limited ventilation (as is common in many developing countries), exposures experienced by household members, particularly women and young children who spend a large proportion of their time indoors, have been measured to be many times higher than World Health Organization (WHO) guidelines and national standards (Bruce et al. 2000; Smith 1987). In this regard, the number of annual deaths in developing countries associated with domestic biomass combustion is estimated at 1,849,000 (SSA, 2005).

In México currently, biofuels represent about 8% of the total energy demand (SE, 1998), 46% of residential energy use (Figure 1) and more than 80% of the energy demand in the rural sector (Masera, 1996b). The three main types of biofuels used in the country are: bagasse, which is used in the sugar cane industry, fuelwood and charcoal. Fuelwood is by far the dominant woodfuel, with charcoal being used mainly in street industries and for barbecues. Total fuelwood use accounts for 3 times the total commercial timber legally harvested in the country (Masera, 1996a). In this regard, in 1990, one of three Mexicans used fuelwood for cooking, including 91 percent of rural residents and 11 percent of urban residents. It is estimated that 25.6 million people used this fuel in their homes in 1993 and that by 2000 this number had declined to 17,256,471. Fuelwood is also used in many small (cottage) industries, like pottery making, "tortilla" making, brick making, and others. The use of fuelwood is concentrated in rural areas and small towns. Fuelwood markets are mostly of a local nature. Most fuelwood comes from forest areas (including here all degraded lands and semi-arid forests); many of the species used are of no commercial value, and the use of agricultural residues and dung is not widespread (Masera, 1996a; Masera et al, 1997). Among the states in México with the highest use of fuelwood are Oaxaca and Chiapas, where it is estimated that 50-60 percent of the population uses this type of fuel (Figure 2). The general pattern is that a higher proportion of people are exposed to fuelwood and charcoal in the southern part of the country. These are largely rural states with some of the poorest populations; thus, exposure to fuelwood-related pollutants is more prevalent here.

The heaviest biomass fuel usage by household is in southern and north central Mexico, where utilization may approach 90 to 100 percent in some localities (Figure 3). These are largely rural states with some of the poorest populations.

The majority of households using solid fuels in México burn them in open fires or simple stoves that release most of the smoke into the home (Figure 4). The resulting indoor air pollution is a major threat to health, particularly for women and young children, who may spend many hours close to the fire. Furthermore, the reliance on solid fuels and inefficient stoves has other, far-reaching consequences for health, the environment, and economic development (Reddy et al., 1997; WHO, 1997).

3. Indoor air pollution

The current dominant pattern of household fuelwood use presents several problems. People depend mostly on

open fires, leading to very high indoor air pollution levels, particularly for women and children. Fuelwood combustion emits a complex mixture of organic compounds and gases, which include carbon monoxide (CO), nitrogen and sulphur oxides (NOx and SOx), aldehydes, polycyclic aromatic hydrocarbons (PAHs), volatile organic compounds (VOCs), chlorinated dioxins, breathable particulate matter (PM) with diameters < 10 microns (PM10),

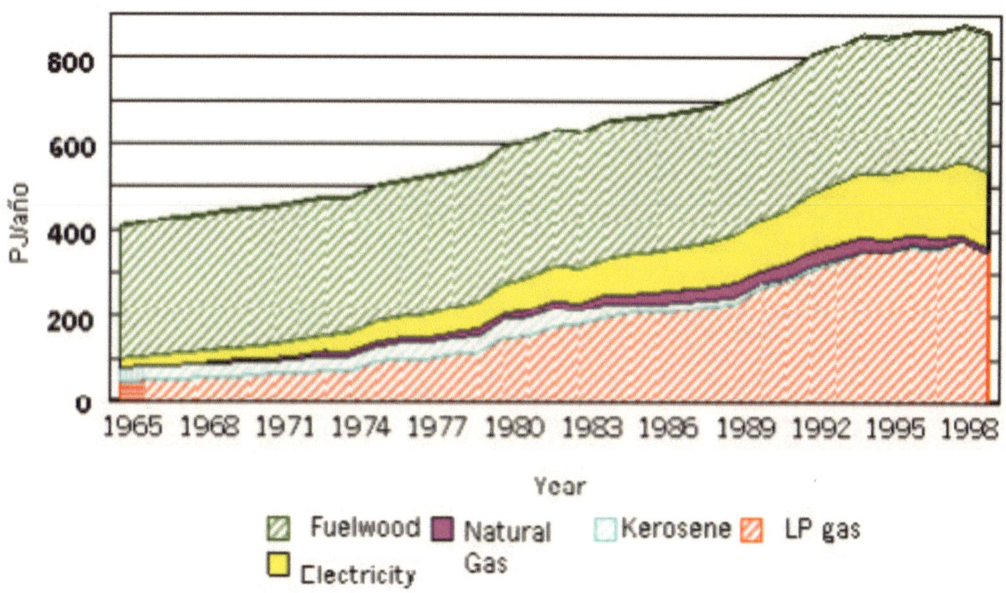

Fig. 1. Residential Energy Use in Mexico 1965-1998. **Source:** Masera O.R, Drigo R., and Trossero M.A. 2003. Woodfuels integrated supply/demand overview mapping. Universidad Autónoma de México, FAO-EC Partnership Programme. Food and Agriculture Organization of the United Nations.

and free radicals (Albalak, 2001; Mishra, 2003; Table 1). Moreover, wood users rely on simple and rustic stoves such as open "three-stone" fires and mud, clay, or metal stoves that result in incomplete and inefficient combustion (Reddy et al., 1997; WHO, 1997).

Compound	Major indoor sources
Carbon monoxide	Fuel/tobacco combustión
Nitrogen oxides	Fuel combustión
Sulphur oxides	Coal combustion
Aldheydes	Furnishing, construction materials, cooking
Polycyclic aromatic hydrocarbons	Fuel/tobacco combustión, fumes from food being cooked.
Volatile and semi-volatile organic compounds	Fuel/tobacco combustión, consumers products, Furnishing, construction materials, fumes from food being cooked.
Chlorinated dioxins	Fuel/tobacco/ coal combustión
Breathable particulate matter	Fuel/tobacco combustión, cleaning, fumes
free radicals	Indoor chemistry

Table 1. Toxic compounds generated during fuelwood combustion.

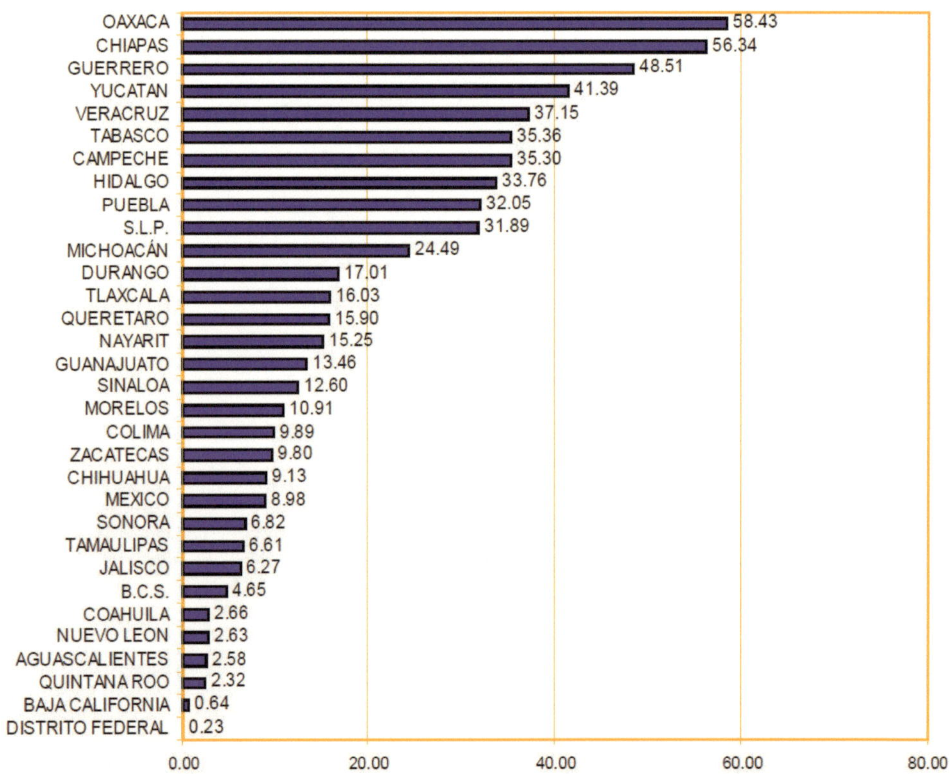

Fig. 2. Percentage of Mexico's General Population Exposed to Biomass Smoke, by Region, 2000. Source: Diagnóstico Nacional de Salud Ambiental y Ocupacional 2002. Dirección General de Salud Ambiental.

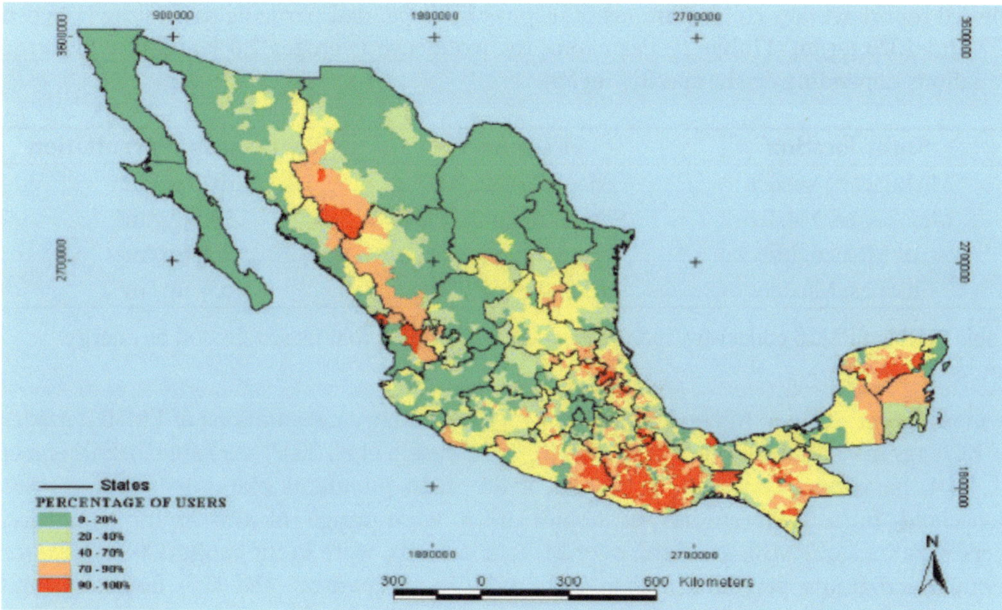

Fig. 3. Percentage of Fuel Wood Users, at the Municipal Level, in Mexico, 2000. **Source:** Masera O.R, Drigo R., and Trossero M.A. 2003. Woodfuels integrated supply/demand overview mapping. Universidad Autónoma de México, FAO-EC Partnership Programme. Food and Agriculture Organization of the United Nations.

Fig. 4. Traditional wood-burning cookstove.

For example, studies carried out in rural households belonging to different regions of Mexico report average concentrations of respirable suspended particulates ranging between 537 and 1020 μg/m³ (Table 2). Per capita fuelwood use averages 2.0 kg/day, with large variations depending on the specific region.

Study location	Reference	PM10 mean concentration
Michoacán, México	Masera et al. 2007	1020 μg/m³
Michoacán, México	Saatkamp et al. 1999	995 μg/m³
State of México, México	Brauer et al. 1996	768 μg/m³
Chiapas, México	Riojas, et al. 2001	537 μg/m³

Table 2. Mean PM10 concentration in Mexican households that use fuelwood as energy source.

A more recent study in Michoacán, México report average concentrations of PM10 particles of 1020 μg/m³ with maxim levels of 4230 μg/m³ (Masera et al. 2007; see Table 2). Saksena et al. 2004, have compiled data on several of the main pollutants associated with various household fuels from studies of homes in a wide range of developing countries. Concentrations of PM10, averaged over 24-hour periods, were in the range 300 to 3,000 (or more) micrograms per cubic meter (μg /m³). By comparison, the U.S. Environmental Protection Agency's annual air pollution standard for PM10 is 50 μg/m³, one to two orders of magnitude lower than levels seen in many homes in developing countries.

With use of biomass, CO levels are generally not as high in comparison, typically with 24-hour averages of up to 10 parts per million (ppm), somewhat below the World Health Organization (WHO) guideline level of 10 ppm for an eight hour period of exposure. Much higher levels of CO have been recorded, however. For example, a 24-hour average of around 50 ppm was found in Kenyan Masai homes (Bruce et al. 2002), and one Indian study reported carboxyhemoglobin (COHb) levels similar to those for active cigarette smokers (Behera et al. 1988). In this regard our group, demonstrated in a rural community in San Luis Potosí, Mexico that use biomass as the principally energy source, mean COHb level of 4.93% and moreover, 55% of those people presented COHb levels above 2.5% considered a safe level (Kleinman, 2000). Moreover, this 55% of people had geometric mean COHb levels of 8.39% (Torres-Dosal et al. 2008). In this study also was monitored exposure to PAH´s in urine of people using 1-hydroxypyrene (1-OHP) as exposure biomarker to PAH´s. Because PAH exposure occurs as a mixture of compounds, and because pyrene is almost always found in this mixture, pyrene and its metabolite 1-hydroxypyrene (1-OHP) are considered appropriate surrogate markers of total PAH exposure (Jacob and Seidel, 2002). Results show that the geometric mean level of 1-OHP ranged between 1.1 to 17.8 μmol/molCr (mean ± SD; 6.71 ± 3.58 μmol/molCr; (Torres-Dosal et al. 2008). The levels of 1-OHP found in this study are higher than the mean urinary 1-OHP concentrations from people throughout the world (range: 0.03–0.76 μmol/mol creatinine; Levine, 1995; Zhao et al., 1992). These values are also higher than levels observed in people living in traditional houses from rural districts in Burundi (1.50 μmol/mol creatinine, range: 0.26–15.62), that use wood as the principal fuel (Viau et al., 2000). We don't know the type of wood burned in Burundi, but it is possible that the difference in exposure between Mexico (our study) and Burundi is the type of wood used. It has been shown that different emission rates of PAHs and other compounds are dependent on fuel and burning conditions (Jenkins et al., 1996; Zhang and

Smith, 1999; Mcdonald et al., 2000). In a more recent study performed in México, our group evaluated PAHs exposure in children in different residential categories (sites included in the study are recognized for their industrial activity, indoor wood combustion, waste disposal and brick manufacturing using different materials as fuel sources). We performed a random sampling in nine communities in Mexico (Table 3). The participant group consisted of 65 children living in communities with brick furnaces (communities of San Vicente and Tercera in San Luis Potosí (SLP), state); 105 children who live in houses where firewood is the domestic fuel (communities of Ramonal in Quintana Roo state, Ventanilla in Oaxaca state, Victoria in Chiapas state and Tancuime in San Luis Potosí, state); 32 children living next to a municipal landfill in San Luis Potosi city with waste combustion (community of Milpillas) and 56 children living in areas next to highways with moderate and high vehicular trafficc (communities of El Centro and Domingo in San Luis Potosí, state, respectively). The highest levels of 1-OHP in this study were found in Victoria, Chiapas (mean, 4.4 ± 3.7 μmol/mol creatinine; 5.9 ± 5.1 μg/L), in this community, children are exposed to biomass combustion. Interestingly, the communities that also are using biomass combustion as Victoria (i.e., Ramonal, Ventanilla and Tancuime) had children with urinary 1-OHP mean levels, similar to those found in Victoria (Martínez-Salinas et al. 2010). When grouped by exposure scenario, (a) moderate vehicular traffic; (b) heavy vehicular traffic; (c) fumes from a municipal landfill; (d) fumes from brick kilns and (e) indoor air pollution by biomass combustion, the highest levels of urinary 1-OHP were found in children exposed to indoor air pollution (approximately one order of magnitude higher than the other scenarios; Table 4). Children living in communities with brick kiln industry, children living in Milpillas (landfill) and children living in Domingo (heavy vehicular traffic) were the next communities in order (Table 4), leaving the community of El Centro (moderate vehicular traffic; Table 4) at the end (Martínez-Salinas et al. 2010). Moreover, Jongeneelen (2001) proposed a benchmark guideline for occupational exposure to PAHs, taking into account urinary 1-hydroxypyrene levels. Following this guideline, the reference value as a 95th percentile in non occupational exposed controls is 0.24 μmol/mol creatinine and 0.76 μmol/mol creatinine for non-smokers and smokers, respectively (first level). A no-biological-effect-level of 1-hydroxypyrene in urine for exposed workers was fixed at 1.4 μmol/mol creatinine. It is the lowest reported level at which no genotoxic effects were found and therefore the estimate for the second level of the benchmark guideline. Finally, two reference values for the third level were proposed 2.3 μmol/mol creatinine and 4.9 μmol/mol creatinine, in two types of industry, coke ovens and primary aluminum production, respectively, and it was designated occupational exposure limit (OEL). When our results in children were compared with this guideline for adult workers, only in El Centro did we find a low risk condition; in the rest of the communities, an important percentage of children were found at risk (Table 5). It is important to note that the guideline values proposed by Jongeneelen (2001) are derived for workers and for adult populations. Thus, our results are more important in terms of public health as we studied children in non-occupational scenarios.

Finally, taking account the above results mentioned assessed in children, our group performed a study to evaluated 1-OHP levels in women (aged 15-50 years) in two exposure scenarios: 1) People living in an indigenous community in La Huasteca in San Luis Potosi state that use fuelwood as the principally energy source and 2) People living in the city of San Luis Potosi that use gas LP as the principally energy source. As expected the levels of 1-OHP in urine were higher in people that use fuelwood (table 6).

Community	Characteristics
Victoria, Chiapas	Rural community with fuelwood combustion
Ventanilla, Oaxaca	Rural community with fuelwood combustion
Ramonal, Quintana Roo	Rural community with fuelwood combustion
Tancuime, SLP	Rural community with fuelwood combustion
San Vicente, SLP	Community with 50 brick kilns using different materials as fuel source
Milpillas, SLP	Community living next to municipal landfill in the city of San Luis Potosí
Tercera, SLP	Community with 75 brick kilns using different materials as fuel source
Domingo, SLP	Community in San Luis Potosí state exposed to heavy vehicular traffic
Centro, SLP	Urban community in the city of San Luis Potosí with moderate vehicular traffic

Table 3. Characteristics of studied sites

Residential Group	Mean ± SD 1-OHP (μmol/mol creatinine)
moderate vehicular traffic	0.08 ± 0.2
heavy vehicular traffic	0.20 ± 0.2
fumes from a municipal landfill	0.30 ± 0.4
fumes from brick kiln	0.35 ± 0.3
indoor air pollution by biomass combustion	2.26 ± 2.8

Table 4. Urine concentrations of 1-OHP in children by exposure scenario in México.

4. Health effects associated with indoor air pollution

The adverse effects on respiratory health of products of incomplete solid-fuel combustion are summarized in Table 7, which also includes some of the known or proposed mechanisms of damage. Exposure to solid fuel smoke can be lifelong, beginning before birth and early infancy, and continuing during adulthood, especially in women, who are traditionally charged with the task of cooking. Exposure is longer in cold communities that require fire-related heating, and may adversely impact lung growth and development, both directly and through an increase in lung infections. Indoor air pollution from indoor burning of solid fuels has been associated with an increased risk of several diseases and health conditions (Table 7). In general, studies are scarce, and show varied health outcomes. Moreover, they commonly lack a quantitative exposure assessment, and rely instead on qualitative or semi-quantitative indicators, such as the use of open fire indoors. The amount of time that children and/or women spend in proximity to fires is the crucial determinant of the health impact of indoor air pollution (Barnes et al. 2005). For other health outcomes, the adverse effects of exposure to solid fuel smoke from coal or biomass is expected, as from exposure to tobacco smoke, but information is lacking or scarce about other consequences such as low birth weight and adverse perinatal outcomes (stillbirth), among others (WHO, 2007) For example, biomass smoke in Guatemalan women has been shown to increase diastolic blood pressure (McCracken et al. 2007) According to World Health Organization estimates, worldwide exposure to solid fuel smoke produces 1.6 million deaths yearly, 693 000 due to COPD and 910 000 due to acute lower respiratory infections (ALRI), as well as 38.5 million disability adjusted life years (DALYs), most due to ALRI, being the eighth overall cause of DALYs in the world and the eleventh cause of death (Smith et al. 2004). This is likely an underestimation, as only diseases with a strong evidence base, i.e., COPD, ALRI, and lung cancer from coal burning, are considered.

Residential Group	<0.24 μmol/mol creatinine (%)	0.24-1.39 μmol/mol creatinine (%)	1.4-2.3 μmol/mol creatinine (%)	>2.3 μmol/mol creatinine (%)
moderate vehicular traffic	95.0	5.0	0.0	0.0
heavy vehicular traffic	53.0	47.0	0.0	0.0
fumes from a municipal landfill	66.0	32.0	2.0	0.0
fumes from brick kiln	51	47.0	2.0	0.0
indoor air pollution by biomass combustion	7.0	14.0	25.0	54.0

Table 5. Percentage of children in each range of 1-OHP urinary levels.

Community	Mean ± SD 1-OHP (μmol/mol creatinine)
La Huasteca in San Luis Potosi state	3.97 ± 5.51
City of San Luis Potosí	0.44 ± 0.70

Table 6. Urine concentrations of 1-OHP in women (aged 15-50 years) in San Luis Potosi state.

In México information regarding health effects associated with indoor air pollution is scarce. However, in a study conducted in the rural village of Solis, composed of 13 small communities, located 200 km northwest of Mexico City and at an elevation of 2,600 m above sea level. Women in Solis have cooked with wood, crop residues, and corn cobs as fuel, but now some of them also use natural gas either alone or supplemented with biomass fuel. When compared with those cooking with gas, current use of a stove burning biomass fuel was associated with increased reporting of phlegm (27 vs. 9%) and reduced FEV1/FVC (79.9 vs. 82.8%). Levels of FEV1 were 81 ml lower and cough wasmore common (odds ratio, 1.7;

95% confidence interval, 1.0–2.8) in women from homes with higher PM10 concentrations. All women found with moderate airflow obstruction were cooking with biomass stoves (Regalado et al. 2006). Similar results were found by Romieu et al. 2009.
In order to study other health effects associated with indoor air pollution, our group evaluated blood carboxyhemoglobin (% COHb) and DNA damage in blood cells (with comet assay) in a community in San Luis Potosi state exposed to wood smoke (Torres-Dosal et al. 2008). Since carbon monoxide (CO) is probably the most important single contaminant emitted during combustion of wood (Viau et al., 2000), A marker use to indirectly assess CO exposure is carboxyhemoglobin (COHb) that reflects binding of CO to the hem portion of hemoglobin. A concentration of COHb <2.5% is currently considered safe (Kleinman, 2000). The lowest level of COHb, at which adverse effects are observed, ranges from 2.9 to 3% (Estrella et al., 2005). Moreover, COHb concentrations >5% are associated with effects, such as neurobehavioral function, impaired visual function, task performance, and maintaining alertness (WHO, 1999; EPA, 2000; Raub and Benignus, 2002). All studied individuals had a geometric mean COHb level of 4.93% and 55% of those people presented COHb levels above 2.5% considered a safe level. While, DNA damage in people exposed was 5.8 ± 1.3 of Tail Moment. In this study also was monitored 1-OHP in studied individuals and significant positive correlation was obtained between urinary 1-OHP and DNA damage in blood cells. Our data suggest that DNA damage in lymphocytes is at least partially related to exposure to wood smoke, as a relationship was found between DNA damage in lymphocytes and levels of 1-OHP in urine of people studied (Table 8).

Compound	Potential health effect
Breathable particulate matter	Wheezing, exacerbation of asthma, respiratory infections, chronic bronchitis, and chronic obstructive pulmonary disease.
Carbon monoxide	Low birth weight, increase in perinatal deaths
Polycyclic aromatic hydrocarbons	Lung cancer, cancer of mouth, nasopharynx and larynx.
Nitrogen oxides	Wheezing, exacerbation of asthma, respiratory infections, reduced lung functions in children.
Sulphur oxides	Wheezing, exacerbation of asthma, Exacerbation of chronic obstructive pulmonary disease, cardiovascular disease.

Table 7. Adverse effects on respiratory health associated with indoor air pollution

5. Interventions programs

Far from a relatively simple problem with fixed technical solutions, the "biofuel problématique" in developing countries has turned out to be a very complex issue with multiple implications. The patterns of household biofuel use are very heterogeneous, as are also the people, the environment and the cultures that depend on these fuels to solve their essential cooking needs. The uses of energy in the home—for example, for cooking and

keeping warm and as a focus of social activities—have important attributes that are specific to the locality, culture, and individual households and are often associated with established traditions and deeply held beliefs. Encouraging the use of cleaner and more efficient energy technologies by populations that are among the poorest in the world has not been easy, but recent years have seen progress being made with respect to suitable technology that meets the needs of households and with respect to the development of supportive policy.
There is limited experience with improved wood-burning cookstove programs in Mexico (Olguín, 1994; Arias and Cervantes, 1994; Navia, 1992; Dutt et al., 1989). The government launched a large-scale effort at the beginning of the '80s, with poor results, and basically abandoned the field (Vargas, 1990). In recent years, a number of initiatives have been started in different regions of Mexico, particularly within the states of Michoacan, Chiapas, Oaxaca and San Luis Potosí, working with a diversity of stove designs and dissemination schemes (RETA, 2004; Cayetano, 1997). Most groups still work with massive Lorena-type improved stoves, but there are groups now disseminating "Justa" and portable "Rocket" stoves in southern Mexico. Most of the improved cookstove programs are local (village- level) or micro-regional in scope, they are generally part of larger initiatives directed at restoration of local forests or conservation of biodiversity, or they are part of the portfolio of well-established peasant organizations. Efforts are still concentrated on the construction of stoves with little or no follow-up; stoves are entirely or mostly subsidized; there is still emphasis on stoves being constructed by users themselves. Some institutions promoting these stove programs have become more aware of users' priorities and needs, resulting in higher stove acceptance rates.

	Before program intervention	**After program intervention**
% COHb (mean ± SD)	4.9 ± 4.3	1.0 ± 0.2
1-OHP (μmol/mol creatinine; mean ± SD)	6.7 ± 3.6	4.8 ± 3.3
Tail moment (comet assay; mean ± SD)	5.8 ± 1.2	2.4 ± 1.3

Table 8. A risk reduction program using biomarkers of exposure (COHb and 1-OHP) and a biomarker of effect (DNA damage).

In this regard, several intervention programs have been implemented. For example, the Grupo Interdisciplinario de Tecnología Rural Apropiada (GIRA) has recently disseminated 4,000 improved Patsari cookstoves, most of them in the Purépecha region of Michoacán state, Mexico. In paired comparisons in a subset of kitchens in a single community before and after installation of an improved Patsari cookstove, 48-hour average kitchen concentrations of carbon monoxide (CO) and fine particulate matter (PM2.5) were reduced by 66 % (n = 32) and 67 % (n = 33), respectively (Masera et al. 2007). The results are excitants; however health concerns are now a major focus of the new initiatives. A randomized study was conducted in the Central Mexican state of Michoacan. Households were randomized to receive the Patsari stove or keep their traditional open fire. A total of 552 women were followed with monthly visits over 10 months to assess stove use, inquire about respiratory and other symptoms, and obtain lung function measurements. Women who reported using the Patsari stove most of the time compared with those using the open fire had significantly lower risk of respiratory symptoms (relative risk [RR], 0.77; 95%confidence interval [CI], 0.62–0.95 for cough and RR, 0.29; 95% CI, 0.11–0.77 for wheezing) adjusted for confounders. Similar results were found for other respiratory symptoms as well as for eye discomfort, headache, and back pain (Romieu et al 2009). Our group performed a study to evaluate a risk reduction program using biomarkers of exposure (COHb and 1-OHP) and a biomarker of effect (DNA damage) in a community in San Luis Potosí state. An initial census survey of the study area was used to identify homes with open fires stoves, indoor soot adhered to roofs and internal walls and dirt floors. Thus, for the intervention we offered 1) removal of indoor soot adhered to roofs and internal walls, 2) paving dirt floors, and 3) an improved stove (the Patzari stove),which is constructed using sand, clay and cement; it also has a metal chimney that expels the smoke outdoors (Masera et al., 2007; Figure 5). The complete 3-stage risk reduction program was applied in all houses studied. We applied a questionnaire to confirm that the improved stove was well accepted by people in the community. The program was introduced in ten indigenous houses during August 2005; we assessed exposure to carbon monoxide, PAHs (1-OHP) and DNA damage previously and one month after the intervention program. Table 8 shows the geometric mean COHb level in all subjects studied before and after intervention. Before intervention all individuals had a geometric mean COHb level of 4.93% and 55% of those people presented COHb levels above 2.5% considered a safe level. Moreover, this 55% of people had geometric mean COHb levels of 8.39%. However, in all the studied individuals the levels of COHb were reduced to below 2.5% (mean level 1.0%) one month after intervention. Urinary 1-OHP was used to assess the exposure to PAHs (Table 8). Results show that the geometric mean level of 1-OHP before intervention was significantly higher in studied people (6.71±3.58 µmol/molCr) than after the program (4.80±3.29 µmol/ mol Cr). The amount of DNA damage in the studied people is depicted in Table 8. When compared, DNA damage in people exposed before the intervention was higher (5.8±1.3 of Tail Moment) than when the program was introduced (2.8±0.9 of Tail Moment). When the studied population was divided by gender or age groups, we did not observe differences in Tail Moment. Usually, risk reduction programs for indoor air pollution caused by biomass combustion only include the installation of a stove with a chimney (Chapman et al., 2005; Smith-Sivertsen et al., 2004; Khushk et al., 2005; Naeher et al., 2000). Those programs were evaluated either through a decrease of indoor air pollution (Riojas-Rodriguez et al., 2001; Naeher et al., 2000; Khushk et al., 2005; Zuk et al., 2007) and/or by a reduction in the incidence of diseases related with this kind of contamination (Riojas-Rodriguez, et al., 2001; Chapman et al., 2005; Boy et al., 2000; Smith-Sivertsen et al., 2004). We believe that our study represents an improvement in this area, as we installed a stove with a metal chimney that expels smoke outdoors, but also indoor soot adhering to roofs and internal walls was removed, and dirt floors were paved (it has been shown that biomass burning might be the major origin of PAHs in rural soil; Zhang et al., 2006). It is important to mention that soot is a sink of several chemicals generated during combustion (Jonker and Koelmans, 2002a, 2002b) and moreover, an extremely slow desorption of PAHs from soot has been demonstrated (Jonker et al., 2005). Therefore, although we do not have information regarding the importance of each stage in risk reduction, the elimination of soot sources (in roofs, walls and dirt floor) is a relevant action considering that this material is a sink of hydrophobic compounds such as PAHs and Dioxins. Children and adults can be exposed to this material, by inhalation (indoor air particles), ingestion (dust particles) or direct dermal exposure. This work was a pilot study and although with the limitation in sample size, the improvement in exposure (1-0HP, and COHb) and effect (comet assay) was so homogeneous that the State Government of San Luis Potosi, using the precautionary principle has expanded the program to different communities. We are following our studies in these sites.

Cookstove dissemination programs are evolving from "projects" narrowly targeting stove construction and fuelwood savings to more integrated "program" approaches looking at providing a set of health, environmental and socio-economic benefits. For them to be successful, systemic and interdisciplinary approaches are needed, including work on technology innovation, users' needs and priorities, market development and innovative financing. Strong and sustained efforts need to be devoted to monitoring stove performance in the field and to ensure the sustainability of the achieved benefits. Being more complex and ambitious, the new approaches come with a new set of challenges. More than ever, an intense cross-fertilization among groups, both North-South and South-South, is urgently needed. Donors and governments need to realize that sustained and longterm efforts rather than the typical two- to three-year projects are needed for setting up the type of integrated approaches described above. There is a need to support continuous technology innovation, getting research institutions to work hand in hand with local organizations on new stove designs that are more robust, cost-effective and clean. Participatory research is also needed to better understand users' priorities and needs and also to better assess the dynamics of multiple fuel use. Innovative and creative financial mechanisms are required that help users

Fig. 5. Patsari wood-burning cookstove.

overcome the investment costs of improved cookstoves. Users – specifically women -- need to have a strong say in program design and implementation. Finally, to be successful in the medium and long term, cookstove programs need to look at the overall policy context and integrate cookstove programs within larger sustainable rural development strategies. The challenge is substantial, but fostering the current partnership and networks on household energy use and learning from each other's experience will surely pave the way to a larger number of success stories in the near future.

Characteristics of Patsari stoves: The Patsari is oriented to mass dissemination; it is built with the help of a mould, and includes several custom-made pieces. The size and geometry of the combustion chamber (primary furnace), the tunnels and the secondary furnaces are designed to increase heat transfer to the pot. Metal flat pans (or comales) are tightly adjusted to the furnaces to prevent smoke leakage. The stove comes with a metal chimney, which is inserted into the stove body with the help of a custom-made metal base built in the stove. Currently, two models of Patsari, one with a single entry for fuelwood and a second with two entries, are being disseminated. In the latter case, both a ceramic comal and a metal comal are used in the primary furnaces (Figure 6).

Fig. 6. Design of Patsari wood-burning cookstove.

6. Conclusion

Indoor pollution will continue to be an essential field for health studies and interventions, because exposure to varied indoor substances will likely increase in coming years. Better studies dealing with genetic susceptibility to indoor pollutants, their carcinogenic effect and their impact on lung growth, lung development and, later on, lung aging, are also required. To answer several of these questions, longitudinal studies are required. A formal evaluation of improved stove programs from many viewpoints is also essential to improve guidance for countries and communities as they implement their own programs. Although local adaptation of programs will always be required, improved biomass stoves will likely be more common, with better community acceptance, reduced burden on forests, and increased spare time for homemakers.

7. Reference

Albalak, R. (2001). Indoor respirable particulate matter concentrations from an open fire, improved cookstove, and LPG/open fire combination in a rural Guatemalan community. Environ Sci Technol. 35:2650–5.

Arias, T.; & Cervantes, V. (1994). Las estufas de barro ahorradoras de leña ¿Una tecnología apropiada para la región de la Montaña de Guerrero?, Programa de Aprovechamiento Integral de Recursos (PAIR)-UNAM, México.

Barnes, B.; Mathee, A.; & Moiloa, K. (2005). Assessing child time-activity patterns in relation to indoor cooking fires in developing countries: a methodological comparison. Int J Hyg Environ Health. 208: 219-225.

Behera, D.; Dash, S.; & Malik, S. (1988). "Blood Carboxyhaemoglobin Levels Following Acute Exposure to Smoke of Biomass Fuel." Indian Journal of Medical Research 88: 522–524.

Boy, E.; Bruce, N.; Smith, K.R.; & Delgado, H. (2000). Fuel efficiency of an improved wood-burning stove in rural Guatemala: implication for health, environment and development. Energy Sustain Dev. 2:21–29.

Brauer, M.; Bartlett, K.; Regalado-Pineda, J.; & Pérez-Padilla, R. (1996). Assessment of particulate concentrations from domestic biomass combustion in rural Mexico. Environ.Sci. Tech, 30(1): 104-109.

Bruce, N.; Pérez-Padilla, R.; & Albalak, R. (2000). "Indoor air pollution in developing countries: a major environmental and public health challenge". Bulletin of the World Health Organization, 78(9): 1078-1092.

Bruce, N. G.; Bates E.; Nguti, R.; Gitonga, S.; Kithinji, J.; & Doig, A. (2002). "Reducing Indoor Air Pollution through Participatory Development in Rural Kenya." In Proceedings of 9th International Conference on Indoor Air Quality and Climate, Monterey, CA, 590–95.

Cayetano, H. (1997). Curso taller de promoción y construcción de estufas rurales en la comunidad de Santa Cecilia Lalana, Asesoría Técnica a Comunidades Oaxaqueñas (ASETECO), Oaxaca, Oax., Mexico.

Chapman, R.S.; He, X.; Blair, A.E.; & Lan, Q. (2005). Improvement in household stoves and risk of chronic obstructive pulmonary disease in Xuanwei, China: retrospective cohort study. BMJ. 331:1050–1056.

Dutt, G.S.; Navia, J.; & Sheinbaum, C. (1989). Cheranátzicurin: tecnología apropiada para cocinar con leña. Ciencias. 15: 43-47.

EPA, U. S. Environmental Protection Agency. (2000). Air quality criteria for carbon monoxide. EPA 600/P-99/001F. Washington, D.C. Office of Research and Development, U. S. Environmental Protection Agency.

Estrella, B.; Estrella, R.; Oviedo, J.; Narváez, X.; Reyes, M.T.; Gutiérrez, M.; & Naumova, E.N. (2005). Acute respiratory diseases and carboxyhemoglobin status in school children of Quito, Ecuador. Environ Health Perspect. 113:607–611.

Jacob, J.; & Seidel, A. (2002). Biomonitoring of polycyclic aromatic hydrocarbons in human urine. J Chromatogr B Analyt Technol Biomed Life Sci 778:31–47.

Jenkins, BM.; Jones, A.D.; Turn, S.Q.; & Williams, R.B. (1996). Emission factors for polycyclic aromatic hydrocarbons from biomass burning. Environ Sci Technol. 30:2462–2469.

Jongeneelen, F.J. (2001) Benchmark guideline for urinary 1-hydroxypyrene as biomarker of occupational exposure to polycyclic aromatic hydrocarbons. Ann Occup Hyg. 45:3–13

Jonker, MT.; Hawthorne, S.B.; & Koelmans, A.A. (2005). Extremely slowly desorbing polycyclic aromatic hydrocarbons from soot and soot-like materials: evidence by supercritical fluid extraction. Environ Sci Techno. 39: 7889–7895.

Jonker, M.T.; & Koelmans, A.A. (2002a) Sorption of polycyclic aromatic hydrocarbons and polychlorinated biphenyls to soot and soot-like materials in the aqueous environment: mechanistic considerations. Environ Sci Technol. 36: 3725–3734.

Jonker, M.T.; & Koelmans, A.A. (2002b). Extraction of polycyclic aromatic hydrocarbons from soot and sediment: solvent evaluation and implications for sorption mechanism. Environ Sci Technol. 36: 4107–4113.

Khushk, W.A.; Fatmi, Z.; White, F.; & Kadir, M.M. (2005). Health and social impacts of improved stoves on rural women: a pilot intervention in Sindh, Pakistan. Indoor Air. 15: 311–316.

Kleinman, M.T. (2000). Carbon monoxide: evaluation of current California air quality standards with respect to protection of children. Prepared for California Air Resources Board, California Office of Environmental Health Hazard Assessment. Irvine, CA: University of California Irvine; [Departament of Community and Environmental Medicine].

Levine, J.O. (1995). First international workshop on hydroxypyrene as a biomarker for PAHexposure inman—summary and conclusions. Sci Total Environ. 163:164–168.

McCracken, J.P.; Smith, K.R.; Diaz, A.; Mittleman, M.A.; & Schwartz J. (2007). Chimney stove intervention to reduce long-term wood smoke exposure lowers blood pressure among Guatemalan women. Environ Health Perspect. 115: 996–1001.

Martínez-Salinas, R.I.; Leal, M.E.; Batres-Esquivel, L.E.; Domínguez-Cortinas, G.; Calderón, J.; Díaz-Barriga, F.; & Pérez-Maldonado, I.N. (2010). Exposure of children to polycyclic aromatic hydrocarbons in Mexico: assessment of multiple sources. Int Arch Occup Environ Health. 83:617–623.

Masera, O. (1996a). Deforestación y degradación forestal en México. Documentos de trabajo N°19, Grupo Intrediciplinario de Tecnología Rural Apropiada (GIRA A.C.) Pátzcuaro, México.

Masera, O. (1996b). Uso y Conservación de Energía en el Sector Rural: El caso de la leña. Documentos de Trabajo N°21, Grupo Interdisciplinario de Tecnología Rural Apropiada (GIRA A.C.). Pátzcuaro, México.

Masera, O.; Navia, J.; Arias, T.; & Riegelhaupt, E. (1997). Patrones de Consumo de Leña en Tres Micro-regiones de México: Síntesis de Resultados. Proyecto FAO/MEX/TCP/4553(A). Grupo Intrediciplinario de Tecnología Rural Apropiada (GIRA A.C.). Pátzcuaro, México.

Masera, O.R; Drigo, R.; & Trossero, M.A. (2003). Woodfuels integrated supply/demand overview mapping. Universidad Autónoma de México, FAO-EC Partnership Programme. Food and Agriculture Organization of the United Nations.

Masera, O.; Edwards, R.; Armendáriz-Arnez, C.; Burrueta, V.; Johnson, M.; Rojas-Bracho, L.; Riojas-Rodríguez H.; & Smith, KR. (2007). Impact of Patsari improved cookstove on indoor air quality in Michoacán, México. Energy Sustain Dev. 9:25–36.

Mcdonald, J.D.; Zielinska, B.; Fujita, E.M.; Sagebiel, J.C.; Chow, J.C.; & Watson, J,G. (2000). Fine particle and gaseous emission rates from residential wood combustion. Environ Sci Technol. 34:2080–2091.
Mishra, V. (2003). Effect of indoor air pollution from biomass combustion on prevalence of asthma in the elderly. Environ Health Perspect. 111:71–78.
Naeher, L.P.; Smith, K.R.; Leaderer, B.P.; Mage, D.; & Grajeda, R. (2000). Indoor and outdoor PM2.5 and CO in high- and low-density Guatemalan villages. J Expo Anal Environ Epidemiol. 10: 544–551.
Navia, J. (1992). Estufas mejoradas, programa de difusi6n en Cheran Atzícurin, Primera Reuni6n Internacional sobre energía y medio ambiente en el sector residencial mexicano, comp. J. Quintanilla.
Olguín, E. (1994). 'Evaluación y optimización del uso de la leña a nivel familiar y de pequeñas industrias rurales. Instituto de Ecología, A.C. Jalapa, Veracruz, Mexico.
Raub, J.A.; & Benignus, V.A. (2002). Carbon monoxide and the nervous system.Neurosci Biobehav Rev. 26:925–940.
Red de Tecnologías Apropiadas (RETA). (2004). Memorias del Taller de Intercambio de experiencias sobre estufas ahorradoras de leña, San Cristóbal de Las Casas, Chiapas, México, 11 pp.
Reddy, AKN.; Williams, RH.; Johansson TB. (1997) Energy after Rio: prospects and challenges. New York: United Nations Development Programme.
Regalado, J.; Perez-Padilla, R.; Sansores, R.; Paramo Ramirez, J.I.; Brauer, M.; Pare, P.; & Vedal, S. (2006). The Effect of Biomass Burning on Respiratory Symptoms and Lung Function in Rural Mexican Women. Am. J. Resp. Crit. Care Med. 174: 901-905.
Riojas, R.H.; Romano, P.; Santosburgoa, C.; &Smith, K.R. (2001). Household firewood use and the health of children and women of Indian communities in Chiapas, México. Int J Occup Environ Health, 7(1): 44-53.
Romieu I.; Riojas-Rodrıguez, H.; Marron-Mares, A.T.; Schilmann, A.; Perez-Padilla, R. & Masera, O. (2009). Improved Biomass Stove Intervention in Rural Mexico. Impact on the Respiratory Health of Women. Am. J. Resp. Crit. Care Med. 180: 649-656.
Saksena, S.; Thompson, L.; & Smith, K.R. (2004). "Indoor Air Pollution and Exposure Database: Household Measurements in Developing Countries." http://ehs.sph.berkeley.edu/hem/page.asp?id=33.
Secretaría de Energía (SE). (1998). Balance Nacional de Energía. Gobierno de México, Secretaría de Energía, México D.F.
Secretariat of Health in Mexico (SSA). (2005). Children Health and the Environment in North America. A First Report on Aviable Indicators and Measures.
Available in http://www.cec.org/Storage/27/1802_CountryReport-Mexico-CHE_en.pdf. Accessed on 18-03-2011.
Smith KR. (1987). Biofuels, air pollution, and health: a global review. New York: Plenum Press.
Smith, K.; Mehta, S.; & Maeusezahl-Feuz, M. (2004). Indoor air pollution from household use of solid fuels. In: Ezzati M, Lopez A, Rodgers A, Murray C, eds. Comparative quantifi cation of health risks. Global and regional burden of disease attributable to s elected major risk factors. Geneva, Switzerland: World Health Organization. pp 1435–1493.

Smith-Sivertsen, T.; Diaz, E.; Bruce, N.; Diaz, A.; Khalakdina, A.; Schei, M.A. et al. (2004). Reducing indoor air pollution with a randomized intervention design—a presentation of the stove intervention study in the Guatemala Highlands. Nor Epidemiol. 14: 137-143.

Saatkamp, B.D.; Masera, O.; & Kammen, D.M. (1999). Energy and health transitions in development: fuel use, stove technology, and morbidity in Jarácuaro, Mexico. Energy for Sustainable Development. IV(2): 7-16.

Torres-Dosal A, Pérez-Maldonado IN, Jasso-Pineda Y, Martínez Salinas RI, Alegría-Torres JA, Díaz-Barriga F (2008) Indoor air pollution in a Mexican indigenous community: evaluation of risk reduction program using biomarkers of exposure and eVect. Sci Total Environ 390:362-368

Vargas, F. (1990). Breve diagnóstico sobre el proyecto de estufas rurales en México, Secretaria de Agricultura y Recursos Hidraúlicos (SARH), Mecanografiado, 8 pp.

Viau, C.; Hakizimana, M.; & Bouchard, M. (2000). Indoor exposure to polycyclic aromatic hydrocarbons and carbon monoxide in traditional houses in Burundi. Int Arch Occup Environ Health. 73:331-338.

World Health Organization. (WHO. 1997). Health and environment in sustainable development. Geneva, Switzerland: World Health Organization.

World Health Organization. (WHO. 1999). Environmental health criteria 213, carbon monoxide. IPCS, International Programme on Chemical Safety. Second Ed.Geneva, Switzerland: World Health Organization; 1999.

World Health Organization. (WHO. 2007). Indoor air pollution from solid fuels and risk of low birth weight and stillbirth: report from a symposium held at the Annual Conference of the International Society for Environmental Epidemiology (ISEE), September 2005, Johannesburg. Geneva, Switzerland.

Zhang, J. (1999). Smith KR. Emissions of carbonyl compounds from various cookstoves in China. Environ Sci Technol. 33:2311-2320.

Zhang, H.B.; Luo, Y.M.; Wong, M.H.; Zhao, Q.G.; & Zhang, G.L. (2006). Distribution and concentrations of PAHs in Hong Kong soils. Environ Pollut. 141:107-114.

Zhao, Z.H.; Quan,W.Y.; Tian, D.H. (1992). The relationship between polynuclear aromatic hydrocarbons in ambient air and 1-hydroxypyrene in human urine. J Environ Sci Health. A27:1949-1966.

Zuk, M.; Rojas, L.; Blanco, S.; Serrano, P.; Cruz, J.; Angeles, F.; Tzintzun, G.; Armendariz, C.; Edwards, R.D.; Johnson, M.; Riojas-Rodriguez, H.; Masera, O. (2007). The impact of improved wood-burning stoves on fine particulate matter concentrations in rural Mexican homes. J Expo Sci Environ Epidemiol. 17:224-232.

18

Epidemiologic and Economic Impact of Natural Gas on Indoor Air Pollution in Colombia 2007

Nelson Alvis Guzmán[1], Luis Alvis Estrada[1] and Fernando De la Hoz Restrepo[2]
[1]Universidad de Cartagena
[2]Universidad Nacional de Colombia
Colombia

1. Introduction

Colombian development since 1990, it can be describe according to evolution of three components of the Human Development Index: internal product, life expectancy and education. As for the gross domestic product, there has been show a slow but steady growth, excluding the 1999 crisis that resulted in a decrease of about four years, from which now is recovering. Life expectancy of Colombian population increased for about 5 years during 90th decade, because important epidemiologic and demographics changes, such as improvements in infant mortality and maternal health services, leading to a gradual recovery of the indicator, with some differences between genders, because of the greater impact than men suffered from the effect of violence. In education, there have had significant advances that although were hampered by the economic crisis, currently with the emphasis in social policies relation to education, there has been considerable improvement in the indicators. Although the overall picture is positive, there are still large differences between population groups, making equity a major challenge to human development in Colombia. Analyses by state, indicators are similar between Bogota and Hungary (country of high human development, 38th in the world) while the Chocó resemble Kenya (country of low human development, since 148 in the world). For Zones, most of the indicators in rural areas are lower than in urban areas and between social groups by deciles of the income distribution is unsatisfactory (Departamento Nacional de Planeacion de Colombia, [DNP], 1991).

In search of a universal sign for development, Colombia and 188 nations agreed at the Millennium Summit in September 2000 within the framework of the General Assembly of the United Nations, eight Millennium Development Goals in the long term, with coordination of UN, World Bank, IMF and OECD. Each country undertook to define national targets, which can be reached in 2015, for each of the Target (DNP, 1991).

However, there are studies that question the possible compliance of Colombia, some of the MDGs, especially related to poverty reduction: "... the scope in terms of poverty reduction is not so encouraging. Although according to international standards (PPP U.S. $ 1 and $ 2 PPP), the country would be very close to achieving the target reduction of 50%, it is clear that when the measuring stick becomes the national poverty line developments indicate that the moderate level of poverty would be only 20% lower than the 2001 (The poverty rate would rise from 55.2% to 44.2%). These results indicate that progress in other areas of society is not enough to boost the capacity of generating household income Colombians and that the strategy should focus on attacking directly additionally other determinants of poverty such as lack of access to labor markets and financial and serve the population in extreme poverty in an integrated way to break the poverty traps facing this population " (Nuñez, 2007).

In this sense, the MDGs by the different countries depends not only on the commitment of states to finance social development and undertake appropriate actions, also the contribution of civil society and the various cooperating agencies are willing to make to achieve more just societies.

Moreover, both programs targeting of social spending by the state as part of national and international, are coherent strategies for progress towards achieving better results in material development and social welfare. However, although the application of resources in development projects and / or involve social opportunities for communities to improve their capacities to address the different risks; the sole decision to implement grant programs to improve access to social services, does not mean that the results are guaranteed. Therefore, it is necessary to impact assessment procedures for specific programs and projects in order to gauge the true impact of the funds on the expected results.

2. Household natural gas as health technology

Living conditions, especially those related to health, an important influence on the economic situation and welfare of everyone in any society, which is particularly true in low income countries (where social safety nets are weak or nonexistent) and for the poor, due to the vicious circle of poverty and disease (Schultz, 1962). Health is considered a capital stock which the product is long lasting health, which can be used to obtain money or wages in the market for "goods" in the home or non-market sector (Grossman, 1972). It is well documented the relationship between health risk and poverty (World Health Organization [WHO], 2001).

On the other hand, the use of solid fuels for cooking and / or heating is the largest source of air pollution inside houses in the world. Nearly half of households in the world continue to use biomass fuels (wood, charcoal, dung and agricultural wastes). When biofuels are used, for example the stoves inefficient combustion, generate high concentrations of pollutants. Exposure to biomass smoke has been associated with low birth weight, acute respiratory infections, infant mortality and recently, anemia and mental retardation in children (Mishra, 2007). These pollutants, called solid fuel smoke (smoke) include respirable particulates, carbon monoxide, nitrogen oxide and sulfur, benzene, formaldehyde, 1,3 butadiene and polyaromatic compounds such as benzo (alpha) pyrene (Smith, 1987; Pandey, 1989). In households with limited ventilation, (both in developing countries and industrialized) exposure to such contaminants is higher in women and children, who stay longer in the house. Their measurements showed that concentrations of these pollutants were above the international standards (WHO) (Smith, 1987; Bruce, 2000).

The combustion of natural gas worldwide is classified as the cleanest among the traditional industrial fuels. In fact, emissions of particulate matter to meet the highest international standards, without having to invest in gas processing equipment. One of the great advantages of natural gas over other fuels, are the low combustion emissions, which can be seen in the following table (Innergy Soluciones Energeticas, 2007).

Fuel	MP Particulado Matter	SOX Sulfur Oxide	NOX Nitric Oxide
Gas Natural	1	1	1
Gas City	3	61	0,5
Gas	1,4	23	2
Kerosene	3,4	269	1,5
Diesel	3,3	1.209	1,5
Fuel Oil N° 5	15	4.470	4
Fuel Oil N° 6	39,4	4.433,	4
Coal	157	5.283	6
Firewood	140	13	2

Table 1. Concentrations of contaminants in relation to alternative fuels different of the NG

Most research on environmental pollution intradomicilliary has been conducted in developing countries and has been associated with the determination of pollutants and their sources (Spengler, 2001). This has been done by estimating the levels of risk factors generated by the combustion of biomass, through the determination of the exposure variable. To do this, Smith et al. 2004, proposes the use of the well established exposure-response epidemiological studies conducted outside the houses and measuring concentrations of pollutants, such as an estimate of what happens inside the houses (Smith, 2004). Another approach is to use studies in developing countries, dividing the population into categories of exposed and unexposed to smoke the product of biomass burning. However, there are gaps in data on exposure of such populations, and variability between groups exposed. In addition, this method can be affected by the differential use of fuels and different mixtures of pollutants, which could prevent establishment of only two categories.
Apart from the use of biomass fuels, another factor to consider is related to the ventilation conditions of housing and particularly the kitchen. This equation can build a better, in which exposure to smoke from solid fuels (biomass) is a function of the number of people using such fuels, multiplied by a factor of ventilation.

$$P_e \cong P_b \times F_V$$

PE = Population exposed to smoke from biofuels
PB = Population using biofuels
FV = Ventilation factor
Based on this model were estimated using solid fuels for subregions in the world (Smith, 2004). For the subregion in which America is Colombia (AMR-B), the population using solid fuels was estimated at 24.6% (CI. 18.8-30.8). This means that in Colombia about a quarter of the population could be using biofuels. Of total DALYs, it is estimated that 2.7% is attributed to diseases associated with smoke from solid fuels (2.5% in men and 2.8% in women). Of this total, 32% occur in Africa (AFR-D), 37% in Southeast Asia (SEAR-D) and 16% in China, Australia and other countries (WPR-B) (Smith, 2004). Surely, that is spread by region and intra-regionally, as far as access to clean fuels is different for each region.
The strength of evidence for each factor was determined based on the structured assessment of causality, using the criteria of Bradford Hill (Hill, 1992). Therefore the specificity, dose-response and experimental evidence is often difficult to assess environmental exposures and their health effects, we used the epidemiological evidence together with the emission

mechanism of exposure and pollution (Smith, 2000; Zodpey, 1999). There are determined three health effects for which there is clear evidence:

1. Acute Lower Respiratory Infection
2. Chronic Obstructive Pulmonary Disease
3. Lung Cancer.

Children under 5 years are the highest exposure group, although levels are unknown globally (Ezzati, 2001; Saksena, 1992). About half of the world's countries cook with solid fuels, including more than 75% of people in India, China and nearby countries, 50 to 75% of people in some countries of South America and Africa. Inadequate ventilation in homes is a common factor in developing countries, which increases the exposure, especially for women and children who stay more time inside the house. In many measurements made by the WHO, has been found that concentrations of pollutants inside homes exceed by a large margin, the measurements on the outside. Multiple studies have shown a consistent relationship between the use of solid fuels and several diseases. These tests show that smoke from solid fuels causes about 35.7% of the IRA, 22% of COPD and 1.5% of cancers tracheae, bronchi and lungs (WHO, 2002).

Health intervention is considered, all that technology into their practices creates possibilities of improving the living conditions of the individual and / or community. Natural gas has become a health intervention as far as a major impact not only the household economy, but can reduce the burden of respiratory diseases in the communities in which it is implemented, especially in children, women and elderly.

3. Estimation of economic and epidemiological impact of household natural gas - analysis plan

To perform the study combined the following studies: a) secondary data analysis (a systematic review of the literature) and analysis of databases provided by the Foundation Promigas, demographic data from DANE 2005 Census, and databases of economic statistical series of the Bank of the Republic. b) Analysis of disease burden estimated from the parameters identified in the previous studies. c) Analysis of the cost burden of illness and the estimated costs avoided then implemented the program natural gas home connections (Arredondo, 1997). d) Cost effectiveness analysis of connections of households to service natural gas home (Drummond, 2005).

The study was conducted in two phases: a) Estimating the burden of disease associated diseases biomass smoke (ARI and COPD) in the population covered by the project further at this stage to combine the results of the literature review; b) Estimating the potential economic impact of the grant program's connection to the Household Natural Gas -HNG.

The estimated population data are from field research conducted in the projected 2005 population census DANE.

We carried out an analysis of cost-effectiveness of introducing natural gas house on morbidity and mortality from respiratory diseases in children and older adults, using a simple decision that assessed the burden and costs of illness in the absence and presence of HNG.

Both share the possible scenarios in which events can be found elsewhere, for example: patients with ARF and / or IRA without COPD and / or COPD. The transition from one event to another is determined by the probability of coverage and efficiency of natural gas as a health technology. For the assessment of the model assumes that: 1) persons living in homes with natural gas which are not protected against respiratory diseases and are equally likely to suffer respiratory infections caused by other causes, 2) the effective protection that natural gas generates against respiratory diseases depends also on housing conditions (ventilation of the kitchen and independence and efficiency of the stove).

The study horizon is 5 years, for both children and adults, with a discount rate of 3% for both costs and outcomes. The parameters for the model shown in the following table.

Model Inputs	Base Case	Plausible Range		Source - Reference
		Lower Limit	Upper Limit	
Demográphics				
Households covered	34.137			
Population covered	143.861	--	--	Current analysis
Population under 5 years	11.940	--	--	Current análysis
Population aged 60	14.675	--	--	Current análysis
Average monthly income per household	174	137	250	Current analysis
Monthly expenditure on fuels other than HNG	9,1	8,0	10,1	Current analysis
Average monthly expenditure without grant HNG	4,0	3,5	4,5	Current análysis
Average monthly expenditure on subsidized HNG	2,1	1,9	2,3	Current análysis
Discount rate	3%			Current análysis
Odds (cumulative incidence)				
Cumulative Incidence of IRA	12,6	11,5	13,5	(Constenla, 2007)
Proportion of hospitalization	30	25	35	(Alvis, 2008)
Proportion of hospital mortality	4	3	5	(Alvis, 2008)
Prevalence of EPOC[1]	8,9	7,3	13,4	(Caballero, 2005)
Proportion of hospitalization	30	25	35	(Alvis, 2008)
Proportion of hospital mortality	8	5	11	(Alvis, 2008)
Effectiveness of Gas to avoid events				
Effectiveness ARI	62,7	55	69	Current analysis
Effectiveness COPD	26,6	22	31	Current analysis
Costos (US$ 2005)				
Average cost of patients hospitalized*	886	867	1 325	(Constenla, 2007)
Average cost of patients Ambulatory ARI*	93	78	221	(Constenla, 2007)
Average cost of patients hospitalized COPD*	5 215	4 800	5 500	(Pérez, 2007)
Average cost of patients ambulatory COPD*	500	450	550	(Pérez, 2007)
Cost of installation of HNG	146	127	170	Current analysis
Weighting of disability				
Disability weighted for ARI	0,280			(Murray, 1996)
Disability weighted for COPD	0,334			(Murray, 1996)

* Includes direct and indirect costs ** Including installation costs and the average annual turnover of the home. Installation costs are deferred for 5 years

Table 2. Parameters for the economic assessment of the impact of the installation program Household Natural Gas

[1] The study by (Caballero, 2005) was performed in 5539 adults of 40 years or more in 5 cities in the country (Bogotá, Bucaramanga, Medellín, Cali and Barranquilla). In conclusion the prevalence of COPD in Colombia is 8.9%, with the Medellín city that was the highest percentage 13.5% and Barranquilla with the lowest 6.2%. The study assessed exposure to wood smoke and found that 60.7% of the population had ever used firewood for cooking. The highest frequency, 67.5%, was observed in Bucaramanga, followed by 62.5%, in Cali, 59.2% in Medellín, 58.1% and 56.3% in Bogota, Barranquilla. The prevalence among men and women were similar: 60.1% in men and 61.0% in women. 64.8% have used the wood for over ten years.

The outcome measures considered or result of the program for the model are: The total costs of the burden of respiratory disease associated with smoke from biomass, the costs of averted disease burden by the grant program to HNG connections, where both ambulatory (medical) and ARI in children hospital and COPD in older adults, deaths from these diseases and the years of disability-adjusted life (DALYs) averted.

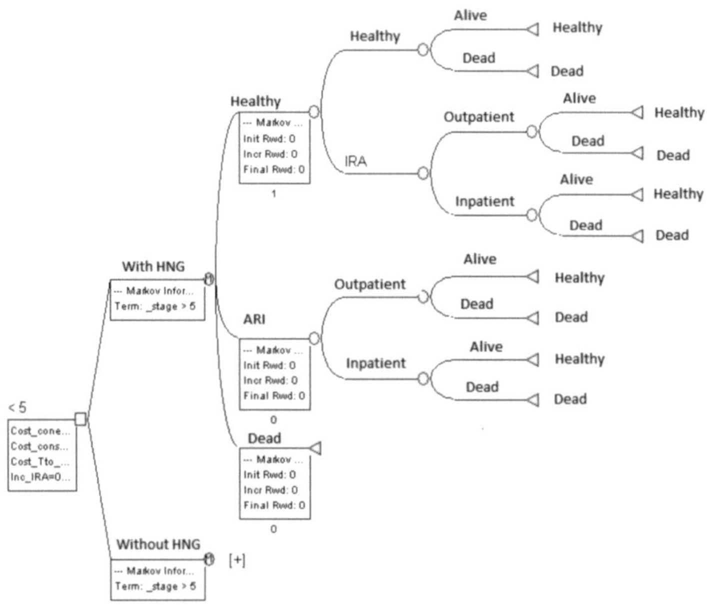

Fig. 1. Choice models for ARI.

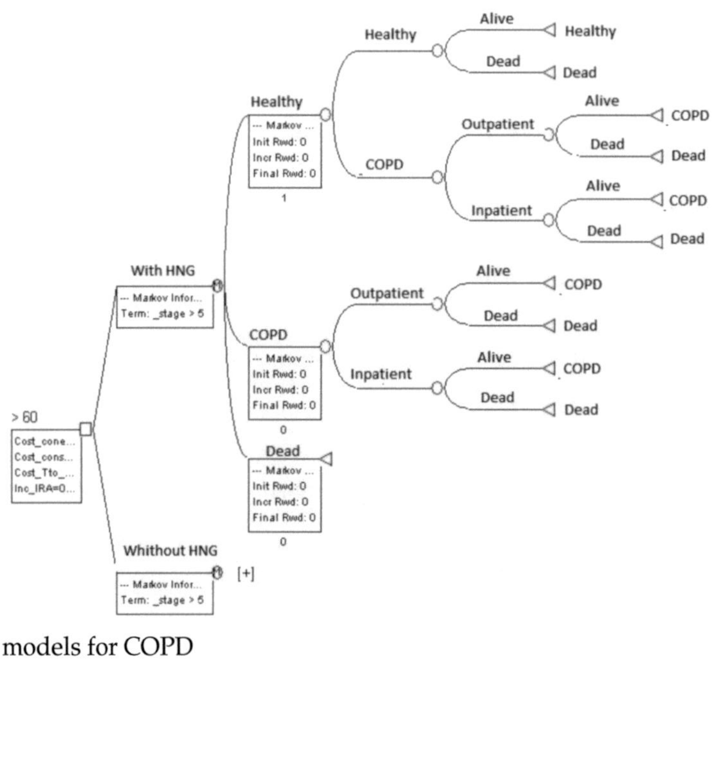

Fig. 2. Choice models for COPD

It also uses the ratio of incremental cost-effectiveness (ICER) for comparing alternatives with and without a program using the following formula:

$$\text{ICER}=\frac{\text{Connections Program Costs to HNG - Avoided costs of the burden of respiratory disease associated with smoke from biomass}}{\text{Consequently avoided (DALYs, deaths avoided, AVS)}}$$

The ICER compares the difference in cost on the difference in health outcomes in different scenarios. For the analysis of cost effectiveness of disease avoided costs are deducted from the cost of implementing the grant program and are divided between health outcomes (life expectancy gained per woman, year of life saved, deaths prevented, etc.). The costs are expressed in dollars and pesos in 2008. All future costs and benefits are discounted at an annual rate of 3%.

3.1 Sensitivity analysis

For the sensitivity analysis using different assumptions for the parameters. Moreover, to assess the profitability of the investment is estimated Internal Rate of Return of the project assuming that: a) all households covered by the project receive an average income between 137 and $ 250 a month, b) the costs of the monthly averages without subsidy bill are $ 2.3 to 1.9, c) the average monthly cost of the subsidy bill is $ 4.5 to 3.5, d) all households receive a subsidy of the cost of installation HNG of $ 146, and e) all households prior to the installation of the HNG, various fuels consumed by an average monthly cost between $ 10.1 and 8.0.Households were followed for 20 years and yields are assumed constant.

4. Burden of disease from ARI and COPD in population

In 84 localities in the absence of the facilities of HNG would occur between 6288 and 7382 outpatient cases of ARI and between two thousand five hundred and five hundred thousand hospitalizations in the 5 years of observation. This could generate between 47 and 129 deaths and between 75 thousand and 127 thousand DALYs per IRA during the period. Also be presented between 5 and 9 thousand cases of outpatient visits for COPD and hospitalizations between 1200 and 3152, just as would occur between 61 and 347 deaths from COPD and between 12 thousand and 35 thousand DALYs.

As shown in Table 3, the installation of Household Natural Gas avoided during the study period, about 4 thousand to 5 thousand cases of outpatient ARI and between 1200 and 2300 for COPD. In addition, about 1,200 hospitalizations avoided by IRA and 500 COPD. Also avoid death between 45 and 170 would represent about 45 thousand to 90 thousand DALYs.

5. Economic impact

The economic costs of the burden of disease from ARI and COPD on stage without HNG would be between US$ 10.7 and US$ 23.6 million. On stage with HNG these costs would be reduced by about 32%, see Table 4.

As for the incremental cost effectiveness of the HNG would be between minus 8 thousand (cost savings) and 18 thousand dollars to avoid a death, and betwen under US$ 16 to US$ 17 avoid a DALY. The net cost would be from under 1.4 million dollars (cost savings) and 834 thousand dollars of 2005.

	In the absence of the installation of Household Natural Gas			In the presence of the installation of Household Natural Gas		
(ARI) Events < 5 years	Base Case	Lower Limit	Upper Limit	Base Case	Lower Limit	Upper Limit
Outpatient cases	6.562	6.288	7.382	2.165	2.075	2.436
Hospitalizations	1.969	1.572	2.584	650	519	853
Deaths	79	47	129	26	16	43
DALYs	96.049	75.697	127.723	31.696	24.980	42.149
(COPD) Events > 60 years						
Outpatient cases	6.049	4.906	9.006	4.476	3.631	6.664
Hospitalizations	1.815	1.227	3.152	1.343	908	2.332
Deaths	145	61	347	107	45	257
DALYs	19.764	12.822	35.711	14.625	9.488	26.426
Total avoided events						
Outpatient cases avoided of ARI				4.068	3.899	4.577
Hospitalizations avoided for ARI				1.220	975	1.602
Deaths avoided for ARI				49	29	80
DALYs avoided for ARI				59.550	46.932	79.188
Outpatient cases avoided of COPD				1.573	1.276	2.341
Hospitalizations avoided for COPD				472	319	820
COPD Deaths				38	16	90
COPD AVAD avoided				5.139	3.334	9.285

Table 3. Burden of disease in the localities. Colombia 2008

	In the absence of the installation of Household Natural Gas			In the presence of the installation of Household Natural Gas		
(ARI) Events < 5 years	Base Case	Lower Limit	Upper Limit	Base Case	Lower Limit	Upper Limit
Outpatient cases	616.810	591.109	693.911	234.388	224.622	263.686
Hospitalizations	1.748.065	1.396.024	2.294.335	664.265	530.489	871.847
Subtotal costs ARI	2.364.874	1.987.133	2.988.246	898.652	755.111	1.135.533
(COPD) Events > 60 years						
Outpatient cases	3.024.324	2.453.063	4.502.883	2.238.000	1.815.267	3.332.133
Hospitalizations	9.279.836	6.272.482	16.119.419	6.867.079	4.641.637	11.928.370
Subtotal costs COPD	12.304.160	8.725.545	20.622.302	9.105.079	6.456.903	15.260.503
Total Cost Burden	14.669.035	10.712.678	23.610.547	10.003.731	7.212.014	16.396.037
Total costos avoided by the project				4.665.304	3.500.664	7.214.511
Total installation costs of GND				4.984.002	4.335.399	5.803.290
Cost/Deaths avoided for the period				3.682	-8.289	18.473
Cost / DALY avoided for the period				5	17	-16
Total neto costs for the period				318.698	-1.411.221	834.735

Table 4. Costs of Burden of disease in the localities. Colombia 2008

Other benefits generated by the HNG connection to households are the savings that are generated by stop using different fuel HNG which is more than twice as costly as paying the monthly bill in the event that households do not receive service subsidies from the State. In the scenario where households receive the state subsidy of these costs can be up to 4 times what the household pays each month for the service. If we assume that these households maintained this condition for about 20 years, it is possible to evaluate returns on those savings can generate the initial investment by the donor (grant connect HNG).

Ítems	Average	Lower Limit	Upper Limit
Total household income	1.096.087.341,3	865.834.808,4	1.569.325.590,3
Spending on different fuel to GND	57.009.359,7	50.367.234,2	63.652.387,1
GND spending without subsidy	25.109.209,4	21.970.558,3	28.247.860,6
GND spending with subsidy	13.182.335,0	11.926.874,5	14.437.795,4
State subsidies	11.926.874,5	10.043.683,8	13.810.065,2
Savings without subsidies	31.900.150,3	28.396.676,0	35.404.526,5
Savings with subsidies	43.827.024,8	38.440.359,8	49.214.591,7
Internal Rate of Return without subsidies	42%	37%	46%
Internal Rate of Return without subsidies	57%	50%	64%

Table 5. Financial return of the project for households for 20 years. Colombia 2008

Adding these savings to the costs avoided by the impact of the HNG on the incidence of respiratory diseases, we can assess the IRR of the project from the perspective of the donor, assuming that their interest is the achievement of maximum benefit to each household subsidized. As shown in Figure 3 the IRR is positive in both scenarios (with and without subsidies) as low as 44% during the first five years and a maximum of 78% if the period of observation is 20 years, this scenario without the subsidy cost of the bill (consumer subsidy). These numbers vary in the same scenario with allowance for consumption of between 65% and 96% when we evaluated the project to 20 years.

6. Sensitivity analysis of the results

As shown in Figure 4, a sensitivity analysis in a way to display the variables that have the greatest impact incremental effect on the cost of connecting homes to the HNG are: a) the prevalence of COPD, b) the costs of HNG installation, c) the effectiveness of the HNG to impact the prevalence of COPD, d) the proportion of hospitalizations for COPD and e) the average costs of outpatient and hospitalization ARI. In this regard, the decision to subsidize the costs of installing the HNG in poor households, allowing for an intervening variable of greatest impact for the evaluation of benefits of installing.

Fig. 3. Internal rate of return of the the installation of Household Natural Gas.

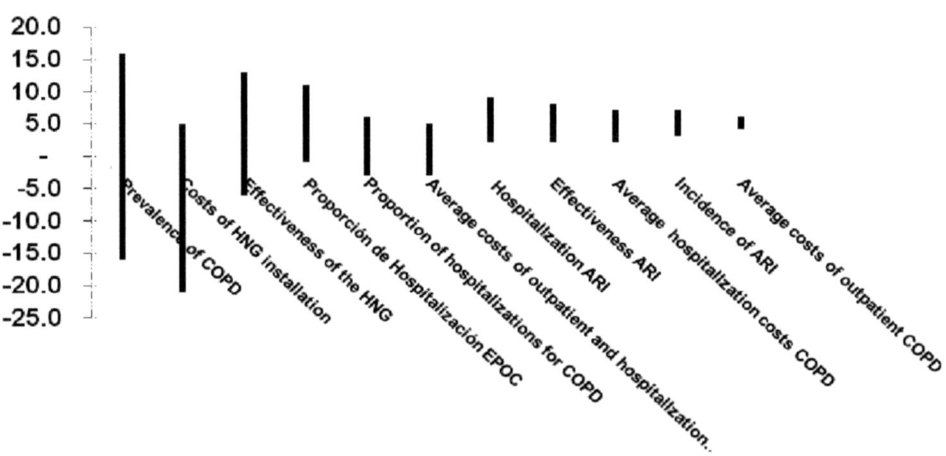

Fig. 4. Tornado Diagram of ICER of Cost/DALYs

7. Conclusion

According to the DANE census of 2005 in Colombia were 9.74 million households of which 3.92 million (40.3%) accessed the service of HNG. For the fourth quarter of 2008 the Ministry of Mines and Energy of Colombia reported 4.93 million homes connected to the HNG (Murray, 1996), which represents a growth of more than one million homes in three years and covers over 50% of all households. These households are concentrated in 462 locations in about 350 municipalities in 22 of the 33 departments in the country. Of these households, 55% belong to strata 1 and 2 (Murray, 1996). Since according to the quality of life survey 2003 (Alzate, 2006) 95.5% of rural households in Colombia belong to the socioeconomic strata 1 and 2, we can say that the coverage in the poorest and most vulnerable populations are not yet adequate.

On the other hand, this demonstrated the increased risk of ARI, COPD and lung cancer in those households who cook with biofuels (Norman, 2007). In our study, the limited information, only measure the disease burden of ARI and COPD which assumes an underestimation of the protective effects of GND, for example, about lung cancer. However, as shown, assuming the WHO criteria (WHO, 2001), the GND is a highly cost effective technology and thus its incorporation into benefits for the social development of poor and marginalized communities, is beyond dispute.

Thus, this study is a good estimator of the potential impact of universal access to GND by the poorest and most vulnerable households, which are precisely those that generate the greatest burden of disease by the pollution. This investigation provides means of support of public policy with a view to providing decision makers to mitigate or eliminate pollution as change or improve the efficiency of the stoves, fuel switching to biomass use GND generating reducing the frequency of cases of ARI, COPD and other respiratory diseases and the reorientation and improvement in the household economy (Larson, 2002).

These findings are consistent in scale, with the referenced document CONPES 3344 through 2005 which set out "guidelines for the formulation of policy on prevention and control of air pollution" (DNP, 2005) and estimated that in Colombia, According to Larsen, is generated per year 6 000 deaths, 7 400 new cases of chronic bronchitis (COPD), 13 000 hospitalizations and 255 000 visits to emergency rooms, with costs between 230 and 600 billion pesos in 2004 ie between about 100 and 270 million dollars.

In Colombia there are already incentives rule (Article 97 of Act 142 of 1994) which facilitate the implementation of the policy of subsidies: "In any case, the cost of household connection, attack and meter of stratas 1, 2 and 3 can be covered by the municipality, the department or the nation through budgetary support, but political will is an important input to achieve the goals for the reduction of pollution by smoke from biomass. However, given the high rates of return for this project (37% and 46% in cases that do not subsidize the consumption of the service and between 50% and 64% in the scenario with subsidies from the state) is very difficult to escape of these choices.

In summary we can state that: 1) Given the high rates of return of the project, both on stage and without subsidies subsidized consumption, for every $ 100 invested by the donor in the year zero of the project, households received net benefits of between 65 and 96 dollars annually. 2) Decisions to invest in subsidizing the connection of poor households to the GND are highly cost effective, to be cost saving, especially in poorer households with higher risks of respiratory diseases (health risk) and lower income (economic risk) is say the most vulnerable. So, with less than $ 20 is spent a year of life saved, which in the worst case is 8 times lower than required to save a year of life using the pneumococcal vaccine, which needs between $ 154 and $ 5252 in different countries of Latin America and the Caribbean (Constenla, 2007; Sinha, 2008). 3) If the state took the decision to subsidize both the connection and GND consumption by households more vulnerable, more could be close to achieving the MDGs, as is clearly demonstrated improvements in quality of life households and high social return on investment in this area, which is essential in the fight against poverty and misery in our country.

8. References

DNP. Departamento Nacional de Planeación.. (2005). Metas y estrategias de colombia para el logro de los objetivos de desarrollo del milenio - 2015, In : *Consejo Nacional de Política Económica y Social República de Colombia, Conpes 091*. Bogotá D.C.

Nuñez, J; Gonzalez, N. & Hernández, G. (2007). ¿Es posible alcanzar los objetivos del Milenio en Colombia? Una evaluación de estrategias de focalización y financiamiento, In : *Archivos de Economía* N° 330. Departamento Nacional de Planeación, Dirección de Estudios Económicos: Bogotá D.C., Colombia.

Schultz, T. (1962). Reflections on investment in man. *Journal of Political Economy*, Vol. 70, pp. 101-109.

Grossman, M. (1972). On the concept of health capital and the demand for health. Journal of Political Economy, Vol. 80, No. 2, (March-April 1972), pp. 223-255.

WHO, World Health Organization (2001). Macroeconomics and health: investing in health for economic development. Report of the Commission on Macroeconomics and Health. In : *Commission on Macroeconomics and Health*, Geneva.

Mishra, V. & Retherford R. (2007). Does biofuel smoke contribute to anaemia and stunting in early childhood? *Int J Epidemiol*, Vol. 36, No. 1, pp. 117-29.

Smith, K. (1987). Biofuels, air pollution and Health: a global review. Plenum, Editor. ISBN: 0-306-42519-X, New York.

Pandey, M ; Smith, K ; Boleij, J. & Wafula, E. (1989). Indoor air pollution in developing countries and acute respiratory infection in children. *Lancet*, Vol. 333, No. 8635, pp. 427-9.

Bruce, N; Perez, R. & Rachel, A. (2000). Indoor air pollution in developing countries: a major environmental and public health challenge. In: *Bulletin of the World Health Organization*, Vol. 78, No. 9, pp. 1067-1071.

Agency, U.S.E.P. (1997). Revisions to the National Ambient Air Quality Standards for Particles Matter. Federal Register. pp. 38651-38701.

Innergy Soluciones Energeticas. (2007). Ventajas Para el Medio Ambiente. Ventajas del Gas Natural, 02.10.2007, Available from: http://www.innergy.cl/ventajas1.htm.

Spengler, J ; Samet, J. & McCarthy, J. (2001). Indoor Air Quality handbook., ISBN: 0-07-445549-4. New York: McGraw Hill.

Smith, K ; Mehta, S. & Maeusezahl, F. (2004). Indoor Air Pollution from Household Use of Solid Fuels. In : *Comparative quantification of health risks: Global and regional burden of disease due to selected major risk factors*, World Health Organization: Geneva.

Hill, A. & Hill, I. (1992). Bradford Hill's Principles of Medical Statistics. *Biometrics*. Vol. 48, No. 2, pp. 659-660.

Smith, K. ; Samet, J ; Romieu, I. & Bruce, N. (2000). Indoor air pollution in developing countries and acute lower respiratory infections in children. *Thorax*. Vol. 55, No. 6, pp. 518- 532.

Zodpey, P. & Ughade, S. (1999) Exposure to cheaper cooking fuels and risk of agerelated cataract in women. *Indian Journal of Occupational and Environmental Medicine*, Vol. 3, No 4,(Oct- Dec 1999) pp. 159-161.

Ezzati, M. & Kammen, D. (2001). Quantifying the effects of exposure to indoor air pollution from biomass combustion on acute respiratory infections in developing countries. *Environmental Health Perspectives*, Vol. 109, No. 5, (May 2001) pp. 481-488.

Saksena, S .; Prasad, R.; Pal, R. & Joshi, V. (1992). Patterns of daily exposure to TSP and CO in the Garhwal, Himalaya. *Atmospheric Environment*, Vol. 26A, No. 11, (August 1992), pp. 2125-2134.

WHO, World Health Organization. (2002). Addressing the links between indoor air pollution, household energy, and human health. In : *Meeting report of the WHO-USAID global consultation on the health impact of indoor air pollution and household energy in developing countries*, Geneva.

Caballero, A; Torres, C. ; Maldonado, D. ; Jaramillo, C. & Guevara, D. (2005). Prevalencia de la Enfermedad Pulmonar Oclusiva Crónica en cinco ciudades colombianas. *Revista Médica Sanitas*. Edición Especial.,.

Alvis, N; Alvis, L. & Orozco, J. (2008). The cost-effectiveness of installing natural gas as a sanitary alternative for rural communities on the Colombian Caribbean coast burning biomass fuels. *Rev. salud pública*, Vol. 10, No. 4, (september 2008) pp. 537 - 549.

Pérez, N ; Murillo, R. ; Pinzón, C. & Hernández, G. (2007). Smoking Attributable Costs of Lung Cancer, COPD, and AMI in Colombia (A PAHO Multicentric Proyect). *Rev Colomb Cancerol*. Vol. 11 No. 4, (December 2007) pp. 241 - 249.

Arredondo, A. & Damian, T. (1997). The economic costs in the production of health services: from the cost of inputs to the cost of case management. *Salud Publica Mex*, Vol. 39, No. 2, pp. 117-24.

Drummond, M. ; O'brien, B. ; Stoddart, G. & Torrance, G. (2005). Methods for the economic evaluation of health care programs. ISBN 0198529457. Third Edition ed., Oxford Medical Publications Oxford GB:.

Murray, C. & Lopez, A. (1996). The Global Burden of Disease. Boston: Harvard University Press, .

Alzate, M. (2006). La estratificación socioeconómica para el cobro de los servicios públicos domiciliarios en Colombia ¿Solidaridad o focalización?. In : *Estudios y Perspectivas 14*, ISBN: 92-1-322968-2, Bogotá D.C..

Norman, R. ; Barnes, B. ; Mathee, A. ; Bradshaw, D. & South African Comparative Risk Assessment Collaborating Group. (2007). Estimating the burden of disease attributable to indoor air pollution from household use of solid fuels in South Africa in 2000. *S Afr Med J*, (August 2007), Vol. 97, No. 8 Pt 2, pp. 764-71.

Larson, B. & Rosen, S. (2002). Understanding household demand for indoor air pollution control in developing countries. *Soc Sci Med*, Vol. 55, No. 4, pp. 571-84.

DNP, Departamento Nacional de Planeación,. (2005). Lineamientos para la formulación de la política de prevención y control de la contaminación del aire. In : *Consejo Nacional de Política Económica y Social República de Colombia, Conpes 3344*. Bogotá D.C.

Constenla D. ; Gomez E.; De la Hoz, F.; O'Loughlin, R.;, Sinha, A.; Valencia, J. & Valenzuela, M. (2007) The Burden of Pneumococcal Disease and the Cost Effectiveness of a Pneumococcal Vaccine in Latin America and the Caribbean: A review of the evidence and a preliminary economic analysis. Available from: http://www.sabin.org.

Sinha, A. ; Constenla D. & O'Loughlin, R.; Gomez E.; De la Hoz, F. ; Valencia, J. ; Valenzuela, M. & De Qadros, C. (2008). Cost-effectiveness of pneumococcal conjugate vaccination in Latin America and the Caribbean: a regional analysis. *Rev Panam Salud Publica*, (November 2008)Vol. 24, No. 5, pp. 304-13.

19

Economic Growth and Air Pollution in Iran During Development Programs

H. Asadikia, R. Oyarhossein, S. Zare and I. Saleh
Department of Agricultural Economics,
University of Tehran
Iran

1. Introduction

Air pollution means combination of air with gases that lead to decrease in quality of the air. Pollutant elements are Carbon dioxide, Monoxide, Sulfur Dioxide and etc.

Estimated annual losses of mortality due to urban air pollution are 640 million dollars i.e. 5100 billion Rials that is equivalent 0.57 percent of GDP. The disease caused of urban air pollution creates 260 million- dollar or 2100 billion- Rial cost (0.023 percent of GDP) for the economy of Iran. (World Bank, 2005).

Grossman and Krueger (1991) has analyzed relationship between air pollution and economic growth in USA. GNP per each person, time and pollution index were input in that model. Results show that there is relationship like "U" between GNP per each person and SO_2 emission.

Lee (1996) has surveyed air pollution in Korea during 1985-92. The effects show that CO and income have relationship as "U" whereas Sulfur Dioxide, NO_2 and Ozone did not have it.

Khan (2002) has perused correlation between revenue in each family and CO, NO and Ozone in USA for the year of 1990. One of the important things is that in this research population, labor force, unemployment, literate people, workmen's, housekeeper, rental house that have an effect on pollution, were Exogenous Variables. Consequences expressed that linkage in equation between income and pollution of gasses is inverse.

Pazhoyan and etal (2007) have examined respect of economic growth and air pollution with Kuznets hypothesis in 67 countries with different income. The effect of economic growth, rural population, provision of environment, number of cars, the ratio of summation of export and import on GNP have been surveyed and results emphasized Kuznets curve for environment in those countries.

This study surveys relationship between air pollution and economic growth and also, evaluation of Economical and Social Development Programs in the field of air pollution reduction. For this purpose, the study used data for 1979-2005 from the Balance Sheet Energy (2009), Iran Department of Environment (2009), Central Bank of Iran (2009) and World Development Reports (2008) has been collected. Calculation and estimates is done by Shazam and SPSS packages.

2. Methodology

The estimation of relationship between economic growth and environmental destruction has done by the use of Grossman and Krueger model with some adjustments (Eq. (1))

$$LNP = \alpha_0 + \alpha_1 LNGDP + \alpha_3 LNGDP^2 + \alpha_4 LNcar + \alpha_5 LO + \alpha_6 LNU + \alpha_7 D_1 + \alpha_8 D_2 + \alpha_9 D_3 + \varepsilon \quad (1)$$

Where LNP: the logarithm of pollution variable (carbon dioxide emissions per capita in thousand tons), LNGDP: the logarithm of GDP per capita in Iran (billion Rials), $LNGDP^2$: the square logarithm of GDP per capita in Iran, LNCAR: the logarithmic number of cars produced and imported, LNO: the logarithm of the degree of openness of an economy (ratio of total exports and imports to GDP), LNU: logarithm of population (thousands), D_1: dummy variable for the function of first Development Program, D_2: dummy variable for the function of second Development Program, D_3: dummy variable for the function of third Development Program (in each dummy variables, the number 1 is related to the program years and the number 0 for the other years). For each of the variables some classes are considered, and then the effect of them is investigated by the find results from SPSS package. Many studies done in the field of air pollution considered carbon dioxide emissions as an index of the air pollution and this gas is one of the most important gases leading to climate change and global warming. Additionally, about 72 percent of greenhouse is related to carbon dioxide, so in this study carbon dioxide emission is a criterion for air pollution.

3. Summary and conclusion

Study data are time series, so stationary test is done and the result is available in Table (1) that shows two lags are optimal and whereas calculated value is more than critical value the model is stationary.

Error term	Description
data level	Variable
-5.835	AIC criteria
2	Lag length
3.5573*	Calculated value
-2.57	Critical value

(Source: research findings)
* 10% significance level

Table 1. Generalized
Dickey - Fuller Unit root test results

Chi-square test shows that there are significant differences in carbon dioxide per capita for various variable levels. Considering the fifth column table (2), when the population is over 60 million people, 100 percent of pollution will be more than five ton. Significant differences exist in different classes of the export and import at one percent level of and when imports increase the level of pollution also increases. It is obvious that the highest pollution level in import and export class is in 40 to 50 billion-dollar level.

variable	variables Classification	1-2 ton (Percent)	3-4 ton (Percent)	3-4 ton (Percent)	More than 5 ton (Percent)	Total	Test Results χ^2
Country population	20-30	25	0	8	0	7	**20.87
	40-60	50	50	0	0	22	
	50-60	25	50	33	0	33	
	More than 60	0	0	59	100	38	
Total export and import (million person)	20000-30000	75	25	0	0	18	**22.02
	40000-50000	25	25	50	0	34	
	50000-60000	0	37.5	16	0	19	
	More than 60000	0	12.5	34	100	29	
GDP in fixed price (billion Riala)	100000-200000	100	37.5	0	0	25	***31.16
	200000-300000	0	62.5	59	0	44	
	300000-400000	0	0	41	60	25	
	More than 400000	0	0	0	33	3	
The numbers of produced and imported cars	8000-28000	50	37.5	0	0	18	**17.54
	28000-48000	0	25	8	0	11	
	48000-68000	0	37.5	25	0	22	
	More than 68000	50	0	67	100	49	

(Source: research findings),** 5%,*** 1% significance level

Table 2. Estimated effects of different factors on carbon dioxide emissions

Considering table (3) the emissions of CO2 during various Development Programs is different. In first and second Development Programs pollution emissions is almost equal, while during third Development Program pollution increase has had significant decrease.
The highest effect on pollution is related to economic growth variable. Since increase in economic growth is as an important goal for all government, it is necessary to this reach this goal while sustainable development is considering and environmental damage is at least.
As an illustration the most important variable that has an effect on pollution is GNP after that population has significant effect that Government must control it with training them to protect the environment.
Although summation of import and export on GNP are not punctual, it explains if importing and exporting is done easier, we can hope that productions with high expenditure will be imported without harmful effect on effect on economy of the Iran.
Instead of that we can export productions with competitive advantages.
Also government must supervise the number of cars and consumption of fuel with pollution must be prohibited, rules in big city should be modified and etc.
The of third Development Program had the remarkable effect on falling the pollution. Authorities should expand it if they want to receive sustainable development I the world.

variables	coefficients	standard deviation	t statistic
The logarithm of GDP in fixed price	-7.7103***	1.729	-4.458
the logarithm of GDP square in fixed price	2.842***	0.5672	5.011
The logarithm of total vehicles number produced and imported	0.0292	0.0204	1.432
Logarithmic degree of openness in a economy	-0.2879	0.0960	0.2998
The logarithm of population	0.3372**	0.1491	2.262
dummy variable for the function of first Development Program	0.200***	0.0394	5.079
dummy variable for the function of second Development Program	0.2447***	0.0532	4.596
dummy variable for the function of third Development Program	0.0794**	0.0436	1.821
Constant coefficient	2.6196	2.249	1.165

R-SQUARE = 0.9494 R-SQUARE ADJUSTED = 0.9270
DURBIN-WATSON = 2.7128
JARQUE-BERA NORMALITY TEST- CHI-SQUARE(2 DF)=1.6833
P-VALUE=0.431

(Source: research findings),** 5%,*** 1% significance level

Table 3. Estimated logarithmic model

4. References

Central Bank of Iran. 2009.

Grossman,G. m. , A. G. Krueger. 1991 .Environmental Impacts of a North American Free Trade Agreement; National Bureau of Economic Research, NBER Working Paper. 3914.

Iran Department of Environment. 2009.

Khanna,N. 2002. The Income Elasticity of Non-point Source Air pollutants: Revisiting the Environmental Kuznets Curve; Economics Letters, 77, 387-392.

Lee, k. and M. Lee. 1996. The Evaluation of the Regional Development Considering Environment, Environmental Economic Review, Vol.5 No.1, pp. 143-188.

Pazhoyan, j. and Moradhasel,N. 2007. Surveying the effect of economic growth on tha air pollution, economic magazine, VOL.7 NO. 4, PP 160-141.

the Balance Sheet Energy. 2009.

UNDP. 2008. Human Development Report 2004 (Energy And The Environment), World Bank 2007b, Baised on DATA from IEA.

www.ttic.ir/new/ Air polution.aspa.

World Bank. 2005. Human Development Report. (www.worldbank.org)

World Development Reports. 2008.

20

Managing Air Pollution: How Does Education Help?

Gaye Tuncer Teksoz
Middle East Technical University
Turkey

1. Introduction

We aware since Tbilisi Declaration (UNESCO, 1977) that, science and technology can no doubt provide solutions to environmental problems, which probably helped to cause, nevertheless, solutions sought should not be short-term ones nor too narrowly conceived. Solutions, on the other hand, have to take into account social and cultural factors which are so often at the root of environmental problems. What is necessary is a close examination of the complex relationships between people and their environment. The equilibrium in the flow of matter and energy through natural ecosystems as well as ecosystems already modified by humanity must be re-established. In addition, models of economic growth, development, environment and culture must be reconsidered. It has become essential to look-over the lifestyles to distinguish between the essentials and luxuries for both the environment and development. This is one of the basis to advocate a holistic approach to the management of environmental problems. Therefore, the recognition of reasons, results and implications of environmental problems must be coupled with an increasing awareness of solidarity among nations. Improved management of the environment should aim reducing existed disparities as pertaining a sustainable use of natural resources and at bringing about international relations based on equity. Environmental Education (EE), therefore, has an evident role to play if the issues are to be grasped and if all concerned are to be provided with the knowledge, skills, and attitudes to modify the existing situation for the better. Building on more than 30 years of experience in environmental education, education for sustainable development (ESD) continues to highlight the importance of addressing the issues of natural resources as part of the broader agenda of sustainable development. In December 2002, the United Nations General Assembly (UNGA) adopted resolution (57/254) to put in place a United Nations Decade of Education for Sustainable Development (DESD), spanning from 2005 to 2014. Teaching society how to behave responsibly towards the environment lies at the core of ESD; the founding value of ESD is respect: for others and respect for the planet and what is provides us with. ESD wants to challenge us all to adopt new behaviours and practices to secure our future, seeks to integrate the principles, values, and practices of sustainable development into all aspects of education, in order to address the social, economic, cultural and environmental problems we face in the 21st century. Air pollution is one of the themes of environmental perspectives of the ESD. Educating about air pollution builds the skills and attitudes needed to question the way we think, the values we hold and the decisions we make in the context of sustainable development. Improving awareness about the sustainability involves issues like the impact of human activities on earth systems, control of green house gases, land and energy use, consumption patterns, pollution and transport. ESD for air pollution can be appreciated as one of the powered tools for air pollution management since it offers innovative ways of framing to make sense in people's daily lives and of translating passive awareness into active concern and behaviour change. EE and ESD for air pollution has been covered in several researches all over the world the major areas of focus being, problems on the implementation of EE, misconceptions of both teachers and students especially related to global warming and ozone layer depletion issues , attitudes towards air pollution problems and solutions and environmental literacy components and environmental responsible behaviour (Cutter, 2002; Michail et al., 2007; Dove, 1996; Gayford, 2000; Özdemir & Çobanoğlu, 2008; Summers, et al., 2000; Michail et al.,2007). The researchers also interested in the regional differences in students' and citizens' attitudes and behaviours toward air pollution issues. All this research have shown that people of all ages has positive attitudes toward air pollution issues; that is to say we all are aware of the threats of population increase, industrialization and consumption patterns on the air pollution problems but, almost none of us can make connections between these threats and individual contributions. Therefore, through ESD, we seek for ways to help people to touch the reality, so to make sustainable decisions. The main idea of this chapter therefore, is to present both reasons and results of air pollution in a wider perspective, i.e. through the point of view of EE and ESD, as well as to present education as a component for air pollution management.

2. Action and response

Hunter-gatherers live in the forest, agriculturalists live adjacent to but within striking distance of the forest, and urban-industrial men live away from the forest. Paradoxically, the more the spatial separation from the forest the greater the impact on its ecology, and the further removed the actors from the consequences of this impact! (Gadgil & Guha, 1992, p.67)

Earliest probable evidence of fire used deliberately to clear forests in the Kalambo Falls site in Tanzania points out 60,000 years before present. (Grove, 1995). Air pollution is concurred with the appearance of humans, continues to grow and as we stand still unconsciously, it will continue to intrude the life on Erath. Signals are clear and actually there are number of evidences for air pollution beginning from ancient civilizations that should not be disregarded:

When Homo sapiens first lighted fire, its smoke provided the first medium of environmental pollution. The burning of fuels for heating and cooking has contributed to indoor air pollution. The walls of caves, inhabited several thousands of years ago, are covered with thick layers of soot. The presence of smoke must have made breathing difficult and must have irritated the eyes in the confined space as well. Most of the lungs of mummified bodies from the Palaeolithic have a black tone. In the first inhabited areas smoke was not driven away (one of the practical reasons might have been protection against mosquitoes) and the people dwelling in these inner areas found shelter in the smoke (McNeill, 2001). Humans, on the other hand, seem to have been living together with this unhealthy form of air pollution for many thousands of years. Following section presents the brief history of air pollution beginning from 13th century till today, based mainly on the "Environmental History Timeline, which originally appeared in Mass Media and Environmental Conflict, a book written by Mark Neuzil and William Kovarik published by

Sage in 1996. The first web publication of the timeline was on 6/18/96, and it was expanded in 1998 and 2001 (www.radford.edu/~wkovarik/envhist).

The Roman Senate introduced a law about 2000 years ago, according to which: 'Polluting air is not allowed'. The Institutes issued under the Roman emperor Justinian in 535 AD were used as a text in law schools. Under the section Law of Things, our right to the air is clear: 'By the law of nature these things are common to mankind - the air, running water, the sea, and consequently the shores of the sea.' (Makra & Brimblecombe, 2004).

In 1257, Queen Eleanor of Provence was forced to leave Nottingham Castle for Tutbury Castle because heavy coal smoke fouls the air.

In 1306, Edward I forbidden coal burning in London, but like many attempts to regulate coal burning, it has little effect.

Between 1560 and1600, rapid industrialization in England leaded to heavy deforestation and increasing substitution of coal for wood. In 1590, Queen Elizabeth was "greatly grieved and annoyed" by coal smoke in Westminster Palace.

In 1661 John Evelyn wrote "Fumifugium, or the Inconvenience of the Air and Smoke of London Dissipated" to propose solutions for London's air pollution problem. These include large public parks and lots of flowers:

"The immoderate use of, and indulgence to, sea-coal in the city of London exposes it to one of the foulest inconveniences and reproaches that can possibly befall so noble and otherwise incomparable City... Whilst they are belching it forth their sooty jaws, the City of London resembles the face rather of Mount Aetna, the Court of Vulcan... or the suburbs of Hell [rather] than an assembly of rational creatures..."

In his diary, Evelyn wrote in 1684 that smoke was so severe "hardly could one see across the street, and this filling the lungs with its gross particles exceedingly obstructed the breast, so as one would scarce breathe."

Abraham Darby of Coalbrookdale, England used coal instead of wood in 1709 for manufacturing iron. British coal production around this time was 3 million tons per year, or five times more than the rest of the world combined.

In 1775 English scientist Percival Pott found that coal was causing an unusually high incidence of cancer among chimney sweeps.

Benjamin Franklin noted in 1784 that the switch from wood to coal had saved what remained of England's forests and he urged France and Germany to do the same.

In 1795 Sir Thomas Percival, a UK physician, leaded group of doctors to supervise textile mills and recommended hours and working conditions, children were only permitted to work 12 hours per day.

Phileppe Lebon became the first to illuminate a public building with gas. The hotel Seignelay in Paris was lit using wood gas, not coal. Project was restricted and then Lebon's untimely death by robbery 1804 ended the experiments.

In 1804, impacts of smoke had begun to be felt in Pittsburgh. The smoke affected the "comfort, health and peace and harmony" of the new city. As in most other cities, the solution was to build higher chimneys.

First gas light introduced in 1812 in London. This "town gas" or manufactured gas would be used in every major US and European city, but residual coal tar would remain an environmental problem well into the 21st century.

In 1819 British Parliamentary committee expressed concern that steam engines and furnaces "could work in a manner less harmful to public health."

James Fenimore Cooper wrote "The Pioneers" in 1823, which contains the idea that humans should "govern the resources of nature by certain principles in order to conserve them.
Jean Baptiste Joseph Fourier wrote the first scientific reference to global warming in 1824, "Remarks on the Temperature of the Terrestrial Globe and Planetary Spaces", in which he proposed the theory that the sun's heat is partially trapped in the earth's atmosphere like a giant glass jar.
In 1835 Alexis de Tocqueville published Journey to England and described the industrial city of Manchester:
"*Thirty or forty factories rise on the tops of the hills...six stories (high). The wretched dwellings of the poor are scattered haphazard around them. Round them stretches land uncultivated but without the charm of rustic nature.,, the fetid, muddy waters stained with a thousand colours by the factories ... Look up and all around this place and you will see the huge palaces of industry. You will hear the noise of furnaces, the whistle of steam. These vast structures keep air and light out of the human habitations which they dominate; they envelope them in perpetual fog; here is the slave, there the master; there is the wealth of some, here the poverty of most.*"
In 1843 House of Commons Select Committee on the Smoke Nuisance recommended all manufacturers be removed to a distance of 5 to 6 miles from city centre.
In 1848, Andrew Jackson Downing, a landscape architect, proposed creation of a 500 acre People's Park in New York, which is now known as Central Park.
Novelist Charles Dickens published his novel "Bleak House" in 1853, with an image of London as a twisted, twilight world of smoke, shadows and wraiths. Dickens wrote:
"*Smoke lowering down from chimney-pots, making a soft black drizzle, with flakes of soot in it as big as full-grown snowflakes -- gone into mourning, one might imagine, for the death of the sun.*"
In 1860, Prof. Augustine Mouchot of Lycee de Tours, France, said: "*One cannot help coming to the conclusion that it would be prudent and wise not to fall asleep regarding this quasi security. Eventually industry will no longer find in Europe the resources to satisfy its prodigious expansion Coal will undoubtedly be used up. What will industry do then?*" Mouchot's answer was to build solar energy machines. In 1874 he built a collector with 54 square feet of reflecting surface for alcohol distillation which worked at the rate of 5 gallons a minute. The machine could also power a 1/2 hp motor and develop 75 psi of steam.
John Tyndall explained the "greenhouse effect" in 1863 in a lecture to the British Royal Society entitled "On Radiation through the Earth's Atmosphere." It was the first confirmation and extension of Joseph Fourier's idea that the earth would be much colder without its atmosphere.
First of a series of "killer fogs" in London occurred in December 1873. Over 1,150 died in three days. Similar incidents happened in 1880, 1882, 1891, 1892 and later.
In 1874 German graduate student Othmar Zeider discovered chemical formula for the insecticide DDT.
Inversions lead to another "killer fog" in London with 700 deaths in January 1880.
In 1881, Chicago became the first American city to create a local ordinance regulating smoke discharges, followed that same year by Cincinnati.
In the same year, in 1881, Norway tracked first signs of acid rain on its western coast.
English writer Edward Carpenter published "**Civilization: Its Cause and Cure**" in 1889, which later had a great influence on Mahatma Ghandi. The following is what Carpenter wrote about the town of Sheffield:
"*Only a vast dense cloud, so thick that I wondered how any human being could support life in it, that went up to heaven like the smoke from a great altar. An altar, indeed, it seemed to me, wherein*

thousands of lives were being yearly sacrificed. Beside me on the hills the sun was shining, the larks were singing; but down there a hundred thousand grown people, let alone children, were struggling for a little sun and air, toiling, moiling, living a life of suffocation, dying (as the sanitary reports only too clearly show) of diseases caused by foul air and want of light -- all for what? To make a few people rich!"

Clarence Kemp, "the father of solar energy in the U.S.", patented first commercial Climax Solar Water Heater in1891. By 1910, the Climax had competition, especially from the Night and Day solar hot water company, which used a secondary loop from the collector to a water tank. By 1920, over 5,000 Night and Day heaters had been sold in California. At the same time, a boom in solar hot water heaters started in Florida, where electricity was a very expensive competitor. About 15,000 units were sold by 1937.

In April, 1896, Swedish chemist Svante August Arrhenius summarized scientific opinion about the effect of carbon dioxide in the atmosphere, predicting a global temperature increase of 8 or 9 degrees Fahrenheit for a doubling of CO_2 in the atmosphere; "On the Influence of Carbonic Acid in the Air upon the Temperature of the Ground".

In 1898, Coal Smoke Abatement Society formed to pressure government agencies to enforce pollution laws in England.

In 1900, the world's leading scientists gathered in Paris to consider new elements with unusual powers discovered by Pierre and Marie Curie. Minerals like thorium, uranium and radium emitted a new kind of light, the Curies had found. A year later, Ernest Rutherford and Frederick Soddy found that thorium was turning itself into radium -- evidence of long-sought transmutation of metals. Tapping the energy within atoms would mean that a future awaited that "would bear as little relationship to the past as a dragonfly does to that of its aquatic prototype." Indeed, he said, the power would allow mankind to "transform a desert continent, thaw the frozen poles, and make the whole world one smiling garden of Eden." (Weart, 1988, p. 6).

Smoke Prevention Association of America founded in Chicago in 1907.

Air pollution lawsuit begun in Supreme Court in 1907. In various decisions through 1915, the Court decided to limit the amount of sulphur and other noxious fumes that can emerge from the Tennessee Copper Co. following a suit by the State of Georgia. The suit involved sulphur dioxide fumes from Copper Basin smelters in Tennessee that were killing forests and orchards and making people sick over the Georgia border. The state of Tennessee refused to move against the copper companies and disputed Georgia's right to interfere. Georgia sued in 1907 and won in 1915 after investigation and attempts to reduce the pollution, including a court-mandated reduction and mandatory inspections by a university professor. The majority opinion was delivered by the Chief Justice:

"*It is a fair and reasonable demand on the part of a sovereign that the air over its territory should not be polluted on a great scale by sulphurous acid gas, that the forests on its mountains should not be further destroyed or threatened by the act of persons beyond its control, that the crops and orchards on its hills should not be endangered.*" - Georgia v. Tennessee Copper Co. and Ducktown Sulphur, Copper & Iron Co, 206 U.S. 230 (1907)

In 1908, Swedish chemist Svante Arrhenius argued that the greenhouse effect from coal and petroleum use is warming the globe. According to his calculations, doubling CO_2 would lead to average temperature increase of 5 to 6 degrees Celsius. Rather than being alarmed, Arrhenius was pleased that people in the future would "*live under a warmer sky and a less harsh environment than we were granted.*" In his book World in the Making, he says that with increased CO_2 "*we may hope to enjoy ages with more equable and better climates, especially as regards the colder regions of the earth, ages when the Earth will bring forth much more abundant crops than at present for the benefit of rapidly propagating mankind."*

Glasgow, Scotland, winter inversions and smoke accumulations kill over 1,000 in 1909.

In October 1921, General Motors demonstrated car powered by 30 percent alcohol-gasoline blend.

In December 1921, General Motors researchers discovered tetraethyl lead as an anti-knock gasoline additive. Despite strong private warnings about its danger and a secret Public Health Service inquiry, the new gasoline went on sale without safety tests 14 months after it was invented, with disastrous consequences.

Leaded gasoline went on sale in February 1923 in Dayton, Ohio at a gas station.

In 1926, the first large scale survey of air pollution occurred in U.S., in Salt Lake City.

Air pollution control begun in 1928 in eastern US cities, reporting sunlight cut by 20 to 50 percent in New York City.

In 1930, Meuse River Valley killer smog incident occurred in Belgium. Three day weather inversion in this industrial valley killed 63, with 6,000 made ill.

The date 1939, October 11 was recorded as St. Louis smog episode. Smog was so thick that lamps were needed during daylight for a week.

Donora, Pennsylvania smog incident occurred in October 30-31, 1948. Twenty people died, 600 hospitalized and thousands stickered in this nationally publicized environmental disaster.

In 1948 a "killer fog" in London caused 600 deaths.

First US conference on air pollution held in 1949.

Four thousand people died in the worst of the London "killer fogs" in Dec. 4-8, 1952. Vehicles used lamps in broad daylight, but smog was so thick that busses run only with a guide walking ahead. By Dec. 8 all transportation except the subway had stopped.

In May 4, 1953, Gilbert N. Plass presented paper on global warming at American Geophysical Union. The Washington Post story (May 5) says:

World Industry, ***pouring its exhausts into the air, may be making the earth's climate warmer,*** *a Johns Hopkins physicist, reported here yesterday. Releases of carbon dioxide from burning coals and oils, said Dr. Gilbert N. Plass, blanket the earth's surface 'like glass in a greenhouse.' So much carbon dioxide has been released in this industrial century that the earth's average temperature is rising 1 1/2 degrees (F) a century, he said. Similar but more naturally caused changes in the air's carbon dioxide content may account for the ice ages and warm intervals in geologic time, he added... Latest experimental and theoretical calculations, he reported, show that* ***doubling the carbon dioxide content of the atmosphere causes surface temperatures to rise four degrees (F)*** *if no other changes occur. But, he added, still other earth warming factors may also be triggered by increased carbon dioxide in the air. It could cause less rainfall by its effect on the clouds and less cloud cover for the earth,' both tending to make the climate warmer and drier,' he said. Dr Plass said the newer calculations bolster the theory first proposed in 1861 that decreases in the carbon dioxide content of the earth's atmosphere caused the ice ages in geologic history. The theory, he said, has not generally been accepted because the effects 'appeared to be too small.' It appears now, he said, that* ***even the physicists supporting the theory underestimated the climate-changing effects of the carbon dioxide*** *content in the earth's atmosphere.* ("Industrial Gasses Warming Up Earth, Physicist Notes Here," Washington Post, May 5, p. 5, probably by Nate Haseltine).

New York smog incident killed between 170 and 260 in November, 1953.

Heavy smog conditions shut down industry and schools in Los Angeles for most of October in 1954.

In 1955 International Air Pollution Congress held in New York City.
Another killer smog occurred in London in 1956; 1,000 died.
British Parliament passed Clean Air Act in 1956.
World's first commercial nuclear electric power plant was opened at Sellafield in the United Kingdom in March 31, 1956.
Chelyabinsk nuclear waste explosion occurred in Kyshtym in Russia in 1957-58. Two million curies spread throughout the region, exposing to radiation over a quarter million people.
Another smog phenomenon in London caused 750 die in 1962.
A reaction took place in 1962 to Silent Spring by Rachel Carson. Some agronomists asked whether Carson is intending to starve people by banning pesticides. By 1970 DDT was banned, but other more toxic chemicals were not. Silent Spring was often seen as a turning point in environmental history because it opened a much stronger national dialogue about the relationship between people and nature.
In January 1970, General Motors president promised "pollution free" cars by 1980 and urged the elimination of lead additives from gasoline in order to allow the use of catalytic converters.
In the seventies, air pollution was cut back dramatically through use of catalytic converters on new cars that use only unleaded gasoline. But the predicted "pollution free car" proved to be imaginary.
Disasters showed the weak and fragile side of industrial technology in 1980's:
Bhopal Disaster was recorded as one, 2 decades after the "Silent Spring": Bhopal is a town in India, was similar to many others, before that night in December 3, 1984. Union Carbide Co. fertilizer plant leaked methyl isocyanide at 5 past midnight in Bhopal. 2000 dead, another 8,000 die of chronic effects, estimated 2000 casualties, 100,000 injuries, and significant damage to livestock and crops. The International Medical Commission on Bhopal estimated that as of 1994 upwards of 50,000 people remained partially or totally disabled.
"India's night of death: (Brelis, 1984).
"The first sign that something was wrong came at 11 p.m. A worker at the Union Carbide pesticide plant on the outskirts of Bhopal (pop. 672,000), an industrial city 466 miles south of New Delhi, noticed that pressure was building up in a tank containing 45 tons of methyl isocyanate, a deadly chemical used to make pesticides. At 56 minutes past midnight, the substance began escaping into the air from a faulty valve. For almost an hour, the gas formed a vast, dense fog of death that drifted toward Bhopal.
The vapour passed first over the shantytowns of Jaiprakash and Chhola, just outside the walls of the plant, leaving hundreds dead as they slept. The gas quickly enveloped the city's railway station, where beggars were huddled against the chill. In minutes, a score had died and 200 others were gravely ill. Through temples and shops, over streets and lakes, across a 25-sq.-mi. quadrant of the city, the cloud continued to spread, noiselessly and lethally. The night air was fairly cool (about 60° F), the wind was almost calm, and a heavy mist clung to the earth; those conditions prevented the gas from dissipating, as it would have done during the day.
A few hundred yards from the chemical plant, M.A. Khan, a farmer, was lying in bed when he heard several thumps at a nearby dairy farm and sensed that his own cows were milling about restlessly. He arose and went outside. Two cows were dead on the ground. A third gave out a loud groan and collapsed as Khan watched. Then the farmer's eyes began to smart painfully. He ran into the darkness. The day after, at Bhopal's Hamidia Hospital, his eyes shut tightly and tears streaming down his cheeks, Khan described his fear: "I thought it was a plague."

Fig. 1. Bhopal Disaster; cover in Time Magazine

By week's end more than 2,500 people were dead in the worst industrial disaster the world has known. At least 1,000 more were expected to die from the fumes in the next two weeks; some 3,000 remained critically ill. In all, 150,000 people were treated at hospitals and clinics in Bhopal and surrounding communities. Most of the dead had succumbed because their lungs had filled with fluid, causing the equivalent of death by drowning. Others had suffered heart attacks. The disaster struck hardest at children and old people, whose lungs were either too small or too weak to withstand the poison. A number of the survivors were permanently blinded; others suffered serious lesions in their nasal and bronchial passages…

Another disaster showing the weak and fragile side of industrial technology happened in 1986: On April 26, 1986, at 1:23 a.m. an explosion and fire occurred in Reactor Number 4 of the Chernobyl Nuclear Power Plant in the former USSR (now Ukraine), located 80 miles north of Kiev. Before engineers and scientists could get it under control, 190 tons of highly radioactive material was released into the atmosphere. The radioactive particles rained down not only on Chernobyl, but all over Ukraine, as well as the neighbouring countries of Belarus and Russia, and drifted over to other European countries such as Poland. Scientists estimate that the amount of particles released was equivalent to the effect of 20 nuclear bombs. The Chernobyl accident remains the largest peacetime nuclear disaster ever. The massive radiation killed 31 people within a short time, mostly plant workers and people close to the accident site who died of radiation sickness. As time passed it became clear that

> *"Good evening, comrades. All of you know that there has been an incredible misfortune -- the accident at the Chernobyl nuclear plant. It has painfully affected the Soviet people, and shocked the international community. For the first time, we confront the real force of nuclear energy, out of control."*
>
> **Soviet President Mikhail Gorbachev**

(http://www.megaessays.com/viewpaper/99365.html)

Fig. 2. Announcement for Chernobyl Disaster

the accident had left a number of serious long-term health problems for the people who lived in the area. These health problems were made worse by the poverty, poor nutrition, and lack of medical care in the region.

In the wake of the Chernobyl disaster, in 1988 Russian scientists form the Ecology and Peace Association, electing as president S.P. Zalyghin, whose astonishing statement -- "Only the people can save nature" -- implied that the government had failed.

At the time when Zalyghin pointed out the people as the saver of the nature, UNEP was working on public awareness raising activities to reduce ozone layer lose. The story of success in ozone layer protection is presented by Stephen O. et al. (2002) and education is dedicated as one of the major contributors.

In 1985, British scientist Joe Farman published discovery of, so called, ozone hole over Antarctica and it was confirmed by US NASA satellite monitoring. Meanwhile, US EPA had begun reconsidering CFC regulations. And the United Nations Environment Program had begun negotiations under the Vienna Convention for the Protection of Ozone which leads to the 1987 Montreal Protocol. The Montreal Protocol international agreement to phase out ozone-depleting chemicals signed by 24 countries, including the US, Japan, Canada and EEC nations in the same year. The treaty calls for phasing out production and consumption of ozone depleting substances. Accordingly, beginning from March 1989, European nations begun ban on ozone - depleting chemicals.

A new report released in September 2010 by the UN, says international efforts to protect the ozone layer are a success and have stopped additional ozone losses. The joint World Meteorological Organization and U.N Environment Program report is the first comprehensive update in four years. World Meteorological Organization Research Department Director Len Barrie says the treaty is working: "It has protected us from severe ozone depletion over the past decade, global ozone, including ozone in the Polar Regions," Barrie said. "It is no longer decreasing, but not yet increasing. One phase for the ozone layer protection work realized by UNEP is related to public awareness raising activities. Apart from the technical reports, UNEP has published myriad of booklets, brochures, posters aimed at the public and the General Assembly of the United Nations designated 16 September as the International Day for the Preservation of Ozone Layer. In most countries, improved servicing was the source of early and substantial emissions reductions. One of the components of the improved servicing, on the other hand, was training. Trade and professional associations, labor unions and private companies primarily undertook training. The objective of the ozone treaties was certainly a difficult one: to persuade the entire world to give up the use of many profitable chemicals. To be persuaded were not only the governments, but also the producers of these chemicals, all major multi- national giants of industrialized countries, and thousands of industries. Behind them were the billions of consumers who wanted and needed the products that contained ozone-depleting chemicals. (Stephen, O et al., 2002).

Therefore, an important role of the success in ozone layer protection can be dedicated to education and training, believing that, re-establishment efforts of the equilibrium of natural systems that have already been distorted by human activities can be satisfactory on the condition that, they include the human itself. Any innovative technology produced to handle an air pollution problem should consider human being who will get the advantage of it.

In the book titled "Earth in Mind", Orr (1994, p.6) wrote, *"If one listens carefully, it may even be possible to hear the Creation groan every year in May when another batch of smart, degree-holding but ecologically illiterate, Homo sapiens who are eager to succeed are launched into the biosphere."* The things on which our future health and prosperity depend are in dire danger and according to Orr (1994) this is not the work of ignorant people, rather it is largely the results of work by people with degrees. Because what was wrong in their education is, it emphasized theories instead of values, concepts rather than human beings, abstraction rather than consciousness, answers instead of questions, ideology and efficiency rather than conscience. And he added (page 8), *"It is not education but education of a certain kind that will save us."* And so, he reported several myths for this kind of education: First one is about the myth that, ignorance is a solvable problem. According to him, ignorance is not a solvable problem; it is rather an inescapable part of the human condition. The advance of knowledge always carried with it the advance of some form of ignorance. As for the case of chlorofluorocarbons (CFCs); *in 1929 the knowledge of what a substance like CFCs would do to the stratospheric ozone and climate stability was a piece of trivial ignorance as the compound had not yet been invented. But in 1930, after the compound discovered, what had been a piece of trivial ignorance became a critical life threatening gap in human understanding of the biosphere. Not until the early 1970's no one did ask "What does this substance to do what? In 1986 we discovered that CFCs had created a hole in the ozone over the South Pole the size of the lower 48 U.S. states; by the early 1990's*

CFCs had created a worldwide reduction of ozone. With the discovery of CFCs, knowledge increased, but like the circumference of an expanding circle, ignorance grew as well (Orr, 1994, p.9). Likewise, one can refer to the above mentioned history of London fog; it was 1257 when Queen Eleanor left Nottingham Castle because heavy coal smoke fouls the air, a series of "killer fogs" occurred in London in 1873, similar incidents happened in 1880, 1882, 1891, 1892 in London, in 1948 another "killer fog" in London caused 600 deaths, New York smog incident killed almost 260 in 1953 and another smog phenomenon in London caused 750 die in 1962 and 4000 people died in the worst of the London "killer fogs" in 1952, 7 centuries later then the first sign. What's more, the disaster caused by a pesticide factory caused unavoidable impacts on the people in 1984, although Rachel Carson's Silent Spring was seen as a turning point in environmental history because it opened a much stronger national dialogue about the relationship between people and nature in 1962. Besides, according to J.Russell (2007), in 2006, coal accounted for 25 percent of world primary energy supply. Due to its high carbon content, coal was responsible for approximately 40 percent of the carbon dioxide emissions from fossil fuels, despite supplying only 32 percent of fossil fuel energy. Management of this plentiful but heavily polluting energy resource has tremendous implications for human welfare, the health of ecosystems, and the stability of the global climate. World coal consumption reached a record 3,090 million tons of oil equivalent (Mtoe) in 2006, an increase of 4.5 percent over 2005. China led world coal use with 39 percent of the total. The United States followed with 18 percent. The European Union and India accounted for 10 percent and 8 percent, respectively.

Another myth Orr (1994) wrote is that, with enough knowledge and technology, we can "manage planet earth". According to Orr, however, what might be managed is us, human desires, economies, politics and communities. But, our attention is caught by those things that avoid the hard choices implied by politics, morality, ethics, and common sense. It makes far better sense to reshape ourselves to fit a finite planet than to attempt to reshape the planet to fit our infinite wants (Orr, 1994, p.9). The global trend on the use of materials, on the other hand, reveals that, we are, in the year 2011, still trying to reshape the planet: Yet, as Gardner (2010), global use of materials (the food, feed, forest products, metals, and minerals that constitute the foundation of modern economies) was up 2.7 percent in 2007, reminding that, materials use is a proxy indicator for environmental impact: the greater the tonnage of virgin materials extracted, processed, consumed, and disposed of, the greater an economy's environmental footprint.

One other myth Orr (1994) presented is related to knowledge, and by implication human goodness, is increasing. According to Orr, rapid increase in data, words and paper, that is an information explosion, should not be mistaken for an increase in knowledge and wisdom, which is not easy to measure. What can be said, as he states, is that some knowledge is increasing while other kinds of knowledge are being lost. As Lopez, B.(1989) says, *"[I am] forced to the realization that something strange, if not dangerous, is afoot. Year by year the number of people with firsthand experience in the land dwindles. Rural populations continue to shift to the cities.... In the wake of this loss of personal and local knowledge, the knowledge from which a real geography is derived, the knowledge on which a country must ultimately stand, has come something hard to define but I think sinister and unsettling."*

All in all, the message is that, we are becoming more ignorant of the things we must know to live well and sustainably on the Earth.

In an Environmental Impact Assessment study, for example, we ask public opinion on the investment. But, how public will assess the investment without having related evaluation skills on the causes and effects? Without having knowledge, positive attitudes, responsibility, skills on the environmental issue, public will act under the influence of economical and cultural pressures and cannot make sustainable choices. An investment even with a very efficient technology on environmental protection has no meaning if the people, who will use and be served by, do not share the responsibility, and responsibility comes with education. Similarly, that efficient technology on environmental protection has no meaning if manager, engineer, technician of the plant, governmental authority responsible for controlling the plant do not feel and share the responsibility. Didn't Chernobyl disaster happen because of such a reason, or what was the reason for the dispersion of the fatal gas in Bhopal? Or how items in the Kyoto Protocol will be effective if people do not understand and feel them? Therefore, education should be a part of any investment. Moreover, we don't need to look for investment proposals for educating people. We can create environmentally literate citizens through several means like integrating EE and ESD into the curriculum of preschools, high schools and universities, offering seminars for women, young, teachers, managers, decision makers, workers etc.

Because, helpless poor in Philippines who has to destroy their own values to survive children living in Africa, people of Bangladesh who do not have any idea about the threat of the global warming, citizens of the developed nations who consume to be happy; decision makers who sign world's biggest investments that discard environmental issues, managers who decide to build new highways instead of constructing new bicycle routes, teachers, students, politicians, women, ...Everybody shall take the responsibility.

Education for Sustainable Development (ESD) may be an answer: It was a quarter of a century ago, that education was described as the "greatest resource" for achieving a just and ecological society. Since then, a series of major international reports have emphasized the critical role education can play in the search for sustainable living. The Brundtland Report, (WCED 1987) argued that teachers had "a crucial role to play in helping to bring about the extensive social changes" (p. xiv) necessary for sustainable development. This message was reiterated by *Caring for the Earth* which identified education's vital role in ensuring that people learn, accept and live by the principle of living sustainably (IUCN, 2002).

Sustainable living must be the new pattern for all levels: individuals, communities, nations and the world. To adopt the new pattern will require a significant change in the attitudes and practices of many people. We will need to ensure that education programs reflect the importance of an ethic for living sustainably.
(IUCN, UNEP & WWF 1991 p. 5)

Unlike most education movements, the inception of ESD was not created by the education community. One major outside thrust for ESD came from international political and economic forums. From the time sustainable development was endorsed in the UN General Assembly in 1987, the parallel concept of education supporting sustainable development was being explored. From 1987 to 1992, the concept of sustainable development matured as committees discussed, negotiated, and wrote the 40 chapters of *Agenda 21*. The initial thoughts concerning ESD were captured in Chapter 36 of *Agenda 21*, "Promoting Education, Public Awareness, and Training" (UNESCO 1992).

Education, including formal education, public awareness and training should be recognized as a process by which human beings and societies can reach their fullest potential. Education is critical for promoting sustainable development and improving the capacity of the people to address environment and development issues. (*Agenda 21*, Chapter 36, p. 3)

3. Millennium goals and hopes for a sustainable air pollution solutions

Does it make a change if people know and feel deeply the difference they make - for their carbon footprint - by choosing to buy locally grown potatoes instead of imported ones?
Does it make a change if people know and deeply feel how an American meal - hamburger, fries and coke -pays for global warming?
Does it make a change if people can realise that we live the lives that "assigned" to us, and as we continue to do so, we'll continue to face with worse problems of air pollution?
It is our choice to make a change...
Former President of Ireland, Mary Robinson (Robinson, 2011) wrote about her visit to Bangladesh in The Hoffington Post: "Travelling by seaplane to Koyra, in the delta area of Bangladesh, was the equivalent of a journey some years into the future, when the devastating effects of climate change will be an accepted reality worldwide. We landed in an area still devastated by cyclone "Aila" which hit Bangladesh in 2009. A huge amount of once cultivated land was still under water, because of daily tidal fluctuations and the fact that some embankments had not been mended in the nearly two years since Aila... A memorable stop on my visit was to a local primary school run by BRAC (a development organization dedicated to alleviating poverty by empowering the poor to bring about change in their own lives). It was organized on the same principles as a BRAC school I had visited the day before in Korail slum, the largest slum in Dhaka. The schools have 30 plus pupils and one teacher, who teaches these children the five year curriculum in four years, .. In the school in Koyra the children enacted with great gusto -- and acting skills -- how climate change may happen. One of the taller boys acted as the tree which the others cut down, even though warned not to. The winds came, and the consequences were played out -- they all knew where the climate shelter was! As I watched with a grandmother's eye, it struck me that every primary school around the world should be beginning to bring home to children what we must all do to change our habits. Reminding that, Bangladesh is a least developed country (LDC) which has become the leading LDC negotiator on climate change issues. Its contribution to the problem of green house gas emissions is negligible, but the additional burden of climate change is already being felt. It is predicted by the officials in Dhaka that 20 million people may have to leave this region if the global temperature increases by more than 2° Celsius and sea levels rise as predicted. There is nothing theoretical about the climate change issue from this local viewpoint. The injustice of a poor LDC country having to bear huge additional costs from climate impacts it did not contribute to be self evident. Every school needs to be a "green school", so that children can educate their parents. For some it will be knowing where the nearest climate shelter is. For others -- in the developed world -- it will be learning to reuse, reduce, recycle, eat less meat, and travel by public transport, among other ideas".
One of the choices to make a change, eating less meat is investigated by Eshel and Martin (2006). Authors compared the energy consumption of animal- and plant-based diets and the range of energetic planetary footprints spanned by reasonable dietary choices. As a result it was demonstrated that the greenhouse gas emissions of various diets vary by as much as the

difference between owning an average sedan versus a sport-utility vehicle under typical driving conditions. Therefore, authors concluded with a brief review of the safety of plant-based diets, and find no reasons for concern. As Walsh (2008) wrote in his article titled "Meat : Making global warming worse", in 2008 the head of the U.N.'s Nobel Prize-winning Intergovernmental Panel on Climate Change, Pachauri advised people around the world to cut back on meat in order to combat climate change. Pachauri is absolutely right, considering the numbers. In a 2006 report, the U.N. Food and Agriculture Organization (FAO) concluded that worldwide livestock farming generates 18% of the planet's greenhouse gas emissions, by comparison, the entire world's cars, trains, planes and boats account for a combined 13% of greenhouse gas emissions. Much of livestock's contribution to global warming come from deforestation, as the growing demand for meat results in trees being cut down to make space for pasture or farmland to grow animal feed. Livestock takes up a lot of space, nearly one-third of the earth's entire landmass. And, there's manure, all that animal waste generates nitrous oxide, a greenhouse gas that has 296 times the warming effect of CO_2. And of course, there is cow flatulence: as cattle digest grass or grain, they produce methane gas, of which they expel up to 200 L a day (full). And as Tung (2010) reported, global meat production increased by 0.8 percent in 2009 to 281.5 million tons, a slowdown from the 2.4 percent growth rate of 2008. But, as he wrote, the increase continued the steady growth of the past decade. Since 2000, global meat production has risen by 20 percent. So, what if people have aware that, giving up average 176 lb. of meat a year is one of the greenest lifestyle changes one can make as an individual. One can drive a more fuel-efficient car, or install compact fluorescent light bulbs or improve insulation, really no green way to get meat — although organic, locally farmed beef or chicken is better than its factory-raised equivalents. If every American reduced meat consumption by just 20%, the greenhouse gas savings would be the same as if we all switched from a normal sedan to a hybrid Prius. Likewise, if a portion of French fries is preferred with meat one shall need to be sure if the potatoes are locally grown or not.

It's a lesser known fact that up to 1/3 of all greenhouse gas emissions (GHGs) can be attributed to the food we eat. Many people think of transportation of food as the big contributor to climate change. But, although transporting food over great distances does contribute to climate change, how food is produced matters just as much. In fact, growing and harvesting, heating and cooling, processing and packing, transportation and storage ALL contribute to the GHGs emitted by our food system. Focusing on the potatoes and a coke accompanying, yet, just simply try to imagine what it takes to consume a favourite American meal (apart from the meat); the potato was dug with a diesel-powered harvester and then trucked to a processing plant where it was dehydrated, sliced, and frozen. The freezing was done by a cooling unit containing hydrofluorocarbons, some of which escaped into the atmosphere and likely contributed to global climate change. The frozen fries were then trucked to a distribution centre, then on to a fast-food restaurant where they were stored in a freezer and then fried in corn oil heated by electricity generated by hydropower. The meal was served in a fast-food restaurant built on what once was originally forest, then farmland, and then converted to commercial/industrial uses as the city expanded. The ketchup in aluminium- foil packets came from Pittsburgh and was made from Florida tomatoes. The salt came from Louisiana River. The high-fructose corn syrup came from Iowa, as did the carbon dioxide used to produce the fizz, which is produced by fermenting corn. The caffeine came from a processing plant that makes decaffeinated coffee. The cola can was made from one-third recycled aluminium and two-thirds bauxite ore strip-mined in Australia. It came to Washington State on a Korean freighter, and was processed into aluminium using an amount of energy equivalent to a quart of gasoline. The energy came from some of the same dams mentioned earlier that have contributed to a 97 percent decrease in the salmon runs of the Columbia Basin. The cola came from a Seattle processing plant. It is made of 90 percent water from the Cedar River. The high-fructose corn syrup came from Iowa, as did the carbon dioxide used to produce the fizz, which is produced by fermenting corn (Ryan & Durning, 1997).

Knowing the theory of global warming, acid rains, photochemical smog is not enough to construct the links, knowledge should be accompanied by feeling the links deep inside and human being has this potential. ESD is the way to take out the potential and help individuals to feel how their presence is connected with nature, the way we arrange our lives in harmony with nature and to respect nature as well as respecting ourselves. The way each individual shape their living styles, on the other hand, is unique. ESD does not mean to outline the styles, but shows the human and environment interdependence considering environmental, economical, social aspects together. Therefore, having environmentally literate citizens promise to take a step in managing air pollution issues. Disinger & Roth (1992) provide a generally accepted 'definition' of environmental literacy pointed out that, Environmental literacy is essentially the capacity to perceive and interpret the relative health of environmental systems and take appropriate action to maintain, restore, or improve the health of those systems (p.2). They go on to explain that environmental literacy draws upon six major components: environmental sensitivity, knowledge, skills, attitudes and values, personal investment and responsibility, and active involvement (Disinger & Roth, 1992). This definition is in line with a growing literature that sees pro-environmental behaviour as a function of environmental literacy and the literature that views environmental behaviour as something that may be learned through increased environmental knowledge (Wilke 2005; Hines et al., 1986)

4. How can education help? How education can be integrated in managing air pollution?

> Education is how we live our lives and how you live with everything around you. Everything in existence teaches us something about life... everything around us educates, how we interact with the land, minerals, trees, sky, animals everything, even our thoughts. Our thoughts too can become a force, for we are in charge of them. (Profeit-Le Blanc 1996, p. 14)

The Intergovernmental Panel on Climate Change (IPCC, 2007) concluded in 2007 that, global warming is inevitable and that human activity is likely to be the main cause. However, according to a survey of the American public in the same year (ABC News 2007), while 33% cited climate change as the world's top environmental issue, and 84% thought it was probably happening today, only 41% of the American public believed that global warming was caused by human activity. Furthermore, while 86% believed global warming would become a serious environmental problem if not corrected, 63% thought it could be reduced, with 62% claiming they knew a moderate amount about global warming. However, reported that, only 18% of the US public agree that every time we use coal or oil or gas, we contribute to the greenhouse effect (Nisbet and Myers, 2007). What these findings imply is the importance of women, young, decision makers, teachers, ... learning about the greenhouse effect in order to understand the arguments and debates about the science of global warming and climate change to promote that is knowledgeable about global warming and climate change, and one which can assume informed responsibility for the management and policymaking decisions facing our planet (Brown 1992; Bybee 1993 quoted in Shepardson et.al. 2011).

Accordingly, ESD for air pollution issues have been covered in several researches all over the world and the major areas of focus are; implementations, cultural differences in attitudes, misconceptions, factors effecting satisfactory results, problems for an effective ESD. And although there are a number of recommendations made by researchers, the targets for ESD still have not been satisfied. But, it is reality that, although research in developing ESD is the task of the education society, increasing environmental literacy is the task for all sectors, including universities, governmental authorities, NGOs, private sector, etc. The following section, therefore, highlights the recent research, focusing on the results, on ESD implementations related to air pollution issues, global warming being the focus of almost all researchers, major challenges and the driving force.

Although air pollution has been a problem since 17th century (please refer to the 1st section of this chapter), it has just become more complex and difficult to manage and control, requiring a growing need for improvement in public understanding of environmental science and policy. Thus, comprehensive and meaningful education is decided to be a promising avenue for equipping members of society in identifying potential solutions to environmental problems in order to protect valuable natural resources. ESD can produce an environmentally literate citizenry able to actively address environmental challenges and problems (Hungerford and Peyton, 1976; UNESCO, 1980; Roth, 1992). Therefore, comprehensive environmental education may be an important mean for societies to meet the increasing need for improved public understanding of environmental issues, trade-offs and other alternatives. The 1972 United Nations-Stockholm Conference (UNEP, 1972) helped articulate a shared outlook and set of principles for inspiring and guiding efforts focused on helping the public learn to pre-serve and enhance healthcare and environment. In 1977, an international assembly of environmental educators developed a set of definitions and principles for environmental literacy and education at the Inter-governmental Conference on Environmental Education in Tbilisi (UNESCO, 1977). A decade later, Hines et al. (1986/87) pointed out that environmental education efforts must go beyond providing simplistic information and move towards providing: knowledge of complex environmental issues, specific knowledge about approaches for addressing such issues and decision-making skills. Hines et al. also called for efforts to change certain effective qualities (attitudes) that result in people caring about and paying more attention to environmental conditions. It seems axiomatic to observe that the current global environmental and natural resources conditions are worse today, the world over than in the past. Therefore, it is imperative that the goals of 21st century educational systems should include environmental education and the formation of an environmentally literate citizenry able to actively participate in solving environmental problems. Disinger and Roth (1992) provide a generally accepted 'definition' of environmental literacy and have pointed out that: Environmental literacy is essentially the capacity to perceive and interpret the relative health of environmental systems and take appropriate action to maintain, restore or improve the health of those systems (p. 2).

Individuals (adults and children) may change their environmental behaviour when their values, beliefs, and pro-environmental norms change (Dietz et al., 2005). Improved environmental education as well as increased environmental literacy may result in such changes. As Clair (2003) pointed out: Environmental literacy for adults means developing and participating in the social practices likely to change the way our societies think about and act upon ecological issues. Literacy is a powerful metaphor that contributes a great deal to thinking through the question of what each of us can contribute for a more just and sustainable way of life for the planetary community (p. 77).

Coyle (2005), in reviewing 10 years of NEETF/Roper research on environmental literacy in the United States, points out that creating more widespread environmental literacy depends on (1) bringing sound environmental education programming into the education realm, and (2) channelling public environmental literacy efforts to focus more on depth rather than accuracy. Thus, environmental literacy is distinct from simple awareness or personal conduct knowledge because of its depth of information and the actual skills (thinking and doing) imparted. Knowledge and attitudes are essential components of environmental literacy, especially if the goal of environmental education is to change behaviour. Individuals' environmental behaviours may change as a result of changes in their values, beliefs, and pro-environmental norms (Dietz et al., 2005).

A research on the environmental literacy of pre-service teachers in Turkey (Tuncer et.al, 2009) revealed that, the least percent (34%) of correct answer for environmental knowledge items were concerned about motor vehicles as the major contributor to carbon monoxide while more than 60% of respondents incorrectly identified factories and businesses as the major source of carbon monoxide. 77% of the teachers correctly answered the item that the ozone layer serves as a protective layer from cancer-causing sunlight. The most frequently answer received for the types of pollution "very concerned', on the other hand, was related to "indoor air pollution" (42%), "ozone depletion and global warming" (38%), while 42% of respondents indicated that they were very concerned about "ozone depletion and global warming," only 19% of them were 'very concerned' about automobile emission. Therefore, the conclusion was that, preservice teachers in this study either do not understand the cause - effect relationship between automobiles and global warming or that their fear of more restrictions on their automobiles motivated them to answer strategically. Besides, gender appeared to play a role in elucidating the variation in the two components of environmental literacy variables along with environmental attitude and uses. Female pre-service teachers tended to have more positive attitudes and have more responsible actions toward the environment than male pre-service teachers, which is in line with other studies (Alp et al., 2006; Berberoglu and Tosunoglu, 1995; Huang and Yore, 2003; Chu et al., 2007; Tikka et al., 2000; Yilmaz et al., 2004; Worsley and Skrzypiec, 1998; Zelezny et al., 2000). For example, according to Tikka et al. (2000), whereas males are more likely to emphasize mastering nature and taking benefits from natural resources, females obtain a more emotional attitude toward nature. Since females, as indicated by the authors, have traditionally been responsible for looking after the home and children, such behaviours could be perceived as a way of taking care of their offspring. Indeed parallel to this explanation, there are two theories, namely socialization-based theories and structural theories proposed in the literature to clarify the gender difference in environmental variables: Socialization-based theory posits (sets) that females are more likely than males to associate themselves with 'caregiver' roles. It is argued that this leadswomen to be more in tune with their locality and the world at large and, consequently, to turn their compassion toward the ecological environment. Women's close affinity with nature is viewed as a result of socialization due to cultural and social-structural forces rather than resulting from biological differences. Structural theories suggest that it is the gendered segmentation of the economy and workplace that frames the perspective of women and men toward the environment. It is argued that although women may be knowledgeable and accepting of the aims of economic growth, they are more prone than men to question the consequences of such growth. The reasoning behind this argument lies in the combination of women's role as caregivers for children and their role in the household, where they do most of the house work, in addition to working in the paid labor force. This role is in direct contrast to men's historical "breadwinner" role. (Weaver, 2002;p.83).

Referring to the above mentioned definition of environmental literacy, it seems to rest on an assumption that individuals have a competent level of environmental knowledge. It is concluded in one of the recent studies realised in Turkey, to determine environmental literacy of preservice teachers (Tuncer et.al, 2009), that, a majority of Turkish pre-service teachers do not possess enough knowledge to be classified as having an acceptable level of environmental knowledge (NEETF and Roper, 2005). Slightly less than half of the pre-service teachers of this study (49%) received a passing grade, based on the NEETF and Roper Starch grading scale. Yet, 66% of the students from Michigan State University [MSU] (Kaplowitz and Levine, 2005) reported to receiving a "passing grade" based on the same grading scale. Interestingly, in that same university, MSU College of Education students had one of the lowest mean correct scores compared with students from other MSU colleges such as agriculture and natural resources (8.84 of 11) and Natural Science (8.48 of 11). One possible explanation for the low level of passing grades for Turkish respondents' environmental knowledge, on the other hand, was explained be the absence of course works, relevant to environmental education in the current teacher education programs in Turkey. Of course, there may be demographic characteristics of students that relate to differences in academic level and environmental knowledge. Despite their low levels of environmental knowledge, Turkish respondents expressed positive attitudes toward the environment as well as high degree of concern about environmental problems. The respondents also expressed feelings of responsibility for environmental problems; expressed the view that environmental problems are one of the most important problems of their lives; and shared their feeling that they are comfortable with their background on environmental issues. Such results beg the question of what it would take to create a critical mass for increased environmental learning throughout the educational system.

In service elementary teachers' knowledge about air pollution in Turkey was the subject of another study (Tuzun et.al, 2008). The study pointed out specific results. One of the questions related to teachers' general knowledge about air pollution sources was that, "Which human activities contributed to air pollution?" The answers for this question were categorized as, individual, societal, and industrial contribution. Approximately 24% of the teachers reported exhaust gases as the main cause of the air pollution. Which, was not a surprising result for the authors, reminding that the study area is one of the crowded cities in Turkey with a number of cars has been increasing and traffic is getting more and more challenging every day. Use of sprays and deodorants was considered as the second main individual source of air pollution by the participants. The reason was explained in relation with not the content of textbooks but with the effect of media. Use of coal at home for heating was declared as the third most important individual source of air pollution by the teachers of the study. Teachers seemed to have knowledge about importance of improper use of other energy sources. Moreover, cigarette, forest fires, education, and trash were found as the other important sources of air pollution declared by the teachers. Contribution of community to air pollution was declared by Turkish teachers as exhaust gases, as the main source, unconscious use of energy especially at homes and forest fires. Besides, awareness toward environmental pollution, trash, and industrial emissions were also considered as the sources of the air pollution by the teachers. When industrial contribution was in consideration, more than 70% of the teachers were stated industrial emissions as the main contributor to air pollution. Whereas, they also indicated that lack of control mechanism is an important aspect for air pollution in Turkey. According to the teachers of this study, emissions from solid waste disposal areas also cause air pollution. Therefore, it was decided by the authors that, teachers participated the study have a considerable knowledge about the reasons of air pollution. However, teachers were asked further questions to get the nature of their knowledge. As a result, teachers reported that, global warming was caused mainly by carbon dioxide (44.3%), methane (28.4 %), chlorofluorocarbons (19.7 %) and ozone (10.4 %). But at the same time, 34.9 % of them declared CO as one of the gases causes global warming, while 18.1 % declared SO2 as so. Teachers declared CFCs (31.7%), spray deodorants (61.7%) and air conditioners and refrigerators (38 %) as the reasons for the ozone layer depletion problem. The interesting result at this point was that, although about 32 % of the teachers indicated CFCs as a reason for the ozone layer depletion, spray deodorants were indicated by 62 %. This result explained with the media effect as; teachers were more familiar with the term rather than the chemical formula. Because, ozone layer depletion has always coincided with the use of spray deodorants in media; especially by the phrase "ozone friendly" in the deodorant advertisements. Moreover, sulphur and nitrogen compounds were stated as the pollutants for acid rain problem by relatively higher percentages. It was inferred as a result that, teachers' knowledge about global warming, ozone layer and acid rain cannot go beyond just defining the concepts. Misconceptions of the teachers of this study about global warming and ozone layer were also obvious by their answers for the multiple choice question (*Which characteristic of the green house gases make them cause green house effect*?): Although 61% of the teachers answered this question correctly, as they absorb *the sunlight emitted from the Earth surface,* 15 % of them answered as they are *found in the upper layers of the atmosphere relative to other gases.* This result points out a common misconception observed in the literature. As was declared by Michail, Stamou and Stamou (2007), Cutter (2002) and Summers, Kruger, and Childs (2000), for example, teachers had confusion about green house effect and ozone layer depletion and they had misconception on these issues. In general terms, the location of ozone layer was confused with that of the greenhouse gases and stratospheric ozone with tropospheric ozone. Teachers in Turkey were also asked about renewable energy question, emphasizing that teachers need to teach renewable energy sources to allow their students to show appropriate actions to prevent air pollution. Just having knowledge about causes of the air pollution will not help them prevent air pollution, they can help prevent air pollution if they know the ways for sustainable use of the sources. As a result, majority of the teachers knew that wind, sun light, wave, and geothermal are renewable energy sources and some of them also indicated that biogas, hydrogen, and biodiesel as renewable energy sources. Similar research conducted with Greek, Australian, and English teachers had also revealed that they did not had enough understanding about air pollutants, green house gases and ozone layer (Cutter, 2002; Michail, Stamou, & Stamou, 2007; Summers et al., 2000).

Shepardson et. Al. (2011) derived five distinct mental models of the greenhouse effect from an inductive analysis of the content of the drawings and explanations of 225 students' from three different schools in the Midwest in the US. Based on the mental models identified it was apparent that students lacked a clear understanding of the greenhouse effect. At best 48% of the students realized that greenhouse gases, whatever they may be, cause the greenhouse effect and that the sun's energy is either 'trapped' by or 'bounced' back to the Earth by the greenhouse gas layer. On the other end of the spectrum, 29% of the students lacked an understanding of the greenhouse effect.

Investigation on the regional differences in 15 year old Turkish students' awareness, perception, optimism and responsibility development toward environmental pollution issues carried out, with 4942 fifteen year-old-students attending 160 schools across 78 provinces and 7 geographical regions (Teksoz, Tekkaya, Erbas, 2009). Results indicated that a minority of students, across seven regions reported as they aware of the increase of greenhouse gases in the atmosphere (9.7%). The percent of students who stated that they never heard about these issues were 26.5% in average. As far as the acid rain issue has been considered, 18.2% stated that they were familiar and 9.2% stated that they never heard about the acid rain issue. Furthermore, while 63.7% of the participants believed that they were familiar with the consequences of clearing forests for other land use, 3.9% declared that they never heard about the issue. The mean scores calculated for environmental perception, environmental awareness and responsibility development components show almost the same trend among the regions. Those for the regions in the West part of the country are higher than those for the Eastern regions. Whereas, mean values for the environmental optimism component show a different pattern between the regions: students living in the Eastern regions seem more optimistic than those in other regions. In the first place, the results of the study provided some evidence that the place where students live had an effect on their environmental awareness, concern, optimism and responsibility for sustainable development. For example, the most noticeable characteristic was that; although the students of the two of the least industrialized regions (Southeast Anatolia and East Anatolia) displayed lower awareness and concern toward environmental issues, they displayed highest degree of optimism concerning the development over the next 20 years of the problems associated with air pollution, clearing of forests for other land use. In fact, the results reflect the transcontinental feature of Turkey. Among the 7 geographical provinces, Marmara having students with comparably higher environmental concern, responsibility but low degree of optimism, distinguishes from the others with its being heavily advanced in industry, commerce, tourism and transportation because of its close location to Europe. Thus the children living in such circumstances are more aware of air pollution problems, are concerned about them and pessimistic about the future state of the problems. The significant feature of Aegean, which has the students with very high environmental awareness, concern, responsibility and comparably higher optimism, is that, most of the population and cities are concentrated on the coast line because of its convenience for sea transportation and tourism and it's also being both industrialized and agriculturalized. Students from one of the most important trading and tourism centre and the rapidly growing port, the Mediterranean region, on the other hand revealed high awareness and concern but lowest optimism toward environmental issues. The students from the Black Sea region, one of the most heavily forested regions with very rich fauna and flora, revealed a similar trend with the former regions. The students from plateau-like heartland of the country, Central Anatolia, revealed comparably higher perception, responsibility and optimism toward environmental issues are considered. The students showing a distinguished feature, lowest awareness, perception, responsibility but highest optimism toward environmental issues, in the current study come from the Eastern Anatolia, where the population and habitat not dense because of the harsh climate and high mountains and has the highest unemployment rate in Turkey and South-eastern Anatolia where a special atmosphere exists throughout, uniquely different from other parts of the country, thus, reflecting a specific life style over its land. Thus, referring the very well known phrase of the environmental studies, *think globally act locally* (UNEP, 1972), efforts to explain environmental perceptions and concern as a function of social structure and socio-demographic characteristics can be combined with the regional features of a country and such a relationship, if any, is valuable for strategy development for developing air pollution perceptions. Thus, as Matthews (1995) reviewed in his study, regional features and culture affect children's behaviour in large-scale environments and it follows that as the life worlds of children from different socio cultural backgrounds differ, the way in which children encounter place and make sense of their everyday worlds are also likely to be at variance. Therefore, it may be concluded that, "area of residence is a silent predictor of responsible environmental behaviour". Therefore, efforts for creating environmentally literate generations need to consider regional socio-economical features as well as the people's perceptions towards environmental issues. Such an evaluation will be very valuable leading the education specialists to establish a national strategy for and will help to make the strategy regional, as suggested in Chapter 36 of Agenda 21 (UNCED, 1992).

Efforts for developing EE and ESD as a tool for managing air pollution therefore continues under the light of the research, several examples of which are summarised above. Although there is are international conventions leading such efforts, regional, national and local strategies are needed to get more effective outputs. Efforts, although concentrated in teachers' education, shall diffuse to all areas, engineering being one of the vital one. Ashford (2004) pointed out the concern as follows: "Scholars and professionals committed to fostering sustainable development have urged a re-examination of the curriculum and the restructuring of research in engineering-focused institutions of higher learning. The focus is on engineering, more than on the natural and physical sciences or on social science, because the activities that drive the industrial state - the activities that implement scientific advance - are generally rooted in engineering. Moreover, engineers are known as 'problem solvers' and if economies are becoming unsustainable because of engineering, it is natural to ask whether engineering as an activity and as a profession can be re-directed toward achieving sustainable transformations. Of course, engineering cannot do it alone; scientific as well as social and legal changes must occur as well" (p.239).

Nevertheless, results of research investigating people awareness, attitudes, behaviour on the air pollution issues have a single common result that, most of the people from different countries with different socio-demographic features and life styles, aware of the natural resources, like air, are vulnerable resources and deserve conservation for the sake of supporting human life or just for the health of itself. Most of the people have an eco-centric approach to the natural sources. But beyond this point, when it comes to individual responsibilities, i.e. making changes in living styles for natural protection, such as preferring public transport instead of private cars or using less energy at home, the approach slides through anthropocentrism. Or, citizens both in the developing and developed world cannot make a relation between, for example the way we eat and the global warming, or the way we shop and acid rain, or the way we consume, we produce and ozone layer depletion, etc, which as a result makes it difficult to make a change. Education is a powerful tool to make people aware of such relations and get rid of the ignorance we have been carrying since 17th century.

Living sustainably depends on a duty to seek harmony with other people and with nature. The guiding rules are that people must share with each other and care for the Earth. Humanity must take no more from nature than nature can replenish. This in turn means adopting lifestyles and development paths that respect and work within nature's limits. It can be done without rejecting the many benefits that modern technology has brought, provided that technology also works within those limits. (IUCN, UNEP and WWF 1991, p. 8)

Knowledge, learning, information, and skilled intelligence are the new raw materials of international commerce and are today spreading throughout the world as vigorously as miracle drugs, synthetic fertilizers, and blue jeans earlier... Learning is the indispensable investment required for success in the "information age" we are now entering. (National Commission for Excellence in Education, 1983, p. 7)

5. References

ABC News (2007). Concern soars about global warming as world's top environmental threat. URL http://abcnews.go.com/images/US/1035a1Environment.pdf.

Alp, E., Ertepinar, H., Tekkaya, C., Yilmaz, A.,(2006). A study on children's environmental knowledge and attitudes: the effect of grade level and gender. *International Research in Geographical and Environmental Education* 15, 210–223.

Andersen, S.O. & Madhava Sarma, K. (2002). Protecting the Ozone Layer, *The United Nations History*, Sinclair,L (Ed.). Earthscan.

Berberoglu, G., Tosunoglu, C.(1995). Exploratory and confirmatory factor analyses of an environmental attitude scale (EAS) for Turkish university students. *The Journal of Environmental Education* 26, 40–44.

Ashford, N., A. (2004). Major challenges to engineering Education for Sustainable Development: What has to change to make it creative, effective and acceptable to the Established Disciplines, *International Journal of Sustainability in Higher Education* 5(3): 239-250.

Aydemir M. (2007). The investigation of teachers with respect to knowledge level on environmental concepts. Ms.Thesis. Middle East Technical University, Ankara.

Chu, H.E., Lee, E.A., Ko, H.R., Shin, D.H., Lee, M.N.,Min, B.M., Kang, K.H. (2007). Korean year 3 children's environmental literacy: a prerequisite for a Korean environmental education curriculum. *International Journal of Science Education* 29, 731-746.

Clair, R. (2003). Words for the world: creating critical environmental literacy for adults. *New Directions for Adult and Continuing Education* 99, 69-78.

Coyle, K.(2005). Environmental Literacy in America. The National Environmental Education Training Foundation, Washington, DC.

Cutter, A. (2002). The value of teachers' knowledge: Environmental education as a key study, ERIC DocumentReproduction Service No. ED 466456.Brelis, D., (1984). Time Magazine, December 17.

Dietz, T., Fitzgerald, A., Shwom, R. (2005). Environmental values. *Annual Review of Environment and Resources* 30, 335-372.Disinger, J. F., Roth, C. E. (1992). Environmental Literacy. ERIC/CSMEE Digest.URL http://eric.ed.gov/ERICDocs/data/ericdocs2/

Dove, J. (1996). Student teacher understanding of the greenhouse effect, ozone layer depletion and acid rain. *Environmental Education Research, 2(1)*, 89-100.Eshel, G., Martin, P.A. (2006). Diet, Energy and Global Warming, *Earth Interactions*, 10:1-17.

Gadgil, M. and Guha, R. (1992). This Fissured Land: An Ecological History of India. Berkeley: University of California in Orr, D., (1994) *Earth in Mind on Education, Environment and the Human Prospect.*

Gardner, G. (2010). Materials Use Up, *Vital Signs 2010*, Worldwatch Institute.Gayford, C. G. (2000). Biodiversity education: A teacher's perspective. *Environmental Education Research, 6:4*, 348-361.

Grove, R.H.(1995). Green Imperialism: Colonial Expansion, Tropical Island Edens and the Origins of Environmentalism, 1600-1860. Cambridge, England: Cambridge University Press, in http://www.radford.edu/~wkovarik/envhist/Envbiblio.html#G

Hines, J.E., Hungerford, H.R., Tomera, A.N.(1986/87). Analysis and synthesis of research in responsible environmental behavior: a meta-analysis. *Journal of Environmental Education* 18, 1-8.

Huang, H., Yore, L.(2003). A comparative study of Canadian and Taiwanese grade 5 children's environmental behaviours, attitudes, concerns, emotional dispositions, and knowledge. *International Journal of Science and Mathematics Education*, 1, 419-448.

Hungerford, H.R., Peyton, R.B.(1976). *Teaching Environmental Education.* Weston Walch, Portland, ME.

IUCN, The World Conservation Union (2002). Education and Sustainability Responding to the Global Challenge, Editors: Daniella Tilbury, Robert B. Stevenson, John Fien, Danie Schreuder.

IUCN, UNEP, WWF (1991). Caring for the Earth: A Strategy for Sustainable Living. IUCN, UNEP, WWF: Gland, Switzerland.

IUCN, UNEP, WWF (1991). Caring for the earth: a strategy for sustainable living. IUCN: Gland, Switzerland. in Education and Sustainability Responding to the Global Challenge (2002). Tilbury, D., Stevenson, R.B., Fien, J., Schreuder, D., (Eds.) IUCN - The World Conservation Union

IPCC (2007). Intergovernmental Panel on Climate Change., *Climate change 2007: The physical science basis.* Geneva ,IPCC.

Lopez, B. (1989), American Geographics, Orion , in Orr, D. (1994), *Earth in Mind on Education, Environment and the Human Prospect,* p.10.

Makra, L., Brimblecombe, P. (2004). Selections from the history of environmental pollution, with special attention to air pollution. Part 1, *J. Environment and Pollution,* 22:6, 641-656.

Matthews, H. (1995). Culture, environmental experience and environmental awareness: making sense of young Kenyan children's views of place. *The Geographical Journal, 161,* 285-295.

McNeill, J.R. (2001). *Something New Under the Sun. An environmental history of the twentiethcentury world,* W.W. Norton & Company.

Michail, S., Stamou, A. G. & Stamou, G. P. (2007). Greek primary school teachers' understanding of current environmental issues: An exploration of their environmental knowledge and images of nature. *Science Education, 91(2),* 244-259.

National Commission for Excellence in Education (1983). *A Nation at Risk.* Washington, DC. in *Education and Sustainability Responding to the Global Challenge* (2002). Tilbury, D., Stevenson, R.B., Fien, J., Schreuder, D., (Eds.) IUCN - The World Conservation Union

Orr, D. W.(1994). *Earth in Mind, On Education, Environment and the Human Prospect,* Island Press.

Özdemir, N. ve Çobanoğlu, E. O. (2008). Türkiye'de nükleer santrallerin kurulması ve nükleer enerji Kullanımı konusundaki öğretmen adaylarının tutumları. *Hacettepe Üniversitesi Eğitim Fakültesi Dergisi, 34,* 218-232.

Profeit-LeBlanc, Louise (1995). "Transferring Wisdom through Storytelling" in Jickling, R. (ed.) *A Colloquium on Environment, Ethics and Education.* Whitehorse:Yukon. in *Education and Sustainability Responding to the Global Challenge* (2002). Tilbury, D., Stevenson, R.B., Fien, J., Schreuder, D., (Eds.) IUCN - The World Conservation Union

Robinson, M. (2011), Climate change: A Glimpse of the Future, The Huffington Post, March 4, 2011. URL http://www.huffingtonpost.com/mary-robinson/post_1734_b_823803.html).

Roth, C.E.(1992). Environmental Literacy: Its Roots, Evolution and Directions in the 1990s. ERIC/CSMEE Publications.

Russell, J. (2007), Coal Use Rises Dramatically Despite Impacts on Climate and Health World Watch Institute Vital Signs 2007.

Ryan, J.C. and Durning, A.T. (1997). *Stuff—The Secret Lives of Everyday Things,* Sightline Institute.

Schumacher, F. (1973). Small is Beautiful: Economics as if People Really Mattered. Abacus:London, in, *Education and Sustainability Responding to the Global Challenge* (2002). Tilbury, D., Stevenson, R.B., Fien, J., Schreuder, D., (Eds.) IUCN - The World Conservation Union

Shepardson, D.P., Soyoung, C., Niyogi, D.and Charusombat, U. (2011). Seventh grade students' mental models of the greenhouse effect, *Environmental Education Research,* 17:1, 1–17.

Summers, M., Kruger, C., Childs, A. & Mant, J. (2000). Primary school teachers' understanding of environmental issues: An interview study. *Environmental Education Research, 6(4),* 293 - 312.

Teksöz, G., C.Tekkaya, A.K.Erbaş (2009). "Geographical Regions as a Silent Predictor of Responsible Environmental Behaviour", *H.U. Journal of Education,* 36, 249-259.

Tikka, P.M., Kuitunen, T.M., Tynys, M.S.(2000). Effects of educational background on students' attitudes, activity levels, and knowledge concerning environment. *The Journal of Environmental Education* 31, 12–19.

Tuncer, G., C. Tekkaya, S. Sungur, J. Cakiroglu, H. Ertepinar, M. Kaplowitz (2009. "Assessing pre-service teachers' environmental literacy in Turkey as a mean to develop teacher education programs," *International Journal of Educational Development,* 29, 426-436.Tung, A. (2010), Vital signs 2010, World Watch Institute.

Tüzün, Ö., Y., G. Teksöz Tuncer, M. Aydemir (2008) "An investigation on the elementary teachers' knowledge about air pollution issues", *H. U. Journal of Education,*35, 374-385.

UNCED, (1992). United Nations Commission for Environment and Development *Agenda 21, the United Nations programme of action from Rio.* New York: United Nations.

UNEP (1972). *United Nations Conference on the Human Environment.* Retrieved from United Nations: URL http://www.unep.org/Documents.Multilingual/Default.asp?DocumentID=97

UNESCO, (1977). Final Report, Intergovernmental Conference on Environmental Education organized by Unesco in co-operation with UNEP, Tbilisi (USSR).

UNESCO, (1992). *United Nations Conference on Environment and Development: Agenda 21.* UNESCO, Switzerland.

UNESCO, (1977). In: International Conference on Environmental Education, Tbilisi, USSR.UNESCO, (1980). In: Environmental Education in the Light of the Tbilisi Conference, Paris, France.

UNEP, (1972). In: Declaration of the United Nations Conference on the Human Environment.

Walsh, B. (2008). Meat : Making global warming worse, Time Magazine, September 10, 2008. URL http://www.time.com/time/health/article/0,8599,1839995,00.html

Weart, S. (1988). *Nuclear Fear: A History of Images,* Harvard University Press.

Weaver, A. (2002). Determinants of environmental attitudes: a five-country comparison nternational Journal of Sociology 32, 77–108.

Wilke, R. (1995). Environmental literacy and the college curriculum. *EPA Journal 21,* 28-30.WCED, (1987). *Our Common Future,* Oxford University Press.

Worsley, A., Skrzypiec, G. (1998). Environmental attitudes of senior secondary school students in South Australia. Global *Environmental Change* 8, 209–255.

Yilmaz, O., Boone, W.J., Anderson, H.O.(2004). Views of elementary and middle school Turkish students toward environmental issues. *International Journal of Science Education* 26, 1527–1546.

Zelezny, L.C., Chua, P., Aldrich, C.(2000). Elaborating on gender differences in environmentalism. *Journal of Social Issues* 56, 443–457.

21

Do Developing or Developed Nations Pollute Air More? An Assessment of Health Consequences

Ramalingam Shanmugam and Attila Hertelendy
School of Health Administration, Texas State University, San Marcos, Texas USA

1. Introduction

Air pollution is a serious threat to the healthy life in society. Nations across the continents in the 21st century need to evaluate the air pollution data to make up fair and implementable global policies, which support healthy life. In this process of examining the data, the following relevant question could be answered. Do developing or developed nations contribute more air pollutions? As this chapter illustrates, the pollutions related data in the public domain WebPages (see [1] through [21] for details) hide the answer.

The developing nations aspire to improve their standard of life by advancing their industrial or agricultural productivities. The developed nations, on the contrary, tend to maintain their higher standard of living with the same or more industrial and agricultural productivities. When it comes to a discussion on who contribute the air pollutions, both the developed and developing nations argue and are willing to make a few policy changes but preach and expect the other side to do more to cut down the global air pollutions. How should this escalated conflict be resolved? What do the relevant data suggest? How a compromise between the developed and developing nations is plausible?

To answer this and related questions, this chapter examines and assesses existing global air pollutions data using statistical techniques. The findings as they are stated in this chapter are interesting and pave ways to formulate equitable global policies to reduce air pollutions and thereby increase the healthy life at large.

2. Motivation

Whether air pollutions are natural or human generated, the damages significantly affect the quality of life. Environmentalists warn that the human generated air pollutions are on the rise to an alarming level in the recent years. Such a rise results in tremendous health impacts. The illness and sufferings among the humans, animals, natural life and vegetations are on the steady increase. The land and water sources are also contaminated because of the air pollutions. A recovery from the damages due to the air pollutions requires enormous resources, efforts and money. Many health hazardous chemicals released by the human initiated activities in day-to-day life, in industrial and agricultural sectors are additional add on to the existing natural air pollutions due to phenomena like volcanoes or ozone leaks (see [23] through [28] for details on natural calamities). Some of these hazardous chemicals are absorbed by the air and stay there as permanent air contaminants. Not only the local

communities where the air pollutions originate but also those reside in nearby or far away communities suffer as the air contaminants are blown through because of the wind. Consequently, many chronic diseases and illnesses increase everywhere in the world.
The World Health Organization (WHO) estimates over 2.4 million people die each year from the air pollutions. Among those, about 1.5 million deaths are due to the indoor air pollutions. About 310,000 Europeans died in year 2005 alone due to the air pollutions. The number of deaths due to the air pollutions is rising and it is more than the automobile accident deaths. The 1984 Bhopal's Union Carbide disaster caused more than 2,000 deaths and over 500,000 injuries. See [11] and [12] for historical details.
The health effects of the air pollutions include breathing difficulties, wheezing, coughing, pulmonary exacerbations, cancer, respiratory and cardiac failures. Consequently, the consumption of medications and the usage of emergency room visits increase. The hospitals and clinics are overwhelmed by the influx of patients for treatments.
The land and water sources are not immune from the air pollutions. When it rains, because of the air contaminants, the rain exhibits acidity. The acidity in rain causes water pollutions. Especially when the drinking water and arable land receive the acidity, the food chains are contaminated and the public health deteriorates. The land becomes less fertile and the crops are of less safe. The food sources are even depleted in some situations. On the health arena, the acid rain contains a high amount of carbon oxides (CO2), sulfur oxides (SO2) and nitrogen dioxides (NO2). The nations across all continents suffer even though they are not the originators of the air pollutions. The neighboring nations constantly complain to the polluting nations. For an example, Canada, in 1980s, complained to the USA that their air and water pollutions originated in USA and wanted the USA to make remedies. Such complaints were also the reasons for the US congress to legislate the US Clean Air Act of 1990. Similar complaints were lodged by some European nations resulted in the famous Helsinki Protocol Agreement of 1985 in which about 21 European nations committed to reduce their sulfates in the air. Much more, of course, need to be done to alert both the developed and developing nations to commit to drastic improvements for pure air. The findings in this chapter based on relevant data analysis are critical and educational for having healthy life.
Incidentally, drinking the distilled water does not remedy the problem of water contaminations. Why is it so? Though the drinking water contains no carbon dioxide (CO2) after the distillation, its average amount of phosphate (pH) is 5.6μ, which is still acidic. Note that an acceptable pH level in vinegar sold in the supermarkets is only 2.4μ on the average. Thus, the acidity in the rain needs to be taken seriously and corrected. However, the acidity in rain arises from the PM10, sulfuric, carbon or nitrogen oxides in the air.
The acid rain was first noticed in Manchester, England 1852. The carbon, nitrogen, PM10 and sulphate in the air are some important sources of acidity in the rain. The carbon dioxide and the sulfur dioxide (SO2) emanate from the auto emissions. The glacial ice and historical lakes attest to the fact, according to the environmentalists, that the volcanoes and environmental factors such as ozone leaks and deforestation contribute to the increased level of sulfur in the air. The coal and petroleum products contain more sulfur and hence, produce sulfur dioxide upon their combustion. The carbon monoxide and dioxide are due to the combustion of cooking or fossil fuels. The particulate matter (PM10) in the air is quite health hazardous (see [22] and [23] for details). The PM10 are tiny particles of solid or liquid in the gas. The ammonia and chlorofluorocarbons in the air are toxic and are so banned by the environmental protection agency (EPA). The principal and significant causes of the acid

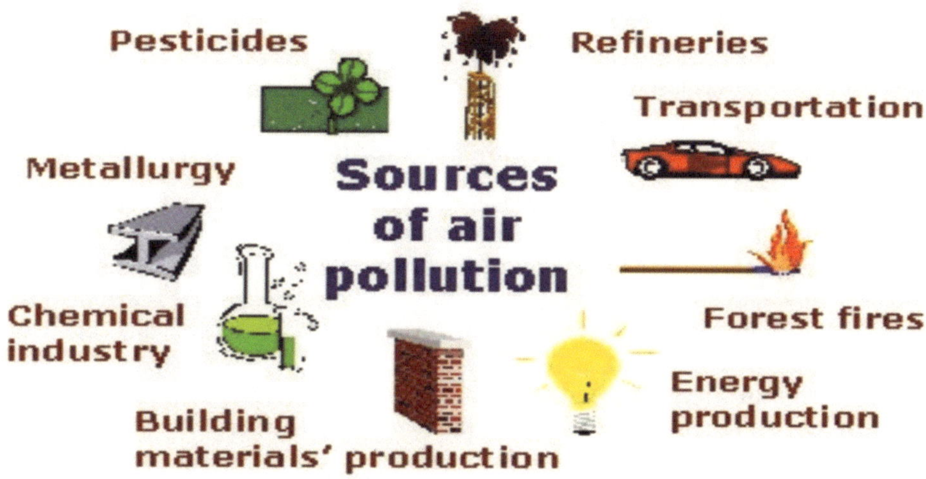

Fig. 1. The group of air pollutants

rain are therefore the chemical compounds emitted by the manufacturing industries or motor vehicles. See Shanmugam et al (2007) for details about the increase in stroke incidences due to the air pollutions. See Figure 1 below for the sources of air pollutions in general.

The developed nations pollute the air by their excessive use of fossil fuels and coal energies in manufacturing industries, oil refineries and the auto emissions in particular. The developing nations aspire to catch up with the developed nations' status and hence add up to the air pollutions by their increased efforts to use excessively the fossil fuels and coal energies in personal and public life. Their use of saltpeter, urea, ammonia (NH_3) and nitric oxides (NO_2) in agricultural activities do contribute to the air pollutions. The farming and agriculture activities in the developing nations are also sources of the air pollutions. Spreading fertilizers and pesticides from the airplanes causes the air pollutions with these solid particles. The soil surface erosions also generate large amounts of particulates with their diameter up to 2.5 μm (the so-called PM 2.5). The particulates are transported in the wind to near and far away communities.

To assess whether the developed and developing nations have parity when it comes to causing the air pollutions, we need to define first who the developed nations are really. How are the developed versus developing nations defined and distinguished? Many definitions are contentious. Could the *per capita income* be an identifier of a developed nation? There might be nations still in developing status though they have a higher per capita income. Could the level of *industrialization* be an indicator of a developed nation? Could the *gross domestic product* (GDP) be an indicator of the developed nations? The list of indicators goes on. The *human development index* (HDI) might be an appropriate and acceptable index according to the discussions in the United Nations, to classify a nation as developed or developing among the members of the United Nations.

If so, what is HDI? The HDI is an aggregate measure of life expectancy, literacy level, education rate, purchasing power and standard of living among the citizens in a nation. The two economists Sen and Haq created the concept and calculating formulas for the HDI during 1990. See [21] for details about the 2010 Human Development Report containing the calculation and interpretation of HDI. In this chapter, the HDI is involved to formulate a list of developed versus developing nations.

2. Do data distinguish developed and developing nations in air pollutions?

To answer this crucial question, an analysis of PM10, sulfuric, carbon or nitrogen oxides emitted to the air by the nations in all continents should be performed. To begin with, for comparisons sake, four groups of nations are formed and considered. All the nations in both the North and South American continental surroundings are grouped together for the analysis. The nations in Asia, New Zealand, Australia and Pacific Oceania are grouped together as part of Asia continental surroundings. The Russia, eastern and western European nations are grouped together as part of European continental surroundings. The Madagascar and all the nations in African continental surroundings are grouped together for the analysis. In all these four groups, there are developed and developing nations. How do these nations perform within the group and across the groups?

The median HDI in each of the four groups of nations is used to classify a nation as a developed or developing in the group. If a nation's HDI is more than the group's median HDI scores, it is a developed nation. See Tables 1 through 4 for the median HDI scores and the status of each nation in the group. Notice that the median HDI scores vary across groups. The median HDI scores are 0.403, 0.695, 0.61 and 0.81 among the 56 African, 51 American, 66 Asian and 54 European nations respectively. Most of the European nations are developed while many of the American nations are developing type. The African and Asian nations are evenly split. The percent of the developed nations are 50, 31, 56 and 61 among African, American, Asian and European continental nations respectively. Hence, no continent contains completely developed nations. A comprehensive, equitable and global air pollution policy is therefore necessary. All members of the United Nations should frame such a policy collectively.

Are the per capita *gross domestic productivities* (GDP) of the developed nations in these four groups close enough? The answer is no as they exhibit a quite heterogeneity according to the Box plots in the Figures 1 through 4 below. The European nations are more homogenous (with the exception of outlier nation Luxemburg). The African nations, American nations (with the exception of outlier nations USA and Canada) and Asian nations (with the exception of outlier nation Qatar) are quite heterogeneous in per capita GDP.

One can not avoid wondering whether the density of the population is a factor in for so much heterogenous per capita GDP or HDI. In fact, it is indeed a factor according to the Figures 5 through 8 which portray the density of the population in millions per square mile. The Asian (with the exception of outlier nations Hong Kong, Taiwan and Singapore) and European (with the exception of outlier nations Belgium, Ireland and Malta) nations have more dense population per square mile than in American (with the exception of outlier nations Bermuda and Barbados) or African nations (with the exception of outlier nations Mauritius, Rwanda and Comoros) in general.

However homogenous or heterogeneous are the nations within the groups, they differ significantly in their contributions to air pollutions. The Figures 9 through 12 portray their contributions to PM10 in the air. The European nations (with the exception of outlier nation Turkey) turn in consistently a low amount of PM10. The Asian nations (with the exception

Countries	HDI	Is it a developed nation?	Countries	HDI	Is it a developed nation?
Algeria	0.677	Yes	Madagascar	0.435	Yes
Angola	0.403	Yes	Malawi	0.385	No
Benin	0.435	Yes	Mali	0.309	No
Botswana	0.633	Yes	Mauritania	0.433	Yes
Burkina Faso	0.305	No	Mauritius	0.701	Yes
Burundi	0.282	No	Morocco	0.567	Yes
Cameroon	0.46	Yes	Mozambique	0.284	No
Cape Verde	0.534	Yes	Namibia	0.606	Yes
Central African Republic	0.315	No	Niger	0.261	No
Chad	0.295	No	Nigeria	0.423	Yes
Comoros	0.428	Yes	Reunion	0	No
Congo (Brazzaville)	0.489	Yes	Rwanda	0.385	No
Congo (Kinshasa)	0.239	No	Saint Helena	0	No
Cote d'Ivoire	0.397	No	Sao Tome	0.488	Yes
Djibouti	0.402	No	Senegal	0.411	Yes
Egypt	0.62	Yes	Seychelles	0	No
Equatorial Guinea	0.538	Yes	Sierra Leone	0.317	No
Eritrea	0.328	No	Somalia	0	No
Ethiopia	0.328	No	South Africa	0.597	Yes
Gabon	0.648	Yes	Sudan	0.379	No
Gambia, The	0.39	No	Swaziland	0.498	Yes
Ghana	0.467	Yes	Tanzania	0.398	No
Guinea	0.34	No	Togo	0.428	Yes
Guinea-Bissau	0.289	No	Tunisia	0.683	Yes
Kenya	0.47	Yes	Uganda	0.422	Yes
Lesotho	0.427	Yes	Western Sahara	0	No
Liberia	0.3	No	Zambia	0.395	No
Libya	0.755	Yes	Zimbabwe	0.14	No
Madagascar	0.435	Yes	Median HDI score	0.403	# developed nations=28

Table 1. Nations in Africa

Countries	HDI	Is it a developed nation?	Countries	HDI	Is it a developed nation?
Antarctica	0	No	Haiti	0.404	No
Antigua-Barbuda	0	No	Honduras	0.604	No
Argentina	0.775	Yes	Jamaica	0.688	No
Aruba	0	No	Martinique	0	No
Bahamas	0.784	Yes	Mexico	0.75	Yes
Barbados	0.788	Yes	Montserrat	0	No
Belize	0.694	No	Netherlands Antilles	0.29	No
Bermuda	0	No	Nicaragua	0.565	No
Bolivia	0.643	No	Panama	0.755	Yes
Brazil	0.699	Yes	Paraguay	0.64	No
Canada	0.888	Yes	Peru	0.723	Yes
Cayman Islands	0	No	Puerto Rico	0.79	Yes
Chile	0.783	Yes	Saint Kitts	0	No
Colombia	0.689	No	Saint Lucia	0	No
Costa Rica	0.725	Yes	Saint Pierre	0	No
Cuba	0	No	Saint Vincent	0	No
Dominica	0.663	No	Suriname	0.646	No
Dominican Republic	0.663	No	Trinidad-Tobago	0.736	Yes
Ecuador	0.695	Yes	Turks Caicos Islands	0	No
El Salvador	0.659	No	United States	0.902	Yes
Falkland Islands	0	No	Uruguay	0.765	Yes
French Guiana	0	No	Venezuela	0.696	Yes
Grenada	0	No	Virgin Islands, U.S.	0	No
Guadeloupe	0	No	Virgin Islands	0	No
Guatemala	0.56	No	Median HDI score	0.695	
Guyana	0.611	No	# developed nations	16	

Table 2. Nations in South and North America

Countries	HDI	Is it a developed nation?	Countries	HDI	Is it a developed nation?
Armenia	0.695	Yes	Macau	0	No
Afghanistan	0.349	No	Malaysia	0.744	Yes
American Samoa	0.414	No	Maldives	0.602	No
Australia	0.937	Yes	Mongolia	0.622	Yes
Azerbaijan	0.713	Yes	Nauru	0.414	No
Bahrain	0.801	Yes	Nepal	0.428	No
Bangladesh	0.469	No	New Caledonia	0.414	No
Bhutan	0.518	No	New Zealand	0.937	Yes
Brunei	0.805	Yes	Niue	0.414	No
Burma (Myanmar)	0.451	No	Oman	0.49	No
Cambodia	0.494	No	Pakistan	0.49	No
China	0.862	Yes	Papua New Guinea	0.431	No
Cook Islands	0.614	Yes	Philippines	0.638	Yes
Fiji	0.669	Yes	Qatar	0.803	Yes
French Polynesia	0.614	Yes	Samoa	0.49	No
Guam	0.614	Yes	Saudi Arabia	0.752	Yes
Hawaiian Trade Zone	0.902	Yes	Singapore	0.846	Yes
Hong Kong	0.863	Yes	Solomon Islands	0.494	No
India	0.519	No	Sri Lanka	0.658	Yes
Indonesia	0.6	No	Syria	0.589	No
Iran	0.702	Yes	Taiwan	0.869	Yes
Iraq	0	No	Tajikistan	0.58	No
Israel	0.872	Yes	Thailand	0.654	Yes
Japan	0.884	Yes	Timor-Leste	0.502	No
Jordan	0.681	Yes	Tonga	0.677	Yes
Kazakhstan	0.714	Yes	Turkmenistan	0.669	Yes
Kiribati	0.614	Yes	U.S. Pacific Islands	0.614	Yes
Korea, North	0	No	United Arab Emirates	0.815	Yes
Korea, South	0.877	Yes	Uzbekistan	0.617	Yes
Kuwait	0.771	Yes	Vanuatu	0.614	Yes
Kyrgyzstan	0.598	No	Vietnam	0.572	No
Laos	0.497	No	Wake Island	0.614	Yes
Lebanon	0	No	Yemen	0.439	No
#developed nations	37		Median HDI score	0.61	

Table 3. Nations in Asia, Australia, New Zealand and Pacific Islands

Countries	HDI	Is it a developed nation?	Countries	HDI	Is it a developed nation?
Albania	0.719	No	Latvia	0.769	No
Austria	0.851	Yes	Lithuania	0.783	No
Belarus	0.732	No	Luxembourg	0.852	Yes
Belgium	0.867	Yes	Macedonia	0.701	No
Bosnia-Herzegovina	0.71	No	Malta	0.815	Yes
Bulgaria	0.743	No	Moldova	0.623	No
Croatia	0.767	No	Montenegro	0.769	No
Cyprus	0.81	Yes	Netherlands	0.89	Yes
Czech	0.841	Yes	Norway	0.938	Yes
Denmark	0.866	Yes	Poland	0.795	No
Estonia	0.812	Yes	Portugal	0.795	No
Finland	0.871	Yes	Romania	0.767	No
France	0.872	Yes	Russia	0.719	No
Georgia	0.698	No	Serbia	0.735	No
Germany	0.885	Yes	Slovakia	0.818	Yes
Gibraltar	0	No	Slovenia	0.828	Yes
Greece	0.855	Yes	Spain	0.863	Yes
Greenland	0	No	Sweden	0.885	Yes
Hungary	0.805	No	Switzerland	0.874	Yes
Iceland	0.869	Yes	Turkey	0.683	No
Ireland	0.895	Yes	Ukraine	0.71	No
Italy	0.854	Yes	United Kingdom	0.849	Yes
#developed nations	33		Median HDI score	0.81	

Table 4. Nations in Europe

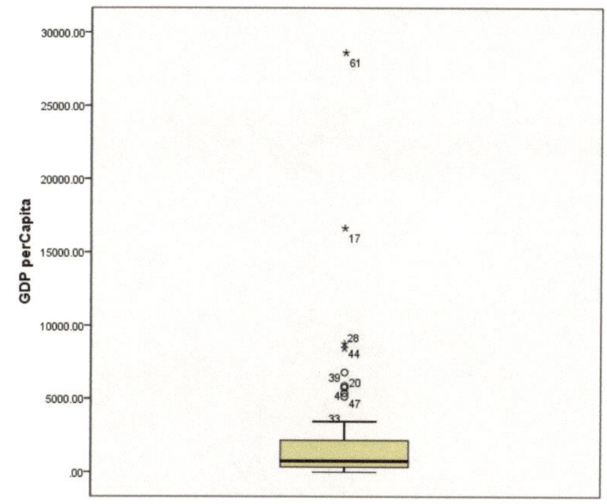

Fig. 1. GDP in Africa

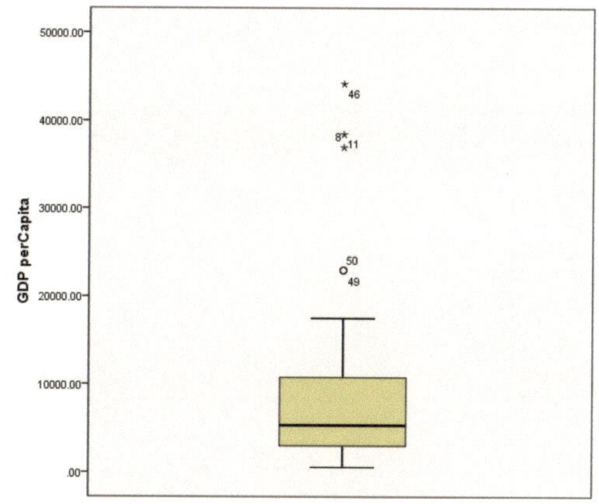

Fig. 2. GDP in America

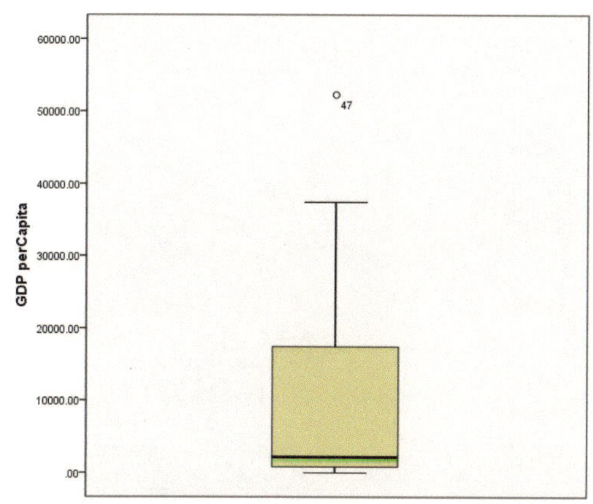

Fig. 3. GDP in Asia

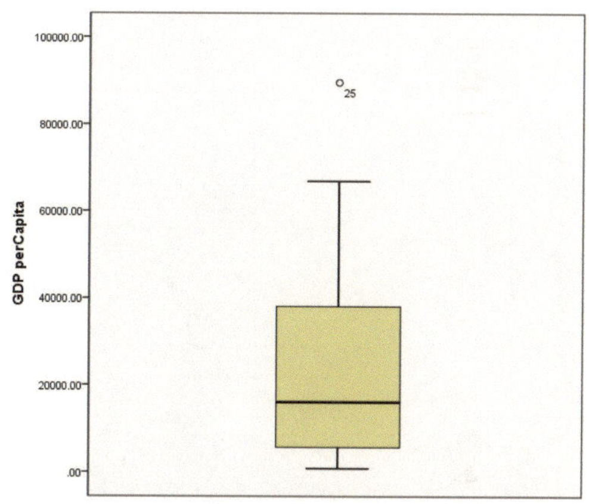

Fig. 4. GDP in Europe

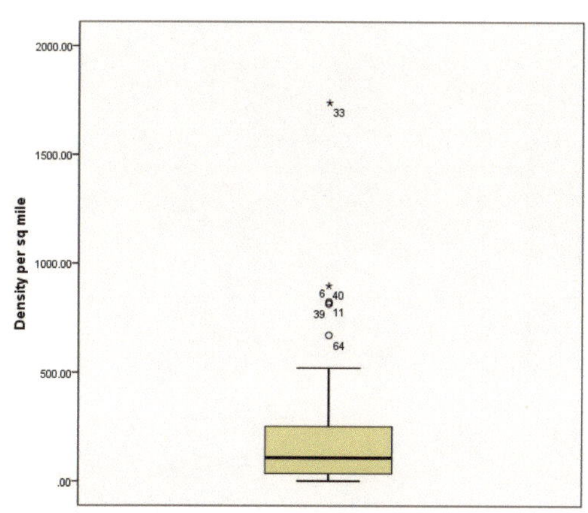

Fig. 5. Density/Square mile in Africa

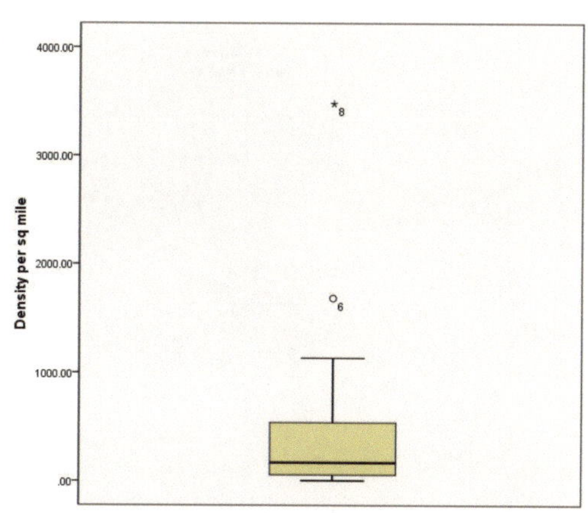

Fig. 6. Density/Square mile in America

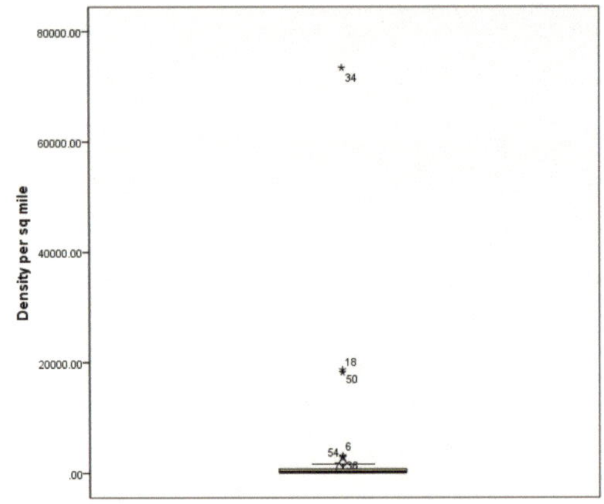

Fig. 7. Density/Square mile in Asia

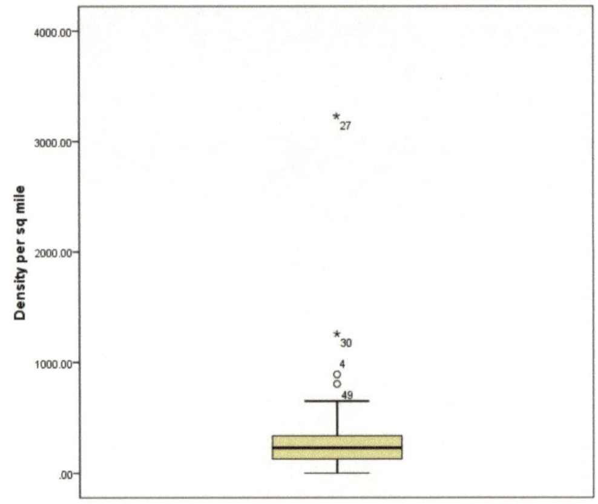

Fig. 8. Density/Square mile in Europe

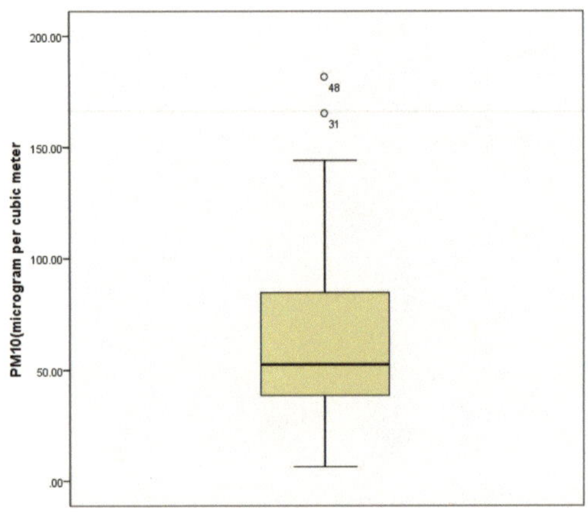

Fig. 9. PM10 in Africa

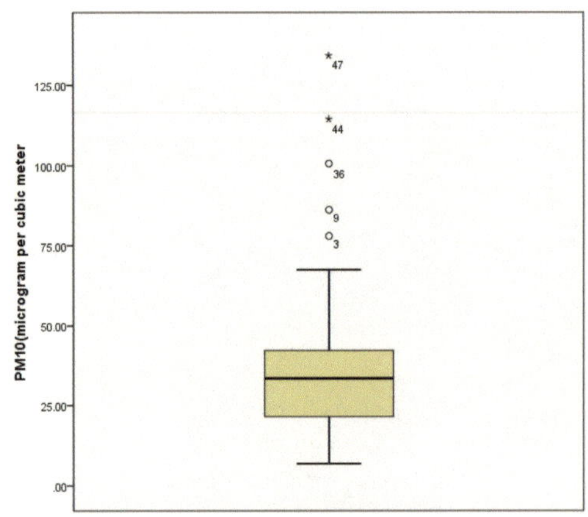

Fig.10. PM10 in America

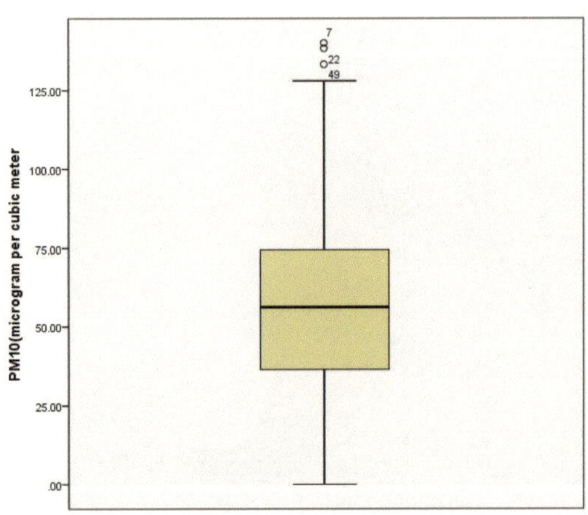

Fig. 11. PM10 in Asia

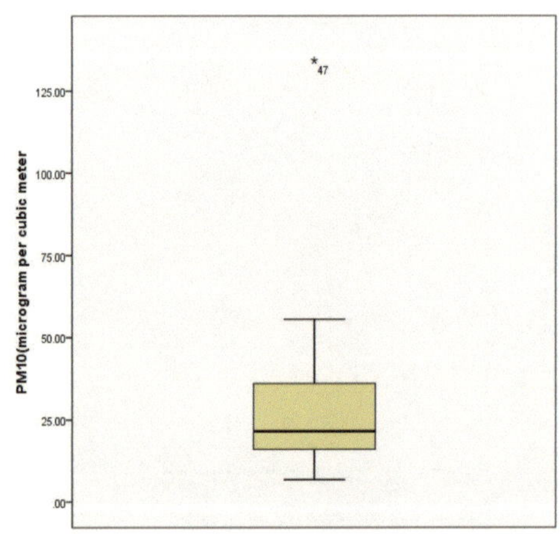

Fig. 12. PM10 in Europe

of outlier nations United Arab, Saudi Arabia, Bangladesh and Iraq) turn in a high amount of PM10. Though the median amount of PM10 is same in both the African and American nations, the nations within each group differ quite a bit. The African nations (with the exception of outlier nations Mali and Sudan) have much different PM10 while there are outlier nations in American continents like Paraguay, Trinidad and Uruguay. It appears that the presence of excessive PM10 in the air is due to industrial and agricultural sectors.

The contributions by the nations to the carbon dioxide (CO2) in the air also differ significantly. The Figures 13 through 16 exhibit the CO2 per capita in metric tons. The per capita CO2 in Asia (with the exception of outlier nations Bahrain, Kuwait, Qatar and United Arab) and in Africa (with the exception of outlier nations Equatorial Guinea, Libya, Seychelles and South Africa) are low. The per capita CO2 in America (with the exception of outlier nations USA, Aruba, Antilles and Trinidad) and in Europe (with the exception of outlier nations Belgium, France and Luxemburg) are high. It appears that the presence of excessive CO2 in the air is due to industrial sectors.

The sulfur dioxide (SO2) in metric tons are illustrated in Figures 17 through 20. The SO2 is low in Asia (with the exception of outlier nations Tajikistan, Australia, China, South and

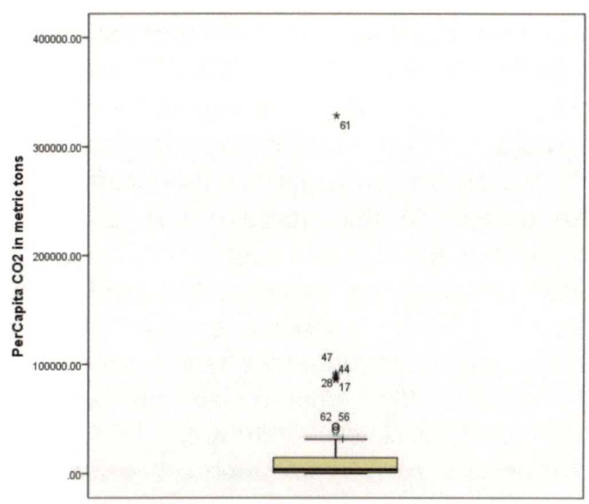

Fig. 13. Per capita CO2 in Africa

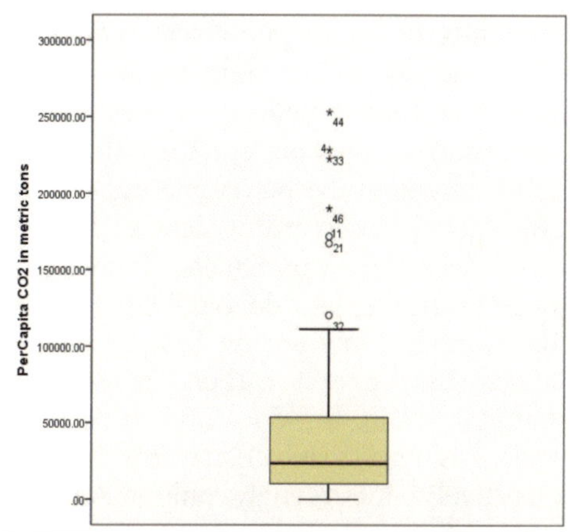

Fig. 14. Per capita CO2 in America

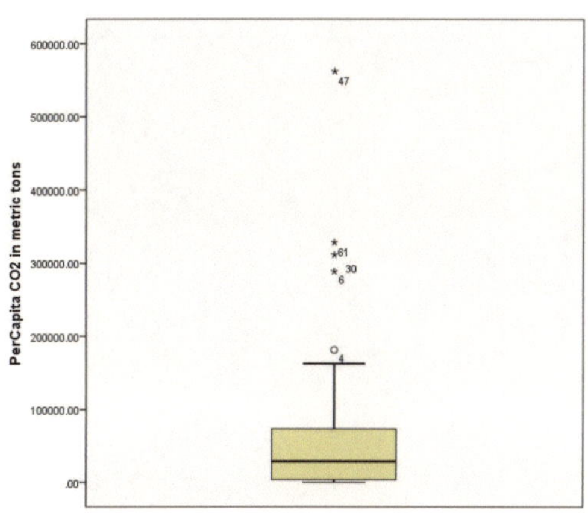

Fig. 15. Per capita CO2 in Asia

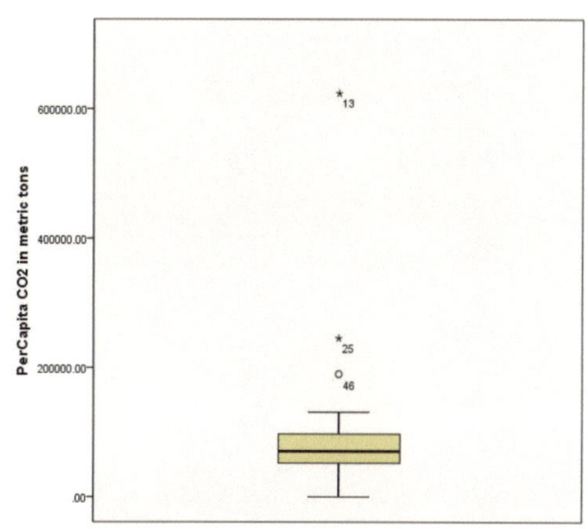

Fig. 16. Per capita CO2 in Europe

North Koreas) and in Americas (with the exception of outlier nations Jamaica, USA and Canada). The SO2 is more in Europe (with the exception of outlier nation Belgium). There are outlier nations in Africa (with the exception of outlier nations Egypt, Libya and South Africa) when it comes to the SO2 level. It appears that the presence of excessive SO2 in the air is due to energy sectors.

The nitric oxide (NO2) level in metric tons are displayed in Figures 21 through 24 for the nations in Africa, America, Asia and Europe. The NO2 is more in Africa (with the exception of the outlier nations like Congo, Ethiopia, Nigeria, Sudan and Tanzania) than in Asia (with the exception of the outlier nations like Cook Island and India). The NO2 in America (with the exception of the outlier nations like USA, Canada and Virgin island) and in Europe (with the exception of the outlier nations like Germany, Russia, Spain and United Kingdom) are low. It appears that the presence of excessive NO2 in the air is due to the agricultural sectors.

3. Clusters of nations based on the principal components

In the previous section, it was noticed that the developing and developed nations contribute differently to the air pollutions. A natural follow up question is then whether the nations within a group pollute more in a single contamination or in all of CO2, NO2, SO2 and PM10. How these contaminations are matching up with the factors such as acidity in the rain, the population density per square mile, the per capita GDP, the percent contribution to the global emission, the per capita expenditure in budget for public health, the incidences of lung cancer and cardiovascular deaths? An answer to this question resides on the significance of their partial correlations after controlling the HDI variable.

For this purpose, let r denote the partial correlation between two factors after controlling the HDI variable. Only among the nations in Africa (but not in American, Asian or European nations), the per capita carbon dioxide (CO2) is significantly correlated with the emission level (that is, $r = 0.8$). In Africa and in Asia, the per capita budget amount spent on the health services is significantly correlated with the per capita CO2 level (that are, $r = 0.9$ and0.9 respectively). Interestingly, only in American nations, the partial correlation between the per capita CO2 and the PM10 is significant (that is, $r = -0.9$). Among the European nations, the per capita CO2 and the nitric oxide (NO2) are significantly correlated (that is, $r = 0.7$).

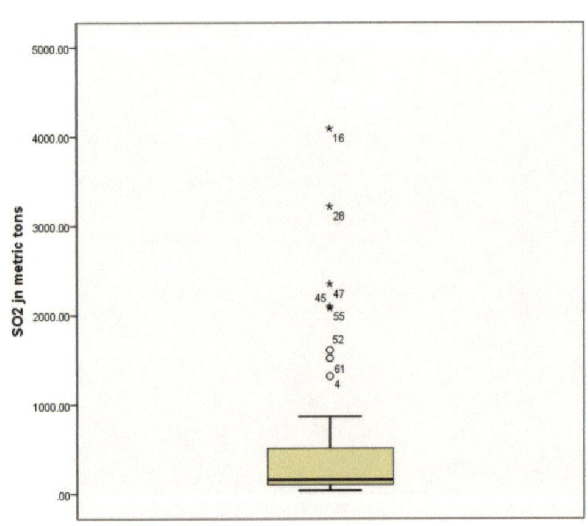

Fig. 17. SO2 in Africa

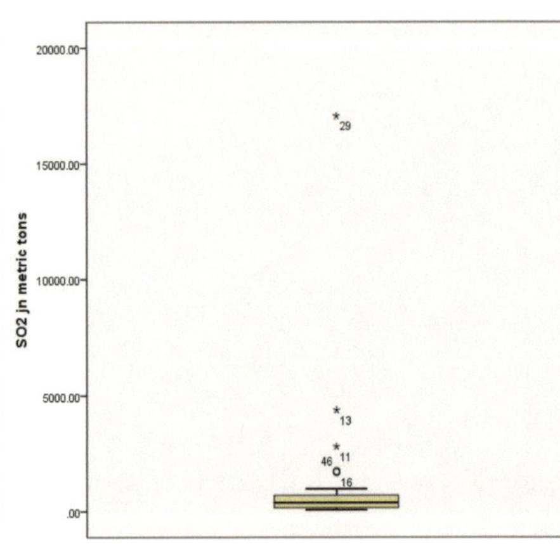

Fig. 18. SO2 in America

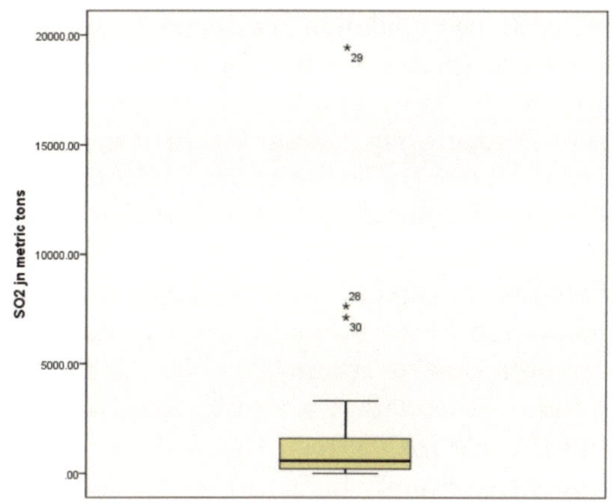

Fig. 19. SO2 in Asia

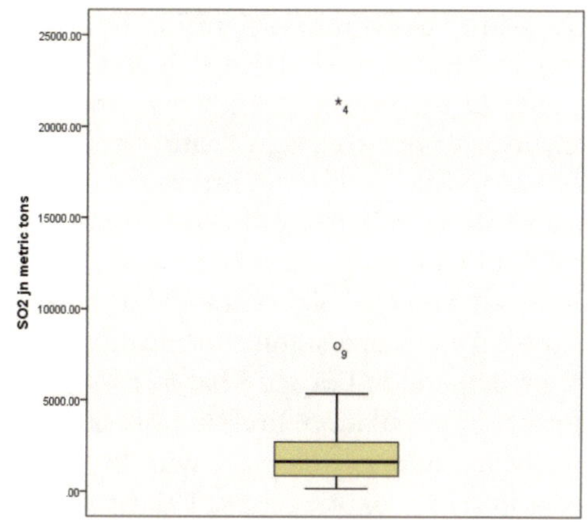

Fig. 20. SO2 in Europe

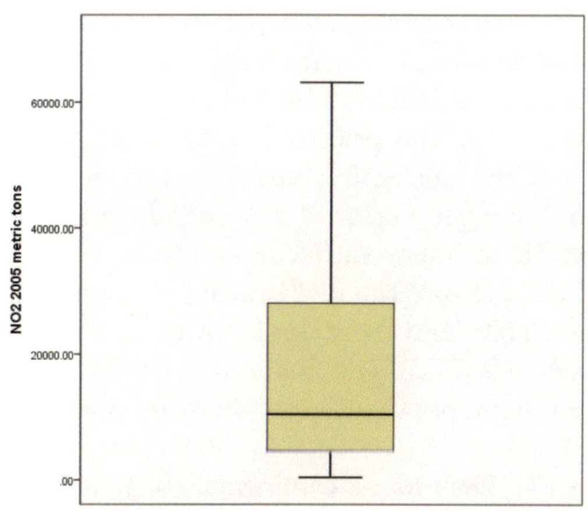

Fig. 21. NO2 in Africa

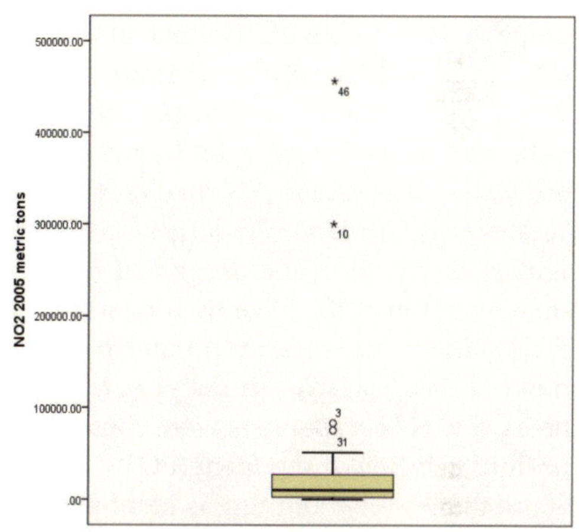

Fig. 22. NO2 in America

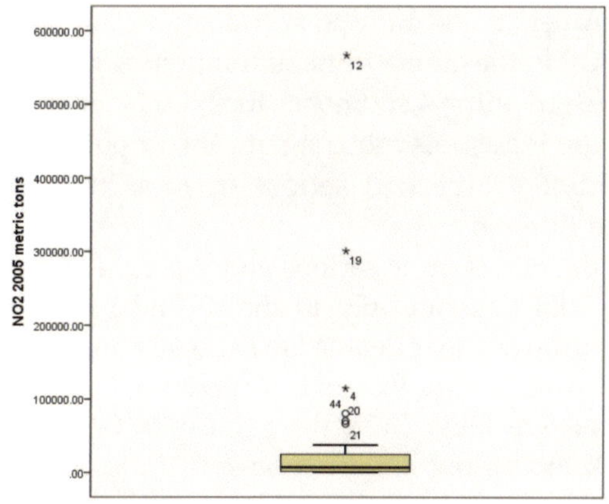

Fig. 23. NO2 in Asia

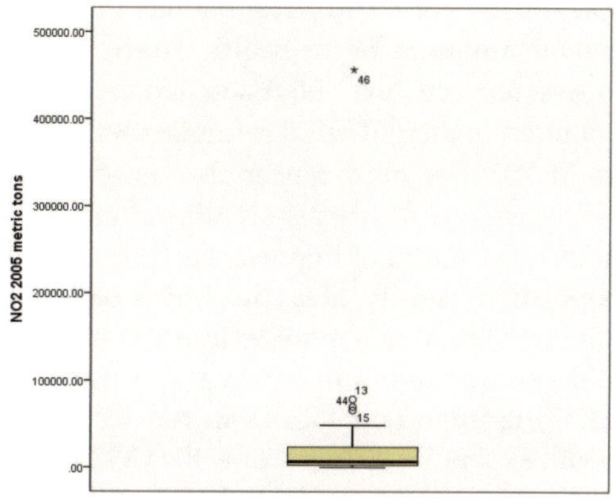

Fig. 24. NO2 in Europe

The acidity in the rain is significantly correlated with the population density per square mile only in African (that is, r = 0.9) and in Asian nations (that is, r = 0.6). In other words, the acidity in the rain is more where more people live in Africa and in Asia. These types of relations do not exist significantly in Americas or Europe. Only in Americas (that is, r = 0.9), the expenditure on health improvement has increased along with the acidity in the rain. The acidity in the rain is significantly more with the higher sulfur dioxide only in Asia (that is, r = 0.7) and in Europe (that is, r = 0.5).

Strangely enough, higher the PM10 means lower the per capita CO2 only in Americas (that is, r = - 0.9). Such a significant relation is not noticed in Africa, Asia or Europe. In Asia (that is, r = 0.5) and in Europe (that is, r = 0.5), more sulfur dioxide occurs when there is a higher density of population. In Asia (that is, r = 0.9) and in Europe (that is, r = 0.7), cardiovascular deaths increase significantly with the higher nitric oxide (NO2) level. This pattern does not exist in Africa or Americas. The emission level is significantly higher in higher density of nations only in Asia (that is, r = 0.9) but not in America, Africa or Europe.

No clear consistent pattern among the developed and developing nations either within or across the groups is emerging. It is, therefore, worthwhile to perform the principal component analysis with the data on CO2 from the coal burning, HDI, per capita GDP, CO2, NO2, SO2, PM10, acidity in the rain, population density per square mile, per capita GDP, percent contribution to global emission, per capita expenditure in budget for public health, incidences of lung cancer and cardiovascular deaths. The principal component analysis formulates three major principal components and the aggregate percent of explained total variations in the data. The larger coefficient in the Eigen vector of a principal component signifies a degree of importance of the factor. These important factors are compiled and summarized in Table 5 below for each of the four groups. The first principal component 1 (PC1) contains the most important pollution variables and these are followed by the next round of data variables in the principal component 2 (PC2). The final round (because only three principal components were chosen to begin with) of pollution variables are selected in the third principal component (PC3).

Notice that the coal burning is identified in the PC1 itself for all continents. The CO2 in the air is serious in Africa and Asia. The emissions cause major air pollution in all the continents. The GDP is quite diversified in Africa and Americas. Only in Asia, the health expenditure is not causing variation in the data. An implication is that the nations in Asia spend more for health. The nations in Africa, America and Europe spend significantly not enough amounts for the health. The SO2 and NO2 in the air are quite different in Africa and in Asia respectively. The Asian nations are more agricultural and hence the NO2 is major air pollution in there. The oil refineries and petroleum industries cause significant air pollutions in Africa. The lung cancer deaths are significantly large and serious in Americas. The cardiovascular deaths are significant in Asia and Europe.

In the next round of importance, the acidity in the rain is quite serious and it occurs in high population density areas in Africa only. The HDI appears only in the second round of importance and it is more so in Asia only. The cardiovascular deaths are seriously important in the second round in Africa and America. Other important factors in America are NO2. In Asia, the important factors in the second round are HDI, CO2, the per capita GDP and health expenditure. In Europe, the CO2 and NO2 are serious air pollutions.

In the third round, the HDI, PM10 and NO2 are important factors in Africa and only PM10 is important in America. The acidity, high population density areas and SO2 are important factors in Asia and only SO2 is an important factor in Europe so far as air pollutions. See Figures 25 through 28 for specifics.

Group	Principal Component 1	Principal Component 2	Principal Component 3	Explained variation
Africa	1. Coal burning 2. CO2 3. GDP 4. Health expenditure 5. SO2 6. Emission	1. Acidity 2. Population density 3. Cardiovascular death 4. Lung cancer	1. HDI 2. PM10 3. NO2	89%
America	1. Coal burning 2. HDI 3. Acidity 4. CO2 5. GDP 6. Health expenditure 7. Lung cancer 8. Emission	1. NO2 2. Cardiovascular deaths	1. PM10	72%
Asia	1. Coal burning 2. NO2 3. Cardiovascular deaths 4. Emission	1. HDI 2. CO2 3. GDP 4. Health expenditure	1. Acidity 2. Populati on density 3. SO2	77%
Europe	1. HDI 2. Health expenditure 3. Cardiovascular deaths 4. Emission	1. CO2 2. NO2	1. SO2	88%

Table 5. Principal factors in all four groups

4. Equitable global air pollution policy for better world health

In this section, the pollutions in both the developed versus developing nations are compared not only within each group but also across the groups. In developed African nations (see Figure 29), in developed America nations (see Figure 31), in developed Asian nations (see Table 33) and in developed European nations (see Figure 35), the per capita carbon dioxide, sulfur dioxide, carbon dioxide from the coal burning, nitric oxide, emissions, acidity and PM 10 are significant pollution variables. In developing African nations (see Figure 30), in developing American nations (see Figure 32), in developing Asian nations (see Figure 34) and in developing European nations (see Figure 36), the carbon dioxide from the coal burning, sulfur dioxide, PM 10, nitric oxide, emission and the carbon dioxide are the pollution variables. There is not much of difference between the developed and the developing nations in African and American continents. All nations are polluting the air significantly.

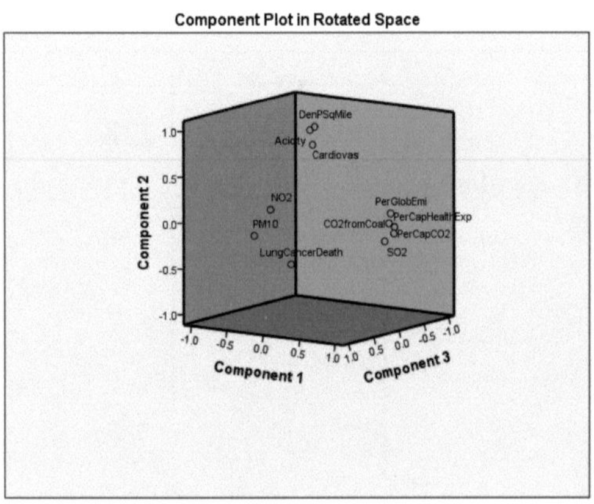

Fig. 25. Significant factors in Africa

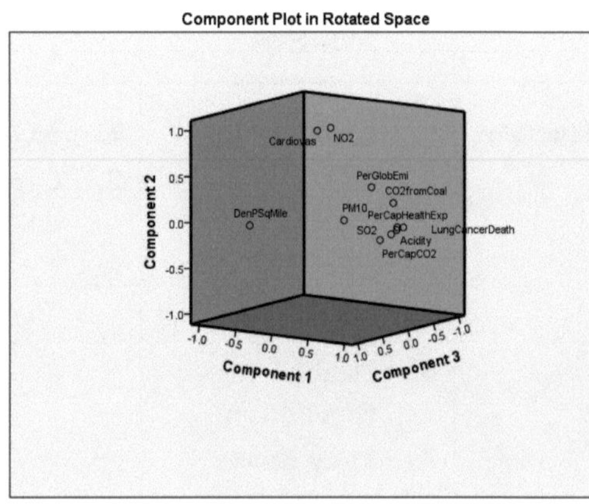

Fig. 26. Significant factors in America

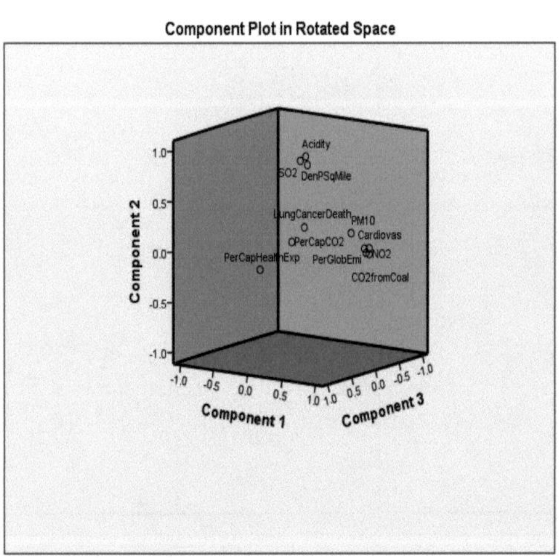

Fig. 27. Significant factors in Asia

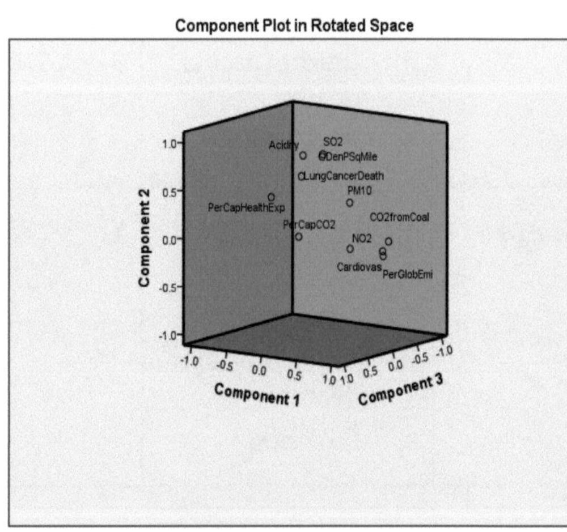

Fig. 28. Significant factors in Europe

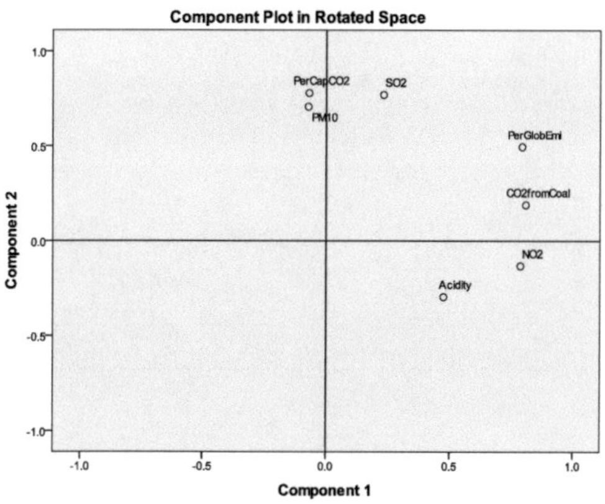

Fig. 29. In developed Africa

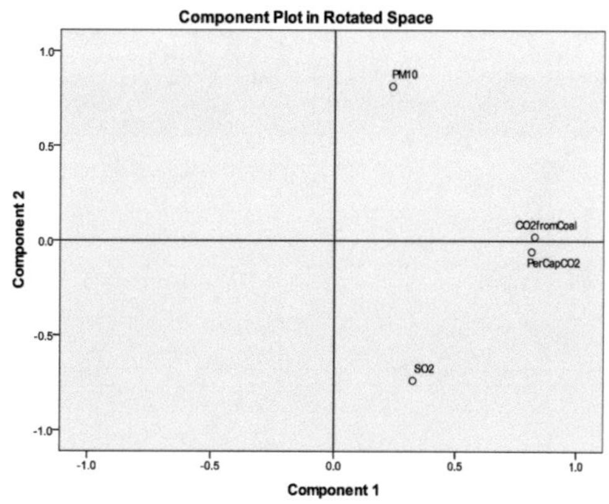

Fig. 30. In developing Africa

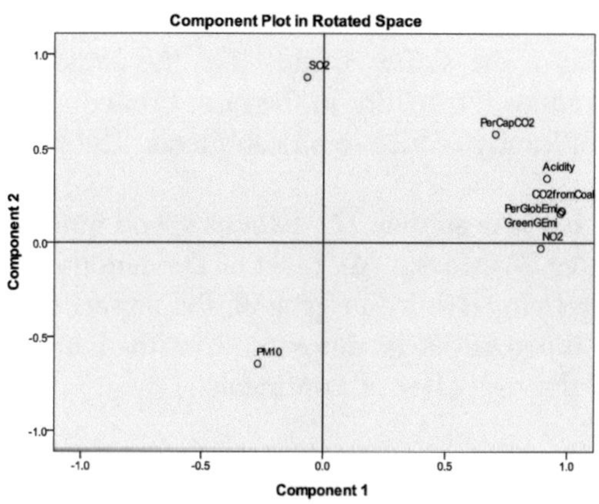

Fig. 31. In developed America

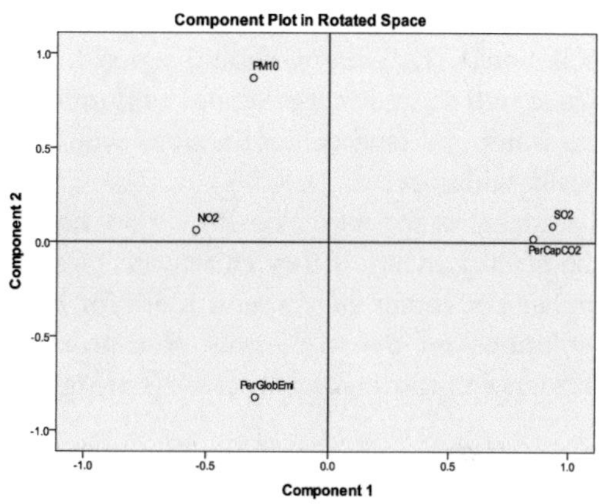

Fig. 32. In developing America

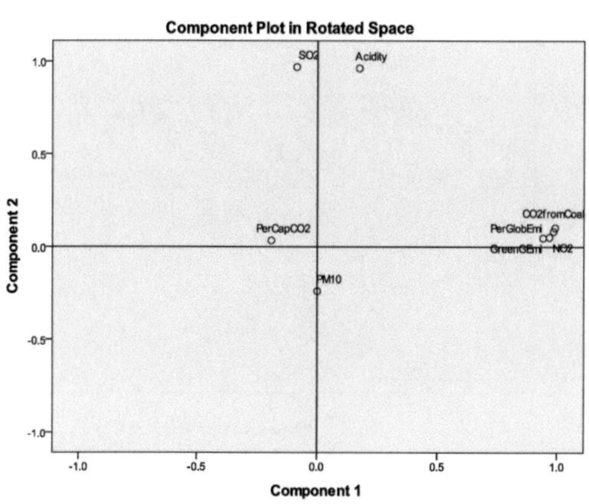

Fig. 33. In developed Asia

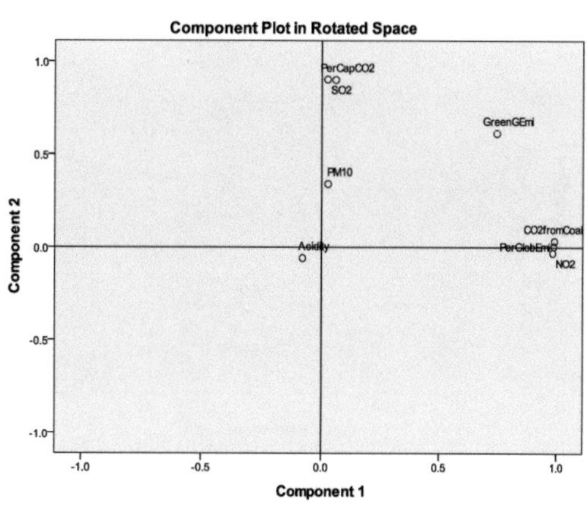

Fig. 34. In developing Asia

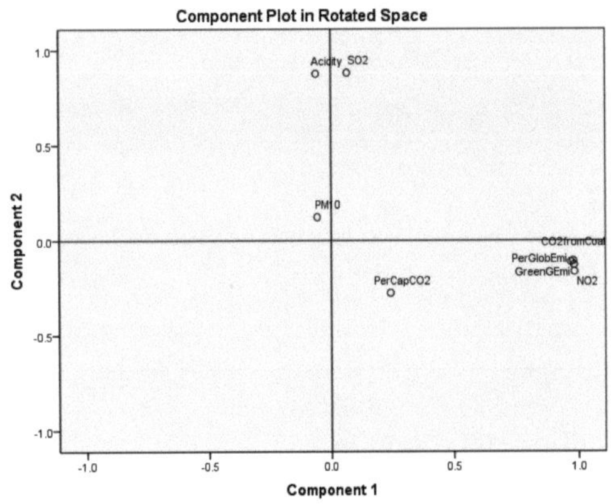

Fig. 35. In developed Europe

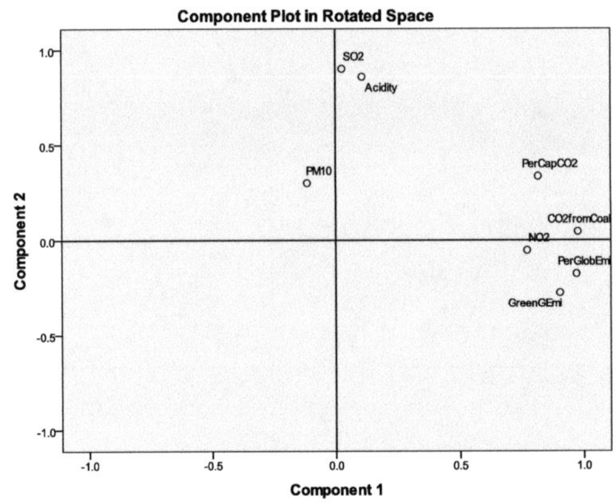

Fig. 36. In developing Europe

The nations in all continents should perhaps collectively create and implement a policy to restrict the coal burning as it produces significant amount of CO2 and PM 10 to the air pollutions. The nations should agree to minimize the sulfur dioxide and the emissions. These two commitments would undoubtedly reduce the acidity in the rain. Consequently, the water and land contaminations would significantly reduce in all continents. The world health will improve.

The status of the world health is now not in what is desirable. The nations spend more for the healthy living, if they can afford (see Figures 37 through 44). Most of the nations with higher per capita GDP spend more for healthy living. Notice in general, the impact of air pollutions on the incidences of cancer and tuberculosis is more positive than on the incidence of cardiovascular cases (see Figure 37 through 44) in all continents.

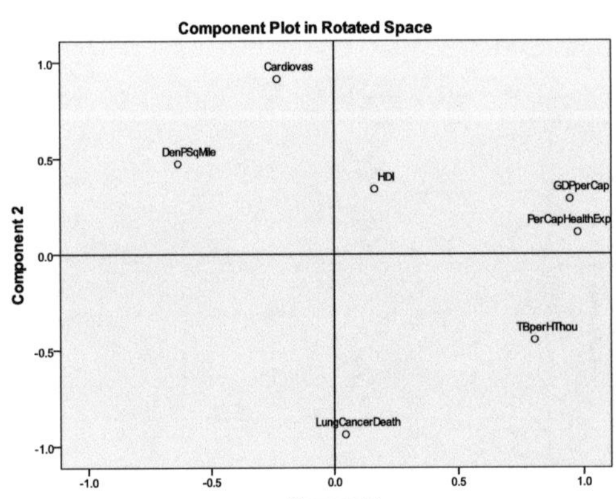

Fig. 37. Health in developed Africa

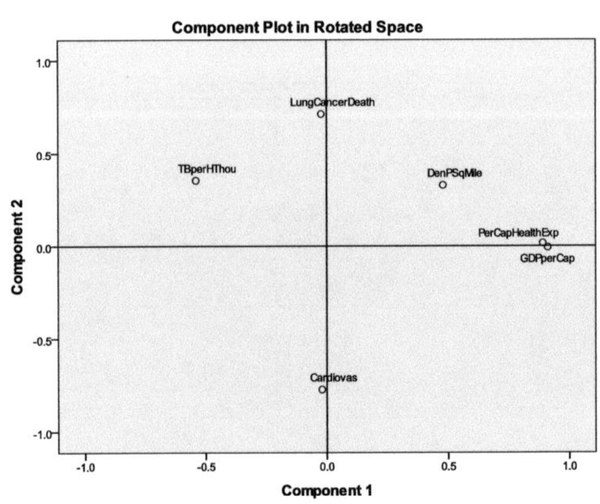

Fig. 38. Health in developing Africa

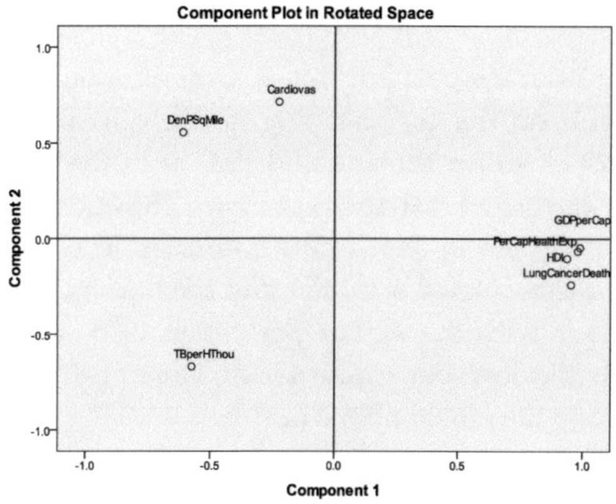

Fig. 39. Health in developed America

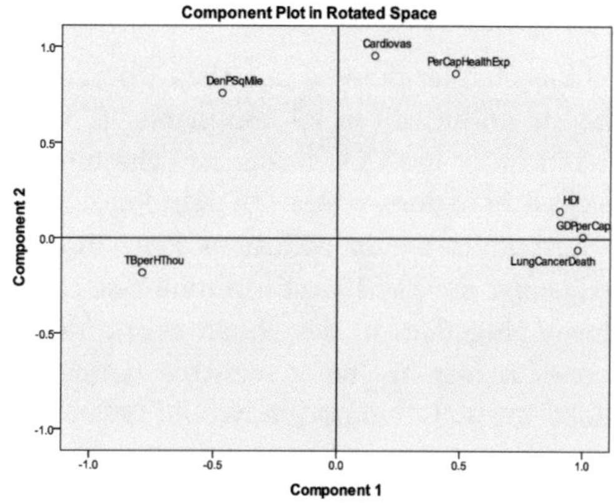

Fig. 40. Health in developing America

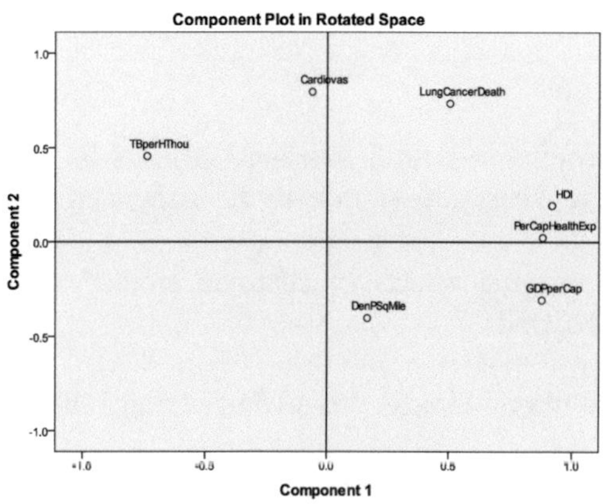

Fig. 41. Health in developed Asia

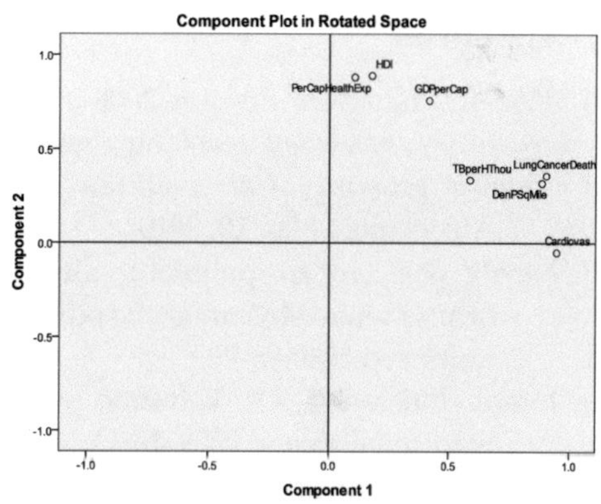

Fig. 42. Health in developing Asia

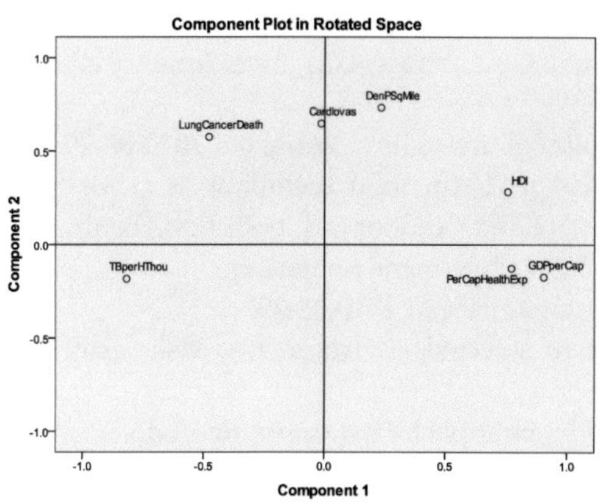

Fig. 43. Health in developed Europe

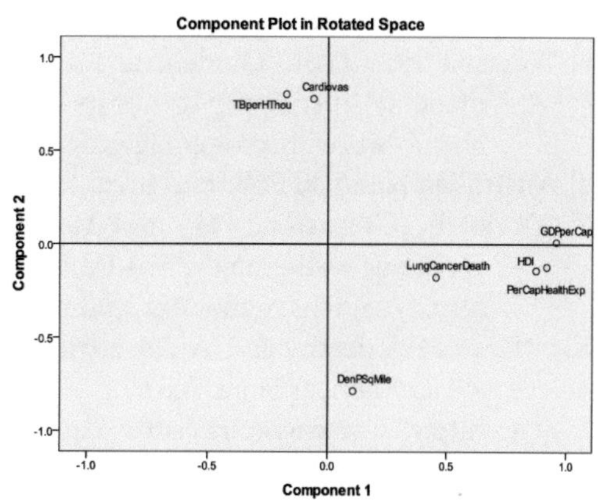

Fig. 44. Health in developing Europe

5. Conclusions

As this chapter pointed out, the carbon dioxide from the coal burning is a serious source of the air pollutions in all continents. A restriction on the coal burning is likely to reduce significantly the CO2 in the air. The abundance of sulfur dioxide and PM 10 in the air is noticed in nations where the petroleum industries related activities are more. The emissions contribute to the air pollutions more in developed than in developing countries. The nitric oxide and the PM 10 in the air are due to agriculture related activities and they are more in developing than in developed countries. Nations with the higher per capita GDP spend more for healthy living than the nations with the low per capita GDP. More restrictive global air pollution policies would help to improve the global public health.

6. Acknowledgement

This research work was supported by a Faculty Research Enhancement leave during the spring 2008 semester from the Texas State University, San Marcos, Texas.

7. References

[1] EPA: Air Pollutants".Epa.gov.2006-06-28 http://www.epa.gov/ebtpages/airairpollutants .html. Retrieved 2010-08-29.

[2] Evidence growing of air pollution's link to heart disease, death // American Heart Association. May 10, 2010

[3] "Newly detected air pollutant mimics damaging effects of cigarette smoke" (PDF). http://www.physorg.com/pdf138201201.pdf. Retrieved 2010-08-29.

[4] "Infant Inhalation Of Ultrafine Air Pollution Linked To Adult Lung Disease". Sciencedaily.com. 2009-07-23. http://www.sciencedaily.com/releases/2009/07/090722123751.htm. Retrieved 2010-08-29.

[5] "United Kingdom's emission factor database". Naei.org.uk. http://www.naei. org.uk/emissions/index.php. Retrieved 2010-08-29.

[6] European Environment Agency's 2005 Emission Inventory Guidebook

[7] "Revised 1996 IPCC Guidelines for National Greenhouse Gas Inventories (reference manual)". Ipcc-nggip.iges.or.jp. http://www.ipcc-nggip.iges.or.jp/public/gl/invs6.htm. Retrieved 2010-08-29.

[8] Australian National Pollutant Inventory Emissions Estimation Technique Manuals[dead link]

[9] "Duflo, E., Greenstone, M., and Hanna, R. (2008) "Indoor air pollution, health and economic well-being". "S.A.P.I.EN.S." '"1"' (1)". Sapiens.revues.org. http://sapiens.revues.org/index130.html. Retrieved 2010-08-29.

[10] "Estimated deaths & DALYs attributable to selected environmental risk factors, by WHO Member State, 2002". http://www.who.int/entity/quantifying_ehimpacts/nationprofilesebd.xls. Retrieved 2010-08-29.

[11] "Study links traffic pollution to thousands of deaths". *The Guardian* (London, UK: Guardian Media Group). 2008-04-15. http://www.guardian.co.uk/society/2008/apr/15/ health. Retrieved 2008-04-15.
[12] Simi Chakrabarti. "20th anniversary of world's worst industrial disaster". Australian Broadcasting Corporation. http://www.abc.net.au/worldtoday/content/2004/s1257352.htm.
[13] http://www.latimes.com/features/health/la-me-pollute13-2008nov13,0,5432723. story, http://www.sfgate.com/cgi-bin/article.cgi?f=/c/a/2008/11/13 /MNQP143CPV.DTL
[14] Christopher H. Goss, Stacey A. Newsom, Jonathan S. Schildcrout, Lianne Sheppard and Joel D. Kaufman (2004). "Effect of Ambient Air Pollution on Pulmonary Exacerbations and Lung Function in Cystic Fibrosis". *American Journal of Respiratory and Critical Care Medicine* 169 (7): 816–821. doi:10.1164/rccm.200306-779OC. PMID 14718248.
[15] Zoidis, John D. (1999). "The Impact of Air Pollution on COPD". *RT: for Decision Makers in Respiratory Care*. http://www.rtmagazine.com/issues/articles/1999-10_06.asp.
[16] Holland WW, Reid DD. The urban factor in chronic bronchitis. Lancet. 1965; I: 445-448.
[17] J. Sunyer (2001). "Urban air pollution and Chronic Obstructive Pulmonary disease: a review". *European Respiratory Journal* 17 (5): 1024–1033. doi:10.1183/ 09031936.01.17510240.PMID 11488305. http://erj.ersjournals.com/cgi/content/abstract/17/5/1024.
[18] "2005 BC Lung Association report on the valuation of health impacts from air quality in the Lower Fraser Valley airshed" (PDF). http://www.bc.lung.ca/pdf/ health_and_air_quality_2005.pdf. Retrieved 2010-08-29.
[19] "World Bank Statistics" http://siteresources.worldbank.org/DATASTATISTICS/ Resources/table3_13.pdf. Retrieved 2010-08-29.
[20] International Carbon Dioxide Emissions and Carbon Intensity Energy Information Administration
[21] "2010 Human development Report". United Nations Development Programme. pp. 148–151. http://hdr.undp.org/en/media/HDR_2010_EN_Complete.pdf. Retrieved 4 November 2010.
[22] Shanmugam, R., Anand, S., & Burke, G. (2007). Mining air pollution Data to Learn Stroke Incidence, *Model Assisted Data Analysis, 2*, 53-58.
[23] World Health Organization 2010 Report at http://www.who.int/2010en/.
[24] The World Bank: Data Factoids at http://go.world.bank.org/.
[25] US Census Bureau. World Population Information at http://census.gov/world.html/.
[26] International Panel on Climate Change at http://www.ipcc.ch/

[27] International Agency for Research on Cancer at http://www.iarc.fr/
[28] US Environemental Protection Agency at http://www.epa.gov/
(Completed on Saturday, March 05, 2011)

Die süddeutsche Städtelandschaft – ein interregionaler Vergleich

Wolfgang Wüst / Klaus Wolf (Hrsg.)

Die süddeutsche Städtelandschaft – ein interregionaler Vergleich

Unter Mitarbeit von Christoph Gunkel

PETER LANG

Bibliografische Information der Deutschen Nationalbibliothek
Die Deutsche Nationalbibliothek verzeichnet diese Publikation in der Deutschen Nationalbibliografie; detaillierte bibliografische Daten sind im Internet über http://dnb.d-nb.de abrufbar.

Umschlagabbildung: Städtischer Brückenschlag in (Bad) Camberg, 2020. Die Gebäudeverbindung steht symbolisch für die Netzwerke einer Städtelandschaft. Konkret richtet sich der Blick auf den Amtshof der hessischen, von 1535 bis 1794 unter gemeinsamer Verwaltung von Kurtrier und der Grafschaft Nassau stehenden Stadt. Der Amtshof zählt zu den größten Fachwerksbauten der frühen Neuzeit in Hessen, der sich aus älteren Gebäudeteilen der Herren von Riedesel, von Nassau sowie derer von Heiden und Metternich zusammensetzt.
Bildnachweis/Foto: Wolfgang Wüst (privat), 2020.

Für die Unterstützung der Drucklegung bedanken wir uns beim Zentralinstitut für Regionenforschung an der Friedrich-Alexander-Universität Erlangen-Nürnberg und bei der Forschungsstiftung Bayerische Geschichte.

ISBN 978-3-631-80117-8 (Print) · E-ISBN 978-3-631-84476-2 (E-PDF)
E-ISBN 978-3-631-84477-9 (EPUB) · E-ISBN 978-3-631-84478-6 (MOBI)
DOI 10.3726/b17945

Peter Lang – Berlin · Bern · Bruxelles · New York ·
Oxford · Warszawa · Wien

Diese Publikation wurde begutachtet.

www.peterlang.com

Inhalt

Sektion 2: Die süddeutsche Städtelandschaft – regional prägend

Sektion 3: Die süddeutsche Städtelandschaft – literarisch, musisch, onomastisch

Sektion 4: Die süddeutsche Städtelandschaft – ökonomisch

Sektion 5: Die süddeutsche Städtelandschaft – urban und stadtprägend

Sektion 6: Die süddeutsche Städtelandschaft – europäisch vergleichend

Verzeichnis der Autoren und Mitarbeiter

Dr. Franz-Rasso Böck, Stadtarchiv Kempten, Hochschule Kempten

Prof. Dr. Ivana Ebelová, Karls-Universität Prag (Tschechien)

Dr. Dorothea Fastnacht, Friedrich-Alexander-Universität Erlangen-Nürnberg / Marloffstein

Prof. Dr. Victor A. Ferretti, Universität Augsburg

Prof. Dr. Helmut Flachenecker, Julius-Maximilians-Universität Würzburg

Prof. Dr. Gerhard Fritz, Pädagogische Hochschule Schwäbisch-Gmünd

Verena Gawert, M.A., Universität Augsburg

Christoph Gunkel, M.A., Friedrich-Alexander-Universität Erlangen-Nürnberg

Dr. Thomas J. Hagen, Kommission für bayerische Landesgeschichte bei der Bayerischen Akademie der Wissenschaften, München

Dr. Senta Herkle, Universität Stuttgart

Prof. Dr. Zdeněk Hojda, Karls-Universität Prag (Tschechien)

Prof. Dr. Franz Körndle, Universität Augsburg

PD Dr. Johannes Lang, Stadtarchiv und Stadtheimatpflege Bad Reichenhall, Universität Salzburg

Dr. Wolfgang Mährle, Hauptstaatsarchiv Stuttgart

Prof. Dr. Ilgvars Misāns, Latvijas Universitāte Riga (Lettland)

Prof. Dr. Alois Schmid, Ludwig-Maximilians-Universität München

Dr. Lina Schröder, Julius-Maximilians-Universität Würzburg

Prof. Dr. Tom Scott, University of St Andrews, School of History (GB)

Prof. Dr. Klaus Wolf, Universität Augsburg

Sabine Wüst, Friedrich-Alexander-Universität Erlangen-Nürnberg / Nürnberg

Prof. Dr. Wolfgang Wüst, Friedrich-Alexander-Universität Erlangen-Nürnberg

Geleitwort des Präsidenten des Deutschen Städtetages

Foto: Deutscher Städtetag.

Städte sind immer gut, wenn sie über ihre Stadtgrenzen hinaus denken und wirken. Die Stadt der Zukunft ist eine agile Stadt. Regionale Initiativen und interkommunale Kooperationen sind dabei kein neues Phänomen. Ihre Bedeutung hat in den letzten Jahren stetig zugenommen. Denn es gibt viele gute Gründe, gemeinsam und partnerschaftlich die vorhandenen Aufgaben zu erfüllen, Standards zu halten oder zu erhöhen. Demografischer Wandel, die Schaffung bezahlbaren Wohnraums, Zuwanderung und Integration, Digitalisierung und eine klimagerechte Stadtentwicklung – für all diese Themenfelder und viele weitere müssen aus kommunaler Sicht dringend Lösungen gefunden werden. Und das in Zeiten knapper finanzieller und personeller Ressourcen sowie eines wachsenden Wettbewerbs der Regionen.

Dabei ist ein Kirchturmdenken schon lange nicht mehr angesagt. Denn es geht schlicht darum, Synergieeffekte zu nutzen und das städtische Leistungsspektrum für die Bürgerinnen und Bürger zu erhöhen, ohne die eigene Identität zu verlieren. Partnerschaftliche Aufgabenerfüllung ist also vielmehr Ausdruck der kommunalen Selbstverwaltung und eines verantwortungsbewussten Umgangs mit personellen und finanziellen Ressourcen. Das ist gerade in Zeiten der Krise wichtiger denn je. Die Steuerschätzungen zeigen: Die Corona-Pandemie verschlechtert auch in den kommenden Jahren dramatisch die Finanzlage der Kommunen. Interkommunale Zusammenarbeit ermöglicht Kosteneinsparungen durch Spezialisierungsvorteile, beispielsweise bei der Beschäftigung von teuren IT-Fachspezialisten, und schafft Größenvorteile bei kommunalen Leistungen. Warum zwei kleine Klärwerke an der jeweiligen Stadtgrenze betreiben, wenn ein großes Klärwerk genauso gut, aber günstiger arbeitet?

Das zeigt, der Aufgabenvielfalt sind kaum Grenzen gesetzt. Gerade für schrumpfende Regionen ist die interkommunale Zusammenarbeit ein wichtiges Instrument, um ihre Leistungen für die Bürgerinnen und Bürger verlässlich zu erbringen und zusätzliche oder erweiterte Infrastruktureinrichtungen bereitzustellen. Auch wachsende Städte kooperieren, etwa bei der Infrastrukturplanung, der Ausweisung von Wohnbauland oder Gewerbegebieten, bei der Wirtschaftsförderung bis hin zu Projekten im Bereich Tourismus oder Regionalmarketing. Ebenso rücken Bürgerdienste, Einkauf, Personaldienstleistungen oder die Informationstechnologie in den Fokus der Städte.

Wir erleben in der kommunalen Praxis eine institutionelle Vielfalt, die von der informellen Zusammenarbeit über gemeinsame Vereinbarungen, Zweckverbände oder Projektgesellschaften bis hin zu zivilgesellschaftlichen Initiativen reichen. Interkommunale Kooperationen sind damit auch flexibler als dauerhafte Aufgabenübertragungen. Gerade für einmalige Projekte, die keine dauerhafte Organisationsstruktur benötigen, sind sie von Vorteil.

Sorgenvoll blicken wir auf die Absicht des Bundes, die interkommunale Zusammenarbeit voll zu besteuern. Das passt so gar nicht zu den Appellen, für mehr Gleichwertigkeit und bessere Daseinsvorsorge die interkommunale Zusammenarbeit zu stärken. Daher sind die Städte dringend darauf angewiesen, dass die Übergangsfrist zur Anwendung des neuen Umsatzsteuerrechts für die Kommunen bis Ende 2022 verlängert wird.

Gemeinsam und partnerschaftlich arbeiten wir auch im Deutschen Städtetag – ein Bündnis, das seit 115 Jahren besteht. Darauf sind wir stolz.

Oberbürgermeister Burkhard Jung, Stadt Leipzig
Präsident des Deutschen Städtetages

Geleitwort des Vorsitzenden des Bayerischen Städtetages

Foto: Stadt Straubing.

Der Bayerische Städtetag ist – mit bald 125 Jahren – der älteste kommunale Spitzenverband in Bayern: Am 1. März 1896 haben sich erstmals auf Einladung des damaligen Münchner Oberbürgermeisters von Borscht 25 Städte getroffen. Die Bürgermeister wollten sich gegen ein Gesetz zur Novellierung des Heimatrechts zur Wehr setzen. Sie kritisierten „aufs Schärfste" den Gesetzentwurf. Dies war die Geburtsstunde des Bayerischen Städtetages – die Bürgermeister erkannten den Wert regelmäßiger Treffen und des engen Erfahrungsaustauschs. Schon ein Jahr später wurde eine organisatorische Form für den freiwilligen Zusammenschluss der bayerischen Städte vereinbart. Die Städte kritisierten bereits vor über einem Jahrhundert die wachsende Fülle von Aufgaben, die der Staat den Kommunen auferlegte. Daher hatten die Stadtväter damals bei der Gründung vereinbart, sich gemeinsam gegen „unberechtigte Ansinnen" des Staates zu wehren. Aus der zunächst informellen Art eines Städtetages konstituierte sich bereits nach wenigen Jahren ein Verband mit einer professionell besetzten Geschäftsstelle, die seit 1909 in München arbeitet.

Jede Stadt in Bayern ist einzig in ihrer Geschichte, in ihrer Stadtgestalt, in ihrer Tradition, sei es als alte Reichsstadt, als Bischofssitz, als Verwaltungssitz, als Handelszentrum oder als Verkehrsachse, als Standort für Handwerk und Industrie. Jede Stadt hat ihre eigenen Stärken, jede Stadt hat ihre ganz eigenen Wurzeln, Ausprägungen und Traditionen – aber in vielen Bereichen ähneln sich unsere Städte. Diese Erfahrungen auszutauschen, Erwartungen und Probleme zu bündeln, ist eine der Aufgaben des Bayerischen Städtetages. Die unterschiedlichen Mitglieder unternehmen es im Bayerischen Städtetag, die Vielfalt der Städte und Gemeinden Bayerns zu einer Einheit zu bringen. Und es geht im Bayerischen Städtetag um Erfahrungsaustausch – die Mitglieder helfen sich selbst mit ihrem jeweiligen Fachwissen. Dies kann durchaus auch den Charakter einer Selbsthilfegruppe annehmen.

In den Mitgliedsstädten des Städtetages wohnen rund 7 Millionen Menschen – das ist mehr als die Hälfte der Bevölkerung Bayerns. Der Bayerische Städtetag ist kein beliebiger Interessenverband. Die Basis des Verbandes sind gewählte Bürgermeister und Stadträte. Somit kann sich der Bayerische Städtetag auf ein allgemeines Mandat der gesamten Bürgerschaft berufen. In den Gremien des Bayerischen Städtetages entscheiden Bürgermeisterinnen und Bürgermeister, Oberbürgermeisterinnen und Oberbürgermeister sowie die Mitglieder von Stadt- und Gemeinderäten. Diese gewählten Mandatsträger bestimmen über die Richtung und die Ziele des Bayerischen Städtetages. Der Bayerische Städtetag vertritt die Interessen von Städten und Gemeinden gegenüber Landtag und Staatsregierung. Er pflegt den Meinungsaustausch und die Interessenvertretung gegenüber anderen landesweiten Verbänden der Wirtschaft, Wohlfahrtsverbänden und gesellschaftlichen Akteuren. Der Städtetag bündelt die Interessen seiner Mitglieder in einer Hand und ist laut Verfassung im Gesetzgebungsverfahren zu beteiligen.

Der Bayerische Städtetag ist der Verband der zentralen Orte Bayerns. Städte und Gemeinden können sich freiwillig zur Mitgliedschaft entscheiden. Neben allen 25 kreisfreien Städten und allen 29 Großen Kreisstädten sind über 200 weitere kreisangehörige Städte, Märkte und Gemeinden Mitglied. Die Bandbreite der über 290 Städtetagsmitglieder reicht von Gemeinden mit knapp 2.000 Einwohnern bis zur Landeshauptstadt München mit 1,5 Millionen Einwohnern.

Die inhaltlichen Kontinuitäten eines Städtetages sind so stark, dass es nicht entscheidend ist, wer nun mit welcher Parteifarbe an der Spitze steht: Diese Überparteilichkeit zeichnet den Städtetag aus. Im Austausch von großen und kleinen Kommunen, von Franken, Altbayern und Schwaben, von Kommunalpolitikern aller Couleur liegt der besondere Reiz des Bayerischen Städtetages. Dies vertreten die Vorsitzenden nach außen: Jeder verkörpert dies auf seine persönliche Art. Vorsitzende wie Hans-Jochen Vogel, Josef Deimer, Hans Schaidinger, Ulrich Maly und Kurt Gribl gaben in den letzten Jahrzehnten dem Bayerischen Städtetag Gesicht und Stimme. Die Bemühungen zur Lösung der vielfältigen Herausforderungen der Gegenwart – von der Corona-Pandemie, Mobilität, Integration, Energieversorgung und kommunaler Daseinsvorsorge bis zu Digitalisierung oder der Finanzausstattung – orientieren sich immer auch an dem Fundament eines reichhaltigen geschichtlichen Erbes unserer Städte. Für den Blick in die Vergangenheit in unterschiedlichen Facetten kann dieses Buch wertvolle Anregungen geben.

Oberbürgermeister Markus Pannermayr, Stadt Straubing
Vorsitzender des Bayerischen Städtetages

Vorwort

Im Frühsommer des Jahres 2020 sollte eigentlich eine interdisziplinäre sowie internationale Tagung im Egloffstein'schen und Wildenstein'schen Palais in der Universitäts-, Plan- und Hugenottenstadt der Frühmoderne, in Erlangen, stattfinden. Wir haben uns die Entscheidung nicht leicht gemacht: Die Situation in Bezug auf „Corona" war damals noch schwer einschätzbar und wir konnten bei Teilnehmern ein erhöhtes Risiko mit schwerem Erkrankungsverlauf nicht ausschließen. Deshalb musste die geplante Tagung, die sich eines sehr regen Zuspruchs erfreute, abgesagt werden.

Um den Wissensdurst um unser Thema *Die süddeutsche Städtelandschaft – ein interregionaler Vergleich* dennoch zeitnah stillen zu können, legten wir den Fokus direkt auf die Drucklegung der mannigfaltigen Forschungsergebnisse. Hervorzuheben ist in erster Linie das Interesse des Peter Lang Verlags, der dank der Initiative seines Leitenden Lektors Herrn Dr. Hermann Ühlein die Publikation der Studien ermöglichte. Großer Dank gilt allen fleißigen Autorinnen und Autoren, die trotz widriger Bibliotheks- und Archivschließungen unter Lockdown-Bedingungen ihre Beiträge zeitnah druckfertig einreichten. Eine stets mit größter Sorgfalt verfolgte Buchredaktion unter Federführung von Herrn Christoph Gunkel M.A. in Erlangen trieb das Projekt in jeder Hinsicht an. Auch Frau Verena Gawert M.A. in Augsburg trug ihren Teil dazu bei. Die Herausgeber arbeiteten in guter kooperativer Zusammenarbeit, die andauernd gegenseitiges Verständnis, Flexibilität und Offenheit voraussetzte, sodass der Band in kurzem Zeitabstand zur erwähnten Tagung veröffentlicht werden konnte. Am Ende gilt unser besonderer Dank nochmals Herrn Christoph Gunkel, der das Orts- und Personenregister wiederum mit Akribie und Sachverstand erstellte.

Finanziell unterstützt wurde der Band – es handelt sich bereits um die siebte Verlagsveröffentlichung der bordeauxroten Erlanger Reihe – von der Forschungsstiftung Bayerische Geschichte in München, der Friedrich-Alexander-Universität Erlangen-Nürnberg (FAU) und der Sektion Franken des Zentralinstituts für Regionenforschung an der FAU. Als verantwortlicher Mitherausgeber und im Namen aller Autorinnen und Autoren danke ich dafür recht herzlich.

Nachhaltigkeit sollte sich in diesen schwierigen Zeiten zu einem positiven Lebensgefühl entwickeln. So wünsche ich dem optisch sowie inhaltlich reizvoll gestalteten Buch eine breite Rezeption in Wissenschaft und Öffentlichkeit. Auf diesem Wege alles Gute für die Zukunft und bleiben Sie gesund!

Erlangen, zum Martinsfest (fränkisch: *Pelzmärtel*) des Jahres 2020

Wolfgang Wüst

Themeneinführung und Zusammenfassung

Wolfgang Wüst

Themeneinführung – Teil 1: Die süddeutsche Städtelandschaft – ein interterritorialer Vergleich

Abstract: As a description of cultural space, the urban landscape took on a pioneering role outside of the geosciences, also in the fields of history, architecture and urbanism, to which other landscape-oriented concepts were committed, such as monastery and sacral landscapes, aristocratic landscapes, landscapes of writing and education, school landscapes, psychiatric landscapes, pleasure landscapes, digital landscapes or the rarely-used village landscapes. The concept of landscape has experienced an almost inflationary spread in parts of the research, so that there has been no lack of criticism and opposition.

1. Bausteine

Im Peter Lang Verlag sind in einer bordeauxroten Reihe die Ergebnisse aus fünf interdisziplinären und zum Teil internationalen Tagungen – sie gründeten auf Initiativen des Erlanger Lehrstuhls für Bayerische und Fränkische Landesgeschichte und des Zentralinstituts für Regionenforschung an der Friedrich-Alexander-Universität – zu folgenden, dem Phänomen des Landschaftsbegriffs zugeschnittenen Themen festgehalten: Dazu zähl(t)en zu Beginn mit Blick auf die Adelslandschaft die im Mai 2016 im Bildungszentrum Kloster Banz – seit 1978 führt dort die CSU-nahe Hanns-Seidel-Stiftung Regie – durchgeführte Tagung zum Thema „Bayerns Adel – Mikro- und Makrokosmos aristokratischer Lebensformen“ (gedruckt: 2017)[1] und das im Februar 2017 in Kloster Ettal mehrtägig veranstaltete Symposium „Adelslandschaften – Kooperationen, Kommunikation und aristokratischer Konsens in Mittelalter, Früher Neuzeit und Moderne“

1 Wolfgang Wüst (Hg.), Bayerns Adel – Mikro- und Makrokosmos aristokratischer Lebensformen. Referate der internationalen und interdisziplinären Tagung. Kloster Banz, Bad Staffelstein, 26.–29. Mai 2016. Unter Mitarbeit von Lisa Bauereisen, Frankfurt am Main u.a. 2017 (380 Seiten, 61 Abbildungen, Orts- und Namensregister).

(gedruckt: 2018).[2] Ebenfalls der räumlichen, sozialen und kulturellen Prägung der adelsnahen Schicht (reichs-)städtischer Patrizier verpflichtet blieben die im Oktober 2016 im Egloffstein'schen Palais zu Erlangen zu hörenden Vorträge zu „Patrizier – Wege zur städtischen Oligarchie und zum Landadel. Süddeutschland im Städtevergleich" (gedruckt: 2018).[3] Es folgten dann im März 2018 im oberbayerischen Ettal „Die süddeutsche Klosterlandschaft – Kultur, Religion, Politik und Umwelt" (gedruckt: 2019)[4] und, in Partnerschaft mit dem Bezirk von Schwaben (Bezirksheimatpflege), im März 2019 in Kloster Irsee die Tagung „Klöster, Kultur und Kunst in Spätmittelalter und früher Neuzeit. Eine Leistungsbilanz zur Sakrallandschaft im deutschen Südwesten" (partiell gedruckt: 2020).[5] Nicht unmittelbar den Landschaftskoordinaten zugeordnet, aber doch auch der räumlichen Dehnung von ökonomischen und ökologischen Entwicklungen verpflichtet, blieb ein weiteres, auf zwei Termine im September 2017 und im März 2018 verteiltes Symposium in Kloster Banz zum Thema „Umwelt-, Klima- und Konsumgeschichte. Fallstudien zu Süddeutschland, Österreich und der Schweiz" (gedruckt: noch 2018).[6]

2. Städtelandschaften als Forschungsfeld

Städtelandschaften sind keine Forschungs- und Publikationsraritäten. Recherchiert man in den bis Ende des Jahres 2015 von der Berlin-Brandenburgischen

---

2 Wolfgang Wüst (Hg.), Adelslandschaften – Kooperationen, Kommunikation und Konsens in Mittelalter, Früher Neuzeit und Moderne. Unter Mitarbeit von Lisa Bauereisen, Berlin u.a. 2018 (285 Seiten, 30 Abbildungen, Orts- und Namensregister).

3 Wolfgang Wüst (Hg.), Patrizier – Wege zur städtischen Oligarchie und zum Landadel. Süddeutschland im Städtevergleich. Referate der internationalen und interdisziplinären Tagung, Egloffstein'sches Palais zu Erlangen, 7.–8. Oktober 2016. Unter Mitarbeit von Marc Holländer, Berlin u.a. 2018 (315 Seiten, 38 Abbildungen, Orts- und Namensregister).

4 Wolfgang Wüst (Hg.), Die süddeutsche Klosterlandschaft – Kultur, Religion, Politik und Umwelt. Unter Mitarbeit von Lisa Bauereisen und Christoph Gunkel, Berlin u.a. 2019 (379 Seiten, 40 Abbildungen, Orts- und Personenregister).

5 Wolfgang Wüst (Hg.), Klöster, Kultur und Kunst – Süddeutsche Sakrallandschaft in Spätmittelalter und früher Neuzeit. Unter Mitarbeit von Lisa Bauereisen und Christoph Gunkel, Berlin u.a. 2020 (374 Seiten, 63 Abbildungen, Orts- und Personenregister).

6 Wolfgang Wüst / Gisela Drossbach (Hg.), Umwelt-, Klima- und Konsumgeschichte. Fallstudien zu Süddeutschland, Österreich und der Schweiz. Unter Mitarbeit von Lisa Bauereisen und Christoph Gunkel, Berlin u.a. 2018 (565 Seiten, 93 Abbildungen, Orts- und Personenregister).

Akademie der Wissenschaften betreuten „Jahresberichte für Deutsche Geschichte"[7] – sie erfassten von 1949 bis 2015 möglichst vollständig[8] das deutsch- und fremdsprachige Schrifttum zur deutschen Geschichte – stößt man unter der einschlägigen Suchanfrage „Städtelandschaft" bereits auf über 40 Treffer. Zur Einstimmung wählen wir aus dieser Datenbank in kaum zu leistender repräsentativer, biographischer Auswahl einige Titel, um das Anliegen unseres Tagungsbandes aus landeshistorischer Sicht zu exemplifizieren.

Wählen wir mit den Jahren 2009, 2010, 2011 und 2013 vier jüngere Stichjahre. Im Jahr 2009 hießen fünf der in den Jahresberichten registrierten Beiträge wie folgt: 1. „Die Entwicklung der Kommunikationsnetze und ihre Beziehung zur europäischen Städtelandschaft" (Ralf Roth),[9] 2. „Über den Zusammenhang von Verkehr, Kommunikation und europäischer Städtelandschaft im 19. und 20. Jahrhundert" (Ralf Roth),[10] 3. „Eine Städtelandschaft im Wandel. Urbanisierungsprozesse im südwestlichen Ostseeraum des 12. Jahrhunderts" (Sven Rabeler),[11] 4. „Einführung: Stadt, Stadtraum, Städtelandschaft – Räume als Analysekategorien der mediävistischen Städteforschung" (Jörg Oberste/Susanne Ehrlich)[12] und schließlich 5. „Die Kulmer Handfeste, das kulmische Recht und die Stadt Kulm. Ein Beitrag zur Gestaltung der Städtelandschaft im Ordensland Preußen" (Roman Czaja).[13]

---

7 Sie standen in der Tradition der von 1880 bis 1916 erschienenen „Jahresberichte der Geschichtswissenschaft", die mit Forschungsberichten zu allen Bereichen der Geschichtswissenschaft von der Historischen Gesellschaft zu Berlin herausgegeben wurden.

8 Es handelt sich um ca. 764.000 Titel von Monographien und Aufsätzen der Erscheinungsjahre 1949 bis 2015.

9 Ralf Roth, Die Entwicklung der Kommunikationsnetze und ihre Beziehung zur europäischen Städtelandschaft, in: Ders. (Hg.), Städte im europäischen Raum. Verkehr, Kommunikation und Urbanität im 19. und 20. Jahrhundert (Beiträge zur Stadtgeschichte und Urbanisierungsforschung 9), Stuttgart 2009, S. 23–62.

10 Ralf Roth, Über den Zusammenhang von Verkehr, Kommunikation und europäischer Städtelandschaft im 19. und 20. Jahrhundert, in: Informationen zur modernen Stadtgeschichte 2009/1, S. 78–87.

11 Sven Rabeler, Eine Städtelandschaft im Wandel. Urbanisierungsprozesse im südwestlichen Ostseeraum des 12. Jahrhunderts, in: Gerhard Fouquet / Gabriel Zeilinger (Hg.), Die Urbanisierung Europas von der Antike bis in die Moderne (Kieler Werkstücke, Reihe E 7), Frankfurt am Main 2009, S. 121–145.

12 Jörg Oberste / Susanne Ehrlich, Einführung: Stadt, Stadtraum, Städtelandschaft – Räume als Analysekategorien der mediävistischen Städteforschung, in: Susanne Ehrlich / Jörg Oberste (Hg.), Städtische Räume im Mittelalter (Forum Mittelalter, Studien 5), Regensburg 2009, S. 7–16.

13 Roman Czaja, Die Kulmer Handfeste, das kulmische Recht und die Stadt Kulm. Ein Beitrag zur Gestaltung der Städtelandschaft im Ordensland Preußen, in: Ders. / Carsten

Im Jahr 2010 publizierte unter anderem Markus Naser als Buch (analog) und Datenbank (digital im Internetportal „Historisches Unterfranken")[14] den Titel „Digitale Karten zur Geschichte der Städte in Franken".[15] 2011 folgten aus der Feder dreier Autoren (Vera Denzer, Anne Dietrich und Haik Thomas Porada) die „Hansetage der Neuzeit. Eine europäische Städtelandschaft als Erinnerungsraum".[16]

2013 wurden schließlich folgende drei Titel zur Städtelandschaft registriert: 1. „‚Stadtlandschaft' – ‚Städtelandschaft' – ‚Städtesystem'. Der Beitrag kleiner Städte zur ‚Urbanisierung' im ‚hansischen Raum' während des späten Mittelalters" (Heidelore Böcker),[17] 2. „Zum Wandel der Städtelandschaft am nördlichen Oberrhein um 1200" (Gerold Bönnen)[18] und 3. „Die Eifel als Städtelandschaft" (Margret Wensky).[19]

Jahnke (Hg.), Städtelandschaften im Ostseeraum im Mittelalter und in der Frühen Neuzeit, Toruń 2009, S. 73–85.

14 Internetportal „Historisches Unterfranken", in: Datenbanken zur Fränkischen Landesgeschichte, hg. vom Lehrstuhl für Fränkische Landesgeschichte an der Julius-Maximilians-Universität Würzburg, URL: http://www.historisches-unterfranken.uni-wuerzburg.de/index.php (letzter Zugriff: 1.10.2020).

15 Markus Naser, Digitale Karten zur Geschichte der Städte in Franken (Mainfränkische Studien 79), Würzburg 2010. Als Datenbank (Link: „Digitale Karten zur Geschichte der Städte in Franken.pdf" – letzter Zugriff: 15.10.2020) mit dem Vermerk: „Zwischen den Jahren 1100 und 1800 können für jeden beliebigen Zeitpunkt und jede Stadt in Franken die Herrschaftsverhältnisse übersichtlich dargestellt und erklärt werden." Internetportal „Historisches Unterfranken" (wie Anm. 14), 2010.

16 Vera Denzer / Anne Dietrich / Haik Thomas Porada, Die Hansetage der Neuzeit. Eine europäische Städtelandschaft als Erinnerungsraum, in: Siedlungsforschung. Archäologie – Geschichte – Geographie 29, 2011, S. 199–218.

17 Heidelore Böcker, „Stadtlandschaft" – „Städtelandschaft" – „Städtesystem". Der Beitrag kleiner Städte zur „Urbanisierung" im „hansischen Raum" während des späten Mittelalters, in: Frank Braun / Stefan Kroll / Kersten Krüger (Hg.), Stadt und Meer im Ostseeraum im 17. und 18. Jahrhundert. Seehandel, Sozialstruktur und Hausbau – dargestellt in historischen Informationssystemen. Beiträge des wissenschaftlichen Kolloquiums in Stralsund vom 8. und 9. September 2005 (Geschichte. Forschung und Wissenschaft 17), Berlin u.a. 2013, S. 30–69.

18 Gerold Bönnen, Zum Wandel der Städtelandschaft am nördlichen Oberrhein um 1200, in: Karsten Igel u.a. (Hg.), Wandel der Stadt um 1200. Die bauliche und gesellschaftliche Transformation der Stadt im Hochmittelalter. Archäologisch-historischer Workshop, Esslingen am Neckar, 29. und 30. Juni 2011 (Materialhefte zur Archäologie in Baden-Württemberg 96), Stuttgart [2]2013, S. 119–130.

19 Margret Wensky, Die Eifel als Städtelandschaft, in: Wolfgang Schmid (Hg.), Die Eifel. Beiträge zu einer Landeskunde. Festschrift 125 Jahre Eifelverein (1888–2013), Bd. 2, Düren 2013, S. 151–174.

Die gewählten biografischen Einträge der Datenbank zu den Jahresberichten konkretisieren bereits einen primär anthropologisch-geowissenschaftlich geprägten Begriff, den der international bekannten Städtelandschaft(en) und ihrer entsprechenden fremdsprachigen Äquivalente als englisch *urban landscape*, italienisch *paesaggio urbano*, französisch *réseau urbain*, *paysage urbain*, spanisch *paisaje urbano*, portugisisch *paisagem urbana*, japanisch 都市景観 oder *last but not least* chinesisch 城市景观. Die Internationalität des Landschaftsbegriffs leistete dann auch Vorschub für grenzüberschreitende Entwicklungsprojekte in der Europäischen Union wie dem von 2003 bis 2006 bestehenden Verbund aus *Sustainable and Accessible Urban Landscapes* (SAUL)[20] als nachhaltigen und zugänglichen Stadtlandschaften. Ziel ist es, in altindustrialisierten Räumen die Lebens- und Wohnqualität zu erhöhen, Räume für Erholung und Freizeit zu schaffen sowie die kulturelle Identifikation im Sinne einer starken und nachhaltigen Wirtschaft zu fördern. Zu den Partnern zählten auf deutscher Seite unter anderem das Institut für Landes- und Stadtentwicklungsforschung und Bauwesen des Landes Nordrhein-Westfalen (ILS NRW) und der Planungsverband Ballungsraum Frankfurt/Rhein-Main.

3. Landschaft als geowissenschaftliches Interessensgebiet?

In der Geografie, als einer Leitwissenschaft in Landschaftsfragen, kann man sich aktuell unter einer Städtelandschaft maßstäblich und räumlich ganz unterschiedliche Ebenen vorstellen, die sich auf einzelne Städte, auf bestimmte Stadtregionen oder selbst auf Metropolregionen oder die Megalopolis übertragen lassen. Die Megalopolis wurde darunter als Landschaftsbegriff erstmals 1918 von dem Philosophen und Kulturhistoriker Oswald Spengler (1880–1936) in seinem Werk „Der Untergang des Abendlandes“ gebraucht.[21] Die Stadtregion muss sich dabei keineswegs auf ihre historisch-territorialen Wurzeln im Sinne reichs-, residenz- oder binnenstädtischer Landterritorien beschränken. Die Stadt und ihr territoriales Umland wurden, auch wenn das Bürgerrecht in und außerhalb der Stadtmauern oft unterschiedliche Freiheitsstufen kannte, wie im Falle der schwäbischen Reichsstadt Ulm um 1715 als kartografische Einheit,

20 Sustainable & Accessible Urban Landscapes (SAUL), in: European Urban Knowledge Network, URL: https://www.eukn.eu/news/detail/sustainable-accessible-urban-landscapes-saul (letzter Zugriff: 1.10.2020).

21 Wolfgang Krebs, Die imperiale Endzeit. Oswald Spengler und die Zukunft der abendländischen Zivilisation, Berlin 2008.

als *„Nova et accurata Territorii Vlmensis cum Dominio Wainensi Descriptio“* abgebildet.

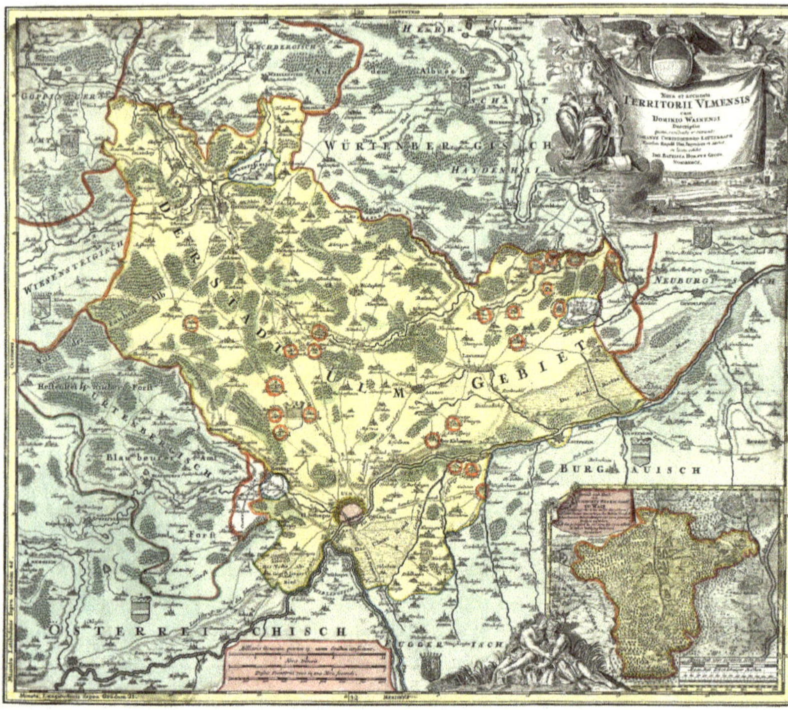

Abb. 1: *„Nova et accurata Territorii Vlmensis cum Dominio Wainensi Descriptio“. Das Territorium der Reichsstadt Ulm mit der Exklave Wain, maßstabsgetreue kolorierte Karte von Johann Baptist Homann (1664–1724), um 1715. Bildnachweis: Wolfgang Wüst (privat).*

Für die ehemalige fränkische Reichsstadt Rothenburg ob der Tauber galt Ähnliches. Der Fokus lag hier allerdings auf der friedensichernden „Landhege“, die im Kartenwerk der frühen Neuzeit wiederholt mit zugehörigen Herrschafts- und Verteidigungslinien gezeichnet und beschrieben wurde.

Abb. 2: *Die „Landhege" der Reichsstadt Rothenburg ob der Tauber ist eine detaillierte, aber nicht unbedingt maßstabsgetreue Darstellung zu dem seit dem Spätmittelalter forcierten Herrschafts- und Verteidigungsausbau im städtischen Umland. Stadtlandschaften inkludierten stets städtische Landterritorien. Karte des 1543 in Würzburg verstorbenen Kartographen und Malers Wilhelm Ziegler, 1537. Bildnachweis: Germanisches National-museum Nürnberg (GNM), Graphische Sammlung, La 4040.*

Abb. 3: *„Geographischer Grundris der in Actis angezogenen und articulierten Sogenannten Rotenburgischen Landwehr in Sachen Comburg contra Rottenburg". Kopie der maßstabsgetreuen Karte im Rahmen eines Reichskammergerichtsverfahrens, 10. April 1776. Bildnachweis: Stadtarchiv Rothenburg, Karten und Pläne, 1/02.*

Der Landschaftsbegriff, der im urbanen Umfeld relativ früh Verwendung fand, geht seitens der Kultur- und Sozialgeographie und der Architekturtheorie auf die frühen 1920er Jahre zurück.[22] Seit dieser Zeit spricht man

22 Werner Durth, Stadtlandschaft im Wiederaufbau, in: Hubertus Fischer (Hg.), Zukunft aus Landschaft gestalten. Stichworte zur Landschaftsarchitektur (CGL-Studies 17),

vermehrt von Städtelandschaft(en), um Kultur- und Siedlungsräume zu umschreiben und sie abzugrenzen. Der in Ostpreußen geborene Geograph, Geologe und Paläontologe Siegfried Passarge (1866–1958), Begründer der Landschaftsgeografie, schärfte den Landschaftsbegriff zu Beginn des 20. Jahrhunderts ebenfalls in mehreren Veröffentlichungen. Die Stadtlandschaft diente Passarge zur Abgrenzung von Naturlandschaften zum Kulturraum.[23] Der Stadtlandschaft fiel als Kulturraumbeschreibung dann außerhalb der Geowissenschaften auch in der Geschichtswissenschaft, der Architektur und der Urbanistik eine Pionierrolle zu, an die sich andere landschaftsorientierte Begriffsbildungen orientierten wie die Kloster- und Sakrallandschaften,[24]

München 2014, S. 217–224; Ders., Stadt und Landschaft. Kriegszerstörungen und Zukunftsentwürfe, in: Jörn Düwel u.a. (Hg.), 1945. Krieg - Zerstörung - Aufbau. Architektur und Stadtplanung 1940–1960 (Schriftenreihe der Akademie der Künste 23), Berlin 1995, S. 127–175.

23 Max Eckert, Die Entwicklung der kartographischen Darstellung von Stadtlandschaften, in: Siegfried Passarge (Hg.), Stadtlandschaften der Erde, Hamburg 1930, S. 1–14.

24 Vgl. zum Landschaftsbegriff in nicht repräsentativer Auswahl: Karl Hausberger, Das Bistum Regensburg als Klosterlandschaft. Ein historischer Abriß, in: Peter Morsbach (Hg.), 1250 Jahre Kunst und Kultur im Bistum Regensburg. Berichte und Forschungen (Kunstsammlungen des Bistums Regensburg. Kataloge und Schriften 7), München/Zürich 1989, S. 177–194; Klaus Wittstadt, Die Klosterlandschaft Frankens heute, in: Frankenland 43, 1991, S. 172–201; Rudolf Renz, Klosterlandschaft Oberschwaben, 2 Bde. (Materialien zur Landeskunde und Landesgeschichte 10 und 11), Tübingen 1994; Bernd Brauksiepe / Anton Neugebauer, Klosterlandschaft Eifel. Historische Klöster und Stifte zwischen Aachen und Bonn, Koblenz und Trier (Große Kunstführer 191), Regensburg 1994; Otto Weiss, Die neue Klosterlandschaft in Bayern, in: Rainer Braun / Joachim Wild (Hg.), Bayern ohne Klöster? Die Säkularisation 1802/03 und die Folgen (Ausstellungskataloge der Staatlichen Archive Bayerns 45), München 2003, S. 483–512; Ewald Wegener, Klosterlandschaft Eifel-Mosel. Praktische Denkmalspflege am Beispiel des Klosters Stuben bei Bremm/Mosel (Rheinische Heimatpflege 48), Neuss 2011; Gert Melville, „Klosterlandschaft". Kritische Bemerkungen zum wissenschaftlichen Wert einer Wortschöpfung, in: Franz J. Felten / Harald Müller / Heidrun Ochs (Hg.), Landschaft(en): Begriffe – Formen – Implikationen (Geschichtliche Landeskunde 68), Stuttgart 2012, S. 195–222; Michael Schmitt, Die Klosterlandschaft des Bistums Würzburg als GIS-Projekt, in: Würzburger Diözesangeschichtsblätter 76, 2013, S. 189–210; Nathalie Kruppa, Die Klosterlandschaft im Bistum Hildesheim im frühen und hohen Mittelalter im Vergleich zu ihren Nachbarbistümern Paderborn, Minden, Verden und Halberstadt, in: Niedersächsisches Jahrbuch für Landesgeschichte 85, 2013, S. 135–189; Hans-Georg Aschoff, Die katholische Klosterlandschaft im Fürstbistum Hildesheim während der Frühen Neuzeit, in: Hansjörg Küster / Joachim Wolschke-Bulmahn (Hg.), Zu den Qualitäten klösterlicher Kulturlandschaften. Geschichte, Kultur, Umwelt und Spiritualität (CGL-Studies 20), München 2014, S. 13–33;

Adelslandschaften,[25] Kulturlandschaften,[26] Schrift- und Bildungslandschaften,[27] Schullandschaften,[28] Psychiatrischen Landschaften,[29] Genusslandschaften,[30]

Joachim Oepen, Das Bergische Land – eine Klosterlandschaft?, in: Stefan Gorißen / Horst Sassin / Kurt Wesoly (Hg.), Geschichte des Bergischen Landes, Bd. 1: Bis zum Ende des alten Herzogtums 1806 (Bergische Forschungen 31), Bielefeld 2014, S. 141–186; Johannes Molitor, Klosterlandschaft – Metten und Niederaltaich, in: Wolfgang Jahn (Hg.), Niederbayerische Donau (Edition Bayern), Regensburg 2015, S. 44–49.

25 Beispielsweise: Václav Bůžek, Die Adelslandschaft der böhmischen Länder im 16. und 17. Jahrhundert, in: Jill Bepler / Helga Meise (Hg.), Sammeln, Lesen, Übersetzen als höfische Praxis der Frühen Neuzeit. Die böhmische Bibliothek der Fürsten Egenberg im Kontext der Fürsten- und Fürstinnenbibliotheken der Zeit (Wolfenbütteler Forschungen 126), Wiesbaden 2010, S. 17–52; Joachim Schneider, Die Bünaus in der wettinischen Adelslandschaft des Spätmittelalters. Gesamtbelehnungen und Wappenführung als Elemente sozialer Strategien zwischen Kernfamilie und Gesamtgeschlecht, in: Martina Schattkowsky (Hg.), Die Familie von Bünau. Adelsherrschaften in Sachsen und Böhmen vom Mittelalter bis zur Neuzeit, Leipzig 2008, S. 167–190.

26 Ein Allroundbegriff mit unterschiedlichen inhaltlichen Zielsetzungen: Herbert Fitzek / Friederike Krause / Alina Kuzborska (Hg.), Kulturlandschaften. Auf Herders Spuren zwischen Mohrungen und Königsberg, Olsztyn 2012; Gerhard Gabel, Vielfalt der Kulturlandschaften Bayerns – Kulturlandschaftliche Gliederung als Beitrag zur Biodiversität, in: Siedlungsforschung. Archäologie – Geschichte – Geographie 29, 2011, S. 311–318; Gerhard Henke-Bockschatz, Kulturlandschaften wahrnehmen und entschlüsseln oder: Welche Kompetenzen brauchen SchülerInnen zur historischen Spurensuche?, in: Heike Düselder / Annika Schmitt / Siegrid Westphal (Hg.), Umweltgeschichte. Forschung und Vermittlung in Universität, Museum und Schule, Köln u.a. 2014, S. 123–141.

27 Hedwig Röckelein, Schriftlandschaften, Bildungslandschaften und religiöse Landschaften des Mittelalters in Norddeutschland (Wolfenbütteler Hefte 33), Wiesbaden 2015; Rolf Kießling, Die Universität Dillingen und die „Bildungslandschaft" Schwaben im Zeitalter der Konfessionalisierung, in: Jahrbuch des Historischen Vereins Dillingen 101, 2000, S. 9–34; Matthias Asche, Der Ostseeraum als Universitäts- und Bildungslandschaft im Spätmittelalter und in der Frühen Neuzeit. Baustein für eine hansische Kulturgeschichte, in: Blätter für deutsche Landesgeschichte 135, 1999, S. 1–20.

28 Helmut Flachenecker / Rolf Kießling (Hg.), Schullandschaften in Altbayern, Franken und Schwaben. Untersuchungen zur Ausbreitung und Typologie des Bildungswesens in Spätmittelalter und Früher Neuzeit (Zeitschrift für bayerische Landesgeschichte, Beihefte B 26), München 2005.

29 Elisabeth Dietrich-Daum (Hg.), Psychiatrische Landschaften. Die Psychiatrie und ihre Patientinnen und Patienten im historischen Raum Tirol seit 1830, Innsbruck 2011.

30 Vgl. hierzu den Beitrag von Sabine Wüst zu europäischen Genusslandschaften oder Foodscapes im interregionalen Vergleich in diesem Band, S. 461–507; Anna Moragues-Faus / Kevin Morgan, Reframing the foodscape. The emergent world of urban food policy, in: Environment and Planning A: Economy and Space 47/7, 2015, S. 1558–1573,

Digitalen Landschaften[31] oder die selten geführten Dorflandschaften.[32] Der Landschaftsbegriff erfuhr in Teilen der Forschung geradezu eine inflatorische Verbreitung, sodass Widerspruch und kritische Stimmen nicht ausblieben.[33]

4. Voraussetzungen und Wirkung

Die Frage nach dem Zentrum und der Zentralität in der jeweiligen Landschaft spielte für viele Themenbereiche als Ergebnis interdisziplinär geführter Überlegungen eine Rolle, keineswegs nur in einer archäologisch fundierten Stadt- und Siedlungsgeschichte.[34] Voraussetzung für die Entstehung der Städtelandschaften im mitteleuropäischen Raum war, wie nicht zuletzt Helmut Flachenecker in seinem Beitrag „Städtelandschaft Franken" (S. 89–116) exemplarisch zeigt, zunächst die Urbanisierung der Fläche. Die Zahl der Städtegründungen war vor allem im Mittelalter hoch, insbesondere in der Zeit zwischen dem 12. und der Mitte des 14. Jahrhunderts. Ein Jahrhundert später, um 1450, zählte man dann bereits circa 5.000 Stadtgründungen im deutschen Sprachraum. Die Chroniken des späten Mittelalters und der frühen Neuzeit, allen voran die Chronik der Bischöfe von Würzburg aus der Feder des fürstbischöflichen Sekretärs Lorenz Fries (1489/91–1550),[35] rekurrierten bereitwillig auf die dynamische Verstädterung einzelner Regionen. In der von den ursprünglich drei Ausfertigungen[36] einzig erhaltenen, mit 176 Miniaturen von dem Würzburger Maler Martin Seger (um 1510/

URL: https://journals.sagepub.com/doi/10.1177/0308518X15595754 (letzter Zugriff: 1.10.2020).

31 Manuel Schramm, Digitale Landschaften (Pallas Athene 30), Stuttgart 2009.

32 Michael A. Hartenstein, Neue Dorflandschaften. Nationalsozialistische Siedlungsplanung in den „eingegliederten Ostgebieten" 1939 bis 1944 (Wissenschaftliche Schriftenreihe Geschichte 6), Berlin 1998.

33 Melville, „Klosterlandschaft" (wie Anm. 24).

34 Bernhard W. Bahn, Zur Frage der Zentralität von Landschaften, Orten und Trassen, in: Gerson H. Jeute / Jens Schneeweiß / Claudia Theune (Hg.), „aedificatio terrae". Beiträge zur Umwelt- und Siedlungsarchäologie Mitteleuropas. Festschrift für Eike Gringmuth-Dallmer zum 65. Geburtstag (Internationale Archäologie. Studia honoraria 26), Rhaden/Westfalen 2007, S. 39–44.

35 Thomas Heiler, Die Würzburger Bischofschronik des Lorenz Fries (gestorben 1550). Studien zum historiographischen Werk eines fürstbischöflichen Sekretärs und Archivars (Veröffentlichungen des Stadtarchivs Würzburg 9), Würzburg 2001.

36 Die Chronik war auf Anordnung des Fürstbischofs Konrad III. von Bibra (1490–1544) in drei Exemplaren hergestellt worden.

1515–1580) illustrierten Handschrift[37] der Würzburger Chronik – sie wurde 1546 fertiggestellt – sind fränkische Residenz- und Reichsstädte wie Bamberg oder Schweinfurt eingebettet in den Kontext ihrer Landschaft.

Abb. 4: *Stadtansicht von Bamberg aus der Chronik von Lorenz Fries, 1546. Bildnachweis: Universitätsbibliothek Würzburg, M.ch.f.760, fol. 34 v, URL: http://franconica.uni-wuerzburg.de/ub/permalink/fries-130.*

Abb. 5: *Stadtansicht von Schweinfurt aus der Chronik von Lorenz Fries, 1546. Bildnachweis: Universitätsbibliothek Würzburg, M.ch.f.760, fol. 300 r, URL: http://franconica.uni-wuerzburg.de/ub/permalink/fries-661.*

37 Heutiger Standort: Stadtarchiv Würzburg, Ratsbuch 412.

Die Stadtgründungswellen des Mittelalters waren Voraussetzung für das Entstehen frühmoderner Städtelandschaften und es gab anschließend keine Epoche mehr, in der sich Ähnliches im urbanen Gründungsbereich vollzog.[38] Städtelandschaften bedurften aber zunehmend auch einer administrativen Effizienz, die bereits während der Städtebündnisse des 15. und 16. Jahrhundert die Zentralität einzelner Orte stärkte. So gewann beispielsweise Ulm unter den über dreißig beteiligten, meist südwestdeutschen Reichsstädten[39] des 1488 auf dem Esslinger Reichstag gegründeten Schwäbischen Bundes als Hauptort eine führende Rolle. Ulm dominierte mit seiner Rats- und Bündnisverwaltung eine weitgestreute Städtelandschaft, die der Friedenssicherung verpflichtet war. Rats- und Amtszentren und die repräsentativen Ratshäuser[40] vergrößerten sich fortlaufend, wenn die territorialen oder bündnisorientierten Verpflichtungen der Städte wuchsen. Ebenso stieg die Frequenz der Ratskorrespondenz an, wenn zusätzliche, oft landschaftsgebundene Koordinierungsaufgaben die Ratsversammlungen beschäftigten. In der hessischen, von 1535 bis 1794 paritätisch durch das Haus Nassau und Kurtrier regierten Stadt Camberg – seit 1981 Bad Camberg – dominiert bis heute ein überdimensionierter Amtshof in Fachwerksausführung die Stadtsilhouette. Gebäudebrücken wurden angesichts stetig steigender

38 Vgl. generell hierzu: Eberhard Isenmann, Die deutsche Stadt im Spätmittelalter 1250–1500. Stadtgestalt, Recht, Stadtregiment, Kirche, Gesellschaft, Wirtschaft, Stuttgart 1988; Helmut Flachenecker / Rolf Kießling (Hg.), Städtelandschaften in Altbayern, Franken und Schwaben. Studien zum Phänomen der Kleinstädte während des Spätmittelalters und der Frühen Neuzeit (Zeitschrift für bayerische Landesgeschichte, Beihefte B 15), München 1999.

39 Dazu zählten im Februar 1488 die Reichsstädte Ulm, Esslingen, Reutlingen, Überlingen, Lindau, Schwäbisch Hall, Nördlingen, Memmingen, Ravensburg, Schwäbisch Gmünd, Biberach an der Riß, Dinkelsbühl, Pfullendorf, Kempten, Kaufbeuren, Isny, Leutkirch, Giengen, Wangen und Aalen. Im April 1488 folgten die Reichsstädte Weil der Stadt und Bopfingen, im November 1488 Augsburg, Heilbronn, Donauwörth und Wimpfen. Vgl. Horst Carl, Der Schwäbische Bund 1488–1534. Landfrieden und Genossenschaft im Übergang vom Spätmittelalter zur Reformation (Schriften zur südwestdeutschen Landeskunde 24), Leinfelden-Echterdingen 2000, S. 61–63.

40 Vgl. für Süddeutschland und Augsburg: Bernd Roeck, Reichsstädtische Rathäuser der frühen Neuzeit und ihre Bildprogramme, in: Rainer A. Müller (Hg.), Bilder des Reiches. Tagung in Kooperation mit der schwäbischen Forschungsgemeinschaft und der Professur für Geschichte der Frühen Neuzeit der Katholischen Universität Eichstätt im Schwäbischen Bildungszentrum Kloster Irsee vom 20. bis 23. März 1994 (Irseer Schriften 4), Sigmaringen 1997, S. 275–295; Ders., Elias Holl. Ein Architekt der Renaissance, Regensburg 2004; Wolfram Baer / Hanno-Walter Kruft / Bernd Roeck (Hg.), Elias Holl und das Augsburger Rathaus, Ausstellungskatalog, Regensburg 1985.

zentralisierender Amtsgeschäfte eine Notwendigkeit, um der Stadt und ihrer Landschaft administrativ gerecht zu werden.

Abb. 6: *Ehemaliger Amtshof in Bad Camberg (Taunus), Südansicht, 2018. Die Gebäudeverbindungen stehen symbolisch für die Netzwerke der Städtelandschaft. Der Blick richtet sich konkret auf die hessische, von 1535 bis 1794 unter gemeinsamer Verwaltung von Kurtrier und der Grafschaft Nassau stehende Stadt. Das Amtshaus zählt zu den größten Fachwerksbauten der frühen Neuzeit in Mitteldeutschland. Es setzt sich aus älteren Gebäudeteilen der Herren von Riedesel, von Nassau sowie derer von Heiden und Metternich zusammen. Bildnachweis/Foto: Wikimedia Commons, Johannes Robalotoff.*

Klaus Wolf

Themeneinführung – Teil 2: Die süddeutsche Städtelandschaft aus sprach- und literaturgeschichtlicher Perspektive

Abstract: Urban landscapes as historically and geographically comprehensible units can also be described on the basis of linguistic and literary-historical criteria. As far as the German-speaking area is concerned, corresponding research by German studies, especially in the last third of the twentieth century, has been carried out with increasing intensity, although this varies greatly from region to region. While, for example, the Upper German area was responsible for the late Middle Ages and the early modern period also because of the culturally tangible Habsburg hegemony and an Holy Roman Emperor Maximilian I, apostrophised as the "Mayor of Augsburg". While the Central German region, including Frankfurt and Weimar, aroused the interest of German studies at an early stage, or the so-called Weimar Classicism in the shape of Goethe in the eighteenth and early nineteenth centuries, which made the region virtually canonical with its cities as literary centers, the Lower German region has remained under-researched.

Städtelandschaften als historisch und geographisch fassbare Einheiten lassen sich auch anhand sprachhistorischer und literaturgeschichtlicher Kriterien beschreiben. Was den deutschsprachigen Raum anbelangt, so sind entsprechende Forschungen von Seiten der Germanistik insbesondere im letzten Drittel des 20. Jahrhunderts mit wachsender, freilich je nach Region durchaus unterschiedlicher Intensität erfolgt. Während etwa der oberdeutsche Raum für das Spätmittelalter und die Frühe Neuzeit auch wegen der kulturell fassbaren habsburgischen Hegemonie und einem als „Bürgermeister von Augsburg" apostrophierten Kaiser Maximilian I. schon früh das Interesse der Germanistik weckte oder etwa der mitteldeutsche Raum mit Frankfurt am Main und Weimar wegen der sogenannten Weimarer Klassik in Gestalt Goethes im 18. und frühen 19. Jahrhundert mit seinen Städten als Literaturorten geradezu kanonisch wurde, ist der niederdeutsche Raum trotz der Bedeutung des Mittelniederdeutschen als *lingua franca* der Hanse bis heute durchaus unterforscht.[1]

1 Vgl. etwa neuerdings die einschlägigen Pionierarbeiten: Jörn Bockmann / Andreas Bieberstedt, Die mittelniederdeutsche Bibliothek. Zur Konzeption und Realisierung einer neuen Buchreihe, in: Korrespondenzblatt des Vereins für niederdeutsche Sprachforschung 126, 2019, S. 31–35.

Im Folgenden gilt es daher, den germanistischen Beitrag insbesondere bezüglich süddeutscher Städtelandschaften zu bilanzieren und im Blick auf Forschungsdesiderate zu perspektivieren.

Sprachgeschichtlich bietet sich hierbei eine auf Städte oder Stadtsprachen bezogene Betrachtungsweise im Grunde erst seit dem späten Mittelalter an, konkret mit dem Durchbruch der Volkssprache auch in städtischen Kanzleien gegenüber dem Lateinischen. Darüber hinaus treten im späten Mittelalter eine Vielzahl von fundamentalen Änderungen in den kommunikativen Verhältnissen zutage, die auch mit der stärkeren wirtschaftlichen Bedeutung der Städte zusammenhängen. Städtische Märkte führen vielfältige Berufsgruppen zusammen, die weniger auf Latein, sondern in der Volkssprache kommunizieren. Meister einer solchen Massenkommunikation sind nicht zuletzt die Bettelorden, welche sich seit dem 13. Jahrhundert in den Städten niederlassen und mittels öffentlicher Volkspredigt breite Schichten erreichen und neue literarische Gattungen anregen. Gerade süddeutsche Städte wie Augsburg und Regensburg waren frühe Einfallstore der Franziskaner und ihrer volkssprachigen Literatur, namentlich auch von frühneuhochdeutschen Gesetzestexten und geistlicher Prosa, einer Gattung, die nicht unwesentlich zur sogenannten Literaturexplosion des Spätmittelalters beitrug. Auch das städtische Schulwesen ist als sprachhistorischer Faktor nicht gering zu schätzen. Für solche Schulen wurden dann deutsche Fibeln verfasst, eine der ersten stammt übrigens aus Augsburg. Die wegen des im Spätmittelalter ausgebauten städtischen Schulwesens zunehmende Lese- und Schreibfähigkeit geht mit einem Ausbau des Kanzleiwesens einher, wobei die Stadtsprachen einerseits regionale Eigenheiten konservieren, andererseits im überregionalen Austausch auch sprachliche Ausgleichstendenzen fassbar werden.[2]

Solche Ausgleichstendenzen werden insbesondere im deutschsprachigen Buchdruck virulent, der einen durchaus süddeutschen Fokus mit Straßburg, Nürnberg und Augsburg aufweist, wobei aber die Pionierrolle und überregionale Wirksamkeit der Augsburger Druckersprache im wahrsten Sinne des Wortes Sprachgeschichte geschrieben hat, nicht zuletzt mit einer Wirksamkeit jenseits der oberdeutschen Sprachlandschaft auch im Mitteldeutschen, denn die Distribution der oberdeutschen Druckerzeugnisse erfolgte über die Frankfurter Buchmessen.[3]

---

2 Vgl. Wilhelm Schmidt, Geschichte der deutschen Sprache. Ein Lehrbuch für das germanistische Studium, Stuttgart [10]2007, S. 105 f.

3 Vgl. Akihiko Fujii, Günther Zainers druckersprachliche Leistung. Untersuchungen zur Augsburger Druckersprache im 15. Jahrhundert (Studia Augustana 15), Tübingen 2007, passim; Hans-Jörg Künast, Getruckt zu Augspurg. Buchdruck und Buchhandel in Augsburg zwischen 1468 und 1555 (Studia Augustana 8), Tübingen 1997, passim.

Damit vermochte der Buchdruck früh zur Ausbildung einer neuhochdeutschen Schriftsprache Entscheidendes beizutragen, einer Ausgleichssprache, die ebenso frühzeitig verschiedene Städtelandschaften durchaus im Sinne einer *lingua franca* verbinden konnte.

Dieser Tendenz zur sprachlichen Überregionalität steht auf der anderen Seite eine in ihren Einzelheiten keineswegs erforschte literaturgeschichtliche Vielfalt in einzelnen Städten und Städtelandschaften gegenüber. Neben methodisch wegweisenden Studien[4] und neueren städtischen Literaturgeschichten[5] wäre auch dem bislang vernachlässigten zwischenstädtischen Austausch verstärkt nachzugehen. Neuere Untersuchungen zeigen in diesem Sinne etwa, dass in der Vormoderne größere süddeutsche Städte, obwohl verschiedenen Sprachlandschaften und mitunter sogar unterschiedlichen Reichskreisen angehörig, in Spätmittelalter und Früher Neuzeit einen bislang unterschätzten lebhaften wechselseitigen literarischen Austausch pflegten, während andererseits eine grundsätzlich fassbare kulturelle Trennlinie zwischen tendenziell süddeutsch-katholischen, romanischem Einfluss offenstehenden Stadt- und Literaturlandschaften einerseits und einem mitteldeutsch-protestantisch geprägten literaturgeographischen Block andererseits beschrieben werden kann.[6] Regionale Literaturgeschichtsschreibung hat hier nicht nur in Buchform noch ein großes Aufgabengebiet vor sich. Denn künftig dürften auch digitale Formen der Literaturgeographie wie etwa digitale Literaturatlanten[7] nicht nur das literarische Leben beispielsweise in einzelnen süddeutschen Städten bis hin zur Überlieferung auf verschiedenen Ebenen detailliert erfassen, sondern auch die geographischen Vernetzungen im Sinne einer Verflechtungsgeschichte digital dokumentieren.

---

4 Hier ist insbesondere Ursula Peters zu nennen mit ihrem Buch: Ursula Peters, Literatur in der Stadt. Studien zu den sozialen Voraussetzungen und kulturellen Organisationsformen städtischer Literatur im 13. und 14. Jahrhundert, Tübingen 1983.

5 Waldemar Fromm / Manfred Knedlik / Marcel Schellong (Hg.), Literaturgeschichte Münchens, Regensburg 2019.

6 Vgl. Klaus Wolf, Bayerische Literaturgeschichte. Von Tassilo bis Gerhard Polt, München 2018, passim.

7 Vgl. URL: https://www.uni-augsburg.de/de/fakultaet/philhist/professuren/germanistik/deutsche-literatur-und-sprache-bayern/forschung/digi/ (letzter Zugriff: 29.10.2020).

Verena Gawert

Zusammenfassung

Der Aufsatz von Wolfgang Wüst (Erlangen-Nürnberg) beleuchtet die frühmodernen „Städtebänke“ als Gremien urbaner Interessensvertreter in Kreis- und Reichstagen. „Städtelandschaft“ soll zunächst die Forschung zur Geschichte der Städte und Märkte auf internationaler und insbesondere auf süddeutscher Ebene beschreiben. Was bedeutet „Städtelandschaft“ eigentlich? Der renommierte Historiker Franz Irsigler versteht unter „Städtelandschaft“ eine mittelgroße räumliche Einheit mit unterschiedlichem Urbanisierungsgrad; und unter „Stadtlandschaft“, dass eine Stadt eine bestimmte räumliche Einheit dominieren kann, dies freilich im diachronen Wandel. In Wolfgang Wüsts Beitrag soll dies im Zusammenhang mit der städtischen Korrespondenz auf den Reichs- und Territorialräten diskutiert werden. Politische Netzwerke werden in der Korrespondenz der Räte transparent, dabei fehlte es nicht an Treffpunkten und Zeiten für die untersuchten Räte. Diese Sitzungen fanden häufig und über lange Zeiträume statt, was den beteiligten Delegationen einen Raum der engen Kommunikation eröffnete. Sie waren ein Ort, an dem zwischenstädtische Reformen beschlossen und Pläne nach besten Kräften korrigiert wurden, wodurch die Städte teilweise finanziell benachteiligt wurden. Diese kommunikative Leistung bei der Konsensfindung wird nicht dadurch geschmälert, dass fast alle Städte kaum durch eigene Abgeordnete vertreten sein konnten. Überdies erreichten Beschlüsse von Reichstag und Kreistag auch die in der bisherigen Forschung oft als „verschlafen“ bezeichneten Reichsstädte.

Tom Scott (St Andrews) betrachtet in seinem Aufsatz die Oberrheinische Städtelandschaft. Der Oberrhein besitzt keine klar definierten territorialen Grenzen; gemeinhin wird er als das von Süden nach Norden verlaufende oberdeutsche Rheintal angesehen. In der Tat sind seine kommerziellen und urbanisierten Merkmale auf einer Ost-West-Achse, die die Westschweiz einbezieht und sich bis in die Franche-Comté erstreckt, ebenso ausgeprägt. Ein großer Teil der Region ist von mittelgroßen oder kleinen Städten geprägt, von denen viele Weinbaugemeinden waren. Weder Basel noch Straßburg haben sich als echte Metropolen entwickelt, die die Wirtschaft ihres Hinterlandes dominieren oder integrieren. Walter Christallers ursprüngliche Zentral-Ort-Theorien, die von einem sechseckigen Marktgebiet ausgingen, funktionieren nicht für den

Oberrhein. Stattdessen sind die Netzwerke, wie er später zugab, als Rhomben zu verstehen, die die Exporte flussauf- und flussabwärts in einer Reihe von Rauten verbinden. Zu unterscheiden sind urbane und verstädterte Landschaften (Franz Irsigler). Ein Vergleich mit dem württembergischen Neckartal und der Westschweiz zeigt, dass weder eine spärliche noch eine dichte Siedlungsverteilung ihren Charakter als Stadtlandschaften ausmacht; nur der integrierte, exportorientierte Weinbau des Elsass kommt dagegen der regionalen Textillandschaft Ostschwabens nahe. Diese mangelnde Integration und Koordination machte die städtischen Siedlungen vor allem im konfessionellen Zeitalter besonders anfällig für ländliche Handwerks- und Marktkonkurrenz sowie für politische Rivalitäten zwischen den Territorialherrschaften.

Helmut Flachenecker (Würzburg) beleuchtet in seinem Beitrag die fränkische Städtelandschaft. Die Geschichte Frankens im Hoch- und Spätmittelalter ist geprägt von einer erstaunlichen Vielfalt territorialer Herrschaftsverhältnisse. Eine Vielzahl mittelgroßer und kleinerer Territorien, die in Franken liegen, prägen eine eindrucksvolle Kulturlandschaft aus Burgen, Städten und Märkten mit massiven – und heute meist romantisierten – Mauersystemen und bedeutenden Klöstern, die jedem Betrachter eine Vorstellung von den einstigen politischen Strukturen noch in der Gegenwart vermittelt. Der Beitrag diskutiert insbesondere den Begriff „Stadtlandschaft“ als einen wichtigen Teil der Kulturlandschaften. Anschließend wird die Entwicklung der Städte in Franken in chronologischer Reihenfolge dargestellt. Da die meisten Städte in dieser Region einem geistlichen Herrscher unterstellt waren, werden die Entwicklungen in den Bistümern Würzburg und Eichstätt näher beleuchtet. Die Bedeutung der Städte für die historische Entwicklung Frankens steht außer Frage, insbesondere ihr Beitrag zur allgemeinen Urbanisierung des Gebietes. Der Forschungsschwerpunkt hängt jedoch von aktuellen regionalgeschichtlichen Ansätzen ab, die Stadtstrukturen fallweise eher ökonomisch, eher als Sonderrechtsbezirk, eher als Zentrum für das Umland oder in ihren Funktionen, die der Erweiterung der Territorien der Stadtherrscher dienen, untersuchen.

Der Beitrag von Gerhard Fritz (Schwäbisch-Gmünd) widmet sich Stadt und Kloster im Herzogtum Württemberg. Der Aufsatz zeigt insbesondere die komplizierten Beziehungen zwischen Städten und Klöstern im Herzogtum Württemberg im Spätmittelalter und in der Frühen Neuzeit. Das Herzogtum bestand aus etwa 40 Grafschaften und 14 separaten Klosterbezirken. Jede Grafschaft und jedes Kloster (aber nur Konvente, keine Nonnenklöster) hatte das Recht, Stellvertreter auf den Landtag in Stuttgart zu entsenden. Obwohl die lutherische Reformation in den Jahren 1534 und 1552 eingeführt wurde und die Klöster als katholische Einrichtungen aufgehoben wurden, blieben die

Klöster als Verwaltungsbezirke bestehen. Jeder Landkreis bestand aus einer zentralen Stadt und mehreren umliegenden Dörfern, Weilern und Scheunenhöfen. In mehreren Klosterbezirken wurden jedoch Städte gegründet, die zum Teil eng mit dem Kloster verbunden waren und nie eine größere Selbständigkeit erlangten wie etwa in Ellwangen oder Murrhardt. Teilweise entwickelten sich die Städte und die Klöster in unterschiedlicher Weise, sodass die Städte absolut unabhängig wurden und ihre eigenen Territorien bildeten wie etwa in Blaubeuren oder Backnang. Noch komplizierter wurde die Situation, als der Herzog von Württemberg seine Souveränität über ein Kloster nicht durchsetzen konnte, wie beispielsweise in Ellwangen, wo Kloster und Stadt katholisch blieben.

Alois Schmid (München) betrachtet in seinem Aufsatz die Städte und Märkte in der Oberpfalz und geht dabei insbesondere auf die Grundzüge ihrer Entwicklung im späten Mittelalter und in der beginnenden Neuzeit ein, wobei die Oberpfalz eine andere Entwicklung als das Herzogtum Bayern aufweist. Die Region war durch kleinteilige Verwaltungsstrukturen gekennzeichnet, die zur Bildung zahlreicher Städte führten. Diese historische Dimension wird bei der Untersuchung des heutigen Regierungsbezirks deutlich. Dabei ist dieser gekennzeichnet durch eine Vielzahl urbaner Formen, die von den vorherrschenden Kleinstädten bis zur Zentralstadt Regensburg als „Mutter" aller Städte Bayerns, Bischofsresidenz und Reichsstadt, reichen. Der Aufsatz untersucht die zeitlichen Schichten der Urbanisierung in der Region, die verschiedenen beteiligten staatlichen Organe, die physische Gestalt der Städte und die Partizipationsmodi innerhalb der sich entwickelnden Regionalparlamente. In der Tat sind die Auswirkungen der historischen Besonderheiten der Stadtlandschaft der Region auch heute noch spürbar.

Im Beitrag von Thomas J. Hagen (München) werden die zwei Pole Konkurrenz und Kooperation in den Blick genommen. Im besonderen widmet er sich dabei den bambergischen Landstädten im Ringen um politische Geltung und wirtschaftliche Prosperität, aufgezeigt am Beispiel Ebermannstadts und Waischenfelds in der Fränkischen Schweiz. Im Gegensatz zu den großen und freien Reichsstädten erscheinen die fast unzähligen kleineren Landstädte, die es weder zum Status einer Residenz noch in die Reichsfreiheit geschafft haben, oft als unbedeutende oder gar machtlose Mitglieder der politischen und wirtschaftlichen Strukturen des Heiligen Römischen Reiches Deutscher Nation. Ausgehend von den beiden Landstädten Ebermannstadt und Waischenfeld im Hochstift Bamberg zeigt dieser Beitrag, dass Städte als eine Art Zwischenstufe zwischen landwirtschaftlich strukturierten Dörfern und großen Wirtschafts- und Handelszentren tatsächlich wichtige administrative und politische Relevanz erlangen konnten.

Klaus Wolfs (Augsburg) Beitrag mit dem Fokus auf Reichsstadt versus Residenzstadt widmet sich vor allem der bislang wenig systematisch erforschten literaturgeschichtlichen Rolle der Städte mit dem Schwerpunkt auf Bayern. Es lässt sich feststellen, dass Residenzstädte das literarische Leben von Anfang an, seit dem frühen Mittelalter, geprägt haben. Dies gilt aber nicht nur für die althochdeutsche, sondern insbesondere für die mittelhochdeutsche Literatur. Andererseits lässt sich zeigen, dass ein eher urbanes literarisches Leben in den Reichsstädten, vor allem in der Frühen Neuzeit, aber bereits auch im Spätmittelalter, den Motor des literaturgeschichtlichen Fortschritts bestimmte. Erst im 18. Jahrhundert wurde die Adelsresidenz, insofern sie wie im Falle Friedrichs des Großen, aber auch am herzoglichen Hof in Weimar letztlich von der Aufklärung geprägt war, zu einem prägenden Merkmal des literarischen Lebens im gesamten deutschsprachigen Raum.

Franz Körndle (Augsburg) bietet in seinem Aufsatz einen Überblick zu den Stadtpfeifern in der süddeutschen Städtelandschaft unter besonderer Berücksichtigung ihrer Professionalisierung. Bereits vor mehr als hundert Jahren machte der Schweizer Musikforscher Karl Nef auf die kultur- und musikpolitische Bedeutung der Instrumentalensembles in den Städten vom ausgehenden Mittelalter bis in die Neuzeit aufmerksam und beschrieb die Anfänge der Stadtpfeifer als fest angestellte Musiker, ihre Pflichten, Berufsbezeichnungen und das Instrumentarium bis zum Übergang der alten Institution Stadtpfeiferei zum Sinfonieorchester im bürgerlichen Konzertbetrieb. Diesen grundlegenden Informationen fügte erst 1982 Heinrich Schwab eine substantielle Studie zur Professionalisierung des Stadtmusikerberufs hinzu. In jüngerer Zeit widmete sich Helen Coffey dem Klang in den Städten während der Zeit Kaiser Maximilians I. Franz Körndles Beitrag hebt auf neuester archivalischer Grundlage die Rolle der freiberuflich tätigen Musiker im Rahmen des urbanen Musiklebens hervor. Dabei nehmen Reisen und Konflikte einen wichtigen Raum ein.

Dorothea Fastnacht (Marloffstein) beleuchtet in ihrem Beitrag das „Historische Ortsnamenbuch von Bayern“ als Grundlage für die Stadtgeschichtsforschung. Um einen bequemen Einstieg in die stadtgeschichtliche Forschung zu Bayern zu finden, eignen sich die Bände der Reihe „Historisches Ortsnamenbuch von Bayern“ besonders. Nicht ohne Grund ist das Projekt bei der Kommission für bayerische Landesgeschichte in München angesiedelt. Denn obwohl in diesen Büchern die Sprachentwicklung der Ortsnamen im Vordergrund steht, werden die Ortsnamen-Dokumente in aussagekräftigen Regesten präsentiert, die sorgfältig erfasst und genau datiert sind. Ein kompetenter Redakteur sorgt für die Verlässlichkeit aller Informationen und Ergebnisse.

Victor A. Ferretti (Augsburg) bietet einen Einblick zur Dimension Augsburgs in Miguel de Cervantes' „Don Quijote". Die Miszelle ist der Evokation des Augsburger Raums im Zweiten Teil (1615) von Miguel de Cervantes' „Don Quijote" als frühneuzeitlicher Ausdruck eines kontingenten sozio-historischen Imaginären im habsburgischen Spanien gewidmet. Dabei geht es nicht zuletzt um die internationale Wahrnehmung des Augsburger Religionsfriedens.

In seinem Beitrag „Monopol – Krise – Marktdynamik. Zur Entstehung hochmittelalterlicher Salinenstädte im Ostalpenraum unter besonderer Berücksichtigung Reichenhalls" betrachtet Johannes Lang (Bad Reichenhall/Salzburg) unter anderem den enormen Einfluss des Rohstoffs Salz auf die Städtelandschaft in Bayern, welcher bereits gut erforscht ist. Dennoch wird kaum wahrgenommen, dass die Salinenstadt Reichenhall ein herausragendes Beispiel für eine dynamische Stadtentwicklung im 12. Jahrhundert ist. Mit Ausnahme der Bischofsstädte verfügt kein anderer Ort auf dem Gebiet des ehemaligen Herzogtums über so gute schriftliche Quellen für das Früh- und Hochmittelalter wie Reichenhall, sodass ein plastisches Bild der Saline und der Stadt rekonstruiert werden kann. Bis zum Ende des 12. Jahrhunderts hatte Reichenhall als einziger leistungsfähiger Salzproduzent der Ostalpen ein überregionales Monopol, während andere Salinen, wie beispielsweise Hall beim Stift Admont, nur von lokaler Bedeutung waren. Während die Saline Reichenhall infolge des allgemeinen Bevölkerungswachstums ab der Mitte des 12. Jahrhunderts an der Kapazitätsgrenze produzierte, erlebte die Stadt mehrere schwere Krisen. Gegen Ende des Jahrhunderts griffen immer mehr Salzproduzenten auf die neu entwickelte Technologie des Laugungsprozesses zurück, was Reichenhall einen Wettbewerbsnachteil brachte. Die Zerstörung der Stadt während des Krieges im Jahre 1196 setzte eine Marktdynamik frei: Sie gab dem Erzbischof von Salzburg die Möglichkeit, die neue Technik auf dem Dürrnberg einzusetzen und mit Hallein eine neue Salinenstadt nach dem Vorbild Reichenhalls zu schaffen. Während Reichenhall mehr als zwei Jahrzehnte lang niedergerissen und ruiniert war, hatte Hallein hervorragende Voraussetzungen, um schnell zum führenden Salzproduzenten im Alpenraum zu werden. Ein weiterer Produzent, Hall in Tirol, entstand fast zeitgleich mit Hallein und verfügte ebenfalls über sehr gute Standortfaktoren, die auch eine bemerkenswerte städtebauliche Entwicklung ermöglichten.

Senta Herkle (Stuttgart) widmet ihr Thema den konkurrierenden Interessen im Handel süddeutscher Reichsstädte im 18. Jahrhundert. Ausgehend von der Prämisse, dass überregionale Handelsbeziehungen vor Ort austaxiert werden müssen, fragt dieser Aufsatz, wie der (Transit-)Handel im 18. Jahrhundert vor Ort, insbesondere in der Reichsstadt Ulm, organisiert war. Die schriftlichen Aufzeichnungen des Steueramtes veranschaulichen Anpassungen der

Transithandelsorganisation an unterschiedliche politische, wirtschaftliche und technische Rahmenbedingungen. Darüber hinaus erlaubt die archivalische Tradition eine Rekonstruktion der Wechselwirkungen zwischen Ulm und anderen Handelsorten des Reiches, deren Organisationsformen je nach Standort ebenfalls unterschiedlich waren. Diese Unterschiede machen sich zum Beispiel im Umgang der Stadt mit dem Kutschergewerbe bemerkbar.

Die Administration des Bildungswesens in süddeutschen Reichsstädten der Frühen Neuzeit nimmt Wolfgang Mährle (Stuttgart) in seinem Beitrag in den Blick. Im 16. Jahrhundert wurden in zahlreichen süddeutschen Reichsstädten, vor allem in reformatorischen oder bikonfessionellen Städten mit protestantischer Mehrheit, sogenannte Scholarchate eingerichtet. Die Scholarchate waren von den Stadträten eingesetzte Kommissionen zur Verwaltung des gesamten Schulwesens, das sowohl deutsche als auch lateinische Schulen umfasste. Der vorliegende Beitrag analysiert vergleichend die Organisation, das Personal und die Funktionen der Scholarchate in den Reichsstädten Schwabens, Frankens und Bayerns. Er zeigt, dass in protestantischen und bikonfessionellen Städten die Geistlichkeit in der Frühen Neuzeit erheblichen Einfluss auf das Schulwesen ausübte, obwohl die Magistrate für das Schulwesen zuständig waren.

Franz-Rasso Böck (Kempten) widmet seinen Beitrag Kempten im Netz der interurbanen Kommunikation. Das Binnenverhältnis zwischen der Herrschaft der Fürstäbte mit der katholischen Stiftsstadt einerseits und der evangelischen Reichsstadt andererseits im „Mikrokosmos“ Kempten ist ein traditioneller Schwerpunkt der lokalgeschichtlichen Forschung. Über die Außenbeziehungen der Reichsstadt Kempten insbesondere zu anderen Reichsstädten, angefangen bei den schwäbischen Nachbarn, ist jedoch wenig bekannt. Dieses Desiderat ergibt sich in der Regel aus der schlechten Quellenlage zur Reichsstadt Kempten, insbesondere angesichts der sehr lückenhaften Aufzeichnungen der Korrespondenz des Stadtrates, für die es keine Eingangs- und Ausgangsbücher in Form von sogenannten Briefbüchern gibt. Die Wege für die Erforschung der Außenbeziehungen der Reichsstadt Kempten führen daher zu Archiven ehemaliger Reichsstädte, von denen bekannt ist, dass sie – wie Augsburg – über eine größere Anzahl von Briefbüchern verfügen, die auch relevante Korrespondenzen zu Kempten abbilden.

Ivana Ebelová und Zdeněk Hojda (Prag) legen einen Beitrag mit dem Thema „Die Metropole als verdichtete Städtelandschaft. Einige Aspekte zur Entwicklung des frühneuzeitlichen Prag“ vor. Prag als politisch-administratives, wirtschaftliches und kulturelles Zentrum des böhmischen Staates stellt einen komplexen Organismus dar, nämlich ein Konglomerat aus mehreren Städten. Ihre ungleichmäßige Entwicklung, die gegenseitigen Beziehungen und das

Zusammenleben ihrer Bewohner rufen ein breites Spektrum von Themen hervor, von den rechtlichen und administrativen Bedingungen der Prager Städte bis hin zu den baulichen und städtebaulichen Besonderheiten ihrer einzelnen Teile oder funktionalen Unterschiede. Zunächst wird das Augenmerk auf die Jüdische Stadt gerichtet, das wichtigste Zentrum des aschkenasischen Judentums in Mitteleuropa. Dabei war die enge Verbindung zwischen den Prager Juden und ihren christlichen Nachbarn während der Zeit des Rudolfinischen Prags, die als die Blütezeit des Prager Ghettos gilt, besonders intensiv. Schließlich wird ein Blick auf die Prager Stadtlandschaft aus der Sicht von Ausländern einbezogen. Der relativ kurze Zeitraum, in dem Prag zu einer der führenden europäischen Hauptstädte wurde, prägte das Bild der Stadt wesentlich mit. Die grundlegenden städtebaulichen Impulse der Rudolfinischen Zeit waren jedoch auch in den folgenden Jahren noch aktiv, bevor sie durch den Dreißigjährigen Krieg abrupt unterbrochen wurden.

Mit der „Städtelandschaft des mittelalterlichen Livlands“ beschäftigt sich Ilgvars Misāns (Riga) in seinem Beitrag, der die spezifischen Merkmale dieser Städtelandschaft anhand der räumlichen und herrschaftlichen Grundlagen der Urbanisierung sowie der wirtschaftlichen Vernetzung und institutionellen Bindungen der Städte untersucht. So zeigt sich, dass in Livland nicht nur die Intensität der Urbanisierung bescheiden und das städtische Leben verhältnismäßig schwach entwickelt war, sondern auch, dass die politische Zersplitterung des Landes, der geringe Zuzug neuer Einwohner aus deutschen Ländern sowie die demographischen und wirtschaftlichen Krisen die volle Entfaltung des Urbanisierungsprozesses angehalten haben. Außerdem dominierten in der Städtelandschaft drei überragende Fernhandelszentren: Riga, Dorpat und Reval, welche sich auf unterschiedliche Fernhandelsrichtungen spezialisierten und landesweite Wirtschaftsräume im Norden, in der Mitte und im Süden unter sich aufteilten. Die drei größeren Städte kontrollierten somit auch die Kleinstädte und Märkte im jeweiligen wirtschaftlichen Hinterland. Eine wichtige koordinierende Rolle für die Politik livländischer Städte spielte seit der Mitte des 14. Jahrhunderts der Städtetag, der sich auf der lokalen Ebene sowohl mit Angelegenheiten der hansischen Gemeinschaft als auch mit Fragen der Landespolitik beschäftigte. Als Bindeglied zwischen den Städten wirkte diese Institution für die Städtelandschaft integrierend.

Sabine Wüst (Nürnberg) nimmt in ihrem Aufsatz die europäischen Genusslandschaften – Foodscapes – interregional in den Blick. Die regionale Ernährungslandschaft kann mit verschiedenen Ansätzen untersucht werden. Beispielsweise ist die regionale Sprache auf Bayern bezogen nicht das einzige Identitätsmerkmal besonderer Lebensmittel oder regionaler Agrarprodukte. Als ein Produkt unter vielen anderen sind „Roude Ranna“ oder Rote Beete auch über die Grenzen Bayerns hinaus zu finden. Ratsdekrete, Ratsbücher, Lebensmittelgesetze, Haushaltsbücher, aber auch Aufzeichnungen aus Klöstern und Krankenhäusern sowie Rezeptbücher geben Auskunft über Essgewohnheiten und Ernährung im Mittelalter und in der Frühen Neuzeit. Die Foodscape variierte und entwickelte sich über viele Jahre. Nach dem Zustrom von Einwanderern, der in den 1960er Jahren begann, hat sich die Ernährungslandschaft Bayerns, Deutschlands oder auch Europas bis zum Ende des Jahrhunderts drastisch verändert. So nehmen heute beispielsweise Einflüsse der mediterranen Küche einen wichtigen Platz in der Lebensmittelszene ein.

Lina Schröder (Würzburg) beleuchtet in ihrem Beitrag die Topographie als Katalysator für die Stadtentwicklung in den „Lage Landen“. Der Aufsatz stellt die „Lage Landen“ in grober Korrelation zu den heutigen Benelux-Ländern als Makroregion dar, die sich auf ihre mittelalterliche und frühneuzeitliche „Stadtlandschaft“ beziehungsweise „Städtelandschaft“ konzentriert. Nach dem derzeitigen Stand der Forschung ist unklar, welcher der beiden Begriffe, die beide von Franz Irsigler im Jahre 1999 geprägt wurden, anwendbar ist. Es ist jedoch offensichtlich, dass die Topographie und der Mangel an königlicher Souveränität beziehungsweise kirchlicher Macht zur Entwicklung einer föderalistisch regierten Kulturlandschaft in der gesamten Region und ihren Provinzen führte, in der die Städte eine bedeutende Rolle spielten. Im Hinblick auf die Topographie werden die folgenden drei Aspekte diskutiert, die zuvor von zeitgenössischen vormodernen Protagonisten genannt wurden: 1. die unterschiedliche Bodenstruktur in der gesamten Region, 2. das Flusssystem des Rhein-Maas-Schelde-Deltas sowie 3. die Tatsache, dass ein Viertel der „Lage Landen“, insbesondere der nordwestliche Teil, unter dem Meeresspiegel lag. Daraus lässt sich ableiten, dass gerade die Städte als Produktions-, Organisations- und Kommunikationszentren angesichts dieser herausfordernden topographischen Bedingungen als entscheidende Drehscheiben dienten, die das Überleben der Menschen innerhalb dieser komplexen Makroregion sichern.

Sektion 1:
Die süddeutsche Städtelandschaft – landschaftsprägend

Wolfgang Wüst

Frühmoderne „Städtebänke“ – Gremien urbaner Interessenvertreter in Kreis- und Reichstagen

Abstract: "Städtelandschaft" is initially intended to describe research into the history of cities and markets on an international and, for our request, on an especially southern German level. What does "Städtelandschaft" actually mean? The renowned historian Franz Irsigler understands "urban landscape" to mean a medium-sized spatial unit with varying degrees of urbanisation; and "cityscape" to mean that the towns and cities dominate specific spatial units. Man has always remained human, but his environment, including the urban landscape, has changed over the centuries. In my contribution, this is to be discussed in the context of urban correspondence at imperial and territorial Councils. Political networks become transparent in the Councils' correspondence. There was no lack of meeting places and times for the councils examined. For example, the number of council sessions ("Kreistage") in Nuremberg in 1783 and 1785 exceeded 2,000! Meetings were held often and for long periods of time, which opened up a space of close communication for the delegations involved. It was a place where inter-urban reforms were decided upon and where plans were corrected to the best of one's ability which put cities at a financial disadvantage. This communicative achievement in seeking a consensus is not diminished by the fact that almost all cities could hardly ever be represented by their own deputies. Resolutions passed by the Imperial Assembly ("Reichstag") and the "Kreistag" also reached the inferior imperial cities ("Reichsstädte"), which were often described as "sleepy" in former research. This group of cities in particular was sometimes more space-defining than the early modern metropolises with their strongly developed self-interests.

1. Grundlagen – Gemeinsames Handeln

„*Wir werden ia beinahe u^e^berall als Stiefkinder behandelt. Alle Fu^e^rsten sperren uns ihre Lande.*“[1] Diese Einschätzung der Nürnberger Marktvorsteher gegenüber dem Rat vom 13. Dezember 1785 war noch vor der 1796 vollzogenen Annexion großer Teile des reichsstädtischen Landgebiets – darunter befanden sich auch die

1 Bayerische Staatsbibliothek München (künftig: BSB), Res. 4° Eur. 399 25: Ludwig Friedrich Grub, Ueber einige Reichsstädte Teutschlands. Ein Wort zu seiner Zeit, [s.l.] 1786, S. 2, VD18 10659595.

Nürnberger Vorstädte Wöhrd und Gostenhof – protokolliert worden. Sie ist Ausdruck des fortgesetzten Mediatisierungsdrucks seitens der Fürstenherrschaft, der sich 1792 mit dem Erbfall der süddeutschen Markgraftümer Brandenburg-Ansbach und Brandenburg-Bayreuth an Preußen und der folgenden Berufung Karl August von Hardenbergs (1750–1822)[2] zum leitenden Minister in Franken erheblich verstärkte. Und es war sicher kein Zufall, dass ausgerechnet ein „Fürstendiener" *in personam* des Thurn-und-Taxis'schen Hofrats Ludwig Friedrich Grub (1760–1818)[3] seiner Darstellung zu den „*teutschen*" Reichsstädten den Stoßseufzer der Nürnberger Marktvorsteher voranstellte, befand sich doch auch die Reichsstadt Regensburg im Vorhof des Fürstenhofes der Thurn und Taxis in St. Emmeram.

Reichsstädte waren in die politische Defensive geraten und ihre Wirtschaftskraft hatte den Zenit im 18. Jahrhundert längst überschritten. Dennoch – oder gerade deshalb – kam es gegen Ende des tintenklecksenden Säkulums unter den nicht armierten Ständen des Alten Reiches zu einer einzigartigen Solidarisierungswelle mit den bedrohten Reichsstädten. Gleichwohl wies Georg Schmidt bereits 1984 in seiner einschlägigen Tübinger Dissertation zum Thema „Städtetag in der Reichsverfassung" darauf hin, dass die territoriale und rechtliche Gemengelage zwischen Städten und Fürstenländern und die Erfahrungen aus den spätmittelalterlichen Städtekriegen die Politik in den frühmodernen Corpora auf den Städtebänken mitbestimmte. Selten stand man deshalb bedrohten reichsständischen Kommunen „demonstrativ und provokativ" bei, sondern wählte diplomatische Formen der Unterstützung, die den Dialog mit expansiven Nachbarländern nicht abbrechen ließen.[4] Das Netzwerk der Reichsstädte verdichtete sich dennoch, wichtige innere Verfassungsreformen wie der Nürnberger Grundvertrag von 1794[5] wurden

---

2 Eberhard Weis, Hardenberg und Montgelas. Versuch eines Vergleichs ihrer Persönlichkeiten und ihrer Politik, in: Jahrbuch des Historischen Kollegs 1997, S. 3–20; Peter G. Thielen, Karl August von Hardenberg, 1750–1822 – Eine Biographie, Köln/Berlin 1967.

3 Volker Grub, Ludwig Friedrich Grub (1760–1818), Thurn und Taxisscher Hofrat. Ein Beitrag zur Geschichte des Hauses Thurn und Taxis in der napoleonischen Zeit, in: Verhandlungen des Historischen Vereins für Oberpfalz und Regensburg 149, 2009, S. 85–166.

4 Georg Schmidt, Der Städtetag in der Reichsverfassung. Eine Untersuchung zur korporativen Politik der Freien und Reichsstädte in der ersten Hälfte des 16. Jahrhunderts (Veröffentlichungen des Instituts für Europäische Geschichte Mainz, Abtl. Universalgeschichte 113 = Beiträge zur Sozial- und Verfassungsgeschichte des Alten Reiches 5), Stuttgart 1984, S. 189.

5 Stadtarchiv (künftig: StadtA) Nürnberg, B 3, B 11 und B 37; Rudolf Endres, Die Rolle der Kaufmannschaft im Nürnberger Verfassungsstreit am Ende des Alten Reiches,

teils auf Veranlassung kaiserlicher Untersuchungskommissionen, teils auf innerstädtische Initiative auf den Weg gebracht und die Reichsgerichte nahmen sich der klagenden Räte an. In den Jahrzehnten vor der sogenannten Napoleonischen Flurbereinigung entstand ein (reichs-)städtischer Kommunikations- und Rechtsraum, den man als süddeutsche Städtelandschaft beschreiben und definieren kann.

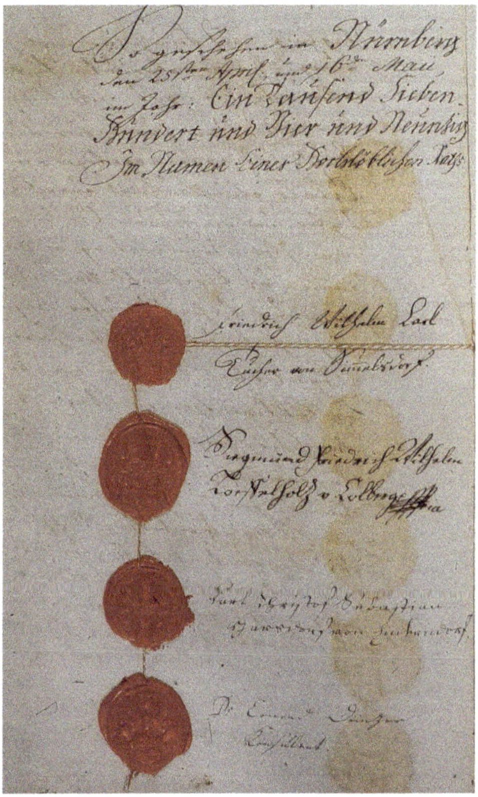
So geschehen in Nürnberg
den 25ten April und 16ten May
im Jahr: Ein Tausend Sieben
Hundert und Vier und Neunzig
Im Namen Eines Hochlöblichen Raths

Abb. 7: *Nürnberger Grundvertrag von 1794, Siegelseite. Bildnachweis: StadtA Nürnberg, B 3; Michael Diefenbacher / Rudolf Endres (Hg.), Nürnberger Stadtlexikon, Nürnberg 1999, S. 390.*

in: Jahrbuch für fränkische Landesforschung (künftig: JfL) 45, 1985, S. 125–167; Franz Buhl, Der Niedergang der reichsstädtischen Finanzwirtschaft und die kaiserliche Subdelegationskommission von 1797–1806, in: Mitteilungen des Vereins für Geschichte der Stadt Nürnberg 26, 1926, S. 111–278.

Die kommunikativ-korporative Geschlossenheit vieler Städte – insbesondere der Reichsstädte – werden wir exemplarisch an der Gremienarbeit, am Stimmverhalten und den Voten ausgewählter Stichjahre im Reichstag, im Kreistag und, soweit es landsässige Gemeinden betrifft, im Landtag untersuchen. Die in ausführlichen Protokollserien festgehaltene Geschlossenheit städtischer Delegierter[6] auf den ihnen zugewiesenen Sitzbänken ist erstaunlich und erklärungsbedürftig.

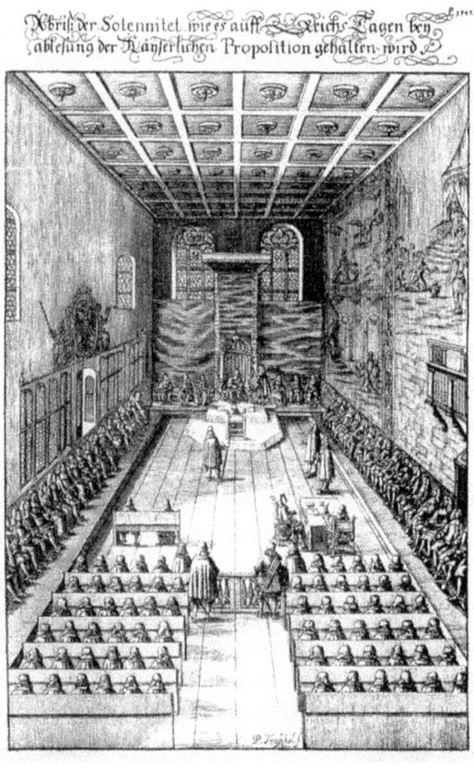

Abb. 8: *Kurien des Reichstages des Heiligen Römischen Reiches („Abriß der Solennitet wie es auff Reichs-Tagen bey ablesung der Kayserlichen Proposition gehalten wird."), Kupferstich von Peter Troschel, Nürnberg 1675. Die Vertreter der Reichsstädte verteilten sich über die zwölf Bänke im Vordergrund. Vor den reichsstädtischen Delegierten saßen die Reichstagsschreiber. Bildnachweis: Wikimedia Commons (gemeinfrei).*

6 Für die Landtage wurden auch die landestypischen altbayerischen Märkte untersucht. Vgl. dazu: Wilhelm Liebhart, Zur spätmittelalterlichen landesherrlichen Marktgründungspolitik in Ober- und Niederbayern, in: Pankraz Fried (Hg.), Bayerisch-schwäbische Landesgeschichte an der Universität Augsburg 1975–1977 (Augsburger

Die vielschichtige äußere Klammer einer Städtelandschaft besteht – so meine These – aus dem gemeinsamen Planen und Handeln unter den in räumlicher Perspektive keineswegs homogenen Sitzgruppen städtischer Gesandter und Abgeordneter in frühmodernen „Parlamenten". Die Schriftwechsel der Gesandtschaften aus den Reichs-, Kreis- und Landtagen mit ihrer jeweiligen Ratsherrschaft zeigen, dass es jenseits lokaler Interessen in der Regel auch um kommunale und urbane Anliegen in genereller Perspektive ging. Dabei konnte es im Reichs- oder Städtetag auch zu Solidaritätsaktionen kommen, die weit entfernte Städte verbanden. So ging es für die Nürnberger Delegation auf dem Wormser Städtetag 1606 um die Unterstützung der Städte Magdeburg und Braunschweig gegen Übergriffe von Herzog Heinrich Julius von Braunschweig-Wolfenbüttel (1564–1613).[7] 1607/08 standen dann – so die Nürnberger Reichstagsabgeordneten – während eines Regensburger Reichstags die Religionsbeschwerden der Stadt Aachen im Mittelpunkt.[8] 1613/14 nahmen Nürnbergs Diplomaten wieder in Worms die Religionsbeschwerden der schwäbischen Reichsstädte Kempten und Biberach an der Riß zur Kenntnis.[9] Kein Widerspruch zur These eines gemeinsamen „Bank"-Interesses ist es, das spezifische Umfeld vor Ort in die Überlegungen zur frühmodernen Städtepolitik aufzunehmen.

Städte hatten wie andere Reichs- und Landstände auch an ihren Grenzen außerhalb der Mauerringe territoriale Anrainer, mit denen zunächst bilaterale Verträge abzuschließen waren. Die Marktlage war selbst für die Gruppe der süddeutschen Handelsstädte keineswegs uniform; Konkurrenten in und außerhalb der Bannmeilen agierten primär im singulären und nicht im landschaftlichen Interesse. Politisches Unheil drohte aber auch in und aus den Gemeinden selbst. Reformkreise gingen gegen ihre oligarchischen Ratsherrschaften vor Gericht. Bischöfe und Ratsregimenter stritten sich über lange Zeit um die Stadtherrschaft; Köln contra Köln,[10] Augsburg contra

---

Beiträge zur Landesgeschichte Bayerisch-Schwabens 1), Sigmaringen 1979, S. 141–152; Wilhelm Liebhart, Kloster, Wallfahrt und Markt in Oberbayern, in: Studien und Mitteilungen zur Geschichte des Benediktinerordens und seiner Zweige 88, 1977, S. 141–152.

7 Staatsarchiv (künftig: StaatsA) Nürnberg, Reichsstadt Nürnberg (Rep. 24/b), Reichstagsakten, Nr. 154, Schriftwechsel zwischen Nürnberg, Magdeburg und Braunschweig, 1604–1606.

8 Ebd., Nr. 165–167, Religionsbeschwerden der Stadt Aachen, Bd. 1: 1560–1581, Bd. 2: 1581/82, Bd. 3: 1583–1608.

9 StaatsA Nürnberg, Reichsstadt Nürnberg (Rep. 24/b), Reichstagsakten, Nr. 184, Korrespondenz 1613/14.

10 Hans-Michel Becker, Köln contra Köln: Von den wechselvollen Beziehungen der Stadt zu ihren Erzbischöfen und Kurfürsten, Köln 1992; Hugo Stehkämper, „Köln contra Köln". Erzbischöfe und Bürger im Ringen um die Kölner Stadtautonomie,

Augsburg[11] oder Weißenburg contra Weißenburg[12] wären hierfür repräsentative Beispiele. Ferner sprechen die vor allem während der Aufklärung publizierten Vorbehalte gegen die „Erstarrung" ursprünglich dynamischer Gemeinwesen, die ihre einstige Geschlossenheit als Folge von Vetternwirtschaft, Korruption, Ämterkauf und Verschwendung aufs Spiel setzten.

Der Chronist Ludwig Friedrich Grub beschrieb 1786 die Kardinalsünden selbstgefälliger städtischer Gesellschaftskreise, die an interstädtischer Kooperation und Vernetzung wenig oder kein Interesse zeigten. Zu den Kritikpunkten zählte auch die in städtischen Policey-Ordnungen[13] stets wiederholte, aber selten befolgte Warnung vor Verschwendung, Ausschweifung, Vorteilnahme und Eigennutz. Grub meinte dazu 1786: „*Der Grundcaracterzug der heutigen Reichsstädter bestehet in dem Hang, in der Liebe zur Schwelgerei und zum Wohlleben. Sterb- und Tauffälle, Bedienstungen, Beförderungen, Rechnungsverhöre, Apothecker-Visitationen, Zehentverleihungen etc. werden nach wohlhergebrachter alter Sitte mit Schmäusen begleitet, wobei meistens auf Rechnung des städtischen Aerarii gezechet wird. In der fränkischen Reichsstadt R. v. d. T. fangen diese Gastereien mit dem Frühlinge an, und währen biß späte in Herbst fort. Selbst das Gut, das iedem heilig seyn soll, das Gut der Armen wird dabei nicht geschont, indem bei der Ablegung des dasigen Hospitalrechnung zwei (oder gar vier – ich erinnere mich der Anzahl nicht mehr genau) Schmäuse gehalten werden, wovon ieder nach*

---

in: Franz-Heinz Hye (Hg.), Stadt und Kirche (Beiträge zur Geschichte der Städte Mitteleuropas 13), Linz 1995, S. 53–82.

11 Wolfgang Wüst, Augsburger Bürgerschaft, Domkapitel und Fürstbischöfe im 17. und 18. Jahrhundert: geistlich-weltliche Allianz oder politisch-ständischer Gegensatz? in: Wolfram Baer / Bernhard Kirchgässner (Hg.), Stadt und Bischof (Südwestdeutscher Arbeitskreis für Stadtgeschichtsforschung 24), Sigmaringen 1988, S. 66–96. An zeitgenössischer Prozessliteratur: Johann Philipp Lupin (Ratskonsulent), Actenmäßiger Conspectus des bey dem hochpreißlichen Kaiserlichen und Reichs-Cammergericht zu Wezlar in prätendirten Mandatssachen Sr. Hochfürstl. Durchl. des Hrn. Bischofs zu Augsburgs und dasigen Hochw. Domkapitels contra die wohllöbl. Reichsstadt Augsburg fürwaltenden Processus die Superioritatem territorialem & inde dependentia jura in denen zur kaiserl. Reichs und Landvogtey Augsburg gehörigen Dörfern Gersthofen und Lankweid betreffend, Wetzlar 1760.

12 Gustav Mödl, „Weissenburg contra Weissenburg". Ein Beitrag zum Verhältnis zwischen Rat und Bürgerschaft, in: Uuizinburc – Weissenburg, 867–1967. Beiträge zur Stadtgeschichte, Weißenburg in Bayern 1967, S. 105–110.

13 Wolfgang Wüst (Hg.), Die „*gute*" Policey im Reichskreis. Zur frühmodernen Normensetzung in den Kernregionen des Alten Reiches, Bd. 7: Policeyordnungen in den fränkischen Reichsstädten Nürnberg, Rothenburg ob der Tauber, Schweinfurt, Weißenburg und (Bad) Windsheim. Ein Quellenwerk, Erlangen 2015.

sicheren Nachrichten wenigstens auf 50. Th[a]*l*[er] *zu stehen kommt.*“[14] Weitere Kritikpunkte an der Reformwilligkeit in der Städtelandschaft war der Sittenverfall mit Blick auf Alkoholkonsum und außerehelichem „*Beischlaf*“, gemessen an der Zahl unehelicher Geburten und der Zunahme von Geschlechtskrankheiten.[15] „*Es ist natürlich*“ – so Ludwig Friedrich Grub – „*daß da wo Bacchus und Ceres verehrt werden, auch der unreinen Venus ein Altar gewidmet sey.*“[16] Es stellt sich nun die Frage, wie ähnlich sich Städte in einer Städtelandschaft waren, ob die von Grub genannten Verfallserscheinungen urbaner Kultur und Bildung weniger singulär als vielmehr gemeinschaftlich zu deuten sind. Fest- und Jahresessen auf Kosten des jeweiligen Amts- oder Stadtsäckels waren europaweit an Neujahr – der Beginn des städtischen Wirtschaftsjahres fiel keineswegs immer auf den 1. Januar –, bei Rechnungsabschlüssen, Namenstagen, Patrozinien oder Jubiläen üblich. In Nürnberg – ein Beispiel unter vielen – wurde den Finanzherren im Losungsamt mit ihren Familien am 13. April 1565 während und nach dem üppigen „*Rechnung Mahl*“ auch noch ordentlich zum Mitnehmen eingepackt. Es handelte sich, obwohl der Tag auf einen Dienstag fiel, um ein traditionelles Fischessen und die kredenzten Speisen konnten alternativ in Geld ausgezahlt werden. „*Erstlich 72 £ Hecht zum Ain sultzen* […]. *Mer 57 £ Karpffen ain £ per 18 d tut fl 60 ort 18 d. Mer 10 ½ £ Esch*[en] […]. *Mer 18 £ Orffen*[17] *zum haimschicken vnd auff die Diesch* […]. *Mer zwen mass Kugelhaubtt*[18] *thut fl 1. Mer dreÿ halbe stidla Grundt*[19] *tut 4 ort 10 d. Item das Gewürtz so zue der Vischsultz vnd andern kumpt als Safferian, Zimett, Pfeffer, Ingwer, Negelein, Muschatnus* […], *auch hauffen Plassen*[20] *tut 9 fl 4 ort 22 d. Mer 21 £ Zucker zue der Vieschsultzen* […]. *Mer 68 £ Venedisch Mandel* […]. *Mer zue der Vieschsultzen ist kumen 4 mas Maluisier* […]. *Mer zum Sultzen 35 mass Rheinisch Wein zue 26 d thutt 3 fl 5 ort 4 d.*“ An weiteren Produkten und Beigaben wurden laut Rechnung für diese Tafel aufgetragen: Schmalz, Weinbeeren, „*Bomerantzen*“, „*Krenn*“, Schnecken, roter und weißer Essig, Salz, „*Fischgestraidt*“ und jede Menge „*Handtwasser*“ zum

14 Ludwig Friedrich Grub, Ueber einige Reichsstädte Teutschlands (wie Anm. 1), S. 12.

15 Wolfgang Wüst, Die Franzosenkrankheit („maladie française“) und policeyliche Vorsorge – Seuchen und Epidemien im frühmodernen Diskurs, in: Michaela Sohn-Kronthaler / Jacques Verger (Hg.), Europa und Memoria/Europe et Mémoire. Festschrift für Andreas Sohn zum 60. Geburtstag/Mélanges offerts à Andreas Sohn à l'occasion de son 60ème anniversaire, St. Ottilien 2019, S. 427–450, 606.

16 Ludwig Friedrich Grub, Ueber einige Reichsstädte Teutschlands (wie Anm. 1), S. 16.

17 Orfen zählen zu den karpfenartigen Fischen.

18 Fischart.

19 Gründling.

20 Brassen[?].

Händewaschen. In der Summe zahlte die Stadt über das Losungsamt über 90 Gulden.[21] Ein weiterer Höhepunkt im Zyklus der annualen Festessen war 1569 in Nürnberg das „*Kaÿsser Mhall*“[22] im reichsstädtischen Waldamt St. Sebald.

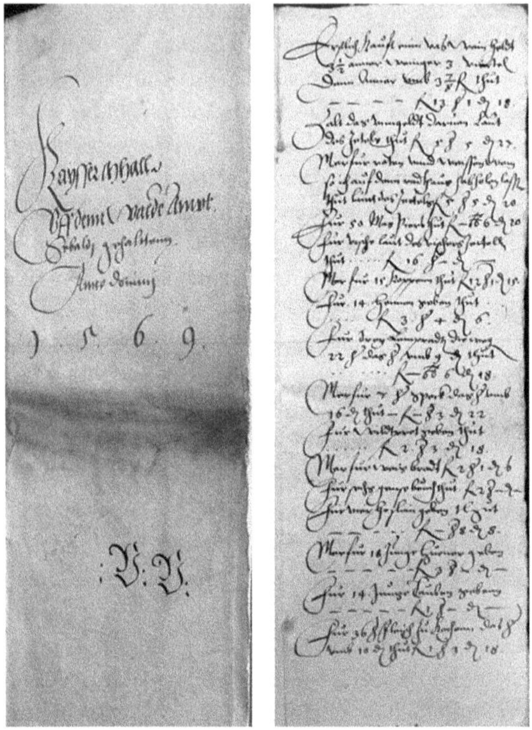

Abb. 9 a und 9 b: *Kaiseressen („Kaÿsser Mhall“) im Nürnberger Reichswald St. Sebald, 1569. Bildnachweis: StaatsA Nürnberg, Reichsstadt Nürnberg, Losungsamt, Stadtrechnungsbelege, Bündel 251.*

---

21 StaatsA Nürnberg, Reichsstadt Nürnberg, Losungsamt, Stadtrechnungsbelege, Büschel 365. Losungsämter als Einnehmer- oder oberste Finanzbehörden waren kein Nürnberger Spezifikum. Vgl. Jaroslav Čechura, Srovnání berních knih města Brna z let 1365 a 1422. Poznámky k metodice studia pramenů hromadné povahy (= Ein Vergleich der Losungsbücher der Stadt Brünn aus den Jahren 1365 und 1422. Notizen zur Methodik des Studiums von Quellen mit Massencharakter), in: Časopis Matice moravské 117/2, 1998, S. 341–352. Für das Zitat: Die Klein- und Großschreibung wurde bei den Tafelprodukten modernisiert.

22 StaatsA Nürnberg, Reichsstadt Nürnberg, Losungsamt, Stadtrechnungsbelege, Bündel 251.

Geladen waren 67 Personen aus den Rats- und Forstgremien in der waldreichen Pegnitzstadt. Auf „*Herrentischen*" wurde in Fülle alles kredenzt, was damals gut und teuer war. An Fleischsorten tischten Küchenhelfer auf gebratene, gekochte oder gesottene Kapaune, junge Tauben, Hühner, Ziegen („*capra*"), Wildbret und Rehschlegel sowie jede Menge an Speck. Dazu aß man Krebse, Schnecken, Fische (Hechte, Orfen, Karpfen), Pistazien, Nüsse, Weiß- und Schwarzbrot sowie Gemüse – vor allem Kichererbsen – aus der Region. Unter den leckeren Nachspeisen gab es neben verschiedenen Käsesorten Erdbeeren („*erper*"), Limetten, gezuckerte Kuchen und Lebkuchen. Auch Rosenwasser durfte am Tisch nicht fehlen. Zum „*Zutrinken*" schenkte man steuerpflichtigen exklusiven Weiß- und Rotwein aus dem mainfränkischen Klingenberg sowie Unmengen an Bier („*50 Mas Piers*") aus.[23] Sabine Wüst ging für unseren Tagungsband der Frage nach, ob und wie sich solche „Foodscapes" insbesondere mit Blick auf die Wurst- und Käseherstellung generierten. Im Kontext historisch gewachsener Städtelandschaften wären urbane Konsum- und Essgewohnheiten äußerst aufschlussreich, zumal sie sich nicht nur in Kochbüchern, sondern vor allem in kommunalen Rechnungen widerspiegeln. Der Landschaftsvergleich städtischer Essens- und Speisegewohnheiten sprengt hier bereits im Mittelalter seine Grenzen, so tischte man im 15. Jahrhundert beispielsweise bei den Empfängen für die Hochmeister des Deutschen Ordens in Preußen ähnliche Fischgerichte wie in Nürnberg und anderen Reichsstädten auf.[24] In meinem Beitrag geht es aber in erster Linie nicht um die Frage, ob Städte im Vergleich ähnliche Lebens-, Ess-, Wohn- und Arbeitsgewohnheiten ihrer Bürger aufweisen. Die zentrale Frage und These lautet vielmehr, ob in Verfassungsgremien, Gerichten und vormodernen Parlamenten (Kreistage, Reichstage, Landtage) Städte gemeinsame politische und steuerökonomische Interessen vertraten.

2. Städte im Kreistag

Über die Teilnahme und die Voten des Städtegremiums in den Versammlungen der Reichskreise ist – im Gegensatz zu den Reichs- und Landtagen – trotz ausgezeichneter archivalischer Überlieferung nur wenig bekannt. Eine Quellengruppe, aus der solidarisches urbanes Stimmverhalten seitens der Reichsstädte

23 Ebd.

24 Andrzej Radzimiński, Küche und Lebensmittelkonsum der Ritter und Geistlichen des Deutschordens in Preußen, in: Sabine Wüst (Hg.), Schätze der Welt aus landeshistorischer Perspektive. Festschrift zum 65. Geburtstag von Wolfgang Wüst, St. Ottilien 2018, S. 313–320, hier: S. 315.

hervorgeht, sind ungezählte Protokollserien[25] der Reichskreise. Insbesondere traf dies für den Schwäbischen und Fränkischen Reichskreis zu, hatten doch während der Kreistage in Schwaben bis zu 35[26] und in Franken fünf[27] Reichsstädte

25 Die dichte Überlieferung der Reichsstadt Nürnberg als ein süddeutsches Beispiel sieht zu den Sitzungsprotokollen für die Kreistage bis 1705 wie folgt aus:
StaatsA Nürnberg, Reichsstadt Nürnberg (Rep. 22), Kreistagsakten, Nr. 1: 1525–154, Nr. 2: 1542–1549, Nr. 3: 1553–1563, Nr. 4: 1564–1567, Nr. 5: 1568–1569, Nr. 6: 1570–1576, Nr. 7: 1577–1591 [fehlt], Nr. 8: 1610–1619, Nr. 9: 1620–1638, Nr. 10: 1638–1639, Nr. 11: 1639–1642, Nr. 12: 1642–1643, Nr. 13: 1643–1644, Nr. 14: 1644–1645, Nr. 15: 1645–1646, Nr. 16: 1645–Dez. 1650, Nr. 17: 1651–1654, Nr. 18: 1663–1664, Nr. 19: 1664, Nr. 20: 1664, Nr. 21: 1664, Nr. 22: 1664, Nr. 23/Tom. 23a: 1672, Nr. 24/Tom. 23b: 1672, Nr. 25: 1673, Nr. 26: 1673, Nr. 27: 1673, Nr. 28: 1674, Nr. 29: 1674, Nr. 30: 1675, Nr. 31: 1675, Nr. 32: 1675, Nr. 33: 1675, Nr. 34: 1676, Nr. 35: 1676, Nr. 36: 1676, Nr. 37: 1677, Nr. 38: 1677–1678, Nr. 39: 1678, Nr. 40: 1678, Nr. 41: 1679, Nr. 42: 1679, Nr. 43: 1679, Nr. 44: 1681, Nr. 45: 1681, Nr. 46: 1682, Nr. 47: 1682, Nr. 48: 1682, Nr. 49: 1683, Nr. 50: 1683, Nr. 51: 1683, Nr. 52: 1684, Nr. 53: 1684, Nr. 54: 1684, Nr. 55: 1684, Nr. 56: 1685, Nr. 57: 1685, Nr. 58: 1686, Nr. 59: 1686, Nr. 60: 1687, Nr. 61: 1687, Nr. 62: 1688, Nr. 63: 1688/89, Nr. 64: 1688/89, Nr. 65: 1688/89, Nr. 66: 1688/89, Nr. 67: 1689, Nr. 68: 1689, Nr. 69: 1689, Nr. 70: 1689, Nr. 71: 1689, Nr. 72: 1690, Nr. 73: 1690, Nr. 74: 1690, Nr. 75: 1690, Nr. 76: 1690, Nr. 77: 1691, Nr. 78: 1691, Nr. 79: 1691, Nr. 80: 1691, Nr. 81: 1691, Nr. 82: 1692, Nr. 83: 1692, Nr. 84: 1692, Nr. 85: 1692, Nr. 86: 1692, Nr. 87: 1692, Nr. 88: 1692, Nr. 89: 1693, Nr. 90: 1693, Nr. 91: 1693, Nr. 92: 1693, Nr. 93: 1693/94, Nr. 94: 1693/94, Nr. 95: 1694, Nr. 96: 1694, Nr. 97: 1694, Nr. 98: 1694, Nr. 99: 1694, Nr. 100: 1694, Nr. 101: 1695, Nr. 102: 1695, Nr. 103: 1695, Nr. 104: 1695, Nr. 105: 1696, Nr. 106: 1696, Nr. 107: 1696, Nr. 108: 1696, Nr. 109: 1697, Nr. 110: 1697, Nr. 111: 1697, [Nr. 112–114 fehlen], Nr. 115: 1698, Nr. 116: 1698, Nr. 117: 1698, Nr. 118: 1699, Nr. 119: 1699, Nr. 120: 1699, Nr. 121: 1700/01, Nr. 122: 1701, Nr. 123: 1701, Nr. 124: 1701, Nr. 125: 1701, Nr. 126: 1701, Nr. 127: 1701/02, Nr. 128: 1702, Nr. 129: 1702, Nr. 130: 1702, Nr. 131: 1702/03, Nr. 132: 1702/03, Nr. 133: 1703, Nr. 134: 1703, Nr. 135: 1703, Nr. 136: 1703, Nr. 137: 1703, Nr. 138: 1703, Nr. 139: 1703/04, Nr. 140: 1704, Nr. 141: 1704, Nr. 142: 1704, Nr. 143: 1704, Nr. 144: 1704, Nr. 145: 1704, Nr. 146: 1704, Nr. 147: 1705, Nr. 148: 1705, Nr. 149: 1705 und Nr. 150: 1705. Berücksichtigt man auch die Kreistage bis zum Ende des Alten Reiches, endet die Serie der Kreistagsakten mit Nr. 345: 1792–1807.

26 Aalen, Augsburg, Biberach, Bopfingen, Buchau, Buchhorn (heutiges Friedrichshafen), Esslingen, Dinkelsbühl, Donauwörth (bis 1607), Gengenbach, Giengen, Heilbronn, Isny, Kaufbeuren, Kempten, Konstanz (bis 1548), Leutkirch, Lindau, Memmingen, Nördlingen, Offenburg, Pfullendorf, Ravensburg, Reutlingen, Rottweil, Schaffhausen (bis 1501/45), Schwäbisch Gmünd, Schwäbisch Hall, St. Gallen (bis 1544), Überlingen, Ulm, Wangen, Weil der Stadt, Wimpfen und Zell am Harmersbach.

27 Nürnberg, Rothenburg ob der Tauber, Schweinfurt, Weißenburg und Windsheim.

Stimmrecht auf der Städtebank.[28] Im reichsstädtereichen Schwaben wurden für eine 1789/90 in Ulm[29] tagende Kreisversammlung gedruckte Steuerlisten in Umlauf gesetzt, auf der man die Städtebank mit der Summe des jeweiligen „*Kreiß Extra-Ordinariums*" als Gruppe führte.

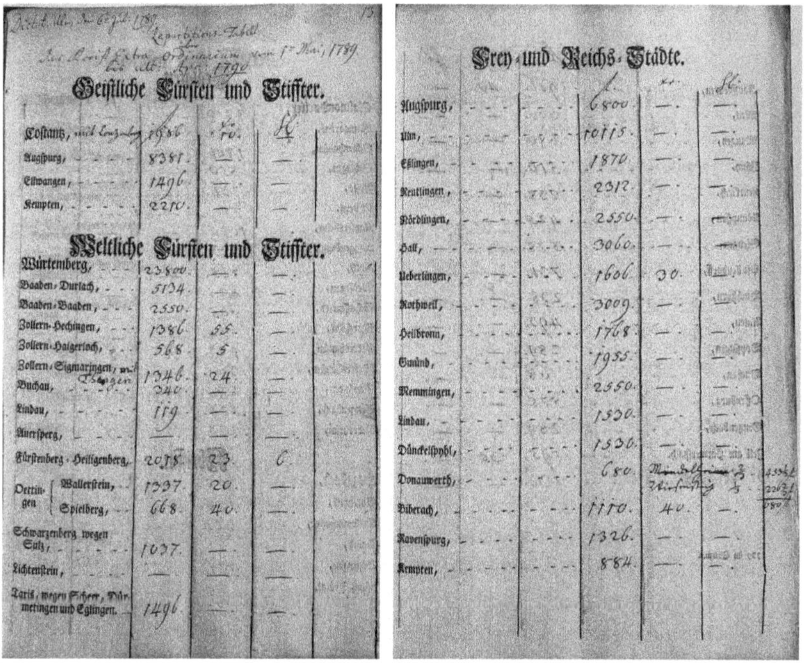

Geistliche Fürsten und Stiffter.

Costantz,	1986	10	
Augspurg,	8381	—	—
Ellwangen,	1496	—	—
Kempten,	2210	—	—

Weltliche Fürsten und Stiffter.

Würtemberg,	23800	—	—
Baaden-Durlach,	5134	—	—
Baaden-Baaden,	2550	—	—
Zollern-Hechingen,	1386	55	—
Zollern-Haigerloch,	568	5	—
Zollern-Sigmaringen,	1346	24	—
Buchau,	340	—	—
Lindau,	119	—	—
Auersperg,	—	—	—
Fürstenberg-Heiligenberg,	2038	23	6
Oettingen Wallerstein,	1337	20	—
Oettingen Spielberg,	668	40	—
Schwarzenberg wegen Sulz,	1037	—	—
Lichtenstein,	—	—	—
Taris, wegen Scheer, Dürmetingen und Eglingen.	1496	—	—

Frey- und Reichs-Städte.

Augspurg,	6800	—	—
Ulm,	10115	—	—
Eßlingen,	1870	—	—
Reutlingen,	2312	—	—
Nördlingen,	2550	—	
Hall,	3060	—	
Ueberlingen,	1606	30	
Rothweil,	3009	—	
Heilbronn,	1768	—	—
Gmünd,	1955	—	—
Memmingen,	2550	—	—
Lindau,	1530	—	—
Dünckelspyhl,	1530	—	—
Donauwerth,	680		
Biberach,	1110	40	—
Ravenspurg,	1326	—	—
Kempten,	884	—	—

Abb. 10 a und 10 b: *„Repartitions-Tabell" über zusätzliche Reichs- und Kreissteuern während des Ulmer Reichskreistages, 6.7.1789. Bildnachweis: StaatsA Augsburg, Reichsstadt Nördlingen, Münchner Bestand, Nr. 693.*

---

28 Vgl. zur Charakterisierung der Reichsstädte des Heiligen Römischen Reiches neuerdings: Peter Bühner, Die Freien und Reichsstädte des Heiligen Römischen Reiches. Kleines Repertorium (Schriftenreihe der Friedrich-Christian-Lesser-Stiftung 38), Petersberg 2019.

29 Hans Eugen Specker, Die Reichsstadt Ulm als Tagungsort des Schwäbischen Reichskreises, in: Wolfgang Wüst (Hg.), Reichskreis und Territorium. Die Herrschaft über der Herrschaft? Supraterritoriale Tendenzen in Politik, Kultur, Wirtschaft und Gesellschaft. Ein Vergleich süddeutscher Reichskreise (Augsburger Beiträge zur Landesgeschichte Bayerisch-Schwabens 7), Stuttgart 2000, S. 179–196.

Aufschlussreich für das landschaftliche Stimmverhalten in der Zeit vom 1. Mai 1789 bis Ende April des Folgejahres ist der Umstand, dass – meist aus Kostengründen – nicht alle Reichsstädte eigene Delegierte nach Ulm schickten. Es waren nun gerade die kleineren Städte, die sich dem Votum des mehrfach beauftragten Kreisdiplomaten fügten. Während Augsburg mit dem Geheimen Rat und Steuereinnehmer Moritz Ludwig von Carl zu Mühlbach und dem Ratskonsulenten Johann Friedrich von Tröltsch zweifach und das gastgebende Ulm mit dem Oberrichter Freiherrn Marcus Theodosius von Welser, mit dem Geheimen Rat Johann August Schneidenlay sowie den Ratskonsulenten Dr. Gottlob Dietrich Miller und Dr. Johann Gottfried Benjamin Härlin vierfach vertreten war, zählten andere auf die Hilfe Augsburgs oder Ulms. Ulm vertrat 1789/90 die Interessen von Aalen, Bopfingen, Esslingen, Reutlingen, Schwäbisch Hall, Heilbronn, Lindau, Kempten, Kaufbeuren, Weil, Wangen, Isny und Leutkirch. Augsburg votierte für Nördlingen, Schwäbisch Gmünd und Memmingen. Die Reichsstadt Überlingen stellte mit seinem Bürgermeister Freiherrn Franz Conrad Lenz von Lenzenfeld einen eigenen Kreisdiplomaten, der auch für Ravensburg, Pfullendorf, Buchau, Offenburg und Zell am Harmersbach votierte. Biberach, Dinkelsbühl – votierte auch für Giengen – und Rottweil leisteten sich einen eigenen Kreisgesandten. Die schwäbische Reichsstadt Wimpfen hatte keine Vertretung; der Protokolleintrag lautete: „*Nemo*“. Keiner der städtischen Kreisdelegierten hatte übrigens ein Mandat für eines der zahlreichen Fürstentümer, Grafschaften, Reichsklöster oder Reichsstifte. Zur Reichsritterschaft konnte es aus formalen Gründen keine direkten Beziehungen geben, da sie außerhalb der Reichskreise stand.[30] So ließ sich das adelige Damenstift Buchau trotz reichsstädtischer Nachbarschaft vom Kreisvertreter der Hohenzollern in Sigmaringen vertreten. Der fürstenbergische Diplomat votierte für das Damenstift in Lindau und ein von der Reichsabtei Ochsenhausen delegierter Herr von Schott vertrat das Reichskloster St. Georg in Isny.[31] Die unmittelbare Grenze zur jeweiligen Reichsstadt führte auf den Kreistagen offenbar zu keinerlei Kooperationen beim Votum – ein Umstand, der für die Städtelandschaft spricht.

---

30 Wolfgang Wüst, Ritterkreis und Reichskreis – Kooperationen für Franken, in: Ders. (Hg.), Adelslandschaften – Kooperationen, Kommunikation und Konsens in Mittelalter, Früher Neuzeit und Moderne. Unter Mitarbeit von Lisa Bauereisen, Frankfurt am Main 2018, S. 97–126.

31 StaatsA Augsburg, Reichsstadt Nördlingen, Münchner Bestand, Nr. 693.

Auch in Franken spiegelte sich die Städtelandschaft in den Kreistagen. So votierten in der 1.901.(!) Sitzung des Nürnberger Kreistags[32] – sie fand am 18. August 1783 statt – fünf fränkische Städte *unisono* für eine Steuerreform. Das „*Directorium*", bestehend aus Bamberg und Brandenburg-Ansbach, gab bekannt: „*Da die in frage stehende Sache nunmehro res Circuli seÿe, so könne auch Rothenburg und Bamberg ohne Anstand mit votiren. Rothenburg, Windsheim, Schweinfurth* [und] *Weißenburg*" stimmten anschließend „*wie Nürnberg*".[33] Das war gewissermaßen der Musterfall bei wichtigen Abstimmungen der Kreisdelegierten; keineswegs eine Einzelentscheidung. Am 24. September 1785 kam es ebenfalls während eines Nürnberger Kreistags zur Frage, ob die Zahlungsrückstände der Kreisstände mit „*Poenalzinnsen*" eingetrieben werden sollten, zu einem geschlossenen reichsstädtischen Votum. Nürnberg: „*In der HauptSache ebenfalls beÿstimmend. Was hingegen illiquide Rückstände betrefe; So könnten von selbigen, selbst nach dem Sinn des Cr*[eis]*-Schlußes de 1752 keine Zinsen angerechnet werden; man müsse sich also auf allen fall dagegen verwahren.*" Rothenburg: „*Man sehe die Abstellung der 3 monatlichen Pönal-Zinnß-Ansäze ebenfalls für gut an, halte aber auch mit Nürnberg dafür, daß die Pönal-Zinnße nur von liquiden Rückständen zu verstehen seÿen. Windsheim; Schweinfurt und Weisenburg wie die vorstimmende Städte, Bamberg wie Würzburg und andere vorstimmende.*"[34]

32 Zu den Kreistagen: Wolfgang Wüst, Windsheim – Kreistage als Erinnerungsorte einer frühmodernen Reichsverfassung in Franken, in: Ferdinand Kramer (Hg.), Orte der Demokratie in Bayern (Zeitschrift für bayerische Landesgeschichte [künftig: ZBLG] 81/1), München 2018, S. 87–104; Wolfgang Wüst, Inhalte, Debatten und Abstimmungen in süddeutschen Kreiskonventen, in: Ders. (Hg.), Mitregieren und Herrschaftsteilung in der Frühen Neuzeit. Beiträge zur Machtfrage im Alten Reich und in Bayern (Franconia 8. Beihefte zum JfL), Erlangen 2016, S. 55–76; Wolfgang Wüst, Entscheidungsnot und Konsenssuche – zur Arbeit frühmoderner Kreiskonvente in Schwaben, Franken und Bayern, in: Ders. / Michael Müller (Hg.), Reichskreise und Regionen im frühmodernen Europa – Horizonte und Grenzen im *spatial turn*. Tagung bei der Akademie des Bistums Mainz, Erbacher Hof, 3.–5. September 2010 (Mainzer Studien zur Neueren Geschichte 29), Frankfurt am Main u.a. 2011, S. 319–339.

33 StaatsA Nürnberg, Reichsstadt Nürnberg, Kreistagsakten, Nr. 450, Protokolleintrag vom 18.8.1783.

34 Ebd., Nr. 453, Sitzungsprotokoll vom 24.8.1785.

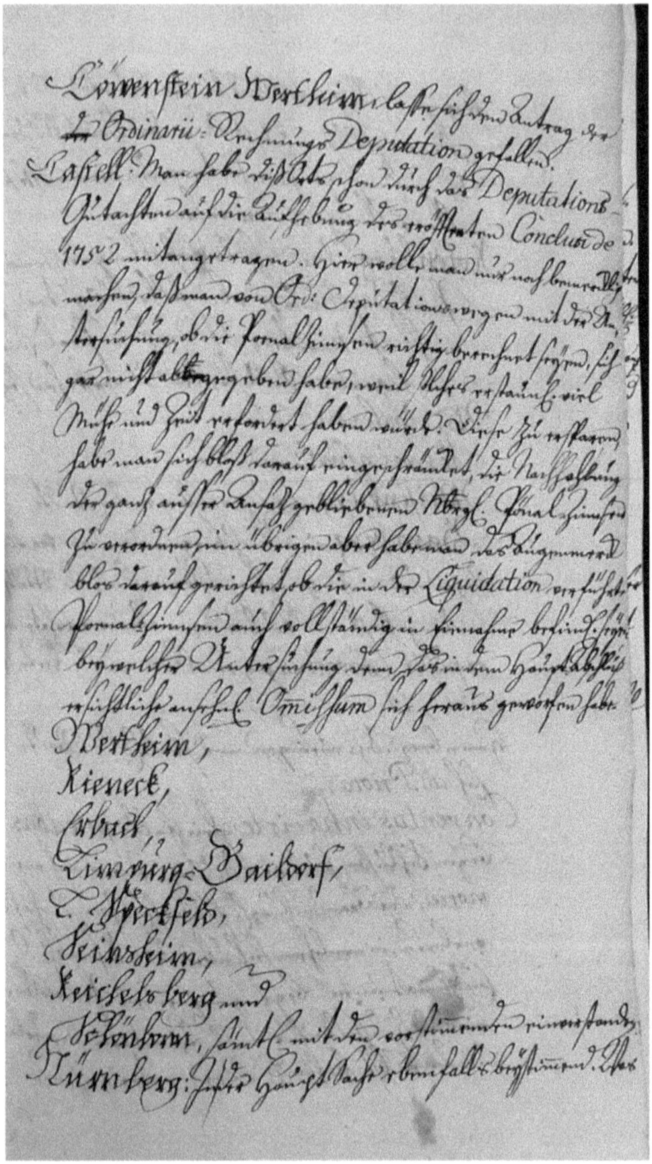

Abb. 11: *Protokollauszug des Nürnberger Kreiskonvents, 1785. Bildnachweis: StaatsA Nürnberg, Reichsstadt Nürnberg, Kreistagsakten, Nr. 453.*

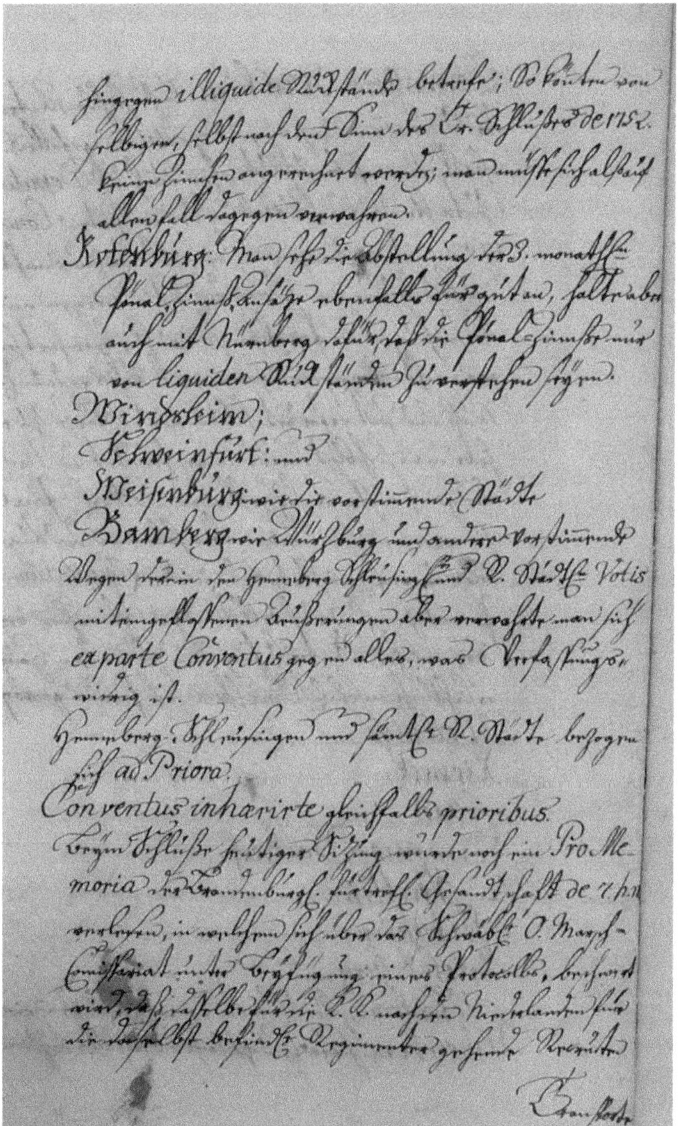

Abb. 12: *Protokollauszug des Nürnberger Kreiskonvents, 1785. Bildnachweis: StaatsA Nürnberg, Reichsstadt Nürnberg, Kreistagsakten, Nr. 453, „Sessio“ 1.999 vom 24.8.1785.*

Am 4. Oktober 1785 verständigten sich auf Antrag des kaiserlichen Ministers Graf Trauttmansdorff die Gesandtschaften „*stando in Circulo*" zu den Friedensbedingungen im Konflikt zwischen Österreich und Holland. Hintergrund waren die wiederholten Pläne Kaiser Josephs II., um Bayern gegen die Österreichischen Niederlande einzutauschen; ein Plan, der am Widerstand Friedrichs II. und des Fürstenbundes scheiterte. Zu dieser Frage äußerten sich auch die kleineren fränkischen Reichsstädte Windsheim und Weißenburg, die sich ansonsten oft durch Nürnberger Kreisdiplomaten vertreten ließen, in eigenen Voten, doch strikt in Anlehnung an andere Reichsstädte. Windsheim: „*In Voraussezung der R*[eichs]*st*[adt] *Nürnbergischen Verbindlichkeit zur unmittelbahren Cassa Vertrettung auf das vergangene, seÿe man gleichermaßen instruirt auf geringen billigen Nachlaß des Cassa-Rückstands und zu societätsmäßiger Assistenz zu deren Indemnisation aus der von* [Den] *Hagenschen Concurs-Massa förderlichst mit zustimmen. Übrigens wie Rothenburg. Schweinfurth und Weißenburg: In Gemäsheit der vorliegenden Grundsäze u*[nd] *in Hoffnung, daß eine Wohll*[öbliche] *R*[eichs]*st*[adt] *Nürnberg zu der unmittelbahren Cassa-Vertrettung auf das Vergangene vorderlichst näher u*[nd] *bestimmter sich erklären werde, seÿe man mit Rothenburg einverstanden.*"[35]

Grenzüberschreitende Koordinationen um das Wohl der Städte im Reichskreis hielten gegen das Bild überfälliger Reformen unter schlafenden Bürgermeistern,[36] selbstgefälligen Räten und arbeitsscheuen verschwenderischen Bürgern. Und das gezeigte Stimmverhalten der Reichsstädte in den 1780er Jahren war keine erkenntnisfördernde Eintagsfliege, sondern es handelt sich vielmehr um die übliche Praxis kommunaler Gremienpolitik. Kreistage[37] – zeitgenössisch

---

35 StaatsA Nürnberg, Reichsstadt Nürnberg, Kreistagsakten, Nr. 453, „*Sessio*" 2.003 vom 4.10.1785.

36 Friedrich von Lupin bemerkte als Sohn eines reichsstädtischen Kanzleidirektors in Memmingen 1794 dazu treffend: „Man bemerkt überall das Festsitzen im Alten und das Festhalten am Alten. In dem Schlafsessel, in welchem der Urgroßvater gestorben, erwartet seine Herrlichkeit, der Herr Bürgermeister v. S. von Tag zu Tag, schon zehn Jahre lang, die letzte Stunde." Vgl. dazu: Friedrich Freiherr von Lupin, Selbst-Biographie, 2 Bde., Weimar ²1847, hier: Bd. 1, S. 296 f.; Rita Huber-Sperl, Memmingen zwischen Zunfthandwerk und Unternehmertum. Ein Beitrag zur reichsstädtischen Gewerbegeschichte 1648–1802 (Memminger Forschungen 5), Memmingen 1995.

37 Peter Claus Hartmann, Die Kreistage des Heiligen Römischen Reiches – Eine Vorform des Parlamentarismus? Das Beispiel des Bayerischen Reichskreises (1521–1793), in: Zeitschrift für Historische Forschung 19, 1992, S. 29–47; Nicola Schümann, Der Fränkische Kreiskonvent im Winter 1790/91. Ein Verfassungsorgan an der Schwelle zur Moderne, in: JfL 62, 2002, S. 231–258; Nicola Humphreys, Der Fränkische Kreistag 1650–1740 in kommunikationsgeschichtlicher Perspektive (Veröffentlichungen der

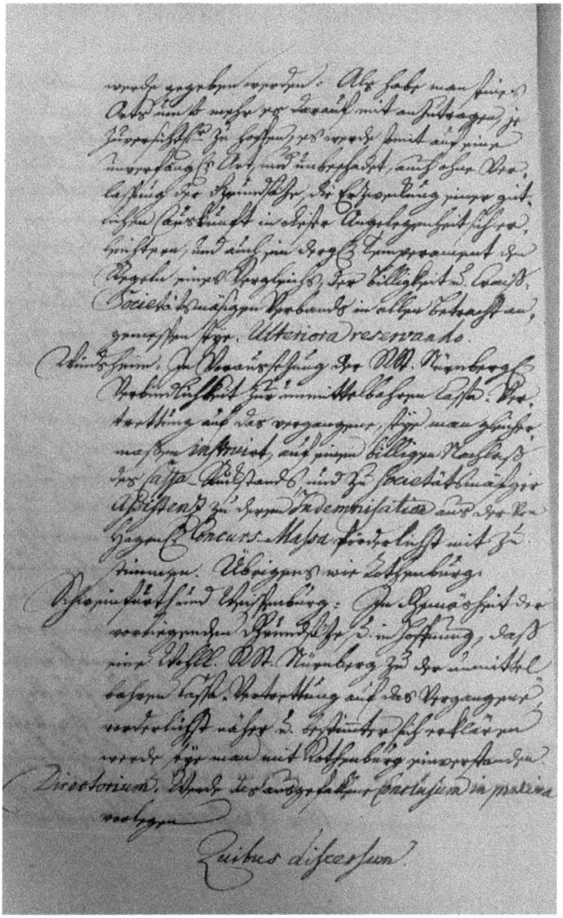

Abb. 13: *Protokollauszug des Nürnberger Kreiskonvents, 1785. Bildnachweis: StaatsA Nürnberg, Reichsstadt Nürnberg, Kreistagsakten, Nr. 453, „Sessio" 2.003 vom 4.10.1785.*

oft als „*Craÿßtage oder Craÿs-Versamblungen*" bezeichnet – fanden insbesondere in süddeutschen Reichskreisen in der Frühmoderne ziemlich regelmäßig, beizeiten sogar fast jährlich statt. Es gab Jahre, in denen sich die Kreisdelegierten

---

Gesellschaft für fränkische Geschichte II/3), Würzburg 2011; Wüst, Entscheidungsnot und Konsenssuche (wie Anm. 32), S. 319–339; Ders., Inhalte, Debatten und Abstimmungen in süddeutschen Kreiskonventen (wie Anm. 32), S. 55–76.

an einem Ort zu unterschiedlichsten Zeiten mehrmals im Jahr trafen. Die (Bad) Windsheimer Stadtchronik[38] berichtete beispielsweise für 1551: „*Sonsten sind in diesem Jahr dreÿ Fränckische Craÿs convent, allhie gehalten worden.*“[39]

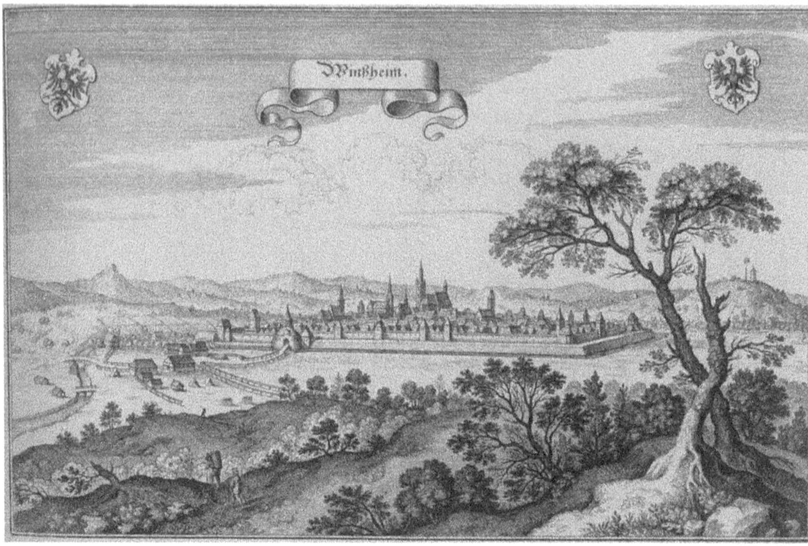

Abb. 14: *Windsheim („Winßheim“) im 17. Jahrhundert nach einem Kupferstich von Matthäus Merian, 21 x 33 cm, vor 1648. Bildnachweis: Topographia Franconiæ. Das ist / Beschreibung, Vnd Eÿgentliche Contrafactur der Vornembsten Stätte / Vnd Plätze des Franckenlandes […], Frankfurt am Main 1648, ²1656.*

3. Städte im Reichstag

Reichsorientierte und städtische Netzwerke als Grundkomponenten einer Städtelandschaft entstanden neben den Kreistagen vor allem im Umfeld der

---

38 Am 20. Juni 1961 wurde der Stadtname offiziell zu Bad Windsheim erweitert. Die Heil- und Solequellen der ehemaligen Reichsstadt in Mittelfranken wurden zuvor oftmals besucht und beschrieben; erstmals 1752 von Wilhelm Simon Carl Hirsching als „*Gesundbrunnen auff dem Kehrenberg*“. Vgl. Wilhelm Volkert (Hg.), Handbuch der bayerischen Ämter, Gemeinden und Gerichte 1799–1980, München 1983, S. 582.

39 StadtA Bad Windsheim, 2a: Georg Andreas Dienst, Chronik der Stadt Windsheim, Windsheim [1666], Bd. 1, fol. 124, 132.

Reichstage. Waren sich viele Städte 1495 in Worms noch nicht einig, ob sie im Reichstag koordiniert an den Beratungen teilnehmen sollten, so forderten sie ihre Teilnahme am Reichstag spätestens zur Mitte des 16. Jahrhunderts förmlich ein, um die Nähe zum Kaiser, den Kurfürsten und den übrigen Reichsständen zu suchen. Ferner bot sich über die Städtekurien die Möglichkeit, Einfluss auf die 1521 festgelegte hohe Besteuerung der 86 Reichsstädte[40] zu nehmen. Beeindruckend ist die äußere Formation der Reichsstädte auf den Reichstagen in der Intention, neben dem Kurfürstenkollegium (Kurfürstenrat) und dem Reichsfürstenrat, größeres Gewicht bei den Beratungen zu den Reichsabschieden zu gewinnen. Die beiden oberen Kurien konnten sich allerdings mit dem Reichsoberhaupt auch ohne das Votum der Städte einigen, denen auf den Reichstagen formal zunächst kein Entscheidungsrecht zustand.[41] Das Reichsstädtekollegium erhielt dann aber mit dem Augsburger Reichstag von 1582 und den dort beschlossenen „*vota decisiva*“ stärkeres politisches Gewicht. Der Westfälische Friede bestätigte das Entscheidungsrecht der Städte, wenngleich die beiden kollektiven Kurienstimmen der Städtebänke keineswegs zwangsläufig den Ausschlag gaben, wenn die Voten der Kurfürsten und Fürsten uneins waren. Die Reichsklöster und Stifte waren im Reichstag über ihre Rheinische und Schwäbische Prälatenbank[42] mit je einer kurialen Stimme ähnlich strukturiert. Die städtischen Voten mussten anschließend dem Kaiser zur Annahme oder Ablehnung des Reichsgutachtens vorgelegt werden.[43] Der städtische Einfluss war insbesondere in Handels- und Wirtschaftsfragen groß, wie die frühe Genese der 1731 implementierten Reichshandwerksordnung zeigte.[44] Inhaltlich ging es um

40 Im Jahr 1792 zählte man nur noch 51 Reichsstädte.

41 Barbara Stollberg-Rilinger, Das Heilige Römische Reich Deutscher Nation. Vom Ende des Mittelalters bis 1806 (Beck'sche Reihe Wissen 2399), München 2006, S. 34.

42 Sarah Hadry, Reichsprälatenkollegium, publiziert am: 13.11.2008, in: Historisches Lexikon Bayerns, URL: http://www.historisches-lexikon-bayerns.de/Lexikon/Reichsprälatenkollegium (letzter Zugriff: 24.8.2019); Ernst Böhme, Das Kollegium der Schwäbischen Reichsprälaten im 16. und 17. Jahrhundert. Untersuchungen zu den Möglichkeiten und Grenzen der korporativen Politik der mindermächtigen Reichsstände, in: Rottenburger Jahrbuch für Kirchengeschichte 6, 1987, S. 267–300.

43 Johannes Burkhardt, Vollendung und Neuorientierung des frühmodernen Reiches 1648–1763 (Handbuch der deutschen Geschichte 11), Stuttgart [10]2006, S. 60.

44 Kristina Winzen, Handwerk – Städte – Reich. Die städtische Kurie des Immerwährenden Reichstags und die Anfänge der Reichshandwerksordnung (Vierteljahrschrift für Sozial- und Wirtschaftsgeschichte, Beiheft 160), Stuttgart 2002. Zur Edition vgl.: Helmut Neuhaus (Hg.), Zeitalter des Absolutismus 1648–1789 (Deutsche Geschichte in Quellen und Darstellung 5), Stuttgart 1997, S. 98–109.

die Fortsetzung der gerade für den städtischen Raum wichtigen Reichspolicey-Ordnungen[45] des 16. Jahrhunderts und um die Meldepflicht mit Geburtsbrief und Lehrbrief für die von Stadt zu Stadt wandernden Handwerksgesellen.[46]

Im Reichstag nahmen die Städte, denen im Gegensatz zu den Fürsten keine Virilstimme zustand, getrennt auf der „*Schwäbischen*" oder „*Rheinischen Bank*" Platz. Die Zahl der Delegierten, die allerdings auf keinem Kreis- und Reichstag vollständig erschienen, blieb getrennt nach Landschaften unterschiedlich. An der gemeinsamen schwäbischen Stimme waren am Ende des 18. Jahrhunderts noch 37, am rheinischen Votum 14 Reichsstädte beteiligt. Für Schwaben votierten in alphabetischer Reihenfolge Aalen, Augsburg, Biberach, Bopfingen, Buchhorn, Dinkelsbühl, Esslingen, Gengenbach, Giengen, Heilbronn, Isny, Kaufbeuren, Kempten, Leutkirch, Lindau, Nördlingen, Nürnberg, Memmingen, Offenburg, Pfullendorf, Ravensburg, Rottweil, Schwäbisch Gmünd, Schwäbisch Hall, Regensburg, Reutlingen, Rothenburg, Schweinfurt, Überlingen, Weißenburg am Nordgau, Ulm, Wangen, Weil, Wimpfen, Windsheim und Zell am Harmersbach.[47] Saßen auf der Schwabenbank abgesehen von den fränkischen Reichsstädten Dinkelsbühl,[48] Nürnberg, Rothenburg, Schweinfurt, Weißenburg und Windsheim ausschließlich südwestdeutsche Ratsdelegierte im Reichstag, so präsentierte sich die Rheinische Bank als ein regional sehr heterogenes

---

45 Matthias Weber (Hg.), Die Reichspolizeiordnungen von 1530, 1548 und 1577. Historische Einführung und Edition (Ius commune, Sonderheft 146), Frankfurt am Main 2002; Wüst, Die „*gute*" Policey im Reichskreis, Bd. 7: Policeyordnungen in den fränkischen Reichsstädten (wie Anm. 13).

46 Johann Joseph Pachner von Eggenstorff (Hg.), Vollständige Sammlung aller von Anfang des noch fürwährenden Teutschen Reichs-Tags de Anno 1663 biß anhero abgefaßten Reichs-Schlüsse [...], Teil 4, Regensburg 1740, S. 374–384, Nr. CXXXVIII.

47 Als Überblick zur Entwicklung der beteiligten Städte vgl. die Bände des Handbuchs der Historischen Stätten: Hans-Michael Körner / Alois Schmid (Hg.), Handbuch der Historischen Stätten. Bayern I: Altbayern und Schwaben, Bayern II: Franken (Kröners Taschenbuchausgabe 324/325), 2 Bde., Stuttgart 2006; Max Miller / Gerhard Taddey (Hg.), Handbuch der Historischen Stätten. Baden-Württemberg (Kröners Taschenbuchausgabe 276), Stuttgart 1980. Zu den Reichsstädten als Reichsstände vgl.: Helmut Neuhaus, Das Reich in der Frühen Neuzeit (Enzyklopädie deutscher Geschichte 42), München 1997, S. 34–36.

48 Dinkelsbühls Zugehörigkeit zu Franken ist ein Ergebnis der Gebietszuteilungen des 19. Jahrhunderts, während die Stadt zuvor politisch in Schwaben verwurzelt war. So schloss sich die Stadt bereits 1378 dem Schwäbischen Städtebund an. Ludwig Schnurrer, Dinkelsbühl, in: Körner / Schmid, Handbuch der Historischen Stätten. Bayern II: Franken (wie Anm. 47), S. 105–108.

Gremium. Von einer räumlich geschlossenen Landschaft kann hier überhaupt nicht die Rede sein, eher von einer Sammelstelle für west-, nord- und mitteldeutsche Reichsstädte. In West-Ost-Richtung erstreckte sich die Gruppe von Aachen und Köln am Rhein bis nach Goslar in Niedersachsen und nach Mühlhausen und Nordhausen in Thüringen. Bremen, Goslar, Hamburg, Lübeck und Wetzlar lagen ebenso wie die thüringischen Reichsstädte fernab der Rheinregion.

Grundsätzlich blieben für die städtische Teilnahme an den Reichstagen zwei Fragen vordringlich. Welche *„freien“* Städte galten als reichsständisch und wie platzierte man sie nach der Wormser Reichsmatrikel? Die Matrikel begann mit den geistlichen Fürsten, die insbesondere über ihre Kur- und Hochstifte präsent waren. Innerhalb der *Germania Sacra* folgte dann die Gruppe der Reichsprälaten,[49] die bis zur Säkularisation korporativ vertreten waren. Den geistlichen Ständen saßen weltliche Fürsten, Grafen und Herren gegenüber, wobei vor allem für den Adel unterhalb des Fürstenstandes eine hohe Fluktuation und Mobilität kennzeichnend war. Matrikelbeiträge und Reichstagsbesuche seitens des Adels variierten demnach erheblich. Schließlich stand für alle Reichsstädte der Reichstag offen. Die Wormser Reformmatrikel nannte unter ihnen zunächst Regensburg, dessen Steuerleistung mit 120 Gulden allerdings weit hinter den Beträgen von Nürnberg, Köln, Ulm, Straßburg oder Augsburg lag. Die Regensburger Finanzkraft war aber vergleichbar mit den Steuerbeiträgen, wie sie für Kempten, Schweinfurt und Konstanz galten. Und die *„H. Röm. Reichs matricull“* legte trotz aller Ungenauigkeiten im Detail innerhalb der städtischen Landschaft die Steuerwertigkeit fest.[50] Der Kreis der immer wieder auch bei Reichstagen über gelehrte und einflussreiche Ratsdelegierte vertretenen Städte war offiziell festgelegt. Als reichsstädtische *„Agenten“* sind sie vor allem in den Stadtrechnungen[51] über die Spesen ihrer Dienstreisen präsent. Ferner sorgten sie als Stadtdelegierte neben den offiziellen Protokollführern für die Mitschriften zu den Verhandlungen und Beschlüssen. Sie baten um Entscheidungsbefugnisse und überführten wie auf dem Regenburger Reichstag 1607/08 aus den Stadtregistraturen zum

49 Armgard von Reden-Dohna, Reichsstandschaft und Klosterherrschaft. Die Schwäbischen Reichsprälaten im Zeitalter des Barock (Institut für Europäische Geschichte Mainz, Vorträge 78), Wiesbaden 1982.

50 Erstellt nach: Hanns Hubert Hofmann (Hg.), Quellen zum Verfassungsorganismus des Heiligen Römischen Reiches Deutscher Nation 1495–1815 (Freiherr vom Stein-Gedächtnisausgabe 13), Darmstadt 1976, S. 40–51.

51 StaatsA Nürnberg, Reichsstadt Nürnberg, Losungsamt, Stadtrechnungsbelege, Bündel 743, 822, 908 und 990. Laufzeiten: 1630–1639, 1640–1649, 1650–1659 und 1660–1669.

Teil ausführliche Handakten, die ihr Votum optimierten.[52] So vertrat beispielsweise Ratsschreiber Dr. Lazarus Spengler (1479–1534) die Reichsstadt Nürnberg mehrmals auf Reichstagen des frühen 16. Jahrhunderts. Spengler hatte in Leipzig die Rechte studiert, übernahm 1507 das Amt des Vordersten Ratsschreibers, wurde 1516 als Genannter Mitglied des Äußeren Rats und war Mitglied der „*Sodalitas Staupiziana*". 1521 vertrat er seine Stadt erstmals auf dem Reichstag zu Worms.[53] Es war ferner Usus, dass größere Reichsstädte wie Augsburg, Frankfurt am Main, Köln, Nürnberg oder Ulm kleinere Mitglieder der Städtekollegien nicht nur auf den Kreistagen, sondern auch in Reichsversammlungen und auf Reichstagen vertraten. So bat 1543 der Rat zu Weißenburg Nürnberg die kleine fränkische Reichsstadt auf einem Städtetag zu Frankfurt zu vertreten.[54] Dies war der Kreis städtischer Gesandter, aus dem auch der Reichstag seine Legitimität schöpfte.

Um die korporative Geschlossenheit in einer zu definierenden Städtelandschaft nachzuweisen, ist der Fokus keineswegs nur auf die politische Bedeutung des „*votum decisivum*" in der Städtekurie[55] des Reichstags zu beschränken. Das städtische Votum wurde auch in der Einschätzung beteiligter Reichsstände unterschiedlich bewertet. 1671 bestätigte der Nürnberger Rat in seiner Korrespondenz mit der Reichsstadt Frankfurt die rechtliche Bedeutungslosigkeit vieler Entscheidungen in der Kurie, an deren Zustandekommen nur wenige Delegierte Anteil hatten: „*Beim Reichstag ist der Ehrbaren Frei- und Reichsstädte Ehr und Stand im merkliche Verachtung gekommen, alldieweil dieselbe in so gar geringer Zahl vorhanden sind und gleichsam das publicum* [...] *beiseits setzen.*"[56] Reichstage

52 StaatsA Nürnberg, Reichsstadt Nürnberg (Rep. 24/b), Reichstagsakten, Nr. 164.

53 Berndt Hamm, Lazarus Spengler. Der Nürnberger Ratsschreiber im Spannungsfeld von Humanismus und Reformation, Politik und Glaube. Mit einer Edition von Gudrun Litz (Spätmittelalter und Reformation, Neue Reihe 25), Tübingen 2004.

54 StaatsA Nürnberg, Reichsstadt Nürnberg, Ratskanzlei, A-Laden, Akten, A 111 Nr. 47.

55 André Krischer, Inszenierung und Verfahren auf den Reichstagen der Frühen Neuzeit: das Beispiel der Städtekurie und ihres politischen Verfahrens, in: Jörg Peltzer / Gerald Schwedler / Paul Töbelmann (Hg.), Politische Versammlungen und ihre Rituale: Repräsentationsformen und Entscheidungsprozesse des Reichs und der Kirche im späten Mittelalter (Mittelalter-Forschungen 27), Ostfildern 2009, S. 181–205; Günter Buchstab, Reichsstädte, Städtekurie und Westfälischer Friedenskongreß. Zusammenhänge von Sozialstruktur, Rechtsstatus und Wirtschaftskraft (Schriftenreihe der Vereinigung zur Erforschung der Neueren Geschichte 7), Münster 1976.

56 Zitiert nach: Klaus-Peter Schroeder, Das Alte Reich und seine Städte. Untergang und Neubeginn. Die Mediatisierung der oberdeutschen Reichsstädte im Gefolge des Reichsdeputationshauptschlusses 1802/03, München 1991, S. 103.

waren aber vielmehr auch Plätze informeller Handlungsabläufe und kommunikativer Netzwerke. Susanne Friedrich hat das Informations- und Kommunikationssystem des Immerwährenden Reichstags für die Zeit um 1700 zum Thema ihrer einschlägigen Augsburger Dissertation gewählt.[57] Das galt sicher auch für den Austausch von Ratssatzungen,[58] Verordnungen zur *„guten"* Policey sowie einer Vielzahl an reichsstädtischen Dekreten, Edikten und Mandaten. Nicht von ungefähr wurden in den Reichsstädten die während der Reichstage erlassenen Reichspolicey-Ordnungen aus den Jahren 1530, 1548 und 1577 hundertfach rezipiert und regional fortgeschrieben.[59] Reichstage galten als Informationsdrehscheiben und Stätten der Begegnung. Das traf insbesondere für süddeutsche Reichsstädte zu, wo man eben entsprechend oft tagte. In Nürnberg fanden im Zeitraum von 1378 bis 1491 mindestens 46 Reichstage beziehungsweise Städte- oder Fürstentage mit königlicher Beteiligung statt, deren Zustandekommen auch durch die zentrale Lage der Stadt und durch die Existenz der Kaiserburg als repräsentativer Herberge[60] auf sicherem Reichsboden gefördert wurde. Hier deckten sich öfter binnenstädtische Kommunikationsstränge mit den Interessen des Reiches und des Königs. Im 15. und 16. Jahrhundert, als Reichs- und Königsstädte von Augsburg über Nürnberg[61] bis Ulm besonders häufig als Standorte

57 Susanne Friedrich, Drehscheibe Regensburg. Das Informations- und Kommunikationssystem des Immerwährenden Reichstags um 1700 (Colloquia Augustana 23), Berlin 2007.

58 Für Nürnberg als frühe Quellenbearbeitung: Werner Schultheiß (Bearb.), Satzungsbücher und Satzungen der Reichsstadt Nürnberg aus dem 14. Jahrhundert (Quellen zur Geschichte und Kultur der Stadt Nürnberg 3), Nürnberg 1965.

59 Matthias Weber, Die Reichspolizeiordnungen von 1530, 1548 und 1577. Historische Einführung und Edition (Ius commune, Sonderheft 146), Frankfurt am Main 2002; Wolfgang Wüst, Normen als Grenzgänger. Policey und Wissenstransfer in Süddeutschland vor 1800, in: Konrad Ackermann / Alois Schmid (Hg.), Staat und Verwaltung in Bayern. Festschrift für Wilhelm Volkert zum 75. Geburtstag (Schriftenreihe zur bayerischen Landesgeschichte 139), München 2003, S. 287–306.

60 Katharina Heinemann (Hg.), Die Nähe zum Reichsoberhaupt – Privilegien und Verpflichtungen der Reichsstadt Nürnberg, in: Kaiser – Reich – Stadt. Die Kaiserburg Nürnberg. Begleitbuch zur Ausstellung „Kaiser – Reich – Stadt. Die Kaiserburg Nürnberg" in Nürnberg, 13. Juli bis 10. November 2013, Petersberg 2013, S. 41–52; Edmund Ries, Die Nürnberger Burg als Herberge der deutschen Kaiser und Könige, in: Das Bayerland 45, 1934, S. 718–720.

61 Nürnberg kann aufgrund der Bestimmungen der Goldenen Bulle, dort jeweils den ersten Reichstag nach der Königswahl auszurichten, zumindest für das 15. und beginnende 16. Jahrhundert als *die* Stadt der Reichstage bezeichnet werden. Vgl. dazu auch: Bernd-Ulrich Hergemöller, Fürsten, Herren, Städte zu Nürnberg 1355/56. Die

für Reichsversammlungen ausgewählt wurden, sorgten städtische Boten für die Verbreitung der Reichstagsladungen.[62] Zudem informierten sie über gemeinsame städtische Interessen im Vorfeld der Reichstage. So empfing auch der Nürnberger Rat vor dem Frankfurter Reichstag 1477 eine Nachricht aus Augsburg, dass man „*im besten vnd gemeinen nutz*" ihnen „*vnnd etlichen anndern vnnsern gutten frunden*" in den Städten schrieb, um vor „*dem hanndel berattenlich zereden vnd zu rathschlagen, was auf dem gemeldten tag zu Frankfurt zehanndeln vnnd fürzenemen sey.*" Die Nürnberger sollten zur Vorbereitung ihre „*ratzbotschaft mit volem gewallt gen Eßlingen schicken, das die auf freytag vor des hailligen crütztag schirist zu nacht daselbst an der herberg sey furter des morgens mit anndern vnnser frewnd der stette vnd vnnsern machtbotten in obgemellter maynung das beste fürzenemen.*"[63] Im gleichen Jahr überbrachte der Nürnberger Ratsdiener Hainz Wonsesser von Innsbruck aus am 17. Juni zur Weiterbeförderung den Augsburger Räten ein Schreiben Erzherzogs Maximilian, das man ihm dort in Abschrift gleich „*nach vesperzeit*" mit auf den Weg gegeben hatte. Vier Tage später traf dieses dann zusammen „*mit eingelegten copeien ains briefs von vnnser allergnedigst hern den Romischen kayser vnd zway* [anderen] *briefe von* [...] *Maximilian herzogen zu Ostenreich etc. vß gangen inhallt ewers briefs*" in Nördlingen und Nürnberg ein. Im August desselben Jahres teilte man ferner Bürgermeistern und Rat zu Kaufbeuren mit, dass in Augsburg aus Nürnberg Schreiben eingetroffen seien, die man „*mit eingelegter copein ettlicher vnnser gnädige heren des reichs kurfürsten vnd fürsten briefe an sy*" in der Reichsstadt Kaufbeuren weiterleiten wolle.[64]

Das Reichstagsgeschehen wurde sorgfältig dokumentiert. Es ist deshalb seit 1858 eines der Langzeitprojekte primär reichsstädtischer Erforschung, die Protokollserien zu edieren, soweit sie sich in staatlichen wie kommunalen Archiven befinden. Jenseits von Ratsversammlungen und Ratsentscheidungen manifestierten sich dort die spannungsgeladenen Beziehungen der Reichsstädte zum Reich. Die Reichstagsakten gelten seit der Initiative Leopold Rankes von 1858 zu ihrer Erschließung anlässlich der Gründung der Historischen Kommission bei der

---

Entstehung der „Goldenen Bulle" Karls IV. (Städteforschung A 13), Köln 1983; Ders., Die Verfasserschaft der „Goldenen Bulle" Karls IV., in: Bohemia 22, 1981, S. 253–299.

62 Wolfgang Wüst, Reichsstädtische Kommunikation in Franken und Schwaben. Nachrichtennetze für Bürger, Räte und Kaufleute im Spätmittelalter, in: ZBLG 62, 1999, S. 681–707.

63 StadtA Augsburg, Reichsstadt, Schätze 105/VII, Nr. 131, 31. August 1477. Der Quellennachweis bezieht sich auch auf die folgenden Zitate.

64 StadtA Augsburg, Reichsstadt, Schätze 105/VII, Nr. 101, 102, Einträge vom 17. und 20. Juni 1477.

Bayerischen Akademie der Wissenschaften als primäre, aber nicht ausschließlich politische Hauptquellen[65] der Zeit. Aus diesem Grund hat sich auch die neuere landeshistorische Forschung[66] dem institutionellen Paar Reichsstadt und Reichstag vor allem für die Zeit des Immerwährenden Reichstags mit interdisziplinären Fragestellungen genähert. Gleiches kann sicher auch für die vorausgegangenen Reichsversammlungen gelten, auch wenn diese nur für kürzere Dauer einberufen wurden. Die Quellen sind für die deutschen Reichstage – diese fanden übrigens vorwiegend in den Reichsstädten[67] statt – dank der drei Editionsreihen zu den „Deutschen Reichstagsakten" in Teilen vorbildlich erschlossen. Die Ältere Reihe umschließt dabei den Zeitraum von 1376 bis 1485; es folgen die Mittlere und Jüngere Reihe für die Jahrzehnte von 1486 bis 1518 und von 1519 bis 1555. Sie belegen demnach die zeitlichen Schwerpunkte reichsstädtischer Forschung bis zur Mitte des 16. Jahrhunderts,[68] doch liegen mittlerweile auch für die Endzeit des

65 Eike Wolgast, Deutsche Reichstagsakten, in: Lothar Gall (Hg.), „... für deutsche Geschichts- und Quellenforschung". 150 Jahre Historische Kommission bei der Bayerischen Akademie der Wissenschaften, München 2008, S. 79–120; Heribert Müller, Die Reichstagsakten (Älterer Reihe) und ihre Bedeutung für die europäische Geschichte, in: Heinz Angermeier / Erich Meuthen (Hg.), Fortschritte in der Geschichtswissenschaft durch Reichstagsaktenforschung. Vier Beiträge aus der Arbeit an den Reichstagsakten des 15. und 16. Jahrhunderts (Schriftenreihe der Historischen Kommission bei der Bayerischen Akademie der Wissenschaften 35), Göttingen 1988, S. 17–46.

66 Beispielsweise: Martin Dallmeier (Hg.), Reichsstadt und Immerwährender Reichstag (1663–1806). 250 Jahre Haus Thurn und Taxis in Regensburg. Beiträge des Regensburger Herbstsymposions zur Kunstgeschichte und Denkmalpflege vom 17. bis 22. November 1998 (Thurn und Taxis-Studien 20), Kallmünz 2001.

67 Dazu zählten im 15. und 16. Jahrhundert: Augsburg (1473, 1474, 1500, 1518, 1525, 1530, 1547/48, 1550/51, 1555, 1559, 1566 und 1582), Eger (1437), Esslingen am Neckar (1488), Frankfurt am Main (1410, 1411, 1427, 1437, 1440, 1454, 1485, 1477, 1486 und 1489), Köln (1505), Konstanz (1408[?], 1415, 1486 und 1507), Lindau (1496/97), Nürnberg (1401, 1421, 1422, 1426, 1438, 1443, 1434/44, 1444, 1447, 1456, 1460, 1461, 1466, 1467, 1470, 1479, 1480, 1481, 1487, 1491, 1522, 1522/23, 1524, 1542 und 1543), Regensburg (1454, 1469, 1471, 1532, 1541, 1546, 1557, 1567, 1576, 1594 und 1597/98) und Ulm (1434).

68 Zuletzt sind in der „Jüngeren Reihe" erschienen: Ursula Machoczek (Bearb.), Der Reichstag zu Augsburg 1547/48 (Deutsche Reichstagsakten unter Kaiser Karl V. 18), 3 Teil-Bde., München 2005; Erwein H. Eltz (Bearb.), Der Reichstag zu Augsburg 1550/51 (Deutsche Reichstagsakten unter Kaiser Karl V. 19), 2 Teil-Bde., München 2005; Rosemarie Aulinger / Erwein H. Eltz / Ursula Machoczek (Bearb.), Der Reichstag zu Augsburg 1555 (Deutsche Reichstagsakten unter Kaiser Karl V. 20), 4 Teil-Bde., München 2009; Rosemarie Aulinger (Bearb.), Der Reichstag zu Augsburg 1525, der Reichstag zu Speyer 1526, der Fürstentag zu Esslingen 1526 (Deutsche Reichstagsakten unter Kaiser Karl V. 5–6), 4 Teil-Bde., Göttingen 2011.

seit 1663 „*immerwährend*" tagenden Reichstags zu Regensburg bahnbrechende Arbeiten vor.[69] Sie sind vielfach quellenorientiert angelegt und mit Blick auf die Beziehungen zum Reich inhaltlich wie methodisch innovativ ausgerichtet.

4. Städte im Landtag

1605 wurde im Herzogtum Bayern zum wiederholten und nicht zum letzten Mal ein Landtag[70] einberufen. Dieser Landtag zog früh das literarische[71] und zu Beginn des 19. Jahrhunderts auch das editorische Interesse auf sich. Insgesamt

---

69 Stellvertretend für kommunikationsorientierte Forschungen: Friedrich, Drehscheibe Regensburg (wie Anm. 57); Johannes Burkhardt, Vorparlamentarische Formen im Heiligen Römischen Reich?, in: Peter Claus Hartmann / Florian Schuller (Hg.), Das Heilige Römische Reich und sein Ende 1806. Zäsur in der deutschen und europäischen Geschichte, Regensburg 2006, S. 23–37; Susanne Friedrich, Beobachten und beobachtet werden. Zum wechselseitigen Verhältnis von gedruckter Zeitung und Immerwährendem Reichstag, in: Volker Bauer (Hg.), Die Entstehung des Zeitungswesens im 17. Jahrhundert. Ein neues Medium und seine Folgen für das Kommunikationssystem der Frühen Neuzeit, Bremen 2011, S. 159–178; Christoph Kampmann, Der Immerwährende Reichstag als „erstes stehendes Parlament"? Aktuelle Forschungsfragen und ein deutsch-englischer Vergleich, in: Geschichte in Wissenschaft und Unterricht 55, 2004, S. 646–662; Karl Härter, Der Immerwährende Reichstag (1663–1806) in der historischen Forschung, publiziert am: 30.1.2013, in: zeitenblicke 11/2, URL: http://www.zeitenblicke.de/2012/2/Haerter/index_html (letzter Zugriff: 30.8.2019), URN: urn: nbn: de:0009-9-34227.

70 In den Jahren 1802 bis 1807 erschienen ohne Autorennennung neun Quellen- und Editionsbände („Der Landtag im Herzogthum Baiern") in großem zeitlichen Abstand zur Landtagsüberlieferung der Jahre 1514 bis 1669. Die chronologisch ungeordnete Reihe begann mit dem Landtag von 1605 und edierte sieben von insgesamt 37 bayerischen Landtagen des 17. Jahrhunderts. 1669 wurde der letzte Landtag einberufen. Als Herausgeber dieser protokollgestützten Landtagsverhandlungen ist die Landschaft selbst und nicht Franz von Krenner (1762–1819) zu vermuten, die sich im Jahrzehnt vor ihrer Auflösung 1808 im politisch-rechtlichen Tagesgeschäft zurückmeldete. Vgl. Joachim Wild, Quellenlage zum Alten Landtag, in: Walter Ziegler (Hg.), Der Bayerische Landtag vom Spätmittelalter bis zur Gegenwart. Probleme und Desiderate historischer Forschung. Kolloquium des Instituts für Bayerische Geschichte am 20. Januar 1995 im Maximilianeum in München (Beiträge zum Parlamentarismus 8), München 1995, S. 127–139. Bei dem hier ausgewerteten Band handelt es sich um: Der Landtag im Herzogtum Baiern vom Jahre 1605. Aus einer gleichzeitigen Handschrift. Erste und zweyte Abtheilung, [s.l.] 1802. Für die Städte und Märktelandschaft: „*Gravamina deß Standß von Stött vnnd Märkhten des Löblichen Fürstenthumbs Bayrn*", S. 323–350. Literatur- und Quellenachweis: BSB, Bavaria 1523–25, URN: urn: nbn: de: bvb:12-bsb00008586-4.

71 Johann Meyer, Dialogus: Ein Gespräch, deß jüngst gehaltenen bayerischen Landtags, in der weitberühmten fürstlichen Hauptstadt München, Anno 1605, München 1606.

hat sich die historische Forschung[72] mit den bayerischen Landtagen zwar intensiv beschäftigt, doch blieben Analysen zur Städtepolitik einzelner Landtage des 17. und 18. Jahrhunderts selten.[73]

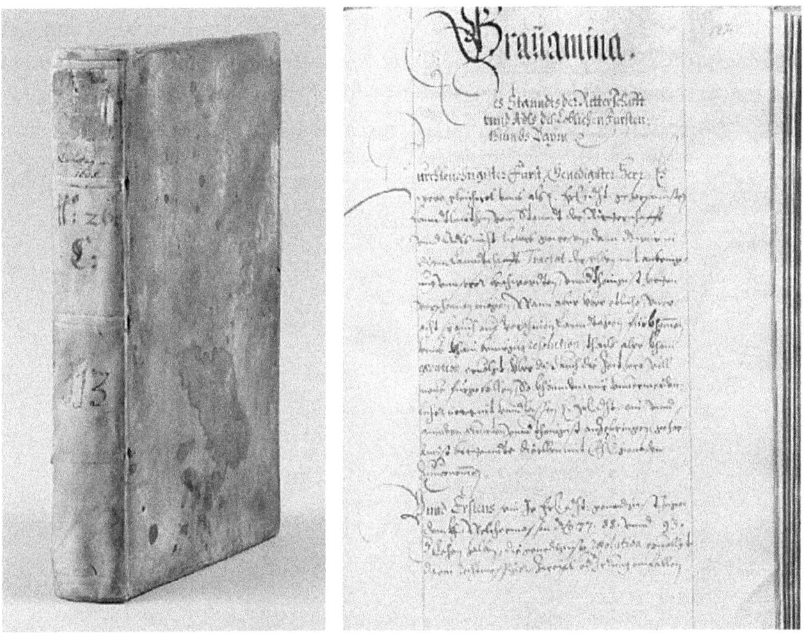

Abb. 15 a und 15 b: *Ausfertigung des Protokolls über den Landtag von 1605 unter Herzog Maximilian I., hier: „Gravamina“ der Ritterschaft und des bayerischen Adels. Bildnachweis: Buch- und Auktionshaus Zisska & Lacher, München, Auktionskatalog Nr. 73, 2019.*

---

72 Zur Landtagsforschung im Alten Reich vgl. insbesondere: Ziegler, Der Bayerische Landtag vom Spätmittelalter bis zur Gegenwart (wie Anm. 70); Rudolf Schlögl, Der bayerische Landtag von 1669. Nach einem Bericht des geheimen Kammersekretärs Corbinian von Prielmair, in: ZBLG 52, 1989, S. 221–254; Alois Schmid, Landtagsforschung in Bayern. Stand, Aufgaben, Perspektiven, in: ZBLG 75, 2013, S. 691–714; Reinhard Heydenreuter, Zur Rechtsstellung des landsässigen Adels im Kurfürstentum Bayern zwischen dem 16. und 18. Jahrhundert, in: Walter Demel / Ferdinand Kramer (Hg.), Adel und Adelskultur in Bayern (ZBLG, Beihefte B 32), München 2008, S. 43–105; Markus Nadler, Die Landstände von Pfalz-Neuburg (1505–1808) – Tragende Säule des Fürstentums und Garant seiner Selbstständigkeit, in: Neuburger Kollektaneenblatt 163, 2015, S. 7–43; Josef Breinbauer, Die Landstände des Hochstifts Passau, in: Passauer Jahrbuch 52, 2010, S. 137–146.

73 Carl Albin Hoffmann, Landesherrliche Städte und Märkte im 17. und 18. Jahrhundert. Studien zu ihrer ökonomischen, rechtlichen und sozialen Entwicklung in Oberbayern (Münchner Historische Studien, Abtl. Bayerische Geschichte 16), Kallmünz/Opf. 1997.

Erkenntnisreich ist hier neben der Analyse zu den kleinen Städten von Helmut Rankl[74] vor allem die Münchner Dissertation von Carl Albin Hoffmann, der 1997 an vier Beispielen[75] – dazu zählten die Residenzstadt München, das Rentamt Burghausen, die Stadt Weilheim und der *„gefreite"* Markt Trostberg – die oberbayerische Städtelandschaft charakterisierte. Der Schulterschluss mit den Landtagsprotokollen erfolgte dort allerdings schwerpunktmäßig mit dem letzten Landtag von 1669,[76] dessen Ausschüsse ausführlich über die schwierige ökonomische Ausgangslage der Städte nach dem Dreißigjährigen Krieg berichteten.[77]

Abb. 16: *Protokoll des Bayerischen Landtags, 1622. Bildnachweis: Buch- und Auktionshaus Zisska & Lacher, München, Auktionskatalog Nr. 73, 2019.*

74 Helmut Rankl, Altbayerische Kleinstädte im Spiegel landesherrlicher Erhebungen des 17. und 18. Jahrhunderts: Erding, Rosenheim, Trostberg und Murnau (Studien zur bayerischen Verfassungs- und Sozialgeschichte 28), München 2011.

75 Ebd., S. 18 f.

76 Franz von Krenner, Der Landtag im Churfürstenthum Baiern vom Jahre 1669, [s.l.] 1802.

77 Carl Albin Hoffmann, Landesherrliche Städte und Märkte im 17. und 18. Jahrhundert (wie Anm. 73), S. 2, 14 f., 33 f., 38, 56 f., 60, 63, 113, 116 f., 128, 235, 248, 250 f., 289, 299, 390, 406 und 442 f.

In der Übersicht der Verhandlungspunkte waren bayerische Märkte und Städte in der Landschaft unter der Rubrik *„Gravamina deß Standts von Stoett vnnd Maerkhten des loeblichen Fuerstenthumbs Bayrn“*[78] als Beschwerden des Bürgerstandes auf allen Landtagen präsent. Folgen wir den 1605 artikulierten Landtagsklagen – es handelte sich um 30 „*Gravamina*“ – altbayerischer Städte und Märkte als *partes pro toto*, dann wird schnell klar, dass sich zu Beginn des 17. Jahrhunderts die Interessen in der Städtelandschaft keineswegs mit der Politik der Landesherren und adeliger Hofmarken deckten. Die städtischen Jahrmärkte wurden zu Tummelplätzen für Landkrämer und migrierende Savoyarden,[79] die weder Markt- noch Zollgebühren respektierten, steuerschädlichen Fürkauf anboten und die importierten Waren keiner städtischen Qualitätsschau vorlegten.[80] Ähnliches galt für die landesherrlichen Wein-, Bier- und Salzhändler, die „*wohlfayle*“ Güter zusehends direkt aus „*Tyrol, Oesterreich vnnd Vlmb*“[81] importierten. Sie schädigten damit die städtischen Tavernen und ihre Schankgerechtigkeit an den Mautstraßen.[82] Der Bauernstand kaufte zum eigenen Vorteil aus „*faulenzerey*“ am Hof nicht verzolltes und auf Umwegen ins Land gebrachtes Getreide auf. Den Kornmärkten der Städte und Märkte entgingen dadurch Einnahmen, während die Bauern „*ettwann in den wirthsheußern ein bessers Sippl als anhaimbs zu essen haben*“.[83] „*Furth, Eschl- Cham vnnd Neuekirchen*“ und andere Gemeinden im Bayerischen Wald führten Klage, da die herzoglichen Brauhäuser gegen die bürgerlichen Brauereien konkurrierten und den Kommunen damit wichtige Ungelder entgingen.[84] In der bayerischen Textilregion verkauften von Haus zu Haus ziehende „*Gey-*“ und Leinenweber „*Rupfen vnnd Zwilch*“ an der Gewerbeaufsicht der Märkte vorbei. Ähnliches galt für die Störschneider, die den lukrativen „*Tuechschnitt*“ aus den Städten und Märkten zusehends auf das Land zogen.[85] Fürkäufer untergruben überall im Herzogtum das Monopol städtischer Jahres- und Wochenmärkte,[86] die seit dem Mittelalter über teure

---

78 Der Landtag im Herzogtum Baiern vom Jahre 1605 (wie Anm. 70), S. 323–350.

79 Martin Zürn, Savoyarden in Oberdeutschland. Zur Integration einer ethnischen Minderheit in Augsburg, Freiburg und Konstanz, in: Carl A. Hoffmann / Rolf Kießlng (Hg.), Kommunikation und Region (Forum Suevicum 4), Konstanz 2001, S. 381–419.

80 Der Landtag im Herzogtum Baiern vom Jahre 1605 (wie Anm. 70), 1. Gravamen.

81 Ulm.

82 Der Landtag im Herzogtum Baiern vom Jahre 1605 (wie Anm. 70), 2. Gravamen.

83 Ebd., 3. Gravamen.

84 Ebd., 5. Gravamen.

85 Ebd., 6. und 8. Gravamen.

86 Wolfgang Wüst, Frühneuzeitliches Marktgeschehen in der Fürsten- und Residenzstadt im Spiegel einer „*Verneuerten Policey= Ordnung*“ von 1678, „*wie solche bey gemeiner*

urkundliche Privilegien geschützt waren.[87] Das Herzogtum Bayern öffnete sich vor allem über den Getreide- und Salzhandel[88] gegenüber seinen Nachbarländern. Infolgedessen strömte zusehends fremdes Geld – gemeint waren die weißen Pfennige – ins Land. Die Fremdwährungen schätzte man in der Summe auf *„vber 200.000 fl.“* In den Städten kamen die Marktleute mit den ungewohnten Münzen nicht mehr zurecht und es wurde schwierig, Falschgeld zu identifizieren. Die kommunalen Märkte wurden so nachhaltig geschädigt, ebenso wie der schlechte Straßenzustand im Land von städtischen Fuhrleuten bisher vergeblich beklagt wurde.[89] Erst der Chausseebau[90] des 18. Jahrhunderts sollte im ruinösen Wegebau in den Reichskreisen Abhilfe schaffen. Die Ärare bayerischer Städte und Märkte litten ferner unter zunehmender Steuer- und Ungeldbefreiung für herzogliche Beamte und Angehörige eingesessener Klöster und Stifte. Die Beamten und Geistlichen könnten *„yederzeith ohne entgelt außgehen vnnd in den vnbefuegter weiß veruhrsachten Vncosten nie conemnirt werden.“*[91] Das Herzogtum griff im 16. und 17. Jahrhundert in das kommunale Wahlrecht ein mit der Verfügung, dass man künftig *„nit mehr Järlich ein Rathswahl halten“* solle.[92] Herzogliche Pfleger, Kastner und Rentmeister messen sich dazu bürgermeisterliche Befugnisse an, indem sie das *„Burgerliche Weßen, polizey vnnd andere Burgerliche sachen“* an sich ziehen.[93]

Stadt Oettingen solle gehalten werden“, in: Markus Würmseher / René Brugger (Hg.), Grenzüberschreitungen zwischen Altbayern und Schwaben. Geschichte, Politik und Kunst zu beiden Seiten des Lechs. Festschrift für Wilhelm Liebhart, Regensburg 2016, S. 119–134.

87 Der Landtag im Herzogtum Baiern vom Jahre 1605 (wie Anm. 70), 10. Gravamen.

88 Martin Ott, Salzhandel in der Mitte Europas. Raumorganisation und wirtschaftliche Außenbeziehungen zwischen Bayern, Schwaben und der Schweiz, 1750–1815 (Schriftenreihe zur bayerischen Landesgeschichte 165), München 2013; Ders., Salzfürsten in Bayern? Die Historiographie zum bayerischen Salz an der Wende zum 19. Jahrhundert, in: Alois Schmid / Hermann Rumschöttel (Hg.), Wittelsbacher-Studien. Festgabe für Herzog Franz von Bayern zum 80. Geburtstag (Schriftenreihe zur bayerischen Landesgeschichte 166), München 2013, S. 545–562.

89 Der Landtag im Herzogtum Baiern vom Jahre 1605 (wie Anm. 70), 16. und 17. Gravamen.

90 Wolfgang Wüst, Chausseen in Franken – Kunststraßen nach französischem Vorbild, in: Altfränkische Bilder NF 9/2014, 2013, S. 22–24.

91 Der Landtag im Herzogtum Baiern vom Jahre 1605 (wie Anm. 70), 20. Gravamen.

92 Ebd., 22. Gravamen.

93 Ebd., 27., 28. und 29. Gravamen, S. 346–348.

Gegen Ende des Landtagsprotokolls von 1605 wurden die Anliegen der südostdeutschen Städtelandschaft gegenüber der Landesherrschaft auf den Punkt gebracht. „*So wir dann ohne das der vnzweifenlichen hoffnung sein, E*[ure] *f*[ü]*rstl*[iche] *D*[urc]*hl*[aucht] *Seyen dißem gehorsamisten Standt in dißen vnd andern beschwerungen hilflich zuerscheinen, vnnd wo mit nur Jammer möglich Stätt vnnd Märkhten, die in etlich wenig iharen in merkhlichem vnd augenscheinlichen abfahl khommen, es seye mit befürderung der Gewerben oder in andere Weeg widerumb abzuhelffen, ohne das vnnd fürnemblich auch darumben, weil bey dißen Laydigen schweren vnnd gefehrlichen Zeitten gemeinen Vatterlandt in aller fürfallender gefahr, daß dißer standt in guetten würden seye: nit weniger daran gelegen, g*[ne]*di*[g]*st vnnd vätterlich gewült.*“[94]

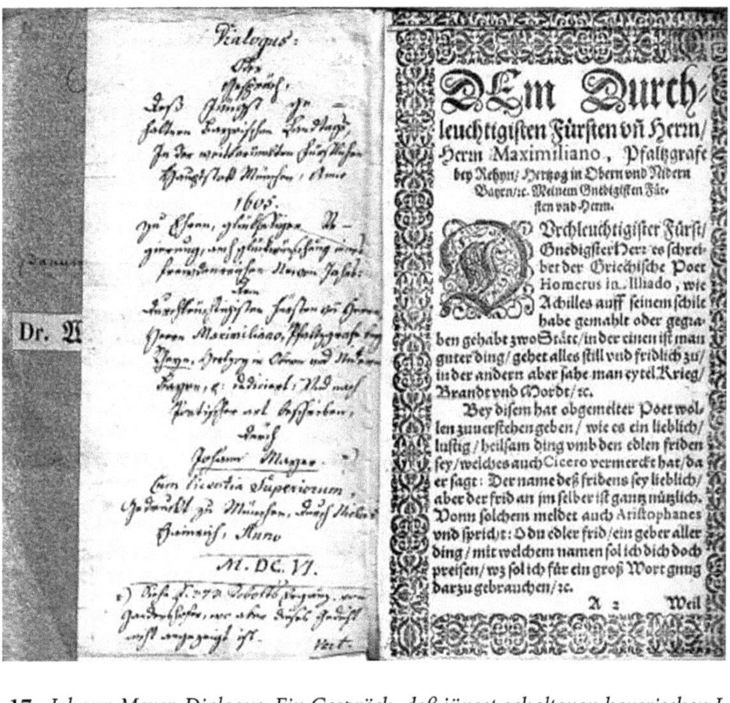
DEm Durchleuchtigisten Fürsten vñ Herrn/ Herrn Maximiliano, Pfaltzgrafe bey Rehyn/ Hertzog in Obern vnd Nidern Bayern/2c. Meinem Gnedigisten Fürsten vnd Herrn.

DVrchleuchtigister Fürst/ Gnedigster Herr es schreibet der Griechische Poet Homerus in Illiado, wie Achilles auff seinem schilt habe gemahlt oder gegraben gehabt zwo Stätt/ in der einen ist man guter ding/ gehet alles still vnd fridlich zu/ in der andern aber sahe man eytel Krieg/ Brandt vnd Mordt/2c.

Bey disem hat obgemelter Poet wollen zuuerstehen geben/ wie es ein lieblich/ lustig/ heilsam ding vmb den edlen friden sey/ welches auch Cicero vermerckt hat/ da er sagt: Der name deß fridens sey lieblich/ aber der frid an jm selber ist gantz nützlich. Vonn solchem meldet auch Aristophanes vnd spricht: O du edler frid/ ein geber aller ding/ mit welchem namen sol ich dich doch preisen/ wz sol ich für ein groß Wort gnug darzu gebrauchen/2c.

A 2 Weil

Abb. 17: *Johann Meyer, Dialogus: Ein Gespräch, deß jüngst gehaltenen bayerischen Landtags, in der weitberühmten fürstlichen Hauptstadt München, Anno 1605, München 1606. Bildnachweis: Antiquariat und Buchhandlung Carl Wegner, Berlin.*

94 Der Landtag im Herzogtum Baiern vom Jahre 1605 (wie Anm. 70), S. 349.

5. Ergebnisse

Die süddeutsche Städtelandschaft – im Plural wären sie präziser zu definieren – stellte sich in den drei zentralen Initiativen der Reichs-, Kreis- und Landtage als „sitzende“ Interessen- und Wertegemeinschaft dar. An Tagungsorten und -zeiten mangelte es den dort zusammengerufenen Konventen nicht; so überstieg die Zahl der Kreistagssessionen („*sessiones*“) zu Nürnberg in den Jahren 1783 und 1785 die Zahl von 2.000! Es wurde oft und lange getagt, wodurch sich für die beteiligten Delegationen ein Raum dichter Kommunikation öffnete. In ihm beschloss man interstädtische Reformen und korrigierte nach Kräften Pläne, die die Städte finanziell benachteiligten. Diese konsenssuchende kommunikative Leistung wird nicht geschmälert durch die Tatsache, dass sich so gut wie nie alle Städte durch eigene Deputierte vertreten ließen. Reichstags- und Kreistagsbeschlüsse erreichten auch die in der älteren Forschung oft als „verschlafen“ verzeichneten mindermächtigen Reichsstädte.[95] Gerade diese Städtegruppe war bisweilen landschaftsprägender als die frühmodernen Metropolen mit stark ausgeprägten Eigeninteressen. Große europäische Reichs-, Residenz- und Handelsstädte handelten nicht immer nach dem gruppenspezifischen Grundsatz: *Unus pro omnibus, omnes pro uno*. Quellen sind dazu in Fülle vorhanden. Endlose Korrespondenzserien auf der einen und zahllose Gravamina auf der anderen Seite spiegeln gleichermaßen – spezifisch und generell – urbane Interessen wider. In der Summe könnten sich daraus Städtelandschaften formiert haben.

---

95 So auch in dem erkenntnisreichen Überblick bei: Axel Gotthard, Das Alte Reich 1495–1806 (Geschichte kompakt), Darmstadt 2003, S. 18.

Tom Scott

Oberrheinische Städtelandschaften*

Abstract: The Upper Rhine lacks clearly defined territorial boundaries; commonly it has been seen as the valley of the Rhine running south to north. In fact, its commercial and urbanised features are just as prominent on an east-west axis, which takes in the Romandie and stretches into the Franche-Comté. Much of the region is stamped by medium-sized or small towns, many of which were wine-growing communities. Neither Basel nor Strasbourg developed as true metropolises, dominating or integrating the economy of their hinterlands. Walter Christaller's original central-place theories, predicated upon a hexagonal market area, do not work for the Upper Rhine; instead, as he later conceded, the networks should be seen as rhomboids, linking exports down- or upriver in a succession of lozenges.

Urban and (thoroughly) urbanised landscapes need to be distinguished (Franz Irsigler); the former comprehended settlements where the distinction between (small) towns and villages gives rise to conceptual imprecision: many small towns were in reality merely circumvallated villages. A comparison with the Neckar valley in Württemberg and with the Romandie shows that neither a sparse nor a dense distribution of settlements accounts for their character as urban landscapes; only the integrated, export-oriented viticulture of Alsace, by contrast, comes close to echoing the regional textile landscape of eastern Swabia. This lack of integration and co-ordination rendered the urban settlements peculiarly vulnerable to rural craft and market competition and to political rivalries between territorial lordships, especially in the confessional age.

Wer den Oberrhein als eine Städtelandschaft analytisch in den Griff bekommen will, sieht sich unmittelbar vor eine zweifache Schwierigkeit gestellt. Zum einen lässt sich die Region im Gegensatz zu Ländern wie Bayern, Franken oder Schwaben kaum territorial erfassen, auch flächenmäßig sind seine Grenzen unscharf: Wo hört nämlich der Oberrhein auf, wo fängt der Mittelrhein an? Wie ist überdies der Hochrhein – der Flusslauf vom Bodensee bis zum Rheinknie bei Basel – einzuordnen?[1] Zum anderen erweisen sich Versuche, die Städte der Region in zentralörtlichen Kategorien gemäß Walter Christallers ursprünglich

---

* Dem Andenken an meinen Kollegen und Freund Rolf Kießling (1941–2020) gewidmet.

1 Grundlegend, mit ausführlichen bibliographischen Hinweisen: Heinz Krieg, Zur Geschichte des Begriffs „historische Landschaft" und der Landschaftsbezeichnung „Oberrhein", in: Peter Kurmann / Thomas Zotz (Hg.), Historische Landschaft – Kunstlandschaft? Der Oberrhein im späten Mittelalter (Vorträge und Forschungen 68), Ostfildern 2008, S. 31–64.

übergreifendem Marktprinzip – das heißt hexagonal wie eine Bienenwabe – zu kartographieren, als außerstande, sie sinnvoll zueinander in Beziehung zu setzen.[2] Vielmehr ist man gezwungen, auf Christallers erst nach dem Zweiten Weltkrieg ausgearbeitetes Verkehrsprinzip zu rekurrieren, wonach sich die hierarchische Verteilung von zentralen Orten rautenförmig gestaltet, das heißt nach längsgezogenen Achsen, die dem auf Export ausgerichteten Wirtschaftsgefüge entlang des Rheins als Hauptverkehrsader entsprechen.[3] Auch dann, wie Franz Irsigler bemerkt hat, bleibt Christallers Modell zu statisch, um geographische beziehungsweise historische Unregelmäßigkeiten zu erklären: Es vermag sowohl Überlappungszonen als auch Leerzonen ohne städtische Durchdringung kaum Rechnung zu tragen.[4]

Es tritt ein zweites Problemfeld hinzu. Gerade im Westen des Reichs sind umfangreiche Forschungsprogramme über das Städtewesen – man denke an das unlängst abgeschlossene Trierer Projekt zur Städtelandschaft zwischen Rhein und Maas[5] – auf das Hindernis gestoßen, einen überzeugenden Katalog von städtischen Merkmalen aufzustellen. Die Trierer Forschungsgruppe von Monika Escher, Alfred Haverkamp und Frank G. Hirschmann hat einen Katalog von 56 Kriterien ausgearbeitet, die in einem additiven Punktesystem den Stadtcharakter einer Siedlung erfassen will, wobei die Autoren die Schwelle zum echten Stadtstatus um die Mitte des 14. Jahrhunderts bei mindestens sieben Punkten ansetzen.[6] Für den linksufrigen Oberrhein hat Bernhard Metz die Trierer Ergebnisse

---

2 Walter Christaller, Die zentralen Orte in Süddeutschland. Eine ökonomisch-geographische Untersuchung über die Gesetzmäßigkeit der Verbreitung und Entwicklung der Siedlungen mit städtischer Funktion, Jena 1933.

3 Walter Christaller, Das Grundgerüst der räumlichen Ordnung in Europa. Die Systeme der europäischen Zentralen Orte (Frankfurter Geographische Hefte 24), Frankfurt am Main 1950.

4 Franz Irsigler, Stadt und Umland in der historischen Forschung. Theorien und Konzepte, in: Neithard Bulst / Jochen Hoock / Franz Irsigler (Hg), Bevölkerung, Wirtschaft und Gesellschaft. Stadt-Land-Beziehungen in Deutschland und Frankreich vom 14. bis zum 19. Jahrhundert, Trier 1983, S. 13–38, hier: S. 17. Er kritisiert Christallers Dispersionsmuster als zu ökonomisch und vom damaligen Geist der im Osten zu gewinnenden Lebensräume getragen.

5 Monika Escher / Frank G. Hirschmann (Hg.), Die urbanen Zentren des hohen und späten Mittelalters. Vergleichende Untersuchungen zu Städten und Städtelandschaften im Westen des Reichs und in Ostfrankreich, 3 Bde. (Trierer Historische Forschungen 50/1–3), Trier 2005.

6 Monika Escher / Alfred Haverkamp / Frank G. Hirschmann (Hg.), Städtelandschaft – Städtenetz – zentralörtliches Gefüge (Trierer Historische Forschungen 43), Mainz 2000, Einleitung, S. 9–53, hier: S. 14–16, 52 f.

dahingehend kritisiert, dass er die entscheidenden Merkmale zwar bei sieben belassen will anstatt der von Trier vorgeschlagenen 56, fragt aber zugleich, warum sich die Trierer Gruppe auf ein Mindestmaß von sieben Punkten festgenagelt hat, anstatt etwa acht, neun oder zehn![7] Zieht man ferner die Stadtgründungen in Württemberg und in der Westschweiz (Romandie) – ich fasse den Oberrhein hier absichtlich weitläufig auf – in Betracht, so treten eklatante Unterschiede auf, die vornehmlich herrschaftlich aber auch bisweilen naturräumlich und verkehrsstrategisch bedingt waren.[8] Eine weitere Differenzierung hat Franz Irsigler vorgenommen, wonach er zwischen Stadtlandschaften und Städtelandschaften unterscheiden will. Diese kennzeichnen Gebiete mittlerer Größe mit variablem Urbanisierungsgrad, wo entweder ein bestimmter Stadttypus oder eine klar erkennbare Städtehierarchie dominiert. Jene weisen eine markante Städtedichte auf, wo über ein Viertel der Bevölkerung in Städten angesiedelt ist, deren Beziehungen zum Hinterland eine Intensität erreichen, dass die ländliche Bevölkerung am eigenen Profil verliert.[9] Solche Stadtlandschaften begegnen am Oberrhein allenfalls in den Vogesenvorbergen, wo sich eine Überzahl an kleinen, dicht beieinander liegenden Siedlungen – in den meisten Fällen eigentlich ummauerte Weinbaudörfer – eine erlahmende Konkurrenz zu liefern scheint, bis man sich vergegenwärtigt, dass ihre Wirtschaft auf Export ausgerichtet war.

---

7 Bernhard Metz, Essai sur la hiérarchie des villes médiévales d'Alsace (1200–1350), in: Revue d'Alsace 128, 2002, S. 49–53. Im Elsass seien, nach Metz, die Stadtrechtsverleihungen ein unzuverlässiges Kriterium (ebd., S. 51). Der zweite Teil des Aufsatzes erschien in: Revue d'Alsace 134, 2008, S. 129–166. Ein abschließender dritter Teil steht noch aus. Ich verzichte hier bewusst auf eine Analyse des Stadtbegriffs, die offenbar kein Ende nehmen will. Entgegen den diesbezüglichen ausufernden Arbeiten zum Spätmittelalter sei auf das jüngst erschienene Werk für den Zeitraum von 1100 bis 1400 verwiesen: Gabriel Zeilinger, Verhandelte Stadt. Herrschaft und Gemeinde in der frühen Urbanisierung des Oberelsass vom 12. bis zum 14. Jahrhundert (Mittelalter-Forschungen 60), Ostfildern 2018, das den derzeitigen Forschungszustand Revue passieren lässt.

8 Tom Scott, Kleine Städte, keine Städte. Das sogenannte „urbane Netz" in Südwestdeutschland im ausgehenden Mittelalter, in: Herbert Knittler (Hg.), Minderstädte – Kümmerformen – gefreite Dörfer. Stufen zur Urbanität und das Märkteproblem (Beiträge zur Geschichte der Städte Mitteleuropas 20), Linz 2006, S. 181–202.

9 Franz Irsigler, Städtelandschaften und kleine Städte, in: Helmut Flachenecker / Rolf Kießling (Hg.), Städtelandschaften in Altbayern, Franken und Schwaben. Studien zum Phänomen der Kleinstädte während des Mittelalters und der Frühen Neuzeit (Zeitschrift für bayerische Landesgeschichte, Beihefte B 15), München 1999, S. 13–38.

Bei näherer Betrachtung lässt sich die Städtelandschaft des Oberrheins in drei Zonen aufteilen: 1. Der Vogesenrand bis an die Ill, deren Verkehrsfunktion bis Straßburg mit dem Rhein effektiv parallel verläuft;[10] 2. Das platte Land zwischen Ill und Rhein, ein sumpfiges, mit Hardten durchsetztes Terrain, fast städteleer, sodann der hügelige Sundgau mit ausgedehntem Ackerbau und mehreren vorderösterreichischen Amtsstädten; 3. Die älteren rechtsrheinischen Gründungen, ob Territorial- oder Landstädte, die Verwaltungs- und Marktfunktionen für ihr Umland wahrnahmen. Nach Norden kam es dank der Städtepolitik der Markgrafen von Baden zu einer Häufung von Kleinstädten, deren Einflusspotential bescheiden war und nie raumbildend wirkte.[11]

Richten wir zum Vergleich unseren Blick auf angrenzende Landschaften, so wird das Bild von einer bevölkerungsreichen, dennoch von mittleren und kleinen Städten geprägten Region bestätigt. In Württemberg entstammten Stadtgründungen nicht selten der Absicht des Fürsten, rivalisierende Siedlungen kaltzustellen, die in unbequemer Nähe nebeneinander standen. Das gleiche gilt für die Romandie,[12] wo Städtchen entlang der Broye oder der Saane feindliche Blicke von beiden Ufern tauschten. Diese Städtchen, die mehrheitlich im 13. und 14. Jahrhundert entstanden, dienten eher dem Repräsentationsbedürfnis

10 Gerrit Jasper Schenk, ‚der straßen halb uff der Yll'. Wasserwege im Elsass als „kritische Infrastruktur" für Wirtschaft und Gesellschaft (1350–1550), in: Kurt Andermann / Nina Gallion (Hg.), Weg und Steg. Aspekte des Verkehrswesens von der Spätantike bis zum Ende des Alten Reiches (Kraichtaler Kolloquien 11), Ostfildern 2018, S. 121–152.

11 Scott, Kleine Städte, keine Städte (wie Anm. 8), S. 185–189; Arnold Scheuerbrandt, Südwestdeutsche Stadttypen und Städtegruppen bis zum frühen 19. Jahrhundert. Ein Beitrag zur Kulturlandschaftsgeschichte und zur kulturräumlichen Gliederung des nördlichen Baden-Württemberg und seiner Nachbargebiete (Heidelberger Geographischen Arbeiten 32), Heidelberg 1972, S. 44 f. Zu den württembergischen Stadtgründungen in nachstaufischer Zeit siehe: Jürgen Sydow, Adlige Stadtgründer in Südwestdeutschland, in: Erich Maschke / Jürgen Sydow (Hg.), Südwestdeutsche Städte im Zeitalter der Staufer (Stadt in der Geschichte 6), Sigmaringen 1980, S. 184–186; Rüdiger Stenzel, Die Städte der Markgrafen von Baden, in: Jürgen Treffeisen / Kurt Andermann (Hg.), Landesherrliche Städte in Südwestdeutschland (Oberrheinische Studien 12), Sigmaringen 1994, S. 89–130.

12 Scott, Kleine Städte, keine Städte (wie Anm. 8), S. 193–197; Hektor Ammann, Die schweizerische Kleinstadt in der mittelalterlichen Wirtschaft, in: Festschrift Walther Merz, Aarau 1928, S. 158–215; Ders., Über das waadtländische Städtewesen im Mittelalter und über landschaftliches Städtewesen im Allgemeinen, in: Schweizerische Zeitschrift für Geschichte 4, 1954, S. 1–87.

von kleineren adligen Herrschaften,[13] als dass sie eine wirtschaftliche oder verkehrsstrategische Funktion erfüllten, geschweige denn eine Territorialpolitik *en miniature* anstrebten. Ihre Zahl schrumpfte bis 1500 zusehends dahin, sie wurden aufgelassen und verschwanden von der Bildfläche.

Am Oberrhein haben wir es also mit einer breiten Palette von städtischen Siedlungen zu tun, in der Weinbaustädte, Marktflecken und Amtsstädte dominierten. Sogar die berühmte elsässische Dekapole fällt hier kaum aus dem Rahmen, abgesehen von ihrem privilegierten Status als Reichsstädte, denn unter ihnen konnten allenfalls Colmar, Schlettstadt und Hagenau ein wirtschaftliches Gewicht für sich beanspruchen. Kein Mitglied der Dekapole erreichte eine Bevölkerungszahl von über 5.000 Einwohnern. Mülhausens von seinem Nachbarn Basel 1516 betriebener Anschluss an die Eidgenossenschaft als zugewandter Ort mag zwar der Reichsstadt zunächst zu größerer Bewegungsfreiheit verholfen haben, sie blieb indes von vorderösterreichischem Territorium umzingelt, sodass sie im Laufe des 16. Jahrhunderts zunehmender katholischer Pression seitens der Habsburger ausgesetzt wurde.[14] Auch die rechtsrheinischen Reichsstädte in der Ortenau, Offenburg, Gengenbach und Zell am Harmersbach, waren politisch und wirtschaftlich von untergeordneter Bedeutung.

Damit sind wir bei den einzigen oberrheinischen Metropolen Basel und Straßburg angelangt. In Basel brachte der Sieg der Handwerkerzünfte 1526 einen allmählichen Rückgang seiner kommerziellen Blüte mit sich, derer sich die Stadt als Schaltstelle zwischen Oberschwaben und Burgund einst erfreute.[15] Der berühmte Basler Kaufmann an der Wende vom 15. zum 16. Jahrhundert Ulrich

---

13 Siehe für Südwestdeutschland das bissige Urteil von: Gerhard Fouquet, Stadt, Herrschaft und Territorium – Ritterschaftliche Kleinstädte Südwestdeutschlands an der Wende vom Mittelalter zur Neuzeit, in: Zeitschrift für die Geschichte des Oberrheins 141, 1993, S. 120: „Die Stadtqualität, das in bzw. über der Stadt liegende Schloss, die Kirchen, vor allem die noch bis ins 16. Jahrhundert hinein recht wirkungsvolle Ummauerung waren für die Ritterschaft Ausweise ihres Standes und Reichtums. Diese Versatzstücke von Urbanität inmitten von stadtähnlichen Dörfern wurden schließlich zu steinernen Zeugen der Reichsunmittelbarkeit dieser Kleinherrschaften."

14 Georges Livet / Raymond Oberlé, Histoire de Mulhouse des origines à nos jours, Straßburg 1977, S. 79 f., 280 f.

15 Hans Füglister, Handwerksregiment: Untersuchungen und Materialien zur sozialen und politischen Struktur der Stadt Basel in der ersten Hälfte des 16. Jahrhunderts (Basler Beiträge zur Geschichtswissenschaft 143), Basel/Frankfurt am Main 1981, S. 272–292. Die Gewerbeordnung wurde 1552 aufgehoben, da sie ihre Wirkung verfehlt hat (ebd., S. 281 f.).

Meltinger hatte bezeichnenderweise keinen Nachfolger.[16] Basels Ausstrahlung auf den Oberrhein blieb weitgehend auf den kulturellen und geistigen Bereich beschränkt, denkt man an die humanistischen Zirkel und die ihnen dienenden Offizinen. Zwar avancierte Basel zum führenden Kreditplatz der Schweiz (und außerdem zum Geldleiher an süddeutsche Adelsfamilien und Gotteshäuser) auf der Basis der 1504 gegründeten Basler Kredit- und Wechselbank, man kann indes schwerlich behaupten, diese finanzielle Vorrangstellung habe Basels Position innerhalb des oberrheinischen Wirtschaftsgefüges wesentlich beeinflusst. Basel gelang es, ein bescheidenes Territorium zu erwerben, seine Rolle im regionalen Handel hat es dafür weitgehend eingebüßt.[17]

Straßburg, trotz seiner herausragenden Funktion als Stapel- und Umschlagplatz für den begehrten elsässischen Wein, hat sich nie zu einer internationalen Handels- oder Messestadt entwickelt. Obwohl Straßburgs Bevölkerung Mitte des 15. Jahrhunderts um die 18.000 Einwohner zählte und kontinuierlich auf 25.000 im Jahre 1580 stieg, um kurz nach 1600 sogar 29.000 zu erreichen,[18] lieferte sie weder den süddeutschen Metropolen Nürnberg, Augsburg und Ulm noch dem niederrheinischen Köln eine ernsthafte Konkurrenz. Straßburgs Beteiligung an der Reichspolitik der Reformationsjahre – vorab die diplomatische Tätigkeit Jakob Sturms – verschaffte der Reichsstadt zwar politischen Ruhm und Einfluss, ihre wirtschaftliche Vorrangstellung in der oberrheinischen Städtelandschaft erfuhr dadurch keine Aufwertung. 1570 wurde zwar der Weihnachtsmarkt in eine wahre Handelsmesse umgewandelt, dies allerdings zu einer Zeit, wo die Besucherfrequenz zurückging und sich die ersten Anzeichen eines wirtschaftlichen Siechtums zu erkennen gaben.[19] Allein als Herberge für französische Religionsflüchtlinge konnte sich Straßburg mit Genf messen.

---

16 Matthias Steinbrink, Ulrich Meltinger. Ein Basler Kaufmann am Ende des 15. Jahrhunderts (Vierteljahrschrift für Sozial- und Wirtschaftsgeschichte, Beiheft 197), Stuttgart 2007.

17 Hans R. Guggisberg, Basel in the sixteenth century. Aspects of the city republic before, during, and after the Reformation, St. Louis/MO 1982, Neudruck: Eugene/OR 2010, S. 27; Martin Körner, Solidarités financières suisses au seizième siècle. Contribution à l'histoire monétaire, bancaire et financière des Cantons suisses et des États voisins (Bibliothèque Historique Vaudoise 66), Lausanne 1980.

18 Zahlen bei: Knut Schulz, Handwerksgesellen und Lohnarbeiter. Untersuchungen zur oberrheinischen und oberdeutschen Stadtgeschichte des 14. bis 17. Jahrhunderts, Sigmaringen 1985, S. 29 f.

19 Jean-Pierre Kintz, La société strasbourgeoise du milieu du XVIe siècle à la fin de la Guerre des Trente Ans 1560–1650. Essai d'histoire démographique, économique et sociale, Paris 1984, S. 380–386.

Das ist ein ernüchterndes Fazit. Um 1600 geriet der einst blühende Oberrhein als eine ausgesprochene Städtelandschaft ins Hintertreffen. Worauf ist dies zurückzuführen? Ein ausdifferenziertes, an Ressourcen reiches und verkehrsstrategisch optimal gelegenes Wirtschaftsgebaren erwies sich längerfristig als unfähig, ein Abgleiten des Oberrheins an die Peripherie der europäischen Wirtschaft aufzuhalten.[20]

An erster Stelle wurden gemeinhin die Handwerkszünfte für eine wirtschaftliche Sklerose verantwortlich gemacht. Ihre Abkapselungsgesinnung und ihre Aversion gegen neue Gewerbepraktiken, die einem freien Arbeitsmarkt ohne Aufnahmebeschränkungen sowie einer freien Marktwirtschaft ohne reglementierte Preistarife und Produktionshöchstgrenzen mit Bedenken und Misstrauen gegenüberstanden, galten lange Zeit als entwicklungshemmend und unternehmensfeindlich. Der Sieg der Basler Handwerkszünfte 1526 und das Bestreben der Freiburger Zünfte um 1500, eine Politik der sogenannten Stadtwirtschaft, das heißt Protektionismus, durchzusetzen, mögen diesem Argument eine gewisse Plausibilität verleihen.[21] Nur ist dabei zu erwägen, ob diese Bemühungen eher als eine Reaktion auf eine wirtschaftliche Schlechtwetterlage aufzufassen sind, als dass sie einen Niedergang mitverschuldet hätten. In den letzten Jahren ist das meist negative Urteil einer positiveren Einschätzung gewichen, die den Zünften innovatorische Fähigkeiten von vornherein nicht abspricht.[22]

Die Drohung für eine florierende Städtewirtschaft, die das Aufkommen von ländlichem Handwerk, Dorfmärkten und Hausieren darstellte, ist am Oberrhein lediglich für den Breisgau eingehend erforscht worden,[23] obgleich das Phänomen im Spätmittelalter in weiten Teilen des Reichs verbreitet war und als Usurpation der ökonomischen Funktion der Städte angesehen wurde. Für den Südwesten sind Beschwerden über ländliche Konkurrenz für Württemberg, Baden, das Elsass, den Breisgau und die Schweiz überliefert.[24] Den städtischen Obrigkeiten

20 Siehe die Überlegungen in: Tom Scott, Regional identity and economic change: The Upper Rhine, 1450–1600, Oxford 1997, S. 287–322.

21 Guggisberg, Basel in the sixteenth century (wie Anm. 17.), S. 27; Tom Scott, Freiburg and the Breisgau. Town-country relations in the age of Reformation and Peasants' War, Oxford 1986, S. 128–141.

22 John Langton / Göran Hoppe, Town and country in the development of early modern western Europe (Historical Geography Research Series 11), Norwich 1983.

23 Scott, Freiburg and the Breisgau (wie Anm. 21), S. 116–118, 157–160.

24 Scott, Regional identity and economic change (wie Anm. 20), S. 104–121; Ders., Bürger, Handwerker und Sondergruppen. Zur verfächerten Sozialstruktur des Dorfs in Südwestdeutschland um die Wende vom Mittelalter zur Neuzeit, in: Kurt Andermann / Oliver Auge (Hg.), Dorf und Gemeinde. Grundstrukturen der ländlichen Gesellschaft

standen zwei Lösungen zur Verfügung: Sie konnten versuchen, mit polizeilichen Maßnahmen gegen die konkurrierenden Handwerker und Märkte vorzugehen, meist ohne ersichtlichen Erfolg, oder sie zwangen die ländlichen Handwerker, den städtischen Zünften beizutreten, wo sie deren Gewerbeordnungen fortan zu gehorchen hatten. So entstanden im Sundgau und in einigen schweizerischen Stadtstaaten Territorialzünfte, die die ländliche Konkurrenz entweder nivellierten oder ausschalteten.

Ebenso wenig ist die herrschaftliche Kleinkammerung des Oberrheins geeignet, als Erklärung für eine Städtelandschaft ohne herausragende Zentren herzuhalten. Nicht die Größe eines Territoriums an sich ist ausschlaggebend, sondern vielmehr die Wirtschaftspolitik des Herrschaftsträgers. Im Zeitalter der Reformation gewann dieser Aspekt an Brisanz, da sich konfessionelle Rivalitäten wirtschaftshemmend oder -störend auswirken konnten: Am rechten Rheinufer sind die langwierigen Streitigkeiten zwischen dem katholischen Vorderösterreich und der evangelischen Markgrafschaft Baden-Durlach zu nennen, deren Herrschaftsgebiete ineinandergriffen, mit entsprechenden gegenseitigen Zoll- und Marktsperren. Diese konnten sogar zwischen den Markgrafschaften selber erfolgen, als Baden-Durlach Blockaden seitens der katholischen Markgrafschaft Baden-Baden ausgesetzt war.[25]

In der Frühneuzeit kam es am Oberrhein zu keiner wirtschaftlichen Stufenleiter, die kleine Manufakturzentren über Umschlagplätze und Zulieferer mit den Handels- und Verteilungsmetropolen in einer wirtschaftsrationalen Hierarchie verbunden hätte, wie Rolf Kießling für die Textillandschaft Ostschwabens mustergültig dargelegt hat.[26] Dieses sogenannten Verlagssystems hat sich seinerzeit der Basler Kaufmann Ulrich Meltinger teilweise bedient, vor allem in der westschweizerischen Textilindustrie, zumal in seinen Beziehungen zu Freiburg im Üchtland, und im ostschweizerischen Eisenerzbergbau. Ansonsten trifft man am Oberrhein selten auf Verlag; lediglich im Norden des Elsass um Weißenburg sind Ansätze zu einer spezialisierten Textilmanufaktur zu beobachten. Die von Hektor Ammann so betitelte „Textilgruppe“, die Weißenburg, Pfaffenhofen, Hagenau, Zabern, Saarburg und Oberehnheim umfasste, hat Tuche für den Regionalhandel hergestellt, die Verleger aus Straßburg in

in Spätmittelalter und Frühneuzeit (Kraichtaler Kolloquien 8), Epfendorf 2012, S. 143–147.

25 Scott, Regional identity and economic change (wie Anm. 20), S. 135–149, 159–162.

26 Rolf Kießling, Die Stadt und ihr Land. Umlandpolitik, Bürgerbesitz und Wirtschaftsgefüge in Ostschwaben vom 14. bis ins 16. Jahrhundert (Städteforschung A 29), Köln/Wien 1989.

Auftrag gaben.[27] Mit der Verbreitung des Verlagssystems, das häufig etwas übertreibend als Motor eines neuen Entwicklungsschubs in der europäischen Wirtschaft angesehen wird, ging in Ostschwaben eine starke Zunahme von Marktortgründungen einher, die sich ohne Rücksicht auf städtische Ratserlasse, Gewerbeordnungen und Rechtsprivilegien entfalten konnten. Mögen damit die Vorrangstellung der mit Stadtrechten ausgestatteten städtischen Siedlungen und somit ihre Zentralität eingeschränkt worden sein, so ist die starke Vermehrung von Marktorten (gelegentlich als „Minderstädte" bezeichnet) ein ungefähres Zeichen für die wirtschaftliche Vitalität solcher Landschaften.[28] Eine ähnliche Entwicklung sucht man am Oberrhein vergebens: Im Breisgau entstammten die zahlreichen Marktgründungen des 15. Jahrhunderts dem Bestreben der Obrigkeiten, in einer herrschaftlich zersplitterten Landschaft Konkurrenz entgegenzuwirken beziehungsweise Umlandsicherung zu bewerkstelligen.[29] Dennoch hat die internationale Forschung jüngst auf die Ankurbelung hingewiesen, die die Marktgründungen für den gesteigerten Regionalbedarf in Europa nach der spätmittelalterlichen Strukturkrise geleistet haben.[30]

Zum Schluss gilt es, die oberrheinische Städtelandschaft in ein zentralörtliches Raster einzuordnen. Hier hat wiederum Rolf Kießling eine Vorreiterrolle gespielt. Er unterscheidet nämlich zwischen drei Modellen: einem

---

27 Hektor Ammann, La place de l'Alsace dans l'industrie textile du Moyen Âge, in: La bourgeoisie alsacienne. Études d'histoire sociale (Publications de la Société Savante d'Alsace et des Régions de l'Est, grandes publications 5), Straßburg 1967, S. 80–89.

28 Rolf Kießling, Kleinstädte und Märkte als regionalpolitische Instrumente. Ostschwaben vom 14. bis zum 16. Jahrhundert, in: Flachenecker / Kießling, Städtelandschaften in Altbayern, Franken und Schwaben (wie Anm. 9), S. 243–288; Carl A. Hoffmann, Der altbayerische Markt in der frühen Neuzeit: eine „Minderstadt" in der bayerischen Städtelandschaft, in: Knittler, Minderstädte - Kümmerformen - gefreite Dörfer (wie Anm. 8), S. 305–323.

29 Tom Scott, Städte und Märkte im spätmittelalterlichen Breisgau zwischen wirtschaftlicher Konkurrenz und Umlandsicherung, in: Ursula Huggle / Thomas Zotz (Hg.), Burgen, Märkte, kleine Städte. Mittelalterliche Herrschaftsbildung am südlichen Oberrhein, in: Das Markgräflerland 2, 2003, S. 164–181.

30 Stephan R. Epstein, Regional fairs, institutional innovation and economic growth in late medieval Europe, in: Economic History Review 47, 1994, S. 459–482. Sein Argument bezieht sich vornehmlich auf die Toskana. Vgl. hierzu: Paolo Malanima, La formazione di una regione economica: la Toscana nei secoli XIII–XV, in: Società e Storia 6, 1983, S. 229–269. Im weiteren Zusammenhang siehe: Stephan R. Epstein, Freedom and growth. The rise of states and markets in Europe, 1300–1750 (Routledge Explorations in Economic History 17), London/New York 2000, S. 73–88.

Territorial-Modell (Altbayern), einem regionalen Hierarchie-Modell (Ostschwaben) und einem regionalen Dispersions-Modell (Oberrhein). In Altbayern hatten die recht zahlreichen Marktorte vornehmlich eine Verwaltungsfunktion wahrgenommen, beschränkten sich daher ökonomisch auf Warenaustauch mit dem Umland und erreichten allenfalls ein stadtähnliches Profil. In dem Textilrevier Ostschwaben nahmen die vielen Marktsiedlungen eine Zuliefererfunktion innerhalb einer monostrukturierten Gewerbelandschaft ein. Von Anfang an waren sie in eine urbane Hierarchie integriert. Am Oberrhein herrschte dagegen keine übergreifende oder einheitliche Gewerbelandschaft vor, sondern, so Kießling, „eine komplexe Verbindung verschiedener Wirtschaftssektoren in kleinräumigen Schwerpunktbereichen, die auf inneren Austausch und partiellen Export orientiert sind".[31] In diesem Modell reichen Mittel- und Kleinstädte aus, die Bedürfnisse der Wirtschaft an Ressourcen, Produktion und Verteilung zu decken; Metropolen mit finanzkräftigen Kaufleuten und verfügbarem Investmentkapital sind kaum gefragt.

Im Mittelalter hat die ausdifferenzierte Wirtschaft des Oberrheins eine Städteblüte herbeigeführt; in der Frühneuzeit, als der Ertrag von den Silberbergwerken versiegte beziehungsweise der Erzabbau zu kostspielig wurde und die Gewinne vom elsässischen Weinbau infolge eines Geschmackswandels auf rheinische und pfälzische Weine oder in Norddeutschland auf gehopftes Bier sanken,[32] verfielen die oberrheinischen Mittel- und Kleinstädte in geruhsame Selbstgenügsamkeit.

---

31 Rolf Kießling, Zwischen Stadt und Dorf? Zum Marktbegriff in Oberdeutschland, in: Peter Johanek / Franz-Joseph Post (Hg.), Vielerlei Städte. Der Stadtbegriff (Städteforschung A 61), Köln/Weimar/Wien 2004, S. 121–143, hier insbes.: S. 140 f.

32 Médard Barth, Der Rebbau des Elsaß und die Absatzgebiete seiner Weine. Ein geschichtlicher Durchblick, Bd. 1, Straßburg/Paris 1958; Tom Scott, Medieval viticulture in the German-speaking lands, in: German History 20, 2002, S. 95–115, hier insbes.: S. 112–114.

Helmut Flachenecker

Städtelandschaft Franken

Abstract: The history of Franconia in the High and Late Middle Ages is characterised by an amazing diversity of territorial dominions. A huge number of middle-sized and smaller territories, situated in Franconia, characterise an remarkable Culture Landscape, made of castles, towns, markets with impressive – and today most romanticised – wall systems and monasteries, which gives every observer's eye an idea of the former political structures still in the present.

In the following, the term "Urban Landscape" is discussed as an important part of Cultural Landscapes. Then the development of towns in Franconia is presented in a chronological way. Because most of the towns in this region were subordinated to a clergy-ruler, the developments in the dioceses of Würzburg and Eichstätt are being examined in more detail. The importance of towns for the historical development of Franconia is out of question, especially their contribution to the general urbanisation of the area. However, the research focus depends on current regional historical approaches, which explore town-structures more economically, or more as a special legal district, or as a centre for surrounding regions, or in its functions serving to expand the territories of the town-rulers.

1. Städtelandschaft – Kulturlandschaften

Zwischen 1100 und 1350 sind in Mitteleuropa in hoher Zahl Städte gegründet worden. Mit diesen neuen Formen gesellschaftlichen Zusammenlebens veränderte sich die Kulturlandschaft auf nachdrückliche Weise. Schätzungsweise sollen es um 1450 an die 5.000 gewesen sein. Anschließend gab es keine Periode mehr in Europa, in der in einer solchen hohen Zahl urbane Zentren unterschiedlichster Größe gegründet worden sind.[1] Diese intensive Phase der Stadtgründung und des Stadtausbaus haben die kulturellen Landschaften in Europa

---

1 Eberhard Isenmann, Die deutsche Stadt im Spätmittelalter, Stuttgart 1988, Köln/Weimar/Wien ²2014, S. 19; Ulrich Knefelkamp, Alltag in der Stadt und auf dem Land im späten Mittelalter, in: Reinhardt Butz / Gert Melville (Hg.), Coburg 1353. Stadt und Land Coburg im Spätmittelalter (Schriftenreihe der Historischen Gesellschaft Coburg 17), Coburg 2003, S. 41–64, hier: S. 56; Karlheinz Blaschke, Kaufmannssiedlungen als Frühformen städtischer Entwicklung, in: Kurt-Ulrich Jäschke / Christian Schrenk (Hg.), Was macht im Mittelalter zur Stadt? Selbstverständnis, Außensicht und Erscheinungsbilder mittelalterlicher Städte, Heilbronn 2007, S. 91–124, hier: S. 94.

nachhaltig verändert, besonders im Bereich Infrastruktur und Wirtschaft, aber auch in den gesellschaftlichen Strukturen. Städte – und nicht mehr allein Klöster und Burgen – bildeten Drehscheiben für kulturelle Innovationen[2] wie auch für Konsumangebote, deren Genüsse nicht nur den adeligen Oberschichten zugutekamen, sondern beim „Massenkonsum" von Textilien, Wein und Getreide praktisch allen Bewohnern.[3] Daneben darf das religiöse „Angebot" einer *civitas sancta* nicht vergessen werden: Der Aufstieg Nürnbergs beispielsweise ist ohne die Wallfahrt zum heiligen Sebald, die Bedeutung der fränkischen Bischofsstädte ohne Kilian, Willibald beziehungsweise dem heiliggesprochenen Kaiserpaar Heinrich und Kunigunde nicht zu erklären.[4]

Die Städte entstanden, zumindest im fränkischen Raum, weniger auf der sprichwörtlichen grünen Wiese, vielmehr dienten bereits bestehende Stifte,[5] Klöster, Burgen, Königspfalzen als präurbane Siedlungskerne, an denen dann Städte mit einem neuen, eigenen Rechtsbereich entstanden.[6] Die Stadtanlage folgte den vorgefundenen geographischen Gegebenheiten, aber auch gestalterischen Interessen: Stadtmauern, Markt und Kirche bildeten die Zentren, daneben traten Mühlen und Brücken, Rathäuser und Hospitäler, aber auch Schulen[7] hinzu. Städte unterlagen damit einem gewissen gestalterischen Ordnungsprinzip, aus dem dann die Stadt als „gebauter Kosmos" erwuchs. Inwieweit dieser antiken philosophisch-mathematischen Vorlagen folgte, soll hier nicht diskutiert werden.[8]

---

2 Wolfgang Wüst (Hg.), Frankens Städte und Territorien als Kulturdrehscheibe. Kommunikation in der Mitte Deutschlands, Neustadt an der Aisch 2008.

3 Wolfgang Wüst (Hg.), Regionale Konsumgeschichte vom Mittelalter bis zur Moderne (Franconia 7. Beihefte zum Jahrbuch für fränkische Landesforschung), Stegaurach 2015.

4 Helmut Flachenecker, Heilige Bischöfe als einheitsstiftende Klammer für mittelalterliche Diözesen, in: Römische Quartalschrift 97, 2002, S. 194–214.

5 Helmut Flachenecker, Geistliche Stifte als Kristallisationskerne für Orte der Herrschaft und Motoren für Urbanität in Franken, in: Ders. / Rolf Kießling (Hg.), Urbanisierung und Urbanität. Der Beitrag der kirchlichen Institutionen zur Stadtentwicklung in Bayern (Zeitschrift für bayerische Landesgeschichte [künftig: ZBLG], Beihefte 36), München 2008, S. 15–48.

6 Helmut Flachenecker / Rolf Kießling (Hg.), Städtelandschaften in Altbayern, Franken und Schwaben. Studien zum Phänomen der Kleinstädte während des Spätmittelalters und der Frühen Neuzeit (ZBLG, Beihefte B 15), München 1999.

7 Helmut Flachenecker / Rolf Kießling (Hg.), Schullandschaften in Altbayern, Franken und Schwaben. Untersuchungen zur Ausbreitung und Typologie des Bildungswesens in Spätmittelalter und Früher Neuzeit (ZBLG, Beihefte B 26), München 2005.

8 Kerstin Geßner, Die Vermessung des Kosmos. Zur geometrischen Konstruktion von urbanem Raum im europäischen Mittelalter, Köln 2020, S. 11.

Somit sind Städte ein bedeutsames Leitfossil für den gesellschaftlichen Wandel in den einzelnen Kulturlandschaften Mitteleuropas. In ihren sich zeitlich immer wieder verändernden Funktionen waren sie Zentren für ein jeweils neu zu definierendes Umland: Sie besaßen und besitzen kulturelle wie raumprägende Funktionen. Bis heute gliedern die urbanen Zentren auch die Kulturlandschaft in Franken in signifikanter Weise und prägen die Lebensumstände der Bevölkerung und deren Identitäten.[9]

Die Vielfalt des fränkischen Städtewesens ist außerdem ein Ergebnis der seit dem 13. Jahrhundert einsetzenden, territorialen Zersplitterung dieser Region. Jeder Landesherr benötigte idealtypisch eine Stadt als Markt- und Handelsort, die mit ihrer Zentralität auf ein mehr oder weniger großes Umland ausstrahlte, neben einer Burg als militärisch-administratives und ein Kloster als ein religiös-memoriales Zentrum. Ab dem 13. Jahrhundert wurden in Franken viele Städte in herrschaftlich umstrittenen Räumen gegründet, um so politische Ansprüche manifest zu machen. Zugleich setzt eine Veränderung der Bevölkerungskonzentration im Raum ein; Dorfwüstungen stehen Städteakkumulationen gegenüber.[10]

Der Begriff Städtelandschaft wird vielgestaltig und in vielen Fällen in einer spielerisch-schimmernden Uneindeutigkeit verwendet, zum Teil ästhetisierend, zum Teil metaphorisch, zum Teil von touristisch-öffentlichen Moden beeinflusst. Er beschreibt aber stets einen Ausschnitt einer generellen Kulturlandschaft. Denn der Mensch hat – pauschal betrachtet – die ihm umgebende naturräumliche Landschaft als seine jeweilige Umwelt gestaltet, in der er leben und überleben konnte und kann.[11] Das führt allgemein zu einer zunehmenden Differenzierung von Natur- und Kulturlandschaften, wobei letztere in ihrem Umfang tendenziell beständig zugenommen haben. Eine Kulturlandschaft historisiert sich gleichzeitig in ihren Einrichtungen, das heißt, es gibt dort auch

9 „Altfränkische" Definitionen im Rahmen von Romantik und Heimatschutzbewegung bis hin zur Identitätsfindung in der Gegenwart werden für die Region Spessart neuerdings diskutiert bei: Anika Magath, Der Spessart als Kulturlandschaft. Blickwinkel auf eine Kulturlandschaft und das Projekt der Europäischen Kulturwege (Veröffentlichungen des Geschichts- und Kunstvereins Aschaffenburg 69), Aschaffenburg 2020.

10 Helmut Flachenecker, Landschafts- und Reichsbindung von Städten in Franken, in: Johannes Merz / Robert Schuh (Hg.), Franken im Mittelalter. Francia orientalis, Franconia, Land zu Franken: Raum und Geschichte (Hefte zur Bayerischen Landesgeschichte 3), München 2004, S. 167–187.

11 Dass dieser Prozess in der Gegenwart andauert, zeigt die aktuelle Diskussion um die Bewohnbarkeit der gegenwärtigen urbanen Zentren im Angesicht der Herausforderung des globalen Klimawandels!

Objekte beziehungsweise spezifisch gestaltete Räume, die nicht mehr notwendig sind und daher verschwinden oder für neue Anforderungen überformt beziehungsweise neueren Bestimmungen zugeführt werden. Die Ergebnisse der sich kontinuierlich veränderten Gestaltung wird von Menschen als Kulturlandschaft deklariert, dahinter steckt also auch ein permanenter Prozess einer bewussten gesellschaftlichen, wirtschaftlichen und kulturellen implizierten Neuausrichtung, die dann Räume mit ihren dortigen Objekten nachträglich zum kulturellen Erbe erklären kann. Damit kommt auch die wissenschaftliche Auseinandersetzung mit einer Landschaft ins Spiel. Diese wurden von sozialen Gruppen und deren Zentren mitbestimmt,[12] worunter neben den Städten auch Klöster und Burgen als herausragende Punkte zu zählen sein dürften. Diese „Reconnecting Culture and Nature"[13] durch Städte soll hier nur knapp angesprochen werden: Die gegenseitige Beeinflussung von „culture and environment"[14] kann damit zwar nicht in ihrer Totalität diskutiert werden, denn die Stadt kann nicht allein für den allumfassenden Begriff „Kultur" gleichgesetzt werden, auch wenn sie eine bedeutsame Einheit in diesem Prozess einer *longue durée* sein dürften.

Eine exakte Definition des Begriffs Kulturlandschaft ist, soviel dürfte klar geworden sein, schwer zu geben und immer vom jeweiligen Untersuchungsgegenstand abhängig. Von daher muss er in seinen Bedeutungsinhalten offen definiert werden, um stets wandelnde Annäherungen an dieses human gestaltete und daher vielfältige Phänomen zuzulassen. In unserem Zusammenhang bedeutet dies: Stets stehen politische, wirtschaftliche, soziale und religiöse Beziehungen zwischen den Städten und deren Umfeld im Mittelpunkt der jeweils nach immer neuen Fragestellungen die Kulturlandschaft analysierenden Untersuchungen. Rücksichtnahmen auf die naturräumlichen Voraussetzungen, also auf Geologie und Hydrologie, müssen stets mit in die Überlegungen einbezogen werden. Unabdingbare Voraussetzung für eine Städtegründung waren etwa ertragreiche Böden, um die Bewohner ernähren zu können. Landwirtschaftliche Prosperität

---

12 Ilke Marschall, Wozu Kulturlandschaftsschutz? Motivationen, mögliche Ziele sowie Möglichkeiten eines lebendigen Kulturlandschaftsschutzes, in: Alfred Toepfer Akademie für Naturschutz (NNA) (Hg.), Umgang mit historischer Kulturlandschaft. NNA-Berichte, 23. Jahrgang, Heft 1, 2010, S. 70–77.

13 Harald Plachter / Mechthild Rössler, Cultural Landscapes. Reconnecting Culture and Nature, in: Bernd von Droste / Harald Plachter / Mechthild Rössler (Hg.), Cultural Landscapes of Universal Value – Components of a Global Strategy, Jena 1995, S. 15–18.

14 Lester B. Rowntree, The Cultural Landscape Concept in American Human Geography, in: Carville Earle / Ken Mathewson / Martin S. Kenzer (Hg.), Concepts in human geography, Lanham/Md. 1996, S. 127–160.

schuf auch die Basis für die Größe der Städte und deren Grad der Urbanisierung. Die Lage am Fluss garantierte den Anschluss an Handelswege, eine Furt beziehungsweise Brücke auch an ein überregionales Straßensystem.

Diese Vielfalt von Raumkonstrukten[15] ist aber keineswegs ein Ausdruck begrifflich-methodischer Schwächen, wie Heinz-Dieter Heimann eindrücklich konstatiert: „Danach adressieren nicht allein die epochenübergreifend arbeitende Landesgeschichts- und Regionalforschung oder sozialgeschichtlich ausgerichtete Kulturhistorische Forschungen, diese freilich intensiv, aus verwandten Raumtheorien und Ansprüchen abgeleitete, aber auch unterschiedlich adressierte Raumkonstrukte. Diese Vielfalt von Räumen ist kein Ausdruck von Schwäche der Geschichtswissenschaft. Sie ist tatsächlich ein Ausweis geschichtswissenschaftlichen Erkenntnisfortschritts und hoher Sensibilität gegenüber reklamierten Raum- und Identitätsansprüchen auch geschichtspolitischer Instrumentalisierungen.“[16] Es geht also, um nochmals Heimann zu zitieren, um die Erfassung von „Codierungen“ in einer Landschaft, deren Ursprünge auch in der städtischen Kultur liegen.[17] Diese müssen von vielfältiger Art sein, in

15 Karl Heinz-Spieß (Hg.), Landschaft im Mittelalter, Stuttgart 2006; Enno Bünz / Werner Freitag, Räume und Grenzen. Traditionen und Konzepte der Landesgeschichte. Epochenübergreifende Sektion auf dem 45. Deutschen Historikertag, in: Blätter für deutsche Landesgeschichte 139/140, 2003/2004, S. 145–338; Jörg Döring / Tristan Thielemann (Hg.), Spatial turn. Das Raumparadigma in den Kultur- und Sozialwissenschaften, Bielefeld 2008; Wilfried Reininghaus / Bernd Walter (Hg.), Räume – Grenzen – Identitäten (Forschungen zur Regionalgeschichte 71), Paderborn 2013; Sigrid Hirbordian / Christian Jörg / Sabine Klapp (Hg.), Methoden und Wege der Landesgeschichte (Landesgeschichte 1), Ostfildern 2015; Holger Gräf / Alexander Jendorff / Piere Monnet (Hg.), Land – Geschichte – Identität (Quellen und Forschungen zur hessischen Geschichte 174), Darmstadt/Marburg 2016; Falk Bretschneider / Christoph Duhamelle, Fraktualität. Raumgeschichte und soziales Handeln im Alten Reich, in: Zeitschrift für Historische Forschung 43, 2016, S. 703–743.

16 Heinz-Dieter Heimann, Landschaften – Musiklandschaften: Wie Raumkonstrukte der Geschichtswissenschaft „kreative Räume“ adressieren, in: Sabine Meine / Arnold Otto / Johannes Süßmann (Hg.), Musiklandschaften zwischen Pader und Rhein. Pluralisierung und Verflechtung entlang des Hellwegs in der Frühen Neuzeit (Musik – Kultur – Geschichte [die Bandnummer steht noch nicht fest]), Würzburg 2020 [in Vorbereitung].

17 Heinz-Dieter Heimann, Klöster und Landschaften, in: Frank Göttmann / Eva Maria Seng (Hg.), Die Landschaft. Kultur, Natur-, Wissenschafts- und Erfahrungsräume [erscheint demnächst], [S. 2]: „Als fernes Ziel solcher Arbeiten könnte ein Atlas monastisch-kultureller Codierungen von Räumen stehen, deren Formate Klöster und Landschaften als kognitive Landschaften, als Identitätsräume ausweisen.“

unterschiedlichen Zeiträumen differierende Wertigkeiten besitzen, aber diese Codierungen sind teilweise bis heute noch in einer Kulturlandschaft nachweisbar.

2. Städte in Franken

Doch wie sieht die Bindung der Städte an einen Raum, hier Franken, in den Quellen des Hoch- und Spätmittelalters aus? Wenn Bischof Gerhard von Würzburg 1387 bei einer Landfriedenseinung zwischen seinen Städten und jenen des Abtes von Fulda von „*vnser oder vnsir Stete zu Franken*“ spricht und dabei zwanzig Orte aufzählt, dann scheint die Landschaftsbindung zu Franken augenfällig und klar.[18] Sie werden von den Orten „*in Fuldischen Lande*“ abgegrenzt – Fulda gehört in dieser Sicht nicht zur Landschaft Franken, obwohl politisch enge Beziehungen bestehen. Die in der Vereinbarung vertretene klare Unterscheidung besticht, hat aber ein großes Manko, denn sie kommt selten in Urkunden und Privilegien vor, um weiter untersucht werden zu können.

Knapp dreihundert Jahre später umschrieb Matthäus Merian (1593–1650) in seiner „Topographia Franconiae“ das Frankenland: „*Es ist aber diß das Franckenland, welches gegen Auffgang den Nortgau und Bayern, gegen Niedergang die Untere= Pfalz, gegen Mittag das Schwabenland und gegen Mitternacht Thüringen angrentzend hat.*“ Die genannten Landschaften sind schillernde Begriffe, um fest begrenzte politische Gebilde – eventuell mit Ausnahme vom Herzogtum Bayern – handelt es sich nicht. Sie sind eher dialektal-kulturell umschrieben. Statt der „*Oberen= Pfalz*“ verwendet er die alte Nordgau-Bezeichnung. Stattdessen nimmt Merian eine mehr naturräumlich-topographische Sicht bei der Binnengliederung Frankens ein: Es sei von acht Flüssen durchzogen, besitze neben großen Waldgebieten auch Regionen mit Waldmangel, wie etwa im Bereich von Ochsenfurt. „*Von Handels= Städten und Emporiis sind Nürnberg, Bamberg, Würtzburg, Schweinfurt und Kitzingen.*“[19] Seine Auswahl umfasst also

---

18 Hennebergisches Urkundenbuch 4, hg. v. Georg Brückner, Meiningen 1861, S. 195–197, Nr. 335 (1387 März 12): „[…] *vnsern Steten mit Wirtzburg, Karlstat, Iphofen, Gerolcshofn, Nuwenstat, Meynungen, Mellerstat, Kunitzshoven, Fladungen, Bischoffsheim, Sesselbach, Ebern, Hasfurt, Eltmeyn, Botingen, Hohenburg, Rotenfels, Gemunde, Aernstein, Bybert* […].“

19 Matthaeus Marian, Topographia Franconiae das ist Beschreibung vnd eygentliche Contrafactur der vornembsten Stätte vnd Plätze des Franckenlandes vnd deren die zu dem hochlöblichen Fränkischen Craiße gezogen werden, Frankfurt 1656, ND Braunschweig 2005, S. 3–5. – Die Texte stammen von Martin Zeiller aus Ulm. Zu Merian und zur Topographie vgl.: Uwe Müller (Hg.), Matthäus Merian d.Ä. Ätzkünstler und Verleger (Veröffentlichungen des Stadtarchivs Schweinfurt 17), Schweinfurt 2003, zur Topographie: S. 115–133.

Bischofs-, Reichs- und Landstädte, Kriterien für diese Nennungen gibt er aber nicht an. Zur weiteren Untergliederung des Raumes ging Merian nun streng nach einer politischen Einteilung vor, nämlich jener des Fränkischen Kreises. Bei jeder Herrschaft nannte er die Ämter, Städte, Festungen und Burgen. An seinen Grenzen gebe es Überschneidungen. Etwa sei das Untere Stift von Eichstätt eigentlich mehr bayerisch, der Bischof gehöre aber zum Fränkischen Kreis.[20] Ähnlich Coburg: „*Darzu wir auch die Coburgische thun wollen; weilen solche nicht allein im Franckenlande gelegen seyn, sondern auch vorzeiten zur Fürstlichen Graffschafft Hennenberg (deren Städte allhie einkommen) gehört haben; wiewol sie die Coburgischen sonsten der Zeit zum Obersächsischen, wie ingleichem andere so dem Lager und Landsart nach Fränckisch seyn zu andern Craysen ausser Lands wegen ihrer Herren* […].“[21]

Ebenso gab es zum Schwäbischen Reichskreis Überschneidungen: Dinkelsbühl, deren Bewohner „*Noricos vnd Portofrancones*“ genannt wurden, weil sie nahe am „*Ost-Franckenland*“ wohnten, sei „*mit der Statt Hall d*[er] *Schwaben Vormawer wider die Francken, gleich wie Rotenburg 5 Meilen von hinnen gelegen, der Franken wider die Schwaben gewesen.*“[22]

Das einmal nach naturräumlichen, das andere Mal nach politischen Kriterien beschriebene Franken kennt nur gelegentlich feste Grenzen. Diese Geschichtslandschaft besitzt nicht eindeutig zu definierende Grenzräume, nach Ernst Schubert „ein Charakteristikum des Regionalismus, der keine Grenzen, sondern Brücken zu anderen Ländern bildet“.[23] Bei dieser Brückenfunktion spielten auch die Städte eine wichtige Rolle. So war die Reichsstadt Rothenburg im 14. Jahrhundert meist Mitglied der fränkischen Landfrieden, aber auch bisweilen schwäbischer Städtebünde.[24] Die niederschwäbischen Städte waren seine unmittelbaren

---

20 Merian, Topographia Franconiae (wie Anm. 19), S. 6, 15.

21 Ebd., S. 14.

22 Mathäus Merian, Topographia Sueviae […], Frankfurt 1643, ND Braunschweig 2005, S. 58. Ein weiterer Ort an dieser Grenze war das bei Hall gelegene Kloster Murrhardt, so Staatsarchiv Würzburg (künftig: StAWü), Standbuch 1012, fol. 496 v [hier wird nach der neuesten, dafür auch durchgängig verlaufenden Blattzählung zitiert!]: „*Murrhart ain closter Benedicter ordens zwo meil hinder der stat Hall am Kochen, in den grenitzen der Franken vnd Schwaben gelegen* […].“

23 Ernst Schubert, Franken als königsnahe Landschaft unter Karl IV., in: Blätter für Deutsche Landesgeschichte 114, 1978, S. 865–890, hier: S. 867.

24 Ludwig Schnurrer, Die Reichsstadt Rothenburg im Zeitalter Karls IV. 1346–1378, in: Blätter für Deutsche Landesgeschichte 114, 1978, S. 563–612, bes.: S. 606–609: Rothenburg war bei den fränkischen Landfrieden von 1340, 1349, 1353, 1358

Nachbarn. Und Weißenburg, etwa in der Mitte zwischen Nürnberg und Nördlingen gelegen, gehörte stets zum Glacis der fränkischen Handels- und Politikmetropole, sein Rat ließ sich durch Nürnberger Gesandten auf den Hof- und Reichstagen vertreten, die Bewohner besaßen jedoch intensive Handels- sowie familiäre Beziehungen zum schwäbischen Nördlingen.[25] Die Stadt wollte dem starken Nürnberger Druck mit einer Anlehnung an schwäbische Reichsstädte im Spätmittelalter zumindest etwas entgehen.

Eine bewusste fränkische Identität ist bei den Städten dieses Raumes schwer direkt nachzuweisen. Lediglich die immer wieder zitierte Bestimmung der Goldenen Bulle (I, 8),[26] wer dem König von Böhmen Geleit bei seinem Zug nach Frankfurt geben solle, enthält die *„civitates Nuremberg, Rotemburg et Windesheim*". Sind sie also die herausragenden fränkischen Städte, verbunden in ihrem Dienst am Reich? Zunächst ist diese Auswahl vom Gegenstand bedingt. So lagen die übrigen Reichsstädte Frankens schlicht nicht auf dem Weg von Prag nach Frankfurt. Zum anderen konnte die Goldene Bulle nicht andere landesherrliche Städte nennen, weil diese nicht direkt vom Reich lehensabhängig waren. Und ein drittes Argument: Nürnberg, Rothenburg und Windsheim sollten auch das Geleit für den Herzog von Sachsen übernehmen (I, 12). Es wird dabei ihre Rolle in Reichsdiensten hervorgehoben, nicht jene in Franken.

Es waren Reichsstädte, die bei den fränkischen Landfrieden des 14. und frühen 15. Jahrhunderts primär beteiligt waren. Nürnberg spielte dabei, was nicht weiter verwunderlich ist, eine führende Rolle. Die Stadt konnte in gewisser Weise auch eigenständig handeln. Am Landfrieden von 1398 beispielsweise waren zunächst nur Nürnberg und Weißenburg von Beginn an dabei. In einem eigenen Paragraphen wurde den Nürnbergern die Vollmacht gegeben, innerhalb einer knappen Frist (zwischen 2. und 31. März) auch noch die Reichsstädte Windsheim und Schweinfurt mit aufzunehmen.[27]

(Landfriede zu Rothenburg), 1368, 1371 beteiligt, gehörte aber 1352 bzw. 1357 zu schwäbischen Städtebünden. Am 17. Mai 1378, wenige Wochen vor Karls Tod, hat es sich „für den harten, antifürstlichen Kurs der schwäbischen Städte entschieden."

25 Ute Jäger, Weißenburg und seine Beziehungen zu den schwäbisch-fränkischen Reichsstädten im Spätmittelalter, in: Flachenecker / Kießling, Städtelandschaften (wie Anm. 6), S. 188–220.

26 Karl Zeumer, Die Goldene Bulle Kaiser Karls IV., 2 Teile (Quellen und Studien zur Verfassungsgeschichte des Deutschen Reiches in Mittelalter und Neuzeit 2/1–2), Weimar 1908/09, hier: Teil 2, S. 11.

27 Gerhard Pfeiffer (Bearb.), Quellen zur Geschichte der fränkisch-bayerischen Landfriedensorganisation im Spätmittelalter, München 1975, Nr. 373 (1398 März 2).

Der Vorrang Nürnbergs lag in seinem ausdifferenzierten diplomatischen Apparat. Dieser vertrat die anderen kleinen Reichsstädte, die sich ein derart kostenintensives Personal nicht leisten konnten. So vertraten die Diplomaten meist die Interessen aller fränkischen Reichsstädte in einer Hand. Dabei kam es in der Regel vor, dass die Nürnberger Botschafter Vollmachten der anderen besaßen beziehungsweise die übrigen Reichsstädte informierten. Ein Beispiel von vielen: Als 1423 ein neuer Landfrieden ausgehandelt wurde, gaben die Rothenburger und Windsheimer den Nürnbergern entsprechende Verhandlungsvollmachten. Weißenburg dagegen wurde vom Nürnberger Rat von den Verhandlungen informiert und erbat deren Meinung.[28]

Gruppenbewusstsein kam vor allem auf, wenn es gegen die alle Reichsstädte bedrohende Gefahr einer Verpfändung durch den Kaiser ging. Deshalb war es in ihrem ureigensten Interesse, in die Landfriedensbestimmungen einen Paragraphen wie den folgenden unterzubringen: „*Ez sol auch niemant fur unsern obgenanten herren den keyser pfenden noch angreiffen, weder dez reichs stet noch anders yemant.*“[29] Reichsstädte als spezielle Glieder des Reiches nutzten aber auch ihre Stellung aus und mussten unter Umständen vom König selbst in die Schranken verwiesen werden. Die in Franken gelegene Gruppe – also Nürnberg, Rothenburg, Schweinfurt, Windsheim, Weißenburg[30] – erhob vor und um 1414 Zölle zur Finanzierung von Landfriedensmaßnahmen. König Sigmund musste ihnen am 1. Oktober 1414 einschärfen, diese Praxis am Ende eines zeitlich befristeten Landfriedens, bei Androhung einer Geldstrafe, zu beenden. Offensichtlich hatten die Reichsstädte, wohl mit Verweis auf den König, die betreffenden Zolleinnahmen auch über das jeweilige Ende eines Landfriedens fortgesetzt: Schlitzohrigkeit oder ein besonderes Reichsbewusstsein, für die Friedenserhaltung in Franken in kontinuierlicher Weise wirken zu müssen?[31]

Der Landfrieden Karls IV. für Franken von 1349 ist insofern für die vorliegende Fragestellung bedeutsam, weil durch die vier regelmäßig im Jahr aufzusuchenden Gerichtsorte, die Städte Würzburg und Bamberg, Neustadt an der Aisch und Nürnberg, die Region wie in einem trapezartigen Netz umfasst wurde. Prägnant hat dies Lorenz Fries in seiner „Hohen Registratur“ zusammengefasst: „*Anno*

28 Ebd., Nr. 727 (1423 Okt. 21).

29 Ebd., Nr. 62 § 28 (1368 Nov. 24). Ediert bei: Ernst Fischer, Die Landfriedensverfassung unter Karl IV., Diss. phil., Göttingen 1883, S. 114.

30 Statt die verschiedenen stadtgeschichtlichen Forschungen im Einzelnen aufzuführen, genüge ein genereller Hinweis auf: Rainer A. Müller (Hg.), Reichsstädte in Franken, 2 Bde., München 1987.

31 Pfeiffer, Quellen (wie Anm. 27), Nr. 694 (1414 Okt. 1).

1349 Sontag nach Michaelis hat kunig Carl der viert ainen Landfriden in dem land zu Francken vfgericht, vnd zu Nurnberg beschlossen auch derwegen im Jare vier gerichte alwegen am sontag nach ainem ieden quatember halten lassen, das erst zu Nurenberg, das ander zu W[ürzburg], *das trit zu Bamberg, das viert zue Newenstat an der Aisch.*“[32]

Diese Auswahl der Tagungsorte für Landfriedensentscheidungen blieb keine Episode. Die genannten vier Städte von 1349 werden auch in den Egerer Landfrieden von 1389 übernommen,[33] bei der Verlängerung des Bundes 1395 durch König Wenzel erneuert,[34] um schließlich beim 1404 von König Ruprecht erlassenen fränkischen Landfrieden wiederum bestätigt zu werden.[35] Der Landfriede König Sigmunds 1423 erwähnte die vier Gerichtsorte nicht,[36] jedoch beschlossen die fränkischen Fürsten und Städte in ihrer Landfriedensvereinbarung, die im Februar 1427 ohne königliche Mitwirkung geschlossen wurde, dass sich der Landfriedensausschuss vierteljährlich „*zu Bamberg, Wirtzburg, Nuremberg oder Newenstat an der Eysch*“ treffen solle.[37] Die burggräfliche Landstadt Neustadt sowie die Bischofsstädte und die Reichstadt Nürnberg wurden so zu häufig besuchten Verhandlungszentren für anliegende Entscheidungen, die nicht nur zu den festen Tagen stattgefunden haben, sondern auch, wenn es die aktuelle Lage darüber hinaus erforderte.[38] Sie bildeten damit über einen längeren Zeitraum hinweg ein „fränkisches Trapez“, das vielleicht einen nicht unerheblichen Beitrag zur Stärkung eines fränkischen Gemeinschaftsbewusstseins geleistet haben könnte.

Fränkische Landstädte treten ansonsten kaum in Landfrieden auf. Lediglich die Bischofsstädte werden 1340 separat genannt. Es kann darüber diskutiert werden, ob damit eine besondere autonome Stellung demonstriert werden sollte oder nicht. Es ist nicht erkennbar, dass sie als dezidiert fränkische Städte auftreten. Dies gilt auch für die Masse der übrigen landesherrlichen Städte. Denn: Reichsstädte bildeten, entgegen der Quellenüberlieferung und der herausragenden Bedeutung in der Forschung, nur eine Minderheit der Städte in Franken, die meist als Landstädte

---

32 StAWü, Standbuch 1012, fol. 378 v (Stichwort „*Lantfrid*“).

33 Pfeiffer, Quellen (wie Anm. 27), Nr. 139 (1389 Mai 5).

34 Ebd., Nr. 309 (1395 Juni 11).

35 Ebd., Nr. 414 (1404 Juli 11).

36 Ebd., Nr. 728 (1423 Nov. 24).

37 Ebd., Nr. 747 (1427 Febr. 5).

38 Zu Neustadt an der Aisch vgl.: Helmut Flachenecker, Die Rolle Neustadts an der Aisch in den Fränkischen Landfrieden, in: Streiflichter aus der Heimatgeschichte 2003, hg. v. Geschichts- und Heimatverein Neustadt an der Aisch, S. 1–6.

bezeichnet wurden. Zum besseren Verständnis muss daher eine chronologische Sicht der Entwicklung eingeschalten werden.

3. Stadtentwicklung in Franken im chronologischen Überblick

Bevölkerungsreiche Orte mit einer unterschiedlichen Anzahl von Zentralitätsfunktionen für eine engere oder weitere Umgebung finden sich im Frühmittelalter nur an den Bischofssitzen.[39] So waren Würzburg und Eichstätt im 8. und Bamberg im beginnenden 11. Jahrhundert jene Orte, in denen die Wandlung präurbaner Siedlungskerne (etwa Klöster, Stifte, Höfe, Befestigungen) zu Städten im mittelalterlichen Sinne begann.[40] Sie sollte im 12./13. Jahrhundert ihren Höhepunkt finden, was die Zahl der Gründungen betrifft. Die damalige Bevölkerungszunahme, einhergehend mit der sogenannten Binnenkolonisation bisher unerschlossener Waldgebiete, erlaubte diesen Schub, der in Franken dazu führte, dass es um 1500 über 150 Städte gab, die auch heute noch die Kulturlandschaft prägen. Viele von ihnen sind an den zentralen Flüssen Main, Fränkische Saale, Regnitz, Pegnitz und Altmühl orientiert. Gerade der Main, der sich als einziger größerer Fluss von Ost nach West orientierte und damit Franken mit den großen Städten am Rhein und in Flandern verband, entwickelte sich als Magnet für Städteneugründungen. Der Gründungsvorgang lag in Händen des jeweiligen Grundherrn beziehungsweise Vogtes, der das zukünftige Areal in Besitz hatte. Eine königliche Bestätigung wurde meist im Nachhinein eingeholt, eigentliche Stadtgründungsurkunden sind wenige bekannt. Stattdessen bestätigen die meisten von ihnen eine bereits vorhandene urbane Entwicklung.

Als Stadtherren traten neben den Bischöfen zunächst die staufischen Könige auf, gefolgt von den Grafen und Herren Frankens. Am Obermain agierten bis zu ihrem Aussterben 1248 die Andechs-Meranier, in weiten Bereichen des heutigen

39 Die folgende Gesamtdarstellung verzichtet auf eine ausführliche Bibliographie; stattdessen soll auf die kartenmäßige Darstellung von Helmut Flachenecker, Städtelandschaft Franken, in: Wolfgang Jahn / Jutta Schumann / Evamaria Brockhoff (Hg.), Edel und Frei. Franken im Mittelalter (Veröffentlichungen zur Bayerischen Geschichte und Kultur 47/4), Augsburg 2004, S. 308–312, verwiesen werden sowie auf Markus Naser, Digitale Karten zur Geschichte der Städte in Franken (Mainfränkische Studien 79), Baunach 2010. – Im Internetportal Historisches Unterfranken befindet sich eine Datenbank zur Geschichte der einzelnen fränkischen Städte: http://www.historisches-unterfranken.uni-wuerzburg.de/staedte/staedte.php (letzter Zugriff: 24.8.2020).

40 Helmut Flachenecker, Der Bischof und sein Bischofssitz: Würzburg – Bamberg – Eichstätt im Früh- und Hochmittelalter, in: Römische Quartalschrift für christliche Altertumskunde und Kirchengeschichte 91, 1996, S. 148–181.

Mittel- und Oberfrankens die Burggrafen von Nürnberg beziehungsweise späteren Markgrafen von Ansbach und Bayreuth. Da die Territorienbildung neben Burgen und Klöstern auch Städte einschloss, finden sich letztere in allen größeren oder kleineren Herrschaften Frankens. Ab der Mitte des 13. Jahrhunderts haben Würzburger und Bamberger Bischöfe viele Städteprivilegien vergeben, in eingeschränkter Anzahl auch die Eichstätter. Von den weltlichen Machtträgern wären zu nennen: die Grafen von Henneberg im Norden, die Herren von Hohenlohe im Südwesten, die Grafen von Wertheim und von Rieneck am Mainviereck, aber auch die Marschälle von Pappenheim an der Altmühl oder die Herren von Heideck im südlichen Mittelfranken besaßen zumindest am Ort ihrer Hauptburg eine städtische Siedlung, die zumeist Marktfunktionen (Jahr-, Wochenmärkte) besaßen. Den Status einer eigenständigen Pfarrei haben sie meist erst in einem zähen Kampf mit dem ursprünglichen, ländlichen Pfarrsitz erwerben können. Mit den Steuerabgaben waren die neuen urbanen Zentren, die bevölkerungsmäßig häufig klein blieben, für ihre Herren eine hochwillkommene monetäre Einnahmequelle, die in Notzeiten auch verpfändet werden konnten. Viele Städte wechselten in solchen Fällen ihre Stadtherren, aber auch nach dem Aussterben der Familien sowie bei Besitzveränderungen im Zuge von Umschichtungen der Machtverhältnisse (Herren von Dürn, Schlüsselberger, Grafen von Castell).

Ab der Mitte des 13. Jahrhunderts konnten sich einige königliche Städte zu Reichsstädten entwickeln, meist in dem Moment, als der Einfluss der Staufer nachließ. An deren Spitze, als einzige „Großstadt" Frankens, stand Nürnberg (Großes Handelsprivileg Friedrichs II. von 1219), gefolgt von Rothenburg und Schweinfurt. Die kleineren Reichsstädte Weißenburg und Windsheim konnten, wie gezeigt, nur unter dem Schirm Nürnbergs überleben. Ein größeres Landgebiet lässt sich im Mittelalter nur bei Rothenburg, zu Beginn des 16. Jahrhunderts auch für Nürnberg nachweisen. Nürnberg wurde daraufhin selbst Stadtherr für Altdorf, Betzenstein, Gräfenberg, Hersbruck, Lauf, Velden, vorübergehend auch in Heideck und Hilpoltstein. Feuchtwangen verlor 1376 einen entsprechenden Status durch Verpfändung an die Zollern. Besonders Würzburg und Nürnberg, weniger Bamberg und kaum Eichstätt konnten von königlichen Hoftagen profitieren.

Im 14. Jahrhundert kam es unter den Königen Ludwig IV. und Karl IV. im Zuge ihrer „Landbrückenpolitik" zur Wetterau beziehungsweise nach Böhmen zu einer hohen Zahl von königlichen Städteprivilegierungen, die meist bestehende Orte noch einmal rechtlich bestätigten. Solche neuen Privilegierungen waren nicht in jedem Falle durchsetzbar, sodass sich – trotz eines entsprechenden Privilegs – die Orte nicht in jedem Falle zu einer Stadt entwickeln konnten (Gailnau, Haldenburgstetten, Hasloch, Hornberg, Mussen, (Markt-)

Bergel, Neubrunn, Thüngen). Häufig lag dies am Widerstand der benachbarten Territorien.

So bestätigte König Ruprecht (1400–1410) dem Erkinger von Seinsheim zum Stephansberg eine Urkunde Karls IV. von 1357, der zufolge die Seinsheimer die Halsgerichtsbarkeit im Dorf Astheim ausüben durften.[41] In problematischen Fällen sollten die Schöffen allein beim Rat der Stadt Schweinfurt Erkundigungen einziehen. Zugleich erlaubte der König, dass Erkinger und seine Nachfolger Astheim zu einem Markt mit Bürgern ausbauen dürften.[42] Derselbe König schließlich erhob den Ort am 16. Juni 1410 zur Stadt.[43] Ferner gelang es Erkinger bereits 1408, im Einvernehmen mit dem damaligen Volkacher Pfarrer sowie mit dem Dekan des Würzburger Domkapitels – der das Patronat über die Volkacher Kirche innehatte –, Astheim zu einer eigenständigen Pfarrei zu erheben. Nur ein Jahr später, 1409, gründete Erkinger ein Kartäuserkloster in Astheim.[44] Allerdings entwickelte sich Astheim, trotz dieser umfangreichen Privilegierungen, zu keiner Stadt. Sicher dürfte das hochstiftische Volkach zu nahe gelegen haben, sodass auf der gegenüberliegenden Mainseite kein zusätzlicher und zugleich konkurrierender Ort mit zentralörtlichen Funktionen entstehen konnte.

Um noch weitere Beispiele hier kurz zu erwähnen: Thüngen entwickelte sich nicht zu einer Stadt, obwohl die Herren von Thüngen 1465 von Kaiser Friedrich III. ein Stadtprivileg erhalten hatten.[45] Der Ort, im Werntal gelegen und damit abseits von der bedeutenderen Mainroute, blieb auf dem Status eines Marktes

41 Regesta Imperii Online, Regesten Pfalzgrafen 2, Nr. 990 (1401 Juni 24 Mergentheim), URL: http://www.regesta-imperii.de/regesten/10-0-0-ruprecht/nr/1400-08-22-1-0-10-0-0-6290-6675.html?tx_hisodat_sources%5B%40widget_0%5D%5BcurrentPage%5D=20&cHash=774dc3f821d16a03cf10c558b94c0888#rinav (letzter Zugriff: 24.8.2020); Joseph Chmel (Hg.), Regesta chronologico-diplomatico Ruperti regis Romanorum, Frankfurt 1834, Nr. 476 mit falscher Ortsangabe „*Offheim*".

42 Regesta Imperii Online, Regesten Pfalzgrafen 2, Nr. 5552 (1408 Oktober 28 Nürnberg), URL: http://www.regesta-imperii.de/regesten/10-0-0-ruprecht/nr/1400-08-22-1-0-10-0-0-6290-6675.html?tx_hisodat_sources%5B%40widget_0%5D%5BcurrentPage%5D=112&cHash=8502a0442cd7f48b6e2fdc47f1d65203#rinav (letzter Zugriff: 24.8.2020); Chmel, Regesta Ruperti (wie Anm. 41), Nr. 2670.

43 Christa Benedum / Karl-Peter Büttner / Gerhard Egert / Franz Pfrang / Werner Stahr, Astheim und seine Kartause, Würzburg 1991, S. 22.

44 Ebd., S. 24; Ulrich Wagner / Walter Ziegler (Hg.), Lorenz Fries – Die Chronik der Bischöfe von Würzburg 742–1495, Würzburg 1999–2002, hier: Bd. 3, S. 130 (mit falschem Datum 1414).

45 Fritz Kugler, Thüngener Heimatbuch, Thüngen 1988, S. 28–32.

stehen.[46] Ein erneuter Versuch im Jahre 1690, die Stadtrechte zu beanspruchen, konnte auch zu diesem Zeitpunkt nicht durchgesetzt werden. Neubrunn scheiterte ebenfalls, da das Stadtprivileg von 1323 – von Ludwig dem Bayern an den Deutschen Orden vergeben – nicht durchsetzbar war.[47] Es waren die Wertheimer Grafen, die eine Stadtwerdung verhinderten. Dasselbe gilt für Hasloch, das von Karl IV. dem Mainzer Erzbischof übertragen und zur Stadt erhoben worden war. Da später von keiner Stadt mehr in den Urkunden die Rede ist, der Ort selbst wieder ganz in die Hände der Grafen von Wertheim gelangte, ist dieser Versuch ebenfalls gescheitert.[48] Ähnliches gilt für die von Karl IV. vorgenommenen Stadtrechtsprivilegierungen für hohenlohische Orte (etwa Gailnau, Haltenbergstetten, Hasloch, Hornberg, Mussen), von denen nur Neuenstein und Kirchberg sich zu Städten mit geringer regionaler Bedeutung entwickeln konnten.[49]

Zu Städtewüstungen kam es kaum: ein Fall wäre das pfalzgräfliche Wallhausen, das den Konkurrenzkampf gegen das mainzische Miltenberg bereits Mitte des 13. Jahrhunderts verloren hatte. In einem anderen Fall entschied eine vergleichbare Konkurrenz Windsheim gegen Lenkersheim, welches auf den Status eines Dorfes zurücksank, für sich. Alzenau schließlich wurde erst 1951 zur Stadt erhoben, obwohl der Mainzer Erzbischof bereits 1401 vom römischen König das Recht erhielt, das damalige Willanzheim zur Stadt auszubauen. Insgesamt fehlte diesen Städten, da erst im 14. Jahrhundert als solche privilegiert, ein größeres Umland, um prosperieren zu können.

Demgegenüber lassen sich in den größeren Städten Frankens jüdische Bevölkerungsgruppen nachweisen (Würzburg, Rothenburg, Nürnberg Bamberg und andere), die ein Indikator für eine überdurchschnittliche Bedeutung des urbanen Zentrums sind, die aber auch zahlreichen Pogromen und Vertreibungen ausgesetzt waren. Gerade beim Pogrom von 1298 werden interessanterweise auch jüdische Gruppen in kleineren Städten sichtbar: unter anderem Aub, Burgkunstadt, Gemünden, Gerolzhofen, Gunzenhausen, Hammelburg, Haßfurt, Kissingen. Ebenso finden sich, vorwiegend im 14. Jahrhundert, Hinweise auf Synagogen (Judenschulen),

46 Zuletzt: Naser, Digitale Karten (wie Anm. 39), S. 202.

47 Der Ort gehörte 1290–1484 dem Deutschen Orden, kam dann an das Erzstift Mainz, ab 1655 an das Hochstift Würzburg. Vgl. Franz L. Brunner, Geschichte der Deutschordenskomthurei und des Marktfleckens Neubrunn, Würzburg 1893.

48 Wilhelm Störmer, Marktheidenfeld (Historischer Atlas Bayern. Teil Franken 10), München 1962, S. 37 f., 66 f.

49 Hohenlohisches Urkundenbuch 3, hg. v. Karl Weller / Christian Belschner, Stuttgart 1912, Nr. 11 (1351 Neuenstein), 344 (1367 Haldenburgstetten), 467 (1373 Kirchberg), 471 f. (1373 Hornberg und Gailnau).

Judengassen und Friedhöfe (Amorbach, Dinkelsbühl, Miltenberg, Münnerstadt, Nürnberg, Rothenburg, Wertheim, Würzburg und andere). Nach der generellen Vertreibung der jüdischen Gemeinden aus den Städten im Spätmittelalter entwickelte sich in der Frühen Neuzeit ein ausgeprägtes Landjudentum besonders im Raum des heutigen Unterfrankens. Wiederansiedlungen von jüdischen Gemeinden waren ab der Mitte des 18. Jahrhunderts möglich, etwa in Bayreuth.

Weit über die Hälfte der Städte Frankens verblieb im Mittelalter in geistlicher Hand: Neben den drei fränkischen Bischöfen und dem Mainzer Erzbischof am Mainviereck waren es vor allem der Deutsche Orden (Mergentheim, Ellingen) und die Domkapitel von Bamberg (Staffelstein) beziehungsweise Würzburg (Ochsenfurt), deren Städte politische und kulturelle Landschaften prägten.[50] Neben den Domsitzen sorgten Klöster (etwa Amorbach, Stadtschwarzach, aber auch die Reichsabtei Fulda) und Stifte (etwa Ansbach, Herrieden) für die Ausprägung zentralörtlicher Funktionen. Lediglich fünf Reichsstädte finden sich in diesem Raum: Rothenburg, Schweinfurt, Weißenburg, Windsheim und, als größte Stadt überhaupt, Nürnberg. Die Markgrafen von Ansbach-Bayreuth führten unangefochten die Städtegründungen von Seiten der weltlichen Adeligen an. Die Städtevielfalt Frankens ist ein Indikator für die territoriale Zersplittertheit, jedoch erreichten viele der Kommunen, wie bereits angedeutet, keine großen zentralen Funktionen. Ihr Umland blieb flächen- und bevölkerungsmäßig gering. Viele Orte, die im Herzogtum Bayern als Märkte bezeichnet werden, sind in Franken Städte; eine eindeutige Grenzziehung zwischen beiden rechtlich definierten Ortsbezeichnungen ist kaum möglich. Von daher ist die oben genannte Gesamtzahl der fränkischen Städte stets diskussionsbedürftig. Je später eine Stadt gegründet wurde, desto geringer waren ihre Entwicklungschancen. Sie dienten meist als Orte der Herrschaftsausübung in politisch umstrittenen Regionen. In Franken war eine Städtelandschaft mit vorwiegend kleinen Städten entstanden; sie haben damit einen regional begrenzten Beitrag zum europaweit zu beobachtenden demographischen, wirtschaftlichen, politischen und sozialen Urbanisierungsprozess geleistet.[51] Durch die dennoch erreichte hohe Raumdichte ging von ihnen ein Verstädterungsprozess aus, der städtische Lebensweisen – wenn auch von überwiegend kleinstädtischem Niveau aus – in die Breite vermittelte. Dazu gehörten Bildungseinrichtungen wie Schulen als Wachstumsphänomene einer Stadt (circa 70 % der nachweisbaren Schulen in Städten), aber auch der Einfluss

50 Wolfram Unger, Grundzüge der Städtebildung in Franken. Träger – Phasen – Räume, in: Jahrbuch für fränkische Landesforschung (künftig: JfL) 59, 1999, S. 57–85.

51 Helmut Flachenecker, Fränkische Städtelandschaften. Anmerkungen zu einem Forschungsdesiderat, in: JfL 59, 1999, S. 87–108.

von neuen Moden (Kleidung, Feste) oder die Beschäftigung mit Geschichte (Chronistik). Rathäuser, Kirchen und Stadttore und Mauern bildeten die hauptsächlichsten Repräsentativbauten, mit denen die Stadt regionale Räume beherrschte. Mit ihren Stadtgerichten und Stadtfrieden unterschieden sie sich vom Umland. Zugleich konzentrierten sie Infrastruktureinrichtungen (Mühlen, Brauhäuser, Brücken et cetera). Schließlich muss die Verbreitung der neuen Tageseinteilung durch Uhren im Zusammenhang mit der Städtedichte gesehen werden; das Verhältnis zwischen Arbeit und Zeit veränderte sich. Universitäten entwickelten sich – wenn auch erst seit der Frühen Neuzeit kontinuierlich – in Würzburg (1402, wiedererrichtet 1582), nach der Reformation in Altdorf (1622) und Bamberg (1649). Lediglich die Waldgebiete des Spessarts, des Steigerwalds und Teile der Rhön blieben städtefrei. Und abschließend: Auch in Franken war die Reformation weitgehend städtisch geprägt. Dies führte unter anderem zu einer konfessionell differenzierten Schullandschaft.

In der Frühen Neuzeit dann kommt es zu einem Konzentrationsprozess der Stadtherrschaften auf die dominierenden Fürsten Frankens.[52] Städtegründungen waren nur noch selten. Zum einen war der Bedarf nach neuen Städten nicht mehr gegeben, zum anderen entsprach kommunale Autonomie nicht mehr den Herrschafts- und Staatsvorstellungen der Frühen Neuzeit. Zwischen 1500 und 1800 sind lediglich sechs Neugründungen zu finden: Ostheim (1596) und Kaltennordheim (1668) durch die Wettiner, Hofheim (1576) durch den Würzburger Bischof sowie das nürnbergische Betzenstein (1533) und die beiden markgräflichen Neugründungen Christian-Erlang (1701) und St. Georgen bei Bayreuth (1705: „*Neue Stadt zu St. Georgen*“).[53] Die beiden letzteren folgten einer strengen rechteckigen Grundrissform, wie sie nicht nur in Europa, sondern auch in Amerika zu beobachten ist. Christian-Erlang wurde für die Neuansiedlung der vertriebenen Hugenotten angelegt.

Der entscheidende Beitrag der Städte in der Territorienausbildung wird in einer zunehmenden herrschaftlichen Instrumentalisierung gesehen. Deswegen waren diese Städte in der Regel Zentren der herrschaftlichen Ämterstruktur, einzelne sogar des spätmittelalterlichen Territoriums beziehungsweise

52 Wolfram Unger, Die fränkischen Städte in der Frühen Neuzeit, in: Frankenland 2005, S. 340–350.

53 Andreas Jakob, Erlangen im 18. Jahrhundert, München 2012; Bernhard Rupprecht, „Gedächtnuß der von Uns angefangenen Neuen Stadt zu St. Georgen am See“. Die Planstadt des Markgrafen Georg Wilhelm, in: Helmut Neuhaus (Hg.), Aufbruch aus dem Ancien régime. Beiträge zur Geschichte des 18. Jahrhunderts, Köln u.a. 1993, S. 135–156.

frühmodernen Staates insgesamt. In besonderen Situationen spielten wirtschaftliche Besonderheiten eine große Rolle, so etwa in Goldkronach der Gold- und Silberbergbau, der jedoch ab der Mitte des 15. Jahrhunderts schon wieder an Bedeutung verlor. Die Stadt als Repräsentations- und Statussymbol für eine Herrschaft verlor auch im 18. Jahrhundert nicht ihre Rolle, allerdings sind neue Stadtgründungen wie etwa das bereits erwähnte St. Georgen große Ausnahmen. Meist hatten die fränkischen Landstädte Mühe, den urbanen Charakter überhaupt aufrechtzuerhalten (etwa Baunach, Hofheim, Teuschnitz, Nordhalben). Der Verlust kommunaler Rechte nach der Niederlage im Bauernkrieg 1525 konnte manche Stadt, wie etwa Ludwigsstadt, in die Gefahr des „Abstiegs" zum Markt bringen. Und die Niederlage des „Gemeinen Mannes" 1525 zeigte auch im Falle Gerolzhofens, um ein weiteres Beispiel zu nennen, schwerwiegende Folgen.[54] Insgesamt gesehen wurden die Städte in der Frühen Neuzeit verstärkt dem obrigkeitlichen Reglement unterworfen, wodurch sie zwangsläufig kommunale Sonderrechte verloren. Die Sonderrolle der Städte wurde in den frühneuzeitlichen Staaten nivelliert.

4. Städte in den Hochstiften Würzburg und Eichstätt

Die Mehrzahl der Städte Frankens besaß, soviel dürfte aus dem generellen Überblick deutlich geworden sein, einen geistlichen Stadtherren, zuvörderst die Bischöfe von Würzburg, Bamberg und Eichstätt. Daneben beanspruchten die Äbte von Fulda mehrere Stadtherrschaften, der Abt von Münsterschwarzach eine. Auf einige Hochstiftsstädte der würzburgischen und eichstättischen Bischöfe soll ein etwas genauerer Blick gerichtet werden.

4.1. Würzburg

„Carlburg vnd Carlstat slos vnd stat haben kunig Pipin vnd seine sune, kaiser Carln. den der erst vnd grosse, vnd kunig Carlmann in vfrichtung vnd widmung des newen bistumbs zu Wirtzburg Sant Burckharten dem bischof vnd seinen nachkomen, neben vnd mit anderen schlossern, stetten vnd flecken des erledigten Hertzogtumbs zu Francken zu aigen gegeben vnd zugestelt. Man heltet auch dafur das gemelte schlos vnd stat Carlburg vnd Carlstat von Carlen Martellen, egedachten kunig Pipins vater, herzog in Braband vnd grosmair der Francken empfangen haben, vnd der alten kunig zu Francken kunigliche schatzkamere diser vnd

54 Helmut Flachenecker, Gerolzhofen im Bauernkrieg, in: Gerolzhofen. Stadtchronik 779–2012, Baunach 2012, S. 293–302.

anderer anstossenden lande gebawet worden vnd iren namen von ime empfangen haben, vf dem gemelten schloß Carlburg gewest sein solle."[55] Mit dieser die frühmittelalterliche Geschichte aufgreifende Erzählung beginnt der bekannte Würzburger Archivar und Berater dreier Fürstbischöfe, Lorenz Fries († 1550), den Eintrag zu Karlstadt in seiner „Hohen Registratur".[56] Ähnlich schreibt er zur Karlburg: „*Carlburg ein dorf, ist vor alten zeiten ain closter gewest vnd durch Kaiser Carln den Grossen an den Stifft Wirtzburg gegeben auch volgends demselbigen von den Kaisern vnd Kenigen bestettigt worden.*"[57] Diese Registratur ist nicht nur ein einfaches Archivrepertorium dieser Tage, sondern, da der Archivar damals in die Politik seines Herrn eingebunden war, ein Herrschaftsinstrument der besonderen Art. Die in der Registratur kopial verzeichneten Urkunden und Schriftstücke konnten bei aktuellen Ereignissen hervorgezogen und in den Auseinandersetzungen eingesetzt werden. Von daher war es für die Würzburger Kanzlei aktuell besonders wichtig, über die Karlstädter Situation informiert zu sein. Deshalb übersprang Fries viele Ereignisse, es fehlt etwa ein Hinweis auf die Stadtgründung durch Bischof Konrad von Querfurt, und setzte zentral wieder im 14. und 15. Jahrhundert ein.

Es war für die würzburgische Verwaltung wichtig zu wissen, wer die Ortsherrschaft in Folge der zahlreichen Verpfändungen jeweils innehatte. Damit wird klar, dass eine Stadtherrschaft kein Selbstläufer war. So versetzte Bischof Johann von Egloffstein die Stadt an das Domkapitel, ebenso wie „*stat vnd ambt Melrichstat*".[58] Derselbe Bischof hatte auch 1405 „*das dorf Carlburg mit allen seinen zu vnd eingehorungen Dietzen von Thungen dem Jungeren fur 950 guldein vf ablosung verschriben vnd nichts vorbehalten dann erbhuldung, volg, steur vnd bete*".[59] Der Thüngen konnte die grundherrschaftlichen und gerichtsrechtlichen Einnahmen in der Zeit der Verpfändung einnehmen, Huldigung, Steuer und

55 StAWü, Standbuch 1011, fol. 141 r.

56 Eine Internet-Edition ist im Aufbau; die derzeitigen Ergebnisse können im Internetportal Historisches Unterfranken abgefragt werden: http://www.historisches-unterfranken.uni-wuerzburg.de/fries/fries.php (letzter Zugriff: 24.8.2020). – Helmut Flachenecker, Die Hohe Registratur des Lorenz Fries († 1550), in: Mathias Thumser / Janusz Tandecki (Hg.), Editionswissenschaftliche Kolloquien 2005/2007. Methodik – Amtsbücher. Digitale Edition – Projekte, Toruń 2008, S. 129–144; Helmut Flachenecker, Lorenz Fries als Historiograph, in: Franz Fuchs / Stefan Petersen / Ulrich Wagner / Walter Ziegler (Hg.), Lorenz Fries und sein Werk. Bilanz und Einordnung (Veröffentlichungen des Stadtarchivs Würzburg 19), Würzburg 2014, S. 1–27.

57 StAWü, Standbuch 1011, fol. 140 v.

58 Ebd., fol. 141 v.

59 Ebd., fol. 140 v.

Heeresfolge blieben beim Bischof. Die Bewohner von Karlburg mussten damit zwei Herren dienen. Das Wissen, dass zu der Burg Burgmänner gehörten sowie Güter, mit deren Erträgen die Besatzung bezahlt werden sollten, war im frühen 16. Jahrhundert noch vorhanden.[60]

Die Bischöfe kümmerten sich auch um die wirtschaftliche Situation der Stadt: Bischof Johann von Brunn hat „*auch der stat Carlstat trei jarmerckt nemlich den ersten vf sontag Quasimodogeniti, den 2. vf sontag vor sant Jacobs tag, den 3 vf sontag nach vnser lieben frawen geburtstag, vnd dann ain wochenmark, allen sambstag, geschehen am triten Ostertag im 1438 jare*". Die Einnahmen aus dem Rathaus, das auch immer als Kaufhaus diente, wurden zwischen Stadt und Bischof geteilt: „*Kauf oder Rathaus zu Carlstat, was man aus den läden vnd sunst fur zins davon hat, stet halb vnserm gnedigen heren vnd halb der stat zu.*" Schließlich musste der Bischof seine Zustimmung geben, dass eine neue Mühle gebaut werden durfte, welche die Schifffahrt auf dem Main nicht beeinträchtigen durfte: „*Bischof Johanns von Brun hat den burgeren zu Carlstat vergont ain muln in Main zubawen, doch dem wasserstrom vnd schiffart on schaden, vnd das die mullere nit zu grosse nutz nemen auch ime vnd seinen nachkomen jherlich XII malter korns zu gulte geben sollen, am dinstag nach Quasimodogeniti anno 1436.*"[61]

Das Karlstädter Beispiel führt mitten hinein in die verwickelte fränkische Geschichte und damit auch in die Stadtgeschichte. Die „Hohe Registratur" vermittelt einen spannenden Einblick auch für andere Landstädte: „*Wie aus dem dorf Dettelbach ain stat gemacht worden ist. Dettelbach ist bis dahere ain dorf gewest, aber aus vergunstigung kaiser Friderichen jn dem jore des heren 1484 zu ainer stat gemacht, vnd der selben stat freihait vnd recht geben worden, wie andern stetn im stift gelegen.*"[62]

Lorenz Fries schildert eine spätmittelalterliche Stadtgründung im Hochstift Würzburg mit einer kaiserlichen Bestätigung. Aber war dem so? Nun, Lorenz Fries hat die Sache etwas verkürzt dargestellt. Dettelbach war schon längere Zeit ein bedeutsamer Ort als Burgsitz der Ritter von Dettelbach, ehemals hohenlohische Ministeriale, die den Ort als würzburgisches Pfand lange Zeit beherrschten.[63] Seit 1331 hatten die Bischöfe in dem Ort Fuß ge-

---

60 Ebd., fol. 145 v: „*Burgmennere vnd burgguter zu Carlburg wer die vor alter gewest vnd itzund sind, ec. davon such im raisbuch.*"

61 Ebd., fol. 145 r.

62 Wagner / Ziegler, Lorenz Fries – Chronik (wie Anm. 44), Bd. 4, S. 263.

63 Hans Bauer, Die kulturlandschaftliche Entwicklung des alten Amtes Dettelbach seit dem 16. Jahrhundert (Mainfränkische Studien 17/1 und 2), Würzburg 1977, S. 8–34.

fasst.[64] Mitte des 15. Jahrhunderts erwarb das Würzburger Benediktinerkloster St. Stephan umfangreichen Besitz, Bischof Rudolf II. von Scherenberg bemühte sich anschließend erfolgreich um eine Wiedereinlösung des Pfandes und der Rückführung Dettelbachs in die bischöfliche Hände. Bereits 1482 hatte er eine „*Ordnung*" („*Policeyordnung*") erlassen.[65] Darin hatte er für das noch Dorf genannte Dettelbach einen inneren und äußeren Rat genehmigt; die Bewohner wurden bereits als Bürger bezeichnet und deren Rechte unter der Aufsicht des bischöflichen Schultheißen geregelt. Die Königsnähe des Bischofs hat Kaiser Friedrich III. wohl dazu bewogen, Dettelbach 1484 formal den Status einer Stadt zu geben,[66] Bischof Rudolf baute die Stadt, die nunmehr so auch in den bischöflichen Urkunden (etwa 1487)[67] bezeichnet wurde, zu einem Amtssitz aus. Den Bürgern wurden alte Gastungsrechte des Bischofs erlassen, mussten dafür einen Teil des Ungelds an ihn abführen.[68] Gleichzeitig gewährte er ihnen drei Jahresmärkte und einen Wochenmarkt.[69] Die

64 StAWü, Standbuch 1011, fol. 167 r: „*Dettelbach daselbst vnd zu Iphoven vnd Reppern-dorff hat etwan Grave Vlrich von Hanaw etliche gutere gehabt vnd dieselben mit iren zugehorenden rechtenehren vnd nutzen gabe Bischof Wolframen vnd seinem stiffte zu kauff geben vnd kaiser Ludwig der 4 solchen kauff besteigt. Sambstag in der pfingstwochen anno 1331. Registrata privilegiorum fo. 52.*"

65 Fritz Mägerlein / Walter Scherzer (Bearb.), Die Archive des Landkreises Kitzingen. Teil 2: Die Stadtarchive, München 1969, S. 1–3: U 1 (1482 Feb. 22). – Analog wie in Ebern siehe: StAWü, Standbuch 1011, fol. 190 r: „*Ordnung im Jare des Heren 1489 gab bischof Rudolf obgedacht den burgermaistern, rathe vnd gemainde der stat Eberen ain policei ordnung. 4 Rudolfi fo. 105 d.*" Ebenso Fladungen: ebd., fol. 228 r: „*Ordnung hat bischof Rudolf der stat Fladungen geben, Montag nach Dorothee, anno 1489, in 4o contractum Rudolfi fo. 112 d.*" – Stadtordnungen für Ebern und Bischofsheim siehe: StAWü, libri diversarum formarum 15, S. 234, 248.

66 Auffälligerweise fehlt ein Hinweis auf diese Urkunde in der Fries'schen Registratur. Alfred Wendehorst (Bearb.), Das Bistum Würzburg 3: Die Bischofsreihe von 1455 bis 1617 (Germania Sacra NF 13), Berlin/New York 1978, S. 38, hat keinen Zweifel an dem Stadtrecht, kann aber auch keinen Fundort für die Urkunde angeben.

67 Mägerlein / Scherzer, Archive des Landkreises Kitzingen (wie Anm. 65), S. 3 f.: U 1 a (1487 Nov. 23).

68 StAWü, Standbuch 1011, fol. 168 r: „*der stiffte Wirtzburg hat vor alter die gerechtikait zu Detelbach gehabt, wan ain Bischof oder seine rathe dahin komen, das gemaine stat dieselbigen aus der herberg lösen mussen, das hat inen bischof Rudolf frei nachgelassen, dargegen haben si dem stifft Witzburg den halbtail am vngelt daselbst vbergeben, vnd ist derwegen ein vngeltordnung da furgenomen dinstag nach vocem Iocunditatis des 1484 jars Registrata in 4 Contract. Rudolfi fo. 299.*"

69 Ebd., fol. 171 r: „*Anno 1484 gibt bischof Rudolff der stadt Dettelbach 3 Jahrmärckht als an sontag nach Purificationi Maria vff S. Maria Magdalenen tag der dritt vff S. Matheus*

pfarrliche Eigenständigkeit folgte. Die Gerichtsrechte von St. Stephan wurden ebenfalls von Bischof Rudolf im selben Jahr abgelöst.[70] Der Bischof hatte einen zusätzlichen zentralen Ort, sowohl wirtschaftlich wie auch administrativ und gerichtlich, im Bereich seines Hochstiftes, also seines weltlichen Herrschaftsbereiches. Ein fürstbischöflicher Amtskeller war politisch wie auch gesetzgeberisch tätig, ein Schultheiß stand sowohl dem Dettelbacher Stadtgericht wie auch den Gerichten in den Dörfern des Amtes vor. Steuern und Zinsen mussten von den Bewohnern in Dettelbach bei der bischöflichen Verwaltung abgeliefert werden. Im ausgehenden 16. Jahrhundert betrug die Bevölkerungszahl im Amt annähernd 1.758 Einwohner, davon 1.281 in Dettelbach selbst. Hinzu kamen noch die Bewohner anderer Herrschaften innerhalb des Amtes, unter ihnen markgräfliche. Mit der Stadtgründung wurde der Ort am Main ein würzburgischer Verwaltungs- und Gerichtsmittelpunkt – Sitz eines Amtes eben. Die Stadtgründung hatte also eine entscheidende Begründung im Herrschaftsausbau des Würzburger Bischofs in der Grenzregion zur Markgrafschaft.

Lorenz Fries schätzte die Bedeutung der Städte sehr hoch ein. In seiner „Hohen Registratur" hat er zahlreiche und ausführliche Einträge zu den einzelnen Hochstiftsstädten verfasst. Dabei hat er in regestartiger Form die wichtigsten, im Würzburger Archiv für diese Städte vorhandenen Urkunden zusammengefasst. Aber auch kleine Ortsgeschichten sind in der Registratur enthalten: Neben Karlstadt ist etwa Neustadt an der Saale zu nennen, dem Fries ebenfalls eine Gründungsgeschichte widmet: *„Es sein etwan zwai dorfer in disem ambt gelegen, vnd das ain Ober das ander Vnder Saltz genant, aber aus dem ainen vnd nemlich Vndersaltz ist ain stat gemacht worden, die an gebewen, leuten, gutern, handel, wandel vnd anderen burgerlichen wesen dergestalt zugenomen vnd gewachsen, das si heut zutag vnder anderen des Stiffts W*[ürzburg] *steten nit die geringste ist. Wann aber solchs beschehen sei davon ist kain gewis vrkund noch antzaigung vorhanden, doch allerlai gelegenhait noch acht ich darfur, es sei in regierung B*[ischof]

---

tag vnd dan auch einen wochenmarckht vff eines jedes montag. Registrata in 7 Contr. Rud. fol. 296."

70 Ebd., fol. 168 r: *„Abbt vnd closter zu Sant Steffan zu Wirtzburg haben ain Gericht zu Detelbach gehabt vnd dasselb vf Montag nach Sant Martins tag in 1484 jare bischof Rudolfen vnd dem stifft Wirtzburg zugestelt dagegen bischof Rudolf inen den sitz vnd weiler Oberndurbach so vor lehen gewest geaignet hat. in 3 Contract. Rudolfi fo. 192 193 Item auch iren hof zu Detelbach ligend so lang der vnverendert bei dem closter bleib gefreiet. ibidem fo. 194."*

Herman des ersten des namens gebornen heren von Lobdeburg beschehen, bei dem ich auch erstlich ain stat genent find anno 1232."[71]

Im besonderen Interesse standen bei Fries immer die rechtlichen Bindungen der Städte zu den Bischöfen. So werden etwa besonders viele Hinweise auf die Stadt Lauda gegeben, weil diese häufig verpfändet war und der Bischof seine Stadtherrschaft gegen Ansprüche der Herren von Hohenlohe beziehungsweise der Pfalzgrafen bei Rhein beständig durchsetzen musste. Dabei wird auf die frühe Inbesitznahme durch Würzburg verwiesen: „*Lauden da sein vor alten zeiten heren gesessen vnd hat dem Stifft W*[ürzburg] *sein leben lang lehen gemacht vnd nach seinem dot zu aigen geben im jare des Heren 1169. Wie schloß vnd stat Lauden an die hertzogen von Bairen komen, davon find ich nichts bei der cantzlei, aber volgender zeit haben die heren von Hohenlohe, bede schlössere vnd stete Lauden vnd Jagsberg (villeicht durch ainen heirat an sich bracht). vnd die bede dem Stifft W*[ürzburg] *zukauffen geben, wie in der rechtvertigung zwischen hertzog Wilhelmen von Bairen (der die widerlosung darauf haben wollen) vnd Bischof Conraten von Thungen ergangen* [...]."[72] Mit dem Argument der frühen würzburgischen Inbesitznahme fundiert Fries in vielen Fällen die Herrschaftsansprüche in seiner Gegenwart. Er ist eben Historiker und Politiker!

Auch Meiningen gehörte seit den Zeiten der Bamberger Bistumsgründung, also seit dem beginnenden 11. Jahrhundert, zu Würzburg: „*Mainingen ain burg stat vnd ambt im Grabfelt gelegen*" sei „*dem Stifft W*[ürzburg]" unterworfen.[73] Oder, um ein weiteres Beispiel zu nennen, Möckmühl soll, so Fries, bereits seit den Zeiten Heinrichs III. immer zu Würzburg gehört haben: „*Meckmülen etwan Megdmülen genant etliche gefelle vnd nutzung daselbst hat kunig Hainrich der trit Bischof Brunen vnd seinem stifft zu aigen gegeben, am anderen tag des jäners anno 1042. So ist sunst das schloß, dergleichen die stat Meckmuln des stifft W*[ürzburg] *aigentumb vnd der herschafft Hohenlohe lehen gewest* [...]."[74]

Intensiv beschäftigte sich der bischöfliche Archivar mit Kitzingen, einer Stadt, deren Stadtherrschaft höchst umstritten war. „*Kitzingen die stat am Main zum merern tail sambt seinen zu vnd eigen des stiffts Wirtzburg aigenthumb zum mererntail vnd der Marggraven zu Onoldbach pfandschaft ist etwan der heren von Hohenlohe zum teil frei aigen zum tail lehen gewest, das si vom reich entpfangen*

71 StAWü, Standbuch 1012, fol. 141 r. – In der Chronik geht Fries nicht darauf ein: Wagner / Ziegler, Lorenz Fries – Chronik (wie Anm. 44), hier: Bd. 2, S. 128–157. – Vielleicht ist unter „*allerlai gelegenhait*" der Erwerb Botenlaubens 1234/42 zu subsummieren.

72 Ebd., fol. 19 r.

73 Ebd., fol. 76 r.

74 Ebd., fol. 104 v.

vnd getragen haben."[75] Mit der für fränkische Städte nicht ungewöhnlichen Situation, mehrere und zum Teil zeitlich wechselnde Stadtherren zu besitzen, beginnt der Registratureintrag. Fries setzt mit dem Kauf des städtischen Anteils der Grafen von Hohenlohe 1336/39 ein. Aber genau hier hat Fries schon ein Verwaltungsproblem: „*Den Kaufbrief hab ich nit gefunden*" – so stellt er resigniert fest, er müsse aber stattgefunden haben, weil die nachfolgende Tauschbelehnung der Hohenlohe mit der würzburgischen Stadt Möckmühl in der Registratur kopial auffindbar wäre; er erschließt also die verlorene Kaufurkunde durch eine vorhandene.

Es war dann die Empörung der Bürger der Bischofsstadt Würzburg im Bündnis mit anderen Hochstiftsstädten, die Bischof Gerhard von Schwarzburg 1399 zwang, seinen Anteil an der Stadt Kitzingen an die Nürnberger Burggrafen Johann und Friedrich zu verpfänden, um diese für ihre Unterstützung bei der Niederwerfung des städtischen Aufbegehrens zu entlohnen. Erneut hat Fries Schwierigkeiten, denn er kann den Umfang des Würzburger Anteils an Kitzingen, der nun verpfändet war, erst aus einer nachfolgenden Urkunde – von 1405 – feststellen: „[...] *lauter meldung davon, das es der tritail sei gewest vf der stat Kitzingen weniger aines sechzehendentails am gantzen wie dan zum tail hievor angezaigt vnd hernach ferner gemeldt wurt.*"[76]

Damit war aber auch ein Jahrhunderte andauernder Streit vorprogrammiert, nämlich wer welche Rechte an der Stadt Kitzingen habe. Es kam 1404 zu einem Schiedsgericht in Bamberg, aber nicht dieses, sondern ein weiteres legte ein Jahr später fest, dass der Burggraf die Stadtrechte des Bischofs an seinem Drittel minus ein Sechzehntel nicht streitig machen dürfe. Die Bürger sollten dem Bischof für seinen Anteil an der Stadtherrschaft auch huldigen, die Stadtsteuer sollte entsprechend geteilt werden; Tore, besonders gegenüber dem Benediktinerinnenkloster, mussten von den Burggrafen offen gehalten werden; Ladungen vor den jeweiligen Landgerichten der beiden Herren mussten formal korrekt angezeigt werden und durften keinesfalls mit Gewalt erfolgen. Für diese burggräflichen Zugeständnisse hob der Bischof im Gegenzug den über die Stadt verhängten Bann auf. Und zugleich trat dann auch noch eine vierte Partei auf, die Grafen von Wertheim. Deren Rechte am Zoll zu Kitzingen sollten in keiner Weise geschmälert werden. Schließlich wurde das Würzburger Recht auf Wiedereinlösung des Pfands von burggräflicher Seite ausdrücklich anerkannt.[77]

---

75 StAWü, Standbuch 1011, fol. 423 r.

76 Ebd., fol. 425 r.

77 Ebd., fol. 425 v–426 v.

Damit waren aber die Fronten nicht für alle Zeiten geklärt. Fries hat erneut Schwierigkeiten, die weitere Entwicklung urkundenmäßig mit der ihm vorliegenden Überlieferung weiter zu verfolgen. Immerhin: 1405 erwarb Würzburg von Hans von Hohenlohe ein weiteres Drittel an der Stadt: „*Vnd ist an disem ort zumercken, das von der gantzen suma ain sechzehend tail erstlich vnd dan der trittail den Burggraven, vnd die anderen zwen dritail dem stifft Wirtzburg zugeburen wie dan solchs aus ainer missiven so Bischof Johanns an burgermaister vnd rathe zu Kitzingen gethan hat, lauter zuvernemen ist.*“[78] Peinlich genau möchte der Archivar die Rechte der Würzburger Bischöfe in Kitzingen belegen können. Für seine Akkuratesse spricht ein kleines Detail am Rande: „*Anno predicto 1406 hat Burggrave Fridrich von Nurenberg die mark zu Brandenburg erkaufft die in gewer vnd geseß gewonnen vnd sich furter davon geschriben.*“ Aber einer gewissen Bosheit und Arroganz kann er sich als politisch agierender Archivar dann doch nicht enthalten. Trotz dieser Klarstellung spricht er erst seit den 1430er Jahren tatsächlich von Markgrafen, zuvor benutzt er den geringeren Titel der Burggrafen.

Die Pfändungsgeschichte wird im gesamten 15. Jahrhundert, von Bischof zu Bischof, penibel verfolgt, auch die mehrmaligen Vereinbarungen zwischen Würzburg und Ansbach über die Aufteilung der Kitzinger Stadtsteuer. Ebenso werden die von jedem neuen Bischof geforderten Huldigungen der Bürgerschaft aufgezeichnet: 1495 vor Lorenz von Bibra, 1519 vor Konrad von Thüngen, 1541 vor Konrad von Bibra, 1546 vor Melchior Zobel.[79]

Fries hält – darauf weisen auch die Karlstädter Eintragungen hin – erneut die wirtschaftlichen Privilegien der Würzburger Hochstiftsstädte fest: „*Anno 1356 hat bischof Albrecht von Hohenlohe zu Melrichstat ain schriftliche freihait geben, das si alle jare vff S. Lorentzen tag ain tag darvor vnd ain tag darnach ainen freien jarmarkt halten mögen.*“[80] Oder für Röttingen: „*Jarmarckt hette die stat Rottlingen alwegen* […]“, und für Münsterschwarzach: „*Jarmarckt zu stat Schwartzach.*“[81]

In diesen wirtschaftlichen Zusammenhang gehört natürlich auch die Vergabe von Rechten und Lehen in einer Stadt. Penibel wird in der Registratur festgehalten, welchem Adeligen oder welchem Kloster der Bischof Lehen vergeben hat,

78 Ebd., fol. 427 v.

79 Ebd., fol. 430 v.

80 StAWü, Standbuch 1012, fol. 111 v.

81 Ebd., fol. 461 r bzw. fol. 489 r.

nicht nur im Falle von Städten, sondern auch in Dörfern, wie etwa im Falle des „*dorffs*" Northeim im Grabfeld den Klosterfrauen von Wechterswinkel.[82]

Schließlich werden die politischen Rahmenbedingungen, allen voran die bischöflichen Stadtrechte verzeichnet. Entsprechende Hinweise gibt es dann für Neustadt an der Saale: „[...] *des ietzgemelten 1435 jars am freitag nach Letare gab bischof Johanns von Brun den burgern zue Newenstat ain gemaine burger oder stat ordnung* [...]" beziehungsweise „[...] *Anno 1478 am sambstag nach Circumcisionis hat bischof Rudolf den burgeren zue Neuenstat ain gemaine statordnung geben.*"[83] Oder für das bereits erwähnte Northeim: „*Northaim vor der Rone weisthumb, vnd was der stift vnd die Gemein alda fur gerechtigkait hat findest du* [...]."[84] Auf weitere städtische Beispiele aus der ersten Hälfte des 16. Jahrhunderts kann hier nur summarisch hingewiesen werden: „[...] *Geroltzhofen statordnung* [...] *Melrichstat statordnung* [...] *Munerstat statordnung durch bede herschaften Wirtzburg vnd Hennenberg* [...] *Volkach der stat ordnung.*"[85]

Landstadt als politische Rechtsform bedeutete nicht automatisch politisches Wohlgefallen der Bewohner gegenüber dem Stadtherrn. Gerade die würzburgischen Hochstiftsstädte haben die Hauptstadt in ihrem Kampf um Unabhängigkeit unterstützt, mussten aber 1400 in der berühmten Schlacht von Bergtheim eine Niederlage einstecken. So heißt es dann in der Registratur unter Neustadt an der Saale: „*Es heten sich Burgermaister, rath vnd gemainde zue Neuenstat mit W*[ürzburg] *vnd anderen steten wider Bischof Gerharten verbünden, wurden aber bei Berchthaim geschlagen vnd von gemelten Bischof Gerharden wider zu gnaden angenomen vnd vertragen.*"[86]

Für Lorenz Fries waren die Städte Herrschaftsmittelpunkte im Hochstift. Sie hatten eine funktionale Aufgabe für die bischöfliche Landesherrschaft. Fries betonte das wirtschaftliche und politische Interesse der Bischöfe an diesen Orten – als Mittel für die Geldleihe mit Hilfe von Verpfändungen oder als Objekte in der Auseinandersetzung und Ausschaltung konkurrierender Machtansprüche anderer Fürsten, Grafen und Herren. Kommunale Freiräume wurden

---

82 Ebd., fol. 147 v: „*Anno 1327 am Montag vor Martini hat B*[ischof] *Wolfram zu W*[ürzburg] *den closterfrawen von Wechterswinkel etliche gefelle vnd nutzung zu Northaim verkaufft vf widerlösung. Registratum Capitulari fol. 232 d.*" Vgl. Heinrich Wagner, Urkunden und Regesten des Klosters Wechterswinkel (Quellen und Forschungen zur Geschichte des Bistums und Hochstifts Würzburg 70), Würzburg 2015, Nr. 179, S. 236.

83 Ebd., fol. 143 r bzw. fol. 144 r.

84 Ebd., fol. 148 r.

85 Ebd., fol. 162 r, zusammengefasst unter dem Titel: „*Von ordnungen vff dem land.*"

86 Ebd., fol. 141 v.

in der ersten Hälfte des 16. Jahrhunderts als antiquiert und überholt angesehen. Die Würzburger Bischöfe versuchten – wie alle anderen Territorialherren der Zeit auch – die Städte als wichtigen Bestandteil eines sich vereinheitlichenden und auf die Hauptstadt mit der dortigen Landesverwaltung sich ausrichtenden frühmodernen Staates zu instrumentalisieren. Würzburg stand mit einem solchen territorialen Ansatz nicht allein, ein knapper Blick auf das Hochstift Eichstätt möge dies illustrieren.

4.2. Eichstätt

Auch im Falle Eichstätts bildete der Bischofssitz die Grundlage für die älteste Stadt im Hochstift. Im 14. Jahrhundert lassen sich viele kleinere Städtegründungen an den Rändern der Territorien beobachten. Kleinstädte waren „im besonderen Maße Instrumente landesherrlicher Politik“ (Peter Johanek).[87] Im 15. Jahrhundert folgten eine Vielzahl von Märktegründungen und Städteausbauten in der Binnenlage der Hochstifte.[88]

Ein Blick auf die hoch- und spätmittelalterlichen Eichstätter Bischofsbiographien zeigt, dass sich praktisch alle Bischöfe mehr oder weniger intensiv um den Ausbau ihrer Städte, Märkte und Burgen kümmerten, um mit deren Hilfe den Landesausbau voranzutreiben.[89] Im Falle Eichstätts lassen sich zwei Verdichtungsphasen festhalten, nämlich um 1300 und in der zweiten Hälfte des 15. Jahrhunderts. Um 1300 bestanden die Bischöfe erfolgreich die Auseinandersetzungen mit ihren beiden mächtigen Hochstiftsvögten (Hirschberg, Oettingen). In dieser veränderten politischen Konstellation erfolgte der bischöfliche Zugriff auf Eichstätt und Berching als Schwerpunkte im Unterstift, denen 1310 dann, durch königliche Übertragung, der Markt Greding hinzugefügt werden konnte. Die Zentren des östlichen Oberstifts wurde aus den Händen der Burggrafen von Nürnberg erworben, nämlich Spalt (1294) und Abenberg (1296),

87 Peter Johanek, Landesherrliche Städte – kleine Städte. Umrisse eines europäischen Phänomens, in: Jürgen Treffeisen / Kurt Andermann (Hg.), Landesherrliche Städte in Südwestdeutschland (Oberrheinische Studien 12), Sigmaringen 1994, S. 9–25.

88 Helmut Flachenecker, Die Städte im Hochstift Eichstätt während des Spätmittelalters, in: Ders. / Kießling, Städtelandschaften (wie Anm. 6), S. 152–187.

89 MGH SS 25, S. 590–609. Zu den Viten siehe: Stefan Weinfurter / Helmut Flachenecker / Monika Fink-Lang / Ernst Reiter / Klaus Kreitmeir, Die Viten der Eichstätter Bischöfe im „Pontifikale Gundekarianum“, in: Andreas Bauch / Ernst Reiter (Hg.), Das Pontifikale Gundekarianum. Kommentarband zur Faksimile-Ausgabe, Wiesbaden 1987, S. 111–147, hier bes.: S. 117–130 [wiederabgedruckt in: Stefan Weinfurter, Eichstätt im Mittelalter. Kloster – Bistum – Fürstentum, Regensburg 2010, S. 155–178].

schließlich 1301 Pleinfeld. Den Oettingern wurden schließlich Herrieden und Ornbau (1310/11) entrissen, die fortan die verwaltungsmäßigen beziehungsweise wirtschaftlichen und bei Herrieden auch religiösen Mittelpunkte des westlichen Oberstifts bildeten. Insgesamt verläuft dieser Prozess im Hochstift Eichstätt fast ein Jahrhundert später als in anderen Hochstiften.

In der zweiten Hälfte des 15. Jahrhunderts erfolgte ein Binnenausbau des Hochstifts. Maßnahmen zur Wirtschaftsförderung wie etwa die Bestätigung und Koordinierung von Jahrmärkten (1482) sowie der Erwerb der Märkte Dollnstein (1440) und Arnsberg (1475) und die Erhebung Arbergs (1454) und Pleinfelds (1483) zu Märkten schlossen sich an. Die mehr oder weniger systematischen Bestätigungen von Handwerkerordnungen gehören ebenfalls in diesen Zusammenhang. Die militärische Sicherung der bischöflichen Gebiete wurde zudem durch einen verstärkten Mauerbau forciert: Nassenfels, Mörnsheim und Dollnstein – alle im Südwesten des Unterstifts an der Grenze zum Herzogtum Bayern gelegen – wurden militärisch stärker gesichert. Auch Ornbau im Oberstift erhielt unter Bischof Wilhelm von Reichenau eine neue Ummauerung. Zu seiner Zeit wurden zudem an mehreren bereits bestehenden Befestigungen (Beilngries, Greding) gearbeitet.

5. Fazit

Die für Franken typische Herrschaftszersplitterung brachte eine große Städtedichte hervor.[90] Damit grenzte sich die Region stark vom bayerischen Herzogtum ab, das wenige Städte, aber relativ viele Märkte kannte. Meist handelte es sich, insbesondere bei den Stadtgründungen im Spätmittelalter, um kleinere beziehungsweise Kleinstorte mit geringen Chancen für eine intensive kommunale wie wirtschaftliche Entfaltung. In Franken dominieren daher, von Nürnberg und den Bischofsstädten abgesehen, Städte mit einer niedrigen Bevölkerungszahl (geschätzt 800–2.000), die sich von jener größerer Dörfer bisweilen kaum unterschied. Städtegründungen des 12. und beginnenden 13. Jahrhunderts betrafen meist wirtschaftlich wie herrschaftlich zentral gelegene präurbane Siedlungen in der Umgebung von Burgen, Klöstern und Stiften, zum Teil mit frühmittelalterlichen Wurzeln. Stadtprivilegierungen des 14. und 15. Jahrhunderts hatten primär eine herrschaftsabgrenzende (Stadt als „Großburg“ in Grenzlage) und nur sekundär eine wirtschaftliche Zielsetzung. Die Lage an einem Fluss (Main,

90 Wilhelm Störmer, Die Gründung von Kleinstädten als Mittel herrschaftlicher Territorialaufbaus, gezeigt an fränkischen Beispielen, in: ZBLG 36, 1973, S. 563–585.

Regnitz, Pegnitz, Altmühl) war eine bevorzugte und stets gesuchte. Als Trend lässt sich *grosso modo* behaupten: Je älter eine Stadt war, desto größer erwies sich die Chance zu einer umfassenderen kommunalen wie wirtschaftlichen Entfaltung.

Die Bedeutung der Städte für die historische Entwicklung dürfte in Franken außer Frage stehen, besonders ihr Beitrag zur generellen Urbanisierung des Raumes. Ihre Bewertungen sind aber auch abhängig von den jeweils gängigen landesgeschichtlichen Forschungsansätzen,[91] die sie einmal mehr wirtschaftlich, einmal mehr in ihrer für den Ausbau der Landesherrschaft dienenden Funktion oder mehr als rechtlichen Sonderbezirk beleuchten. Die allgemeine stadtgeschichtliche Forschung hat sich naturgemäß eher auf die Großstädte wie Nürnberg oder Augsburg, Köln oder Straßburg gestürzt, weil hier die Quellenlage ungleich besser ist als etwa für Langenzenn, Arzberg oder Thüngen, um willkürlich drei Beispiele aus den fränkischen Regierungsbezirken heranzuziehen. Von der Geographie kommend wurde zunehmend der Begriff der Zentralität in der Geschichtsforschung heimisch. Damit wird die Stadt in ihrer Funktionalität für das sie umgebende Umland gesehen und bewertet. Gerade in Franken mit seiner ausgeprägten Vielfalt von Herrschaftsträgern und -formen ist dieser Ansatz weiterführend, um die damit entstandene Städtelandschaft besser analysieren und verstehen zu können.

91 Helmut Flachenecker, Die Darstellung der Stadt im Historischen Atlas von Bayern, Teil Franken, in: ZBLG 73, 2010, S. 761–778.

Sektion 2: Die süddeutsche Städtelandschaft – regional prägend

Gerhard Fritz

Stadt und Kloster im Herzogtum Württemberg

Abstract: The essay shows the complicated relations between cities and monasteries in the Duchy of Württemberg in the later Middle Ages and the early Modern Times. The duchy consisted of about 40 counties and 14 separate monastery districts. Every county and every monastery (but only friaries, not nunneries) had the right to send deputies to the Duchy's Diet in Stuttgart. Although the Lutheran reform was introduced in 1534 and 1552 and the monasteries as catholic institutions were suspended, the monasteries continued to exist as administrative districts. Every county consisted of a central town (and often some smaller towns) and several surrounding villages, hamlets and barnyards. The structure of a monastery district was quite different. There were villages, hamlets and barnyards as in the normal counties, but there was no town at all. However, in several monastery districts cities were founded. Sometimes these cities were closely connected with the monastery and never gained a greater independence as in Ellwangen or Murrhardt. Sometimes the cities and the monasteries developed in different ways and the cities became absolutely independent and formed their own counties as in Blaubeuren or Backnang. The situation even became more complicated when the Duke of Württemberg was not able to accomplish his sovereignty over a monastery as in Ellwangen, where the monastery and the city remained catholic and were able to escape from Württemberg and developed their own semi-independent territory.

1. Stadt, Amt und Kloster

Das Herzogtum Württemberg bestand in der Frühen Neuzeit aus etwa 40 Ämtern und – je nach Zeitpunkt – aus 13 oder 14 Klosterämtern.[1] Jedes Amt

---

1 Die Mannsklöster und nach der Reformation die Klosterämter waren: 1. Adelberg, 2. Alpirsbach, 3. Anhausen, 4. Bebenhausen, 5. Blaubeuren, 6. Denkendorf, 7. Herrenalb, 8. Hirsau, 9. Königsbronn, 10. Lorch, 11. Maulbronn, 12. Murrhardt, 13. Reichenbach, 14. St. Georgen. Die viel zahlreicheren Frauenklöster, deren Besitz den Mannsklöstern deutlich nachstand, bildeten in der Regel keine eigenen Klosterämter. Eine Zwitterstellung hatten verschiedene Männer- und Frauenstifte, die zwar manchmal eigene Ämter bzw. Kloster- und Stiftsverwaltungen bildeten, aber keine Mitgliedschaft in der Landschaft hatten. Eine Sonderstellung hatte das Klosteramt Lichtenstern, das aus einem Zisterzienserinnenkloster hervorgegangen war und ein Klosteramt bildete. Da es verständlicherweise keinen Abt besaß, war Lichtenstern auch nicht in der Landschaft vertreten.

wiederum bestand aus der Amtsstadt sowie den sie umgebenden Dörfern, Weilern und Höfen. Manchmal waren neben der Amtsstadt auch weitere, in der Regel kleinere Städte vorhanden, die hin und wieder – um sie von der Amtsstadt zu unterscheiden – inoffiziell bis halboffiziell als „Städtle" bezeichnet wurden.[2] Die Klosterämter waren insofern ein württembergischer Sonderfall, als trotz der 1534 erstmals von Herzog Ulrich und nach dem Interim seit 1552 endgültig von Herzog Christoph eingeführten Reformation die alten Klöster nominell weiterbestanden. Sogar das Amt des Abtes existierte noch bis zum Ende des Alten Reiches, aber selbstverständlich waren die Äbte der nachreformatorischen Zeit protestantische Geistliche. Nach wenigen Jahrzehnten war es üblich geworden, dass diese evangelischen Äbte verdiente alte Pfarrer waren, denen der Herzog gleichsam für ihren Lebensabend als wohldotierte Altersversorgung ein Kloster und das Amt des Abtes übertragen hatte. Um die alten Herren ein wenig von der geistlichen Alltagsarbeit zu entlasten, wurde ihnen in der Regel ein Diaconus (manchmal auch deren zwei oder drei) beigegeben. Der Diaconus war keineswegs das, was man heute unter einem Diakon versteht, sondern es handelte sich um ganz normal ordinierte, meist jüngere Pfarrer.

Im Prinzip gliederte sich um ein Kloster ein Klosteramt, das heißt, auf den ersten Blick schien das Kloster in den Klosterämtern dieselbe Position einzunehmen wie die Stadt in einem (weltlichen) Amt. Aber es gab auch wesentliche Unterschiede zwischen einem Klosteramt und einem weltlichen Amt. Weltliche Ämter, das heißt, Amtsstädte, eventuell vorhandene Städtle und die Dörfer samt all ihren Bewohnern unterstanden der vollen Landesherrschaft des Herzogs. In den Lagerbüchern wurde das seit dem 16. Jahrhundert mit Formulierungen wie den folgenden ausgedrückt: Der Herzog sei rechter und alleiniger Herr, natürlicher Landesfürst, habe Stab, Gebot und Verbot, Geleit und alle andere Obrigkeit und Herrlichkeit, Gericht, Zwang, Unrecht, Strafen, Rügungen und Bußen. Zudem genossen alle Bewohner der weltlichen Ämter, also die Bürger der Städte und die Bauern der Dörfer, Weiler und Höfe, wichtige Rechte, die der Tübinger

2 Vgl. etwa das Beispiel der Städte Weilheim und insbesondere Owen im Amt Kirchheim unter Teck. Owen als nachgeordnete Stadt stand im 16. Jahrhundert, weil es sich gegenüber der Amtsstadt aus unterschiedlichsten Gründen benachteiligt fühlte, in der zweiten Hälfte des 16. Jahrhunderts in langem Streit mit Kirchheim; dazu: Georg M. Wendt, Legitimation durch Vermittlung. Herrschaftsverdichtung und politische Praxis in Württemberg am Beispiel von Kirchheim/Teck, Schorndorf und Steinheim/Murr (1482–1608) (Schriften zur südwestdeutschen Landeskunde 79), Ostfildern 2018, S. 38–42, 158–162.

Vertrag von 1514 gewährte.[3] Dieser oft als württembergisches Staatsgrundgesetz bezeichnete Vertrag gewährte allen Bürgern und Bauern das Recht des freien Zuges, also die Freizügigkeit. Damit war die Leibeigenschaft, die formal durchaus weiterbestand, in einem entscheidenden Punkt durchlöchert. Alle württembergischen Untertanen, also nicht nur die Bürger, sondern auch die Bauern, profitierten auch von der im Tübinger Vertrag festgelegten Rechtsgarantie: In Fragen, die Ehre, Leib und Leben betrafen, war ein ordentliches Gerichtsurteil erforderlich und somit herzogliche Willkürentscheidungen ausgeschaltet.

In einem wichtigen Punkt unterschied sich die Rechtsstellung der Klosteruntertanen zunächst von der der Untertanen weltlicher Ämter. Während die Bewohner weltlicher Ämter unstrittig in den Genuss der im Tübinger Vertrag gewährten Freizügigkeit gekommen waren, war dies bei den Untertanen der Klosterämter jahrzehntelang strittig und wurde erst 1551 auch für diese eingeführt.[4]

Die Bürger der Städte genossen durch den Tübinger Vertrag weitere Rechte. So entsandte jede Stadt ihre Vertreter in die „Landschaft", das heißt die württembergische Ständeversammlung. Die Bezeichnung „Landschaft" war die allgemein übliche, obwohl es sich um nichts anderes als den Landtag handelte.[5] Durch ihre Mitgliedschaft in der Landschaft hatten die Städte dank der Regelungen des

---

3 Die Edition des Vertragstextes bei: Eugen Schneider (Hg.), Ausgewählte Urkunden zur Württembergischen Geschichte (Württembergische Geschichtsquellen 11), Stuttgart 1911, S. 87–93; Götz Adriani / Andreas Schmauder (Hg.), 1514. Macht. Gewalt. Freiheit. Der Vertrag zu Tübingen in Zeiten des Umbruchs, Ostfildern 2014; Andreas Schmauder, Der Tübinger Vertrag und die Rolle Tübingens beim Aufstand des Armen Konrad 1514, in: Sönke Lorenz / Volker Schäfer (Hg.), Tubingensia: Impulse zur Stadt- und Universitätsgeschichte. Festschrift für Wilfried Setzler, Ostfildern 2008, S. 187–208; Andreas Schmauder / Wilfried Setzler, Vor 500 Jahren: Württemberg im Aufstand. Der Arme Konrad und der Tübinger Vertrag von 1514, in: Schwäbische Heimat 1, 2014, S. 15–23.

4 Walter Grube, Der Stuttgarter Landtag 1457–1957. Von den Landständen zum demokratischen Parlament, Stuttgart 1957, S. 200; Otto Hering, Leibbuch, Leibrecht, Leibeigenschaft im Herzogtum Württemberg, in: Zeitschrift für Württembergische Landesgeschichte 11, 1952, S. 157–188, hier: S. 170–173; Dieter Stievermann, Landesherrschaft und Klosterwesen im spätmittelalterlichen Württemberg, Sigmaringen 1989, S. 165 f.

5 Vgl. zum württembergischen Landtag: Grube, Stuttgarter Landtag (wie Anm. 4); Ders., Stände in Württemberg, in: Günther Bradler / Franz Quarthal (Red.), Von der Ständeversammlung zum demokratischen Parlament. Die Geschichte der Volksvertretungen in Baden-Württemberg, Stuttgart 1982, S. 51–61.

Tübinger Vertrags maßgeblichen Einfluss auf die Finanzen, insbesondere die Steuererhebung und auf die militärischen Belange des Herzogtums. Kriege, die über die rein defensive Landesverteidigung hinausgingen, konnte der Herzog nur „mit Wissen und Willen" der Landschaft führen, bei Verteidigungskriegen musste er die Landschaft informieren und ihren Rat einholen. Anders gesagt: Die herzogliche Wehrhoheit war entscheidend eingeschränkt.

Anders als in vielen anderen Territorien des Reiches bestand die Landschaft – außer in den Jahrzehnten nach dem ersten Landtag 1457 – nicht aus drei Ständen (Geistliche, Adel, Bürger), sondern nur aus zwei Ständen. Der Adel war in Württemberg nie landsässig geworden. Die Adligen waren fast durchweg niederadlige Ritter. Diese standen zwar in vielfacher Abhängigkeit vom Herzog, sie lebten aber als nominell reichsunmittelbare Reichsritter in ihren meist winzigen Herrschaften, die häufig nur über einen Teil der Hoheitsrechte verfügten und oft als Enklaven mitten im württembergischen Territorium lagen. Als Reichsritter hatten sie zwar als häufige Mitglieder der Regierung in Stuttgart politisches Gewicht, sie waren aber nicht Teil der Landschaft. Dort waren nur die Vertreter der Städte und die Äbte der Klöster beziehungsweise der Klosterämter vertreten. Da die Städte gegenüber den Klöstern weit in der Mehrheit waren, waren sie in der Landschaft tonangebend. Durch die Aktivitäten in der Landschaft waren die Führungsschichten der Städte landesweit eng miteinander vernetzt, was sich nicht zuletzt auch in zahlreichen Heiratsverbindungen ausdrückte. Mit dieser als württembergische „Ehrbarkeit"[6] bezeichneten städtischen Führungsschicht, die, wie die über Jahrhunderte ständig gleichen Familiennamen zeigen, recht stabil war, hatte sich jeder Herzog auseinanderzusetzen, und die „Ehrbarkeit" wusste ihre im Tübinger Vertrag verbrieften Rechte mit Erfolg zu verteidigen. Wenn ein Herzog, was immer wieder vorkam, absolutistische Tendenzen zeigte und an der Landschaft vorbei regieren wollte, begaben sich deren Vertreter, meist solche aus den Städten, im äußersten Falle zum Kaiser und erinnerten diesen daran, dass im Jahr 1514 Kaiser Maximilian den Tübinger Vertrag mit ausgestellt hatte, dass also jeder Eingriff des Herzogs in die Rechte des Tübinger Vertrags ein Eingriff in die kaiserlichen Garantien wäre. Letztlich musste der Herzog dann klein

---

6 Gabriele Haug-Moritz, Die württembergische Ehrbarkeit. Annäherungen an eine bürgerliche Machtelite der Frühen Neuzeit (Tübinger Bausteine zur Landesgeschichte 17), Ostfildern 2009; Nina Kühnle, Richter, Vögte, Landschaftsvertreter. Die Ehrbarkeit im spätmittelalterlichen Württemberg, in: Elisabeth Gruber (Hg.), Mittler zwischen Herrschaft und Gemeinde, Innsbruck u.a. 2013, S. 217–241; Martin Goll, Die „Ehrbarkeit" im Staate Beutelsbach, in: Südwestdeutsche Blätter für Familien- und Wappenkunde 32, 2014, S. 53–114.

beigeben, und in der Tat blieb das fein austarierte System des Tübinger Vertrags bis 1806 erhalten, ja das im württembergischen Kollektivbewusstsein verankerte „gute alte Recht" des Tübinger Vertrags war über das Ende des Reiches hinaus wirkmächtig. Noch die Diskussion über die 1819 in Kraft gesetzte württembergische Verfassung mit ihren für die Verhältnisse des Deutschen Bundes ziemlich liberalen Regelungen wurde stark von den Vorstellungen des „guten alten Rechts" und den Partizipationsmöglichkeiten der Bürger geprägt.

So vorteilhaft die Rechtsposition für die württembergischen Städte (genauer: für deren Führungsschicht, das heißt die Ehrbarkeit) war, so defizitär war die Rechtsposition der nichtstädtischen Einwohner eines Amtes. Sie waren in der Landschaft nicht vertreten. Lange Zeit gab es einen gewissen Ausgleich durch die Vögte, also die führenden herzoglichen Verwaltungsbeamten der einzelnen Ämter, die ebenfalls der Landschaft angehörten und bis zu einem gewissen Grad die Interessen der Außerstädtischen vertraten. Aber seit 1629 waren die Vögte aus der Landschaft ausgeschlossen, und die aus der Amtsstadt stammenden Landschafts-Mitglieder eines Amtes sahen sich nur bedingt als Interessenvertreter der Dörfer und Weiler. Die Nicht-Vertretung der Landbewohner in der Landschaft war 1514 die Konsequenz der Vorgeschichte des Tübinger Vertrags gewesen. Dem Tübinger Vertrag unmittelbar vorausgegangen war der Aufstand des Armen Konrad, der ganz Württemberg erfasst hatte und der maßgeblich von den nichtstädtischen bäuerlichen Schichten getragen worden war.[7] Herzog Ulrich hatte sich seinerzeit aus der für seine Herrschaft bedrohlichen Lage herausmanövriert, indem er dem führenden städtischen Bürgertum mit weitreichenden Zugeständnissen entgegengekommen war und dafür im Gegenzug die Unterstützung der Städte erhalten hatte. Im Grunde war der Tübinger Vertrag ganz entscheidend ein Bündnis des Herzogs mit den städtischen Führungsschichten gegen die bäuerliche Bevölkerung. Die Konsequenz war die jahrhundertelange Benachteiligung beziehungsweise Nicht-Vertretung der Bauern in den Tagungen der Landschaft.

Lange Zeit wurde das von den Bauern hingenommen. Im 17. und 18. Jahrhundert wuchs allerdings der Einfluss der Bauern allmählich. In den 1790er Jahren waren, auch beeinflusst durch die Französische Revolution, bäuerlicher und

7 Aus der umfangreichen Literatur über den Armen Konrad sei nur die umfassendste Darstellung erwähnt: Andreas Schmauder, Württemberg im Aufstand – der Arme Konrad 1514. Ein Beitrag zum bäuerlichen und städtischen Widerstand im Alten Reich und zum Territorialisierungsprozeß im Herzogtum Württemberg an der Wende zur Frühen Neuzeit (Schriften zur südwestdeutschen Landeskunde 21), Leinfelden-Echterdingen 1998.

allgemeiner Unwille dann so gut koordiniert, dass sich die Verhältnisse änderten, und die Bauern konnten auf dem von 1797 bis 1799 tagenden Reformlandtag grundsätzlich ihre Mitvertretung in der Landschaft thematisieren, während die nicht mit Volksrepräsentation zu vereinbarende Anwesenheit der Äbte qua Amt immer heftigerer Kritik ausgesetzt war. Durchschlagenden Erfolg hatte der Reformlandtag freilich bis zum Ende des Alten Reiches nicht.[8]

In diesem kurz skizzierten württembergischen Verwaltungssystem und in der Landschaft waren die Klöster beziehungsweise Klosterämter zwar zahlenmäßig gegenüber den Vertretern der Städte in der Minderheit. Andererseits kam ihnen durchaus ein respektables Gewicht zu. Sie waren aber zunächst einmal Interessenvertreter ihrer eigenen Sache, also nicht primär Interessenvertreter der Einwohner ihres Klosteramts und erst recht nicht Interessenvertreter der Städte – auch wenn sich selbstverständlich immer wieder Interessenidentitäten ergaben. Solche Interessenidentitäten mit den Städten ergaben sich auch, weil die meisten evangelischen Äbte denselben ehrbaren Familien angehörten wie die Landschaftsmitglieder aus den Städten.

Die Rechtstitel, mit denen der Herzog von Württemberg seine Herrschaft in den Klosterämtern ausübte, unterschieden sich insofern von den Rechtstiteln in den weltlichen Ämtern, als die württembergische Herrschaft in allen Klöstern und den im Zuge der Reformation aus ihnen entstehenden Klosterämtern aus der Schutzvogtei beziehungsweise „Schutz und Schirm" und/oder Kastvogtei erwachsen war. Die Begriffe werden in der Forschung manchmal auseinandergehalten. Der Kastvogtei schreibt man eher ökonomische Aspekte zu, der Zusammenhang zwischen Schutzgewährung und Abgabenverpflichtung wird betont. „Schutz und Schirm" (*defensio, tuitio, protectio*) stamme dagegen ursprünglich aus dem zisterziensischen Bereich und betone die reine Schutzfunktion ohne materielle Gegenleistung. Allerdings hatten sich spätestens im Spätmittelalter die Begriffe vermengt und wurden meist synonym verwendet.[9]

Nominell waren ursprünglich jeweils Abt und Konvent die Inhaber der eigentlichen Klosterherrschaft gewesen, und der Vogt hatte nur die Aufgabe gehabt, das jeweilige Kloster militärisch zu schützen und im Vogtgericht Kapitalverbrechen abzuurteilen, die auf Klostergebiet vorgefallen waren. Die Pflichten der Klöster gegenüber dem Vogt waren ursprünglich überschaubar: Beherbergungs- und

8 Vgl. zur Entwicklung bis zum Reformlandtag: Grube, Stuttgarter Landtag (wie Anm. 4); Ders., Stände in Württemberg (wie Anm. 5), jeweils passim.

9 Vgl. die Zusammenfassung der Diskussion um die Begriffe bei: Stievermann, Landesherrschaft und Klosterwesen (wie Anm. 4), S. 15–71, insbesondere: S. 20–24, 118–127.

Atzungspflicht des Schutzvogts und seines Gefolges während des Gerichts, Transportdienste im Kriegsfalle und – etwas seltsam – die Pflicht, Hunde für den Schutzvogt zu halten. Aber die Vogtei war im Laufe des 15. und frühen 16. Jahrhunderts nach und nach immer mehr erweitert und in Richtung einer alle Rechte umfassenden Landesherrschaft ausgedehnt worden. Das war in den einzelnen Klöstern zu unterschiedlichen Zeiten und mit unterschiedlicher Intensität geschehen, aber dieser Prozess war im Wesentlichen schon vor den Zeiten Martin Luthers abgeschlossen. Die anderthalb Jahrzehnte, in denen Herzog Ulrich von 1519 bis 1534 wegen allen möglichen Übergriffen, nicht zuletzt wegen seines Überfalls auf die Reichsstadt Reutlingen, aus seinem Land ins hessische Exil vertrieben war, und in denen die Habsburger Württemberg verwalteten, änderten an den Rechtsverhältnissen der Klöster nichts.

Ulrich führte zwar bei seiner Rückkehr 1534 überall die Reformation ein, die Klöster als Wirtschafts- und Verwaltungseinheiten tastete er aber ebenso wenig an wie sein Sohn Christoph, der seit 1552 die Reformation endgültig durchsetzte. Aber noch bis Mitte des 16. Jahrhunderts war es üblich, dass die Untertanen der Klosterämter ihrem jeweiligen Abt huldigten und nicht dem Herzog. Erst seit 1550 änderte Herzog Christoph das und ließ die Klosteruntertanen ihm selbst huldigen. Das führte anfangs durchaus zu Irritationen der Klosteruntertanen, für die bisher immer der Abt der „gnädige Herr" gewesen war und nicht der Herzog. Immerhin wich in den Lagerbüchern auch noch in der zweiten Hälfte des 16. und 17. Jahrhunderts die formelhafte Aufzählung der württembergischen Rechte in den Klosterämtern von der Aufzählung der Rechte in den weltlichen Ämtern ab: Zwar wurden dieselben Rechte wie bei den weltlichen Ämtern genannt (der Herzog als rechter und alleiniger Herr, natürlicher Landesfürst, Inhaber von Stab, Gebot und Verbot, Geleit und aller anderen Obrigkeit und Herrlichkeit, Gericht, Zwang, Unrecht, Strafen, Rügungen und Bußen), aber dem war vorausgestellt, dass der Herzog Patron, Kastvogt, Schutz- und Schirmherr des Klosters war.[10] Insofern war den Juristen und Verwaltungsbeamten auch dieser späten Zeit durchaus noch klar, dass die Wurzel aller herzoglichen Rechte in den Klöstern und Klosterämtern eine ganz andere war als in den weltlichen Ämtern.

10 Vgl. Gerhard Fritz (Hg.), Das Lagerbuch der Murrhardter Weiler von 1575 (Schriftenreihe des Instituts für Gesellschaftswissenschaften der Pädagogischen Hochschule Schwäbisch Gmünd 7), Schwäbisch Gmünd 2017, S. 11 (Bl. 4 a), URN: urn: nbn: de: bsz:752-opus4-708.

Mit den Klöstern und Klosterämtern gab es im Dreißigjährigen Krieg nochmals gravierende Probleme. Mit dem Restitutionsedikt von 1629 ordnete Kaiser Ferdinand II. an, alle nach dem Passauer Vertrag von 1552 beziehungsweise nach dem Augsburger Religionsfrieden von 1555 protestantisch gewordenen Klöster wieder zu rekatholisieren und erneut mit Mönchen zu besetzen. Die in Württemberg dann 1630 realisierte Rekatholisierung blieb Episode, weil im Westfälischen Frieden von 1648 alle Klöster wieder an Württemberg zurückgegeben und damit wieder evangelisch wurden. Die Reibereien zwischen Katholiken und Protestanten während der Zeit der Restitution waren aber Anlass für das Entstehen von mancherlei Schriftsätzen, die Einblick in die rechtlichen Verhältnisse der Klöster und Klosterämter erlauben.

Normalerweise besaß ein Klosteramt keine Amtsstadt. Das Kloster nahm deren Funktionen wahr. Zwar gab es in unmittelbarer Nachbarschaft aller 14 württembergischen Klöster eine kleinere oder größere dörfliche oder dorfähnliche Ansiedlung. Soweit diese Ansiedlungen und Dörfer Stadtrechte erhielten, geschah dies meist erst im 19. Jahrhundert.[11] In zwei Fällen entstand aber bereits im Mittelalter direkt neben einem Kloster eine Stadt. Das war zum einen der Fall in Blaubeuren, zum anderen in Murrhardt. Aber die beiden Fälle waren durchaus nicht vergleichbar. Denn während in Blaubeuren rund um die Stadt zusätzlich zum Klosteramt ein vollkommen separates weltliches Amt entstand, waren die Verhältnisse in Murrhardt ganz anders. Hier gab es zwar eine Stadt, aber die hing nicht nur eng mit dem Kloster zusammen, sondern war auch vollkommen von diesem abhängig. Ein weltliches Amt Murrhardt ist nie entstanden. Die Fälle Blaubeuren und Murrhardt sollen näher betrachtet werden.

2. Stadt und Kloster Blaubeuren

Um die unterschiedliche Entwicklung von Blaubeuren und Murrhardt zu verstehen, muss auf die Frühgeschichte beider Städte zurückgegriffen werden. Die Stadt Blaubeuren ging auf eine dorfähnliche Ansiedlung neben dem 1085 gegründeten (beziehungsweise nach Blaubeuren verlegten) Benediktinerkloster zurück.[12] Um die Mitte des 13. Jahrhunderts wurde die Ansiedlung von den

---

11 Alpirsbach, Herrenalb, Lorch, Maulbronn und St. Georgen erhielten erst in der zweiten Hälfte des 19. Jahrhunderts Stadtrechte. Vgl. dazu: Gerhard Fritz, Stadt und Kloster Murrhardt im Spätmittelalter und in der Reformationszeit (Forschungen aus Württembergisch Franken [künftig: FWFr] 34), Sigmaringen 1990, S. 143 f.

12 Die folgenden Ausführungen zur Stadt Blaubeuren grundsätzlich nach: Immo Eberl, Die Stadt Blaubeuren im Spätmittelalter. Zur Entwicklung einer landesherrlichen

Pfalzgrafen von Tübingen, die zugleich auch Vögte des Klosters waren, zur Stadt ausgebaut. 1267 wird Blaubeuren erstmals als „*civitas*" erwähnt. Das Kloster war über die Aufwertung der Siedlung zur Stadt anfangs offenbar nicht glücklich, wurde aber 1267 unter anderem damit entschädigt, dass klösterliche Eigenleute in der Stadt von Abgaben, Diensten und Wachen befreit waren. Auch was Heiraten der klösterlichen Leibeigenen anging, kam der Pfalzgraf dem Kloster entgegen. Nur mit Erlaubnis des Abtes waren Heiraten von Klosterleuten erlaubt. Weitere Zugeständnisse sollten das Kloster offenkundig mit der Stadtgründung aussöhnen.

Nach diversen Turbulenzen der Pfalzgrafen von Tübingen fielen Stadt und Kloster Blaubeuren 1273 an die Grafen von Helfenstein, die Stadt und Kloster an die Habsburger als Lehensherren auftrugen und sich von diesen umgehend damit belehnen ließen. Unter ihnen war das Verhältnis zwischen Stadt und Kloster, soweit sich das bei der mäßigen Quellenlage erkennen lässt, abwechselnd mal ausgeglichen, mal eher gespannt. Insgesamt wurden aber bis um 1300 die rechtlichen Strukturen der Stadt immer deutlicher erkennbar, das heißt auch vom Kloster unterscheidbar. Anfangs standen an der Spitze der Stadt zwei Amtmänner – ein gräflicher und ein klösterlicher – und ein Gericht aus zwölf Richtern. Für eine gewisse Eigenständigkeit der Stadt spricht auch das Vorhandensein einer eigenen Schule spätestens Mitte des 14. Jahrhunderts, die vermutlich sogar eine Lateinschule war.

Was das städtebauliche Aussehen Blaubeurens anging, so nimmt man an, dass die Stadt schon 1267 ummauert gewesen sei, wobei die Mauer zu Beginn aber durchaus ein Holz-Erde-Wall mit Türmen gewesen sein dürfte.

Kleinstadt, in: Hansmartin Decker-Hauff / Immo Eberl (Hg.), Blaubeuren. Die Entwicklung einer Siedlung in Südwestdeutschland, Sigmaringen 1986, S. 177–219; zum Kloster: Otto-Günter Lonhard, Das Kloster Blaubeuren im Mittelalter. Rechts- und Wirtschaftsgeschichte einer schwäbischen Benediktinerabtei (Veröffentlichungen der Kommission für geschichtliche Landeskunde in Baden-Württemberg B 25), Stuttgart 1963; Klaus Schreiner, Mönchtum im Geist der Benediktregel. Erneuerungswille und Reformstreben im Kloster Blaubeuren während des hohen und späten Mittelalters, in: Decker-Hauff / Eberl, Blaubeuren (wie oben), S. 93–176; zusammenfassend: Immo Eberl, Blaubeuren, in: Wolfgang Zimmermann / Nicole Priesching (Hg.), Württembergisches Klosterbuch. Klöster, Stifte und Ordensgemeinschaften von den Anfängen bis in die Gegenwart, Ostfildern 2003, S. 155 f.

Abb. 18: *Kloster Blaubeuren. Bildnachweis/Foto: Wikimedia Commons (gemeinfrei).*

Wesentlich für die Entwicklung der Stadt waren die seit der zweiten Hälfte des 14. Jahrhunderts immer deutlicher hervortretenden Finanzprobleme der Grafen von Helfenstein. Die Stadt vermochte dies in den 1380er Jahren auszunutzen und eine ganze Reihe von Rechten von den Grafen zu bekommen. Infolge der pestbedingten Krise und den anzunehmenden schweren Bevölkerungsverlusten bemühten sich die Helfensteiner um die Anwerbung von Neubürgern bei gleichzeitigen Abwanderungsverboten.

In den beiden letzten Jahrzehnten gerieten Stadt und Kloster in erhebliche Wirrungen, weil auf die Grafen von Helfenstein im Zuge verwickelter Verpfändungs- und Veräußerungsvorgänge kurzfristig die Herren von Landau, die Reichsstadt Ulm und andere regionale Adelsgeschlechter folgten, bevor sich dann Ulm 1397 als Pfandinhaber etablierte. Stadt und Kloster gingen dabei nicht immer gleiche Wege, da sich das Kloster 1398 für die Grafen von Württemberg als neue Vögte entschied. Dauerhaft unter württembergische Herrschaft kamen Kloster und Stadt aber erst 1447, da in den Jahren um 1400 auch die Helfensteiner nochmals ihre Herrschaftsansprüche geltend machten. In den turbulenten letzten Jahrzehnten des 14. Jahrhunderts lässt sich auch, was typisch für spätmittelalterliche Städte ist, ein Bettelordenskloster nachweisen. Dessen Gründung geschah zwar nicht direkt in den Mauern der Stadt Blaubeuren, aber im unmittelbar benachbarten Ort Weiler, wo sich eine seit 1382 nachgewiesene Eremitinnenklause zu einem Franziskaner-Terziarinnen-Kloster entwickelte. Die führenden Bürgerfamilien brachten dort im 15. Jahrhundert wiederholt ihre Töchter als Nonnen unter.[13]

Wesentlich für die erste Hälfte des 15. Jahrhunderts war weiterhin, dass im Kloster immer häufiger städtische Familien als Mönche und Äbte auftauchten. Wichtig war weiterhin, dass in der Stadt auf Initiative verschiedener Bürger ein Spital gegründet wurde, das zu erheblichem Besitz kam und eine ausgedehnte Erwerbspolitik betrieb.[14] Der Stadt gelang es auch, auf die Besetzung der verschiedenen in der Blaubeurer Pfarrkirche gestifteten Pfründen maßgeblichen Einfluss auszuüben. Überhaupt ging es der Stadt wirtschaftlich gut. Um 1440 beliefen sich ihre jährlich an die Grafen von Helfenstein bezahlten Abgaben auf etwa 430 lb h.[15]

Der eigentliche Verkauf Blaubeurens von Helfenstein an Württemberg fand 1447 erst statt, nachdem es in den Jahren zuvor ein langes Hin und Her zwischen den verschiedenen Mitgliedern des Hauses Helfenstein gegeben hatte. Die Integration Blaubeurens nach Württemberg lief offenbar rasch und unproblematisch, was nicht zuletzt mit der ähnlichen Struktur der Bürgerschaft mit einer starken Ehrbarkeit zu erklären sein dürfte. Württemberg trennte von Beginn an die Stadt Blaubeuren samt den zu ihr gehörenden Dörfern und Weilern vom Kloster Blaubeuren ab und schuf ein separates, weltliches Amt.

13 Außer Eberl, Stadt Blaubeuren (wie Anm. 12), S. 197 f., insbesondere: Otto-Günter Lonhard, Das Kloster Weiler, in: Das Blaumännle 24–26, 1976; zum Erwerb der Klostervogtei durch Württemberg: Stievermann, Landesherrschaft und Klosterwesen (wie Anm. 4), S. 111.

14 Vgl. außer Eberl, Stadt Blaubeuren (wie Anm. 12), S. 195 f., insbesondere: Rudolf Seigel, Spital und Stadt in Alt-Württemberg. Ein Beitrag zur Typologie der landstädtischen Spitäler, Tübingen 1966, S. 34; Otto-Günter Lonhard, Das Spital zum Heiligen Geist in Blaubeuren, in: Ulm und Oberschwaben 39, 1970, S. 22–24.

15 Eberl, Blaubeuren (wie Anm. 12), S. 200 f., nach: Max Huber, Ein Einkünfteregister der Grafschaft Kirchberg-Kirchberg von 1379/1438, in: Ulm und Oberschwaben 40/41, 1973, S. 27–68, hier: S. 36 f.

3. Stadt und Kloster Murrhardt im Vergleich zu Blaubeuren

Die Aufhellung der Verhältnisse in Murrhardt ist durch eine noch schlechtere Quellenlage als in Blaubeuren erschwert. Das Kloster in Murrhardt war wesentlich älter als das in Blaubeuren und geht mindestens auf das frühe 9., wenn nicht auf das späte 8. Jahrhundert zurück. Neben dem Kloster existierte jedenfalls schon früh eine Ansiedlung. Ein als Bauholz verwendeter Balken konnte dendrochronologisch in ottonische Zeit datiert werden. Die Ansiedlung lag offenbar neben dem Gelände, in dem die Zivilsiedlung eines von etwa 150 bis 260 existierenden römischen Kohortenkastells lag. Nur in Randbereichen scheinen sich die mittelalterliche Ansiedlung und der römische Vicus überschnitten zu haben.[16]

Die Umwandlung der neben dem Kloster gelegenen Siedlung in eine Stadt erfolgte nach allem, was sich erschließen lässt, im späten 13. Jahrhundert, am ehesten um 1287, als die Grafen von Löwenstein auch die Siedlung unter ihrer namengebenden Burg zur Stadt erhoben.[17] Bei den Grafen von Löwenstein handelte es sich um das zweite Geschlecht dieses Namens, nachdem das erste 1277 seine Grafschaft erst an das Bistum Würzburg verkauft hatte, von wo diese 1281 an das Reich und König Rudolf von Habsburg kam. Der verlieh Löwenstein 1282/83 an seinen illegitimen Sohn Albrecht von Schenkenberg, der mit der Verleihung zum Grafen von Löwenstein aufstieg. Albrecht betrieb eine Politik ganz im Sinne seines Vaters, das heißt, er baute seine Herrschaft mit Frontstellung gegen die aufsteigenden Grafen von Württemberg durch Städtegründungen aus. Das Kloster Murrhardt wurde von den Löwensteinern bevogtet, und offenbar sah Albrecht es als sinnvoll an, die Ansiedlung neben dem Kloster zur Stadt auszubauen. In später Überlieferung aus dem 16. Jahrhundert ist davon die Rede, dass der damalige Abt aus der Familie von Hohenstein die Ansiedlung neben dem Kloster „*zu einem marckt umbgraben*" habe. Die Ansiedlung muss im Übrigen schon lange vor dem späten 13. Jahrhundert als Markt fungiert haben, denn es ist eine aus dem 12. Jahrhundert stammende spezifisch murrhardtische Münzprägung nachgewiesen, die nur einen Sinn ergibt, wenn es hier schon zu diesem Zeitpunkt einen nicht ganz unbedeutenden Warenumschlag gab.

---

16 Gerhard Fritz, Murrhardt, in: Zimmermann / Priesching, Württembergisches Klosterbuch (wie Anm. 12), S. 338–340; zur Frühzeit des Klosters: Ders., Kloster Murrhardt im Früh- und Hochmittelalter. Eine Abtei und der Adel an Murr und Kocher (FWFr 18), Sigmaringen 1982; Ders., Die Anfänge des Klosters Murrhardt, in: Württembergisch Franken (künftig: WFr) 100, 2016, S. 7–26.

17 Gerhard Fritz, Zum Zeitpunkt der Stadterhebung von Murrhardt, in: WFr 69, 1985, S. 281–285.

Von der frühen Stadt weiß man kaum etwas. Ausdrücklich als solche dokumentiert ist sie erstmals 1318 und dann 1328.[18] Informationen über ihre innere Verfassung liegen lange Zeit kaum vor. Klar war nur, dass Kloster und Stadt unmittelbar neben einander lagen und eine gemeinsame Umfassungsmauer hatten. Klar ist weiter, dass - und hier liegt eine Parallele zu Blaubeuren - das Geschlecht der Klostervögte, die Grafen von Löwenstein, in der zweiten Hälfte des 14. Jahrhunderts (als Auswirkungen der Pest-Epidemien?) in erhebliche finanzielle Probleme gerieten. Schon 1365 machten sowohl Württemberg als auch Baden fürs Erste noch gescheiterte Anstalten, Murrhardt zu übernehmen. 1388 wurde dann der damalige junge Erbe der Grafschaft Löwenstein in der Schlacht von Döffingen erschlagen, in der die Reichsstädte gegen den Grafen von Württemberg standen. Jetzt übernahm Graf Eberhard der Greiner von Württemberg sofort - unter Verdrängung der jüngeren Brüder des getöteten Löwensteiner Grafen - 1388 erst die Stadt; 1393/95 folgte die Übernahme des Klosters Murrhardt. Der rechtliche Hebel, die Stadt in die Hand zu bekommen, war die Mitgift der löwensteinischen Gräfin Udelhild, die Württemberg der finanzschwachen Gräfin vorgestreckt hatte. Udelhild war auf die Stadt Murrhardt als Widerlegung angewiesen, und da sie ihre Schulden an Württemberg nicht zurückzahlen konnte, griff sich Eberhard der Greiner ersatzweise die Stadt.

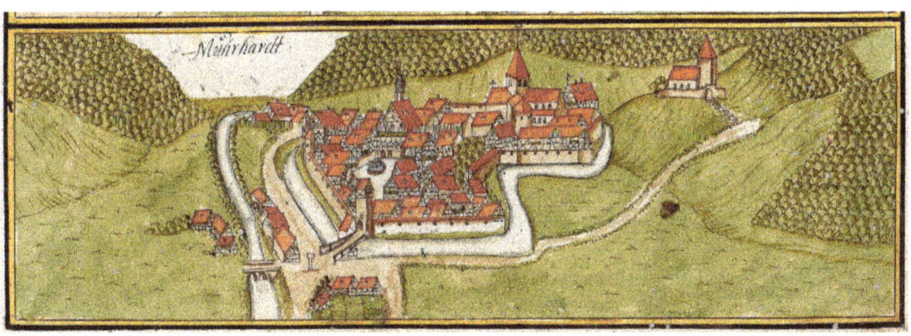

Abb. 19: *Murrhardt um 1685, Abbildung aus dem Kieserschen Forstlagerbuch. Bildnachweis: HStA Stuttgart, H 107/14, Bd 6, Bl. 6.*

18 Gerhard Fritz, Die Geschichte der Grafschaft Löwenstein und der Grafen von Löwenstein-Habsburg vom späten 13. bis zur Mitte des 15. Jahrhunderts (FWFr 29), Sigmaringen 1986, Regesten, S. 265 f., Nr. 11; S. 271, Nr. 137.

Während die Stadt schon 1388 die württembergische Herrschaft anerkannte und sich zu einer offenbar für sie akzeptablen jährlichen Zahlung von 20 lb h und 20 Scheffel Hafer an den neuen Stadtherrn verpflichtete, konnte sich das Kloster, das von Abt Eckhard, einem ausgesprochenen Parteigänger der Löwensteiner, geführt wurde, nicht so schnell mit dem Gedanken an eine württembergische Vogtei anfreunden. Erst nach über einem Jahr willigte Eckhard 1389 in die neuen Verhältnisse ein. Man kann aus der zeitlich versetzten Übernahme eigentlich nur sehen, dass man Stadt und Kloster zu dieser Zeit als juristisch einigermaßen getrennte Objekte ansah, die man sich separat aneignen musste. Die noch sehr jungen löwensteinischen Grafen, die Brüder des 1388 bei Döffingen Gefallenen, unternahmen zusammen mit ihrer Mutter Udelhild bis 1393/95 noch einige Versuche, unter anderem vor dem Hofgericht in Rottweil, Stadt und Kloster von Württemberg zurückzubekommen, aber alle Anläufe, die alten Verhältnisse wiederherzustellen, scheiterten.[19]

Insgesamt war also – trotz der löwensteinischen Rekuperationsversuche – der Übergang von Stadt und Kloster Murrhardt an Württemberg wesentlich einfacher als im Falle Blaubeurens, wo es ja ein langes Hin und Her zwischen den Grafen von Helfenstein, dem Ritter von Landau und der Reichsstadt Ulm und dann nochmals den Grafen von Helfenstein gegeben hatte, bevor Stadt und Kloster unter württembergische Herrschaft kamen.

Für das 14. Jahrhundert erfährt man nur die Namen einiger Bürger sowie die bemerkenswerte Tatsache, dass der Pfarrer von Murrhardt, der offenbar kein Mönch war, 1338 siegelführend auftritt.[20] Anders als im Falle Blaubeurens bleibt bei Murrhardt auch im 15. Jahrhundert die Quellenlage zur Stadt miserabel.[21]

---

19 Fritz, Löwenstein (wie Anm. 18), S. 53–69.

20 1338 VIII 24, in einer Urkunde des Murrhardter Abtes Heinrich und des Konvents als Zeugen: „*Burglin und Heinrich Battenburg, burger ze Murrhart*" (Hauptstaatsarchiv [künftig: HStA] Stuttgart, A 469, U 138); 1349 IV 25: Hartmann der Obermann, Bürger zu Murrhardt (Fritz, Löwenstein (wie Anm. 18), S. 281, Regest Nr. 184); 1388 III 10: „*Chuntz Ekhart von Murrhart*"; derselbe auch: 1389 VI 30 (Friedrich Pietsch, Die Urkunden des Archivs der Reichsstadt Schwäbisch Hall, 2 Bde. (Veröffentlichungen der staatlichen Archivverwaltung Baden-Württemberg 21, 22), Stuttgart 1967–1972, S. 256, U 854; S. 260, U 878, 879).

21 Fritz, Stadt und Kloster Murrhardt (wie Anm. 11), S. 138–223. Zu erwähnen sind lediglich zwischenzeitlich neu gefundene Murrhardter: „*Cunrat Wisshar*", der 1404 I 14 Besitz bei Schmiedelfeld hat (StA Ludwigsburg, B 113 I, U 425); 1432 V 14: „*Betz Schryner zu Murhart*" (Hospitalarchiv Schwäbisch Hall, U 723); 1445 II 23 noch derselbe (ebd., U 672); 1469 II 9 ist derselbe tot, seine Ehefrau lebt (ebd., U 564); 1457 VI 10 und 17: der Bruch in Murrhardt (HStA Stuttgart, A 602, U 2614 und 2615); 1467

Die schriftliche Überlieferung nennt lange Zeit so gut wie gar nichts. Bekannt ist nur, dass die Äbte Wilhelm Egen und Johannes Schradin in den letzten Jahren des 15. Jahrhunderts große Baumaßnahmen an Stadt und Kloster durchführten. Vielleicht fand damals auch eine (in ihrem Umfang bescheidene) Vergrößerung der Stadtanlage statt. Die Stadt erscheint jedenfalls völlig als Anhängsel des Klosters, Ansätze zu irgendeiner Selbstständigkeit oder eigenständigem Handeln sind nicht zu erkennen. Als Organe der Stadt werden 1406 Schultheiß, Richter und „*burger gemeinlich*" genannt.[22]

Die nachgeordnete Position der Stadt gegenüber dem Kloster und die grundsätzlich geringe Bedeutung der Stadt wird auch durch die Tatsache unterstrichen, dass in Murrhardt – anders als in Blaubeuren – nie ein städtisches Spital entstand. Ebenso wenig entstand in Murrhardt ein Bettelordenskloster wie in (beziehungsweise unmittelbar bei) Blaubeuren das Franziskanerinnenkloster Weiler.

Die wirtschaftliche Leistungsfähigkeit der Städte Blaubeuren und Murrhardt war extrem unterschiedlich. Es ist zwar nicht ganz unproblematisch, die Murrhardter Abgaben an die Grafen von Württemberg von 1388 mit den Blaubeurer Abgaben um 1440 zu vergleichen, aber auch wenn man den Vergleich nicht 1:1 durchführen kann, wird doch deutlich, dass den Murrhardter Abgaben von 20 lb h und 20 Scheffel Hafer Blaubeurer Abgaben von etwa 430 lb h gegenüberstanden.

Anders als in Blaubeuren, wo sich auf der Grundlage dieser relativ großen wirtschaftlichen Leistungsfähigkeit eine bildungs- und aufstiegsbewusste Ehrbarkeit herausgebildet hatte (man schickte seine Söhne immer wieder zum Studium auf verschiedene Universitäten), fehlt eine solche überlokal vernetzte Ehrbarkeit in Murrhardt im 15. Jahrhundert und auch später völlig. Die Blaubeurer Ehrbarkeit wuchs durch ihre Teilnahme an den seit 1457 stattfindenden württembergischen Landtagen rasch in die gesamtwürttembergische Ehrbarkeit hinein. Den einigermaßen wohlhabenden Murrhardter Familien, die ja keinerlei Teilnahme am Landtag hatten, fehlte diese landesweite Vernetzung. Sie standen immer im Schatten des Abtes, der allein der Landschaft angehörte. In Blaubeuren hatten die Bürger der Stadt und zusätzlich und separat von ihnen der Abt des Klosters die Landstandschaft inne.

III 12: „*Michael Metzler von Murhart*" (Pietsch, Urkunden (wie Anm. 20), S. 320, U 2546); ein Gleichnamiger auch 1466 VI 4 als „*von Plintheim*", Laie der Würzburger und Augsburger Diözese (ebd., S. 318, U 2521); 1491 III 24: Bürgermeister Hans Miller (HStA Stuttgart, A 508, U 8).

22 HStA Stuttgart, A 508, U 2.

Aus dem städtischen Archiv ist nur eine Zinsliste von 1471, ein Stadtrecht von 1502 und eine Metzlerordnung von 1522 erhalten.[23] Die Zinsliste, die 40 Namen nennt, mutmaßlich meist Bürger, zeigt ausgesprochen bescheidene Einkünfte der Stadt. Das Stadtrecht bezieht sich in erster Linie auf Erbangelegenheiten und zeigt dabei wenig Parallelen zu anderen württembergischen Städten. Zusätzlich wird erkennbar, dass die Stadt 1502 über Bürgermeister, Gericht und Rat verfügte, bemerkenswerterweise nur über einen Bürgermeister, während in den anderen württembergischen Städten in der Regel deren zwei vorhanden waren. Als eigentliches (und einziges) Fachpersonal lässt sich neben dem ehrenamtlich tätigen Bürgermeister samt Gerichts- und Ratsmitgliedern ein Stadtschreiber namhaft machen. Außerdem wird der Rechtszug bei Appellationsfällen beschrieben. War ein Bürger mit einem Urteil des Stadtgerichts nicht zufrieden, konnte er ab einem Streitwert von mehr als 6 lb h und nach zehn Tagen Bedenkzeit beim Abt appellieren. Der Abt war also Appellationsinstanz, was die enge Abhängigkeit der Stadt vom Kloster unterstreicht, die viel enger war als in Blaubeuren. Die gegenüber dem Kloster nachgeordnete Position der Stadt wird auch durch die Metzlerordnung von 1522 unterstrichen: Grundsätzlich musste beim Fleischverkauf zuerst der Abt bedient werden, dann schwangere Frauen und erst dann andere Käufer. Anzunehmende weitere Handwerksordnungen neben der Metzlerordnung sind nicht erhalten.

1503/06 lassen sich einige Konflikte zwischen der Stadt Murrhardt und den Weilern des Amtes feststellen, bei denen sich erstaunlicherweise das Kloster, das doch die Stadt so eng an sich gebunden hatte, außerstande sah, diese Streitigkeiten, die Jahrzehnte zurückreichten, erfolgreich zu schlichten.[24]

Weitere Aussagen zum Murrhardter Stadtrecht stammen von 1565, geben aber zweifellos ältere Verhältnisse wieder. Es geht darin um Fischrechte, Frondienstpflichten, Schatzung, herzogliche Ablösungshilfe, Steuer und Schlossgeld, Beholzungsrechte, Besetzung von Gericht und Rat sowie die Frage, ob ein altes Statutenbuch vorhanden sei (was verneint wird). Die Frondienstpflichten waren für Bürger der Stadt äußerst bescheiden und beschränkten sich auf wenige Tage, wenn das Kloster einen „Hauptbau" errichtete, also bei großen Baumaßnahmen. Solche Baumaßnahmen kamen aber selten vor. Für die Bewohner der außerhalb der Stadt gelegenen Weiler, die kein Bürgerrecht hatten, waren die Fronpflichten

---

23 Gerhard Fritz, Murrhardter Stadtrechtsquellen aus dem 15. und 16. Jahrhundert, in: WFr 94, 2010, S. 38–63.

24 Gerhard Fritz, Untersuchungen zur Sozial- und Wirtschaftsgeschichte des Schwäbisch-Fränkischen Waldes im 15. und frühen 16. Jahrhundert, in: WFr 96, 2012, S. 59–111.

nicht ganz so günstig. Weiter wird 1565 darauf hingewiesen, dass man bisher mit dem Abt und dem Vogt gut zusammengearbeitet habe. Zugleich wird aber deutlich, dass zu diesem Zeitpunkt – die endgültige Reformation des Klosters lag eben anderthalb Jahrzehnte zurück – das Herzogtum sich zunehmend in den Vordergrund geschoben hatte und die Gewichte zwischen Kloster und Stadt sich dadurch geändert hatten.[25]

Ein nach dem Bauernkrieg von der Stadt an das Kloster geleisteter Kredit, der dann vom Kloster nicht bedient wurde, führte zu Spannungen, die 1537 in einen Bürgeraufstand mündeten, der aber vom Herzogtum – nicht mehr vom eben 1534 reformierten Kloster – durch Inhaftierung der Rädelsführer rasch niedergeschlagen werden konnte. Ein weiterer Bürgeraufstand folgte 1564/65, der auf dieselbe Weise beendigt wurde. Diesmal war es darum gegangen, dass etliche Bürger meinten, ihnen werde die Verlesung eines Buches mit ihren Rechten verweigert. Beide Male – 1537 und 1564/65 – handelte es sich nicht um einen Konflikt der Stadt gegen das Kloster oder gegen das Herzogtum, sondern primär um einen Konflikt unter den Bürgern, die mit ihren Bürgermeistern, Gerichts- und Ratsherren nicht zufrieden waren.[26] Das Kloster seinerseits war in beiden Fällen (wie schon 1503/06) nicht in der Lage, den Konflikt alleine in den Griff zu bekommen. Das gelang übrigens auch dem Vertreter Württembergs in Murrhardt, dem Vogt, nicht. Dazu war jeweils das Herzogtum erforderlich, das mit seinen Räten unmittelbar eingriff.

Wie eng die Stadt vom Kloster auch in nachreformatorischer Zeit abhing, wird deutlich, wenn man die Gerichtssitzungen, namentlich die mehrfach jährlich stattfindenden Rug- und Vogtgerichte betrachtet. Neben den eigentlichen Gerichtsmitgliedern waren grundsätzlich auch der Abt und der Vogt anwesend.[27] Die Autonomie des städtischen Gerichts war also gering. Allerdings wandelte sich die Gewichtsverteilung allmählich: Der evangelische Abt Etzel beschwerte sich 1593, dass der Vogt die Besetzung des städtischen Gerichts vornehme, wo dies doch dem Abt zustehe.[28] Dieser Prozess, der mit der Huldigung an den Herzog statt an den Abt 1550 begonnen hatte, war offenbar auch in der Folge nicht aufzuhalten.

---

25 Fritz, Murrhardter Stadtrechtsquellen (wie Anm. 23), S. 54–60.

26 Gerhard Fritz, Murrhardter Bürgeraufstände des 16. Jahrhundert, in: WFr 67, 1983, S. 55–71.

27 Gerhard Fritz, Murrhardter Sozialgeschichte von der Mitte des 16. Jahrhunderts bis zum Beginn des Dreißigjährigen Krieges (ca. 1559 bis 1620) (historegio 10), Remshalden 2020, S. 37, 176.

28 Ebd., S. 39.

Viel mehr Eigenständigkeit gewann die Stadt dadurch nicht. Zwar erwähnten Bürgermeister, Gericht und Rat 1565, dass sie ein neues Rathaus errichtet hätten, aber dieser Ausdruck bürgerlichen Selbstbewusstseins wurde getrübt durch die Tatsache, dass sich die Stadt dadurch und durch andere Baumaßnahmen finanziell völlig übernommen hatte.[29] Die geringe Eigenständigkeit der Stadt Murrhardt gegenüber dem Kloster wird bis in die Gegenwart hinein von der dialektalen Bezeichnung „Städtle" statt Stadt unterstrichen.

Ein wichtiger Aspekt städtischer Autonomie war die Frage der Blutgerichtsbarkeit. Diese soll vor 1550 – also so lange dem Abt und nicht dem Herzog gehuldigt worden war – vom Stadtgericht ausgeübt worden sein. Zwar ist die Beweislage nicht ganz klar, aber immerhin weist eine heute noch am Flurnamen „Galgen" erkennbare Hinrichtungsstätte darauf hin, dass es in Murrhardt ein Hochgericht und damit eine tatsächlich praktizierte Blutgerichtsbarkeit gegeben haben muss. Nach 1550 war die Hochgerichtsbarkeit aber auf jeden Fall nicht mehr vorhanden und wurde nun auf Murrhardter Klostergebiet vom Gericht in Marbach ausgeübt.[30]

Immerhin sind in nachreformatorischer Zeit verschiedene, unabhängig vom Kloster gemachte Eingaben von Bürgermeister, Gericht und Rat beziehungsweise Gemeinde an den Herzog beziehungsweise an die herzoglichen Räte nachgewiesen, entweder in akuten Krisensituationen oder bei wichtigen Baumaßnahmen. In diese direkte Korrespondenz mischte sich gegebenenfalls auch der herzoglich Vogt ein, dessen positive oder negative Stellungnahme ein Anliegen der Stadt maßgeblich fördern oder hemmen konnte.[31] Überhaupt hat man den Eindruck, dass das Kloster in nachreformatorischer Zeit alte Traditionen, die die Verbundenheit zu den Bürgern jahrhundertelang wenigstens symbolisch unterstrichen hatte, nun allmählich einschlafen lassen wollte. Hier wäre der sogenannte Martinswein zu nennen, zu dem das Kloster alljährlich seine Bürger einlud, über dessen Abschaffung sich die Bürger aber 1576 beim Herzog beschwerten und damit auch Erfolg hatten.[32]

---

29 Ebd., S. 48 f.

30 Ebd., S. 233.

31 Vgl. ebd., S. 64, eine Eingabe wegen der akuten Hungersnot 1571; S. 72, wegen des Baus der (vom Kloster unabhängigen) Burgermühle 1583; HStA Stuttgart, A 508, Bü 6, Briefwechsel zwischen Stadt, Vogt und herzoglichem Oberrat, d.h. der Regierung 1630/31 mit negativer Stellungnahme des Vogtes.

32 Ebd., S. 196.

Selbstverständlich waren auch die katholischen Äbte der Restitutionszeit auf den Sitzungen von Gericht und Rat anwesend und bestimmten dort den Kurs.[33] Aber ihre Linie war eher noch härter als die ihrer evangelischen Vorgänger, wie Eingriffe in städtische Rechte zeigen. Der seit 1635 in Murrhardt etablierte Benediktinerabt Emmerich Fünkler hatte zwar versprochen, die althergebrachten Rechte der Stadt nicht zu beeinträchtigen. Als er sein Versprechen nicht einhielt, wandte sich die Stadt an den Bischof von Würzburg (Kontakt zum alten Landesherrn, dem Herzog von Württemberg, war den Bürgern von Fünkler verboten worden) und zählte in einer langen Liste die gebrochenen Versprechen des Abtes auf. Besonders wichtig war die Frage der Huldigung: Fünkler hatte hier wieder die Verhältnisse von vor 1550 eingeführt. Ihm wurde 1635 von den Bürgern gehuldigt, nicht mehr dem Herzog, der laut Restitutionsedikt im Kloster ja nichts mehr zu schaffen hatte. Fünkler hatte den Bürgern ausdrücklich verboten, etwa nach Stuttgart zu reisen und dort dem Herzog zu huldigen. Verschiedene Geldeinkünfte, die nach einem komplexen System zwischen Stadt und Kloster aufgeteilt waren, hatte Fünkler ins Kloster gezogen und die Fronpflichten erhöht. Die Dokumente, auf denen die Rechte der Stadt aufgeschrieben waren, hatte Fünkler an sich genommen und hielt sie unter Verschluss.[34] Das hatten seine evangelischen Vorgänger nicht getan.

4. Geistliche Institutionen, die keine Vertreter zur Landschaft entsandten, und die dazugehörigen Städte

Wieder anders waren die Verhältnisse, wenn sich kein in der Landschaft vertretenes Klosteramt herausgebildet hatte, sondern wenn es bei Kloster- beziehungsweise Stiftsverwaltung geblieben war. Hier waren – anders als bei den meisten Klöstern – durchaus Städte vorhanden. Die Kloster- und Stiftsverwaltungen waren zwar, was den Haushalt und die Buchführung anging, von den jeweiligen Städten getrennt, aber die Kloster- und Stiftsverwaltungen traten an Bedeutung und öffentlicher Wahrnehmung gegenüber den Klöstern in den Klosterämtern merklich zurück.

33 Stadtarchiv Murrhardt, Vermögensbuch, fol. 222 b, Sitzung vom 11.10.1647; künftig in derselben Studie.

34 Vgl. zur Supplik von 1638 an den Bischof: StA Würzburg, Gebrechenamt II, M 92. Zum gesamten Sachverhalt ist eine Studie des Verfassers in Vorbereitung.

Abb. 20: *Backnang um 1685, Abbildung aus dem Kieserschen Forstlagerbuch. Bildnachweis: HStA Stuttgart, H 107/14, Bd 6, Bl. 5.*

Beispiele gibt es mehrere. In Backnang existierte ein um 1116 gegründetes, von den Markgrafen von Baden bevogtetes reguliertes Augustiner-Chorherrenstift, neben dem um 1220/30 eine von den Badenern gegründete Stadt entstanden war. Beides, Stift und Stadt, gingen um 1300 an Württemberg über,[35] das heißt, die Integration in die Grafschaft und das spätere Herzogtum hatte fast 100 Jahre vor Murrhardt und etwa 150 Jahre vor Blaubeuren begonnen. Die einigermaßen wirtschaftsstarke Stadt war zwar in vielfacher Weise mit dem Stift verflochten,[36] aber eigentlich hätte man bei dieser Konstellation eine ähnliche Entwicklung erwartet wie in Blaubeuren, nämlich, dass sowohl ein weltliches Amt, das sich um die Stadt gruppierte, als auch ein Klosteramt entstanden wären. Dazu kam es nicht. Der entscheidende Grund dafür war die 1477 durchgeführte Umwandlung des regulierten Chorherrenstifts in ein Säkularstift. Diese Umwandlung hing maßgeblich mit der Gründung der Universität Tübingen zusammen, für die das Vermögen des Säkularstifts Sindelfingen verwendet worden war. Zum Ausgleich für die Pfründen dieses Stifts wurden im nunmehrigen Säkularstift Backnang neue Pfründen für Säkularkanoniker geschaffen. Die Augustiner-Chorherren erhielten für den Verlust Backnangs als eine Art Trostpflaster ein neues, allerdings sehr kleines Stift in Sindelfingen, wo man einige wenige Teile

35 Gerhard Fritz, Backnanger Geschichte vom 11. bis 13. Jahrhundert, in: Backnanger Jahrbuch 25, 2017, S. 71–80.

36 Gerhard Fritz, Wirtschaftsgeschichte des Stifts Backnang vom späten 14. bis zur Mitte des 15. Jahrhunderts. Auswertung des Lagebruchs von 1393 und der Gültverzeichnisse von 1450/53, in: Backnanger Jahrbuch 18, 2010, S. 56–94.

des vorherigen Säkularstifts in ein Augustiner-Chorherrenstift der Windesheimer Kongregation umgewandelt hatte.[37]

Obwohl das Backnanger Stift zwischen 1477 und der Reformation des Herzogs Ulrich 1534 durchaus eine auch in geistlicher und geistiger Hinsicht bedeutende Rolle spielte – es war unter dem Propst Peter Jakobi von Arlon (1492–1509) ein Zentrum des Humanismus – und obwohl das Stift durchaus auch wirtschaftlich auf festen Grundlagen stand, dachte offenbar niemand daran, einen Backnanger Propst auf einen Landtag einzuladen. Warum das so war, wird nirgends gesagt, aber offenbar hing das mit dem Status als Säkularstift zusammen. Während das alte Augustiner-Chorherrenstift ständig als Kloster angesehen wurde und sich selbst in etlichen Urkunden auch als solches bezeichnete, sah man ein Säkularstift nicht mehr als landtagsfähiges Kloster an. Das alte Augustiner-Stift hatte gegenüber Württemberg auch dieselben Leistungen zu erbringen wie ein Kloster, also Transportpflichten im Kriegsfall sowie Beherbergungs- und Atzungspflicht für den Grafen und sein Gefolge, wenn dieser zur Erfüllung seiner Schutz- und Kastvogts-Aufgaben in Backnang weilte. Das neue Säkularstift wurde 1477 von diesen Aufgaben befreit, was mit eine Rolle beim Verlust der Landstandschaft gespielt haben könnte – obwohl der finanziell klamme Herzog Ulrich später all diese Pflichten wieder einführte.[38] Zumindest anfangs wird das Stift das Fehlen in der Landschaft kaum als Defizit empfunden haben, denn es scheint, dass unter dem ersten Propst des Säkularstifts noch eine enge Verbindung zu den landständischen Äbten und anderen Prälaten bestand, und der gebildete Propst Jakobi gehörte zum unmittelbaren Beraterkreis des Herzogs Ulrich, übte also durchaus reale politische Macht aus.[39]

Im selben Maße wie das Stift an politischer Bedeutung einbüßte, schob sich die Stadt Backnang nach vorne. Sie war rasch Zentralort eines der größten Ämter des Herzogtums mit den beiden Unterämtern Reichenberg und Unterweissach. Ihre Vertreter spielten in der württembergischen Landschaft eine durchaus

---

37 Vgl. zur Umwandlung: Sabine Beate Reustle, Stift und Stadt Backnang im 16. Jahrhundert (Backnanger Forschungen 2), Backnang 1996, S. 25–53; Dies., Backnang, in: Zimmermann / Priesching, Württembergisches Klosterbuch (wie Anm. 12), S. 191 f.; Oliver Auge, Stift und Herrschaft. Eine Studie über die Instrumentalisierung von Weltklerus und Kirchengut für die Interessen der Herrschaft Württemberg [...] (Veröffentlichungen des Stadtarchivs Sindelfingen 4), Sindelfingen 1996.

38 Gerhard Fritz / Sabine Reustle, Backnang, in: Sönke Lorenz / Oliver Auge / Sigrid Hirbodian (Hg.), Handbuch der Stiftskirchen in Baden-Württemberg, Ostfildern 2019, S. 79–86.

39 Reustle, Stift und Stadt Backnang im 16. Jahrhundert (wie Anm. 37), S. 50 f.

angemessene Rolle. Dessen ungeachtet, hatte das Stift noch in der gesamten Frühen Neuzeit umfangreichen Grundbesitz innerhalb der Stadt, wo die Mehrzahl aller Gebäude vom Stift zu Lehen rührten.[40]

In Sindelfingen, wo 1477 das Einkommen des dortigen Stifts zur Finanzierung der neu gegründeten Universität Tübingen verwendet worden war, bestand erst recht kein Grund, weshalb das Stift, das nur noch als kleine Rumpf-Institution weiter existierte, einen Vertreter in die Landschaft hätte entsenden sollen. Sindelfingen trat nur als Stadt samt dazugehörigem weltlichem Amt auf.[41]

Noch eindeutiger als in Sindelfingen war die Situation in Herrenberg. Dort war das erst 1439 gegründete weltliche Chorherrenstift, das zwischendurch von 1481 bis 1517 ein Stift der Brüder vom gemeinsamen Leben war, wieder zum Säkularstift geworden. Niemals stand zur Debatte, dass das Stift ein eigenes Amt gebildet hätte. Hier war allein die Stadt und ihr weltliches Amt maßgeblich.[42] Die Reihe mit Stiften und Städten, in denen ähnliche Verhältnisse herrschten, wo also nur die Stadt mit ihrem Umland ein weltliches Amt bildeten und wo das Stift nie in die Lage kam, ein eigenes stiftisches Amt zu bilden und den Propst in die Landschaft zu entsenden, lässt sich beliebig fortsetzen. Das gilt auch für zeitweilig besonders wichtige und reiche Stifte, die nach der Reformation als dauerhaft eingerichtete „Kloster"-Schule (obwohl dieser Begriff für ein Stift nicht richtig passt) große Bedeutung gewannen, wie etwa das Amandus-Stift in Urach.[43]

Für Frauenklöster und -stifte war es sowieso völlig außerhalb jeglicher Vorstellung, dass diese parallel zu den Städten, in denen sie lagen, ein eigenes Klosteramt mit Landstandschaft bildeten. Nirgendwo finden sich auch nur Ansätze, die in diese Richtung gehen.[44]

---

40 Gerhard Fritz (Hg.), Die Backnanger Lagerbücher von 1587, in: Carsten Kottmann / Bernhard Trefz (Hg.), Glaube – Bildung – Gesellschaft. Leben in der Frühen Neuzeit. Beiträge der Backnanger Tagung vom 21. Juni 2003 (Backnanger Forschungen 7), Backnang 2006, S. 85–178.

41 Oliver Auge, Sindelfingen, in: Lorenz / Auge / Hirbodian, Handbuch der Stiftskirchen (wie Anm. 38), S. 599–603. Zur Stadtgeschichte liegt nur eine ältere Arbeit vor: Helmut Graessle, Sindelfingen. Dorf, Stadt und Stift bis zur Mitte des 16. Jahrhunderts, Sindelfingen 1954.

42 Roman Jansen, Herrenberg, in: Lorenz / Auge / Hirbodian, Handbuch der Stiftskirchen (wie Anm. 38), S. 288–195.

43 Roland Deigendesch, Bad Urach, in: Lorenz / Auge / Hirbodian, Handbuch der Stiftskirchen (wie Anm. 38), S. 112–120.

44 Vgl. etwa bei Wendt, Legitimation durch Vermittlung (wie Anm. 2), S. 154–157, das Dominikanerinnenkloster Kirchheim unter Teck.

Abb. 21: *Steinheim an der Murr um 1685, Abbildung aus dem Kieserschen Forstlagerbuch. Bildnachweis: HStA Stuttgart, H 107/14, Bd 6, Bl. 14.*

Das galt auch für Sonderfälle wie Steinheim an der Murr. Dort gab es ein eigenes Unteramt, in dem es ein seltsames Spannungsverhältnis zwischen dem örtlichen Dominikanerinnenkloster und dem stadtähnlichen Dorf Steinheim gab. Steinheim hatte von 1298 bis ins späte 14. Jahrhundert sogar Stadtrechte besessen und den Status einer Reichsstadt gehabt. Diese herausgehobene Position war allerdings gegen Ende des 14. Jahrhunderts verloren gegangen, und der Ort existierte nur noch als Dorf weiter, allerdings ein Dorf mit einer Stadtmauer, städtischer Infrastruktur und einer bemerkenswerten stadtähnlichen Verfassung mit erheblichen Selbstverwaltungsrechten, von denen „württembergische Kommunen nur träumen konnten". Die Grafen von Württemberg waren um 1250 zwar an der Gründung des Dominikanerinnenklosters Marienthal in Steinheim beteiligt gewesen, waren dann aber für längere Zeit aus Steinheim verdrängt worden. Das Kloster kam 1422 unter hohenlohische Vogtei, und es dauerte bis zur Mitte des 16. Jahrhunderts, bis es Württemberg gelang, Hohenlohe nach und nach hinauszudrängen. Steinheim wurde dann als Unteramt dem benachbarten Amt Marbach angegliedert. Die Konsequenz war, dass weder das stadtähnliche

Dorf Steinheim noch das zwischen 1559 und 1582 reformierte Kloster einen Vertreter in die Landschaft entsandte.[45]

5. Städte und Klöster, bei denen die Integration nach Württemberg scheiterte: Ellwangen und Zwiefalten

Die Städte Blaubeuren und Murrhardt unterschieden sich in ihrer Wirtschaftskraft und im Maß ihrer Abhängigkeit vom jeweiligen Kloster. Diejenigen Städte, die neben sich oder in ihren Mauern nicht landschaftsfähige Stifte oder Frauenklöster beherbergten, waren wieder ein anderer Fall. Aber alle hatten insofern eine Gemeinsamkeit, als sie fest und auf Dauer ins politische System der Grafschaft und seit 1495 des Herzogtums Württemberg integriert waren.

Eine dritte Variante lässt sich bei den Klöstern Ellwangen und Zwiefalten beobachten. Dort beanspruchte Württemberg zwar ebenfalls lange Zeit die Vogtei, aber, anders als in den späteren Klosterämtern, gelang es dort nie, die Vogtei zur vollen Landesherrschaft auszubauen.

Ellwangen[46] war eines der ältesten Klöster im südwestdeutschen Raum und Ende des 8. oder Anfang des 9. Jahrhunderts entstanden. Als Reichskloster spielte es über die Jahrhunderte eine bedeutende Rolle. Seit 1370 hatte Württemberg Schutz und Schirm über das Kloster inne. Der wohl pestbedingte Niedergang des 14. Jahrhunderts führte zu einer lang andauernden Krise, die auch von Württemberg veranlasste Reformversuche zu Beginn des 15. Jahrhunderts nicht bewältigen konnten. Auf Wunsch der Mönche wurde das Kloster 1460 in ein Säkularstift umgewandelt. Diese hatte eine allmähliche Stabilisierung der wirtschaftlichen Lage zur Folge. Die Position Württembergs war anders als in den übrigen württembergisch bevogteten Klöstern, weil Ellwangen nominell reichsunmittelbar war und dies auch dauerhaft durchsetzen konnte.

---

45 Wendt, Legitimation durch Vermittlung (wie Anm. 2), S. 43 f. Ein knapper Überblick zu Steinheim bei: Matthias Untermann, Steinheim, in: Zimmermann / Priesching, Württembergisches Klosterbuch (wie Anm. 12), S. 463 f.

46 Hans Pfeifer, Ellwangen, in: Zimmermann / Priesching, Württembergisches Klosterbuch (wie Anm. 12); Franz Brendle, Ellwangen, in: Lorenz / Auge / Hirbodian, Handbuch der Stiftskirchen (wie Anm. 38), S. 181–190. Fast die gesamte umfangreiche Literatur zu Ellwangen befasst sich mit dem Kloster und Stift; die Stadt ist kaum einmal Gegenstand der Untersuchungen. Zum Erwerb der Schutzherrschaft 1370 auch: Stievermann, Landesherrschaft und Klosterwesen (wie Anm. 4), S. 108 f.

Neben dem Kloster entstand – und zwar offensichtlich auf Initiative des Klosters – noch vor 1200 eine Stadt,[47] die schon 1201 zusammen mit dem Kloster einem Brand zum Opfer fiel. Baulich bildeten Kloster und Stadt eine Einheit. In rechtlicher Hinsicht war der Abt beziehungsweise nach 1460 der Propst der Stadtherr. Ihm mussten, wie im Zinsbuch von 1337 überliefert ist, die Bürger schwören, und er verlieh das Amt des Schultheißen und besetzte das Gericht. Außerdem schuldete die Stadt ihm eine jährliche Steuer von 100 lb h, zu denen weitere 36 lb h kamen, die von der Pfarrei an den Abt zu entrichten waren. Für die Verleihung des Schultheißenamtes kassierte der Abt jeweils 26 lb h. Außerdem bestand ein Mühlbann des Abtes; die Bürger mussten in seiner Mühle mahlen lassen. Erbenloser Besitz fiel an das Kloster. Zusätzlich schuldeten die Bürger dem Abt für ihre Häuser und Grundstücke Zinsen und Frondienste, die jeweils zu Martini zu entrichten waren. Die völlige Abhängigkeit der Stadt vom Kloster im Allgemeinen und dem Stift im Besonderen wird aus verschiedenen Quellen des 14. und 15. Jahrhunderts sichtbar, wo Eide zur Bürgeraufnahme oder erbrechtliche Regelungen ein eindeutiges Bild ergeben. Immerhin hatte die Stadt in der Frühen Neuzeit und zweifellos auch schon im Spätmittelalter eine eigene Rechnungsführung, aus der deutlich wird, wieviel Geld ans Stift abzuführen war. Die Umwandlung des Stifts brachte für die Stadt keinerlei Änderung der Verhältnisse.

Anders als im Falle von Murrhardt und Blaubeuren tritt die Stadt Ellwangen im Umgang mit den Vögten nie als separate Größe hervor. Württemberg hat nie, wie das bei Murrhardt und Ellwangen war, Verträge mit der Stadt Ellwangen abgeschlossen. Die Anbindung der Stadt an Kloster beziehungsweise Stift war derart eng, dass es dafür offenbar keinerlei Ansatz gab. Nachdem Württemberg seinen Einfluss in Ellwangen angesichts von dessen starker reichsunmittelbarer Position nicht halten und erst recht nicht zur Landesherrschaft ausbauen konnte, brach auch die letzte Möglichkeit weg, einen Keil zwischen Kloster beziehungsweise Stift einerseits und Stadt andererseits zu treiben – und Württemberg scheint dies auch nie versucht zu haben.

---

47 Die folgenden Ausführungen zur mittelalterlichen und frühneuzeitlichen Stadt Ellwangen folgen grundsätzlich: Eugen Weis, Bürger zu Ellwangen unter Abt und Propst, in: Viktor Burr (Hg.), Ellwangen 764–1964. Beiträge und Untersuchungen zur Zwölfhundert-Jahrfeier, 2 Bde., Ellwangen 1964, hier: Bd. 1, S. 168–178. Die von Weis herangezogene Quelle von 1337 ist mittlerweile ediert: Hubert Häfele (Bearb.), Das älteste Urbar der Abtei des *gotzhuses* zu Ellwangen von 1337 (Veröffentlichungen der Kommission für geschichtliche Landeskunde in Baden-Württemberg A 52), Stuttgart 2008, S. 15–238.

Anders als in Württemberg wurde in Ellwangen die Reformation nie eingeführt. Aber nicht einmal die konfessionellen Differenzen des 16. Jahrhunderts boten einen Ansatz, etwa die Stadt Ellwangen gegen das Stift auszuspielen. Württemberg hielt trotzdem fast das ganze 16. Jahrhundert über an seinem Anspruch auf Schutz und Schirm über Ellwangen fest, obwohl dieser je länger, desto weniger mit Substanz zu füllen war. Erst 1589 gab es diesen endgültig auf.[48]

Neben Ellwangen schaffte es Württemberg auch im Kloster Zwiefalten nicht, seine Vogteirechte zur Landesherrschaft auszubauen.[49] Württemberg hatte die dortige Vogtei 1365/74 von Rudolf von Österreich erhalten, dessen Familie sie ihrerseits seit 1303 innehatte.[50] Mit Zwiefalten hatten die Württemberger insofern Schwierigkeiten, als das Kloster 1478 sogar einmal einen ihm nicht genehmen württembergischen Untervogt absetzte – ein Vorgang, den ein Kloster sonst kaum je riskierte. Zwiefalten ging noch weiter und setzte es durch, dass Württemberg nur noch dem Kloster genehme Untervögte einsetzte. Dies führte zur Entwicklung von „Binnenstrukturen […], die das Überleben der Abtei im Reformationszeitalter wesentlich förderten".[51] Württemberg hatte, obwohl längst evangelisch, nominell die Klostervogtei über das katholische Kloster bis 1750(!) inne, als Zwiefalten sie gegen eine Geldzahlung endlich ablöste.[52] Zur Landesherrschaft konnte Württemberg die Vogteirechte aber nie verdichten. Da aber beim Kloster Zwiefalten nie eine eigene Stadt entstand, sondern da die dortige Ansiedlung immer dörflichen Charakter hatte, sind weitere Ausführungen hier entbehrlich.

6. Fazit

Es gab viele Klöster in Städten, insbesondere die kaum zu zählenden Bettelordensklöster in Reichsstädten, zum Teil auch in Landstädten. Aber diese Klöster beherrschten nie, wie das in Murrhardt und Ellwangen der Fall war, die Stadt.

---

48 Dieter Stievermann, Fürstabtei Ellwangen, in: Meinrad Schaab / Hansmartin Schwarzmaier (Hg.), Handbuch der baden-württembergischen Geschichte, 2. Bd.: Die Territorien im Alten Reich, Stuttgart 1995, S. 526–536.

49 Vgl. zu Zwiefalten allgemein: Hermann Josef Pretsch, Zwiefalten, in: Zimmermann / Priesching, Württembergisches Klosterbuch (wie Anm. 12), S. 525–529; ausführlich: Wilfried Setzler, Kloster Zwiefalten. Eine schwäbische Benediktinerabtei zwischen Reichfreiheit und Landsässigkeit, Sigmaringen 1979; Hermann Josef Pretsch (Hg.), 900 Jahre Benediktinerabtei Zwiefalten, Ulm 1989.

50 Stievermann, Landesherrschaft und Klosterwesen (wie Anm. 4), S. 107 f.

51 Ebd., S. 126 f.

52 Pretsch, Zwiefalten (wie Anm. 49), S. 527.

Grundsätzlich war bei diesen Bettelordensklöstern auch die Stadt zuerst da. Erst als die Stadt vorhanden war, entstanden die Bettelordensklöster.

Im Falle von Murrhardt und Ellwangen war dagegen die Stadt später entstanden als das Kloster, und das Kloster hatte jeweils Anteil an der Stadtgründung, wenn diese nicht – wie im Falle Ellwangens – überhaupt vom Kloster alleine ausging. Auch außerhalb von Württemberg war die in Murrhardt und Ellwangen sichtbare Konstellation einer Stadt unter Klosterherrschaft selten. Man hat hier auf Städte unter der Klosterherrschaft Radolfzell verwiesen, auf den Markt des Klosters Reichenau am Nordufer des Bodensees, auf Säckingen am Hochrhein und auf Ottobeuren. Überall, wo es diese Konstellation gab, kamen die Klosterstädte nie „über das Stadium der Kleinstadt hinaus".[53] Überall fehlte der Fernhandel, der bei erfolgreichen, größeren Städten die Grundlage von Wachstum und Wohlstand war. Das war zwar auch in Blaubeuren der Fall, aber offenbar hatte die Stadt aufgrund ihres nicht ganz geringen wirtschaftlichen Potentials die Chance, sich autonomer zu entwickeln, als dies in Murrhardt und Ellwangen der Fall war, und so Zugang zu den Landtagen zu gewinnen.

Insgesamt ist es erstaunlich, dass das Thema „Stadt und Kloster" – sei es in der seltenen Konstellation wie in Murrhardt, Ellwangen und Blaubeuren, sei es in der häufigen Konstellation mit Bettelordensklöstern in größeren Städten – noch nicht Gegenstand einer der seit etlichen Jahrzehnten durchgeführten Tagungen des südwestdeutschen Arbeitskreises für Stadtgeschichtsforschung war. Ebenso wenig war bisher einer der mittlerweile über 50 einschlägigen Tagungs- und Themenbänden diesem Inhalt gewidmet.[54] Für die Zukunft wäre – über die genannten württembergischen Beispiele und über den südwestdeutschen Raum hinaus – das Verhältnis von Stadt und Kloster auf breiterer geographischer Basis zu untersuchen.

---

53 Jürgen Sydow, Die Klein- und Mittelstadt in der südwestdeutschen Geschichte, in: Hans-Peter Becht (Hg.), Pforzheim im Mittelalter. Studien zur Geschichte einer landesherrlichen Stadt (Pforzheimer Geschichtsblätter 6), Pforzheim 1983, S. 9–38, hier: S. 20; Johannes Waldschütz, Tagungsbericht: Kloster und Stadt am südlichen Oberrhein im späten Mittelalter und in der frühen Neuzeit. Historisches Seminar, Abteilung Landesgeschichte der Universität Freiburg und der Stadt Neuenburg am Rhein, 5. und 6.3.2010, in: Informationen zur modernen Stadtgeschichte 1, 2010, S. 112–116.

54 Vgl. die Zusammenstellung der Tagungen und der Bände der Reihe „Stadt in der Geschichte" und der verschiedenen stadtgeschichtlichen Publikationen der „Veröffentlichungen der Kommission für geschichtliche Landeskunde in Baden-Württemberg, Reihe B", URL: http://www.stadtgeschichtsforschung.de/publikationen.htm (letzter Zugriff: 4.7.2020).

Alois Schmid

Städte und Märkte in der Oberpfalz. Grundzüge ihrer Entwicklung im späten Mittelalter und in der beginnenden Neuzeit

Abstract: In the late Middle Ages and the early Modern Times the Upper Palatinate developed differently than the Duchy of Bavaria. The region was characterised by small governance structures which led to the formation of numerous little towns. This historical dimension becomes apparent when investigating today's administration district. It is characterised by a multitude of urban forms reaching from the prevalent small towns to the central city of Regensburg, the "mother" of all towns in Bavaria, episcopal residence and imperial city. The study examins the temporal strata of urbanisation in the region, the different government bodies involved, the physical appearance of the towns and modes of participation within the developing regional parliaments. As a matter of fact, the effect of the historical peculiarities of the region's urban landscape are still tangible today.

Die Oberpfalz stellt für die aus durchaus einsichtigen Gründen ohnehin nicht besonders ausgeprägte Stadtgeschichtsforschung in Bayern einen weithin weißen Fleck dar. Wohl liegen für viele Städte und Märkte zahllose Lokalstudien, meist älteren, vereinzelt aber auch neueren Datums – oftmals innerhalb des „Historischen Atlasses von Bayern" – vor,[1] doch haben diese noch kaum eine Zusammenfassung oder breitere Auswertung unter überörtlich ansetzenden, systematisierenden Fragestellungen erfahren. Zu Recht stellte Wilhelm Volkert in seinem Beitrag über die Oberpfalz zum „Handbuch der bayerischen Geschichte" 1995 fest: Eine „zusammenfassende Untersuchung

1 Maßgebliche Zusammenstellungen: Erich Keyser / Heinz Stoob (Hg.), Bayerisches Städtebuch, 2 Bde. (Deutsches Städtebuch 5/1–2), Stuttgart u.a. 1971/1974; Karl Bosl (Hg.), Bayern (Handbuch der historischen Stätten Deutschlands 7), ³1981; Hans-Michael Körner / Alois Schmid (Hg.), Handbuch der historischen Stätten. Bayern I: Altbayern und Schwaben, Stuttgart 2006. – Geringfügig überarbeitete Fassung des Erstdrucks in: Helmut Flachenecker / Rolf Kießling (Hg.), Städtelandschaften in Altbayern, Franken und Schwaben. Studien zum Phänomen der Kleinstädte während des Spätmittelalters und der Frühen Neuzeit (Zeitschrift für bayerische Landesgeschichte [künftig: ZBLG], Beihefte B 15), München 1999, S. 113–151.

über das Städtewesen fehlt."[2] In diese Richtung sind noch nicht mehr als erste Forschungsansätze entwickelt worden, nachdem die für Altbayern angefertigten Untersuchungen von Ludwig Rothenfelder,[3] Klaus Fehn,[4] Wilhelm Liebhart[5] oder Karl Bosl[6] nur wenig über die Donau nach Norden ausgriffen. Die Verfasser der wichtigsten diesbezüglichen Studien sind Maria Recknagel,[7] Ernst Klebel,[8] Reinhard H. Seitz,[9] Konrad Ackermann[10] und Erwin

---

2 Wilhelm Volkert, Die staatliche Organisation, in: Max Spindler (Begr.) / Andreas Kraus (Hg.), Handbuch der bayerischen Geschichte, Bd. III/3, München [3]1995, S. 150.

3 Ludwig Rothenfelder, Die Wittelsbacher als Städtegründer in Bayern von Otto I. dem Großen bis auf Ludwig IV. den Bayern, in: Verhandlungen des Historischen Vereins für Niederbayern 47, 1911, S. 1–106.

4 Klaus Fehn, Die zentralörtlichen Funktionen früher Zentren in Altbayern. Raumbindende Umlandbeziehungen im bayerisch-österreichischen Altsiedelland von der Spätlatènezeit bis zum Ende des Hochmittelalters, Wiesbaden 1970.

5 Wilhelm Liebhart, Die frühen Wittelsbacher als Städte- und Märktegründer in Bayern, in: Hubert Glaser (Hg.), Die Zeit der frühen Herzöge. Von Otto I. zu Ludwig dem Bayern. Beiträge zur Bayerischen Geschichte und Kunst 1180–1350, Bd. 1 (Wittelsbach und Bayern I/1), München/Zürich 1980, S. 307–317. Siehe: Anm. 15.

6 Karl Bosl, Die mittelalterliche Stadt in Bayern (ZBLG, Beihefte B 6), München 1974; Ders., Der Osten Bayerns. Kräfte, Formen, Ergebnisse geschichtlicher Entwicklung, Passau 1986; Ders., Die bayerische Stadt in Mittelalter und Neuzeit. Altbayern – Franken – Schwaben, Regensburg 1988. Zur Thematik weiterhin in größerem Rahmen: Carl A. Hoffmann, Landesherrliche Städte und Märkte im 17. und 18. Jahrhundert. Studien zu ihrer ökonomischen, rechtlichen und sozialen Entwicklung (Münchener Historische Studien, Abt. Bayerische Geschichte 16), Kallmünz/Opf. 1997 (der die Opferpfalz freilich nicht berührt); Alois Schmid, Die Stadt in Bayern – Stationen der Entwicklung bis zu Ludwig dem Bayern, in: Mitteilungen des Verbandes bayerischer Geschichtsvereine 27, 2016, S. 11–34; Peter Wolf u.a. (Hg.), Stadt befreit. Wittelsbacher Gründungsstädte. Katalog zur Bayerischen Landesausstellung, Augsburg 2020.

7 Maria Recknagel, Die Städte und Märkte des baierischen Donaugebietes, in: Mitteilungen der Geographischen Gesellschaft München 20, 1927, S. 1–118.

8 Ernst Klebel, Die Städte und Märkte des baierischen Stammesgebietes in der Siedlungsgeschichte, in: ZBLG 12, 1939/40, S. 37–93; Ders., Städte, Burgen und Siedlungen in der Oberpfalz, in: Oberpfälzer Heimat 2, 1957, S. 21–34.

9 Reinhard H. Seitz, Zum Problem Markt und Stadt im Spätmittelalter in der Oberpfalz, in: Emil Meynen (Hg.), Zentralität als Problem der mittelalterlichen Stadtgeschichtsforschung (Städteforschung A 8), Köln u.a. 1979, S. 272–283. Dieser Autor hat zahlreiche weitere kleinere Studien vorgelegt, von denen mehrere im Folgenden genannt werden.

10 Konrad Ackermann, Die Oberpfalz. Grundzüge ihrer geschichtlichen Entwicklung, München 1987, S. 20–33; Ders., Zur Entwicklung der oberpfälzischen Städte vom Spätmittelalter zur frühen Neuzeit, in: ZBLG 50, 1987, S. 441–462; Ders., Die Oberpfalz als Städtelandschaft, in: Arnika 3, 1989, S. 125–131.

Herrmann.[11] Die Frage nach den besonderen architektonischen Grundmerkmalen der oberpfälzischen Stadt hatte schon 1926 Bernhard Hermann Röttger thematisiert.[12] Für die herrschafts- und verfassungsgeschichtlich ausgerichtete Städteforschung bleibt die Oberpfalz jedoch weithin erst zu entdecken. Sie wird eines der Grundprobleme sein, das es bei der anstehenden Auswertung des „Historischen Atlasses von Bayern“ zu formulieren gilt.[13]

Dazu will die folgende Studie einen vorbereitenden Beitrag leisten. Ihr Ziel ist, die bemerkenswert breit besetzte, vielgestaltige, alle Typen und Hierarchien von Städten aufweisende Städtelandschaft der Oberpfalz vorzustellen und die bestimmenden Grundlinien ihrer Entwicklung aufzuzeigen. Notgedrungen muss sie sich auf ausgewählte Aspekte konzentrieren. Im ersten Abschnitt möchte sie den Aufbau der oberpfälzischen Städtelandschaft durch die Aufdeckung der wichtigsten zeitlichen Schichten verdeutlichen. Dieser Einstieg soll mit der Frage nach den entscheidenden Trägern und den von ihnen jeweils bevorzugten Räumen weitergeführt werden. In einem dritten Abschnitt werden dann einige typologische Grundmerkmale herausgearbeitet. Abschließend soll nach der Rolle der Städte und Märkte in den oberpfälzischen Ständevertretungen bis hin zur Wiedereingliederung des Fürstentums der Oberen Pfalz nach Bayern durch Kurfürst Maximilian I. von Bayern in den Jahren 1623/28 gefragt werden. Zeitlich werden die Erörterungen somit die drei Jahrhunderte zwischen dem Hausvertrag von Pavia 1329 und der Regierungszeit Maximilians I. im Blick haben. Am ehesten in dieser Epoche des Spätmittelalters und der beginnenden Neuzeit hatte die Oberpfalz ihre eigene Geschichte, weswegen auch wirklich von einer oberpfälzischen Städtelandschaft gesprochen werden kann. Natürlich kann das angesprochene Fragenbündel hier weithin nur angerissen und nicht auch erschöpfend abgehandelt werden. Es muss bei der Absteckung eines Untersuchungsfelds und der Aufdeckung einiger Grundlinien bleiben, das einmal in einer Dissertation auf der erforderlichen breiten archivalischen Grundlage in Verbindung historischer und kulturgeographischer Fragestellungen aufgearbeitet werden sollte.

---

11 Erwin Herrmann, Zur Stadtentwicklung in Nordbayern, in: Archiv für Geschichte von Oberfranken 53, 1973, S. 31–79.

12 Bernhard Hermann Röttger, Die oberpfälzische Stadt (Alte Kunst in Bayern 12), Augsburg 1926. Für die bayerische Ebene ist in dieser Hinsicht noch immer anregend: Alois Elsen, Wie entstand die altbayerische Stadt?, in: Das Bayerland 46, 1935, S. 481–512.

13 Vgl. Wilhelm Volkert, Die Oberpfalz im Historischen Atlas von Bayern, in: Andreas Kraus (Hg.), Land und Reich – Stamm und Nation. Probleme und Perspektiven bayerischer Geschichte. Festgabe für Max Spindler zum 90. Geburtstag (Schriftenreihe zur bayerischen Landesgeschichte 78), München 1984, S. 35–54.

1. Der Aufbau der Städtelandschaft der Oberpfalz

Die Erörterungen haben von der Exposition des Untersuchungsmaterials ihren Ausgang zu nehmen: der Vorstellung der Städtelandschaft Oberpfalz. Doch bereits hier beginnen die Schwierigkeiten: Auf welchen Raum sollen die Untersuchungen aufgebaut werden? Angesichts der starken Aufsplitterung der Oberpfalz in eine Reihe kleinerer und kleinster Herrschaften während des Untersuchungszeitraums muss zweckmäßigerweise auf den modernen Regierungsbezirk Oberpfalz zurückgegriffen werden. Ungeachtet der zahlreichen durch die Zugrundelegung einer erst 1838 begründeten modernen Verwaltungseinheit für eine historische Untersuchung heraufbeschworenen neuen Probleme wird nur auf diesem Wege eine ausreichend breite und damit wirklich aussagekräftige Arbeitsbasis geschaffen. Die einzelnen Kleinterritorien geben für sich alleine kaum eine eigenständige Städtelandschaft ab. Da sich die historische Städtelandschaft der Oberpfalz natürlich keinesfalls mit dem modernen Regierungsbezirk deckt, muss vereinzelt darüber hinaus ausgegriffen werden.

Zu dieser ersten grundlegenden Vorentscheidung kommt als zweite, dass die Städte und Märkte nicht getrennt, sondern gemeinsam betrachtet werden sollen. Eine Trennung hat wenig Sinn, weil die Übergänge gerade in der Oberpfalz im Untersuchungszeitraum ohnehin fließend blieben und die Zusammenschau aussagekräftigere Ergebnisse erbringt.

Eine erste Bestandsaufnahme der Städte und Märkte in der Oberpfalz am Ausgang des Mittelalters haben Gertrud Diepolder und Sebastian Hiereth für den „Bayerischen Geschichtsatlas“ erstellt.[14] Sie haben für den Raum des Regierungsbezirks Oberpfalz insgesamt 86 Ortseinträge gemacht. Diese Arbeitsgrundlage wird nach den Ergebnissen zusätzlicher Quellenstudien infolge der Ausweitung des Untersuchungszeitraums geringfügig auf 91 Orte erweitert. Davon betreffen 37 Siedlungen mit Stadtrang. Diese Zahl ist in Verbindung zu den anderen Regierungsbezirken zu setzen. Für Oberbayern und Niederbayern werden von Diepolder und Hiereth jeweils 75 Orte mit Stadt- beziehungsweise Marktrang, für Schwaben insgesamt 73 Orte vermerkt. Dagegen wurden für die drei fränkischen Regierungsbezirke zusammen über 200 Städte und Märkte eingetragen. Bereits dieser statistische Befund ergibt, dass die Oberpfalz zu den städtereicheren Teilen des modernen Staatsbayern gehört und von der Struktur ihrer Siedlungen her mehr auf Franken als auf Altbayern und Schwaben hin ausgerichtet ist. Sie ist zu

---

14 Gertrud Diepolder / Sebastian Hiereth, Städte und Märkte im Mittelalter, in: Max Spindler (Hg.) / Gertrud Diepolder (Red.), Bayerischer Geschichtsatlas, München 1969, Karte 22–23a. Siehe: Abb. 22.

den städte- und märktereicheren historischen Landschaften innerhalb des heutigen Staatsbayern zu zählen. Das System der Zentralorte ist hier ähnlich strukturiert wie in Franken. Das ist eine erste wichtige Feststellung.

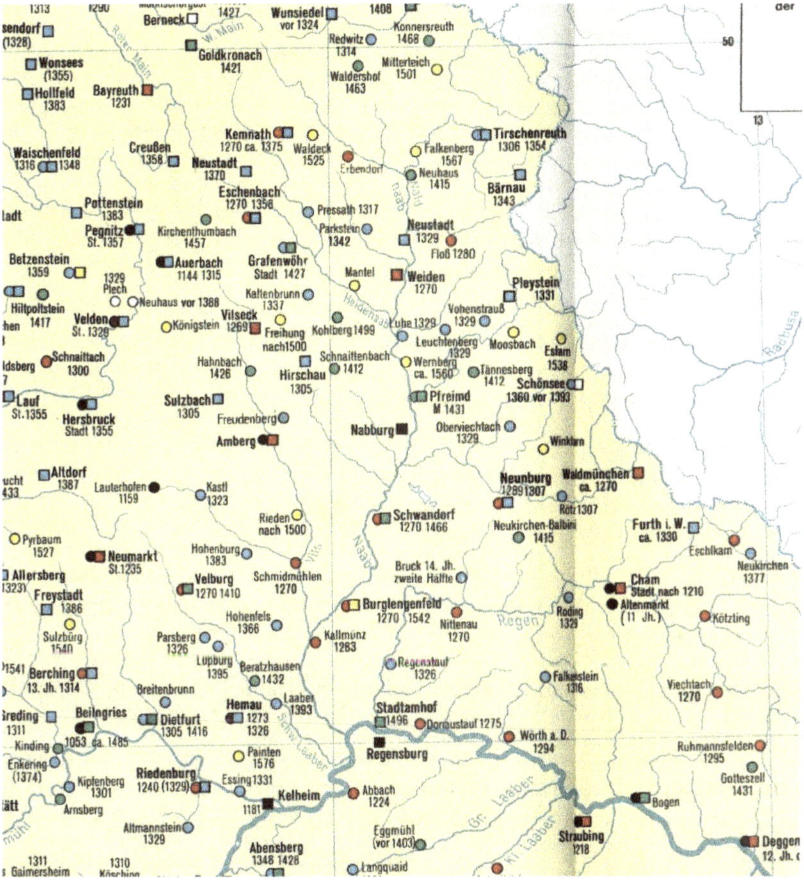

Abb. 22: *Städte und Märkte in der Oberpfalz um 1500. Bildnachweis: Max Spindler (Hg.) / Gertrud Diepolder (Red.), Bayerischer Geschichtsatlas, München 1969, Karte 22–23a (Ausschnitt).*

Die nächste Frage soll der Entstehung dieser Städtelandschaft gelten. Wenn das für die angeführte kartographische Erfassung zugrunde gelegte und schon von den Bearbeitern selber als wenig befriedigender Notbehelf erkannte

Einteilungsprinzip nach Jahrhunderten beibehalten wird, dann ergibt sich folgende Verteilung auf die hier zur Betrachtung anstehenden Jahrhunderte:

Ortstyp	**vor 1200**	**13. Jh.**	**14. Jh.**	**15. Jh.**	**16. Jh.**	**17. Jh.**
Städte	2	7	20	6	2	–
Märkte	5	20	24	12	14	1

Die Auswertung dieser Aufstellung erbringt mehrere weitere wichtige Feststellungen:

- Die Städteentwicklung erreicht in der Oberpfalz ihre Blütezeit vergleichsweise spät erst im 14. Jahrhundert. Das ist ein wichtiger Unterschied zum benachbarten Herzogtum Bayern.
- Die Kurven für die Städte und die Märkte erreichen ihren Höhepunkt gleichzeitig jeweils eindeutig im 14. Jahrhundert. Doch verläuft sie bei den Märkten anders als bei den Städten. Lediglich im 14. Jahrhundert kommt die Anzahl der Städte in etwa an die Anzahl der Märkte heran, während im vorausgehenden 13. und dann wieder im folgenden 15. und 16. Jahrhundert das Übergewicht viel stärker bei den Märkten liegt.
- Im 15. Jahrhundert fällt die Kurve der Städtegründungen steil ab, um im 16. Jahrhundert gänzlich auszulaufen.
- Dagegen dauert die Gründung von Märkten bis ins 17. Jahrhundert hinein an. Rund ein Drittel der oberpfälzischen Märkte entsteht in einer Zeit, als kaum noch Städte gegründet werden.

Die Kurven der Städte- und Märkteerhebungen verlaufen also für die Oberpfalz in keiner Weise parallel, sondern haben ein gänzlich unterschiedliches Aussehen. Sie nehmen auch einen deutlich anderen Verlauf als im wittelsbachischen Altbayern, wo die Phase intensiver Städtegründungen bereits zwischen 1200 und 1270 liegt und um 1270 recht abrupt beendet wird, um in eine Epoche der nicht minder ausgeprägten landesherrlichen Märktepolitik überzugehen.[15] Auch diese Eigenheit entspricht den fränkischen Gegebenheiten, die einen vergleichbaren Bruch nicht kennen und von ähnlicher Kontinuität bestimmt sind.

15 Wilhelm Liebhart, Zur spätmittelalterlichen landesherrlichen Marktgründungspolitik in Ober- und Niederbayern, in: Pankraz Fried (Hg.), Bayerisch-schwäbische Landesgeschichte an der Universität Augsburg 1975–1977. Vorträge – Aufsätze – Berichte (Augsburger Beiträge zur Landesgeschichte Bayerisch-Schwabens 1), Stuttgart 1979, S. 141–152.

Die statistische Auflistung der Stadt- und Markterhebungen nach Jahrhunderten macht gewiss Grundzüge der Entwicklung deutlich. Dennoch verwischt sie andererseits die entscheidenden geschichtlichen Einschnitte. Die vorgelegte Aufstellung bedarf deswegen der Interpretation unter verstärkter Einbeziehung historischer Gesichtspunkte. Es lassen sich noch genauere zeitliche Schichtungen herausarbeiten. Insgesamt können vier Hauptphasen unterschieden werden.

1.1. Die Frühzeit bis 1294

In der Epoche, als in Altbayern die meisten Städte entstehen, sind in der Oberpfalz erst wenige Städte festzustellen. Am Anfang steht Regensburg, das aber in mehrfacher Hinsicht, als Römerstadt, als Pfalzstadt, als Bischofsstadt und als Reichs- und Freistadt, gänzlich aus der Reihe fällt und sicherlich die Mutterstadt des oberpfälzischen Raumes schlechthin ist. Regensburg ist innerhalb der Städtelandschaft der Oberpfalz ein gänzlich „singuläres Phänomen“.[16] Auf Regensburg folgen als wittelsbachische Stadterhebungen des Hochmittelalters nur die aus ottonischen Reichsburgen herauswachsenden Städte Cham[17] und Nabburg.[18] Die Frühzeit der salierzeitlichen Königsgründung Amberg ist noch immer mit vielen Fragezeichen belastet; der Erstbeleg als „*civitas*“ liegt im Jahre 1242.[19] Ähnliche Verhältnisse sind in Sulzbach gegeben, das sich aus einer Burg heraus ebenfalls in der zweiten Hälfte des 13. Jahrhunderts zur Stadt auswuchs.

---

16 Fehn, Frühe Zentren (wie Anm. 4), S. 157–159, 68–171.

17 Johann Brunner, Geschichte der Stadt Cham, Cham 1919, S. 15–19; Karl Bosl, Cham. Die Geschichte der Stadt und ihres Umlandes in 1200 Jahren (Bayerische Städtebilder), Stuttgart 1989, S. 22–24; Alois Schmid, Die Städtepolitik der frühen Wittelsbacher: Die Gründungsstadt Cham, in: ZBLG 77, 2014, S. 833–855.

18 August Scherl, Verfassung und Verwaltung der Stadt Nabburg bis zum Ausgang des 16. Jahrhunderts, in: Verhandlungen des Historischen Vereins für Oberpfalz und Regensburg 96, 1955, S. 93–276; Elisabeth Müller-Luckner, Nabburg (Historischer Atlas von Bayern. Altbayern 50), München 1981, S. 106–113.

19 Michael Schwaiger, Chronica oder kurze Beschreibung der churfürstlichen Stadt Amberg in der obern Pfalz, neu hg. v. Felix Joseph Lipowski, Amberg 1818, S. 1–3; Amberg 1034–1984. Aus tausend Jahren Stadtgeschichte (Ausstellungskataloge der Staatlichen Archive Bayerns 18), München 1984, S. 45; Alois Schmid, Amberg im Hohen Mittelalter: Villa – forum – oppidum – civitas, in: Johannes Laschinger (Hg.), Aus Ammenberg wird Amberg. Historische Vorträge aus 975 Jahren Amberger Geschichte, Amberg 2010, S. 71–89. Zur allgemeinen Entwicklung: Hans Bungert / Franz Prechtl (Hg.), Ein Jahrtausend Amberg (Schriftenreihe der Universität Regensburg 11), Regensburg 1985.

Die wittelsbachische Städtepolitik, die in der Frühzeit stark auf Regensburg ausgerichtet war, griff in dieser frühen Phase kaum über die Donau in den Nordgau aus. Dennoch weist sie bereits in dieser ersten Phase auf das Gerüst der späteren Zentralorte voraus.[20] Die ältesten Städte sind dann auf Dauer die bedeutendsten geworden.

Die wittelsbachische Städtepolitik, die von Süden in den Nordgau hineinwirkte, stieß auf den Gegendruck der nicht minder zielstrebigen, von Westen in den Raum ausgreifenden staufischen Städtepolitik. Im Bestreben, einen möglichst geschlossenen Königslandgürtel von Franken ins Egerland und thüringische Mitteldeutschland aufzubauen, betätigten sich auch die Staufer als aktive Städtegründer, deren zukunftweisende Städtepolitik den oberpfälzischen Raum einerseits im Süden im Donaubogen um Regensburg,[21] andererseits am Nord- und Westrand erfasste. Neumarkt[22] und Weiden[23] sind die wichtigsten staufischen Gründungen geworden, die beide vorübergehend sogar den Rang von Königsstädten beanspruchten, aber nicht behaupten konnten. Im Übrigen ist noch bei einer Reihe weiterer Städte mit staufischen Anfängen zu rechnen: Auerbach,[24] Erbendorf,[25] Vilseck,[26]

---

20 Vgl. Erich Götz, Die Münzprägung der Oberpfalz. Geschichte und Katalog, Nürnberg 1992.

21 Alois Schmid, Regensburg. Reichsstadt – Fürstbischof – Reichsstifte – Herzogshof (Historischer Atlas von Bayern. Altbayern 60), München 1995, S. 79–108. Zur staufischen Städtepolitik allgemein: Erich Maschke, Die deutschen Städte der Stauferzeit, in: Reiner Hausherr (Hg.), Die Zeit der Staufer: Geschichte – Kunst – Kultur. Ausstellungskatalog, Bd. 3: Aufsätze, Stuttgart 1977, S. 59–73.

22 Karl Ried, Neumarkt in der Oberpfalz, Neumarkt 1960, S. 3–6; Bernhard Heinloth, Neumarkt (Historischer Atlas von Bayern. Altbayern 16), München 1967, S. 31–57; Alois Schmid, Zwischen Bayern und dem Reich. 850 Jahre Neumarkt i.d.OPf., in: Jahresbericht des Historischen Vereins für Neumarkt i.d.OPf. und Umgebung 26, 2011, S. 1–31.

23 Heribert Sturm, Neustadt an der Waldnaab – Weiden (Historischer Atlas von Bayern. Altbayern 47), München 1978, S. 21; Konrad Ackermann, Weiden – Tor und Brücke zu Böhmen (Bayerische Städtebilder), Stuttgart 1992, S. 12–15.

24 Fritz Schnelbögl, Auerbach in der Oberpfalz. Aus der Geschichte der Stadt und ihres Umlandes, Auerbach 1976.

25 Sturm, Neustadt an der Waldnaab – Weiden (wie Anm. 23), S. 20 f., 58; Josef Höser, Geschichte der Stadt Erbendorf, Erbendorf 1967. Vgl. Heribert Sturm, Tirschenreuth (Historischer Atlas von Bayern. Altbayern 21), München 1970, S. 22–54.

26 Georg Blössner, Die Bamberger Fürstbischöfe in ihren Beziehungen zur Oberpfalz, in: Verhandlungen des Historischen Vereins für Oberpfalz und Regensburg 80, 1930,

Vohenstrauß;[27] auch die heute mittelfränkischen Städte Lauf[28] und Hersbruck[29] sind in diesen Kontext zu stellen. Doch ist der Stadtcharakter bei diesen Stauferorten zunächst nur wenig ausgeprägt und oftmals auch kaum mehr nachzuweisen, weil die staufische Periode in jedem Fall nur von kurzer Dauer war und bald von den folgenden Entwicklungsstufen gänzlich überlagert wurde.

Das war die Lage zu Beginn der neunziger Jahre des 13. Jahrhunderts. In der Zeit, als in Altbayern das Gros der wittelsbachischen Städte entstand, wurde der Raum der späteren Oberpfalz nur am Rande von diesem zukunftweisenden Vorgang der Städtegründungen betroffen. Am ehesten ist Cham in diese Reihe zu stellen. Um 1290 war der alte Nordgau erst wenig mit Städten durchsetzt.

1.2. Die Epoche Ludwigs des Bayern

In die Phase Ludwigs des Bayern von seinem Herrschaftsantritt im Teilherzogtum Oberbayern 1294 bis zu seinem Tod 1347 fällt die Blütezeit des oberpfälzischen Städtewesens. Dem wittelsbachischen Teilherzog, König und schließlich Kaiser sowie seinem Bruder Rudolf I. verdankt eine Reihe oberpfälzischer Orte ihren Stadtstatus. Die Oberpfalz gehört zu den Räumen, die von ihrem Bruderzwist profitierten. Beide versuchten ihre Position gerade in dieser Randzone auszubauen und nahmen in diesem Sinne eine intensive Städtepolitik auf. Den Anfang machte dabei Rudolf I., der seit seinem Herrschaftsantritt im Jahre 1294 geradezu ein Füllhorn von Privilegien über die sich verfestigende Bürgerschaft in den Kommunen auch auf dem alten Nordgau ausschüttete.[30] Vor allem Amberg,[31]

S. 201; Joseph Kopf, Die Vogtei und das bambergische Amt Vilseck, in: Verhandlungen des Historischen Vereins für Oberpfalz und Regensburg 81, 1931, S. 51–117; Eugen Hierold, Das bambergische Amt Vilseck, in: Oberpfälzer Heimat 10, 1966, S. 101–115.

27 Volker Wappmann, Der Markt Vohenstrauß, in: Oberpfälzer Heimat 25, 1981, S. 114–124; Dieter Bernd, Vohenstrauß: Pflegamt Tännesberg - Treswitz, Amt Vohenstrauß, Pflegamt Pleystein, Landgrafschaft Leuchtenberg, Herrschaft Waldthurn (Historischer Atlas von Bayern. Altbayern 39), München 1977, S. 33–39.

28 Fritz Schnelbögl, Lauf - Schnaittach. Eine Heimatgeschichte, Lauf an der Pegnitz 1941, S. 60–66.

29 Georg Schmidt, Stadt und Landkreis Hersbruck. Ein Heimatbuch, Hersbruck 1951, S. 14; Wilhelm Schwemmer / Gustav Voit, Lauf - Hersbruck (Historischer Atlas von Bayern. Franken I/14), München 1967, S. 9–12.

30 Volkert, Die staatliche Organisation (wie Anm. 2), S. 150 f.

31 Original: Stadtarchiv Amberg, Urkundensammlung, 1. Druck: Johannes Laschinger (Hg.), Denkmäler des Amberger Stadtrechts, Bd. I: 1034–1450 (Bayerische Rechtsquellen 3), München 1994, S. 4–6, Nr. 4, Regest: Alfons Sprinkart, Kanzlei, Rat und

Nabburg,[32] Schwandorf[33] und Sulzbach[34] erhielten von ihm für die Stadtentwicklung entscheidende Urkunden; dieser Gruppe müssen auch Lauf[35] und das damals gerade niederbayerische Cham zugerechnet werden, dessen Stadtrecht von 1293 am Beginn dieses Entwicklungsschubs steht.[36]

Rudolfs I. Bruder, Konkurrent und schließlicher Bezwinger Ludwig der Bayer griff dieses Beispiel auf und verteilte noch mehr Stadtprivilegien. Vor allem Ludwig der Bayer ist der große Förderer der oberpfälzischen Städte geworden. Er hat zwar keine einzige Stadt neu gegründet, aber zahllose großzügig gefördert, weil er sich auch von deren Wirtschaftskraft eine Besserung seiner misslichen finanziellen Lage erhoffte. Ihm verdankt eine lange Reihe von oberpfälzischen Kommunen die Erhebung in den Stadt- beziehungsweise Marktrang oder zumindest die weitere Ausgestaltung der Privilegierung.[37] Kein Wittelsbacher hat mehr für

---

Urkundenwesen der Pfalzgrafen bei Rhein und Herzöge von Bayern 1294 bis 1314 (1317). Forschungen zum Regierungssystem Rudolfs I. und Ludwigs IV. (Forschungen zur Kaiser- und Papstgeschichte des Mittelalters 4), Köln/Wien 1986, S. 438, Nr. 173. Vgl. Carl Fuchs, Die Entwicklung des Amberger Stadtrechtes, Diss. jur. masch., Erlangen 1926, S. 2 f.; Rudolf Rubenbauer, Die Verfassung der Stadt Amberg 1294–1808, Diss. jur. masch., Erlangen 1955, S. 12–20, Teil 2, S. 2–5, Nr. 3; Rudolf Fritsch, Ambergs Stadtverfassung und städtische Verwaltungsorgane, in: Amberg 1034–1984 (wie Anm. 19), S. 45–47.

32 Sprinkart, Urkundenwesen (wie Anm. 31), S. 448, Nr. 284. Vgl. Scherl, Nabburg (wie Anm. 18), S. 235–237.

33 Sprinkart, Urkundenwesen, S. 460, Nr. 417. Vgl. Alois Schmid, Schwandorf: Der lange Weg vom Dorf zur Stadt, in: Jahrbuch für fränkische Landesforschung 66, 2006 (= Festschrift für Werner K. Blessing), S. 19–33; wieder in: Jahrbuch zur Kultur und Geschichte im Landkreis Schwandorf 18/19, 2007/08, S. 86–105.

34 Sprinkart, Urkundenwesen, S. 482, Nr. 681. Vgl. Eisenerz und Morgenglanz. Geschichte der Stadt Sulzbach-Rosenberg, hg. v. der Stadt Sulzbach-Rosenberg, Amberg 1999, bes.: S. 191–205.

35 Sprinkart, Urkundenwesen, S. 460, Nr. 413. Vgl. Schnelbögl, Lauf – Schnaittach (wie Anm. 28), S. 63.

36 Stadtarchiv Cham (Depositum im Staatsarchiv Amberg), Urk. 6. Vgl. Ludwig Schnurrer, Urkundenwesen, Kanzlei und Regierungssystem der Herzöge von Niederbayern 1255–1340 (Münchener Historische Studien, Geschichtliche Hilfswissenschaften 8), Kallmünz 1972, S. 370, Nr. 338; Bosl, Cham (wie Anm. 17), S. 28 f.

37 Vgl. Scherl, Nabburg (wie Anm. 18), S. 237 f.; Wilhelm Volkert, Amberg und Ludwig der Bayer, in: ZBLG 43, 1980, S. 29–44; Wilhelm Nutzinger, Neunburg vorm Wald (Historischer Atlas von Bayern. Altbayern 52), München 1982, S. 233 f. Das Thema Ludwig der Bayer und die oberpfälzischen Städte bedarf noch der systematischen Bearbeitung.

die Städte- und Märktelandschaft der Oberpfalz getan als der Kaiser. Diesen Zustand bringt bereits der Hausvertrag von Pavia 1329 zum Ausdruck, der für das Viztumsamt Burglengenfeld neben 38 Burgen schon 8 Städte und immerhin 26 Märkte als entscheidende Herrschaftsstützpunkte aufführt.[38] Diese Nennungen machen die Verdichtung des Netzes der Zentralorte bereits vor der Abtrennung der Oberen Pfalz vom Herzogtum Bayern deutlich. Der große Städtefreund hat seine breite Städtepolitik[39] auch in der Oberpfalz zur Geltung gebracht. Die pfälzischen Wittelsbacher haben sie nach dem Hausvertrag von Pavia fortgesetzt.[40]

Am meisten profitierte von dieser angestrengten wittelsbachischen Städtepolitik sicherlich Amberg. Die wissenschaftliche Bearbeitung des Amberger Stadtrechts bietet bis 1450 nicht weniger als 82 für die Stadtentwicklung grundlegende Urkunden, die fast ausschließlich von Wittelsbachern ausgestellt wurden;[41] darunter finden sich immerhin 18 Privilegien Ludwigs des Bayern.[42]

1.3. Die Frühzeit Kaiser Karls IV.

Der nächste große Förderer der oberpfälzischen Kommunen wurde dann in einer dritten Phase der luxemburgische Kaiser Karl IV. Ihm verdankt eine nicht minder große Anzahl von oberpfälzischen Städten ihren Status, der ihnen im Rahmen des Aufbaus des neuböhmischen Territoriums verliehen wurde. Die Reihe ist lang: Lauf,[43] Freystadt, Hirschau, Erbendorf, Floß,[44] Neustadt an der Waldnaab,[45] Neustadt am

---

38 Hans Rall (Hg.), Wittelsbacher Hausverträge des späten Mittelalters. Die haus- und staatsrechtlichen Urkunden der Wittelsbacher von 1310, 1329, 1392/93, 1410 und 1472 (Schriftenreihe zur bayerischen Landesgeschichte 71), 1987, S. 41–174. Vgl. Karl-Otto Ambronn / Rita Maria Sagstetter (Bearb.), Das Fürstentum der Oberen Pfalz. Ein wittelsbachisches Territorium im Alten Reich (Ausstellungskataloge der Staatlichen Archive Bayerns 46), München 2004, S. 13.

39 Pankraz Fried, Die Städtepolitik Kaiser Ludwigs des Bayern, in: ZBLG 60, 1997, S. 105–114.

40 Scherl, Nabburg (wie Anm. 18), S. 238–253; Laschinger, Amberger Stadtrecht (wie Anm. 31), S. 21–93, Nr. 23–82.

41 Laschinger, Amberger Stadtrecht (wie Anm. 31), S. 1–93, Nr. 1–82.

42 Ebd., S. 6–20, Nr. 5–22. Vgl. auch: Schnelbögl, Auerbach (wie Anm. 24), S. 68.

43 Wilhelm Schwemmer / Wilhelm Kraft, Kaiser Karls IV. Burg und Wappensaal zu Lauf (Schriftenreihe der Altnürnberger Landschaft 7), Nürnberg 1960.

44 Adolf Wolfgang Schuster, 1000 Jahre Floß, Floß 1976, S. 50–57.

45 Heinrich Ascherl, Neustadt WN unter Karl IV., in: Oberpfälzer Heimat 23, 1979, S. 65–72; Ders., Geschichte der Stadt und Herrschaft Neustadt an der Waldnaab, Neustadt an der Waldnaab 1982, S. 217–219 u.ö.; Karl-Otto Ambronn, Neustadt an der Waldnaab. Anfänge und frühes Werden einer oberpfälzischen Haupt- und Residenzstadt, in: Oberpfälzer Heimat 28, 1984, S. 162–181.

Kulm,[46] Grafenwöhr,[47] Auerbach,[48] Kemnath,[49] Pleystein,[50] Bärnau,[51] Vohenstrauß,[52] Luhe,[53] wohl auch Eschenbach,[54] Pressath[55] oder Weiden.[56] Nicht in allen Fällen werden die Eingriffe des Luxemburgers, der mehrfach auf entsprechende Anträge der regionalen Herrschaftsträger hin tätig wurde, in Deutlichkeit ersichtlich, weil die Urkunden zum Teil verschollen sind. Dennoch ist Karl IV. wegen seiner zahlreichen Privilegien ein weiterer wichtiger Förderer der oberpfälzischen Städte geworden. Diesen Sachverhalt verdeutlicht die beigegebene Kartenskizze, auf der die ungewöhnliche Massierung der privilegierten Städte in der nördlichen Oberpfalz deutlich ersichtlich ist. Vor allem Sulzbach erlebte als Mittelpunkt des neuböhmischen Territoriums damals seine Glanzzeit, deren Spuren im Stadtbild noch heute sichtbar sind.

1.4. Die Folgezeit

Mit Karls IV. recht abrupter Abwendung von seinem neuböhmischen Territorium im Jahre 1373 kam die Blütezeit der landesherrlichen Städtepolitik weithin zum Abschluss. Im späteren 14. und während des gesamten 15. Jahrhunderts klingt sie nur mehr aus. Zum Teil setzt nun sogar eine gegenläufige Entwicklung ein, indem einzelne der privilegierten Orte ihren Status nicht umzusetzen

---

46 Keyser / Stoob (Hg.), Bayerisches Städtebuch II (wie Anm. 1), S. 480–483.

47 Hermann Schenkl, Geschichte der Stadt Grafenwöhr, Grafenwöhr ^{2}1961, S. 39–42; Reinhard H. Seitz, Zur Entstehung von Markt und Stadt Grafenwöhr, in: Oberpfälzer Heimat 14, 1970, S. 62–74.

48 Schnelbögl, Auerbach (wie Anm. 24), S. 43–45, 73 f.

49 Heribert Sturm, Kemnath (Historischer Atlas von Bayern. Altbayern 40), München 1975, S. 16–21; Reinhard H. Seitz, Zur Entwicklung der Stadt Kemnath. Von der „Kemenathe" zu Markt und Stadt, in: Oberpfälzer Heimat 15, 1971, S. 97–112; Anton Reger, Aus der Geschichte der Stadt Kemnath, Kallmünz 1981, S. 26–34.

50 Sigfried Pablotzki, Geschichte der Herrschaft, der Stadt und der Pfarrei Pleystein, Pleystein 1980, S. 23 f.

51 Heribert Sturm, Entstehung des Pflegamts Bärnau, in: Oberpfälzer Heimat 10, 1966, S. 60–78; Christine Brunner-Hastreiter, Bärnau – ein Heimatbuch, Bärnau 1972, S. 45–56.

52 Bernd, Vohenstrauß (wie Anm. 27), S. 79–81, 141–143.

53 Ebd., S. 186–188.

54 Reinhard H. Seitz, Die Stadt Eschenbach und ihre Bannmeile, in: Oberpfälzer Heimat 17, 1973, S. 100–113.

55 Reinhard H. Seitz, Der spätmittelalterliche Markt Pressath, in: Oberpfälzer Heimat 18, 1974, S. 72–89.

56 Sturm, Neustadt an der Waldnaab – Weiden (wie Anm. 23), S. 29–36.

vermochten und andere Städte sich nicht mehr weiterentwickelten, sodass sie zu „Minder-" oder „Kümmerstädten" in der Terminologie von Heinz Stoob[57] wurden oder aber gar ihren Stadtstatus nicht behaupten konnten. Im Unterschied zur stagnierenden Städtepolitik lebten aber die Markterhebungen in dieser Zeit noch einmal mächtig auf. Das ist zumindest zum Teil das Werk der Pfalzgrafen im 1505 neugegründeten Fürstentum Pfalz-Neuburg,[58] die zum Zwecke des Herrschaftsausbaus an diesem herkömmlichen Mittel festhielten und ihrer Hauptstadt auf dem Nordgau Burglengenfeld noch 1542 den Stadtrang verschafften.[59] Burglengenfeld sollte die letzte Stadterhebung auf Jahrhunderte hin bleiben. Die Markterhebungswelle dagegen setzt sich in beachtlicher Intensität fort und kommt erst im 17. Jahrhundert mit Windischeschenbach 1605 und Waldsassen – zusammen mit Kastl der einzige Klostermarkt – 1693 zum Abschluss.[60] Freilich haben diese letzten Ausläufer nichts mehr mit den Wittelsbachern zu tun.

2. Die Träger

Die Untersuchung der entscheidenden Schichten der Städtelandschaft Oberpfalz ist untrennbar verbunden mit der Frage nach den Trägern.

Der frühere Münchner Kulturgeograf Hans Fehn hat als erster den Zusammenhang zwischen Herrschaftsstruktur und Stadtverteilung erkannt. Er konnte schon in den 1930er Jahren deutlich machen, dass die Städtelandschaft unserer Gegenwart ihre entscheidende Wurzel in den herrschaftlichen Verhältnissen des Spätmittelalters hat, weil die Städtepolitik ein wichtiges Mittel des Aufbaus und des Ausbaus der Territorien war.[61] Nach Fehn ist das System der urbanen

57 Heinz Stoob, Minderstädte. Formen der Stadtentstehung im Spätmittelalter, in: Vierteljahrschrift für Sozial- und Wirtschaftsgeschichte 46, 1959, S. 1–28; Ders., Minderstädte. Formen der Stadtentstehung im Mittelalter, in: Ders., Forschungen zum Städtewesen in Europa: Räume, Formen und Schichten der mittelalterlichen Städte. Eine Aufsatzfolge, Köln/Wien 1970, S. 225–245; Ders., Über frühneuzeitliche Stadttypen, in: ebd., S. 246–284.

58 Dieser Aspekt der Geschichte des Fürstentums Pfalz-Neuburg verdient eine eigene Behandlung.

59 Ludwig Brandl, Heimat Burglengenfeld. Geschichte einer Stadt, Burglengenfeld 1968, S. 39–58; Margit Berwing, Burglengenfeld. Die Geschichte der Stadt und ihrer Ortsteile, Regensburg 1996, S. 32–34.

60 Sturm, Tirschenreuth (wie Anm. 25), S. 118.

61 Hans Fehn, Das Siedlungsbild des niederbayerischen Tertiärhügellandes zwischen Isar und Inn, in: Mitteilungen der Geographischen Gesellschaft München 28, 1935, S. 5–94;

Zentralorte weithin in der Herrschaftsgeschichte des Mittelalters begründet. Er entwickelte seine Theorie an der Kulturlandschaft Ober- und Niederbayerns. Sie hat zweifellos auch Gültigkeit für die Oberpfalz, wo sie angesichts der starken herrschaftlichen Aufsplitterung sogar noch deutlicher fassbar wird.

Der alte Nordgau war vor allem durch den Landesausbau der Herzöge von Bayern von Süden her erschlossen worden. Im frühen 12. Jahrhundert geriet er aber zudem aus Westen in den Wirkungskreis der staufischen Königslandpolitik. Nach deren Zusammenbruch vermochten die Wittelsbacher die Vorherrschaft im Raum zu erringen. Doch wurde dieser im beginnenden 14. Jahrhundert unter die beiden wittelsbachischen Hauptlinien aufgeteilt. Während der Süden beim Teilherzogtum Bayern-München verblieb, wurde durch den Hausvertrag von Pavia (1329) die mittlere Oberpfalz der Pfalzgrafschaft am Rhein zugeschlagen und später als Fürstentum der Oberen Pfalz bezeichnet. Dazu kam ab 1505 das Fürstentum Pfalz-Neuburg mit beträchtlichen Territorialanteilen nördlich der Donau.[62] Vor allem die Wittelsbacher haben in ihren verschiedenen Linien auf die Städtelandschaft der Oberpfalz prägenden Einfluss ausgeübt. Für sie war, in Weiterführung des staufischen Vorbilds, Städtepolitik das entscheidende Mittel der Raumerfassung, das nun das frühere Zentrum der Burg abzulösen begann.[63] Sie strebten vornehmlich die Ausbildung eines flächendeckenden Netzes von Administrationsmittelpunkten an. Der Aufbau eines leistungsfähigen Verwaltungssystems wurde auch hier in erster Linie auf Städte und Märkte ausgerichtet. Freilich behaupteten vereinzelt Burgorte ihre herkömmliche Zentralität als Sitze von Pflegämtern: Wetterfeld, Parkstein, Treswitz. Im Allgemeinen aber erhielten die neuen Städte die Zentralortfunktionen in administrativer, wirtschaftlicher, militärischer und oft auch kirchlicher Hinsicht. Stadt und Markt als neue Herrschaftsmittelpunkte – das war die Hauptaufgabe, die den neuen Vororten zugedacht wurde.

Unter diesen mehrschichtigen Aufgaben war von besonderer Bedeutung der militärische Funktionsbereich: die Stadt als wichtiger Stützpunkt der Landesverteidigung. Die Oberpfalz war Grenzland, an deren Rand sich die Grenze nach Böhmen hin immer mehr ausbildete. Deswegen musste vor allem hier

---

Ders., Städte und Märkte, in: Leonhard Rückert (Hg.), Oberbayern – Land und Leute, Essen 1960, S. 110–135.

62 Zur Geschichte des oberpfälzischen Raumes: Wilhelm Volkert, Die politische Entwicklung der Pfalz, der Oberpfalz und des Fürstentums Pfalz-Neuburg bis zum 18. Jahrhundert, in: Spindler / Kraus, Handbuch der bayerischen Geschichte III/3 (wie Anm. 2), S. 3–141. Siehe: Karte 30–31 in: Spindler, Bayerischer Geschichtsatlas (wie Anm. 14).

63 Klebel, Städte, Burgen und Siedlungen in der Oberpfalz (wie Anm. 8).

die Landesverteidigung nach außen organisiert werden. Damit war bereits in ottonischer Zeit mit der Anlage der Reichsburgen begonnen worden. Der entscheidende Burgengürtel der ottonisch-salischen Zeit war, zunächst sehr weitmaschig rückwärts versetzt, in Cham, Nabburg, Neunburg, Sulzbach und Eger positioniert gewesen.[64] Im Zeitalter der Städtebildung wurde er weiter an die Grenze vorgeschoben und enger geflochten. Die Landesverteidigung erfolgte nun von Grenzstädten aus, die sich überwiegend ebenfalls an Burgen anlehnten. Die Kette dieser Grenzstädte reicht lückenlos von Furth im Wald[65] über Waldmünchen,[66] Oberviechtach,[67] Schönsee,[68] Eslarn,[69] Pleystein,[70] Tirschenreuth[71] letztlich hinauf bis nach Eger.[72] Der Blick auf die Karte zeigt die Konzentration der Städte und Märkte gerade am Ostrand der Oberpfalz sehr deutlich. Das hängt mit der Grenzsituation zusammen. Die Grenzstadt ist ein zweiter wichtiger Stadttypus für die Oberpfalz.

---

64 Karl Bosl, Die Markengründungen Kaiser Heinrichs III. auf bayerisch-österreichischem Boden, in: ZBLG 14, 1943/44, S. 177–247; wieder in: Ders. (Hg.), Zur Geschichte der Bayern (Wege der Forschung 60), Darmstadt 1965, S. 364–442; erneut in: Konrad Ackermann / Erich Lassleben (Hg.), Oberpfalz und Oberpfälzer. Geschichte einer Region, Kallmünz 1978, S. 47–99; Mathias Hensch, Burg Sulzbach in der Oberpfalz. Archäologisch-historische Forschungen zur Entwicklung eines Herrschaftszentrums des 8. bis 14. Jahrhunderts in Nordbayern, 3 Bde. (Materialien zur Archäologie in der Oberpfalz 3/1–3), Büchenbach 2005.

65 Josef Pongratz, Furth im Walde, Furth im Wald 1959; Reinhold Macho / Siegi Wild (Red.), Furth im Wald 1332–1982, Furth im Wald 1982; Werner Perlinger, Die Grenzhauptmannschaft „Furtt vor dem Oberen Behaimer Waldt“, in: Oberpfälzer Heimat 42, 1998, S. 134–140.

66 Josef Kraus, Waldmünchen. Ein Heimatbuch, Waldmünchen 1968; Emma Mages, Waldmünchen. Die Pflegämter Waldmünchen und Rötz (Historischer Atlas von Bayern. Altbayern 56), München 1991, S. 74–81.

67 Emma Mages, Oberviechtach (Historischer Atlas von Bayern. Altbayern 61), München 1996, S. 80–86.

68 Teresa Guggenmoos, Stadt Schönsee, Schönsee 1981; Mages, Oberviechtach (wie Anm. 67), S. 86–90.

69 Hans Schlemmer, Geschichte des Marktes Eslarn in der Oberpfalz, Weiden 1960; Josef Hanauer, Heimatbuch der Marktgemeinde Eslarn, Eslarn 1975.

70 Pablotzki, Pleystein (wie Anm. 50), S. 304–309.

71 Sturm, Tirschenreuth (wie Anm. 25).

72 Heribert Sturm, Eger. Geschichte einer Reichsstadt, Bd. 1, Augsburg ^{2}1960; Bd. 2, Augsburg 1952. Vgl. Ders., Districtus Egranus. Eine ursprünglich bayerische Region (Historischer Atlas von Bayern. Altbayern II/1), München 1981.

Angesichts dieser Feststellung lohnt sich der vergleichende Blick in die westliche Oberpfalz. Auch hier kam es im Rahmen des Aufbaus der Herrschaftswelt in Franken zur Abgrenzung. Doch erfolgte diese nicht in gleichem Ausmaß mithilfe von Städten oder Märkten. Der Raum um Neumarkt und Hersbruck bis hinauf nach Auerbach ist von einer auffallenden Städte- und Märkteleere gekennzeichnet, auf die ausdrücklich aufmerksam gemacht sei. Die Abgrenzung gegenüber dem aufsteigenden Nürnberg und der Hohenzollernherrschaft[73] ist wesentlich weniger ausgeprägt als die Abgrenzung nach Osten. Freystadt und Allersberg sind hier derartige, freilich nicht besonders bedeutsame Grenzstädte geworden.[74] Die Oberpfalz ist jedenfalls in Bezug auf die Städtedichte von einem deutlichen Strukturunterschied zwischen dem Osten und dem Westen gekennzeichnet. Die nämliche Städteleere bestimmt den Süden, in dem offensichtlich das dominierende Regensburg dem Aufstieg weiterer Orte zur Stadt deutlich im Wege stand. Die Hauptorte im Hochstift der Bischöfe zu Donaustauf und Wörth erlangten nur den Marktstatus. Der wittelsbachische Amtssitz wurde sogar in ein gänzlich unbedeutendes Dorf, nach Haidau, verlegt.

Mit dem Blick auf diese Orte ist auf den dritten Typus der Gegenstädte verwiesen, die bewusst gegen andere Herrschaftsträger aufgebaut werden. Ein beträchtlicher Anteil der Städte ist als Gegenstädte gegründet worden. Sie liegen deswegen bevorzugt in Grenzpositionen. Das deutlichste Beispiel dafür ist Stadtamhof[75] vor der Reichsstadt Regensburg, das schon durch seinen Namen auf den Bezug zum bisherigen Herzogshof verweist, wenn der bis heute nicht erklärte ungewöhnliche Name richtig interpretiert wird.[76] Als ähnliche Gegenstädte gegen die Reichsstadt Nürnberg waren Altdorf,[77] Lauf[78] und

73 Zum Befestigungswesen der Markgrafen von Brandenburg: Daniel Burger, Die Landesfestungen der Hohenzollern in Franken und Brandenburg im Zeitalter der Renaissance (Schriftenreihe zur bayerischen Landesgeschichte 128), München 2000. Die Dissertation zeigt, dass auch die Hohenzollern in diesem Raum keine besonderen Aktivitäten entwickelten.

74 Reinhard H. Seitz, Die Grenzstadt Freystadt, in: Oberpfälzer Heimat 20, 1976, S. 50–59.

75 Diethard Schmid, Regensburg, Bd. I: Das Landgericht Stadtamhof, die Reichsherrschaften Donaustauf und Wörth (Historischer Atlas von Bayern. Altbayern 41), München 1976; Ders., Die Ausbildung der wittelsbachischen Landesherrschaft im Raum Regensburg, in: Verhandlungen des Historischen Vereins für Oberpfalz und Regensburg 124, 1984, S. 313–332; Statt am Hoff – Stadtamhof, hg. vom Heimatverein Stadtamhof, Regensburg 1981.

76 A. Schmid, Regensburg (wie Anm. 21), S. 98 f.

77 Keyser / Stoob, Bayerisches Städtebuch I (wie Anm. 1), S. 33–39.

78 Schnelbögl, Lauf – Schnaittach (wie Anm. 28), S. 60–66.

Hersbruck[79] konzipiert, die freilich nach langem Kampf schließlich von der aufsteigenden Reichsstadt vereinnahmt und damit auf Dauer entschärft werden konnten.[80] Eine ähnliche Gegenstadt gegen das Hochstift Eichstätt ist Dietfurt,[81] das dem fürstbischöflichen Beilngries entgegengesetzt wurde.[82] Der Markt Schnaittenbach war eine wittelsbachische Gegengründung gegen das zunächst hirschbergische und später neuböhmische Hirschau.[83] Schwandorfs Aufstieg zur Stadt hängt mit seiner Lage an der Grenze der wittelsbachischen Herrschaftsbereiche zusammen.[84] Auch der Typus der Gegenstadt begegnet im Oberpfälzischen oftmals, was angesichts der komplizierten Binnenstruktur nicht verwundern darf. Er wird in besonderer Deutlichkeit im Gebiet der nur wenige Kilometer voneinander entfernten fünf Städte Kemnath, Eschenbach, Neustadt am Kulm, Grafenwöhr und Creußen fassbar, das darüber hinaus mit den drei Märkten Kirchenthumbach, Waldeck und Pressath besetzt wurde. Eine ähnliche Konzentration von Zentralorten auf engstem Raum im Städtedreieck Burglengenfeld, Maxhütte-Haidhof und Teublitz ist allerdings erst ein Ergebnis späterer Entwicklungen. Noch heute streiten sich Amberg und Weiden um den Vorrang in der mittleren und nördlichen Oberpfalz; diese Rivalität hat jedoch ihre historische Dimension in der Zugehörigkeit zu unterschiedlichen wittelsbachischen Teillinien.

Das waren die wesentlichen Beobachtungen zur Rolle der Wittelsbacher als Träger der Städte- und Märktepolitik in der Oberpfalz. Einer grundsätzlich anderen Leitlinie folgte die Städtepolitik des luxemburgischen Kaisers Karls IV. Ihm ging es während der nur zwei Jahrzehnte seiner besonderen Aktivitäten auf dem Nordgau mehr um Raumerfassung als um Grenzsicherung oder

---

79 Schmidt, Hersbruck (wie Anm. 29), S. 11–15.

80 Heinz Dannenbauer, Die Entstehung des Territoriums der Reichsstadt Nürnberg (Arbeiten zur deutschen Rechts- und Verfassungsgeschichte 7), Stuttgart 1928, S. 169–207; Schwemmer / Voit, Lauf – Hersbruck (wie Anm. 29), S. 28 f.

81 Keyser / Stoob, Bayerisches Städtebuch II (wie Anm. 1), S. 144–146.

82 Michael Schattenhofer, Beilngries. Chronik zur Neunhundert-Jahrfeier der Marktverleihung, Kallmünz/Opf. 1953, S. 14–32.

83 Heribert Batzl, Geschichte der Stadt Hirschau, Hirschau 1968, S. 9–14, 40–46; Reinhard H. Seitz, Hirschau – eine Stadtgründung Kaiser Karls IV. an der Goldenen Straße?, in: Oberpfälzer Heimat 16, 1972, S. 69–82.

84 Joseph Pesserl, Chronik und Topographie der Stadt Schwandorf, in: Verhandlungen des Historischen Vereins für Oberpfalz und Regensburg 24, 1866, S. 163–586; Hans Schneider, Eine alte Landesgrenze. Geschichte einer politischen Naht bei Schwandorf, in: Oberpfälzer Heimat 3, 1958, S. 92–101; Schwandorf in Geschichte und Gegenwart. Stadtchronik, hg. v. der Stadt Schwandorf, Bd. 1, Schwandorf 2001, S. 73–114 [Hans Schneider].

Territorienabgrenzung. Karl IV. hat stattdessen versucht, quer durch den oberpfälzischen Raum entlang der „Goldenen Straße“ von Nürnberg nach Prag den Riegel einer neuböhmischen Landbrücke von Eger nach Nürnberg und weiter nach Frankfurt zu legen, um auf diesem Wege eine Territorialverbindung zwischen diesen für ihn besonders wichtigen Reichsstädten aufzubauen. Zusammen mit Burgen waren Städte die entscheidenden Stützpunkte, deren Hauptziel die Bildung einer herrschaftlichen Landbrücke war. Dabei waren nicht alle Stadterhebungsversuche des Luxemburgers erfolgreich; das Stadtprivileg für das Dorf Sindlbach, zehn Kilometer nördlich von Neumarkt gelegen, kam aus unersichtlichen Gründen nicht zur Ausführung.[85] Natürlich hatte die Städtepolitik Karls IV. in der Nordoberpfalz[86] eine prinzipiell antiwittelsbachische Stoßrichtung und beschwor ihrerseits wittelsbachische Gegengründungen herauf. Die besonderen Aktivitäten der Wittelsbacher in der Oberpfalz sind auch in diesen politischen Gegebenheiten begründet. Diese führten zum unverkennbaren Städtereichtum gerade der nördlichen Oberpfalz; er hängt hauptsächlich mit der vor allem in dieser Berührungszone ausgetragenen Rivalität zwischen Wittelsbachern und Luxemburgern zusammen. Der Blick auf die Karte macht das Neuböhmische Territorium Karls IV. als städtischen Verdichtungsraum in besonderer Eindringlichkeit deutlich.

---

85 Es ist überliefert in: Bayerisches Hauptstaatsarchiv München (künftig: BayHStA), KU Gnadenberg, 1370 Februar 8. Kurzer Hinweis bei: Heinloth, Neumarkt (wie Anm. 22), S. 225; Michael Odorfer, Sindlbacher Lesebuch über Burg Haimburg, Heimatdorf und Heimatkirche, Sindlbach 1981, S. 37 f. mit Faksimile der Urkunde.

86 Zur Politik Karls IV. in der Nordoberpfalz: Hanns Hubert Hofmann, Karl IV. und die politische Landbrücke von Prag nach Frankfurt am Main, in: Zwischen Frankfurt und Prag, hg. vom Collegium Carolinum, München 1963, S. 51–74; Josef Staber, Die Oberpfalz und Niederbayern im Kulturprogramm Kaiser Karls IV., in: Verhandlungen des Historischen Vereins für Oberpfalz und Regensburg 109, 1969, S. 51–62; Lenka Bobková, Nordbayern in den territorialen Plänen Karls IV., in: Verhandlungen des Historischen Vereins für Oberpfalz und Regensburg 131, 1991, S. 281–286; Volkert, Staatliche Organisation (wie Anm. 2), S. 55–67; Bernhard Fuchs, Die Städte und Märkte in der nördlichen Oberpfalz unter Karl IV. (Regensburger Beiträge zur Regionalgeschichte 16), Regensburg 2012. Siehe: Abb. 23.

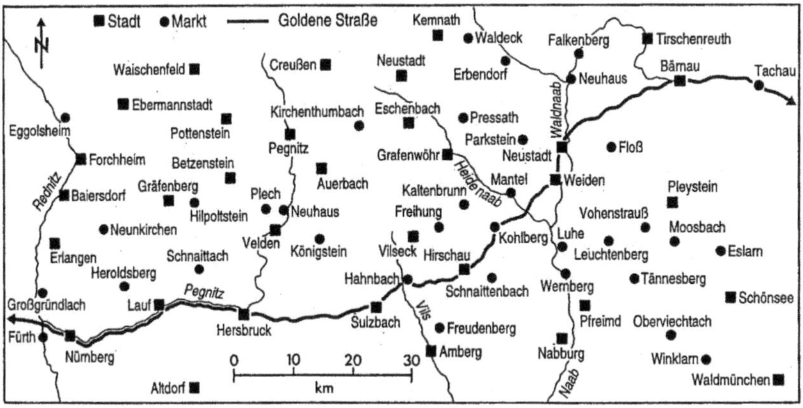

Abb. 23: *Städte und Märkte in der nördlichen Oberpfalz entlang der „Goldenen Straße". Bildnachweis: Helmut Flachenecker / Rolf Kießling (Hg.), Städtelandschaften in Altbayern, Franken und Schwaben. Studien zum Phänomen der Kleinstädte während des Spätmittelalters und der Frühen Neuzeit (ZBLG, Beihefte B 15), München 1999, S. 125.*

Neben den vorgestellten wichtigsten Herrschaftsträgern bildete sich gerade an den Rändern der Oberpfalz noch eine Reihe weiterer kleiner Herrschaften aus. Auch sie nahmen im Rahmen ihrer begrenzten Möglichkeiten eine abgestufte Städte- beziehungsweise Märktepolitik in Angriff. Auch diese Kleinterritorien wurden planvoll mit Kleinstädten oder Märkten besetzt.

Das gilt etwa für die Landgrafschaft Leuchtenberg. Die sehr wirkungsvolle Städtepolitik der Leuchtenberger Reichsgrafen im niederbayerischen Raum ist schon Erich Stahleder aufgefallen.[87] Sie lässt sich auch im Raum nördlich der Donau verfolgen, wo ihnen Pfreimd als Hauptstadt ihres Zwergterritoriums den Stadtrang verdankt.[88] Pfreimd erhielt das Stadtrecht zwischen 1366 und 1372 von Landgraf Ulrich II. Weitere Leuchtenberger Städte oder Märkte sind Pegnitz,

87 Erich Stahleder, Landgraf Johann I. von Leuchtenberg als Städtegründer in Niederbayern, in: Verhandlungen des Historischen Vereins für Niederbayern 96, 1970, S. 41–60. Vgl. auch: Illuminatus Wagner, Geschichte der Landgrafen von Leuchtenberg, Bd. II, Kallmünz 1950, S. 130–180.

88 Reinhard H. Seitz, Zu Entstehung und Entwicklung, Privilegien und Recht der leuchtenbergischen Residenzstadt Pfreimd, in: Verhandlungen des Historischen Vereins für Oberpfalz und Regensburg 111, 1971, S. 27–56, bes.: S. 31, 45, Regest Nr. 1; Bernd, Vohenstrauß (wie Anm. 27), S. 156 f., 180–182.

Luhe,[89] Pleystein,[90] Grafenwöhr[91] und Schönsee.[92] Den aus dem Altmühltal in den oberpfälzischen Raum hineinwirkenden Grafen von Hirschberg verdankt die nach ihnen benannte Stadt Hirschau ihre Anfänge.

Im Übrigen war der oberpfälzische Raum gerade an den Rändern nach Franken hin stark aufgesplittert. Und auch hier hat sich jede Herrschaft ihr Zentrum in Form einer Stadt oder eines Marktes eingerichtet. Die kleine Reichsherrschaft Laaber verlieh ihrem gleichnamigen Vorort 1393 das Marktrecht.[93] Die nordwestlich angrenzende Reichsherrschaft Ernfels stattete ihren Vorort Beratzhausen ebenfalls schon um 1300 mit dem Marktrecht aus.[94] Mit den Herren von Laaber stand die kleine Herrschaft Lupburg in Verbindung, deren herrschaftliche Zugehörigkeit lange umstritten war. Nach dem Anfall an Bayern 1392 errichteten die Wittelsbacher auch hier sofort 1395 einen Markt.[95] Das benachbarte Parsberg, 1205 als *„castrum"* belegt, das allerdings zunächst nicht weiter ausgebaut wurde, erhielt von König Ludwig dem Bayern als Oberherrn 1326/34 entscheidende Gerichtsprivilegien, aber ebenfalls nicht das volle Stadtrecht.[96] In gleicher Weise wurde der Vorort der Reichsherrschaft Breitenegg Breitenbrunn nur mit dem Marktrecht ausgestattet.[97] Auch in den beiden Landesteilen der Doppelherrschaft Sulzbürg-Pyrbaum wurden die Vororte Sulzbürg und Pyrbaum ebenfalls erst spät 1527 beziehungsweise 1540 von den Reichsgrafen von Wolfstein mit Marktrechten begabt.[98] Die Reichsherrschaft Störnstein richtete ihren Vorort in Neustadt an der Waldnaab ein, das nach dem Übergang an die Fürsten von Lobkowitz auch deren Residenzstadt wurde. Heinrich von Plauen verlieh Rötz, dem Zentrum der ererbten Kleinherrschaft Schwarzenburg, um 1500 das Stadtrecht. Eine bambergische Enklave in der nördlichen Oberpfalz war Vilseck, das 1330 wohl vom Bischof von Bamberg das Stadtrecht erhielt, dem auch

---

89 Bernd, Vohenstrauß (wie Anm. 27), S. 186–188.
90 Pablotzki, Pleystein (wie Anm. 50), S. 23 f., 34–39.
91 Schenkl, Grafenwöhr (wie Anm. 47), S. 35–42; Seitz, Grafenwöhr (wie Anm. 47).
92 Guggenmoos, Schönsee (wie Anm. 68), S. 53–55; Mages, Oberviechtach (wie Anm. 67), S. 86–90.
93 Manfred Jehle, Parsberg (Historischer Atlas von Bayern. Altbayern 51), München 1981, S. 125–152.
94 Robert Dollinger, Elfhundert Jahre Beratzhausen in der ehemaligen reichsfreien Herrschaft Ernfels, Regensburg 1966, S. 75 f.; Jehle, Parsberg (wie Anm. 93), S. 152–191.
95 Jehle, Parsberg (wie Anm. 93), S. 191–220.
96 Ebd., S. 359–382.
97 Ebd., S. 342–359.
98 Heinloth, Neumarkt (wie Anm. 22), S. 73–111.

Auerbach sein Marktrecht verdankt.[99] Damit sind auch diese fränkischen Bischöfe in der Städtelandschaft der Oberpfalz tätig geworden. Bischöflich-regensburgische Herrschaften waren Hohenburg, Donaustauf und Wörth. Hohenburg[100] erhielt das Marktrecht 1383 offensichtlich ebenso von den Bischöfen wie Wörth[101] und Donaustauf[102] zu einem nicht genau fixierbaren Zeitpunkt; auch hier richteten die Bischöfe bezeichnenderweise nur Märkte und keine Städte ein. Im Hochstift Regensburg war keine einzige Stadt gelegen. Am Nordrand der Oberpfalz vermochte schließlich das Zisterzienserstift Waldsassen sein ausgedehntes Stiftland aufzubauen, dessen Vorort Tirschenreuth das Stadtrecht 1364 vom Abt des Klosters erhielt.[103] Der Klosterort Waldsassen selber beschließt die Reihe der Markterhebungen in der Oberpfalz 1693 überhaupt.[104]

Zu welchen Komplikationen die herrschaftliche Zerteilung der Oberpfalz führen konnte, wird am deutlichsten am Beispiel der Stadt Windischeschenbach, die an der Berührungszone von drei Herrschaftsbereichen vorübergehend wirklich dreigeteilt war, wobei der Marktrang nur an einem Teil haftete.[105]

Die vorgestellte Übersicht über die wichtigsten Träger der Stadt- und Marktentwicklung belegt in Eindringlichkeit, dass auch in der Oberpfalz die Verteilung der Städte gewiss mit der historischen Herrschaftsstruktur zu tun hat. Die Feststellung von Hans Fehn lässt sich auch in diesem Raum verifizieren. Vor allem hängt die Vielzahl der Städte und Märkte in der Oberpfalz mit der Vielzahl der Herrschaftsträger zusammen. Die unverkennbare Konzentration der Städte gerade an

---

99 Blössner, Bamberger Fürstbischöfe (wie Anm. 26); Manfred Winkler, Stadt und Burg Vilseck, in: Oberpfälzer Heimat 1, 1956, S. 56–58. Vgl. Schnelbögl, Auerbach (wie Anm. 24), S. 38–40.

100 Hans Dachs, Das Marktrecht von Hohenburg auf dem Nordgau, in: Verhandlungen des Historischen Vereins für Oberpfalz und Regensburg 84, 1934, S. 3–86; Wilhelm Volkert, Hohenburg auf dem Nordgau, in: Die Oberpfalz 50, 1962, S. 159–163, 187–191; Jehle, Parsberg (wie Anm. 93), S. 382–404.

101 D. Schmid, Regensburg I (wie Anm. 75), S. 236–240; Josef Fendl (Hg.), Wörth – Stadt zwischen Strom und Berg, Regensburg 1979, bes.: S. 30 f.

102 D. Schmid, Regensburg I (wie Anm. 75), S. 224–226; Josef Fendl, Die Burg Donaustauf, „das beste und festeste Haus der Regensburger Kirche“, in: Die Oberpfalz 65, 1977, S. 7–12, 51–53; Hermann Hage (Hg.), Donaustauf – Moderne Marktgemeinde mit großer Vergangenheit, Regensburg 1994.

103 Sturm, Tirschenreuth (wie Anm. 25), S. 77–83; Max Gleissner, Inselstadt Tirschenreuth, in: Oberpfälzer Heimat 26, 1982, S. 72–81.

104 Sturm, Tirschenreuth (wie Anm. 25), S. 118, 305.

105 Sturm, Neustadt an der Waldnaab – Weiden (wie Anm. 23), S. 335, 374.

zwei Schwerlinien in der mittleren und nördlichen sowie in der östlichen Oberpfalz ist Folge dieser herrschaftlichen Verhältnisse; die Verstädterung hatte ihre Schwerpunkte in Grenz- und Konkurrenzräumen. Als weitere Grundlinie zeichnet sich ab, dass kleine Dynastien Stadt- und vor allem Markterhebungen länger durchführten als die großen Territorien. Auch hier zeigt sich, dass sie in der Wahl ihrer Herrschaftsmittel hinter den großen Territorien hinterherhinkten.

3. Typologische Merkmale

An der oberpfälzischen Stadt lassen sich mehrere typologische Merkmale feststellen. Ein erster Blick soll dem Vorgang der Stadtgründung gelten. Die oberpfälzischen Städte gehen auf recht unterschiedliche Wurzeln zurück. Die Ausnahme stellt auch hier die Stadtgründung auf der grünen Wiese aus wilder Wurzel dar. Ein Beispiel dafür ist das mit einem diesbezüglich redenden Namen belegte Freystadt, über Jahrhunderte hinweg eine sehr bewusst platzierte Grenz- und Gegenstadt; ihr Plancharakter ist noch heute deutlich erkennbar. Freystadt wurde vor 1305 vom adeligen Hiltpolt von Stein wohl gleich als Stadt gegründet.[106] Ein irgendwie gearteter Vorgänger ist nicht zu erkennen. In ähnlicher Weise wurde Neustadt am Kulm zwischen zwei Geländeerhebungen auf die grüne Wiese gestellt. Doch das sind Sonderfälle.

Der Normalfall ist die Stadtgründung in Anlehnung an einen älteren präurbanen Kern. Für die Oberpfalz ist besonders typisch die Anlehnung an eine Burg. Hier begegnet auffallend oft der Typus der Burgstadt beziehungsweise des Burgmarktes.[107] Beispiele dafür sind: Sulzbach, Burglengenfeld, Neunburg, Velburg, Parsberg, Hohenburg, Waldmünchen, Donaustauf, Wörth. Der Kern der Burg, der noch heute vielfach sichtbar ist, klingt bei einem Drittel der oberpfälzischen Städte sogar im Stadtnamen an, der oftmals auf das Grundwort „*-burg*" gebildet wird.[108] Das ursprüngliche Veldorf wurde in städtischer Zeit schließlich in Velburg umbenannt.[109] Die aus einer Hirschberger Wasserburg erwachsene Stadt Hirschau nahm in ihrer Benennung unverkennbaren Bezug auf das Gründergeschlecht. An die in frühere Jahrhunderte zurückreichende, oftmals mit der

106 Röttger, Die oberpfälzische Stadt (wie Anm. 12), S. IX; Seitz, Freystadt (wie Anm. 74).

107 Klebel, Städte, Burgen und Siedlungen in der Oberpfalz (wie Anm. 8); Andreas Boos, Burgen im Süden der Oberpfalz. Die früh- und hochmittelalterlichen Befestigungen des Regensburger Umlandes (Regensburger Studien und Quellen zur Kulturgeschichte 5), Regensburg 1998.

108 Klebel, Städte und Märkte (wie Anm. 8).

109 Jehle, Parsberg (wie Anm. 93), S. 232–269.

Ausbildung der Grenze nach Böhmen hin zusammenhängende Burg lehnte sich eine Burgsiedlung an, die im Verlaufe des Spätmittelalters mit dem Markt- oder auch Stadtrecht begabt wurde. Das war vor allem dann der Fall, wenn der Ort Behördensitz wurde. Die Anfänge der Burg und die Erhebung zum Markt beziehungsweise zur Stadt liegen oftmals Jahrhunderte auseinander. Es sei auf das Beispiel Nabburg verwiesen. Der Erstbeleg der Burg findet sich in einem Königsdiplom Heinrichs I. 929,[110] während der Ort das Stadtrecht erst um 1270 von Herzog Ludwig II., dem Strengen, erhielt.[111]

Neben den Typus des in der Oberpfalz auffallend häufigen Burgorts tritt der landesherrliche Markt oder die landesherrliche Stadt, die neben eine ältere Dorfsiedlung gesetzt wurde. Auch dieser Vorgang ist mehrfach noch heute in der Ortsbenennung fassbar. Von den insgesamt 18 Orten im deutschen Sprachraum mit dem Namen Neustadt finden sich immerhin zwei in der nördlichen Oberpfalz: Neustadt an der Waldnaab und Neustadt am Kulm.[112] Zu dieser Gruppe darf man sicher auch Neumarkt rechnen.[113] In diesen Fällen kommt also die Neugründung neben einer Altsiedlung auch in der Benennung zum Ausdruck. Nach der Errichtung der Neugründung erhielt der präurbane Kern verschiedentlich das Attribut „Alt“ als Ortsnamen; es wird von Altenstadt, Altendorf oder ähnlich gesprochen. Neustadt an der Waldnaab wird unmittelbar neben Altenstadt gestellt.[114] Ein anderes Altenstadt liegt vor der Stadt Vohenstrauß.[115] In gleicher Weise wird die wittelsbachische Gründungsstadt Cham zwischen die Vorgängerorte Altenstadt und Altenmarkt platziert.[116] Es gibt noch viele weitere Beispiele für diesen aussagekräftigen Namenparallelismus (Weiden,[117]

110 MGH DD H I., Nr. 19; Kurt Reindel, Die bayerischen Luitpoldinger 893–989. Sammlung und Erläuterung der Quellen (Quellen und Erörterungen zur bayerischen Geschichte NF 11), München 1953, S. 147 f., Nr. 75.

111 Müller-Luckner, Nabburg (wie Anm. 18), S. 106–113.

112 Sturm, Neustadt an der Waldnaab – Weiden (wie Anm. 23), S. 367.

113 Heinloth, Neumarkt (wie Anm. 22), S. 46–51.

114 Heinrich Ascherl, Frühgeschichte von Neustadt an der Waldnaab, in: Oberpfälzer Heimat 6, 1961, S. 22–32; Sturm, Neustadt an der Waldnaab – Weiden (wie Anm. 23), S. 370 f.

115 Volker Wappmann, Zur Geschichte von Altenstadt bei Vohenstrauß, in: Oberpfälzer Heimat 22, 1978, S. 85–90; Bernd, Vohenstrauß (wie Anm. 27), S. 141–143.

116 Bosl, Cham (wie Anm. 17), S. 14–25.

117 Adolf Schuster, Verfassungsgeschichte der Stadt Weiden im Mittelalter und in den ersten Jahrhunderten der Neuzeit, in: Verhandlungen des Historischen Vereins für Oberpfalz und Regensburg 92, 1951, S. 49–160, hier: S. 57 f.; German Weiss, Das „Alte Dorf“ bei Weiden, in: Oberpfälzer Heimat 4, 1959, S. 83–94; Sturm, Neustadt an der Waldnaab – Weiden (wie Anm. 23), S. 5.

Pegnitz,[118] Erbendorf,[119] Bayreuth[120]). Auch in der Oberpfalz lässt sich das Phänomen der Alt[en]- und Neustädte gut verfolgen, auf das schon Karl Bosl mit Nachdruck aufmerksam gemacht hat.[121]

In mehreren Fällen erwuchs die neue Stadt aus einem älteren Markt. Dass eine Stadt oftmals die Weiterentwicklung eines Marktes ist, lässt sich auch im Oberpfälzischen durch mehrere Beispiele belegen. Eine beachtliche Reihe der oberpfälzischen Städte ist aus Märkten hervorgegangen. Auch dieser Vorgang hat sich in besonderer Deutlichkeit im Falle Neustadts im Ortsnamen niedergeschlagen. Aus dem früheren Neumarkt an der Waldnaab wurde nach der Stadterhebung Neustadt.[122] Doch ist umgekehrt nicht in jedem Fall das Vorstadium eines Marktes nachzuweisen, wie sich andererseits nicht jeder Markt zu einer Stadt auswachsen konnte. Der präurbane Kern von Schönsee Stadlern, der seinen Namen vom Deminutiv Städtlein herleitet, ist gar zum Dorf abgesunken.[123] Auch hier wurde der Zentralort in die Neugründung Schönsee gelegt. Gerade Karl IV. versuchte, den Stadtstatus auch vielen Dörfern ohne Marktrang zukommen zu lassen. Der Zusammenhang von Stadt und Markt oder einem anderen präurbanen Kern ist offensichtlich sehr vielschichtig.

Der präurbane Kern war in den meisten Fällen unmittelbar neben der Neugründung gelegen. In Einzelfällen konnte der Abstand aber durchaus auch mehrere Kilometer betragen. Die staufische Gründungsstadt Neumarkt liegt immerhin sechs Kilometer vom Vorgänger- und Bezugsort, dem Königshof Berngau, entfernt.[124] In etwa dieselbe Entfernung trennt Oberviechtach von der vorausgehenden Burganlage Obermurach. Für die Positionierung der Gründungsstadt waren im Einzelnen unterschiedliche Motive maßgeblich, in diesem

118 Heinrich Bauer, Geschichte der Stadt Pegnitz und des Pegnitzer Bezirks, Pegnitz ^{2}1938; Ludwig Büttner, Geschichte der Stadt Pegnitz. Zur 600-Jahrfeier, Pegnitz 1955.

119 Sturm, Neustadt an der Waldnaab – Weiden (wie Anm. 23), S. 145–149, 414 f.; Höser, Erbendorf (wie Anm. 25), S. 32 f., 55 f.

120 Klaus Peter Dietrich, Territoriale Entwicklung, Verfassung und Gerichtswesen im Gebiet um Bayreuth bis 1603 (Schriften des Instituts für fränkische Landesforschung an der Universität Erlangen 7), Erlangen 1958, S. 43 f.; Peter Segl, Bayreuth im Mittelalter, in: Rudolf Endres (Hg.), Bayreuth. Aus einer 800-jährigen Geschichte (Bayreuther Historische Kolloquien 9), Köln u.a. 1995, S. 65–97.

121 Karl Bosl, Alt[en]stadt und Neustadt als Typen in Bayern, in: Civitatum Communitas. Studien zum europäischen Städtewesen. Festschrift für Heinz Stoob, Bd. I (Städteforschung A 21/1), Köln/Wien 1984, S. 158–180.

122 Sturm, Neustadt an der Waldnaab – Weiden (wie Anm. 23), S. 118 f.

123 Mages, Oberviechtach (wie Anm. 67), S. 86 f.

124 Heinloth, Neumarkt (wie Anm. 22), S. 43–51.

Fall die wesentlich günstigere Verkehrslage an einer an Bedeutung gewinnenden Fernstraße. Doch gibt es auch Städte, die aus den Vorgängersiedlungen unmittelbar förmlich herausgewachsen sind. Ein besonders bezeichnendes Beispiel dafür ist Weiden, dessen Altes Dorf im Stadtkern aufgegangen ist.[125] Die ursprünglichen Verhältnisse wirkten besonders in der Pfarreiorganisation oftmals lange nach. In Parsberg wurde der Pfarrsitz erst 1444, in Vohenstrauß im späteren 15. Jahrhundert, in Cham erst nach dem Dreißigjährigen Krieg 1674 aus dem älteren Chammünster in die Stadt,[126] in Neustadt an der Waldnaab schließlich gar erst 1929 von Altenstadt in die Neustadt verlegt. Ähnliche Verhältnisse sind in Nabburg oder im Markt Falkenstein gegeben.[127]

Das entscheidende Mittel der Stadt- oder Markterhebung war die zu diesem Anlass ausgestellte Urkunde. Sie ist in jedem Fall vorauszusetzen, jedoch nur selten nachzuweisen. Aus der Oberpfalz ist vor allem die Stadterhebungsurkunde von Grafenwöhr (1361) im Original bekannt.[128] Ein weiterer Indikator sind sodann die in jedem Fall ausgestellten Stadtrechtsverleihungsurkunden. Da sie für die weitere Entwicklung der jeweiligen Kommune von großer Bedeutung waren, sind sie sehr sorgsam aufbewahrt und oftmals auch mehrfach abgeschrieben worden. Sie sind uns demzufolge in größerer Anzahl erhalten, und zwar in recht unterschiedlichen Formen: sei es original, sei es kopial, sei es in Transsumpten, sei es in Konfirmationen aus späterer Zeit. Als Beispiel sei auf die abschriftlich überlieferte Marktrechtsverleihung für Kaltenbrunn verwiesen.[129] Doch liegen durchaus nicht für alle Städte und Märkte derartige Privilegien vor; vor allem bei den Märkten bleibt der Vorgang der Marktwerdung – wie in Freudenberg – bisher völlig im Dunkeln.

125 Weiss, Das „Alte Dorf" bei Weiden (wie Anm. 117); Sturm, Neustadt an der Waldnaab – Weiden (wie Anm. 23), S. 5.

126 Bernd, Vohenstrauß (wie Anm. 27), S. 27, 69. – 1250 Jahre Chammünster. Festschrift der Pfarrei, hg. vom Pfarrgemeinderat Chammünster 1989, S. 82.

127 Josef Pieler / Heinrich Ascherl, Chronik der Pfarrei Altenstadt an der Waldnaab und Neustadt an der Waldnaab, Neustadt an der Waldnaab [2]1977, S. 14 f.; Gerhard Zückert, Urpfarrei Altenstadt WN, in: Oberpfälzer Heimat 19, 1975, S. 53–62. – Josef Heigl, Geschichte von Falkenstein/Opf. und Umgebung mit besonderer Berücksichtigung der Kirchengeschichte, Falkenstein o.J., S. 113–119.

128 Schenkl, Grafenwöhr (wie Anm. 47), nach S. 36: Faksimile der Urkunde.

129 Gottfried Glockner, Kaltenbrunn. Die Gründungsurkunde vom Jahre 1344, in: Oberpfälzer Heimat 16, 1972, S. 92–113. – Zu Freudenberg: Georg Leingärtner, Geschichte der Herrschaft Freudenberg, Kallmünz 1980, S. 134 f.

Bei den Stadtrechten zeichnet sich auch in den oberpfälzischen Städten oftmals die Anlehnung an das Stadtrecht von Vororten ab. Für den oberpfälzischen Raum werden vor allem zwei Stadtrechtsfamilienkreise greifbar, die sich freilich in keinem Fall an die Reichsstadt Regensburg anlehnten, dessen eigene Entwicklung als Reichsstadt sich auch auf diesem Gebiet zeigt.[130] Der größte der oberpfälzischen Stadtrechtskreise gruppiert sich um die Stadt Amberg. Vom Ende des 13. bis zum Beginn des 15. Jahrhunderts sind folgende Anlehnungen an das Stadtrecht der Hauptstadt Amberg nachzuweisen: Nabburg 1294, Lauf 1298, Schwandorf 1299, Sulzbach 1305, Kastl 1323, Neustadt an der Waldnaab 1339, Nittenau 1345, Neunburg vorm Wald 1345, Neumarkt 1354, Kemnath 1375, Pressath 1398, Altdorf 1400, Bärnau 1405, Velburg 1410.[131]

Eine zweite Gruppe von Städten lehnte sich in ihrem Stadtrecht an die benachbarte, durch vielfache wirtschaftliche Verflechtungen mit dem Oberpfälzer Raum eng verbundene Reichsstadt Nürnberg an.[132] Das Nürnberger Stadtrecht wurde verbindlich etwa für Neumarkt 1361, Schönsee 1354, Pleystein 1331, Grafenwöhr 1361, Pfreimd oder vor allem Weiden.[133] Schließlich gruppieren sich um die Städte Weiden[134] und Sulzbach[135] kleinere eigene Stadtrechtsfamilien. In Einzelfällen ist aber auch die Übernahme des Egerer Stadtrechts (Bärnau[136]), des Tachauer Stadtrechts (Eschenbach), des Ingolstädter

---

130 Zum Stadtrecht von Regensburg: Heinrich Gottfried Philipp Gengler, Die Quellen des Stadtrechts von Regensburg aus dem XIII., XIV. und XV. Jahrhundert (Beiträge zur Rechtsgeschichte Bayerns 3), Erlangen 1892; A. Schmid, Regensburg (wie Anm. 21), S. 108–120.

131 Scherl, Nabburg (wie Anm. 18), S. 223–229; Josef Klose, Das Nittenauer Marktrecht, in: Der Regenkreis 7, 1967, S. 29–36, 63–67, 113–126; Fritsch, Ambergs Stadtverfassung (wie Anm. 31); Wilhelm Volkert, Das spätmittelalterliche Städtewesen, in: Max Spindler (Begr.) / Andreas Kraus (Hg.), Handbuch der bayerischen Geschichte, Bd. II/2, München ²1988, S. 590.

132 Werner Schultheiss, Die Einwirkung Nürnberger Stadtrechts auf Deutschland, besonders Franken, Böhmen und die Oberpfalz (Der Nürnberger Stadtrechtskreis), in: Jahrbuch für fränkische Landesforschung 2, 1936, S. 18–54, bes.: S. 26–37.

133 Schenkl, Grafenwöhr (wie Anm. 47), S. 42; Adolf Schuster, Rechtsbeziehungen zwischen Weiden und Nürnberg (Weidner Heimatkundliche Arbeiten 13), Weiden 1969.

134 Schuster, Verfassungsgeschichte der Stadt Weiden (wie Anm. 117), S. 63 f.

135 Keyser / Stoob, Bayerisches Städtebuch II (wie Anm. 1), S. 102: 1414 erhält Bärnau das Sulzbacher Stadtrecht.

136 Adalbert Busl, 650 Jahre Stadtrechte Bärnau, in: Oberpfälzer Heimat 38, 1994, S. 43–55. Vgl. Brunner-Hastreiter, Bärnau (wie Anm. 51), S. 46 f.

Stadtrechts (Riedenburg[137]) oder selbst des Münchner Stadtrechts (Hemau[138]) festzustellen. Doch konnte ein einmal verliehenes Stadtrecht durchaus später durch ein anderes ersetzt werden; derartiger Wechsel ist an mehreren Orten der Oberpfalz zu konstatieren.

Die Anlehnung erfolgte in sehr unterschiedlichen Formen. Das Stadtrecht konnte entweder ganz oder nur teilweise übernommen werden. Stadtrechte konnten in entsprechend modifizierter Form sogar auf Märkte transferiert werden. Üblicherweise gewährte die Stadtrechtsmutterstadt der Tochter bei Bedarf auch weiterhin konkrete Rechtshilfe. Sie übernahm Beratertätigkeiten für die Stadträte und Stadtgerichte. Die Zusammenarbeit der Rathäuser wurde also auch nach der Übertragung der Rechte fortgesetzt. Viele oberpfälzische Städte haben derartigen Rechtsbeistand gerade bei der Reichsstadt Nürnberg in Anspruch genommen: Sulzbach, Neumarkt oder selbst die Hauptstadt Amberg. Weiden taucht nicht weniger als 143 Mal mit der Bitte um entsprechende Unterstützung in den Nürnberger Ratsverlässen auf.[139] Andererseits waren Amberg, Nabburg, Weiden oder Sulzbach selber begehrte Ansprechpartner der kleineren oberpfälzischen Städte in Rechtsangelegenheiten.[140]

Weitere Rechtsdokumente der Stadtgemeinden sind Stadtwappen[141] und Stadtsiegel,[142] die verschiedentlich die herrschaftlichen Zusammengehörigkeiten

---

137 Keyser / Stoob, Bayerisches Städtebuch II (wie Anm. 1), S. 599.

138 Johann Nepomuk Müller, Chronik der Stadt Hemau, Regensburg 1861, S. 37; Karl Dachs, Die Entstehung der Stadt Hemau „auf dem Tangrintel", in: Verhandlungen des Historischen Vereins für Oberpfalz und Regensburg 90, 1940, S. 125–162, hier: S. 129.

139 Staatsarchiv Nürnberg, Rep. 60a: Reichsstadt Nürnberg, Verlässe des Inneren Rates 1-4456. Druck: Die Nürnberger Ratsverlässe I. Heft 1: 1449–1450, bearbeitet von Irene Stahl; Heft II: 1452–1471, bearbeitet von Martin Schieber (Schriften des Zentralinstituts für fränkische Landeskunde und Allgemeine Regionalforschung an der Universität Erlangen-Nürnberg 23/1, 2), Neustadt an der Aisch 1983/1995. Weiterhin: Schuster, Verfassungsgeschichte der Stadt Weiden (wie Anm. 117), S. 85–90.

140 Schuster, Verfassungsgeschichte der Stadt Weiden, S. 70 f., 89 f.; Ders., Rechtsbeziehungen zwischen Weiden und Nürnberg (wie Anm. 133), S. 18–26.

141 Adolf Schuster, Das Wappen Weidens, in: Oberpfälzer Heimat 1, 1956, S. 23 f.; Anton Neckermann, Das Wappen der Stadt Neunburg, in: Oberpfälzer Heimat 5, 1960, S. 102–104; Johann Baptist Fröhlich, Der böhmische Löwe in Weiden und Neustadt, in: Oberpfälzer Heimat 11, 1967, S. 101–103; Batzl, Hirschau (wie Anm. 83), S. 9; Klemens Stadler, Deutsche Wappen: Bundesrepublik Deutschland, Bd. 6: Die Gemeindewappen des Freistaates Bayern, Teil II: M–Z, Nachträge zu Bd. 4 und 6, Bremen 1968, S. 29 (Neustadt an der Waldnaab); S. 91 (Weiden).

142 Vgl. Scherl, Nabburg (wie Anm. 18), S. 197–201; Rubenbauer, Stadt Amberg (wie Anm. 31), S. 71 f.; Fritsch, Ambergs Stadtverfassung (wie Anm. 31), S. 57.

zum Ausdruck bringen. Weiden, Eschenbach, Kemnath und Moosbach führen gemeinsam in der oberen Hälfte des Wappenschilds den pfälzischen Löwen und die bayerischen Rauten zur Versinnbildlichung der beiden wichtigsten Herrschaftsträger; nur die untere Hälfte ist jeweils verschieden gestaltet. Vereinzelt wird – wie im Auerbach der luxemburgischen Zeit – auch in der Siegelumschrift auf die herrschaftliche Zuordnung Bezug genommen.

Als entscheidendes Signum des Stadtranges wurde lange Zeit die Wahrnehmung des Befestigungsrechts angesehen. Vor allem die Stadtmauer galt als das Hauptmerkmal einer Siedlung mit Stadtrang. Freilich bereitete die Anwendung dieser Feststellung auf die Oberpfalz immer große Mühe. Denn bei der Überprüfung an den Städten dieses Raumes wurde schon bald erkannt, dass es hier – wie im Falle Pressath oder Grafenwöhr[143] – sogar sehr aussagekräftige Belege für die Richtigkeit dieser These gibt, dass ihr aber andererseits verschiedene Beobachtungen deutlich widersprechen. Mit Sicherheit verfügten in der Oberpfalz auch Orte über eine Ummauerung, die nie oder aber auch erst später den Stadtrang erhielten, während andererseits auch Stadtgemeinden ohne das Attribut einer Stadtmauer begegnen. Gerade diese Beobachtungen waren der Ausgangspunkt der wichtigen Studie von Reinhard H. Seitz, die ergab, dass die Stadtmauer zumindest hier nicht als derart exklusives Unterscheidungsmerkmal betrachtet werden kann.[144] Der Vergleich mit Altbayern zeigt, dass hier wohl ein oberpfälzisches Spezifikum vorliegt. Für diesen Raum ist eine besonders ausgeprägte Befestigungstätigkeit kennzeichnend, die wohl vor allem mit den Hussiteneinfällen in Verbindung zu bringen ist. Oberpfälzer Märkte mit Mauer sind etwa Pressath, Oberviechtach, Tännesberg oder Bruck. Gerade diese äußere Bedrohung bedingte, dass hier auch Märkte verschiedentlich eine Stadtmauer errichteten. Der Zusammenhang mit dieser äußeren Bedrohung wird auch dadurch nahegelegt, dass diese Orte über eine derartige Befestigung nicht von Anfang verfügten, sondern diese erst zu einem späteren Zeitpunkt erhielten. In der Oberpfalz hat also die Grenzlage zu besonderen Verhältnissen geführt, die nicht in jeder Hinsicht in das gewohnte Systematisierungsmuster passen.

143 Seitz, Markt Pressath (wie Anm. 55), S. 86. – Schenkl, Grafenwöhr (wie Anm. 47), nach S. 36: […] *„daß sie den margt zu Grauenwird mit mauren, pforten, turnen, graben vnd anders bestenen mugen,* […] *als man ander stete pfleget zu besten.“*

144 Seitz, Oberpfalz (wie Anm. 9).

Mit dem Blick auf das Befestigungswesen ist die Frage nach dem Verhältnis von Stadt und Markt angesprochen. Die Abgrenzung dieser beiden Ortstypen gegeneinander ist in der Oberpfalz offensichtlich schwierig. Dafür ist vor allem Schwandorf ein aussagekräftiges Beispiel; hier hat man in der entscheidenden Urkunde, die den Stadtstatus verfügte, auffallender Weise das Attribut Markt weiter beibehalten. Die Bezeichnungen Stadt und Markt werden für mehrere Orte durcheinander verwendet, ohne dass dafür ein System erkenntlich würde. Doch ist gleiches in Altbayern etwa für Aichach, Friedberg oder Vohburg festzustellen.[145]

Aus diesem Befund erwächst die Frage, inwieweit aus dem Wechsel in den Bezeichnungen der Quellen auf einen Wechsel in der realen Rechtsqualität der betroffenen Orte geschlossen werden darf. Das Problem hat der Lokalgeschichte vielerorts großes Kopfzerbrechen bereitet. Doch sollte die Diktion der Quellen nicht überbeansprucht werden. Es ist jeweils exakt die Quellengattung in Rechnung zu stellen. Aber selbst in hochrangigen Verwaltungsdokumenten erfolgt die Verwendung der Begriffe Stadt und Markt nicht immer in der wünschenswerten Klarheit. Als Beispiel sei auf die wittelsbachischen Hausurkunden des späten Mittelalters[146] verwiesen, die auch den oberpfälzischen Raum betreffen. Wer die Verwendung der Bezeichnungen Stadt und Markt in dieser Quellengruppe genau analysiert, wird feststellen, dass sie einem merkwürdigen Wechsel unterworfen ist. Es gibt Orte, die zunächst als Städte, später aber als Märkte bezeichnet werden; das gilt etwa für Riedenburg.[147] Es gibt aber auch Orte, die umgekehrt zunächst als Märkte und später als Städte bezeichnet werden (zum Beispiel Auerbach,[148] Eschenbach,[149] Hersbruck,[150] Hirschau,[151] Kemnath,[152] Lauf[153] oder Velburg[154]). Einzelne Orte sind sogar einem mehrfachen Wechsel in der Einstufung unterworfen. Es geht sicher zu weit, aus jedem Wechsel in der

145 Alois Schmid, Die Anfänge der Stadt Vohburg, in: ZBLG 60, 1997, S. 491–520, bes.: S. 515–517.

146 Rall, Hausverträge (wie Anm. 38).

147 Ebd., S. 89, 192, 198.

148 Ebd., S. 84, 251.

149 Ebd. Vgl. Anton Fischer, Eschenbach in Vergangenheit und Gegenwart, Eschenbach 1950.

150 Rall, Hausverträge (wie Anm. 38), S. 84, 214, 250.

151 Ebd., S. 85, 214, 251. Batzl, Hirschau (wie Anm. 83), geht auf diese Probleme nicht ein.

152 Rall, Hausverträge (wie Anm. 38), S. 84, 245.

153 Ebd., S. 84, 214.

154 Ebd., S. 89, 192, 198, 250.

Begrifflichkeit auch einen Wechsel im Verfassungsstatus ableiten zu wollen, was die Lokalgeschichte oftmals getan hat. Im Spätmittelalter war die Unterscheidung von Stadt und Markt noch nicht in der späteren Deutlichkeit festgelegt. Für diese Epoche ist noch mit einer gewissen Offenheit in der Terminologie auch der Rechtssprache zu rechnen, die den Kanzlisten noch einen gewissen Formulierungsspielraum beließ. Zudem gingen Eigen- und Fremdbezeichnung – wie in Beilngries im 15. Jahrhundert lange Zeit – vereinzelt bewusst auseinander. Hier hat erst der Verwaltungsstaat der Neuzeit größere Genauigkeit durchgesetzt. Vor allem die Errichtung der Städtekurie in den landständischen Vertretungen zwang zur Präzisierung des verfassungsrechtlichen Status.

Die Stadtbefestigung hat den Blick auf den Bereich der Architektur gelenkt. Hier stellt sich als entscheidende Frage: Hat die oberpfälzische Stadt eine typische Form etwa in Gestalt eines bezeichnenden Stadtgrundrisses oder in den Bauformen, die man neben die altbayerische[155] oder auch die böhmische[156] Stadt stellen könnte? Die Frage ist gerade in der Oberpfalz gut zu untersuchen, weil die oberpfälzischen Städte ihren ursprünglichen Grundriss oftmals in unsere Gegenwart hinein gerettet haben. Diese Tatsache hat den Kunsthistoriker Röttger zu seiner bereits genannten Monographie über die oberpfälzischen Städte veranlasst.[157] Er stellt als Grundmuster der oberpfälzischen Stadt die an einer Fernstraße aufgereihte langgestreckte Siedlung vor, die sich in der Mitte zum Stadtplatz weitet, auf dem das Rathaus platziert ist. Besonders bezeichnende Beispiele dafür sind Lauf an der Pegnitz, auch der Ort Pegnitz selber, Kemnath, Freystadt oder Weiden.[158]

---

155 Klaus Kratzsch, Wittelsbachische Gründungsstädte. Die frühen Stadtanlagen und ihre Entstehungsbedingungen, in: Glaser, Die Zeit der frühen Herzöge I (wie Anm. 5), S. 318–337.

156 Für sie ist bezeichnend die ausgedehnte Stadtplatzanlage. Vgl. Hans Planitz, Die deutsche Stadt im Mittelalter. Von der Römerzeit bis zu den Zunftkämpfen, Köln/Wien [5]1992, S. 176 f.

157 Röttger, Die oberpfälzische Stadt (wie Anm. 12).

158 Das machen noch die Blätter der Urkatasteraufnahme deutlich: Bayerische Städte im jungen Königreich. Ortsblätter des bayerischen Flurkartenwerks im 19. Jahrhundert, hg. vom Bayerischen Landesvermessungsamt, München 1983. Vgl. Röttger, Die oberpfälzische Stadt (wie Anm. 12), S. XI.

Abb. 24: *Katasterblatt von Neumarkt in der Oberpfalz (1831). Bildnachweis: Bayerische Städte im jungen Königreich. Ortsblätter des bayerischen Flurkartenwerks im 19. Jahrhundert, hg. vom Bayerischen Landesvermessungsamt, München 1983.*

Freilich begegnen derartige Schemata auch andernorts, sodass man sie kaum als typisch oberpfälzisch ansprechen kann. Ein auch anderwärts festzustellender Grundtyp der Stadt ist gerade in der Oberpfalz so häufig anzutreffen, dass keinesfalls gänzlich zu Unrecht schon einmal vom „plangerechten Oberpfälzer Straßenmarkt“ gesprochen wurde.[159]

Ähnliches gilt bezüglich der Hofformen. Die oberpfälzische Stadt ist im Normalfall eine Ackerbürgerstadt und deswegen oftmals von deren Bauformen gekennzeichnet. Typische Elemente sind die großen Zufahrtstore in die durch eine Ummauerung von der Hauptstraße abgetrennten Wirtschaftsgebäude mit den zugehörigen Innenhöfen. Auch diese Eigenheiten verbinden sie mehr mit der fränkischen als der bayerischen Stadt und können somit gewiss nicht als oberpfälzisches Spezifikum bezeichnet werden.

4. Die Städte und Märkte in der Landschaft

Nach der Errichtung des Fürstentums der Oberen Pfalz 1329 bildeten sich hier eigene Landstände. Ihnen hat Klaus Köhle eine erste Darstellung gewidmet.[160] Er stellt vor allem einen Punkt heraus: Die oberpfälzische Landschaft orientierte sich in ihrer Entwicklung in den entscheidenden Punkten mit zeitlicher Verzögerung am Vorbild des Herzogtums Bayern. Sie nahm mehrfach auf das dortige Grunddokument der „Ottonischen Handfeste“ von 1311 Bezug und organisierte sich ebenfalls nach dem Dreikurienschema in die Korporationen des Adels, der Prälaten sowie der Städte und Märkte. Auch hier gelang es also den Städten und Märkten, ihre Vertretung in einer eigenen Korporation auf den Landtagen durchzusetzen. Ansätze zum Zusammenschluss in Form des späteren Städtekorpus sind erstmals zum Jahre 1454 in den Quellen angesichts des Amberger Aufruhrs von 1453 fassbar.[161]

Der Vorgang wiederholte sich nach der Einrichtung des Fürstentums Pfalz-Neuburg im Jahre 1505. Die pfalz-neuburgische Landschaft hat Michael Cramer-Fürtig in einer profunden Untersuchung bearbeitet.[162] Sie nahm ihre Tätigkeit noch im Jahre 1505 auf und wies von Anfang an eine Städte- und Märktekurie

159 Hans Scherzer (Hg.), Gau Bayerische Ostmark. Land, Volk und Geschichte, München 1940, S. 247 f. Ein schönes Beispiel ist Neumarkt in der Oberpfalz (Abb. 24).

160 Klaus Köhle, Landesherr und Landstände in der Oberpfalz von 1400–1583. Sozialstruktur und politische Repräsentanz eines frühneuzeitlichen Territoriums (Miscellanea Bavarica Monacensia 16), München 1969.

161 Köhle, Landstände, S. 22–27, 51–53; auch: Scherl, Nabburg (wie Anm. 18), S. 166–169; Rubenbauer, Stadt Amberg (wie Anm. 31), S. 31 f.

162 Michael Cramer-Fürtig, Landesherr und Landstände im Fürstentum Pfalz-Neuburg. Staatsbildung und Ständeorganisation in der ersten Hälfte des 16. Jahrhunderts (Schriftenreihe zur bayerischen Landesgeschichte 100), München 1995.

auf. Dieser gehörten mit einer eigenen Organisation „Nordgau" die Städte Burglengenfeld, Sulzbach, Hemau, Schwandorf, Erbendorf, Vohenstrauß sowie die Märkte Floß, Kohlberg, Kaltenbrunn, Allersberg, Kallmünz, Schmidmühlen, Regenstauf und Laaber an.[163]

Für viele Städte wurde die regelmäßige Teilnahme an den Aktivitäten der Landschaft bald eine Belastung, die sie auf Dauer nicht tragen konnten und nicht tragen wollten. Deswegen zogen sie sich rasch aus dem Städtekorpus zurück und überließen die Vertretung immer mehr den großen Kommunen. Der Vorgang ist vor allem im Fürstentum der Oberen Pfalz gut zu verfolgen. Hier kristallisierte sich ab 1526 eine kleine Gruppe von acht Städten heraus, welche die Vertretung des Bürgertums in der Landschaft zu ihrer Aufgabe machten. Die Entwicklung weist auf die Ausschüsse der Folgezeit voraus. Diese Vororte wurden als „Gezirkstädte" bezeichnet. Zu ihnen gehörten folgende Orte: Amberg, Neumarkt, Nabburg, Auerbach, Kemnath, Weiden, Cham und Neunburg vorm Wald. Diese acht Gezirkstädte nahmen zusammen mit dem zur gleichen Zeit gebildeten Ausschuss der Märkte (Pressath und Bruck) die Vertretung auf dem Landtag sowie die Abwicklung der sonstigen anfallenden Geschäfte, vor allem die Einziehung der Steuern, wahr.

Die städtischen Selbstverwaltungsrechte wurden mit dem Eintritt in die Neuzeit für die Territorialfürsten immer mehr ein Stein des Anstoßes. Auf breiter Front begannen sie, im Rahmen ihrer ständefeindlichen Politik, sie zurückzudrängen. Die Städtepolitik des sich auf den Weg zum Frühabsolutismus begebenden Territorialstaats ist von zunehmender Repression und Restriktion gekennzeichnet. Das gilt auch für die Oberpfalz. Hier spielt – anders als in Franken – der Bauernkrieg keine diesbezügliche Rolle; es ist kein Beispiel bekannt, dass in diesem Rahmen eine Kommune ihren Stadtstatus eingebüßt hätte.

Dagegen wurde die Reformation sehr wohl in dieser Hinsicht ausgenutzt. Auch die kurpfälzischen Landesherren benutzten die Glaubensfrage, um die bürgerliche Autonomie einzuschränken. Dafür boten gerade die mehrfachen Konfessionswechsel in der Oberpfalz vielfältige Ansatzpunkte. Den entscheidenden Vorstoß trug der pfälzische Kurfürst im Jahre 1592 im Rahmen der weiteren Durchsetzung des Calvinismus vor.[164] Diesen erneuten Konfessionswechsel

---

163 Ebd., S. 350–353. Zum bayerischen Rahmen vgl.: Heinz Lieberich, Der Bürgerstand in der baierischen Landschaft, in: Mitteilungen für die Archivpflege in Oberbayern 24, 1945, S. 633–665.

164 Johann Baptist Götz, Die religiösen Wirren in der Oberpfalz von 1576 bis 1620 (Reformationsgeschichtliche Studien und Texte 66), Münster 1937.

wollten mehrere oberpfälzische Städte nicht mitmachen; sie verharrten beim Luthertum. Als der Landesherr seiner Forderung Nachdruck zu verleihen suchte, kam es an mehreren Orten zu Widerstand und Auseinandersetzungen mit den Vertretern der Landesherrschaft. In Nabburg und Tirschenreuth waren in diesem Zusammenhang sogar Tote zu beklagen. In Nabburg wurde der calvinische Prediger Sebastian Breitschedl tot geprügelt.[165] Auch in Tirschenreuth wurde der kurpfälzische Stiftshauptmann Valentin Winsheim erschlagen.[166] Diese Ausschreitungen nahmen die Kurfürsten zum Anlass, um in Breite gegen die bürgerliche Autonomie vorzugehen. Sie erlegten den betroffenen Städten nicht nur harte Geldstrafen auf. In Nabburg, Tirschenreuth, Amberg und Neumarkt wurden auch die städtischen Freiheiten aufgehoben und das Ratsregiment der Bürger auf Jahrzehnte suspendiert.[167] Vor allem wurden die Wahlrechte beseitigt und Amtshandlungen so strenger landesherrlicher Aufsicht unterstellt, dass in Vohenstrauß und Erbendorf schließlich nicht mehr klar war, ob die Orte nun Stadt- oder Marktrecht hatten.[168] In Erbendorf gebrauchte die Kommune zwar weiterhin das Stadtattribut, das ihr von dem Regierungsbehörden vorenthalten wurde, sodass Eigen- und Fremdbezeichnung auch hier nicht mehr übereinstimmten.[169] Auch in den Städten der Oberpfalz wurde also die Glaubensfrage benutzt, um die landesherrliche Position auf Kosten der Kommunen auszubauen. Die pfälzischen Kurfürsten haben ihre gezielte Städtepolitik in die Oberpfalz getragen und im Rahmen der Reformation die Selbstverwaltung der Städte und Märkte faktisch aufgehoben.

Mehrere Orte haben durch diese Eingriffe ihren Stadtstatus auf lange Zeit verloren, zum Beispiel Rötz[170] oder Eschenbach.[171] Auch Vohenstrauß erscheint nach der letzten Nennung 1449 als Stadt in der Folgezeit bis ins 19. Jahrhundert

---

165 Karlheinz Kirch, Selbstverwaltung der Stadt Nabburg, in: Oberpfälzer Heimat 21, 1977, S. 105–114, bes.: S. 112 f.; Müller-Luckner, Nabburg (wie Anm. 18), S. 102–104.

166 Johann Brunner, Die Ermordung des kurfürstlichen Stiftshauptmanns Valentin Winsheim zu Tirschenreuth am 24. Febr. 1592, in: Verhandlungen des Historischen Vereins für Oberpfalz und Regensburg 80, 1930, S. 173–199; Ackermann, Die oberpfälzischen Städte (wie Anm. 10), S. 448 f.

167 Scherl, Nabburg (wie Anm. 18), S. 230–234.

168 Bernd, Vohenstrauß (wie Anm. 27), S. 141; Sturm, Neustadt an der Waldnaab – Weiden (wie Anm. 23), S. 58; Höser, Erbendorf (wie Anm. 25), S. 69.

169 Ackermann, Die oberpfälzischen Städte (wie Anm. 10), S. 449–453.

170 Rötz. Unsere Heimat in Geschichte und Gegenwart, hg. zur 475 Jahr-Feier der Stadt Rötz, Rötz 1981.

171 Keyser / Stoob, Bayerisches Städtebuch II (wie Anm. 1), S. 181–185. Vgl. Anm. 149.

hinein nur mehr als Markt;[172] ähnliches gilt für Luhe.[173] Das waren die gravierendsten Folgen der repressiven Städtepolitik der frühneuzeitlichen Territorien. In jedem Fall aber dauerte die Kassierung der kommunalen Selbstverwaltungsrechte bis zum Ausgang der pfälzischen Zeit in der Oberpfalz an.[174] Sie wurde nicht mehr rückgängig gemacht.

Als nun die Obere Pfalz ab 1621 an Kurbayern zurückgeführt wurde, war die Frage für die neuen Landesherren, wie sie sich in dieser heiklen Angelegenheit verhalten sollten.[175] Doch hatten sie in ihrem Herzogtum bereits bisher eine durchaus parallele Politik betrieben. Sie hatten hier ebenfalls ein strenges Aufsichtsrecht über die Städte und Märkte durchgesetzt. Dieses bekam 1618 vor allem Riedenburg zu spüren, als nach einer Kontrolle alle kommunalen Funktionsträger wegen Verfehlungen streng bestraft wurden. Riedenburg verlor in dieser Epoche seinen früheren Stadtrang auf Jahrhunderte und wurde scharfer landesherrlicher Aufsicht unterstellt.[176] Die rigorose Städtepolitik der Wittelsbacher ist im Rahmen ihres besonders entschlossenen Kampfes gegen die Landstände zu sehen. Die Münchner Zentrale legte auch für die oberpfälzischen Städte umfassende und detaillierte Sammlungen der Stadtrechte an und unterwarf sie einer eingehenden Prüfung. Die Entscheidung über die Anerkennung der einzelnen Privilegien ging immer von der Voraussetzung aus, dass es allein in der Kompetenz des Kurfürsten „*liege, den Stätten und Märckhten die privilegien zu ertheiln*" oder nicht.[177] Denn sie wurden betont als Äußerungen der Gnade des Landesherrn hingestellt. Die landesherrlichen Behörden haben auch die bürgerlichen Vorrechte allenthalben mit Entschiedenheit bekämpft. Die Privilegienbestätigungen der Städte ließen zum Teil lange auf sich warten.

---

172 Bernd, Vohenstrauß (wie Anm. 27), S. 141 f.

173 Ebd., S. 186–188.

174 Christian Reinhardt, Fürstliche Autorität versus städtische Autonomie. Die Pfalzgrafen bei Rhein und ihre Städte 1449 bis 1618: Amberg, Mosbach, Nabburg und Neustadt an der Haardt, Stuttgart 2012.

175 Zur Politik gegenüber den oberpfälzischen Ständen: Sigmund von Riezler, Geschichte Baierns, Bd. VI (Geschichte der europäischen Staaten 20/6), Gotha 1903, S. 34 f.; Karl-Otto Ambronn / Achim Fuchs (Bearb.), Die Oberpfalz wird bayerisch. Die Jahre 1621 bis 1628 in Amberg und der Oberpfalz (Ausstellungskataloge der Staatlichen Archive Bayerns 10), Amberg 1978.

176 Reinhard Heydenreuter, Der landesherrliche Hofrat unter Herzog und Kurfürst Maximilian I. von Bayern (1598–1651) (Schriftenreihe zur bayerischen Landesgeschichte 72), München 1981, S. 257–260.

177 BayHStA, Oberpfalz, Lit. 244: „*Privilegien und Freyheiten der Städte und Märkte in der Obern Pfalz*". Das Zitat: „*Allerunterthenigstes Anfiegen*" vom 3.10.1652.

Maximilian I. hat sie unter Berufung auf das Kriegsgeschehen erst nach vieljährigem Abwarten ausgesprochen. Verschiedentlich erfolgte sie aber auch erst unter seinem Sohn und Nachfolger Ferdinand Maria.[178] Zu einer Wiederaufnahme der landständischen Politik ist es in der Oberpfalz mit Sicherheit nicht gekommen, weil der Landtag noch mehr als in Altbayern in seinen Kompetenzen beschnitten wurde. Zu einem spürbaren Faktor des politischen Entscheidungsprozesses sind die Städte und Märkte hier nicht mehr geworden. Die Kommunalautonomie wurde untergraben, wo immer es nur ging. Das Kurfürstentum hat die Städte und Märkte in die Rentmeisterumritte einbezogen und so seine Aufsicht deutlich ausgedehnt.[179]

Aus diesem Grund hatte der Übergang an Bayern für die oberpfälzischen Städte insgesamt eine deutliche Minderung ihres Rechtsstatus zur Folge. Entscheidend war die Neudefinition ihrer Rechtsbasis. Auch hier wurde bestritten, dass die Städte über alte oder gar autogene Rechte verfügten, indem alle Privilegien auf Verleihung durch den Landesherrn zurückgeführt wurden. Zugleich wurde deutlich gemacht, dass diese Verleihungen jederzeit widerrufen werden könnten. Der Vorgang entspricht der Neubestimmung der Rechtslage der Klöster[180] und steht in Zusammenhang mit der Ausrichtung aller Gewalt im Staat auf den absolutistischen Landesherrn.

An diesen Verhältnissen sollte sich in der bayerischen Folgezeit bis zum Ende des Kurfürstentums nichts mehr ändern. Je mehr sich der Münchner Zentralismus durchsetzte, umso mehr gerieten die Städte in dieser der aufsteigenden kurbayerischen Haupt- und Residenzstadt München besonders fernen Randprovinz ins Abseits. Für deren wirtschaftliche Erneuerung haben die Kurfürsten kaum etwas getan. Die Oberpfalz erholte sich von ihrer wirtschaftlichen Rezession seit dem beginnenden 16. Jahrhundert nie mehr. Hauptleidtragende dieses wirtschaftlichen Niedergangs waren die Städte, die nun oftmals zu Landstädten verkümmerten. Vom Sonderfall Regensburg abgesehen, konnten sich allein

178 Zum Beispiel Auerbach: BayHStA, MInn 30, Nr. 806. Vgl. Schnelbögl, Auerbach (wie Anm. 24), S. 74.

179 Druck der Rentmeisterinstruktion von 1669: [Franz Xaver Alois Wiguläus Freiherr von Kreittmayr,] Sammlung der neuest- und merkwürdigsten churbaierischen Generalien und Landesverordnungen, München 1771, S. 547–557.

180 Alois Schmid, Die Klöster der Oberpfalz von der Wiedererrichtung bis zur Säkularisation 1802/03, in: Bernhard Löffler / Karsten Ruppert (Hg.), Religiöse Prägung und politische Ordnung in der Neuzeit. Festschrift für Winfried Becker zum 65. Geburtstag, Köln u.a. 2006, S. 113–135.

Amberg[181] und Sulzbach[182] sowie Neustadt an der Waldnaab[183] und Pfreimd[184] als kleine Residenzstädte einen zumindest bescheidenen Anteil am Glanz des höfischen Zeitalters verschaffen. Im Übrigen ist die Stadtentwicklung des späteren 17. und des 18. Jahrhunderts von breiter Stagnation und Niedergang gekennzeichnet.

Diese Zustände dauerten auch nach dem Eintritt in die Neueste Zeit an. Die Oberpfalz nahm an den entscheidenden Vorgängen der sozialen und wirtschaftlichen Entwicklung, der Industrialisierung und Urbanisierung, nur verhältnismäßig begrenzten Anteil.[185] Hier kam es nicht zu dem breiten Aufbruch des Städtewesens im 19. Jahrhundert, der für die anderen Landesteile kennzeichnend ist. Eher ist Stagnation zu konstatieren, die die Abwanderung beträchtlicher Bevölkerungsüberschüsse in die aufsteigenden Ballungsräume zur Folge hatte. Und manche oberpfälzische Stadt hat an diesem Aufschwung der urbanen Welt im 19. Jahrhundert überhaupt keinen Anteil genommen, sondern ist weithin auf dem um 1800 erreichten Zustand stehen geblieben. Die Oberpfälzer Städte sind so ein sehr aussagekräftiges Beispiel für den von Heinz Stoob[186] als „Minderstädte" und „Kümmerstädte" bezeichneten Typus von Städten geworden. Die Kulturgeographie hat dafür den Begriff der oberpfälzischen „Zwergstädte" geprägt; Schönsee, Rötz, Pleystein oder Bärnau sind bezeichnende Beispiele für diesen Stadttypus.[187] Gerade in diesen oberpfälzischen Zwergstädten wird noch heute in Deutlichkeit das Phänomen der bayerischen Landstadt der vorindustriellen Zeit greifbar.

---

181 Wilhelm Volkert, Amberg und die Kurfürsten von der Pfalz, in: Amberg 1034–1984 (wie Anm. 19), S. 61–74.

182 Geschichte der Stadt Sulzbach-Rosenberg (wie Anm. 34).

183 Heinrich Ascherl, Neustadts Schlösser und ihr verborgener Kunstschatz, in: Oberpfälzer Heimat 1, 1956, S. 46–51; Ders., Gemälde im Neustädter Schloß freigelegt, in: Oberpfälzer Heimat 9, 1964, S. 73–75; Wilhelm Volkert, Neustadt an der Waldnaab und die Fürsten Lobkowitz, in: Verhandlungen des Historischen Vereins für Oberpfalz und Regensburg 100, 1959, S. 175–194; Sturm, Neustadt an der Waldnaab – Weiden (wie Anm. 23), S. 367–378.

184 Seitz, Residenzstadt Pfreimd (wie Anm. 88).

185 Wilhelm Kaltenstadler, Bevölkerung und Gesellschaft Ostbayerns im Zeitalter der frühen Industrialisierung (1780–1820), Kallmünz 1977.

186 Stoob, Minderstädte (wie Anm. 57).

187 Einprägsame Beispiele in: Topographischer Atlas von Bayern, hg. vom Bayerischen Landesvermessungsamt, München 1968, Karten 72–79; Hans Fehn (Hg.), Luftbildatlas Bayern. Eine Landeskunde in 72 farbigen Luftaufnahmen, München 1973, S. 51–67, Nr. 16–23.

Thomas J. Hagen

Zwischen Konkurrenz und Kooperation: Bambergische Landstädte im Ringen um politische Geltung und wirtschaftliche Prosperität. Aufgezeigt am Beispiel Ebermannstadts und Waischenfelds in der Fränkischen Schweiz

Abstract: In contrast to the large and free imperial cities, the almost innumerable smaller country cities, which have neither made it to government seats nor to imperial freedom, often appear as insignificant or even powerless members of the political and economic structures of the Holy Roman Empire of the German Nation. Based on the two episcopal Bamberg country towns Ebermannstadt and Waischenfeld, this article shows that country towns, as a kind of intermediate stage between agriculturally structured villages and large economic and trading centers, could actually acquire important administrative and political relevance.

1. Einleitung

Im Schatten großer Residenz- und Reichsstädte erscheinen die schier unzähligen kleineren Landstädte, die es weder zu Regierungssitzen noch zu reichsunmittelbarer Freiheit geschafft haben, oftmals als unbedeutende oder gar ohnmächtige Glieder der politischen und wirtschaftlichen Strukturen des Alten Reiches. Dass diese – als eine Art Zwischenstufe zwischen agrarisch strukturierten Dörfern und großen Wirtschafts- und Handelszentren – aber eine durchaus wichtige verwaltungs- und herrschaftspolitische Relevanz erlangen konnten, soll hier am Beispiel zweier ehemals bischöflich-bambergischer Städte aufgezeigt werden: Ebermannstadt und Waischenfeld.

Was diese beiden Orte auszeichnet und gleichzeitig verbindet, ist deren Lage in der Mittelgebirgsregion „Fränkische Schweiz" und deren Entstehungsgeschichte. Letztere reicht noch in vorbambergische Zeit zurück – einer Epoche, in der die Region während einer allgemeinen Phase der Herrschaftsbildung von machtpolitischen Auseinandersetzungen zwischen kleineren Territorialherren und größeren Reichsfürsten geprägt war. Als militärisches Zünglein an der Waage war es den aus Adelsdorf stammenden Reichsherren von Schlüsselberg im 13. Jahrhundert gelungen, auf der Fränkischen Alb eine Art Pufferstaat zwischen den Territorien des Fürstbischofs von Bamberg und des Burggrafen von

Nürnberg zu schaffen. Allerdings mangelte es in diesem, vom Flüsschen Wiesent durchzogenen Gebiet noch an einem städtischen Zentrum, auf das sich die Schlüsselberger Macht hätte wirtschaftlich stützen können. Schließlich gelang es Konrad II. von Schlüsselberg im Zuge seiner Militärhilfe für Ludwig den Bayern Anfang des 14. Jahrhunderts, die erwünschten königlichen Stadtprivilegien zu erhalten. Nicht zuletzt, um die Dominanz der im Norden, Osten und Westen gelegenen altbambergischen Marktzentren Hollfeld, Pottenstein, Eggolsheim und Forchheim zu brechen, schienen die verkehrsgünstig an den Handelswegen zwischen Nürnberg und Kulmbach gelegenen Orte Ebermannstadt und Waischenfeld als neue städtische Mittelpunkte besonders geeignet.[1]

2. Die Stadt Waischenfeld

2.1. Die Entstehung der Stadt Waischenfeld

Erstmals erwähnt wird das inmitten des Karstgebirges der Frankenalb gelegene Waischenfeld 1122 als befestigter Ansitz eines „*Wirint de Wischenvelt*".[2] In den Besitz der Herren von Schlüsselberg gelangte diese Burg samt dem dortigen Centgericht (Hochgericht) - vermutlich im direkten Erbgang - kurz nach 1216. Es war wohl der um 1277 geborene Konrad II. von Schlüsselberg selbst, der die zu Füßen dieser Burg liegende Talsiedlung gleichen Namens gründete. Aus der Burgkapelle entstand

---

1 Vgl. hierzu und zum Folgenden: Gustav Voit, Die Schlüsselberger. Geschichte eines fränkischen Adelsgeschlechtes (Schriftenreihe der Altnürnberger Landschaft 37), Nürnberg 1988, S. 72–130; Paul Österreicher, Der Reichsherr Gottfried von Schlüsselberg. Ein geschichtlicher Abriß, Bamberg 1821, S. 6; Paul Österreicher, Neue Beiträge zur Geschichte der ehemaligen Reichsherrschaft Schlüsselberg, Bamberg 1823; Rudolf Endres, Konrad von Schlüsselberg, in: Gerhard Pfeiffer (Hg.), Fränkische Lebensbilder. Neue Folge der Lebensläufe aus Franken, Bd. 4 (Veröffentlichungen der Gesellschaft für fränkische Geschichte [künftig GffG] VII A/4), Würzburg 1971, S. 27–48; vgl. zum Thema „Städtegründung" allgemein: Wolfram Unger, Grundzüge der Städtebildung in Franken. Träger, Phasen, Räume, in: Jahrbuch für fränkische Landesforschung (künftig: JfL) 59, 1999, S. 57–85, hier insbes.: S. 70–73; vgl. zu den Märkten Hollfelds, Pottensteins, Eggolsheims und Forchheims die Literaturangaben unter Anm. 140; vgl. zu den besonderen naturräumlichen Bedingungen, Geologie und Klima der „Fränkischen Schweiz" u.a.: Philipp Hümmer / Cyrus Samimi, Wasser auf der nördlichen Frankenalb - ein kostbares Gut, in: Mitteilungen der Fränkischen Geographischen Gesellschaft 49, 2002, S. 111–146.

2 Gemäß anderen Überlieferungen („Miracula" von ca. 1200) dürfte das „*castello Wiskenvelt*" bereits vor 1033 bestanden haben. Vgl. Dorothea Fastnacht, Ebermannstadt. Ehemaliger Landkreis Ebermannstadt (Historisches Ortsnamenbuch von Bayern. Oberfranken 4), München 2000, S. 324. Vgl. zur geologischen Besonderheit des Karstgebirges: Hümmer / Samimi, Wasser auf der nördlichen Frankenalb (wie Anm. 1), S. 111–125.

später die Stadtpfarrkirche. In zwei Privilegien von 1315 und 1316 erhielt Waischenfeld von König Ludwig dem Bayern mit der Freiung und dem Stadtrecht auch die Befugnis, Juden *„zu behausen"* (letzteres Schutzrecht wurde 1343 wieder entzogen). Im Privileg von 1316 verbot der König jedweden Handel innerhalb einer halben Meile um die Stadt und sicherte damit dem Ort das Handelsmonopol einschließlich des Marktrechts. 1322 wurde schließlich königlicherseits festgelegt, dass Konrad II. bei Übertretung dieses Gebots 10 Pfund Haller als *„Buße"* pfänden durfte.[3]

2.2. Das Bamberger Amt Waischenfeld

Nach dem Tode Konrads II. kam Waischenfeld 1348 durch Kauf an das Hochstift Bamberg (bestätigt 1376). Waischenfeld wurde bischöfliches (Cent- und Vogtei-)Amt und war nachweislich seit 1400 Sitz eines bedeutenden hochstiftischen Kasten-, später auch Steuer- und Forstamts. Diese Ämter umfassten ungewöhnlich weite Gebiete. So zogen sich die Grenzen des Waischenfelder Cent- beziehungsweise Hochgerichtsbezirks von Plankenfels bis Behringersmühle und von Draisendorf bis Hintergereuth. Der Grundverwaltung des Kastenamts unterstanden erbzinspflichtige Lehenleute nicht nur im Waischenfelder Amtsbezirk und bis 1780 in drei weiteren, diesem zugeordneten Vogteiamtssprengeln (Gößweinstein, Pottenstein und Leienfels), sondern auch zahlreiche Lehenstücke in den daran angrenzenden bambergischen, markgräflichen und nürnbergischen Ämtern. Zwischen 1438 und 1562 war die Stadt zusammen mit dem bischöflichen Schloss und Amt immer wieder an verschiedene Adelige verpfändet.[4] Die zeitweise Verpfändung von Ämtern (mit den zugehörigen Burgen, teils auch Städten) an Niederadelige, auch die Veräußerung bestimmter Herrschaftsrechte „auf Wiederkauf", hatte für das bischöfliche Ärar den Vorteil, dass anstatt des aufwändigen und wenig lukrativen Einzugs der Naturalabgaben

3 Vgl. Voit, Die Schlüsselberger (wie Anm. 1), S. 77, 128; Hellmut Kunstmann, Die Burgen der östlichen Fränkischen Schweiz (Veröffentlichungen der GffG IX/20), Würzburg 1965, S. 92; Reinhard Löwisch, Die drei Waischenfelder Stadtprivilegien und ihre Auswirkungen, in: Kurt Neuner (Hg.), Stadt Waischenfeld. 700 Jahre Stadtrecht 1315–2015. Jubiläumsschrift, Waischenfeld 2015, S. 20–29, hier: S. 21, 25–29. Stimmt die Annahme J. Andraschkes, dass die Ansiedlung von Juden und der Sitz eines Centgerichts auf das Vorhandensein eines Marktes hindeuten, könnte sich in Waischenfeld bereits vor der Stadtgründung ein Markt befunden haben. Vgl. Joachim Andraschke, Geschichte des Marktes zu Eggolsheim. Entstehung und Entwicklung einer fränkischen Marktgemeinde, Bamberg 2011, S. 34 f., 69–73.

4 Vgl. Fastnacht, Ebermannstadt (wie Anm. 2), S. 326 f.; Kunstmann, Die Burgen der östlichen Fränkischen Schweiz (wie Anm. 3), S. 93–97, 128–130; Johann Caspar Bundschuh, Geographisches Statistisch-Topographisches Lexikon von Franken, Bd. 6, Ulm 1804, Sp. 134–136.

erhebliche Bargeldsummen, die der Pfandnehmer zu zahlen hatte, eingenommen werden konnten. Das Besatzungsrecht („Öffnungsrecht") an der Burg im Kriegsfalle und bestimmte herrschaftliche Rechte (wie zum Beispiel Bargeldsteuern) behielt sich der Bischof freilich vor. Erst im Laufe des 16. Jahrhunderts verschwand die Praxis der Ämterverpfändung, so auch in Waischenfeld, wodurch wieder reguläre bischöfliche Amtmänner die Verwaltung übernahmen.[5] Ob nun verpfändet oder unter direkter hochstiftischer Verwaltung – die am Ort agierenden, einen unverhältnismäßig großen Bezirk umfassenden bischöflichen Ämter trugen wesentlich zur Entwicklung und zum Ansehen Waischenfelds als regionales Zentrum bei.

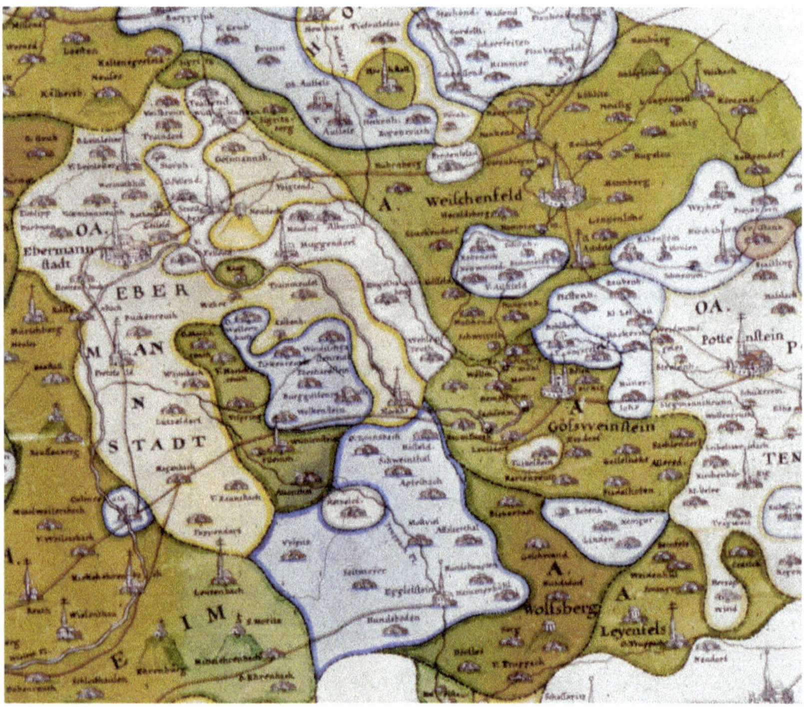

Abb. 25: *Vogteiamtsbezirke „im Gebirg" (Fränkische Schweiz), Ausschnitt aus der Karte „Geographische Vorstellung des Kaiserlichen Hochstifts und Fürstenthums Bamberg", Handzeichnung von Johann Baptist Roppelt, 1796. Bildnachweis: StABa, A 240, Nr. T584.*

5 Vgl. allgemein: Johannes Köhler, Territorialisierung im Hochstift Würzburg. Etappen zur frühmodernen Staatlichkeit, in: Mitteilungen des Verbandes bayerischer Geschichtsvereine 28, 2018, S. 77–84, hier: S. 80 f.

2.3. Beschreibung und Entwicklung der Stadt Waischenfeld

Waischenfeld besaß zwei durch die Wiesent geteilte Stadtteile („Stadt" und „Vorstadt"), die durch eine Brücke verbunden waren. 1514 wurde am Ort infolge einer Stiftung des Domherrn Eberhard von Rabenstein ein Spital erbaut. Eine Schule ist dort bereits 1422 erwähnt.[6] Waischenfeld verfügte seit dem Mittelalter über eine mehr als 1.700 Meter lange Stadtmauer mit vier Stadttoren (Afterntor, Brucktor, unteres Tor, Bamberger Tor) und erfüllte damit allgemeine Ansprüche einer „echten" Stadt. Dennoch musste der unterhalb der bischöflichen Burg gelegene Ort im Laufe der Zeit schwere Zerstörungen hinnehmen, so während des Hussitenkrieges 1430, im Zweiten Markgräflerkrieg 1553 und im sogenannten Schwedenkrieg 1632.[7] 1553 wurde Waischenfeld, so ein zeitgenössischer Bericht, „*me*[h]*renteilß*" in Asche gelegt, nämlich „*über 80 Heuser, auch das Rathauß, sampt den Preuheusern* [Brauhäusern], *Thor*[e]*n, Thürm*[en] *und Wehren*".[8] Um 1555 lebten in der „Stadt" noch 65 und in der „Vorstadt" 46 bischöfliche Untertanen. 1632 wurde nahezu die gesamte Stadt (außer Kirchen und Schloss) mit 149 Häusern und 78 Stadeln eingeäschert.[9] Nach dem Dreißigjährigen Krieg wird Waischenfeld beschrieben als eine Stadt „*mit einem fürstlich bambergischen Amtsschloß, einer Vogtei, dem als Kastnerhaus benutzten Mönchshof, einem Forstmeister, Bürgermeister und Rat, Spital, Gotteshaus St. Johannes Baptista, öder Lorenzikapelle, öder Siechhaushofstatt, Almosenamt, Zoll und Bürgermeisterrechnung*".[10] Wie hieraus zu entnehmen ist, hatte der Ort trotz seiner relativen ländlichen Abgeschiedenheit auch über ein funktionierendes Armen- und Gesundheitswesen verfügt.

Nach einer Phase der Erholung finden sich in Waischenfeld im 18. Jahrhundert neben dem bereits erwähnten Amtsschloss mit herrschaftlichem

---

6 Vgl. Fastnacht, Ebermannstadt (wie Anm. 2), S. 326 f.

7 Vgl. Kunstmann, Die Burgen der östlichen Fränkischen Schweiz (wie Anm. 3), S. 97; vgl. zur mittelalterlichen Stadtmauer und den vier Stadttoren: Reinhard Löwisch, Geschichtliches und Sagen um Denkmäler der Gemeinde Waischenfeld (Arbeitskreis Heimatkunde im Fränkische-Schweiz-Verein), Erlangen 1990, Kap. 21 (mit umfangreicher Literaturangabe).

8 Eintrag einer Klarissennonne zu den Schäden im Markgrafenkrieg (7.7.1553), zitiert nach: Löwisch, Die drei Waischenfelder Stadtprivilegien (wie Anm. 3), S. 23.

9 Vgl. Löwisch, Die drei Waischenfelder Stadtprivilegien (wie Anm. 3), S. 23; Kunstmann, Die Burgen der östlichen Fränkischen Schweiz (wie Anm. 3), S. 97; Fastnacht, Ebermannstadt (wie Anm. 2), S. 326 f.

10 Beschreibung aus dem Jahre 1692, zitiert nach: Fastnacht, Ebermannstadt (wie Anm. 2), S. 328.

Getreideboden, dem Kastenhof mit Nebengebäuden, der Pfarrkirche zum Hl. Johannes und dem St.-Elisabethen-Spital, auch ein Oberamtshaus mit Ökonomiegebäuden, ein Forsthaus mit Nebengebäuden, eine Kapelle „zur Hl. Anna", die wieder aufgebaute Stadtkapelle St. Michael, ein Pfarrhaus, ein Schulhaus, ein Rathaus, ein Badhaus, ein Hirtenhaus, ein Gemeindebrauhaus, ein Wirtshaus und 109 andere Häuser mit 17 außerhalb der eigentlichen Stadt gelegenen Stadeln und 16 Hofstätten. Hinzu kamen eine „alte" Pfarrhaushofstatt mit einem kleinen Haus, eine (Stadt-)Mühle mit vier Mahlgängen und einem Schlaggang sowie zwei Hammermühlen (Obere und Untere).[11]

In Waischenfeld waren die meisten Häuser freieigener Besitz der Stadtbürger, nur zwei „Häuslein", das Gemeindebrauhaus und das Hirtenhäuslein, waren Lehen des städtischen Rates, dem auch die Stadtkapelle „St. Laurentius und Michael" gehörte; über geringen Grundbesitz verfügten daneben das Waischenfelder Spital (ein großes Haus) und das bischöfliche Kastenamt (die Amtsgebäude, sechs Häuser und das örtliche Badhaus).[12]

Waischenfeld verfügte über eine offensichtlich gut funktionierende städtische Selbstverwaltung, wofür aus den Jahren 1500 und 1610 Gemeindestatuten belegt sind. Die Stadt besaß einen Bürgermeister, einen Stadtschreiber und zehn „Rathsverwandte" (Ratsherren), die unter dem Vorsitz des (Ober-)Amtmanns oder, im Falle von dessen Abwesenheit, des Amtsvogts tagten. Innerhalb des Stadtrayons („*in dem Städtchen und soweit außerhalb gewisse Markzeichen oder sogenannte Schranken gehen*") verfügten Bürgermeister und Stadtrat über die Vogteilichkeit (niedere Gerichtsbarkeit).[13]

---

11 Vgl. Johann Baptist Roppelt, Historisch-topographische Beschreibung des Kaiserlichen Hochstifts und Fürstenthums Bamberg, Zweyte Abtheilung: südlicher Theil, Nürnberg 1801, S. 460–463; vgl. auch: Löwisch, Die drei Waischenfelder Stadtprivilegien (wie Anm. 3), S. 24.

12 Vgl. Roppelt, Historisch-topographische Beschreibung (wie Anm. 11), S. 460–463; Fastnacht, Ebermannstadt (wie Anm. 2), S. 105, 328 f.; Staatsarchiv Bamberg (künftig: StABa), Hochstift Bamberg, Vogt-, Kasten- und Steueramt Waischenfeld, Nr. 8.

13 Zitiert nach: Fastnacht, Ebermannstadt (wie Anm. 2), S. 328; vgl. auch: Löwisch, Die drei Waischenfelder Stadtprivilegien (wie Anm. 3), S. 23 f.; Roppelt, Historisch-topographische Beschreibung (wie Anm. 11), S. 460–463. Nach J. G. Biedermann waren der Waischenfelder Bürgerschaft hingegen vier Bürgermeister und acht Ratsverwandte vorgesetzt. Vgl. Johann Gottfried Biedermann, Topographische Beschreibung des Fürstenthums Bamberg (1752), Handschrift, aus: Staatsbibliothek Bamberg, J.H. Msc. hist 123, fol. 132, Nr. 702.

Abb. 26: *Waischenfeld um 1802, nach einem Gemälde von Sebastian Förtsch. Bildnachweis: Wikimedia Commons (gemeinfrei).*

Von der Waischenfelder Bürgerschaft ist uns eine ganze Reihe von Namen bereits aus der Zeit um 1400 bekannt, darunter die Reiterer und Brunsteiger (1399), Liphart, Schad, Schatz und Schuwartt (1409/15); in ihrer Eigenschaft als Burghüter, als Pfandinhaber von „Stadt, Schloss und Amt" oder als bischöfliche Amtmänner sind nachgewiesen unter anderem die aus der Region stammenden Adelsfamilien derer von Rüssenbach, Wannbach, Streitberg, Aufseß, Groß, Modschiedler (seit der zweiten Hälfte des 14. Jahrhunderts), Wichsenstein, Plankenfels und Königsfeld (seit Anfang des 15. Jahrhunderts). Ab 1492 saßen die Lochner „zu Waischenfeld", im 16. Jahrhundert tauchen die Stiebar zu Rabenstein, die Lüchau, Groß von Christanz, Seckendorff, Breitenstein, Künsberg und Eyb dort auf.[14]

14 Vgl. Fastnacht, Ebermannstadt (wie Anm. 2), S. 104, 264 f., 325–328. Im Jahre 1547 war Liborius von Wannbach *„Amtmann, Kastner und Bürger zu Waischenfelt"* (ebd., S. 327).

2.4. Die kirchenrechtlichen Verhältnisse Waischenfelds

Die katholische Pfarrei Waischenfeld, deren Kirche ursprünglich zur Urpfarrei Nankendorf gehörte, geht auf die von den Schlüsselbergern erbaute Burgkapelle St. Johannis Baptistae mit einem herrschaftlichen Frühmesser zurück. Frühmesspfründe sind 1384 und 1402 erwähnt. Nach einem Heiltumsverzeichnis aus dem Jahre 1558 war Konrad II. von Schlüsselberg Stifter der ersten („Pfarr-")Kirche. Tatsächlich hatte bereits Gottfried I. von Schlüsselberg 1308 deren Mutterkirche in Nankendorf mit einer Dotation bedacht. Die auf Königsgut entstandene Kirche von Nankendorf zählt mit ihrem Martinspatrozinium zu den frühesten Kirchen der ostfränkischen Kolonisation. Als Urpfarrei umfasste ihr Pfarrsprengel anfangs praktisch die gesamte Cent Waischenfeld. Anfang des 15. Jahrhunderts schien den Waischenfelder Stadtherren die kirchliche Zuordnung zum kleineren, nördlich gelegenen Nankendorf wohl nicht mehr standesgemäß, infolgedessen sich die Stadtkirche von Waischenfeld aus ihrem ursprünglichen Pfarrverband herausschälte. Belegt wird dies durch die 1406 und 1415 erfolgte Nennung eines eigenständigen Waischenfelder Pfarrers, der als „Oberpfarrer" auch die Pfarrei Nankendorf mit versah. Nicht lange nach dieser faktischen Separierung galt Waischenfeld (spätestens seit 1520) als übergeordnete, sogenannte „untere" Pfarrei der nun als eine Art Filialkirche betrachteten „oberen Pfarrei" Nankendorf.[15] Damit war „die Mutterpfarrei Nankendorf in ihrer Tochterpfarrei untergegangen" und das Filialverhältnis hatte sich *de facto* umgekehrt.[16] Zwar behauptete weiterhin, wie schon seit 1301, das Bamberger Domkapitel das Patronatsrecht über die beiden Pfarreisprengel „*in turno*", das Ernennungsrecht der Benefiziaten an der städtischen St. Michaelskapelle behielt sich aber der Stadtrat von Waischenfeld vor.[17]

15 Vgl. Edeltraud Herlitz, Die Pfarrei Waischenfeld in Geschichte und Gegenwart. Eine historische Darstellung mit Beispielen für die unterrichtliche Umsetzung, Zulassungsarbeit, Bayreuth 1991, S. 3–5; Fastnacht, Ebermannstadt (wie Anm. 2), S. 40* (insbes. Anm. 149), 46 f., 187–189 und 325–327; Erich Freiherr von Guttenberg / Alfred Wendehorst, Das Bistum Bamberg 2. Die Pfarreiorganisation (Germania Sacra, Zweite Abteilung, Die Bistümer der Kirchenprovinz Mainz 1, Das Bistum Bamberg), Berlin 1966, S. 14, 140–143; Bernhard Schweßinger, Geschichte der Pfarrei Waischenfeld, in: Stadt und Pfarrei Waischenfeld (Hg.), 500 Jahre Bischof Friedrich Nausea. Aus der Geschichte der Stadt und Pfarrei Waischenfeld, Waischenfeld 1987, S. 65–90, hier: S. 65–77; vgl. ausführlich auch: Kurt Neuner, 250 Jahre Pfarrkirche „St. Martinus" Nankendorf. Festschrift, Nankendorf 1998.

16 Herlitz, Die Pfarrei Waischenfeld in Geschichte und Gegenwart (wie Anm. 15), S. 5 f.

17 Vgl. ebd., S. 7 f.; StABa, Markgraftum Brandenburg-Bayreuth, Geheimes Hausarchiv Plassenburg, Akten und Bände, Nr. 8988, Register des Einkommens der Pfarrei und

2.5. Die wirtschaftliche Entwicklung Waischenfelds

Von Waischenfeld aus führten wichtige Altstraßen in nördlicher Richtung über Nankendorf nach Plankenfels (mit Anschluss an bedeutende Geleit- und Handelsstraßen), im Süden vorbei an der Burg Schlüsselberg und Oberailsfeld nach Tüchersfeld beziehungsweise Pottenstein und im Osten über den Burgstuhl nach Muthmannsreuth, Glashütten, Mistelgau und Donndorf nach Bayreuth.[18] Waischenfeld war Schnitt- und Kontrollpunkt weiterer Verkehrsadern, die aus dem Pegnitz- beziehungsweise Ahorngrund kommend, nach Westen und hier insbesondere zum uralten Königshof Königsfeld führten.[19] Eine nicht unwesentliche Einnahmequelle für die landesfürstlichen und städtischen Finanzen stellten daher Zölle dar. In Waischenfeld waren gleich zwei fürstliche Zolleinnahmen zugegen, die betreffende Abgaben erhoben. Der „Stadtzoll" wurde von Bürgermeister und Rat, der „Amtszoll" hingen vom bischöflichen Kasten- und Vogteiamt eingehoben.[20] Ferner war das in der weiter südöstlich gelegenen Stadt Pottenstein angesiedelte bischöfliche Zollamt dem Waischenfelder Kastner zugeordnet.[21] Weitere Waischenfelder respektive bambergische Zollstätten sind unter anderem am Kaupersberg (bei Nankendorf) belegt, wo bis etwa 1720 eine sogenannte „*Zolltafel*" (Zollsäule) stand.[22] Das Geleit „*gein Werschennfeld*" gehörte

der drei Pfründen Frühmesse, Engelmesse und Spitalmesse zu Waischenfeld, 1552; vgl. auch: Guttenberg / Wendehorst, Das Bistum Bamberg 2 (wie Anm. 15), S. 140–142.

18 Vgl. Hellmut Kunstmann, Die Burgen der westlichen und nördlichen Fränkischen Schweiz, 1. Teil: Der Südwesten. Unteres Wiesenttal und Trubachtal (Veröffentlichungen der GffG IX/28,1), Würzburg 1971, Karte im Anhang.

19 Vgl. Kunstmann, Die Burgen der östlichen Fränkischen Schweiz (wie Anm. 3), S. 13 f. Zu den Altstraßen und deren wirtschaftliche Bedeutung vgl. auch weiter unten, Kap. 3.5.

20 Vgl. StABa, Hochstift Bamberg, Vogt-, Kasten- und Steueramt Waischenfeld Nr. 32, Aufstellung über die Zolleinnahmen der Stadt und des Amts Waischenfeld (Verzeichnisse des Amtszolls 1757–1766 und des Stadtzolls 1767–1772); Fastnacht, Ebermannstadt (wie Anm. 2), S. 328.

21 Verwaltet wurde das dortige Zollamt vom Pfleger bzw. Vogt von Pottenstein, der daraus auch einen Teil seines Salärs bezog. Vgl. Friedrich Wachter, Pottenstein. Geschichte des ehemaligen Pflegamtes und der Pfarrei Pottenstein sowie der Filiale Kirchenbirkig und des Herrschaftssitzes Kühlenfels. Nach urkundlichen Quellen bearbeitet, Bamberg 1895, S. 15–22.

22 Diese wurde offenbar später nach Breitenlesau verlegt. Vgl. StABa, Markgraftum Brandenburg-Bayreuth, Geheime Landesregierung, Nr. 3396, Amtsbeschreibung Kastenamt Streitberg 1752.

hingegen gemäß einer Ämterbeschreibung aus dem Jahre 1568 zu den obrigkeitlichen Rechten des bischöflichen Amtes Forchheim.[23]

Die Stadt besaß ein eigenes Getreidemaß, es wurden dort zunächst aber „*vielerlei Maß gebraucht: Waischenfelder, Hollfelder, Pottensteiner und Ebermannstadter Maß*".[24] Um diesen Verwirrung stiftenden Zustand zu ändern, wurde 1740 ein neues (einheitliches) Getreidemaß für das gesamte Kastenamt Waischenfeld eingeführt.[25] Zu wiegen und messen gab es viel, insbesondere auf den in der Stadt abgehaltenen Märkten. Gemäß einer Beschreibung aus dem Jahre 1752 war Waischenfeld mit sechs Jahrmärkten privilegiert, daneben hielt es Vieh- und Wochenmärkte ab.[26]

Wie wir heute wissen, hat in den Städten die Entwicklung hin zu spezialisierter Produktion schon seit dem Hochmittelalter zum Bau und Betrieb von Mühlen geführt, die in der Regel nicht nur die eigene Bürgerschaft, sondern mehrere Ortschaften versorgten. Hierin machten auch die – für den Betrieb günstig am Wiesentfluss gelegenen – Schlüsselbergischen Stadtneugründungen keine Ausnahme. Schon 1307 erfolgte die erste urkundliche Erwähnung der nahe Waischenfelds gelegenen „Schlüsselmühle". Nach der Stadtgründung tauchen mit der Nennung der „*Götelmanns Mül*" (Göckermühle) zu Ebermannstadt 1371, der Waischenfelder Hammermühle 1399 und der „Stadtmühle" von Waischenfeld 1450 auch innerstädtische Mühlenbetriebe auf.[27]

Tatsächlich entwickelte sich Waischenfeld aufgrund seiner Stellung als Amtsmittelpunkt, seiner Lage am Schnittpunkt wichtiger Altstraßen und mit Hilfe seines Marktrechts einschließlich des privilegierten Handels- und Bierbraumonopols im Laufe der Zeit zu einem florierenden kleinen Handwerker- und

---

23 Vgl. Fastnacht, Ebermannstadt (wie Anm. 2), S. 327. Zum Wesen und zur Bedeutung des Geleitrechts und der Geleitstraßen vgl. weiter unten, Kap. 3.5.

24 Zitiert (gemäß einer obrigkeitlichen Festsetzung des Maßes aus dem Jahre 1577) nach: Johannes E. Bischoff, Die mittelalterlichen Stadtsiegel von Hollfeld (1326–1554), in: Stadt Hollfeld (Hg.), 650 Jahre Stadt Hollfeld 1329–1979, Hollfeld/Bayreuth 1979, S. 11–23, hier: S. 13.

25 Vgl. StABa, Hochstift Bamberg, Vogt-, Kasten- und Steueramt Waischenfeld, Nr. 37, Einführung eines neuen Getreidemaßes im Kastenamt Waischenfeld, 1740.

26 Vgl. Biedermann, Topographische Beschreibung des Fürstenthums Bamberg (wie Anm. 13), fol. 132, Nr. 702.

27 Vgl. zu diesen und anderen Mühlen der Fränkischen Schweiz: Fastnacht, Ebermannstadt (wie Anm. 2), S. 11, 16, 40, 51, 79, 86, 91, 104, 106, 110, 127, 145, 157, 161, 185, 188, 218, 253, 266, 276, 315, 358 und 362; vgl. auch: Johann-Bernhard Haversath, Mühlen in der Fränkischen Schweiz, Erlangen 1987, S. 25–28.

Handelszentrum mit über 60 „*Professionisten*".[28] Die wirtschaftliche Bedeutung von Stadt und Amt Waischenfeld belegen Aufzeichnungen über die jährlichen Einkünfte des bischöflichen Ärars. So nennt ein Urbar aus der zweiten Hälfte des 14. Jahrhunderts an Waischenfelder Einnahmen – ohne die Einkünfte aus Stadt- und Landgericht – fast 228 Sumer Roggen, 45 Sumer Weizen, 105 Sumer Hafer, 1.068 Käse, 267 Hühner und Geldzins, woraus sich ein Gesamtwert von etwa 400 Pfund Hallern errechnete.[29] Noch 1805 berichtet eine Landbeschreibung, dass der „*Hauptnahrungsstand der Einwohner*" Waischenfelds „*Gewerbschaften*" seien, „*indem der Ort über 60 Handthierungen*" zähle.[30] Die meisten Gewerbeleute betrieben allerdings im Nebenerwerb Landwirtschaft, da ihr Handwerksbetrieb nur in den wenigsten Fällen ein ausreichendes Einkommen liefern konnte. So sei auch „*der Feldbau auf dem steinigten Boden nicht zu vergessen*", die „*vielen Hutweiden, deren kurzes Gras sehr nahrhaft und gesund*" sei, ernährten Schafe, Ziegen und anderes Vieh. Übel sei allerdings, dass „*man von Bamberg her des schlechten Weges halber nicht ohne Gefahr in das Städtchen fahren*" könne.[31] Freilich hielten sich die landwirtschaftlichen Erträgnisse aufgrund der ungünstigen geologisch-klimatischen Bedingungen im Vergleich zu anderen Gegenden des Hochstifts in engen Grenzen, weswegen zwangsläufig Handel und Gewerbe die wichtigsten Einnahmequellen der Stadt und ihrer Bürger blieben.[32]

---

28 Vgl. StABa, Hochstift Bamberg, Obereinnahme/Hofkriegsrat, Nr. 296, Steuerkataster für Stadt und Amt Waischenfeld, 1731; Löwisch, Die drei Waischenfelder Stadtprivilegien (wie Anm. 3), S. 23 f.; Roppelt, Historisch-topographische Beschreibung (wie Anm. 11), S. 460–463.

29 Vgl. Fastnacht, Ebermannstadt (wie Anm. 2), S. 326.

30 Johann Georg Prändel, Erdbeschreibung der gesammten pfalzbairischen Besitzungen mit steter Hinsicht auf Topographie, Geschichte, physische Beschaffenheit, Land- und Staatswirthschaft, 2. Abt., Amberg 1806, S. 206.

31 Ebd.

32 Noch Mitte des 19. Jahrhunderts bestanden die Bodenerträgnisse hauptsächlich aus Korn, Hafer und Kartoffeln, in geringerem Maße aus Weizen, Gerste, Erbsen, Linsen, Buchweizen, Hanf, Flachs und Hopfen. Die geringe Obstbaumkultur beschränkte sich auf Zwetschgen, Äpfel und Birnen. Während eine eher unbedeutende Viehzucht vorzufinden war, lieferten die Flüsse und Bäche „*gute Forellen*". Vgl. „*Bericht des königl. Bayerischen Rentbeamten Kiefhaber*" vom März 1855 (Amtsbeschreibung des Rentamts Waischenfeld), nach: Haversath, Mühlen in der Fränkischen Schweiz (wie Anm. 27), S. 109 f., hier: S. 110.

3. Die Stadt Ebermannstadt

3.1. Die Entstehung der Stadt Ebermannstadt

Ebermannstadt tritt erstmals im Jahre 981 aus dem Dunkel der Geschichte. In jenem Jahr verlieh Kaiser Otto II. den Einwohnern des Ortes „*Ebermarestad*" Immunität gegenüber der Gerichtsbarkeit des Radenzgaugrafen Hasso. Vorher hatte ein gewisser Razzo die dortigen Güter dem Klosterstift St. Peter zu Aschaffenburg geschenkt. Zwischen 1059 und 1064 übertrugen der Domdiakon Theodor und sein Bruder Dompropst Hermann der „Bamberger Kirche" ihre Besitzungen in Ebermannstadt, 1109 befand sich ein Teil des Ortes im Eigentum des neu gegründeten Bamberger Kollegiatstifts St. Jakob. Die Nennung eines „*Thimo und Cunradus de Ebermarstat*" im Jahre 1194 legt das Bestehen eines Edelmannssitzes in oder bei Ebermannstadt nahe.[33]

Ebenso wie für Waischenfeld, erwiesen sich die Herren von Schlüsselberg als Förderer für die Entwicklung Ebermannstadts als überaus wichtig. Das Dorf Ebermannstadt gelangte wohl spätestens im Zuge der Verpfändung der Burg und des Amtes Neideck um 1248/49 als Burgzubehör in den Besitz Eberhards V. von Schlüsselberg. Auch der Bau einer Kirche dürfte auf das aus Adelsdorf stammende Geschlecht zurückgehen – jedenfalls wurde ein Ebermannstädter Gotteshaus erstmals in Verbindung mit Gottfried I. von Schlüsselberg erwähnt, aus dessen Vermögen im Jahre 1308 eine Stiftung von zwei Pfund Hallern getätigt wurde. 1323 gelang es dem Schlüsselberger Konrad II., von König Ludwig dem Bayern das Stadtrecht für Ebermannstadt zu erlangen, samt dem städtischen Handelsmonopol im Umkreis von einer Meile.[34] Bemerkenswert scheint der Umstand, dass in der königlichen Gründungsurkunde ausdrücklich auf das kurz vorher zur Stadt erhobene Waischenfeld Bezug genommen wird. So habe der Schlüsselberger Landesherr ebenso wie die Bürger Ebermannstadts bei Zuwiderhandlungen gegen das Handelsmonopol „*Gewalt zu pfänden um zehn Pfund*" Haller, wie dies bereits Waischenfeld erlaubt worden sei. Während jedoch Waischenfeld die gleichen Freiheiten und Rechte, wie sie der Stadt Bamberg zugebilligt worden waren, erhalten hatte, musste und durfte sich Ebermannstadt

---

33 Vgl. Anton Girsig, Ebermannstadt von der Gründung des Ortes bis ins 20. Jahrhundert, in: Georg Förtsch / Anton Girsig / Adolf Schön (Hg.), Ebermannstadt. Ein Heimatbuch, Ebermannstadt/Bamberg 1973, S. 56–97, hier: S. 58–60; Fastnacht, Ebermannstadt (wie Anm. 2), S. 50 f.

34 Vgl. Voit, Die Schlüsselberger (wie Anm. 1), S. 105.

nach dem Nürnberger Recht richten.[35] Die befestigte, durch zwei Flussarme geschützte Stadt Ebermannstadt wurde damit – neben der nahe gelegenen Festung Neideck und der Stadt Waischenfeld – zum dritten Zentrum des schlüsselbergischen Juraterritoriums.

3.2. Burg und Amt Neideck und die Verbindung zu Ebermannstadt

Neben Ebermannstadt war auch die Burg Neideck, samt der damit zusammenhängenden Hochgerichtsbarkeit („Cent"), 1248/49 vom Hochstift Bamberg per Pfandvertrag an Eberhard V. von Schlüsselberg übergeben worden. 1285 durch Ulrich V. um zwei Vorburgen erweitert, wurde Neideck zur Veste ausgebaut und zum Schlüsselberger Amtssitz erhoben. Als Amtmann ist uns ein „*Waltherus dictus Stubech officiatus in Neydecke*" aus dem Jahre 1312 überliefert.[36] 1347 kam es zwischen Konrad II. einerseits und den Bischöfen von Bamberg und Würzburg sowie den Burggrafen von Nürnberg andererseits zu einem erbitterten Streit über eine von ersterem errichtete Mautstelle am Talübergang zwischen den Burgen Neideck und Streitberg. Im Zuge der Belagerung Neidecks durch die Burggrafen verlor der dort weilende Konrad II. sein Leben. Mit ihm starben die Schlüsselberger im Mannesstamm aus, die Veste Neideck wurde von Bamberg und Würzburg besetzt. Endgültig verzichtete Konrads Verwandter Johann von Schwarzburg 1376 auf den Besitz, nachdem die übrigen Erben ihre diesbezüglichen Rechte bereits 1348 an den Bischof von Bamberg verkauft hatten. Das an der Herrschaft Neideck hängende Ebermannstadt war nach dem Tode Konrads II. zunächst in den gemeinsamen Besitz der Bischöfe von Bamberg und Würzburg gelangt, bis letzterer seine Hälfte im Jahre 1384 ans Hochstift Bamberg veräußerte.[37]

Ähnlich wie im Falle Waischenfelds erwies sich auch für Ebermannstadt dessen Funktion als herrschaftlicher Amtssitz als maßgeblich für die Entwicklung zu einem regionalen Zentrum. Schon 1332 residierte dort ein schlüsselbergischer Vogt; seit 1347 befand sich am Ort ein bischöfliches Vogteiamt, das in der

35 Vgl. Girsig, Ebermannstadt von der Gründung des Ortes bis ins 20. Jahrhundert (wie Anm. 33), S. 62 f.

36 Zitiert nach: Voit, Die Schlüsselberger (wie Anm. 1), S. 72.

37 Vgl. hierzu und zur Territorialpolitik der Schlüsselberger: Österreicher, Reichsherr Gottfried von Schlüsselberg (wie Anm. 1), S. 6; Voit, Die Schlüsselberger (wie Anm. 1), S. 72–75, 103–130; vgl. zum Ende der schlüsselbergischen Herrschaft ausführlich: ebd., S. 85–95; vgl. allgemein auch: Gustav Voit, Der Adel am Obermain. Genealogie edler und ministerialer Geschlechter vom 11. bis 14. Jahrhundert („Die Plassenburg". Schriften für Heimatforschung und Kulturpflege in Ostfranken 28), Kulmbach 1969, S. 301 f.

Regel von Bürgerlichen versehen wurde. In der Nürnberger Gebietsbeschreibung „*Gelegenhait der Landschaft*“ wird Ebermannstadt 1504 ausdrücklich als Stadt „*mit eigenem Halsgericht*“ beschrieben. Allerdings waren die kasten- und steueramtlichen Befugnisse Forchheim zugeordnet. Als Meilenstein dürfte gelten, dass Ende des 16. Jahrhunderts der Neidecker (Cent-)Amtmann, der ab dem 17. Jahrhundert den Titel „Oberamtmann“ führte, in Ebermannstadt seinen Sitz nahm. Nachdem im Zweiten Markgräflerkrieg seine Amtsburg in schwere Mitleidenschaft gezogen worden war, hatte er in der Stadt Aufnahme gefunden.[38] Auch dessen, aus dem alten Amt Neideck hervorgegangener Verwaltungs- und Gerichtskörper hatte zunächst nur das Centamt und das Vogteiamt umfasst, während sich das zuständige Steuer- und Kastenamt in Forchheim befand. Um 1625 erhielt Ebermannstadt dann ein eigenständiges Steueramt, die Kastenamtsgeschäfte wurden ihm im Zuge der Schönborn'schen Reformen im Jahre 1742 übertragen. Von nun an übernahm der Ebermannstädter „Stadt- und Amtsvogt“ auch die Geschäfte eines Kastners für den gesamten Ebermannstadt-Neidecker Amtssprengel.[39]

3.3. Beschreibung und Entwicklung der Stadt Ebermannstadt

Wenngleich von den bischöflich-bambergischen Vögten dominiert, entwickelte sich in Ebermannstadt doch früh ein starkes städtisches und bürgerliches Selbstbewusstsein. Relativ eigenständige Ober- und Unterbürgermeister sind seit 1623 belegt, waren aber sicher schon früher als solche tätig. Im 18. Jahrhundert bestand die Stadtverwaltung aus einem dem Stadtvogt unterstellten Bürgerrat mit einem Oberbürgermeister, der auch „*Auffschläger*“ (Einnehmer von Biersteuern und Zöllen) war, einem Unterbürgermeister, einem Stadtschreiber sowie acht Ratsverwandten und vier Viertelmeistern. Die „*Stadt-Seßionen*“ (Ratssitzungen) fanden unter dem Vorsitz des bischöflichen Oberamtmanns statt. Mitte des 16. Jahrhunderts besaß das Hochstift Bamberg in Ebermannstadt 99 Untertanen, 1623/27 waren es 86 „*haussässige*“ Bürger und 18 „*Beständner*“ (Pächter), die alle dem Bischof erbhuldigungspflichtig waren. Von den circa 100

38 Vgl. Fastnacht, Ebermannstadt (wie Anm. 2), S. 51 f.; Christoph Beck, Ebermannstädter Heimatbuch, Ebermannstadt 1926, S. 97 f.; Reinhold Glas, Forchheim. Stadt und Bürgerschaft zwischen Obrigkeit und Selbstverwaltung vom Mittelalter bis zum Übergang an Bayern (1802/03), Nürnberg 2008, S. 127.

39 Vgl. Glas, Forchheim. Stadt und Bürgerschaft (wie Anm. 38), S. 127; Reinhold Glas, Pretzfeld II. Die Entwicklung der Ortsteile im Spiegel der Geschichte seiner Anwesen und Bewohner, Erlangen 2007, S. 929.

Wohnhäusern waren fast alle Lehen des Rates der Stadt, was dessen Dominanz über das Gemeinwesen unterstrich; nur zwei Häuser gehörten dem Kasten Forchheim und eines dem Bamberger Bürger Wolf Hofmayster. Schweren Schaden erlitt die Stadt 1430 durch den Hussiteneinfall und während des Dreißigjährigen Krieges, als 1633 die Schweden alles brandschatzten. Im 18. Jahrhundert befanden sich im Ort etwa 112 Häuser, darunter *„viele Wirthsschenken"*, ein Badhaus und vier Mühlen, zwei Gemeindebrauhäuser, ein Amtshaus, ein Rathaus *„auf dem Markt"*, ein Schulhaus (*„Schulmeister zu Ebermannstadt"* sind seit 1464 belegt) und zwei katholische Kirchen. Fast alle Einwohner bekannten sich zur katholischen Religion.[40] Ein wichtiges Standbein für die städtischen Finanzen stellten die Einnahmen aus dem vermutlich noch aus Schlüsselbergischer Zeit stammenden Stadtwald am sogenannten Schlüsselstein dar.[41]

Zwar besaß Ebermannstadt drei Tore beziehungsweise Torhäuser, im Übrigen scheint der Ort jedoch nicht durch Mauern, sondern lediglich durch die ihn umgebenden Fluss- und Bachläufe, Gräben, Erdwälle und einen Palisadenzaun geschützt gewesen zu sein. Auch die 1452 vom Fürstbischof genehmigte Errichtung einer befestigten *„Behausung"* und die 1511 beurkundete Abtretung von Grundstücken zum Festungsbau führten nicht – wie anderswo üblich – zum Bau einer festen Stadtmauer. Jedenfalls zeigen die seit 1625 überkommenen Stadtansichten lediglich einige, an die äußeren Häuserfronten angebaute Torhäuser, Wälle und

40 Vgl. hierzu ausführlich: Georg Kraus, Ebermannstadt-Breitenbach, I. Teil, Hagen-Haspe 1960; Georg Kraus, Die Häuser Nr. 1–122 von Ebermannstadt, Hagen-Haspe 1959; vgl. auch: Girsig, Ebermannstadt von der Gründung des Ortes bis ins 20. Jahrhundert (wie Anm. 33), S. 68–77; Fastnacht, Ebermannstadt (wie Anm. 2), S. 52 f.; Beck, Ebermannstädter Heimatbuch (wie Anm. 38), S. 125 f.; Roppelt, Historisch-topographische Beschreibung (wie Anm. 11), S. 535; Johann Caspar Bundschuh, Geographisches Statistisch-Topographisches Lexikon von Franken, Bd. 1, Ulm/Stettin 1799, Sp. 670; Josef Elias Seyfried, Statistische Nachrichten über die ehemaligen geistlichen Stifte Augsburg, Bamberg, Constanz, Eichstätt, Freisingen, Passau, Regensburg, Salzburg und Würzburg, nebst einer historisch-politischen Uebersicht der gesamten säcularisirten teutschen Kirchen-Staaten, Landshut 1804, S. 67, 82 f.; Michel Hofmann, Die Außenbehörden des Hochstifts Bamberg und der Markgrafschaft Bayreuth, in: JfL 3, 1937, S. 52–97, hier: S. 82; Biedermann, Topographische Beschreibung des Fürstenthums Bamberg (wie Anm. 13), fol. 33 f., Nr. 112; Günther Hofmann, Märkte und Warenverkehr, in: Stadt Hollfed (Hg.), 1000 Jahre Hollfeld. Stadt und Land 1017–2017, Hollfeld 2017, S. 53–56, hier: S. 55; StABa, Hochstift Bamberg, Vogt-, Kasten- und Steueramt Ebermannstadt, Nr. 6 und 8; vgl. zu dem 1464 genannten Schulmeister: Beck, Ebermannstädter Heimatbuch (wie Anm. 38), S. 102.

41 Vgl. Beck, Ebermannstädter Heimatbuch (wie Anm. 38), S. 69.

Palisadenzäune; noch in einem Mandat Kaiser Leopolds vom Mai 1691 wird anlässlich eines Streits um das städtische Handels- und Biermonopol moniert, dass Ebermannstadt – trotz diesbezüglicher Vorgaben Kaiser Ludwigs des Bayern – noch immer nicht „*zu einer* [wirklichen] *Statt gemacht*" worden sei, „*sondern dieser orth noch itzo ein offener flecken, von keinem burgerlich- oder Stättischen Gewerb und nahrung seye*".[42]

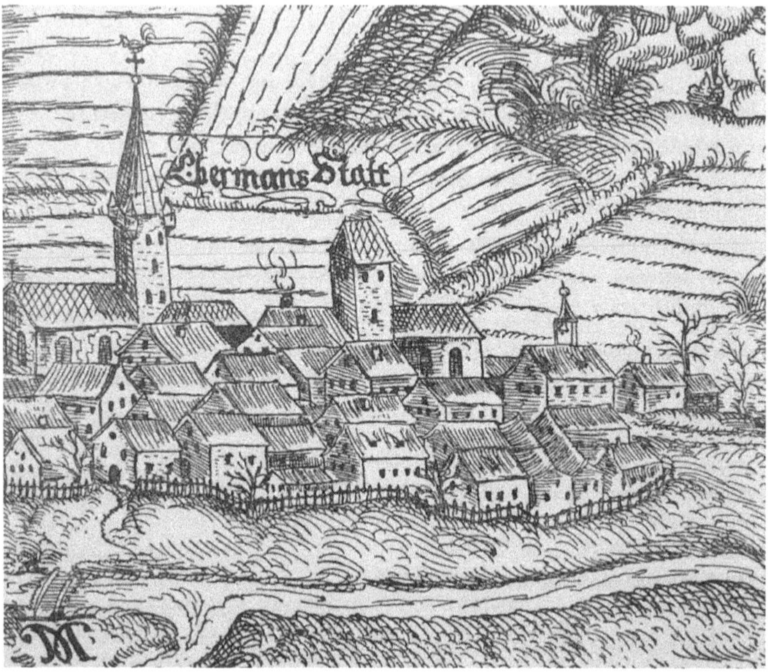

Abb. 27: *Eine der frühesten Darstellungen Ebermannstadts, zeitgenössisch um 1625. Bildnachweis: Christoph Beck, Ebermannstädter Heimatbuch, Ebermannstadt 1926, S. 85.*

42 Mandat Kaiser Leopolds vom 28. Mai 1691, zitiert nach: Georg Kraus, Pretzfeld, Hagen-Haspe 1964, S. 107 f.; vgl. auch: Löwisch, Die drei Waischenfelder Stadtprivilegien (wie Anm. 3), S. 24; Beck, Ebermannstädter Heimatbuch (wie Anm. 38), S. 117 f.; vgl. auch mit Abbildung der entsprechenden Stadtansichten: Girsig, Ebermannstadt von der Gründung des Ortes bis ins 20. Jahrhundert (wie Anm. 33), S. 64 f.; Fastnacht, Ebermannstadt (wie Anm. 2), S. 52.

Geprägt war das städtische Leben vom Wirken einer stolzen Bürgerschaft. Noch aus dem 14. Jahrhundert sind uns die ersten Namen Ebermannstädter Bürgerfamilien bekannt. Hierzu zählen unter anderem die Mol, Schmitt, Engeltraut, Eppenauer, Treyndorf, Gensel und Oberein (allesamt beurkundet 1332), außerdem die Gast und Hollfelder (1371), Klüglein, Schreiber und Körnlein (1377), Sporlein, Kiener, Lüglein und Übellein (1379), Scholle, Noltz und Putteler (1384), Weygelein und Strügelein (1389), Techlein, Schnauder und Peter (1391), Drummer (1394), Sartz und Klauß (1399).[43] Im Jahre 1453 stiftete der im Umland reichbegüterte Ebermannstädter Untervogt Eberhard Eppenauer (erwähnt 1446 bis 1466) das dortige Mittelmessbenefizium, wofür er sich und seinem Geschlecht das Präsentationsrecht vorbehielt. Zur Vornehmheit des Landadels gehörte es im auslaufenden Mittelalter, sich einen Wohnsitz in der Stadt zu nehmen. Daher hatte, ebenso wie im Falle Waischenfelds, auch in Ebermannstadt eine ganze Reihe von Adelsfamilien Haus- und Grundbesitz. So findet sich hier mit den Ebermannstädter Bürgern und Unter-Richtern Ulrich Lochner (1332) und Heinz Lochner (1394–1417) auch das in Waischenfeld anzutreffende Adelsgeschlecht der Lochner.[44] Weitere Mitglieder aus Ministerialen- und Adelsgeschlechtern, die zeitweise zur Bürgerschaft Ebermannstadts zählten, waren unter anderem die Stübig (1371/1450), Ochs (Dietrich Ochs, 1372 Vogt von Ebermannstadt; 1399/1450), Modschiedler (1400/14), Tockler (1421), Hetzelsdorfer (1428) und Drosendorfer (1467/89).[45]

---

43 In der ersten Hälfte des 15. Jahrhunderts treten u.a. hinzu die Futterer, Neubauer, Kleienpeutel, Bot, Hofmann, Stieber, Krauß, Zech, Stolz, Roßmort, Stengler, Schütz, Jahn, Ramer, Rußwurm, Herzog, Heym, Ortlein, Hetzelsdorf, Rechsel und Rat. In der zweiten Hälfte des 15. Jahrhunderts werden erstmals genannt die Mulner, Rohleder, Fleischach, Wilhelm, Nagel, Schuster, Prater, Knorrlein, Oheim (der Schulmeister), Seybot, Genslein, Thayler, Dorn, Nöt, Tockler, Clas, Löhner („*Loner*"), Fiderlein, Schneyder, Feulner, Frank, Rot (in Breitenbach), Fleischmann, Püttner, Muckner, Kaufer, Heller, Scheffer, Eyban (am Breitenbach), Kuner, Pautler, Rüger, Striegel, Rest, Schmitzer, Weisel und Gebhardt. Vgl. hierzu und zu weiteren Bürgernamen auch des 16. Jahrhunderts: Beck, Ebermannstädter Heimatbuch (wie Anm. 38), S. 101–103; Fastnacht, Ebermannstadt (wie Anm. 2), S. 51.

44 Vgl. Beck, Ebermannstädter Heimatbuch (wie Anm. 38), S. 85–87, 98 und 101; Glas, Forchheim. Stadt und Bürgerschaft (wie Anm. 38), S. 413.

45 Vgl. Fastnacht, Ebermannstadt (wie Anm. 2), S. 51 f.; Beck, Ebermannstädter Heimatbuch (wie Anm. 38), S. 102.

3.4. Kirchliche Verhältnisse Ebermannstadts

Bis zum Ende des Mittelalters gehörte die erstmals 1308 erwähnte Kirche von Ebermannstadt zum Sprengel der – wohl noch aus dem 9. Jahrhundert stammenden – würzburgischen Urpfarrei Pretzfeld. Spätestens seit Anfang des 14. Jahrhunderts besaßen die Schlüsselberger das Patronatsrecht über diese Pfarrei samt Zugehörungen als Würzburger Lehen. Nach dem Tode Konrads II. von Schlüsselberg übertrug 1350 der Würzburger Bischof das Patronat an das Kloster Schlüsselau. Als Filiale der Pfarrei Pretzfeld war für Ebermannstadt der Pretzfelder Pfarrer zuständig, der hierfür einen besonderen „Pfarrgesellen" beziehungsweise Kaplan abzustellen hatte. Im Laufe des 15. Jahrhunderts versuchte das immer mächtiger gewordene Ebermannstadt sich von seiner Mutterpfarrei zu lösen. Als Vorbild diente offenbar Waischenfeld, wo bekanntlich um 1400 eine selbstständige Stadtpfarrei entstanden war. 1469 veranlasste schließlich Bischof Georg von Bamberg die Äbtissin des Klosters Schlüsselau, der Separation Ebermannstadts von Pretzfeld zuzustimmen. Von nun an bildete Ebermannstadt mit den umliegenden Dörfern eine eigene Pfarrei, wenngleich noch bis ins 20. Jahrhundert „Separationsgelder" an Pretzfeld gezahlt werden mussten und die Mutterpfarrei das Präsentationsrecht für die städtische Pfarrstelle (bis 1827) behielt. Die Besetzung der vor 1453 gestifteten Stelle eines örtlichen „*Mittelmeßbenefiziaten*" behielt sich hingegen die Stifterfamilie, wie bereits erwähnt, vor.[46] Streit gab es allerdings um das 1489 vom Bischof genehmigte Engelmessbenefizium, wegen dem der örtliche Pfarrer im Jahre 1595 gegen den Rat der Stadt Ebermannstadt klagte, weil dieser ihm das „*Engelmeßhaus und die zugehörigen Besitzungen*" entziehen wollte.[47]

3.5. Die wirtschaftliche Entwicklung Ebermannstadts

Sich auf die königliche Stadtgründungsurkunde von 1323 berufend, behauptete Ebermannstadt durchwegs das Zoll-, Markt-, Waag- und Bierverlagsrecht, Privilegien, welche die Stadt – wie wir noch sehen werden – auch rigoros durchzusetzen entschlossen war. Nach dem Ebermannstädter Maß- und Gewichtssystem,

---

46 Vgl. Glas, Pretzfeld II (wie Anm. 39), S. 37 f.; Josef Pieger, Aus der Pfarrgeschichte von Ebermannstadt, in: Förtsch / Girsig / Schön, Ebermannstadt (wie Anm. 33), S. 156–163, hier: S. 156 f.; Fastnacht, Ebermannstadt (wie Anm. 2), S. 51, 227–229; Guttenberg / Wendehorst, Das Bistum Bamberg 2 (wie Anm. 15), S. 124–128. Zum Eppenauer'schen Mittelmessbenefizium vgl. u.a.: StABa, Hochstift Bamberg, Geistliche Regierung, Akten und Bände, Nr. 1704; StABa, Hochstift Bamberg, Geistliche Regierung, Akten und Bände, Nr. 1713.

47 StABa, Hochstift Bamberg, Geistliche Regierung, Akten und Bände, Nr. 1702.

das sich nach „*altem Herkommen*“ am „*Nürnberger Rats Stadt maß und gewicht*“ orientierte, mussten sich alle umliegenden Orte richten.[48] 1514 bestätigte der bischöfliche Landesherr „*seiner Stat Ebermanstat*“ das Recht, montags einen Wochenmarkt abzuhalten, im Jahre 1608 genehmigte er den Ebermannstädtern zu ihren zwölf Jahrmärkten jährlich vier Viehmärkte abzuhalten (bestätigt 1655).[49] Eine Beschreibung aus dem Jahre 1752 attestiert der Stadt, sie sei mit sieben Jahrmärkten privilegiert und halte daneben „*gute Viehmärkte*“ ab.[50] Wie in vielen anderen Städten, hatte auch in Ebermannstadt das Handwerk goldenen Boden. Aus dem 18. Jahrhundert kennen wir von dort eine ganze Reihe von Zunftordnungen beziehungsweise Meisterbüchern, so von den Zimmerleuten, Müllern, Maurern, Steinhauern und Schneidern.[51]

Für den Standort Ebermannstadt zentral erwies sich dessen Lage an der für die Region wohl wichtigsten hoch- und spätmittelalterlichen Altstraße, die von Nürnberg aus über Erlangen kommend quer durch die Fränkische Schweiz nach Kulmbach beziehungsweise Bayreuth führte. Die Straße war Teil des Handelsnetzes, das Nürnberg für seinen überaus wichtigen Ost-West-Handel nach Polen und nach den Ostseegebieten nutzte. Dabei handelte es sich um eine sogenannte Geleitstraße, auf der die Handelswagen von einem bewaffneten, regional zuständigen herrschaftlichen Geleitzug geschützt wurden. Für diese Dienste erhob der zuständige „Geleitsmann“ eine Gebühr, in der auch der festgesetzte Zoll enthalten war. Hatte das Handelsgeleit (ebenso wie das Zollregal) ursprünglich zu den Rechten des Königs gehört, galt es spätestens seit der Zeit König Rudolfs I. von Habsburg (1273–1291) als landesherrlich und frei übertragbar.[52] Wenngleich sich die Burggrafen von Nürnberg vom Prinzip her schon während des Meranischen Erbfolgestreits (1250–1260) alle diesbezüglichen Rechte gesichert hatten, übten zunächst doch die Schlüsselberger das Geleit im Ebermannstädter Grund

48 Ersuchen Ebermannstadts an den Nürnberger Rat um die Zusendung eines neuen Zentnergewichtes aus dem Jahre 1516, zitiert nach: Beck, Ebermannstädter Heimatbuch (wie Anm. 38), S. 116.

49 Vgl. Hofmann, Märkte und Warenverkehr (wie Anm. 40), S. 55; Fastnacht, Ebermannstadt (wie Anm. 2), S. 52–54.

50 Biedermann, Topographische Beschreibung des Fürstenthums Bamberg (wie Anm. 13), fol. 32, Nr. 112.

51 Vgl. Beck, Ebermannstädter Heimatbuch (wie Anm. 38), S. 114 f.

52 Vgl. Rudolf Endres, Burgen und Burgenverfassung in der Fränkischen Alb, in: Franz Tichy / Rainer Gömmel (Hg.), Die Fränkische Alb (Schriften des Zentralinstituts für Fränkische Landeskunde 28), Neustadt an der Aisch 1989, S. 161–193, hier: S. 176–178; Rudolf Endres, Ein Verzeichnis der Geleitstraßen der Burggrafen von Nürnberg, in: JfL 23, 1963, S. 107–138, hier: S. 108–110.

und auf der daran anschließenden Albhochflächenstraße in Richtung Hollfeld beziehungsweise Bayreuth aus. Schmerzlich zu spüren bekamen dies die Nürnberger Kaufleute, als im Jahre 1347 Konrad II. von Schlüsselberg bei Streitberg eine zusätzliche Zollsperre errichtete. Die Folge war bekanntlich die Belagerung der gegenüberliegenden schlüsselbergischen Hauptburg Neideck durch Würzburg, Bamberg und die Burggrafen, die mit der Tötung Konrads II. endete. Nach dem Ende der Schlüsselbergischen Territorialherrschaft beanspruchten und erhielten nun die Nürnberger Burggrafen auch faktisch das Geleitsrecht auf diesem Weg durch die Alb, obwohl er überwiegend durch bischöfliches Territorium führte.[53] Im sogenannten Forchheimer Vertrag von 1538 wurde schließlich vereinbart, dass das umstrittene Geleitsrecht auf der Straße zwischen dem bei Baiersdorf in die Regnitz mündenden Kreuzbach über Ebermannstadt, Hollfeld und Kasendorf nach Kulmbach fortan Bamberg und Brandenburg gemeinsam gehören sollte.[54] In Ebermannstadt, durch das der Weg mitten hindurchführte, befand sich eine bedeutende und einträgliche Zollstation, woran der alte Ebermannstädter Flurname „*Zollstab*" erinnert.[55] Allerdings entnehmen wir einer Beschreibung aus dem Jahre 1752 über den Zustand der Landstraße, dass „*in dem Städtlein* [Ebermannstadt] *selbst* [...] *lauter Morast, und darinnen nicht*

---

53 Das burggräfliche Geleit zog sich von Bayreuth aus über Schönfeld bzw. Hollfeld, Plankenfels, Wüstenstein, die Streitberger Steige hinab, Streitberg, Niederfellendorf, Rothenbühl, Ebermannstadt (Wiesentübergang), Pretzfeld, Kirchehrenbach, Wiesenthau, Kersbach, Baiersdorf, Erlangen, Tennenlohe, Buch und Thon bis hin zum Nürnberger Stadttor. Vgl. Endres, Burgen und Burgenverfassung (wie Anm. 52), S. 176–178; Kunstmann, Die Burgen der westlichen und nördlichen Fränkischen Schweiz, 1. Teil (wie Anm. 18), S. 5, 8.

54 Die Geleitsmänner hatten von nun an eine Büchse mit beiden Wappen zu führen. Vgl. Forchheimer Vertrag vom 1. Juli 1538, in: Gerhard Rechter (Bearb.), Die Archive der Grafen und Freiherren von Seckendorff. Die Urkundenbestände der Schloßarchive Obernzenn, Sugenheim, Trautskirchen und Unternzenn, Bd. 2 (Bayerische Archivinventare 46), München 1993, S. 523–526; vgl. hierzu auch: Endres, Verzeichnis der Geleitstraßen (wie Anm. 52), S. 137 f.

55 Zum Ebermannstädter Zollwesen, über das sich 1453/55, 1509 und 1607 Nürnberger Kaufleute beschwerten und das aufgrund stattgehabter „Unregelmäßigkeiten" in den „*Hochfürstlichen Verordnungen für das Municipal Städtlein Ebermannstadt*" 1740 und 1776 einer strengen Regelung unterzogen wurde, vgl. weiter unten. Vgl. auch: Staatsarchiv Nürnberg (künftig: StAN), Reichsstadt Nürnberg, Ratskanzlei, A-Laden, Akten A 64, Nr. 4, „*Beschwerde* [der Reichsstadt Nürnberg] *wider die Amtleute zu Forchheim, Höchstatt und Ebermannstadt, daß dieselben sich unterstehen, die nürnbergischen Fürkheufel von Viktualien, der Zoll abgewandert auch samt mit Spaten niederlassen des Fannens gehindert werden*", 1509; Girsig, Ebermannstadt von der Gründung des Ortes bis ins 20. Jahrhundert (wie Anm. 33), S. 76 f. (Hochfürstliche Verordnungen 1740/1776); Beck, Ebermannstädter Heimatbuch (wie Anm. 38), S. 114, 116.

einmahl zu fueß fort zu kommen, auch bös zu reiten" sei, weswegen dringend eine Pflasterung empfohlen wurde, um den Weg „*in paßablen Stand*" zu setzen.[56]

Die Lage Ebermannstadts hatte weitere, für die Entwicklung des Städtchens durchaus nennenswerte Vorteile. So testieren noch Landbeschreibungen aus den Jahren 1795 und 1805 Ebermannstadt „*eine der angenehmsten* [Lagen] *wegen des herrlichen Wiesgrundes und Wiesentflusses, worin unter andern Fischen die größten Forellen angetroffen*" würden.[57] Tatsächlich hatte der Fischfang in den klaren Bächen und kleinen Flüssen der Fränkischen Schweiz mit einer Vielfalt von 15 Fischarten früher einzelne Fänge mit bis zu fünf Zentnern erlaubt.[58] Außerdem sei der Ebermannstädter Hanfanbau, so zeitgenössische Berichte, „*besonders einträglich, und die Viehmastung sehr gut*".[59] Auch nährten sich die Bürger „*vorzüglich durch Bierbrauen*", die „*Viehmärkte daselbst*" würden „*von Ausländern stark besuchet*".[60] Auf das recht einträgliche Biergeschäft und den Viehhandel werden wir im Anschluss noch näher eingehen. Dass indes der Hanfhandel beziehungsweise -anbau für die Stadt auch seine Schattenseiten hatte, wird aus einer „*Anzeige des k*[öniglichen] *Phisikats*" vom September 1840 deutlich. So sei „*das Aufstellen des Hanfes an den Häusern wegen Ausdünstung desselben* [und wohl wegen der daraus resultierenden Rauschwirkung] *für die Gesundheit nachteilig*", weshalb „*dieses bisher übliche Verfahren untersagt*" wurde.[61] Ob ein ausgiebiger Bierkonsum oder die Ausdünstungen des Hanfes Nebenwirkungen auch bei der städtischen Elite zeitigten, ist nicht überliefert, wenngleich manche Verhaltensweisen im Umgang mit Nachbarorten – wie wir gleich sehen werden – einen Zusammenhang durchaus nahelegen.

---

56 „*Grund-Riß über Die auf Bambergischen Territorio durch die beyden Hochfürstl. Ämterr Vorchheim und Ebermannstadt gefurte Landstraß Von Erlang nacher Bayreuth*" (1752), zitiert nach: Girsig, Ebermannstadt von der Gründung des Ortes bis ins 20. Jahrhundert (wie Anm. 33), S. 81.

57 Prändel, Erdbeschreibung (wie Anm. 30), S. 203.

58 Nach einem Tagebucheintrag des Johann Michael Füssel vom Ende des 18. Jahrhunderts, in: Ursula Pfistermeister, Fränkische Schweiz. Hersbrucker Schweiz, Nürnberg ^{4}1989, S. 134 f.

59 Prändel, Erdbeschreibung (wie Anm. 30), S. 203.

60 Geographie des Fürstenthums Bamberg von J. G. Wehrl (1795), zitiert nach: Girsig, Ebermannstadt von der Gründung des Ortes bis ins 20. Jahrhundert (wie Anm. 33), S. 82; vgl. auch: Biedermann, Topographische Beschreibung des Fürstenthums Bamberg (wie Anm. 13), fol. 32, Nr. 112.

61 Schreiben des königlichen Landgerichts vom 17. September 1840, zitiert nach: Girsig, Ebermannstadt von der Gründung des Ortes bis ins 20. Jahrhundert (wie Anm. 33), S. 82.

Abb. 28: *Das Wappen von Ebermannstadt auf den „Bamberger Vasallentafeln“ von 1603 (darin: Konterfei Kaiser Ludwigs des Bayern, Schlüssel als Wappen der Schlüsselberger, Eber als Wappentier Ebermannstadts). Bildnachweis: Christoph Beck, Ebermannstädter Heimatbuch, Ebermannstadt 1926, S. 16 [b].*

4. Behauptung wirtschaftlicher Interessen und Auswüchse kleinstädtischer Machtvollkommenheit am Beispiel Ebermannstadts

4.1. Die Ebermannstädter „Bier- und Handelskriege“

Bekanntlich hatte sich Konrad II. von Schlüsselberg gleichzeitig mit dem Stadtrecht auch das Privileg des exklusiven Handels seiner beiden Städte Waischenfeld und Ebermannstadt innerhalb eines bestimmten Rayons vom König urkundlich zusichern lassen. Unter diesem Privileg wollten später die städtischen Ratsherren gerne auch das Recht des „Bierverlags“, das heißt des Bierhandels, verstanden wissen. Dabei war es keineswegs unumstritten, inwieweit der in der Stadtgründungsurkunde niedergelegte Vermerk, dass eine „*Meil umb*“ die Stadt „*herumb niemand einen feilen Kauff haben solle*“, auch das exklusive Recht der Kommune für das Brauen und Ausschenken von Bier umfasste.[62] Nichtsdestotrotz hielten nach dem Ende Konrads II. und der Verteilung seines Erbes unter den Hochstiftern Bamberg und Würzburg und den hohenzollerischen Burggrafen die nun bambergischen Städte an ihren vom König garantierten Rechten, insbesondere was das Bierverlagswesen anging, fest.[63]

Naturgemäß waren es enge persönliche Verflechtungen zwischen den Ratsmitgliedern, den Händlern und Brauern, die das Interesse an einem möglichst hohen Ertrag der ansässigen kommunalen und privaten Gewerbe wachhielten. Die in direkter und indirekter Verbindung zu den städtischen Fiskaleinnahmen stehenden fürstlichen Steuerflüsse sicherten den Bamberger Landstädten, so auch Ebermannstadt, die diesbezügliche Unterstützung des Hochstifts respektive des mächtigen Bamberger Fürstbischofs. So ließ sich der Rat Ebermannstadts im Laufe der Jahrhunderte mehrmals die *anno* 1323 vom König zugesicherten Handels- und Bierverlagsprivilegien von der Bamberger Hofkanzlei schriftlich bestätigen.[64] Es ging dabei nicht nur um wirtschaftliche und fiskalische Interessen der Stadt und des Fürstbistums, sondern auch um eine politische Machtdemonstration gegenüber fremden Grundherren, wie den Reichsritterschaften, die sich seit Beginn der Frühen Neuzeit als kleine Territorialherren innerhalb des hochstiftischen Gebiets etablierten.[65]

Trotz der königlichen Privilegien kam es immer wieder zu Streitigkeiten zwischen den Städten und den umliegenden Dorf- und Ritterschaften, um das Recht Handel zu treiben, zu mälzen und zu brauen. Dabei führte das gemeinsame Interesse von Ebermannstädter Stadträten, Kaufleuten und Bierbrauern zu eisernen Allianzen, die weder vor langwierigen juristischen noch bewaffneten Auseinandersetzungen zurückschreckten, um das seit dem 14. Jahrhundert reklamierte

62 Zitiert nach: Kraus, Pretzfeld (wie Anm. 42), S. 108. Die Urkunde über die Verleihung des Stadtrechts an Ebermannstadt durch Ludwig den Bayern findet sich vollständig abgedruckt bei: Beck, Ebermannstädter Heimatbuch (wie Anm. 38), S. 70 f.

63 Vgl. Voit, Die Schlüsselberger (wie Anm. 1), S. 85–87.

64 So in den Jahren 1510, 1653/54 und 1711. Vgl. Girsig, Ebermannstadt von der Gründung des Ortes bis ins 20. Jahrhundert (wie Anm. 33), S. 71 f.

65 Vgl. hierzu und zum Folgenden: Thomas J. Hagen, Oberfrankens Braustätten und Bierkonsum im Fadenkreuz frühneuzeitlicher Wirtschafts- und Fiskalpolitik, in: Mitteilungen des Verbandes bayerischer Geschichtsvereine 28, 2018, S. 111–130, hier: S. 112–125; Thomas J. Hagen, Konfliktpotential alltäglicher Konsumgüter: Der Streit ums Biermonopol im frühneuzeitlichen Oberfranken, in: Wolfgang Wüst / Gisela Drossbach (Hg.), Umwelt-, Klima- und Konsumgeschichte. Fallstudien zu Süddeutschland, Österreich und der Schweiz. Unter Mitarbeit von Lisa Bauereisen und Christoph Gunkel, Berlin u.a. 2018, S. 45–73, hier: S. 58–73.

Handelsmonopol durchzusetzen. Waren es während des Bauernkrieges vor allem *„Flecken und Dörfer"*, die sich *„mit dem bräuen und schenken neuigkeit gesucht vnnd furgenommen"* hatten, setzten sich danach immer öfter auch kleinere Rittergüter über das von Ebermannstadt beanspruchte Biermonopol hinweg.[66] Abnehmer fanden diese reichlich in Form des „gemeinen Mannes". Daran waren die städtischen Bierbrauer nicht ganz unschuldig, hatten diese doch mitunter geglaubt, ihr Monopol missbrauchen zu können, indem sie den für die Landbevölkerung bestimmten Gerstensaft panschten und dadurch ihre Gewinne in die Höhe trieben. So war es möglich, dass sich manch eine Ritterherrschaft zu einer ernstzunehmenden Konkurrenz im Biergeschäft entwickelte.[67] Im benachbarten Hetzelsdorf beispielsweise, wo das Bierbrauen schon eine alte Tradition hatte, genehmigte der Markgraf von Brandenburg 1623 – ohne Rücksicht auf die Ebermannstädter Stadtprivilegien – den Bau eines neuen Bräuhauses; in Wannbach sah sich 1681 der Bischof von Bamberg genötigt, die Errichtung eines von der Ritterherrschaft betriebenen Bräuhauses samt Mälzerei zu erlauben; im nahegelegenen Pretzfeld, das mehreren Grundherren, darunter den Rittern von Stiebar, unterstand, wurde bereits vor 1510 mehr oder weniger eigenmächtig Bier gebraut.[68]

Dass es hierbei zwangsläufig zu heftigen Auseinandersetzungen um das Biermonopol kommen musste, liegt auf der Hand. Die Reihe der über die Jahrhunderte erfolgten Reibereien ist lang. So hatten die Ebermannstädter Zank mit den Freiherren von Wiesenthau um das Schankrecht in Hagenbach, mit den Herren von Seinsheim gewaltsame Auseinandersetzungen um den Bierverlag in Wannbach. Um den Bierausschank in Unterleinleiter stritt Ebermannstadt gleich mit mehreren Herrschaften, so mit den Herren von Egloffstein, den Herren von Streitberg, den Herren von Dienheim, den Herren von Lüchau und den Herren von Schlammersdorf. Selbst vor einem Streit mit den im Hochstift einflussreichen Geschlechtern der Schenk von Stauffenberg und der Voit von Rieneck schreckten die Ebermannstädter nicht zurück. So gab es Reibereien mit ersteren wegen des Bierhandels in Trainmeusel, Burggrub und Wohlmannsgesees, mit letzteren wegen des Ausschanks in Burg- und Windischgaillenreuth. Nicht selten arteten die Bierpfändungen in

---

66 Zitiert nach: Georg Fehn, Chronik von Kronach, Bd. 4, Kronach 1969, S. 228; vgl. hierzu: ebd., S. 231; vgl. auch: Günter Dippold, Kleinstädte im Barock. Studien am Beispiel bambergischer Amtsstädte, in: Dieter J. Weiß, (Hg.), Barock in Franken (Bayreuther Historische Kolloquien 17), Dettelbach 2004, S. 197–215, hier: S. 205.

67 Vgl. mit einer genauen Aufstellung der von Bamberg festgelegten Bierpreise zwischen 1520 und 1711: Dominikus Kremer, Maineck. Geschichte eines hochfürstlich-bambergischen Dorfes, Bamberg/Münsterschwarzach 1983, S. 560–562.

68 Vgl. Glas, Pretzfeld II (wie Anm. 39), S. 310, 1039.

bewaffnete Überfälle der Ebermannstädter aus, so im Jahre 1685 auf das Wirtshaus in Unterleinleiter, 1747 in Morschreuth oder 1779 in Wannbach.[69]

Wollte das Bamberger Landstädtchen Ebermannstadt die Aushöhlung seines Monopols verhindern, musste es zwangsläufig zu drastischen Maßregeln greifen. Dabei wurde es zur Gewohnheit, dass unmittelbar nach Bekanntwerden des „Bierfrevels" eine Abordnung von Bürgern unter Führung einer Amtsperson ausrückte, um den Beschuldigten zur Rede zu stellen, die betreffenden Krüge und Fässer mitsamt dem Inhalt zu pfänden und entweder an Ort und Stelle auszutrinken oder mitzunehmen. Nicht selten wurde den Delinquenten eine empfindliche Strafzahlung aufgebürdet. Delikater Weise profitierten die Stadtoberen selbst am meisten von den Pfändungen, da sie nicht nur am „Freibiertrinken" beteiligt waren, sondern oftmals auch kostengünstig in den Besitz der beschlagnahmten Zapfhähne und Bierbehältnisse kamen.[70] Gleich ob bei Dorffesten, Richtfesten, Hochzeiten, Taufen oder Leichentrünken – überall rückten Ratsmitglieder, Viertelmeister und Amtsknechte aus, um das städtische Bierverlagsmonopol kompromisslos durchzusetzen. Nur selten kam es vor, dass die Städter „Gnade vor Recht" ergehen ließen, wie im Falle des Hans Och in Unterleinleiter, dem die Ebermannstädter zwar ein Fässchen fremden Bieres *„abgepfändet, das Bier aber seiner Tochter alß einer Kinds Betterin"* überlassen haben.[71]

Wenngleich es in der großen Mehrzahl ritterschaftliche Braustätten waren, mit denen es Auseinandersetzungen gab, machte der Ebermannstädter Monopolanspruch doch auch vor anderen Konkurrenten nicht halt. Fast unvermeidlich kam es daher mit benachbarten bambergischen Amtsstädten zu Rangeleien; so in den Jahren 1700 bis 1756, als es im Streit mit Waischenfeld und Gößweinstein um die Herausgabe von Bier für eine Hochzeitsfeier im Grenzort Windischgaillenreuth ging oder um ähnliche Ansprüche in den Dörfern Wichsenstein, Urspring und Morschreuth.[72] Selbst mit dem Landalmosenamt der mächtigen

---

69 Gräflich und freiherrlich von und zu Egloffstein'sches Archiv Kunreuth, Alt-Sign. 662 und 649; Landesarchiv (künftig: LA) Baden-Württemberg, Abt. Staatsarchiv (künftig: StA) Sigmaringen, Gesamtarchiv Schenk von Stauffenberg, Akten Burggrub-Greifenstein, Dep. 38 T 5, Nr. 64/18 und 157; Stadtarchiv Ebermannstadt (künftig: StadtA Ebermannstadt), A 10/050, A 10/052–053, A 10/070, A 10/100–101, A 10/104–105 und A 10/110.

70 Vgl. u.a.: Kraus, Die Häuser Nr. 1–122 (wie Anm. 40), S. 6–14; vgl. auch die zahlreichen diesbezüglichen Akteneinträge in: StadtA Ebermannstadt, A 10/050–057, A 10/070 und A 10/100–111.

71 *„Actum Ebermannstadt den 30. Octobris 1764"*, in: Kraus, Die Häuser 1–122 (wie Anm. 40), S. 13 f.

72 StadtA Ebermannstadt, A 10/105, A 10/107–110.

Reichsstadt Nürnberg legten sich die Ebermannstädter im 18. Jahrhundert mehrmals an, indem sie es wagten, bei dessen Untertanen im nahegelegenen Zaunsbach gewaltsame Bierpfändungen durchzuführen.[73]

Ein regelrechter, über mehrere Jahrhunderte dauernder „Bierkrieg" fand indes mit dem alten Pfarrdorf Pretzfeld statt. Fast die gesamte Frühe Neuzeit über focht Ebermannstadt mit diesem Nachbarort, dessen Pfarrer bekanntlich bis 1469 die kirchliche Oberhoheit über die Stadt besessen hatte, einen teils erbitterten Kampf um die Einhaltung seines Biermonopols aus.[74] Die erste tätliche Auseinandersetzung zwischen beiden Orten ist uns aus dem Jahre 1510 überliefert. Das reklamierte Biermonopol Ebermannstadts ignorierend, brauten und verkauften zu jener Zeit die Einwohner Pretzfelds, deren Dorf eigentlich innerhalb der Ebermannstädter Bierbannmeile lag, ihren eigenen Gerstensaft. Der Pretzfelder Ortsvorsteher verteidigte das fortgesetzte Braugeschäft mit den sehr schleppenden, manchmal gänzlich ausbleibenden Bierlieferungen aus Ebermannstadt. Außerdem, so seine Klage, sei das Ebermannstädter Bier so schwach eingebraut, *„dass an der Kirchweih nicht ein einziges Mal gerauft wurde"*.[75] Den Pretzfeldern kam entgegen, dass die gerichtliche Zugehörigkeit des Ortes umstritten war. Zwar behauptete der Bischof von Bamberg die Territorial-, Hochgerichts- sowie prinzipiell die Dorf- und Gemeindeherrschaft. Aufgrund des grundherrschaftlichen Gerangels zwischen den Rittern von Wiesenthau, den Stiebar von Buttenheim, der Reichsstadt Nürnberg, dem Markgrafentum Kulmbach-Bayreuth und dem Hochstift Bamberg erlangte die Dorfgemeinde aber eine vergleichsweise starke Selbstverwaltungsbefugnis.[76] Aus diesem Grund war es dem bischöflichen Vogteiamt nicht ohne weiteres möglich, gegen Pretzfeld vorzugehen – zum Verdruss der um ihr Monopol bangenden Ebermannstädter.

Nachdem weder Drohungen des Ebermannstädter Magistrats noch Beschwerden beim Bischof die Pretzfelder von ihren *„unbefugten Handirungen"*

73 StadtA Ebermannstadt, A 10/106; StAN, Reichstadt Nürnberg, Landalmosenamt, Akten I, Nr. 2892.

74 Vgl. hierzu und zum Folgenden auch: Hagen, Konfliktpotential alltäglicher Konsumgüter (wie Anm. 65), S. 58–73; Erich Arneth, Der Bierkrieg zwischen Ebermannstadt und Pretzfeld, in: Stadt Ebermannstadt (Hg.), Ebermannstadt. Heimatkundliche Beiträge. Folge 1, Bayreuth 1981, S. 141–155.

75 Zitiert nach: Reinhard Löwisch, Von der „Bierseele" Frankens, veröffentlicht am: 29.3.2016, in: inFranken.de, URL: http://www.infranken.de/regional/forchheim/Von-der-Bierseele-Frankens;art216,1745269 (letzter Zugriff: 21.9.2016).

76 Vgl. Reinhold Glas, Pretzfeld [I]. Häuser- und Familienchronik eines Marktortes in der Fränkischen Schweiz, Erlangen/Jena 1994, S. 30–32.

beim Bierbrauen abhielten, eskalierte der Streit.[77] Im Morgengrauen des 16. August 1510 zog eine schwer bewaffnete Heerschaar von Ebermannstädtern unter der Führung des Stadtvogts und des Bürgermeisters nach Pretzfeld. Dort angekommen, drang das aufgebrachte Bürgerheer in die Ortschaft ein, um alle Biervorräte zu vernichten, und zwar in der Weise, indem man alles, was nicht sofort von der wütenden Menge ausgetrunken werden konnte, ausgoss, um schließlich die Fässer in Stücke zu schlagen. Selbst die Zuber in den Badestuben wurden zerstört, um den Pretzfeldern jegliche Möglichkeit zu nehmen, neu gebrautes Bier einzulagern. Die als Anstifter für den Bierfrevel geltenden Dorfmeister wurden gefangengesetzt und in Forchheim eingekerkert.[78]

Abb. 29: *Kolorierter Holzschnitt des Stiebar'schen Schlosses in Pretzfeld, gegen das ein Heer aufständischer Bauern mit Trommler und Fahnenträger zieht (aus der Zeit des Bauernkriegs 1525). Bildnachweis: Reinhold Glas, Pretzfeld [I]. Häuser- und Familienchronik eines Marktortes in der Fränkischen Schweiz, Erlangen/Jena 1994, S. 133.*

---

77 Vgl. Girsig, Ebermannstadt von der Gründung des Ortes bis ins 20. Jahrhundert (wie Anm. 33), S. 70; Erich Arneth, Der Ebermannstädter Bierkrieg, in: Förtsch / Girsig / Schön, Ebermannstadt (wie Anm. 33), S. 264–268, hier: S. 264 f.

78 Vgl. ebd.

Verständlicherweise hatte dieser kriegerische Gewaltakt ein Nachspiel, da sich die Pretzfelder nun ihrerseits beim Bamberger Fürstbischof als oberstem Herrn der Aggressoren beschwerten. Erwartungsgemäß kam das 1513 ergangene, sogenannte „*Compromiß-Urtheil*“ des Fürstbischofs den Ebermannstädtern sehr entgegen, indem es den Pretzfeldern zwar den Betrieb einer Reihe von Gewerben, wie das Fleischer- und Bäckerhandwerk, erlaubte, ihnen jedoch künftig untersagte, „*zu mälzen und zu bräuen*“.[79] Zum Ärger Ebermannstadts hielten sich die Pretzfelder nicht an das bischöfliche Urteil, weswegen sich die Stadt 1569 erneut genötigt sah, gegen ihren Nachbarn vorzugehen.[80] Auch während des Dreißigjährigen Krieges setzte sich der Streit um das Bierverlagsrecht fort. Verstärkt durch die konfessionellen Konflikte zwischen dem bischofstreuen katholischen Ebermannstadt und den protestantisch gewordenen Pretzfelder Rittergutsherren von Stiebar, wandte sich nun der Argwohn gegen deren, „*nach altem herkommen*“ betriebene Schlossbrauerei. Auf dem Höhepunkt des Konflikts drangen die Ebermannstädter Anfang der 1630er Jahre neuerlich in Pretzfeld ein und pfändeten dem Gutsbesitzer ein „*ganzes Gebräu*“ Bier.[81] Zwar klagte der Stiebar'sche Vogt daraufhin erfolgreich auf Schadensersatz.[82] Aufgrund der Kriegswirren gelang es jedoch den Ebermannstädtern, die Auszahlung der festgesetzten Schadenssumme zu verhindern. Umso mehr sah sich nun der Schlossverwalter im Recht, nicht nur weiter Bier zu brauen, sondern dieses auch an ritterschaftliche Untertanen auszuschenken. Wenige Jahre nach dem Ende des Krieges versuchten daher die Ebermannstädter erneut, der Pretzfelder Schlossverwaltung das Bierbrauen und den Verkauf ihres Malzgebräus streitig zu machen. Mit Hilfe eines fürstbischöflichen Dekrets wurde den Pretzfelder Adeligen 1661 zum wiederholten Male jedweder Handel mit ihrem Gebräu untersagt.

---

79 Zitiert nach: Kraus, Pretzfeld (wie Anm. 42), S. 98; vgl. auch: Glas, Pretzfeld [I] (wie Anm. 76), S. 569, 968. Vgl. zu dem ganzen vorausgegangenen Schlichtungsvertrag: StadtA Ebermannstadt, A 10/020, „*Hauptbrieff oder Vertrag deren von Ebermannstadt und Pretzfeldt*“ vom Januar 1511.

80 Vgl. Kraus, Pretzfeld (wie Anm. 42), S. 108; StadtA Ebermannstadt, A 10/021, Aufforderung an den Bürgermeister und Rat der Stadt Ebermannstadt durch den Forchheimer Schultheißen Georg Groß von Pfersfeld, zu der Pretzfelder Klage Stellung zu nehmen.

81 Vgl. zur Geschichte Pretzfelds und Ebermannstadts während des Dreißigjährigen Krieges: Hans Haas, Geschichte des Pfarrdorfes und Marktes Pretzfeld, Erlangen 1969, S. 37–40.

82 „*Schreiben Conrad Galsterern an Ebermannstadt*“ vom 4./14.10.1631, in: Kraus, Die Häuser Nr. 1–122 (wie Anm. 40), S. 98. Vgl. auch die wechselseitigen Amtsschreiben vom 14.10.1631 bis 23.6.1632, in: ebd., S. 99–102.

Lediglich für den eigenen Hausbedarf sollte es ihnen gestattet sein, Bier herzustellen. Die übrigen Einwohner Pretzfelds aber, so die bischöfliche Anordnung, hätten ihr Bier in Ebermannstadt abzuholen und zu versteuern.[83]

Da sich die Pretzfelder Ritter auch an diesen landesfürstlichen Befehl nicht hielten, kam es zu neuerlichen Auseinandersetzungen. So fielen die Ebermannstädter im Mai 1681 mehrmals in die nahe der Stadt gelegene Stiebar'sche „*Erbschenkstatt zu Moggast*" gewalttätig ein, um das dort vorgefundene Pretzfelder Bier zu pfänden.[84] Dem Wirt wurde dabei nach „*vielen verbal- und real-injurien*" unter Androhung von „*straff und fernern gewalts*" unmissverständlich klar gemacht, dass er sein Bier künftig „*nirgend anderswo, denn zu Ebermannstatt zukauffen und abzuholen*" habe. Da half es auch nichts, dass die Reichsritterschaft des Kantons Gebirg den Stiebar zu Hilfe kam, indem sie den Fürstbischof offiziell aufforderte, „*solch violentes, anbey aigenmächtiges beginnen und dieß orts unerhörte Neuerung*" künftig zu unterbinden.[85]

Zwar sah sich im Dezember 1682 die Regierung des Hochstifts der Form halber genötigt, einen allgemeinen Erlass „*wegen vorgefallener Gewalttätigkeiten*" herauszugeben.[86] In diesem wurde den Bamberger Städten und Marktflecken verboten, „*ihre Rechte, namentlich wegen des Bierverlags, durch eigenmächtige Ausfälle und Pfändungen zu handhaben*" und befohlen, ihr „*Recht am gehörigen Orte zu suchen*". Allerdings blieben die Bürger weiterhin berechtigt, ja sogar verpflichtet, „*wenn andere Herrschaften und Adelsuntertanen dem hergebrachten Bierverkauf oder anderen Hantierungen Eintrag tun*", sich „*gleichwohl selbst bei ihren Rechten durch ‚gewehrlichen außfall und Pfandungen' unter Vermeidung von Exzessen* [zu] *schützen, wenn sie solches vorher ihren Beamten angezeigt*" hätten.[87]

83 Vgl. zur Auseinandersetzung zwischen den Stiebar und der Bamberger Obrigkeit: Heinrich Joachim Jäck, Geschichte der Provinz Bamberg vom Jahre 1006 bis 1803, 1. Teil, Bamberg 1809, S. 114–116; Haas, Geschichte des Pfarrdorfes und Marktes Pretzfeld (wie Anm. 81), S. 37–39; Kraus, Pretzfeld (wie Anm. 42), S. 107 f.; vgl. auch: StadtA Ebermannstadt, A 10/024, Dekret des Bamberger hochfürstlichen Rats vom 3.8.1661; StadtA Ebermannstadt, A 10/024, Kopie eines Schreibens des Fürstbischofs Philipp Valentin an Stiebar vom 18.2.1662; „*Extractus Protocolli uff der fürstlichen Rathstube*" vom 7./17.4.1663, in: Kraus, Die Häuser Nr. 1–122 (wie Anm. 40), S. 104; StadtA Ebermannstadt, A 10/024, Teil-Abschrift einer Antwortnote (Februar 1662) an den Bamberger Fürstbischof.

84 StadtA Ebermannstadt, A 10/026.

85 Ebd.

86 Bamberger Verordnung vom 12.12.1682, zitiert nach: Fehn, Chronik von Kronach (wie Anm. 66), S. 239.

87 Ebd.

In der Folge wandten sich die Pretzfelder schließlich ans Reichsoberhaupt nach Wien, um Schutz vor den Ebermannstädter „Ausfällen" zu suchen. In der Tat verfasste der den Reichsritterschaften wohlgesonnene Kaiser im Juli 1690 ein kaiserliches „*Rescriptum*" wider die Stadt Ebermannstadt, in dem er anordnete, dass diese die „*Gemeind zu Pretzfeld* [...] *unter anderen auch in dem freyen Bierkauff*" künftig nicht mehr „*beschwehren*" dürfe.[88] Nichtsdestotrotz sahen sich der Ebermannstädter Bürgermeister und Magistrat im Recht und marschierten am 6. September 1690 „*sambt der gantzen und völligen Burgerschafft zu Ebermannstatt* [...] *nebst etlichen Zimmerleuten mit ihren Hacken und Hebeisen* [...] *unter offen Drommelschlag gleichsamb als in einem rechten Krieg*" nach Pretzfeld.[89] Dort angekommen, stürmten sie das Schloss und das Brauhaus, um sämtliches Brauzeug „*in kleine stück und trümmer*" zu schlagen und das dort vorgefundene Bier auszuschütten.[90] Wie später dem Kaiser berichtet wurde, habe das wütende Kriegsvolk den Stiebar'schen Schlossverwalter, den herrschaftlichen Jäger und einen Bräugehilfen „*mit stossen und schlägen grausambtlich tractirt, und zu schand und spott gefänglich mit* [...] *nacher Ebermannstatt geschleppet*", wo diese gleich den „*ärgsten Maleficanten* [...] *in ein böses schimpfliches Gefängnis geworffen*" wurden.[91] Dem Schlossvogt wurde unverhohlen angedroht, dass man „*bey wiederanstellendem Breuwerck bald wiederkommen und übel ärger machen*" werde.[92] Freilich blieben dieser „Kriegszug" und die Androhung weiterer Gewalt nicht ohne Folgen. So erließ Kaiser Leopold im Mai 1691 ein Mandat gegen die Ebermannstädter, in welchem er ihr „*hochmüttiges unternehmen*" als schweren Landfriedensbruch anprangerte und die Stadtoberen vor den Wiener Reichshofrat zitierte.[93]

88 Zu finden im Mandat Kaiser Leopolds vom 28.5.1691, zitiert nach: Kraus, Pretzfeld (wie Anm. 42), S. 111.

89 „*Brief des Adam Ernst Marschalk von Ebert in Wildenberg an den Consulenten und Advocaten der Reichsritterschaft aller Sechs Orth in Franken in Nürnberg, Johan Friedrich Schober*", vom 6.9.1690; Mandat Kaiser Leopolds vom 28.5.1691, beides abgedruckt bei: Kraus, Pretzfeld (wie Anm. 42), S. 105, 109.

90 Mandat Kaiser Leopolds vom 28.5.1691, abgedruckt bei: Kraus, Pretzfeld (wie Anm. 42), S. 109.

91 Ebd., S. 109 f. und „*Brief des Adam Ernst Marschalk von Ebert in Wildenberg* [in seiner Eigenschaft als Stiebarscher Vormund] *an den Consulenten und Advocaten der Reichsritterschaft aller Sechs Orth in Franken, Johann Friedrich Schober*", vom 6.9.1690, abgedruckt bei: Kraus, Pretzfeld (wie Anm. 42), S. 105.

92 Mandat Kaiser Leopolds vom 28.5.1691, abgedruckt bei: Kraus, Pretzfeld (wie Anm. 42), S. 110.

93 Mandat Kaiser Leopolds vom 28.5.1691, abgedruckt bei: Kraus, Pretzfeld (wie Anm. 42), S. 111 f.

Scheinbar völlig unbeeindruckt von der kaiserlichen Anordnung rückten die Ebermannstädter bereits zwei Wochen später erneut in Pretzfeld ein. Nur mit Mühe konnte der Pretzfelder Vogt die ritterschaftlichen Untertanen daran hindern, sich den Eindringlingen mit Waffen entgegenzustellen. Unweigerlich, so berichteten Zeugen, wäre sonst „*ein großer tumult* [...] *und leicht Mord und Todtschlag*“ entstanden, da „*das ganze Ebermannstatt mit Knechten und Buben, alle bewehrt in Pret*[z]*feld eingestürmet*“ seien, „*dabey mit Rohren viel Schuß gethan*“.[94]

Zwar versuchte der Bamberger Bischof den politischen Schaden einzugrenzen, indem er kurz darauf die „*Burgerschaft zu Ebermannstadt*“ eindringlich ermahnte, sich künftig der „*allerhöchsten Verordnung*“ des Kaisers seiner Intention gemäß zu beugen.[95] Doch diesmal waren die Ebermannstädter zu weit gegangen. Aufgrund des erneuten Bruchs des Landfriedens und der Missachtung des kaiserlichen Mandats bekam die Stadt den Zorn Kaiser Leopolds zu spüren. Dieser ließ die Ortschaft kurzerhand von kaiserlichen Truppen besetzen, um die Ratsoberen langen Verhören zu unterziehen und den Bürgermeister wegen Majestätsbeleidigung anzuklagen. Am Ende mussten die Bürger hohe Bußzahlungen entrichten.[96] Dennoch vergingen nur wenige Jahre, bis die Ebermannstädter Strafaktionen und Bierpfändungen gegen umliegende Dorfschaften wieder wie gewohnt einsetzten.[97] Resignierend notierte 1729 der Stiebar'sche Steuereinnehmer in Pretzfeld, dass das „*Umgeld von denen dermahligen dreyen Wirthen*“ nach wie vor „*nicht in Ansatz gebracht werden* [könne], *weil dieselben gleich denen anderen, das benötigte Bier in Ebermannstadt zu nehmen gezwungen*“ würden.[98]

Was für Ebermannstadt galt, lässt sich – wenngleich weniger dramatisch – auch von anderen bambergischen Landstädten, wie der Stadt Waischenfeld,

---

94 StadtA Ebermannstadt, A 10/027, Zeugenvernehmungen des Stadtrichters zu Nürnberg, Tobias Tucher von Simmelsdorff und Winterstein, vom 3.10.1691 und des markgräflich streitbergischen Amtmanns Christian Ratiborsky von Sechzebus vom 13.10.1691; vgl. auch: Kraus, Pretzfeld (wie Anm. 42), S. 117 f., 121–123.

95 StadtA Ebermannstadt, A 10/028, Schreiben des Fürstbischofs Marquard Sebastian [Schenk von Stauffenberg] zu Bamberg an „*Unßeren lieben getreuen, Burgermeister und Rath, dan gesambter Burgerschafft zu Ebermanstadt*“ vom 29.8.1691, gegengezeichnet in Ebermannstadt am 31.8.1691.

96 Vgl. Arneth, Ebermannstädter Bierkrieg (wie Anm. 77), S. 268; Kraus, Pretzfeld (wie Anm. 42), S. 130.

97 Aktenkundig wieder nachweisbar ab ca. 1700. Vgl. StadtA Ebermannstadt, A 10/100 und A 10/106–111.

98 Zitiert nach: Glas, Pretzfeld [I] (wie Anm. 76), S. 569, Anm. 31; vgl. auch: ebd., S. 52.

berichten. So führte letztere in den Jahren 1669 bis 1737 unter anderem ein Streitverfahren mit den Rittern von Rabenstein wegen des Bierverlags in Kirchahorn und Umgebung.[99]

Wie im Falle Ebermannstadts und Waischenfelds führten gewaltsame Übergriffe und Pfändungen auch anderer Bamberger Landstädte auf ihr Umland nicht selten zu längeren juristischen Auseinandersetzungen, die vor dem Reichskammergericht oder dem Reichshofrat ausgetragen wurden.[100] Im Verlaufe dieser Auseinandersetzungen entschloss sich die Reichsritterschaft des fränkischen Kantons Gebirg, gemeinsam gegen Waischenfeld, Ebermannstadt und fünf weitere hochstiftische Munizipalstädte (Pottenstein, Hollfeld, Weismain, Burgkunstadt und Kupferberg) juristisch vorzugehen.[101] Vorher hatte sich der Bamberger Fürstbischof Lothar Franz von Schönborn unzweideutig auf die Seite seiner Stadträte und Bürgermeister gestellt.[102] Die klagenden Reichsritter hatte er 1717 wissen lassen, dass in Sachen der *„Ritterschaft Orts Gebürg“* gegen die Stadt Ebermannstadt, *„den Bierverlag betreffend* [...] *von dem Bischöflich Bambergischen Gericht, Hofrathspräsidenten, Kanzlern, Vicekanzlern und weltlichen Räthen zurecht erkannt* [worden sei], *daß eine Meilweegs Rings umb Ebermannstatt denen Klägern das Recht des Bier-Verlags ab- denen Beklagten aber zuzusprechen sey, inmaßen dann ein solches recht bemelten Klägern ab- und denen beklagten Ebermannstättern hiermit zugesprochen wird*“.[103] Mit diesem und ähnlichen Urteilen mochte sich die Korporation

---

99 Vgl. StABa, Hochstift Bamberg, Hofkammer, Akten und Bände, Nr. 2247, Reichskammergerichtsprozess zwischen der Stadt Waischenfeld bzw. dem Hochstift Bamberg und den von Rabenstein wegen des Bierverlags in den Rabenstein'schen Orten.

100 So prozessierte der Bürgermeister und Rat zu Hollfeld u.a. mit den Rittern von Eisenheim und den Herren von Aufseß. Die Bamberger Stadt Pottenstein stritt mit den Rittern Groß von Trockau wegen des Bierverlags und des Umgelds in Kohlstein, Rackersberg und Trockau. Vgl. zur Streitsache zwischen Hollfeld und Friedrich Hector von Aufseß (1688) wegen des Bierverlags zu Wohnsdorf: StABa, Archiv der Freiherrn von und zu Aufseß, Sign. 2805; zu Pottenstein (nachgewiesen u.a. für das Jahr 1698) vgl.: StABa, Hochstift Bamberg, Hofkammer, Akten und Bände, Nr. 1440.

101 Vgl. die Streitschriften in: StadtA Ebermannstadt, A 10/032; LA Baden-Württemberg, Abt. StA Sigmaringen, Gesamtarchiv Schenk von Stauffenberg, Akten Burggrub-Greifenstein, Dep. 38 T 5, Nr. 157.

102 Zum im absolutistischen Stil regierenden Fürstbischof Lothar Franz von Schönborn vgl. u.a. die Beiträge in: Erich Schneider / Dieter J. Weiß (Hg.), 300 Jahre Schloss Weißenstein ob Pommersfelden. Wissenschaftliches Symposium der Gesellschaft für fränkische Geschichte am 15. und 16. September 2011 (Veröffentlichungen der GffG VIII/17), Würzburg 2014.

103 LA Baden-Württemberg, Abt. StA Sigmaringen, Gesamtarchiv Schenk von Stauffenberg, Akten Burggrub-Greifenstein, Dep. 38 T 5, Nr. 157, Brief vom Hochstift

der Reichsritter nicht abfinden, weswegen sie zunächst beim Kreisgericht, dann beim kaiserlichen Reichshofrat in Wien dagegen appellierte.[104] Die Unsummen an Prozesskosten verschlingenden Verhandlungen zogen sich bis in die Mitte der 1730er Jahre hin und führten beinahe zum finanziellen Ruin Ebermannstadts. Immerhin konnte hierdurch ein klares Urteil gegen den Stadtmagistrat wegen seines gewaltsamen Vorgehens gegen das Umland verhindert werden.[105]

4.2. „Zollkrieg" mit Nürnberg und Handelskonkurrenz durch ritterschaftliche „Schutzjuden"

Bekanntlich verfügte die Stadt Ebermannstadt seit ihrer Gründung über das Markt- und Zollrecht. Auch hier sind Auseinandersetzungen mit außerstädtischen Institutionen bekannt. So klagten die Nürnberger Kaufleute respektive der Rat der Reichsstadt Nürnberg 1453, 1455 und 1509 mehrmals über die von den Ebermannstädtern erhobenen, ihrer Auffassung nach ungerechtfertigten Zölle.[106] Die Nürnberger beschwerten sich, dass ihnen beim Besuch der Jahr- und Wochenmärkte zu Ebermannstadt *„wider altes Herkommen"* ein Zoll von Hennen, Hühnern, Vögeln, Eiern, Schmalz, Käse und anderen dergleichen *„essenden Dingen, so sy desselben ennde zu faylen marckt kauffen"* wollten, auferlegt würde.[107] Es komme sogar vor, so die Klage, dass man den Nürnbergern

---

Bamberg an *„den Reichsfrey-Hochwohlgebohrnen Herrn Philipp Adam Sigmund Schenk Freyherrn von Stauffenberg und Greifenstein* [...]", vom 28. Juli 1717.

104 Vgl. LA Baden-Württemberg, Abt. StA Sigmaringen, Gesamtarchiv Schenk von Stauffenberg, Akten Burggrub-Greifenstein, Dep. 38 T 5, Nr. 157, Actum Burggrub, 3. Aug. 1755; vgl. auch: StadtA Ebermannstadt, A 10/030 und A 10/031.

105 Vgl. hierzu die zahlreichen Akteneinträge aus den Jahren 1700 bis 1736 im Ebermannstädter Stadtarchiv: StadtA Ebermannstadt, A 10/030–032. Allein für die Jahre 1733–1735 betrugen die Kosten für den von Ebermannstadt bestellten Agenten am Wiener Reichshofrat 95 Gulden 34 Kreuzer. Vgl. StadtA Ebermannstadt, A 10/030, Aufstellung der Prozesskosten vom 16. Juli 1735. Im Jahre 1700 waren für Ebermannstadt bereits 248 Gulden 4 Groschen Anwaltskosten fällig gewesen. Vgl. Kraus, Pretzfeld (wie Anm. 42), S. 130. Vgl. allgemein auch: Rainer Hofmann / Bärbel Kerkhoff-Hader (Hg.), Bierkultur – „Rund um's Bier in Franken", Tüchersfeld 2007, S. 28.

106 Vgl. Girsig, Ebermannstadt von der Gründung des Ortes bis ins 20. Jahrhundert (wie Anm. 33), S. 77. Vgl. zur 1509 genannten Zollerhebung durch den Ebermannstädter Amtmann: StAN, Reichsstadt Nürnberg, Ratskanzlei, A-Laden, Akten A 64, Nr. 4. Vgl. im Übrigen auch: Beck, Ebermannstädter Heimatbuch (wie Anm. 38), S. 114, 116.

107 Beschwerde des Nürnberger Rates über Übergriffe der Ebermannstädter den Nürnberger Kaufleuten gegenüber (1509), zitiert nach: Beck, Ebermannstädter Heimatbuch (wie Anm. 38), S. 116.

überhaupt nicht mehr gestatten wolle, etwas zu kaufen. Ein offensichtliches Ziel der Ebermannstädter war es dabei, den billigen Aufkauf ihrer Waren durch die Nürnberger zu verhindern. Andererseits versuchte die Stadt fremde Konkurrenz von ihren Märkten und ihrem Einzugsgebiet fernzuhalten, weswegen sich der Nürnberger Rat im Jahre 1607 beim Bischof erneut, diesmal über den Viehzoll, der in Ebermannstadt den Nürnberger Kaufleuten auferlegt wurde, beschwerte.[108] 1681 plante die Markgrafschaft Bayreuth gemeinsam mit dem Nürnberger Landalmosenamt gegen Ebermannstadt vorzugehen, da es dort ihren Bäckern verboten worden war, Brot auf den öffentlichen Markttagen zu verkaufen.[109] Einem Muggendorfer Bäcker war sogar das Brot weggenommen worden, nachdem man ihn vorher das Standgeld hatte bezahlen lassen. Die Pretzfelder Bäcker vermieden es hingegen von vornherein, die Ebermannstädter Märkte aufzusuchen, da sie sonst, so eine schriftliche Klage des Nürnberger Landalmosenamtes vom August 1681, „*unfehlbar zu besorgen gehabt hätten, daß man ihnen dort das Brot wegnehmen werde, da sie nicht dahin, sondern nach Gräfenberg zünftig seien*".[110] Konflikte gab es bis zum Ende des Alten Reiches auch wegen der Zollerhebung in Pretzfeld. Dort hatte die bischöfliche Regierung Ebermannstadt erlaubt, eine „*Zolltafel*" aufzustellen und Zoll zu erheben. Dies widersprach jedoch dem dort bestehenden Kondominium, worüber sich der markgräfliche Amtmann in Streitberg beschwerte.[111]

Eine weitere Konkurrenz, gegen die sich das Bamberger Landstädtchen Ebermannstadt zur Wehr setzte, waren Viehmärkte fremder Herrschaften. Letztere befanden sich in der Regel in Händen jüdischer Händler, die, historisch bedingt, außerhalb der Städte wirtschaftlich Fuß gefasst hatten. Hatten die fränkischen Juden in ältester Zeit nur in größeren Städten gelebt, so siedelten sie seit dem 12., insbesondere aber seit dem 13. Jahrhundert vermehrt auch in Burgsiedlungen und Marktorten auf dem Land. Veranlasst worden war diese Entwicklung in erster Linie durch Adelige, die sich von der Ansiedlung von Judengemeinden wirtschaftliche Vorteile erhofften. Eine wichtige wirtschaftliche Grundlage für das Entstehen von Judengemeinden waren neben Handel und Gewerbe in größeren Orten vor allem der Vieh- und Hopfenhandel sowie das Webereihandwerk. Auch der Geldverleih stellte eine wichtige Einnahmequelle jüdischer Bewohner dar. Für Gegenden, in denen sich Juden mit ihren guten Handelsbeziehungen

---

108 Vgl. Beck, Ebermannstädter Heimatbuch (wie Anm. 38), S. 116.

109 Vgl. hierzu und zum Folgenden: ebd., S. 117.

110 Zitiert nach: ebd.

111 Vgl. StABa, Markgraftum Brandenburg-Bayreuth, Geheime Landesregierung, Nr. 3396, Amtsbeschreibung Kastenamt Streitberg 1752, Rubrik „Pretzfeld".

und ihrer Finanzkraft angesiedelt hatten, bedeutete dies nicht nur Vorteile für den Waren- und Geldverkehr, sondern auch für das örtliche und regionale Kreditwesen.[112]

Mehrfache Pogrome, in denen die Juden aus den großen Städten vertrieben wurden – so etwa anlässlich des Kreuzfahrer-Pogroms von 1096 oder des sogenannten Rintfleisch-Pogroms von 1298 –, führten zu deren Flucht aufs Land. Gemäß einer urkundlichen Überlieferung aus dem Jahre 1343 gab es gegen Ende des 13. Jahrhunderts Judengemeinden in Waischenfeld, Ebermannstadt, Hollfeld, Eggolsheim und Forchheim.[113] Besonders infolge des Pogroms von 1298, während dem auch in Ebermannstadt Verfolgungen stattfanden, hatten sich Juden vermehrt unter dem Schutz des Adels, insbesondere der Schlüsselberger, in Märkten und Dörfern des Fränkischen Jura angesiedelt. Verstärkt wurde dieser Trend durch weitere Pogrome, die in Franken Mitte des 14. Jahrhunderts im Zusammenhang mit der Pest stattfanden. Ihren wirtschaftlichen Vorteil durch die Ansiedlung der Juden erkennend, hatten sich beispielsweise die Herren von Stiebar im Jahre 1403 vom bischöflichen Amt Eggolsheim einen Judenschutzbrief ausstellen lassen. Zu ritterschaftlichen Orten, in denen sich solchermaßen geschützte jüdische Untertanen der Stiebar niederließen, zählten die nahe Ebermannstadts gelegenen Dörfer Weilersbach, Pretzfeld, Rüssenbach und Hagenbach.[114] Auch in anderen umliegenden Gemeinden mit ritterschaftlichem Grundbesitz wurden Juden im Laufe der Zeit sesshaft, so in Aufseß, Heiligenstadt oder Wannbach.[115]

---

112 Vgl. Andraschke, Geschichte des Marktes zu Eggolsheim (wie Anm. 3), S. 37; Glas, Pretzfeld II (wie Anm. 39), S. 107–109.

113 Vgl. Andraschke, Geschichte des Marktes zu Eggolsheim (wie Anm. 3), S. 34–37. Vgl. hierzu und zum Folgenden insbes. auch: Gerhard Philipp Wolf / Walter Tausendpfund, Obrigkeit und jüdische Untertanen in der Fränkischen Schweiz, in: Fränkische-Schweiz-Verein (Hg.), Jüdisches Leben in der Fränkischen Schweiz (Die Fränkische Schweiz - Landschaft und Kultur 11), Erlangen 1997, S. 79–149, hier: S. 79–113.

114 Jüdische Einwohner sind belegt in Pretzfeld durchgehend seit 1566 (vorher schon 1324 und 1452–1461), in Hagenbach seit 1638. Vgl. Andraschke, Geschichte des Marktes zu Eggolsheim (wie Anm. 3), S. 36 f.; Glas, Pretzfeld [I] (wie Anm. 76), S. 931–936; Fastnacht, Ebermannstadt (wie Anm. 2), S. 229–231; Glas, Pretzfeld II (wie Anm. 39), S. 103–110. Vgl. zu den Stiebar'schen Schutzjuden in Rüssenbach: Josef Seitz, Pretzfeld, in: Fränkische-Schweiz-Verein, Jüdisches Leben in der Fränkischen Schweiz (wie Anm. 113), S. 343–392, hier: S. 367. Vgl. zur Judenverfolgung in Ebermannstadt u.a.: Fastnacht, Ebermannstadt (wie Anm. 2), S. 51.

115 Belegt in Wannbach seit 1593, in Heiligenstadt seit 1574. Vgl. Glas, Pretzfeld II (wie Anm. 39), S. 946–951; Dieter Zöberlein, Gemeindechronik Markt Heiligenstadt i.OFr. – Herausgegeben anläßlich der Feier des 450-jährigen Jubiläums der Verleihung

Im Jahre 1422 verabredeten die Hochstifte Würzburg und Bamberg mit dem Markgraftum Brandenburg, die Juden aus ihren Territorien auszuweisen. Betroffen hiervon war unter anderem die jüdische Gemeinde im nördlich von Waischenfeld gelegenen Städtchen Hollfeld, wo sich die Spur der jüdischen Bewohner im dritten Viertel des 15. Jahrhunderts völlig verliert. Vermutlich galt dies auch für Ebermannstadt und Waischenfeld. In letzterem wird 1433 letztmals ein dort wohnender Jude namentlich erwähnt, in Ebermannstadt verließen spätestens 1504 die letzten Juden den Ort.[116] Nach dem Hochstift Bamberg, wo die Ausweisungen zwischen 1464 und 1478 vollzogen wurden, machten auch die umliegenden Territorialmächte ernst mit den Judenvertreibungen. So wurden ab 1499 in Nürnberg keine Juden mehr geduldet, 1515 verfügte der Markgraf von Brandenburg-Kulmbach die vollständige Ausweisung der Judenschaft, desgleichen geschah 1560. Infolgedessen finden sich insbesondere seit den 1560er Jahren immer mehr „Schutzjuden" in ritterschaftlichen Orten der Fränkischen Schweiz, darunter insbesondere als Hintersassen und Beständner auf Lehen der Herren von Streitberg, Aufseß, Egloffstein, Wiesenthau und Stiebar. Auf bischöflichen Lehen tauchen hingegen erst seit dem Ende des Dreißigjährigen Krieges wieder Juden als Hausbesitzer auf, was durch den Heimfall oder Ankauf von Rittergütern bedingt war.[117] Aus jener Zeit kennen wir eine landesfürstliche Verordnung, die das (Berufs-)Leben der in den Bamberger Amtsbezirken lebenden Juden streng reglementierte. Darin wurde bestimmt, *„daß in denen jenig Städten Flecken und Dorffschafften, wo bishero kein Jud gewohnet, auch fürs künfftig keiner eingelassen, dann in denen jenig Orthen, wo vor alters nur eine gewisse anzahl geduldet werde, sothane anzahl nicht vermehret sondern allenfalls dahin wiederumb nach und nach reduciret werd solle"*.[118]

---

des Marktrechtes an Heiligenstadt i.OFr., Heiligenstadt 1995, S. 159 f., 181, 194–203, 220–233 und 237.

116 Vgl. Gerhard Philipp Wolf / Walter Tausendpfund, Schnaittach – Landesrabbinat, in: Fränkische-Schweiz-Verein, Jüdisches Leben in der Fränkischen Schweiz (wie Anm. 113), S. 632–687, hier: S. 632–636; Germania Judaica II/1, S. 367 f. und Germania Judaica III/1, S. 571 f. (Hollfeld); Germania Judaica II/1, S. 182 f. (Ebermannstadt); vgl. hierzu und zur Nennung eines *„Abraham, Jude von Weyschenfelt"* im Jahre 1433 ausführlich: Löwisch, Die drei Waischenfelder Stadtprivilegien (wie Anm. 3), S. 20–29, insbes.: S. 26–28.

117 Vgl. Glas, Pretzfeld [I] (wie Anm. 76), S. 931–936; Glas, Pretzfeld II (wie Anm. 39), S. 103 f., 946 f.; Seitz, Pretzfeld (wie Anm. 114), S. 345–347; Wolf / Tausendpfund, Schnaittach (wie Anm. 116), S. 632–636.

118 *„Hohe Landsherrliche Bamberg*[ische] *Verordnung"* (hier für die Stadt und das Amt Waischenfeld), um 1680, zitiert nach: Löwisch, Die drei Waischenfelder Stadtprivilegien (wie Anm. 3), S. 26–28.

Trotz der zeitweise sehr judenfeindlichen Politik der größeren Territorien blieben die fränkischen Landjuden für den regionalen Handel und das Kreditgeschäft – soweit sie nicht von Pogromen und Vertreibungen betroffen waren – eine wichtige Personengruppe im Wirtschaftsleben. Von anderen Berufszweigen bis 1813 ausgeschlossen, waren sie als Krämer, Hausierer und Geldverleiher, aber auch als erfolgreiche Viehhändler tätig. In Hagenbach erhielten die israelitischen Viehhändler beispielsweise 1698 das Recht, gegen eine Pachtgebühr von einem Reichstaler ihr Handelsvieh den Sabbat über auf den gemeindlichen Weideflächen zu hüten und zu weiden. Dort, in Wannbach und Pretzfeld, wurde ihnen gegen die Zahlung von Zöllen seit den 1730er Jahren die offizielle Befugnis des Viehhandels zugestanden. In Wannbach, wo sie hierfür eigens ein Viehhaus unterhielten, durfte seit 1739 an jedem Freitag vor dem Ebermannstädter Markttag ein regulärer Viehmarkt abgehalten werden.[119] Als sogenannte „*Marktjuden*" erschienen jüdische Händler auch auf anderen Jahrmärkten der Fränkischen Schweiz, wie unter anderem für 1785 in Muggendorf belegt.[120]

Eine Folge dieser Entwicklung war, dass es zwar auf dem Land rund um Waischenfeld und Ebermannstadt reges jüdisches (Handels-)Leben gab, die betreffenden Städte aber seit dem Ende des Mittelalters diesbezüglich sehr eingeschränkt waren.[121] So kann es nicht verwundern, dass es auch mit den jüdischen Händlern zu teils erbitterten Auseinandersetzungen kam, sobald die städtischen Eliten ihre Rechte und monetären Interessen bedroht sahen. Einen besonderen Zankapfel stellten hierbei die Wannbacher Viehmärkte dar, gegen deren Abhaltung sich die Ebermannstädter mehrmals – wenngleich wenig erfolgreich – wehrten. Das Dörfchen Wannbach lag zwar innerhalb der Ebermannstädter Bannmeile, war jedoch von der örtlichen Ritterschaft dominiert. So sind dort bereits seit 1430 Gastwirte, seit dem 17. Jahrhundert auch andere Gewerbe, wie Bäcker, Metzger, Schuster, Schneider, Weber, Zimmerleute und Maurer, belegt.[122] Zunächst war es 1779 in einem Bierverlagsstreit zu einem „*Überfall*"

---

119 Vgl. Glas, Pretzfeld II (wie Anm. 39), S. 105, 107 f., 948–950 und 958.

120 Vgl. ebd., S. 952.

121 Zwar berichtet J. G. Biedermann 1752 in seiner Beschreibung des Hochstifts Bamberg, dass auch in Waischenfeld und Ebermannstadt Juden wohnten; in Ebermannstadt gebe es sogar „*eine nahmhafte Anzahl* [von] *Schutzjuden*". Doch scheint deren tatsächliche Anzahl und ihr Einfluss auf das örtliche Wirtschaftsleben recht unbedeutend gewesen zu sein, da hierüber aus anderen Quellen kaum etwas überliefert ist. Vgl. Biedermann, Topographische Beschreibung des Fürstenthums Bamberg (wie Anm. 13), fol. 32, Nr. 112 (Ebermannstadt) und fol. 132, Nr. 702 (Waischenfeld).

122 Vgl. Glas, Pretzfeld II (wie Anm. 39), S. 956 f.

durch die Ebermannstädter gekommen, wogegen die Wannbacher beim Hochstift Beschwerde *„wegen Anmaßung von Rechten"* eingereicht hatten.[123] Zwischen 1790 und 1800 stritt dann Ebermannstadt mit den Wannbachern um den am Ort betriebenen Viehhandel, wobei es dabei zu *„landfriedensbrüchigen Störungen der Viehmärkte"* kam.[124] Selbst die Kritik des Ebermannstädter Vogts Möhrlein am Vorgehen gegen die Wannbacher Märkte ließ den Bürgerrat zu Ebermannstadt unbeeindruckt. Im Gegenteil legte dieser gegen die Ermahnung des eigenen Vogts im Jahre 1800 sogar noch Beschwerde ein.[125] Ähnlich wie der „Bierkrieg" mit Pretzfeld, zogen auch diese Auseinandersetzungen weite Kreise in der Öffentlichkeit. So berichtete Johann Caspar Bundschuh in seinem 1804 erschienenen Standardwerk „Geographisch Statistisch-Topographisches Lexikon von Franken" über Wannbach, dass sich dort zwar zahlreiche Juden vom Viehhandel nährten und es einen regelmäßigen Viehmarkt gebe. Jedoch wolle die *„gemeine Stadt Ebermannstadt* […] *ein kaiserl*[iches] *Privilegium haben, vermöge dessen sie das Zoll- und Marktregale zu Wannbach in Anspruch nehmen will, weil gedachter Ort in ihrer Bannmeile läge"*.[126] Ungeachtet dessen, dass die Grafen von Seinsheim 1764 mit dem Rittergut Wannbach belehnt worden seien, habe sich das Amt Ebermannstadt *„vermöge besonderen Vertrags ausdrücklich das Zoll- und Aufschlagsregale"* vorbehalten.[127] Konnten die Ebermannstädter die jüdische Konkurrenz im Wannbacher Viehhandel nicht gänzlich unterbinden, so wollte man diese doch durch Störungen zurückdrängen oder zumindest daran mitverdienen.

Wie diese Beispiele zeigten, schafften es die Ebermannstädter und – wenngleich durch die dürftigere Quellenlage nicht so deutlich belegbar – offenbar auch die Waischenfelder meist, ihren Monopolanspruch und damit ihre wirtschaftlichen Interessen auf Grundlage der Stadtprivilegien König Ludwigs des Bayern durchzusetzen; und dies gegen nicht unerhebliche Widerstände in der Landbevölkerung und beim Adel. Weder langwierige Gerichtsprozesse noch die Ungnade des den Landfrieden schützenden Kaisers hielten die beiden Kleinstädte davon ab, ihre Dominanz über die mindermächtigen, jedenfalls politisch schwächeren Nachbardörfer und Rittersitze zu behaupten – wenn nötig, auch mit fragwürdiger Gewalt.

---

123 StadtA Ebermannstadt, A 10/070, Bierverlagsstreit mit Wannbach.

124 StadtA Ebermannstadt, A 10/071–073.

125 StadtA Ebermannstadt, A 10/073.

126 Bundschuh, Geographisches Statistisch-Topographisches Lexikon von Franken, Bd. 6 (wie Anm. 4), Sp. 102.

127 Ebd.

Dies war durchaus keine Selbstverständlichkeit, wie uns das Beispiel der 20 Kilometer nördlich von Waischenfeld gelegenen Ortschaft Wonsees nahelegt. Dieser, von alters her mit einem Vogteiamt ausgestattete Centgerichtsort war 1290 vom Nürnberger Burggrafen Friedrich III. erworben worden. 1328 hatte der in jenem Jahr zum Kaiser gekrönte Ludwig der Bayer auch diesem Dorf das Recht erteilt, „*eine Stadtbefestigung zu bauen*", dort auf Dauer „*die Blutgerichtsbarkeit auszuüben und einen Wochenmarkt abzuhalten*".[128] Gleichzeitig verlieh er der „*Stadt*", ebenso wie vorher schon Ebermannstadt, alle Freiheiten, „*wie sie Nürnberg*" besaß. 1355 bestätigte Kaiser Karl IV. die städtischen Privilegien mit der Erlaubnis, sechs Jahrmärkte abzuhalten und den Ort mit einer Mauer zu umgeben; in Urkunden des Jahres 1392 wird Wonsees als eigenständige Pfarrei aufgeführt. Doch während die Ortschaft zu jener Zeit noch als Stadt bezeichnet wird, taucht sie in den Quellen ab 1497 nur mehr als „*Markt*" Wonsees auf. Ob Verheerungen des Hussitenkrieges von 1430 hierfür mitverantwortlich sind, ist mangels Quellen nicht zu klären. Einer der Gründe für den Verlust seiner Stellung als „Stadt nach Nürnberger Recht" dürfte aber sicher der fehlende Anschluss an eine der (Fern-)Handelsstraßen gewesen sein. In einer Beschreibung aus dem letzten Viertel des 17. Jahrhunderts wurde dem Ort bescheinigt, dass mangels einer „*Straße*" durch den Marktflecken das dortige Gewerbe nur schlecht liefe. Außerdem stand Wonsees in dauerhafter Konkurrenz zum in unmittelbarer Nähe gelegenen burg- beziehungsweise markgräflichen Kastenamt Zwernitz, das sich Stück für Stück Wonseeser Gerichtsrechte anmaßte. Während der Burg Zwernitz, die mit ihrem Felsengarten Sanspareil zu einem fürstlichen Jagd- und Lustschloss ausgebaut wurde, zollerischerseits immer größeres Augenmerk geschenkt wurde, verlor das abseits gelegene Wonsees an Bedeutung. Hinzu kam die gewerbliche Konkurrenz der in direkter Nachbarschaft gelegenen bambergischen Stadt Hollfeld, die mit ihren florierenden Märkten und Handwerken ein Aufkommen des an der Bayreuther Peripherie gelegenen Wonsees zu verhindern wusste.[129]

128 Zitiert nach: Fastnacht, Ebermannstadt (wie Anm. 2), S. 371.

129 Vgl. F[riedrich] G[ottlob] Leonhardi, Erdbeschreibung der Fränkischen Fürstenthümer Bayreuth und Anspach, Halle 1797, S. 114 f.; Joachim Heinrich Jäck, Der Ober-Mainkreis im Königreich Bayern, historisch, statistisch, topographisch und geographisch beschrieben von vielen Gelehrten, 1. Bd., Augsburg 1830, S. 579 f.; Gustav Voit, Geschichtlicher Abriß, in: Gustav Voit / Brigitte Kaulich / Walter Rüfer (Hg.), Vom Land im Gebirg zur Fränkischen Schweiz. Eine Landschaft wird entdeckt, Erlangen 1992, S. 11–31, hier: S. 16, 26; Rüdiger Barth, Kulmbach. Stadt und Altlandkreis (Historischer Atlas von Bayern, Teil Franken, Reihe I, Heft 38/I und II), München 2012,

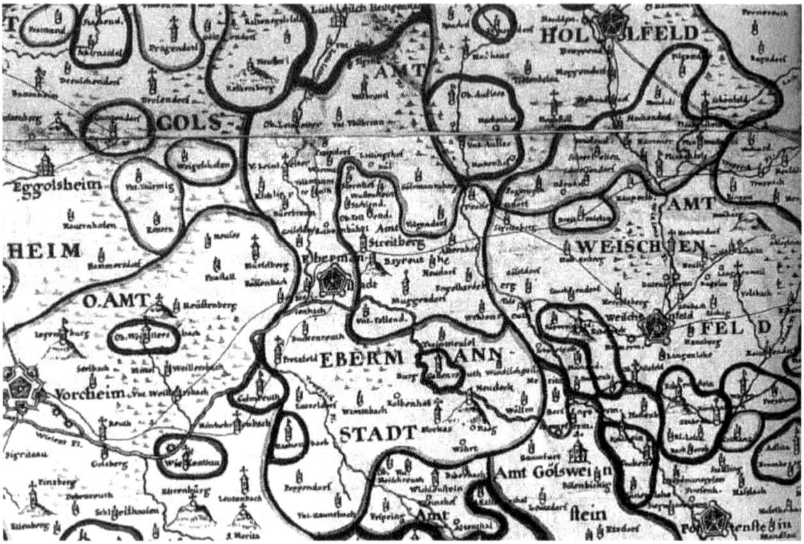

Abb. 30: *Die Städtelandschaft der Fränkischen Schweiz im 18. Jahrhundert, Ausschnitt aus: „Das Hochstift Bamberg" von Johann Baptist Roppelt, 1799. Kolorierte Übersichtszeichnung mit Ämtereinteilung und Angabe der Anrainer. Bildnachweis: Hans Vollet, Weltbild und Kartographie im Hochstift Bamberg („Die Plassenburg". Schriftenreihe für Heimatforschung und Kulturpflege in Ostfranken 47), Kulmbach 1988, Abb. 1.*

5. Höhepunkt und Ende der städtischen Überheblichkeiten

Insofern ist es durchaus bemerkenswert, dass es Waischenfeld und insbesondere dem im Dunstkreis der Königs- und Festungsstadt Forchheim gelegenen kleinen Landstädtchen Ebermannstadt gelang, die von Ludwig dem Bayern verliehene Stellung nicht nur zu behaupten, sondern im Laufe der Jahrhunderte auch wirtschaftlich und machtpolitisch auszubauen.[130] Fast schien es sogar, als könne

S. 112–114; Fastnacht, Ebermannstadt (wie Anm. 2), S. 371 f. Vgl. zur Beschreibung von Wonsees nach dem Dreißigjährigen Krieg (1685): Hellmut Kunstmann, Die Burgen der westlichen und nördlichen Fränkischen Schweiz, 2. Teil: Der Nordwesten und Norden. Leinleitertal, Aufsesstal und oberes Wiesenttal und Randgebiete (Veröffentlichungen der GffG IX/28,2), Würzburg 1972, S. 233.

130 Zur herausragenden Bedeutung Forchheims für das südliche Fürstbistum Bamberg und zu dessen Stadtgeschichte vgl. insbes.: Glas, Forchheim. Stadt und Bürgerschaft (wie Anm. 38).

keine Macht, nicht einmal der als oberster Richter auftretende Kaiser, dem anmaßenden Verhalten der kleinstädtischen Führungseliten auf Dauer Einhalt gebieten. Die vermeintliche wirtschaftliche und politische Machtposition führte aber schließlich zu Überheblichkeiten, die, wie im Falle Ebermannstadts, ein jähes Ende der städtischen Selbstherrlichkeit brachten. Von den Erfolgen, die mit den teils gewaltsamen Übergriffen erzielt worden waren, bestärkt, legte sich die Bamberger Kleinstadt und der dort residierende Vogt am Ende mit einem – wie sich zeigen sollte – zu großen Gegner an.

Schon seit Mitte des 18. Jahrhunderts war es mehrmals zu Auseinandersetzungen zwischen Ebermannstadt und dem nahe gelegenen Markt Muggendorf gekommen. Was die Angelegenheit besonders brisant machte, war, dass Muggendorf zwar innerhalb der städtischen Bannmeile, jedoch außerhalb des Bamberger Territoriums in der markgräflichen Exklave Streitberg lag. Nichtsdestotrotz hatte sich der Rat Ebermannstadts in den Jahren 1748 und 1755 genötigt gesehen, in mehreren Ortschaften Bierfässer gewaltsam pfänden zu lassen, die entgegen dem beanspruchten Monopol aus dem „*fremdländischen*“ Muggendorf bezogen worden waren.[131] Nach weiteren Reibereien mit dem ungeliebten protestantischen Nachbarn kam es schließlich 1786/87 zu einer ernsthaften Auseinandersetzung, die in die spätere Literatur als „*Schauspiel eines eignen Kriegs en miniature, herbeigeführt durch die eifersüchtige Spannung der durch Regierung und Religion geschiedenen Bewohner*“ Eingang fand.[132] Diesmal ging es nicht um wirtschaftliche Interessen, wie den Viehhandel oder Bierverlag, sondern um den Anspruch auf die Hochgerichtsbarkeit über einen zwischen Muggendorf und Streitberg gelegenen Weiler namens Wöhr. Dort beanspruchte, wie es hieß, „*seit undenklichen Zeiten*“ der Markgraf von Brandenburg-Bayreuth die hohe Gerichtsbarkeit, was allerdings die Ebermannstädter bestritten. Im Verlauf des Streits fiel im Mai 1786 eine Schaar Ebermannstädter unter Führung des

131 Vgl. Kraus, Die Häuser Nr. 1–122 (wie Anm. 40), S. 6, 8 f.; vgl. auch: LA Baden-Württemberg, Abt. StA Sigmaringen, Gesamtarchiv Schenk von Stauffenberg, Akten Burggrub-Greifenstein, Dep. 38 T 5, Nr. 157, Actum Burggrub, 3. Aug. 1755 und 9. Sept. 1755. Vgl. zum bereits in den Jahren 1680–1683 schwelenden Streit um das Verbot des Bierverlagsgeschäfts eines Georg Trautner aus Muggendorf in Gößmannsberg: LA Baden-Württemberg, Abt. StA Sigmaringen, Gesamtarchiv Schenk von Stauffenberg, Akten Burggrub-Greifenstein, Dep. 38 T 5, Nr. 64/18.

132 L[orenz] Kraußold / G[eorg] Brock, Geschichte der Fränkischen Schweiz oder Muggendorfs und seiner Umgebungen mit einem kurzgefassten vollständigen Wegweiser für solche, welche die Gegend besuchen, Nürnberg 1837, S. 143.

bischöflichen Stadtvogts in Muggendorf ein, um zwei ihrer *„eifrigsten Gegner"* zu verhaften.[133]

Nachdem kurze Zeit später vom Bamberger Landvolk auch noch ein bayreuthischer Husar misshandelt worden war, griff die hohenzollerische Obrigkeit in das Geschehen ein und verhaftete ihrerseits die Schuldigen, um diese in Steitberg einzukerkern. In Ebermannstadt, so ein zeitgenössischer Bericht, sann man nun auf Rache und wollte die verhafteten Landsleute aus dem Streitberger Gefängnis befreien. Im Gegenzug sollten die vermeintlichen fremden Rädelsführer abgeführt werden. Am Morgen des 19. Januar 1787 rückten der Bamberger Stadtvogt samt Bürgermeister und 700 Mann, *„der ganzen waffenfähigen Mannschaft"* des Distriktes, gegen Streitberg vor. Obwohl die Streitberger nur rund 30 mit Gewehren bewaffnete Männer aufbieten konnten, die durch ebenso viele Mitstreiter aus Muggendorf verstärkt worden waren, konnten sie die Angreifer abwehren. Während des Feuergefechts wurde ein Ebermannstädter Bürger tödlich in die Brust getroffen und mehrere andere verwundet. Besonders, so berichten mündlich überlieferte Quellen, soll sich ein Schneider aus Streitberg im Kampf ausgezeichnet haben, der, als sein Blei verschossen war, von zu Hause eilig *„gläserne Hemdknöpfe holte und nun diese unter die Feinde verschoss"*.[134] Nach kurzer Zeit begannen sich die Angreifer zurückzuziehen und als auch noch *„die Doppelhaken des Streitberger Schlosses sie begrüßten"*, verwandelte sich der Rückzug in die wildeste Flucht.[135] Bayreuther Militär schützte von nun an die markgräflichen Untertanen, der Streit um Wöhr musste auf diplomatischem Wege gelöst werden. Der Ebermannstädter Vogt aber war gezwungen, seinen Hut zu nehmen. Lange Zeit, so berichteten Zeitgenossen, *„konnten die Ebermannstädter ihre Niederlage nicht verschmerzen, so wie sie den Mut ihrer kampffertigen Gegner hob"*.[136]

Nicht lange nach dem Streitberger Desaster gipfelte schließlich die Selbstherrlichkeit der städtischen Elite Ebermannstadts in deren Entmachtung durch die Staatsgewalt. Schon in den Jahren 1740 und 1776 hatte es wegen dem *„sehr unrichtig verwaltet*[en] [...] *Zollwesen und dessen Berechnung zu Ebermannstadt"* einer strengen Maßregelung in zwei *„Hochfürstlichen Verordnungen für das*

---

133 Vgl. ebd., S. 143–145.

134 Ebd., S. 143–145. Hierzu und zum Folgenden vgl. auch: Andreas Neubig, Die Fehde zwischen den Streitbergern und Ebermannstädtern im Jahre 1787, in: Archiv für Geschichte und Altertumskunde von Oberfranken 6/3, 1856, S. 59–71.

135 Kraußold / Brock, Geschichte der Fränkischen Schweiz (wie Anm. 132), S. 145.

136 Ebd.

Municipal Städtlein Ebermannstadt" bedurft.[137] Neben den ungenügend angefertigten Stadtrechnungen wurde insbesondere das „*Amtiren der Burgermeistern*" gerügt. So sei „*solches zeithero zur Ungebühr mißbrauchet worden*", und zwar in der Weise, als „*daß dieselben entweder alleinig oder auch mit dem Stadtschreiber die mehrsten Klag-Händel außer dem versammelten Rath zu Hauß vorgenommen und abgetan*" hätten.[138] Aufgrund des uneinsichtigen Verhaltens der Stadtoberen folgte schließlich 1801 die obrigkeitliche Suspendierung der Führungselite. Wie aus einer Urkunde des Jahres 1802 hervorgeht, führte von nun an ein besonderes Gremium unter Zuhilfenahme eines Rechtsgelehrten die Geschäfte der Stadt, da der „*eigentlich und von undenklichen Zeiten her aus zwey Burgermeistern, zehn Rathsgliedern und einem Stadtschreiber*" bestehende Bürgerrat „*seit einem Jahre aus besonderen hochfürstlichen Verordnungen, und hauptsächlich wegen übelgeführter Wirtschaft einstweilen und bis nach gelegten gründlichen gemeiner Stadt Rechnungen*" seiner Zuständigkeit enthoben war.[139] Die einst glorreiche, durch wirtschaftliche Stärke und kleinstädtische Machtdemonstration begleitete Politik hatte im Taumel von Korruption und Amtsanmaßung ihr jähes Ende gefunden. Erst nach der Einverleibung ins Kurfürstentum respektive Königreich Bayern konnte wieder zu geregelten Verhältnissen in der Stadtverwaltung zurückgekehrt werden; die einst erbittert verteidigten wirtschaftlichen Sonderprivilegien waren nun aber für immer dahin.

6. Ein Resümee

Mit Hilfe ihrer militärischen Stärke und Dank kaiserlicher Klientelpolitik hatten es die Herren von Schlüsselberg an der Wende vom 13. zum 14. Jahrhundert geschafft, einen Großteil des heute als Fränkische Schweiz bekannten Gebiets unter ihre Kontrolle zu bringen. Um seinen Machtanspruch zu behaupten, kam dem Schlüsselberger Konrad II. die zu jener Zeit in Mode gekommene Gründung von Städten sehr gelegen – zumal sein königlich-kaiserlicher Herr, Ludwig der Bayer, aus machtpolitischen Erwägungen heraus von diesbezüglichen Privilegienerteilungen gerne Gebrauch machte. Innerhalb weniger Jahre erhielten so die unter schlüsselbergischer Herrschaft stehenden Dörfer Waischenfeld und

137 Zitiert nach: Girsig, Ebermannstadt von der Gründung des Ortes bis ins 20. Jahrhundert (wie Anm. 33), S. 77; vgl. auch: Beck, Ebermannstädter Heimatbuch (wie Anm. 38), S. 114, 116.

138 Zitiert nach: Girsig, Ebermannstadt von der Gründung des Ortes bis ins 20. Jahrhundert (wie Anm. 33), S. 77.

139 Zitiert nach: ebd., S. 90.

Ebermannstadt das Markt- und Stadtrecht sowie weitreichende Handelsfreiheiten. Trotz Konkurrenz zu den umliegenden altbambergischen Marktzentren Hollfeld, Pottenstein, Eggolsheim und Forchheim konnten beide Städte ihre Privilegien behaupten, auch nachdem sie infolge des Erlöschens der Schlüsselberger 1347 in den Besitz des Hochstifts Bamberg gelangt waren.[140] Mehr noch, wuchsen beide Orte zu kleinen Handwerks- und Handelszentren der mitunter recht kargen Gebirgsregion heran, wenngleich nicht nur die umliegenden altbambergischen Märkte, sondern später auch die reichsritterschaftlichen Gewerbeleute und „Schutzjuden" sowie die markgräflichen Nachbarn eifrig Konkurrenz betrieben. Selbst Versuchen der mächtigen Nürnberger Kaufmannsgilde, im Wiesentgebiet marktpolitisch Fuß zu fassen, konnte früh durch eine geschickte (wenn auch umstrittene) Zollpolitik erfolgreich begegnet werden.

Einen wesentlichen Anteil an der Entwicklung beider Städte hin zu regionalen Zentren leisteten die am Ort angesiedelten landesfürstlichen Ämter, deren Amtsleute teils aus der Bürgerschaft rekrutiert wurden. Im Gegenzug profitierten die Ämter und damit die bischöfliche Regierung vom Schutz und Rückhalt, den die wehrhaften Städte mit ihren gut ausgebauten Strukturen und ihrem Einfluss auf das Umland den fürstlichen Beamten boten – von den durch die Bemühungen der Magistrate garantierten regelmäßigen Steuereinnahmen, insbesondere dem für den Bierkonsum erhobenen Umgeld, gar nicht zu reden.

Parallel hierzu nahmen das bürgerliche Selbstbewusstsein und der damit einhergehende Selbstverwaltungsanspruch ernstzunehmende Formen an. Die Herausbildung eigener Pfarreien durch die Herauslösung aus den altehrwürdigen Kirchenstrukturen war nur ein Ausdruck dieses städtischen Selbstverständnisses. Meist zogen beide, durch Geschichte und gemeinsame Landesherren verbundene Landstädte an einem Strang, besonders wenn es gegen wirtschaftliche Konkurrenten aus der Reichsritterschaft ging. Zwischen den beiden Protagonisten selbst kam es hingegen nur im Ausnahmefall zu Streitigkeiten, obwohl deren „Bannmeilen" und Amtsbezirke doch recht dicht beieinander lagen. Trotz zahlreicher, vor allem durch kriegerische Zerstörungen bedingter Rückschläge

140 Vgl. zu Forchheim: Glas, Forchheim. Stadt und Bürgerschaft (wie Anm. 38); vgl. zu Hollfeld: Stadt Hollfeld, 650 Jahre Stadt Hollfeld (wie Anm. 24); Stadt Hollfed, 1000 Jahre Hollfeld (wie Anm. 40); vgl. zu Pottenstein: Wachter, Pottenstein (wie Anm. 21); vgl. zu Eggolsheim, das zwar bis 1347 zur Hälfte an die Schlüsselberger verpfändet, im Übrigen aber im Besitz und unter der Verwaltung des Hochstifts Bamberg blieb: Andraschke, Geschichte des Marktes zu Eggolsheim (wie Anm. 3); Hans Puff, Zur Verfassungsgeschichte des fürstbischöflich-bambergischen Amtes Senftenberg-Eggolsheim, Diss., Erlangen 1939, S. 7–10.

brachte erst die Einverleibung des Hochstifts Bamberg ins moderne Bayern das Ende des alten kaiserlichen Sonderstatus für beide Gemeinden. Freilich handelten sich insbesondere die Ebermannstädter mit ihrer zur Überheblichkeit neigenden Art im Laufe der Jahrhunderte auch viel Ärger ein, die dem Stadtrat schon vor Auflösung der alten staatlichen Verhältnisse zum Verhängnis wurde.

Abb. 31: *Ansicht des Stadtzentrums von Waischenfeld, Lithografie, nach Originalzeichnung von Carl Kaeppel. Bildnachweis: Theodor Rothbarth / Carl Kaeppel, Die Fränkische Schweiz. Cyclus der interessantesten Punkte aus der Umgegend von Muggendorf und Streitberg, Nürnberg 1840, [o.S.].*

Sektion 3:
Die süddeutsche Städtelandschaft – literarisch, musisch, onomastisch

Klaus Wolf

Reichsstadt versus Residenzstadt. Zum literaturgeschichtlichen Profil oberdeutscher Städte vor 1800

Abstract: The following essay deals with the literary-historical role of cities with a focus on Bavaria. It can be stated that royal residence cities have shaped literary life from the very beginning, since the early Middle Ages. But this is true not only for old high German literature, but especially for middle high German literature. On the other hand, it can be shown that a more urban literary life in imperial cities, especially in the early modern period, but already beginning in the late Middle Ages, determined the motor of progress. It was not until the eighteenth century that the aristocratic residence, insofar as it was ultimately influenced by the Enlightenment as in the case of Frederick the Great, but also at the Ducal Court in Weimar, became a defining feature of literary life.

Die Stadt als Literaturort ist scheinbar erst eine moderne Erscheinung. Bert Brecht verfasst Gedichte für Städtebewohner.[1] Ein älterer Schwabe, er stammte aus Kaufbeuren beziehungsweise aus dem schwäbischen Welden, Ludwig Ganghofer nämlich, dessen 100. Todestag sich am 24. Juli 2020 ausmachen lässt, schien sich im Gegensatz zu Bert Brecht eher auf die ländliche Literatur spezialisiert zu haben. Viele seiner Romane, Erzählungen und Theaterstücke spielen ja in den Alpen. Hierin folgte er einer literarischen Mode um 1900, die längst zuvor durch Ludwig Steub, durch Maximilian Schmidt, genannt Waldschmidt, durch Josef Ruederer, durch Ludwig Thoma, durch Lena Christ und viele andere zum Erfolg geführt worden war.[2] Gerade dieser Rückzug auf ein idealisiertes bayerisches Landleben geschah aber zu einer Zeit, als auch in Bayern Industriestädte ihren Ruß in den Himmel schleuderten. Dies gilt für Nürnberg, Augsburg und München gleichermaßen.[3]

1 Bertolt Brecht, Gedichte für Städtebewohner, herausgegeben und mit einem Nachwort von Franco Buono, Frankfurt am Main 1988.

2 Vgl. die alpine Literaturexplosion in der Dokumentation von: Ulrich Hohoff, Voralpenland und bayerische Alpen in Erzählungen und Romanen. Bibliographie der Jahre 1850–1920 (Editio Bavarica VI), Regensburg 2018.

3 Vgl. Wolfgang Wüst / Tobias Riedl (Hg.), Industrielle Revolution. Regionen im Umbruch: Franken, Schwaben, Bayern (Franconia 6. Beihefte zum Jahrbuch für fränkische Landesforschung), Erlangen 2013.

Ludwig Ganghofer, der selbst übrigens sehr viel Zeit im urbanen München verbrachte, wo er einen mondänen literarischen Salon unterhielt, und der literarisch nicht zuletzt als Feuilletonist in der pulsierenden Metropole Wien geprägt war, stilisierte sich auf zahlreichen Fotografien als Landmensch in Lederhose. Seine alpine Literatur wie die seiner Zeitgenossen Ludwig Thoma und Georg Queri markierte vielmehr eine literarische Mode als Antwort auf krisenhafte Erscheinungen von Industrialisierung, Landflucht und Verstädterung.

Abb. 32: *Ludwig Ganghofer, Veranstalter mondäner literarischer Salons in München, posiert in ländlicher Tracht. Bildnachweis/Foto: Markt Welden.*

Demgegenüber ist festzuhalten, dass die deutsche Literatur von Anfang an städtisch war. Dies will ich bei einem Durchgang durch die Gattungen und Jahrhunderte vom Frühen Mittelalter bis ins ausgehende 18. Jahrhundert exemplifizieren und mich dabei besonders auf den oberdeutschen Raum und speziell das Territorium des heutigen Freistaats Bayern konzentrieren. Dabei wird es in einer

Feinabstimmung um die Differenzierung verschiedener Städtetypen gehen. Als Antipoden seien daher Reichsstädte und Residenzstädte genannt.

Beginnen wir mit Bayerns ältester Dynastie, den Agilolfingern. Diese waren alles andere als germanische Krieger. Sie entstammten wohl selbst teilweise römischen Vorfahren, sie sollen nämlich senatorische Ahnen gehabt haben und versippten sich erfolgreich nicht zuletzt mit den langobardischen Königen. So ist der Tassilokelch mit seiner entsprechenden Umschrift als königliche Allianz zu lesen: „*tassilo dux fortis liutpirc virga regalis*". Selbstredend waren die Langobardenkönige städtisch orientiert. Ihre Residenzen etwa in Monza waren mit römischer Kunst prachtvoll ausgestattet. Nicht weniger urban repräsentierten auch die Agilolfinger, die an städtischen Residenzen ihre Macht und ihr kulturelles Leben entfalteten. Hier sind zuvorderst die alten Römerstädte Regensburg und Salzburg zu nennen. Für Regensburg preist Arbeo von Freising die „*muros maximos*", die gewaltigen römischen Mauern der Stadt. In der alten Römerstadt Salzburg sollte mit dem Virgildom als „*capella regia*" eine königliche Kirche errichtet werden, die durchaus mit dem karolingischen Saint-Denis zu vergleichen war. Nur Freising war kein alter Römerort, gleichwohl hatte Freising städtischen Charakter. An diesen städtischen Residenzen der Agilolfinger gab es ein blühendes literarisches Leben, eine elitäre lateinische Geistigkeit, die sich in prachtvoll illuminierten Pergamentcodices manifestierte.[4]

In späterer Zeit waren die Welfen als bedeutendes bayerisches Herrschergeschlecht nicht weniger urban. Mit der Stadt Regensburg, in der sie gerne residierten, sind zentrale mittelhochdeutsche Werke wie die „Kaiserchronik", aber auch das „Rolandslied" zu verbinden.[5] Nicht weniger urban auch in ihrem Literaturgeschmack und ihrem literarischen Leben waren die großen Konkurrenten der Welfen. Die Staufer nämlich, denen die Wittelsbacher als aufsteigende Dynastie treu dienten. Berühmt für die Staufer ist literaturgeschichtlich Kaiser Heinrich VI., welcher als Minnesänger sogar den *Codex Manesse* mit seiner Miniatur eröffnet:

4 Vgl. Klaus Wolf, Gab es eine Literaturpolitik der Agilolfinger? Ein Beitrag zur regionalen Literaturgeschichtsschreibung, in: Archiv für das Studium der neueren Sprachen und Literaturen 161, 2009, S. 283–292.

5 Vgl. Klaus Wolf, Bayerische Literaturgeschichte. Von Tassilo bis Gerhard Polt, München 2018, S. 67–70.

Abb. 33: *Der Stauferkaiser Heinrich VI. als Minnesänger, in: Große Heidelberger Liederhandschrift (Codex Manesse), Zürich, ca. 1300 bis ca. 1340. Bildnachweis: Universitätsbibliothek Heidelberg, Cod. Pal. germ. 848, fol. 6 r, URL: https://digi.ub.uni-heidelberg.de/diglit/cpg848/0007.*

Aufgeführt wurde Minnesang der Stauferzeit aber nicht zuletzt auf städtischem Terrain. Ein gutes Beispiel dafür ist das Mainzer Hoffest von 1184, situiert in einer großen Stadt, wo sich neben der ritterlichen auch die künstlerische Elite, die großen Dichter und Minnesänger Europas getroffen haben.[6] Und die Wittelsbacher als getreue Parteianhänger der Staufer, die diese auch zu ihren Zügen nach Italien begleiteten, wo Städte wie Mailand ein prächtiges urbanes, den höfischen Künsten gewogenes Milieu abgaben, schauten sich die literarischen Gepflogenheiten der Staufer ab. Im Umfeld der Staufer sangen nämlich die *Troubardours*, die *Trouvères* und die Minnesänger natürlich vor allem im städtischen Milieu. Und es ist kein Wunder, dass die Wittelsbacher an ihre neue Städtegründung Landshut gleich den damals kühnsten und stellenweise sogar Walther von der Vogelweide überflügelnden Minnesänger Neidhart etablierten. Neidhart brachte es auf den Punkt: „*dû sage ze Landeshuote, / wir leben alle in hôhem muote, / niht unvruote!*"[7]

Die Wittelsbacher nutzten aber nicht nur Landshut als Ort ihrer literarischen Repräsentation, sondern selbstredend München. Nach Ludwig dem Bayern, wo vor allem eine lateinische Geistigkeit mit Wilhelm von Ockham, Michael von Cesena und Marsilius von Padua gepflegt wurde, waren es im 15. Jahrhundert vor allem die Herzöge Albrecht III. und Albrecht IV., die München zu einer auch literarisch glanzvollen Residenz ausbauten.[8] Daneben ist aber im frühen 15. Jahrhundert Ingolstadt keineswegs zu vernachlässigen. Ludwig im Bart, dessen Schwester Isabeau de Bavière in Frankreich die Regentschaft ausübte, lebte lange Jahre in Paris und wollte nun auch die literarischen Eindrücke aus Frankreichs Hauptstadt nach Ingolstadt verpflanzen. Hierzu gehört ein wichtiges literarisches Unternehmen, nämlich eine gewaltige Bibelübersetzung mit anspruchsvollen Miniaturen. Ein Verfahren, das Ludwig VII. in Paris durch

6 Vgl. Joachim Heinzle, Das Mittelalter in Daten. Literatur, Kunst, Geschichte. 750–1520, München 1993, S. 138.

7 Vgl. Klaus Wolf, „Du sage zu Landeshuote, wir leben alle in hôhem muote!" Neidhart, Wolfram und andere Dichter an städtischen Höfen der Wittelsbacher, in: Peter Wolf / Evamaria Brockhoff / Richard Fischer / Sarah Schormair / Marco Veronesi (Hg.), Stadt befreit. Wittelsbacher Gründerstädte. Katalog zur Bayerischen Landesausstellung 2020. Friedberg, Wittelsbacher Schloss. Aichach, FeuerHaus (Veröffentlichungen zur Bayerischen Geschichte und Kultur 69), Regensburg 2020, S. 170–174.

8 Vgl. Klaus Wolf, Literatur am Münchener Hof im 15. Jahrhundert, in: Wolfgang Wüst (Hg.), Bayerns Adel – Mikro- und Makrokosmos aristokratischer Lebensformen. Unter Mitarbeit von Lisa Bauereisen, Frankfurt am Main 2017, S. 227–239.

persönliche Bekanntschaften mit Jean Gerson, dem berühmten Professor der Sorbonne kennengelernt hatte.[9]

Abb. 34: *Der Minnesänger Neidhart, der am wittelsbachischen Hof in Landshut auftrat, in: Große Heidelberger Liederhandschrift (Codex Manesse), Zürich, ca. 1300 bis ca. 1340. Bildnachweis: Universitätsbibliothek Heidelberg, Cod. Pal. germ. 848, fol. 273 r, URL: https://digi.ub.uni-heidelberg.de/diglit/cpg848/0541.*

9 Vgl. Klaus Wolf, München leuchtete! Zur Rolle der Wittelsbacher im deutsch-romanischen Kulturtransfer des Spätmittelalters, in: Bernd Bastert / Sieglinde

Abb. 35: *Ludwig VII. mit französischer Bartpracht. Bild über dem Eingang zum Südwestturm im Liebfrauenmünster in Ingolstadt. Bildnachweis/Foto: Kurt Scheuerer; URL: www.ingolstadt.de.*

So wurden im späten Mittelalter die Wittelsbacher Residenzen München und Ingolstadt zu wichtigen Literaturorten in Oberdeutschland. Im Gegensatz zu dieser Literatur an festen Residenzen ist auf Ludwig den Bayern zu verweisen, der tatsächlich eher als Reisekaiser vor allem die Reichsstädte frequentierte. Sein Itinerar zeigt ihn eben besonders häufig in Städten wie Nürnberg und Frankfurt am Main. Er tat dies natürlich des Geldes willen, denn seine Geldgeber Konrad Groß in Nürnberg und Jakob Knoblauch der Reiche in Frankfurt am Main sicherten den monetären Nachschub. Besonders in Frankfurt am Main gab es auch ein entsprechendes literarisches Leben, hier sind „Frankfurter Dirigierrolle" und „Frankfurter Osterspielfragment" zu nennen, die in Zeiten des Interdikts auf der Bühne machtvoll die Einsetzung der Eucharistie durch Jesus Christus selbst thematisieren. Kaiser Ludwig der Bayer lässt also in diesem öffentlichkeitswirksamen Passionsspiel auf einem großen Markt- und Versammlungsplatz in dieser und in anderen Reichsstädten den Ungehorsam gegen das päpstliche Interdikt massenwirksam inszenieren. Er tut dies ähnlich oder lässt dies ähnlich in der Reichsstadt Worms aufführen, wo sogar gegen den von Avignon protegierten

Hartmann (Hg.), Romania und Germania. Kulturelle und literarische Austauschprozesse in Spätmittelalter und Früher Neuzeit (Jahrbuch der Oswald von Wolkenstein-Gesellschaft 22), Wiesbaden 2019, S. 412–423.

Bischof Salmann Cleman agitiert wird. Auf der Bühne nämlich gibt es einen „*Salman Phariseus*", also einen unbiblischen Pharisäer Salman, womit niemand anderes als der franzosenfreundliche Bischofsaspirant gemeint ist.[10]

Neben den Wittelsbachern sind aber auch andere Adelige als Akteure des literarischen Lebens in den Blick zu nehmen. Schon im 12. Jahrhundert ist hier der Burggraf von Regensburg aus dem Geschlecht der Riedenburger zu nennen. Die Riedenburger hatten nämlich das Burggrafenamt in Regensburg inne. Tatsächlich werden dort in der Stadt ihre Minnelieder erklungen sein. Das zeigt sich auch in der Tatsache, dass das sogenannte Budapester Fragment tatsächlich der Regensburger Buchmalerschule entsprang. Auch darin sind die Lieder des Burggrafen von Regensburg enthalten. Und dass der Minnesang nicht zuletzt eine städtische Literatur- und Kunstgattung darstellte, erweisen auch die Regensburger Neidhartfresken ebenso wie die Neidhartfresken in Wien.[11]

Überhaupt darf man nicht Richard Wagner auf den Leim gehen, der es scheinbar eindeutig auf die Opernbühne gebracht hatte, wonach die Meistersinger von Nürnberg vor allem treudeutsche Handwerker waren und der Sängerkrieg auf der Wartburg – also jenseits von städtischem Milieu – stattfand. Beides ist zu falsifizieren. Minnesang fand nicht zuletzt häufig in Städten statt und eben nicht auf zugigen Ritterburgen; und noch die Wartburg mit ihrem berühmten, eher legendären Sängerkrieg, thronte ja oberhalb der Stadt Eisenach, wo man auch viel besser Feste feiern konnte. Für höfische Feste bedurfte es nämlich einer gewissen Infrastruktur, die nur durch eine Stadt zu gewährleisten war. Und die Meistersinger von Nürnberg sind der Ausnahmefall mit tatsächlich mehrheitlich handwerklichen Angehörigen, dagegen sind die Meistersinger von Memmingen oder die Meistersinger von Augsburg durchaus auch von Akademikern geprägt.[12]

---

10 Vgl. Klaus Wolf, König Ludwig IV., genannt „der Bayer". Ein Herrscher zwischen Tradition und Innovation, in: Franz Niehoff (Hg.), Das Goldene Jahrhundert der Reichen Herzöge. Publikation zur Ausstellung der Museen der Stadt Landshut in der Spitalkirche Heiliggeist (Schriften aus den Museen der Stadt Landshut 34), Landshut 2014, S. 14–29.

11 Vgl. neuerdings zur biographischen Identität: Klaus Wolf, Dietmar von Aist: Minnelieder, in: Günter Butzer / Katja Sarkowsky / Hubert Zapf (Hg.), Große Werke der Literatur. Band XV. Eine Ringvorlesung an der Universität Augsburg 2017/18, Tübingen 2020, S. 9–23.

12 Vgl. Klaus Wolf, Die Meistersinger von Memmingen, in: Dietmar Schiersner / Georg Seiderer (Hg.), Schwaben und Franken. Regionalgeschichte im Vergleich (Forum Suevicum. Beiträge zur Geschichte Ostschwabens und der benachbarten Regionen 13), Konstanz 2020, S. 95–106.

Abb. 36: *Der Burggraf von Regensburg, in: Große Heidelberger Liederhandschrift (Codex Manesse), Zürich, ca. 1300 bis ca. 1340. Bildnachweis: Universitätsbibliothek Heidelberg, Cod. Pal. germ. 848, fol. 318 r, URL: https://digi.ub.uni-heidelberg.de/diglit/cpg848/0631.*

Auch für die Gattung Drama führen die Anfänge im Mittelalter nur teilweise an ländliche Höfe. Zwar ist der Tegernseer „*ludus de antichristo*“, das Tegernseer Spiel vom Antichrist, im 12. Jahrhundert und in der Stauferzeit tatsächlich mit dem Kloster und vielleicht sogar mit der Klosterkirche selbst als Bühnen- und Aufführungsort[13] in Zusammenhang zu bringen, andererseits sind die schon erwähnten Passionsspiele, nämlich Frankfurter Dirigierrolle, das Wormser Passionsspiel des 14. Jahrhunderts und das Frankfurter Passionsspiel des ausgehenden 15. Jahrhunderts tatsächlich in großen Reichsstädten zu lokalisieren, wo etwa auf dem Frankfurter Römerberg die Bühne des Frankfurter Passionsspiels stand, wobei die patrizischen Zuschauer es sich auf dem Römer selbst und auf dem zinnenbewährten Dach der Nicolai-Kirche mit Rotwein und Kirschen gemütlich machen konnten.[14] Es bedurfte also gerade für große Passionsspielaufführungen eines städtischen – ja reichsstädtischen – Forums, um alleine schon die anspruchsvolle Logistik der hölzernen Bühnenbauten zu verwirklichen.

Aber schon eine dezidiert ehrwürdige mittelalterliche Gattung wie das Heldenepos bedurfte durchaus ebenso eines urbanen Milieus. Und während die Literaturwissenschaft schon lange weiß, dass das Epos „Kudrun“ vermutlich im Regensburg des 13. Jahrhunderts entstand und mit der Kaufmannslist Horants und Fruotes sowie Wates auf die Kaufleute von Regensburg als Publikum letztlich anspielt,[15] ist es selbst für das archaische Nibelungenlied nicht von der Hand zu weisen, dass dem Autor beziehungsweise der sogenannten Nibelungenwerkstatt, die im 12. Jahrhundert und im frühen 13. Jahrhundert wohl in Passau das Werk verschriftlichte, ein urbaner Hintergrund vor Augen stand. So erfahren wir in der ersten Aventiure des Nibelungenlieds mit aller wünschenswerten Deutlichkeit etwas von der städtischen Residenz der Nibelungen.

> „*Ze Wormez bi dem Rîne // si wonten mit ir kraft.*
> *in diente von ir landen // vil stolziu ritterscaft*
> *mit lobelîchen êren // unz an ir endes zît.*
> *Si stúrben sît jáemerlîche // von zweier edelen frouwen nît.*“[16]

Während mit Worms die Heimatstadt von Kriemhild, Gunther, Gernot und Giselher bezeichnet wird, kommt Siegfried ebenfalls aus städtischem Milieu, wie wir mit wünschenswerter Klarheit am Beginn der zweiten Aventiure erfahren.

---

13 Vgl. Wolf, Bayerische Literaturgeschichte (wie Anm. 5), S. 60 f.

14 Vgl. Klaus Wolf, Kommentar zur „Frankfurter Dirigierrolle“ und zum „Frankfurter Passionsspiel“, Tübingen 2002, passim.

15 Vgl. Wolf, Bayerische Literaturgeschichte (wie Anm. 5), S. 38.

16 Helmut de Boor / Roswitha Wisniewski (Hg.), Das Nibelungenlied (Deutsche Klassiker des Mittelalters), Wiesbaden 1979, S. 4.

> *„Do wuohs in Niderlanden // eins edelen küneges kint,*
> *des vater der hiez Sigemunt, // sîn muoter Sigelint,*
> *in einer rîchen bürge // wîten wol bekannt,*
> *nidene bî dem Rîne: // diu was ze Sántén genant.*"[17]

„*Sántén*" oder Xanten meint natürlich die städtische Residenz von Siegfrieds Eltern; und der berühmte Streit der Königinnen Kriemhild und Brünhild fand ebenfalls im städtischen Milieu, nämlich vor dem Wormser Dom, statt und der Nibelungenlieddichter hatte sicher das staufische Nordportal, wo auch die Könige traditionell einzogen, im Blick, als er das Aufeinandertreffen der Königinnen schildert:

> *„Hie stuont vor dem münster // daz Guntheres wîp,*
> *dô hete kurzewîle // vil maniges ritter lîp*
> *mit den schoenen vrouwen, // der si dâ nâmen war.*
> *dô kom diu vrouwe Kiremhilt // mit maniger hêrlîchen schar.*"[18]

In der Nibelungenklage wird schließlich Bischof Pilgrim von Passau zum Urheber des Nibelungenlieds und seiner Verschriftlichung selbst.

> *„Von Pazowe der biscof Pilgerin*
> *durh liebe der neven sîn*
> *hiez scrîben ditze maere,*
> *wie ez ergangen ware,*
> *in latînischen buochstaben.*"[19]

Erst in Passau wurde aus den „*alten mären*", die wohl von umherziehenden Sängern im Vortrag relativ frei als improvisierte Heldenlieder vorgetragen wurden, ein Buchepos. Die Literaturwerdung des Nibelungenlieds ist also nur im urbanen Milieu, in der Residenz- und Bischofsstadt Passau vorstellbar. Die Verstädterung und Verbürgerlichung des Nibelungenlieds – man könnte von einer *adaptation bourgeois* sprechen – fand ihren Höhepunkt im sogenannten „Augsburger Nibelungenlied", welches das heldenepische Geschehen in eine Augsburger Patrizierstube der Gossembrots verlegte.[20] Kriemhilds Falkentraum wird vom Illustrator der Handschrift in ein stadtbürgerliches Schlafzimmer verlegt und der Maler vergisst nicht einmal den Nachttopf unter Kriemhilds Bett.

---

17 Ebd., S. 7.

18 Ebd., S. 140 f.

19 Karl Bartsch (Hg.), Diu Klage. Mit den Lesarten sämtlicher Handschriften, Leipzig 1875 (Nachdruck: Darmstadt 1964), S. 221 f.

20 Vgl. Michaela Eser, Augsburger Nibelungenlied und -klage. Edition und Untersuchung der Nibelungen-Handschrift b (Editio Bavarica II), Regensburg 2015.

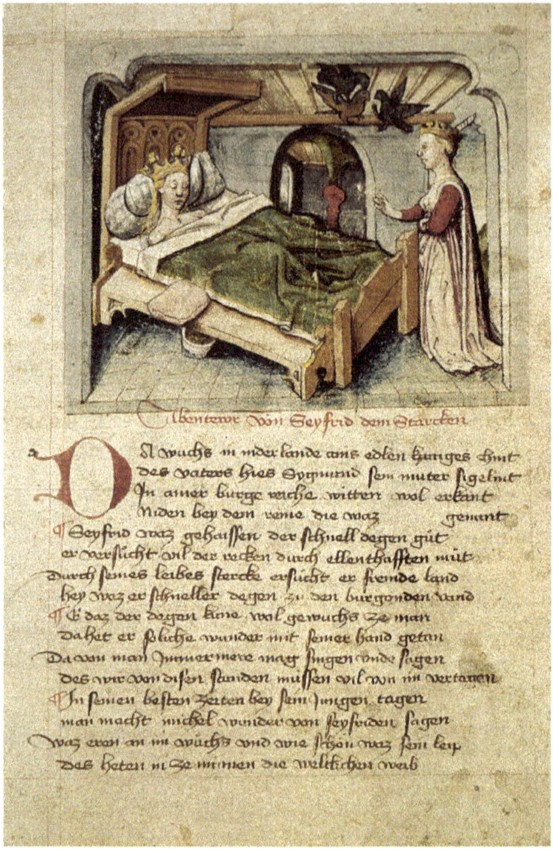

Abb. 37: *Kriemhilds Falkentraum im Augsburger Nibelungenlied. Bildnachweis: Staatsbibliothek zu Berlin – Preußischer Kulturbesitz; prometheus.uni-koeln.*

Die nibelungischen Helden sind im 15. Jahrhundert bei den Gossembrots, die teilweise ein Universitätsstudium hatten, in einem reichen, aber hochgebildeten humanistischen Milieu angelangt, in der Reichsstadt Augsburg eben. Überhaupt gibt es ja nicht wenige Beispiele, dass das Patriziat versuchte, die Lebensweise des Adels zu imitieren, erinnert sei an die Nürnberger Patrizier, die Tucher und wie sie alle hießen, die auf Burgen und Schlössern um die Reichsstadt herum ein rittermäßiges Leben zu pflegen suchten. Dies galt übrigens auch für Fugger und andere zu Geld gekommene Kaufleute und Vertreter des Patriziats, die um

Augsburg herum in Rittergütern und Jagdschlössern, wie etwa Mickhausen, ein rittermäßiges Leben führen wollten. Gleichwohl spielte die literarische Musik im Spätmittelalter vornehmlich in den Städten.

Die Dominanz des Urbanen gilt sogar für eine Gattung, die man mit dem Rittertum schlechthin identifiziert, für den Minnesang. Denn die berühmteste Minnesangsammlung des Mittelalters, der *Codex Manesse*, gehört tatsächlich nach Zürich in das Bürgergeschlecht Manesse, sogar die Handschrift selbst dokumentiert dies ausführlich. So waren es wohl Stadtbürger, die den alten ritterlichen Minnesang am Leben erhielten. Der Fund des Erfurter Minnegürtels zeigt übrigens, dass auch jüdische Städtebewohner – hinzuweisen wäre auch auf den „*dukus horant*" in hebräischen Buchstaben – die ritterliche Literatur pflegten.[21] Zumindest seit dem 14. Jahrhundert wurden vermeintlich ritterliche Gattungen wie Heldenlied und Minnesang vornehmlich in den Städten gepflegt und am Leben gehalten. Dies gilt übrigens auch für die vollständigste Sammlung mit Neidhartliedern, nämlich die sogenannte Handschrift c, welche aus Nürnberg stammte. Das Beispiel Minnesang und Neidharts insbesondere zeigt, dass einige Gattungen zwischen Stadt und Land munter mäandern. Und während der Minnesang letztlich in der Stadt – nicht zuletzt in den Neidhartfresken – eine spätmittelalterliche Heimat findet, zieht eine andere literarische Gattung längerfristig von der Stadt aufs Land. Die Rede ist vom Passionsspiel, das *per se* keine ortsgebundene Gattung ist. In die Reichsstadt führen Frankfurter Passionsspiel oder Augsburger Passionsspiel, wobei auch die Residenzstadt Wien ein mit dem Augsburger Spiel vergleichbares Passionsspiel aufzuweisen hatte, freilich im Kontext der Residenz, nämlich – wie wir durch einen Aufführungsbeleg wissen – „*in castro ducis*". Im Innenhof der habsburgischen Residenz hatte man eine Passion und eine Marienklage inszeniert. Je länger, je mehr – besonders im Gefolge der Aufklärung – gerät jedoch die Gattung Passionsspiel in die Krise, besonders die Aufklärer versuchen sie abzuschaffen. So konnte im ausgehenden 18. und im frühen 19. Jahrhundert die Gattung Passionsspiel sich nur noch im alpinen Rückzugsraum halten, und das berühmteste Beispiel bis auf den heutigen Tag ist wohl Oberammergau, auch die Passionsspiele von Erl in Tirol wären in diesem Kontext zu nennen.[22]

Gegen den bisherigen Befund einer städtischen Dominanz, was das literarische Leben anbelangt, lässt sich wenigstens für Altbayern der Hinweis auf die Klöster dingfest machen. Tegernsee, schon im frühen Mittelalter gegründet,

21 Vgl. Wolf, Dietmar von Aist (wie Anm. 11), passim.

22 Vgl. Wolf, Bayerische Literaturgeschichte (wie Anm. 5), passim.

erfasste im 15. Jahrhundert eine gewaltige literarische Blüte, es wurde geradezu vor Gründung der Universität Ingolstadt 1472 zu einen *thinktank* für eine gehobene Geistigkeit in lateinischer und deutscher Sprache. Und noch im 18. Jahrhundert fassen wir dieses ländliche literarische Milieu im *parnassus boicus* einer Aufklärervereinigung, deren Anhänger eigentlich Prälaten in ländlichen Klöstern waren. Das literarische Leben ist in Altbayern also bis ins 18. Jahrhundert hinein durchaus städtisch dominiert, aber das Land spielt keine unwichtige Rolle, insbesondere in den Klöstern mit ihren beeindruckenden Bibliotheken und ausgebauten Schulen.[23] Anders dagegen in Neubayern, also Franken und Schwaben, dort sind es vor allem die Reichsstädte, und so kann Nürnberg im 15. und 16. Jahrhundert beinahe als eine Literaturhauptstadt des Reichs gewertet werden, was die deutschsprachige Literatur anbelangt.

Die literarische Dominanz der Reichsstädte vor allem in Neubayern beruht auf ihrem Charakter als Druckerstädte. Die Erfindung des Buchdrucks durch Johannes Gutenberg reüssierte dann weniger in Mainz, sondern in oberdeutschen Städten wie Ulm, Straßburg, Nürnberg, Augsburg – allesamt freie Reichsstädte. In Augsburg etwa blühte daneben das geistliche Spiel. Hinzuweisen ist hierbei insbesondere auf das Augsburger Georgspiel als Drachenstich vor dem Wertachbruckertor, wo die Stadt selbst Sujet der Spielhandlung wird. Der heilige Georg befreit nämlich eine Stadt von einem Drachen. Die Bewohner der Stadt sind in der Logik des Spiel Moslems, die nach der Befreiung vom Drachen zum Christentum konvertieren. Auch dies ist eine städtische Thematik, denn die Türkenkriegssteuern wurden vor allem bei den reichen Kaufleuten der Reichsstädte eingetrieben, auch bei den Juden, so sie noch nicht vertrieben waren. Die Türkenkriegspublizistik gehört ebenfalls in die Reichsstädte, und da ist Augsburg im 15. bis zum 18. Jahrhundert durchaus als Druckerstadt enorm wichtig.[24] In große Städte führen auch die Buchmessen, so Frankfurt am Main mit der Buchmesse seit dem 15. Jahrhundert, das später durchaus von Leipzig bis um 1800 überflügelt wurde. Eine rein städtische Gattung, die sich zäh halten konnte, ist auch der Meistersang. Man könnte von freiem Singen für freie Bürger

---

23 Vgl. Klaus Wolf, Zwischen Melker Reform und Aufklärung. Der literaturgeschichtliche Beitrag oberdeutscher Köster, in: Wolfgang Wüst (Hg.), Klöster, Kultur und Kunst – Süddeutsche Sakrallandschaft in Spätmittelalter und früher Neuzeit. Unter Mitarbeit von Lisa Bauereisen und Christoph Gunkel, Berlin u.a. 2020, S. 215–224.

24 Vgl. Klaus Wolf, Kruzitürken! Das Türkenthema in der vormodernen schwäbischen Literatur nebst methodischen Überlegungen zur schwäbischen Literaturgeschichtsschreibung, in: Zeitschrift des Historischen Vereins für Schwaben 107, 2015, Mit einer Festgabe für Georg Kreuzer zum 75. Geburtstag, S. 109–121.

sprechen, eine durchaus reichsstädtische Gattung, die sich in Memmingen sogar bis ins 19. Jahrhundert hinein hielt, und neben der Aufführung von Meisterliedern auch das Theaterspielen vorsah. Theaterspielen war nämlich immer ein Politikum und unter dem Schutz einer Meistersingergesellschaft konnten auch revolutionäre Stücke wie Schillers „Räuber" schon früh in Memmingen aufgeführt werden. Eine städtische Bewegung ist schließlich auch zusammen mit dem ihr folgenden Literaturboom die Reformation. Die gedruckten deutschen Texte entstammten eben den reichsstädtischen Druckorten. Die fromme Revolte war nicht zuletzt eine urbane Bewegung. Urban schließlich auch die wohl bedeutendste literarische Gesellschaft des 17. Jahrhunderts, die Nürnberger Pegnitzschäfer, in deren Reihen es sogar bedeutende Dichterinnen gab. Hier bot die Reichsstadt einen Schutzraum der Freiheit und der künstlerischen Entfaltung nach den schrecklichen Ereignissen des Dreißigjährigen Krieges. Das 18. Jahrhundert führte dagegen teilweise zu einem literarischen Aufschwung ländlicher Residenzen. Hier sind nicht zuletzt die Prälatenresidenzen, die Klöster, zu nennen. Für Süddeutschland ist hier unbedingt Pater Sebastian Sailer wichtig, nicht nur mit seiner „Schwäbischen Schöpfung", sondern mit seinen Dichtwerken, welche in den Lustschlössern der Prälaten gelesen wurden. Im Gegensatz zur eher ländlichen Prälatenliteratur ist das dezidiert städtische bürgerliche Trauerspiel zu sehen. Die dort als tragische Helden inszenierten Figuren wie Emilia Galotti oder Luisa Millerin sind unschwer nach stadtbürgerlichen Vorbildern gezeichnet. Auch scheinbare Historiendramen wie „Die Bernauerin" des Grafen Törring sind in Wirklichkeit bürgerliche Trauerspiele, denn Graf Törring rezipiert die Massengattung Bürgerliches Trauerspiel und macht aus der bürgerlichen Agnes Bernauer eine Emilia Galotti des Spätmittelalters. Der Fürst, den sie begehrt, wird sie nicht bekommen und sie endet tragisch durch Hinrichtung, durch Ertränkung als Hexe. Das scheinbar mittelalterliche Historienspektakel ist in Wirklichkeit ein Bürgerliches Trauerspiel und insofern letztlich in städtischem Milieu gegründet. Im 18. Jahrhundert wird aber neben der Reichsstadt ein anderer Städtetypus als literarischer Schauplatz stilbildend. Hier ist unbedingt die Residenzstadt Weimar zu nennen. Denn Goethe hatte ja seine Heimatstadt, die freie Reichsstadt Frankfurt, gerne verlassen, die ihm erstarrt schien. Schon Goethes Vater, ein Anhänger der Aufklärung, zog sich resigniert und verzweifelt aus dem öffentlichen Leben zurück. Im absolutistischen Weimar hingegen konnte Goethe einen Musenhof finden, der einer damals modernen Literatur Unterstützung bot, während die Reichsstädte wie Frankfurt oder Regensburg ja im ausgehenden 18. Jahrhundert literarisch eher rückständig waren. Tradition und Fortschritt hatten also ein wenig die Plätze getauscht. Bis ungefähr zum Ausbruch des Dreißigjährigen Krieges boten die Reichsstädte den

Motor des literarischen Fortschritts. Nach dem Dreißigjährigen Krieg kamen zunehmend die Adelsresidenzen in den Genuss einer fortschrittlichen Literatur. Erinnert sei an Sanssoucis, wo Voltaire mit Friedrich dem Großen literarisch geprägte Gespräche führte, und natürlich an den Weimarer Musenhof. Beinahe anachronistisch mutet „München leuchtete“ an, so der Ausruf in Thomas Manns „Gladius dei“, aber es ist nicht nur die Verlagsstadt, die Stadt der literarischen Zeitschriften, die München damals um 1900 so fortschrittlich erscheinen ließen, sondern es ist letztlich auch eine tolerantere Haltung etwa des Prinzregenten Luitpold, der die ganze Epoche prägte, sodass die Zensur auch in München kaum wütete, während ein Organ wie der „Simplicissimus“ in Berlin nie hätte erscheinen können. Der literarische Fortschritt war um 1900 also bayerisch, man könnte sagen: Der Fortschritt sprach Bairisch![25]

Damit aber sind wir mit unserem Ausblick zu den Literaturorten Stadt versus Land an das Ende gelangt. Als Fazit lässt sich ziehen, dass Residenzstädte von Anfang an, schon seit dem frühen Mittelalter, für das literarische Leben prägend sind, im Gegensatz zu zugigen Ritterburgen auf dem Land. Auf der anderen Seite lässt sich zeigen, dass ein mehr stadtbürgerliches literarisches Leben in Reichsstädten vor allem in der Frühen Neuzeit, aber schon beginnend im Spätmittelalter, den Motor des Fortschritts bestimmte. Erst im 18. Jahrhundert wird die Adelsresidenz, sofern sie letztlich aufklärerisch geprägt war wie bei Friedrich dem Großen, aber auch am Herzogshof in Weimar, für das literarische Leben prägend.

---

25 Vgl. Wolf, Bayerische Literaturgeschichte (wie Anm. 5), passim.

Franz Körndle

Stadtpfeifer in der süddeutschen Städtelandschaft. Professionalisierung, Konflikte, Reisen

Abstract: It was only at the beginning of the twentieth century, when musicological scholarship drew attention to the cultural and musical-political importance of instrumental ensembles in cities from the end of the Middle Ages to modern times and described the beginnings of the "Stadtpfeifer" as permanently employed musicians, their duties, job titles and the range of instruments until the transition of the old institution "Stadtpfeiferei" to a symphony orchestra in the bourgeois concert business. This contribution is devoted to urban musical life by examining the process of professionalisation with a view to the freelance musicians in a city. Travel and conflicts play an important role.

Vor mehr als hundert Jahren machte der Schweizer Musikforscher Karl Nef auf die kultur- und musikpolitische Bedeutung der Instrumentalensembles in den Städten vom ausgehenden Mittelalter bis in die Neuzeit aufmerksam und beschrieb die Anfänge der Stadtpfeifer als fest angestellte Musiker, ihre Pflichten, Berufsbezeichnungen und das Instrumentarium bis zum Übergang der alten Institution Stadtpfeiferei zum Sinfonieorchester im bürgerlichen Konzertbetrieb.[1] Diesen, auf knappem Raum gegebenen, aber grundlegenden Informationen fügte erst 1982 Heinrich Schwab eine substantielle Studie zur Professionalisierung des Stadtmusikerberufs hinzu.[2] In jüngerer Zeit widmete sich Helen Coffey dem Klang in den Städten während der Zeit Kaiser Maximilians I.[3] Während diese Beiträge sich relativ eng auf die berufliche Stadtmusik selbst konzentrierten, gilt der hier vorgelegte Beitrag dem urbanen Musikleben mit einem

---

1 Karl Nef, Die Stadtpfeiferei, in: Die Schweiz. Schweizerische illustrierte Zeitschrift 7, 1903, S. 540–544.

2 Heinrich W. Schwab, Die Anfänge des weltlichen Berufsmusikertums in der mittelalterlichen Stadt. Studie zu einer Berufs- und Sozialgeschichte des Stadtmusikantentums (Kieler Schriften zur Musikwissenschaft 24), Kassel/Basel/London 1982.

3 Helen Coffey, City Life and Music for Secular Entertainment during the Reign of Maximilian I, in: Ian Biddle / Kirsten Gibson (Hg.), Cultural Histories of Noise, Sound and Listening in Europe, 1300–1918, London/New York 2017, S. 171–185.

erweiterten Fokus, indem zum Prozess der Professionalisierung die freiberuflich tätigen Musiker in einer Stadt hinzugenommen werden. Dabei nehmen Reisen und Konflikte einen wichtigen Raum ein.

Die Erkenntnisse aus den älteren Studien schufen zwar erste Einblicke in die Musikkultur der Städte am Beginn der Neuzeit, sie vermochten allerdings noch nicht einmal in der weiteren Fachwelt die Vorurteile auszuräumen, die mit dem Bild des Stadtmusikanten verbunden waren – und es bis heute sind. Offenbar nicht selten, geprägt durch wenig sorgfältig recherchierte historische Romane,[4] werden Stadtpfeifer gerne mit Spielleuten gleichgesetzt. Auf diese Art und Weise wird ein Spielmann dann durchaus zum Vorläufer des späteren Kammermusikers.[5] Unklarheit darüber, was denn einen Stadtpfeifer ausmache, bestand zweifellos seit den frühesten Zeiten der Instrumentalmusik selbst und hält sich bis in die Gegenwart. Ein eher schauerliches Beispiel hierzu bot die Darstellung eines Hexenprozesses, von dem Johann Reiche im Jahre 1703 berichtet: „*An einem Ort haben in der Marter viel Personen nach einander bekennet, daß sie in einer Nacht zugleich auff dem Hexen-Tantze gewesen; Die erste sagte, die Spielleute waren Geiger: die andere, die Spielleute waren Zinckenbläser und Posauner: die dritte, die Spielleute waren Pfeiffer: die vierdte, die Spielleute waren Leyrmänner.*“[6]

Ganz offensichtlich wollte der juristische Berichterstatter mit seinem Text Widersprüche in den Aussagen der Beschuldigten darstellen. Man darf allerdings den Wert einer solchen Beweisführung mit guten Gründen bezweifeln, da Menschen nicht einfach unter Androhung der Folter zu Organologen werden können, selbst wenn die Schreckung (Territion) mit den Worten befohlen würde, ihnen die Instrumente zu zeigen.[7] Daher bemerkt man die erstaunliche Kompetenz, mit der in dem Bericht fast das komplette Instrumentarium vorgestellt wird, über das eine Stadtpfeiferei in der frühen Neuzeit verfügte. Diese Vielfalt machte es den Zeitgenossen schwer, zu unterscheiden, ob es sich um einen Zinken (auch *Cornetto* genannt) oder eine Pfeife handelte. Ohne genauere Kenntnis der jeweiligen

4 Etwa: Oliver Pötzsch, Der Spielmann: Die Geschichte des Johann Georg Faustus, Berlin 2018; oder: Ingrid Ganß, Der Spielmann, Laubach 2019.

5 Karlheinz Weber, Vom Spielmann zum städtischen Kammermusiker. Zur Geschichte des Gürzenich-Orchesters, Teil 1, Kassel 2009.

6 Johann Reiche, Unterschiedliche Schrifften von Unfug des Hexen-Proceßes, zu fernerer Untersuchung der Zauberey, Magdeburg 1703, S. 498.

7 Vgl. Bertolt Brecht, Leben des Galilei. Schauspiel, Berlin [77]2016, S. 108; Arthur Kaufmann (Hg.), Die Peinliche Gerichtsordnung Kaiser Karls V. von 1532 (Carolina), Stuttgart [6]1975, S. 18 f.; Hans-Jürgen Wolf, Hexenwahn und Exorzismus. Ein Beitrag zur Kulturgeschichte, Kriftel/Taunus 1980, S. 282.

Tonerzeugung konnte man sich lediglich darauf beziehen, was man sah oder hörte. Zudem umfasst der Begriff „Pfeife" ohnehin nicht nur die Flöte, sondern im weiteren Sinne so gut wie das gesamte Blasinstrumentarium. Trotzdem gehörte schon früh das Spiel auf Geigen oder Fiedeln ebenfalls zum Tätigkeitsbereich.

Die beschriebenen terminologischen Unschärfen kann man aus heutiger Sicht nicht nur bei den Instrumenten, sondern bei der Bezeichnung der Musiker selbst beobachten. „Musikant", „Musicus", „Musicus instrumentalis", „Plaser", „Pfeifer" und dergleichen mehr konnten nebeneinander und synonym gebraucht werden. Die Frage nach den Bezeichnungen erlaubt es jedoch nicht, die eigentliche Problematik der urbanen Musikkultur in der frühen Neuzeit zu erfassen, im Gegenteil, sie ist Teil des Problems. In dem lang andauernden Prozess der Loslösung einer professionellen städtischen Musikgruppe von den Türmern, die das Musikinstrument hauptsächlich zur Signalgebung einsetzten, entstand erstmals eine berufliche Ausrichtung, die sich nicht nur dem Erzeugen von Klang zu widmen, sondern musikalische Aufgaben und Funktionen zu erfüllen hatte. Dazu gehörten städtische Veranstaltungen ebenso wie das Mitwirken bei der Kirchenmusik und das Aufspielen beim Tanz.[8] Mit der festen Anstellung und regelmäßigen Bezahlung von Musikern bei einer Stadt kam freilich eine soziale Trennlinie in die Welt der Instrumentalisten, die seither nicht mehr verschwunden ist.[9] Vereinfacht könnte man davon sprechen, dass aus den Bereichen der oftmals fahrenden sowie frei tätigen Spielleute einerseits und den musikalisch Ambitionierten bei den Türmern andererseits ein neuer Berufsstand, der der Stadtmusiker hervorging. Nicht überall wurde jedoch eine klare Unterscheidung zwischen Türmern und Stadtpfeifern vorgenommen. Manche Städte übertrugen einfach einigen ihrer Türmer zusätzlich musikalische Aufgaben, andere verzichteten ganz auf eine professionelle Musikergruppe.

Eine Abbildung in der „*Zürcher- und Schweizerchronik*" des Gerold Edlibach, entstanden ab 1485/86, zeigt einen Turmbläser sowie drei Musiker mit Instrumenten auf den Mauern einer Stadt; erkennbar sind eine S-förmig gebogene Trompete und zwei Schalmeien oder Pommern, die Bläser tragen Gewänder mit sichtbarer Heraldik der Stadt Zürich. Den aus den Instrumenten hervorgehenden Schall hat der Illustrator mit kleinen Strichlein angedeutet. Im „*Kopienband zur Zürcher- und Schweizergeschichte bis zum 16. Jahrhundert*" wird zur identischen historischen Situation unter der Rubrik „*Wie die eigenossen das lütten vnd pfiffen herten das in der stat beschach dess sy sich ser verwunderten*" eine

8 Armin Brinzing, Studien zur instrumentalen Ensemblemusik im deutschsprachigen Raum des 16. Jahrhunderts, Bd. 1: Darstellung, Göttingen 1998, S. 11.

9 Schwab, Berufsmusikertum (wie Anm. 2), S. 33–73.

ähnliche Ikonographie angeboten. Zwar ist die Heraldik auf ein einzelnes Züricher Wappen reduziert, doch die Anzahl der Bläser auf den Mauern hat sich verdoppelt.[10] Karl Nef hatte die beiden Bilder in Nachzeichnungen reproduziert. In den modernen Digitalisaten der Handschriften ist dagegen deutlicher zu erkennen, wie der Kopienband die gewachsene Bedeutung des urbanen Klangs mit einer Vermehrung der Musiker anzuzeigen gewillt war.[11]

Abb. 38: *Turmbläser und Musiker auf den Mauern Zürichs, dargestellt in: Gerold Edlibach u.a., Zürcher- und Schweizerchronik, [Zürich, zwischen 1485 und 1532]. Bildnachweis: Zentralbibliothek Zürich, Ms A 75, S. 119.*

10 Zentralbibliothek Zürich, Ms A 75: Gerold Edlibach / Ludwig Edlibach / Johann Martin Usteri, Zürcher- und Schweizerchronik, [Zürich, zwischen 1485 und 1532], S. 119; ebd., Ms A 77: Gerold Edlibach / Johann Martin Usteri, Kopienband zur Zürcher- und Schweizergeschichte bis zum 16. Jahrhundert, [Zürich, zwischen 1506 und 1566], fol. 68 v; Regula Schmid, Schweizer Chroniken, in: Gerhard Wolf / Norbert H. Ott (Hg.), Handbuch Chroniken des Mittelalters, Berlin/Boston 2016, S. 267–300, hier bes.: S. 286.

11 Vgl. Nef, Stadtpfeiferei (wie Anm. 1), S. 541.

Abb. 39: *Turmbläser auf den Mauern Zürichs, dargestellt in: Gerold Edlibach u.a., Kopienband zur Zürcher- und Schweizergeschichte bis zum 16. Jahrhundert, [Zürich, zwischen 1506 und 1566]. Bildnachweis: Zentralbibliothek Zürich, Ms A 77, fol. 68 v.*

Die städtische Obrigkeit musste ein Augenmerk darauf haben, dass das in den neuen Planstellen auszugebende Geld auch gut investiert war. Die Stadtpfeifer gelangten nicht nur in den Genuss einer Dienstwohnung und bezogen Gehalt, sondern wurden mit Naturalien wie Brennholz, Getreide und anderen Lebensmitteln, dazu Wachslichtern und Kleidung ausgestattet.[12] Als Repräsentanten der eigenen Stadt war diese in den heraldischen Farben gestaltet.[13] Im Vertrag mit

12 Sabine Žak, Musik als „Ehr und Zier" im mittelalterlichen Reich. Studien zur Musik im höfischen Leben, Recht und Zeremoniell, Neuss 1979, S. 113, 121–168.

13 Vgl. Moritz Kelber, Die Musik bei den Augsburger Reichstagen im 16. Jahrhundert (Münchner Veröffentlichungen zur Musikgeschichte 79), München 2018, S. 87.

dem „*Stattpfeuffer*“ Marte Sommer legte der Rat von Rothenburg ob der Tauber auch die „*klaydung, Nemblich genauer Statt farb Rott und Weyß wie von allter herkomen*“, fest.[14] Die Augsburger Stadtpfeifer trugen rote Hosen, grünes Wams sowie weiße Ärmel und Halskrause, zu sehen etwa an den Musikerfiguren des sogenannten Babylonischen Turms, den die Goldschmiedewerkstatt des Hans Schlottheim um 1602 in Augsburg fertigte.[15] Der Kleidung kam insbesondere bei Reisen eine Bedeutung zu, wenn die Musiker städtische Delegationen zu begleiten hatten. Mit Erkennungsmelodien zeigten sie der empfangenden Stadt akustisch ihre Herkunft an.[16] Dabei war es keineswegs eine Selbstverständlichkeit, dass eine städtische Musikergruppe mit dem Privileg zum Blasen von Trompeten ausgestattet war.[17] Beim Lesen von historischen Dokumenten ist allerdings durchweg Vorsicht geboten, da Rechnungsschreiber oder Chronisten nicht immer über die entsprechende Kenntnis und Beobachtungsgabe verfügten.

Aufenthalte auswärtiger Bläsergruppen sind dann in städtischen Rechnungsbüchern dokumentiert, wenn die Musiker irgendwelche Verehrungen oder Geschenke erhielten. Die früh im 15. Jahrhundert ansetzenden Stadtrechnungen der Reichsstadt Dinkelsbühl erwähnen schon 1437 nicht nur einen Pfeifer „*von Werde*“ (Donauwörth), sondern auch aus Memmingen, „*vnsers herren marggrafen Pfeiffer*“, „*vnsers herren von ottingen Pfeiffer*“, und weiterhin Pfeifer aus Nürnberg und Nördlingen unter den (all-)gemeinen Ausgaben.[18] Die Stadt selbst bezahlte damals einen einzigen eigenen Musiker, nämlich Peter Hutzing, der im Jahr 19 Gulden erhielt.[19] In der Stadt Mindelheim bekamen am 27. März 1597 „*dreyen Thrummetern von Chur*“ 16 Kreuzer, am 26. April zahlte man einem „*Thrumeter von Kirchhaim an der Tegg*“ drei Kreuzer sowie „*Ainem Thrummeter von Schwabisch Gmündt*“ ebenfalls drei Kreuzer und am 9. August „*ainem Thrummeter von Eßlingen*“ drei Kreuzer.[20] „*Ainem Trummeter*

14 Stadtarchiv (künftig: StadtA) Rothenburg, A 1319, fol. 220 r.

15 Staatliche Kunstsammlungen Dresden, Grünes Gewölbe, Inv.-Nr.: V 140: Kugellaufuhr.

16 Vgl. Wilhelm Ehmann, Tibilustrium: Das geistliche Blasen. Formen und Reformen, Kassel/Basel 1950, S. 51.

17 Timothy A. Collins, Of the Differences between Trumpeters and City Tower Musicians. The Relationship of Stadtpfeifer and Kammeradschaft Trumpeters, in: The Galpin Society Journal 53, 2000, S. 51–59. Vgl. auch: Helen Green, Defining the City “Trumpeter”: German Civic Identity and the Employment of Brass Instrumentalists, c. 1500, in: Journal of the Royal Musical Association 136, 2011, S. 1–31.

18 StadtA Dinkelsbühl, Stadtkämmerei 69, Steuerrechnungen 1437, fol. 80 r, 81 r und 82 r.

19 Ebd., fol. 183 r.

20 StadtA Mindelheim, Inv.-Nr.: II G 6 (1597), fol. 121 r. Für die Exzerpte sei dem Stadtarchivar in Mindelheim, Herrn Andreas Steigerwald M.A., sehr herzlich gedankt.

von Weissenburg" wurden vier Kreuzer zugeeignet, dann „*Vier Musicis*" (ohne Ortsangabe) sechs Kreuzer verehrt. Im Folgejahr erhielten drei Trompeter von Oettingen und Wallerstein 20 Kreuzer, zwei Bläser von Dillingen und Höchstädt 10 Kreuzer, zwei Trompeter von Nördlingen acht Kreuzer, zwei Trompeter von Gundelfingen zwölf Kreuzer, ein Trompeter von Pfaffenhofen in Bayern acht Kreuzer.[21] Anhand der Zahlungen in weiteren Jahren können Trompeter und/oder Stadtpfeifer aus Würzburg, Ulm, Eichstätt, Ravensburg, Aichach, Monheim, Weißenhorn, Dinkelsbühl, Ingolstadt, Günzburg und Straubing als Gäste in Mindelheim nachgewiesen werden.[22]

Mindelheim selbst verfügte über keine eigene Stadtpfeiferei, auch dem Türmer waren keine musikalischen Aufgaben übertragen, was aus den Bestallungsurkunden für die Jahre 1580 und 1591 deutlich hervorgeht.[23] Wegen der Übersichtlichkeit der Stadt genügte es, wenn einer der Türme, nämlich der des unteren Tors („Plaserturm") mit einem Mann besetzt war, der die nötigen Signale geben konnte. Das oberste Geschoss war als Wohnung ausgebaut. Wenn eine Stadt wie Mindelheim damit erhebliche Kosten für Musiker einsparen konnte, soll dies allerdings nicht bedeuten, dass man gerne darauf verzichten wollte. Zu besonderen Gelegenheiten, etwa dem Anblasen des Neuen Jahres, ließ man sich die Stadtpfeifer aus Landsberg oder Memmingen kommen.[24] Ob bei einem solchen Anlass möglicherweise beide Ensembles miteinander spielten, ist aus den Dokumenten allerdings nicht abzulesen.

Die städtischen Arbeitgeber schützten ihre Angestellten mit Erlassen und sicherten ihnen das Privileg zu, bei privaten Anlässen, etwa Hochzeiten, für die Tanzmusik engagiert werden zu müssen.[25] Die Nürnberger Hochzeitsordnung von 1603 setzt das Engagement von wenigstens zwei Stadtpfeifern bei den Trauungsgottesdiensten als selbstverständlich voraus – eine Bezahlung für die dabei zu verrichtenden Tätigkeiten ist explizit geregelt.[26] Extrakosten für das Einüben

21 Ebd., fol. 122 r.

22 StadtA Mindelheim, Inv.-Nr.: II G 6 (1596), (1598) und (1599).

23 StadtA Mindelheim, Inv.-Nr.: U-0456: „*Bestallung Michel Bairs, des Turners*", 1580; Inv.-Nr.: U-0474: „*Hanns Schnebelins des Plasers zue Mindlhaim Bestallung*", 1591.

24 StadtA Mindelheim, Inv.-Nr.: II G 6 (1594) und (1597), fol. 122 r; auch: (1598) und (1599).

25 Schwab, Berufsmusikertum (wie Anm. 2), S. 72.

26 Verneuerte Ordnung, wie es hinfüro auff den Erbarn vnd verlegten Hochzeiten in Nürmberg gehalten werden soll, Nürnberg 1603, S. 57–61. Vgl. Wolfgang Wüst (Hg.), Die „*gute*" Policey im Reichskreis. Zur frühmodernen Normensetzung in den Kernregionen des Alten Reiches. Ein Quellenwerk, Bd. 7: Policeyordnungen in den

von Musikstücken anzusetzen, war bei Strafe verboten: „*Ebenmässig soll den Stattpfeiffern oder Organisten bey obbegriffener Straff verbotten sein, hinfüro über obbestimpte Besoldung, vnter dem schein, also daß sie die Gesang zuvor probiert hetten, weiter etwas zubegeren oder zu nehmen.*“[27] Andere Musiker waren von der Mitwirkung ausgeschlossen: „*Vnd sollen hinfüro zu der Erbarn Hochzeiten Handschlägen vnnd Heimladungen, keine andere Spilleut, als eines Ehrnvesten Raths Stattpfeifer, vnd die Organisten, so in den Kirchen die Orgel zu schlagen pflegen, jedoch obbegriffener Ordnung gemäß, gebraucht werden.*“[28]

In Rothenburg ob der Tauber waren die Honorarsätze für die Musiker bei Hochzeiten pauschal mit 36 bis 40 Kreuzer pro Tag festgesetzt. Die Veranstaltungen durften auch nicht länger als zwei Tage dauern.[29] In Ulm konnte man für die Mitwirkung bei der „*Copulation*“ 30 Kreuzer sowie „*wegen der music bey der mahlzeit und dem tanz jeweils 1 Gulden und 30 Kreuzer*“ bekommen.[30] In (Bad) Windsheim bekam von den drei Stattpfeifern bei einer Hochzeit jeder einen Gulden, sollten die Tänze bis lange in die Nacht dauern oder „*wenn mann sie aber sonderlich die jungegesellen, uff der gassen gebrauchen wollten, so soll ihnen ein besondere Verehrung desswegen zubegehren, oder zu nehmen erlaubt*“ sein.[31] Mit den zusätzlichen Einnahmen konnte ein städtischer Musiker seine Einkünfte bei entsprechendem Engagement nicht nur verdoppeln, sondern sogar vervielfachen.

Aus den bisher vorgestellten Informationen sind die Grundlagen für die permanenten Konflikte zwischen Stadtpfeifern und freiberuflichen Musikern leicht abzulesen. Was für die einen eine bequeme und weitgehend durch Privilegien geschützte Zusatzeinnahme darstellte, bildete für die anderen das Existenzminimum. Voraussetzung dafür war allerdings erst einmal, dass bei einer

---

fränkischen Reichsstädten Nürnberg, Rothenburg o.d.T., Schweinfurt, Weißenburg und (Bad) Windsheim, Erlangen 2015, S. 385–396.

27 Ebd., S. 60.

28 Ebd., S. 60 f.

29 Erneuerte Policey-Ordnung deß Heil[igen] Röm[ischen] Reichs Statt Rotenburg ob der Tauber, Rothenburg ob der Tauber 1685, S. 13, 16 f. Vgl. Wolfgang Wüst, Die „*gute*“ Policey im Reichskreis. Zur frühmodernen Normensetzung in den Kernregionen des Alten Reiches, Bd. 2: Die „*gute*“ Policey im Fränkischen Reichskreis, Berlin 2003, S. 167–187.

30 Steffen Schure, Die Geschichte des Stadtmusikantentums in Ulm (1388–1840), Ulm/Stuttgart 2007, S. 181 f.

31 StadtA Bad Windsheim, Stadtmusik [ohne Signatur], Musicanten Bestallung, o.O. [1670], fol. 79 r.

Veranstaltung keiner der Stadtpfeifer verfügbar war. In der Regel durften erst dann andere Musiker engagiert werden. Wenn sich unmittelbar nach Bekanntgabe einer bevorstehenden Hochzeit eine Gruppe für die musikalische Gestaltung anbot, war freilich für den Hochzeiter als Auftraggeber oft nicht erkennbar, dass er nahe daran war, sich strafbar zu machen. Reisen von Stadtpfeifern wurden gern gesehen, wenn man freiberuflich tätig war.

Als sich am Beginn des 16. Jahrhunderts noch einige wenige Städte professionelle Ensembles leisteten, hatten diese oft sogar überregional zu tun. Die Rothenburger Stadtmusik bediente etwa auswärtige Hochzeiten im Umkreis von circa 25 bis 50 Kilometern. Manche Auftritte lagen im unmittelbaren Einzugsbereich anderer Reichsstädte oder Fürstenhöfe. 1505 holte Dyonisius Geßner „*ettliche pfeyffer unnd spilleut*" nach Schwäbisch Hall,[32] 1506 bat Michel Stoll um Musiker für eine Hochzeit, „*di weyll wir dan hie zu Dinckelpühel mit Spilleutth nit gefast sein, bit ich ewer weyßheyt* [...] *Ewer stat pfeufer zu vergonnen dan sie auff negstes suntag zu nacht hie zu dinckelspühel sein mogt*[en] *vnd sp*[ä]*ter zwen tag bey vns beleyben.*"[33] 1515 spielten zwei „*Hiesige StadtPfeiffer*" auf der Hochzeit des Adam Bernhamer in Ansbach.[34] Im Laufe des 16. Jahrhunderts veränderte sich die Situation zusehends. Immer mehr Städte und auch die Höfe verfügten über eigene Musiker. Auch wenn die Gruppen zunächst klein waren, sorgten sie selbst für den akustischen Rahmen bei den allfälligen Tänzen. Und überall, wo es professionelle Musiker gab, zog man eigenen Nachwuchs heran, der nach Lehrjahren am Ort sich anderweitig als Gehilfe verdingen wollte. Das gelang freilich nicht immer. Der Spielraum für gut ausgebildete Musiker, die aus den Stadtpfeiferfamilien kamen, wurde enger.

Im nahen Wertheim am Main waren die Bedingungen etwas komplizierter als in Rothenburg. Der Graf von Löwenstein, dem die Stadt gehörte, baute erst zu Beginn des 17. Jahrhunderts um den Schlosstürmer herum ein kleines Musikensemble auf. Weil die Mittel für weitere Gesellen fehlten, musste ein freiberuflicher Instrumentalist als Daueraushilfe herhalten. Um an lukrative Auftritte bei Tänzen zu kommen, schlossen sich andere Musiker aus der Stadt mit genau diesem Mann zusammen und bildeten eine „*Companey*". Mit juristischen Mitteln

32 StadtA Rothenburg, A 1319, fol. 190 r.

33 Ebd., fol. 192 r.

34 Ebd., fol. 196 r.

musste der mehr und mehr ausgeschlossene eigentliche Stadtmusiker sein Mitwirkungsrecht erzwingen.[35]

Da die Berufsbezeichnung „Musicus" oder „Musikant" nicht geschützt war, konnte sich jeder Instrumentalist in dieser Weise kennzeichnen. Besonders elegant stellte es 1591 Hans Dilger (Dillinger) in Rothenburg an. Er heiratete die Witwe des Stadtpfeifers Melchior Blaicher und beantragte das Bürgerrecht unter dem Versprechen, den Stadtpfeifern kein Geschäft wegzunehmen. Außerdem gab er seine Bereitschaft an, bei Bedarf mitzuhelfen und was „*in Kirchen vnd Schulen bedrifft, bin Jch ebenmeßig anpurtig, den Concentum mit allerlay Instrumenten nach meinem besten Vermügen, helffen zu stercken vnd zu zieren.*" Am Ende seiner Eingabe formulierte er jedoch die Bitte, „*die wöllen mich mit einem geringen järlichen Honorarii gunstiglich bedenken.*"[36] Die Unterschrift trägt die Bezeichnung „*Musicus Jnstrumentalis*". Da sich 1614 ein Hans Dillinger jun. um eine Stadtpfeiferstelle bewarb und dabei seinen Vater erwähnt,[37] könnte man annehmen, dieser sei in der Zwischenzeit regulär bestellt worden. 1622 sind Vater und Sohn Dilger tatsächlich bei den Stadtpfeifern aufgeführt.[38] In einem Inventar der Rothenburger Stadtpfeiferei wird mit Georg Dillger vermutlich ein weiteres Mitglied dieser Musikerdynastie genannt als Spieler von Fagott, Flöte und Krummhorn jeweils in der Altlage sowie des zweiten Tenor-Pommers und einer Diskantgeige.[39]

Nicht immer konnten Übergänge bei freien Stellen und das Zusammenwirken unterschiedlicher Musiker so reibungslos gestaltet werden. Im Folgenden geht es um den Stadtpfeifer Heinrich Bronner (oder Brunner) und Mitglieder der Familie Kern, ebenfalls in Rothenburg. Von den älteren Brüdern Kern war Simon als Stadtpfeifer, sein Bruder Heinrich als Turmbläser angestellt. Im September 1645 strengte der Stadtpfeifer Heinrich Bronner (oder Brunner) eine Klage gegen dessen Söhne Hans und Conrad Kern an.[40] Bronner hatte mit deren Onkel Simon

---

35 Vgl. hierzu: Franz Körndle / Moritz Kelber, Die Wertheimer Stadtmusik, in: Wertheimer Jahrbuch 2016/2017, hg. v. Historischen Verein Wertheim in Verbindung mit dem Staatsarchiv Wertheim, Wertheim 2018, S. 95–112, hier bes.: S. 106–109.

36 StadtA Rothenburg, A 1319, fol. 287 r. Für wertvolle Hinweise auf die Informationen in den Rothenburger Stadtpfeiferakten danke ich Frau Katharina Preller M.A. (DFG-Projekt „Urbane Musikkultur in der frühen Neuzeit" an der Universität Augsburg, URL: https://www.uni-augsburg.de/de/fakultaet/philsoz/fakultat/musikwissenschaft/forschung/dfg-projekte-laufend/urbane-musikkultur/).

37 Ebd., fol. 312 r.

38 Ebd., fol. 330 v.

39 StadtA Rothenburg, AA 135, fol. 3 r.

40 StadtA Rothenburg, A 1319, fol. 354 r–356 v (daraus auch die folgenden Zitate).

gemeinsam bei einer Hochzeit „*vffgewartet*". Am folgenden Tag ließ er den Lehrjungen an seiner statt musizieren. Und offenbar hatte auch Simon Kern sich vertreten lassen, und zwar durch seine beiden Neffen. Hans und Conrad Kern, so der Bericht Bronners, hätten den Lehrjungen „*alsobalden außgelacht, verhönt, verspottet*", denn „*er sey nicht so gut, das Er neben ihnen auffwarte, vnd wann er schon außgelernet, so lassen Sye ihn doch nicht Passieren*". Am folgenden Tag kam es erneut zu einem Vorfall. Die Brüder Hans und Conrad Kern waren spät im Wirtshaus Schwarzer Adler erschienen und ließen sich „*ein maß Wein einschencken*". Obwohl die Wirtin bereits zu Bett gehen wollte, zogen „*doch beede Gebrüder vnter dem Mantel die Geigen herfür, vnd vertrieben also denn bede den Pfisterer, vndt meinen LehrJungen aus* [dem] *Hauß*!" Bronners bis hierher formulierter Vorwurf betrifft die praktizierte Übertretung der Regeln, dass nämlich Heinrich Kern seinen Posten als Türmer verlassen hätte, um mit seinen Söhnen im Wirtshaus aufzuwarten, weshalb ihm (Bronner) „*alß besteltem Musicanten, das Brodt vor dem Maul abgekürzt worden.*" Nicht genug damit, erdreistete sich offenbar Hans Kern, die Hälfte des Lohns für das Musizieren bei der besagten Hochzeit zu fordern. An einem anderen Tag verschafften sich Hans und Conrad Kern in Abwesenheit Bronners Zugang zu seinem Haus und konfrontierten die Ehefrau mit Äußerungen wie: „[Bronner] *könnte keinen Bauren Dieners Geigen, waß* [er] *dann für ein Meister sein sollte.*" An die alte Mutter und die Magd seien Beleidigungen gerichtet worden, Bronner „*gehör ins Siechhauß*". Einen Zinken, der auf der Hobelbank gelegen habe, habe Hans Kern an sich genommen, die Magd geschändet und bei seinem Abgang gerufen, er würde „*bey des Teüffels treü, die selbe nacht* [Bronner] *den Halß brechen*".

Hans und Conrad Kern gaben vor Gericht eine Gegendarstellung ab. Hans habe nach der Sonntagsvesper den Lehrjungen vor Bronners Haus angetroffen. Da Bronner nicht da gewesen sei, habe er dessen Frau angesprochen, warum er für das „*schuldige Kopffstück die begerte Seyten*"[41] nicht geschickt habe. (Dabei könnte es sich um Noten für eine Kantate oder einen Psalm als Hauptstück im Gottesdienst handeln, denkbar wären aber auch Saiten für eine Theorbe oder einen Chitarrone, dessen Wirbelkasten für die Basssaiten auch als Kopfstück bezeichnet werden kann.) Sie habe ihn daraufhin geschmäht, „*man wisse wohl, was wir für ein loses leichfertiges thürmers vnnd Landfahrers gesindtlein seyen*".[42] Er habe sich infolgedessen zu der „*geclagten injurien sachen zum höchsten gemüssiget.*" Der Prozess zog sich bis ins Frühjahr 1646 hin. Am 8. April machte

41 Ebd., fol. 357 r.
42 Ebd.

Dr. David Frisch den Vorschlag eines Vergleichs, nach dem Hans Kern eine Strafe von 30 Pfund zu zahlen hatte und die Gerichtskosten zu teilen waren. Da Kern damit einverstanden war, wurde der Fall am 24. April 1646 abgeschlossen.[43] Mit dem Abstand von mehr als 370 Jahren kann kaum entschieden werden, welcher der Rothenburger Musiker als unangenehmer zu empfinden gewesen wäre. Allerdings spricht manches Argument dafür, Hans Kern für einen schwierigen Zeitgenossen zu halten. 1655 gab der Lehrjunge bei den Türmern, Hans Beier, bekannt, dass er sich bei Graf Siegfried von Hohenlohe-Gleichen zum Kriegsdienst verpflichten wolle, denn sein Lehrmeister Hans Kern habe ihn *„geschlagen wie einen Esel"*.[44] Obwohl dieser Hans Kern auch seinen Dienst als Rathaustürmer sorglos ausübte und im September 1655 eine Feuersbrunst in Schandhof übersah,[45] brachte ihn das beruflich kaum in Verlegenheit. Erst als er 1659 für sich eine sechste Musikantenstelle einforderte mit dem Erpressungsversuch, er würde ansonsten seinen Dienst quittieren, sah der Rat der Stadt eine günstige Gelegenheit. Zum 14. März 1659 wurde Hans Kern verabschiedet.[46]

Über eine andere Variante pädagogischer Vorgehensweisen gibt ein ebenfalls vor Gericht ausgetragener Streit in Wertheim Auskunft. 1668 hatte der Lehrjunge Johann Bach seinen Lehrherrn Michael Senkeisen verklagt. Er sei zwar nach der mehrjährigen Ausbildung in der Lage gewesen, die anspruchsvollen Partien bei Aufführungen übernehmen zu können, jedoch sei sein Zeugnis so miserabel ausgefallen, dass er damit bei Bewerbungen an keiner anderen Stelle habe erfolgreich sein können. Ganz offensichtlich hatten die Fähigkeiten des jungen Mannes den Lehrer dazu verführt, ihn langfristig als billige Arbeitskraft an die Stadtmusik zu binden. Sollte Johann Bach damit der Weg auf eine anderweitig feste Anstellung verbaut worden sein, wäre so der Weg in den Niedriglohnbereich eines Freiberuflerdaseins geebnet worden.[47]

Fast zur gleichen Zeit eskalierten die alten Streitereien unter den Musikern in Rothenburg. Es ging weiterhin um die Frage, wer bei Hochzeiten und bei Wirtshaustänzen aufspielen durfte. Auch durfte der eigentlich verabschiedete Hans Kern aus Gründen der familiären Bindung weiterhin mitwirken. Das ging dem Kollegen Heinrich Bronner und dem neu als Stadttürmer eingestellten Michel Bernhardt zu weit. Sie holten sich Rat aus den Städten Nördlingen und Ulm,

43 Ebd., fol. 366 r–369 v.
44 Ebd., fol. 380 r.
45 Ebd., fol. 381 r.
46 Ebd., fol. 385 r–387 r.
47 Körndle / Kelber, Wertheimer Stadtmusik (wie Anm. 35), S. 105 f.

von wo sie Informationen erhielten, wie man dort mit ähnlichen Problemen umzugehen pflegte.[48] Der ebenfalls neu nach Rothenburg gekommene Abraham Martini griff schließlich zum Äußersten und bat die Stadtpfeifer aus Eger, Graßlitz, Adorf, Hof und Windsheim um eine öffentliche Stellungnahme, dass Simon Kern nicht als Stadtpfeifer tragbar wäre. Er habe wie „*ain Birfiedel geiger gelernet, vndt* [sei] *nun also ain geraume Zeit vff ain Schein alß ain wohl erlernter Musicant* [...] *Passiret worden*", Vater und Sohn würden „*der Edlen Music ein solchen Despect anthun, daß sich die gantze Companij wie auch andre Erliche liebhabende, ihrer Schämen müßen.*"[49] Am 4. Oktober 1667 musste der Rat der Stadt wieder einmal eingreifen. Dabei wurde Simon Kern auf seiner Stelle bestätigt, desgleichen sein Sohn Leonhard und auch der Sohn Hans Georg Dillingers mit Namen Hans Leonhard. Ebenso bestätigt wurden die alten – aber nie eingehaltenen – Regeln zur Verteilung der Gelder aus dem Ertrag bei Hochzeiten.[50] Wenig später bat Michel Bernhardt um seinen Abschied,[51] 1669 verließ auch Abraham Martini die Stadt.[52]

Auch wenn in den zahlreichen Eingaben und Klageschriften der Musiker in allen Reichs- und Landstädten mehrfach von fremden Instrumentalisten die Rede ist, sind meistens eben doch die freiberuflich tätigen „*Bierfiedler*", „*Hochzeitsgeiger*", „*Stümpler*" und „*Dorfpfeifer*" gemeint, die sich längst niedergelassen hatten, aber ihren Lebensunterhalt mit Gelegenheitsgeschäften verdienen mussten. Neben dem Aufspielen in Wirtshäusern, sofern sie nicht von den Stadtpfeifern bedient wurden, gehörten vor allem die Hochzeiten zu den begehrten Anlässen, um die es oftmals Streitereien gab. Es ist leicht vorzustellen, was geschah, wenn die angestellten Stadtmusiker mit einer offiziellen Delegation über mehrere Tage auf Reisen waren. Die ortsansässigen, freiberuflichen Musikanten bekamen in solchen Zeiten freien Zugriff auf die finanziell einträglichen Gelegenheitsgeschäfte, da sie von den Stadtpfeifern nicht bedient werden konnten. Würde man einen Vergleich ziehen von den Anfängen bei der Professionalisierung im urbanen Musikleben zu Berufsausbildung und -ausübung in der Gegenwart, dürften die Parallelen in beeindruckender Weise ins Auge fallen.

---

48 StadtA Rothenburg, A 1319, fol. 429 r (Johann Heinrich Hetsch d. Ältere an Heinrich Bronner); A 1319, fol. 430 r und 430 v („*Ulmischer Vergleich der Herrn Musicanten*").

49 Ebd., fol. 433 r und 433 v.

50 Ebd., fol. 426 r–428 r.

51 Ebd., fol. 437 r.

52 Ebd., fol. 444 r–445 r.

Dorothea Fastnacht

Das Historische Ortsnamenbuch von Bayern – Grundlage für die Stadtgeschichtsforschung

Abstract: To find a comfortable entry point to urban history research of Bavaria, the volumes of the series "Historisches Ortsnamenbuch von Bayern" are suitable. Not without reason it is located at the "Kommission für bayerische Landesgeschichte" in Munich. Because although the language development of the place names is in the foreground in these books, the place name documents are presented in meaningful regesta, that are carefully recorded and exactly dated. A competent editor ensures reliability of all informations and results.

1. Vorbemerkungen

Dass die Veranstalter bei der Planung dieser Vortragsreihe beziehungsweise des Sammelbandes von Aufsätzen zum Thema „süddeutsche Städtelandschaft" an das Reihenwerk „Historisches Ortsnamenbuch von Bayern" (HONB) gedacht haben, liegt auf der Hand: Die Bände von den bearbeiteten Landkreisen tragen schon im Titel den Namen einer Stadt. Oft kommen in einem Band zwei Städte vor, denn die 1862 aus den Landgerichten älterer Ordnung gebildeten Bezirksämter entstanden meist durch Zusammenfügung von zwei Ämtern.[1] 1939 wurde dafür die Bezeichnung „Landkreis" eingeführt. Das 1947 gestartete Unternehmen HONB ist an diesem historisch gewachsenen Gebietsstand orientiert. Man nennt sie jetzt im Hinblick auf die großen Umstrukturierungen bis hin zur Auflösung eines Teils dieser Landkreise in der Gebietsreform von 1972 „Altlandkreise". In den von mir bearbeiteten Bänden sind dies die ehemaligen Landkreise Ebermannstadt (2000)[2] mit den Städten Ebermannstadt, Waischenfeld und Hollfeld, Staffelstein (2007)[3] mit Staffelstein und Seßlach, Erlangen

1 Vgl. Wilhelm Volkert (Hg.), Handbuch der bayerischen Ämter, Gemeinden und Gerichte 1799–1980, München 1983, S. 451.

2 Dorothea Fastnacht, Ebermannstadt. Ehemaliger Landkreis Ebermannstadt (Historisches Ortsnamenbuch von Bayern [künftig: HONB], Oberfranken 4), München 2000.

3 Dorothea Fastnacht, Staffelstein. Ehemaliger Landkreis Staffelstein (HONB, Oberfranken 5), München 2007.

mit Erlangen und Baiersdorf (2015)[4] sowie das Manuskript vom Stadtkreis Nürnberg.[5]

Zwar gehören Aspekte wie die Stadtwerdung nicht zu den Grundanliegen dieser Reihe. Dort wird als eigentliches Ziel verfolgt, die Ortsnamen in ihrer Primärbedeutung mit ihrer sprachlichen Entwicklung herauszuarbeiten sowie die räumliche und zeitliche Entwicklung des jeweiligen Ortsnamentyps und die damit im Zusammenhang stehende Siedlungsgeschichte darzulegen – und zwar von der Wüstung bis zur Großstadt. Auf diesem Weg wird versucht, mit den Mitteln der Ortsnamenforschung die (früh-)mittelalterlichen Verhältnisse zu erhellen, wobei Einblicke in die Zeiträume, in denen die Namen geprägt wurden, und in die Entwicklung eines spezifischen Sprachraums gewonnen werden. Es geht um das Aufspüren der Anfänge von etwas Neuem – in unserem Fall die ortsgebundene Besiedlung, die einen besonderen Reiz ausüben kann – fast einen stärkeren als die Darstellung des Vollentwickelten.[6]

Wenn also aus einer Siedlung später eine Stadt wird, so ist das eine außersprachliche Entwicklung, wofür der Ortsname nicht mehr relevant ist. Dass Nürnberg den Namen einer Stellenbezeichnung „Nürnberg" trägt, spielt bei der Weichenstellung, den Platz zu fördern, sodass bald eine städtische Siedlung entsteht, keine Rolle mehr. Nicht im Gründungsdatum, sondern im überörtlichen Marktgeschehen liegt ja der Keim jeder Stadtwerdung.

Historiker und Ortsnamenforscher treffen sich aber bei der Quellenarbeit.

2. Die korrekte Information

Jeder, der sich in die bayerische „Städtelandschaft" einarbeiten und den Wissensstand von der Stadtwerdung der genannten Landstädtchen und bald der Reichsstadt Nürnberg abrufen will, kann die Recherchen in den verfügbaren Bänden des HONB als Einstieg nutzen – tragen sie doch unter dem Schirm der Kommission für bayerische Landesgeschichte und der kompetenten Redaktion[7]

4 Dorothea Fastnacht, Erlangen. Ehemaliger Stadt- und Landkreis (HONB, Mittelfranken 7), München 2015.

5 Alle in diesem Aufsatz vorkommenden Zitate sind den Ortsartikeln in diesen Bänden entnommen.

6 Vgl. Hermann Schreibmüller, Otnand, der erste große Reichsministeriale in Franken, in: Günther Schumann (Hg.), Hermann Schreibmüller, Franken in Geschichte und Namenwelt. Ausgewählte Aufsätze (mit einem Schriftenverzeichnis) zum 80. Geburtstag des Verfassers, Würzburg 1954, S. 12–28, hier: S. 12.

7 Auch dieser Aufsatz verdankt einige Anregungen dem Redakteur Dr. Wolfgang Janka, dem ich die Arbeit vorlegen durfte.

ein Gütesiegel für verlässliche Information. In den sogenannten „Ortsartikeln" von jeder Siedlung im Untersuchungsgebiet werden die vorhandenen Quellenbelege mit den quellenkritischen Angaben in chronologischer Reihenfolge dargeboten. Von großem Wert dürfte dabei die Vorgabe sein, dass in der Belegsammlung auch die wichtigsten historischen Fakten im Kurzregest erscheinen. So werden die überlieferten Nachrichten jeder Siedlung, Wüstung und – mit größeren Beleganfall – auch der Städte[8] erfasst. Vor allem die jüngeren Bände bieten eine reiche Grundlage des überlieferten Wissens von einer Siedlung – von Indizien auf die primäre Bedeutung des Siedlungsplatzes bis hin zu den neuesten statistischen Angaben zum Größenwachstum.

Bei den Machern von Homepages der Städte oder in den Wikipedia-Einträgen davon im Internet für die schnelle Information scheint das HONB noch weitgehend unbekannt zu sein. Von Ebermannstadt ist zu lesen: „Das Dorf wurde 531 vom Stamm der Thüringer gegründet." Derartige als Fakt ausgegebene Vermutungen wird man in der Belegreihe zu Ebermannstadt vergeblich suchen. Indirekt und über das Register gut auffindbar werden aber die möglichen weiträumigen Verbindungen der alten „*-stat*"-Namen mit Thüringen in der „Siedlungsgeschichtlichen Einleitung" angesprochen. Der tatsächliche Erstbeleg von 981 kommt in einer Kopie des 12. Jahrhunderts vor. Richtig steht in der Quelle „*villa Ebermarestat*" geschrieben.

Für die Stadtwerdung von Relevanz dürfte der 1109 im Ausstattungsgut des Stifts St. Jakob befindliche bambergische Herrenhof sein, der seit 1194 bezeugte Ortsadel und die zu 1298 gegebene Nachricht von einer verfolgten Judenschaft. Dem treuen Bannerträger in der Schlacht von Mühldorf und Ampfing, Konrad II. von Schlüsselberg, freit König Ludwig der Bayer 1323 sein Dorf Ebermannstadt, verleiht ihm alle Rechte und Freiheiten, wie sie die Reichsstadt Nürnberg besitzt, und gestattet, es mit einer Mauer zu umziehen. Doch das Format Nürnbergs wird Ebermannstadt nie erreichen, das Warum ist aus den Belegen herauszulesen. Vierzig Jahre nach dem Tod des Stadtherrn wird das Städtchen bischöflichbambergisch und firmiert zuweilen auch wieder als Dorf oder als Markt. Erst im Laufe der Jahrhunderte gewährt der bischöfliche Stadtherr dem Städtchen von

---

8 „Stadt" bedeutet in dem Zusammenhang Stadt. Das ahd./mhd. Appellativum „*stat*" „Ort, Platz, Stelle, Stätte" nahm im 13. Jahrhundert deutlicher die Bedeutung „Wohnstätte, Stadt" an und löste in der Funktion das ältere „*Burg*" ab. Das Wort „Bürger" erinnert noch an die (ehemals) bei der Siedlung vorhandene Burg oder auch einen burgartig befestigten Herrenhof der Karolingerzeit, der zur Siedlung erweitert werden konnte. Vgl. Manfred Niemeyer (Hg.), Deutsches Ortsnamenbuch, Berlin/Boston 2012, S. 103, 606.

gleichbleibender Größe die Separierung von der Pfarrei Pretzfeld, unterstützt den Bau einer neuen Stadtmauer, genehmigt einen Wochenmarkt und fast 100 Jahre später vier Viehmärkte. Im letzten ausführlich zitierten Beleg von Roppelt ist festgehalten: „*Alle landesherrliche Gerechtsame gehoͤren dem Hochstift, auch ist ein herrschaftliches Amthaus daselbst. Der Rath* [...] *uͤbt die Vogteilichkeit in dem Staͤdtchen aus.*“

Durch die konsequente Durchforstung der in den einschlägigen Archiven lagernden Archivalien konnten zahlreiche Datierungen aktualisiert werden. In der im Bayerischen Geschichtsatlas von Max Spindler zu findenden, von Sebastian Hiereth und Gertrud Diepolder bearbeiteten Karte „Städte und Märkte im Mittelalter“,[9] die erklärtermaßen nur „eine erste, grobe Bestandsaufnahme“ hergibt, sind nach den von mir vorgenommenen Recherchen folgende Datierungen zu ersetzen: Staffelstein war bereits 1290 „*opidum*“; Hollfeld war circa 1285 schon „*forum*“ und verwendete 1326 „*der Stat Insigel zu Holvelt*“; König Ludwig privilegierte das Dorf Waischenfeld 1315 mit den Rechten und Freiheiten, wie sie der Stadt Bamberg verliehen worden sind; bei Erlangen erfolgte die Marktverleihung und erste Stadtnennung im Jahr 1326. Bezüglich des zum Markt abgesunkenen Wonsees gestattete Kaiser Ludwig dem Nürnberger Burggrafen Friedrich bereits 1328, „*in Wunnsez*“ eine Stadtbefestigung zu bauen, die Blutgerichtsbarkeit auszuüben und einen Wochenmarkt abzuhalten, und verlieh der Stadt alle Freiheiten, wie sie die Stadt Nürnberg besitzt.

3. Die ausführliche Information

In Staffelstein, dessen hohes Alter durch die Verbindungen mit dem Adel des Grabfeldgaus und mit Kloster Fulda im 9. Jahrhundert belegt werden kann, wurde 1130 speziell den Bamberger Domkanonikern von König Lothar Marktrecht erteilt. Zwischenzeitlich war diesen Domkanonikern in der Person Adalberts von Truhendingen, der von Friedrich von Rothenburg die würzburgischen Lehen in Staffelstein erhalten hatte, ein Störfaktor im Marktgeschehen erwachsen. Den darüber entbrannten Rechtsstreit „*de mercato Staphelstein*“ entschied Kaiser Friedrich 1165 zugunsten des Domkapitels: Nur von seinen Leuten „*habitantes in bonis sancti Kyliani Staphelstein*“ dürfen „*in cymiterio plebane ecclesie* [...], *que iure fundi ad ecclesiam Wirceburgensem spectabat*“, weiterhin Kaufbuden betrieben werden. Die ausführliche Belegkette schließt mit dem Eintrag von 1801 in der historisch-topographischen Beschreibung des kaiserlichen

9 Max Spindler (Hg.) / Gertrud Diepolder (Red.), Bayerischer Geschichtsatlas, München 1969, S. 22 f. mit Kommentar auf S. 81–83.

Hochstifts und Fürstentums Bamberg von Johann Baptist Roppelt ab.[10] Weil dieser das Format der schmucken Stadt Staffelstein sehr informativ widerspiegelt, soll er in vollem Umfang wiedergegeben werden:

„1801 Staffelstein. Eine schoᵉne in einer angenehmen Gegend gelegene Stadt mit Mauern, Thuᵉrmen und Graben umfaßt. Die Lauther umgiebt einen Theil derselben, treibt 7 Muᵉhlen und ergieset sich in einem herrlichen Wiesgrunde unterhalb Nedensdorf in den Maynfluß. Da die Chaussee durch die Stadt geht, so ist uᵉber diesen Bach, der bey großem Wasser oft sehr reißend wird, eine hoᵉlzerne Bruᵉcke erbaut worden. Die Landeshoheit gehoᵉrt dem Hochstift, die hohe Zent dem Domkapitel zu Bamberg und ist der Richtplatz vor der Stadt an der Landstrasse. Der Stadtrath besteht aus 4 Buᵉrgermeistern, einem Stadtschreiber und 8 Rathsverwandten. Er ist sowohl im Ganzen als im Einzelnen der Gerichtsbarkeit des Amtsverwesers unterworfen, dem auch im Steuer und Zollwesen, dann in Zentsachen die Direction zustehet, so wie er auch die niedere Gerichtsbarkeit zu verwalten hat. Die Stadt hat die Zunftgerechtigkeit uᵉber die mehresten Gattungen der Handwerker. Es werden 8 Jahrmaᵉrkte gehalten. Auch ist eine Post daselbst und wird von Bamberg bis dahin 1 1/2 Poststationen gerechnet. Die Stadt hat eine Pfarrkirche ad St. Kilianum, welche ehemals ein Filial der Wuᵉrzburgischen Pfarrey Moᵉrschbach gewesen ist [?], *eine St. Georgen Kapelle, eine St. Anna Kapelle und außer der Stadt eine Kreutzkapelle, ferner einen schoᵉnen neugebauten Pfarrhof, ein Kaplaneyhaus mit 2 abgesonderten Wohnungen, ein Schulhaus, ein Kirchnershaus, ein Domkapitlisches Amtsverwesershaus mit Stadel, Stallung und Nebengebaᵉuden, ein Rathhaus, ein Stadtbraᵉuhaus, ein Spital und ein Siechhaus. Das uᵉbrige der Stadt wird in 4 Viertel eingetheilt. Im ersten Viertel sind, 1 Gasthaus, 44 andere Haᵉuser, 24 Staᵉdel und 4 Hofstaᵉtte; im zweyten Viertel 1 Gasthaus, 45 andere Buᵉrgershaᵉuser, 3 Muᵉhlen, eine Badstube, 23 Staᵉdel und 6 Hofstaᵉtte; im dritten 2 Gasthaᵉuser, 41 andere Haᵉuser, 4 Muᵉhlen, eine Badstube, 17 Städel und eine Hofstatt; im vierten Viertel 1 Gasthaus, 59 andere Haᵉuser, 35 Staᵉdel und 4 Hofstaᵉtte. Die Geistlichkeit ist Bambergisch und die Pfarrey sowohl als auch beyde Kaplaneyen werden von dem Domkapitel vergeben. Der große Stadtzehend ist in 9 Theile getheilt, davon 6 Theile dem Domcustos zu Bamberg, die 3 andern dem Grafen von Schoᵉnborn und Herrn Zoᵉllner auf dem Brand gehoᵉren. Der Sturjenzer Zehend gehoᵉrt zur Haᵉlfte dem Veitspfarrer im Domstifte zu Bamberg, zur andern Haᵉlfte einem Burger zu Staffelstein, und hat auch der dasige Pfarrer einen besonderen Zehend. Das Kloster Banz hat in der Staffelsteiner Flur viele lehenbare Grundstuᵉcke, auch in der Stadt*

10 Johann Baptist Roppelt, Historisch-topographische Beschreibung des Kaiserlichen Hochstifts und Fürstenthums Bamberg […], Nürnberg 1801.

2 haeusliche Gueter, welche Rittermannlehen und demselben nach Absterben der Zoellnerischen Familie wieder heimgefallen sind; es hat auch allda einen besonderen Lehensverwalter aufgestellt.“

Vergleichbar vermitteln die mit den Roppelt'schen Angaben von 1801 die Belegreihe abschließenden Schlusszitate in den Ortsartikeln Ebermannstadt, Hollfeld und Waischenfeld eine Vorstellung vom tatsächlichen Entfaltungsumfang dieser ansehnlichen Städtchen im Bistum Bamberg, ihrer Organisation und den Besonderheiten des Wirtschaftslebens, was aus den zahlreichen Belegen davor und auch aus dem Wortlaut der Stadterhebungsurkunde nicht hervorgegangen wäre.

4. Die besondere Information

Wenn der Ortsnamenforscher die Archivbestände systematisch auswertet, wird er möglicherweise diesen und jenen wichtigen Beleg finden, der bei zielgerichtetem Suchen nicht gefunden wird. Ein Indiz für die künftige Eignung eines Burgortes oder eines Dorfes zur Stadterhebung könnte neben dem vorhandenen Herrschaftspunkt der Finanzstatus gewesen sein. Die in Geldgeschäften erfahrenen Juden haben vor der Mitte des 12. Jahrhunderts in Nürnberg Zuflucht gefunden. Durch das „*Martyrologium*“ des Nürnberger Memorbuches werden zu 1298 die Opfer der Pogrome in Nürnberg, in Ebermannstadt, Hollfeld und Waischenfeld bezeugt.[11] Als König Ludwig der Bayer sich bereits 1314 veranlasst sah, seinen Bannerführer Konrad II. von Schlüsselberg für dessen Einsatz in der Schlacht von Gammelsdorf zu ehren, erhob der König das Dorf Waischenfeld 1315 zur Stadt und erlaubte ausdrücklich, dass er Juden behausen dürfe. Erst später werden auch in Staffelstein (1405), Erlangen (1408) und Baiersdorf (1425) Juden genannt.

Eine ganz spezielle Adelung einer Dorfsiedlung, lange bevor sich diese zum Markt und meist auch zur Stadt entwickelt hat, ist das Vorkommen im Wunderkatalog der Heiligen Kunigunde. Das etwa 1200 erstellte Verzeichnis der „*Miracula*“ war Vorbedingung für die Heiligsprechung. Es berichtet von Wunderheilungen in Waischenfeld („*an einer edlen Dame ‚de castello‘, das ‚Wiskenvelt‘ heißt*“), Hollfeld („*an einer Frau ‚de Holevelt‘*“), Heiligenstadt („‚*in Haldenstat‘ wurde ein blindes Mädchen zweimal erleuchtet*“), Seßlach („‚*De Sezzelach‘ wurde eine Frau von langer Krankheit befreit*“), Rattelsdorf („*blinde Frau ‚de Ratelsdorf* [...]‘ *sehend gemacht*“), Baiersdorf („*Mann ‚de Beierstorf‘ und viele andere aus*

---

11 Siegmund Salfeld (Hg.), Das Martyrologium des Nürnberger Memorbuches (Quellen zur Geschichte der Juden in Deutschland 3), Berlin 1898.

Nbg. haben beschworen, daß sich die Erde, die sie vom Grab der hlg. Kunigunde fortgetragen hatten, wahrhaftig in Getreide verwandelt hätte") und Nürnberg („*Ein Mädchen mit einem verbildeten Fuß und Knie ‚de burgo, qui dicitur Nurenberc‘, wurde von Kaiserin Cunigunde geheilt*").

5. Die richtigstellende Information

Da und dort kann ein gut recherchiertes HONB wohl eine Lücke schließen, wo der Historische Atlas von Bayern (HAB) oder entsprechende Arbeiten fehlen oder veraltet sind, wo die Quellenkritik nicht auf der Höhe der Zeit ist, wo der Historiker anfechtbare Thesen vertritt und der Germanist alias Historiker in der Lage ist, dies zu widerlegen. In diesen Fällen könnte das HONB die Grundlage für die Geschichtsforschung zu einer ganz bestimmten Stadt sein.

Die Kernkompetenz der Ortsnamenforschung in Bayern ist die Befähigung, mit sprachlichen Mitteln – wie die Kombination von „*Schwab-*" und „*-wang*"-Namen mit deren häufiger Neigung zum „*w*"-Schwund – die frühmittelalterlichen Verhältnisse zu erhellen. Gerade bei der Bearbeitung des HONB Erlangen konnten auf dem Weg neue Horizonte eröffnet werden und vor allem die Urkunde mit dem Erstbeleg neu interpretiert werden, sodass sich nun die Abläufe stimmig ineinander fügen. Auch der in größere räumliche und zeitliche Zusammenhänge gestellte Ortsname konnte mit größtmöglicher Sicherheit gedeutet werden. Von Vorteil ist dabei das Herangehen an die geschichtliche Materie ohne Vorgaben durch veraltete Theorien wie etwa die Prämisse von der Gaugrenze als mittelalterlicher Siedlungsgrenze, wo in den urkundlichen Angaben lediglich grundsätzliche Orientierungen auf einen Gau angesprochen werden.

Desgleichen kann der Historiker durchaus von den Belegsammlungen der Bände des HONB profitieren, wenn der Ortsnamenforscher den Ehrgeiz hat, entgegen bisher vertretenen Darstellungen die richtigen Kontinuitäten herauszufinden, das heißt, gleich und ähnlich klingende Belege mit der richtigen Siedlung zu verbinden.

Vor allem bei der Bearbeitung des HONB Nürnberg hat sich dafür ein reiches Betätigungsfeld geboten, galt es doch, gegenüber einer Vielzahl von einmal eingebürgerten Meinungen, einen durch Quellenarbeit gestützten Standpunkt dagegenzustellen. So schlugen die phantasiereichen Vermutungen des Architekten und Fotografen Friedrich August Nagel manchen in ihren Bann. Zum Beispiel findet man seine in den 1950er Jahren[12] formulierte Hypothese, in

12 Friedrich August Nagel arbeitete von ca. 1908 bis 1959 an seinen oft undatierten Ortsgeschichten. Die meisten schloss er in den 1950er Jahren ab.

Steinbühl habe es neben dem Ansitz der Breitenstein einen älteren Vorgängersitz gegeben, den der Reichsbutigler Heinrich von Stein um 1260 erbaut habe, ungeprüft im Stadtlexikon Nürnberg.[13] Der „älteste Turmhausbau Nürnbergs" sei dieser gewesen. Aufgrund ergiebiger Belege wird man im Band Nürnberg den alten Bürgersitz – ein Weiherhaus, das Breitenstein'sche Eigenlehen – mit seinen zum Schloss gehörigen Gebäuden deutlich von einem vormals breitensteinischen Hof unterscheiden können und die Bestätigung im Grundsteuerkataster finden. Danach „geht aus dem Salbuch von 1688 klar hervor, dass im vorliegenden Fall zum einen die ehemalige Zugehörigkeit zum Sitz und die Lage unmittelbar am Schlossgarten (wie auch beim Söldengut: ‚*Alter Schloß= hoff. Ein groß Hauß und Stadel unter einem Dach, mit Wald- und Zweÿen Feuerrechten, samt einer Hoffraith und einfahrt im Dorff am Schloßgraben- und garten gelegen*') irreführende Bezeichnungen generiert hat, zum anderen, daß das zweite Stockwerk (‚*mit einem aufgesetzten gaden, daß Schlößlein genand*') das Hofgebäude gewissermaßen schloßähnlich gemacht hat." Einen zweiten Sitz hat es aber nie gegeben.

6. Landesstädte in Franken

Die oben genannten sieben Städte waren einst sogenannte „Landesstädte" – landläufig auch „Landstädte", „Klein- oder Minderstädte" genannt –,[14] die einem Landesherrn unterstanden, der nach Belieben über sie verfügen konnte. Auch wenn sie nicht mit den Begriffen „landsässige Städte" oder „Munizipalstädte" benannt werden, erkennt man sie später als Sitz eines Amtes. Erlangen bietet noch die Besonderheit, dass es im Rahmen seiner „Neu-Böhmen"-Politik 1361 von Kaiser Karl IV. gekauft und 1374 als Stadt mit Marktrechten ausgestattet wurde („*Statt zw Erlanng vnnd der purger gemeinlichenn doselbst* zwei Jahrmärkte *zw Erlanngenn* […] *Inn allen freÿheÿttenn Rechtenn vnnd guttenn gewonnheÿttenn, als das Inn anndern stettenn gewonnheÿtt ist,* desgleichen für ewige Zeiten einen Wochenmarkt"). Der Sohn Wenzel beurkundete 1398 noch ausdrücklich den Schöffen, Rat und Gemeinde daselbst, zum größeren Nutzen „*Vnnsers marckts zu Erlang,* […] *das sÿ furpaß Ewigklichenn Burgerrecht vnnd Statrecht habenn vnnd der geprauchenn sollen vnnd mogenn Inn aller der maß*

13 Helmut Beer, Art. „Steinbühl", in: Michael Diefenbacher / Rudolf Endres (Hg.), Stadtlexikon Nürnberg, Nürnberg ^{2}2000, S. 1038.

14 Vgl. Wilhelm Störmer, Die Gründung von Kleinstädten als Mittel herrschaftlichen Territorienaufbaus, gezeigt an fränkischen Beispielen, in: Zeitschrift für bayerische Landesgeschichte 36, 1973, S. 563–585.

als vnnsere Burger vnnd statt zw Awerbach haben vnnd geprauchenn". Darüber hinaus erlaubt der König, dass sie in derselben Stadt ein Kaufhaus an einem passenden Ort „*mit Maurn annhebenn vnnd pawenn mogen*". Der Gewinn soll „*zu der selbenn vnnser Statt zw Erlanng nutz vnnd frommen, Die zubefestenn mit mawrnn thurnnen vnnd grebenn*" verwendet werden. Von den Kurfürsten 1400 als deutscher König abgesetzt, verlor der seit 1381 durchaus zum Wohle Erlangens tätige König Wenzel Neu-Böhmen größtenteils an die Wittelsbacher, während er Erlangen 1402 seinem Schwager, dem Nürnberger Burggrafen Johann, verpfänden musste. Seit 1686 erfahren wir von der Anlage einer neuen Stadt, „*la Nouvelle Ville d'Erlang, que nous voulons nommer Christianstatt*". Diese direkt an das bestehende Erlangen um den Martin-Luther-Platz grenzende, als Residenzstadt konzipierte Neustadt mit der „*Alt= Stadt Erlang*" zu vereinen, war eine schwierige Aufgabe. Bis 1812 gab es zwei Ratskollegien und zwei Magistrate. Der Ortsartikel „19a. Erlangen" im HONB Erlangen bietet noch zahlreiche weitere Details, an denen sich die ganz individuelle Ausprägung jeder Stadtgeschichte vor Augen führen lässt.

7. Die Quellengrundlage für die Entfaltung einer Reichsstadt

Im Gegensatz zu den landsässigen Städtchen der weltlichen oder geistlichen Fürsten, die im späten Mittelalter aus ihrem dörflichen, von Grundherrschaft und Pfarreisprengel vorbestimmten Wirkungskreis herausgehoben wurden, um wirtschaftliche, herrschaftliche und kulturelle Bedürfnisse zu befriedigen, trägt Nürnberg den Charakter einer königlichen Gründungs- oder Gründerstadt, die um 1040 auf Reichsland angelegt wurde, auch wenn die archäologischen Funde vom Gelände vom Bergfelsen zum Pegnitzufer auf ein gewisses Vorleben schließen lassen, das am ehesten mit den Einrichtungen einer alten Wald- und Wegeaufsicht in Verbindung zu bringen ist. Der archäologische Forschungsstand wird dabei berücksichtigt, auch wenn für die in den vergangenen Jahren gemachten Funde die abschließende Bewertung noch aussteht. Mit den gebotenen Vorbehalten zur Vorab-Datierung der neuesten Keramikfunde in den Medien bietet der Band Nürnberg das Material für das allmähliche Entstehen einer der modernen Salierstädte von den Anfängen bis zur heutigen Großstadt.

In einem Augenblick der Schwäche des Bistums durch den Tod Bischof Eberhards 1040 geschah die bewusst anmutende Gründung Nürnbergs durch Heinrich III. als Folge der Gründung des Bistums Bamberg im Jahr 1007. Dahinter standen wohl politische Erwägungen – teils, um Reichsland zurückzugewinnen, teils, um in dem verschenkten Gebiet durch eine Marktsiedlung von überörtlichem Charakter Terrain zu erobern, teils, um an dem Weg nach Osten einen

Stützpunkt zu schaffen. „Das Besondere Nürnbergs liegt in der Verflechtung seines Ursprungs und seiner ganzen Entwicklung mit dem Reichsgeschehen."[15] Der König selbst hat widerrechtlich das Fürther Marktrecht auf den auf unbekannte Weise akquirierten Platz auf und um den Nürnberg, den „Felsberg" oder möglicherweise den „Berg des **Nuoro*", verlegt und damit den Kristallisationspunkt für das Werden einer Reichsstadt bestimmt, der 1105 unter Heinrich IV. von einer Burg aus verteidigt werden konnte. Als die Staufer 1183 auf dem weithin sichtbaren Felsen ein „*palacium*" errichtet haben und um der wohlfeilen Versorgung des Hofes willen die Kaufleute Zoll- und Geleitsfreiheit im ganzen Reich genießen lassen, hatte sich dort ein leistungsfähiges Wirtschaftsleben entwickelt.

Die im Laufe der Zeit hinzugekommenen Indikatoren für den Aufstieg sind den Regesten der Belegsammlung zu entnehmen: Reichswald – lokaler Markt, Markt mit Fernhandel und Gewerbe, Zollfreiheit, Münzstätte, Juden, Zollrecht – Kirchen, Wallfahrtsort zum Grab des Lokalheiligen, Klöster, Schule – Burg, Stadtmauer, Pfalz, Schanzen – Freiheitsbrief, Stadtgemeinde, Stadtsiegel, Stadtwappen, Rat – Spitäler, Siechkobel – Gewerbestätten und Viehställe vor der Stadtmauer – europaweites Unternehmertum, Reichtum, Bildung, Kunst, Herrscheraufenthalte, Reichsgeltung. Die Punkte wären um viele weitere zu ergänzen.

Das HONB Nürnberg (Stadtkreis) wird viele Nachweise, Belege und Indizien für die Entfaltung der wirtschaftlich europaweit florierenden Stadt in der Fläche und für den Reflex der Organisation des Lebens in städtischer Enge enthalten. In den archivalischen Nachrichten zu den vor dem Stadtring entstehenden Siedlungen geht es um das Stemmen der Bäche, die Anlagen der Gleißhämmer, die Gründung von Siechkobeln an den vier Hauptausfallstraßen, die ersten Wasserleitungen, um Mühlen auf engstem Raum, um die Bleichen der Tucher, das Amt des Fallmeisters und die vor den Toren liegenden Schweineställe oder Ziegelhütten. Sehr viele Informationen vermitteln die Akten der Waldämter und die Privatarchive der Patrizier. Die Literatur zu den Einzelthemen ist schier unüberschaubar. Sicherlich konnte nicht alles in der Bearbeitungszeit berücksichtigt

15 Internet-Beitrag „Stadtrecht Nürnberg", Teil I. Der Königsmarkt 1040/50 bis 1138/40, auf der Grundlage von Werner Schultheiß, Geschichte des Nürnberger Ortsrechtes, Nürnberg 1957 (u.a.), durchgesehen von Hartmut Frommer, URL: https://www.nuernberg.de/internet/stadtrecht/koenigsmarkt.html (letzter Zugriff: 15.6.2020). Die Fortsetzung („Nürnberg stellt die erste Marktgründung eines deutschen Königs dar, die auf dem Reich gehörigen Rodungsland im heutigen Nordbayern aus politischen Erwägungen heraus neu angelegt worden ist.") muss allerdings modifiziert werden. Der Forst zwischen Erlanger Schwabach und Pegnitz war ja 1021 ohne Einschränkung der Bamberger Kirche geschenkt worden.

werden. Einen Überblick vermitteln der Artikel „Nürnberg" von Werner Schultheiß,[16] der Artikel „Reichsstadt" von Walter Bauernfeind[17] oder die Internetbearbeitung zum Nürnberger Stadtrecht von Hartmut Frommer nach Werner Schultheiß.[18]

Natürlich spielte auch bei der Darstellung der Frühgeschichte Nürnbergs die Bewertung der historischen Quellen (Quellenkritik) eine große Rolle. Auf diesem Wege konnten über das HONB Nürnberg noch neue Sichtweisen in die Materie eingebracht werden.

Als Beispiel sei die Winzelbürg ausgewählt. Der Hof am Nordhang des Schübelsbergs – des heutigen Rechenbergs – wird erstmals 1467 aktenkundig. Der „*bürg*"-Name wird als Spottname gedeutet. Viel Verwirrung um dieses Anwesen hat der von Sebald Rech im Zusammenhang mit dem „Fraischprozess" um die Rechtmäßigkeit des circa 1524 mit Quadersteinen und Schießscharten erbauten Bürgersitzes Rechenberg um 1536 unternommene Versuch angerichtet, von einer älteren Bebauung auf der Höhe des Berges zu berichten, die – ganz gegen die schriftliche Überlieferung – „*vor alters die Winselburg*" genannt worden sei („*die Wintzburg* [...] *hatt vor Jaren ein Burgers Hauß oder Sitz auff der Höhe gehabt*"). Der Wahrheitsgehalt dieser Aussage an sich (mittelalterliche Bebauung des Schübelsbergs?) wird seitdem unterschiedlich beurteilt. Dabei ist auch von einem gewaltigen Fundament und von Quaderstücken die Rede, die auf „*ein gewaltig stark Gebäu*" deuteten, und ein Teil der Nürnberger Stadthistoriker war bereit, „*Winselburg*" für den Namen der mittelalterlichen Bebauung des Bergs zu halten, worauf Sebald Rech um 1524 seine Kemenate erbaut hatte. Da dies der einzige Platz ist, der mit der Nachricht von Mauerresten und einem Abzugskanal im Untergrund Anhaltspunkte für ältere Bebauung lieferte, wurden auf Initiative des Nürnberger Stadtarchivars Ernst Mummenhoff Grabungen durchgeführt, in denen aber nur der von Sebald Rech genannte „*Thol*" („Wasserleitung") bestätigt werden konnte. Alles andere war ja nach der Bedrohung durch den Markgrafen 1553 bis auf den Grund abgetragen, die Quader hinweggeführt und der Berg eingeebnet worden. Ohne jede Kenntnis der Vorgeschichte des auf der Nordseite des Schübelsbergs gelegenen Anwesens Winzelbürg, wie sie im Ortsartikel „Winzelbürg" zusammengetragen worden sind, und ohne kritische

---

16 Werner Schultheiß, Nürnberg, in: Handwörterbuch zur deutschen Rechtsgeschichte 3, 1984, Sp. 1114–1119.

17 Walter Bauernfeind, Art. „Reichsstadt", in: Diefenbacher / Endres, Stadtlexikon Nürnberg (wie Anm. 13), S. 877.

18 Siehe oben, Fußnote 10.

Hinterfragung des sich in Wahrheit auf den Rechenberg beziehenden Belegs von circa 1536, folgerte Mummenhoff 1904: „Winzelbürg und Kühberg oder Schübelsberg sind auch hier nicht, wie auch sonst so häufig, auseinander gehalten worden. [...] Die Erinnerung an diese Befestigung auf dem Rechenberg war damals schon vollständig erloschen. Daraus ist zu schließen, dass sie schon vor sehr langer Zeit abgebrochen worden war und in eine frühe Zeit der Nürnberger Geschichte zurückreicht." 1956–1957 hat Friedrich August Nagel auf der Basis der Angaben in eben jenem Beleg ein Phantasiegebäude von einer gewaltigen militärischen, von Kaiser Heinrich II. befohlenen Sicherungsanlage gegen den Bischof von Eichstätt errichtet, die noch in der heimatkundlichen Literatur von 1968 aufgegriffen worden ist, obwohl sie keiner näheren Prüfung standhält.

8. Grenzen der HONB

Wie der Stadt-Status letztendlich mit Leben erfüllt worden ist, stellt sich in den Belegen meist nur bedingt dar. Das mag zum Beispiel im Fall Ebermannstadt daran liegen, dass die Masse der Belege mit Immobilientransaktionen befasst ist, die fast immer im Belieben des bischöflichen Stadtherrn liegen. Ganz selten mischt sich einmal ein später Beleg wie *„1678 Rat und Bürgerschaft ‚zu Ebbermanstatt'"* darunter. Um noch mehr vom überörtlichen Marktgeschehen und dem Ermessensspielraum des Rates als Indikatoren für die Entfaltung städtischen Gemeinwesens zu erfahren, müssten noch die Stadtbücher, Ratsbücher oder Gerichtsbücher durchgesehen werden. Was die bischöflichen Landstädtchen betrifft, so ist allerdings nur mit einem begrenzten Spielraum der Bürgerschaft zu rechnen. Man war den Neuerungen von Haus aus abhold und praktizierte bei Bedarf die Verpfändung. Auch Nürnberg befand sich 1269 bis 1273 nominell im Besitz der Wittelsbacher, wusste aber der Gefahr durch seinen Status als privilegierte Königsstadt (1219) zu entkommen. Die Darstellung des Weges zur Selbstverwaltung wäre allerdings noch mit Leben zu erfüllen. Im HONB erschöpft sie sich an Begriffen wie „*universitas civium*" (1240), Bezeichnung der Nürnberger Bürger als „*cives*", die Nennung des Stadtsiegels und Wappens, das Auftreten von „*consules et universitas civium*" (1256) neben dem Reichsschultheißen. Zur „Freien" Reichsstadt wurde Nürnberg im 14. Jahrhundert durch den Erwerb des Reichsschultheißenamts (endgültig 1427).

Desgleichen wird die städtische Bevölkerung nicht wirklich deutlich. Die führenden Städter haben nur selten ihre Wurzeln im näheren Umland wie Reimar „*officiatus*" von Mögeldorf, der 1213 als „*provisor*" von Nürnberg fungiert, oder der Reichsschultheiß Giselher von Erlenstegen oder der Reichsministeriale Leupold von Gründlach. Nur wenn die Deutung des Ortsnamens dies erfordert (zum Beispiel Hallerweiherhaus), wird der Zuzug solcher unternehmerischer Familien nach Nürnberg unter die Lupe genommen.[19] Auch die Organisation des Alltagslebens in der städtischen Enge findet eher einmal am Rande Berücksichtigung, wie in dem modern anmutenden Bericht des Nürnberger „Fallmeisters" über die Ausübung seiner Pflichten als „Hundsschlager" oder „Nachtjäger" auf streunende Hunde im Ortsartikel „Fallhütte": „1724 Der Wasenmeister Joh. Dieterich Behringer gibt zu Protokoll, dass er in der vergangenen Nacht mit Hilfe des Wasenmeisters zu Altdorf und dessen Söhnen, des Fallmeisters von Lauf und eines Fallknechts von Hersbruck unter Bedeckung durch zwei Stadtknechte ‚*nach Hunden gestöbert*' habe. Dabei sei ein so großer Tumult von Leuten entstanden, dass sie, ehe sie die ganze Stadt verjagt hätte, aufhören mussten. Falls dem allzu großen Volksauflauf in Zukunft nicht gesteuert werde, könnten sie das Stöbern nicht mehr verrichten."

9. Akzeptanzprobleme

Obwohl die in jahrelanger Recherche erarbeiteten Bände mit ihrem reichhaltigen Inhalt in einer feierlichen Veranstaltung im Kreis von Herausgeber, Betreuer, Redakteur, wissenschaftlich Interessierten und daneben Landrat und Oberbürgermeister vorgestellt werden und eine pauschale Würdigung in der Presse finden, so tauchen die HONB auf keiner Homepage der von mir bearbeiteten Städte und auch nicht in ihrem Wikipedia-Eintrag auf. Ist es das dröge Linguistendeutsch, für dessen Feinheiten man sich Zeit nehmen muss? Verhindert das fehlende Wissen vom althochdeutschen Sprachschatz, von der deutschen Sprachentwicklung und ihren mundartlichen Ausformungen das Verständnis?

Nicht jeder Historiker ist bereit, anzuerkennen, dass es vorkommen kann, dass der Germanist alias Historiker in der Lage ist, seine „Thesen" zu widerlegen. Da kann ein bestens recherchiertes HONB wohl einmal eine Lücke schließen, aber die Akzeptanz ist bei den städtischen Kulturschaffenden sogar in der „Universitätsstadt Erlangen" gleich null. Aus dem Rathaus war nie ein irgendwie gearteter Kommentar unter anderem auf die begründete Identifizierung des Erstbelegs von 1002 mit der rezenten Stadt Erlangen oder die mit Hilfe von analogen Namen-Konstellationen in Franken erfolgte Zuweisung des Ortsnamens zu den „*wang*"-Namen zu hören, als gäbe es das HONB Erlangen nicht. Der Archäologe Jochen Haberstroh hat als Ursache, dass bei Archäologen und

---

19 Siehe: Peter Fleischmann, Rat und Patriziat in Nürnberg. Die Herrschaft der Ratsgeschlechter vom 13. bis zum 18. Jahrhundert (Nürnberger Forschungen 31/1–3), Bd. 2: Ratsherren und Ratsgeschlechter, Nürnberg 2008.

Historikern „der Restzweifel an den Ergebnissen der Namenkunde […] virulent ist“, das ihnen „unbekannte Methodengerüst“ genannt. Gerade diejenigen Historiker, denen ein eigenes Urteil zu den Erkenntnissen der Sprachwissenschaft, zu deren Methoden und Möglichkeiten durch Unwissen verschlossen ist, neigen ja dazu, die ganze Branche links liegen zu lassen oder sich sogar abfällig über den Wert der Ortsnamenforschung zu äußern. Gerade sie zeigen mitunter aber auch große Anfälligkeit, ihren Beifall abstrusen Lösungen zu zollen.

Mit solchem Gebaren haben die Bände des HONB nichts zu tun. Es ist zu hoffen, dass dieser Beitrag das Augenmerk auch der Historiker auf das Reihenwerk mit seinem Reichtum an vielfältig brauchbaren Informationen gelenkt hat.

Victor A. Ferretti

Augusta-Miszelle zu Cervantes' „Don Quijote"

Abstract: The miscellany is dedicated to the evocation of the Augsburg area in the Second Part (1615) of Miguel de Cervantes' "Don Quixote" as an Early Modern expression of a contingent sociohistorical imaginary in Habsburg Spain.

Es ist sicher kein Zufall, dass die punktuelle Erwähnung Augsburgs in Miguel de Cervantes' zweiteiligem Meisterwerk „Don Quijote" zu Beginn des 17. Jahrhunderts in einem religiös-ökonomischen Kontext steht. So trifft in Kapitel 54 des Zweiten Teils Don Quijotes „Knappe" – der mittlerweile zum „Gouverneur" beförderte – Sancho Panza auf eine Gruppe von Pilgern, worunter sich ein alter Bekannter von ihm, ein Moriske namens Ricote, befindet, der ihm das unverhoffte Wiedersehen mit folgenden Worten hinlänglich erklärt: „Wohl weißt du, Sancho Panza, mein Nachbar und lieber Freund!, wie sehr uns der Aufruf Seiner Majestät, der öffentliche Erlass, den er gegen die von meiner Herkunft verhängt hat, in Angst und Schrecken versetzte. Zumindest bei mir saßen sie so tief, dass ich den Eindruck hatte, schon vor der Zeit, da wir Spanien verlassen sollten, würden ich und meine Kinder die Härte der Strafe leibhaftig spüren. So hielt ich es für klug, Vorkehrungen zu treffen, denn wer weiß, dass man ihm dann und dann sein Dach über dem Kopf wegnehmen wird, der sucht sich ein anderes, weshalb ich also vorsorglich allein, ohne meine Familie, das Dorf verließ, um zu sehen, wohin ich sie bringen könnte […], denn ich begriff sehr wohl, wie all die Älteren von uns, dass die Aufrufe keine leeren Drohungen waren, wie manche behaupteten, sondern eherne Gesetze, die zur festgelegten Zeit vollstreckt würden […]. [§] Ich verließ also unser Dorf und ging nach Frankreich, und wenn man unsereins dort auch gut aufnahm, wollte ich etwas von der Welt sehen. Ich zog weiter nach Italien und kam schließlich nach Deutschland, wo wir, wie mir schien, freier würden leben können, denn die Leute dort sind nicht allzu zimperlich, jeder lebt, wie er mag, nach eigenem Gusto und Gewissen. Ich mietete ein Haus in einem Dorf in der Nähe von Augsburg [Augusta im Original] und schloss mich dieser Gruppe von Pilgern an, von denen viele Jahr für Jahr nach Spanien kommen, um dessen Heiligtümer zu besuchen […]. Sie durchreisen fast das ganze Land, und kein Dorf gibt es, das sie nicht mit gesalbtem Magen und geölter Kehle, wie man sagt, verlassen, im Beutel wenigstens einen Real in kleinen Münzen, sodass sie am Ende ihrer Reise einen Gewinn von über hundert Escudos beisammen haben, die machen sie zu Gold und verstecken sie entweder in ihren hohlen Pilgerstäben, nähen sie in die Flicken ihrer Pilgerkleider oder schmuggeln sie durch welche List auch immer aus dem Königreich und in ihr Heimatland, so sehr man an den Zoll- und Grenzstationen oder Häfen auch aufpasst, dass sie kein Geld ausführen. [§] Meine Absicht ist es nun, Sancho, den Schatz auszubuddeln, den ich vergraben habe, was ich ohne Gefahr tun kann, da er sich außerhalb des Dorfes befindet, und dann schreibe oder fahre ich von Valencia aus nach Algier, wo meine Tochter[1] und meine Frau sind, wie ich weiß, und suche nach Mitteln und Wegen, sie zu einem französischen Hafen zu bringen und von dort aus nach Deutschland, wo wir uns Gottes Willen anvertrauen wollen."[2]

Der Christ muslimischer Abkunft bezieht sich hier auf die Morisken-Vertreibung ab 1609 unter Philipp III. von Spanien, deren Auguren ihn also bis in das Augsburger Umland getrieben hätten, wohin ihm nun seine Familie nachfolgen solle. Als „Exulant" kehrt er sonach im Pilgerhabit in seine alte Heimat zurück,[3] um noch Barschaft aus dem Land zu schaffen, wobei er sich bei dieser „Kapitalflucht" gleichsam in guter Pilgertradition weiß. Als Zufluchtsort fungiert das unmittelbare Umfeld der *Pax Augustana*, worauf, wie Armstrong-Roche erwähnt,[4] nicht von ungefähr angespielt wird. Denn diese symbolisiert – in idealisierender Art und Weise –[5] eine Ablösung von einer absoluten „Territorialreligion" zugunsten einer – und hier ist der Blick in den Originaltext unabdingbar – partikularistischen „*libertad de conciencia*".[6] Nun ist schon

1 In den Kapiteln 63 bis 65 des Zweiten Teils tritt Ricotes Tochter Ana Félix noch in Aktion, womit, wie Güntert hervorstreicht, auch der Barceloner Abtritt Don Quijotes eingeläutet wird. Vgl. Georges Güntert, Der Diskurs der Minderheiten: Sancho und der Moriske Ricote, in: Christoph Strosetzki (Hg.), Miguel de Cervantes' *Don Quijote*. Explizite und implizite Diskurse im *Don Quijote*, Berlin 2005, S. 83–97, hier: S. 95–97.

2 Miguel de Cervantes, Don Quijote von der Mancha, übersetzt und hg. v. Susanne Lange, 2 Bde., München 2008, Bd. 2, Kap. 54, S. 469–471.

3 Bereits im Jahre 1604 hatte Lope de Vega einen Abenteuerroman mit dem Titel „El peregrino en su patria" („Der Pilger in seiner Heimat") veröffentlicht, in dem jedoch eine romantische Liebeskonstellation leitend ist.

4 Vgl. Michael Armstrong-Roche, Cervantes' Epic Novel. Empire, Religion, and the Dream Life of Heroes in *Persiles*, Toronto 2009, S. 64.

5 Mit Neuschäfer ließe sich diese Wunschvorstellung bezüglich des Religionsfriedens in der Fiktion als ein „Als-ob der Toleranz" fassen. Vgl. dazu: Hans-Jörg Neuschäfer, Sancho Panza bietet seinem Herrn Paroli, veröffentlicht am: 29.8.2015, in: Neue Zürcher Zeitung 199, S. 51, URL: https://www.nzz.ch/feuilleton/buecher/sancho-panza-bietet-seinem-herrn-paroli-1.18603628? (letzter Zugriff: 30.10.2020).

6 Miguel de Cervantes, Don Quijote de la Mancha, hg. v. Francisco Rico, Barcelona 2001, S. 1073.

einiges zu dieser cervantinischen Fügung *anno* 1615 geschrieben worden,[7] die ja heutigentags einen konstitutiven Anspruch moderner Gesellschaften darstellt. Und vielleicht hallt die frühneuzeitliche Rede von einer Gewissensfreiheit gerade deswegen so frappierend nach, da besagte „*libertad*" im Spanischen ebenso ein Fehlen von Gewissen, gar eine Ruchlosigkeit indizieren könnte,[8] was wiederum eine gegenreformatorische Volte darstellte. Ganz gleich nun, zu welcher Lesart man selbst tendiert, der Text wahrt hier eine Doppeldeutigkeit, die gerade deswegen so reflektiert, da sie recht unvermittelt, ganz *en passant* von einem „Exulanten" ins Feld geführt wird. Dabei scheint für den iberischen Geflüchteten mehr als das Gewissen ein „Frei-Sein" im Kontext des Augsburger Reichsfriedens Determinans zu sein; etwas, was man ihm in seiner alten Heimat Spanien als Morisken ethnizistisch nicht mehr gestattet. Die bei Augsburg verortete Freiheit des Gewissens annonciert demnach die Chance, dass jemand, der „[…] mehr von einem Christen als von einem Mauren […]"[9] habe und sich überdies „[…] wie ein echter Kastilier […]"[10] zu verständigen verstehe, ebendort obrigkeitsgemäß als Christenmensch da sein und bleiben dürfe. Das „Augsburger Bekenntnis" des sogenannten „Neuchristen" sagt sich sonach zuvörderst in einer Art von christianischer Vergewisserung aus.

Dass nun die deutsche Pilgergruppe, unter der sich Ricote befindet, so geschäftstüchtig wie geschickt wallzufahren weiß, bringt noch eine zweite Dimension von Augsburg ins Spiel, die im Roman schon etwas früher zum Ausdruck gelangt, als Don Quijote im wohl „kühnsten" Kapitel des Zweiten Teils von der wundersamen „*Cueva de Montesinos*" (II, 23) Kunde gibt. So sei ihm in besagter abenteuerlicher Höhlenwelt auch seine Dulcinea erschienen, die ihn über eine Begleiterin etwas unverhofft um ein Darlehen gebeten habe, worauf der ehrenwerte Junker seiner Dame habe ausrichten lassen, „dass ihre Drangsal mir in der Seele brennt und ich wünschte, ich wäre ein Fugger, um sie zu lindern".[11] Wie der Herausgeber Francisco Rico zu dieser Stelle kommentiert,[12] fungieren die Fugger, die zu Zeiten Cervantes' auch eine Niederlassung in Almagro (im heutigen Kastilien-La Mancha) hatten, hier als vormoderne Synekdoche für einen „hyperbolischen" Reichtum (und vielleicht ja auch für ein wohltätiges

---

7 Vgl. dazu kompendiös: José Antonio Rodríguez García, Cervantes y la libertad de conciencia: recapitulando, in: Laicidad y libertades 17, 2017, S. 211–234.

8 Vgl. Cervantes, Don Quijote (wie Anm. 6), S. 1073, Anm. 36.

9 Cervantes, Don Quijote (wie Anm. 2), S. 471.

10 Ebd., S. 469.

11 Ebd., S. 212.

12 Vgl. Cervantes, Don Quijote (wie Anm. 6), S. 827, Anm. 50.

Engagement im Geiste der Fuggerei). In jedem Fall scheint auch Augsburger Geschäftstüchtigkeit auf die deutsche Pilgerschaft bei Cervantes auszustrahlen, in deren Gefolge sich der „exulierte" Moriske Ricote nun wieder in die alte Heimat wagt, um seine verscharrte Habe außer Landes Richtung Augsburg zu bringen. Dabei lohnt es, eine Perspektive von Güntert einzubeziehen, der zum exilierten Morisken bei Cervantes anmerkt: „Ricote ist der neue, fast schon liberal denkende Mensch, der gleichzeitig mehreren Kulturen angehört und staatenlos wird, wodurch er mehr denn je auf seine einzige existenzielle Sicherheit, das Geld, angewiesen ist."[13] Fernab affluenter Klischees dient die historische Handels- und Reichstagsstadt Augsburg in diesem Sinne auch als frühneuzeitlicher *τόπος* einer Form von Ubiquität in prekär-migrationsdynamischen Zeiten.

Cervantes' Augusta-Evokation ruft in jedem Fall ein Imaginäres zu Beginn des 17. Jahrhunderts hervor, das sowohl durch den nach ihr benannten Religionsfrieden als auch als Heimstatt der katholischen Fugger konditioniert ist und in dem der „[…] brave Ricote, der sich von einem Morisken in einen Deutschen […] verwandelt hatte",[14] sogar *in puncto* Alkoholkonsum bereits integriert zu sein scheint, zumal „*Spanioli und tudeschi, tutto uno: gut compagno*".[15]

Wenn also in Kapitel 54 des Zweiten Teils die (gelehrte) Rede von „*Augusta*" ist,[16] unterstreicht das die beträchtliche imaginäre Dimension Augsburgs, die neben der religionspolitischen Ausstrahlung ihrer Reichstage unter Karl V. noch erweitert wird, berücksichtigt man einen augusteischen Aspekt. So gemahnt das signifikante Aufeinandertreffen von Exiliertem und Dagebliebenem bei Cervantes an Vergils famose erste Ekloge, in der der vertriebene Hirte Meliboeus auf den „akkommodierten" Hirten Tityrus trifft, der Ersteren zum Schluss anbietet, in bukolischen Fluren zu verweilen. Im „Quijote", dessen Held gegen Ende sogar erwägt, bukolischer Hirte zu werden,[17] kehrt nun ein „Exulant" in seine geliebte Heimat zurück und trifft auf einen alten Dorfnachbarn. Die frühneuzeitliche Inversion zeigt sich darin, dass bei Cervantes der vertriebene Ricote dem dagebliebenen Sancho eine größere Freiheit im Exil eröffnet. „Heimat" ist im Falle

13 Güntert, Der Diskurs der Minderheiten (wie Anm 1.), S. 94.

14 Cervantes, Don Quijote (wie Anm. 2), S. 468.

15 Ebd.

16 Wiewohl Covarrubias in seinem zeitgenössischen „Tesoro" von 1611 (fol. 104 r, s.v.) auf die zahlreichen städtischen „*Augustas*" verweist, bezieht sich Cervantes mit seiner „ehrwürdigen" Stadt in Deutschland (nicht auf Trier, sondern) auf Augsburg, was durch den explizit religionspolitischen Kontext, in dem diese genannt wird, verdeutlicht wird.

17 Vgl. dazu: Victor A. Ferretti, Bucolica mente. Zu Diskursivität und Medialität romanischer Bukolik in früher Neuzeit (Studia Romanica Chiloniensia 1), Kiel 2017, S. 413 f.

des Morisken bereits stärker an ein Frei- denn an das *In-situ*-Sein gekoppelt – nach der Losung: „Wo man dich sein lässt, da lass' dich nieder." Der Verlust eines Heimatlandes wird dieserart durch den Hinzugewinn von „Aufenthaltsberechtigung" relativiert. Und so sehr in Ricote noch eine nostalgische Vertriebenensaite *à la* Meliboeus nachschwingt,[18] so sehr bleibt er, was die kommende Zeit angeht, Pragmatiker. Was ihn wiederum dem Tityrus annähert – und noch mehr dem Moeris aus der neunten Ekloge Vergils,[19] der sein Vertriebenenlos mehr nüchtern als ernüchtert mit einem „[...] *quod nunc instat agamus*"[20] angenommen zu haben scheint.[21]

Wie nun in bukolischen Gefilden eine goldene Aura am Horizont gleißt, wahrt „Augustanisches" bei Cervantes für einen Vertriebenen in Früher Neuzeit etwas von einer „arkadischen Freiheit", die Erich Köhler im Kontext von Cervantes so treffend abgeschritten hat[22] und die mehr als ein *Laissez-faire* solchermaßen eine Form von vormoderner Liberalität verkündigt. Dass Ricote Letzteres nicht sicher weiß, sondern dass es ihm lediglich derart zu sein scheint, weist noch einmal auf die Kontingenz eines vormodernen Augsburger Imaginären hin, zu dem nun neben einer „*libertad de conciencia*" und repräsentativem Reichtum dann auch ein gewisser „arkadischer" Liberalismus gehörte, worin sich angesichts der Komplexität von heterogener Gesellschaft und entborgener Subjektivität, wie sie nicht zuletzt Don Quijotes „Selbstverritterlichung" repräsentierte,[23] die

---

18 „Wo immer wir sind, wir weinen um Spanien, schließlich ist es unsere Wiege, unsere angestammte Heimat. [...] Wir haben nicht gewusst, was für einen Schatz wir hatten, bis wir ihn verloren, und so übermächtig ist bei fast allen der Wunsch, nach Spanien zurückzukehren, dass die meisten, die die Sprache so gut sprechen wie ich, und das sind viele, hierher zurückkommen und ihre Frauen und Kinder drüben im Stich lassen: so groß ist ihre Liebe zu diesem Land, und jetzt weiß ich aus eigener Erfahrung, was es bedeutet, wenn man sagt, süß ist die Liebe zur Heimat." Cervantes, Don Quijote (wie Anm. 2), S. 470.

19 Vgl. dazu: Helmut Seng, Vergils Eklogenbuch (Spudasmata 72), Hildesheim u.a. 1999, S. 40 f.

20 Verg. ecl. IX, 66.

21 Vgl. zu dieser komplexen Ekloge den Kommentar in: Vergil, Bucolica. Hirtengedichte. Studienausgabe. Lateinisch/Deutsch, übersetzt, kommentiert und hg. v. Michael von Albrecht, Stuttgart 2001, S. 199–212.

22 Vgl. Erich Köhler, Wandlungen Arkadiens: die Marcela-Episode des *Don Quijote* (I, 11–14), in: Ders. (Hg.), Esprit und arkadische Freiheit. Aufsätze aus der Welt der Romania, München 1984, S. 302–327, hier bes.: S. 312–315.

23 Vgl. dazu *en bref*: Victor A. Ferretti, Don Quijote und das sanchoeske Imaginäre, in: Jörg Dünne / Kurt Hahn / Lars Schneider (Hg.), Lectiones difficiliores – Vom Ethos der Lektüre (Orbis Romanicus 11), Tübingen 2019, S. 403–408, hier: S. 406.

Freiheitsgrade zwischen einem Selbst und seiner Um-Welt ausagieren. Augusta als ebensolcher frühneuzeitlicher Projektionsraum erlangt in dieser Hinsicht einen weltliterarischen, die Jahrhunderte überdauernden Platz, den ihr auch das nachmoderne Augsburg, wie es in einem der bedeutendsten Romane jüngerer Zeit, nämlich Roberto Bolaños „2666", apperzipiert wird,[24] nicht mehr nehmen kann. Und akzeptiert man, dass sich literarische Fiktion nicht mit einer platonischen Mimesis von etwaiger Wirklichkeit begnügen muss, sondern gerade die aristotelische Poiesis von Auch-anders-Möglichkeit ins Blickfeld rücken kann, birgt „Augustanisches" in Cervantes' Meta-Fiktion, wie auch Thomas Manns „Meerfahrt mit Don Quijote" bezeugte,[25] eine weit über die Frühe Neuzeit und die Region Augsburg hinausstrahlende Aktualität.

24 Vgl. Roberto Bolaño, 2666, New York 2009, S. 24.

25 Vgl. dazu eingehend: Klaus Meyer-Minnemann, Zur Entstehung, Konzeption und Wirkung des *Don Quijote* in der europäischen Literatur, in: Tilmann Altenberg / Klaus Meyer-Minnemann (Hg.), Europäische Dimensionen des *Don Quijote* in Literatur, Kunst, Film und Musik, Hamburg 2007, S. 11–46, hier: S. 39–41.

Sektion 4:
Die süddeutsche Städtelandschaft – ökonomisch

Johannes Lang

Monopol – Krise – Marktdynamik. Zur Entstehung hochmittelalterlicher Salinenstädte im Ostalpenraum unter besonderer Berücksichtigung Reichenhalls

Abstract: The enormous influence of salt as a commodity on the urban landscape in Bavaria is well known in science. Nevertheless, it is hardly received that the saltworks town of Reichenhall is an outstanding example of dynamic urban development during the twelfth century. With the exception of the episcopal cities, no other place in the territory of the former tribal duchy has such good written sources for the early and high Middle Ages as Reichenhall, so that it is possible to reconstruct an approximate picture of the salt works and the town. Until the end of the twelfth century Reichenhall held the monopoly as the only efficient salt producer in the Eastern Alps, whereas other salt works, such as Hall near the Admont Monastery, were only of local importance. While the salt works of Reichenhall were producing at the limit of its capacity as a result of a general increase in population from the middle of the twelfth century, the town experienced several difficult crises. At the end of the century, more and more salt producers resorted to the newly developed technology of the leaching process, which put Reichenhall at a competitive disadvantage. The destruction of the town during the war in 1196 unleashed a market dynamic: It gave the Archbishop of Salzburg the opportunity to use the new technology on the Dürrnberg, near Reichenhall, and, with Hallein, to create a new salt-works town based on the Reichenhall model. While Reichenhall was razed and ruined for more than two decades, Hallein had excellent prerequisites to quickly become the leading salt producer in the Alpine region. Another producer, Hall in Tyrol, was created almost at the same time as Hallein and also had very good location factors that allowed an urban development too. For the still young "Länder" of Tyrol and Salzburg, the towns gained enormous importance because they made them independent of salt imports. Reichenhall was rebuilt after 1196 by the Bavarian Duke and the saltworks continued at great expense to ensure the Duchy's salt supply over a long term.

Im Jahre 1930 veröffentlichte der französische Wirtschaftshistoriker Henri Hauser (1866–1946) einen Aufsatz, worin er die Bedeutung des Salzes für die vergangene Menschheitsgeschichte mit dem Stellenwert des Rohöls für den Menschen der technisierten Gegenwart verglich.[1] Beide Rohstoffe seien, so Hauser, das

1 Henri Hauser, Le sel dans l'histoire, in: Ders., Les Origines historiques des prblemès économiques actuels, Paris 1930, S. 53–69.

Ausgangsmaterial für bestimmte wichtige Verfahren und Erzeugnisse. Ihre Vorkommen seien weltweit ungleich verteilt und nicht selten knapp, weshalb das unbedingte Angewiesensein bestimmter Arbeitszweige oder Gesellschaften darauf zu Abhängigkeiten führe. Jean-François Bergier (1931–2009) erblickte dagegen einen wesentlichen Unterschied darin, dass Rohöl heute der Industriegesellschaft einen hohen Standard zu sichern vermag, während Salz in der Vergangenheit als Konservierungsmittel dem bloßen Überleben der Menschheit diente.[2] Das Wissen um den überragenden Wert des Salzes und der damit verbundene Machtanspruch müssen folglich am Beginn einer jeden Salz- und Salinengeschichte stehen.

Als Joseph Ernst Ritter von Koch-Sternfeld (1778–1866) im Jahre 1836 eine vergleichende Studie zu den deutschen Salinenorten vorlegte, brachte er genau diesen Anspruch zum Ausdruck, indem er die ungewöhnliche Menge an frühmittelalterlichen Schriften anführte, an denen sich die Bedeutung der Saline Reichenhall ablesen lasse: Mit der Ankunft des heiligen Rupertus „müssen die verschiedenen Quell-Salzwerke des Landes überhaupt in einem gesegneten, den großen inländischen Bedarf und die ebenso starke Nachfrage im Osten und Westen, im Norden und Süden zugleich deckenden Betrieb gestanden haben".[3] Über die Einschätzung, wonach die dortige Saline das gesamte Frühmittelalter über bis zum ausgehenden 12. Jahrhundert im Ostalpenraum und darüber hinaus eine singuläre Stellung einnahm, da sie als einzige exportorientiert zu arbeiten im Stande war, besteht in der Geschichtswissenschaft seit Langem allgemeiner Konsens.[4]

2 Jean-François Bergier, Die Geschichte vom Salz, Frankfurt am Main 1989, S. 15.

3 Joseph Ernst Ritter von Koch-Sternfeld, Die deutschen, insbesondere die bayerischen und österreichischen Salzwerke zunächst im Mittelalter, München 1836, S. 31.

4 Herbert Klein, Zur älteren Geschichte der Salinen Hallein und Reichenhall, in: Gesellschaft für Salzburger Landeskunde (Hg.), Festschrift für Herbert Klein, Salzburg 1965, S. 385–409, hier: S. 396: „Die Bedeutung Reichenhalls als Salzproduktionsstätte durch das frühe und hohe Mittelalter bis zur Neuerschließung der Bergwerke ist kaum zu überschätzen." Heinrich Wanderwitz, Studien zum mittelalterlichen Salzwesen in Bayern (Schriftenreihe zur Bayerischen Landesgeschichte 73), München 1984, S. 2, spricht davon, dass Reichenhall für das 12. Jahrhundert die quellenmäßig am besten belegte Stadt in Bayern sein dürfte. Fritz Koller, Salzgewinnung und Salzhandel unter den Erzbischöfen, in: Heinz Dopsch (Bearb.), Salz. Salzburger Landesausstellung. Hallein, Pernerinsel, Keltenmuseum, 30. April bis 30. Oktober 1994, Salzburg 1994, S. 128–149, hier: S. 128; Herwig Wolfram, Grenzen und Räume. Geschichte Österreichs vor seiner Entstehung, Wien 1995, S. 364; Johannes Lang, Geschichte von Bad Reichenhall, Neustadt an der Aisch 2009, S. 82–99.

Abb. 40: *Früheste Darstellung Reichenhalls durch Hans Donauer, 1584. Bildnachweis: Johannes Lang, Geschichte von Bad Reichenhall, Neustadt an der Aisch 2009, S. 361.*

In der bayerischen Stadtgeschichtsforschung wurde zwar wiederholt die Bedeutung des Reichenhaller Salzes für die Entstehung der Städtelandschaft Bayerns hervorgehoben. Doch der Rückschluss, wonach für das Florieren einer Großsaline dieses Formats ein urbanes Umfeld zwingend erforderlich gewesen sein muss – gerade wenn man an die arbeitsteilig notwendige Konzentration von Menschen sowie die Versorgung der umfangreichen Arbeiterschaft denkt –, wurde überwiegend außer Acht gelassen.[5] Dass zum Zeitpunkt der Übertragung des Herzogtums Bayern an den Pfalzgrafen Otto I. von Wittelsbach im Jahre 1180 in Bayern nur die Bischofssitze Regensburg, Freising, Passau und Salzburg, die mit Ausnahme von Freising auf römische Vergangenheit rekurrierten, urban

5 Eine von mehreren Ursachen liegt wohl in der kartographischen Darstellung der Städte und Märkte im Mittelalter in Altbayern, siehe: Max Spindler / Gertrud Diepolder, Bayerischer Geschichtsatlas, München 1969, S. 22 f. Als Ausnahme sei genannt: Fitz Koller, Die Anfänge der Salzburger Städte. Civitas und verwandte Begriffe in den Salzburger Städten, in: Mitteilungen der Gesellschaft für Salzburger Landeskunde (künftig: MGSL) 128, 1988, S. 5–32, hier: S. 15–16.

geprägt gewesen seien, ist eine ebenfalls wiederholt getätigte Annahme.[6] Gerade im Falle der Stadtwerdung Reichenhalls muss die Klassifizierung nach Bischofsstädten, Burgstädten, wittelsbachischen Gründungsstädten und dergleichen unwirksam bleiben, da die urbane Entwicklung des Salzproduktionsortes eine singuläre Erscheinung darstellt, die zudem, ähnlich wie beim Salzumschlagplatz Laufen an der Salzach, nicht von den Bayernherzögen, sondern von den Salzburger Erzbischöfen maßgebliche Impulse erfuhr. Schon Ernst Klebel (1896–1961) sprach von einer „alten Selbstverwaltung" Reichenhalls, das zudem im 12. Jahrhundert mit den Begrifflichkeiten der „*cives*" sowie der „*civitas*" aufwarte.[7] Fritz Koller konnte feststellen, dass der Begriff „*civitas*" im Verlaufe vom 8. bis zum 12. Jahrhundert einem Bedeutungswandel unterworfen war, dass die Konzentration des nicht-agrarischen Wirtschaftssektors das wichtigste Kriterium auf dem Weg zur Stadt darstellte und dass das Vorhandensein von Bürgern darüber entschied, ob aus einer Siedlung eine Stadt hervorging. Daher könne, so Koller, erst nach einer ersten Blüte des Bergbaus und des Handels überhaupt vorbehaltlos von einer Stadt gesprochen werden, da die beiden Hauptkriterien – die wirtschaftliche Dynamik und, darauf aufbauend, der korporative Zusammenschluss der Wirtschaftstreibenden – erkennbar würden.[8] So etwa tauchen in den Salzburger Quellen des 12. Jahrhunderts für das Herzogtum Bayern nur die Orte Salzburg, Reichenhall, Freising und Regensburg als „*civitates*" auf.[9]

Während für das um 1131 als „*urbs*" bezeichnete Burghausen, das 1157/58 als „*forum*" urkundlich erwähnte München sowie das 1162 als „*castrum*" genannte Landsberg – die Orte stehen übrigens mit dem Reichenhaller Salzhandel in

---

6 Wilhelm Liebhart, Städte und Märkte in Altbayern (Mittelalter/Frühe Neuzeit), veröffentlicht am: 2.2.2015, in: Historisches Lexikon Bayerns, URL: https://www.historisches-lexikon-Bayerns.de/Lexikon/St%C3%A4dte_und_M%C3%A4rkte_in_Altbayern_(Mittelalter/Fr%C3%BChe_Neuzeit) (letzter Zugriff: 13.10.2020); Alois Schmid, Die Städtepolitik der frühen Wittelsbacher. Die Gründungsstadt Cham, in: Zeitschrift für bayerische Landesgeschichte (künftig: ZBLG) 77/3, 2014, S. 833–855, hier: S. 834; Ders., Die Anfänge der Wittelsbacherstadt Straubing. Örtliche Befunde und überörtliche Bezüge, in: ZBLG 82/1, 2019, S. 1–41, hier: S. 11. Hubert Raab, Städte als Burgen. Der Fall Friedberg, in: Peter Wolf / Evamaria Brockhoff / Richard Fischer u.a. (Hg.), Stadt befreit. Wittelsbacher Gründerstädte (Katalog zur Bayerischen Landesausstellung 2020), Augsburg 2020, S. 188–198, hier: S. 188, zählt zu den bayerischen Bischofsstädten auch Eichstätt dazu.

7 Ernst Klebel, Die Städte und Märkte des baierischen Stammesgebietes in der Siedlungsgeschichte, in: ZBLG 12, 1939, S. 37–93, hier: S. 52.

8 Koller, Salzburger Städte (wie Anm. 5), S. 17.

9 Ebd., S. 9.

Verbindung – nur sehr vage Aussagen über einen Urbanisierungsgrad für das 12. Jahrhundert gemacht werden können, lassen die schriftlichen und archäologischen Quellen für das hochmittelalterliche Reichenhall ein bemerkenswert plastisches Bild zu. Dieses Bild beginnt sich seit dem ausgehenden 7. Jahrhundert zu verdichten und hält bald nach der Mitte des 12. Jahrhunderts sämtliche Merkmale einer Stadt bereit, wie im Folgenden zu zeigen sein wird.

Eine weitere darzulegende Entwicklung wurde in der Geschichtswissenschaft bislang übersehen: Erst die schweren krisenhaften Erscheinungen, mit denen Reichenhall in der zweiten Hälfte des 12. Jahrhunderts zu kämpfen hatte, führten im Ostalpenraum zu einer Marktdynamik und damit zur Entstehung weiterer Salinenstädte, die sich in mehrerlei Hinsicht am älteren Reichenhaller Vorbild orientierten. Diese Dynamik wurde noch beschleunigt durch neue Methoden der bergmännischen Salzgewinnung, die damit einhergehende Möglichkeit zur Etablierung salinarischer Großbetriebe sowie einen schärfer werdenden Wettbewerb infolge der Territorialisierung Mitteleuropas.

1. Die Reichenhaller Saline als Monopolist

Im Gegensatz zu den benachbarten Bergbauorten Berchtesgaden und Dürrnberg spielte die bergmännische Salzgewinnung in Reichenhall keine Rolle. Mit der nacheiszeitlichen Klimaentwicklung, dem Ansteigen der Waldgrenze und einem allgemeinen Absinken des Wasserspiegels könnte es im 3. Jahrtausend v. Chr. zur Auffindung der Reichenhaller Solequellen gekommen sein.[10] Noch heute deuten der Verlauf der Saalach und die Abbruchkante am sogenannten Gruttenberg darauf hin, dass der Gebirgsfluss ursprünglich nahe den Solequellen das Tal durchflossen haben muss.[11] Dass er nun in der Form einer großen

10 Karl Brunnacker / Jürgen Freundlich / Manfred Menke / Hans Schmeidl, Das Jungholozän im Reichenhaller Becken (Eiszeitalter und Gegenwart 27), Öhringen 1976, S. 159–173, hier: S. 166; Josef Maurer, Die Reichenhaller Solequellen und der prähistorische Mensch (Teil 3), in: Heimatblätter (Beilage zum „Reichenhaller Tagblatt") 23, 1924, S. 89 f., hier: S. 89. Thomas Stöllner, Archäologisches zu Salz, Salzgewinnung, Salzhandel, in: Reallexikon der Germanischen Altertumskunde 26, 2004, S. 357–379, 364, geht ohne nähere Erklärung von einer Reichenhaller Salzgewinnung im 16./17. Jh. v. Chr. aus. Lang, Geschichte von Bad Reichenhall (wie Anm. 4), S. 26.

11 Vgl. Brunnacker u.a., Jungholozän im Reichenhaller Becken (wie Anm. 10), S. 163–168.

Doppelkurve an der Stadt vorbeiströmt, ist überwiegend das Ergebnis künstlicher Flussbettverlegungen sowie aufwendiger Uferverbauungen.[12]

Während im Weichbild der Stadt mehrere Siedlungsstätten seit der Übergangszeit bekannt sind, fehlen Hinweise auf das eigentliche Stadtgebiet sowie den engeren Salinenbereich.[13] Im Gegensatz zu anderen Salinenstätten, wie beispielsweise Bad Nauheim, sind systematische archäologische Untersuchungen hier bislang ausgeblieben. Die archäologischen Lesefunde im unmittelbaren Salinenareal – zwei Beile – sind für den Nachweis einer prähistorischen Existenz ungeeignet, zumal sich deren Fundumstände als zweifelhaft erweisen.

Auch wenn vieles für eine Nutzung der Reichenhaller Saline in prähistorischer Zeit spricht, beginnt die dokumentierbare Ausbeutung der natürlichen Solequellen mit dem Einsetzen der schriftlichen Quellen im 8. Jahrhundert. Wie dominant Reichenhall als Salzerzeugungsort damals bereits gewesen ist, zeigt nicht nur die geringfügig später zu datierende „Raffelstetter Zollordnung" von 903/906, worin die Handelsware Salz beim Handel auf der Donau eine herausragende Rolle einnimmt, sondern auch das signifikante Auftreten des Namens „*Hall*". Entgegen der veralteten Ansicht, „Hall" sei keltischer Herkunft, besteht heute kein Zweifel mehr darüber, dass es sich um ein althochdeutsches Wort des frühen Mittelalters handelt.[14] Zunächst scheint der Begriff in den Bergbaugebieten von Halle an der Saale und Halle in Westfalen entstanden zu sein, allerdings noch nicht im Zusammenhang mit der Salzgewinnung. Erst in Reichenhall erfolgte die Gleichsetzung von „*Hall*" = „Salzerzeugungsstätte", wovon noch die Rede sein wird.

Der fränkische Glaubensbote Rupertus, der wohl auf Bitten des Herzogs Theodo II. nach Bayern kam, und um das Jahr 696 in Iuvavum eintraf, erhielt zur Gründungsausstattung einer von Rupertus zu errichtenden Kirchenstruktur vom Herzogssohn Theodebert ein Drittel der Soleschüttung an der Saline

---

12 Adolph Bühler, Reichenhall in vorchristlicher Zeit. Beilage zum Grenzboten vom 18.2.1904, o.S., weist auf eine 1879 bei Tiefbauarbeiten in der „Solereserve 4" etwa 60 Meter nordwestlich der Solequellen entdeckte Streichwehr hin. Vgl. Wanderwitz, Salzwesen (wie Anm. 4), S. 23 f.

13 Johannes Lang, Ad Salinas. Zur Frage nach der Existenz einer vor- und frühgeschichtlichen Saline in Bad Reichenhall, in: Felix Lang / Raimund Kastler / Wilfried Kovacsovics / Stefan Traxler (Hg.), Colloquium Iuvavum 2012 (Archäologie in Salzburg 8), Salzburg 2014, S. 225–245, hier: S. 226–228.

14 Jürgen Udolph, Hall in Ortsnamen, in: Reallexikon der Germanischen Altertumskunde 13, 1999, S. 433–442, hier: S. 438; Ders., Die Ortsnamen Hall, Halle, Hallein, Hallstatt und das Salz, Bielefeld 2014, S. 9–18.

Reichenhall, dazu 20 Salzsiedeanlagen, verbunden mit spezialisierten Salinenarbeitern. Die Bedeutung des Salinenortes sollte sich noch vor 750 in Iuvavums deutscher Bezeichnung „Salzburg", das seinen Namen dem Salz aus Reichenhall verdankt, niederschlagen.[15]

Rupertus erhielt außerdem den von der Reichenhaller Bevölkerung gereichten Salzzehent, ebenso ein Zehntel der bis dahin ausschließlich dem Herzog zugestandenen Steuerabgabe.[16] Die Salzburger Güterverzeichnisse lassen für die Saline zu diesem frühen Zeitpunkt ein bereits eingespieltes System erkennen, was nur durch deren langfristige Etablierung erklärbar ist. Reichenhall präsentiert sich zu diesem Terminus als eine in wesentlichen Belangen vollendete Ansiedlung, bei der die Konzentration von Arbeitskräften mit der Wirtschaftskraft des Standortes einherging. Dies und die Stellung Reichenhalls als einzige leistungsfähige Großsaline des gesamten Ostalpenraums[17] sowie weiter Teile Mitteleuropas bedingen zweifellos lange Erfahrungswerte. Eine dem Monopolisten Reichenhall ähnliche Bedeutung für die Salzerzeugung besaß im 12. Jahrhundert vermutlich nur noch die weit entfernt liegende Salinenstadt von Halle an der Saale (Ummauerung der Stadt im 12. Jahrhundert), mit Abstrichen eventuell Lüneburg (Stadtrecht 1247), während die Salinen von Schwäbisch Hall, Nauheim und Salins (Burgund) in jener Zeit deutlich zurückstanden. Das in Chioggia hergestellte venezianische Salz dürfte im Alpenraum kaum Bedeutung besessen haben.

Das Areal unterhalb des im felsigen Steilgelände aufragenden Gruttenbergs war bereits zur Zeit des heiligen Rupertus beherrscht von vermutlich aus Holz errichteten Hütten, in denen die einzelnen Siedepfannen untergebracht waren. In den Güterverzeichnissen aus dem Ende des 8. Jahrhunderts werden diese als „*loca fornacium*" (Herdstätten) bezeichnet, die mit „*patellae*" (Pfannen) ausgestatten waren.[18] Damals mögen dies rund 60 Sudhäuser gewesen sein, da Rupertus vom Bayernherzog 20 Sudhäuser mit den dazugehörigen Pfannen erhielt, was dem

---

15 Wanderwitz, Salzwesen (wie Anm. 4), S. 6 f.

16 Fritz Lošek, Notitia Arnonis und Breves Notitiae. Die Salzburger Güterverzeichnisse aus der Zeit um 800: Sprachlich-historische Einleitung, Text und Übersetzung, in: MGSL 139, 1990, S. 5–192, hier: S. 80 (Notitia Arnonis [künftig: NA] 1,3: „*Simulque etiam tradidit iam dictus dux in eodem pago in loco, qui vocatur Salinas, fornaces XX et totidem patellis et terciam partem de putiatorio ibidem concessit, quod barbarice dicitur galgo. Insuper et in iam dicto loco concessit decimam de sale et de teloneo, quod datur in censo dominico*"; Breves Notitiae [künftig: BN] 2,5).

17 Koller, Salzgewinnung und Salzhandel (wie Anm. 4), S. 128.

18 Lošek, Salzburger Güterverzeichnisse (wie Anm. 16), S. 108 (BN 4,5), 116 (BN 9,6).

dritten Teil der gesamten Soleschüttung entsprach.[19] Die Sole befand sich offenbar in einem einzigen großen „*putiatorium*“ beziehungsweise „*puteus*“,[20] worin man offenbar keinen ausgeschachteten Brunnen zu erblicken hat, sondern eine weitläufige Grube, an deren oberem Rand „*galgo*[nes]“ (Schöpfgalgen) platziert waren,[21] weshalb man eventuell von einer Galgbrunnen-Anlage sprechen könnte.[22]

Abb. 41: *Schematische Darstellung des Solebrunnens in der „Cosmographia“ des Sebastian Münster, 1544. Bildnachweis: Johannes Lang, Geschichte von Bad Reichenhall, Neustadt an der Aisch 2009, S. 83.*

Auf dem Grund des großen Grubenareals, das ab dem 12. Jahrhundert in den Schriftquellen auch als „*salinarius fons*“ (Salzbrunnen) auftaucht,[23] bestanden zahlreiche kleinere Schächte, die als Schöpfstellen dienten und deren wohl quadratisch angelegte Grundrisse zumindest so groß gewesen sein müssen, dass jeweils neben einem Auffanggefäß, zum Beispiel einem Holzfass, auch zwei Arbeiter, sogenannte Vaher (Schöpfknechte), darin Platz fanden. Die Schächte waren mit Brettern nach unten hin abgeteuft, sodass sie mit Hilfe weiterer „Werkbretter“ beliebig tief geführt werden konnten, um zusätzliche Sole führende Schichten

19 Ebd., S. 80 (NA 1,3), 104 (BN 2,5).

20 Ebd.

21 Ebd., S. 118 (BN 11,3): „[…] *ad illum puteo hoc, quod dicitur galgo*“ (dt.: „An jener Grube das, was Galgen genannt wird“. Das Adverb *ad* macht deutlich, dass *puteus* und *galgo* nicht identisch waren.

22 Jakob und Wilhelm Grimm, Deutsches Wörterbuch, Bd. 4, Leipzig 1878, Sp. 1166.

23 Salzburger Urkundenbuch, Bd. I: Traditionscodices (künftig: SUB I), hg. v. Willibald Hauthaler, Salzburg 1910, S. 380, Nr. 243.

anzuschneiden. (Als weiterer *terminus technicus* war der Begriff „*Schrot*" in Verwendung, worunter man einen abgeschnittenen Holzstamm verstand.) Allerdings scheint dies – um die Quellschüttung nicht zu beeinträchtigen – eine sehr diffizile Angelegenheit gewesen zu sein, weshalb für das weitere Abteufen mittels Werkbrettern offenbar eine obrigkeitliche Zustimmung des Hallgrafen beziehungsweise des Kaisers, später auch des Stadtrichters vonnöten war.[24]

An einer bestimmten Stelle gelangten die Vaher über Treppen oder Leitern in die Grube hinunter, um dort ihrer Arbeit als Schöpfknechte nachgehen zu können.[25] Da auch die Bezeichnung einer Schöpfstelle „*an dem steg*"[26] auftaucht, ist davon auszugehen, dass die über die gesamte Grube – wohl eher vergleichbar mit der Baugrube einer modernen Großbaustelle – verteilten Soleschächte durch labyrinthartige Wege und Stege miteinander verbunden waren. Eine Überdachung der Grube ist zwar wohl aus Gründen der Statik auszuschließen, aber am Grubenboden muss es dennoch ein Bauwerk gegeben haben – möglicherweise die Konstruktion eines über den Grubenrand hinausragenden Schöpfwerks –, denn in den Schriftquellen werden „*colummnae*" (Pfeiler) genannt, zwischen denen sich ebenfalls Sole sammelte.[27]

Die unterschiedlichen Bezeichnungen der zahlreichen Soleschächte erfolgte einerseits nach deren Eigentümern. So etwa ist das „*Tekkendorferbret*[t]"[28] nach den Herren von Deggendorf, einem mit den niederösterreichischen Grafen von Pernegg identischen sowie den Babenbergern eng verwandten Adelsgeschlecht benannt.[29] Das „*Abbatsbret*[t]"[30] stand ursprünglich im Eigentum des Abtes von St. Peter in Salzburg. Einige Soleschächte lassen namentlich einstige Besitzverhältnisse erkennen, so etwa der „*Eichstätter*",[31] ursprünglich wohl dem Bistum

24 Darauf deutet eine Urkunde Kaiser Friedrich Barbarossas hin, der das Einrichten eines neuen Werkbretts genehmigte (MGH DD F I., Tom. X, Pars III, S. 26, Nr. 560). Vgl. Wanderwitz, Salzwesen (wie Anm. 4), S. 203, Anm. 1234.

25 Dies geht aus einer Textstelle des herzoglichen Urbars von 1301/07 hervor: „[…] *an dem stain ze nahst da di trager auz dem prunn gent*", zitiert nach: Wanderwitz, Salzwesen (wie Anm. 4), S. 177.

26 Ebd., S. 176.

27 SUB I, S. 394, Nr. 270 a.

28 SUB I, S. 380, Nr. 243: „*Tekkindorfariprét*".

29 Johann Gruber, Mittelalterliche Herren- und Dienstmannensitze im Umland von Deggendorf, in: Deggendorfer Geschichtsblätter 8, 1987, S. 105–120, hier: S. 107 f.

30 SUB I, S. 451, Nr. 366: „*Abbatspret*".

31 Bayerisches Hauptstaatsarchiv München, KL Benediktbeuern, Nr. 9, fol. 40 r: „*Aisteter*".

Eichstätt zugehörig, das aber bereits im 12. Jahrhundert keine Anteile an der Reichenhaller Saline mehr hatte, oder aber der „*Schultheißer*",[32] worunter man den Gemeindevorsteher verstand. Das „*Bischofsloch*"[33] dürfte ehedem im Besitz des Salzburger Erzbischofs gewesen sein.

Andererseits wurden die Schächte nach ihren Eigenschaften benannt, beispielsweise die „*Ruzenlache*" (mhd. *ruzen* = rauschen; *lache* = Pfütze), oder aber nach ihrer Lage, zum Beispiel „*an dem hohen Stein*" (vermutlich der Felsen des Gruttenbergs). Da es im 12. Jahrhundert kaum mehr einen Soleschacht gab, der einem Besitzer allein zu Eigen war, musste die geschöpfte Sole in einem bestimmten Verhältnis aufgeteilt werden, wobei einige Namen, so etwa „*Vierteiler*", „*Fünfteiler*", „*Siebenteiler*", „*Neunteiler*" oder „*Elfteiler*", die prozentuale Aufteilung dieser Schöpfstellen erkennen lassen. Auf diese Weise ergaben sich sogar auf Achtzehntel-Teile aufgegliederte Schächte.

Damit die theoretisch beanspruchten Salzwasserteile auch in der Realität zugewiesen werden konnten, wurde die Sole mittels Schöpfeinrichtungen – Galgbrunnen – nach oben befördert. Ursprünglich bestand zu jedem Schacht ein Galgen, der auch danach benannt war. So etwa findet sich in den Schriftquellen zunächst neben dem „*Watzmannbrett*" auch ein „*Watzmannsgalgen*";[34] später ist nur mehr die Rede von einer Betriebseinheit: dem „*Watzmanner*".[35] Der waagrechte und in einem großen Radius schwenkbare Jochbalken – der eigentliche Galgen – einer solchen Brunnenkonstruktion befand sich am oberen Rand der großen Grube. Die unten im Schacht stehenden Vaher schöpften mit Hilfe von kleinen Ledereimern die am Boden zusammenfließende Sole in das hölzerne Fass, das anschließend von der Bedienmannschaft des Galgbrunnens nach oben gezogen und dessen Inhalt auf ein offenes Gerinne geschüttet wurde. Um eine durch Vertrag und Gewohnheitsrecht geregelte Aufteilung zu gewährleisten, schloss sich an den jeweiligen Galgen eine Vorrichtung zum Ausgießen der Sole, der sogenannte „Angoss",[36] der mittels Rohrleitungen nur eine bestimmte Schüttung und somit exakte Zuteilungen zuließ. Da der Angoss aus Holz bestand und ohne größere Schwierigkeiten um weitere Ausgüsse vergrößert oder aber verkleinert werden konnte, waren anteilsmäßige Veränderungen jederzeit möglich.

---

32 Monumenta Boica, Bd. 3 (künftig: MB 3), hg. v. der Bayerische Akademie der Wissenschaften, München 1764, S. 561, Nr. 27: „*Schultheizere*".

33 SUB I, S. 477, Nr. 412: „*Bischolvesloch*".

34 Oberösterreichisches Urkundenbuch/Urkunden-Buch des Landes ob der Enns, Bd. II (künftig: OÖUB II), bearb. v. Erich Trinks, Wien 1856, S. 151, Nr. 51: „*Wazimannsgalgen*".

35 SUB I, S. 477, Nr. 412: „*Wazmannære*".

36 Vgl. Wanderwitz, Salzwesen (wie Anm. 4), S. 204.

Vom Angoss aus lief die Sole durch ein Rohrleitungssystem im natürlichen Gefälle in die nahe gelegenen Sudhäuser, wo sich die Herdstätten mit den zur Verdampfung des Wassers erforderlichen Pfannen befanden.[37] Diese auch als „Pfannhäuser" bezeichneten Gebäude, die zuweilen dieselben Namen trugen wie die zugehörigen Soleschächte und Schöpfgalgen, konnten, ähnlich der Soleschüttung, auf unterschiedliche Besitzer aufgeteilt sein, ebenso die umgebenden Grundstücke, wo die zur Endverarbeitung des Salzes notwendigen Gerätschaften standen. Zu einer salinarischen Betriebseinheit gehörten Anteile an der Sole, am Pfannhaus, an Stätten der Weiterverarbeitung, an Wäldern sowie an Holzlagerplätzen an der Saalach. Sie bildeten die unbedingt erforderlichen betriebswirtschaftlichen Voraussetzungen.

Während sich eine Reichenhaller Schiffsanlegestelle[38] wohl in der Nähe der um 1050 erstmals urkundlich erwähnten Saalachbrücke (an der Stelle der heutigen Luitpoldbrücke) befand,[39] gab es in deren Nähe Verladestätten, von wo aus die Befrachtung von Spezialkähnen vorgenommen werden konnte.[40] Noch vor Ort erfolgte sowohl für den Land- als auch für den Wassertransport eine erste Zollabgabe (Maut),[41] bevor die Salzschiffe über die Saalach auf die Salzach und, nach einer Umladung in größere Kähne, von dort auf den Inn und weiter auf die Donau gelangten.[42] In Passau, dem zentralen Ort des Flusssystems Saalach–Salzach–Inn–Donau und einem der wichtigsten Salzumschlagplätze überhaupt, wurde das „Weiße Gold" teilweise auf Wagen gepackt, um damit über den sogenannten „Goldenen Steig" bei Prachatitz den böhmischen Raum zu erreichen, während das übrige Salz weiter Donau abwärts geführt wurde.[43] Auf dem Landweg gingen Wagenladungen durch die Furt der Enns in Richtung Osten. Aber auch gegen den Strom wurden die Salzschiffe – vermutlich mittels

37 Vgl. ebd., S. 200.

38 SUB I, S. 477, Nr. 412: „*Sceflende*".

39 SUB I, S. 234, Nr. 7: „*ultra pontem in Salinis*".

40 Fritz Koller, Die Salzachschiffahrt bis zum 16. Jahrhundert, in: MGSL 123, 1983, S. 1–126, hier: S. 7 unter Hinweis auf die 973 genannten „*loca onustaria*" (MGH DD Ko I. / DD H I. / DD O I., Tom. I, S. 584, Nr. 431).

41 Wanderwitz, Salzwesen (wie Anm. 4), S. 25, A 112, geht von einer Wasser- wie auch einer Weg-Maut aus, die in Reichenhall erhoben wurden.

42 Ebd., S. 209 f.

43 Indirekt wird der „Goldene Steig" im Jahre 1010 genannt. Vgl. Paul Praxl, Der Goldene Steig, Grafenau 1976, S. 5; Zdeněk Šimeček, Salz aus dem Alpenraum in Konkurrenz mit dem Salz aus dem Norden in den böhmischen Ländern, in: Jean-Claude Hocquet / Rudolf Palme (Hg.), Das Salz in der Rechts- und Handelsgeschichte, Innsbruck 1991, S. 135–145, hier: S. 136 f.; Wolfram, Grenzen und Räume (wie Anm. 4), S. 365.

Menschenkraft – getreidelt, um die großen Salzstadel von Regensburg sowie der anderen Machtzentren an der Donau zu füllen. Sogar bis zur Mündung der Iller und diese wiederum aufwärts bis ins Allgäu wurde das Reichenhaller Salz geführt, um in die Speicherräume der Abtei Kempten zu gelangen.

2. Bürgertum und Patriziat im 12. Jahrhundert

Vor dem Hintergrund der reichhaltigen Schriftquellen des 12. Jahrhunderts gestaltete sich die Saline von Reichenhall als ein durchaus komplexes Gefüge, das einen großen Personalstand voraussetzte. Für das frühe 15. Jahrhundert sind mehr als 700 Personen anzunehmen, die direkt in der Salzerzeugung tätig waren.[44] Aufgrund der Monopolstellung, die Reichenhall im Salzwesen bis zu seiner Zerstörung und Schleifung im Jahre 1196 besaß, muss man für das 12. Jahrhundert sogar von deutlich mehr Menschen ausgehen: Rund 60 Sudhäuser und damit verbundene Betriebseinheiten sind konstant in der Zeit vom ausgehenden 7. bis zum Ende des 12. Jahrhunderts anzunehmen,[45] währenddessen die Saline zeitweise an der Kapazitätsgrenze arbeitete. Ein späteres Schriftstück spricht beispielsweise von 436 Vahern, die allein zur Förderung der Solequellen herangezogen worden sein sollen.[46] Dies wären im Durchschnitt sieben Vaher pro Betriebseinheit gewesen, was durchaus realistisch erscheint, wenn man in Betracht zieht, dass die mit dem Soleschöpfen und der Bedienung der Galgbrunnen Beschäftigten in einer Tag- und Nachtschicht arbeiteten. Hinzu kamen die Arbeiter an den Salzpfannen, die mit dem Heranschaffen des Brennholzes Beschäftigten, die Küfer mit ihren Gesellen, eine große Zahl weiterer Werkleute, die auf der Stoßstatt Tätigen – darunter zahlreiche Frauen („Pfieseldirnen“) –, die für den Transport zuständigen Aufleger, die für die Schiffsfrachten auf der Saalach Verantwortlichen, das große Heer an Triftarbeitern sowie sonstige Salinenknechte. Man wird die Zahl der Werktätigen an der Saline folglich mit über 1.000 veranschlagen müssen.

---

44 Lang, Geschichte von Bad Reichenhall (wie Anm. 4), S. 268.

45 Das ergibt einerseits die in den Güterverzeichnissen genannte Zahl von 20 Sudhäusern, was einem Drittel der Soleschüttung entsprach, und andererseits die Auflistung der kirchlichen und adeligen Anteilsnehmer bei: Wanderwitz, Salzwesen (wie Anm. 4), S. VII–IX.

46 Nach: Wolfgang Jahn, Die Saline Reichenhall vom Spätmittelalter bis zum Beginn des 19. Jahrhunderts, in: Manfred Treml / Wolfgang Jahn / Evamaria Brockhoff (Hg.), Salz macht Geschichte, Augsburg 1995, S. 83–92, hier: S. 84.

Analog zu den vielfältig spezialisierten Arbeitsschritten beim Produktionsprozess herrschte eine klare Hierarchie. Dass die Vaher ein geringes soziales Ansehen genossen, liegt auf der Hand.[47] Der Pfannmeister hingegen hatte die Oberaufsicht über die Vorgänge im Pfannhaus inne. Darüber hinaus muss es jemanden gegeben haben, der nicht nur das Pfannhaus, sondern auch die Logistik, den Vertrieb und die Zusammenarbeit mit den anderen Salzsiedebetrieben im Überblick hatte. Dabei ging es, ähnlich einem Wirtschaftskartell, auch um gemeinsame betriebswirtschaftliche Ziele, Absatzmärkte und Preisabsprachen – letztlich um Unternehmenspolitik. Aus einer derart komplexen Gemengelage an Aufgaben erwuchs noch im 11./12. Jahrhundert eine unternehmerische Elite: das Salzpatriziat.[48]

Die Patriziatsforschung der letzten Jahre hat neue und grundsätzliche Aspekte beleuchtet, doch für die meisten Fallbeispiele setzte die Forschung hierzu – quellenbedingt – erst im Spätmittelalter ein.[49] Im Falle Reichenhalls hingegen lassen sich dazu bereits im Frühmittelalter betriebswirtschaftliche Strukturen und im Hochmittelalter damit einhergehende gesellschaftliche Entwicklungen erkennen. Zunächst machen die Quellen deutlich, dass die an der Saline tätigen Eigenleute von ihren Grundherren zusammen mit Salzpfannen, Gütern oder Wäldern verschenkt, verkauft oder vertauscht werden konnten; gleichzeitig müssen die im Bereich der Betriebsführung Tätigen schon bald eine Sonderrolle eingenommen haben.[50] Die Erkenntnis, auf die Expertise einer bestimmten Gruppierung angewiesen zu sein, führte automatisch zu deren Privilegierung, gesetzt den Fall, dass diese Abhängigkeit auch der Gruppierung selbst bewusst war und diese daher ihre Verhandlungsposition zu nutzen wusste.

---

47 Johannes Lang, Arbeit zwischen Tradition und Rationalisierung – Wirtschafts- und Überlebensstrategien an der mittelalterlichen Saline Reichenhall, in: Verband Österreichischer Historiker und Geschichtsvereine / Österreichisches Staatsarchiv (Hg.), Tagungsband des 23. Österreichischen Historikertages in Salzburg 2002, Salzburg 2003, S. 614–625, hier: S. 617.

48 Heinz Dopsch / Johannes Lang, Bürgerliche Eliten in Salinen- und Salzhandelsstädten des Ostalpenraums im Mittelalter und der frühen Neuzeit, in: Janez Mlinar / Bojan Balkovec (Hg.), Meste elite v srednjem in zgodnjem novem veku med Alpami, Jadranom in Panonsko nižino (Urban Elites in the Middle Ages and the Early Modern Times between the Alps, the Adriatic an the Pannonian Plain), Ljubljana 2011, S. 101–131, hier: S. 102–110.

49 Siehe z.B.: Michael Hecht, Patriziatsbildung als kommunikativer Prozess. Die Salzstädte Lüneburg, Halle und Werl in Spätmittelalter und Früher Neuzeit (Städteforschung A 79), Köln/Weimar/Wien 2010; Peter Fleischmann, Rat und Patriziat in Nürnberg. Die Herrschaft der Ratsgeschlechter in der Reichsstadt Nürnberg vom 13. bis zum 18. Jahrhundert (Nürnberger Forschungen 31/1–3), Neustadt an der Aisch 2008.

50 Lang, Geschichte von Bad Reichenhall (wie Anm. 4), S. 114.

Allein die auf über 300 eigentumsrechtliche Einzelanteile aufgesplitterte Soleschüttung muss für Außenstehende diffus gewirkt haben, zumal im 12. Jahrhundert die an der Reichenhaller Saline begüterten Bistümer, Klöster und Adeligen in der Regel Anteile an unterschiedlichen Soleschächten innehatten. Die Pfannhäuser erhielten ihre Sole folglich nicht mehr aus einer, sondern aus mehreren Schöpfstellen, weshalb die Soleverteiler zunächst wohl immer wieder umständlich ummontiert werden mussten. Man war plötzlich auf die Vaher anderer Schöpfstellen angewiesen und musste sich fremder Schöpfgalgen bedienen, was zwangsläufig Fragen der Bezahlung sowie des Bauunterhalts aufwarf. Selbst wenn die Salzproduktion an der Reichenhaller Saline eine Zeit lang in dieser komplizierten Form erfolgt sein sollte, war dies auf längere Sicht nicht praktikabel.

Man sah sich also regelrecht dazu gezwungen, die Eigentumsverhältnisse wie auch die Organisationsform an der Reichenhaller Saline zu vereinfachen. Träger dieser Reform waren jene, die über Herrschaftswissen verfügten und um die Abläufe beim Salzproduktionsprozess am besten Bescheid wussten. Auf sie waren die Grundherren und Eigentümer der Salinenanteile angewiesen, was ihren Marktwert steigen ließ: Bereits im 11. Jahrhundert begann die Klientel der Pfannenverwalter damit, bestimmte Salineneinheiten nicht mehr nur zu verwalten, sondern als selbstständige Unternehmer auf der Grundlage von Pachtverhältnissen zu bewirtschaften. Dem vorausgegangen waren nicht selten Konflikte der Grundherren mit ihren Eigenleuten, die teilweise eigenmächtig und ohne Rücksprache Besitz an der Saline veräußert hatten. Ein gewisser „*Pabo de Curia*“ (vom Hofe) beispielsweise, dem eines der wenigen Steinhäuser in Reichenhall gehörte, dürfte ab den 1120er Jahren zu den mächtigsten Akteuren an der Saline gezählt haben. Obwohl er ursprünglich zu den Eigenleuten des Salzburger Domkapitels gezählt hatte, erwarb er selbstständig mehrere Anteile am Solebrunnen und weigerte sich beharrlich, die vom Grundherrn geforderten Abgaben zu entrichten.[51] Interessant erscheint aber auch der Umstand, dass in der darüber ausgestellten Aufzeichnung als Verkäufer weitere Eigenleute des Domkapitels vor Ort auftreten. Offenbar hatten sich die Rechtsverhältnisse derart verselbstständigt, dass Anteile an der Saline ohne obrigkeitliche Zustimmung den Besitzer wechselten und es für die Grundherren schwierig wurde, Besitzentfremdungen entgegenzutreten. Zugleich betrachteten sich genau jene einst von den Grundherren vor Ort eingesetzten Verwalter immer stärker als die eigentlichen Leistungsträger an der Saline und suchten untereinander den korporativen Schulterschluss.

51 SUB I, S. 613, Nr. 58.

Die seit der Mitte des 12. Jahrhunderts urkundlich belegbare bürgerliche Spitzengruppe war für die Entwicklung der Stadt offensichtlich so prägend, dass einzelne Salzpfannen bis in das ausgehende Mittelalter nach ihnen benannt wurden.[52] Den kirchlichen und adeligen Eigentümern flossen nur mehr Pachterträge zu,[53] die jedoch im 12. Jahrhundert noch erheblich gewesen sein müssen. Ihre Macht aber erfuhr deutliche Einschränkungen, denn die „Geschäftsführer" vor Ort agierten weitgehend unumschränkt, ohne einer nennenswerten Kontrolle unterworfen zu sein, und damit bis zu einem gewissen Grad rechtlich unabhängig. Wenn Wissen und Können vom Vater auf den Sohn übergingen, dann neigten auch die Rechte dazu, erblich zu werden.[54]

Abb. 42: *Das von Erasmus Grasser errichtete Brunnhaus der Saline, 2. Hälfte 18. Jahrhundert. Bildnachweis: Stadtarchiv Bad Reichenhall.*

52 Lang, Geschichte von Bad Reichenhall (wie Anm. 4), S. 120.

53 Fritz Koller, Die Abtei St. Peter als Salzproduzent und Montanunternehmer, in: Studien und Mitteilungen zur Geschichte des Benediktinerordens und seiner Zweige 93, 1982 (= Festschrift Erzabtei St. Peter in Salzburg 582–1982, Salzburg 1982), S. 159–186, hier: S. 161.

54 August Funke, Die Reichenhaller Saline bis zur Begründung des herzoglichen Produktions-Monopols (ca. 1520), Meppen 1911, S. 22; Hugo Schröder, Entwicklungsgeschichte der Reichenhaller Sieden vom Ausgang des Mittelalters bis in das erste Viertel des neunzehnten Jahrhunderts. Mit besonderer Berücksichtigung der Siedeherren und deren Familiengeschichte, Typoskript, München 1942, S. 1.

Dies führte im Laufe der Zeit zu einer Aushöhlung der alten Eigentumsverhältnisse, denn die Eigentümer waren üblicherweise weit weg vom Geschehen. Die vor Ort tonangebenden „Geschäftsführer" schlossen sich zu einer Organisation zusammen, die am ehesten mit der Form eines Syndikats vergleichbar ist, ging es doch um eine gemeinsame Produktionsstätte, gleichgeartete Vertriebsstrukturen und einen monopolisierten Absatz. Man legte Preise sowie Produktionsmengen fest, vereinbarte die rechtlichen Bedingungen für den Verkauf und schuf mit der Etablierung eines Stadtrichters in den 60er Jahren des 12. Jahrhunderts ein Schiedsgericht.[55]

Innerhalb der Reichenhaller Bevölkerung begann sich mit dem Salzpatriziat eine gesellschaftliche Oberschicht herauszukristallisieren, die selbstbewusst den Stadtwerdungsprozess Reichenhalls um die Mitte des 12. Jahrhunderts vorantrieb.[56] Waren bislang nur die Vertreter des Adels und der Kirche an dem Salinenort tonangebend gewesen, so verstand es diese neue bürgerliche Schicht, immer größere Machtbefugnisse an sich zu ziehen. Mit dem untrüglichen Instinkt für wirtschaftlichen Erfolg ausgestattet, streckte man die Fühler dorthin aus, wo finanzieller Gewinn zu gewärtigen war: Laufen an der Salzach, wo sich ein wichtiger Umschlagplatz des Reichenhaller Salzes befand und der Fernhandel auf Kapital und Münzgeld angewiesen war, erhielt eine Münzprägestätte.[57] Schon früh scheinen die Erzbischöfe beim Ausbau Laufens mit den Reichenhaller Patriziern zusammengearbeitet zu haben, was sich spätestens seit dem 12. Jahrhundert in sehr engen verwandtschaftlichen Beziehungen und Heiratsverbindungen zwischen beiden Orten niederschlug. Selbst bis nach dem Weinbauzentrum Krems – immer dem Weg des Salzes auf dem Wasser folgend – reichten die Verflechtungen der Reichenhaller Oberschicht in jener Zeit. Im Falle von Laufen beabsichtigten die Patrizier wohl, neben der Salzproduktion in der Saline, auch den lukrativen Handel und Transport unter ihre Kontrolle bringen zu können.

55 Karl August Muffat, Schenkungsbuch der ehemaligen Gefürsteten Probstei Berchtesgaden, in: Franz Michael Wittmann / Karl August Muffat (Hg.), Schenkungsbücher bayerischer Klöster (Quellen und Erörterungen zur bayerischen und deutschen Geschichte 1), München 1856 (Neudruck: Aalen 1969), S. 225–364, hier: S. 328, Nr. 152: „*Liuthwinus iudex hallensis*".

56 Lang, Geschichte von Bad Reichenhall (wie Anm. 4), S. 114–123.

57 Heinz Dopsch, Die Anfänge der Siedlung und die Entwicklung zur Stadt, in: Ders. / Hans Roth (Hg.), Laufen und Oberndorf. 1250 Jahre Geschichte, Wirtschaft und Kultur an beiden Ufern der Salzach, Laufen/Oberndorf 1998, S. 47–60, hier: S. 52 f.

3. Die Stadt im Spiegel der Quellen des 12. Jahrhunderts

Dem Zuwachs an Macht und Möglichkeiten der Reichenhaller Bevölkerung und ihres Patriziats suchte man auch formal Rechnung zu tragen, zumal der Salinenort im frühen 12. Jahrhundert in einem raschen Wachstum begriffen war. Zur bereits seit dem 8. Jahrhundert bestehenden St.-Johannes-Kirche kamen in jener Zeit St. Ägidius (um 1159) und St. Nikolaus (um 1181) hinzu, außerdem das rund 2,5 Kilometer von der Saline entfernt liegende Augustiner-Chorherrenstift St. Zeno, das ab den 1120er Jahren über rund ein Jahrhundert zu einem der größten romanischen Gotteshäuser im gesamten bayerischen Raum ausgebaut werden sollte. Auffälliger Initiator dieses Brückenkopfs der Salzburger Kirche war Erzbischof Konrad I. von Abenberg (reg. 1106–1147),[58] der bereits im Falle des in Kärnten liegenden Ortes Friesach für ein städtebauliches Programm gesorgt hatte.[59] Konrad, der 1120 aus dem Exil in Sachsen in seine Residenzstadt Salzburg zurückgekehrt war, dürfte denn auch bald darauf zur treibenden Kraft für die Urbanisierung Reichenhalls geworden sein.

Der Umfang an Profanbauten, insbesondere Wohnhäusern, war bereits im 12. Jahrhundert beträchtlich. Aufgrund der Aufzeichnungen zu einer Brandkatastrophe aus dem Jahre 1037 wissen wir, dass allein das Kloster Niederaltaich über rund 50 Eigenleute in Reichenhall verfügte, die ein einziges großes Haus bewohnten.[60]

Dies deutet auf die Existenz überdurchschnittlich großer Wohngebäude hin, die zum Teil nicht nur aus Holz, sondern bemerkenswerterweise auch schon aus Stein errichtet waren, wie die Erwähnung eines steinernen Hauses („*lapidea domus*“) des erwähnten Pabo vom Hofe aus der ersten Hälfte des 12. Jahrhunderts zeigt.[61]

58 Johannes Lang, St. Zeno in Reichenhall. Geschichte des Augustiner-Chorherrenstifts von der Gründung bis zur Säkularisation (Studien zur bayerischen Verfassungs- und Sozialgeschichte 22), München 2009, S. 27–36; Ders., Das Augustiner-Chorherrenstift St. Zeno in Reichenhall (Germania Sacra. Dritte Folge 9: Das Erzbistum Salzburg 2), Berlin/Boston 2015, S. 108–123.

59 Heinz Dopsch, Die Länder und das Reich. Der Ostalpenraum im Hochmittelalter, Wien 1999, S. 314.

60 Lang, Geschichte von Bad Reichenhall (wie Anm. 4), S. 102.

61 SUB I, S. 394, Nr. 270 a.

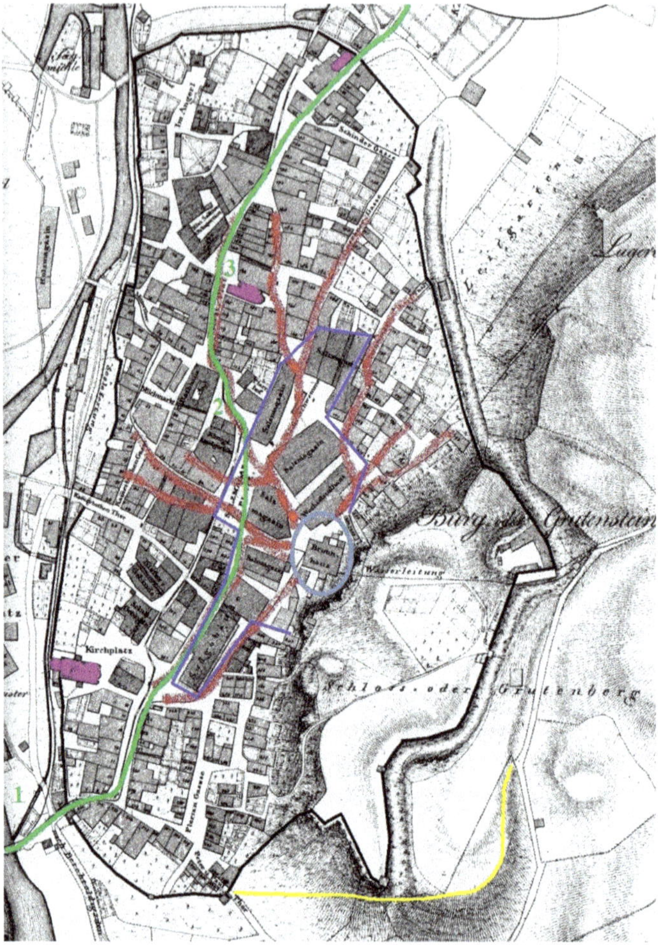

Abb. 43: *Im Reichenhaller Stadtplan von 1817 lassen sich die amorphen städtebaulichen Strukturen des 12. Jahrhunderts ablesen. Es ist erkennbar, dass die Solequellen den Ausgangspunkt für die Stadtwerdung darstellten: Von der Sole-Brunngrube (hellblau) führten einzelne Gassen (rot) sternförmig zunächst durch das Areal (dunkelblau) der im Hochmittelalter über 60 Sudhäuser, woran sich im weiteren Verlauf die Wohnbebauung anschloss. Für den Fernverkehr zog sich die Reichsstraße (grün) der Länge nach durch die Stadt und passierte dabei die Schiffsverladestätten an der Saalach (1), das „forum“ (2) des 12. Jahrhunderts und den neuen Marktplatz (3) des 13. Jahrhunderts. Für das 12. Jahrhundert lassen sich innerhalb der Stadt drei Kirchen (violett) nachweisen, außerdem eine Stadtmauer (schwarz), deren ursprünglich veränderter Verlauf (gelb) archäologisch nachgewiesen werden konnte. Bildnachweis: Entwurf und Ausführung: Johannes Lang, Bad Reichenhall.*

In jeder Hinsicht beherrschte das der Gerichtsbarkeit des Hallgrafen unterworfene und im 12. Jahrhundert ein Sechstel der städtischen Gesamtfläche einnehmende Salinengelände das Gepräge Reichenhalls. Bekannt sind einige zwischen den Pfannhäusern verlaufende Gassen, wie etwa die in jener Zeit erwähnte „*Huntgasse*", „*Dietmarsgasse*", „*Weite Gasse*", „*Oheimer Gasse*" oder „*Milchgasse*", die im weiteren Verlauf in die Wohnviertel der Stadt führten, ihre Namen teilweise bis in das 19. Jahrhundert bewahren konnten und noch im heutigen Stadtplan abzulesen sind. Das Prinzip der an der Saline beginnenden Gassen gibt einen Hinweis darauf, dass die Werkssiedlung an den Sudhäusern sternförmig ihren Ausgang nahm. Da sich im Osten der anstehende Fels des Gruttenbergs erhob und im Westen der Verlauf der Saalach beziehungsweise des Triftgeländes eine Barriere darstellte, breitete sie sich gleichermaßen gegen Süden wie auch gegen Norden aus.

Unmittelbar am Gelände der Saline entlang verlief, Reichenhall in der Länge durchschneidend, die um 1180 als „*Richenstraze*" bezeichnete Reichsstraße,[62] die als Fernstraße Salzburg mit Innsbruck verband. Im Gegensatz zu den mit dem Wortbestandteil „Gasse" gebildeten Bezeichnungen finden wir nur in diesem Fall die Benennung „Straße", was auf den Rang dieser Verkehrsverbindung hindeutet.

Außerhalb des Salinenareals sind nur wenige Bezeichnungen bekannt, so der um 1231 genannte „*Alte Markt*",[63] dessen Benennung bereits damals die Anlage eines neuen Marktes voraussetzt. Die exakte Lage dieses um die Mitte des 12. Jahrhunderts erstmals urkundlich erwähnten älteren Marktes („*forum*") ist nicht bekannt,[64] doch könnte er unmittelbar westlich an das Salinengelände angegrenzt haben. An der Ägidikirche (heute Poststraße) etablierte sich später der neue Markt. Das Marktrecht, dessen Ursprünge wir nicht kennen und das vielleicht sogar noch ins 11. Jahrhundert zurückreicht, bildete eine wichtige wirtschaftliche Voraussetzung für die Entstehung der mittelalterlichen Stadt.

62 SUB I, S. 693, Nr. 225.

63 Ingrid Heeg-Engelhart, Das älteste bayerische Herzogsurbar. Analyse und Edition (Quellen und Erörterungen zur bayerischen Geschichte NF 37), München 1990, S. 9: „*an dem alten Marct*".

64 Muffat, Schenkungsbuch Berchtesgaden (wie Anm. 55), S. 284, Nr. 88: „[...] *et iuxta forum locum sartaginis et in loco petre* [...]".

Neben dem Regensburger Pfennig kursierte in Reichenhall seit der Mitte des 11. Jahrhunderts Salzburger, Passauer, Freisinger, Brixener, österreichisches und steirisches Geld, sowohl in Form von bischöflichen, herzoglichen als auch von kaiserlichen Prägungen.[65] Während die in Reichenhall im Umlauf befindlichen Salzburger Münzen hauptsächlich aus der Münzstätte Laufen stammten, womit einmal mehr die sehr engen Verflechtungen zwischen dem Salzproduktionsort Reichenhall und dem Salzumschlagplatz Laufen belegt sind, deuten vor allem die Freisinger, Passauer, Regensburger und österreichischen Münzen den Weg des „Weißen Goldes“ in den Münchener Raum sowie entlang der Donau an. Der Tauschhandel gehörte im 12. Jahrhundert weitgehend der Vergangenheit an; eingenommenes Geld diente hauptsächlich der Thesaurierung und Kapitalisierung des Patriziats. Nur so ist es zu erklären, dass man in Reichenhall und dessen Umgebung wiederholt regelrechte Münzhortfunde aus dem hohen Mittelalter gemacht hat.[66]

Mit seiner Lage zwischen der Saalach und dem Fuße des Gruttenstein bot die Natur dem Ort einen nur unzulänglichen Schutz, weswegen mit teils enormem Aufwand eine Ringmauer errichtet wurde. Die heute immer noch in Teilen existente „*rinchmower ze Halle*“ lässt sich in den Schriftquellen zwar erst im Jahre 1285 fassen,[67] 2006 jedoch entdeckte man außerhalb des bekannten

---

65 Vgl. Joseph Eucharius Obermayr, Historische Nachricht von Bayerischen Münzen oder muthmaßliche Erklärung derer zu Reichenhall ausgegrabenen und in dem XI. und XII. Jahrhundert geschlagenen Münzen, Regensburg 1763, S. X–XII; Christoph Mayrhofer, Die Münzstätte Laufen und ihr Gepräge, in: Das Salzfass 2, 1998, S. 73–83, hier: S. 78 f.

66 Dazu gehört der Münzhortfund von 1753 in Reichenhall (vgl. Obermayr, Historische Nachricht von Bayerischen Münzen (wie Anm. 65), S. III f.) und jener in Karlstein von 1851 (vgl. Heinrich Buchenau, Der Münzfund von Karlstein bei Reichenhall (1851) mit Beiträgen zur Münzkunde der beiden ersten Baiernherzöge aus dem Hause Wittelsbach, in: Mitteilungen der Bayerischen Numismatischen Gesellschaft 32/33, 1915, S. 77–114, hier: S. 77 f.) sowie zuletzt ein um das Jahr 2012 vermutlich in Bad Reichenhall oder Umgebung gemachter Fund mit ca. 1.600 Münzen, die bis in die 1180er Jahre reichen (vgl. Johannes Lang, Nehmen wir einmal an … oder: Unglaubliches aus dem Leben eines Heimatpflegers, in: Pulverturm. Mitteilungen des Vereins für Heimatkunde und Umgebung e.V. 2019, S. 52–54; Tilmann Mittelstrass, Drei Münzgefäße aus Österreich, Bayern und Slowenien, in: Beiträge zur Mittelalterarchäologie in Österreich 30, 2014, S. 211–222).

67 Salzburger Urkundenbuch, Bd. III: Urkunden von 1200–1246 (künftig: SUB III), bearb. v. Willibald Hauthaler / Franz Martin, Salzburg 1918, S. 145, Nr. 125 a.

Stadtmauerverlaufs Überreste einer älteren Quadermauer, die stellenweise eine Mindeststärke von 2,40 Meter aufwies und vom Bayerischen Landesamt für Denkmalpflege dokumentiert wurde.[68] Die beiden im Zusammenhang mit dieser archäologischen Untersuchung aufgefundenen teils mächtigen Brandschichten wurden einer ^{14}C-Beprobung unterzogen und – unter Vorbehalt der üblichen Kalibrierungsschwankungen – auf die Jahre um 830 sowie 1063 datiert, wobei die untersuchten Holzkohlen den Wachstumszeitpunkt des beprobten Holzes widerspiegeln.[69] In der Folge konnte der weitere Verlauf dieser als älter einzustufenden Stadtmauer mittels Prospektion festgestellt werden, was zu einem Nachtrag in der Denkmalliste führte: Offenbar dehnte sich die Stadt über das Terrain der heute bekannten mittelalterlichen Stadt aus und schloss auf diese Weise eventuell auch den höchstgelegenen Punkt der Umgebung, den sogenannten Streitbichl mit dem dortigen Burgstall, mit ein. Sowohl die Lage als auch die ^{14}C-Datierung als auch die Ereignisgeschichte deuten darauf hin, diese ältere Befestigung mit einer im Jahre 1196 zerstörten Stadtmauer in Verbindung zu bringen, wovon noch die Rede sein wird. Diese ältere Ummauerung Reichenhalls muss folglich im Verlaufe des 12. Jahrhunderts erfolgt sein, vielleicht bereits in den 1120er Jahren.

Die für das 12. Jahrhundert archäologisch nachweisbare Ausdehnung der Stadt erstreckte sich über eine maximale Länge von circa 750 Metern sowie eine Breite von 550 Metern. Damit besaß Reichenhall die ungefähren Ausmaße des damaligen Freising, das – wie eingangs erwähnt – in den Salzburger Quellen als einer von nur vier bayerischen Orten mit dem „*civitas*"-Begriff bedacht wurde, wogegen andere mit urbanen beziehungsweise präurbanen Strukturen für das 12. Jahrhundert in Frage kommende Orte (Laufen an der Salzach, Wasserburg, Mühldorf, Burghausen, Landsberg am Lech) zu jener Zeit teilweise deutlich kleiner ausfielen.

Das für die über zwei Kilometer lange Reichenhaller Stadtmauer benötigte Steinmaterial bezog man vom nah gelegenen Kirchholz, wo sich das leicht zu bearbeitende Konglomerat oberflächennah findet und noch heute große

68 Dazu der interne Untersuchungsbericht durch den Grabungstechniker des Bayerischen Landesamtes für Denkmalpflege, Ulf Händler, vom 12.4.2006 (Wohnhaus Glück im Winkel 3, Flst.-Nr. 833/7, Gmkg. und Gmde. Bad Reichenhall, Lkr. Berchtesgadener Land).

69 Die Beprobung erfolgte durch das AMS-Labor Erlangen, Physikalisches Institut an der Friedrich-Alexander-Universität Erlangen-Nürnberg, Dipl.-Phys. Andreas Scharf am 17. Juli 2006.

Abbauflächen erkennbar sind. Auf dem Streitbichl stand die nach dem darunter liegenden Ort benannte Burg, die in einer päpstlichen Urkunde Lucius' II. im Jahre 1144 als *„castrum Halle“*,[70] 1193 dann in der deutschen Übersetzung als *„Halburch“* (Hallburg) bezeichnet wurde.[71] Mit ihrem ovalen Grundriss (längsovaler Durchmesser: circa 100 Meter) erscheint die Anlage fast geometrisch und in der Tradition einer ab etwa 1050 üblichen Motte. Diese verfügte über einen inneren wie auch äußeren Ringwall und war wohl nur durch Erdwerke – Wälle und Gräben – sowie Holzpalisaden gesichert.[72] Die Hallburg diente dem seit 1077 urkundlich auftretenden Hallgrafen aus dem Geschlecht derer zu Wasserburg, wenn er zu Rechtgeschäften in Reichenhall weilte, als Residenz und Verwaltungssitz.[73] In seiner Abwesenheit allerdings trug ein Kastellan beziehungsweise Burggraf, wie er 1159 erstmals urkundlich erwähnt wird,[74] die Verantwortung für den Wehrbau, zu dem auch eine Burgkapelle gehörte.

Zu Füßen der Hallburg lag die durch Mauern und Gräben bewehrte und in unterschiedliche Arbeits-, Handels- und Wohnbereiche gegliederte Siedlung, die um 1150 als *„oppidum“* Erwähnung findet,[75] worunter ein befestigter Ort zu verstehen ist. Der um die Mitte des 12. Jahrhunderts in den Urkunden erkennbare Wechsel unterschiedlicher mittellateinischer Begrifflichkeiten für das Wesen Reichenhalls macht deutlich, dass der Ort in jener Zeit einer hohen städtebaulichen Dynamik unterworfen war. Während der Salzburger Erzbischof Konrad I. als äußerer Impulsgeber anzusprechen ist, ging der innere Impuls von den Bewohnern des Salinenortes aus, denn im Falle Reichenhalls trafen zwei maßgebliche Kriterien zusammen: einerseits ein rasches wirtschaftliches Wachstum auf der Grundlage des hier gewonnenen Rohstoffes und, darauf aufbauend, andererseits der Zusammenschluss aller Wirtschaftstreibenden und Werktätigen zu einer gewerblich orientierten Gemeinde.[76] Mit der Nennung aller Reichenhaller – sie werden gemäß dem mitunter auftauchenden Ortsnamen *„Salinae“* (für Reichenhall) als *„salinarii“* bezeichnet – anlässlich der im Jahre 935 in Reichenhall

---

70 Germania Pontificia, Bd. 1: Provincia Salisburgensis et episcopatus Tridentinus, hg. v. Albert Brackmann, Berlin 1911, S. 66; MB 3, S. 530, Nr. 3: *„capellam in castro Halle“*.

71 Salzburger Urkundenbuch, Bd. II: Urkunden von 790–1199 (künftig: SUB II), bearb. v. Willibald Hauthaler / Franz Martin, Salzburg 1916, S. 688, Nr. 509.

72 Johannes Lang, Reichenhaller Burgenweg. Führer zu den Burgen und Schlössern im Reichenhaller Raum, Reichenhall 2004, S. 22 f.

73 Vgl. zu den Hallgrafen: Lang, Geschichte von Bad Reichenhall (wie Anm. 4), S. 100–106.

74 SUB II, S. 479, Nr. 342; S. 489, Nr. 349: *„Rudegerus castellanus Hallensis“*.

75 SUB I, S. 422, Nr. 313 c.

76 Koller, Salzburger Städte (wie Anm. 5), S. 17.

erfolgten Huldigung des bayerischen Herzogssohnes Eberhard wird nicht nur die Bedeutung des Salinenortes für die luitpoldingische Macht unterstrichen, sondern es lässt sich auch bereits eine gildenartig ausgeprägte Gemeinschaft erahnen, die eine Vorform des späteren Bürgertums darstellte.[77] Das Auftreten sämtlicher Bevölkerungsschichten zeugt schon im frühen 10. Jahrhundert von einem bemerkenswerten Selbstbewusstsein.

Im Gegensatz zu anderen Orten, die ihren natürlichen Mittelpunkt im kaufmännisch-handwerklichen Markt fanden, bildete in Reichenhall die industriell anmutende Salzproduktion und der Handel mit dem „Weißen Gold" die einigende Grundlage aller mit dem Salz Beschäftigten. Die Gruppe wurde nun nicht mehr, wie zuvor, nur als „*salinarii*" – als „Reichenhaller" – angesprochen, sondern als Bürger („*cives*").[78] Tatsächlich erlangte der Begriff ab der Mitte des 12. Jahrhunderts eine immer häufigere Anwendung – zu einem Zeitpunkt also, als das Werden einer Bürgerschaft und ihres Siedlungsmittelpunktes weitgehend abgeschlossen war. Alle Gewerbetreibenden dieses Ortes galten fortan als Bürger.[79]

Anno 1153 weilte der Salzburger Erzbischof Eberhard I. in Reichenhall und bezeichnete in einer Urkunde die Bewohner Reichenhalls als Bürger einer befestigten Siedlung („*oppidi cives*").[80] Im Jahre 1158, als München als „*forum*" erstmals erwähnt wird und dessen städtische Ausformung noch in einiger Ferne liegt, findet sich in einer Salzburger Urkunde mit Heinrich Ferre erneut ein Reichenhaller Bürger („*Hallensis civis*").[81] Knapp anderthalb Jahre später, in einem

---

77 Wilhelm Störmer, Vergesellschaftungsformen des Meliorats und des Handwerks in den Städten des bayerisch-österreichischen Raumes (Vorträge und Forschungen 29), Sigmaringen 1985, S. 337–375, hier: S. 343; Lang, Geschichte von Bad Reichenhall (wie Anm. 4), S. 96 f.

78 Theodor Bitterauf, Die Traditionen des Hochstifts Freising (Quellen und Erörterungen 5/2), München 1909 (Neudruck: Aalen 1967), S. 339, Nr. 1491 c: „*cives de Halla*", betrachtet das Jahr 1098 (Todesjahr Bischof Meginwards) als *terminus ante quem* für die Nennung eines gewissen Gerold, der in den Traditionen als „*cives* [sic!] *de Halla*" genannt wird. Tatsächlich muss es sich um einen nachträglichen späteren Eintrag (vermutlich Mitte des 12. Jahrhunderts) handeln.

79 Hans Planitz, Die Deutsche Stadt im Mittelalter. Von der Römerzeit bis zu den Zunftkämpfen, Graz 1954, S. 101.

80 SUB II, S. 425, Nr. 304.

81 SUB II, S. 460, Nr. 331; Birgit Gruber-Groh, Bad Reichenhall (Historischer Atlas von Bayern, Altbayern 57), München 1995, S. 110, macht eine erste „*civis*"-Nennung zum Ende des 11. Jahrhunderts aus, den Traditionen von Freising folgend (Bitterauf,

Diplom vom 16. November 1159, bezeichnete der Salzburger Erzbischof Eberhard I. Reichenhall erstmals als Stadt („*civitas*").[82] Mit dieser Nennung ist Reichenhall vorbehaltlos als vollendete Stadt mit einem ausgeprägten Bürgertum zu betrachten.[83] In den Urkunden aus der zweiten Hälfte des 12. Jahrhunderts findet sich Reichenhall wiederholt mit diesem Attribut, so etwa in einem 1163 auf der Burg in Steyr ausgestellten Diplom als „*bawarica civitas Halla*".[84] Um das Jahr 1193 wandte sich der bayerische Herzog Ludwig I. brieflich an „*burgenses omnes in Halle*".[85]

In der Reihe jener, die den oben erwähnten Urkundentext von 1159 bezeugten, findet sich die Spitzengruppe des damaligen Bürgertums. Ihre wirtschaftliche Potenz verlieh diesen Familien ein ausgeprägtes Selbstbewusstsein, das zur Bildung einer politisch bevorrechteten Elite und damit zur Entstehung eines auf Reichtum gegründeten Ratswesens in der Stadt führte. Diese etwa dreißigköpfige Gruppe kartellartig zusammengeschlossener Salinenbetreiber und Siedeherren stellte fortan den Rat der Stadt und traf die Entscheidungen in erster Linie nach persönlichen Interessen.

Eine geplante Stadtgründung, formale Stadterhebung oder Stadtrechtsverleihung gab es, wie wir gesehen haben, in Reichenhall nicht. Vielmehr deutet bei der langwierigen Entwicklungsgeschichte und Stadtwerdung alles auf einen älteren Stadttypus hin, der sich seine besonderen Rechte über die Zeiten hinweg erworben hatte. Altes Herkommen schuf Rechte: Reichenhall besaß eine steinerne Ummauerung, Türme und Tore. Seine Bürger waren persönlich frei und setzten sich vor allem aus Salzproduzenten, Kaufleuten, Fernhändlern und Handwerksmeister zusammen. Als verbindliches Stadtrecht galt jenes mündlich überlieferte Abkommen, das sich an dem Ort etabliert hatte und das in einer

---

Traditionen des Hochstifts Freising (wie Anm. 78), S. 339, Nr. 1491 c. Tatsächlich wird der „*cives*" (sic!) Gerolt in einer späteren Tradition (S. 491, Nr. 1673 c) ohne „*civis*"-Bezeichnung genannt. Die Verlässlichkeit der Quelle ist somit fraglich.

82 SUB II, S. 478, Nr. 342: „[…] *in ipsa civitate*".

83 Koller, Salzburger Städte (wie Anm. 5), S. 17.

84 OÖUB II, S. 328, Nr. 224.

85 Zitiert nach: Heinrich Wanderwitz, Die frühen wittelsbachischen Herzöge und das bayerische Salzwesen (1180–1347), in: Hubert Glaser (Hg.), Wittelsbach und Bayern. Die Zeit der frühen Herzöge. Von Otto I. zu Ludwig dem Bayern, Bd. I/1, München 1980, S. 338–348, hier: S. 344, Anm. 16 und S. 339 (Übersetzung aus dem Lateinischen: Paul Rauscher). Die von Wanderwitz vorgenommene Datierung mit „vor 1191 IV 14" lässt sich meines Erachtens nicht aufrechterhalten, sondern dürfte wohl um 1193 anzusetzen sein.

Urkunde Kaiser Heinrichs VI. aus dem Jahre 1194 – in dem Diplom ausdrücklich abgegrenzt vom Landrecht („*ius provinciale*") – als „*ius hallensium*" bezeichnet wurde.[86] Darüber zu wachen hatte der bereits erwähnte Richter, der sowohl 1193 in einem herzoglichen als auch ein Jahr später in einem kaiserlichen Brief direkt namentlich angesprochen wird.[87]

Äußeres Hoheitszeichen und Symbol städtischen Selbstverständnisses bildete das Siegel. Reichenhalls ältestes erhaltenes Stadtsiegel stammt von 1279, dürfte in der Form aber schon früher bestanden haben: ein mit der Siegesfahne schreitendes Gotteslamm, dessen Haupt vom Strahlenkranz umgeben ist, im äußeren Kreis die Umschrift „*SIGILLUM CIVIUM DE HALLE*".[88] In dieser christlichen Symbolik ist jener Schirmherr zu erkennen, den die junge Stadt noch vor dem heiligen Nikolaus (Patrozinium der Stadtpfarrkirche) zu ihrem ersten Schutzpatron erkoren hatte: den heiligen Johannes den Täufer. Denn zum Zeitpunkt der Entstehung des Siegels gab es wohl innerhalb der Stadtmauern nur die aus dem Frühmittelalter stammende Kirche des heiligen Johannes. Um ihr Engagement bei der Errichtung der Stiftskirche St. Zeno dauerhaft in Erinnerung zu rufen, ließen die Bürger an der Wende vom 12. zum 13. Jahrhundert das Lamm Gottes in reliefierter Form im Bogenfeld des romanischen Eingangsportals anbringen – angesichts der Dimensionen dieses Gotteshauses ein Beweis für die Finanzkraft der Reichenhaller Bürgerschaft.

4. Salinen im Ostalpenraum während des 12. Jahrhunderts

Ein ab der Mitte des 12. Jahrhunderts anzunehmendes Klimaoptimum mit dem damit verbundenen Bevölkerungswachstum führte auch im Umfeld Reichenhalls zu einer neuerlichen Rodungs- und Besiedelungswelle. Es entstanden mehrere markante Rodungsinseln, wie sie heute beim Kugelbachbauern oder als gegen das Staufengebirge vorgeschobene Siedlungsstellen beim Preindler und bei Pading anzutreffen sind. Dies bedeutete für das Reichenhaller Tal nicht nur einen Zugewinn an Siedlungsboden und landwirtschaftlicher Nutzfläche, sondern war – wie im gesamten Alpenraum – verbunden mit einem Anstieg der

86 MB XXXI/1, S. 456, Nr. CCXXXVII; vgl. Erich Keyser / Heinz Stoob (Hg.), Bayerisches Städtebuch, 2. Teil (Deutsches Städtebuch. Handbuch städtischer Geschichte V/2), Stuttgart 1974, S. 87.

87 Wanderwitz, Wittelsbachische Herzöge (wie Anm. 85), S. 344, Anm. 16; MB, XXXI/1, S. 455, Nr. CCXXXVI.

88 Vgl. Johannes Lang, Die Geschichte des Bad Reichenhaller Stadtwappens, in: Adreßbuch 1999 – Bad Reichenhall, Bayerisch Gmain, Piding, München 1999, o.S.

Abb. 44: *Holzlagerplatz für die Reichenhaller Saline in Unken, frühes 20. Jahrhundert. Bildnachweis/Foto: Hohenwarter, Unken.*

Bevölkerung. In den Alpen entstanden Schwaigen – Bauerngüter, die sich auf Schaf- und Viehhaltung sowie auf die Herstellung von Käse spezialisiert hatten. In der Umgebung Reichenhalls, in Weißbach, findet um das Jahr 1180 ein „*svaichof*" urkundliche Erwähnung.[89] Um die gleiche Zeit begann ein Ministeriale der Salzburger Kirche, Rudolf von Schnellenberg, am Fuße des Untersbergs mit einer Schwaigenkolonie,[90] während der jenseits des Reichenhaller Tales gelegene Jochberg von den Grafen von Plain landwirtschaftlich erschlossen und mit Schwaigen ausgestattet wurde.[91]

89 SUB I, S. 693, Nr. 225.

90 Das Großgmainer Klingerbauerngehöft erscheint urkundlich vom 14. bis zum 18. Jahrhundert als „*Schwaig Schnellnberg*" (vgl. Johannes Lang / Max Schneider, Auf der Gmain. Chronik der Gemeinden Bayerisch Gmain und Großgmain, Bayerisch Gmain/Großgmain 1995, S. 101). Rudolf von Schnellenberg („*Snellenberc*") tritt in den Salzburger Urkunden von 1188 bis 1193 auf (SUB I, S. 489, Nr. 436; S. 715, Nr. 274; S. 720, Nr. 283; SUB II, S. 612, Nr. 451; S. 656, Nr. 482).

91 Regesta sive Rerum Boicarum [...], Bd. II, bearb. v. Karl Heinrich Ritter von Lang, München 1822, Nr. 390.

Auf diese Weise wurde der Bedarf an Käse und später an Butterschmalz in dem bevölkerungsreichen Salinenort durch die Schwaigen der umliegenden Gebirgszüge gedeckt. Die zu jener Zeit im gesamten Ostalpenraum zu beobachtende Schwaighofkolonisation ließ wiederum die Nachfrage nach Salz sprunghaft ansteigen, da sowohl für die Käseherstellung als auch für die intensive Viehwirtschaft enorme Mengen an Käserei- und Viehsalz nötig waren. In Reichenhall führte diese Entwicklung, verbunden mit einem allgemeinen Bevölkerungsanstieg, zu dem oben geschilderten bemerkenswerten Wirtschaftswachstum.[92]

Von der verstärkten Nachfrage nach Salz profitierte nicht nur die etablierte Saline von Reichenhall, sondern es gab im Ostalpenraum schon seit Längerem einige Klein- und Kleinstsalinen, wie etwa jene im heutigen Bad Hall (Oberösterreich). Sie findet bereits im Zusammenhang mit der Stiftung des Klosters Kremsmünster 777 durch den bayerischen Herzog aus dem Hause der Agilolfinger, Tassilo III., als *„salina, que ad sulzibach est"* urkundliche Erwähnung.[93] In der bezeichneten Textstelle ist die Rede von *„drei Männern, die dort leben und Salz kochen"*. Allein daran lässt sich deren geringer Umfang im frühen Mittelalter ermessen. Vermutlich zum Ende des 10. Jahrhunderts interpolierte man die Gründungsurkunde um einen Besitztitel *„in salina vero maiore"* – die *„größere Saline"*.[94] Heinrich Wanderwitz hat diese nicht näher erläuterte Örtlichkeit mit Reichenhall in Verbindung gebracht,[95] was in Ermangelung geeigneter anderer Gleichsetzungen eine hohe Wahrscheinlichkeit besitzt. Dass man im weit entfernten Kremsmünster von Reichenhall als der „größeren Saline" sprach, wirft einerseits ein Schlaglicht auf die salinarische Vormachtstellung Reichenhalls und verdeutlicht andererseits die wirtschaftlich untergeordnete Bedeutung der anderen inneralpinen Salinen zu jener Zeit. Mit der Schaffung des Herzogtums Österreich im Jahre 1156 finden sich unterschiedliche Ortsbenennungen für Hall: vorerst *„Herzogenhall"*, zeitweise auch *„Halle inferior"*, was man im Deutschen mit „Niederhall" übersetzt hat.[96]

---

92 Vgl. Klein, Geschichte der Salinen Hallein und Reichenhall (wie Anm. 4), S. 400; Bergier, Geschichte vom Salz (wie Anm. 2), S. 53.

93 Rudolf Palme, Rechts-, Wirtschafts- und Sozialgeschichte der inneralpinen Salzwerke bis zu deren Monopolisierung, Frankfurt am Main 1983, S. 10.

94 OÖUB II, S. 4, Nr. 2: *„In salina vero maiore unum hominem salem coquentem concessimus"*. Dabei scheint es sich um einen gefälschten Zusatz (vermutlich Ende 10. Jahrhundert) eines im jahre 777 von Tassilo III. für Kremsmünster ausgestellten Diploms zu handeln.

95 Wanderwitz, Salzwesen (wie Anm. 4), S. 112.

96 Oberösterreichisches Urkundenbuch/Urkunden-Buch des Landes ob der Enns, Bd. I (künftig: OÖUB I), bearb. v. Erich Trinks, Wien 1852, S. 242, Nr. 68.

Seit dem frühen 10. Jahrhundert existierte in Hall bei Admont (Steiermark) eine weitere Saline, die ihr Salz – wie in Reichenhall – aus am Fuße des Pitzbergs, des Lärchecks sowie des Leichenbergs entspringenden Solequellen generierte.[97] Allerdings lassen die aus dem hochmittelalterlichen Schrifttum herauszulesenden Verhältnisse erkennen, dass die Salzerzeugung dort im vergleichsweise kleinen Stil erfolgte und sich auf circa elf Salzpfannen beschränkte.[98] Die Anzahl der im 12. Jahrhundert dort begüterten Grundherren war mit den Bistümern Gurk und Bamberg sowie den Klöstern Admont, St. Lambrecht, Steiergarsten und St. Georgen am Längsee überschaubar, verglichen mit den über 60 Anteilsnehmern an der Reichenhaller Saline. Über eine dörfliche Ansiedlung von Salinenarbeitern wuchs Hall nicht hinaus, was mehrere Gründe hatte: Zuvorderst machte die nur geringe Grädigkeit der im 12. Jahrhundert etwa sechs vorhandenen Solequellen einen Großbetrieb unmöglich. Da das zum Teil steile Gelände der unterschiedlichen Quellorte eine unmittelbare Platzierung der Salzpfannen nicht erlaubte, musste die Sole mittels Rohrleitungen herangeschafft werden. In Ermangelung geeigneter Eigengründe war das Hochstift Bamberg sogar gezwungen, die Sole mit Traggeschirren zu seinem Pfannhaus bringen zu lassen. Eine Konzentration des Salinenareals war auf diese Weise nicht möglich, worunter die Wirtschaftlichkeit zweifellos gelitten haben muss. Der Holzreichtum der Umgebung sowie die zahlreichen triftbaren Bäche und der Fluss Enns erlaubten zwar eine ausreichende Brennholzzufuhr, die im Flurnamen „*Grieß*“ (Triftgelände) ihren Niederschlag erfuhr. Problematisch erwies sich aber die Schiffbarkeit der Enns und damit der großzügige Transport des Salzes auf dem Wasserweg. Diese war nur flussaufwärts möglich, da flussabwärts hinter Hall das schluchtenartige „Gesäuse“ beginnt.

Zu nennen ist schließlich die starke grundherrschaftliche Dominanz an der Saline, insbesondere des nur anderthalb Kilometer nah gelegenen Klosters Admont, was die Herausbildung einer korporativen Gemeinschaft an Werktätigen – Grundlage eines künftigen Bürgertums – nicht beförderte. Im Gegenteil vereinigte das Kloster, das ab 1093 die Gerichtsbarkeit über die gesamte Saline innehatte, bis zum Jahre 1303 sämtliche Salinenanteile in Hall unter seiner Verwaltung.[99] Dementsprechend entwickelte sich eine marktähnliche Ansiedlung nicht in Hall, sondern rund um das Kloster Admont, das erst 1443 das Marktrecht förmlich verliehen bekam.[100] Der von der aufblühenden Saline Aussee ausgehende wirtschaftliche Druck führte

---

97 Erstmalige Erwähnung im Jahre 931: SUB I, S. 80, Nr. 13.

98 Albert von Muchar, Geschichte des Herzogthums Steiermark, 3. Teil, Graz 1846, S. 99 f.

99 Ebd., S. 99.

100 Waltraus Krassnig / Ulrich Faust, Die benediktinischen Mönchs- und Frauenklöster in Österreich und Südtirol, Teil 1, St. Ottilien 2000, S. 174.

bereits im 13. Jahrhundert zu einem Niedergang der Admonter Salzproduktion, die 1540 auf landesfürstliches Geheiß zum Erliegen kam.[101]

Eine weitere Solequelle, die in ihrer Bedeutung zurückstand, befand sich im steirischen Halltal bei Mariazell und wurde 1025 urkundlich erwähnt.[102] Der Name des unmittelbar vorbeifließenden Flusses Salza, der sich für die Transportschifffahrt nicht geeignet haben dürfte, weist auf eine vermutlich weiter zurückreichende Tradition dieser bescheidenen Salzerzeugungsstätte hin. Im Jahre 1103 schenkte der Herzog von Kärnten, Heinrich III. von Eppenstein, die wohl nur aus einer Handvoll Sudpfannen bestehende Saline an das Benediktinerkloster St. Lambrecht,[103] das die vermutlich schwachgrädigen Quellen ausbeutete. Nur die Abgeschiedenheit des Ortes, die eine Salzversorgung der Umgebung durch auswärtige Händler schwierig machte, bescherte der kleinen Saline eine bis in das 17. Jahrhundert hineinreichende Produktivität.

Darüber hinaus existierten im Ostalpenraum bis zur Mitte des 12. Jahrhunderts einige zusätzliche kleine Solequellen, die nur lokale Bedeutung besaßen, so etwa jene in Altaussee, Weißenbach an der Enns (Steiermark) oder an dem über Spital am Pyhrn aufragenden Berg Gulch. Entsprechend den unbedeutenden Erträgen blieb es dort lange Zeit nur bei einer weilerartigen Siedlungsstruktur.

Ungeachtet der oben geschilderten Anlagen von meist nur lokaler Bedeutung behauptete Reichenhall um die Mitte des 12. Jahrhunderts bei der Produktion des überregional gehandelten Salzes seit inzwischen einem halben Jahrtausend überlegen die Marktführerschaft. Zur namentlichen Unterscheidung der zunächst noch zum Herzogtum Kärnten gehörenden Salinen im steirischen Ennstal sowie den teils weiter entfernten Anlagen in der Mark Österreich erfuhr die Reichenhaller Saline noch zum Ende des 10. Jahrhunderts namentliche Unterscheidungen: Dem Rang und der Bedeutung der Saline gemäß findet sich der Name der *„größeren Saline"*; in ortsfremden Quellen ist von der *„bayerischen Saline"* die Rede,[104] um 1120 vom *„bayerischen Hall"*,[105] womit dessen Lage und

101 Franz Stadler, Salzerzeugung, Salinenorte und Salztransport in der Steiermark, in: Wilhelm Rausch (Hg.), Stadt und Salz, Linz 1988, S. 89–165, hier: S. 101.

102 Ebd., S. 98 f.

103 Imma Waid, Schwaigen im Gebiet von Mariazell, in: Blätter für Heimatkunde 45, 1971, S. 97–99, hier: S. 97.

104 Josef Widemann, Die Traditionen des Hochstifts Regensburg und des Klosters St. Emmeram (Quellen und Erörterungen 8), München 1943 (Neudruck: Aalen 1969), S. 191, Nr. 211: „[…] *Salinam bauvariensem, quam vulgo comprovinciales Hal solent nuncupare* […]".

105 OÖUB I, S. 135, Nr. 25: *„ad bawaricum halle"*.

Unterstellung unter den Herzog von Bayern angedeutet wird. Ab der Mitte des 12. Jahrhunderts bürgerte sich zeitweise der Begriff „*Paierhalle*“ („*paier*“ = bayerisch) ein.[106]

5. Krise und aufkommende Konkurrenz

Vor dem Hintergrund einer immer größer werdenden Nachfrage nach Salz dürfte die Reichenhaller Saline ab der Mitte des 12. Jahrhunderts an der Kapazitätsgrenze gearbeitet haben.[107] Wenig später, ab den 1160er Jahren, stand die Salinenstadt im Zusammenhang mit dem „Alexandrinischen Schisma“ im Brennpunkt schwerer Auseinandersetzungen zwischen Kaiser Friedrich I. (reg. 1152/55–1190) und dem Salzburger Erzbischof Konrad II. von Babenberg (reg. 1164–1168). Einem Hoftag Friedrich Barbarossas in Laufen folgte 1169 die Zerstörung Salzburgs durch kaiserliche Parteigänger, die Grafen von Plain. Nur zwei Jahre später ging Reichenhall, wo der Salzburger Erzbischof maßgeblichen Einfluss ausübte, in Flammen auf, wobei die schriftlichen Quellen von einer fast vollständigen Zerstörung sprechen.[108] Ob ein Versehen vorlag oder eine Brandstiftung ebenfalls kaiserlicher Adlaten die Ursache war, ist ungewiss. Einiges spricht jedenfalls dafür, dass die Grafen von Peilstein ihre Hände dabei im Spiel hatten. Mit der Brandzerstörung der Stadt 1171 erfolgte die Vernichtung der inmitten des Ortes gelegenen Saline, was einen längeren Produktionsausfall zur Folge gehabt haben muss. Allein der Wegfall der für die Verfeuerung der Sole notwendigen Brennholzlager bedeutete für die Salzerzeugung vorerst ein abruptes Ende. Unbeeindruckt davon bewegte sich freilich die Nachfrage nach Salz weiterhin auf hohem Niveau.

Der auf diese Weise entstandene Nachfrageüberhang führte zu einer fieberhaften Suche nach neuen Salzlagerstätten nicht nur in der Umgebung Reichenhalls, sondern im gesamten Ostalpenraum. Man hat diese Suche bisher mit einem allgemein gesteigerten Salzbedarf zu erklären versucht und dabei die allgemeine politische und spezielle lokale Situation außer Acht gelassen.[109] Sowohl das krisenhafte Schisma als auch der 1171 erfolgte Stadtbrand verschärften den

---

106 MGH SS XI, Vita Gebhardi et successorum eius, S. 39.

107 Heinrich Wanderwitz, Zur Technik der Reichenhaller Solegewinnung im 12. Jahrhundert, in: MGSL 123, 1983, S. 143–147.

108 Lang, Geschichte von Bad Reichenhall (wie Anm. 4), S. 130–138.

109 Otto Volk, Salzproduktion und Salzhandel mittelalterlicher Zisterzienserklöster, Sigmaringen 1984, S. 44.

ohnehin angespannten Salzmarkt und bildeten die unmittelbaren Auslöser für die Erkundung neuer Ressourcen.

Dabei wurde man unweit von Reichenhall, an der sogenannten Pfannhauswand bei Unken fündig, als man in den 80er Jahren des 12. Jahrhunderts ebenfalls eine Solequelle entdeckte, die für einige Jahrzehnte von der Abtei St. Peter ausgebeutet wurde.[110] Der im Gelände noch heute sichtbare Solebrunnen bestand – darauf deutet eine Beschreibung des im Jahre 1666 wiederentdeckten Brunnens hin – im Hochmittelalter aus einem rund 12 Meter tiefen mit Holz verzimmerten Schacht, wobei die Sole eine Grädigkeit von etwa 10 Prozent aufwies.[111] Lediglich eine in den Schriftquellen genannte Sudpfanne – möglicherweise an Stelle der später als „Pfannhausgüter“ bezeichneten Gehöfte – diente der Versiedung der Sole, was einen Hinweis auf die offenbar nur geringe Schüttung im Solebrunnen gibt. Obwohl der im Tal liegende Bereich des Flusses Saalach für einen Salinenstandort günstigere Bedingungen für die Produktion, die Brennstoffzufuhr und den Weitertransport des Salzes geboten hätte, verzichtete man auf eine Ableitung der Sole im natürlichen Gefälle dorthin, was möglicherweise mit den zu erwartenden Kosten für die Errichtung und den Unterhalt einer rund 500 Meter langen Rohrleitung zusammenhängt; vermutlich standen diese nicht im Verhältnis zum Ertrag. Obwohl weitere schriftliche Unterlagen nicht vorliegen, ist davon auszugehen, dass die im Jahre 1210 letztmalig erwähnte Saline schon im Verlaufe des 13. Jahrhunderts aufgelassen wurde. Die einstige Salzerzeugungsstätte lebte lediglich im Namen eines Weilers weiter.

Anhand des Unkener Beispiels lassen sich die Determinanten herausfiltern, die für den Erfolg eines Salinenstandortes ausschlaggebend waren: Sole, Holz und Transportmöglichkeiten.[112] Neben dem Vorhandensein menschlicher Arbeitskraft bildeten im Hochmittelalter der Rohstoff Sole sowie der Energieträger Holz die Hauptkriterien für die Erzeugung von Salz aus Quellsole. An Holz bestand in Unken, dessen umliegende Waldungen für die Saline in Reichenhall gewidmet waren und dessen Siedlungsraum durch Schlägerungen für die dortige Saline überhaupt erst entstanden war, kein Mangel. Die Nähe zum Fluss

110 Erwähnung im Jahre 1186 (SUB II, S. 607, Nr. 445); erneut wird diese Pfanne 1210 erwähnt (SUB III, S. 131, Nr. 631).

111 Lugg Hohenwarter, Zeitspuren aus dem Unteren Saalachtal am Beispiel von Schloss Oberrain. Von der St. Petrischen Grundherrschaft auf den Pfannhausgütern und dem Seidelgut/Oberrain bis zur heutigen Nutzungsform an der Schwelle zum 21. Jahrhundert (Kniepaß-Schriften 26), Unken 2013, S. 16–19.

112 Vgl. Fritz Koller, Hallein im frühen und hohen Mittelalter, in: MGSL 116, 1976, S. 1–116, hier: S. 27.

Saalach stellte einen unübersehbaren Standortvorteil dar, da sich das Gewässer einerseits zur Holztrift eignete und auf diese Weise das erforderliche Brennmaterial am Oberlauf des Flusses geschlägert und hierher geschwemmt werden konnte. Andererseits diente der Fluss als Transportweg für das erzeugte Salz. Mit Spezialkähnen – überliefert ist beispielsweise der abgekommene Flurname „*Scheflent*"[113] (Schiffländе) – wurde das erzeugte Salz flussabwärts geführt. Allerdings bot sich die Saalach wohl schon damals nur bedingt für die Transportschifffahrt, was vermutlich nur mit Hilfe von Fahrrinnen und in Zeiten hohen Wasserstands möglich war. Auch wenn Arbeitskräfte in großer Anzahl sowie die für die Versorgung erforderliche Infrastruktur vorhanden gewesen wären, so fehlte offensichtlich ein wichtiges Kriterium, das den Unkener Salinenstandort letztlich scheitern ließ: eine in ausreichend großer Schüttung und hoher Grädigkeit vorhandene Solequelle.

Diese Beobachtung von im Vergleich zu Reichenhall stark determinierten Standorten trifft auch auf die oben geschilderten älteren Salzerzeugungsorte im Ostalpenraum zu, die sämtlich auf natürliche Solequellen rekurrierten. In diesem Reigen bildete die Reichenhaller Saline eine Ausnahme, verfügte sie doch sowohl über einen in reichem Maße vorhandenen Rohstoff als auch über die notwendige Brennstoffzufuhr. Sie lag an der Saalach, die flussaufwärts zur Holztrift aus einem großen geographischen Einzugsbereich diente und flussabwärts als Wasserweg für den Salztransport in Frage kam. Das optimale Zusammentreffen aus Rohstoff, Holz sowie dem Standort an einem schiff- und triftbaren Fluss konnte zur erfolgreichen Etablierung einer Saline führen. Abhängig war die Größe einer Saline freilich auch vom menschlichen Arbeitspotenzial, womit die Schaffung der erforderlichen Infrastruktur einhergehen musste. Die Intensität dieses Prozesses war letztlich entscheidend dafür, ob es bei einer weilerartigen Bebauung blieb oder ob sich an der Saline ein Dorf, ein Markt oder schließlich eine Stadt entwickeln konnte.

Solange die Salzproduktion auf der Grundlage natürlicher Solequellen erfolgte, wie bis weit ins Hochmittelalter üblich, hatte die Reichenhaller Saline keine Konkurrenz zu gewärtigen. Sollte sich hingegen die Gewinnung des Rohstoffs verändern und sollten sich die oben beschriebenen, unbedingt erforderlichen Rahmenbedingungen künstlich generieren lassen, so war der Weg frei für die Schaffung neuer Salinenstandorte, die man bis dahin nicht ins Kalkül gezogen hatte.

113 Lang, St. Zeno in Reichenhall (wie Anm. 58), S. 559.

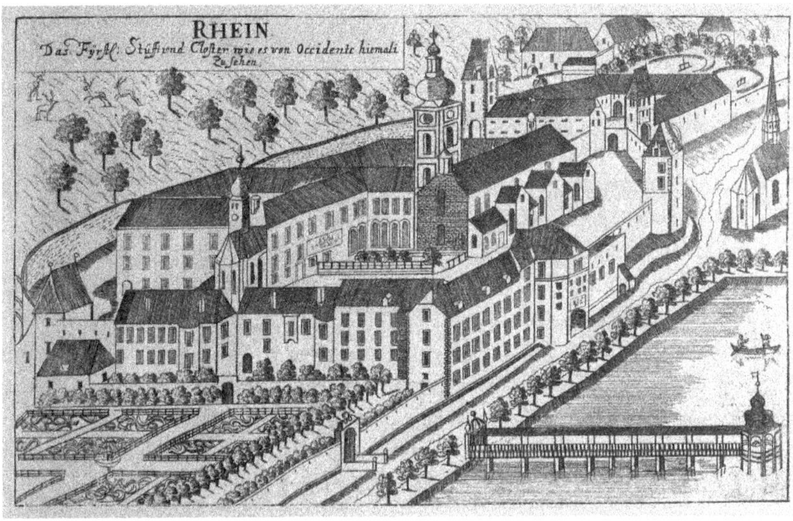

Abb. 45: *Aus dem Zisterzienserkloster Rein kamen die entscheidenden Impulse für die neue bergmännische Salzgewinnungstechnik des Laugwerkverfahrens im Ostalpenraum. Kupferstich von Georg Mattaeus Vischer, 1681. Bildnachweis: Wikimedia Commons (gemeinfrei).*

Die wohl wichtigste Rolle bei dieser grundlegenden Veränderung in der salinarischen Bergwerkstechnik kam dem Orden der Zisterzienser zu, wobei für die Entwicklung im Ostalpenraum von der Zisterze Rein (bei Graz) entscheidende Impulse ausgingen. Zunächst nur auf den lokalen Bedarf beschränkt, erhielten die Mönche im Jahre 1147 aus der Hand des steiermärkischen Markgrafen Otakar III. († 1164) zwei Sudpfannen „*in Enstal apud Mahorn*",[114] im Gebiet um Altaussee, im späteren Salzkammergut gelegen. Ausgangspunkt waren auch hier Solequellen.[115] Die räumliche Diskrepanz zwischen den Solequellen östlich des Dietrichkogels und den Pfannenstandorten am Augstbach deutet darauf hin, dass die Sole vermutlich schon damals mittels Rohrleitungen zu den eigentlichen Salzverarbeitungsstätten geführt wurde. Daneben scheinen die Zisterzienser in dem Gebiet aber auch – erstmals im Ostalpenraum – noch zum Ende des 12. Jahrhunderts Versuche einer 1211 erstmals urkundlich belegbaren bergmännischen

114 Rudolf Werner Soukup, Chemie in Österreich. Bergbau, Alchemie und frühe Chemie von den Anfängen bis zum Ende des 18. Jahrhunderts, Wien 2007, S. 73.

115 Stadler, Salzerzeugung, Salinenorte und Salztransport (wie Anm. 101), S. 105–116.

Erschließung von Salzvorkommen unternommen zu haben.[116] Dabei griff man auf die Bergwerkskenntnisse des gut vernetzten Ordens zurück, zumal die Äbte sämtlicher Zisterzienserklöster einmal jährlich in Cîteaux zusammen kamen und dort auch Erfahrungen austauschten.[117] Immerhin verfügten die meisten Zisterzen über Anteile an den lothringischen Salinen im Tal der Seille.

Dieses bergmännisch gewonnene Salz basierte auf der Technik des sogenannten Laugwerkverfahrens, wobei Süßwasser in künstlich geschlagene Kavernen des salzführenden Haselgebirges geleitet wird, sich dort mit Salz anreichert und als Sole wieder herausgepumpt werden kann. Auf diese Weise lassen sich sowohl Schüttung als auch Grädigkeit frei bestimmen, wogegen man bei der älteren Salzgewinnung aus natürlichen Solequellen auf die Launen der Natur angewiesen war. Diese revolutionäre technische Innovation, verbunden mit der Erkenntnis, mittels Rohrleitungen den Rohstoff Sole über beliebig lange Distanzen vom Bergwerk zum idealtypischen Salinenstandort verbringen zu können, musste zwangsläufig Auswirkungen nicht nur auf die Möglichkeit neuer Salinen haben, sondern auch auf die damit verbundene urbane Entwicklung.

Während im Falle von Altaussee dank der neuen Technik zwar sowohl der Rohstoff als auch der Energieträger umfangreich zur Verfügung standen, verhinderte doch das Fehlen eines nahen schiffbaren Flusses die Entstehung einer größeren Saline. Stattdessen war man beim Transport des Salzes überwiegend auf die kleinteilige Säumerei angewiesen.[118] Erfolgreicher gestalteten sich dagegen Versuche auf Salzbergbau in der weiteren Umgebung Reichenhalls: Um das Jahr 1190 begann die Propstei Berchtesgaden, selbst nicht unwesentlich an der Saline von Reichenhall beteiligt, unweit ihrer Stiftsbauten – am Berchtesgadener Gollenbach, später am Gutratsberg, dem sogenannten Tuval – mit der Herstellung bergmännisch gewonnenen Salzes.[119] Dem Beispiel folgte wenig später auch die Abtei St. Peter im Bereich des Dürrnbergs,[120] wozu sich schließlich auch der

---

116 Stadler, Salzerzeugung, Salinenorte und Salztransport (wie Anm. 101), S. 105–116; Klein, Geschichte der Salinen Hallein und Reichenhall (wie Anm. 4), S. 401 f., Anm. 78; Volk, Zisterzienserklöster (wie Anm. 109), S. 34–36.

117 Werner Rösener, Das Zisterzienserstift Salem und das Erzstift Salzburg, in: Peter F. Kramml (Hg.), Stadt, Land und Kirche. Salzburg im Mittelalter und in der Neuzeit (= Festschrift für Heinz Dopsch), Salzburg 2012, S. 67–84.

118 Klein, Geschichte der Salinen Hallein und Reichenhall (wie Anm. 4), S. 402.

119 Koller, Hallein (wie Anm. 112), S. 38; Ders., Das Salzwesen, in: Walter Brugger / Heinz Dopsch / Peter F. Kramml (Hg.), Geschichte von Berchtesgaden, Bd. I, Berchtesgaden 1991, S. 737–842, hier: S. 741 f., 748 f.

120 Koller, Hallein (wie Anm. 112), S. 39 f.

Salzburger Erzbischof gesellte. Dass der Metropolit 1201 die Zisterze Salem und 1207 die Zisterzienser von Raitenhaslach als Mitsieder an der dortigen Saline verpflichtete, macht den Innovationsschub deutlich, der von dem Orden ausging.

Das bergbauliche Engagement der Klöster Berchtesgaden und St. Peter sowie der Salzburger Kirche löste zwischen den Bürgern von Reichenhall und dem Metropoliten eine schwere politische Krise aus, deren detaillierte Darstellung hier nicht Gegenstand sein kann.[121] Am Ende einer an Dramatik reichen Entwicklung stand 1196 die vollständige Zerstörung Reichenhalls durch den Erzbischof Adalbert III. (reg. 1168–1177/1183–1200), wie sie in den Annalen von Admont und Kremsmünster zum Ausdruck kommt: Demnach wurde die gesamte Stadt („*civitas*") mit den Salinenanlagen, Häusern und Kirchen durch ein Gefecht und eine Feuersbrunst vernichtet. Der Umstand, dass man in Gesprächen zwischen dem Salzburger Metropoliten und dem Bayernherzog nach mehr als zwei Jahrzehnten, im Jahre 1218, eine Verlegung der Stadt von ihrem ursprünglichen Standort ernsthaft in Erwägung zog,[122] unterstreicht eindrucksvoll die Existenzkrise, in welcher sich Reichenhall seit dem Jahre 1196 befand.

Die brutale Vernichtung der Salinenstadt durch den Salzburger Metropoliten erinnert an das bedingungslose Vorgehen Kaiser Friedrichs I. gegen missliebige Städte wenige Jahre zuvor – genannt seien die vollständigen Zerstörungen beziehungsweise Schleifungen von Tortona (1155), Crema (1160), Mailand oder Mainz (beide 1163) –, das eine fatale Signalwirkung auf die Fürsten des Reiches erzeugt haben mag. So etwa ließ Herzog Heinrich der Löwe (reg. 1142–1180) im Jahre 1189 die durch den Handel mit Lüneburger Salz zu Wohlstand gelangte Stadt („*oppidum*") Bardowick nach einer längeren Belagerung mit Ausnahme der Kirchen und Kapellen dem Erdboden gleichmachen.[123] Flankiert wird das Bild der offenbar vollständig geschleiften Stadt Reichenhall durch die in den letzten Jahren durchgeführten archäologischen Kampagnen: Offenbar wurden die Mauern abgebrochen, sodass sich das Bett der Saalach zumindest vorübergehend wieder innerhalb der einstigen Stadtmauern ausbreiten konnte. Unterhalb einer mächtigen Schwemmsandschicht fanden sich in einem ins 12. Jahrhundert zu datierenden Planum mehrere menschliche Skelette, die Spuren von

121 Lang, Geschichte von Bad Reichenhall (wie Anm. 4), S. 147–166.

122 SUB III, S. 253, Nr. 732: „*Antiqua civitas in suo loco permaneat numquam transponenda*".

123 Walter Jarecki, Die Verdener Bischofsgeschichte des Elard von der Hude, in: Andreas Bihrer / Elisabeth Stein (Hg.), Nova de veteribus. Mittel- und neulateinische Studien für Paul Gerhard Schmidt, München/Leipzig 2004, S. 927–961, hier: S. 944.

Gewaltanwendung aufwiesen.[124] Zudem lässt sich erkennen, dass man bei dem nach 1218 anzunehmenden Wiederaufbau Reichenhalls den älteren Stadtmauerverlauf in Teilbereichen verließ, um die neue Stadt in verkleinertem Umfang wieder zu errichten, was einen Hinweis auf die reduzierte Einwohnerzahl liefert.

6. Die Dynamik des Marktes

Die Etablierung einer neuen bergmännischen Salzgewinnungstechnik hatte, fast zeitgleich, ihren Schlusspunkt in der Zerstörung Reichenhalls gefunden, das symbolhaft für die ältere Form der Erzeugung aus Solequellen stand. An eine Reichenhaller Salzproduktion im großen Stil war über Jahre, vermutlich sogar Jahrzehnte nicht zu denken. Der bis dahin vom Monopolisten beherrschte Salzmarkt war nun einer freien Dynamik ausgesetzt, die sich auf lange Sicht auch städtebaulich niederschlagen musste. Am Fuße des Dürrnbergs – rund 15 Kilometer Luftlinie von Reichenhall entfernt –, an der stets wasserreichen und durchgängig schiffbaren Salzach, erwuchs bald nach 1191 der Idealtyp einer Saline, die sich in vielerlei Hinsicht an dem Vorbild von Reichenhall orientierte. Die neue Siedlung war in erster Linie von den salinarischen Erfordernissen geprägt, entstand als Salinen-Großbetrieb gewissermaßen auf der „grünen Wiese" und orientierte sich weder an möglichen Verkehrskreuzungspunkten noch an landwirtschaftlichen Anbauflächen. Möglicherweise waren es sogar jene in der Salzherstellung kundigen Leute aus Reichenhall gewesen, die man für den Aufbau der neuen Produktionsstätte abgeworben hatte.[125] Eine frühe Quelle nennt sie die *„neue Saline bei Salzburg"*.[126] Der Ortsname lautete vorerst auf Mühlbach, was eine ursprünglich wohl weilerartige Siedlungsstruktur an einem für Mühlen geeigneten Bachlauf, dem heutigen Kotbach, andeutet.[127] Nachdem das alte Reichenhall – damals Hall genannt – längst zum Inbegriff einer Salinenstätte geworden war, bezeichnete man den neuen Ort ab den 1210er Jahren gleichermaßen, wählte aber die Deminutivform und nannte ihn fortan „Hallein". Die Standortkriterien dieses „kleinen Hall" waren freilich prädestiniert, um den bisherigen Monopolisten, Reichenhall, in den Schatten zu stellen: Der enorme Salzstock

124 Thomas Beckh, Turm und Tote „Im Angerl". Mittelalterlicher Turmbau zu Bad Reichenhall, in: Das Archäologische Jahr in Bayern 2017, S. 103–105.

125 Vgl. Fritz Koller, St. Peter als Salzproduzent, in: Salzburger Landesregierung (Hg.), St. Peter in Salzburg. Das älteste Kloster im deutschen Sprachraum, Salzburg 1982, S. 104–108, hier: S. 105 f.

126 MGH SS XVII, S. 525: *„nova salina apud Salzburgam"*.

127 Koller, Hallein (wie Anm. 112), S. 32.

Abb. 46: *Die Salinenstadt Hallein an der Salzach im 17. Jahrhundert. Bildnachweis: Franz Dückher von Haßlau, Saltzburgische Chronica, Salzburg 1666, S. 13.*

mit der, dank Laugwerkverfahren, jederzeit in Schüttung und Grädigkeit steuerbaren, künstlich hergestellten Sole bildete einen klaren Wettbewerbsvorteil. Mittels Rohrleitungstechnik war es möglich, die Sole vom Bergwerk, dem Ort der Rohstoffgewinnung, zu einem Ort mit optimalen logistischen Voraussetzungen zur Verarbeitung zu führen. Die Lage der Saline unmittelbar an der Salzach wies sowohl für die Frachtschifffahrt als auch für die Holztrift herausragende Bedingungen auf, zumal auch das Holzeinzugsgebiet ein größeres war als jenes für Reichenhall.[128]

Herbert Klein (1900–1972) ging 1952 noch von einer überlegenen Marketingpolitik – dem Instrument der Valorisation – des Metropoliten aus, womit die neue Saline Hallein das ältere Reichenhall wirtschaftlich überflügelt habe und zum bedeutendsten Salzproduzenten im Ostalpenraum während des 13. Jahrhunderts aufgestiegen sei.[129] Doch angesichts der inzwischen gewonnenen Kenntnisse bedurfte es offensichtlich keines besonderen Marketings, um innerhalb kürzester Zeit die Marktführerschaft zu erringen, da die Chancen für Reichenhall, im Bereich des Salzwesens wieder konkurrenzfähig zu werden, zumindest mittelfristig gering waren. Nicht nur dass die Stadt ruiniert war und an ihrem zielgerichteten Aufbau gehindert wurde, auch der Wasserweg auf Saalach und Salzach wurde über große Strecken vom Erzbischof kontrolliert, und die wichtige Verladestation Laufen geriet, ebenso wie die Verfügungsgewalt über die Salzschiffe, zusehends

128 Koller, Salzachschiffahrt (wie Anm. 40), S. 106.
129 Klein, Geschichte der Salinen Hallein und Reichenhall (wie Anm. 4), S. 385 f.

in erzbischöfliche Hand. Damit wurde das Reichenhaller Salz von den einstigen Hauptabsatzgebieten an der Donau ausgesperrt.[130]

Dagegen war die dynamische Entwicklung Halleins in diesen ersten Jahrzehnten, ausgehend von den dortigen Pfannhäusern beiderseits des Mühlbach, enorm. Im Jahre 1206 wird die von der Abtei St. Peter in Salzburg errichtete und pastorierte Pfarrkirche St. Antonius urkundlich erwähnt.[131] Fritz Koller nimmt im Bereich des Kornsteinplatzes noch für die erste Hälfte des 13. Jahrhunderts eine erste Ortsbefestigung mit Wall und Graben an,[132] ehe sich die Siedlung weiter gegen Norden ausdehnte. Vor dem Jahre 1232 spricht der Salzburger Erzbischof die Gesamtheit der Bewohner von Hallein – gemeinsam mit jenen von Salzburg und Laufen – als „*cives*" an,[133] wobei schon zuvor einzelne als Bürger belegt sind. Im Vergleich zu dem fast einhundert Jahre zuvor erfolgten Stadtwerdungsprozess in Reichenhall fällt auf, dass im Falle von Hallein in den Schriftquellen Begriffe wie „*oppidum*", „*urbs*" oder „*civitas*" bis zur erstmaligen Überlieferung eines Stadtsiegels („*S. CIVIUM DE SA*[LINA]") 1275[134] fehlen. Im Zusammenhang mit der Nennung von Stadtmauern findet sich der Begriff „*civitas*" erstmals 1293,[135] im Zusammenhang mit einem Richter jener Begriff der „*stat*" 1304,[136] nachdem zuvor der Name „*Salina*" zumindest gleichwertig neben „Hallein" rangiert hatte. Sowohl die Eigen- als auch die Fremdwahrnehmung des Ortes reduzierte sich vorerst in erster Linie auf die Salzproduktionsstätte, wogegen das urbane Gepräge offenbar sekundär erschien.

Die Technik des Laugwerkverfahrens und die zielgerichtete Anwendung von Soleleitungen ermöglichten ab der Wende zum 13. Jahrhundert die Etablierung weiterer Salinenstandorte, wobei die wenigsten eine überregionale Bedeutung erlangten. Zu den genannten Standortdeterminanten gesellte sich das Problem des im Bergbau erforderlichen Arbeiterpotenzials, was entsprechendes Kapital

---

130 Koller, Salzachschiffahrt (wie Anm. 40), S. 109.

131 Koller, Hallein (wie Anm. 112), S. 62.

132 Ebd.

133 SUB III, S. 235, Nr. 719.

134 Salzburger Urkundenbuch, Bd. IV: Ausgewählte Urkunden 1247–1343, bearb. v. Franz Martin, Salzburg 1933, S. 86, Nr. 82; Koller, Hallein (wie Anm. 112), S. 71.

135 Archiv der Erzabtei St. Peter, Archiv der Erzabtei (1005–1981), Urk. Nr. 225 – 1293 IV 24, in: monasterium.net, URL: https://www.monasterium.net/mom/AT-StiASP/Urkunden/Urk_Nr_225-1293_IV_24/charter?q=Hallein (letzter Zugriff: 25.10.2020); Koller, Hallein (wie Anm. 112), S. 109 f.

136 Haus-, Hof- und Staatsarchiv Salzburg, Erzstift (798–1806), AUR 1304 VIII 24, in: monasterium.net, URL: https://www.monasterium.net/mom/AT-HHStA/SbgE/AUR_1304_VIII_24/charter?q=Hallein (letzter Zugriff: 25.10.2020).

voraussetzte. Die von der Propstei Berchtesgaden unterhaltene Saline in Schellenberg blieb in ihren Erträgen deutlich hinter dem Halleiner Standort zurück; über die Erlangung des Marktrechts ging die Siedlungsqualität nicht hinaus. Ähnlich verhielt es sich mit den Orten Aussee und Hallstatt, die zwar im Verbund wichtige Salzproduzenten darstellten, denen aber der Aufstieg zu Städten verwehrt blieb.

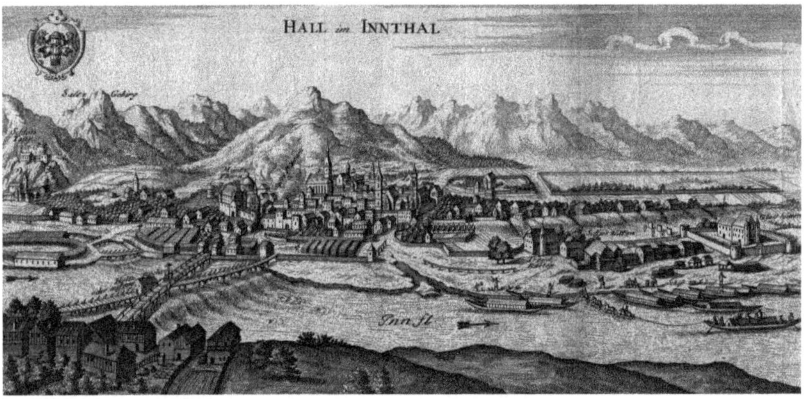

Abb. 47: *Die Salinenstadt Hall im Inntal. Kupferstich von G. Bodenehr, Augsburg um 1720. Bildnachweis: Wikimedia Commons (gemeinfrei).*

Während die Halleiner Saline in den ersten Jahrzehnten des 13. Jahrhunderts den überregionalen Salzmarkt fast nach Belieben dominierte, befand sich eine andere Saline, jene im Inntal, gerade in einem Anfangsstadium. Im Jahre 1232 als eine dem Grafen Albert III. von Tirol gehörige Saline „*iuxta Thaur*" erstmals urkundlich genannt,[137] dürfte die Anlage zunächst noch im Halltal nahe einer Solequelle platziert gewesen sein.[138] Ab der Jahrhundertmitte kam es wegen der besseren Holzbringungs- und Transportmöglichkeiten zu einer Verlegung der Betriebseinheiten ans Ufer des Inn. Zur Steigerung der Rohstoffgewinnung begann man ab den 70er Jahren des 13. Jahrhunderts mit der bergmännischen Aufschließung des dortigen Salzgebirges, wobei die aus Aussee und Hallein bekannte Laugwerktechnik zur Anwendung kam. Ähnlich wie im Falle

137 Rudolf Palme, Rechtliche Probleme bei der mittelalterlichen Salzgewinnung in Österreich, in: Wilhelm Rausch (Hg.), Stadt und Salz, Linz 1988, S. 69–87, hier: S. 69.

138 Rudolf Palme, Geschichte des Salzbergbaues und der Saline von Hall, in: Tiroler Landesmuseum (Hg.), Silber, Erz und Weißes Gold. Bergbau in Tirol, Innsbruck 1990, S. 206–221, hier: S. 206.

Dürrnberg/Hallein lagen die Orte der Gewinnung und der Verarbeitung nun voneinander getrennt: Vom Halltal führte man eine rund zehn Kilometer lange Soleleitung bis zu einer hochwassergeschützten Örtlichkeit, die 1256 als „*Halle*" urkundlich auftaucht.[139] Analog zu Hallein spielten bei der Wahl des Salinenstandortes verkehrs- oder handelspolitische Überlegungen keine Rolle,[140] was darauf hindeutet, dass die sich an dieser Stelle entwickelnde Siedlung vordergründig salinarische Zwecke verfolgte. Die Errichtung eines quer zum Innstrom verlaufenden Triftrechens begünstigte den Bau einer Hafenanlage.

Auch die städtebauliche Dynamik Halls ähnelt jener von Hallein, wobei in Hall zum salinarischen Betrieb – der insgesamt günstigen Verkehrslage geschuldet – in verstärktem Maße gewerbliche Faktoren hinzukamen. Ein Marktrecht ist um das Jahr 1286/88 indirekt erwähnt,[141] könnte aber älter und möglicherweise bald nach der Mitte des 13. Jahrhunderts entstanden sein.[142] So ist es zu erklären, dass sich eine Ummauerung des Marktfleckens für die Zeit um 1280/85 aus schriftlichen und archäologischen Hinweisen ergibt. 1303 erhielt Hall, das offenbar längst die äußeren Merkmale einer Stadt besaß, die Stadtrechte aus der Hand des damals regierenden Landesfürsten, Herzog Otto von Kärnten-Tirol, förmlich verliehen, wobei die für Innsbruck geltenden Bestimmungen zu Grunde gelegt wurden.[143] In der darüber ausgestellten Urkunde findet sich interessanterweise die namentliche Unterscheidung zwischen dem befestigten Ort („*oppidum*"), welcher zur Stadt („*civitas*") erhoben werden sollte. Damit verbunden war das Niederlagsrecht für Salz sowie das Recht zur Errichtung einer neuen Innbrücke, womit sich für die noch junge Stadt – zum Nachteil Innsbrucks – neue handelspolitische Möglichkeiten, insbesondere für den Italien- und Venedig-Handel, eröffneten.

Die Entwicklung von der Salinenstätte zur voll ausgebildeten Stadt dauerte im Falle Halleins rund drei Generationen und erfolgte in Hall im Inntal in einem sogar noch kürzeren Zeitraum. Zurückgreifen konnten beide Orte auf die Erfahrungen

---

139 Franz-Heinz Hye, Hall in Tirol. Gründung und Werdegang einer Salzstadt, in: Rausch, Stadt und Salz (wie Anm. 137), S. 233–245, hier: S. 236 f.

140 Ebd., S. 234 f.

141 Ebd., S. 238 f.

142 Heinz Moser, Die städtebauliche Entwicklung von Hall in Tirol, in: Alexander Zanesco / Romedio Schmitz-Esser (Hg.), Forum Hall in Tirol. Neues zur Geschichte der Stadt, Bd. 1, Hall in Tirol 2006, S. 44–61, hier: S. 44, verweist auf Urkundeninhalte aus der zweiten Hälfte des 14. Jahrhunderts, wonach Hall bereits um die Mitte des 13. Jahrhunderts das Marktrecht erhalten habe.

143 Andreas Faistenberger, Das Haller Stadtrecht von 1303, in: Zanesco / Schmitz-Esser, Forum Hall in Tirol (wie Anm. 142), S. 34–43.

beim „Prototypen" der Salinenstadt des Ostalpenraums, Reichenhall, wo sich die gleiche Entwicklung über eine Spanne von mehreren Jahrhunderten erstreckt hatte. Diese Vorbildwirkung fand ihren augenscheinlichsten Niederschlag in der Namensgebung der neuen Orte, die mit dem Wortstamm „Hall" gebildet wurden. Jürgen Udolph konnte nachweisen, dass es im Alpenraum keine älteren Belegstellen für Hall-Namen gibt als jene Reichenhalls, denn die meisten anderen Salzproduzenten mit dem Begriff „Hall" im Ortsnamen kommen zu einem Zeitpunkt auf, als Reichenhall im Verlaufe des 12. und 13. Jahrhunderts massive Konkurrenz erfährt.[144] Bis zum Monopolbruch durch Hallein um 1190 hatte Reichenhall über mehrere Jahrhunderte eine unangefochtene Position inne, sodass Reichenhall, damals nur „Hall" genannt, *pars pro toto* als Muster eines Salzerzeugungsortes galt (sogenanntes Appellativum). Udolph spricht von der Verselbstständigung des Namens „Hall" im Sinne von Saline. (Man denke beispielsweise an den belgischen Kurort Spa, der bei englischen Kurgästen derart bekannt war, dass der Name in den allgemeinen Wortschatz einging, um damit jegliche Badeorte zu bezeichnen.)

7. Ausblick

Mit der prädestinierten Lage an einem ganzjährig schiffbaren und ein großes Waldgebiet erschließenden Fluss erlangte der Aspekt der Logistik im Verlaufe des 13. Jahrhunderts zentrale Bedeutung für den überregionalen Erfolg einer Saline und deren urbane Entwicklung. Die perfekte Kombination sämtlicher erforderlicher salinarischer Faktoren blieb so auf einige mögliche Standorte beschränkt. Erst aufgrund der technischen Innovation pumpenbetriebener Pipelines überwand man ab dem frühen 17. Jahrhundert die bis dahin erzwungene geographische Nähe von Rohstoffquelle und Verarbeitungsbetrieb, wodurch Salinen fernab der Salzgebirge – wie etwa in Traunstein (1619) und Rosenheim (1810) – überhaupt entstehen konnten.

Dass Reichenhall über das Jahr seiner Schleifung 1196 hinaus als Salinenstadt bestehen blieb, muss in diesem Zusammenhang als durchaus ungewöhnlich betrachtet werden. Denn die schon zuvor für die Schifffahrt nur bedingt geeignete Saalach kam aus politischen Gründen als Transportweg für das Salz ab dem Jahre 1218 nicht mehr in Betracht. Vielmehr war man ausschließlich auf den Landweg angewiesen. Im Hinblick auf den Rohstoff konnte die Reichenhaller Saline auch weiterhin nur auf die den Launen der Natur unterworfenen Solequellen zurückgreifen und hatte keine Möglichkeit einer bergmännischen

144 Udolph, Hall in Ortsnamen (wie Anm. 14), S. 65 f.

Gewinnung. Schließlich erwies sich auch der Umfang der zur Saline gewidmeten Wälder als deutlich geringer als jener am Oberlauf von Salzach und Inn, zumal die überwiegende Lage dieser Reichenhaller Salinenwaldungen auf dem Territorium des Erzstifts Salzburg immer wieder für politische Spannungen sorgte.

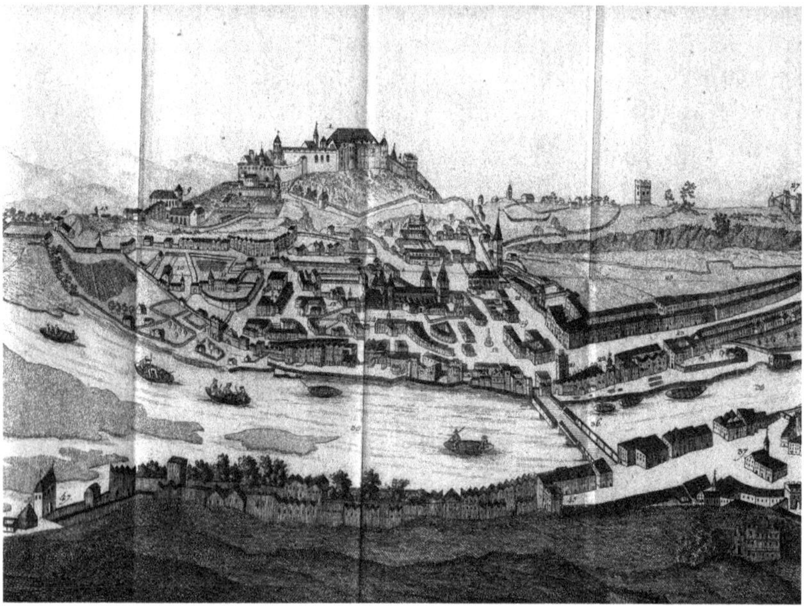

Abb. 48: *Salzburg, das seinen Namen dem „Weißen Gold" aus Reichenhall verdankt, mit den auf der Salzach verkehrenden Salzschiffen aus Hallein, 18. Jahrhundert. Bildnachweis: Lorenz Hübner, Beschreibung der hochfürstlich-erzbischöflichen Haupt- und Residenzstadt Salzburg und ihrer Gegenden, 2. Bd., Salzburg 1793, Beilage.*

Wäre das 1196 zerstörte Reichenhall dauerhaft unter salzburgischen Einfluss geraten, so wäre eine erneute Inbetriebnahme der Saline im Hinblick auf das aufblühende Hallein aus protektionistischen Gründen wohl unterblieben. So aber stand hinter der Wiederaufnahme der Reichenhaller Salzproduktion und dem Wiederaufbau der Stadt in erster Linie der Bayernherzog, der in der wichtigen Frage der Versorgung seines Herzogtums mit Salz nicht in Abhängigkeiten geraten wollte.[145] Dass Salzburg, Tirol und Bayern an eigenen Standorten festhielten

145 Lang, Geschichte von Bad Reichenhall (wie Anm. 4), S. 162.

und diese nach Kräften förderten, wirft ein Bild auf die in jener Zeit sich im vollen Gange befindliche Territorialisierung, denn während sich Tirol unter Meinhard II. als „Land" 1282 formal vom Mutterland Bayern löste, erfolgte dieser Schritt in Salzburg vor der Mitte des 14. Jahrhunderts. Mit seinem Engagement an der Saline Altaussee im Jahre 1211, als er die Zisterze Rein als Betreiberin verdrängte,[146] deutete der österreichische Herzog Leopold VI. die Wichtigkeit landeseigener Salzproduktionsstätten an. Insofern gingen der Landwerdung im Ostalpenraum die umwälzenden Prozesse im Salzwesen – neue Techniken der Gewinnung sowie der Bruch des Reichenhaller Monopols – zeitlich nur geringfügig voraus. Zur Bildung landesfürstlicher Salzmonopole war es nur mehr ein kleiner Schritt.[147]

Die Saline Hall in Tirol stellte 1967, jene in Hallein 1989 ihren Betrieb ein. Verblieben ist – und dies erscheint vor dem Hintergrund der oben geschilderten Widrigkeiten sogar verblüffend – die Reichenhaller Saline, die seit ihrer Ersterwähnung im 8. Jahrhundert bis zum heutigen Tag immer noch am selben Standort Salz aus natürlicher Quellsole erzeugt. Dieses Kontinuum führte unter anderem dazu, dass das allgemein geläufige „Bad Reichenhaller" seit Langem als bekannteste Salzmarke Deutschlands gilt und aufgrund der globalen Distribution inzwischen auch weltweit über einen hohen Markenbekanntheitsgrad verfügt.[148]

146 Stadler, Salzerzeugung, Salinenorte und Salztransport (wie Anm. 101), S. 110.

147 Jean-Claude Hocquet, Weißes Gold. Das Salz und die Macht in Europa von 800 bis 1800, Stuttgart 1993, S. 111–122.

148 Vgl. u.a.: Südwestdeutsche Salzwerke AG (Hg.), Kompetenz in Salz. Unternehmensbroschüre, Heilbronn 2019, S. 20.

Senta Herkle

„... ist wohl zu bedenken, daß Ulm ringsum mit Städten und Plätzen umgeben ist, welche einen ausgebreiteten Handel treiben, und ihre Fuhrwerk haben.“ Konkurrierende Interessen im Handel süddeutscher Reichsstädte im 18. Jahrhundert

Abstract: Based on the premise that cross-regional trade relations had to be negotiated locally, this essay asks how (transit) trade in the eighteenth century was organised on site, especially in the imperial city of Ulm. The written records of the tax office illustrate adaptions of transit trade organisation to different political, economic and technical framework requirements. Furthermore, the archival tradition allows for a reconstruction of the interactions between Ulm and other (imperial city) trade locations, whose forms of organisation also varied depending on the different locations. Those differences become noticeable for instance in how the city dealt with coachmen or in the characteristics of urban facilities.

Im Zentrum verschiedener bedeutender Handelswege gelegen, kam Ulm spätestens seit dem späten Mittelalter große Bedeutung als Umschlagplatz insbesondere für Wein, Eisen und Salz zu.[1] Die Produktion und der Vertrieb von Barchent, einem Mischgewebe aus Baumwolle und Leinen – seit dem 17. Jahrhundert verstärkt auch Leinwand – machte die Reichsstadt zudem zu einem wichtigen Standort für den Export von Textilprodukten.[2] Besonders ausgeprägt waren die Handelsbeziehungen nach Venedig und Genua in Italien und nach Lyon in

---

1 Vgl. Hans Eugen Specker, Ulm. Stadtgeschichte, Ulm 1977, S. 56–62, 177. Ulm verfügte direkt nach Nürnberg über das zweitgrößte reichsstädtische Territorium. Dieses große Territorium war für Ulm durch Geld- und Naturalabgaben als Versorgungsquelle der Stadt wichtig; die Hälfte des Gesamtsteueraufkommens im 18. Jahrhundert machten durchgängig die Landsteuern aus. Vgl. Kurt Rothe, Das Finanzwesen der Reichsstadt Ulm im 18. Jahrhundert (Forschungen zur Geschichte der Stadt Ulm 21), Ulm 1991, S. 164; Reinhard Hildebrandt, Der Fernhandel als städtischer Wirtschaftsfaktor (1500–1650), in: Bernhard Kirchgässner / Hans-Peter Becht (Hg.), Stadt und Handel (Stadt in der Geschichte 22 = Veröffentlichungen des Südwestdeutschen Arbeitskreises für Stadtgeschichtsforschung 32), Sigmaringen 1995, S. 49–61, hier: S. 52.

2 Vgl. Senta Herkle, Reichsstädtisches Zunfthandwerk. Sozioökonomische Strukturen und kulturelle Praxis der Ulmer Weberzunft (1650–1800) (Forschungen zur

Frankreich, Ulmer Händler besuchten aber auch regelmäßig die Messen in Nördlingen und Frankfurt, standen in engen Handelsbeziehungen mit Augsburg und Nürnberg sowie mit den benachbarten Territorien Württemberg und Bayern.[3] Im 18. Jahrhundert waren sowohl die Textilproduktion, der Textilexport als auch der Speditionsverkehr rückläufig, wobei die größten bürgerlichen Vermögen der Reichsstadt in der Hand der Ulmer Speditions- und Leinwand-Händler lagen.[4]

Nach dem Dreißigjährigen Krieg lässt sich ein ständiger Anstieg von Personen- und Güterverkehr verzeichnen, der mit dem Ausbau von Infrastruktur und mit technischen Neuerungen im Transportwesen zusammenhing. Reisen wie auch der Güterverkehr wurden damit schneller, zuverlässiger, sicherer und letztlich auch kostengünstiger.[5] Ausgehend von der These, dass überregionale Handelsbeziehungen lokal ausgehandelt werden mussten, wird im Folgenden danach gefragt, wie Handel im 18. Jahrhundert vor Ort – und hier besonders in Ulm – organisiert war. Zentral war dabei die Rolle des Güterbestätters oder Gutfertigers, der alle ein- und abgehenden Waren in der Reichsstadt abfertigte, dem Steueramt unterstand und als Vermittler zwischen Händlern und Fuhrleuten auftrat. Dabei lässt sich nachzeichnen, wie sich die Organisationsformen des Handels über ein Jahrhundert hinweg an unterschiedliche politische,

---

Geschichte der Stadt Ulm 34), Stuttgart 2014. Zwar blieben die Bestrebungen Ulms insbesondere im 15. Jahrhundert, die Stadt als Messestandort zu etablieren, erfolglos, allerdings hatten sich besonders der Veitsmarkt, aber auch der Nikolausmarkt im 17. und vor allem 18. Jahrhundert zu Märkten mit überregionalem Austausch entwickelt. Vgl. Heinrich Steinmeyer, Die Entwicklung der Ulmer Sommermesse (des späteren Veitsmarktes) und ihre Einordnung in das süddeutsche Handelssystem bis zum Ende der Reichsstadtzeit, in: Vierteljahrschrift für Sozial- und Wirtschaftsgeschichte (künftig: VSWG) 77/3, 1990, S. 323–349.

3 Zur dadurch entstandenen Handelskorrespondenz zwischen Ulm, der Schweiz und Italien ist ein Projekt der Verfasserin in Kooperation mit dem Stadtarchiv Ulm in Arbeit.

4 Vgl. Rothe, Das Finanzwesen der Reichsstadt Ulm (wie Anm. 1), S. 375–379; Friedrich Nicolai, Ulm betreffend. Beschreibung von Ulm und dessen Gebiete …, in: Ders., Beschreibung einer Reise durch Deutschland und die Schweiz im Jahre 1781. Nebst Bemerkungen über Gelehrsamkeit, Industrie, Religion und Sitten. Neunter Band, Berlin/Stettin 1795, Beilage, S. 7, 20 f.

5 Vgl. Rolf Peter Stieferle, Transport und wirtschaftliche Entwicklung, in: Ders. (Hg.), Transportgeschichte (Der Europäische Sonderweg 1), Berlin 2008, S. 34; Wolfgang Behringer, Im Zeichen des Merkur. Reichspost und Kommunikationsrevolution in der Frühen Neuzeit (Veröffentlichungen des Max-Planck-Instituts für Geschichte 189), Göttingen 2003, hier bes.: S. 436–485.

wirtschaftliche und technische Rahmenbedingungen anpassten. Es zeigt sich, dass der Handel wesentlich von den jeweiligen Bedingungen abhängig war, die die lokale Organisation des Transportwesens prägten.

Verschiedentlich erforscht ist der (Fern-)Handel von einzelnen Reichsstädten – einen Schwerpunkt bilden dabei Kaufleute und hier besonders die Forschungen zu Augsburg.[6] Obwohl Ulm zweifellos zur Trias der großen Reichsstädte im Süden gehörte, ist diese Stadt bislang, sieht man von Ausnahmen ab,[7] kaum Gegenstand handelsgeschichtlicher Untersuchungen geworden. Die

---

6 Vgl. exemplarisch: Mark Häberlein, Die Fugger. Geschichte einer Augsburger Familie (1367–1650), Stuttgart 2006; Mark Häberlein / Christof Jeggle (Hg.), Praktiken des Handels. Geschäfte und soziale Beziehungen europäischer Kaufleute in Mittelalter und früher Neuzeit (Irseer Schriften NF 6), Konstanz 2010; Johannes Burkhardt (Hg.), Augsburger Handelshäuser im Wandel des historischen Urteils (Colloquia Augustana 3), Berlin 1996; Mark Häberlein, Brüder, Freunde und Betrüger. Soziale Beziehungen, Normen und Konflikte in der Augsburger Kaufmannschaft um die Mitte des 16. Jahrhunderts (Colloquia Augustana 9), Berlin 1998; Christiane Reves, Vom Pomeranzengängler zum Großhändler? Netzwerke und Migrationsverhalten der Brentano-Familien im 17. und 18. Jahrhundert (Studien zur Historischen Migrationsforschung 23), Paderborn u.a. 2012; Markus A. Denzel, Unternehmen, Handelshäuser und Wirtschaftsmigration im neuzeitlichen Alpenraum. Einführung, Forschungsaufriss und konzeptionelle Überlegungen, in: Marie-Claude Schöpfer / Markus Stoffel / Françoise Vannotti (Hg.), Unternehmen, Handelshäuser und Wirtschaftsmigration im neuzeitlichen Alpenraum (Veröffentlichungen des Forschungsinstituts zur Geschichte des Alpenraums 13), Visp 2014, S. 1–24; Hildebrandt, Der Fernhandel als städtischer Wirtschaftsfaktor (wie Anm. 1), S. 49–61; Michael Diefenbacher, Handel im Wandel: Die Handels- und Wirtschaftsmetropole Nürnberg in der frühen Neuzeit (1550–1630), in: Kirchgässner / Becht, Stadt und Handel (wie Anm. 1), S. 63–81.

7 Vgl. Eugen Nübling, Ulms Weinhandel im Mittelalter, Ulm 1893; Senta Herkle, Obrigkeitliche Strategien zur Förderung der Leinwandproduktion und des Leinwandhandels im deutschen Südwesten nach dem Dreißigjährigen Krieg, in: Zeitschrift für Württembergische Landesgeschichte 75, 2016, S. 129–153; Hans Heinrich Vangerow, Handel und Wandel auf der Donau von Ulm bis Wien in den Jahren 1583 bis 1651, in: Ulm und Oberschwaben 57, 2011, S. 115–168; Otto Ströhle, Ulms Handel im 16. und 17. Jahrhundert mit Frankreich und Italien, Diss. phil. [masch.], Würzburg 1922; Max Erich Feuchtinger, Die Schiffahrt auf der Oberen Donau Ulm–Regensburg im Wandel der Zeiten, in: Süddeutsche Wasserstraßen 12, 1936, S. 52–57; Ernst Neweklowsky, Die Schiffahrt und Flößerei im Raume der Oberen Donau, 3 Bde. (Die Schiffahrt und Flößerei im Raume der oberen Donau 5, 6, 16), Linz 1952–1964.

praktische Organisation des Handels ist bislang überhaupt weitgehend ein Forschungsdesiderat geblieben.[8]

Als Quellengrundlage dieser Untersuchung dienen vor allem die Akten des Ulmer Steueramts,[9] das für die Organisation des reichsstädtischen Handels zuständig war. Dem Steueramt waren die Greth, das örtliche Waag-, Steuer- und Lagerhaus und deren Personal, etwa Grethmeister und Güterbestätter, zugeordnet. Dementsprechend sind hier die Dienstanweisungen, die Instruktionen[10] der Güterbestätter, die im 18. Jahrhundert ihren Dienst verrichteten, überliefert, wie auch Ordnungen und Akten, die Auseinandersetzungen um die Organisation des Handels dokumentieren. In diesen Akten finden sich außerdem Korrespondenzen mit anderen Reichsstädten, die Aufschluss über die jeweilige Situation vor Ort geben.

Im Folgenden werden zunächst die Initiativen des Schwäbischen Kreises um den (Aus-)Bau von Straßen im 18. Jahrhundert und deren spezifische Folgen für Ulm erörtert. Anschließend erfolgt ein Aufriss über die Ulmer Ordinari-Fuhrleute und darauffolgend werden die Güterbestätter als Schaltstelle des Warenverkehrs, die Organisation des (Transit-)Handels und die damit

---

8 Eine Dissertation aus dem Jahr 1932 gibt Hinweise auf die Organisation von Handel in Basel: Robert Frey, Das Fuhrwesen in Basel von 1682 bis 1848 mit besonderer Berücksichtigung der schweizerischen Fuhren. Ein Beitrag zur Verkehrsgeschichte der Schweiz, Basel 1932. Die Augsburger Verhältnisse sind in der 1984 erschienenen Stadtgeschichte kurz abgehandelt: Hermann Kellenbenz, Wirtschaftsleben der Blütezeit, in: Gunther Gottlieb / Wolfram Baer / Josef Becker u.a. (Hg.), Geschichte der Stadt Augsburg von der Römerzeit bis zur Gegenwart, Stuttgart 1984, S. 258–301.

9 Stadtarchiv (künftig: StadtA) Ulm, A [2826], Güterbestäterei in der Gräth: Eid, Ordnungen, Instruktionen und Anstellung der Gräthmeister, Gräthknechte, Ballenbinder und Güterbestäter 1584–1802; A [2827], Beschwerde des Handelsstandes gegen den Güterbestätter; Vorhaben zur Aufstellung eines neuen Faktors oder Spediteurs 1794–1796; A [2897], Niederlage der Güter inner- und außerhalb der Gräth 1774–1804; B 121/20, Nr. 001: Ordinarifuhrleute, Landboten, Überlandfuhrleute 1807–1892.

10 Zur Quellengattung der Instruktionen vgl. das Projekt „Zu Diensten Ihrer Majestät", bearbeitet von Martin Scheuz, das sich mit Instruktionsbüchern des Wiener Hofs auseinandersetzt. URL: https://www.univie.ac.at/hoforganisation/projekt/projekt.html (letzter Zugriff: 16.9.2020). Vgl. auch: Anita Hipfinger / Josef Löffler / Jan Paul Niederkorn u.a. (Hg.), Ordnung durch Tinte und Feder? Genese und Wirkung von Instruktionen im zeitlichen Längsschnitt vom Mittelalter bis zum 20. Jahrhundert (Veröffentlichung des Instituts für Österreichische Geschichtsforschung 60), Wien/München 2012; Jakob Wührer / Martin Scheutz (Hg.), Zu Diensten Ihrer Majestät. Hofordnungen und Instruktionsbücher am frühneuzeitlichen Wiener Hof (Quelleneditionen des Instituts für Österreichische Geschichtsforschung 6), Wien/München 2011.

verbundenen innerstädtischen Interessenkonflikte thematisiert. Diese Interessenkonflikte hatten Korrespondenzen mit anderen Reichsstädten zur Folge und geben damit Einblick in die jeweilige Organisation des Handels.

1. Die Initiativen des Schwäbischen Kreises um den Ausbau des Straßennetzes und die Folgen für den Ulmer Handel

Der Chausseebau im Alten Reich war geprägt von einem „Südwest-Nordost-Gefälle“:[11] Im deutschen Südwesten begannen die Landesherren bereits im ersten Drittel des 18. Jahrhunderts Chausseen anzulegen,[12] Preußen erst im späten 18. Jahrhundert.[13] Im internationalen Vergleich allerdings fällt das Alte Reich insgesamt hinter den entsprechenden Infrastruktur-Initiativen Frankreichs und Englands zurück.[14]

Mittels des südwestdeutschen Chausseebaus sollte vor allem der Transit- und Fernhandel gestärkt werden und hier besonders der Handel zwischen den Niederlanden und der Schweiz respektive aus Oberdeutschland durch die Schweiz in Richtung Italien.[15] Im 18. Jahrhundert kritisierten die Stände des Schwäbischen Kreises die schlechte Beschaffenheit der Straßen, wiesen darauf hin, dass deren Qualität für einen (reibungslosen) Warenverkehr wichtig sei und forderten eine

---

11 Marcus Popplow, Europa auf Achse – Innovationen des Landtransports im Vorfeld der Industrialisierung, in: Stieferle, Transportgeschichte (wie Anm. 5), S. 79–142, hier: S. 117.

12 Vgl. zum Südwesten: Bernd Wunder, Das Chausseestraßennetz des schwäbischen Kreises im 18. Jahrhundert, in: Zeitschrift für die Geschichte des Oberrheins 147, 1999, S. 515–535; Bernd Wunder, Der Chausseebau in Württemberg während des 18. Jahrhunderts, in: Wolfgang Schmierer u.a. (Hg.), Aus südwestdeutscher Geschichte. Festschrift für Hans Martin Maurer, Stuttgart 1994, S. 526–538; Bernd Wunder, Der Kaiser, die Reichskreise und der Chausseebau im 18. Jahrhundert, in: Zeitschrift für Neuere Rechtsgeschichte 18, 1996, S. 1–22.

13 Vgl. zu Preußen: Herbert Liman, Preußischer Chausseebau, Meilensteine in Berlin (Berliner Hefte 5), Berlin 1993; Friedrich-Wilhelm Henning, Die Entwicklung der Verkehrsinfrastruktur in Brandenburg/Preußen als Teil der Staatsbaukunst von 1648 bis 1850, in: Forschungen zur Brandenburgischen und Preußischen Geschichte, NF 7, 1997, S. 211–232; Wilhelm Treue, Wirtschafts- und Technikgeschichte Preußens (Veröffentlichungen der Historischen Kommission zu Berlin 56), Berlin/New York 1984, S. 211 f.

14 England nahm dabei eine Vorreiterrolle ein. Popplow, Europa auf Achse (wie Anm. 11), S. 135.

15 Wunder, Das Chausseestraßennetz (wie Anm. 12), S. 518.

Verbesserung der Situation.[16] Vor allem Württemberg und die badischen Markgrafen verfolgten die Chaussierung der Straßen zur besseren Kommunikation und Förderung des Warentransfers. Dabei wurden nicht nur die bestehenden Straßen ausgebaut, sondern ebenfalls neue Straßen angelegt, um Routen zugunsten neuer Stationen anzulegen, wie das württembergische Beispiel deutlich zeigt.[17] Die territoriale Vielgestaltigkeit des Gebietes erforderte eine Zusammenarbeit der einzelnen Territorien des Schwäbischen Kreises und darüber hinaus auch der vorderösterreichischen und reichsritterlichen Gebiete, die sich aber bei einem ersten Vorstoß des Schwäbischen Kreises 1737 – der damit gleichsam als erster Kreis im Alten Reich eine Chausseepolitik entwickelt hatte –[18] nicht beteiligen wollten.[19] Diese obligatorische Zusammenarbeit war zwar von Erfolgen gekrönt, allerdings wurde der Bau und Unterhalt von Straßen durchgängig von Streitigkeiten wegen der Kostenübernahme begleitet.[20] Spätestens ab 1748 erfolgte ein gemeinschaftlicher, aber durchaus von konkurrierenden Interessen geprägter Chausseebau, der einer „überregionalen Konzeption"[21] der beteiligten Kreisstände folgte und von dem besonders Württemberg profitierte, indem es dort durch den Ausbau alter Verkehrswege und den Bau neuer Straßen gelang, Cannstatt als „Mittelpunkt aller Überlandstraßen im nördlichen Kreisgebiet"[22] zu etablieren.

Für Ulm waren die Straßenbauarbeiten ebenfalls vorteilhaft, denn die Reichsstadt lag an der neu ausgebauten Straße Cannstatt–Augsburg, der Memminger Straße, die über Kempten und Füssen nach Tirol führte, an der Strecke über Biberach und Lindau in Richtung Graubünden sowie donauaufwärts an der neu

---

16 Vgl. dazu: Wunder, Das Chausseestraßennetz (wie Anm. 12); Rothe, Das Finanzwesen der Reichsstadt Ulm (wie Anm. 1), S. 86 f.; StadtA Ulm, A [2901], Akten der Commerciendeputation 1748–1755: Gutachten *„auf was Arth das in Schwaben in Zerfall gekommene Commercium hinwiederum etabliert werden könnte"*.

17 Wunder, Das Chausseestraßennetz (wie Anm. 12), S. 534.

18 Ebd., S. 517.

19 Der Chausseebau im Jahr 1737 scheiterte an der fehlenden Genehmigung des Kaisers hinsichtlich der Beteiligung der vorderösterreichischen und reichsritterschaftlichen Gebiete. Vgl. Wunder, Das Chausseestraßennetz (wie Anm. 12), S. 520.

20 Vgl. dazu die Akten des Herrschaftspflegamts in Ulm: StadtA Ulm, A 3016–A 3062 sowie A 3063–3073.

21 Wunder, Das Chausseestraßennetz (wie Anm. 12), S. 533.

22 Württemberg engagierte sich auch im Ausbau der Straßen innerhalb des Territoriums (sogenannte Kommunikations- und Vizinalstraßen), sodass Wunder von einer „flächendeckenden Erschließung des Landes von der Hauptstadt her" spricht. Wunder, Das Chausseestraßennetz (wie Anm. 12), S. 534.

ausgebauten Straße in Richtung Schaffhausen und Basel.[23] Zudem war Ulm Ausgangspunkt der neu ausgebauten Nürnberger Straße über Ellwangen[24] und der Strecke über Riedlingen, das Höllental und Freiburg nach Straßburg.[25] Neben den Kooperationen mit Württemberg schloss Ulm zusätzlich Verträge mit Bayern (Scheibenstraße von Donauwörth über Niederstotzingen nach Ulm) und mit Österreich (Ausbau der Schweizer Straße) ab.[26]

Anhand der Torzölle Ulms, die Kurt Rothe für das 18. Jahrhundert ausgewertet hat, lässt sich eine Präferenz bestimmter Wege ablesen, denn der Torzoll wurde für Fahrzeuge, Waren und Vieh – mit wenigen Ausnahmen – pauschal erhoben.[27] Zwischen 1719/20 und 1803 nahm die Reichsstadt insgesamt 89.947 Gulden Torzoll ein. Mit 32,9 % Anteil am Gesamtvolumen des erhobenen Torzolls war Rothes Erhebung zufolge das Herdbrucktor an der Herdbrücke besonders stark frequentiert, an dem die Handelswege aus Oberschwaben, Venedig und Lyon zusammentrafen. An zweiter Stelle (32,2 %) lag das Frauentor, durch das die Verkehrswege Richtung Nördlingen und Nürnberg wie auch Esslingen–Cannstatt–Speyer und Heilbronn führten. Platz drei (20,3 %) nahm das Gögglinger Tor ein, durch das der Verkehr aus Biberach–Ravensburg–Schweiz und Urach–Tübingen–Kehl–Straßburg floss. Im Gegensatz zu diesen stark frequentierten Toren fielen die Anteile des Neutors (13,3 %) in Richtung Nordwesten und des im Osten gelegenen Gänstors (1,3 %) deutlich geringer aus.[28] Zur

23 Vgl. Wunder, Das Chausseestraßennetz (wie Anm. 12), S. 535.

24 Ebd.

25 Auf dieser sogenannten „Dauphinestraße“ reiste der Brautzug der Marie Antoinette 1769. Vgl. Meinrad Schaab, Siedlung, Gesellschaft, Wirtschaft von der Stauferzeit bis zur Französischen Revolution, in: Ders. / Hansmartin Schwarzmaier (Hg.), Handbuch der baden-württembergischen Geschichte, Bd. 1, 2. Teil: Vom Spätmittelalter bis zum Ende des Alten Reiches, Stuttgart 2000, S. 457–587, hier: S. 573.

26 Vgl. Albrecht Rieber, Ulm im Schnittpunkt der Verkehrswege (Hefte zur Ulmer Stadtgeschichte 4), Ulm 1971, S. 5.

27 Demnach wurde ein vollbeladener Wagen mit 15 Kreuzern und ein halbbeladener mit 7 Kreuzern verzollt, ein Güterkarren vollbeladen mit 7 Kreuzern und halbbeladen mit 5 Kreuzern. Fuhren mit Eisen, Wein, Hafnergeschirr und Hausrat wurden gesondert berechnet; zum Beispiel kostete ein Wagen vollbeladen mit Eisen 10 Kreuzer. Vgl. Rothe, Das Finanzwesen der Reichsstadt Ulm (wie Anm. 1), S. 339 f.

28 Rothe, Das Finanzwesen der Reichsstadt Ulm (wie Anm. 1), S. 339–341. Die Torzölle entwickelten sich erst positiv, als ab 1773/74 das Torsperrgeld eingeführt wurde, das heißt, das Herdbrucktor und das Frauentor waren nachts durchgängig geöffnet. Vgl. ebd., S. 341–343. Rothe zeigt außerdem, dass die Ausgaben der Reichsstadt Ulm für die Unterhaltung bereits bestehender Straßen innerhalb der Stadtmauern deutlich höher ausfielen als deren Neubau. Vgl. ebd., S. 87.

Beschleunigung des Verkehrsflusses vereinfachte die Reichsstadt außerdem das Zollwesen in ihrem Territorium, wie die stark frequentierte Strecke von Ulm über Geislingen nach Württemberg zeigt. Mitte des 18. Jahrhunderts legte der Magistrat fest, dass der „*helffensteinische*" Zoll, der in Geislingen kassiert wurde, bereits in Ulm beim Güterbestätter bezahlt werden konnte und die Fuhrleute damit keinen Aufenthalt mehr in Geislingen einplanen mussten.[29]

Die Initiativen der Stände des Schwäbischen Kreises hinsichtlich des Straßenbaus wirkten sich insgesamt durchaus positiv auf die Infrastruktur Ulms aus, indem besonders die von Ulm präferierten Straßen Richtung Nördlingen und Richtung Italien ausgebaut wurden.

Auch zeitgenössisch wurde die Initiative Ulms und des Schwäbischen Kreises im Ausbau des Verkehrsnetzes positiv vermerkt, wie das Beispiel Friedrich Nicolais zeigt, der aus Preußen kommend den deutschen Süden per Kutsche bereiste: „*Die Reichsstadt Ulm hat das Verdienst um die Nachbarschaft, daß sie sich den Chausseebau in der umliegenden Gegend eifrig hat angelegen seyn lassen. Die Chaussee auf der Straße nach Nürnberg fand viele Schwierigkeiten wegen der mannichfaltigen umliegenden Herrschaften, durch deren Gebiet die Wege gehen mußten. Endlich ward in den achtziger Jahren auf der Kreisversammlung ein Mittel gefunden, sich desfalls unter Herzoglich Wirtembergischer Vermittelung zu vereinigen. Das Reichsstift Elchingen, der Deutsche Orden, das fürstl. Haus Oettingen-Wallerstein, die Reichsstifter Salmansweil und Kaisersheim, nebst den dabey interessirten Reichsrittern des Kantons an der Donau verstanden sich zu einem bestimmten Beytrage an Gelde; und die Stadt Ulm übernahm es den weitern Bau der Chausseen zu endigen, den sie vorher schon auf ihrem eignen Gebiete bis Albeck, zwei Stunden weit, angefangen hatte.*"[30]

Aber nicht nur die zeitgenössische Wahrnehmung des Chausseebaus zeigt den positiven Effekt dieser Infrastrukturmaßnahmen. Auch historische Untersuchungen kommen zu dem Ergebnis, dass sich durch die Chaussierung in England die Reisezeit halbierte, während sich das Ladungsgewicht der Fuhren verdoppelte[31] und Wolfgang Behringer zeigt am Beispiel der Chaussee

---

29 Vgl. StadtA Ulm, A [2826], Nr. 253, 254: Instruktionen 1756, 1757. Das Zollwesen kann hier nur am Rande miteinbezogen werden, es verdient aber eine nähere Untersuchung.

30 Nicolai, Ulm betreffend (wie Anm. 4), S. 56 f.

31 Vgl. Uwe Müller, Art. „Chaussee", veröffentlicht: 2019, in: Enzyklopädie der Neuzeit Online. Im Auftrag des Kulturwissenschaftlichen Instituts (Essen) und in Verbindung mit den Fachherausgebern hg. v. Friedrich Jaeger, URL: https://referenceworks.brillonline.com/entries/enzyklopaedie-der-neuzeit/*-SIM_251120 (letzter Zugriff: 16.9.2020).

Koblenz–Montabaur–Limburg–Frankfurt auf, dass sich der Postverkehr auf dieser Strecke auf 21 Stunden halbierte.[32] Allerdings schätzt Klaus Gerteis den Chausseebau – trotz erheblicher Verbesserungen im Straßenverkehr – dennoch nicht als Teil einer „Verkehrsrevolution" ein, denn immer wieder waren schlechte Passagen im Straßennetz zu passieren.[33] Dieses Verkehrsnetz war allerdings ausschlaggebend für den Warenverkehr – wenn nur ein Streckenteil nicht gut befahrbar war, hatte dies Folgen für die ganze weiterführende Transportkette. Zwar wies schon Sombart auf die im hohen Mittelalter aufgekommene Einrichtung von regelmäßig verkehrenden Liniendiensten – den Ordinari-Boten oder Schiffen – hin, die zu einer besseren Kalkulierbarkeit im Warenverkehr und dessen Zunahme geführt hätten.[34] Es ist aber zu bedenken, dass beispielsweise im Alpenraum Ordinari-Boten zwar durchaus eingerichtet worden waren,[35] die Witterungsverhältnisse in den Alpen aber immer wieder zu erheblichen Verspätungen – die mitunter eine Woche und mehr betrugen – führen konnten.[36] Hier zeigt sich nun ein Spannungsverhältnis von intendiertem Verkehr nach Plan, seiner praktischen Durchführung und dem daraus resultierenden Umgang mit den Abweichungen. Deutlich wird dieses Spannungsverhältnis bei den Schilderungen des Ulmer Güterbestätters Holl aus dem Jahr 1794, der in einer Stellungnahme an den städtischen Magistrat von der regelmäßigen Verspätung des

---

32 Vgl. Behringer, Im Zeichen des Merkur (wie Anm. 5), S. 541.

33 Klaus Gerteis, Das „Postkutschenzeitalter". Bedingungen der Kommunikation im 18. Jahrhundert, in: Karl Eibl (Hg.), Entwicklungsschwellen im 18. Jahrhundert (Aufklärung 4/1), Hamburg 1990, S. 55–78, hier: S. 69.

34 Werner Sombart, Der moderne Kapitalismus. Historisch-systematische Darstellung des gesamteuropäischen Wirtschaftslebens von seinen Anfängen bis zur Gegenwart, Bd. 2/1: Das europäische Wirtschaftsleben im Zeitalter des Frühkapitalismus vornehmlich im 16., 17. und 18. Jahrhundert, München/Leipzig 1916, S. 352 f.

35 Das bekannteste Beispiel ist der Lindauer oder Mailänder Bote. Vgl. Werner Dobras, Der Mailänder oder Lindauer Bote – eine zuverlässige Transporteinrichtung zwischen Lindau und der Lombardei, in: Bündner Monatsblatt 5, 1989, S. 339–355; Paul Senger-Weiss / Heidi Senger-Weiss (Hg.), Das WeissBuch. Mehr als 500 Jahre unternehmerische Tätigkeit einer Familie im Transport, Wien 2010.

36 15 Tage war der Lindauer oder Mailänder Bote unterwegs, die Wetterbedingungen wie auch die Strecken durch Schluchten und über Bergpässe machten die Wege gefährlich und die Reisedauer stieg mitunter an. Vgl. Senger-Weiss / Senger-Weiss, Das WeissBuch (wie Anm. 35), S. 60 f. Zur Route des Lindauer oder Mailänder Boten: ebd., S. 48–56; Dobras, Der Mailänder oder Lindauer Bote (wie Anm. 35).

Schaffhausener Ordinari-Boten Welti berichtete, die bis zu einer Woche reichen konnten, womit die Fuhren nach Schaffhausen schwerer planbar seien.[37]

2. Die Ulmer Ordinari-Fuhrleute

Der Schaffhausener Ordinari-Bote war dabei einer von insgesamt 18 Fuhrleuten, die Johann Herkules Haid in seiner Chronik aus dem Jahr 1774 aufführte. Allein neun Fuhrleute verkehrten regelmäßig – zumeist wöchentlich – zwischen Ulm und den Reichsstädten Augsburg, Biberach, Esslingen, Frankfurt, Giengen, Kempten, Lindau, Nördlingen und Nürnberg.[38] Am häufigsten frequentiert waren dabei Augsburg und Nürnberg. Regelmäßig von den Fuhrleuten angesteuert wurden außerdem mit Kuchen und Geislingen Orte im Ulmer Territorium, die vorderösterreichischen Städte Ehingen und Riedlingen, die württembergischen Städte Heidenheim und Stuttgart sowie Schaffhausen und Zürich in der Eidgenossenschaft und Ichenhausen, im Besitz der Herren von Stain und in der Nähe von Günzburg gelegen.[39] Günzburg selbst wurde jede

---

37 Vgl. StadtA Ulm, A [2827], Nr. 50: Stellungnahme des Güterbestätters Holl vom 14. Januar 1795.

38 Sonntags und mittwochs steuerten „*fahrende Botten*", montagmittags ein Fuhrmann Augsburg an. Nach Biberach brachte ein Fuhrmann, der im Ulmer Mohren logierte, samstagabends Waren. Lieferungen nach Esslingen wurden vom Stuttgarter Fuhrmann übernommen. Der Handel mit Frankfurt wurde durch die sogenannte „*Guldische Fuhr*" abgedeckt, die unregelmäßig in Ulm eintraf und dort im Schwanen einstellte. Nach Frankfurt fuhr außerdem alle 14 Tage ein Ulmer Fuhrmann. Ein weiterer in Ulm ansässiger Ordinari-Fuhrmann belieferte sonntagvormittags die Reichsstädte Kempten und Memmingen. Lindau wurde von einem dritten Ulmer Fuhrmann angesteuert, der mittwochvormittags aufbrach. Sonntagvormittags machte sich ein Fuhrmann, der ebenfalls im Schwanen einstellte, auf den Weg nach Nördlingen und zum selben Zeitpunkt gingen zwei Fuhren, die von Fuhrleuten aus Giengen und Nördlingen geführt wurden und ebenfalls im Schwanen untergebracht waren, nach Nürnberg. Mit welcher Regelmäßigkeit ein Fuhrmann mit Giengen verkehrte, ist durch Haid nicht überliefert, lediglich, dass dieser Fuhrmann im Schwanen unterkam. Vgl. Johann Herkules Haid, Ulm mit seinem Gebiete, Ulm 1786, S. 277 f.

39 Der Fuhrmann, der Ehingen belieferte, logierte im Goldenen Adler, derjenige, der Heidenheim anfuhr, auf der Hohen Schule, der Ichenhausener Fuhrmann stellte im Kreuz ein. Auch bei diesen drei Fuhren ist die Abfahrtszeit mit Haid nicht überliefert. Nach Kuchen führten fünf Fuhrleute, die auf dem Kreuz einstellten, ihre Waren jeweils samstagmittags, Geislingen wurde samstagmittags sowie mittwochs angeliefert. Samstags brach ein Fuhrmann nach Riedlingen auf, der ebenfalls im Kreuz logierte. Sonntagvormittags wurden Schaffhausen und Stuttgart angefahren, der Stuttgarter Fuhrmann stellte dabei im Goldenen Rad ein. Zürich wurde von den Jungischen Fuhren beliefert,

Woche samstags durch einen Meister der Fischerzunft per Ordinari-Schiff über die Donau angesteuert. Ebenfalls wöchentlich – freitags oder samstags – ging ein Ordinari-Schiff nach Wien. Hier bestimmte ein Los-Verfahren, welcher Meister aus der Schiffer- und Fischerzunft dasselbe steuern durfte.[40] Diese Aufstellung zeigt die überregionalen (Handels-)Beziehungen Ulms mit anderen Reichsstädten, aber auch mit Vorderösterreich, Wien und dem benachbarten Herzogtum Württemberg.

Die Fuhrleute, die regelmäßig ab und nach Ulm verkehrten, waren nur zum Teil Bürger der Reichsstadt, wie sowohl Haid bereits in Einzelfällen als auch eine Gesamt-Aufstellung der Ordinari-Boten des Handelsstandes aus dem Jahr 1826 zeigen.[41] Der oben angesprochene Schauffhausener Fuhrmann Welti, dessen Nachfahren des gleichen Namens die Fuhrmannstätigkeit übernommen hatten, stammte beispielsweise aus Mengen, der Heilbronner Fuhrmann aus Heilbronn. Diejenigen in Ulm ansässigen Fuhrleute scheinen dabei vom Handelsstand eingesetzt und – ähnlich dem Güterbestätter – per Instruktion in den Dienst aufgenommen worden zu sein.[42]

Eine wichtige Voraussetzung für den Fuhrmannsdienst formulierte der Tübinger Jurist Ferdinand Christoph Harpprecht im Jahr 1706: „*Demnach das Fuhrwerck eine nicht gemeine Vorsichtigkeit, Tag und Nacht unermüdeten Fleiß, nicht weniger auch starcke Gliedmassen und einen daurhafften Leib erfordert, so müssen auch die Persohnen, welche zu dieser Profession sich begeben wollen, mit solchen Beschaffenheiten versehen seyn, daß sie der erforderten Aufsicht, auch offtmahls benöhtigten schwehren Arbeit gewachsen seyn, damit nicht durch sie so wol die aufgegebene Güter verwarloset, als auch Wagen und Pferde in Gefahr gesetzet werden.*“[43]

---

die dienstags in Ulm eintrafen und im Schwanen einstellten. Vgl. Haid, Ulm (wie Anm. 38), S. 277 f.

40 Haid, Ulm (wie Anm. 38), S. 278.

41 Ebd.; StadtA Ulm, B 121/20, Nr. 001: Ordinarifuhrleute, Landboten, Überlandfuhrleute 1807–1892. Die Aufstellung zeigt außerdem, dass sich die Ordinari-Fuhren im frühen 19. Jahrhundert geändert haben mussten. 1826 listete der Handelsstand nur noch sieben Ordinari-Fuhren auf: Augsburg, Kempten, Heilbronn, Schaffhausen, Nürnberg, Friedrichshafen und Tübingen.

42 Leider ist bislang nur eine Instruktion aus dem Jahr 1826 aufgefunden worden. Die Überlieferung legt aber nahe, dass bereits im 18. Jahrhundert ähnliche Instruktionen vorhanden waren. StadtA Ulm, B 121/20, Nr. 001: Ordinarifuhrleute, Landboten, Überlandfuhrleute 1807–1892.

43 Ferdinand Christoph Harpprecht, Das Recht der Fuhr-Leuthe, Zelle 1706, S. 5.

Neben der körperlichen Konstitution war aber auch ein gewisser Wohlstand notwendig, um als Ordinari-Fuhrmann in den Dienst treten zu können. Die Lindauer Boten mussten nämlich bei Dienstantritt 1.000 Gulden Kaution hinterlassen[44] – die Überlieferung aus dem Jahr 1826 deutet auch in Ulm auf eine Kaution derselben Höhe für den Nürnberger Ordinari-Fuhrmann hin.[45] Auch für das fränkische Frammersbach ist nachgewiesen, dass das Einkommen der dort ansässigen Fuhrleute weit über dem durchschnittlichen Einkommen der übrigen Dorfbewohner lag.[46]

3. „Des Fuhrmanns Assistent": Der Güterbestätter als Schaltstelle des Warenverkehrs

Das frühneuzeitliche Speditionswesen umfasste eine Reihe von Berufen, die nicht oder nur schwer voneinander abzugrenzen sind. Spätestens im 16. Jahrhundert etablierten sich die Berufe der Gutfertiger und Güterbestätter,[47] deren Hauptaufgabe in der Abfertigung der ein- und ausgehenden Waren lag. Regional konnte der Güterbestätter oder Gutfertiger auch eine eigene Spedition betreiben, das heißt, ein oder mehrere Fuhrwerke unterhalten, andernorts war ihm das vollständig untersagt. Eine semantische Differenz der beiden Ausdrücke scheint es zeitgenössisch nicht gegeben zu haben, sondern eher einen unterschiedlichen regionalen Gebrauch, wie das Ulmer Beispiel nahelegt. Wesentliches Differenzkriterium dieser beiden Berufe zu einem Spediteur war aber offenbar, dass der Güterbestätter/Gutfertiger (zumindest an einigen Orten) ein vereidigtes Amt darstellte, wohingegen der Spediteur ein Privatunternehmer war, der zumeist im Auftrag von Händlern agierte.[48] Beide grenzten sich ihrerseits von Faktoren und

44 Senger-Weiss / Senger-Weiss, Das WeissBuch (wie Anm. 35), S. 46.

45 StadtA Ulm, B 121/20, Nr. 001: Ordinarifuhrleute, Landboten, Überlandfuhrleute 1807–1892.

46 Peter Moser, Mittel- und Nordeuropäischer Landtransport. Die Frammersbacher Fuhrleute und ihr Beitrag zur Transportgeschichte (15.–19. Jahrhundert), Diss., Bamberg 1990, S. 96.

47 Vgl. zu Augsburg: Kellenbenz, Wirtschaftsleben in der Blütezeit (wie Anm. 8), S. 266 f. Vgl. zu Nürnberg: Karl Hald, Alt-Nürnberger Verkehrswesen, Nürnberg 1929; Art. „Güterbestätterei", in: Michael Diefenbacher / Rudolf Endres (Hg.), Nürnberger Stadtlexikon, [2]2000. Vgl. zu Basel: Robert Frey, Das Fuhrwesen in Basel von 1682 bis 1848 mit besonderer Berücksichtigung der schweizerischen Fuhren, Basel 1832, S. 89 f.

48 Vgl. Friedrich Albert Wengler, Beiträge zu der Lehre vom Speditionsgeschäfte für Juristen und Kaufleute, Chemnitz 1860, S. 26–29; Art. „Spediteur", in: Johann Georg Krünitz, Oekonomische Encyklopädie oder allgemeines System der Staats- Stadt- Haus u. Landwirthschaft, in alphabetischer Ordnung,

Kommissionären in dem Sinne ab, dass letztere beiden für die kaufmännische Abwicklung von Geschäften zuständig waren und Spediteure oder Fuhrleute mit dem Transport und der Lagerung von Waren beauftragten.[49]

Die fünf überlieferten Dienstanweisungen (Instruktionen) der Ulmer Güterbestätter oder Gutfertiger,[50] die zwischen 1699 und 1773 entstanden sind, gewähren Einblicke in deren Aufgabenstellungen, aber auch in deren Beziehungen zu den Fuhr- und Kaufleuten. Änderungen und Ergänzungen, die in

---

URL: www.kruenitz1.uni-trier.de (letzter Zugriff: 16.9.2020). Zur auch zeitgenössisch fehlenden Trennschärfe der Begrifflichkeiten: In der „Oekonomischen Encyklopädie" von Johann Georg Krünitz findet man den Gutfertiger unter dem Begriff „Fertigen". Hier heißt es: *„Im Oberdeutschen sind Fertiger, oder Gutfertiger, solche Personen, welche die Waaren anderer durch Schiffer oder Fuhrleute fortschaffen, Güterbestäter, Spediteur*", URL: www.kruenitz1.uni-trier.de (letzter Zugriff: 16.9.2020). Im Grimm'schen Wörterbuch wird der Güterbestätter als oberdeutsch und *„verfrachter von versandgütern, frachtmakler*" beschrieben. Vgl. URL: http://woerterbuchnetz.de/cgi-bin/WBNetz/wbgui_py?sigle=DWB&mode=Vernetzung&lemid=GG32592#XGG32592 (letzter Zugriff: 16.9.2020). Vgl. Art. „Spediteur", in: Johann Heinrich Zedler, Grosses vollständiges Universal-Lexicon aller Wissenschaften und Künste, Bd. 38, Sp. 1391–1392. Werner Sombart sieht den Spediteur als die Weiterentwicklung des Güterbestätters. Vgl. Sombart, Der moderne Kapitalismus (wie Anm. 34), S. 334.

49 Art. „Factor", in: Zedler, Grosses vollständiges Universal-Lexicon (wie Anm. 48), Bd. 9, Sp. 64 f.; Art. „Factor", in: Krünitz, Oekonomische Encyklopädie (wie Anm. 48).

50 Anhand der Überlieferung lassen sich folgende Güterbestätter oder Gutfertiger ausfindig machen: Bereits im 17. Jahrhundert tätig und mit der ersten Instruktion der Bestätter (1699) in Ulm überhaupt versehen war Johann (Hans) Conrad Neubronner, der aus der Handelsfamilie Neubronner stammte, von der zwei Familienmitglieder (Marx Tobias Neubronner und dessen gleichnamiger Sohn) im 17. Jahrhundert in das Ulmer Patriziat aufgenommen wurden. Vgl. Stefan Lang, Die Patrizier der Reichsstadt Ulm. Stadtherren, Gutsbesitzer und Mäzene, Ulm 2011, S. 12. Ihm folgte im Jahr 1726 Marx von Capol, aus dem der Churer Linie der Adelsfamilie von Capol entstammenden Ulmer Zweig, der ebenfalls im Handel tätig war. Nach dessen Verhaftung wurde Elias Albrecht Müller im Jahr 1757 vom Ulmer Handelsstand gewählt, der bis zu seinem Tod im Jahr 1773 im Amt verblieb. Der letzte Güterbestätter des 18. Jahrhunderts war Johann Adam Holl, der Handelsfamilie Holl – Septimus Holl war im 18. Jahrhundert Leinwandhändler, vgl. Herkle, Reichsstädtisches Zunfthandwerk (wie Anm. 2), S. 201 f. – zugehörig. Als Johann Adam Holl im Jahr 1802 seinen Dienst niederlegte, folgte ihm sein Sohn Johann Martin Holl, der ebenfalls vom Handelsstand gewählt wurde, und dem sein jüngerer Bruder Johann Jacob Holl assistierte. Vgl. StadtA Ulm, A [2826], Güterbestäterei in der Gräth 1584–1802; A [2827], Beschwerde des Handelsstandes gegen den Güterbestätter, Vorhaben zur Aufstellung eines neuen Faktors oder Spediteurs 1794–1796.

die Instruktionen eingeflossen sind, hingen mit dem jeweiligen persönlichen Hintergrund der einzelnen Bestätter[51] wie auch mit den sich verändernden wirtschaftlichen Verhältnissen zusammen. Der Text der Instruktionen nahm im Laufe der Zeit im Umfang um das Dreieinhalbfache zu – die wirtschaftliche Lage bedingte dabei eine immer weiter ausdifferenzierte Beschreibung des Berufs. Mit Ausnahme des ersten Güterbestätters Johann Conrad Neubronner, der das Amt von Großvater und Vater geerbt hatte, wählte der Handelsstand, der sich aus Mitgliedern der Kaufleute- und Kramer-Zunft zusammensetzte,[52] die Güterbestätter des 18. Jahrhunderts. Zugeordnet waren die Güterbestätter allerdings dem Steueramt – ihre Amtsstube, für die jährlich ein Zins von vier Gulden entrichtet werden musste, befand sich im örtlichen Zoll-, Waag- und Lagerhaus, der sogenannten Greth.[53] Aus der Überlieferung geht hervor, dass die Voraussetzung für das Amt des Ulmer Güterbestätters oder Gutfertigers ein gewisser Wohlstand war, denn zumindest für zwei der Bestätter ist eine Kaution von jeweils 400 Gulden nachweisbar.[54] Außerdem entstammten alle Amtsinhaber dem Handelsstand. Spätestens seit der Amtszeit des Güterbestätters Elias Albrecht Müller (1765/67) stand dem Güterbestätter ein Assistent zur Seite. Der Kreiskontingentssoldat Johannes Eisen, der seinen Dienst „*verhandglübdet*" antrat, blieb auch in der Amtszeit von Johann Adam Holl im Dienst.[55]

---

51 Beispielsweise wurde in die Instruktion für Johann Adam Holl aus dem Jahr 1773 aufgenommen, dass er sein Warenlager abverkaufen und keine weiteren Waren mehr annehmen soll. Vgl. StadtA Ulm, A [2827], Instruktion 1773.

52 Vgl. dazu, insbesondere zu den Leinwandhändlern: Herkle, Reichsstädtisches Zunfthandwerk (wie Anm. 2), S. 191–194. Zu den Ulmer Kaufleuten ist ein Aufsatz der Verfasserin in Vorbereitung.

53 Der Zins sollte dabei in zwei Raten auf Michaelis und Ostern bezahlt werden. Dieser Zins wurde dem Güterbestätter Holl – wie auch dessen Amtsvorgänger Müller – erlassen. Vgl. StadtA Ulm, A [2826], Nr. 249: „*Additamenta zu deß neuangenommenen GütherBestätters Johann Adam Hollen Instruction*".

54 Elias Albrecht Müller und sein Amtsnachfolger Johann Adam Holl mussten jeweils 400 Gulden Kaution hinterlegen. Dem Assistenten beider Güterbestätter – Johannes Eisen – wurden 200 Gulden Kaution auferlegt. Vgl. StadtA Ulm, A [2826], Nr. 246: Extrakt Steueramt, Protokoll vom 12. Juli 1773.

55 Eisen wird im Jahr 1765 bestätigt. Vgl. StadtA Ulm, A [2826], Nr. 250: Extrakt Steueramt, Protokoll vom 20. Mai 1765. In der Instruktion von 1773 wird erstmals auch der Assistent Eisen genannt. Hier wurde vermerkt, dass der Güterbestätter dem Assistenten die Instruktion zur Kenntnis geben und mit ihm eine sinnvolle Arbeitsteilung absprechen solle. Vgl. StadtA Ulm, A [2826], Nr. 248: Schreiben der Stadtrechner vom 1. Juli 1773, Nr. 247: Dekret vom 2. Juli 1773; A [2827], Instruktion 1773.

Die Verschriftlichung der Dienstanweisung für den Güterbestätter oder Gutfertiger in Ulm erfolgte im Jahr 1699 vor dem Hintergrund, dass die Geleitgelder[56] für die Waren von und nach Nürnberg[57] sowie Frankfurt nach Ansicht des Handelsstandes nicht ordentlich durch den Gutfertiger Neubronner eingezogen worden seien[58] und dieser kein ordentliches Register über die ein- und ausgehenden Waren führe.[59] Dieser reagierte auf den Vorwurf, in dem er darauf hinwies, dass er „*keinen Ayd, also auch keine Verantwortung*" habe.[60] Die darauf vom Magistrat eingeforderte Stellungnahme des Steueramts kam zu dem Schluss, „*daß zu möglicher abwendung izt vorsehender inconvenienzien, welche das Publicum sowohl alß die privatos treffen köndten*" der Gutfertiger oder Güterbestätter an eine Instruktion gebunden werden solle.[61] Eine „*Unmaßgebliche Instruction*" hing der

56 Vgl. zum Messegeleit in Ulm: Rolf Freitag, Das Geleit der Reichsstadt Ulm, in: Ulm und Oberschwaben 37, 1964, S. 85–131. Vgl. zur Frankfurter Messe: Rainer Koch (Hg.), Brücke zwischen den Völkern – Zur Geschichte der Frankfurter Messe, 3 Bde., Frankfurt 1991; Nils Brübach, Die Reichsmessen von Frankfurt am Main, Leipzig und Braunschweig. 14. bis 18. Jahrhundert, Stuttgart 1994, S. 374.

57 Mit Hilfe der Geleitverträge versuchte Nürnberg den Warenverkehr der Handelsleute zu erleichtern. Vgl. Hektor Ammann, Die wirtschaftliche Stellung der Reichsstadt Nürnberg im Spätmittelalter (Nürnberger Forschungen. Einzelberichte zur Nürnberger Geschichte 13), Nürnberg 1970, S. 20–44.

58 In diesem Fall setzte sich der Handelsstand dafür ein, dass nur die Kaufleute die Geleitgelder kassieren dürfen. Vgl. StadtA Ulm, A [2826], Nr. 279: Ratsdekret vom 17. März 1699; Nr. 280: Stellungnahme des Steueramts vom 15. März 1699. Ob diese Regelung eingeführt wurde, muss offen bleiben. Sicher ist allerdings, dass der Güterbestätter ab dem Jahr 1756/57 die Geleitgelder wieder selbst einziehen sollte. Vgl. StadtA Ulm, A [2826], Nr. 253, 254: Instruktion 1756, 1757. Insgesamt erwies sich das Ulmer Geleitswesen im 18. Jahrhundert als ein Zuschussgeschäft für die Stadt, denn die Straßen wurden sicherer und die Kaufleute nutzten für sich und die Warenbeförderung immer weniger das Geleit, das Gebühren verursachte. Für 1743 ist das Geleitgeld überliefert: Ein Kaufmann, der auf die Frankfurter Messe reiste, musste sechs Kreuzer entrichten. Ein vierrädriger Güterwagen kostete 15 Kreuzer, ein Karren fünf Kreuzer. Für die Beibehaltung des Geleits sprach Rothe zufolge lediglich, dass sonst das Geleitrecht an umliegende Territorien abgegeben wurde. Vgl. Rothe, Das Finanzwesen der Reichsstadt Ulm (wie Anm. 1), S. 71–74. Vgl. zum Geleitwesen: Manfred Straube, Geleitwesen und Warenverkehr im thüringisch-sächsischen Raum zu Beginn der Frühen Neuzeit (Veröffentlichungen der Historischen Kommission für Thüringen, Kleine Reihe 42), Köln/Weimar/Wien 2015.

59 Vgl. StadtA Ulm, A [2826], Nr. 279: Ratsdekret vom 17. März 1699; Nr. 280: Stellungnahme des Steueramts vom 15. März 1699.

60 Ebd., fol. 278: Extrakt aus der Anzeige des Handelsstandes vom 7. Mai 1699.

61 Vgl. ebd., Nr. 277: Stellungnahme des Steueramtes vom 11. Mai 1699.

Stellungnahme an, die der Magistrat ohne Einschränkung annahm.[62] Sie richtete sich an den „*Güter-Bestätter oder Guthfertiger*“ und bestand aus vier Punkten, die zunächst die Führung eines Registers nach Nürnberger Vorbild anmahnte, in dem „*mit bemerkung des gewichts, zeichens und numern*“ alle ein- und ausgehenden Waren verzeichnet werden sollten. Der Fuhrmann sollte daraufhin eine „*ordentliche Cartam oder designation über solche seine Ladung*“ erhalten. Die Registrierung der Ware sollte besonders dem Ausweis aber auch und vor allem dazu dienen, die Gebühren einfordern zu können. Weiter war es die Aufgabe des Güterbestätters oder Gutfertigers, die Ware anhand des vom Fuhrmann mitgeführten Frachtbriefes zu kollationieren und bei etwaigen Beschädigungen direkt und vor Ort mit dem Fuhrmann eine Lösung zu suchen. Waren, die als Transitgut und ohne örtliche Kommission in der Ulmer Greth ankamen und teilweise direkt von einem Fuhrmann auf den nächsten übertragen wurden, sollte der Güterbestätter ebenfalls kontrollieren und registrieren. Dies alles war mit Fleiß und Sorgfalt durchzuführen und in strittigen Fällen gegebenenfalls der Rat des Handelsstandes einzuholen.[63] Eine Ausnahme bildeten zu diesem Zeitpunkt die Ordinari-Fuhrleute aus Augsburg und Memmingen, Dolp und Rietmeyer, die auch weiterhin selbständig und ohne Kontrolle durch den Güterbestätter Ware annehmen und befördern durften.[64]

Der Konfliktfall zwischen dem Güterbestätter und dem Handelsstand bewirkte eine Prägung des Berufsbilds, das nun mit Instruktion und Register über eigene – überprüfbare – Medien verfügte. Gleichsam zeigt sich, dass infrastrukturelle Einrichtungen, Dirk van Laak folgend, „keineswegs das Ergebnis einer kohärenten Planung waren“, sondern sich vielmehr im Kontext von Konflikten entwickelten.[65] Mit der Verschriftlichung der Dienstanweisung stand nun gleichermaßen beiden Seiten ein Instrument zur Verfügung, auf das im Konfliktfall Bezug genommen werden konnte, wie folgendes Beispiel zeigt.

Eine wichtige Ergänzung der Tätigkeitsbeschreibung erfolgte in der Instruktion aus dem Jahr 1726 für Marx von Capol,[66] als diesem Güterbestätter untersagt wurde, selbstständig mit Ware zu handeln: „*vor sich aber* […] *alle Spedition*

62 Vgl. ebd., Nr. 275: „*Unmaßgebliche Instruction nach welcher sich der Güter-Bestätter od. Guthfertiger alhier zu richten haben solle*“, 11. Mai 1699.

63 Vgl. ebd.

64 Vgl. ebd., Nr. 274: Ratsdekret vom 12. Mai 1699.

65 Dirk van Laak, Alles im Fluss. Die Lebensadern unserer Gesellschaft – Geschichte und Zukunft der Infrastruktur, Frankfurt am Main 2018, S. 8.

66 Die Instruktion für den im Jahr 1726 eingesetzten Marx von Capol wuchs insgesamt auf sieben Punkte an. Vgl. StadtA Ulm, A [2826], Nr. 267: Instruktion 1726.

und Handlung unter waß praetext alß immer seyn kann, sich gänzlich enthalten."[67] Hier wird die neutrale Position betont, die das Amt vorsah, indem eigennütziges Verhalten des Güterbestätters verhindert werden sollte. Diese Erweiterung der Instruktion hinderte von Capol allerdings nicht daran, vier Fässer nicht näher bestimmten Inhalts „*auf dem Wasser*" anzunehmen, um mit ihnen zu handeln.[68] Aktenkundig wurde dieser Vorfall, weil der Grethmeister Johann Ulrich Kindervatter seinen Mitarbeiter auf frischer Tat ertappte und zur Rede stellte. Zwar betonte von Capol, dass „*er von denen Herren Handthierenden dependire, und von ihme Gräthmeister keinen Befehl anzunehmen hätte.*" Allerdings wurde ihm vom Steueramt auf die Klage des Grethmeisters hin bedeutet, dass er qua Dienstanweisung keine eigene Spedition betreiben dürfe.[69] Durch diesen Vorfall, wie auch andere „*zerschiedene Fehler*" während seiner Amtszeit und weil er die Rechnungen für die Fuhrleute nicht ordnungsgemäß ausgestellt hatte, handelte sich Marx von Capol im Jahr 1756 letztlich eine Turmstrafe von acht Tagen ein.[70] Ob er damit auch sein Amt verlor, muss zwar offen bleiben, allerdings deutet die Einsetzung eines Amtsnachfolgers im Jahr 1756 darauf hin.[71]

Weitere wesentliche Ergänzungen der Instruktionen des 18. Jahrhunderts betrafen vor allem den Umgang des Güterbestätters mit den Fuhrleuten. Im Jahr 1726 legte das Steueramt fest, dass die Frachtbriefe, die der Fuhrmann mitführte, vom Güterbestätter eingefordert und an die entsprechenden Handelsleute übergeben werden sollten, keinesfalls aber durften die Fuhrleute die Frachtbriefe den Wirten vorlegen, bei denen sie logierten.[72] Die Konkurrenz der Wirte, die hier ausschlaggebend war, sollte auch dadurch minimiert werden, dass die Instruktion vorsah, die Fracht in der Greth auf und abzuladen und nicht im Wirtshaus. Die vermittelnde Funktion des Güterbestätters wird hier deutlich: „*Sollte sich ereignen, daß um ein oder andere Ursachen willen ein Handthierender mit dem Fuhrmann, in Disputen komme, so mag zwar der Guthfertiger des Furhmanns Assistent seyn, jedoch aber mit Bescheidenheit und raison bey dem Handthierenden sprechen.*" Der abnehmende Transitverkehr in Ulm war der Grund für

67 StadtA Ulm, A [2826], Nr. 267: Instruktion 1726.
68 Vgl. ebd., Nr. 266: „*Extract Protocolli*" vom 29. November 1734.
69 Vgl. ebd.
70 Vgl. ebd., Nr. 261: Schreiben der Stadtrechner vom 10. Mai 1756.
71 Vgl. ebd., Nr. 254: Instruktion 1756.
72 Vgl. zu den Frachtbriefen respektive zur Handelskorrespondenz als der zeitgenössisch angewandten Betriebswirtschaftslehre: Monika Dommann, Verbandelt im Welthandel: Spediteure und ihre Papiere seit dem 18. Jahrhundert, in: WerkstattGeschichte 58, 2011, S. 29–48.

folgende Ergänzung aus dem Jahr 1756, laut welcher der Güterbestätter dafür zu sorgen hätte, „*die hieher kommende Fuhrleuth durch willige dienste und freundliches bezeug*[en] *dahin zu bewegen suchen, daß Sie die hiesige Lad-Statt nicht verlassen, sondern solche fernerhin willig frequentiren möchten.*“[73] Der letzten reichsstädtischen Instruktion aus dem Jahr 1773 ist schließlich zu entnehmen, dass ein festgelegter Fuhrmannslohn eingeführt wurde, dessen Art und Höhe allerdings in der Instruktion nicht näher bestimmt ist.[74]

Überblickt man die geschilderte Entwicklung der Aufgaben der Güterbestätter oder Gutfertiger, so lässt sich konstatieren, dass er als Vermittler zwischen Kauf- und Fuhrleuten auftrat und gleichzeitig die Interessen der Obrigkeit hinsichtlich des Zolls und der Niederlage der Handelsgüter in der Greth vertrat. In seiner Funktion als Amtmann in der städtischen Greth war der Güterbestätter dazu verpflichtet, Waren abzufertigen. Das heißt vor allem, zu registrieren, zu wiegen, zu kollationieren und die dafür erhobenen Gebühren einzunehmen. Er war aber auch dafür zuständig, den Weitertransport der Güter an den Bestimmungsort zu organisieren und im Falle der nicht adressierten Güter geeignete Weiterbeförderungsmöglichkeiten aufzutun. Selbstständig Warenhandel zu betreiben oder zu spedieren, war dem Güterbestätter dagegen untersagt. Die Instruktionen der Güterbestätter nahmen in Umfang und Inhalt während des 18. Jahrhunderts deutlich zu, was für eine zunehmende Konkretisierung des Berufsbilds in Ulm spricht. Dabei ist bemerkenswert, dass sich die Instruktionen besonders hinsichtlich der Fuhrleute veränderten, die an den Ulmer Standort gebunden werden sollten. Die Kaufleute wollten bewirken, dass die Güterbestätter selbst keine Spedition betrieben, in dem Sinne, dass sie einen bezahlten Zwischenhandel betrieben, vermutlich aber vor allem, dass sich die Bestätter nicht zu einer Konkurrenz zu den oft großen Speditionen der Ulmer Händler entwickelten. Diese Problematik zeigt sich nochmals anhand eines Konflikts zwischen den Güterbestättern und dem Handelsstand.

4. Güterbestätter gegen Handelsstand: der Streit um das Transitgut

Transitgut hatte in der Ulmer Greth besondere Konditionen. Es wurde gewogen, registriert und durch den Güterbestätter ohne zwischengeschalteten Spediteur direkt auf den anschließenden Fuhrwagen umgeladen. Insgesamt acht Kreuzer

73 Vgl. StadtA Ulm, A [2826], Nr. 254: Instruktion 1756.
74 Vgl. StadtA Ulm, A [2827], Instruktion 1773.

hatte der Fuhrmann für das Auf- und Abladen in der Greth zu bezahlen.[75] Diese Begünstigung sollte vor allem die Fuhrleute an die Reichsstadt Ulm binden, der damit Einnahmen durch den Transitverkehr zukamen. Allerdings witterte der Handelsstand gerade hier eine Möglichkeit, das eigene Einkommen zu erhöhen, indem ein eigens eingesetzter Spediteur – gegen ordentliche Vermittlungsgebühr – das Transitgut an die einheimischen Händler zur Weiterbeförderung übergab.

Die Problematik des Transitguts wurde im Jahr 1736 erstmals aktenkundig, als die Ulmer Kaufleute die Aufstellung eines eigenen Faktors forderten und der Grethmeister Johann Ulrich Kindervatter daraufhin zu bedenken gab, dass die Fuhrleute dann einen anderen (günstigeren) Handelsplatz wählen würden.[76] Die Koalitionskriege des späten 18. Jahrhunderts und die damit verbundenen Einschränkungen im Fernhandel ließen diesen Konflikt aber erst richtig aufbrechen.[77]

Als der Ulmer Handelsstand den Güterbestätter Johann Adam Holl im Jahr 1794 beschuldigte, die Transitgüter vor den einheimischen Waren bei der Weiterbeförderung zu bevorzugen und gleichzeitig die Aufstellung zweier Faktoren forderte, die im Auftrag des Handelsstandes agieren sollten, verlangte der Ulmer Magistrat eine Stellungnahme Holls. Dieser schilderte auf ganzen 22 Seiten die Lage des Ulmer Speditionswesens und seiner eigenen Profession. Dieses Dokument erlaubt einen Einblick in die infrastrukturellen Verhältnisse als wesentliche Voraussetzungen für einen funktionierenden Warenverkehr zwischen Ulm und anderen Handelsplätzen. Über die transporttechnische Vernetzung mit anderen Städten heißt es da: „[…] *ist wohl zu bedenken, daß Ulm ringsum mit Städten und Pläzen umgeben ist, welche einen ausgebreiteten Handel treiben, und ihre Fuhrwerk haben. So fahren Fuhren von Nürnberg nach Schaffhausen, und von da wieder Retour von Augspurg nach Straßburg, Hailbronn, Frankfurt, von Kempten und Lindau gleichfalls. Alle diese Städte brauchen den hiesigen Plaz nicht allein nicht*

75 Davon drei Kreuzer Transit-Zoll, dem Güterbestätter zwei Kreuzer, den sechs Ballenbindern für das Abladen und Aufladen zwei Kreuzer und nochmals einen Kreuzer an die beiden Spanner. Vgl. StadtA Ulm, A [2827], Nr. 94: Eingabe des Grethmeisters an den Magistrat vom 21. November 1794.

76 Vgl. StadtA Ulm, A [2826], Nr. 264: „*Extractus Steur. Amtl. Protocolli*" vom 18. Januar 1736.

77 Vgl. Peter Schaller, Zur Wirtschaftsgeschichte Ulms, in: Hans-Eugen Specker (Hg.), Ulm im 19. Jahrhundert. Aspekte aus dem Leben der Stadt (Forschungen zur Geschichte der Stadt Ulm, Reihe Dokumentation 7), Stuttgart 1990, S. 105–168, hier: S. 108–110.

nothwendig, sondern unsere Fuhrleute müssen die dortigen Kaufleute sogar um Ladungen bitten, sonst geben sie solche den directe gehenden fuhren auf."[78]

Diese Passage spiegelt die problematische Lage Ulms im Speditionsgeschäft des 18. Jahrhunderts wider. Denn nicht zuletzt durch den verstärkten Straßenbau waren Bedingungen geschaffen worden, die Zwischenstationen wie Ulm im Speditionsverkehr obsolet machten. Dies zeigt auch die ebenfalls vom Magistrat eingeforderte Stellungnahme des Grethmeisters, der nicht nur die Abwanderung der Fuhrleute befürchtete, sondern auch, dass ein Faktor „*das ganze Waaggeschäfft zertrümmern werden würde*".[79] Seine Befürchtung untermauerte eine Beschwerde aus Nördlingen, die beklagte, dass die zur Weiterführung nach Schaffhausen bestimmte Nördlinger Ware viel zu lange (nämlich ganze vier Wochen) in Ulm im Waaghaus liege, ohne weiterbefördert zu werden.[80] Das Nördlinger Schreiben drohte nun an, eine eigene „*ordinare nach Lindau und Schaffhausen*" aufzustellen.[81] Als den Schuldigen für diese missliche Situation identifizierte Grethmeister Gumpel den Handelsstand, da jener den Güterbestätter angewiesen habe, dem Schaffhausener Fuhrmann Welti so lange keine fremden Güter aufzuladen, bis alle heimischen aufgeladen seien. Offenbar hatte der Handelsstand eigenmächtig auf einer Konferenz in der Unteren Stube[82] entschieden, die Fuhrleute, die Transitgüter nach Ulm brachten, an einen eigens beauftragten Spediteur zu binden.[83] Zunächst wollte der Magistrat den Fall auf sich beruhen lassen, weil der Handelsstand keine neuen Vorstöße in diese Richtung unternommen habe.[84] Wenige Tage später allerdings wandte sich Grethmeister

---

78 Vgl. StadtA Ulm, A [2827], Nr. 50: Stellungnahme des Güterbestätters Holl vom 14. Januar 1795.

79 Vgl. ebd., Nr. 94: Eingabe des Grethmeisters an den Magistrat vom 6. November 1794.

80 Vgl. ebd., Nr. 88: Beschwerde aus Nördlingen vom 5. November 1794.

81 Vgl. zum Handel in Schaffhausen: Markus A. Denzel, Die Geschäftsbeziehungen des Schaffhauser Handels- und Bankhauses Ammann 1748–1779. Ein mikroökonomisches Fallbeispiel, in: VSWG 89, 2002, S. 1–40.

82 Die „Untere Stube" war der Versammlungsort der Kaufleute. Vgl. Rothe, Das Finanzwesen der Reichsstadt Ulm (wie Anm. 1), S. 379–383.

83 Grethmeister Gumpel berichtet von einem Gespräch, in dem der Kaufmann Seeger („*noch kein Jahr allhier sesshaft*"), den Nördlinger Ordinari-Fuhrmann Feldmeyer anwies, „*seine anhero brengende fremde Transito Güther nicht mehr dem Güther Bestätter sondern einem hiesigen Spediteur übergeben müsse*". Feldmeyer entgegnete ihm, dass er „*zu seinen fremden Güthern keinen hiesigen Spediteur nöthig, und das weitere werde sich deswegen schon zeigen*". Vgl. StadtA Ulm, A [2827], Nr. 93: Eingabe des Grethmeisters Gumpel an den Magistrat vom 10. November 1794.

84 Vgl. StadtA Ulm, A [2827], Nr. 89: Extrakt Steueramt, Protokoll vom 13. November 1794.

Gumpel wiederum an den Rat, denn er hatte *„in sichere Erfahrung gebracht"*, dass der Handelsstand eigenmächtig den Güterbestätter Holl samt Assistent vorgeladen und angewiesen habe, künftig die Fuhrleute, die Transitgüter nach Ulm brachten, nicht mehr selbst abzufertigen, sondern sie abwechslungsweise an die ansässigen Händler zur Weiterbeförderung zu übermitteln. Auf den Einwand des Güterbestätters, dass einerseits die Marktwoche bevorstünde und andererseits er und sein Assistent dem Steueramt unterstünden, entgegneten die Kaufleute, dass Güterbestätter wie auch die Fuhrleute vielmehr dem hiesigen Handelsstand verpflichtet seien. Zudem wollten die Kaufleute diese Änderung direkt an die Presse geben und an anderen Handelsorten bekannt machen. Der Grethmeister gab nochmals zu bedenken, dass weitere Kosten entstünden und – wie Nördlingen bereits angedroht hatte – künftig andere Routen, um Ulm herum, bevorzugt werden würden. *„So müssten am Ende samtliche hiesige Ordinarien, /: weilen die Proprio Güther allein vor Sie keine hinlängliche Ladungen geben: / zu grunde gerichtet und das ganze hiesige Fuhrweesen Seinem gänzlichen Ruin ausgesezet werden."*[85] Der Handelsstand wiederum beklagte, dass der Güterbestätter und sein Assistent die fremden Transitfuhren vor den heimischen bevorzugten, was aber dem Handelsstand *„zum grösten Nachtheil"* gereiche.[86]

Der Magistrat wies zunächst Stellungnahmen des Güterbestätters und des Grethmeisters an.[87] Wegen der Aufstellung eines Faktors des Handelsstandes entschied er, ein Zirkular an die Reichsstädte Lindau, Kempten, Nürnberg, Frankfurt, Nördlingen und Esslingen zu senden, um den dort üblichen Umgang mit Transitgut zu erfragen. Bis auf weiteres sollte es bei der alten Regelung bleiben. Die Entscheidung aber, ob ein Faktor aufgestellt werde oder nicht, so betonte der Magistrat, treffe der *„Hochlöbliche Magistrat, nicht aber der Löbliche Handelsstand"*.[88]

Die ausführliche Stellungnahme des Güterbestätters ist aus verschiedenen Gründen aufschlussreich. Als wesentliches Problem erkannte Holl – auch kriegsbedingt – nicht ausreichende Kapazitäten im Fuhrmannsbetrieb, sodass Transitware häufig zu lange in Ulm gestapelt bleibe. Sein Lösungsansatz, nämlich einen zweiten Nördlinger Ordinari-Fuhrmann aufzustellen, sei aber vom

85 Vgl. ebd., Nr. 87: Eingabe des Grethmeisters Gumpel an den Magistrat vom 20. November 1794.

86 Vgl. ebd., Nr. 81: Pro Memoria des Handelsstandes, unterzeichnet von Johann Joseph Kindervatter, Vorgesetzter der Kaufleutezunft, Conrad Heinrich Magirus, Vorgesetzter der Kramerzunft am 18. Dezember 1794.

87 Vgl. ebd., Nr. 64: Stellungnahme des Grethmeister Gumpels, 12. Januar 1795.

88 Vgl. ebd., Nr. 73: Ratsdekret vom 29. Dezember 1794.

Handelsstand verworfen worden. Die Aufstellung eines Faktors aber sei *„für den hiesigen Plaz sehr nachtheilig und schädlich* [...] *da bej Ausgang des gegen wärtigen Kriegs die fremde güther bejnahe alle wegfallen – und alle andern kaufmännische Geschäffte zu treiben uns verbotten ist.“* Er begründete dies durch einen Vergleich Ulms mit Lindau und Kempten. In diesen beiden Städten seien die Verhältnisse andere, denn in Lindau sei es unabdingbar, wegen des Bodensees abzuladen – ein Umweg über Land wäre letztlich für die Fuhren zu teuer. Kempten handele besonders mit Tirol, hier seien die Fuhren mehrheitlich mit teuren und leicht verderblichen Waren bestückt (Baumöl[89] oder Zitronen). *„So ist es bej ihnen nothwendige Folge, daß diese Waaren jemand zur Spedition übergeben werden, der die genaueste vigilance darüber trägt, bis sie wieder weiters verschikt werden können.“* Aus diesem Grund sei in diesen beiden Städten ein Faktor *„nöthig und rathsam. Es haben also beede nicht zu fürchten, daß wegen den Speditions Speesen die Transito Güther von ihrem Plaz vertrieben werden.“* In Ulm allerdings sei es schon häufiger vorgefallen, dass Fuhrleute aufgrund der Gebühren(-erhöhungen) nicht abladen wollten und zu einem anderen Ort wechselten (beispielhaft nannte er Erbach, Gögglingen, Walheim, Donaustetten), *„bloß um die hiesigen Waag Unkosten zu ersparen“*. Wenn nun aber die Fuhrleute Ulm nicht mehr ansteuerten, so schade dieser Umstand dem *„löblichen Publico“*, denn *„Professionisten und Handthierende“*, wie der Wirt, der Sattler, der Wagner oder der Merzler, verlören ihre Einkünfte. Güterbestätter Holl wies sämtliche Anschuldigungen von sich: *„Uns aber, die wir die uns anvertraute obliegende Geschäffte mit Fleiß und accuratesse immerdar redlich und uneygennüzig verrichten, bej unsern Rechten als Bürger und Kramer zu schützen, um so mehr, als es uns äusserst schmerzt, daß uns der löbl. Handelsstand ohne alle erhebl. und rechtl. Ursache mit Absezung droht, gleichsam als wenn wir unter der Despotesten Abhängigkeit von denenselben stünden.“*

Das Ulmer Zirkular[90] in dieser Sache erbrachte Antworten aus allen angeschriebenen Reichsstädten, die Auskunft über die Situation vor Ort gaben. Aus Lindau wurde berichtet, dass Transitgut durch einen dortigen Spediteur weitertransportiert werde, der von dem Fuhr- oder Schiffmann selbst ausgewählt werden könne.[91] *„Dienstbereitwillige Bürgermeister und Rath der Reichs Stadt Eßlingen“* antworteten, dass *„die Übergabe solcher umdisponirten Güter lediglich*

---

89 Baumöl = Olivenöl.

90 Vgl. StadtA Ulm, A [2827], Nr. 71: Zirkular an die Reichsstädte Nürnberg, Frankfurt, Lindau, Kempten, Nördlingen und Esslingen vom 30. Dezember 1794.

91 Vgl. ebd., Nr. 62 f.: Antwortschreiben aus Lindau vom 6. Januar 1795.

als eine Sache der Fuhrleute unter sich betrachet, weniger dißfalls ein eigener Spediteur oder Factor aufgestellt worden." Sie merkten aber auch an: „*Übrigens geschieht dahier selten eine solche Übergab.*" Sollte eine solche dennoch einmal vorkommen, so erteile der Fuhrmann dem Wirt, bei dem er logiere, den Auftrag, einen geeigneten Fuhrmann zu engagieren. Die Antwort aus Esslingen weist außerdem auf die Praxis im benachbarten Herzogtum Württemberg hin, dort sei „*ein besonderer Factor*".[92] In der Tat hatte die wirtschaftliche Strategie des benachbarten Territoriums dazu geführt – wie bereits beim Straßenbau angeklungen ist –, dass der Güterverkehr im 18. Jahrhundert deutlich zunahm und aus diesem Grund im Jahr 1782 eine Niederlage unter dem Stuttgarter Rathaus eingerichtet und ein eigener Güterbestätter bestellt worden war.[93]

Aus Kempten wurde vermeldet, dass diejenigen Transitgüter, die nicht an ein hiesiges Handelshaus adressiert waren, zur Weiterbeförderung an die im Jahr 1771 eingerichteten „*Speditions-Compagnie*" übergeben werden müssten – zum Missfallen der Fuhrleute. Begründet war dies im Schreiben damit, dass die zur „*Sicherstellung des Zollweesens zum besten des Handelsstandes getroffene Verfügung möglichst aufrecht erhalten wird*". Die Kemptener Antwort beinhaltete den Wunsch, dass sich diejenigen „*Schwäbischen Handelsstädte, wo förmliche Güter Niederlaagen vorhanden sind*" – also Waaghäuser –, in Hinblick auf das Speditionswesen absprächen, damit so den „*sonst ganz über Hand nehmenden Eigenschafften der Fuhrleute in das Speditionsweesen abgeholfen und vorgebogen werden könnte.*"[94] Darüber hinaus zeigt die Antwort aus Kempten ein lokales Problem auf, nämlich die vergleichsweise mächtige Position der Fuhrleute, vermutlich aus dem Tiroler Raum.[95] Die besondere Kemptener Situation hinsichtlich der Fuhrleute hatte bereits Güterbestätter Holl angemerkt.

In Nördlingen war es den Fuhrleuten ebenfalls gestattet, die Ware selbstständig zur Weiterbeförderung an einen anderen Fuhrmann zu übergeben oder aber das Transitgut wurde im örtlichen Waaghaus niedergelegt, um dort vom Waagmeister einem Fuhrmann zugeteilt zu werden. Allerdings waren in Nördlingen weder Güterbestätter noch eigener Faktor aufgestellt.[96]

---

92 Vgl. ebd., Nr. 48: Antwortschreiben aus Esslingen vom 8. Januar 1795.

93 Vgl. Rudolph Friedrich von Moser, Beschreibung des Stadtdirections-Bezirkes Stuttgart (Beschreibung des Königreichs Württemberg 36), Stuttgart 1856, S. 279.

94 Vgl. StadtA Ulm, A [2827], Nr. 47: Antwortschreiben aus Kempten vom 12. Januar 1795.

95 Vgl. ebd.

96 Vgl. ebd., Nr. 45: Antwortschreiben aus Nördlingen vom 14. Januar 1795. Die im späten Mittelalter bedeutende Nördlinger Messe hatte im 18. Jahrhundert deutlich an Relevanz verloren. Vgl. Wilfried Sponsel, Art. „Nördlinger Messe", veröffentlicht

Nürnberg hatte qua Ordnung aus dem Jahr 1711 zwar den Fuhrleuten verboten, Transitgut ohne Adressat in Nürnberg selbstständig an einen weiteren Fuhrmann weiterzureichen und stattdessen einen örtlichen Spediteur oder Kaufmann damit zu beauftragen. Allerdings berichtet der Nürnberger Rat an Ulm, dass ein Verstoß gegen diese Ordnung „*dann und wann*" und vor allem vor dem Hintergrund eines geringen Fuhrmannslohns toleriert werde. Allerdings hätte es ganz aktuell wieder Klagen der Kaufleute gegeben, denen man nun nachgehe.[97]

Ähnlich wie in Nürnberg stellte sich die Situation in Frankfurt dar; auch hier sollten die Fuhrleute nach Auskunft der Deputierten des Rechnungs- und Fuhrenamts keine „*direkte oder indirekte Spedition*" betreiben. Allerdings war man der Ansicht, dass gegen die fremden Fuhrleute „*nichts verfügt werden könne*", obwohl auch hier die Kaufleute eine striktere Regelung einfordern wollten.[98] In Nürnberg wie in Frankfurt waren außerdem mehrere Güterbestätter mit der Abwicklung des Warenverkehrs betraut.

Nachdem alle Antworten eingetroffen waren, erging am 16. Dezember 1795 letztlich das Dekret in der Sache Handelsstand gegen Güterbestätter. In Hinblick auf die Anschuldigungen des Handelsstands, dass Holl die fremden Güter bevorzuge, urteilte der Magistrat, den Güterbestätter anzuweisen, „*ganz unpartheyisch zu Werke* [zu] *gehen*" und die einkommenden Güter nach Dringlichkeit zu versorgen. Die Forderung des Handelsstandes zwei Faktoren aufzustellen, lehnte der Rat ab und wies darauf hin, dass die Güter weiterhin durch den Güterbestätter an Fuhrleute weitergegeben werden sollen. Der Güterbestätter allerdings „*solle sich der Spedition jener Güther womit nur die mindeste Correspondenz verknüpft ist, auch ferner gänzlich* […] *enthalten.*"[99]

Zwar legte der Handelsstand im Januar 1796 nochmals sein Veto ein und wies darauf hin, dass er die erste Instanz für den Güterbestätter sei[100] – der Magistrat

---

am: 18.7.2012, in: Historisches Lexikon Bayerns, URL: https://www.historisches-lexikon-bayerns.de/Lexikon/N%C3%B6rdlinger_Messe (letzter Zugriff: 16.9.2020); Rolf Kießling, Die Nördlinger Pfingstmesse im 15./16. Jahrhundert. Aufstieg und Strukturwandel eines süddeutschen Wirtschaftszentrums, in: Jahrbuch des Historischen Vereins für Nördlingen und das Ries 29, 1999, S. 69–95.

97 Vgl. StadtA Ulm, A [2827], Nr. 43: Antwortschreiben aus Nürnberg vom 20. Februar 1795.

98 Vgl. ebd., Nr. 31: Antwortschreiben aus Frankfurt vom 16. Juli 1795. Als im Juli 1795 noch immer keine Antwort aus Frankfurt eingetroffen war, ließ der Ulmer Magistrat dort erneut anfragen. Vgl. StadtA Ulm, A [2827], Nr. 38: Schreiben des Ulmer Magistrats an Frankfurt vom 1. Juli 1795.

99 Vgl. StadtA Ulm, A [2827], Nr. 8: Dekret vom 16. Dezember 1795.

100 Vgl. ebd., Nr. 5: Pro Memoria des Handelsstandes vom 15. Januar 1796.

begegnete diesem von ihm als Anmaßung empfundenen Einwand des Handelsstands deutlich, in dem er dekretierte, *„daß das Steueramt die erste oder gewöhnliche Instanz eines Gutfertigers sey"*. Der Handelsstand dürfe auch weiterhin den Gutfertiger wählen, aber dieser stehe im Dienst des Steueramts. Ähnlich verhielt es sich mit der selbstständigen Aufstellung zweier Faktoren durch den Handelsstand: „[...] *gegen diese Anmassung"* – so hieß es im Ratsdekret – *„müsse man* [...] *protestieren."* Sollte ein Faktor künftig doch nötig sein, so werde man aber darauf achten, dass dieser aus dem örtlichen Handelsstand sei.[101]

Die langwierige Klage des Handelsstandes gegen Güterbestätter Holl in den Jahren 1794 bis 1796 macht die innerstädtischen Interessenkonflikte deutlich sichtbar, die sich aus der Organisation der Abwicklung des Transitguts ergaben. Die amtliche Seite, durch das Personal in der Greth vertreten, befürchtete eine mit der Abwanderung der Fuhrleute verbundene Herabsetzung Ulms als Umschlagplatz, wenn ein Faktor durch den Handelsstand eingerichtet werde. Zudem waren die Fuhrleute auch für die Stadtkasse relevant, indem sie beispielsweise Waaggeld entrichten mussten. Das Interesse der Kaufleute lag in der Optimierung des eigenen Handels. Dabei ist bemerkenswert, dass die Kaufleute nicht nur in Ulm, sondern auch in anderen Reichsstädten versuchten, Einfluss auf den Warenhandel zu nehmen. Unterschiedlich waren indes die Reaktionen auf das Verhalten der Kaufleute – in Kempten beispielsweise scheinen die Kaufleute als Speditionsunternehmer begünstigt worden zu sein, während in Ulm an einer amtlichen Regelung des Speditionsverkehrs festgehalten wurde.

Die Ulmer Handelsorganisation bedingte aber zugleich auch Konflikte mit anderen Reichsstädten, wie das Nördlinger Beispiel zeigt. Diese Konfliktlage bot dann den Ausgangspunkt für einen Vergleich mit anderen Reichsstädten, der vor allem die Unterschiedlichkeit der Formen von Handelsorganisation sichtbar macht. Topographische Gründe – wie in Lindau oder Kempten – konnten dabei eine Rolle spielen, aber auch die Größe und Bedeutung einer Stadt als Handelsstandort – wie Frankfurt oder Nürnberg. Der Vorschlag aus Kempten zu einer Absprache derjenigen Handelsorte mit einer *„förmlichen Güter Niederlaage"* über die Organisation von Transithandel war demnach schwer umsetzbar.

5. Fazit

Der Blick auf die lokale Organisation des Warenhandels zeigt, dass diese sich – wie auch die damit in Verbindung stehenden Berufsfelder – permanent äußeren

101 Vgl. ebd., Nr. 1: Dekret vom 27. Januar 1796.

Faktoren anpassen musste und wiederum je nach Standort unterschiedlich ausgeprägt war. Am Beispiel der Konfliktfelder um die Ulmer Güterbestätter, die zwischen Händler und Fuhrmann vermitteln sollten und gleichzeitig die Interessen des Steueramtes vertraten, wurden die unterschiedlichen innerstädtischen Interessenlagen sichtbar. Während das Steueramt in Vertretung des Güterbestätters versuchte, die Fuhrleute an die Stadt zu binden, um einen Einbruch im Waaggeschäft zu verhindern und Ulm als Speditionsstandort zu stabilisieren, war dem Handelsstand an einem Ausbau der eigenen Geschäfte gelegen. Die Problematik des Transithandels zeigt in diesem Zusammenhang auch der innerstädtische Konflikt um die Entscheidungsbefugnisse von Kaufleuten und Magistrat, den der Magistrat in diesem Fall für sich entschied.

Der Ausbau der Infrastruktur durch den Straßenbau hat einigen Nutzen gebracht, indem die Verkehrswege schneller und sicherer wurden. Ulm profitierte von den Investitionen und Initiativen des Schwäbischen Kreises hinsichtlich des Straßennetzes. Allerdings bedingte dieser Ausbau auch einen Rückgang des Speditionsgewerbes, denn nun waren direkte Warentransporte immer häufiger. Die Ulmer Situation lässt sich dabei nicht auf andere Reichsstädte übertragen, wie die eingeforderten Berichte anderer Handelsplätze nahelegen. Während die Organisationsformen in Esslingen und Nördlingen weniger ausgeprägt waren, was darauf hindeutet, dass diese beiden Reichsstädte im 18. Jahrhundert offenbar keine relevanten Speditionsstandorte waren, wiesen Nürnberg und Frankfurt stark ausgebildete städtische Einrichtungen (wie beispielsweise das Frankfurter Fuhrenamt) und Regelungen auf, die ihre Bedeutung im (Transit-) Handel zum Ausdruck bringen. Bemerkenswert scheint hier auch der Vergleich zum Herzogtum Württemberg zu sein, das sich durch seine Wirtschaftspolitik und den damit verbundenen strategischen Straßenbau im 18. Jahrhundert zu einem Wirtschaftsstandort entwickelte, sodass dort Ende des 18. Jahrhunderts in der Residenzstadt Stuttgart eine Niederlage sowie das Amt eines Güterbestätters etabliert wurde.

Die sehr unterschiedliche Ausprägung der Handelsorganisation von Reichsstädten zeigt sich auch im Falle der Fuhrleute, deren Situation ebenfalls standortabhängig war und die zum Teil einflussreiche Positionen einnehmen konnten, wie der Kemptener Bericht zeigt. Die Relevanz der Fuhrleute für den Transithandel lässt sich am Ulmer Beispiel ablesen, indem sie in die Instruktionen der Güterbestätter aufgenommen wurden, um an den Ulmer Standort gebunden zu werden.

Die Korrespondenz zeigt aber nicht nur die Vielgestalt der Handelsorganisation in den jeweiligen Handelsstädten; die Bedeutung der Korrespondenz als solche spiegelt sich auch darin, dass sie in den Entscheidungsfindungsprozess des Magistrats einfloss. Als im Juli 1795 noch immer keine Antwort aus Frankfurt eingetroffen war, ließ der Ulmer Magistrat dort erneut anfragen, um erst dann eine Entscheidung zu fällen.[102] Die Ulmer Handelsorganisation orientierte sich demnach weniger an merkantilen Theorien als vielmehr an den Erfahrungen und Gepflogenheiten anderer Handelsstädte und an der sich verändernden Situation vor Ort – wie das beispielsweise auch die Anpassung der Dienstanweisungen für die Güterbestätter nahelegt.[103]

Über diese Ergebnisse hinaus wurden Themen angesprochen, die weitere Aufmerksamkeit erfordern: Hier zu nennen wäre etwa das Messegeleit, das sich – zumindest im Ulmer Fall – im 18. Jahrhundert als Minusgeschäft entwickelt hatte, an dessen Recht aber festgehalten wurde. Darüber hinaus verdient das nur am Rande angerissene Zollwesen mehr Beachtung, gerade in Hinblick auf die Beschleunigung des Verkehrs.

---

102 Vgl. ebd., Nr. 38: Schreiben des Ulmer Magistrats an Frankfurt vom 1. Juli 1795.

103 Zu ähnlichen Ergebnissen kommt auch Heinrich Steinmeyer hinsichtlich der Entwicklung der Ulmer Sommermesse. Vgl. Steinmeyer, Ulmer Sommermesse (wie Anm. 2), S. 336 f., 349.

Sektion 5:
Die süddeutsche Städtelandschaft – urban und stadtprägend

Wolfgang Mährle

„... ist fleissige ἐπισκοπή und obacht insgemein über das gantze schulwesen anbefohlen." Die Administration des Bildungswesens in den süddeutschen Reichsstädten in der Frühen Neuzeit

Abstract: In the sixteenth century, so-called scholarchates were established in numerous southern German imperial cities, especially in those that had embraced the Reformation or that were bi-confessional with a Protestant majority. The scholarchates were commissions set up by the city magistrates to administer the entire school system, which included both German and Latin schools. This article analyzes comparatively the organisation, personnel, and functions of the scholarchates in the imperial cities of Swabia, Franconia, and Bavaria. It demonstrates that in Protestant and bi-confessional cities, even though magistrates were responsible for the school system, the clergy exercised considerable influence over education in the early modern period.

Eine vergleichende Untersuchung der seit dem 16. Jahrhundert in den süddeutschen Reichsstädten bestehenden Schulverwaltungen gilt seit langer Zeit als Desiderat.[1] Im Folgenden soll versucht werden, die bestehende Forschungslücke zu schließen. Ziel des vorliegenden Aufsatzes ist es, die Grundstrukturen der süddeutsch-reichsstädtischen Administration des Bildungswesens in der Frühen Neuzeit zu rekonstruieren. Durch den komparatistischen Ansatz sollen die jeweiligen lokalen Gegebenheiten in einen breiteren Interpretationsrahmen gestellt werden. Die Studie soll gleichzeitig eine solide Basis für zukünftige Spezialstudien zu einzelnen reichsfreien Städten bieten.

Einbezogen in die Analyse werden die insgesamt 37 Reichsstädte des Schwäbischen, Fränkischen und Bayerischen Reichskreises, die ihre Immediatstellung bis zum Beginn des 19. Jahrhunderts bewahren konnten.[2] Diese Städte wiesen

---

1 Ernst-Wilhelm Kohls, Die Schule bei Martin Bucer in ihrem Verhältnis zu Kirche und Obrigkeit, Heidelberg 1963, S. 204, Anm. 119; Anton Schindling, Humanistische Hochschule und freie Reichsstadt. Gymnasium und Akademie in Straßburg 1538–1621, Wiesbaden 1977, S. 82.

2 Die frühneuzeitlichen Reichsstädte sind aufgelistet bei: Ulrich Rosseaux, Städte in der Frühen Neuzeit, Darmstadt 2006, S. 29 f. Zu den Reichskreisen und ihren Mitgliedern vgl. zusammenfassend: Winfried Dotzauer, Die deutschen Reichskreise

in demografischer, politischer, wirtschaftlicher, sozialer und religiöser Hinsicht große Unterschiede auf. So umfasst die Untersuchungsgruppe unter anderem die damaligen Großstädte Nürnberg und Augsburg, aber auch zahlreiche ackerbürgerliche Kleinstädte wie Zell am Harmersbach, Wangen oder Giengen. 21 der 37 untersuchten Städte waren evangelisch, fünf bikonfessionell (davon vier seit dem Westfälischen Frieden paritätisch), elf katholisch.

Der Forschungsstand zu den Schulverwaltungen in den süddeutschen Reichsstädten ist sehr unbefriedigend. In der vorliegenden bildungs- beziehungsweise stadtgeschichtlichen Literatur finden sich lediglich punktuell Informationen zu dieser Thematik. Die verfügbaren Angaben sind in der Regel wenig detailliert; sie nehmen zudem häufig auf einen bestimmten Zeitpunkt oder auf ausgewählte Zeitabschnitte der frühneuzeitlichen Epoche Bezug.

Der lückenhafte Forschungsstand zur Schulverwaltung in den süddeutschen Reichsstädten spiegelt in vielen Fällen eine disparate Quellenlage. Häufig fehlen Zeugnisse, in denen die Aufgabenwahrnehmung im Bereich der reichsstädtischen Schuladministration explizit niedergelegt ist. Schriftliche Dienstanweisungen für schulische Verwaltungs- und Aufsichtsorgane sind nur selten überliefert – vielfach hat es sie wohl auch nicht gegeben. In den Schulordnungen der reichsunmittelbaren Städte werden zwar verschiedentlich Organisation und Aufgaben der Schulverwaltungen erwähnt, doch nehmen derartige Angaben eher eine periphere Rolle ein. Die Ordnungen regeln schwerpunktmäßig die Organisation des Lehrbetriebs, die schulische Wissensvermittlung, Disziplinarfragen sowie Angelegenheiten des Stipendienwesens.

---

in der Verfassung des Alten Reiches und ihr Eigenleben (1500–1808), Darmstadt 1989. Ein Gesamtüberblick über die Entwicklung der süddeutschen Reichsstädte in der Frühen Neuzeit fehlt. Vgl. aber zu den Reichsstädten im heutigen Bundesland Baden-Württemberg: Hans Eugen Specker, Die Geschichte der Reichsstädte im Überblick, in: Meinrad Schaab / Hansmartin Schwarzmaier (Hg.), Handbuch der baden-württembergischen Geschichte, 2. Bd.: Die Territorien im Alten Reich, Stuttgart 1995, S. 649–661 sowie die nachfolgenden Städteporträts. Grundlegende, allerdings zum Teil nicht mehr dem heutigen Forschungsstand entsprechende Daten zur Stadtgeschichte finden sich in den von Erich Keyser u.a. herausgegebenen Städtebüchern: Erich Keyser (Hg.), Badisches Städtebuch, Stuttgart 1959; Ders. (Hg.), Württembergisches Städtebuch, Stuttgart 1962; Ders. / Heinz Stoob (Hg.), Bayerisches Städtebuch, 2 Bde., Stuttgart u.a. 1971/1974. Zu den im heutigen Bundesland Bayern gelegenen Reichsstädten vgl. zudem die einschlägigen Artikel im „Historischen Lexikon Bayerns“, URL: https://www.historisches-lexikon-bayerns.de.

1. Die Entstehung reichsstädtischer Schulkommissionen im Reformationszeitalter

Im Laufe des Spätmittelalters erlangten die Magistrate der meisten süddeutschen Reichsstädte die Aufsicht über das kommunale Schulwesen.[3] Sie nahmen eine Kontrollfunktion über die „*Teutschen Schulen*" wahr, die zum Teil städtische Gründungen waren, in der Regel aber von Privatunternehmern geführt wurden.[4] Daneben ging mehr und mehr auch das Patronat über die von kirchlichen Einrichtungen getragenen Lateinschulen (Pfarrschulen, Klosterschulen) auf die Stadtmagistrate über; diese Schulen wurden damit zu „Ratsschulen". Die Stadträte entschieden insbesondere über die Besetzung der Rektoren- und Lehrerstellen in den Lateinschulen, über die Einkünfte des Lehrpersonals und über die schulischen Räumlichkeiten. Verschiedene Schulordnungen aus der Zeit um 1500 belegen ferner, dass die weltlichen Obrigkeiten in mehreren süddeutschen Reichsstädten (beispielsweise in Nürnberg, Nördlingen

---

3 Knapper Überblick hierzu bei: Arno Seifert, Das höhere Schulwesen. Universitäten und Gymnasien, in: Notker Hammerstein (Hg.), Handbuch der deutschen Bildungsgeschichte, Bd. 1: 15. bis 17. Jahrhundert. Von der Renaissance und der Reformation bis zum Ende der Glaubenskämpfe, München 1996, S. 197–374, hier bes.: S. 223–226. Instruktive Überblicksdarstellungen über die Entwicklung des Bildungswesens in den heute bayerischen Gebieten im Untersuchungszeitraum finden sich im „Handbuch der bayerischen Geschichte". Vgl. Dieter Wölfel, Das evangelische Schulwesen, in: Max Spindler (Begr.) / Andreas Kraus (Hg.), Handbuch der bayerischen Geschichte, Bd. 3/1: Geschichte Frankens bis zum Ausgang des 18. Jahrhunderts, München [3]1997, S. 1193–1216; Hans-Michael Körner, Das höhere und niedere Schulwesen, in: Max Spindler (Begr.) / Andreas Kraus (Hg.), Handbuch der bayerischen Geschichte, Bd. 3/2: Geschichte Schwabens bis zum Ausgang des 18. Jahrhunderts, München [3]2001, S. 686–717. Vgl. daneben auch den Überblick von: Rolf Kießling, Ansatzpunkte und Entwicklungstendenzen in den spätmittelalterlichen Schullandschaften Schwabens, in: Helmut Flachenecker / Rolf Kießling (Hg.), Schullandschaften in Altbayern, Franken und Schwaben. Untersuchungen zur Ausbreitung und Typologie des Bildungswesens in Spätmittelalter und Früher Neuzeit, München 2005, S. 247–279. Relativ knapp sind die Abschnitte zum Schulwesen im „Handbuch der baden-württembergischen Geschichte" gehalten. Vgl. Meinrad Schaab, Spätmittelalter (1250–1500), in: Ders. / Hansmartin Schwarzmaier (Hg.), Handbuch der baden-württembergischen Geschichte, Bd. 1/2: Allgemeine Geschichte. Vom Spätmittelalter bis zum Ende des Alten Reiches, Stuttgart 2000, S. 1–145, hier: S. 110–121; Eike Wolgast, Reformationszeit und Gegenreformation (1500–1648), in: ebd., S. 145–306, hier: S. 276–285; Dieter Stievermann, Absolutismus und Aufklärung (1648–1806), in: ebd., S. 307–456, hier: S. 446–451.

4 Vgl. einführend: Rudolf Endres, Handwerk – Berufsbildung, in: Notker, Handbuch der deutschen Bildungsgeschichte (wie Anm. 3), S. 375–424, hier: S. 377–379.

und Memmingen) bereits in vorreformatorischer Zeit einen bestimmenden Einfluss auf das Lehrangebot der Lateinschulen hatten.[5] Ebenfalls der Kontrolle des jeweiligen Magistrats unterlagen die in einigen größeren Städten (Nürnberg, Regensburg) um 1500 gegründeten Poetenschulen.[6] Diese elitären Bildungseinrichtungen boten in Konkurrenz zu den bestehenden Lateinschulen eine humanistisch profilierte Lehre an.

---

5 Vgl. die Abdrucke einer mutmaßlich in die 1490er Jahre zu datierenden Nürnberger Schulordnung, der Memminger Schulordnung von 1515 und der Nördlinger Schulordnungen von 1499, 1512 und 1522, in: Johannes Müller (Hg.), Vor- und frühreformatorische Schulordnungen und Schulverträge in deutscher und niederländischer Sprache, 2. Abt., Zschopau 1885/1886, hier: Abt. 1, S. 119–123; Abt. 2, S. 145–159, 169–175, 180–188, 212–228. Zur Datierung der Nürnberger Schulordnung vgl. zuletzt mit Verweis auf die ältere Literatur: Antonia Landois, Gelehrtentum und Patrizierstand. Wirkungskreise des Nürnberger Humanisten Sixtus Tucher (1459–1507), Tübingen 2014, S. 149–151. Vgl. zur Entwicklung des Bildungswesens in Nürnberg um 1500: Heinrich Wilhelm Heerwagen, Zur Geschichte der Nürnberger Gelehrtenschulen in dem Zeitraume von 1485 bis 1526, Nürnberg 1860; Rudolf Endres, Das Schulwesen in Franken im ausgehenden Mittelalter, in: Bernd Möller / Hans Patze / Karl Stackmann (Hg.), Studien zum städtischen Bildungswesen des späten Mittelalters und der frühen Neuzeit, Göttingen 1983, S. 173–214, hier bes.: S. 175–184; Ders., Nürnberger Bildungswesen zur Zeit der Reformation, in: Mitteilungen des Vereins für Geschichte der Stadt Nürnberg (künftig: MVGN) 71, 1984, S. 109–128; Martin Schieber, Wissenspolitik. Die Tätigkeit des Nürnberger Rates für die Entwicklung von Unterricht und Forschung, in: Martial Staub / Klaus A. Vogel (Hg.), Wissen und Gesellschaft in Nürnberg um 1500, Wiesbaden 1999, S. 36–47. Vgl. zu Nördlingen: Daniel Eberhard Beyschlag, Versuch einer Schulgeschichte der Reichsstadt Nördlingen, 6 Teile, Nördlingen 1793–1797; Hans Ockel, Die lateinische Schule der Reichsstadt Nördlingen. Auf Grund der Schulordnungen dargestellt, in: Zeitschrift des Historischen Vereins für Schwaben und Neuburg 34, 1908, S. 133–145, hier bes.: S. 133–136; Ders., Geschichte des höheren Schulwesens in Bayerisch-Schwaben während der vorbayerischen Zeit, Berlin 1931, S. 120–147, hier: S. 120–126; Friedrich Seßler, Die Auswirkungen von Reformation und Aufklärung auf die Entwicklung der Nördlinger Lateinschule, Augsburg 1985, S. 19–48; Dietmar-Henning Voges, Das Schulwesen, in: Ders., Die Reichsstadt Nördlingen. 12 Kapitel aus ihrer Geschichte, München 1988, S. 175–196, hier bes.: S. 177–181. Vgl. zu Memmingen: Ockel, Geschichte des höheren Schulwesens (wie oben), S. 147–171, hier: S. 148–154; Herbert Schallhammer, Das Schulwesen der Reichsstadt Memmingen von den Anfängen bis 1806 (Memminger Geschichtsblätter, Jahresheft 1962), Memmingen 1963, S. 8–12. Vgl. daneben zu den schwäbischen Städten insbesondere: Körner, Das höhere und niedere Schulwesen (wie Anm. 3), S. 688–693.

6 Wolfgang Mährle, Poetenschulen, veröffentlicht am: 11.1.2010, in: Historisches Lexikon Bayerns, URL: http://www.historisches-lexikon-bayerns.de/artikel/artikel_45323 (letzter Zugriff: 27.9.2020). Vgl. zur Nürnberger Poetenschule bes.: Gustav Bauch, Die Nürnberger Poetenschule 1496–1509, in: MVGN 14, 1901, S. 1–64; ergänzend zuletzt: Landois, Gelehrtentum und Patrizierstand (wie Anm. 5), S. 196–199.

Der spätmittelalterliche Übergang des Schulpatronats, der eingebettet war in übergreifende Kommunalisierungsprozesse,[7] vollzog sich in den einzelnen Reichsstädten Süddeutschlands in sehr unterschiedlicher Weise. Ein entscheidender Faktor für die jeweilige Entwicklung der Schuladministration war die Stellung der kirchlichen Institutionen.[8] In Städten ohne bedeutende kirchliche Funktionsträger (zum Beispiel Nördlingen, Memmingen) fiel es den Obrigkeiten leichter, die Kontrolle über die städtischen Lateinschulen sukzessive zu erlangen; in diesen Kommunen war am Ende des Mittelalters das „Schulregiment" des Rats ausgebildet. In Städten mit starken kirchlichen Einrichtungen kam es hingegen zum Teil zu größeren Auseinandersetzungen um die jeweiligen Kompetenzen im Bildungsbereich. So gab beispielsweise in Kempten das Kloster sein Monopol im Bereich der höheren Bildung erst nach längeren Konflikten auf.[9] In Augsburg verhinderte der bischöfliche Einfluss sogar, dass die Stadt das Schulpatronat erlangen konnte. Stattdessen übte bis in die Reformationszeit hinein der Domscholasticus die Aufsicht über die bestehenden fünf Lateinschulen aus.[10]

---

7 Zum „Kommunalismus" an der Wende vom Mittelalter zur Neuzeit existiert eine breite wissenschaftliche Literatur. An dieser Stelle sei lediglich genannt: Peter Blickle, Kommunalismus. Skizzen einer gesellschaftlichen Organisationsform, 2 Bde., München 2000.

8 Vgl. Peer Friess, Das Schulwesen der oberschwäbischen Reichsstädte im 16. Jahrhundert, in: Flachenecker / Kießling, Schullandschaften in Altbayern, Franken und Schwaben (wie Anm. 3), S. 303–329, hier: S. 308 f.

9 Ockel, Geschichte des höheren Schulwesens (wie Anm. 5), S. 223 f.; Margit Bauer / Daniela Sibbe-Fischer, Das Schulwesen im Mittelalter, in: Volker Dotterweich / Karl Filser / Pankraz Fried u.a. (Hg.), Geschichte der Stadt Kempten, Kempten 1989, S. 150–156, hier: S. 152–155.

10 Ockel, Geschichte des höheren Schulwesens (wie Anm. 5), S. 14; Körner, Das höhere und niedere Schulwesen (wie Anm. 3), S. 668; Norbert Hörberg, Domschule, in: Stadtlexikon Augsburg, URL: https://www.wissner.com/stadtlexikon-augsburg/artikel/stadtlexikon/domschule/3596 (letzter Zugriff: 27.9.2020). Zum spätmittelalterlichen Schulwesen in Augsburg vgl. auch: Martin Kintzinger, „Ich was auch ain schueler". Die Schulen im spätmittelalterlichen Augsburg, in: Johannes Janota / Werner Williams-Krapp (Hg.), Literarisches Leben in Augsburg während des 15. Jahrhunderts, Tübingen 1995, S. 58–81; Rolf Kießling, Pfarrschulen, in: Stadtlexikon Augsburg, URL: https://www.wissner.com/stadtlexikon-augsburg/artikel/stadtlexikon/pfarrschulen/5000 (letzter Zugriff: 27.9.2020); Ders., Lateinschulen, in: Stadtlexikon Augsburg, URL: https://www.wissner.com/stadtlexikon-augsburg/artikel/stadtlexikon/lateinschulen/4572 (letzter Zugriff: 27.9.2020).

Im Laufe des 16. Jahrhunderts fiel in allen Reichsstädten, in denen sich die evangelische Bewegung durchzusetzen vermochte, den Magistraten die uneingeschränkte Verfügungsgewalt sowohl über das Kirchenwesen als auch über das Schulwesen zu.[11] In vielen größeren und mittelgroßen sowie in einigen kleineren Reichsstädten Süddeutschlands wurden bereits kurze Zeit nach dem Übergang zur Reformation eigene Schulkommissionen beziehungsweise -deputationen geschaffen, in deren Hände die administrative Betreuung des deutschen und des lateinischen Schulwesens dauerhaft gelegt wurde. Diese Kommissionen wurden – zumeist erst in späterer Zeit – als „Scholarchate" bezeichnet. Ihre Mitglieder führten den Titel eines „*Schulherrn*" beziehungsweise eines „*Scholarchen*"; in Dinkelsbühl war, zumindest zeitweise, die Amtsbezeichnung „*Schulregent*", in Reutlingen „*Schulvogt*" gebräuchlich.[12]

Die erste reichsstädtische Schulkommission wurde in Süddeutschland im Jahr 1528 in der Bodenseestadt Lindau gebildet.[13] In den beiden folgenden Jahrzehnten bis zum Interim entstanden dann in zahlreichen Kommunen entsprechende Einrichtungen, so 1531 in Ulm,[14] 1533 in

---

11 Vgl. zusammenfassend: Seifert, Das höhere Schulwesen (wie Anm. 3), bes.: S. 279–282; vgl. zu den Verhältnissen in den Reichsstädten Straßburg, Nürnberg und Frankfurt am Main: Anton Schindling, Die Reformation in den Reichsstädten und die Kirchengüter – Straßburg, Nürnberg und Frankfurt im Vergleich, in: Jürgen Sydow (Hg.), Bürgerschaft und Kirche, Sigmaringen 1980, S. 67–88. Zuweilen bestanden in der Frühen Neuzeit noch Patronatsrechte, die aus vorreformatorischer Zeit herrührten. Sie stellten indes die städtische Schulhoheit nicht mehr prinzipiell in Frage. Vgl. zu Weißenburg in diesem Kontext: Rainer A. Müller, Strukturen des Schul- und Bildungswesens der Reichsstadt Weißenburg, in: Villa nostra. Weißenburger Blätter für Geschichte, Heimatkunde und Kultur von Stadt und Weißenburger Land 1/2002, S. 5–20, hier: S. 11.

12 Vgl. Paul Warmbrunn, Zwei Konfessionen in einer Stadt. Das Zusammenleben von Katholiken und Protestanten in den paritätischen Reichsstädten Augsburg, Biberach, Ravensburg und Dinkelsbühl von 1548 bis 1648, Wiesbaden 1983, hier: S. 299; vgl. auch die Reutlinger Schulordnung von 1566 bei: Franz Votteler, Geschichte der Lateinschule der Reichsstadt Reutlingen, in: Geschichte des humanistischen Schulwesens in Württemberg, 3 Bde., hg. v. der Württembergischen Kommission für Landesgeschichte, Stuttgart 1912–1928, hier: Bd. 2/1, Stuttgart 1920, S. 327–383, hier: S. 337–341.

13 Ockel, Geschichte des höheren Schulwesens (wie Anm. 5), S. 174.

14 Johannes Greiner, Geschichte der Ulmer Schule, in: Geschichte des humanistischen Schulwesens in Württemberg (wie Anm. 12), Bd. 2/1, Stuttgart 1920, S. 1–90 (bereits zuvor in: Ulm – Oberschwaben. Mitteilungen des Vereins für Kunst und Altertum in Ulm und Oberschwaben 20, 1914), hier: S. 25; Hans Eugen Specker, Das Gymnasium academicum in seiner Bedeutung für die Reichsstadt Ulm, in: Erich Maschke / Jürgen Sydow (Hg.), Stadt und Universität im Mittelalter und in der früheren Neuzeit, Sigmaringen 1977, S. 142–160, hier: S. 144.

Heilbronn,[15] 1534 in Esslingen[16] und Memmingen,[17] 1534/35 in Augsburg,[18] 1541 in Regensburg,[19] 1542 in Dinkelsbühl,[20] (vermutlich) 1542/44 in Kempten,[21] 1543 in Nördlingen[22] und 1546 in Ravensburg.[23] Ebenfalls in der Reformationsepoche begegnen Scholarchen im fränkischen Windsheim.[24]

---

15 Gustav Lang, Geschichte des Gymnasiums der Reichsstadt Heilbronn, in: Geschichte des humanistischen Schulwesens in Württemberg (wie Anm. 12), Bd. 2/1, Stuttgart 1920, S. 91–203, hier: S. 95; Alfred Kolbeck (Bearb.), 350 Jahre Gymnasium in Heilbronn. Festschrift zum Jubiläum des Theodor-Heuss-Gymnasiums, hg. v. der Schulleitung des Theodor-Heuss-Gymnasiums in Heilbronn, Heilbronn 1971, hier: S. 14.

16 Karl Pfaff, Geschichte der Reichsstadt Esslingen. Nach Archivalurkunden und andern bewährten Quellen, Bd. 1, Esslingen 1840, S. 742; Otto Mayer, Geschichte des humanistischen Schulwesens in der freien Reichsstadt Esslingen 1267–1803, in: Geschichte des humanistischen Schulwesens in Württemberg (wie Anm. 12), Bd. 2/1, Stuttgart 1920, S. 204–326, hier: S. 258 f.; Tilman Matthias Schröder, Das Kirchenregiment der Reichsstadt Esslingen. Grundlagen – Geschichte – Organisation, Esslingen 1987, S. 350 f.

17 Ockel, Geschichte des höheren Schulwesens (wie Anm. 5), S. 147–171, hier: S. 156; Schallhammer, Das Schulwesen der Reichsstadt Memmingen (wie Anm. 5), S. 13.

18 Ludwig Greiff, Beiträge zur Geschichte der deutschen Schulen Augsburgs. Aus urkundlichen Quellen gesammelt, Augsburg 1858, S. 8 f.; Ockel, Geschichte des höheren Schulwesens (wie Anm. 5), S. 18; Karl Köberlin, Geschichte des humanistischen Gymnasiums bei St. Anna in Augsburg von 1531 bis 1931, Augsburg 1931, S. 20 f.

19 Christian Heinrich Kleinstäuber, Ausführliche Geschichte der Studien-Anstalten in Regensburg 1538–1880. Erster Teil: Geschichte des evangelischen reichsstädtischen Gymnasii poetici, in: Verhandlungen des historischen Vereins von Oberpfalz und Regensburg 35, 1880, S. 1–152 und 36, 1882, S. 1–142, hier: S. 36 bzw. S. 77. Vgl. auch: Karl Ernst Maier, Die Schulverhältnisse in der Reichsstadt Regensburg, in: Max Liedtke (Hg.), Handbuch der Geschichte des bayerischen Bildungswesens, Bd. 1: Geschichte der Schule in Bayern von den Anfängen bis 1800, Bad Heilbrunn/Oberbayern 1991, S. 447–455, hier: S. 450.

20 Warmbrunn, Zwei Konfessionen in einer Stadt (wie Anm. 12), S. 299.

21 Margit Bauer / Volker Dotterweich / Daniela Sibbe-Fischer, Das Schulwesen von der Reformation bis zur Säkularisation, in: Dotterweich / Filser / Fried u.a., Geschichte der Stadt Kempten (wie Anm. 9), S. 289–303, hier: S. 290.

22 Voges, Das Schulwesen (wie Anm. 5), S. 186.

23 Alfons Dreher, Geschichte der Reichsstadt Ravensburg und ihrer Landschaft von den Anfängen bis zur Mediatisierung 1802, Bd. 2, Weißenhorn 1972, hier: S. 770; Warmbrunn, Zwei Konfessionen in einer Stadt (wie Anm. 12), S. 295.

24 Christian Wilhelm Schirmer, Geschichte Windsheims und seiner Nachbarorte, Nürnberg 1848, S. 130; Monika Baumüller-Rappold, „Schöne Pflanzgärtlein und Werkstätte aller Gottseligkeit, Ehrbarkeit, Tugenden und guter nuzer Lehr". Fünf Reichsstädte in Franken und ihre Schulen, in: Rainer Albert Müller (Hg.), Reichsstädte in Franken, Bd. 2: Aufsätze (Veröffentlichungen zur Bayerischen Geschichte und Kultur 15/2), München 1987, S. 284–297, hier: S. 290.

In weiteren Reichsstädten Süddeutschlands, die sich der Reformation angeschlossen hatten oder die als „bikonfessionelle“ Städte evangelische Bevölkerungsmehrheiten aufwiesen, erfolgte die Einsetzung von Schulkommissionen erst nach dem Einschnitt des Interims beziehungsweise nach dem Augsburger Religionsfrieden. Scholarchen lassen sich in Rothenburg ob der Tauber seit 1557 nachweisen,[25] in Kaufbeuren seit 1562,[26] in Reutlingen seit 1566,[27] in Nürnberg seit 1576/82, in Schweinfurt seit 1578[28] und in Schwäbisch Hall seit 1584.[29] In Biberach markiert die 1588/90 erfolgte Einsetzung je eines evangelischen und eines katholischen Schulvisitationsausschusses, dem jeweils auch Ratsherren angehörten, die Gründung einer dauerhaften, bereits konfessionell getrennten Schulverwaltung.[30] Nicht mehr datieren lässt sich die Einsetzung von Scholarchen in Weißenburg.[31]

Auffallend ist die späte Formierung einer speziellen Schulbehörde im bevölkerungsreichen Nürnberg. In der fränkischen Metropole oblag die Aufsicht

25 Walter Bauer, Die Reichsstadt Rothenburg und ihre Lateinschule, Rothenburg 1979, S. 56.

26 Joseph Sieber, Beiträge zur Schulgeschichte des Landkapitels und der Reichsstadt Kaufbeuren bis zum Jahre 1803, Kaufbeuren [1921], S. 133 f.

27 Votteler, Geschichte der Lateinschule der Reichsstadt Reutlingen (wie Anm. 12), S. 337–341.

28 Valentin Völcker, Geschichte der Studienanstalt Schweinfurt. 1. Teil: Geschichte der lateinischen Schule zu Schweinfurt bis zum Jahr 1634, Schweinfurt 1882, hier: S. 18.

29 Vgl. Wilhelm Kolb, *Schola latina* und *Gymnasium illustre* in Schwäbisch-Hall, in: Geschichte des humanistischen Schulwesens in Württemberg (wie Anm. 12), Bd. 2/1, Stuttgart 1920, S. 490–587, hier: S. 522; Andreas Maisch / Daniel Stihler, Schwäbisch Hall. Geschichte einer Stadt, Schwäbisch Hall 2006, S. 145.

30 Warmbrunn, Zwei Konfessionen in einer Stadt (wie Anm. 12), S. 303; Andrea Riotte, „Diese so oft beseufzte Parität“. Biberach 1649–1825. Politik – Konfession – Alltag, Stuttgart 2017, hier: S. 207.

31 Vgl. Leonhard Götz, Beiträge zur Geschichte der Lateinschule zu Weissenburg a. S. als Erinnerungsschrift an deren 350jährigen Bestand, Weißenburg 1886, S. 13 f.; Gustav Mödl / Karlheinz Trabert, Die Lateinschule von der Reformation bis zum Ende der reichsstädtischen Zeit, in: Gustav Mödl (Hg.), nostra uuizzinburgi. Häuser – Menschen – Bücher, Weißenburg 1977, S. 12–29, hier: S. 18; Baumüller-Rappold, „Schöne Pflanzgärtlein“ (wie Anm. 24), S. 290; Müller, Strukturen des Schul- und

über die zahlreichen städtischen Schulen mehrere Jahrzehnte lang dem Kirchenpfleger Hieronymus Baumgartner (1498–1565), einer herausragenden Figur der reichsstädtischen Politik.[32] Ein Scholarchat als kollegiales Gremium entstand erst zwischen 1576 und 1582 im Zuge des Aufbaus der Hochschule in Altdorf.[33]

Im Unterschied zu den großen und mittelgroßen evangelischen beziehungsweise bikonfessionellen Reichsstädten wurden in den katholischen reichsfreien Städten Süddeutschlands – mit einer Ausnahme – keine speziellen Kommissionen für Schulangelegenheiten gebildet. Belegt sind Scholarchen seit 1563 in Rottweil, das stark von den Traditionen der Straßburger Schulgeschichte geprägt war.[34] Hingegen verfügten die beiden anderen mittelgroßen katholischen Reichsstädte Süddeutschlands, die höhere Schulen unterhielten, Schwäbisch Gmünd und Überlingen, über keine Scholarchate. Die städtische Schuladministration unterlag im katholischen Umfeld bis zu einem gewissen Grad anderen Voraussetzungen als in den evangelischen Gemeinwesen. Die dortigen

---

Bildungswesens der Reichsstadt Weißenburg (wie Anm. 11), S. 11; Ders., Strukturen des Schul- und Bildungswesens der fränkischen Reichsstadt Weißenburg in Spätmittelalter und früher Neuzeit, in: Flachenecker / Kießling, Schullandschaften in Altbayern, Franken und Schwaben (wie Anm. 3), S. 203–224, hier: S. 213. Zuletzt machte Reiner Kammerl darauf aufmerksam, dass sich in Weißenburg erst seit Beginn des 17. Jahrhunderts Scholarchen nachweisen lassen. Vgl. Reiner Kammerl, 675 Jahre Lateinschule. Die Anfänge des Gymnasiums in Weißenburg, in: Dieter Thelsinger (Hg.), 1337–2012. 675 Jahre Gymnasium in Weißenburg. Festschrift. Darf's ein bisschen mehr sein?, S. 12–19, hier: S. 17.

32 Vgl. zuletzt ausführlich: Wolfgang Mährle, Hieronymus I. Paumgartner (1498–1565) und das nachreformatorische Kirchen- und Bildungswesen der Reichsstadt Nürnberg, in: Michael Diefenbacher / Olga Fejtová / Zdzisław Noga (Hg.), Stadt und Reformation. Krakau, Nürnberg und Prag (1500–1618), Praha 2019, S. 369–400 (mit weiterer Literatur). Die Schreibweise Baumgartner/Baumgärtner/Paumgartner/Paumgärtner variierte bereits im 16. Jahrhundert.

33 Wolfgang Mährle, Academia Norica. Wissenschaft und Bildung an der Nürnberger Hohen Schule in Altdorf (1575–1623), Stuttgart 2000, S. 110–145; Peter Fleischmann, Das Verhältnis des Nürnberger Rats zur Universität Altdorf, in: Hanns Christof Brennecke / Dirk Niefanger / Werner Wilhelm Schnabel (Hg.), Akademie und Universität Altdorf. Studien zur Hochschulgeschichte Nürnbergs, Köln/Weimar/Wien 2011, S. 7–28.

34 Vgl. Johannes Greiner, Geschichte der Schule in Rottweil am Neckar, in: Geschichte des humanistischen Schulwesens in Württemberg (wie Anm. 12), Bd. 2/1, Stuttgart 1920, S. 384–465, hier: S. 398 f.

Lateinschulen wurden zum Teil von geistlichen Orden geführt.[35] Am Beispiel Rottweils, wo die Leitung der städtischen Lateinschule seit dem 17. Jahrhundert nacheinander in den Händen mehrerer religiöser Gemeinschaften (Dominikaner, Jesuiten, Benediktiner) lag, lässt sich sehr gut nachvollziehen, dass die Beteiligung von Orden am Schulwesen oft langwierige Aushandlungsprozesse zwischen Stadt und Ordensvertretern notwendig machte, bevor man zu verbindlichen administrativen Regelungen gelangte.[36]

Keine speziellen Kommissionen für Schulangelegenheiten gab es auch in zahlreichen kleineren Reichsstädten, die zum größeren Teil ebenfalls katholisch geblieben waren.[37] In diesen Gemeinwesen war in der Regel keine Lateinschule vorhanden, sondern allenfalls eine „vermengte" Schule, in der Latein zusätzlich zum deutschsprachigen Unterricht angeboten wurde. Eine personell fest umrissene Schulkommission war zur Verwaltung einer derartigen Bildungseinrichtung nicht erforderlich. Die Administration der vorhandenen Schule(n) erfolgte durch den Rat. Insgesamt lässt sich in 15 der untersuchten 37 reichsfreien Städte Süddeutschlands kein Scholarchat nachweisen.[38]

2. Organisation der reichsstädtischen Schulverwaltungen

Die nachreformatorischen Schulkommissionen stellten ein neues Element in der Verwaltung der süddeutschen Reichstädte dar. Die Bildung dieser kollegialen Schulbehörden konnte allerdings an vorreformatorische Entwicklungen

---

35 In Schwäbisch Gmünd existierten zwei Lateinschulen: eine städtische und eine des Franziskanerordens. In Überlingen führten die Franziskaner seit 1658 die zuvor städtische Lateinschule.

36 Vgl. hierzu die Ausführungen bei: Greiner, Geschichte der Schule in Rottweil am Neckar (wie Anm. 34), bes.: S. 421–423, 434–449.

37 Vgl. zum Schulwesen in diesen kleineren Städten: Adolf Diehl, Das humanistische Schulwesen in den Reichsstädten Biberach, Ravensburg, Isny, Leutkirch, Wangen, Buchau, Buchhorn, Weilderstadt, Aalen, Bopfingen und Giengen, in: Geschichte des humanistischen Schulwesens in Württemberg (wie Anm. 12), Bd. 2/1, S. 588–614.

38 Aalen, Bopfingen, Buchau, Buchhorn, Gengenbach, Giengen, Isny, Leutkirch, Offenburg, Pfullendorf, Schwäbisch Gmünd, Überlingen, Wangen, Weil der Stadt, Zell am Harmersbach. Zum Personalbestand der jeweiligen reichsstädtischen Verwaltungen vgl. auch die Angaben in den Adressbüchern des Schwäbischen Kreises aus dem späten 18. Jahrhundert: Des hochlöbl. Schwäbischen Crayses vollständiges Staats- und Addreß-Buch […], Geislingen 1766, 1771, 1773, 1777, 1785 und 1786 sowie Staats- und Adreß-Handbuch des hochlöbl. schwäbischen Kreises […], Ulm 1791.

anknüpfen.[39] Ähnliche Einrichtungen hatte es seit dem 14. Jahrhundert in italienischen und seit dem 15. Jahrhundert auch in mitteleuropäischen Universitätsstädten (Basel, Köln) gegeben. Dort waren jeweils spezielle Ratsausschüsse für das Universitätswesen gebildet worden, die zum Teil umfassende Leitungsfunktionen wahrnahmen und in diesem Kontext unter anderem für die Einstellung von Professoren verantwortlich waren. Einen Anstoß für die Einsetzung von Schulkommissionen in den süddeutschen Reichsstädten, insbesondere in den zahlreichen schwäbischen Kommunen, in denen die Reformation zwinglianischen Mustern folgte (so etwa in Augsburg, Esslingen, Lindau, Memmingen und Ulm), dürfte die Einrichtung eines aus drei Ratsherren bestehenden Scholarchats in der Reichsstadt Straßburg im Jahr 1525 gegeben haben.[40] Allerdings wurden die Schulkommissionen in fast allen Reichsstädten des oberdeutschen Raumes anders ausgestaltet, als dies in Straßburg der Fall war.

Organisation und personelle Zusammensetzung der Scholarchate, die in den süddeutschen Reichsstädten im 16. Jahrhundert gebildet wurden, unterschieden sich erheblich. Wie die Schulkommissionen organisiert wurden, war in erster Linie von den bestehenden verfassungsrechtlichen Rahmenbedingungen, daneben von der Stellung kirchlicher Institutionen sowie von personellen Konstellationen abhängig.[41] Hingegen spielte bei der Einrichtung der Scholarchate das Vorbild anderer Kommunen zumeist keine entscheidende Rolle. Die große Variationsbreite an Organisationsformen in der Schulverwaltung der reichsunmittelbaren Städte Süddeutschlands erstaunt zum einen angesichts der Befunde

39 Vgl. zur Vorbildwirkung der im Spätmittelalter für das Universitätswesen gebildeten Ratsausschüsse bei der Konstitution des Straßburger Scholarchenkollegiums: Schindling, Humanistische Hochschule und freie Reichsstadt (wie Anm. 1), S. 80–82.

40 Vgl. ebd., S. 80. Vgl. zur Vorbildwirkung Straßburgs auf Ulm und Esslingen: Greiner, Geschichte der Ulmer Schule (wie Anm. 14), S. 24 f.; Schröder, Das Kirchenregiment der Reichsstadt Esslingen (wie Anm. 16), S. 350–352. Zur Reformationsgeschichte der schwäbischen Reichsstädte vgl. zusammenfassend: Wilfried Enderle, Ulm und die evangelischen Reichsstädte im Südwesten, in: Anton Schindling / Walter Ziegler (Hg.), Die Territorien des Reichs im Zeitalter der Reformation und der Konfessionalisierung. Land und Konfession 1500–1650, Bd. 5: Der Südwesten, Münster 1993, S. 194–212; Ders., Rottweil und die katholischen Reichsstädte im Südwesten, in: ebd., S. 214–230; Herbert Immenkötter / Wolfgang Wüst, Augsburg, Freie Reichsstadt und Hochstift, in: Schindling / Ziegler, Die Territorien des Reichs im Zeitalter der Reformation und der Konfessionalisierung (wie oben), Bd. 6: Nachträge, Münster 1996, S. 8–35.

41 Friess weist darauf hin, dass den Scholarchaten der Bischofsstädte Straßburg, Konstanz und Augsburg keine Geistlichen angehörten. Vgl. Friess, Schulwesen der oberschwäbischen Reichsstädte (wie Anm. 8), S. 317.

in anderen Bereichen der Bildungspolitik – zum Beispiel bei der Lehrplangestaltung.[42] Sie erstaunt zum anderen aber auch angesichts der dichten Kommunikation, die zwischen den Reichsstädten bestand und deren Auswirkungen in zahlreichen Politikfeldern greifbar sind.[43]

Drei Typen von Scholarchaten lassen sich in den süddeutschen Reichsstädten abgrenzen: In einigen Städten – in Nördlingen, Nürnberg, Schweinfurt und Windsheim – bildeten lediglich die mit den Schulangelegenheiten befassten (zwei bis vier) Ratsmitglieder das Scholarchat. Andere in der Schulverwaltung engagierte Personen, zum Beispiel Geistliche, die als Schulvisitatoren wirkten, nahmen ihre Aufgaben im Auftrag der Scholarchen wahr. In anderen Städten – in Augsburg, Biberach, Regensburg, Rothenburg ob der Tauber und Ulm – wurden außer Ratsmitgliedern ausgewählte geistliche Funktionsträger oder Angehörige der städtischen akademischen Elite zu Scholarchen ernannt.[44] Auch in diesen Städten umfasste das Scholarchat lediglich einen kleinen Personenkreis und agierten weitere Personen im Auftrag der schulischen Oberbehörde. In den meisten reichsfreien Gemeinwesen – in Dinkelsbühl, Esslingen, Heilbronn, Kaufbeuren, Kempten, Lindau, Memmingen, Ravensburg, Reutlingen, Rottweil, Schwäbisch Hall, Weißenburg und Wimpfen – wurden hingegen im Scholarchat alle in der Schulverwaltung an verantwortlicher Stelle tätigen Persönlichkeiten vereint. Die Zahl der Scholarchen war in diesen Kommunen deutlich höher als

---

42 Vgl. Wolfgang Mährle, Straßburg als Vorbild. Das akademische Gymnasium Johannes Sturms und das evangelische höhere Bildungswesen in Süddeutschland (1540–1620), in: Historisches Jahrbuch 133, 2013, S. 167–224.

43 Vgl. zur reichsstädtischen Kommunikation im Spätmittelalter und im frühen 16. Jahrhundert: Wolfgang Wüst, Reichsstädtische Kommunikation in Franken und Schwaben. Nachrichtennetze für Bürger, Räte und Kaufleute im Spätmittelalter, in: Zeitschrift für bayerische Landesgeschichte 62, 1999, S. 681–707; Ders., Kommunikation im Bündnis. Zur Rolle oberdeutscher Reichsstädte in den Bauernunruhen 1524/25, in: Zeitschrift des Historischen Vereins für Schwaben 92, 1999, S. 7–30; Ders., Reichstage und Reichsstädte – Netzwerke in Süddeutschland im ausgehenden Mittelalter und in beginnender Neuzeit, in: Thomas Lau / Helge Wittmann (Hg.), Kaiser, Reich und Reichsstadt in der Interaktion (Studien zur Reichsstadtgeschichte 3), Petersberg 2016, S. 61–84.

44 Vgl. zu Augsburg: Ockel, Geschichte des höheren Schulwesens (wie Anm. 5), S. 21; Köberlin, Geschichte des humanistischen Gymnasiums (wie Anm. 18), S. 29 f.; vgl. zu Biberach: Warmbrunn, Zwei Konfessionen in einer Stadt (wie Anm. 12), S. 303; vgl. zu Regensburg: Kleinstäuber, Ausführliche Geschichte der Studien-Anstalten in Regensburg (wie Anm. 19), 36, 1882, S. 77 f., 82–85; vgl. zu Rothenburg: Bauer, Reichsstadt Rothenburg (wie Anm. 25), S. 56 f.; vgl. zu Ulm: Greiner, Geschichte der Ulmer Schule (wie Anm. 14), S. 30 f., 41 f.

in den Städten der beiden zuerst genannten Gruppen. So amtierten beispielsweise in Heilbronn und Memmingen je zehn, in Kempten bis zu 13, in Esslingen bis zu 15 Scholarchen.[45]

Das Scholarchenkollegium war in allen Städten hierarchisch gegliedert. In der Regel ergab sich eine Rangordnung durch die Amtsfunktion der jeweiligen Mitglieder in der Stadtverwaltung. In Kaufbeuren, Nördlingen, Nürnberg und Regensburg begegnet die Position eines Protoscholarchen, der den Vorsitz im Gremium innehatte.[46] Als Protoscholarch amtierte stets ein Ratsmitglied; falls ein Bürgermeister im Scholarchat vertreten war, fiel ihm die Position des Protoscholarchen zu. Ansonsten wurde der ranghöchste städtische Funktionsträger mit der Leitung der Schulbehörde betraut. In Augsburg und Ulm unterschied man zwischen Ober- und Unterscholarchen. In Augsburg waren die Oberscholarchen stets Ratsmitglieder, die Unterscholarchen Gelehrte.[47] In Ulm amtierten seit etwa 1600 die Pfarrkirchenbaupfleger als Oberscholarchen.[48] Sie waren dem Superintendenten und zwei weiteren Akademikern (einem Juristen und einem Arzt) vorgesetzt, die als „einfache" Scholarchen fungierten.

In den Reichsstädten Kempten, Memmingen, Nördlingen und Ulm findet sich neben dem jeweiligen Scholarchat die Institution eines Schulkonvents. Diese Einrichtung stellte in den genannten Kommunen (im Unterschied etwa zu Straßburg) kein Versammlungsgremium des Lehrpersonals dar, sondern bildete ein Organ der Schulverwaltung.[49] In Kempten gehörten dem Schulkonvent

---

45 Vgl. zu Heilbronn: Lang, Geschichte des Gymnasiums der Reichsstadt Heilbronn (wie Anm. 15), S. 95; Kolbeck, 350 Jahre Gymnasium in Heilbronn (wie Anm. 15), S. 14; zu Memmingen: Schallhammer, Das Schulwesen der Reichsstadt Memmingen (wie Anm. 5), S. 13; vgl. zu Kempten: Bauer / Dotterweich / Sibbe-Fischer, Das Schulwesen von der Reformation bis zur Säkularisation (wie Anm. 21), S. 290; vgl. zu Esslingen: Mayer, Geschichte des humanistischen Schulwesens (wie Anm. 16), S. 259.

46 Vgl. zu Kaufbeuren: Sieber, Beiträge zur Schulgeschichte (wie Anm. 26), S. 133; vgl. zu Nördlingen: Stadtarchiv Nördlingen, R 39 F 13, Nr. 9: Schulreformvorschläge 1747 (freundliche Mitteilung von Dr. Wilfried Sponsel, Stadtarchiv Nördlingen); vgl. zu Nürnberg: Mährle, Academia Norica (wie Anm. 33), S. 134, Anm. 91; vgl. zu Regensburg: Kleinstäuber, Ausführliche Geschichte der Studien-Anstalten in Regensburg (wie Anm. 19), 36, 1882, S. 78. In Schwäbisch Hall begegnet ab 1614 der Titel „Direktor" für den Vorsitzenden im Scholarchat. Vgl. Kolb, *Schola latina* und *Gymnasium illustre* (wie Anm. 29), S. 522.

47 Ockel, Geschichte des höheren Schulwesens (wie Anm. 5), S. 21; Köberlin, Geschichte des humanistischen Gymnasiums (wie Anm. 18), S. 29 f.

48 Greiner, Geschichte der Ulmer Schule (wie Anm. 14), S. 30 f., 41 f.

49 Vgl. zu Straßburg: Schindling, Humanistische Hochschule und freie Reichsstadt (wie Anm. 1), S. 102–104.

diejenigen Scholarchen an, die gleichzeitig das Visitatorenamt innehatten.[50] In Ulm bildeten Schulherren, Visitatoren und der Rektor des Gymnasiums den Schulkonvent, dessen Aufgabe unter anderem in der Beratung der Ergebnisse von Schulvisitationen bestand.[51] Ähnlich lagen die Verhältnisse wohl in Memmingen und Nördlingen.[52]

Die Scholarchate, die sich im Laufe des 16. Jahrhunderts in den süddeutschen Reichsstädten etablierten, wiesen insgesamt eine hohe institutionelle Stabilität auf. Nach einer Konsolidierungsphase, die in einzelnen Städten (zum Beispiel in Ulm) mehrere Jahrzehnte umfasste und in der die Zusammensetzung und die Aufgabenwahrnehmung zum Teil öfters modifiziert wurden, bestanden die Schulverwaltungsbehörden in den meisten Kommunen bis zum Beginn des 19. Jahrhunderts ohne grundlegende Veränderungen fort. In einigen Städten erfolgte eine Zusammenführung von Scholarchat und Konsistorium beziehungsweise Kirchendirektorium: in Rothenburg ob der Tauber 1581 oder 1602,[53] in Esslingen 1656/57[54] sowie in Regensburg 1780.[55] In Schwäbisch Hall scheinen Scholarchat und Konsistorium hingegen von Beginn an identische Gremien gewesen zu sein.[56] Veränderungen, die in zahlreichen Städten während des Dreißigjährigen Krieges aus konfessionellen Gründen erzwungen worden waren, blieben zumeist ohne langfristige Folgen. Lediglich in den nach 1648 paritätischen Städten bewirkte der Krieg strukturelle Korrekturen; hier wurde die konfessionelle Trennung des Bildungswesens festgeschrieben, die in den Jahrzehnten vor 1618 in unterschiedlichem Umfang verwirklicht worden war. So setzte der Rat der Reichsstadt Augsburg nach Kriegsende konfessionell

50 Bauer / Dotterweich / Sibbe-Fischer, Das Schulwesen von der Reformation bis zur Säkularisation (wie Anm. 21), S. 290.

51 Greiner, Geschichte der Ulmer Schule (wie Anm. 14), S. 41 f.; Ders., Die Ulmer Gelehrtenschule zu Beginn des 17. Jahrhunderts und das akademische Gymnasium. Darstellung und Quellenmaterial, Ulm 1912 (zugleich erschienen in: Ulm – Oberschwaben. Mitteilungen des Vereins für Kunst und Altertum in Ulm und Oberschwaben 18, 1912), hier S. 58 f.; Hans Eugen Specker, Ulm. Stadtgeschichte, Ulm 1977, S. 156; Ders., Gymnasium academicum (wie Anm. 14), S. 145 f.

52 Vgl. zu Memmingen: Ockel, Geschichte des höheren Schulwesens (wie Anm. 5), S. 156; vgl. zu Nördlingen: ebd., S. 137; Seßler, Auswirkungen von Reformation und Aufklärung (wie Anm. 5), S. 224 f.

53 Bauer, Reichsstadt Rothenburg (wie Anm. 25), S. 56 f.

54 Schröder, Das Kirchenregiment der Reichsstadt Esslingen (wie Anm. 16), S. 275–278.

55 Kleinstäuber, Ausführliche Geschichte der Studien-Anstalten in Regensburg (wie Anm. 19), 36, 1882, S. 78.

56 Kolb, *Schola latina* und *Gymnasium illustre* (wie Anm. 29), S. 522.

geschiedene Schulaufsichtsbehörden ein.[57] Derartige Einrichtungen gab es bereits in Biberach sowie im ebenfalls bikonfessionellen, wenngleich nicht unter die Paritätsbestimmungen des Westfälischen Friedens fallenden Kaufbeuren.[58] Des Weiteren ist zu beobachten, dass in einigen Reichsstädten neben dem Scholarchat vereinzelt temporäre „Sonder"-Kommissionen für das Schulwesen gebildet wurden, die neben das Scholarchat traten oder in denen die regulären Schuldeputationen aufgingen. So setzten die Reichsstädte Ulm und Reutlingen in den Jahren 1609 bis 1613 (Ulm) beziehungsweise 1668 (Reutlingen) Schulreformkommissionen ein.[59] Diese Einrichtungen blieben jedoch ephemer. Dies gilt auch für die zeitweise Verlagerung von Kompetenzen, die sich in Augsburg und Rothenburg ob der Tauber feststellen lässt: In der Lechmetropole wurde dem Rektor des Gymnasiums bei Sankt Anna, Hieronymus Wolf, von 1557 bis 1562 die Aufsicht über das Schulwesen übertragen.[60] In Rothenburg amtierte von 1592 bis 1607 Andreas Libavius als „*inspector scholae*"; der Rat übertrug Libavius umfassende Befugnisse im Bildungswesen, sodass die Schulherren in dieser Zeit in den Hintergrund traten.[61]

Schwieriger zu beantworten ist die Frage, inwieweit sich in einzelnen Reichsstädten bei gleichbleibender institutioneller Struktur im Laufe der Zeit die Kompetenzen zwischen den verschiedenen, an der Schulaufsicht beteiligten Institutionen veränderten. In Rothenburg ob der Tauber, dessen Bildungsgeschichte

---

57 Vgl. Greiff, Beiträge zur Geschichte der deutschen Schulen Augsburgs (wie Anm. 18), S. 57; Leonhard Lenk, Augsburger Bürgertum im Späthumanismus und Frühbarock (1580–1700), Augsburg 1968, S. 122.

58 In Biberach wurden 1588/90 vermutlich zwei Scholarchate gebildet. Vgl. Warmbrunn, Zwei Konfessionen in einer Stadt (wie Anm. 12), S. 303; Riotte, „Diese so oft beseufzte Parität" (wie Anm. 30), S. 207. In Kaufbeuren existierte seit 1620 (evtl. bereits seit etwa 1600) neben der evangelischen eine eigene katholische Schulbehörde. Vgl. Sieber, Beiträge zur Schulgeschichte (wie Anm. 26), S. 133.

59 Greiner, Ulmer Gelehrtenschule (wie Anm. 51), S. 16 f.; Votteler, Geschichte der Lateinschule der Reichsstadt Reutlingen (wie Anm. 12), S. 350–364; Heinrich Betz, Von der mittelalterlichen Lateinschule zum Friedrich-List-Gymnasium. 700 Jahre Schulgeschichte, in: Ernst-Reinhard Beck (Red.), Festschrift des Friedrich-List-Gymnasiums Reutlingen 1987. 700 Jahre Lateinschule. 100 Jahre Gymnasium, Reutlingen 1987, S. 15–66, hier: S. 26–34.

60 Ockel, Geschichte des höheren Schulwesens (wie Anm. 5), S. 30 f.; Köberlin, Geschichte des humanistischen Gymnasiums (wie Anm. 18), S. 64.

61 Bauer, Reichsstadt Rothenburg (wie Anm. 25), S. 73. Zu Andreas Libavius vgl. einführend: Friedemann Rex, Libavius, Andreas, in: Neue Deutsche Biographie 14, 1985, S. 441 f.

gut erforscht ist, gibt es – abgesehen von der Libavius-Zeit – deutliche Hinweise auf Zuständigkeitsveränderungen. So hat der Rat ab dem ausgehenden 17. Jahrhundert seine Befugnisse gegenüber dem beim Konsistorium angesiedelten Scholarchat erweitert.[62] Ähnliche Kompetenzänderungen könnte es auch in anderen Städten gegeben haben; zu diesem Problembereich bedarf es indes noch vertiefender Forschungen.

3. Funktionen in der Schuladministration

Die reichsstädtischen Scholarchate nahmen die Administration des Bildungswesens im Auftrag und in Vertretung der Obrigkeit, also des Stadtmagistrats, wahr. In allen reichsunmittelbaren Städten delegierte der Rat nur einen Teil der ihm obliegenden Aufgaben. Eine präzise Kompetenzabgrenzung zwischen der politischen Stadtspitze und dem jeweiligen Scholarchat fehlte allerdings in den meisten süddeutschen Reichsstädten. Sofern die Aufgaben der Schulverwaltung schriftlich fixiert wurden, geschah dies in der Form einer – mehr oder weniger vollständigen – Aufzählung von Zuständigkeiten, des Öfteren in einer Schulordnung.

Zu den Angelegenheiten von grundsätzlicher Tragweite, deren Entscheidung sich die Ratsherren in den meisten Städten vorbehielten, zählten insbesondere der Erlass normativer Regelungen (Schulordnungen), die Durchführung von Schulreformen, die Planung und Ausführung von Schulneubauten, disziplinarische Angelegenheiten, die den Rahmen des üblichen schulischen Alltags sprengten, sowie – nicht immer, aber doch häufig – auch die Besetzung von Leitungsfunktionen im Bildungswesen, zum Beispiel von Rektorenstellen.[63] Sofern die Scholarchate Entscheidungen innerhalb ihres Zuständigkeitsbereichs nicht selbst treffen konnten, bereiteten sie Beschlüsse der Stadtmagistrate vor. Auf diese Weise hatten sie auch in Angelegenheiten, in denen sie im Auftrag agierten, einen erheblichen Einfluss. Insgesamt prägten die Schulkommissionen damit sowohl direkt als auch indirekt die Administration des Schulwesens sowie – mit Einschränkungen – die städtische Bildungspolitik.

---

62 Bauer, Reichsstadt Rothenburg (wie Anm. 25), S. 135–140.

63 Vgl. in diesem Kontext exemplarisch den Umgang des Nürnberger Rates mit Albrecht von Wallenstein, der im Studienjahr 1599/1600 an der Hohen Schule in Altdorf als Student eingeschrieben war: Wolfgang Mährle, Der gewalttätige Student. Wallenstein an der Hohen Schule in Altdorf, in: Birgit Emich / Dirk Niefanger / Dominik Sauerer / Georg Seiderer (Hg.), Wallenstein. Mensch – Mythos – Memoria, Berlin 2018, S. 179–209.

Ein instruktives Beispiel für eine normative Festlegung von Kompetenzen des Scholarchats bildet die Ulmer Schulordnung von 1613, die den Titel „*Des ganzen lateinischen schulwesens allhier in Ulm reformiert ernewerte ordnung und bestellung. 1613*" trägt.[64] Der erste Abschnitt des zweiten Teils dieser Ordnung ist den „*oberen und fuerstehern der schuel*" gewidmet. Dort werden neben der Zusammensetzung des Scholarchats auch dessen Aufgaben geregelt. Den Scholarchen waren demnach „*fleissige ἐπισκοπή und obacht insgemein über das gantze schulwesen anbefohlen*".[65] Konkret überwachten sie den Lehr- und Prüfungsbetrieb, waren für die Schuldisziplin verantwortlich und nahmen repräsentative Aufgaben wahr, insbesondere „*bei den probationibus, progressionibus und anderen actibus*". Das wichtigste Instrument für die „Qualitätssicherung" der Lehre war – in Ulm wie andernorts – die Durchführung von Schulvisitationen. Die Evaluationen des Lehrbetriebs, die in den einzelnen Reichsstädten unterschiedlich häufig stattfanden, waren ein neues Element im Schulwesen der Frühen Neuzeit und ein wichtiger Baustein der im 16. Jahrhundert einsetzenden städtischen Schulaufsicht.[66] Zu den in der Ulmer Schulordnung explizit genannten Zuständigkeiten der Scholarchen kamen in zahlreichen Städten noch die Verantwortung für die Schulfinanzen sowie für die Schulgebäude.

Kaum erforscht ist bislang die konkrete Arbeitsweise der reichsstädtischen Scholarchate. In vielen Reichsstädten sind keine Geschäftsordnungen und keine Protokollbände von Sitzungen der Schulkommissionen überliefert.[67] Häufigkeit

---

64 Greiner, Ulmer Gelehrtenschule (wie Anm. 51), S. 44–75, hier: S. 58–61. Zur Aufgabenwahrnehmung des Ulmer Scholarchats im konfessionellen Zeitalter vgl. auch: Angelika Schlenkrich, „Man kann nicht jedermann schulmeister geben." Das Schulwesen der Reichsstadt Ulm im Spiegel der Visitationen im Zeitalter von Reformation und Konfessionalisierung, in: Flachenecker / Kießling, Schullandschaften in Altbayern, Franken und Schwaben (wie Anm. 3), S. 281–302.

65 Ebd., S. 58 (auch das folgende Zitat).

66 Vgl. zu den Anfängen der Schulaufsicht: Friedrich Paulsen, Geschichte des gelehrten Unterrichts auf den deutschen Schulen und Universitäten vom Ausgang des Mittelalters bis zur Gegenwart, Bd. 1, Leipzig ³1919, S. 331.

67 In der bildungsgeschichtlichen Literatur finden sich vereinzelt Hinweise auf die fragmentarische Überlieferung von Protokollbänden der reichsstädtischen Scholarchate. Vgl. zu Heilbronn: Kolbeck, 350 Jahre Gymnasium in Heilbronn (wie Anm. 15), S. 14; vgl. zu Schwäbisch Hall: Walther Ludwig, Deklamationen und Schuldramen im 17. Jahrhundert – das Beispiel des Gymnasiums der Reichsstadt Schwäbisch Hall, in: Bianca-Jeanette / Jens-Peter Schröder (Hg.), Studium declamatorium. Untersuchungen zu Schulübungen und Prunkreden von der Antike bis zur Neuzeit, München/Leipzig 2003, S. 335–372, hier: S. 337; vgl. zu Biberach: Diehl, Das humanistische Schulwesen (wie Anm. 37), S. 592.

und Ablauf der Zusammenkünfte der jeweiligen Gremien können daher im Detail oft nicht mehr rekonstruiert werden.

Mit der Durchführung der Schulvisitationen wurden eigene Kommissionen beauftragt, deren personelle Zusammensetzung in den einzelnen Städten differierte und die sich bisweilen auch innerhalb einer Reichsstadt in wechselnden Konstellationen formierten.[68] Den Visitationskommissionen gehörten in den reichsfreien Städten Süddeutschlands Personen aus drei Funktionsgruppen an: Erstens Vertreter der politischen Führung (das heißt ins Scholarchat berufene Ratsherren), zweitens Repräsentanten der Geistlichkeit (darunter zumeist die jeweiligen leitenden Geistlichen) sowie drittens Angehörige des Lehrpersonals (darunter in der Regel der Schulrektor). Betrachtet man die vorliegenden Informationen über die Durchführung der Visitationen, so zeigt sich, dass die Geistlichkeit in nahezu allen Reichsstädten an der Evaluation des Lehr- und Prüfungsbetriebs beteiligt war.[69] In vielen Kommunen hatte der leitende Geistliche (Primarius/Senior/Superintendent) bei der Schulaufsicht eine Schlüsselrolle inne; er nahm innerhalb der Visitationskommission eine Führungsrolle ein oder trug als „Schulinspektor“ gegenüber dem Rat die Verantwortung für die Durchführung der Lehrkontrollen. Seine Funktion kam damit einer fachlichen Oberaufsicht über die Schulen gleich.[70] Demgegenüber nahmen die mit der Schuladministration betrauten Ratsmitglieder, die in der Regel mindestens

---

68 So etwa in Nördlingen (freundliche Auskunft von Dr. Wilfried Sponsel, Stadtarchiv Nördlingen, vom 20.8.2020).

69 Dies gilt etwa auch für das patrizisch regierte Nürnberg. Vgl. Günther Petsch, Das Nürnberger protestantische Kirchenrecht der reichsstädtischen Zeit. Eine kirchenrechtsgeschichtliche Studie, Düren 1933, hier: S. 38. Auch in der katholischen Reichsstadt Überlingen lässt sich eine Beteiligung des Stadtpfarrers an den Schulvisitationen nachweisen. Vgl. Wilfried Enderle, Konfessionsbildung und Ratsregiment in der katholischen Reichsstadt Überlingen (1500–1618) im Kontext der Reformationsgeschichte der oberschwäbischen Reichsstädte, Stuttgart 1990, hier: S. 230.

70 Eine solche Konstellation findet sich beispielsweise in Esslingen, Nördlingen, Regensburg, Reutlingen und Schweinfurt, d.h. insbesondere in Reichsstädten mittlerer Größe. Vgl. zu Esslingen: Mayer, Geschichte des humanistischen Schulwesens (wie Anm. 16), S. 259; vgl. zu Nördlingen: Dietmar-Henning Voges, Die Eingliederung der Reichsstadt Nördlingen in den bayerischen Staat 1803, in: Ders., Nördlingen seit der Reformation. Aus dem Leben einer Stadt, München 1998, S. 178–210, hier: S. 204; vgl. zu Regensburg: Kleinstäuber, Ausführliche Geschichte der Studien-Anstalten in Regensburg (wie Anm. 19), 36, 1882, S. 82; vgl. zu Reutlingen: Eugen Wendler, Reutlingen. Geschichte und Gegenwart einer lebendigen Stadt, Reutlingen [6]2017, S. 193; vgl. zu Schweinfurt: Völcker, Geschichte der Studienanstalt Schweinfurt (wie Anm. 28), S. 17.

einen Vertreter in die Visitationskommissionen entsandten, in erster Linie eine politische Oberaufsicht über das Schulwesen wahr. Einen Sonderfall stellte die Schuladministration im bikonfessionellen Augsburg dar. Dort wurden von den 1540er Jahren bis zur Einführung der Parität Geistliche nicht für Schulangelegenheiten verwendet.[71] Auf diese Weise sollten konfessionelle Kontroversen vermieden werden.

4. Mehrere Funktionsträger – unterschiedliche Kompetenzen: ein Fazit

In bildungsgeschichtlichen Darstellungen wird üblicherweise herausgestellt, dass im 16. Jahrhundert in den evangelischen Territorien und Städten die Verantwortlichkeit für das Schulwesen von geistlichen Einrichtungen auf staatliche beziehungsweise kommunale Herrschaftsträger überging. Diese Transformation wurde von Teilen der älteren Forschung als Schritt auf dem Weg zur modernen staatlichen Schulaufsicht positiv gedeutet.[72] Die vorliegende Analyse von Organisation, Personal und Aufgabenwahrnehmung der Schuladministrationen in den süddeutschen Reichsstädten legt demgegenüber eine modifizierte Sichtweise und Interpretation nahe.

Die Untersuchung hat gezeigt, dass in fast allen größeren und mittelgroßen Reichsstädten Süddeutschlands, die sich der Reformation zugewandt hatten und in denen die Zuständigkeit für das Bildungswesen daher vollständig auf den jeweiligen Stadtmagistrat übergegangen war, bereits unmittelbar oder kurze Zeit nach dem religiösen Umbruch sogenannte Schulkommissionen beziehungsweise Scholarchate, also spezielle Verwaltungsorgane für das Bildungswesen, formiert wurden. Die personelle Struktur dieser Kommissionen war in den einzelnen Städten unterschiedlich, zum Teil gehörten den Scholarchaten nur Ratsmitglieder an, zum Teil jedoch auch andere Personen, vor allem Geistliche und städtische Akademiker (Juristen, Ärzte). Wichtig für die weitere Entwicklung war, dass unabhängig von der konkreten Organisationsform der Scholarchate in nahezu allen reichsfreien Städten Süddeutschlands neben den Ratsmitgliedern relativ rasch weitere Personen, insbesondere die städtische Geistlichkeit, mit verantwortungsvollen Aufgaben bei der Verwaltung der Schulen betraut wurden. Die Kirchendiener – an der Spitze der jeweilige Primarius/Senior/Superintendent – erlangten in vielen Städten bereits kurze Zeit nach der Etablierung des

71 Ockel, Geschichte des höheren Schulwesens (wie Anm. 5), S. 28 f.

72 Vgl. bes.: Wilhelm Kahl, Zur Geschichte der Schulaufsicht, Leipzig/Berlin 1913.

Scholarchats eine Schlüsselfunktion bei der Durchführung der für die Entwicklung des reichsstädtischen Bildungswesens sehr wichtigen Schulvisitationen. Vielfach war dem ersten Geistlichen der Kommune die Leitung der Evaluation des Lehr- und Prüfungsbetriebs übertragen. Auf diese Weise kamen die bei den Geistlichen angesiedelten Kompetenzen in zahlreichen Reichsstädten einer – in ihrer Bedeutung kaum zu überschätzenden – fachlichen Oberaufsicht über das Schulwesen gleich. Hinzu kommt, dass gerade die Geistlichkeit aufgrund ihrer Expertise und ihrer herausgehobenen Position in der Schuladministration Einfluss nehmen konnte auf Entscheidungen des Stadtmagistrats, bei dem die politische Verantwortlichkeit für das Bildungswesen lag. Eine derartige Möglichkeit zur politischen Einflussnahme galt unabhängig davon, ob das Pfarrpersonal dem Scholarchenkollegium angehörte oder nicht. Insbesondere bei Lehrplanreformen oder bei Berufungen von Professoren beziehungsweise Schulrektoren waren die Empfehlungen und der Rat der in den reichsunmittelbaren Kommunen wirkenden Geistlichen, gegebenenfalls auch weiterer Akademiker, kaum verzichtbar.

Aufgrund dieses Befunds ergibt sich ein differenziertes Bild der süddeutschreichsstädtischen Schuladministration in der Frühen Neuzeit. Unstrittig stellte der Übergang der politischen Zuständigkeit für das Bildungswesen in den seit dem 16. Jahrhundert evangelischen Reichsstädten eine Zäsur dar. Doch gilt es über diese Feststellung hinaus, stärker als dies die ältere Forschung bisweilen tat, die administrative Realität im Blick zu behalten. Analysiert man die Praxis des Verwaltungshandels in den protestantischen Reichsstädten Süddeutschlands in der Zeit vom 16. bis zum 18. Jahrhundert, zeigt sich in allen Kommunen ein – unterschiedlich organisiertes – Zusammenwirken von Obrigkeit und städtischen Akademikern, insbesondere den Kirchendienern.[73] Aufgrund ihrer Kompetenzen bei der fachlichen Aufsicht über die Bildungseinrichtungen übten auch im evangelischen Bereich die Geistlichen einen erheblichen Einfluss auf die schulische Bildung aus. Die Administration des Bildungswesens in den süddeutschen Reichsstädten in der Frühen Neuzeit stand demzufolge stärker in der Tradition des Spätmittelalters, als dies von der Forschung üblicherweise angenommen wird; von einer „weltlichen" Schulaufsicht kann in diesen Kommunen nur mit Einschränkungen gesprochen werden.

---

73 So bereits: Paulsen, Geschichte des gelehrten Unterrichts (wie Anm. 66), S. 331–333.

Franz-Rasso Böck

Kempten im Netz interurbaner Kommunikation

Abstract: The internal relationship between the reign of the prince-abbots with the Catholic monastery town on one hand and the Protestant imperial city on the other hand in the "microcosm" Kempten is a traditional focus of local historical research. However, little is known about the external relations of the imperial city of Kempten in particular with other imperial cities, starting with its Swabian neighbors. This desideratum is generally due to the poor source situation concerning the imperial city of Kempten, especially considering the very fragmentary records of the correspondence of the city council, for which there are no documents about incoming and outgoing letters in the form of so-called "Briefbücher" ("letter books"). The paths for research on the foreign relations of the imperial city of Kempten therefore lead to archives of former imperial cities, of which it is known that, like Augsburg, they have a larger amount of "letter books", which also depict relevant correspondence on Kempten.

1. Sonderfall Kempten

„Was die Geschichte der Reichsstadt Kempten […] von den meisten anderen Reichsstädten unterscheidet […], ist, daß es ihr erst so spät gelang, die längst fällige ‚Abnabelung' vom Stadtherrn und vielfachen Grundherrn durchzusetzen."[1] In der Tat muss auch die verzögerte verfassungsgeschichtliche Entwicklung der Reichsstadt im Zusammenhang mit dem langwierigen Ablöseprozess von der Abtei gesehen werden.[2] Trotz des königlichen Freiheitsbriefes von 1289 übte die tatsächliche Macht in der Stadt bis zum „Großen Kauf" 1525 weiterhin der Abt als oberster Grundherr, Vogt und Marktherr aus. Aufgrund der Dominanz des Fürstabtes über die Stadt und ihrer vergleichsweise bescheidenen Wirtschaftskraft konnten die Kemptener Bürger erst dann alle Rechte einer Freien

1 Vgl. Gerhart Nebinger, Das Bürgertum. Anfänge und Entwicklung. 13.–15. Jahrhundert, in: Volker Dotterweich / Karl Filser/ Pankraz Fried (Hg.), Geschichte der Stadt Kempten, Kempten 1989, S. 98.

2 Vgl. grundlegend: Bettina Löffeler, Kempten: Der Ablösungsprozess von geistlicher Stadtherrschaft im Spätmittelalter, in: Allgäuer Geschichtsfreund 103, 2003, S. 99–149.

Reichsstadt vom Fürstabt ablösen, die ihnen urkundlich von König Rudolf I. bereits 1289 verbrieft und zugesichert worden waren.

2. Außenbeziehungen

Nach wie vor ist die Forschung noch auf das Innenverhältnis zwischen Reichsstadt und Fürststift im „Mikrokosmos" Kempten fokussiert. Die reichhaltigen Dimensionen der Außenbeziehungen beider Herrschaftsträger sind bisher nur ansatzweise und exemplarisch erforscht und dargestellt worden. Eine erste Studie zu den wirtschaftlichen und politischen Außenbeziehungen der Reichsstadt Kempten lieferte Wolfgang Wüst im Rahmen der Geschichte der Stadt Kempten 1989.[3] Eine Darstellung Kemptens aber etwa im Verbund der benachbarten oberschwäbischen Reichsstädte oder der Rolle Kemptens im Schwäbischen Reichskreis steht noch aus, was nicht zuletzt auf die lückenhafte Quellenlage zurückzuführen ist: „Kaum etwas in Schwaben ist nämlich interessanter als die Geschichte Kemptens."[4] Allerdings fehlen aufgrund der schlechten Quellenlage unter anderem Darstellungen für die mittelalterliche und frühneuzeitliche Verfassungsgeschichte sowie für die Geschichte des Kemptener Patriziats. Gerhart Nebinger (1911–1997), der sich um die Erforschung des Kemptener Patriziats mit einer Studie zu den Jenisch in Kempten[5] verdient gemacht hat, spricht von

3 Vgl. Wolfgang Wüst, Bürgertum, Handel, wirtschaftliche und politische Außenbeziehungen der Reichsstadt, in: Dotterweich / Filser / Fried, Geschichte der Stadt Kempten (wie Anm. 1), S. 202–222. In jüngerer Zeit vgl.: Franz-Rasso Böck, Die Außenbeziehungen der Reichsstadt Kempten im Spiegel ihres Fernhandels, in: Wolfgang Wüst / Georg Kreuzer / David Petry (Hg.), Grenzüberschreitungen. Die Außenbeziehungen Schwabens in Mittelalter und Neuzeit (Zeitschrift des Historischen Vereins für Schwaben [künftig: ZHVS] 100), Augsburg 2008, S. 515–520; Ders., Transfer in Politik und Wirtschaft, Wissenschaft und Kunst. Aspekte der italienischen Beziehungen Kemptens, in: Wolfgang Wüst / Peter Fassl / Rainhard Riepertinger (Hg.), Schwaben und Italien. Zwei europäische Kulturlandschaften zwischen Antike und Moderne (ZHVS 102), Augsburg 2010, S. 173–184.

4 Zitiert nach: Franz-Rasso Böck, Patriziat und Stadtherrschaft – das reichsstädtische Regiment in Kempten um 1600, in: Thomas Weiß (Hg.), In Bronze gegossen. Zur Geschichte des Kemptener Rathausbrunnens (Kataloge und Schriften der Museen der Stadt Kempten 9), Kempten 1993, S. 23; Ders., Das Patriziat der Reichsstadt Kempten, in: Wolfgang Wüst (Hg.), Patrizier – Wege zur städtischen Oligarchie und zum Landadel. Süddeutschland im Städtevergleich. Unter Mitarbeit von Marc Holländer, Berlin u.a. 2018, S. 189–196.

5 Gerhart Nebinger, Die Jenisch in Kempten, in: Allgäuer Geschichtsfreund 83/84, 1984, S. 140–255.

großen „Problemen hinsichtlich einer Aufhellung der Geschichte der Kemptener Geschlechter“[6] und bemerkt, „es gibt kaum eine schwäbische Stadt, welche in dieser Hinsicht so quellenarm ist“.[7] Dieser Umstand ist auch auf den Verlust des Archivs des Patriziats zurückzuführen, „von dem man natürlich nicht weiß, wie groß seine Verluste bei der Kemptener Katastrophe von 1633 gewesen sind“.[8] Dies betrifft in besonderem Maße auch Geschichte und Funktion des städtischen Rates, denn wir verfügen für Kempten im Vergleich zu anderen Reichsstädten nur sehr fragmentarisch über Anhaltspunkte zur Korrespondenz des Rates, da keine Eingangs- und Ausgangsbücher vorliegen. Natürlich könnte man schon beim Innenverhältnis von Reichsstadt und Fürststift Kempten von „Außenbeziehungen“ der Reichsstadt sprechen, denn unabhängig von einer gar nicht gegebenen geographischen Distanz befanden sich die Reichsstädter *extra muros* bereits unmittelbar im stiftischen Territorium und damit im Ausland. Doch geht es hier nicht um das Kemptener Innen- wie Außenverhältnis der beiden Kontrahenten untereinander, sondern um die faktischen Außenbeziehungen der Reichsstadt Kempten zu anderen Reichsstädten, wofür aufgrund der geschilderten Quellenlage nur wenige Pinselstriche gesetzt und erste Beispiele gewissermaßen rekonstruiert werden können.

3. Briefbücher

Sofern vorhanden, bietet die archivalische Überlieferung des städtischen Rates die Grundlage für die Erforschung der Außenbeziehungen einer Reichsstadt zu benachbarten oder überhaupt anderen Reichsstädten, seien es Korrespondenzen oder auch Gesandtschaften. An erster Stelle sind die „Briefbücher“ oder „Missivbücher“ zu nennen, in welche die Entwürfe der nach außen gesandten Briefe geschrieben wurden. Sie haben sich, so die jüngste Studie zum Thema am Beispiel Nürnbergs,[9] in mehreren Freien oder Reichsstädten erhalten. Was heißt „mehrere“? Oder: In wie vielen Städten haben sie sich – so wie in Kempten – nicht erhalten? So ist Kempten wohl doch kein Einzelfall und es wäre statistisch wie inhaltlich aufschlussreich, mit welchen Städten Kempten das Fehlen

6 Nebinger, Das Bürgertum (wie Anm. 1), S. 108.
7 Ebd.
8 Ebd.
9 Vgl. Hiromi Abe-Kosaka, Korrespondenz und Gesandtschaften des Nürnberger Stadtrats im 15. Jahrhundert. Eine Analyse aufgrund des Jahrs 1460, in: Sabine Wüst (Hg.), Schätze der Welt aus landeshistorischer Perspektive. Festschrift zum 65. Geburtstag von Wolfgang Wüst, St. Ottilien 2018, S. 483–493.

dieser Quellengattung gemeinsam hat und aus welchen Gründen. Die Quellengattung der Rechnungsbücher, in denen auch Kosten von Gesandtschaften vermerkt wurden, hat sich laut genannter Studie zu Nürnberg ebenso in mehreren Städten erhalten, für die es im Gegensatz zu den Briefbüchern auch für Kempten eine gewisse Überlieferung gibt, die Anhaltspunkte für die weitere Forschung geben könnten. Brief- und Rechnungsbücher werden für verschiedene Fragestellungen zwar schon seit langem in Untersuchungen einbezogen, aber erst seit kurzem selbst als Forschungsgegenstand betrachtet. Für Kempten hat wiederum Wolfgang Wüst zumindest indirekt erste Schneisen geschlagen, der sich am Beispiel Augsburgs mit seiner reichen Überlieferung an Briefbüchern besonders mit der Korrespondenz der schwäbischen und fränkischen Städte beschäftigt hat.[10] Um für Kempten statistisch und inhaltlich mehr bringen zu können als in dieser Notiz mit einem bereits intensiver klingenden Arbeitstitel, gilt es sozusagen, das „Pferd von hinten aufzuzäumen" und von Korrespondenz-Partnern wie Augsburg her die Kemptener Verhältnisse zu beleuchten und zu rekonstruieren. Im Stadtarchiv Kempten selbst verweisen zunächst zwei Beispiele aus der Reformationszeit auf durchaus vorhandene Außenbeziehungen der Reichsstadt Kempten.

4. Schreiben nach Nürnberg, Augsburg und Straßburg

Nach Einführung der Reformation[11] durch den Rat 1527 unterzeichnete die Reichsstadt Kempten 1529 die *Protestatio* von Speyer und 1530 die *Confessio Augustana*, womit der Weg zu einer Reformation lutherischer Prägung beschritten schien. Als jedoch 1529 der gemäßigte alte Pfarrer Sixtus Rummel (1507 der letzte noch vom Fürstabt eingesetzte Geistliche an der reichsstädtischen Pfarrkirche) starb, gewann Jakob Haystung, der bereits 1527 zum ersten Prediger an St. Mang bestellt worden war, mit der zwinglianischen Partei rasch die Oberhand und beeinflusste die Entscheidungen des Rates. Wie in den meisten

10 Vgl. Wolfgang Wüst, Reichsstädtische Kommunikation in Franken und Schwaben. Nachrichtennetze für Bürger, Räte und Kaufleute im Spätmittelalter, in: Zeitschrift für bayerische Landesgeschichte 62, 1999, S. 681–707; Ders., Süddeutsche Reichsstädte als Informationsdrehscheibe, in: Ders. / Kreuzer / Petry, Grenzüberschreitungen (wie Anm. 3), S. 305–326.

11 Vgl. Franz-Rasso Böck, Die Reichsstadt Kempten im Zeitalter der Reformation, in: Wolfgang Wüst / Georg Kreuzer / Nicola Schümann (Hg.), Der Augsburger Religionsfriede 1555. Ein Epochenereignis und seine regionale Verankerung (ZHVS 98), Augsburg 2005, S. 201–206.

oberdeutschen Reichsstädten setzte sich auch in Kempten die zwinglianische Partei mit allen Konsequenzen durch. Doch wurde die Entscheidung des Rates für Zwingli letztlich nicht im Schnellverfahren „von oben" aufgezwungen: Zuvor wurde zwischen den Parteien lange und offen diskutiert und nicht nur mit Worten gerungen, es wurden auswärtige Vermittler bemüht und Gutachten aus Nürnberg, Augsburg und Straßburg[12] angefordert, ob man sich denn nun lutherisch, gemäßigt oder radikal zwinglianisch verhalten sollte – kein Wunder, dass Kempten dreierlei Ratschläge erhielt. Nur das Schreiben nach Nürnberg die „*Protestation wider den kayserlichen Reichsabschied betreffend*" vom 6. August 1529 hat sich in einer Abschrift erhalten, die es noch vergleichend zu überprüfen und zu transkribieren gilt.

Mochte sich durch den späten Beitritt Kemptens zum Schmalkaldischen Bund 1536 außenpolitisch langfristig der allmähliche Übergang zum Luthertum abzeichnen, so blieben doch die Kirchenordnung und das Leben in der Stadt noch lange im Sinne der reformierten Lehre zwinglianisch geprägt.

5. Kempten auf dem Reichstag in Augsburg

Besonders dem Wirken des slowenischen Reformators Primus Truber (1508–1586), den der Rat 1553 zur grundlegenden Neuordnung des Kirchenwesens nach Kempten berufen hatte, war es zu verdanken, dass für die noch junge evangelische Kirche Kemptens nach äußeren Bedrohungen und innerprotestantischen Auseinandersetzungen eine Phase der Kontinuität und Stabilität einsetzte. In der Tat fand Truber in der Reichsstadt Kempten ein Kirchenwesen vor, das noch immer stark an der zwinglianischen Reformation orientiert war. Doch war das konfessionelle Klima in der Reichsstadt nach dem Augsburger Religionsfrieden 1555 spürbar offener geworden und nicht mehr eindeutig auf die nicht in den Frieden einbezogene „schweizerische" Richtung festgelegt. Dies kam Trubers Haltung entgegen, in dogmatischen Fragen zwischen der lutherischen und der reformierten Linie zu vermitteln. Truber wollte nicht den radikalen Bruch mit der zwinglianischen Tradition, sondern diese behutsam abschwächen und der lutherischen Praxis anpassen. 1561 übernahm Truber auf Ruf der krainischen Stände das Amt eines Kirchenvorstehers und Superintendenten in Laibach – der

---

12 Vgl. Stadtarchiv Kempten, Akten A VI 84, Religions- und kirchliche Zustände; A VI 85, Religionssachen 1525, 1529, 1530; Herbert Immenkötter, Stadt und Stift in der Reformationszeit, in: Dotterweich / Filser / Fried, Geschichte der Stadt Kempten (wie Anm. 1), S. 168–183; Paul Warmbrunn, Evangelische Kirche und Kultur in der Reichsstadt, in: ebd., S. 273–289.

Abschied wurde für beide Seiten nicht leicht, die Kemptener Stadtväter ließen den beliebten Prediger nur ungern ziehen. In seiner kurzen Kemptener Zeit aber hat Truber viel erreicht und fruchtbare Spuren hinterlassen. Seine ausgleichende Haltung hat der evangelischen Kirche Kemptens den Weg zu einem – durch die Bestimmungen des Augsburger Religionsfriedens vorgeschriebenen – Luthertum auf der Grundlage der *Confessio Augustana* geebnet.

Auf dem Augsburger Reichstag 1555 wurde die Reichsstadt Kempten durch ihren Stadtschreiber, Archivar und Chronisten Bartholomäus Holdenriedt (1525–1599, Amtszeit 1554–1598/99)[13] vertreten, auf den auch eine erste Ordnung des reichsstädtischen Archivs von 1574 und eine städtische Chronik für den Zeitraum 1543 bis 1598/99 zurückgehen. Über seine Teilnahme in Augsburg berichtet aber ein anderer großer Kemptener Chronist: Christoph Schwarz,[14] 1601 bis 1615 Erster Prediger an der St.-Mang-Kirche, dem Zentrum des gemeindlichen und geistig-wissenschaftlichen Lebens der seit 1527 evangelischen Reichsstadt, verfasste eine von den Anfängen von Stift und Stadt bis ins Jahr 1595 reichende Chronik, die aufgrund ihrer Zuverlässigkeit und Reichhaltigkeit als wertvolle Quelle für die Stadt- und besonders auch die Reformationsgeschichte gilt – nicht zuletzt wegen fehlender anderer Quellen, für die die Chronik einen gewissen, wenn auch letztlich begrenzten Ersatz bieten kann. Schwarz schrieb: „*Auf disem Reichstag, welicher sich biß in den September verzog, wurde der Passauische Vertrag bestettigt, das die beede Religionen, Bäpstlicher und Augspurgischer Confessions Verwandte friedlich nebeneinander lauffen und kein teil den anderen vergewaltigen solle. War auf den Reichstag geschickht Bartholome Holderriedt, genannt Schmidt, Stattschreiber.*“[15]

Nur kleine Ansätze und Hinweise, die zeigen, dass auch die Reichsstadt Kempten im Kontext der Reichs- und Kirchengeschichte außenpolitische Wirksamkeit entfaltet hat, auch wenn es sich mangels eigener Quellen schwierig gestaltet, das Beziehungsgeflecht zu anderen Reichsstädten zu rekonstruieren – Wege der Forschung weisen in die Archive benachbarter Reichsstädte.

---

13 Vgl. Franz-Rasso Böck, Studien zur Geschichte des Stadtarchivs Kempten. Erster Teil: Das Archiv und seine Archivare, in: Allgäuer Geschichtsfreund 94, 1994, S. 111–129, hier: S. 113, 118 f.

14 Vgl. Franz-Rasso Böck, Das Stadtarchiv Kempten und seine Bestände. Teil III: Die Chroniken, in: Der Altstadtbrief 18, 1997, Nr. 24, S. 16 f.

15 Stadtarchiv Kempten, B 31, „Schwarz'sche Chronik“, fol. 141 r.

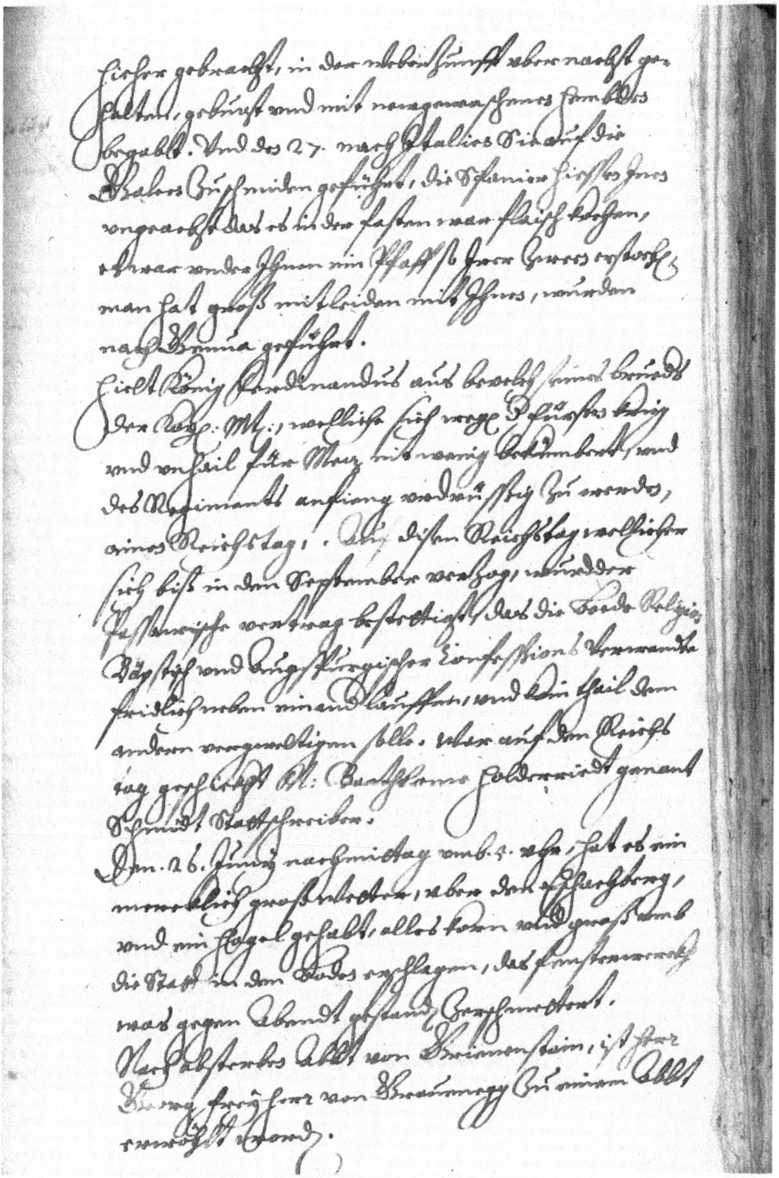

Abb. 49: *Abschnitt zum Augsburger Reichstag 1555 aus der Chronik von Christoph Schwarz. Bildnachweis: Stadtarchiv Kempten, B 31, „Schwarz'sche Chronik", fol. 141 r.*

Sektion 6:
Die süddeutsche Städtelandschaft – europäisch vergleichend

Ivana Ebelová / Zdeněk Hojda

Die Metropole als verdichtete Städtelandschaft. Einige Aspekte zur Entwicklung des frühneuzeitlichen Prag*

Abstract: Prague, as a political-administrative, economic, and cultural center of the Czech state, represents a complex organism, a conglomerate of several cities. Their uneven development, mutual relations and coexistence of their inhabitants evoke a wide range of topics, from the legal and administrative conditions of Prague cities, to the construction and urban specifics of their individual parts or functional differences.

The present study emphasises the areas representing the city on three different, seemingly disparate levels, with the aim of showing the Prague conurbation in its uniqueness. These selected passages are preceded by a text outlining the basic features of the legal and administrative development of Prague's cities, necessary for the integration of the thematic areas into the broader context.

First, the attention is paid to the Jewish Town, the most important center of Ashkenazi Jews in Central Europe. Sovereign privileges were the cornerstone from which the lives of (not only) Prague Jews were derived. The Prague Jews were directly subordinate to the monarch, to whom they regularly paid considerable funds for "protection". Members of the Prague community played an important role in the area of trade, not only financially. Their abilities and wide business contacts, based abundantly on family ties, were enjoyed by members of both, the highest class and the townspeople. The close connection between Prague Jews and their Christian neighbors was particularly intense during the Rudolfinian Prague period, considered to be the heyday of the Prague ghetto.

The second topic is a treatise on an extremely dynamic period when, after the accession of Rudolf II to the throne in 1576, Prague became an imperial residence, a Renaissance metropolis, a thriving and dynamic cultural metropolis for more than three decades.

Finally, a view of the Prague urban landscape through the eyes of foreigners is included, which reflects their ideas, sometimes confronted with immediate impressions, feelings, and experiences. The view from the outside, however superficial it may sometimes be, often illuminates unexpected connections or parallels.

---

* Diese Studie entstand an der Karls-Universität in Prag im Rahmen des Programms „PROGRES Q09: Geschichte – Der Schlüssel zum Verständnis der globalisierten Welt". Wir danken Herrn Christoph Gunkel M.A. von der Friedrich-Alexander-Universität Erlangen-Nürnberg herzlich für Korrekturen und redaktionelle Durchsicht.

The three thematic chapters, which we have selected from many possible problem areas, certainly cannot exhaust such a complex topic as the development of the Prague "Städtelandschaft". A number of equally interesting problems have not been encountered, and many of them still lack basic research.

The relatively short period during which Prague became one of the leading European capitals was significantly inscribed in the image of the city. However, the fundamental city-building impulses of the Rudolfinian period were still active in the following years, before they were brutally interrupted by the Thirty Years' War. For the Czech lands, Prague continued to play an important role not only as an administrative center, the seat of provincial authorities, the archbishop, and the oldest Central European university. Its importance also lies in its symbolic role.

1. Einleitung

Das Thema „Städtelandschaft" bietet ein breites Spektrum von Fragen, welche die Problematik des Stadtraumes samt seiner Veränderungen im Laufe der Jahrhunderte betreffen – eines Stadtraumes, der unter anderem aus geographischer, urbanistischer oder architektonischer Sicht betrachtet werden kann. Gleichzeitig ist es aber nötig, diesen Raum als eine lebhafte Szene der sozialen Kommunikation wahrzunehmen, also als einen Raum des politischen, wirtschaftlichen und kulturellen Lebens.

Prag, seit dem letzten Viertel des 9. Jahrhunderts das politische, wirtschaftliche und kulturelle Verwaltungszentrum des böhmischen Staates, stellt einen komplexen Organismus dar – ein Konglomerat einiger selbstständiger Städte, deren ungleichmäßige Entwicklung, gegenseitige Beziehungen und das Zusammenleben ihrer Bewohner eine Menge Themen hervorruft; angefangen mit den Rechts- und Verwaltungsverhältnissen der Prager Städte bis zu den Bau- und Stadtspezifika der einzelnen Stadtteile und ihren funktionellen Unterschieden.

Es gehört nicht zu unserer Aufgabe (und es wäre weder möglich noch zweckmäßig), in der vorliegenden Studie – sei es nur in ihren Grundrissen – alle Aspekte zu behandeln, die an der Bildung der Prager Konurbation direkt oder indirekt beteiligt waren. Im Gegenteil: Wir bemühen uns, diejenigen Themen zu akzentuieren, welche die Stadt auf drei unterschiedlichen – auf den ersten Blick inhomogenen – Ebenen darstellen, jedoch etwas gemeinsam haben. Mit anderen Worten: Wir wollen das Prager Ballungsgebiet in seiner Einzigartigkeit vorstellen, was allerdings nicht möglich wäre, ohne die ausgewählten Passagen in den breiteren Kontext der Rechts- und Verwaltungsentwicklung der Prager Städte einzubeziehen.

Der Leser wird im Rahmen des ersten Abschnitts nur wenig über die Prager Judenstadt erfahren, da Prag als „Stadt und Mutter Israels“ (wie damals die bedeutenden jüdischen Zentren genannt wurden), eines der bedeutendsten Zentren der aschkenasischen Juden Mitteleuropas samt ihres bekanntesten Rabbis Judah Löw (Jehuda ben Bezal'el Löw), in einem eigenen Kapitel thematisiert wird. Von großer Bedeutung waren die königlichen Privilegien, die das Leben (nicht nur) in der Prager Judenstadt bestimmten. Die Prager Juden waren direkt dem Herrscher unterstellt, dem sie für den „Schutz“ regelmäßig nicht unerhebliche finanzielle Summen leisteten. Eine wesentliche Rolle bekleideten die Vertreter der Judengemeinde im Bereich des Handels, und das nicht nur in finanzieller Hinsicht. Ihre Fähigkeiten und weiten Kontakte, die nicht selten auf Familienbeziehungen beruhten, wurden nicht nur von Seiten der Bürger, sondern auch von den höchsten adeligen Kreisen genutzt. Die enge Verbindung mit den Prager Juden war unter der Regierung Rudolfs II. besonders intensiv, die als eine Blütezeit des Prager Ghettos angesehen wird.

Der zweite Themenbereich behandelt die besonders dynamische Zeit, in der Prag nach dem Regierungsantritt Rudolfs II. im Jahre 1576 zu einer kaiserlichen Residenz wurde – zu einer wahren Renaissance-Stadt, einer prosperierenden und kulturellen Metropole.

Abschließend wird dem Leser Prag aus der Sicht der Ausländer nahegebracht, die ihre Vorstellungen – konfrontiert mit den unmittelbaren Eindrücken, Gefühlen und Erfahrungen – widerspiegelt. Die Sicht von außen, obgleich diese manchmal oberflächlich erscheint, kann oft unerwartete Zusammenhänge und Parallelen bieten.

2. Die frühneuzeitliche Stadtentwicklung

2.1. Von der Gründung der Prager Städte bis zur Stärkung ihrer Selbstverwaltung an der Schwelle zum 16. Jahrhundert

Bei der Vorstellung von Prag denkt man an das malerische Bild einer Stadt an beiden Ufern der Moldau – an das linke Ufer mit der Kleinseite (Malá Strana) und dem Hradschin (Hradčany) einschließlich der Prager Burg sowie an das rechte Ufer mit dem Wyschehrad (Vyšehrad), sämtlichen romantischen Gassen der Altstadt (Staré Město) und der großzügig angelegten Neustadt (Nové Město).[1] Die Prager Burg stellte einen der wichtigsten Stützpunkte dar, um den

1 Es gibt eine umfangreiche Menge an Publikationen über Prag, seien es Monographien, kunsthistorische Arbeiten, Bildstudien oder kürzere Beiträge. Im Kontext des behandelten Themas wird lediglich die Synthese von Václav Ledvinka / Jiří Pešek, Praha (= Prag), Praha 2000, S. 129, genannt, deren Werk ebenfalls eine thematische Bibliographie älterer Literatur zur Geschichte der Hauptstadt beinhaltet.

herum und darunter sich zunächst eine Siedlung konzentrierte. Der ursprünglichen Siedlung am linken Ufer des Flusses kann jedoch keine große Bedeutung beigemessen werden. Der Prozess der Verwandlung der Siedlungsstruktur am rechten Ufer in einen Stadtorganismus begann etwa in der Mitte des 12. Jahrhunderts – in Bezug auf Aspekte wie Siedlungskonzentration, Baucharakter, Erfüllung der Verwaltungs-, Religions- und anderer Funktionen wies dieses Gebiet gewisse Stadtzüge höchstwahrscheinlich schon früher auf –, verlief allmählich und wurde irgendwann in den frühen 30er Jahren des 13. Jahrhunderts abgeschlossen. Eine wichtige Rolle spielte dabei auch der Bau der Stadtmauern (einschließlich der ursprünglichen jüdischen Siedlung am linken Ufer der Moldau), die den Raum für die Entwicklung des Stadtlebens abgrenzte.

Je mehr das Gebiet am linken Ufer zunächst in seiner Entwicklung dem rechten Ufer nachstand, desto schneller war dann seine Verwandlung in eine mittelalterliche Stadt. Diese fand am Ende der 50er Jahre des 13. Jahrhunderts statt, als König Ottokar II. Přemysl die Stammbevölkerung vertrieb und 1257 fremde Kolonisten aus Norddeutschland einlud, die darauf die Kleinere Stadt Prag (später Kleinseite genannt) gründeten. In der zweiten Hälfte des 13. Jahrhunderts existierten so nebeneinander die Größere Stadt Prag (seit 1282 Prager Altstadt genannt) und die Kleinere Stadt Prag. Eine ganze Reihe von kleineren Siedlungen, sowohl am linken als auch am rechten Ufer des Flusses, blieb außerhalb der Stadtmauern.

Einen Meilenstein im Prozess der Verwandlung der Prager Altstadt, die dem süddeutschen (Nürnberger) Stadtrecht folgte,[2] in eine Gemeinde vollberechtigter Bürger, stellte das Jahr 1338 dar, in dem einerseits der König eine Genehmigung zum Bau eines neuen Rathauses erteilte und andererseits die Vertreter der alten Patrizierfamilien dazu verpflichtet wurden, den neuen Ratsherren in Anwesenheit von Vertretern der Stadtgemeinde ihre Geldmittel und Rechnungen offenzulegen. Die Finanzen wurden hiermit zu einer öffentlichen Angelegenheit der allmählich institutionalisierten Gemeinde, der alle vollberechtigten Bürger angehörten. Die Stadt hatte ihre eigene Verwaltung, wobei das Gerichtswesen den wichtigsten Bestandteil darstellte.

2 An die Stadt Nürnberg, die als höchste juristische Autorität galt, wurden ursprünglich auch Berufungen der Prager Altstadt gerichtet; und es handelte sich gerade um die Prager Altstadt, die zum Sitz des Apellationsgerichts für alle Städte Böhmens, die dem Nürnberger Recht folgten, unter anderem auch für die von Karl IV. gegründete Neustadt, wurde.

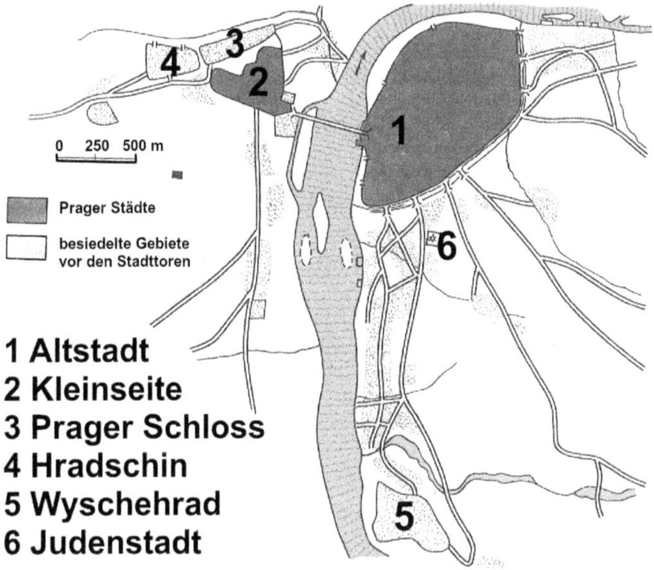

Abb. 50: *Prager Städte vor der Gründung der Neustadt durch Karl IV. Bildnachweis/Copyright: Matyáš Ebel.*

Während die Prager Altstadt direkt dem Herrscher unterstellt war, blieb die Kleinseite, die dem Magdeburger Stadtrecht folgte,[3] in der Verwaltung des königlichen Unterkämmerers.

Irgendwann um das Jahr 1330 kam zu den beiden genannten Städten die neu gegründete Stadt Hradschin hinzu – das Hinterland der Prager Burg, das dem Obersten Burggrafen unterstellt war.

Zum entscheidenden Moment in der Stadtentwicklung – in der Verwandlung in eine dem Anspruch des römisch-deutschen Kaisers entsprechende Residenz –, wurde die Gründung der Prager Neustadt durch König Karl IV. im Jahre 1348. Diese wurde in der Nachbarschaft der Altstadt, östlich und südlich von ihren Stadtmauern, auf dem Gebiet der dort liegenden Vorstadtsiedlungen errichtet.

3 Die Apellationstelle für Städte, die sich nach dem Magdeburger Stadtrecht richteten, hatte später seinen Sitz in Leitmeritz (Litoměřice).

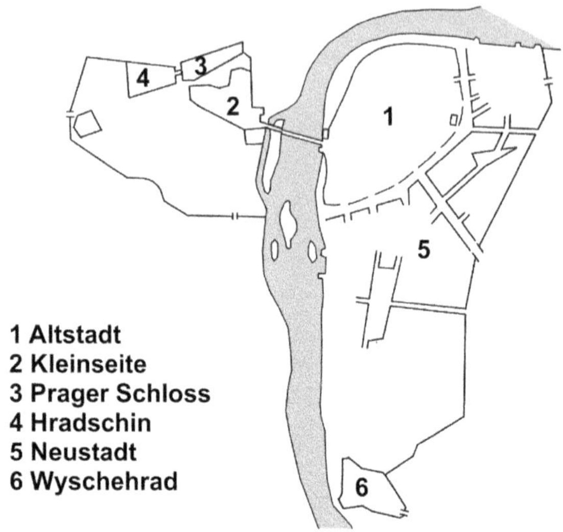

Abb. 51: *Die Prager Städte mit der neu gegründeten Neustadt, nach 1348. Bildnachweis/ Copyright: Matyáš Ebel.*

Der Bau einer neuen Stadt bedeutete eine deutliche Erleichterung zu Gunsten der übervölkerten Altstadt, aus der vor allem eine Reihe von Handwerkern in die Neustadt umzog. Im Vergleich dazu war die Vertretung der in der Neustadt ansässigen Geschäftsleute viel niedriger.

Genauso wie die Prager Altstadt war auch die Neustadt nicht dem Unterkämmerer unterstellt. Bei Konflikten zwischen der neu gegründeten Stadt und ihrem älteren Nachbarn sollten sich die Neustadt-Bewohner mit ihren Rechtsanliegen an das Gericht der Altstadt wenden.[4] Eine wichtige Rolle in der Beziehung beider Städte spielte auch die Tatsache, dass die Funktion des Richters der Neustadt bis zum Ende des 14. Jahrhunderts von den Altstadt-Bürgern ausgeübt wurde. Es ist daher kein Wunder, dass zwischen den Städten rasch eine Rivalität entstand. Kaiser Karl IV. versuchte diese Streitigkeiten mit der Vereinigung beider Städte am rechten Ufer im Jahre 1367 zu lösen. Diese war allerdings nicht von langer Dauer. Wachsende Probleme, von denen wir heute nicht genug gesicherte

4 Die Bedeutung der gesamten Stadt Prag – vor allem die besondere Stellung der Altstadt – stieg mit der Gründung der Universität im Jahre 1348, die über eine selbstständige Verwaltung verfügte.

Informationen zur Verfügung haben, führten schließlich nach zehn Jahren zu einer wiederholten Trennung beider Städte.[5]

Die Auseinandersetzungen unter den einzelnen Prager Städten, jedoch insbesondere die religiösen Konflikte und sozialen Probleme, die alle böhmischen Länder betrafen, steigerten sich seit Mitte der 80er Jahre des 14. Jahrhunderts. Nach dem Tode Wenzels IV. im Jahre 1419 eskalierte dieser Konflikt mit der Hussitenbewegung, die ein selbstständiges, stürmisches Kapitel in der Geschichte der böhmischen Länder darstellt. Olga Fejtová gibt an, die Hussitenbewegung habe gleichzeitig den Beginn des Emanzipationsprozesses der böhmischen königlichen Städte (an deren Spitze die Prager Städte standen), den Beginn des Trennungsprozesses unter den königlichen Städten und ihrer Obrigkeit, sprich dem Herrscher, sowie den Anfang von deren Funktionsdifferenzierung bedeutet.[6]

Die kurzzeitige Regierung Sigismunds von Luxemburg (1368–1437) brachte zwar den Prager Städten eine Bestätigung deren älterer Privilegien und die erneute Unabhängigkeit der Neustadt von der Altstadt (mit Ausnahme der Justiz),[7] jedoch keine allgemeine Beruhigung der herrschenden Verhältnisse. Die andauernde politische Krise vertiefte sich im Zusammenhang mit der Frage der Nachfolgerschaft noch mehr. Prag wurde zum politischen Kampfplatz in den Verhandlungen über den neuen König. Es gehört jedoch nicht zur Aufgabe dieser Studie – im Übrigen ist es wegen des Platzmangels auch nicht möglich – die Situation in den Prager Städten in den folgenden Jahrzehnten nach dem Tode

5 Zur Verwaltungsentwicklung der Prager Städte siehe ausführlicher: Václav Vojtíšek, O vývoji samosprávy pražských měst (= Zur Entwicklung der Selbstverwaltung der Prager Städte), Praha 1927; Jaroslava Mendelová, Vývoj městské správy a samosprávy od 13. století do roku 1784 (= Die Entwicklung der städtischen Verwaltung und Selbstverwaltung vom 13. Jahrhundert bis 1784), in: Václav Ledvinka (Hg.), Osm století pražské samosprávy, Praha 2000, S. 13–29; Olga Fejtová, Formování správy raně novověkých měst pražských jako „demokratizační" proces (= Die Bildung der Verwaltung der frühneuzeitlichen Prager Städte als ein „Demokratisierungs"-Prozess), in: Český časopis historický 108, 2010, S. 615–636. Sämtliche Studien zur Stadtverwaltung (insbesondere der Neustadt Prag) aus den Jahren 1981–2004 von Jaroslava Mendelová wurden in einem Band herausgegeben: „Obecné dobré velebiti, zlé tupiti a lidské spravedlnosti fedrovati …". Výbor statí Jaroslavy Mendelové k dějinám městské správy (= „Das Gemeinwohl rühmen, das Böse schmähen und menschliche Gerechtigkeit fördern …". Eine Auswahl der Studien von Jaroslava Mendelová zur Geschichte der Stadtverwaltung), Praha 2012.

6 Fejtová, Formování správy (wie Anm. 5), S. 616.

7 Eine Ausnahme bildete die Justiz, da das Altstädter Gericht den obersten Gerichtshof in Prag darstellte.

des letzten Luxemburgers auf dem böhmischen Thron im Detail zu betrachten. Erinnert sei in diesem Zusammenhang zumindest an die Tatsache, dass die Prager Städte (und mit ihnen auch die weiteren Städte der böhmischen Länder und ganz Mitteleuropa) zu dieser Zeit den Prozess der Veränderung der Beziehung Stadt – Adel – Herrscher durchmachten beziehungsweise den Prozess der Veränderung von Stadteliten.[8] Im Jahre 1507 wurde die Kleinere Stadt von der Macht des Unterkämmerers befreit; 1514 wurde von Seiten des Herrschers eine neue Wahlordnung herausgegeben, die die Wahl der Schöffen festlegte. Den Höhepunkt der Stärkung der städtischen Autonomie stellte schließlich der sogenannte St.-Wenzels-Vertrag von 1517 dar – ein Kompromissabkommen zwischen dem Adel und den Städten.

2.2. Von der Vereinigung der Altstadt mit der Neustadt bis zur Aufhebung ihrer Unität

Der Abschluss des St.-Wenzels-Vertrags, der die Stellung der Städte als selbstständigen Stand bestätigte, öffnete den Weg zur Einigung der Prager Städte am rechten Moldau-Ufer. Zum sichtbarsten Zeichen der Einigkeit beider Städte wurde zwar die Zuschüttung des Altstädter Grabens, die wesentlichen Änderungen betrafen jedoch die Stadtverwaltung. Ihr Zentrum war das Altstädter Rathaus, während im Neustädter Rathaus niedrigere Ämter ihren Sitz hatten. Die Besetzung der reformierten Stadtbehörden sollte unter den altstädtischen und den neustädtischen Bürgern gleichmäßig aufgeteilt werden.[9] Die Praxis sah jedoch von Anfang an anders aus; die Realität des beiderseitigen Zusammenlebens konnte von einem idyllischen Bild nicht weiter entfernt sein. Die Ursache hierfür war nicht nur die Konkurrenz zwischen den beiden Städten, sondern auch die Rivalität innerhalb der altstädtischen Machteliten, die die höchsten Posten in der Verwaltung der vereinigten Stadt bekleideten. Das Hauptproblem für Konflikte lag hier weniger an verschiedenen Auffassungen über die Bedürfnisse der Stadt, sondern eher auf der politisch-religiösen Ebene: Gegeneinander standen nämlich Anhänger zweier Konfessionen[10] – die progressivere lutherische

---

8 Ledvinka / Pešek, Praha (wie Anm. 1), S. 282.

9 Mendelová, Vývoj městské správy (wie Anm. 5), S. 24; Fejtová, Formování správy (wie Anm. 5), S. 626.

10 Fejtová, Formování správy (wie Anm. 5), S. 628; Ledvinka / Pešek, Praha (wie Anm. 1), S. 284–287.

Kirche (mit Jan Hlavsa von Liboslav an der Spitze)[11] und die konservativen Utraquisten, die von Jan Pašek von Vrat[12] angeführt wurden.

Zu einer grundsätzlichen Wende kam es im Jahre 1526, in dem die Habsburger den böhmischen Thron bestiegen. Der junge und pragmatische Herrscher Ferdinand I. war sich bewusst, dass es notwendig sei, die Macht der geeinigten Stadt zu brechen, um seine Macht durchzusetzen und eine feste Position erreichen zu können. Dies gelang ihm relativ bald nach seiner Krönung, als er 1528 die einheitliche Stadtverwaltung abschaffte und erneut einen selbstständigen Altstadt- und Neustadt-Rat bestätigte. Hiermit öffnete sich gleichzeitig der Weg zur politischen Machtschwächung der protestantischen königlichen Städte und zu ihrer Unterordnung unter den katholisch gesinnten Herrscher.

Bereits im Jahre 1532 kam es zum Verbot des Zusammenrufens der Stadtgemeinden; die Prager Städte verloren das Privileg einer freien Stadtratswahl. Die Kleinere Stadt wurde erneut in die Verwaltung des Unterkämmerers zurückversetzt. Der Konflikt zwischen den Städten und dem König eskalierte 1547, als die böhmischen Städte (an deren Spitze die Prager Städte standen) es ablehnten, einer königlichen Verordnung zu folgen und dem Herrscher militärische Hilfe im Kampf gegen die protestantische Seite im Schmalkaldischen Krieg zu leisten. Nachdem dieser Aufstand von Seiten des Königs unterdrückt wurde, kam es zu einer rasanten Bestrafung der Städte und der einzelnen Bürger. Die Selbstverwaltung der Prager Städte wurde begrenzt, indem sie ihre Privilegien abgeben mussten und die Renten samt den Stadtgütern verloren. Es wurden hohe Geldstrafen auferlegt. Die Aufsicht über politische und öffentliche Angelegenheiten aller Prager Städte führte nun ein vom König eingesetzter Hauptmann (aus dem Adelsstand). Die königlichen Interessen im Rat wurden durch einen Königsrichter beaufsichtigt, der aus dem Milieu der zuverlässigen Bürger stammte. Das Altstädter Stadtgericht verlor das Recht der Apellation für die anderen Städte; das königliche Appellationsgericht wurde neu eingerichtet. Im wirtschaftlichen Bereich fielen die Städte in die Kompetenz der Böhmischen Kammer.[13]

---

11 Er war Prager Bürger, 1497–1499 Schreiber der sogenannten Landtafeln und 1515–1517 Prager Primator (Bürgermeister). Mehr dazu bei: Jiří Pešek, Die intelektuellen Eliten der Stadt Prag und ihre Beziehung zur Macht im 16. Jahrhundert, in: Michael Diefenbacher / Olga Fejtová / Zdzisław Noga (Hg.), Krakau – Nürnberg – Prag: Die Eliten der Städte im Mittelalter und in der frühen Neuzeit. Herkunft, Nationalität, Mobilität, Mentalität, Praha 2016, S. 471–481.

12 Er war nobilitierter Patrizier, Gebildeter, Pädagoge und konservativer Politiker. Vgl. ebd.

13 Mendelová, Vývoj městské správy (wie Anm. 5), S. 25; Fejtová, Formování správy (wie Anm. 5), S. 630–632; Ledvinka / Pešek, Praha (wie Anm. 1), S. 287–292.

Besonders schmerzhaft war für die Städte die Aufhebung ihrer Zünfte, die als bedeutende Zentren der städtischen, antiferdinandeischen Opposition zur Zeit des Aufstandes fungierten.[14] Die Städte verloren praktisch alle Privilegien, die sie während und nach der Hussitenzeit erlangt hatten. Schließlich kam noch die Steuerreform von 1567 hinzu, welche die Städte finanziell zusätzlich belastete.

Eine weitere deutliche Veränderung für die Prager Städte beruhte darin, dass die Stadträte zum ersten Mal seit den Hussitenkriegen konfessionell gemischt waren, das heißt, dass auch die Katholiken zugelassen waren. Eingriffe des Königs in die Zusammensetzung des jeweiligen Stadtrates sowie die zeitlich beschränkte Ausübung des Amtes trugen zur Verhinderung einer Oligarchisierung der Räte bei. Im Hinblick auf die soziale Zusammensetzung war die Struktur der Räte also heterogen.

2.3. Vom Böhmischen Ständeaufstand bis zur Vereinigung der Prager Städte im Jahre 1784

Noch am Ende seiner Regierung gab Ferdinand I. (1503–1564) den Prager Städten einige der weggenommenen Privilegien zurück. Gleichzeitig kam es zur Wiederaufnahme der Zünfte; es entstanden auch einige subdelegierte Ämter und Gerichte als Hilfsorgane des Stadtrates. Als problematisch zeigte sich immer mehr der Ankauf von Bürgerhäusern durch den Adel. Diese wurden infolgedessen von Abgaben befreit und dadurch dem städtischen Schoß entzogen, was vor allem die Kleinere Stadt und den Hradschin betraf und zu deren Verarmung führte.[15]

Die zweite Hälfte der 70er Jahre des 16. Jahrhunderts wurde zu einem entscheidenden Moment in der frühneuzeitlichen Geschichte Prags. 1575 wurde Rudolf II. (1552–1612) in Prag zum böhmischen König gekrönt, wobei die Stadt bald nach der Krönung zu seiner Hauptresidenz wurde. Für die Prager Stadtverwaltung war die folgende Zeit besonders einträglich, da neue Instruktionen die Kompetenzen ihrer einzelnen Bestandteile genauer regelten.

Eines der brisantesten Probleme, die zu dieser Zeit zu lösen waren, stellte die Frage der Religionsfreiheit dar. Die Ständeopposition und die Prager Städte

---

14 Zdeňka Karlová, Změny v cechovním zřízení měst pražských po roce 1547 (= Die Veränderungen in der Zunfteinrichtung der Prager Städte nach 1547), in: Sborník příspěvků k dějinám hlavního města Prahy 8, 1938, S. 58–78.

15 Hradschin wurde 1592 zu einer königlichen Stadt erhoben. Dieser Akt, urkundlich im Jahre 1598 bestätigt, wurde später vom Obersten Burggrafen (als ehemaligen Inhaber) ohne Erfolg in Frage gestellt. Mendelová, Vývoj městské správy (wie Anm. 5), S. 27.

nutzten den dynastischen Konflikt zwischen Rudolf II. und seinem Bruder Matthias aus, um die Ausstellung des sogenannten rudolfinischen Majestätsbriefs im Juni 1609 zu erzwingen. In den Prager Städten entstanden zwei lutherische Kirchen, bei der Salvatorkirche in der Altstadt wurde sogar eine deutsch-lutherische Schule gegründet, mit dem Ziel, diese zu einer künftigen lutherischen Akademie weiterzuentwickeln.[16] Diese sollte neben den existierenden Hochschulen, der utraquistischen (hussitischen) Universität und der jesuitischen Akademie, fungieren. Dieser Plan gelang jedoch aus Zeitgründen nicht. Infolge der Veränderungen im Verwaltungs-, Politik- und Religionsbereich nahm die königliche Macht über die Prager Städte ab. Der bereits schwerkranke und vor allem geistig gestörte König versuchte die Oberhand mit Hilfe der Armee des Passauer Bischofs Erzherzog Leopold zurückzugewinnen. Der militärische Einfall im Februar 1611 führte zwar zur kurzzeitigen Besetzung und Ausplünderung der Kleinseite und des Hradschins, politisch handelte es sich jedoch um einen Misserfolg.[17]

Mit der Abdikation Rudolfs II. im Jahre 1611 und der Thronbesteigung seines Bruders Matthias – er wurde im Mai desselben Jahres als Matthias II. zum böhmischen König gekrönt – endete eine mehr als drei Jahrzehnte dauernde Periode, in der sich Prag von bürgerlich geprägten Städten in eine blühende kaiserliche Residenz verwandelt hatte.[18]

Weit schlimmere Auswirkungen als der „bloße" Bedeutungsverlust der Prager Städte nach der Umsiedlung des Hofes zurück nach Wien 1612 hatte die Wahl des Nachfolgers von Kaiser Matthias im Jahre 1617 – die Wahl des eifrigen Katholiken Ferdinand II. Er lehnte nicht nur die religiösen Zugeständnisse seiner Vorgänger ab, sondern es war gerade er, der mit der Rekatholisierung der böhmischen Länder unter dem Motto „*cuius regio, eius religio*" begann.[19]

Die politischen Eliten und Intellektuellen der Prager Städte lavierten lange Zeit herum, ohne sich auf der Seite der sich radikalisierenden adeligen Opposition zu engagieren. 1617 wurde (noch von Seiten Matthias') für die Prager Städte

16 Ledvinka / Pešek, Praha (wie Anm. 1), S. 322.

17 Der Konflikt wurde schließlich politisch geschlichtet, indem Peter von Rosenberg die Passauer Armee auszahlte und diese dann das Land verließ.

18 Dies gilt insbesondere für die beiden Prager Städte am linken Ufer der Moldau, für Kleinseite und Hradschin. Ledvinka / Pešek, Praha (wie Anm. 1), S. 320.

19 Zur Frage des religiösen Lebens in den böhmischen Ländern (inklusive Prag) siehe Jiří Mikulec, Náboženský život a barokní zbožnost v Českých zemích (= Religiöses Leben und barocke Frömmigkeit in den Böhmischen Ländern), Praha 2013, wo auch weitere Quellen zu diesem Thema zu finden sind.

eine sehr kontroverse, gegen die Städte gezielte und gleichzeitig auch gegenreformatorisch orientierte Instruktion für die kaiserlichen Richter ausgestellt. Die sich stets erhöhende Spannung zwischen den Anhängern des Herrschers einerseits und den protestantischen Ständen andererseits endete mit dem Fenstersturz der kaiserlichen Statthalter Wilhelm Slawata von Chlum und Koschumberg und Jaroslav Bořita von Martinitz am 23. Mai 1618. Am folgenden Ständetag im Karolinum nahmen die Prager Städte aus Furcht vor potenziellen Repressionen nicht teil. Nach diesem Ereignis konnten sie jedoch nicht neutral bleiben und schlossen sich dem Aufstand an. Die von Ferdinand II. eingesetzten Stadträte wurden durch solche ersetzt, die mit dem Ständeaufstand sympathisierten und diesen unterstützten. Die Prager Bevölkerung ging stürmischen Zeiten entgegen.

Der Aufstand mündete schließlich in der unrühmlichen Schlacht am Weißen Berg (bei Prag) am 8. November 1620. Genauso gut bekannt sind auch die katastrophalen Konsequenzen. Es folgten die Ausplünderung der Stadt durch das bayerische Heer und die Hinrichtung der 27 Anführer des Aufstandes auf dem Altstädter Ring am 21. Juni 1621, von denen 17 zum Bürgerstand gehörten. Nach dem Erlass des Rekatholisierungsdekrets folgte eine Konfiskationswelle; die Exulanten wurden gezwungen, ihren Besitz für symbolische Summen an die treuen Anhänger des Herrschers zu verkaufen.[20] Das Zusammentragen eines solchen Besitzes durch einige adelige Familien bildete die eigentliche Grundlage der späteren barocken Neugestaltung der Prager Städte.

Nach der Niederlage am Weißen Berg durften in den Stadträten der Prager Städte ausschließlich katholische Vertreter tätig sein.[21] Die „Verneuerte Landesordnung" schwächte zwar die politische Macht der Städte weitgehend, glich aber die Stellung der einzelnen Prager Städte untereinander aus. Die Kleinseite wurde der Zuständigkeit des Unterkämmerers entzogen; die Kleinseitner konnten nun wieder die freien Güter kaufen.

Im letzten Jahr des Dreißigjährigen Krieges hielt Prag einem schwedischen Angriff stand, wobei die Kleinseite und der Hradschin besetzt und erneut geplündert wurden, während sich die Altstadt und die Neustadt erfolgreich verteidigten und von Ferdinand III. neue Privilegien und verbesserte Stadtwappen erhielten.

---

20 Allein in Prag betraf die Konfiskation etwa 400 Häuser. Ledvinka / Pešek, Praha (wie Anm. 1), S. 326. Siehe zu den Konfiskationen nach der Schlacht am Weißen Berg im Detail: Tomáš Václav Bílek, Dějiny konfiskací v Čechách po roce 1618 (= Geschichte der Konfiskationen in Böhmen nach 1618), 2 Bde., Praha 1882/93.

21 Dies wurde mit der Ausstellung der „Verneuerten Landesordnung" in den Jahren 1627 und 1628 bestätigt.

In der zweiten Hälfte des 17. Jahrhunderts versuchte der Herrscher die nicht allzu gute wirtschaftliche Situation der Prager Städte durch die Stärkung des Einflusses der landesherrlichen Behörden sowie durch einzelne Maßnahmen zu verbessern. Den an finanziellen Mangel leidenden Städten sollte die 1704 neu gegründete „Kommission zur Verbesserung der königlichen Städte" zum Abbau der Schulden verhelfen. Die Maßnahmen waren jedoch wenig erfolgreich und führten nicht zu den gewünschten Ergebnissen.

Die 20er Jahre des 18. Jahrhunderts wurden zu einem weiteren Wendepunkt in der Verwaltungsentwicklung der Prager Städte, insbesondere was die Erneuerung der Stadträte angeht. Im Jahre 1749 wurden die Stadträte auf acht Personen reduziert; die Räte mussten höher qualifiziert sein. Die politische und steuerliche Verwaltung wurde der Aufsicht der Kreishauptleute unterstellt. Von der ursprünglichen Selbstverwaltung der Stadt blieb daher nur ein Torso.

Die Verteilung Prags auf vier selbstständige Städte – Altstadt, Neustadt, Kleinseite und Hradschin – mit eigener Verwaltung wurde aber immer mehr zum Haupthindernis der weiteren Entwicklung. Das Problem löste erst Joseph II. mit seiner radikalen Stadtreform, die im Jahre 1784 mit der Vereinigung der Prager Städte endete.

3. Die Judenstadt

„Die Stadt hatte Häuser, Straßen, Ringplätze und Tempel. Sie war gleich wie die anderen und doch unterschiedlich. Ihre Häuser standen dicht, eines neben dem anderen, […] ihre Straßen haben sich gedreht und gewunden auf einem mittelalterlichen Grundriss. […] So wie die Gassen in Wirklichkeit eher Gässchen waren, war für die Ringplätze eher die Bezeichnung ‚Ringplätzchen' zutreffend: enge Räume und schmale Häuser […], oft nur dank eines Wasserbrunnens und kleinen Geschäften einem wirklichen Ringplatz ähnlich. Die Stadt hatte dagegen einen großen Garten, der bis zum Moldau-Ufer herunterreichte und voll von grünen Anlagen und hohen Bäumen war. Anstatt von Blumenbeeten war der Garten jedoch voll von Reihen oder dichten Anhäufungen von Steintafeln […]. Und Tempel? Ein Dutzend davon stand da zwischen niedrigen und hohen Häusern. Sie waren umgeben von Misstrauen und Neugierde der Menschen, die in diese Stadt gekommen sind, doch in dieser Stadt nicht gelebt haben. Sie waren geschützt vom Glauben und von der Ergebenheit der Menschen, die darin Trost im Gebet und Zuflucht in Zeiten der Bedrohung fanden, da die Tempel Synagogen waren. Der große Garten war ein Friedhof."[22]

---

22 Jiřina Šedinová / Eva Kosáková, Golem kráčí Židovským Městem (= Der Golem schreitet durch die Jüdische Stadt), Praha 1994, S. 9 f.

Eine bessere Charakteristik der Judenstadt kann man sich nur schwer vorstellen als durch das Zitat, mit dem die Publikation „Golem kráčí Židovským Městem" („Der Golem schreitet durch die Jüdische Stadt") eingeleitet ist, an der die führende tschechische Hebraistikerin und Historikerin Jiřina Šedinová mitwirkte. Obwohl dieser Text die Prager Judenstadt im 19. Jahrhundert beschreibt, fängt er ihre Atmosphäre auch für die vorangegangenen Jahrhunderte ein, was auch die erhaltenen historischen Quellen belegen.[23]

Die Judenstadt stellte seit dem Mittelalter ein selbstständiges, geschlossenes Gebiet dar, das von der übrigen christlichen Welt getrennt war. Sie lag in der unmittelbaren Nachbarschaft der Altstadt, von der sie durch einige Tore getrennt wurde (*portae Judaerorum*). In den Quellen werden insgesamt sechs Tore angegeben.[24] Diese steckten das Gebiet ab, hinter dessen Grenzen die Juden nur ausnahmsweise ein Haus kaufen konnten. Das Gebiet der Judenstadt, welches bis zur Mitte des 19. Jahrhunderts praktisch unverändert blieb, war also eine Welt für sich, deren Bewohner, die als Kammerdiener (*servi camerae*) dem Herrscher direkt unterstellt waren, Sicherheit garantiert wurde. Dass es sich hierbei jedoch um eine eher relative Sicherheit handelte, zeigt vor allem der größte Pogrom in Prag, der zu Ostern des Jahres 1389 stattfand und in dessen Rahmen es zur Ermordung eines großen Teils der Bewohner der Judenstadt (es werden über 3.000 Opfer angegeben) kam.[25] Den Impuls hierfür sollen Ausschreitungen jüdischer Jungen gegeben haben, die einen christlichen Priester beleidigten und ihn mit Steinen bewarfen, als er durch das Ghetto ging, um einem Sterbenden die letzte Ölung zu geben. Die Schuldigen wurden zwar gefangengenommen, doch die Prager Bevölkerung, die mittels antisemitischer Predigten aus den altstädtischen Kirchen aufgestachelt worden war, nahm das Gesetz in die eigenen Hände.

---

23 Ibrahim ibn Jakub erwähnt die Prager Juden in seinem Bericht über den Besuch der Stadt im Jahre 966. Allerdings ist unklar, ob es sich um die in Prag ansässigen Juden oder um die durch die Stadt pendelnden Kaufleute handelte. Vgl. Dušan Třeštík, Cesta Ibrahima ibn Jakuba do Prahy v roce 966 (= Ibrahim ibn Jakubs Reise nach Prag im Jahre 966), in: Dějiny a současnost 14/5, 1992, S. 9–13. Nach den Angaben der „*Chronica Boemorum*" des Cosmas von Prag übten die Juden im 11. Jahrhundert die Rolle der Finanziers am Hof der Přemysliden aus. Vgl. Josef Emler / Václav Vladivoj Tomek (Hg.), Fontes rerum Bohemicarum, Bd. II, Praha 1874/75, S. 138.

24 Milada Vilímková, Židovské Město pražské (= Die Prager Judenstadt), Praha 1990, S. 16 f.

25 Tomáš Pěkný, Historie Židů v Čechách a na Moravě (= Die Geschichte der Juden in Böhmen und Mähren), Praha ^{2}2001, S. 38–42.

Auch die Stadtmauer, die auf Befehl König Wenzels IV. kurz davor gebaut worden war, konnte die Bewohner des Ghettos nicht beschützen.

Obwohl dieses Ereignis in einer ganzen Reihe von Chroniken und in manchen literarischen Quellen erwähnt wird, bietet wahrscheinlich Rabbi Avigdor ben Isaak Kara, einer der wenigen Überlebenden des Pogroms, die beste Beschreibung der Ereignisse: „*Der letzte Tag des Festes Pesach, des Erlösungsfestes, ist Zeuge eines Blutvergießens geworden. An diesem bitteren Tage haben sich gegen uns Sünder, schmutziger Pöbel ohne Namen, verschworen* […]. *Sie haben die Tore in Angriff genommen, auch Mauern haben sie durchbrochen, sie haben Schare und Horden gebildet, Lieder gesungen, gefeiert, Blut von Unschuldigen vergossen, Zerstörung und Gewalt gebracht* […]. *Dann sind sie in die Synagogen eingebrochen, in die Alte und die Neue,* […] *sie haben unsere heiligen Bücher verbrannt und zerrissen, unsere Tora* […], *auch den Friedhof, den Ort der ewigen Freiheit, wo die Gebeine meiner großen Vorfahren ruhen, haben sie geplündert, ihre Gebeine ausgegraben, die Grabsteine zerstört* […]."

Sogar der Krakauer Stadtschreiber verzeichnete 1403 dieses Ereignis ungewöhnlichen Ausmaßes: „[…] *item dernoch ist her gefallin an unsir Juden und Kamerknechte die in unserin ganczin Konriche* [sic!] *czu Behemen sint gesessin und hat si geschaczt, abgenomyn, geplundirt und gancz und gar vortervit junge und alde Weyben und Kinder bis aff den lecztin pfennyng und ir gesegilte briefe eingenomyn hat. Und di unsirn* […] *hat wedirgegebin di selbin Juden, wir uns habin offt gehaldin mit demselbin gute czu unserin erern. Und wir den und wen uns des not wurden und gescheen, des si uns mit dem sewm gute mochten gehülfen han.*"[26]

Der König, der eine (oft kritisierte) liberale Einstellung zu den Juden hatte – sei es aus ideellen oder eher pragmatischen Gründen –, ließ dieses Ereignis nicht unbeantwortet. Die jüdische Bevölkerung, die dem Massaker entkommen war, nahm er in Schutzhaft und ordnete gleichzeitig an, alle Wertgegenstände, die im Ghetto gestohlen worden waren, aufzuspüren und im Altstädter Rathaus abzugeben. Die Altstadt musste für die verursachten Schäden aufkommen und eine Strafe in Höhe von 20.000 Schock Groschen bezahlen. Unter Berücksichtigung des Privilegs von Ottokar II. Přemysl aus dem Jahre 1254 (*Statuta Judaeorum*) endete diese Summe schließlich in der königlichen Kasse als Kompensation für den Angriff auf den Besitz der königlichen Kammer. Nichtsdestotrotz kam es bald zu einem Wiederaufbau und einer Neubesiedlung des Ghettos. 1410

26 Zitiert nach: Šedinová / Kosáková, Golem (wie Anm. 22), S. 10 f.

bestätigte Wenzel IV. das Privileg für den Friedhof in der Neustadt (Judengarten genannt),[27] inklusive einiger in der Nähe liegender Häuser (12–14).[28]

Von den sozialen und politischen Veränderungen nach dem Tode Wenzels IV. war auch die jüdische Bevölkerung betroffen. Die Hussitenbewegung, die hinsichtlich ihrer Ideologie das Alte Testament betonte, wurde von Seiten der Juden behutsam, aber mit gewissen Erwartungen empfangen. Während die Prager Katholiken aus der Stadt vertrieben wurden, konnten die Juden bleiben. Die Illusion eines ruhigen Zusammenlebens währte aber nicht lange. Aus dem Jahre 1422, kurz nach der hussitischen Eroberung Wyschehrads, stehen uns Berichte über die Ausplünderung einiger Häuser im Ghetto zur Verfügung. Angriffe gegen die jüdischen Stadtbewohner gingen auch in der Folgezeit weiter. So stammt die nächste Nachricht über Plünderungen in der Judenstadt bereits aus dem Jahre 1448. Auch die folgenden Jahre verliefen im Geiste einer ambivalenten Beziehung seitens des Königs, der Vertreter der Stadtverwaltung und der übrigen Bürger. Eine besondere Feindseligkeit fanden die Juden bei den Handwerkern und Kaufmännern, welche die Juden als unwillkommene Konkurrenz ansahen. Diese feindliche Einstellung hielt auch in den folgenden Jahren an. Und es waren gerade die Vertreter einiger Zünfte, die auf den König Druck ausübten, um die Juden aus der Stadt zu vertreiben. In der gefälligen Einstellung des

---

27 Nach dem Erlass König Vladislavs II. von 1478 kam es zur Aufhebung des Friedhofs und darauf folgenden Aufteilung des Grundstücks auf einzelne Parzellen. Während der archäologischen Forschungen in den Jahren 1997 bis 2000 wurden hier 400 Gräber ausgegraben. Vgl. Bývalý židovský hřbitov zvaný Židovská zahrada – pohřebiště (= Ehemaliger jüdischer Friedhof namens Jüdischer Garten – Grabstätte), veröffentlicht am: 6.9.2010, in: Odbor archeologie NPÚ Praha. Národní památkový ústav v Praze, URL: http://www.archeopraha.cz/byvaly-zidovsky-hrbitov-zvany-zidovska-zahrada-pohrebiste (letzter Zugriff: 23.10.2020). Der Friedhof auf dem Gebiet der Prager Neustadt war älter als der heute bekannte jüdische Friedhof in der ehemaligen Prager Judenstadt, der bis zum Verbot der Bestattung innerhalb der Stadtmauern von Seiten König Josefs II. fungierte.

28 Pěkný, Historie Židů (wie Anm. 25), S. 35 f.; Vilímková, Židovské Město pražské (wie Anm. 24), S. 14. Irgendwann in den letzten zwei Jahrzehnten des 15. Jahrhunderts bekamen die Juden für eine kurze Zeit das Recht, sich auch auf der Kleinseite anzusiedeln. Die jüdische Siedlung innerhalb der Neustadt ist vermutlich irgendwann nach 1502 verloren gegangen, als die Neustädter die Vertreibung der hier ansässigen Juden verlangten. Auf die (am Ende nicht realisierten) Überlegungen über eine Erneuerung der jüdischen Siedlung innerhalb der Neustadt stößt man nach dem großen Brand des Prager Ghettos im Jahre 1689.

Königs den Juden gegenüber kann jedoch kein Altruismus gesehen werden, da er bloß die eigenen ökonomischen Interessen verfolgte.

Die vergeblichen Bestrebungen der Prager Handwerker um die Vertreibung der lästigen Konkurrenz in den ersten zwei Jahrzehnten des 16. Jahrhunderts führten erst Anfang der 40er Jahre zum Erfolg. Zu dieser Zeit wurde gegen die Juden die Anklage erhoben, sie hätten in Prag Spionage zugunsten der Türken betrieben. Einem Landtagsbeschluss zufolge mussten daraufhin alle Juden bis Anfang November 1541 die Stadt verlassen. Angesichts der kurzen Frist wurde ihnen aber die Aufenthaltserlaubnis in Prag mehrmals verlängert. Am Ende blieben in der Stadt – mit Rücksicht auf eine besondere Genehmigung des Königs – nur 15 vornehme Juden, um alle verbliebenden Forderungen auflösen zu können.[29]

Die Wende zum Guten, nicht nur für die Prager, sondern im Allgemeinen für alle Juden in den böhmischen Ländern, begann erst im Zusammenhang mit der Krönung Maximilians II. zum böhmischen König im Jahre 1564. Dieser bestätigte den Juden bereits 1567 alle ihre Privilegien und den in der Stadt ansässigen Juden erlaubte er, Handel zu treiben.

Nach den uns bekannten Informationen[30] lebten in der Judenstadt zu dieser Zeit etwa 1.000 Bewohner. Dies bedeutet, dass sich die Einwohnerzahl seit dem Pogrom Ende der 80er Jahre des 14. Jahrhunderts (siehe oben) praktisch nicht verändert hatte – vor dem Pogrom werden für das Prager Ghetto etwa 4.000 Bewohner angegeben.[31]

Von der Besiedlungsentwicklung des Ghettos, dessen Religions- und Verwaltungszentrum in der Nähe der Altneu-Synagoge zu finden war, wo sich auch ein kleiner Markt für die Bedürfnisse der Ghettobewohner befand, sowie von den Häusern, die in Händen jüdischer Inhaber waren,[32] zeugen die ältesten

---

29 Lenka Matušíková, Pražské Židovské Město a jeho obyvatelé v 17. století (= Die Prager Judenstadt und ihre Bewohner im 17. Jahrhundert), in: Dies., K dějinám Židů v Českých zemích, Praha 2015, S. 131–138, hier bes.: S. 131.

30 Eine Übersicht (basierend auf älterer Literatur) bietet: Pěkný, Historie Židů (wie Anm. 25), S. 637.

31 Ziemlich exakte Angaben stehen erst für das 17. und dann besonders für das 18. Jahrhundert zur Verfügung, und zwar auf Grund der Listen der jüdischen Familien und Familianten. Siehe: Ivana Ebelová, Census Lists of Jewish Families in Bohemia, 1723/1724–1811, in: Judaica Bohemiae XLVIII/1, 2013, S. 79–91.

32 In diesem Zusammenhang soll aber die Unvollständigkeit der Quellen in Betracht gezogen werden. Die Liste der Häuser, bei denen wir wissen, dass diese in jüdischen Händen waren, muss nicht vollständig sein.

altstädtischen Stadtbücher. Seit 1577 wurden diese durch die neu erstellten, sogenannten Jüdischen Weißbücher (*Libri albi Judaeorum*) ersetzt.[33] Beide Büchersorten bieten zugleich Informationen darüber, wie sich die Eigentumsbeziehungen in den Grenzen beider Städte (Altstadt und Judenstadt) änderten. Obwohl die Inhaber der Immobilien in den Grenzen wechselten, das Eigentum von der Hand der Altstadt-Bewohner in die Hand der jüdischen Bewohner überging und umgekehrt, blieben der Umfang und die Struktur der jüdischen Besiedlung im Rahmen des Ghettos (mit gewissen kleinen Änderungen) bis zum Anfang des 19. Jahrhunderts bewahrt.

Die Wahl Prags zur Residenzstadt durch Rudolf II. spielte nicht nur für die gesamte Stadt, sondern auch für die jüdischen Bewohner eine wesentliche Rolle. Die Einwohner des Prager Ghettos bekamen neben der Bestätigung älterer Privilegien ebenfalls eine Versicherung, dass sie aus Prag nicht vertrieben werden. Der Herrscher öffnete den Juden den Weg zur Weiterentwicklung des Handels sowie ihrer Unternehmen und ermöglichte ihnen den Zugang nicht nur zu den böhmischen und mährischen Märkten, sondern auch zu denen im Ausland.[34] Die Judenstadt unterlag direkt der Jurisdiktion der Böhmischen Kammer und der Statthalterei. Die wirtschaftliche Prosperität der Judenstadt verstärkte Rudolf II. durch die Befreiung der Juden von der Zollpflicht in den Prager Städten; er genehmigte die Ausübung einer ganzen Reihe der bis dahin verbotenen Handwerke (siehe weiter unten). Die günstigen Bedingungen spiegelten sich auch in der Einwohnerzahl des Ghettos wider, das mit seinen etwa 3.000 Bewohnern am Anfang des 17. Jahrhunderts zu den größten jüdischen Gemeinden nicht nur in Mitteleuropa, sondern in ganz Europa zählte.[35]

Die Regierung Rudolfs II. hatte eine gute Auswirkung nicht nur auf die Wirtschaftsführung der Stadt und die Verstärkung ihrer ökonomischen Stabilität, sondern auch auf die Entwicklung des Kulturlebens und den Aufschwung von Bauaktivitäten, die jedoch ohne eine starke finanzielle Unterstützung einiger vornehmer Persönlichkeiten nie so groß hätte ausfallen können. Zu erwähnen sind zuvörderst der erfolgreiche Kaufmann, Hoflieferant, Investor und jüdische

---

33 Die *Libri albi Judaeorum* sowie die Altstädter Stadtbücher werden heute im Archiv der Hauptstadt Prag (Archiv hlavního města Prahy) aufbewahrt.

34 Zum Handel vor der Schlacht am Weißen Berg (inklusive der Handelstätigkeiten der Prager Juden) siehe: Marie Buňatová, Pražští kupci na cestách. Předbělohorská Praha a středoevropské trhy (= Prager Kaufleute unterwegs. Prag und die mitteleuropäischen Märkte vor der Schlacht am Weißen Berg), Praha 2013.

35 Matušíková, Pražské Židovské Město (wie Anm. 29), S. 131 f.

Primus Mordechaj Maisel (1528–1601) sowie Jakob Bassevi von Treuenberg (1570–1634), der 1622 sogar in den Adelsstand erhoben wurde.

Mehr als zuvor wirkte sich zu dieser Zeit in beträchtlichem Maße der kosmopolitische Charakter der jüdischen Bevölkerung aus. Dies zeigte sich insbesondere in der erhöhten Handelsaktivität, die vor allem auf den Import der ausländischen Luxus-Ware zielte, ob es sich nun um Textil- oder Kürschnerware auf der einen oder um Kolonialware – vor allem Gewürze – auf der anderen Seite handelte. Im Laufe der Zeit erweiterte sich die Skala der angebotenen Ware immer mehr, samt der Genehmigung, an den christlichen (Jahr-)Märkten teilzunehmen.[36] Den Juden wurde es ermöglicht, einige spezifische Handwerke zu betreiben, inklusive des Goldschmiedehandwerks. Der jüdische Goldschmied Josef Coref bekam in der Folgezeit sogar Aufträge von Seiten der Prager Burg. Vertreter einiger Kunsthandwerke (insbesondere Perlhefter oder Perlsticker) fertigten dann Aufträge nicht nur für die Prager Bürger, sondern auch für kirchliche Institutionen ab.

Hand in Hand mit den erweiterten Möglichkeiten und Gelegenheiten, welche die Prager Juden Ende des 16. Jahrhunderts genossen, gingen trotz des Platzmangels innerhalb der Judenstadt auch die außergewöhnlich großen Bauaktivitäten. Aus eigener Initiative privater Bauherren wurden unter anderem einige Bethäuser und Synagogen errichtet. Dabei handelte es sich zunächst um die Pinkas-Synagoge, gegründet ursprünglich als eine private Synagoge (erwähnt zum ersten Mal im Jahre 1492) im Hause des Rabbis Israel Pinkas. Der Bau einer neuen Synagoge wurde 1535 begonnen; nach 1607 wurde das Objekt erweitert. Hinzu kam außerdem die Hohe Synagoge, die mit dem jüdischen Rathaus verbunden war. Ihr Bauherr war der bereits erwähnte Mordechaj Maisel, der unter anderem auch den Bau der größten Renaissance-Synagoge in Prag – der sogenannten Maisel-Synagoge – finanzierte. Weitere Synagogen, die zu dieser Zeit erbaut wurden, insbesondere die Zigeuner- (Cikánova) und Großhof- (Velkodvorská) Synagoge, wurden zum Opfer eines Brandes im Jahre 1689.[37]

Nicht nur die große Bauaktivität in der unmittelbaren Nähe der Altstadt, sondern auch der Zuzug von fremden Juden und ihre Niederlassung in der Stadt ohne Genehmigung, die wachsenden Handelsaktivitäten der Juden, die

---

36 Gerade in der Erweiterung der Handelsmöglichkeiten sowie der Handwerksproduktion lag die Hauptquelle der Feindseligkeit der christlichen Kaufmänner gegenüber der jüdischen Konkurrenz.

37 Zu den Synagogen und ihrer Bauentwicklung siehe: Pavel Kalina / Jiří Koťátko, Praha 1437–1610. Kapitoly o pozdně gotické a renesanční architektuře (= Prag 1437–1610. Kapitel von der spätgotischen und Renaissance-Architektur), Praha 2011, S. 171 f.; speziell: Vilímková, Židovské Město pražské (wie Anm. 24), S. 102–122.

wirtschaftliche Prosperität der Judenstadt sowie das Aufkaufen von christlichen Häusern in der nächsten Umgebung des Ghettos, all dies wurde zum Dorn im Auge vieler Bewohner der Prager Altstadt.

Die Niederlage der böhmischen Stände in der Schlacht am Weißen Berg im Jahre 1620 brachte den Prager Städten und ihren Bewohnern (inklusive der Judenstadt) viele Änderungen. Im Zusammenhang mit den folgenden Kriegsereignissen[38] sind diese eher negativ zu betrachten – Senkung der Einwohnerzahl, Verarmung, Schulden. Im Vergleich zu den übrigen Prager Städten zielten die hiesigen Bauaktivitäten nach dem Westfälischen Frieden im Jahre 1648 insbesondere auf Reparaturen, An- und Umbauten. Praktisch die ganze Bautätigkeit ging jedoch auf Grund eines großen Brandes 1689 verloren. Der Brand breitete sich von der Prager Altstadt aus und vernichtete fast das gesamte Ghetto samt seiner 318 Häuser und zwölf Synagogen. In der Altstadt selbst sind schließlich insgesamt 382 Häuser abgebrannt.

Der Wiederaufbau der Judenstadt nach 1689 spielte sich bereits im Lichte der völlig veränderten Bedingungen des sogenannten bürokratischen Antisemitismus ab, zu dessen Kennzeichen eine Reihe von staatlichen Restriktionen gehörte – vor allem die Bemühungen um die Beschränkung der jüdischen Bevölkerungszahl oder gar deren Vertreibung.[39]

4. Die Prager Städte als kaiserliche Metropole

Die Verlegung der kaiserlichen Residenz nach Prag, die der Inthronisierung Rudolfs II. im Jahre 1576 folgte – der Hof und die Hofbehörden siedelten bis 1583 über –, beeinflusste deutlich das Leben der Prager Konurbation für mehr als drei Jahrzehnte. Prag avancierte zu den bedeutendsten Metropolen Europas und seine Entwicklung beschleunigte sich gewaltig.[40] Die Präsenz des kaiserlichen

38 Von großer Bedeutung war die aktive Teilnahme der Bewohner der Judenstadt an der Verteidigung der Prager Altstadt zur Zeit der schwedischen Belagerung Prags im Jahre 1648.

39 Ivana Ebelová, Aspekty migrace a emigrace židů z Českých zemí se zřetelem k období tzv. úředního antisemitismu 18. století a vypovězení pražských židů ve čtyřicátých letech (= Aspekte der Migration und Emigration der Juden aus den Böhmischen Ländern hinsichtlich des sogenannten amtlichen Antisemitismus des 18. Jahrhunderts und der Vertreibung der Prager Juden in den 1940er Jahren), in: Sborník archivních prací (im Druck).

40 Vgl. Ledvinka / Pešek, Praha (wie Anm. 1), S. 293–323; Josef Janáček, Rudolf II. a jeho doba (= Rudolf II. und seine Zeit), Praha 1987; Eliška Fučíková, Praha v době vlády Rudolfa II. (= Prag in der Regierungszeit Rudolfs II.), in: Eliška Fučíková / Beket Bukovinská / Ivan Muchka (Hg.), Umění na dvoře Rudolfa II. (= Die Kunst am Hofe Rudolfs II.), Praha 1988, S. 29–59; Jaroslava Hausenblasová / Michal Šroněk, Urbs

Hofes beeinflusste die Konurbation auf vielerlei Weise. In Bezug auf die Wirtschaft verstärkte sie die Handelsbeziehungen und dadurch den Charakter Prags als Verbraucherzentrum, in sozialer Hinsicht verstärkte sie eine vermischte Nationalstruktur und die sozialen Unterschiede, bezüglich der Demographie führte sie zum massiven Anstieg der Einwohnerzahl.[41] Der Hof lockte sowohl Wissenschaftler als auch Künstler und dazu noch geschickte Handwerker und fremde Kaufleute an.[42] Das alles brachte neue kulturelle Impulse mit sich, die das Prager Bürgertum – und bestimmt nicht nur die amtierenden Eliten – bereitwillig annahm. Die Konfrontation mit kultureller, sprachlicher oder konfessioneller Andersartigkeit führte zur Durchsetzung eines offeneren Kulturmodells;[43] der Lebensstil veränderte sich. Dies fand seinen Niederschlag in der baulichen und urbanistischen Transformation der Prager Städte und natürlich

---

aurea. Prague of Emperor Rudolf II, Praha 1997. Vgl. auch: Jiří Pešek, Prag auf dem Weg zur kaiserlichen Residenz (1483–1583), in: Evamaria Engel / Karen Lambrecht / Hanna Nogossek (Hg.), Metropolen im Wandel. Zentralität in Ostmitteleuropa an der Wende vom Mittelalter zur Neuzeit, Berlin 1995, S. 213–223; Zdeněk Hojda, Prag um 1600 als multikulturelle Stadt: Hof – Adel – Bürgertum – Kirche, in: ebd., S. 225–232.

41 Vgl. zur Prager Einwohnerzahl: Zikmund Winter, Kolik obyvatel měla Praha v XVI. věku (= Wie viele Einwohner hatte Prag im 16. Jahrhundert?), in: Zvon 4, 1904, S. 541 f.; Václav Líva, Kolik obyvatelů měla Praha před třicetiletou válkou a po ní (= Wie viele Einwohner hatte Prag vor und nach dem Dreißigjährigen Krieg?), in: Český časopis historický 42, 1936, S. 332–347. Für die genauere Bestimmung der Einwohnerzahl sind die Verleihungen des Bürgerrechts in den dazugehörigen Stadtbüchern sowie die älteren Kirchenmatrikeln von großem Nutzen. Vgl. Eva Semotanová, Knihy měšťanských práv – významný pramen předstatistického období (= Die Bürgerrechtsbücher – eine wichtige Quelle der vorstatistischen Zeit), in: Historická demografie 10, 1986, S. 73–115; Jaroslava Mendelová, Knihy měšťanských práv na Novém Městě pražském 1582–1657 (= Die Bürgerrechtsbücher der Neustadt Prag 1582–1657), Praha 2012 (Edition in Tabellen); Dies., Svatojindřišská matrika 1584–1600 jako demografický pramen (= Die Kirchenmatrikel der Pfarrei St. Heinrich aus den Jahren 1584–1600 als demographische Quelle), in: Documenta Pragensia 7, 1987, S. 340–355.

42 Vgl. zu Prag als Handelszentrum im 16. Jahrhundert: Josef Janáček, Dějiny obchodu v předbělohorské Praze (= Geschichte des Handels in Prag vor 1620), Praha 1955; Buňatová, Pražští kupci na cestách (wie Anm. 34), passim; Dies., Hedvábí, sklo a koření. Obchod mezi Prahou a Itálií (1500–1620) (= Seide, Glas und Gewürz. Der Handel zwischen Prag und Italien (1500–1620)), Praha 2019.

43 Zum Thema der offenen Kultur siehe: Robert J. W. Evans, Bílá Hora a kultura českých zemí (= Die Schlacht am Weißen Berg und die Kultur der böhmischen Länder), in: Československý časopis historický 17, 1969, S. 845–862. Von großer Bedeutung für die wegbereitende Thematisierung einer breit verstandenen Kulturatmosphäre der Zeit bleibt: Robert J. W. Evans, The Making of the Habsburg Monarchy, Oxford 1979, hier: S. 3–67.

auch in der Verstärkung ihrer Besonderheiten und gegenseitigen Unterschiede. Daher lohnt sich ein Rundgang durch diese dynamische Stadtlandschaft.

Das Prager Schloss entsprach beim Eintreffen Rudolfs II. bei weitem nicht den Ansprüchen einer kaiserlichen Residenz. Der alte Königspalast diente seit langem nur noch den Landesbehörden: Er war Schauplatz des Landtags, das Landtafelamt hatte hier seinen Sitz, hier tagte das Landgericht. Mit dem Eintreffen des Kaisers stieg der Einfluss der Böhmischen Hofkanzlei, die durch die Verbindung der älteren Prager Königskanzlei und ihrer Wiener Expedition entstand. Die enge Koexistenz der landesherrlichen und ständischen Institutionen, die zentralen kirchlichen Institutionen, die Kathedrale, die Adelspaläste und das Gewirr der Kleinhäuser für die Höflinge und Diener, das alles schuf das besondere Milieu des Residenzbezirks.[44] Es fehlten „bloß" geeignete Wohnräume für den langfristigen Aufenthalt des Herrschers. Daher mussten für Rudolf II. im westlichen Teil des Schlosses schrittweise neue Wohn- und Repräsentationsräume geschaffen werden, zu denen auch Pferdeställe und Säle für die Kunstkammer und Gemäldegalerie gehörten. Für die Reichsbehörden, Zentralbehörden und Kollegien, die (mit Ausnahme des Kriegshofrats) von Wien nach Prag umgezogen waren, mussten ebenfalls geeignete Orte gefunden werden.[45] Bereits unter Rudolfs Vorgänger Maximilian II. wurde in Hradschin das Kaiserspital errichtet.[46]

Immer auffälliger drang die einheimische und fremde Aristokratie in das Leben der Prager Städte ein. Infolge ihrer Amtspflichten sowie aus politischen

---

44 Ivan Muchka, Rudolf II. jako stavebník (= Rudolf II. als Bauherr), in: Fučíková / Bukovinská / Muchka, Umění na dvoře Rudolfa II. (wie Anm. 40), S. 179–212; Eliška Fučíková, Prague Castle under Rudolf II, His Predecessors and Successors, 1530–1648, in: Dies. (Hg.), Rudolf II and Prague. The Imperial Court and Residential City as the Cultural and Spiritual Heart of Central Europe, Prague/London/Milan 1997, S. 2–71, zur Bautätigkeit im Schloss bes.: S. 15–20; Pavel Vlček, Renesanční proměny Pražského hradu a Hradčan (= Die Renaissance-Veränderungen des Prager Schlosses und Hradschins), in: Ders. (Hg.), Umělecké památky Prahy. Pražský Hrad a Hradčany (= Die Kunstdenkmäler Prags. Prager Schloss und Hradschin), Praha 2000, S. 31–36. Auch die Beschreibung der sich im Schloss befindlichen Personen von 1620 zeigt viel von den Verhältnissen der vorangegangenen Jahre. Archiv des Prager Schlosses (Archiv Pražského hradu), Hofbauamt, Inv.-Nr. 398, I, fol. 1–13; II/1, fol. 1–26 („*Beschreibung aller heuser und zimmer im kuniglichen Prager schloß*").

45 Für die Reichshofkanzlei wurde 1602 ein Haus in der heutigen Loretogasse (Loretánská, Nr. 178/V und 179/IV) gekauft, 1607 mit dem benachbarten Renthaus (Sitz des Rentamts der Böhmischen Kammer) verbunden und zu einem repräsentativen Sitz umgebaut. Vlček, Pražský hrad a Hradčany (wie Anm. 44), S. 404 f.

46 Ebd., S. 283.

und gesellschaftlichen Gründen verbrachten die Adeligen einen immer größeren Teil des Jahres in der Stadt. Dieser Umstand erforderte den Bau von genügend repräsentativen Stadtresidenzen samt zugehöriger Wirtschaftsbasis mit Erholungsorten in der Nähe Prags. Der heimischen Nobilität schlossen sich ab den 70er Jahren des 16. Jahrhunderts die Reichsgeschlechter und österreichische Aristokraten an.[47] Die ersten Adelspaläste wurden im Vorfeld des Schlosses bereits kurz nach dem großen Brand von Kleinseite und Hradschin im Jahre 1541 gebaut – die Paläste von Jan Popel von Lobkowitz[48] und Florian Griespeck von Griespach.[49] Die Anzahl der Häuser der Adelsgeschlechter stieg hauptsächlich in der Epoche Rudolfs II. rasant an.

Im frühen 17. Jahrhundert entstanden in beiden Städten am linken Ufer eine Art von „Adelszonen".[50] Das gilt insbesondere für Hradschin. Für die Situation in der „oberen" Stadt war kennzeichnend, dass sich das Rathaus nicht einmal am Hauptplatz befand und sich mit einer bescheidenen Bauparzelle in einer Nebengasse begnügen musste.

Im östlichen Bereich der Prager Burg bauten ihre Paläste ab den 40er Jahren zwei damals prominente Adelsfamilien im Königreich, die Herren von Rosenberg[51] und die Herren von Pernstein.[52] Direkt im Vorfeld der Burg am Hradschiner

47 Vgl. Jaroslava Hausenblasová, Prager Elitenwandel um 1600: Böhmischer und höfischer Adel zur Zeit Rudolfs II., in: Marina Dmitrieva / Karen Lambrecht (Hg.), Krakau, Prag und Wien. Funktionen von Metropolen im frühmodernen Staat, Stuttgart 2000, S. 173–184.

48 Marius Winzeler, Palác Lobkowiczů jako renesanční residence (= Das Palais Lobkowitz als eine Renaissance Residenz), in: Marius Winzeler (Hg.), Schwarzenberský a Salmovský palác (= Schwarzenberg-Palais und Palais Salm), Praha 2018, S. 106–145.

49 Tomáš Kleisner, The Gryspek Palace on a Drawing by Pieter Stevens, in: Studia Rudolphina 9, 2009, S. 138 f.

50 Im Folgenden bieten wir ein verdichtetes Gesamtbild der Stadt Rudolfs II. ohne konkrete Daten der Fertigstellung der Häuser oder ihres Erwerbs durch die genannten Besitzer. Dieses Bild entspricht etwa der Situation im letzten Jahrzehnt der Regierung Rudolfs II. Die mit den genannten Adeligen angegebenen Ämter und Titel sind immer die höchsten, die sie unter dieser Herrschaft erreichten. Zur Bautätigkeit des Adels zusammenfassend: Jan Muk, K stavební činnosti šlechty v Praze v 16.–18. století (= Zur Bautätigkeit des Adels in Prag vom 16. bis zum 18. Jahrhundert), in: Documenta Pragensia 10, 1991, S. 205–214.

51 Jarmila Krčálová, Palác pánů z Rožmberka (= Das Palais der Herren von Rosenberg), in: Umění 18, 1970, S. 469–485; Ivan Muchka, Rožmberský palác na Pražském hradě (= Das Rosenberg-Palais in der Prager Burg), in: Václav Bůžek (Hg.), Život na dvoře a v rezidenčních městech posledních Rožmberků (= Das Leben am Hof und in den Residenzstädten der letzten Rosenberger), České Budějovice 1993, S. 353–361.

52 Petr Chotěbor / Tomáš Durdík, Lobkovický palác – renesance a manýrismus (= Palais Lobkowitz – Renaissance und Manierismus), in: Umění 43, 1995, S. 411–424.

Platz befand sich an der Nordseite das erzbischöfliche Palais[53] und in seiner Nachbarschaft das Haus des Obersthofmeisters Christoph Popel von Lobkowitz, ein paar Häuser weiter stand der Sitz des Präsidenten der Böhmischen Kammer Joachim Nowohradský von Kolowrat und des Hofmarschalls Jaroslav von Martinitz. Die Westseite des Platzes und die Südseite am Rande der Erhebung waren ebenfalls den aristokratischen Residenzen vorbehalten. Am stattlichsten von ihnen war das Palais Lobkowitz, der erste Prager Palastbau im Stile der Renaissance, der bereits im Jahre 1567 fertig gebaut war.[54] Ab 1600 besaß ihn Peter Wok von Rosenberg. Genau hier fand im Oktober 1601 das Festessen statt, bei dem der kaiserliche Astronom Tycho Brahe infolge akuter Urämie verstarb. Der benachbarte Palast gehörte zunächst dem Präsidenten der Reichshofkanzlei Paul Sixt Trautson, nach 1600 dem Oberstlandkämmerer Wolf Nowohradsky von Kolowrat.

Vom Hradschiner Platz nach Westen gingen zwei Gassen. Die kleinere führte zum Kaiserspital, zur Pagenschule und an deren Ende zur Residenz des spanischen Gesandten. Die Hauptausfallstraße in Richtung Strahover Tor und weiter zur Reichsstraße säumten links kleinere Häuser des Adels und die Reichshofkanzlei, rechts das von Herzog Heinrich Julius von Braunschweig-Wolfenbüttel[55] (Mäzen und Freund des Kaisers) gemietete Haus und – ein Stück weiter – das im Jahre 1599 gegründete Kapuzinerkloster. Vor dem Strahover Tor, unter dem Hang mit der St.-Rochus-Kirche, die ab 1603 als Exvoto für ein Ende der Pestepidemie von 1599 erbaut wurde, stand das Haus des Reichsvizekanzlers Jacob Kurz von Senftenau, das nach seiner Ankunft in Prag der bereits erwähnte dänische Astronom zur Verfügung hatte.[56]

---

53 Vgl. Jiří Kropáček, The Archbishop's Palace in Prague, Kostelní Vydří [2]2014.

54 Vgl. Lucie Večerníková, Majitelé (= Die Besitzer), in: Winzeler, Schwarzenberský a Salmovský palác (wie Anm. 48), S. 54–65. Im Jahre 1600 gab der Kaiser diesen, von der Familie Lobkowitz konfiszierten Palast Peter Wok von Rosenberg.

55 Hilda Lietzmann, Herzog Heinrich Julius zu Braunschweig und Lüneburg (1564–1613). Persönlichkeit und Wirken für Kaiser und Reich, Braunschweig 1993; Jochen Luckhardt, Kunst im Dienste des Reiches: Herzog Heinrich Julius von Braunschweig-Lüneburg und Prag, in: Lubomír Konečný / Beket Bukovinská / Ivan Muchka (Hg.), Rudolf II, Prague and the World. Papers from the International Conference. Prague, 2–4 September, 1997, Prague 1998, S. 81–88; Václav Bůžek, Heinrich Julius von Braunschweig-Wolfenbüttel am Prager Kaiserhof, in: Werner Arnold / Brage Bei der Wieden / Ulrike Gleixner (Hg.), Herzog Heinrich Julius zu Braunschweig und Lüneburg (1564–1613): Politiker und Gelehrter mit europäischem Profil (Quellen und Forschungen zur Braunschweigischen Landesgeschichte 49), Braunschweig 2016, S. 42–56.

56 Václav Brož, O domě říšského místokancléře Jakuba Kurze ze Senftenavy na Hradčanech a o zřízené při něm hvězdárně Tychona Brahe (= Über das Haus des Reichsvizekanzlers

Zwischen dem Burgtor und der Kleinseite, unterhalb der Schlosssteige, konstruierte man auf der prekär abfallenden Baustelle ab Mitte des Jahrhunderts den stattlichen Palast der Herren von Neuhaus, den 1602 Wilhelm Slawata von Chlum und Koschumberg kaufte.[57] Direkt unter den Wallanlagen der Prager Burg, in der heutigen Sněmovní-Straße, hatten die Mitglieder des Reichshofrats Ferdinand und Hans Friedrich Hofmann von Grünbühel ihren Palast.[58] Ferdinand pflegte Kontakte zu interessanten Persönlichkeiten aus Wissenschaft und Kunst (Johannes Kepler, Giuseppe Arcimboldo) und baute eine hervorragende Bibliothek auf.

Die Adelsresidenzen auf der Kleinseite[59] konzentrierten sich vor allem an der Nordseite des Kleinseitner Rings[60] am Fuße des Schlossbergs. Das prächtigste Palais ließ hier Sigismund Smiřický von Schmiritz errichten, einer der reichsten

---

Jacob Kurz von Senftenau in Hradschin und über die Sternwarte Tycho Brahes darin), in: Sborník příspěvků k dějinám královského hlavního města Prahy 1, 1907, S. 61–91.

57 Vgl. Václav Ledvinka, Rezidence feudálního velmože v předbělohorské Praze. Pražské sídlo pánů z Hradce ve 2. polovině 16. století (= Die Residenz eines Feudalherren in Prag vor der Schlacht am Weißen Berg. Der Prager Sitz der Herren von Neuhaus in der zweiten Hälfte des 16. Jahrhunderts), in: Documenta Pragensia 9, 1991, S. 113–134.

58 Vgl. František Spurný, Hans Friedrich Hofmann z Grünbüchlu (= Hans Friedrich Hofmann von Grünbühel), in: Pocta Josefu Polišenskému. Sborník prací moravských historiků k 80. narozeninám univerzitního profesora PhDr. Josefa Polišenského, DrSc. (= Festschrift für Josef Polišenský zum 80. Geburtstag), Olomouc 1996, S. 151–156. Zu seinem Haus (Nr. 170/III) siehe: Pavel Vlček (Hg.), Umělecké památky Prahy. Malá Strana (= Kunstdenkmäler Prags. Kleinseite), Praha 1999, S. 269.

59 Vgl. zur historischen sowie architektonischen Entwicklung der Kleinseite: Václav Vojtíšek, Hradčany a Malá Strana v 2. polovině 16. století (= Hradschin und Kleinseite in der zweiten Hälfte des 16. Jahrhunderts), in: Ders. (Hg.), Z minulosti naší Prahy (= Aus der Vergangenheit unseres Prags), Praha 1919, S. 41–74; Milada Vilímková, Vývoj na levém vltavském břehu do nástupu Rudolfa II. (= Die Entwicklung am linken Ufer der Moldau bis zur Ankunft von Rudolf II.) und Období rudolfinské (= Die rudolfinische Zeit), in: Emanuel Poche (Hg.), Praha na úsvitu nových dějin (= Prag zu Beginn der neueren Geschichte), Praha 1988, S. 50–68 und 68–94; Pavel Vlček, Proměny města v době renesance a za vlády Rudolfa II. (= Die Veränderung der Stadt [Kleinseite] in der Renaissance und in der rudolfinischen Zeit), in: Vlček, Malá Strana (wie Anm. 58), S. 34–38.

60 Vgl. zum Kleinseitner Ring: Jana Doktorová, Malostranské náměstí v Praze (1400–1750). Metamorfózy veřejného prostoru a jejich hybatelé (= Der Kleinseitner Ring in Prag (1400–1700). Metamorphosen des öffentlichen Raums und deren Antriebskräfte), in: Kateřina Jíšová / Jan Hrdina (Hg.), Radnice, rynek a tržiště: místa setkávání, obchodu a reprezentace ve středověkých a raně novověkých městech (= Rathaus, Ring und Marktplatz: Orte der Begegnung, des Handels und der Repräsentation in mittelalterlichen und frühneuzeitlichen Städten) (Documenta Pragensia. Supplementa VIII), Praha/Ústí nad Labem/Dolní Břežany 2017, S. 215–269, für die Jahre 1576–1620: S. 230–251.

Aristokraten im Lande.[61] In diesem Haus fand auch im Mai 1618 das Treffen der Radikalen unter den Ständen am Vorabend der Defenestration der katholischen Statthalter statt. Weitere Adelspaläste befanden sich an der Westseite des Rings, auch hinter dem Kleinseitner Rathaus, wo ein stattliches (nicht erhaltenes) Gebäude dem Oberstlandhofmeister (und Onkel Wallensteins) Adam von Waldstein gehörte. Um 1600 findet man auf der Kleinseite aber noch nicht die zahlreichen Gartenpaläste, die erst der Barock nach der Mitte des 17. Jahrhunderts mit sich brachte.

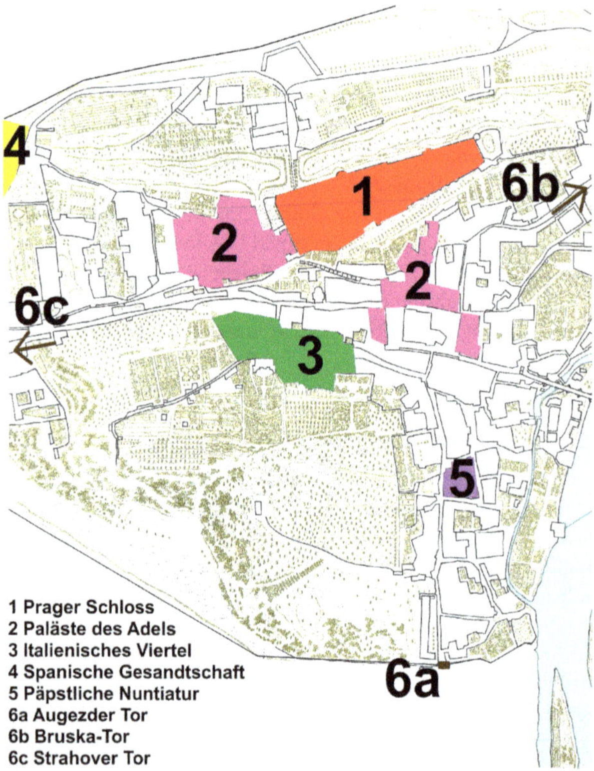

Abb. 52: *Die Kleinseite mit Hradschin in den Jahren 1576–1618: Die überwiegend vom Adel bewohnten Stadtteile – Welsches Viertel – Gesandtenhäuser. Bildnachweis/Copyright: Matyáš Ebel.*

61 Jaroslav Čechura / Veronika Kucrová / Zuzana Vlasáková, Smiřičtí. Krátké dějiny úspěšného rodu (= Die Smirziczky. Eine kurze Geschichte eines erfolgreichen Geschlechts), Praha 2018.

Die hohen Hof- und Landesbeamten und Räte der Regierungskollegien stellten zwar den einflussreichsten, doch zahlenmäßig kleinsten Bestandteil des Hofstaates. Dieser komplizierte Organismus[62] wuchs lediglich während der Herrschaft Rudolfs II. von circa 600 auf etwa 1.000 Personen an, wobei die Ausgaben fast um die Hälfte auf nahezu 300.000 (Rheinische) Gulden jährlich anstiegen. Dazu kamen hauptsächlich Personen mit Ehrenfunktionen und Diener, die Rudolfs persönliche Liebhabereien und Bedürfnisse befriedigten. Hierzu zählten Kammerdiener, Ärzte, Künstler und Kunsthandwerker sowie das Personal in den Pferdeställen. Obwohl ein kleinerer Teil des Kaiserhofs in Wien zurückblieb, wird die Anzahl der Höflinge[63] mit ihren Familien auf etwa 2.500–3.000 Personen geschätzt, das heißt rund fünf Prozent der Prager Bevölkerung. Genauso wie die Spitzen am Hofe suchten auch die Personen in niedrigeren Funktionen Wohnungen in der Nähe des Schlosses. Auf der Kleinseite stellten sie etwa ein Drittel der Bevölkerung dar, in der Stadt Hradschin war ihr Anteil noch höher. Ungefähr die Hälfte von ihnen erwarb ein eigenes Haus, die anderen mieteten Wohnungen.[64] Diese große Gruppe trug gewiss zum sonderbaren Kolorit der Prager Städte am linken Moldau-Ufer bei. Im Vergleich dazu machten die Höflinge in der Prager Altstadt lediglich einen Bruchteil der Bevölkerung aus. Nach der Mieterliste von 1608, die auf Anordnung des Stadtrates erstellt wurde, entsprach ihre Zahl etwa fünf Prozent.

Die meisten Höflinge waren Fremde; eine markante Gruppe unter ihnen bildeten etwa die Italiener. Sie dominierten unter den Sängern und Kammermusikern

---

62 Die personelle Zusammensetzung des Hofes ergibt sich aus den Hofstaatsverzeichnissen, siehe: Jaroslava Hausenblasová, Der Hof Kaiser Rudolfs II. Eine Edition der Hofstaatsverzeichnisse 1576–1612 (Fontes historiae artium 9), Prag 2002.

63 Vgl. Jaroslava Hausenblasová, Die „Höflinge" Kaiser Rudolfs II. Beitrag zur Begriffsbestimmung, in: Konečný / Bukovinská / Muchka, Rudolf II, Prague and the World (wie Anm. 55), S. 239–243.

64 Diese Daten beruhen auf der Analyse des Quartierbuchs für Hradschin und Kleinseite aus den Jahren 1608–1612, die Jaroslava Hausenblasová und der Autor dieser Studie zur Edition vorbereiten. Die Handschrift wird in der Nationalbibliothek der Tschechischen Republik aufbewahrt: Národní knihovna ČR, sign. XXIII D 57, 1608–c. 1612, IV + 776 pp. Vgl. auch: Jaroslava Hausenblasová / Zdeněk Hojda, Pražský rudolfínský dvůr mezi Hradem a městem (= Der Prager Hof Rudolfs II. zwischen Schloss und Stadt), in: Bůžek, Život na dvoře a v rezidenčních městech (wie Anm. 51), S. 115–136; Zdeněk Hojda, Der Hofstaat Rudolfs II. in Prag. Wo und wie wohnten die Hofleute, insbesondere die Künstler?, in: Prag um 1600. Beiträge zur Kunst und Kultur am Hofe Rudolfs II., Freren 1988, S. 118–123.

(Familie Mosto, Caspar, Ardesi),[65] andere finden wir unter den spezialisierten Handwerkern und Künstlern – sogar sechs Steinschneider, unter ihnen zwei aus der Familie Miseroni und Giovanni Castrucci, ein bekannter Hersteller von Steinmosaiken „Florenzer Art". Zu den gut bezahlten Spezialisten zählte man die Brüder Scaiola, die in der Kaiserküche tätig waren. Der Pasteten-Hersteller Filip Scaiola bezog ein Jahresgehalt in Höhe von 168 (Rheinische) Gulden, sein Bruder Lucas, der Leibkoch des Kaisers, verdiente sogar 240 (Rheinische) Gulden. Daneben gab es auch andere hochgeschätzte Fachleute, wie den Paretmacher (Hersteller von Kopfbedeckungen) Ferrante Castell oder den Tanzmeister (Festball-Organisator) Alfonso Pasetti, dessen Einkommen ebenfalls auf 240 (Rheinische) Gulden geschätzt wurde.

Eine besondere Gruppe unter den Ausländern in Prag stellten die Diplomaten und ihr Personal dar. Die Römische Kurie, Spanien, Venedig und Toskana, ab 1608 auch Frankreich, hatten in Prag ihre ständigen Gesandten, die anderen nur Sondergesandten, die sogenannten Agenten. Die diplomatischen Missionen aus Italien (Venedig, Mantua, Florenz), die Gesandten aus Anhalt und Frankreich siedelten in der Prager Altstadt – meistens mieteten sie ein ganzes Haus –,[66] doch die beiden politisch einflussreichsten Botschaften von Rom und Spanien befanden sich am linken Ufer.

Die Apostolische Nuntiatur befand sich im Südteil der Kleinseite im ehemaligen Magdalenerinnenstift (Nr. 459/III), der spanische Botschafter nutzte einen geräumigen Komplex von Häusern mit einem Turm und Gärten in Hradschin (Nr. 72/IV), den er von den Erben des Bürgers Hans Tucher von Schoberau mietete.[67] Gerne besichtigt wurden auch die Ankünfte der fremden Gesandtschaften. Im Jahre 1600 kamen in Prag zum Beispiel die Botschafter des französischen Königs Heinrich IV. unter Führung des Marschalls Bois-Dauphin an;[68]

65 Vgl. Zdeněk Hojda, Hudebníci Rudolfova dvora v ubytovací knize Malé Strany a Hradčan z roku 1608 (= Die Hofmusiker im Quartierbuch von Kleinseite und Hradschin von 1608), in: Hudební věda 24, 1987, S. 162–167.

66 Vgl. Jaroslav Čechura / Zdeněk Hojda / Martina Novozámská, Nájemníci na Starém Městě pražském roku 1608, Rekonstruovaná edice shořelého rkp. 324 z Archivu hl. m. Prahy podle opisu uloženého v Archivu Národního muzea (= Die Mieter in der Altstadt Prag im Jahre 1608. Die rekonstruierte Edition der verbrannten Handschrift 324 aus dem Archiv der Hauptstadt Prag nach der im Archiv des Nationalmuseums aufbewahrten Abschrift), Praha 1997.

67 Vgl. Tomáš Černušák / Pavel Marek, Gesandte und Klienten. Päpstliche und spanische Diplomaten im Umfeld von Kaiser Rudolf II., Berlin/Boston 2020.

68 Eliška Fučíková / Josef Janáček (Hg.), Tři francouzští kavalíři v rudolfínské Praze (= Drei französische Kavaliere im rudolfinischen Prag), Praha 1989, S. 38–73.

noch größere Aufmerksamkeit weckten exotische Gesandte aus Russland (1595) und Persien (1600).[69]

Diese Angaben belegen, dass das höfische Element einen ausschlaggebenden Einfluss auf die Demographie und Sozialtopographie der Kleinseite hatte. Unter den Hausbesitzern waren die Höflinge am häufigsten in der unmittelbaren Vorburg vertreten. Entlang der Schlosssteige gehörten zum Hofstaat beinahe drei Viertel der Hausbesitzer (73 Prozent). Was die Dichte der höfischen Besiedlung angeht, entsprach die Zahl der Häuser am Hang der Prager Burg in etwa dem Gebiet um die St.-Thomas-Kirche des Augustinerstiftes, das die Fremden aus katholischen romanischsprachigen Ländern anzog, und dem „Welschen Viertel" unter dem Laurenziberg, das man auch „Kleinitalien" nannte. Die Wohnsitze der Höflinge, die Quartiere mieteten, waren verständlicherweise eher mehrstöckige, das heißt stattlichere Häuser, deren Belegung mit der Entfernung von der Burg nachließ. Am meisten wurden hier die Wohnungen an der Schlosssteige, in der Sporrengasse und am Hohlweg genutzt, häufig wohnte das Hofpersonal in der Umgebung des Kleinseitner Rings und in der Brückengasse. Der Preis einer Wohnung war abhängig von ihrer Größe, ihrer Lage und auch von der Bauart und Ausstattung des Hauses. Die durchschnittliche Jahresmiete belief sich auf 10 bis 40 (Rheinische) Gulden.[70]

Dank der Anwesenheit des Hofes und der mit ihm verknüpften Möglichkeiten wurde Prag immer attraktiver für neue Anwärter des Bürgerrechts. Obschon die demographische Bilanz auch in früheren Jahrzehnten eindeutig positiv war, verstärkte sich der Strom der Neubürger dank der Konjunktur in den 80er Jahren beträchtlich. Im Vergleich zu der Mitte des 16. Jahrhunderts siedelten sich in der Kleinseite circa zweieinhalb mal so viele und in der Prager Altstadt etwa zweimal so viele an. Auch in der Prager Neustadt weisen die herausgegebenen Bürgerrechtsbücher insbesondere in den 80er Jahren hohe Zahlen auf.[71] Der

69 Siehe zuletzt: Eva Lukášová, Kryštof mladší Popel z Lobkowicz a památky z návštěv perských diplomatických poselstev v Praze na počátku 17. století (= Christoph (der Jüngere) Popel von Lobkowitz und die Andenken an die Besuche der persischen diplomatischen Gesandtschaften in Prag Anfang des 17. Jahrhunderts), in: Petr Pavelec / Martin Gaži / Milena Hajná (Hg.), Ve znamení Merkura. Šlechta českých zemí v evropské diplomacii (= Im Zeichen des Merkur. Der Adel aus böhmischen Ländern in der europäischen Diplomatie), České Budějovice 2020, S. 156–168.

70 Siehe Anm. 64.

71 Für die Altstadt siehe: Josef Teige, Seznamy měšťanů pražských (= Die Bürgerlisten von Prag), in: Almanach královského hlavního města Prahy 1901–1905. Vgl. außerdem die in Anm. 41 angeführte Literatur.

Zuzug von denjenigen, die kein Bürgerrecht beantragten und es nicht einmal erwerben konnten, war zweifelsohne noch größer. Die Schätzungen der Einwohnerzahl der Stadt, extrapoliert aus zugänglichen Indikatoren wie zum Beispiel der Häuserzahl, deuten an, dass die Population Prags während der Regierungszeit Rudolfs II. von circa 20.000–25.000 auf etwa 60.000 Einwohner anstieg.[72]

Namentlich auf der Kleinseite stellten die Fremden einen beträchtlichen Anteil dar.[73] Meistens stammten sie aus deutschsprachigen Gebieten (Kaufleute und Handwerker) und aus Italien (Steinmetze, Maurer, aber auch Kaminfeger). Unter den Kaufleuten begegnet man üblicherweise den Niederländern,[74] weniger häufig den Personen aus französischsprachigen Gebieten. Die Spanier finden wir im Umkreis des Hofes und natürlich des spanischen Gesandten oder im Jesuitenkolleg.[75] Der wachsende Anteil der Fremden veränderte die konfessionellen Verhältnisse Prags. Die Italiener[76] verstärkten die katholische Gemeinschaft und

---

72 Siehe Anm. 41.

73 Vgl. zu den Ausländern in Prag um 1600: Olga Fejtová, Das Verhältnis zwischen den Nationalitäten in den Prager Städten an der Wende vom 16. zum 17. Jahrhundert. Die Prager Neustadt als ein Bestandteil des Prager Stadtkomplexes, in: Berichte und Beiträge des Geisteswissenschaftlichen Zentrums Geschichte und Kultur Ostmitteleuropas 1999, S. 38–59; Dies., Integrace příchozích cizinců do raně novověké pražské společnosti (= Die Integration ankommender Ausländer in die frühneuzeitliche Prager Gesellschaft), in: Český časopis historický 116, 2018, S. 459–485; Jaroslav Pánek, Italové, Nizozemci a Němci v rudolfinské Praze – některé formy a problémy soužití (= Italiener, Niederländer und Deutsche im rudolfinischen Prag – einige Formen und Probleme des Zusammenlebens), in: Documenta Pragensia 19, 2001, S. 67–74. Einen sehr nützlichen Forschungsbericht bietet: Jiří Pešek, Národnostní skupiny, menšiny a cizinci ve městech: Praha středověku i novověku ve středoevropském poznání (= Ethnische Gruppen, Minderheiten und Fremde in den Städten: Prag in Mittelalter und Neuzeit im mitteleuropäischen Vergleich), in: Documenta Pragensia 19, 2001, S. 9–16.

74 Marianne Wünsch, Die Niederländer am Hofe Rudolfs II., in: Franz Höller (Hg.), Prager Jahrbuch, Prag 1943, S. 60–64; Nicolette Mout, Bohemen en de Nederlanden in de zestiende eeuw, Diss., Leiden 1975. Siehe auch die in Anm. 42 aufgeführte Literatur zu den fremden Kaufleuten in Prag.

75 Pavel Štěpánek, Španělé v Praze 16. století (= Die Spanier im Prag des 16. Jahrhunderts), in: Documenta Pragensia 19, 2001, S. 59–65.

76 Zu den Italienern in Prag: Josef Janáček, Italové v předbělohorské Praze (1526–1620) (= Die Italiener in Prag in der Zeit vor der Schlacht am Weißen Berg (1526–1620)), in: Pražský sborník historický 16, 1983, S. 77–118; Pavel Preiss, Italští umělci v Praze (= Italienische Künstler in Prag), Praha 1986, hier: S. 56–143; Gernot Heiss, Mezi ochranou a ohrožením. Italská a židovská obec v Praze a ve Vídni v 16. a 17. století

konzentrierten ihre religiösen und anderen „Minderheits-Aktivitäten“ zuerst auf die Welsche Kapelle am Jesuitenkolleg in der Prager Altstadt (fertig gebaut im Jahre 1600); im Jahre 1617 wurde die Hl. Karl Borromäus-Kapelle im neugebauten Spital im „Welschen Viertel“ unterhalb von Strahov (Nr. 335/III) geweiht.[77] Aus den deutschen Reichsstädten kamen Personen lutherischer Konfession; ein Teil der Niederländer, zum Beispiel der Hofkünstler Jacob Hoefnagel, engagierte sich an der Seite der Calvinisten[78] und den ihnen nahestehenden Böhmischen Brüdern. Für die lutherischen Gemeinschaften wurden nach dem Erlass des rudolfinischen Majestätsbriefes im Jahre 1609, der ein offizielles Bekennen weiterer protestantischer Denominationen, neben den bis dahin erlaubten, und in Prag den mehrheitlichen Utraquismus ermöglichte, rasch zwei neue lutherische Kirchen in der Kleinseite und Prager Altstadt errichtet,[79] wobei der Baumeister im Falle der Kleinseitner Dreifaltigkeitskirche wahrscheinlich der Hofarchitekt Giovanni Maria Filippi war!

Infolge der Einwanderung stieg die Belegung der Häuser. Während man vor 1583 mit einer durchschnittlichen Belegung eines Hauses mit acht bis zehn Personen rechnete, wuchs diese Zahl um 1600 auf etwa 15–18 Personen an.

---

ve zprávách jezuitů (= Zwischen Schutz und Bedrohung. Italienische und jüdische Gemeinden in Prag und Wien im 16. und 17. Jahrhundert in den Berichten der Jesuiten), in: Olga Fejtová (Hg.), Barokní Praha – Barokní Čechie. 1620–1740, Praha 2004, S. 177–189.

77 Zdeněk Kristen, Storia della Congregazione italiana di Praga, Praha 1949; La Congregazione italiana di Praga. Luoghi e memorie dell'Istituto Italiano di Cultura, a cura di Angela Trezza Cabrales, Kutná Hora 2003; Vlček, Malá Strana (wie Anm. 58), S. 435.

78 Thea Vignau-Wilberg, Neues zu Jacob Hoefnagel, in: Studia Rudolphina 10, 2010, S. 196–211; Evelyn Reitz, Der Protestantismus rudolfinischer Künstler im Lichte der „Theutschenn orthodoxischen Gemein“, in: Pražský sborník historický 44, 2016, S. 7–41.

79 Vgl. zu Prager „Kirchenlandschaft“: Štěpán Vácha, „Sub utraque, sub una“. Eine Quelle zur sakralen Topographie des rudolfinischen Prag (zum Jahr 1618), in: Studia Rudolphina 12/13, 2013, S. 116–133; Anna Ohlidal, Kirchenbau in der multikonfessionellen Stadt. Zur konfessionellen Prägung und Besetzung des städtischen Raums in den Prager Städten um 1600, in: Vera Isaiasz / Ute Lotz-Heumann / Monika Mommertz / Matthias Pohlig (Hg.), Stadt und Religion in der frühen Neuzeit. Soziale Ordnungen und ihre Repräsentationen, Frankfurt/New York 2007, S. 67–81; zur Kleinseitner lutherischen Kirche: Hilda Lietzmann, Die Deutsch-Lutherische Dreifaltigkeits-, die spätere Ordenskirche St. Maria de Victoria der Kleinen Seite zu Prag, in: Zeitschrift für Kunstgeschichte 30, 1977, S. 205–226. Vgl. auch: Reitz, Der Protestantismus rudolfinischer Künstler (wie Anm. 78), passim.

Dieser Zuwachs war durch die vielen Mieter, Zwischenmieter und Untermieter bedingt, denn laut der Auflistung von 1608 gab es in der Prager Altstadt nur 1.617 solcher Personen in 907 Häusern.[80] In Kleinseite und Hradschin zählte man nach dem Quartierbuch etwa 180 Personen (meistens Bürger) und weitere direkt im Areal des Prager Schlosses, deren genaue Zahl allerdings unbekannt ist.[81] Es handelt sich hierbei um die niedrigsten Zahlen, wobei die Menschen in den Gasthäusern, Obdachlose et cetera nicht mitgezählt wurden. Der Zuzug neuer Einwohner – freilich auch neue Anforderungen bezüglich der Wohnqualität – übte Druck auf das Bauen neuer Häuser und den Umbau der bestehenden Wohngebäude aus.

Das alles führte natürlich sowohl zur Erweiterung als auch zur Verdichtung des Stadtareals; beide Vorgänge kann man dabei fast wie im Labor am Beispiel der Kleinseite beobachten.[82] Von der erheblichen Bauaktivität zeugen unter anderem zahlreiche Kalköfen und Ziegelbrennereien – im Besitz der Gemeinden oder des Kaisers – am Stadtrand. Wenn wir auf das Prager Prospekt von Michael Peterle aus dem Jahre 1562 mit der Vedute, die Aegidius Sadeler 1606 herausgab,[83] blicken, stellen wir fest, dass die Stadt am linken Ufer in die Höhe wuchs.

80 Čechura / Hojda / Novozámská, Nájemníci na Starém Městě pražském (wie Anm. 66). Die tatsächliche Zahl der Mieter war noch höher, denn in dem Verzeichnis fehlen die adeligen und kirchlichen Häuser sowie die Gemeinde-Häuser.

81 Die in Anm. 44 angegebene Quelle ist hier nicht zu gebrauchen, denn sie stellt die Situation im Jahre 1620 dar, als sich die Zusammensetzung der im Schloss wohnenden Personen sicher schon verändert hatte.

82 Zur Bau- und urbanistischen Entwicklung der Kleinseite siehe die in Anm. 59 erwähnte Literatur sowie: Cyril Merhout, O Malé Straně. Její stavební vývoj a dávný život (= Über die Kleinseite. Ihre Bauentwicklung und vergangenes Leben), Praha 1956, S. 31–34; Václav Hlavsa / Jiří Vančura, Malá Strana (= Kleinseite), Praha 1983, S. 59–86 (kap. „Proměna v palácové a rezidenční město“ [= Kapitel „Die Umwandlung in eine Palast- und Residenzstadt“]).

83 Markéta Lazarová / Jiří Lukas, Praha. Obraz města v 16. a 17. století. Soupis grafických pohledů (= Prag. Das Bild der Stadt im 16. und 17. Jahrhundert. Verzeichnis der graphischen Veduten), Praha 2002, Bild 16.

Abb. 53: *Ansicht der Kleinseite. Detail aus der Radierung: Philip van den Bossche / Johannes Wechter / Aegidius Sadeler, Prag im Jahre 1606. Bildnachweis: Archiv der Altstadt Prag, Grafikensammlung.*

Auf einer älteren Stadtansicht sehen wir Fachwerkhäuser noch auf schmalen gotischen Parzellen, nur selten höher als ein Geschoss. Die Kleinseite ist umgeben von einem breiten Streifen mit Gärten und Weinbergen. Eine ganz andere Sicht bietet die um kein halbes Jahrhundert jüngere Vedute Sadelers. In allen Stadtteilen verzeichnet sie relativ weiträumige Gebäude, oft durch Giebel verbunden oder sogar mit der breiten Stirnseite zur Straße orientiert; zweistöckige Häuser sind üblich, es kommen auch dreistöckige vor. Man bemerkt auch einige Lustschlösser (*Casini*), Loggien, Altane und Belvederi auf den Dächern.[84] Immer noch gibt es genug Gärten, vor allem im Bereich der Insel Kampa an der Moldau; im Jahre 1596 kaufte jedoch die Gemeinde der Kleinseite zwei große Gärten am flachen Hang unter dem Laurenziberg auf und verkaufte die Parzellen überwiegend an die Italiener. Dadurch entstanden zwei parallele Straßen (Wälsche Gasse und Neue Gasse, tschechisch Vlašská und Břetislavova), das oben erwähnte „Kleinitalien". Diesen Stadtteil sehen wir auf dem Prospekt von 1606 wegen der perspektivischen Deformation nicht sehr gut, dafür können wir ein großes Objekt im Renaissancestil unweit vom Augezder Tor detailliert erkennen. Dieses von einem Garten umgebene Haus wird in den Quellen als „Lustschloss" bezeichnet und gehörte Jan Vchynský (Kinský). Es ist eines der ersten Beispiele der später zahlreicheren Palastbauten, die typisch für die Kleinseite waren und in sich die Funktion eines Stadtpalastes und Gartenpalastes verknüpften.[85]

Zur Verdichtung des Stadtraumes trugen auch einige Klöster mit ihren „Nebenrechten", das heißt Enklaven, die außerhalb der Kompetenz der Stadtverwaltung standen, bei. Die Klostergärten in Parzellen aufzuteilen und sie zu vermieten oder zu Bauzwecken zu verkaufen, war eine willkommene Quelle zur Finanzierung des Johanniterordens oder der Augustiner-Barfüßer. Eine neue Siedlung entstand auch vor dem ehemaligen Sandtor, als die Altstädter, denen

---

84 Petr Uličný, Belvederes and Loggias in Prague: Two Faces of the Leisure Architecture of the Imperial City, in: Studia Rudolphina 14, 2014, S. 30–50.

85 Ein gleichartiger, architektonisch sehr wertvoller und von einem nicht-adeligen Bauherrn unternommener Bau war der Palast, den sich in der Tržiště-Straße nach 1609 der Finanzmann Lazarus Henckel von Donnersmarck bauen ließ. Vgl. Josef Kallbrunner, Lazarus Henckel von Donnersmarck, in: Vierteljahrschrift für Sozial- und Wirtschaftsgeschichte 24, 1931, S. 142–156; Petr Uličný, Henckelův palác na Malé Straně (= Henckels Palast auf der Kleinseite), in: Petr Uličný u.a. (Hg.), Architektura Albrechta z Valdštejna. Italská stavební kultura v Čechách v letech 1600–1635 (= Architektur von Albrecht von Wallenstein. Italienische Baukultur in Böhmen in den Jahren 1600–1635), Bd. 2, Praha 2017, S. 953–966.

die Zolleinnahmen aus den Kleinseitner Toren gehörten, das neue Tor bis unter den Bach Bruska am Nordrand der Stadt verschoben und dadurch die Erweiterung der Kleinseite auf dem Gebiet ermöglichten, wo einige Jahre später Albrecht von Wallenstein sein großartiges Bauprojekt begann.

Das Primat unter den Städten Prags behielt jedoch weiterhin die Prager Altstadt. Sie war die dominante ökonomische Kraft der Konurbation und natürliches Zentrum des städtischen Lebens. Angesichts ihrer Lage – von allen Seiten umgeben von ihrem neustädtischen Nachbarn – konnte die Altstadt ihr Stadtareal nicht erweitern, das bereits vor der Epoche der Renaissance dicht bebaut war.[86] Die Prager Altstadt änderte daher ihr Aussehen vor allem durch Umbau und allgemeine Modernisierung. Da wir uns in diesem Fall nicht allzu sehr auf das Zeugnis der ikonographischen Quellen verlassen können – zum Beispiel auf die erwähnte Vedute von 1606, auf denen (wegen des Standorts des Zeichners) nur die Umgebung der Brücke und das anliegende Moldau-Ufer gut sichtbar sind –, stützen wir unsere Erkenntnisse über die Transformation der Altstadt in der Epoche Rudolfs II. hauptsächlich auf die erhaltenen Renaissance-Denkmäler und auf die bauhistorische Forschung.[87]

Als Ergebnis lässt sich festhalten: In der Prager Altstadt änderte sich vor allem die Höhe der Häuser; nicht nur der Straßentrakt wurde erhöht, sondern auch und besonders häufig die hinteren Hoftrakte, die nun als vollwertige Wohnungen dienten. Oft finden wir einen geschlossenen Hof mit Arkaden vor, der italienische Einflüsse verrät, allerdings Parallelen zur zeitgenössischen Transformation der österreichischen oder süddeutschen Städte zeigt. Ein bescheidenerer „Ersatz" der Arkaden waren eine Loggia oder offene Pawlatsche. Eine vollkommene Verschönerungskur erfuhren die Straßenfassaden, die mit bossierten und oft auch plastisch reich verzierten

86 Die Anzahl der Häuser änderte sich daher fast nicht. Vgl. František Dvorský, O počtu domů v Praze a v královských městech v Čechách v 16.–19. století (= Über die Anzahl der Häuser in Prag und anderen königlichen Städten in Böhmen vom 16. bis zum 19. Jahrhundert), in: Časopis Musea Království českého 55, 1881, S. 478–494.

87 Zur stadtbürgerlichen Architektur siehe: Dobroslav Líbal, Bürgerliche Architektur in Prag zur Zeit Rudolfs II., in: Gouk, Prag um 1600 (wie Anm. 64), S. 171–175; Pavel Vlček, Architecture in Prague, 1550–1650, in: Fučíková, Rudolf II and Prague (wie Anm. 44), S. 345–352. Zur Architektur der Altstadt siehe: Dobroslav Líbal, Renesance a manýrismus, in: Pavel Vlček (Hg.), Umělecké památky Prahy. Staré Město a Josefov (= Die Kunstdenkmäler Prags. Altstadt und Josefstadt), Praha 1996, S. 27–30.

Portalen versehen wurden (zum Beispiel das Haus „Zu den zwei goldenen Bären“, Nr. 475/I).[88]

Abb. 54: *Portal des Hauses „Zu den zwei goldenen Bären“, um 1575. Bildnachweis/ Foto: Zdeněk Hojda (privat).*

---

88 Josef Mayer, Dům u dvou zlatých medvědů (= Haus Zu den zwei goldenen Bären), in: Umění 6, 1958, S. 389–396; Vlček, Staré Město a Josefov (wie Anm. 87), S. 328 f.; Dobroslav Líbal / Jan Muk, Staré Město pražské. Architektonický a urbanistický vývoj (= Die Altstadt von Prag. Architektur- und Stadtentwicklung), Praha 1996, S. 249–259.

Man gliederte sie durch plastisch umsäumte Fenster und verzierte sie durch Sgraffitoputz (zum Beispiel den Stadtpalast des Zollamtsverwalters Jakob Granovský von Granov in der Prager Altstadt, Nr. 639–640/I).[89] Eine ausgesprochen dekorative Funktion hatten die Hausgiebel, die oft im damals modernen „Venediger Stil" erbaut wurden. Mit den wachsenden Anforderungen des Wohnens änderten sich auch die Innenräume. Als Beispiel hierfür können neue Kommunikationselemente dienen wie eingebaute, gewölbte Treppenhäuser oder die Vervollkommnung des Heizsystems durch neue „Welsch-Kamine", die mit der Schwarzküche verknüpft waren. Im Haus „Zum Goldenenen Engel" (Nr. 149/I) wurde sogar eine gemalte Kassettendecke entdeckt.[90]

Die prächtigsten Häuser in der Prager Altstadt finden wir am Altstädter Ring, in den Hauptstraßen (Zeltner- und Dominikanergasse, heute Celetná und Karlova) und ebenfalls im sogenannten Galliviertel, wo sich größtenteils der Handel konzentrierte. An dieser Stelle sollen nur ein paar Beispiele genannt werden: Den dreistöckigen Stadtpalast mit fünf Fensterachsen (Nr. 465/I) ließ sich vor 1615 die Familie Nerhoff von Holterberg (auch Holtenperk) erbauen, die bereits in den 60er Jahren des 16. Jahrhunderts von Hattingen im Ruhrgebiet nach Prag übersiedelte und von bescheidenen Anfängen zu den größten Großhändlern und Finanziers der Stadt avancierte.[91] Ihr Haus „Zu den fünf Kronen" erinnert an die Prestigebauten des Augsburger Stadtarchitekten Elias Holl.[92]

---

89 Vlček, Staré Město a Josefov (wie Anm. 87), S. 431 f. Die in der Technik des Chiaroscuro ausgeführte Verzierung hat wahrscheinlich der Hofkünstler Francesco Terzio schon um 1560 gemalt.

90 Ebd., S. 183; Líbal / Muk, Staré Město pražské (wie Anm. 88), S. 266–270. Der Umbau des Hauses wird auf das Jahr 1600 datiert.

91 Buňatová, Pražští kupci na cestách (wie Anm. 34), S. 35. Der erste Nerhoff in Prag war Johann Nerhoff der Ältere aus Hattingen im Ruhrgebiet, der in den 60er Jahren als Diener des niederländischen Kaufmanns Georg Kramer in die Stadt kam. 1571 erhielt er das Bürgerrecht in der Altstadt. Nach Prag zogen dann allmählich auch seine Brüder, die einen Großhandel mit Krämerware aufbauten und auch als Finanziers tätig waren. Peter Nerhoff, Bauherr des Hauses „Zu den fünf Kronen", kam im Jahre 1578 an die Moldau; 1602–1608 war er Altstädter Ratsherr. Das repräsentative Haus wurde wahrscheinlich 1615 (nach dem Datum am Tor des Hauses) fertiggestellt. Vgl. dazu: Vlček, Staré Město a Josefov (wie Anm. 87), S. 321 f.; Líbal / Muk, Staré Město pražské (wie Anm. 88), S. 275.

92 In Prag waren die Bauten des Augsburgers Elias Holl wohl gut bekannt. Ein indirektes Zeugnis darüber stellt auch die Präsenz von Holls Zeichnungen in den Sammlungen Rudolfs II. dar. Vgl. Jürgen Zimmer, Aus den Sammlungen Rudolfs II.: Baurisse von Elias Holl, in: Studia Rudolphina 11, 2011, S. 7–25.

Abb. 55: *Das Haus „Zu den fünf Kronen“, vor 1615. Bildnachweis/Foto: Zdeněk Hojda (privat).*

Im nahen ehemaligen Altstädter Richterhaus (Nr. 404/I), das dem Patrizier Václav Krocín von Drahobejle gehörte – für den Altstädter Ring ließ er in den 90er Jahren des 16. Jahrhunderts einen prächtigen, plastisch geschmückten Brunnen anfertigen –, entstand sogar eine Art Kaufhaus mit Geschäften, einem

Weinschank und einer Apotheke.[93] Dass die Prager Altstadt immer noch eine gute Adresse selbst für „die Leute vom anderen Moldau-Ufer" darstellte, bezeugt zum Beispiel das geräumige Objekt mit einem durchgängigen Arkadenhof (Nr. 188/I) nahe der Karlsbrücke. Seinen Stadtpalast „Zur französischen Krone" ließ hier in den Jahren 1596–1600 Doktor der Rechte und Appellationsrat Mathias Kremer von Grünau bauen.[94] In den Jahren 1607–1612 war in seinem Haus Johannes Kepler untergebracht. Der Verwalter der kaiserlichen Sammlungen (Antiquar) Ottavio Strada kaufte das Haus „Zur goldenen Krone" direkt gegenüber dem Turm des Altstädter Rathauses (Nr. 482/I) und baute es um.[95]

Die Bau- und Wirtschaftskonjunktur unter Rudolf II. hinterließ Spuren in der Infrastruktur der Prager Städte. Die ambitionierte Politik der Städte, die sich bereits von den Strafmaßnahmen des Herrschers nach dem ersten Ständeaufstand im Jahre 1547 erholten, zeigte sich bei der Neugestaltung der Rathäuser. Zu ihrem Umbau führten nicht nur die Repräsentationsansprüche der selbstbewussten Stadträte, sondern vor allem die Betriebsanforderungen einer wachsenden Großstadt mit immer stärker strukturierter Stadtverwaltung inklusive vieler subdelegierter Behörden.

Auch der öffentliche Raum musste höheren Kriterien, insbesondere der Hygiene entsprechen. Aus den Plätzen wurden ungepflegte Wasserbecken, Schuttabladeplätze, verschiedene provisorische Bauten und Läden verdrängt. Auf diesen Stellen entstanden zum Beispiel modernere gemauerte Fleischbänke, die die Kleinseite im Jahre 1599 errichtete. Zumindest die Hauptwege und Plätze wurden gepflastert.[96] Zu den Schlüsselelementen der öffentlichen Hygiene gehörte die Wasserversorgung durch die Wasserleitungen. Alle Prager Städte bauten damals neue Wassertürme, die bis heute das Stadtbild prägen.[97] Die meisten

93 Vlček, Staré Město a Josefov (wie Anm. 87), S. 285 f.; Kalina / Koťátko, Praha 1437–1610 (wie Anm. 37), S. 152.

94 Vlček, Staré Město a Josefov (wie Anm. 87), S. 197; Líbal / Muk, Staré Město pražské (wie Anm. 88), S. 262.

95 Vlček, Staré Město a Josefov (wie Anm. 87), S. 334. Vgl. auch: Annemarie C. van der Boom, The Life and Works of Ottavio Strada, in: Prag um 1600 (wie Anm. 64), S. 20.

96 Der englische Reisende Fynes Moryson, der die Stadt im Jahre 1592 besuchte, urteilte äußerst skeptisch über die öffentlichen Räume Prags: „[…] *and except the stinch of the streetes drive backe the Turkes,* […] *there is small hope in the fortifications thereof. The streets are filthy* […]." Fynes Moryson, An itinerary […], I. Part, I. Book, London 1617, S. 14. Digitalisat unter URL: https://quod.lib.umich.edu/cgi/t/text/text-idx?c=eebo;idno=A07834.0001.001 (letzter Zugriff: 5.11.2020).

97 Jaroslav Jásek / Jaroslav Beneš, Pražské vodní věže (= Die Prager Wassertürme), Praha 2000; Jaroslav Jásek / Michal Fiala, Šítkovská vodárna a Karel Mělnický (=

wurden nach den nahegelegenen Mühlen benannt, die bei dieser Gelegenheit auch modernisiert wurden. Schon 1562 wurde das Petržílek-Wasserwerk vollendet, das die Kleinseite versorgte. Der Altstädter Wasserturm in der Nachbarschaft der Brücke ersetzte den älteren Holzbau aus dem Jahre 1577. Der Šítek-Wasserturm für die obere Neustadt wurde 1591 gebaut, im Jahre 1606 dann der Neumühlen-Wasserturm für den unteren Teil der Neustadt.

Die fundamentale Gesundheitspflege, aber auch das soziale Rettungsnetz stellten die städtischen Spitäler dar.[98] Die Kleinseite etwa baute das neue Gemeinde-Spital unter dem Laurenziberg (Petřín), in der Nachbarschaft der kleinen Jan-Hus-Kirche in der heutigen Karmelitergasse (Karmelitská). Ein weiteres Spital „für kranke Höflinge des Kaisers, oder Bürger und ihr Gesinde" wurde im Haus in Újezd errichtet, das der Kaiser der Kleinseitner Gemeinde im Jahre 1599 schenkte. 1602 wurde das System der Spitäler auf der Kleinseite durch das bereits erwähnte „Welsche Spital"[99] für die Prager Italiener ergänzt.

Indem die Prager Städte immer kosmopolitischer wurden und unbändig wuchsen, entwickelte sich ein nicht gerade erwünschtes Phänomen: Die Großstadt zog verstärkt Leute vom Rande der Gesellschaft an. Die Kleinkriminalität oblag der städtischen *Policey*, weswegen kleinere Delikte nicht vor das Stadtgericht kamen. Der darauf folgende Quellenmangel führt zu einer ungenügenden Bearbeitung dieses Themas.

Unterschiedlichste Landstreicher, Betrüger, Kleindiebe, Falschspieler und andere strömten in die Peripherie der Prager Konurbation, wo sich Kneipen, Weinbrennereien und ähnliche Einrichtungen konzentrierten. Man konnte sie entlang wichtiger (Handels-)Wege, vor der Stadtmauer, am Rande der Neustadt und an den Moldau-Ufern finden.

Jacques Esprinchard, ein französischer Besucher Prags im Jahre 1597, erwähnt Gaststätten und Kegelbahnen auf den Prager Inseln. Derselbe Reisende bemerkte

---

Šítek-Wasserwerk und Karel Mělnický), Praha 2004. Mělnický war ein Baumeister, der sich auf Wasserwerke spezialisiert hatte.

98 Vgl. Rudolf Secký, Staropražské špitály (= Altprager Spitäler), Praha 1928; Petr Svobodný / Ludmila Hlaváčková, Pražské špitály a nemocnice (= Prager Spitäler und Krankenhäuser), Praha 1999, S. 16–18.

99 Petr Svobodný, Vlašský špitál na Malé Straně a jeho místo v péči o matku a dítě do josefinských reforem (1602–1789) (= Das italienische Spital auf der Prager Kleinseite und sein Platz in der Sorge um Mutter und Kind bis zu den josefinischen Reformen (1602–1789)), in: Documenta Pragensia 7/1, 1987, S. 113–133; Ders., L'Ospedale italiano come struttura socio-sanitaria (1602–1789), in: La congregazione italiana (wie Anm. 77), S. 26–38.

große Mengen von Landstreichern und Bettlern in den Straßen Prags.[100] Vom Kolorit der damaligen Unterwelt zeugt auch der Antrag eines gewissen Jacob Prossel aus Nürnberg, der in Prag „Glückstopf" – eine Art Lotterie – betreiben wollte, oder das Privilegium auf das „*Betreiben des freien Kartenspiels*" für Abraham Dicz.[101]

Ganz bestimmt ist es kein Zufall, wenn Niclaus Ulenhart, der Autor der Novelle (im Geiste Cervantes') „*Historia von Isaac Winckelfelder und Jobst von der Schneidt*", die Handlung gerade nach Prag lokalisierte. „Die Protagonisten" der Novelle, ein Falschspieler und ein Beutelschneider, begegnen sich in einem Ausschank vor dem Strahover Tor: „[…] *zu was End ich diesem Keyserlichen Hof* […] *nachzeuch; dann ich bin gänzlich der Meinung, daß uns das Glück an diesem Orth nicht umsonst, sondern mit Fleiß zusammengezielt, ja ich halte davor, daß wir beede inskünfftig einander gute Gesellschafft leisten*",[102] ist der ältere von ihnen zuversichtlich. Und auch der Autor erklärt deshalb mit eigenen Worten sein Vorhaben: „*Noch anderes Intent, und zwar das vürnembst, hat der Autor dieser History gehabt: daß er diejenigen, welche frembde Länder zu bauen und zu besuchen Vorhabens seyn, sonderlich diejenige, welche* […] *sich offtermals in großen Städten (inmaßen hie zu einem Exempel Prag gesetzt, sonst aber man zu Pariß, Venedig, Neapoli, Madrit, Seviglia, Lissabon, Brüssel und London, wol noch größere Gefahr, als hie von Prag vermeldt, unterworffen) aufzuhalten haben,* […] *mit diesem Tractat gleichsam verständiger werden, was ihnen vor Unglück an solchen Orten und von solch verdrackten Gesind, da sie nit wohl Achtung auff sich selbst geben, zu gewarten.*"[103]

5. Die Prager Städte mit den Augen ihrer Besucher gesehen

In den Berichten und Beschreibungen der Reisenden spiegelt sich die Sichtweise der Stadtbesucher wider, deren Aufenthalt in Prag oft nicht länger als ein paar Tage dauerte. Ihre Beobachtungen reflektieren deswegen einerseits die direkten

100 Fučíková / Janáček, Tři francouzští kavalíři (wie Anm. 68), S. 34.

101 Nationales Archiv (Národní archiv), Stará manipulace, S 160/1, fol. 10–12 (21.5.1585 und 7.9.1596).

102 Siehe dazu die neue Edition: Miguel de Cervantes Saavedra / Niclaus Ulenhart, Historia von Isaac Winckelfelder und Jobst von der Schneidt, hg. v. Gerhart Hoffmeister, Leipzig 1983, S. 46. Ulenhart erzählte die Handlung einer Novelle von Cervantes aus der Sammlung „*Novelas ejemplares*" neu, versetzte das Geschehen von Sevilla nach Prag und passte die erwähnten Gegebenheiten dementsprechend an.

103 Ebd., S. 40.

Eindrücke und Erfahrungen, andererseits jedoch auch die im Voraus durch das Studium von zeitgenössischen Handbüchern und Reiseführern geschöpften Kenntnisse und Vorstellungen, aber offenbar auch diejenigen von lokalen Ciceronen. Dadurch vermischen sich oft ungleichartig die Stereotypen mit spontanen Beobachtungen, wobei beide Auskunftsarten für uns wertvoll sind.

Die Reisenden loben oft die Lage der Stadt, die sich, von Hügeln umgeben, im weiten Moldau-Tal erstreckt, und finden diesen Standort reizend oder malerisch. Die Einleitung zur Beschreibung von Prag in Georg Brauns „*Civitates orbis terrarum*“ klingt wie ein humanistisches Städtelob: „*Die Metropole des Königtums Böhmen im herrlichen, in alle Richtungen offenen Tal, zwischen Hügeln und Weinbergen.*“[104] Ähnlich erschien die Topographie der Prager Konurbation auch Giovanni Botero in seinem Werk „*Relazioni universali*“.[105] Bei diesen Beschreibungen können wir natürlich nicht die Szenerien auf den ersten Veduten von Prag, dem sogenannten Breslauer Prospekt von 1562 und dem nach ihm gestochenen Holzschnitt von Franz Hogenberg, der Brauns Text begleitete, außer Acht lassen.[106]

Hundert Jahre später war der Zisterzienser und Universalgelehrte Mauritius Vogt in seinem topographischen Werk „*Das jetzt-lebende Königreich Böhmen*“ konkreter. Laut ihm ist Prag „*theils in einem breiten Tal, theils auf sieben Bergen*“ verteilt.[107] Dies ist zwar eine eindeutige Anspielung auf die sieben Hügel Roms, doch Vogt benennt die Hügel Prags konsequenterweise danach.

Neben dem Gesamteindruck von der Stadt in einem von Erhebungen umgebenen Kessel weckte auch die Moldau die Aufmerksamkeit der Reisenden, die eine klare Trennlinie darstellt. Für John Taylor, der im September 1620 den Hof des „Winterkönigs“ und seiner englischen Gemahlin besuchte, war Prag eine Stadt mit beinahe rundem Grundriss, die in der Mitte durch die Flusslinie halbiert wird.[108] Die dominante Rolle des Flusses, der die Mitte der Stadt durchfließt, betonte in seinem polnischen, mit Latein reichlich gespicktem Tagebuch auch

104 „*PRAGA, Regni Bohemiae metropolis,* […] *in amoenissima & patentissima valle, inter colles & vineta* […].“ Georg Braun / Franz Hogenberg (Hg.), Civitates orbis terrarvm, Bd. 1, Köln 1593, fol. 29. Digitalisat unter URL: https://digi.ub.uni-heidelberg.de/diglit/braun1593bd1/0083 (letzter Zugriff: 5.11.2020).

105 „*La metropoli è Praga, divisa in tre parti, che giacciono tutte in una valle amena e spatiosa.*“ Giovanni Botero, Le relationi universali di Giovanni Botero Benese, divise in sette parti, Venetia 1618, S. 77.

106 Lazarová / Lukas, Praha. Obraz města v 16. a 17. století (wie Anm. 83), Bild 3 f.

107 Mauritius Vogt, Das jetzt-lebende Königreich Böhmen, Frankfurt/Leipzig 1712, S. 118.

108 Fynes Moryson / John Taylor, Cesta do Čech (= Reise nach Böhmen), Praha 1977, S. 191.

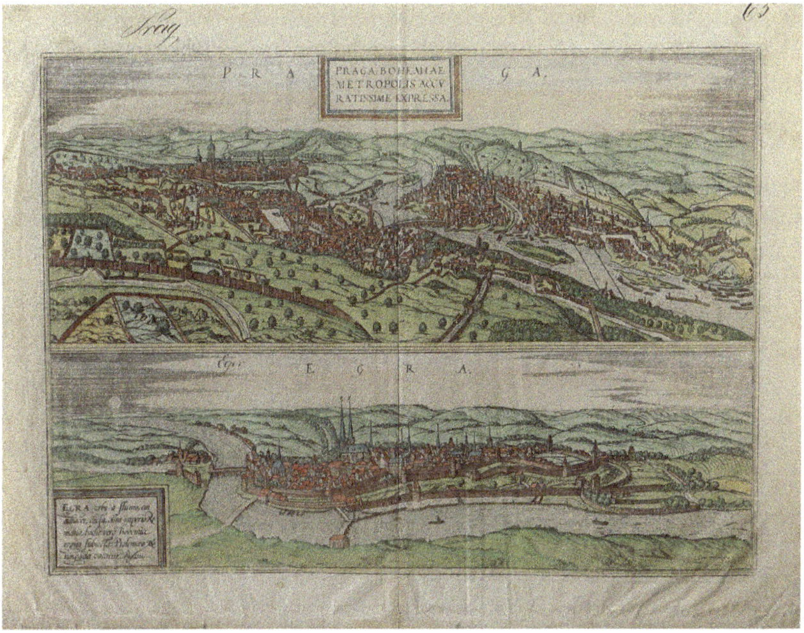

Abb. 56: *Franz Hogenberg, Dartsellung Prags, aus: Ders. / Georg Braun (Hg.), Civitates orbis terrarum, 1572. Bildnachweis: Archiv der Hauptstadt Prag, Grafikensammlung.*

Teodor Billewicz, der 1677 in die Metropole Böhmens mit der Gesandtschaft des Fürsten von Polen-Litauen, Michael Kasimir Radziwiłł, gelangte: „[Die Stadt] *ist durch einen gewaltigen Fluss in zwei Teile getrennt, der inmitten des Ortes fließt.*“[109]

Keine der Beschreibungen und Berichte verzichtete auf die einführende Belehrung des Enea Silvio Piccolomini über die Dreiteilung Prags: „*Praga,* [...] *tres in partes divisa, quis nomina indidere Parvam Pragam, Veterem ac Novam.*“[110]

---

109 „[...] *in duas dividitur partes fluvio potentissimo transfluente środkiem tych miast.*“ Teodor Billewicz, Diariusz podróży po Europie w latach 1677–1678 (= Tagebuch der Reisen durch Europa in den Jahren 1677–1678), hg. v. Marek Kunicki-Goldfinger, Warszawa 2004, S. 122.

110 Dana Martínková / Alena Hadravová / Jiří Matl (Hg.), Aeneae Silvii Historia Bohemica, Pragae 1998, S. 10. Die „*Historia Bohemica*“ wurde im Jahre 1457 verfasst, die erste Druckausgabe erschien 1475. Dieses Werk wurde für eine lange Zeit zur ausschlaggebenden Quelle zur Geschichte Böhmens für das internationale humanistische Publikum.

Die Stadt galt in der Regel als ein aus mehreren Teilen bestehendes Ganzes. Die urbanistische und häufig auch soziale und kulturelle Andersartigkeit der Teile wurde mehr oder weniger treffend geschildert, je nachdem, wie belesen der Autor war und wie lange er sich in der Stadt aufhielt.

Man konnte die Grenzen zwischen den Prager Städten ohne weiteres wahrnehmen und die zugeschütteten Gräben zwischen der Altstadt und der Neustadt waren nicht zu übersehen. Trotzdem finden wir bei einigen Autoren überraschende Angaben über bis zu sieben Prager Städte, wie sie Casimir Freschot im Jahre 1704 schildert: „*Es ist keine einheitliche Stadt, es sind drei, einige meinen sogar sieben Städte, das heißt, zwischen den bedeutenderen Stadtteilen gibt es dermaßen wesentliche Trennlinien, dass man jeden Teil als eine Stadt bezeichnet.*“[111]

Einigermaßen konfus wirkt der Bericht des französischen Chirurgen Charles Patin, den die Stadtbesichtigung im Jahre 1674[112] sehr ermüdete: „*Man sagte mir, dass ich mich nicht über die Stadtgröße wundern soll, denn sie schließe sieben unterschiedliche Städte ein; das steigerte nur meine Verwunderung, denn ich hätte gedacht, dass sie über hundert Städte einschließt.*“[113] Man kann nicht ausschließen,

111 „*La ville n'est pas seule, on y en conte trois, & quelques-uns même sept, c'est à dire qu'entre les parties les plus considérables de la Ville on voit des séparations assez grandes, pour donner à chacune d'elles le nom de Ville.*“ Casimir Freschot, Remarques historiques et critiques, faites dans un voyage d'Italie, en Hollande dans l'année 1704. Contenant les mœurs, interêts, & religion, de la Carniole, Carinthie, Bavière, Autriche, Bohême, Saxe, & des Electorats du Rhin. Avec une relation des differens qui partagent aujourd'hui les Catholiques romains dans les Païs-Bas, Cologne 1705, S. 127 f.

112 Sein Reisebericht – in Form von vier Briefen an verschiedene Reichsfürsten adressiert – „*Relations historiques curieuses de voyages en Allemagne, Angleterre, Hollande, Bohême, Suisse etc.*“ ist zum ersten Mal in Basel („*Bâle*“) gedruckt und dann noch mehrmals herausgegeben (Lyon 1674, Rouen 1676 und Amsterdam 1695) worden. Prag beschreibt er im vierten Brief von 1673. Vgl. Zdeněk Kalista, Charles Patin a jeho návštěva v Čechách let sedmdesátých XVII. v. (= Charles Patin und sein Besuch in Böhmen in den 70er Jahren des 17. Jahrhunderts), in: Český časopis historický 37, 1931, S. 354–361; Jaroslava Kašparová, Čechy očima pařížského lékaře a numismatika 17. století (= Böhmen in den Augen eines Pariser Arztes und Numismatikers des 17. Jahrhunderts), in: Miroslava Hejnová (Hg.), Dr. Bedřišce Wižďálkové. Přátelé a spolupracovníci k významnému životnímu jubileu (= Festschrift für Bedřiška Wižďálková von Freunden und Kollegen für ein bedeutendes Lebensjubiläum) (Příspěvky ke Knihopisu 11), Praha 1996, S. 97–106.

113 „*On me dit que je ne me devois pas étonner de sa grandeur puis qu'elle contenoit sept villes différentes; cela augmenta mon é tonnement, car j'aurois crû qu'elle en contenoit plus d'un cent.*“ Charles Patin, Relations historique et curieuses de voyages, en Allemagne, Angleterre, Hollande, Boheme, Suisse, etc., Rouen 1676, S. 219.

dass die ab und zu erscheinenden Angaben über sieben Städte etwa der falsch verstandenen Schilderung der Prager Ciceronen entnommen wurden, die vielleicht die sieben Hügel meinten.

Die Einsicht in die Verwaltungsgliederung der Prager Städte blieb jedoch den meisten Gelegenheitsbesuchern verborgen. Eine seltene Ausnahme ist Jacques Esprinchard, ein gebildeter französischer Hugenotte aus La Rochelle, der Prag im Jahre 1597 besuchte. „*Prag besteht aus drei Städten* [...]. *Jede dieser Städte verfügt über eine eigene Rechtssprechung und einen eigenen Stadtrat.*“[114] Für den französischen Besucher stellte diese Situation zweifellos eine der Prager Sehenswürdigkeiten dar.

Prag wurde allgemein für eine außergewöhnlich hübsche Stadt gehalten. So lieferte auch ein dem Herausgeber bekannter Anonymus im mehrbändigen „*Lo stato presente de tutti i paesi e popoli del mondo*“ eine Beschreibung Prags: „*Tre sono le parti o sien le Città di qui ella è composta* [...] *ma e l'une e l'altre compariscono assai adornate di superbi Palazzi e di sontuosi Edifizii.*“[115] Jedoch gab es zwischen den einzelnen Städten beträchtliche Unterschiede. Unter den Prager Städten galt natürlich die am dichtesten bevölkerte und reichste Stadt als bedeutendste: die Prager Altstadt. Die detaillierten Angaben verdankte die Altstadt der Tatsache, dass sie die Universität innerhalb ihrer Mauern beherbergte. Ansonsten sind die Informationen zum Stichwort sehr knapp, sowohl bei Enea Silvio Piccolomini als auch in Schedels „*Liber chronicarum*“ oder in Münsters „*Cosmographia*“ – beide zuletzt genannten Werke folgen übrigens genau der Beschreibung Prags in der „*Historia Bohemica*“ des Enea Silvio Piccolomini. Die Universität erwähnten regelmäßig wohl alle Autoren und einige von ihnen erinnerten gar an ihre Spaltung im Jahre 1409, meistens mit stark überschätzten Zahlen von Studenten, die damals Prag verlassen sollten.

Von den weiteren Sehenswürdigkeiten der Altstadt wurden wiederholt das Rathaus und die Teynkirche genannt. Die Vorzüge der Prager Altstadt fasste der bereits erwähnte Mauritius Vogt prägnant zusammen: „*Die Alt-Stadt reich an Volck, eng von Gassen, hoch von Häusern, voll von Kirchen, versehen von Kauffleuten, liegt in einer Ebene, redet böhmisch und teutsch.*“[116] Das Jesuitenkolleg Clementinum wurde manchmal auch als eine besondere Enklave verstanden.

114 Fučíková / Janáček, Tři francouzští kavalíři (wie Anm. 68), S. 29.

115 Thomas Salmon (Hg.), Lo stato presente de tutti i paesi e popoli del mondo naturale, politico e morale. Con nuove osservazioni e correzioni degli antichi e moderni viaggiatori, vol. IX: Dei Regni di Boemia e d'Ungheria, dell'Imperio Germanico in generale ed in particolare, Napoli 1740, S. 4.

116 Vogt, Das jetzt-lebende Königreich Böhmen (wie Anm. 107), S. 123.

Esprinchard spricht sogar von einer Kleinstadt in der Stadt: „*A la fin du pont de Prague en la vieille ville, on voit la magnifique maison des Jesuittes avec leur Eglise et leur eschole, qui contient une petite ville en longeur et largeur.*“[117]

Eine außerordentliche Aufmerksamkeit galt schon immer dem „exotischen“ Judenviertel. Es wurde ebenfalls mit einer selbstständigen Stadt verglichen, namentlich bei Giovanni Botero: „*In questa parte i Giudei vi hanno un Ghetto, che par una città.*“[118] Auch der englische Reisende Fynes Moryson schaute im Jahre 1592 unterwegs von der Steinbrücke in die Richtung der Moldauer Schleife: „[…] *to which as you goe, on the left side is a little City of the Iewes.*“[119] Das Prager Ghetto als eine selbstständige urbane Einheit zu verstehen, war in der Epoche seiner Blütezeit unter Rudolf II. letztendlich völlig berechtigt. Am Anfang des 18. Jahrhunderts, in der Periode der allmählich ansetzenden Regulation der jüdischen Bevölkerung seitens des Staates, war es mit der Konjunktur vorbei. Mit einer unabdingbaren Dosis an zeitgenössischem Antisemitismus schreibt Mauritius Vogt über die Schädlichkeit der jüdischen Konkurrenz: „*In dieser Alt-Stadt ist am Wasser die Juden-Stadt, welche neun Synagogen hat; die Judenschaft aber, welche den Christen die Kauffmannschafft sehr verderbet, mit alt und jung in vielen Tausenden* […].“[120] Auch Casimir Freschot bemerkte wohl die immer noch zahlreiche jüdische Bevölkerung, doch nahm er zugleich ihre schlechte soziale Situation wahr: „*Ces Juifs, qui à Prague habitent seuls une de trois Villes, sont fort nombreux & par conséquent tous assez misérables, les occasions de gain y étant rares, à cause du peu de commerce qui y regne.*“[121]

Die Prager Neustadt unterschied sich wesentlich von ihrem Nachbarn. Es ist wieder Enea Silvio Piccolomini, der diesen Stadtteil wesentlich charakterisierte: „*Haec quoque civitas ampla est et ad coles usque protenditur.*“[122] Mit den Hügeln waren die Erhebungen am Südrand der Stadt nahe der Stadtmauer

---

117 Léopold Chatenay, Vie de Jacques Esprinchard Rochelais et Journal de ses voyages au XVIe siècle, Paris 1957, S. 169.

118 Botero, Le relazioni universali (wie Anm. 105), S. 77.

119 Moryson, An itinerary (wie Anm. 96), S. 14. Einen ausführlichen Kommentar zu Morysons Prager Reisebericht liefern: Josef Giesen / Wilhelm Pfeifer, Der Bericht des Engländers Moryson über seine Reise durch Böhmen, in: Mitteilungen des Vereins für Geschichte der Deutschen in den Sudetenländern 80, 1943, S. 44–76. Zuletzt widmete sich den Reisenden aus England: Hana Ferencová, Čechy a Morava očima anglických cestovatelů (= Böhmen und Mähren in den Augen der englischen Reisenden), Olomouc 2018.

120 Vogt, Das jetzt-lebende Königreich Böhmen (wie Anm. 107), S. 123.

121 Freschot, Remarques historiques et critiques (wie Anm. 111), S. 130 f.

122 Martínková / Hadravová / Matl, Aeneae Silvii Historia Bohemica (wie Anm. 110), S. 10.

gemeint: der Karlshof (Karlov) mit der Kirche St. Mariä Himmelfahrt und St. Karl der Große, der Windberg (Větrov) mit dem Kollegiatstift St. Apollinaris und die Burg Vyšehrad, die zu jener Zeit in Ruinen lag. Fast alle Besucher verwunderte die Prager Neustadt durch die Weite ihres Areals sowie durch dessen „Leere". Abbadino, der Sekretär des Herzogs von Mantua Federico II. Gonzaga, schrieb zum Beispiel im Jahre 1531: „*Il circuito de tutta la città è grandissimo, ma gli è vacuo dentro assai e non fa però in tutto più che sessanta millia persone o siano anime.*"[123] Diese Bemerkung betraf zwar gesamt Prag, ganz sicher aber bezog sie sich besonders auf die Prager Neustadt. Die Stadt schien ihm leer zu sein, jedoch überschätzte er die Prager Population gewaltig; diese Einwohnerzahl erreichte Prag erst gegen Ende der Herrschaft Rudolfs II.

Beide Ufer von Prag, namentlich die Prager Altstadt mit der Kleinseite, verband bereits ab dem 14. Jahrhundert eine Steinbrücke – die eine noch ältere aus dem 12. Jahrhundert ersetzte –, über die der eben erwähnte Abbadino urteilte, sie sei „*die schönste in der ganzen Welt.*"[124] Der nach der Kathedrale beispielsweise am meisten bewunderte Prager Bau von Peter Parler erlebte eine kuriose Schätzung seiner Länge in der Angabe des polnischen Adeligen Billewicz: „*Wir fuhren über den Fluss auf einer Brücke, die so lang war wie vier Bogenschüsse.*"[125]

Weil die Besucher von Westen aus nach Prag kamen, verwundert es nicht, dass sie zuerst die Prager Kleinseite (auch Prager Kleinstadt genannt) beschrieben. Fynes Moryson nahm sogar an, dass dieser Stadtteil bloß eine Vorstadt am linken Moldau-Ufer war: „*On the West side of Molda is the Emperours Castle, seated on a most high Mountaine, in the fall whereof is the Suburbe called Kleinseit, or little side. From this Suburbe to goe into the City, a long stone bridge is to be passed over Molda, which* [...] *divides the suburbe from the City.*"[126] Sein Landsmann, der englische Arzt Edward Brown, der Prag 80 Jahre später besuchte, fand hier schon eine große Zahl an aristokratischen Palästen vor und bezeichnete die Prager Kleinseite als eine reizvolle Stadt, voll von schönen Bauten: „*The Kleine Seitten, or lesser Prague, far exceeds the other for pleasantness and beauty of Buildings and fair Palaces.*"[127] In diesem

123 Rudolf Knott, Ein Bericht über Prag und seine Bewohner aus dem Jahre 1531, in: Mitteilungen des Vereins für Geschichte der Deutschen in Böhmen 38, 1900, S. 155–162, hier: S. 158.

124 „[...] *il fiume Voltava, qua ha sopra un ponte di pietra viva, il più bello che credo sia hoggi dì al mondo.*" Ebd., S. 158.

125 „*Jechaliśmy przez most na tej rzece, który jest tak długi jako na czworo z łuku strzelania.*" Billewicz, Diariusz podróży po Europie (wie Anm. 109), S. 122.

126 Moryson, An itinerary (wie Anm. 96), S. 14.

127 Edward Brown, A brief account of some travels in divers parts of Europe, London 1673, S. 163.

Punkt kam mit ihm sein französischer Fachkollege Charles Patin überein, der schrieb, dass „*die Kleinseite viel schöner sei* [als die Städte am rechten Moldauufer, versteht sich, Anm. ZH], *es gibt dort lauter Paläste,* […] *über dreihundert an der Zahl.*“[128] Die Kleinseite durch die Optik der neu gebauten Paläste wahrzunehmen, war also keine Ausnahme. Trotzdem gelang es dem Zisterzienser Vogt zur Charakterisierung der Stadt eine weitere Facette hinzuzufügen, nämlich eine hohe Zahl an sogenannten Sonderrechten, das heißt Enklaven, die der Kleinseitner Verwaltung entglitten waren und meistens von einer Klosterobrigkeit verwaltet wurden: „*Kleine Seiten* […]*; sie ist durch unterschiedliche Rechte gleichsam in unterschiedliche Städte ausgetheilt.*“[129]

Nur selten bemerkten einige Besucher auch die vierte und kleinste der Prager Städte, nämlich Hradschin (Hradčany). Es verwundert nicht, dass diesen Stadtteil fast nur diejenigen Autoren beachteten, die Prag zu jener Zeit besuchten, als die Prager Burg dem Kaiserhof als Residenz diente. Denn damals erlebte auch ihr Vorfeld Hradschin eine lebhafte Epoche. Jacques Esprinchard wusste im Jahre 1597, dass „*es hinter der Burg noch einen Stadtteil Prags – Hradschin genannt – gibt, der über eigene Rechtsprechung ähnlich wie andere Städte verfügt*“;[130] und sein Landsmann Pierre Bergeron schrieb im Jahre 1603: „*Auf dem Hügel erhebt sich die Burg und es gibt noch eine Vorstadt, äußere Hradschin genannt.*“[131] Im späten 17. Jahrhundert zog die Prager Burg die Reisenden nicht mehr an und die Paläste auf der Anhöhe von Hradschin verschmolzen visuell mit den anliegenden Kleinseitner Gartenresidenzen. Für den aufmerksamen Engländer Brown gab es weiterhin auch „*Oberprag*“, doch zählte er die Paläste mit denen der Kleinseite zusammen: „*A great part of this Town lies high, and upon the Hill stands Hratschin or Upper Prague* […]. *There are also many Houses of the Nobility in this part of the City.*“[132]

Ein Fremder betrachtet in der Regel die Stadt anders als ein Einheimischer. Seine Perspektive unterscheidet sich deshalb. In seinen unmittelbaren Notizen oder späteren Textformulierungen dominiert eine Mischung aus eigenem Erleben, Stereotypen, Irrtümern und Informationen aus zweiter Hand, die oft durch

---

128 „*Cette partie de la ville qui est au de là de la riviére & qu'on appelle la petit côté, est bien plus charmante, on n'y voit que des Palais: comme la retraite de ce qu'il y a de riche en Boheme & dans le pays héréditaires de S.M.I. on y en compte plus de trois cent.*“ Patin, Relations historique et curieuses de voyages (wie Anm. 113), S. 222.

129 Vogt, Das jetzt-lebende Königreich Böhmen (wie Anm. 107), S. 121.

130 Fučíková / Janáček, Tři francouzští kavalíři (wie Anm. 68), S. 31.

131 Ebd., S. 78.

132 Brown, A brief account of some travels (wie Anm. 127), S. 163 f.

Ungenauigkeit belastet sind. Falls der Autor beabsichtigt, die künftigen Stadtbesucher zu belehren und zu informieren, unterscheidet sich seine Methode freilich von den Texten, die „lediglich" der Memoria dienen.

Wir waren bemüht, in diesen Absätzen zu zeigen, wie die Prager Konurbation in ihrer Gesamtheit, im Rahmen der Landschaft, aber auch in ihrer Struktur und in der Andersartigkeit ihrer Bestandteile wahrgenommen wurde. Unser Kriterium bei der Auswahl waren unter anderem die Distanz und die Erfahrungen der Schreiber, die manchmal zum Vergleich mit anderen europäischen Städten führten. So wurde der Vergleich von Prag mit Florenz vom Humanisten Enea Silvio Piccolomini zum Klassiker: „*Praga* […] *neque minor neque ignobilior Etrusca Florentia.*"[133]

Zwei Jahrhunderte später verglich beide Städte – mit Hinweis auf Enea Silvio Piccolomini – Edward Brown gründlich miteinander. Sein Ergebnis war, dass Prag größer und volkreicher ist, die Paläste und schönen Häuser über verglaste Fenster verfügen, während man in Florenz hässlichen und zerrissenen Fenstern aus Papier begegnet. Nicht einmal den toskanischen Fluss Arno kann man mit der Moldau in Prag vergleichen. Doch in Prag fand er keine fein gepflasterten Straßen, keine Stadtkirche mit Marmorbelag und vor allem keine Kunstgalerie voll von prächtigen Kunstwerken und anderen Kostbarkeiten.[134] Derselbe Edward Brown verglich interessanterweise auch Prag mit Lyon. Diese etwas überraschende Parallele basiert einerseits auf dem Stadtrelief – auch Lyon steht teils am Hang, teils in einer Ebene –, andererseits auf zahlreichen unbebauten Grundstücken, die sich innerhalb der Stadtmauer befanden.[135] Nicht selten wird

---

133 Martínková / Hadravová / Matl, Aeneae Silvii Historia Bohemica (wie Anm. 110), S. 10.

134 „*The same Author* [d.h. Enea Silvio Piccolomini] *compares Prague unto the City of Florence in Tuscany, wherefore having seen both places, I cannot omit to say something, I had a view of the City of Florence from the top of the Domo, or Cathedral, and of Prague from the Church of St Veit, upon the Hill in the lesser Town: Prague seemed to my eye to contain a far greater Circuit than Florence; it seemed also more populous, and to exceed it very much in the number of People, the Streets larger, and the Windows of Palaces and fair Houses being of Glass, looked not so tatterdly as the ragged Paper Windows of Florence. The River Arno, which runs through Florence, is not to be compared with the Muldau at Prague,* […]. *But as for the well-paved Streets of Florence, the Domo, or Cathedral with black and white Marble, with a Cupola, second only to that of St. Peters of Rome, for the incomparable Chappel of St. Laurence, and the Dukes Gallery and Rarities, I must confess I saw not any thing in Prague which answered them.*" Brown, A brief account of some travels (wie Anm. 127), S. 165.

135 „*The Walls of this City* [Prag] *seem to enclose the greatest Circuit of ground of any I have seen in Germany but the Hills and void spaces within it take up a large Tract, and*

Prag mit großen Metropolen konfrontiert, doch geht es hier um eher allgemeine Bemerkungen. Pierre Bergeron urteilte, dass Karl IV. die Prager Neustadt, beeinflusst durch seine Jugenderlebnisse, gegründet hatte, da er am französischen Hof erzogen worden war.[136] Giovanni Botero verglich sogar das Gelände von allen drei Prager Städten insgesamt mit dem besiedelten Gebiet innerhalb der Stadtmauern Roms.[137] Interessanter scheint die Bemerkung des sächsischen Diplomaten Karl Ludwig Freiherr von Pöllnitz über die Gesellschaft Prags zu sein, der – mit Ausnahme von Rom, London und Paris – sonst nirgendwo eine so pompös und verschwenderisch auftretende Aristokratie gesehen habe.[138]

6. Schlusswort

Drei Themenschwerpunkte, die wir aus vielen möglichen Problembereichen ausgewählt haben, können kaum ein dermaßen komplexes Thema erschöpfen, welches die Entwicklung der Prager Städtelandschaft darstellt. Eine Reihe von nicht minder interessanten Problemen blieb unbeachtet, bei manchen mangelt es immer noch an Grundlagenforschung.

Eine relativ lange Periode, während der sich Prag unter die ersten Metropolen Europas eingliederte, hinterließ in ihrem Stadtbild deutliche Spuren. Doch die grundsätzlichen urbanistischen Impulse der Epoche Rudolfs II. wirkten noch in den folgenden Jahren nach, bevor sie der Dreißigjährige Krieg brutal unterbrach. Der 400. Jahrestag der Schlacht am Weißen Berg in diesem Jahr erinnert

---

therein it is like the City of Lyon in France.“ Ebd., S. 162 f. Auf Brown basiert offensichtlich auch der anonyme Autor der italienischen „*Weltenzyklopädie*“ aus dem Jahre 1740: „*Capitale non solo della Provincia, ma la Metropoli di tutto il Regno di Boemia. Ella e situata quasi nella stessa figura che la Città di Lion nella Francia, cioè una parte fabbricata sulle falde del Monte, e l'altra nel Piano, rinserrato però da altre Montagne, che sono di là del Fiume Muldau.* Salmon, Lo stato presente de tutti i paesi e popoli del mondo (wie Anm. 115), S. 4.

136 Fučíková / Janáček, Tři francouzští kavalíři (wie Anm. 68), S. 78.

137 „*Se stima che tutte tre queste parti insieme girino non meno, che l'habitato di Roma.*“ Botero, Le relazioni universali (wie Anm. 105), S. 77.

138 „*Excepté Rome, Paris, & Londres, il n'y a point de Ville, où il ait plus de Noblesse, & une Noblesse plus riche. Tout le Monde fait ici de la dépense. Il n'y a point ailleurs de Seigneurs plus magnifiques, & qui se fassent plus honneur de leur bien.* […] *Il n'y a point de Seigneurs au monde plus porté à la dépense que ceux-ci.*“ Memoires de Charles-Louis baron de Pöllnitz, contenant les observations qu'il a faites dans ses voyages, et le caractere des personnes qui composent les principales cours de l'Europe, Tom. I, Liège 1734, S. 274, 277.

an das Trauma, das sich in die Überlegung nicht nur über die Prager Geschichte projiziert. Aus der Perspektive der Geschichte Prags geht es dabei um ein Problem, das grundlegende Fragen zur Kontinuität und Diskontinuität des historischen Prozesses stellt.

Prag verlor zwar das Statut einer kaiserlichen und königlichen Residenzstadt und wurde aus dieser Sicht zu einer Provinzstadt. Doch bis zur zweiten Wiener Türkenbelagerung im Jahre 1683 war Prag als Residenzstadt mehr als eine ernsthafte Konkurrenz zu Wien; jedoch ist das Verhältnis zwischen Zentrum und Peripherie in der Habsburgermonarchie bisher nicht genügend erforscht.[139] Für die böhmischen Länder spielte Prag weiterhin eine wichtige Rolle nicht nur als Verwaltungszentrum, Sitz der Landesämter, des Erzbischofs und der ältesten mitteleuropäischen Universität. Wesentlich blieb weiterhin auch seine symbolische Rolle.

Bis heute ist der Streit um das innerlich kontroverse, „lange" barocke Jahrhundert zwischen den Jahren 1620–1740 nicht völlig beigelegt. Gerade der Barock hinterließ in Prag – zusammen mit dem Gründungswerk Karls IV. – eine dermaßen deutliche Spur, dass diese auch das Tourismus-Marketing von heute nutzt. So könnte unsere Studie mit Überlegungen zur Relativität der Krisen und Konjunkturen fortsetzen, die man durch das Prisma der Entwicklung einer Stadtlandschaft betrachten kann. Dadurch geraten wir jedoch weit über die Grenze der zulässigen Länge dieser Studie hinaus, sodass wir uns diese Anreize für eine andere Gelegenheit aufsparen.

---

139 Siehe dazu: Jiří Pešek, Praha a Vídeň v baroku – centrum a periferie v proměnách barokní kultury? (= Prag und Wien im Barock – Zentrum und Peripherie), in: Fejtová, Barokní Praha – Barokní Čechie (wie Anm. 76), S. 135–150.

Ilgvars Misāns

Städtelandschaft des mittelalterlichen Livlands

Abstract: In order to uncover the specific characteristics of the Livonian urban landscape, the article examines the spatial and political conditions of urbanisation as well as the economic networks and institutional ties of the cities. It has been demonstrated that in Livonia the intensity of urbanisation was modest and urban life was relatively underdeveloped. The political fragmentation of the province, the low influx of new inhabitants from German territories, and demographic and economic crises have stopped the urbanisation process from unfolding fully.

The urban landscape was dominated by three outstanding long-distance trade centres: Riga, Dorpat and Reval. They specialised in different long-distance trade directions and divided among themselves economic areas in the north, centre and south. The three largest cities controlled the small towns and markets in their respective economic hinterlands. From the middle of the fourteenth century onwards, an important coordinating role in the policy of Livonian towns was played by the Diet of Towns ("Städtetag"), which dealt with both Hanseatic community affairs and regional policy issues at the local level. As a link between the cities, this institution had an integrating effect on the urban landscape.

Lange Zeit hat sich die Städteforschung des mittelalterlichen Livlands[1] vornehmlich auf drei größere Kommunen – Riga, Reval (estn. Tallinn) und Dorpat (estn. Tartu) – konzentriert. Einerseits ergibt sich dieser Umstand aus der herausragenden Bedeutung der genannten drei Städte für das wirtschaftliche, politische und kulturelle Leben des Landes, andererseits aus der Tatsache, dass sich vor allem Reval und Riga im Mittelpunkt der livländischen Städteforschung befinden,[2]

1 Livland war eine politische Gemeinschaft geistlicher Landesherrschaften der Bischöfe und des Deutschen Ordens (seit 1237), die vom 13. Jahrhundert bis Mitte des 16. Jahrhunderts auf dem Territorium des gegenwärtigen Estlands und Lettlands existierte. Dabei ist zu beachten, dass Nordestland bis 1346, als es an den Deutschen Orden verkauft wurde, der dänischen Krone unterstellt war.

2 Juhan Kreem, Was ist Stadtgeschichte in Estland?, in: Heidi Hein-Kircher / Ilgvars Misāns (Hg.), Stadtgeschichte des Baltikums oder baltische Stadtgeschichte? Annäherung an ein Forschungsfeld zur baltischen Geschichte (Tagungen zur Ostmitteleuropaforschung 33), Marburg [2]2016, S. 61–73, hier: S. 69–71; Ilgvars Misāns, Die Stadt als Forschungsobjekt in der Geschichtsschreibung Lettlands, in: ebd., S. 75–94, hier bes.: S. 88–92.

was der spezifischen Situation der Quellenüberlieferung zugrunde liegt.[3] Erfreulicherweise hat im letzten Jahrzehnt das Interesse der Forschung auch an der Geschichte der kleineren Städte Livlands beträchtlich zugenommen.[4] Auf die neuesten Forschungsergebnisse stützend, kann man komplexere Fragestellungen als bisher entwickeln, nicht zuletzt die Urbanität im livländischen Raum im Sinne einer Städtelandschaft mit ihren charakteristischen Merkmalen untersuchen.[5]

Obwohl das Konzept der Städtelandschaften schon seit einiger Zeit Wissenschaftler/-innen sowohl in Deutschland als auch darüber hinaus beschäftigt,[6] sind die theoretischen Aspekte dieses Forschungsansatzes bisher nicht ausführlich erörtert worden. Selbst der Inhalt des Begriffs „Städtelandschaft(-en)" wird verschiedenartig gedeutet und in der Literatur existieren unterschiedliche Anhaltspunkte, anhand der sich die Städte einer historisch gewachsenen

---

3 Mehr darüber: Anti Selart, Die kleinen Städte Livlands und ihre Rolle in der oder für die Hanse, in: Juhan Kreem / Jürgen Sarnowsky (Hg.), „Hansisch" oder „nicht-hansisch". Das Beispiel der kleinen Städte und Livlands in der Hanse (Hansische Studien 27), Wismar 2019, S. 23–42, hier: S. 28 f.

4 Als umfassende Darstellungen zu diesem Thema sind besonders zwei Beiträge estnischer Wissenschaftler (mit einschlägiger Literatur) hervorzuheben: Inna Põltsam-Jürjo, Grundzüge des livländischen Städtewesens im Mittelalter, in: Hein-Kircher / Misāns, Stadtgeschichte (wie Anm. 2), S. 117–135; Selart, Die kleinen Städte Livlands (wie Anm. 3), S. 23–42. In Lettland beschäftigt sich derzeit Edgars Plētiens mit den kleineren Städten Livlands. Siehe v.a.: Edgars Plētiens, Livonijas mazpilsētas: kas tās bija un kāda bija to ietekme? (= Die Kleinstädte Livlands: Was waren sie und welchen Einfluss hatten sie?), in: Latvijas Vēstures Institūta Žurnāls 3, 2015, S. 5–44.

5 Eine erste „Vermessung des Forschungsfeldes" für die Beschäftigung mit den Städtelandschaften in Livland habe ich im Jahre 2008/2009 versucht zu unternehmen, als die Geschichte der livländischen Kleinstädte noch ein deutliches Forschungsdesiderat darstellte. Siehe: Ilgvars Misāns, Die Städte als zentrale Orte im mittelalterlichen Livland und die Anfänge ihrer Vernetzung bis zur Mitte des 14. Jahrhunderts, in: Roman Czaja / Carsten Jahnke (Hg.), Städtelandschaften im Ostseeraum im Mittelalter und in der Frühen Neuzeit, Toruń 2009, S. 87–102, hier bes.: S. 88.

6 Siehe v.a.: Helmut Flachenecker / Rolf Kießling (Hg.), Städtelandschaften in Altbayern, Franken und Schwaben. Studien zum Phänomen der Kleinstädte während des Spätmittelalters und der Frühen Neuzeit (Zeitschrift für bayerischen Landesgeschichte, Beihefte B 15), München 1999; Monika Escher / Alfred Haverkamp / Frank G. Hirschmann (Hg.), Städtelandschaft – Städtenetz – zentralörtliches Gefüge. Ansätze und Befunde zur Geschichte der Städte im hohen und späten Mittelalter (Trierer Historische Forschungen 43), Mainz 2000; Holger Thomas Gräf / Katrin Keller (Hg.), Städtelandschaft – Réseau urbain – Urban Network. Städte im regionalen Kontext in Spätmittelalter und früher Neuzeit (Städteforschung A 62), Köln/Weimar/Wien 2004.

Landschaft beschreiben und analysieren lassen.[7] Um die Besonderheiten der Städtelandschaft speziell in Livland in konzentrierter Form zu charakterisieren, wollen wir zwei Aspekte in den Blick nehmen, nach denen dieser Beitrag auch strukturiert wird: 1. Räumliche und herrschaftliche Grundlagen der Urbanisierung des Landes, 2. Wirtschaftliche Vernetzung und institutionelle Bindungen der Städte.

1. Räumliche und herrschaftliche Grundlagen der Urbanisierung

Das mittelalterliche Livland erschien auf der Landkarte Europas im 13. Jahrhundert, als aus Norddeutschland und Skandinavien stammende Kreuzfahrer und Umsiedler einen Teil des von baltischen und finnougrischen Stämmen im östlichen Ostseeraum bewohnten Territoriums[8] in die von Rom geprägte Ökumene einbezogen haben. Nicht nur die natürlichen Voraussetzungen (Ostseeküste, Binnengewässer und menschenleere Urwälder), sondern auch die Missionierung und Kreuzzüge waren für die räumliche Ausdehnung Livlands von grundlegender Bedeutung. Die äußeren Grenzen des neuen lateinisch-christlichen Landes mit den benachbarten orthodoxen Russen im Osten und heidnischen Litauern im Süden wurden dort gesetzt, wo sich die Eroberung erschöpfte.[9] Parallel zu

7 Siehe ausführlicher: Ellen Widder, Südwestdeutsche Städtelandschaften im Vergleich. Chancen, Grenzen und Probleme eines Forschungsansatzes, in: Sigrid Hirbodian / Peter Rückert (Hg.), Württembergische Städte im späten Mittelalter. Herrschaft, Wirtschaft und Kultur im Vergleich (Tübinger Bausteine zur Landesgeschichte 26), Ostfildern 2016, S. 11–36, hier: S. 12–19; Franz Irsigler, Städtelandschaften und kleine Städte, in: Flachenecker / Kießling, Städtelandschaften in Altbayern, Franken und Schwaben (wie Anm. 6), S. 13–38, hier: S. 30–38; Monika Escher / Alfred Haverkamp / Frank G. Hirschmann, Städtelandschaft – Städtenetz – zentralörtliches Gefüge. Einleitung, in: Dies., Städtelandschaft – Städtenetz – zentralörtliches Gefüge (wie Anm. 6), S. 9–53, hier: S. 18–25; Eberhard Isenmann, Die deutsche Stadt im Mittelalter 1150–1550. Stadtgestalt, Recht, Verfassung, Stadtregiment, Kirche, Gesellschaft, Wirtschaft, Köln/Weimar/Wien [2]2014, S. 678 f.

8 Mehr dazu: Michael North, Geschichte der Ostsee. Handel und Kulturen, München 2011, S. 26–28; Manfred Hellmann, Die baltischen Völker zwischen Slaven, Ugrofinnen und Skandinaviern, in: Gli Slavi occidentali e meridionali nell'alto Medioevo, 15–21 aprile 1982 (Settimane di studio del Centro italiano di studi sull'Alto Medioevo 30), Spoleto 1983, S. 515–530.

9 Matthias Thumser, Das Baltikum im Mittelalter. Strukturen einer europäischen Geschichtsregion, in: Jahrbuch des baltischen Deutschtums 58, 2011, S. 17–30, hier: S. 19 f.; Carsten Goehrke, Das Baltikum um 1200 im Spannungsfeld der

den Missionskriegen und Seite an Seite mit der militärischen Unterdrückung der autochthonen Einwohner verlief die Integration des Landes in die römisch-katholische Welt und stabilisierte das neue Herrschaftssystem, das aus geistlichen Landesherrschaften bestand.[10] Es braucht hier im Einzelnen nicht näher dargelegt zu werden, wie sich dieser Prozess vollzog. Einige neugeschaffene geistige Strukturen und dazu gehöhrende Landesherrschaften waren von kurzer Dauer. Manche ursprüngliche Bischofsitze wurden im Laufe der späteren Umstrukturierungen aufgelöst, und der im Jahre 1202 gegründete sogenannte Orden der Schwertbrüder (*Fratres milicie Christi de Livonia*) und sein Besitz wurde 1237 nach einer vernichtenden Niederlage gegen die Litauer in den Deutschen Orden inkorporiert.[11] Erst nachdem um die Mitte des 13. Jahrhunderts das Erzbistum Riga als Metropolitankirche für die Kirchenprovinz Preußen-Livland[12] geschaffen wurde, stabilisierte sich das kirchliche Verwaltungswesen, womit auch die weltlichen Machtbefugnisse zwischen den Kirchenfürsten aufgeteilt wurden. Als Territorialherrschaften etablierten sich in Livland das Erzbistum Riga, die Bistümer von Dorpat, Ösel-Wieck und Kurland sowie das Herrschaftsgebiet des Deutschen Ordens. Seine endgültige territoriale Gestalt fand das Land nach

---

umgebenden Mächte: ein Bild unserer Gegenwart?, in: Ders. / Jürgen von Ungern-Sternberg (Hg.), Die baltischen Staaten im Schnittpunkt der Entwicklungen. Vergangenheit und Gegenwart, Basel 2002, S. 9–24, hier: S. 9–14.

10 Anti Selart, Die Eroberung Livlands (12. und 13. Jahrhundert), in: Karsten Brüggemann / Detlef Henning / Konrad Meier / Ralph Tuchtenhagen (Hg.), Das Baltikum. Geschichte einer europäischen Region, Bd. 1: Von der Vor- und Frühgeschichte bis zum Ende des Mittelalters, [Stuttgart] 2018, S. 159–209; Nils Blomkvist, The Discovery of the Baltic. The Reception of a Catholic World-System in the European North (AD 1075–1125) (The Northern World 15), Leiden 2005, S. 570–572, 623–656; Tiina Kala, The Incorporation of the Northern Baltic Lands into the Western Christian World, in: Alan V. Murray (Hg.), Crusade and Conversion on the Baltic Frontier 1150–1500, Burlington 2001, S. 3–20; Muntis Auns, Acquisition of the Acquired: The Establishing of a Real Administration in Livonia, in: Nils Blomkvist (Hg.), Culture Clash or Compromise? The Europeanisation of the Baltic Sea area 1100–1400 AD. Papers of the XIth Visby Symposium held at Gotland Centre for Baltic Studies, Gotland University College, Visby, October 4th–9th, 1996 (Acta Visbyensia 11), Visby 1998, S. 259–267.

11 Klaus Militzer, Die Geschichte des Deutschen Ordens, Stuttgart [2]2012, S. 115; Bernhart Jähnig, Der Deutsche Orden und die livländischen Bischöfe im Spannungsfeld von Kaiser und Papst, in: Nordost-Archiv NF 7/1, 1998, S. 47–63, hier: S. 47–56.

12 Dabei ist anzumerken, dass das in von Dänen eroberte Nordestland liegende Bistum Reval das ganze Mittelalter rechtlich zur Kirchenprovinz Lund gehörte. Siehe: Richard Hasselblatt, Die Metropolitanverbindung Revals mit Lund, in: Mitteilungen aus dem Gebiete Liv-, Est- und Kurlands 14, 1890, S. 462–466.

dem der Orden im Jahre 1346 Nordestland von Dänemark kaufte und der Hochmeister wenig später die neuerworbene Landschaft dem livländischen Meister beziehungsweise dem livländischen Ordenszweig zur Verwaltung übertrug. Mit 67.000 km^2 ragte nun das Gebiet des Deutschen Ordens in Livland noch stärker als zuvor unter den übrigen Territorialherrschaften hervor – der Erzbischof und drei Bischöfe zusammen verfügten über circa 41.000 km^2.[13]

Bereits während der Eroberung und territorialen Ausdehnung der neuen Herrschaftsstrukturen trat eine Absonderung der Herrschaften und Grenzabsteckung zutage, sodass sich allmählich ein neues räumliches Gefüge festigte.[14] Als Mittelpunkte für Schutz und Verwaltung der Territorialherrschaften errichteten die Landesherren und ihre Lehensleute Burgen und Residenzen.[15] Eine eigene Rolle im Herrschaftskonzept der Territorialfürsten spielten auch die Städte. Ursprünglich, während der Missionskriege, traten die von den Territorialfürsten gegründeten deutschrechtlichen Städte – an Ordens- oder Bischofsburgen angelehnt – nicht nur in der Rolle der Wirtschaftszentren auf, sondern sie übten auch eine Basisfunktion als militärisch-strategische Hauptorte und als Zentren der lateinischen Kirche aus. Erst mit der Zeit gewannen die Bürgergemeinden eine

---

13 Heinz von zur Mühlen, Livland von der Christianisierung bis zum Ende seiner Selbständigkeit (etwa 1180–1561), in: Gert von Pistohlkors (Hg.), Deutsche Geschichte im Osten Europas. Baltische Länder, Berlin 1994, S. 25–172, hier: S. 38–78; Norbert Angermann, Die Baltischen Länder 1350–1650, in: Hermann Kellenbenz (Hg.), Handbuch der europäischen Wirtschafts- und Sozialgeschichte, Bd. 3, Stuttgart 1986, S. 1097–1110, hier: S. 1098.

14 Rafal Simiński, Die lokalen Grenzen in Livland im 13. und 14. Jahrhundert. Entstehung und Funktion, in: Ulrich Knefelkamp / Kristian Bosselmann-Cyran (Hg.), Grenze und Grenzüberschreitung im Mittelalter. 11. Symposium des Mediävistenverbandes vom 14. bis 17. März 2005 in Frankfurt an der Oder, Berlin 2007, S. 93–106.

15 Friedrich Bennighoven, Die Burgen als Grundpfeiler des spätmittelalterlichen Wehrwesens im preußisch-livländischen Deutschordensstaat, in: Hans Patze (Hg.), Die Burgen im deutschen Sprachraum. Ihre rechts- und verfassungsgeschichtliche Bedeutung, Teil 1 (Vorträge und Forschungen 19), Sigmaringen 1976, S. 565–601, hier bes.: S. 567–569, 575–578; Manfred Hellmann, Das Lettenland im Mittelalter. Studien zur ostbaltischen Frühzeit und lettischen Stammesgeschichte, insbesondere Lettgallens (Beiträge zur Geschichte Osteuropas 1), Münster/Köln 1954, S. 222–224; Klaus Neitmann, Die Auswahl von Residenzorten. Methodische Bemerkungen zur spätmittelalterlichen geistlichen Residenzbildung, in: Ders. / Heinz-Dieter Heimann (Hg.), Spätmittelalterliche Residenzbildung in geistlichen Territorien Mittel- und Nordostdeutschlands (Studien zur brandenburgischen und vergleichenden Landesgeschichte 2/Veröffentlichungen des Museums für Brandenburgische Kirchen- und Kulturgeschichte des Mittelalters 3), Berlin 2009, S. 41–88, hier bes.: S. 53–64.

tatsächliche oder (im Falle der kleineren Kommunen) eher vermeintliche Autonomie von den Stadtherren.[16] Immerhin ist in Livland die Städtegründung als Instrument einer Verwaltungsstrategie der Landesherren erkennbar. Aus Städten als Mittelpunkte der Territorialherrschaften organisierten und kontrollierten sie ihre Herrschaftsräume.[17] Die Bischöfe, um den Forderungen des kanonischen Rechts nach dem Residieren in der Stadt zu folgen, waren verpflichtet, dafür zu sorgen, dass in ihrem Zuständigkeitsbereich mindestens eine Residenzstadt existiert.[18] Gleichwohl gründeten die Bischöfe von Dorpat und Kurland jeweils nur eine Stadt; der Bischof von Ösel-Wieck zwei, eine etwas größere Anzahl von Städten gab es in Livland nur in den Herrschaftsgebieten des Erzbischofs von Riga und des Deutschen Ordens.[19] Infolgedessen hatten es die Kommunen in Livland nicht nur mit einem Herrscher zu tun, sondern mussten Rücksicht auf die Interessen verschiedener Landes- und Stadtherren unterschiedlicher Größenordnung nehmen.

Über die Zahl der Städte im mittelalterlichen Livland gibt es unterschiedliche Schätzungen. Zumeist werden in der Literatur 19 oder 20 Städte genannt.[20]

---

16 Die kleineren Städte konnten keine wichtige Entscheidung ohne Zustimmung des Stadtherrn oder seines Vertreters treffen. Plētiens, Livonijas mazpilsētas (wie Anm. 4), S. 41.

17 Siehe ausführlicher: Misāns, Die Städte als zentrale Orte (wie Anm. 5), S. 91–93.

18 Es sei hier auf die neueste, umfassende Publikation zur Residenzenforschung in Livland mit den Artikeln über Dorpat, Hapsal (estn. Haapsalu), Kokenhusen (lett. Koknese), Pilten (lett. Piltene), Riga, Ronneburg (lett. Rauna) sowie der Residenz des livländischen Ordensmeisters Wenden (lett. Cēsis) verwiesen: Harm von Seggern (Hg.), Residenzstädte im Alten Reich (1300–1800). Ein Handbuch, Teil 1: Nordosten, Ostfildern 2018.

19 Põltsam-Jürjo, Grundzüge des livländischen Städtewesens (wie Anm. 4), S. 121. Unerwähnt darf an dieser Stelle nicht bleiben, dass auch einige Großvasallen als Städtegründer auftraten, auch wenn sich nur eine von diesen Initiativen – Roop (lett. Straupe) – als nachhaltig erwies (ebd.).

20 Põltsam-Jürjo, Grundzüge des livländischen Städtewesens (wie Anm. 4), S. 117. Die Zahl hängt davon ab, ob Leal (estn. Lihula) als Stadt angesehen wird. Zum urbanen Potential dieses Ortes siehe: Selart, Die kleinen Städte Livlands (wie Anm. 3), S. 30 f. In der von Edgars Plētiens zusammengestellten Liste wurden sogar 21 Städte aufgenommen, was damit zu erklären ist, dass er im Unterschied zu anderen Autoren auch die Ortschaft Bauske (lett. Bauska) zu den Städten zählt (Plētiens, Livonijas mazpilsētas (wie Anm. 4), S. 23). Plētiens beruft sich dabei alleine auf den Auszug aus einem Lehensbrief aus dem Jahr 1509, in dem Bauske („*Bawske*") als „*stede up deme schilde*" bezeichnet wird (ebd., S. 28). Aus dem Kontext der Quelle ergibt sich jedoch, dass mit dem Wort „*stede*" nicht eine Stadt, sondern eine Stätte gemeint ist.

So lässt sich Livland als eines der am wenigsten urbanisierten Gebiete Europas bezeichnen.[21] Wie dünn das Städtenetz Livlands war, veranschaulicht ein Vergleich mit dem Ordensland Preußen, wo es annähernd 100 Städte gab, von den mehr als 3.000 Städten auf dem Territorium Deutschlands ganz zu schweigen.[22]

Während der Eroberungszeit war es im Interesse der Herrscher, sichere Anlaufpunkte für Kaufleute aus Lübeck und Gotland einzurichten, um sowohl die Überführung der Kreuzfahrer zu erleichtern als auch um neuerworbene Territorialgebiete wirtschaftlich zu stärken. So erteilte Bischof Albert als Anerkennung der Verdienste der Kaufleute im Jahre 1211 den Händlern aus Gotland – vornehmlich aus Norddeutschland stammende Kaufleute, die von dieser Insel regelmäßig nach Livland segelten – ein Bündel von Privilegien, was sich „folgenschwer für den Stadtbildungsprozeß in Riga" erwies, weil sich bald ein Teil von ihnen in der Dünastadt, aber auch im später gegründeten Dorpat und in Reval sowie einigen kleineren Städten dauerhaft als Bürger niederließen.[23] Auch für die Kaufleute zahlte sich trotz mancher anfänglicher Rückschläge die Kooperation mit den neuen Machthabern aus – die Stadt- und Landesherren sicherten die Grundlagen für ihre Handelstätigkeit. Außerdem, wie im Mittelalter üblich, entwickelte sich als zweite Säule der städtischen Wirtschaft neben dem Handel auch das Handwerk, wenn auch seine Bedeutung in Livland die lokalen Grenzen nicht überschritt. Der Wohlstand der livländischen Städte stütze sich in erster Linie auf den Handel.

Für die gezielte Kolonisierung mit Einwanderern aus dem zivilisierten Teil Europas machten die neuen Landesherren von Städtegründungen wenig Gebrauch. Hier ist weder eine gezielte Anlockung der Umsiedlungswilligen noch eine ausgeprägte Privilegierung der Ansässigen erkennbar.[24] Letztendlich

Ohne zusätzliche Begründung ist es voreilig, Bauske den Städten des mittelalterlichen Livlands zuzurechnen.

21 Inna Põltsam-Jürjo, Die Städte: Alltag, soziale Schichten, Handel und Gewerbe, in: Brüggemann / Henning / Meier / Tuchtenhagen (Hg.), Das Baltikum (wie Anm. 10), S. 296–340, hier: S. 297.

22 Norbert Angermann, Einführung in das Thema des Seminars, in: Ders. (Hg.), Städtisches Leben im Baltikum zur Zeit der Hanse. Zwölf Beiträge zum 12. Baltischen Seminar (Baltische Seminare 10), Lüneburg 2003, S. 7–20, hier: S. 12.

23 Hansisches Urkundenbuch, Bd. 1, hg. v. Konstantin Höhlbaum, Halle an der Saale 1876, Nr. 88, S. 38; Detlef Kattinger, Die Gotländische Genossenschaft. Der frühhansisch-gotländische Handel in Nord- und Westeuropa (Quellen und Darstellungen zur hansischen Geschichte NF 47), Köln 1999, S. 175 f.; Peter Johanek, Die mittelalterlichen Fernhändler, in: Wilfried Schlau (Hg.), Sozialgeschichte der baltischen Deutschen, Köln 1997, S. 31–45, hier: S. 40–44.

24 Põltsam-Jürjo, Städte (wie Anm. 21), S. 297 f.

stiegen zu bedeutenden Knotenpunkten des Fernhandels in Livland nur Riga, Dorpat und Reval auf (alle drei Kommunen erhielten die Stadtrechte noch vor der Mitte des 13. Jahrhunderts), eine dichte Urbanisierung fand nicht statt und das Städtenetz blieb großmaschig. Inna Põltsam-Jürjo stellt in diesem Zusammenhang drei Faktoren heraus, die eine intensivere Entwicklung der städtischen Lebensform Livlands behinderten: Erstens ein schwacher Zuzug der Stadtbewohner aus den deutschen Gebieten und der damit verbundene Nachwuchsmangel bei deutschen Stadtbürgern, zweitens das Desinteresse der Landesherren an der Verbreitung der städtischen Freiheiten, weil sie eine Erstarkung der Städte scheuten, und drittens der Widerstand Revals und Rigas gegen die Stärkung der weiteren Städte, um potentielle Konkurrenz zu unterdrücken.[25] Während die zweite Behauptung mithilfe zusätzlicher Forschung noch zu überprüfen wäre, kann man den anderen zwei Argumenten zustimmen. Auch Anti Selart schließt sich der Meinung an, dass die demographische Krise negative Folgen auf die Entwicklung der Kleinstädte im 14. und 15. Jahrhundert hinterließ, und bemerkt in Bezug auf den dritten Punkt, dass wirtschaftlich die kleinen Städte von den großen „tot konkurriert" wurden.[26]

Es gibt Gründe, die Dynamik des Urbanisierungsprozesses in Livland in zwei Zeitabschnitte – vor und nach dem 14. Jahrhundert – einzuteilen. Bis ungefähr 1300 und einige Jahrzehnte danach ist ein Wachstum der städtischen Strukturen und eine Stärkung sowie Ausbau des Städtenetzes zu beobachten. Nicht nur die drei großen Handelszentren Riga, Dorpat und Reval, sondern auch einige kleinere Städte nahmen ihre Nischen im System des Fernhandels und im lokalen Wirtschaftsleben des Landes ein. Wenn nach dem Ende der Eroberungsperiode manche in der Zeit der Kreuzzüge wichtige Zentren untergingen, wurden einigen anderen Ortschaften Stadtrechte verliehen.[27] Nicht nur die größeren, sondern sogar mehrere kleinere Städte verfügten im 14. Jahrhundert schon über genügend Ressourcen, um sich eine Mauer leisten zu können.[28] Ihre Verteidigungsanlagen stärkten das Verteidigungssystem des jeweiligen Ortes, weil sie

25 Põltsam-Jürjo, Grundzüge des livländischen Städtewesens (wie Anm. 4), S. 121–123.

26 Selart, Die kleinen Städte Livlands (wie Anm. 3), S. 41.

27 Siehe dazu: ebd., S. 34 f.

28 Rivo Bernotas, Estonian Small Towns in the Middle Ages: Archaeology and the History of Urban Defense, in: Ajalooline Ajakiri 3, 2013, S. 265–297; Anti Selart, Where was the Home of the Livonian Merchant? Early Urban Mobility in the Baltics, in: Zapiski Historyczne 1, 2019, S. 43–66, hier: S. 47. Selart bezeichnet den raschen Fortschritt der Urbanisierung im 13. Jahrhundert als „a real boom of urbanisation and economic development".

eine Verlängerung der Burgmauer bildeten. Einige von diesen kleineren Städten gehörten somit zum Typus der „Vorburg-Stadt".[29] Nach 1300 sahen ihre Entwicklungsperspektiven nicht schlecht aus. Mit der Zeit lenkte jedoch eine Kombination von ungünstigen inneren und äußeren, wirtschaftlichen und demographischen Faktoren die Entwicklung des urbanen Lebens in Livland in eine Richtung, in der die kleineren Städte in Stagnation und Bedeutungslosigkeit verharrten. Vor allem der Organisationswandel des Handels im gesamten hansischen Wirtschaftsraum und die Pestepidemien verursachten Veränderungen zu ihren Ungunsten. In Livland konnten die kleineren Städte mit niedriger Bevölkerungsdichte die Menschenverluste mithilfe des Zuzugs der bäuerlichen Bevölkerung nicht ausreichend ersetzen. Generell wurden die kleineren Orte aus dem überregionalen Handel zunehmend verdrängt und den größeren Städten untergeordnet, ihre Autonomie, soweit vorhanden, wurde zugunsten der Landesherren und seiner Amtspersonen deutlich eingeschränkt und ihre selbstständige Handlungsfähigkeit geschwächt. Von der wirtschaftlichen und demographischen Krise Livlands in der ersten Hälfte des 15. Jahrhunderts haben sich die Kleinstädte nicht mehr erholt.[30]

Da es in Livland nur sehr wenige Kleinstädte mit Stadtrecht gab, wohnte hier eigentümlicherweise die Mehrheit der Städter nicht in den kleineren Kommunen, wie es gewöhnlich in anderen Ländern Europas der Fall war, sondern in nach lokalen Maßstäben großen Städten.[31] In allen livländischen Kleinstädten mit Stadtrecht waren zusammengenommen weniger Menschen wohnhaft als in den „Großstädten" Riga, Reval und Dorpat.[32]

29 Von zur Mühlen, Livland (wie Anm. 13), S. 102.

30 Siehe ausführlicher: Selart, Die kleinen Städte Livlands (wie Anm. 3), S. 35–41.

31 Riga hatte um die Mitte des 16. Jahrhunderts schätzungsweise 10.000–15.000, Reval 7.000–8.000 und Dorpat 5.000–6.000 Einwohner. Vgl. Norbert Angermann, Die Stellung der livländischen Städte in der hansischen Gemeinschaft, in: Hansische Geschichtsblätter 113, 1995, S. 111–125, hier: S. 112. Im europäischen Vergleich sind sie nach diesem Kriterium den mittelgroßen Städten zuzuordnen.

32 Põltsam-Jürjo, Städte (wie Anm. 21), S. 297. Da es an entsprechenden Quellen mangelt, kann die Einwohnerzahl jeder einzelnen Stadt nur in groben Zügen geschätzt werden. So geben verschiedene Autoren unterschiedliche Zahlen an. Die Tendenz lässt sich wie folgt zusammenfassen: Die Grenze von 1.000 Einwohner hat von den kleineren Städten wahrscheinlich nur Neu-Pernau (estn. Uus-Pärnu) überschritten (zwischen 1.000 und 1.400 Bewohner), ca. 800 bis 1.000 Einwohner hatten Fellin (estn. Viljandi), Wenden und wohl auch Wolmar (lett. Valmiera), es folgte Narwa (estn. Narva) mit 600–700 Einwohnern. In Alt-Pernau wohnten nicht mehr als 300, in Goldingen (lett. Kuldīga) 400 Menschen. Vgl. Põltsam-Jürjo, Grundzüge des livländischen Städtewesens (wie

Wie bereits erwähnt, erreichten nur diese drei größeren Kommunen eine wirkliche Autonomie und nur sie besaßen genügend wirtschaftliche und politische Stärke, um eine Achtung vom jeweiligen Landesherrn zu erlangen. Jede Stadt unterhielt dabei mit ihrem Landesherrn ein anderes Verhältnis, das aus ihrer spezifischen Geschichte und den lokalen Bedingungen hervorging. Reval bekam schon während der dänischen Herrschaftszeit keine starke Hand einer höheren Macht zu spüren und auch nach dem Übergang der Stadt an den Deutschen Orden im Jahre 1346 kam die Kommune gut mit ihrem neuen Stadtherrn zurecht. Der Orden war nicht bestrebt, die rechtlich-politische Autonomie Revals einzuschränken, erwartete aber Loyalität seitens der Stadtbehörden und sah die Stadt als einen politischen Verbündeten und einen Stützpfeiler für den Aufbau seiner Landesherrschaft. Auch Reval legte seinerseits Wert auf gute Kontakte und ein freundschaftliches Verhältnis mit dem Orden. Ebenso Dorpat, soweit bekannt, hatte bis zur Reformation mit seinem Bischof gute Beziehungen gepflegt und bei Bedarf seinen Stadt- und Landesherrn unterstützt. Komplizierter war die Situation Rigas. 1201 von Bischof Albert gegründet, wurde die Stadt an der Düna ab der zweiten Hälfte des 13. Jahrhunderts mit der Bestrebung des Deutschen Ordens konfrontiert, die größte Kommune Livlands unter seine Gewalt zu bringen. Der Orden sah die Stadtherrschaft über Riga als ein Mittel im Kampf um eine vollständige und uneingeschränkte Landesherrschaft in Livland an und hatte mehrmals Versuche unternommen, die Kommune unter seine Gewalt zu bringen. Riga leistete ihm Widerstand, aber da sich die Stadt auch ständig der Zunahme der Herrschaftsrechte des Erzbischofs widersetzte, konnte sie in bestimmten Phasen eine Schaukelpolitik zwischen den beiden größten politischen Territorialgewalten des Landes betreiben. Insgesamt hatte Riga das ganze Mittelalter hindurch ein kompliziertes und gespanntes Verhältnis zum Erzbischof und dem Landmeister des Deutschen Ordens, das von einer Mischung aus Loyalität, Neutralität und Konfrontation geprägt war.[33]

Anm. 4), S. 132; von zur Mühlen, Livland (wie Anm. 13), S. 106; Angermann, Einführung (wie Anm. 22), S. 10; Roman Czaja, Livonian Towns from the 13th to the 16th century, in: Ders. / Andrzej Radzimiński (Hg.), The Teutonic Order in Prussia and Livonia. The political and ecclesiastical Structures 13th–16th century, Toruń 2015, S. 225–252, hier: S. 247.

33 Roman Czaja, Der Deutsche Orden als Stadtherr im Reich, in Preußen und in Livland, in: Ordines Militares. Colloquia Torunensia Historica 14, 2007, S. 127–139, hier: S. 131–135; Ders., Livonian Towns (wie Anm. 32), S. 230–241; Manfred Hellmann, Der Deutsche Orden und die Stadt Riga, in: Udo Arnold (Hg.), Stadt und Orden. Das Verhältnis des Deutschen Ordens zu den Städten in Livland, Preußen und im Deutschen Reich (Quellen und Studien zur Geschichte des Deutschen Ordens 44/

Trotz der Zugehörigkeit zu unterschiedlichen Territorien kooperierten die drei größeren livländischen Städte sowohl im wirtschaftlichen als auch im politischen Bereich eng miteinander. Riga, Dorpat und Reval spannten weitreichende Handelsverbindungen und bildeten über die landesherrschaftlichen Grenzen hinweg eine Städtelandschaft mit eigenen Netzwerken, zu der auch kleinere Städte und Marktorte gehöhrten.

2. Wirtschaftliche Vernetzung und institutionelle Bindungen der Städte

Ähnlich wie im südlichen Ostseeraum hat man auch in Livland während der Entstehung und Stabilisierung des urbanen Lebens nicht nur auf bereits bestehende Siedlungen und Raumstrukturen zurückgegriffen. Viel wichtiger für die Ausdehnung der Städtelandschaft war die Einbindung einzelner Ortschaften in die wirtschaftlichen und damit im Grunde genommen personell bestimmten Netzwerke, welche die urbane Entwicklung maßgeblich beeinflussten und die gesamte Stadtkultur prägten.[34] Ausschlaggebend in Livland ist in diesem Zusammenhang das Stichwort Hanse. Die Rekonstruktion der Verflechtung der Verbindungen der livländischen Kaufleute und Städte erleichtert die von Carsten Jahnke unlängst vorgeschlagene Einteilung der zur Hanse gehörenden Städte in vier Gruppen, der Kern-, Portal-, Regional- und Kleinstädte, von denen in Bezug auf Livland die letzten drei ins Gewicht fallen. Zu den Portalstädten werden Kommunen

---

Veröffentlichungen der Internationalen Historischen Kommission zur Erforschung des Deutschen Ordens 4), Marburg 1993, S. 1–33, hier bes.: S. 15–31; Reinhard Vogelsang, Reval und der Deutsche Orden. Zwischen Autonomie und landesherrlichen Gewalt, in: ebd., S. 34–58; Juhan Kreem, The Town and its Lord. Reval and the Teutonic Order (in the fifteenth century), Tallinn 2002; Thomas Brück, Riga in der ersten Hälfte des 15. Jahrhunderts. Das Verhältnis der Stadt zum Orden, zum Erzbischof und zur Hanse, in: Angermann, Städtisches Leben (wie Anm. 22), S. 43–92, hier bes.: S. 65–76.

34 Karsten Igel, Stadtbildung und Netzwerke im südwestlichen Ostseeraum. Überlegungen zu Greifswald und Stralsund, in: Sunhild Kleingärtner / Gabriel Zeilinger (Hg.), Raumbildung durch Netzwerke? Der Ostseeraum zwischen Wikingerzeit und Spätmittelalter aus archäologischer und geschichtswissenschaftlicher Perspektive, Beiträge des am 28. und 29. Oktober 2010 in Kiel veranstalteten internationalen Workshops (Zeitschrift für Archäologie des Mittelalters, Beiheft 23), Bonn 2012, S. 79–94, hier bes.: S. 80; Alexander Drost, Historische Grenzräume im Ostseeraum: Livland. Borderland-Theorien und kulturelle Schichten, in: Ders. / Michael North (Hg.), Die Neuerfindung des Raumes. Grenzüberschreitungen und Neuordnungen, Köln/Weimar/Wien 2013, S. 113–141, hier: S. 123 f.

gezählt, deren Wohlstand sich im Besonderen auf den Außenhandel in fremde Gebiete stützte, die im überregionalen Handel der Hanse führend und in überregionale Konflikte verwickelt waren, Rechte im Ausland nutzten und sich häufig an den hansischen Tagfahrten beteiligten. Der Anschluss an den Fernhandel der Regionalstädte war dagegen eingeschränkt und ihr Interesse an Privilegien und Konflikten im Ausland dementsprechend geringer. Die Kaufleute dieser Städte spezialisierten sich primär auf den Verkauf der aus der Umgebung stammenden Waren, aber sie dienten auch als regionale Verteilerzentren importierter Güter. Auf diese Art und Weise waren auch diese Zentren in die Rechts- und Netzwerkstrukturen der Hanse eingebunden, ihre Kaufmänner konzentrierten sich jedoch nicht auf den Handel im Ausland, sondern vorwiegend auf den Warenaustausch und -verkehr im innerhansischen Raum. Die Teilhabe am hansischen Handel der in kleineren Städten sesshaften Kaufleute trat dadurch in Erscheinung, dass sie für den Austausch mit Waren und Informationen zwischen den Portal- und Regionalstädten und den lokalen Märkten sorgten.[35]

Gemäß der von Carsten Jahnke vorgeschlagenen Einteilung gehörten zu den Portalstädten in Livland Riga, Dorpat und Reval,[36] die sich nicht zuletzt dank der verkehrsgünstigen Lage in der nordosteuropäischen Region als wichtige Hauptorte im östlichen Teil des hansischen Raumes mit Handelsverbindungen bis nach Russland, Flandern und England etablierten.[37] In die Kategorie der hansischen Regionalstädte mit weit ausgedehntem Handel wird (Neu-)Pernau – im livländischen Maßstab eine mittelgroße Stadt – eingestuft.[38] Dennoch waren auch in kleineren Städten Livlands Händler zu finden, die einen Zugang zum

---

35 Carsten Jahnke, Hansisch oder nicht hansisch? Die kleinen und die großen Hansestädte, in: Kreem / Sarnowsky, „Hansisch" oder „nicht-hansisch" (wie Anm. 3), S. 1–22, hier: S. 17–21 und S. 7, Abb. 4.

36 Ebd., S. 18.

37 Norbert Angermann, Die Bedeutung Livlands für die Hanse, in: Ders. (Hg.), Die Hanse und der deutsche Osten, Lüneburg 1990, S. 97–115; Ivar Leimus / Anu Mänd, Reval (Tallinn): A city emerging from maritime trade, in: Wim Blockmans / Mikhail Krom / Justyna Wubs-Mrozewicz (Hg.), The Routledge Handbook of Maritime Trade Around Europe, 1300–1600: Commercial Networks and Urban Autonomy, London/New York, S. 273–291.

38 Jahnke, Hansisch oder nicht hansisch? (wie Anm. 35), S. 18. Zur Charakteristik und Bewertung des Pernauer Handels siehe: Selart, Die kleinen Städte Livlands (wie Anm. 3), S. 34; Heinrich Laakmann, Geschichte der Stadt Pernau in der Deutsch-Ordenszeit (bis 1558) (Wissenschaftliche Beiträge zur Geschichte und Landeskunde Ost-Mitteleuropas 23), Marburg an der Lahn 1956, S. 145–173.

Fernhandelsnetz hatten und gelegentlich im großen Stil Handel betrieben.[39] Wenn auch hauptsächlich für die wirtschaftliche Tätigkeit der Einwohner dieser Städte die lokale Ebene maßgebend war, waren auch sie mit Portal- und Regionalstädten verbunden und gehörten zum hansischen Handelsnetz. Als Basis des regionalen Markt- und Verteilersystems „waren sie und haben sie sich auf ihrer Ebene als Teil der Hanse in der Region verstanden".[40] In diesem Zusammenhang werden in der aktuellen Hanseforschung nicht primär die institutionellen Bindungen zwischen den Städten, sondern stärker als früher die Rolle der personellen Netzwerke einzelner Kaufleute beachtet. Wenn in der älteren Literatur als maßgebliches Kriterium der Zugehörigkeit zur Hanse dieser oder jener Kleinstadt ihre Beteiligung an den regionalen städtischen Tagfahrten gegolten hat,[41] so lassen Erkenntnisse der modernen Netzwerkforschung[42] in der Betrachtung dieser Frage eine größere Flexibilität zu.[43] Dafür sprechen auch die Erkenntnisse von Anti Selart, wenn er feststellt, dass „not only were 'big towns' connected

39 Friedrich Benninghoven, Ein Osnabrücker Fernhändlergeschlecht im Livlandhandel des 13. Jahrhunderts, in: Hamburger mittel- und ostdeutsche Forschungen 4, 1963, S. 157–189; Selart, Where was the Home of the Livonian Merchant? (wie Anm. 28), S. 45 f.

40 Jahnke, Hansisch oder nicht hansisch? (wie Anm. 35), S. 18 f.

41 Paul Johansen, Die Bedeutung der Hanse für Livland, in: Hansische Geschichtsblätter 65/66, 1940/41, S. 1–55, hier: S. 33 f.; Angermann, Die Stellung der livländischen Städte (wie Anm. 31), S. 113; Ilgvars Misāns, Alt-Livland und die Hanse, in: Goehrke / von Ungern-Sternberg, Die baltischen Staaten (wie Anm. 9), S. 35–45, hier: S. 39.

42 Carsten Jahnke, Handelsnetze im Ostseeraum, in: Gerhard Fouquet / Hans-Jörg Gilomen (Hg.), Netzwerke im europäischen Handel des Mittelalters (Vorträge und Forschungen 72), Ostfildern 2010, S. 189–212; Ders., Moderne Netzwerkforschung in der regionalen Hansegeschichte: Möglichkeiten, Gefahren und Perspektiven, in: Oliver Auge (Hg.), Hansegeschichte als Regionalgeschichte (Kieler Werkstücke A 37), Frankfurt am Main 2014, S. 47–58; Jahnke, Hansisch oder nicht hansisch? (wie Anm. 35), hier bes.: S. 21 f.

43 Deutlich veranschaulicht diese Veränderungen die letzte Neuausgabe des langjährigen Standardwerks zur Hansegeschichte von Philippe Dollinger, wo die Liste der Hansestädte aus Livland im Vergleich mit früheren Auflagen wesentlich gekürzt worden ist. Laut der ursprünglichen Version des Buches galten als Hansestädte Dorpat, Fellin, Goldingen, Kokenhusen, Lemsal, Pernau, Reval, Riga, Roop, Wenden, Windau und Wolmar. Vgl. Philippe Dollinger, Die Hanse, Stuttgart [2]1976, S. 571. In der letzten Auflage sieht die Aufzählung der livländischen Hansestädte schon deutlich knapper und ungenauer aus: Riga, Reval, Dorpat, Fellin (?), Pernau (?), Wenden (?) und Wolmar (?). Vgl. Philippe Dollinger, Die Hanse. Neu bearbeitet von Volker Henn und Nils Jörn, Stuttgart [6]2012, S. 593.

to the whole Hanseatic region but personal networks based on family and/or commercial ties included places on different levels of the settlement hierarchy".[44]

Die Ursprünge des hansischen Netzwerks in Livland reichen bis zur Missions- und Kreuzzugszeit zurück, als sich die Einwanderer aus norddeutschen Gebieten in den vorhandenen Siedlungsorten niederließen, wo sie Seite an Seite mit den autochthonen Einwohner wohnten.[45] Jedoch nur der kleinste Teil dieser Ansiedlungen entwickelte sich mit der Zeit zu Rechtstädten. Als Knotenpunkte des wirtschaftlichen Netzwerkes waren auch kleinere Ortschaften mit ein paar Dutzend Häusern – Marktflecke, Weichbilder oder Hakelwerke –, wenn auch diese die städtische Qualität nicht erreichten, ein Teil des Urbanisierungsprozesses. Ähnlich wie die kleineren Städte blieben diese Zentren in erster Linie auf die Erfüllung der wirtschaftlichen Bedürfnisse der unmittelbaren Umgebung ausgerichtet, was jedoch die Teilhabe einzelner Kaufleute am Großhandel nicht ausschließt. Als charakteristisches Beispiel ist das südestnische Odenpäh (estn. Otepää) zu erwähnen, das Anfang des 13. Jahrhunderts wohl bedeutender als Dorpat oder Fellin war, aber sich nicht zu einer Rechtstadt fortentwickelte und im 14. Jahrhundert nicht mehr als eine zweitrangige Ortschaft war.[46] In den schriftlichen Quellen tauchen die 50–52 nachweisbaren Hakelwerke/Weichbilder des mittelalterlichen Livlands nur fragmentarisch auf und ihre faktische wirtschaftliche Bedeutung sowie ihr rechtlicher Status ist schwer einzuschätzen. Dennoch lassen die überlieferten Quellen zumindest in einigen Hakelwerken Elemente einer Autonomie und eines kleinstadtähnlichen rechtlichen Raums erkennen.[47]

---

44 Selart, Where was the Home of the Livonian Merchant? (wie Anm. 28), S. 51. Über die (Nicht-)Hansezugehörigkeit einzelner kleinerer livländischer Städte siehe: Selart, Die kleinen Städte Livlands (wie Anm. 3), S. 25–28.

45 Misāns, Die Städte als zentrale Orte (wie Anm. 5), S. 90–93; Selart, Where was the Home of the Livonian Merchant? (wie Anm. 28), S. 46 f. In einer anderen neuen Publikation unterstreicht Anti Selart, dass für die einheimische nicht-deutsche Einwohnerschaft gerade die Städte insbesondere in der Zeit der raschen wirtschaftlichen Entwicklung im 13. und 14. Jahrhundert eine Aufstiegschance boten. Vgl. Anti Selart, Post hoc oder propter hoc? Eroberung, Einwanderung und Gesellschaftswandel im mittelalterlichen Livland, in: Blätter für deutsche Landesgeschichte 154, 2018, S. 429–453, hier: S. 441 f. Das Zusammenleben der Menschen von verschiedener Herkunft und Sprache ist ein charakteristisches Merkmal des urbanen Lebens des mittelalterlichen Livlands.

46 Selart, Die kleinen Städte Livlands (wie Anm. 3), S. 31.

47 Siehe dazu: Põltsam-Jürjo, Grundzüge des livländischen Städtewesens (wie Anm. 4), S. 127–130.

Als Portalstädte spielten Riga, Dorpat und Reval in der Städtelandschaft Livlands eine bestimmende Rolle. Dabei ist nicht zu beobachten, dass Riga als wirtschaftlich und einwohnermäßig stärkste Stadt des Landes eine Vorzugsstellung eingenommen hätte oder sich gar an der Spitze der Hierarchie der livländischen Städte befand. Die drei livländischen „Großstädte" waren gleichwertige Partner, die alle auf gleicher Augenhöhe kommunizierten und ihre Beziehungen beruhten auf dem Interessenausgleich und der gegenseitigen Anerkennung der jeweiligen Interessen- und Einflusssphären.[48]

Um jede größere livländische Stadt als Mittelpunkt herum bildeten sich allmählich weitere oder nähere Handelsräume, die aus Unterzentren, kleineren Städten, Weichbildern und Marktorten bestanden.[49] Reval – der wichtigste Umschlagplatz für den hansischen Handel mit Nowgorod – übte eine zentralörtliche Funktion im Norden des Landes aus. Zu Revals Einflussbereich in Livland gehörten der nördliche Teil der Insel Ösel (estn. Saaremaa) und auf dem Festland die dänischen Gebiete, die nach dem Herrscherwechsel im Jahre 1346 der Deutsche Orden übernahm.[50] Überdies hat die Verbindungen Revals zu den Kleinstädten Wesenberg (estn. Rakvere) und Narwa (estn. Narva) auch das lübische Recht gestärkt, das an diese ebenso wie an Reval verliehen wurde (die sonstigen livländischen Städte übernahmen das Rigische Recht).

Im mittleren Teil des mittelalterlichen Livlands (das heißt in Südestland) ragte Dorpat als das bedeutendste Zentrum heraus, dessen Handel auf die benachbarte russische Stadt Pleskau (russ. Pskow) und weiter nach Nowgorod ausgerichtet war.[51] Aufgrund der dürftigen Quellenlage lässt sich die Handelsorganisation Dorpats nicht genau rekonstruieren, sodass man die Umrisse des Handelsbezirks im inneren des Landes nur hypothetisch skizzieren kann. Vilho Niitemaa setzt das wirtschaftliche Hinterland Dorpats mit den Grenzen des gleichnamigen Stifts gleich.[52] Andere Autoren bemerken, dass Dorpat auf dem

48 Ilgvars Misāns, Riga – ein Vorort der livländischen Städte im Mittelalter?, in: Ders. / Horst Wernicke (Hg.), Riga und der Ostseeraum. Von der Gründung 1201 bis in die Frühe Neuzeit (Tagungen zur Ostmitteleuropa-Forschung 22), Marburg 2005, S. 169–179, hier bes.: S. 176–179.

49 Vilho Niitemaa, Der Binnenhandel in der Politik der livländischen Städte im Mittelalter (Annales Academiae Scientiarum Fennicae B 76/2), Helsinki 1949, S. 49.

50 Ebd., S. 50.

51 Norbert Angermann, Deutsche Kaufleute im mittelalterlichen Novgorod und Pleskau, in: Hans Rothe (Hg.), Deutsche im Nordosten Europas, Köln/Wien 1991, S. 59–86, hier: S. 80 f.

52 Niitemaa, Binnenhandel (wie Anm. 49), S. 50.

Flussweg mit Pernau verbunden war und seine Kaufleute die Waren über den dortigen Hafen umschlagen konnten.[53] Es ist wohl in der Tat dem wirtschaftlichen Hinterland Dorpats, in mittlerer Reichweite auch Neu-Pernau und ebenfalls Fellin zuzurechnen. Gleichzeitig ist Neu-Pernau in diesem vergrößerten Wirtschaftsraum teilweise als Stadt mit einem selbstständigen Einzugsgebiet zu betrachten, deren Hinterland sich mit den Orten Burtneck (lett. Burtnieki), Rujen (lett. Rūjiena), Karkus (estn. Karksi), Fellin, Weissenstein (estn. Paide), Rappel (estn. Rapla), Fickel (estn. Vigala) und Leal umschreiben lässt[54] – eine Überschneidung der Handelsbezirke mit einzelnen größeren beziehungsweise mittelgroßen Städten ist gut vorstellbar. Einiges deutet darauf hin, dass um die Mitte des 14. Jahrhunderts in Pernau – damals einer der drei Stapelhäfen Livlands – noch reger Fernhandel herrschte, später jedoch der Warentransit des Ost-West-Handels auf diesem Weg ins Stocken geriet. Im 15. Jahrhundert hat Dorpat seine Waren größtenteils über Reval ausgeführt oder bezogen und die Straße Dorpat–Fellin–Pernau verödete. Neu-Pernau konzentrierte sich zu dieser Zeit wiederum auf die Ausfuhr der Agrarprodukte aus dem Binnenmarkt, insbesondere des Flachses. Sein Hafen wurde zunehmend vom Deutschen Orden für die Ausfuhr der Agrarprodukte aus den estländischen Besitzungen benutzt.[55]

Am größten war der von Riga kontrollierte Wirtschaftsraum. Die günstige geographische Lage an der Mündung der Düna in die Ostsee brachte der livländischen Metropole Vorteile im Handel mit Polozk und dem umliegenden weißrussisch-litauischen Gebiet, den Riga völlig beherrschte. Als Stützpunkt auf dieser Fernhandelsroute fungierte die Kleinstadt Kokenhusen, in der sich Handelspartner der Rigaer Kaufleute niederließen. Außerdem erstreckten sich binnenländische Handelsverbindungen Rigas nach Mittel- und Südlivland sowie in den südlichen Teil von Ösel. Eine Vorstellung von der Ausdehnungsweite des Rigaer Handelsnetzes in die östliche Richtung um 1300 vermittelt die häufige Erwähnung im Rigaer Schuldbuch der Kreditgeschäfte der Kaufleute der Dünastadt mit den Händlern aus den Kleinstädten und Hakelwerken Wenden, Roop,

53 Von zur Mühlen, Livland (wie Anm. 13), S. 102 f.; Põltsam-Jürjo, Städte (wie Anm. 21), S. 314. Zum genauen Verlauf dieser Straße siehe: Friedrich Bruhns / Hugo Weczerka, Hansische Handelsstrassen. Textband (Quellen und Darstellungen zur hansischen Geschichte 13/2), Weimar 1967, S. 760 f.

54 Laakmann, Geschichte der Stadt Pernau (wie Anm. 38), S. 145 f.

55 Selart, Die kleinen Städte Livlands (wie Anm. 3), S. 33, 37; Laakmann, Geschichte der Stadt Pernau (wie Anm. 38), S. 169–171; Põltsam-Jürjo, Städte (wie Anm. 21), S. 314, 321.

Walk (estn. Valga; lett. Valka) und Odenpäh (sie liegen auf der Straße nach Dorpat und weiter nach Pleskau).[56]

Wie intensiv die Beziehungen Rigas zu den Städten und Marktorten Kurlands waren, ist schwer zu bewerten. Es bleibt offen, ob Riga für sie als Portalstadt anzusehen ist. Windau, Goldingen, Hasenpoth (lett. Aizpute) und Pilten waren verkehrsmäßig miteinander gut vernetzt und bildeten anscheinend eine eigene Städtegruppe, deren Außenhandel über kurländische Häfen verlief und die mit Riga nur schwach verbunden war.[57] Kurland und seine Städte schalteten sich nur selten in das politische und wirtschaftspolitische Geschehen Livlands ein, wobei alle Hinweise, die solche Aktivitäten dokumentieren, aus dem 15. Jahrhundert stammen. Die Aufgabe der weiteren Forschung wäre es, zu überprüfen, wie die wirtschaftlichen Beziehungen der kurländischen Städte funktionierten und inwieweit man über Kurland als einen separaten Teilraum der Städtelandschaft des mittelalterlichen Livlands reden kann.

Zwischen den livländischen Städten entwickelte sich mit der Zeit eine enge Kooperation. Auch wenn ihre Kontakte auf der institutionellen Ebene bis zur Mitte des 14. Jahrhunderts nicht belegt sind und wohl nur schwach vorhanden waren,[58] schafften seit den 1350er Jahren die Städtetage (Zusammenkünfte der Delegierten der livländischen Städte), die mehr oder weniger regelmäßig durchgeführt wurden, einen konstanten organisatorischen Rahmen für ihre Zusammenarbeit. Die Tätigkeit dieser Institution ist im Zusammenhang sowohl mit den hansischen Angelegenheiten und dem Hansetag als auch mit der livländischen Territorialpolitik zu betrachten. Die livländischen Städte traten innerhalb der hansischen Gemeinschaft als eine regionale Gruppe auf; häufig wurden auf diesen Tagfahrten die Positionen ihrer Kaufmannschaft zu verschiedenen Fragen der hansischen Politik abgestimmt. Außer der hansischen Thematik wurden zum Gegenstand der Beratungen auf den Städtetagen auch die Fragen der livländischen Innen- und Außenpolitik. Der Meinungsaustausch und die Koordination erleichterte nicht zuletzt das Auftreten der Städte auf dem Landtag – der

56 Niitemaa, Binnenhandel (wie Anm. 49), S. 50–54; Benninghoven, Ein Osnabrücker Fernhändlergeschlecht im Livlandhandel (wie Anm. 39), S. 157–189; Selart, Die kleinen Städte Livlands (wie Anm. 3), S. 32.

57 Vilnis Pāvulāns, Satiksmes ceļi Latvijā XII–XVII gs. (= Verkehrsstraßen in Lettland vom 12. bis zum 17. Jahrhundert), Riga 1971, S. 119 f., 123 f.

58 Ilgvars Misāns, Die späten Anfänge städtischer Zusammenarbeit in Alt-Livland, in: Ortwin Pelc / Gertrud Pickhan (Hg.), Zwischen Lübeck und Novgorod. Wirtschaft, Politik und Kultur im Ostseeraum vom frühen Mittelalter bis ins 20. Jahrhundert. Norbert Angermann zum 60. Geburtstag, Lüneburg 1996, S. 89–98.

allgemeinlivländischen Ständeversammlung, die in den 1420er Jahren ins Leben gerufen wurde. Nicht zufällig fand ein Teil der Städtetage vor, nach oder während der Landtage statt. Die Tagesordnung der livländischen Städtetage bestimmten Riga, Dorpat und Reval, welche die Versammlung völlig dominierten. Nur in einzelnen Fällen erschienen auch Ratssendeboten (abgeordnete Ratsherren) der kleineren Städte Pernau, Fellin, Wenden, Wolmar, Lemsal, Kokenhusen, Windau, Goldingen und Roop auf den Tagfahrten oder wurden mit Vollmacht vertreten. Obwohl sich die kleineren Städte bei den Begegnungen der städtischen Vertreter, wenn man den Quellen Glauben schenken kann, passiv verhalten haben, wurden auch sie durch die Beteiligung an der Arbeit des Städtetages in das Informationsnetz der hansischen Gemeinschaft einbezogen und mit der städtischen Sicht auf Ereignisse der lokalen Landespolitik bekannt gemacht. Gleichwohl wurden die kleineren Städte am Ende des 15. Jahrhunderts aus dem Städtetag verdrängt. Die Institution verengte sich zur Tagfahrt der Ratssendeboten Rigas, Dorpats und Revals, was sich aus der wirtschaftlichen Schwäche und Unbedeutendheit der Kleinstädte ergab. Trotzdem darf man die verbindende Rolle des Städtetages nicht zu niedrig einschätzen. In Livland, dass in mehrere Territorialherrschaften zersplittert war, hat er institutionell die Verbindungen zwischen den Städten gestaltet und den Zusammenhalt der livländischen Städtelandschaft begünstigt.[59]

3. Fazit

Als Resümee ist festzuhalten, dass in Livland die Intensität der Urbanisierung bescheiden und das städtische Leben verhältnismäßig schwach entwickelt war. Politische Zersplitterung des Landes, geringer Zuzug neuer Einwohner aus den deutschen Ländern sowie die demographischen und wirtschaftlichen Krisen haben die Entfaltung des Urbanisierungsprozesses angehalten.

---

59 Speziell mit der Geschichte des livländischen Städtetages habe ich mich in den folgenden Publikationen beschäftigt: Ilgvars Misāns, Der Städtetag. Eine hansische und territoriale Institution in Alt-Livland, in: Nordost-Archiv NF 7/1, 1998, S. 81–96, Ders., Zusammenarbeit und Konkurrenz: Riga, Dorpat und Reval auf den livländischen Städtetagen, in: Nils Jörn / Detlef Kattinger / Horst Wernicke (Hg.), Genossenschaftliche Strukturen in der Hanse (Quellen und Darstellungen zur hansischen Geschichte NF 47), Köln/Weimar/Wien 1999, S. 273–285; Ders., Der Städtetag als Instrument der hansischen Politik der livländischen Städte, in: Hansische Geschichtsblätter 119, 2001, S. 85–102. Bei den neueren Veröffentlichungen ist auf den Aufsatz von Anti Selart hinzuweisen: Selart, Die kleinen Städte Livlands (wie Anm. 3), hier bes.: S. 24 f.

Man kann sich die Städtelandschaft des mittelalterlichen Livlands als eine Struktur vorstellen, in der die drei überragenden Fernhandelszentren Riga, Dorpat und Reval jeweils in einzelnen Landesteilen wirtschaftlich dominierten. Jede Großstadt hatte eigene Prioritäten im Handel mit den benachbarten russischen beziehungsweise weißrussischen Städten, die gegenseitig akzeptiert wurden – für Reval hatte Nowgorod eine besondere Bedeutung, für Dorpat Pleskau und für Riga Polozk. Die Einflusssphären einzelner Städte innerhalb des Landes passten sich diesen Handelsrichtungen an. Als Portalstädte kontrollierten Riga, Dorpat und Reval die Kleinstädte und Marktflächen ihres Hinterlandes. Die Grenzen des Einflussbereichs der jeweiligen „Großstadt" waren dabei nicht streng gezogen und schlossen Überschneidungen nicht aus. Eine existierende politische Einteilung war für die Ausdehnung der Wirtschaftsräume einzelner Städte nicht maßgebend, obwohl im Falle Revals und Dorpats der wirtschaftliche Aktionsradius mit den entsprechenden Landesherrschaften übereinstimmte. Innerhalb der einzelnen, von den größeren Städten kontrollierten Wirtschaftsräume konnten sich auch kleinere Teilräume (beispielsweise Kurland und der Handelsbereich von Neu-Pernau) behaupten.

Eine wichtige koordinierende Rolle für die Politik der livländischen Städte spielte seit der Mitte des 14. Jahrhunderts der Städtetag, der sich auf der lokalen Ebene sowohl mit den Angelegenheiten der hansischen Gemeinschaft als auch mit den Fragen der Landespolitik beschäftigte. Obwohl Riga, Dorpat und Reval alles auf diesen Zusammenkünften bestimmten und die kleineren Städte nur die Rolle der Statisten innehatten, diente diese Institution als Bindeglied zwischen den Städten und übte für die Städtelandschaft eine integrierende Funktion aus.

Sabine Wüst

Europäische Genusslandschaften: Foodscapes – interregional

Abstract: The regional foodscape on southern German and other European territories can be investigated by a variety of approaches. For example, the Bavarian language is not the only identity feature of special food or regional agricultural products. As one product among many others, "Rouge Ranna" (*Roude Ranna*) or beetroot can also be found beyond the borders of Bavaria. Council decrees, council books, food laws, budget books, but also records from monasteries and hospitals as well as recipe books provide information about eating habits and food in the Middle Ages and early Modern Times. The Foodscape varied and developed over many years. Following the influx of immigrants, which began in the 1960s, the foodscape of Bavaria, Germany or even Europe has drastically changed by the end of the century. Influences of Mediterranean cuisine are taking an important place in the food scene. For today, bestselling author Brie Arthur summed it up in his book "The Foodscape Revolution: Finding a Better Way to Make Space for Food and Beauty in Your Garden" (Pittsburgh, PA, USA, 2017).

Die moderne Gesellschaft – wie ernährt sie sich? Wir bewegen uns auf der einen Seite in einer regelrechten „McDonaldisierung", wie es der US-amerikanische Soziologe George Ritzer treffend formulierte.[1] Demgegenüber steht ein richtiggehender Bio-Boom durch vermeintliche Ökoprodukte. Ende 2007 bewirtschafteten bereits 18.700 Bio-Bauern rund 865.000 Hektar Land nach ökologischem Landbau. Das waren rund fünf Prozent aller Äcker und Wiesen in Deutschland. Zehn Jahre später, Ende 2017, waren es dann knapp 45.000 Bauern, die rund 1,4 Millionen Hektar Fläche nach ökologischen Prinzipen bewirtschafteten. Der Anteil an der Gesamtfläche beträgt nun über acht Prozent. Bio ist auf dem Vormarsch, denn fast zehn Milliarden Euro gaben die Deutschen im vergangenen Jahr für ungespritztes Obst und Biofleisch aus.[2] Und das Wachstum hält

1 George Ritzer, Die McDonaldisierung der Gesellschaft, Lizenzausgabe, Frankfurt am Main 1997.

2 Die Begriffe „Bio" und „Öko" sind allgegenwärtig, doch was bedeuten sie eigentlich?: Almut Röhrl / Tobias Aufmkolk, Bio-Lebensmittel, veröffentlicht am: 22.7.2019, in: Planet Wissen (online), URL: https://www.planet-wissen.de/gesellschaft/lebensmittel/bio_lebensmittel/index.html (letzter Zugriff: 11.8.2019). Auch der „Merkur" stellt in einem Artikel von 2017 den Bio-Boom vor, denn nach einer aktuellen Studie der Gesellschaft für Konsumforschung (GfK) hat sich der Bioanteil beim Lebensmittel- und

an. Dadurch entsteht das Image einer gesunden versus einer ungesunden Foodscape.[3] Parallel zur Fastfood-Kultur kommt es zu einer Renaissance oder gar zur Neuerfindung regionaler Kulinarik.

Zunächst fragt sich jedoch an dieser Stelle sicherlich der geneigte Leser, was mit Foodscape gemeint sein könnte. Eine als Metapher ausgedrückte Landschaft von Nahrungsmitteln, dem Essen oder den verschiedenen Speisen, ihre jeweilige Herstellung beziehungsweise Produktion sowie die kulturellen Verknüpfungen. Am deutlichsten zeigt sich die Foodscape in der künstlerischen Umsetzung bei Carl Warner.[4] Vor rund zehn Jahren kam dem in London lebenden Fotografen die Idee, seine Leidenschaft für Stillleben- und Landschaftsfotografie miteinander zu verbinden. Auf den ersten Blick wirken sie wie ganz normale Landschaftsbilder. Aber bei Warner bestehen die Wolken aus Blumenkohl, der Wald aus Brokkoli und die Berge aus Brot. In seinen surrealen Lebensmittel-Landschaften ist alles essbar – ein besonderes Bilderlebnis. Die Schweizer „Urchuchi"[5] setzt nicht nur der 2012 erschienene Artikel „Kulinarisches Erbe der Schweiz. Topi-

---

Getränkeeinkauf in den vergangenen zehn Jahren nahezu verdoppelt – von knapp drei auf fast sechs Prozent. Art. „Bio-Boom: ‚Der erfolgreichste Trend überhaupt'", veröffentlicht am: 19.10.17, in: Merkur.de, URL: https://www.merkur.de/leben/genuss/bio-boom-erfolgreichste-trend-ueberhaupt-8787656.html (letzter Zugriff: 11.8.2019).

3 Norah MacKendrick, Foodscape, in: Contexts 13/3, 2014, S. 16–18, URL: https://journals.sagepub.com/doi/pdf/10.1177/1536504214545754 (letzter Zugriff: 29.8.2019). Weiterführende Literatur: Gunther Hirschfelder, Europäische Esskultur, Frankfurt am Main 2001; Hans Jürgen Teuteberg / Gerhard Neumann / Alois Wierlacher (Hg.), Essen und kulturelle Identität, Berlin 1997; Ulf Matthiesen, Esskultur und Regionale Entwicklung – unter besonderer Berücksichtigung von „Mark und Metropole", in: Beate Binder / Astrid Deuber-Mankowsky (Hg.), Die Botschaft der Botschaften (Berliner Blätter, Heft 34), Berlin 2004, S. 111–145.

4 Harry N. Abrams, Carl Warner's Food Landscapes, New York 2010. Das Buch beinhaltet eine Sammlung Warners früher Arbeiten seit dem Jahr 1998. Es gibt einen Einblick hinter die Kulissen von Warner und seinem Team und enthält Details darüber, was die einzelnen Foodscapes inspiriert hat und wie sie von den ersten Skizzen bis hin zu den in der Fotografie verwendeten Zutaten entstanden sind.

5 In einer Übersetzung würde man „Urküche" dazu sagen, was sich allerdings dahinter versteckt, ist in Martin Weiss, Urchuchi-Rezepte. Das Schweizer Kochbuch, Zürich 2009, nachzulesen. Sieben Jahre lang hat der Autor die Schweiz bis in die hintersten Bergtäler erforscht und regionale Rezepte gesammelt – bei Bauernfrauen und Spitzenköchen. Entstanden ist ein Standardwerk, das die bekanntesten Gerichte aus dem kulinarischen Erbe der Schweiz in einer einzigartigen Sammlung zusammenfasst. „Urchuchi" ist ein von Weiss kreierter Dialektbegriff, verstanden als bodennahe, ehrliche Küche mit Respekt vor der Natur und den Traditionen.

nambur, Pastinake, roter Mangold, Wollschwein, Evolener Rind, Engadinerschaf, Stiefelgeiss, Schweizerhuhn, Pfauenziege, Diepolzer Gans und vieles mehr. Mit zwei Rezepten des kürzlichen erschienenen Buches ‚Kochen mit ProSpecieRara: Blaue Schweden, Grüne Zebra, Roter Feurio' "[6] in den Fokus, sondern auch das Foto von Carl Warner: „Foods of Switzerland".

Abb. 57: *„Foods of Switzerland" von Carl Warner. Bildnachweis: URL: http://www.carlwarner.com/creative/foods-of-switzerland/.*

Der Erwerb von Lebensmitteln, die Zubereitung, die unterschiedlichsten Diskurse darüber und, obgleich auch mannigfaltige Sorten unterschiedlicher Nahrungsmittel in das Feld der Foodscape fallen, auch die Identitätsstiftung – das alles beschreibt das weite Gebiet einer „Essenslandschaft".[7]

Innerhalb der süddeutschen und europäischen Städtelandschaft existieren demzufolge unterschiedlichste Foodscapes, die es näher zu betrachten gilt. Alleine Altbayern, Schwaben und Franken divergieren innerhalb der jeweils regionaltypischen Genussvielfalt relativ stark. Weitet man das Blickfeld auf Europa aus, so wird rasch klar, wie facettenreich die Welt der Nahrungsmittel und Speisen ausfällt.

Die Auswahl der analysierten Regionalküchen beziehungsweise Städteküchen wurde anhand der süddeutschen und europäischen Städtelandschaft getroffen. Der These, dass es früher richtige Foodscapes als regionaltypische und

---

6 Den Artikel findet man online, URL: https://www.foodaktuell.ch/kulinarisches-erbe-der-schweiz/ (letzter Zugriff: 11.8.2019). Details in: Martin Weiss / Albi von Felten, Blaue Schweden, Grüne Zebra, Roter Feurio. Alte Sorten neu entdeckt. Das ProSpecieRara Kochbuch, Aarau/München 2012.

7 Da Foodscape keine Beveragescape meint, ist in diesem Beitrag ausschließlich vom Essen die Rede und sämtliche Getränke, wie zum Beispiel der Frankenwein, welcher für eine regionale Identität spräche, bleiben außen vor.

traditionelle Landschaften gab, die heute im Zuge der Globalisierung ihr Gleichgewicht verlieren, gehen wir im Folgenden auf den Grund.

1. Das Knoblauchsland oder die gesunde Vorspeise

Kleinere Bauernhöfe standen innerhalb der Stadt – größere Gehöfte befanden sich außerhalb der Stadtmauern, wie unter anderem auf einer Nürnberger Postkarte von 1900 zu lesen war:

> *„Auf dem Marktplatz ist es schön,*
> *Wo bei Veilchen Zwiebeln stehn,*
> *Wo die Düfte weisser Rosen*
> *Und vom Topfkäs dich umkosen,*
> *Wo Madame fein im Glanz*
> *Stiehlt dem Bauer eine Gans,*
> *Wo sich Altdorfer Bauernweib*
> *Schnurrbart dreht zum Zeitvertreib.*
> *Wo der Fremde sprachlos höret,*
> *Wie man bayrisch schimpft u. schwöret.*
> *Hier bei Zwiebel, Obst, Gemüse,*
> *Denk ich Dein u. schick Dir Grüsse.“*

Abb. 58: *Postkarte „Gruss aus Nürnberg“, 1900, mit Hauptmarkt und Marktfrauen. Lithographie: Karl Büttner Buchhandlung, Anstalt Rosenblatt Frankfurt. Bildnachweis: Historische Ansichten der Stadt Nürnberg und Umgebung aus dem Archiv der Altstadtfreunde e.V. in Nürnberg, ausgewählt und bearbeitet von Richard Svandrlik, URL: https://www.flickr.com/photos/altstadtfreunde-nuernberg/31680170760/in/album-72157675380862032/.*

Abb. 59: *Postkarte zur Leipziger Markthalle, um 1900. Bildnachweis: Stadtgeschichtliches Museum Leipzig, Inv.-Nr. PK 2325; Wikimedia Commons (gemeinfrei), URL: https://upload.wikimedia.org/wikipedia/de/d/d2/Leipzig_Markthalle_Postkarte.jpg.*

Abb. 60: *Postkarte „Gruss aus Düsseldorf", um 1900, mit Marktplatz und Rathaus. Bildnachweis: Privat; oldthing, URL: https://oldthing.de/Duesseldorf-Marktplatz-Rathaus-Marktfrau-Gedicht-Kat-Duesseldorf-0023114527.*

Abb. 61: *Postkarte „Gruss aus Wesel" (am Niederrhein), um 1900, mit Großem Markt. Bildnachweis: Privat; akpool GmbH, URL: https://static5.akpool.de/images/cards/198/1984661.jpg.*

Schon dieser Reim legt ein Augenmerk auf die heimische Zwiebel. Knoblauch- beziehungsweise Zwiebelzucht hatte diesem Gebiet zu seinem Namen verholfen. Das Wachstafelzinsbuch des Burggrafentums Nürnberg von 1425[8] gibt eine Antwort darauf, warum das Knoblauchsland nach der Knolle benannt wurde. Eine Erklärung zum „*Knoblochsland*" finden wir beim Nürnberger Stadtschreiber Johannes Müllner (1565–1634): „*Und diese bishero erzählte Flecken und Dörfer liegen alle zwischen den Wald und der Rednitz gegen der linken Hand, welcher Tractus insgemein das Knoblochsland genennet wird, aus Ursach, daß viel Zwiefel und Rubsamen dies Orts gebauet und von hinnen in frembde Land, als Welschlandt, Franckreich, Spanien verführt und damit große Hantierung getrieben wird* […]."[9]

---

8 Siehe dazu ausführlicher: Gustav Voit / Werner Schultheiß / Werner Sprung (Bearb.), Das Wachstafelbuch der Reichsveste zu Nürnberg von etwa 1425 und das Reichslehenbuch der Herren von Berg aus dem Jahre 1396, Nürnberg 1967.

9 Gerhard Hirschmann (Hg.), Johannes Müllner, Die Annalen der Reichsstadt Nürnberg von 1623, Teil I: Von den Anfängen bis 1350, Nürnberg 1972, S. 324. Zu dieser Erklärung siehe auch: Johannes Müllner, Kurzgefaßte Reformations-Geschichte der freyen Reichs-Stadt Nürnberg, oder Johann Müllners, ehemaligen Raths Syndici, Bericht von

1562/63 setzte der Kartograf Georg Nöttelein (1525–1567) eine wenige Jahre zuvor vom Mathematiker und Geografen Johann Schöner (1477–1547) handgezeichnete Karte der Nürnberger Wälder Sebaldi und Lorenzi als Kupferstich um. Die Exaktheit ist bei einer großmaßstäbigen Darstellung kaum zu überbieten.

Abb. 62: *Jörg Nöttelein, Große Wald- und Fraißkarte von Nürnberg, s.l. [vermutlich Nürnberg] 1562/63. Bildnachweis: BSB, Mapp. XI,302 ab, URN: urn:nbn:de:bvb:12-bsb00014430-6.*

Aus dem Reim der Postkarte von 1900 wird deutlich, dass das Knoblauchsland[10] in intensiven Wechselbeziehungen zu den angrenzenden städtischen Räumen stand. Meist ist die Rede von Bayerns größtem Gemüsegarten. Diese

Aenderung der Religion und Abschaffung des Papstthums in der Stadt Nürnberg, und was deswegen vorgegangen, Nürnberg 1770, S. 67.

10 In Auswahl: Hermann Rusam, Knoblauchsland. Romantik einer altfränkischen Gemüsebauernlandschaft inmitten des Städtedreiecks Nürnberg, Fürth, Erlangen, Nürnberg 1989; Bertold Frhr. Haller von Hallerstein, Art. „Knoblauchsland", in: Christoph Friedrich / Bertold Frhr. Haller von Hallerstein / Andreas Jakob (Hg.), Erlanger Stadtlexikon, Nürnberg 2002, S. 426.

traditionelle Region innerhalb des Städtedreiecks Reichsstadt Nürnberg, Planstadt Erlangen und Fürth – der Ort hatte lange Zeit durch die Nähe zu Nürnberg ein bäuerlich geprägtes Gemeinwesen – diente der besonderen Lage zufolge in weiten Teilen der Versorgung der Stadtbevölkerung. Die Stadt mit der Kaiserburg behielt durch ihre Kirchenstiftungen und Geschlechter die Macht über das Agrarland. So standen die Bauern des Knoblauchslandes in grundherrschaftlichen Verhältnissen. Schon 1427 verkauften die Burggrafen einen Großteil ihrer Rechte an Nürnberg. Die Stadt wiederum schützte die agrarischen Bereiche in Kriegszeiten und ermöglichte den Landwirten und Einwohnern Zuflucht. Um den Bedarf der Stadtbevölkerung zu decken, waren die Anbauflächen von Natur aus nicht genug nährstoffreich. Daher wurden durch die jahrhundertelange Einbringung von Biomasse (Laub- und Nadelstreu) die Böden aufgebessert.[11] Die Städte und das dörfliche Umland hatten also einen reziproken Nutzen voneinander. Den Städtern verschaffte das Umland eine Nahrungsmittelgrundlage – den Erzeugern brachte es in Krisenzeiten Sicherheit im Zeichen der Kaiserburg. Auch die zahlreichen Herrensitze der Patrizier,[12] wie etwa das ehemalige Wasserschloss der Kreß von Kressenstein in Neunhof, profitierten von dem regionalen Anbau. Daran wird unmissverständlich klar, wie wertvoll eine ertragreiche Fläche für die Entwicklung der regionalen Städtelandschaft ist. Heute besteht das Knoblauchsland aus einer Anbaufläche von etwa 1.900 Hektar Freiland sowie ungefähr 95 Hektar Unterglasanbau.[13] Um eine typische Region ausmachen zu können, stellt sich allerdings die Frage, was denn angebaut wurde?

Bereits Hans Sachs (1494–1576) schrieb 1560: „*Bayerland hat die freyheit, isst kraut mit Löffeln alle zeit, all tag zwey kraut, mach ein jahr siben hundert kraut, darzu dreißig.*“[14]

---

11 Siehe dazu auch: Bayerisches Landesamt für Umwelt (Hg.), Entwurf einer kulturlandschaftlichen Gliederung Bayerns als Beitrag zur Biodiversität, Nr. 21: Ballungsraum Nürnberg – Fürth – Erlangen, Stand: 2011, URL: https://www.lfu.bayern.de/natur/kulturlandschaft/gliederung/doc/21.pdf (letzter Zugriff: 30.8.2019).

12 Siehe dazu u.a.: Peter Fleischmann, Rat und Patriziat in Nürnberg. Die Herrschaft der Ratsgeschlechter in der Reichsstadt Nürnberg vom 13. bis zum 18. Jahrhundert, 3 Bde. (Nürnberger Forschungen 31/1–3), Nürnberg 2008; Michael Diefenbacher, Stadt und Adel – Das Beispiel Nürnberg, in: Zeitschrift für die Geschichte des Oberrheins 141, 1993, S. 51–69.

13 Art. „Knoblauchsland Franken“, in: Unsere Bayerischen Bauern, URL: https://www.unsere-bauern.de/knoblauchsland/ (letzter Zugriff: 27.8.2019).

14 Zu finden im Beitrag „Kraut“, in: Johann Andreas Schmeller, Bayerisches Wörterbuch, Bd. 1 (7. Nachdruck der von G. Karl Frommann bearbeiteten 2. Ausgabe, München 1872/75, mit einer wissenschaftlichen Einleitung von Otto Maußer, Leipzig 1939), München 2008, Sp. 1385.

Der ehemalige Nürnberger Archivdirektor Ernst Mummenhoff[15] (1848–1931) und der Gymnasiallehrer Dr. Alfred Koeberlin[16] (1861–1902) haben sich eingehend mit den Quellen zum Knoblauchsland befasst. Diese beiden Historiker untersuchten handschriftliche Archivalien ebenso wie gedruckte Quellen und Literatur. Dabei beschrieben sie auch die Entwicklungsgeschichte der Nutzpflanzen:[17]

Jahreszahl der Erstnennung	**Name der Nutzpflanze**
911	Wein
1109	Hirse
es folgt	Roggen, Gerste, *Heidel*, Hafer, Kichererbse, Lein, Hanf
1262	Mohn
1350	Zwiebeln und Kraut
1434	Melonen
1437	Weichseln und Amarellen
1438	Gelbe Rübensamen, *Pfebensamen*
1440–1464	*Meerrettig* (Bayersdorf)
1536	Süssholz (bei Bamberg)
1557	Spargel

15 Ernst Mummenhoff, Geschichtliches über die landwirtschaftlichen Verhältnisse der Umgegend von Nürnberg, in: Festschrift gewidmet den Teilnehmern an der 32. Wanderversammlung Bayerischer Landwirthe, Nürnberg 1895, S. 1–58.

16 Alfred Koeberlin, Zur historischen Gestaltung des Landschaftsbildes um Bamberg. Wissenschaftliche Beilage zu dem Jahresbericht des neuen Gymnasiums in Bamberg, Bamberg 1893.

17 Vgl. dazu auch: Abhandlungen der Naturhistorischen Gesellschaft zu Nürnberg, 10. Bd., mit einer Beilage der Flora der Umgegend von Nürnberg-Erlangen, Nürnberg 1897, S. 189 f.: „*§ 39. Nachtrag zu Kapitel III: Die floristische Litteratur über die Gefässpflanzen und Fortsetzng von 1892 bis 1897 unter Erweiterung auf die Litteratur über Nutzpflanzenanbau im Gebiete.*“ Die Untersuchungen betreffen folgende Gebiete: Nürnberg, Spitalhof, Schäfhof, Weigelshof, Schoppershof, Fürth, Burgfarrnbach, Cadolzburg, Langenzenn, Laubendorf, Herzogenaurach, Höchstadt an der Aisch, Schwabach, Pyras, Mörsdorf, Sulzbürg, Niederrorenstadt, Altdorf, Ochenbruck, Schwarzenbruck, Feucht, Egensbach, Hersbruck, Heroldsberg, Eltersdorf, Uttenreuth, Marloffstein, Hausen, Forchheim, Serlbach, Ebermannstadt, Gößweinstein, Pottenstein, Schnaid bei Hirschaid, Bamberg und Umgebung, Erlau, Burgebrach, Scheßlitz, Rattelsdorf, Banz, Weismain und Kloster Langheim.

Jahreszahl der Erstnennung	**Name der Nutzpflanze**
1563	Wermut, Dill, *Spik*, Basilicum, Lavendel, Majoran, Rosmarin, Kamillen und v.a. Gewürzpflanzen
1624	*Kardobenediktenkraut*
1695	Runkelrüben und *Skorzoneren*
1735	Kartoffel
1778	Mais
1790	*Esparsette*, Luzerne und Tobinampur

Ob sich diese Erstnennungen auf den agrarischen Anbau im großen Stile oder lediglich auf eine erste Erwähnung des jeweiligen Produkts beziehen, bleibt ein Desiderat.

Historische Rezepte finden sich ebenfalls im Nürnberger Kochbuch von 1691, wenn es heißt: „*Ein gefülltes weisses Kraut/ noch auf eine andere Weise*" oder „*Einen Spinat oder grünes Kraut zu kochen/ auf andere Art*".[18] In der obigen Tabelle lässt sich die regionale Vielfalt der Nutzpflanzen für die besagte Städtelandschaft gut erkennen.

Eine heute nahezu ausgestorbene Gemüsesorte, die „Bayerische Rübe", taucht erstmals 1276 im Augsburger Stadtbuch auf.[19] Hier wird sie jedoch nicht explizit als solche bezeichnet: „*Rübe unde obez. Bringet ein uzman Rübe her.*"[20] Ein Ausmann beziehungsweise ein Fremder, wohl ein Bauer, bringt hier Speiserüben, vielleicht die Bayerische Rübe, nach Augsburg. Johann Fischart vergleicht „*die bayrischen Ruͤblein, die Koͤstlichkeit halber den galilaͤischen Feigen*".[21] Außerdem berichtet ein Reisender: „*Ein eygentlicher bericht eines gewanderten gesellen, der*

18 Susanne Maria Endter, [Vollständiges Nürnbergisches Koch-Buch] Der aus dem Parnasso ehmals entlauffenen vortrefflichen Köchin/ Welche bey denen Göttinnen Ceres, Diana und Pomona viel Jahre gedienet/ Hinterlassene und bißhero/ Bey unterschiedlichen der Löbl. Koch-Kunst beflissenen Frauen zu Nürnberg/ zerstreuet […], Nürnberg 1691, u.a.: Nr. 175, 180, 185, S. 572–580. Digitalisat der Wolfenbütteler Digitalen Bibliothek, URL: http://diglib.hab.de/drucke/6-oec/start.htm (letzter Zugriff: 28.8.2019).

19 Siehe dazu u.a.: Peter Johanek, Augsburger Stadtbuch, in: Burghart Wachinger (Hg.), Die deutsche Literatur des Mittelalters. Verfasserlexikon, 11. Bd., Berlin u.a. ^{2}2002, Sp. 181–185.

20 Siehe: Johann Andreas Schmeller, Bayerisches Wörterbuch, Bd. 2/1, München 1985, S. 11: „Die Rueb." (Übersetzung: Dr. J. Denz, Kommission für Mundartforschung, Bayerische Akademie der Wissenschaften, München).

21 Ebd.

ein geborner Frantzoß war und wol käß kennen kundt. Es waren Parmasan-käß, die man in wälschen landen (214) zumachen pflegt und für die besten helt, einem handelßman gen Augspurg vonn Potzen auß der graffschafft Tyrol zugeschickt, die so gute sein, das man sie nicht schneyden noch brechen darf, sonder schaben, wie die bayerischen rüblein, die so kostlich sein, daß sie den galileischen feygen verglichen werddn. Von denselbigen käsen redeten wolgewanderte gesellen, undter welchen auch war der geborne Frantzoß […].“[22] Damit wurde nun die Wertigkeit der Bayerischen Rübe deutlich. Diese wird oftmals auch in Beziehung zur Herbstrübe oder Mairübe gesetzt.

Eine regionale Rezeptur zur Bayerischen Rübe findet man im Nürnberger Kochbuch von 1691: „*50. Ein Hecht in Bayerischen= oder Pfetter=*[23]*blein. Schabe die Ruͤblein/ schneide Plaͤtzlein daraus/ und wasche sie schoͤn; darnach thue ein klein wenig Schmalz in eine Pfannen/ schuͤtte die Ruͤblein darein/ und laß so lang schweissen biß sie etwas weich werden; alsdann thue sie in einen Hafen/ gieß warme Fleisch= Bruͤhe daran/ pfeffers/ und laß sieden biß sie recht seyn; thue zuletzt ein wenig Butter mit hinein/ den Hecht aber siede und blaͤttere/ wie oben gemeldet/ lege dann auch eine Lag Ruͤblein in eine Schuͤssel/ mache wieder eine andere Lage von dem Hecht: als= dann Muscatbluͤhe und Butter/ und so immer fort/ gieß ein wenig Fleischbruͤhe daran/ und laß es auf einer Kohlen aufsieden.*“[24]

Noch bis Ende des 19. Jahrhunderts durfte in keinem Kochbuch ein Rezept mit dieser Rüben-Art fehlen. Ab dem Jahr 2001 wurde sie – nach langen Jahren einer Ruhepause – wiederentdeckt und von H. Haseloff durch eine Diplomarbeit untersucht.[25]

---

22 Michael Lindener, Rastbüchlein und Katzipori, hg. v. Franz Lichtenstein, Tübingen 1833, Nr. 77, S. 131, URL: https://archive.org/details/michaellindener02lindgoog/page/n138 (letzter Zugriff: 28.8.2019).

23 Auch als „Regensburger“ oder „Fett-Rübe“ bezeichnet, siehe: M. D. Funk, Neues Fränkisches Kochbuch oder deutliche und bewährte Anweisung zur vortheilhaftesten und schmackhaften Zubereitung der Speisen und Getränke, des Backwerks und der Konfituren, wie auch Früchte einzumachen und zu trocknen, verschiedene Essige zu verfertigen, Pöckelfleisch zu machen, Schinken einzusalzen und zu sieden, Fleisch zu räuchern u.s.w., nebst vielen andern zur Koch- und Haushaltungskunst nützlichen Vorschriften, Ansbach ^{2}1816, Nr. 180, S. 108.

24 Endter, Vollständiges Nürnbergisches Koch-Buch (wie Anm. 18), Nr. 50, S. 124: „*Ein Hecht in Bayerischen- oder Pfetter-Rüblein*“.

25 H. Haseloff, Die Bayerische Rübe (Brassica rapa L. spp. rapa) – eine fast ausgestorbene Speiserübe, dargestellt nach alten Literaturquellen, Diplomarbeit, München/Weihenstephan 2002.

Heute hat sich das Anbauspektrum deutlich erweitert: Unter anderem findet man im Knoblauchsland Blumen, wie Sonnenblumen oder Gladiolen zum Selbstschneiden, Kartoffeln, Kohl, Kohlrabi, Lauch, Radieschen, Rettich, Spargel, Spinat, Tabak, Weizen und Wirsing sowie Zwiebeln, Zucchini, Auberginen, Tomaten und viele Salatsorten.

Wie ernährte man sich im spätmittelalterlichen Nürnberg? Fleisch stand auf der Beliebtheitsskala über dem Gemüse.[26] Das von Conradus Celtis handschriftliche Kochbuch „*Kuchemaistrey*"[27] (ca. 1490) beinhaltet dennoch mannigfache Rezepte, die Obst oder Gemüse als Ingredienzien benötigen. Deshalb ist anzunehmen, dass die Bevölkerung der Reichsstadt diese Produkte auch in größeren Mengen benötigte. So lesen wir zum Beispiel für die Stadt Augsburg: „*Gemüse aller Art ist auf dem Gemüsemarkt am Obstmarkte durch Landleute verkäuflich. Die hießigen Gärtner verkaufen dieses gleichfalls an den Markttagen sowohl, als ausser der Marktzeit, in den sogenannten geschlossenen Gärtnerständen, deren es in verschiedenen Stadttheilen giebt; sie geben Kräutelwaare aber auch in ihren Gärten ab.*"[28] In Nürnberg übernahm der Hauptmarkt die Funktion des Milchmarkts und diente als „Grüner Markt" wesentlich dem Verkauf landwirtschaftlicher Produkte.[29] Bauern aus dem näheren Umland – speziell dem Knoblauchsland – versorgten die Reichsstadt.

Bei den zur Burggrafenburg Nürnberg gehörenden Konstruktionen handelt es sich um den Fünfeckturm, das bereits durch das salische Zeitalter geprägte Burgamtmannshaus und die Reste des alten Vestnertores.[30] Bei Ausgrabungen

---

26 August Jegel, Ernährungsfürsorge des Altnürnberger Rates, in: Mitteilungen des Vereins für Geschichte der Stadt Nürnberg (künftig: MVGN) 37, 1940, S. 73–199.

27 Conradus Celtis, Kuchemaistrey, Nürnberg 1490. Digitalisat der Wolfenbütteler Digitalen Bibliothek, URL: http://diglib.hab.de/inkunabeln/276-quod-2/start.htm (letzter Zugriff: 28.8.2019). Obst- und Gemüserezepte: „*Byrenmuß*", „*Bramber wein*", „*Cumpost bald gut zu machen*", „*Duͤrbyren dienen wol in fuͤlle*", „*Duͤr ruben*", „*Erbesen da von ein hofflich essen*", „*Frucht vber iar zu behalten*", „*Holdermuß*", „*Hyrsen lumelbraten*", „*Knoblauch salssen mach also*", „*Kraͤuter zubehalten vber iar zu salssen*", „*Kraut essen mach also*", „*Maulbermuß*", „*Obß wie man es lang frisch behalten sol*", „*Opffel muß*", „*Opffel gebachen mit seltzsamen farben*", „*Prambermuß*", „*Schwartz gemuͤß*", „*Salssen von bomerantzen*", „*Welsch nuß oder haßelnuß vber iar frisch*", „*Weinber vber iar*".

28 Neuestes Taschenbuch von Augsburg. Oder: Topographisch-statistische Beschreibung der Stadt und ihrer Merkwürdigkeiten, mit Beziehung auf die ältern geschichtlichen Ereignisse. Ein Handbuch für Fremde und Einheimische, Augsburg 1830, S. 258.

29 Zur Quellenrecherche: Stadtarchiv Nürnberg, A 4, F 2.

30 Martin Nadler, Die Ausgrabung 1986 im Burgamtmannsgebäude der Nürnberger Burg, in: MVGN 75, 1988, S. 227–240.

und genaueren archäologischen Untersuchungen fand man Erbsen,[31] Linsen und Schwarzwurzel. Mineralisierte Pflanzenreste waren zudem aus der Latrinenprobe aus der Irrerstraße zu erwarten. Hier waren die Funde jedoch auf Rote Rüben, Melonen oder Gurken beschränkt.[32] Bei Ausgrabungen in der Gaststätte „Zum Rudolf" gab es weitere Entdeckungen: Äpfel, Birnen, Süßkirschen, Pflaumen, Erdbeeren, Hagebutten – diese sprechen für das wohlbekannte typisch fränkische Hiffenmark im Krapfen –, Himbeeren, Brombeeren, Heidelbeeren sowie Schlehen. Heidelbeeren, Schlehen und Pflaumen fand man bereits im Burgamtmannshaus.[33] Beeren wie Heidel-, Johannis-, Erd- oder Himbeeren gehören gegenwärtig zum Knoblauchsland und können an manchen Stellen sogar selbst geerntet werden.[34] Die Konsekration von Feigen und Weinkernen sprach für die Exklusivität dieses Wirtshauses.[35] Darüber hinaus belegte das Burgamtmannshaus auch Haselnüsse. Weitere nennenswerte Gemüsesorten der fränkischen Reichsstadt waren Kohl, Kraut, Petersilie, Rettiche, Rüben, Salat und Zwiebeln, wobei dem aus Kohl zubereiteten und lange haltbaren Sauerkraut eine besondere Rolle zugeschrieben wird.[36] Wildwuchs steht bei den ganzen genannten Sorten der Kultivierung der Pflanzen gegenüber. Im jüdischen Viertel Nürnbergs entstand ein Obstmarkt im Nordosten des Areals und der spätere Hauptmarkt, der zunächst „Grüner Markt" beziehungsweise „Großer Markt" hieß. Ferner gab es *inter alia* einen Mehlmarkt, einen Heringsmarkt, einen Kälbermarkt, einen Weintraubenmarkt, einen Milchmarkt, einen Salzmarkt sowie einen Gänsemarkt.

Für den ehemaligen Landkreis Scheinfeld wird die Regionalität noch dramatischer festgehalten. Ein Hofgärtner sollte auf Schloss Schwarzenberg angestellt

31 Die Erbse galt als Hauptnahrungsmittel der ärmeren Bevölkerung im mittelalterlichen Nürnberg. Vgl. Else Kraft, Fränkische Kochkunst im späten Mittelalter (15. Jahrhundert), in: MVGN 44, 1953, S. 231–237.

32 Hansjörg Küster, Kultur und Nutzpflanzen in Konstanz, in: Marianne Flüeler / Niklaus Flüeler (Hg.), Stadtluft, Hirsebrei und Bettelmönch: die Stadt um 1300, Katalog zur Ausstellung „Stadtluft, Hirsebrei und Bettelmönch – die Stadt um 1300" des Landes Baden-Württemberg und der Stadt Zürich, Zürich, im Hof des Schweizerischen Landesmuseums, 26. Juni bis 11. Oktober 1992, Stuttgart u.a. 1993, S. 292 f.

33 Nadler, Die Ausgrabung 1986 im Burgamtmannsgebäude (wie Anm. 30).

34 Siehe dazu: Art. „Pflücken ohne Bücken", in: Beerenland, URL: https://cafebeerenland.de/unsere-beeren/ (letzter Zugriff: 28.8.2019).

35 Claudia Frieser, Zwei spätmittelalterliche Wirtshäuser in Nürnberg. Kleinfunde aus der Irrerstraße. Mit einem Beitrag von Franz Herig (Arbeiten zur Archäologie Süddeutschlands), Büchenbach 1999.

36 Jegel, Ernährungsfürsorge des Altnürnberger Rates (wie Anm. 26).

werden, weshalb die Besoldung näher thematisiert wurde: „*Dreÿ Sümmern Korn Nürnberger Maßes, Ein halb Malter Dinckhel, zweÿ Metz Erbsen, zweÿ Metz Linsen, sechs Eimer Bier, davon halb Nachbier, dreÿ Schwein jedes von einem Centner, zweÿ rauhe wildt seüe, zwölf Mass Schmalz* […].“ Das war ein Halbjahreslohn „*Von Walburgis bis Michaelis*“.[37] Da es sich bei einer Besoldung um ständig anfallende Kosten handelt, rückten bei Naturalien besonders regionale Produkte in den Fokus. Noch deutlicher wird die Heimatverbundenheit aber, als dem Hofgärtner nun noch folgende Aufgabe zuteilwurde: „*gute Obstarten, ohne solche mit größeren Kosten im Auslande holen zu müßen, verschaffen zu können.*“ Es sollten also unter Einbezug der Untertanen und Bauern Obstkulturen landwirtschaftlich angebaut werden, wozu der Hofgärtner beratend tätig wurde, „*daß die Unterthanen, welche hochstämige Aepfel und Biernbäum von guter und auserleßener Obstart wünschen, sich alljährlich im Herbst und frühjahr beÿ dem hiesigen Schloßgärtner Persch zu melden hätten*“.[38]

Nach dieser fränkischen Vorspeise geht es im Folgenden um die Wurst.

2. Historische Wurstküchen

Reisen wir zuerst in das westliche Nachbarland Frankreich. Die Lyoner – jeder kennt sie – ob als Wurstsalat, Fleischsalat et cetera. *Lugdunum*, wie Lyon[39] lateinisch auch heißt, ist seit langer Zeit eine Hauptstadt für Wurstwaren. Lyon besitzt unzählig viele Spezialitäten: Hartwurst, Rosette, Jésus, Fleischwurst und so weiter. Doch die Lyoner blieb nicht vor Ort, sondern wanderte auch nach Deutschland und in die Schweiz.

---

37 Siehe dazu die Akten der Kanzlei Schwarzenberg vom 1. März 1646: Staatsarchiv Nürnberg (künftig: StAN), Herrschaft Schwarzenberg, Registratur, 1431/11.

38 Siehe dazu die Akten der Regierungskanzlei Schwarzenberg vom 7. November 1794: StAN, Herrschaft Schwarzenberg, Registratur, 1431/11.

39 Zur Stadtgeschichte von Lyon: Alain Ferdière, Gallia Lugdunensis. Eine römische Provinz im Herzen Frankreichs (Antike Welt. Sonderband. Zaberns Bildbände zur Archäologie), Mainz 2011; zur Urbanisierung der Peripherie: Susanne Rau, The Urbanization of the Periphery: A Spatio-Temporal History of Lyon since the Eighteenth Century, in: Historical Social Research 38/3, 2013, S. 150–175; Dies., Räume der Stadt. Eine Geschichte Lyons (1300–1800), Frankfurt am Main 2014, S. 128–160; Dies., Imaginierte Räume. Zur Geschichte nicht-realisierter Stadtplanungsprojekte, in: Informationen zur modernen Stadtgeschichte 2, 2015, S. 129–147; speziell zur Wurstgeschichte: Erich Lissner, Wurstologia oder Es geht um die Wurst. Eine Monographie über die Wurst, Frankfurt am Main 1939.

Ursprünglich kommt diese Wurstsorte, wie schon ihr Name vermuten lässt, aus der Stadt Lyon, wo sie allerdings als *Cervelas*[40] bezeichnet wird. Hergestellt wird diese Wurst aus gepökeltem Schweinefleisch, teils auch mit Rindfleisch vermischt, und Rückenspeck. Dazu kommen verschiedene Gewürze, wie Pfeffer, Kardamom, Kurkuma, Muskat, Koriander, Knoblauch und Ingwer. Bereits im Kochbuch der Augsburgerin Sabina Welser von 1553 steht ein Rezept: „*Wie man zerwùlawirstlach machen soll. Erstlich nempt 4 pfúnd schweinflesch vom zepfflin/ vnnd 2 pfúnd speck, das last klainhacken vnnd thiet 6 lott saltz darain/ ain pfúnd geriben kesß, 3 lott pffeffer, 3 lott jmber, wen es gehackt jst, so knetten das als darein, rerlach*[41] *3 lott, ain 1/2 lott negellach, ain halb lott múscatnúsß, zwaý lott zúcker, die derm músß man saúbermachen vnd nachmals gilben, darf man nit gar ain 1/2 lott saffera, man músß sý binden aúff baiden seitten, aúch vnngeferlich ain qúertlin frisch wasser darangiessen, man músß aúch das saltz, jmber, pfeffer nit gar darainthon, soll es vor versúchen vnnd darnach machen, man soll sý sieden vnngefarlich als 2 air, das gewirtz vnnd saltz músß man darein ton nach aines gúten gedoncken, man músß zuúor versúchen.*“[42] Daran lässt sich erkennen, dass es sich um eine Luxuswurst handelte, denn man verwendete nur beste – teils exotische – Ingredienzien. Auch auf italienischem Terrain lassen sich Nachweise für diese Wurstspezialität finden. So liest man beim wohl bedeutendsten Gastrosophen des 15. Jahrhunderts und zudem dem ersten prominent gewordenen Koch der Welt, Martino de' Rossi (1430–1470): „*Per fare mortadelle di carne de vitello. Piglia de la carne magra de la cossa et battila con un pocho di lardo o di bon grasso de vitello como se batte la carne dei pastelli. Dapoi togli petrosello et maiorana battuta ben menuta et un rosso d'ova con un pocho di caso grattato più et mancho secundo la quantità che vol fare et spetie et zafrano; et mescola tutte queste cose con la ditta carne; et dapoi togli di rete de porcho o di castrone, o d'altra bestia pur che sia bona et lega molto bene la ditta mescolanza in questa tal rete, facendone pezzi di grossezza d'uno ovo o simile;*

40 Mehrere Schreibweisen sind möglich: *Cervelas* oder *Cervelat*, nicht zu verwechseln mit der in Deutschland bekannten schnittfesten Rohwurst, der Zervelatwurst. Während in der welschen Schweiz die *Cervelas* verwendet wird, ist die Deutschschweizer *Cervelat* gemäß „Le Robert. Dictionnaire Historique“ älter. Erstmals erscheint diese Schreibweise 1552 bei François Rabelais. Sie geht auf das milanesische Dialektwort *zervelada* und das daraus entstandene italienische *cervelato* zurück. Siehe dazu auch: Le Matin Dimanche, 27. Januar 2008, S. 25.

41 Editionsfehler: „*berlach*“ (Bärlauch).

42 Hugo Stopp (Hg.), Das Kochbuch der Sabina Welserin, mit einer Übersetzung von Ulrike Gießmann (Germanische Bibliothek NF 4, Texte), Heidelberg 1980, in der Online-Ausgabe unter URL: http://www.staff.uni-giessen.de/gloning/tx/sawe.htm (letzter Zugriff: 11.8.2019), hier: S. 30, Rezeptnummer 24.

et ponila ad rostire nel speto et che si cocha ad agio et che non sia troppo cotta."[43] Er beschreibt zwar die Herstellung der italienischen Mortadella, doch weitere Belege für die *Cervelat* findet man bei François Rabelais im Romanzyklus „*Gargantua und Pantagruel*", dessen fünf Bände 1532, 1534, 1545, 1552 und 1564 erschienen.[44] Das von Rabelais verwendete Vokabular bezieht sich ausdrücklich auf die Küche. Einer Parodie gleich, lässt uns der Autor am Trojanischen Krieg teilhaben. Er verwendet gezielt einen ganzen Katalog von Wurstsorten in seinem Werk. Wenn die Figur des Turners einer Andouille (auch: Andouillette)[45] gegenübersteht, handelt es sich um „*un gros cervelas*", das in zwei Stücke geschnitten werden muss.[46]

Während bei uns in Süddeutschland eher die kalte Variante der Wurstsorte auf den Tisch kommt, findet man auf französischen Speisekarten das warme Gericht: *Cervelas chaud à la beaujolaise.*[47] Auffällig an der ganzen

---

43 Auch genannt: „*Maestro Martino*" oder „*Martino de Rubeis*", Chef im römischen Palazzo des päpstlichen Kaplans (Camerlengo), dem Patriarchen von Aquileia. Die Bewunderung seiner Zeitgenossen zeigt sich auch in seiner Grabinschrift: „Prinz der Köche". Hier: Maestro Martino, Libro de arte coquinaria (sec. XV), Arte della cucina. Libri di ricette, testi sopra lo scalco, i trinciante e i vini. Dal XIV al XIX secolo. A cura di Emilio Faccioli, Bd. 1, Mailand 1966, S. 115–204, hier: S. 128, URL: http://www.staff.uni-giessen.de/gloning/tx/martino2.htm (letzter Zugriff:11.8.2019). Übersetzung des Zitats: „Wie Sie Kalbsmortadella zubereiten: Nehmen Sie mageres Fleisch ‚de la Cossa' und geben Sie ein wenig Schmalz oder Kalbsfett hinzu, anschließend das Ganze gut durchmischen. Geben Sie dann fein gehackte Petersilie und Majoran dazu; ferner ein rohes Ei und geben Sie nach und nach geriebenen Käse, den Rest der Gewürze und Safran hinzu. Mengen Sie die Masse unter die Fleischmischung. Befüllen Sie damit ein Netz aus Schweine- oder Pferdefett, oder wahlweise andere Tierfette. Anschließend schneiden Sie das ins Netz gefüllte Brät in eigroße Stücke o.ä. Danach legen Sie die Stücke in den Gewürzsud und lassen sie sieden, ohne zu kochen."

44 Die beiden Protagonisten, Pantagruel, ein junger Riese, und sein Vater Gargantua, sind heute vor allem noch durch die französischen Redewendungen „avoir un appétit pantagruélique" („einen pantagruelischen Appetit haben") und „un repas gargantuesque" („ein gargantuesker Schmaus") bekannt. Vgl. Les Œuvres De M. François Rabelais, Docteur en Medecine, Bd. 2, s.l. 1663, Buch IV, Kap. XLI, S. 621 u.a.

45 Andouillette: Außer in Lyon gibt es diese Wurstsorte oft in und um Troyes und Cambrai. Sie besteht aus Schweinedarm und Schweinemagen. In Deutschland sagt man „Gekrösewurst", was so viel meint wie Innereienwurst. Anduillien gelten als Trierer und moselländische Spezialität.

46 Les Œuvres De M. François Rabelais (wie Anm. 44), S. 621.

47 Hier das Rezept dazu: Zutaten: 1 grobe Lyoner Wurst, frisch (ca. 750 g), 3 Schalotten, 50 g Butter, 0,25 l Beaujolais, nicht zu jung, Salz, Pfeffer, dazu Pellkartoffeln; Zubereitung: Schalotten fein hacken und zusammen mit Butterflöckchen auf dem Boden einer feuerfesten Form verteilen. Die Lyoner Wurst einige Male fein einstechen und

Fleischwurstgeschichte, die in einer Wurst namens Lyoner endet, ist, dass die Wurst den Namen einer UNESCO-Weltkulturerbe-Stadt[48] trägt und vor Ort aber *Cervelas* heißt. Der Ursprung liegt im Lateinischen *cerebrum*, was so viel wie eine Verniedlichung von Gehirn bedeutet. Das Schweinehirn war ein wesentlicher Bestandteil der Wurstherstellung. Meist wird diese recht einfache Wurst mit Pistazien oder Trüffel verfeinert, was wir eher als italienische Mortadella kennen. Die *Cervelas de Lyon* wird – anders als bei uns in Süddeutschland – roh verkauft und muss deshalb vor dem Verzehr 20 Minuten in siedendem Wasser ziehen.

Die Lyoner oder Leoni – in München die Fleischwurst im Ring – mutierte in den verschiedenen Regionen. Durch die Globalisierung ist jedoch der Ursprung kaum mehr gegenwärtig – außer im Namen.

Die sogenannte „saarländische Lyoner“[49] ist durch den Zusatz regional typisiert und außerdem geschützt. In den Weinbaugebieten Pfalz, Rheinhessen, Hessische Bergstraße und Rheingau ist eine weit verbreitete Mahlzeit das „WWW“ – ausgeschrieben: „Weck, Worscht un Woi“.[50] Anders als in der Literaturwissenschaft, in der mit WWW keine geistige Zufriedenstellung gemeint ist, soll hier im Gegenzug zur Mundartliteratur der regionale Bezug eine Relevanz darstellen. Diese Art „Mainzer Frühstück“ dürfte wohl dem „Münchner Weißwurstfrühstück“ – bestehend aus Weißbier, Weißwürsten, einer „Brezn“ und nicht zu vergessen dem süßen Senf – am nächsten kommen. Patentrechtlich nicht geschützt, kann die Erfindung der Münchner Weißwurst auch nur auf folgender Legende basieren: Gasthaus „Zum ewigen Licht“, Rosenmontag 1857: Durch eine Fehlkalkulation bei den Därmen füllte der Metzger, Sepp Moser, sein Brät in Schweinedärme. Um das Platzen beim Kochen zu verhindern, siedete er diese nur in heißem Wasser.[51]

---

daraufsetzen. Wein zugießen, die Form mit Alufolie abdecken und die Wurst im vorgeheizten Backofen bei guter Mittelhitze (220° C) ca. 30–40 Minuten (je nach Dicke der Wurst) braten. In Lyon serviert man dazu heiße Pellkartoffeln, in dicke Scheiben geschnitten, mit Salz und Pfeffer bestreut und mit Öl und Essig beträufelt. Oder es gibt warme Linsen dazu, die mit dem Rotweinsud, in dem die Wurst zubereitet wurde, gewürzt werden. URL: http://www.rezepte-guru.de/show.php/46388_Lyoner_Wurst_in_Beaujolais.html (letzter Zugriff: 12.8.2019).

48 Insgesamt sind 500 Hektar der Innenstadt als UNESCO-Weltkulturerbe ausgezeichnet.

49 Die Lyoner ist im Saarland maskulin! Günther Klahm, Der Lyoner – Eine saarländische Spezialität mit Kultstatus, Gudensberg 2006.

50 Uwe Herrmann, WWW. Die Abenteuer von Weck, Worscht & Woi. Der Wegweiser zur Pfälzer Lebensart, Offenbach/Queich 2012.

51 Diese Legende widerlegte u.a.: Dieter Wunderlich, Münchner Weißwurst. Geschichte der Weißwurst, veröffentlicht: 2008, in: Dieter Wunderlich. Buchtipps und mehr, URL: https://www.dieterwunderlich.de/weisswurst_muenchen.htm (letzter Zugriff:

Der Münchner Stadthistoriker Richard Bauer fand einen Stich aus dem Jahr 1814, der Münchner in einem Bockbierkeller beim „Zuzeln" der Weißwürste zeigt. Zudem las er in einem Metzgerhandbuch von der „Maibockwurst", welche der Weißwurst anscheinend ähnlich oder gar gleich ist. Daraus ergab sich das Resultat, dass die Weißwurst wohl eine Weiterentwicklung der Maibockwurst ist.[52] Festzuhalten ist an dieser Stelle, dass, was nicht rechtlich markengeschützt ist, keinen Anspruch auf seinen regional tradierten Namen hat.

Als pejorative Bezeichnung begegnet uns der mittelniederdeutsche „*hans myst*" schon 1519 in Sebastian Brants (1457–1521) spätmittelalterlicher Moralsatire „*Das Narrenschyff*".[53] Schon Martin Luther schrieb 1541 über den sogenannten „*Hans Worst*", um „*gottlose*[n] *Tyrannen*" ein Schimpfwort zu geben.[54] Die Reformation, die sich besonders auf die großen Städte fokussierte, hatte mit der Wurst mehr zu tun, als man im ersten Moment annimmt. So schrieb der bedeutende Reformator Ulrich Zwingli im Jahr 1522: „*Von Erkiesen und Fryheit der Spysen: Von Ergernus und Verböserung: Ob man Gwalt hab die Spysen zuo etlichen Zyten verbieten / Meynung Huldrichen Zuinglis zuo Zürich geprediget jm M.D.XXII. Jar.*" Das „Zürcher Wurstessen" beziehungsweise das „Froschauer-Wurstessen" (Drucker Christoph Froschauer) fand 1522 am ersten Sonntag der Fastenzeit in Zürich statt. Demgemäß gab es also einen Bruch des Abstinenzgebots gegenüber der Wurst während der Fastenzeit.

Ein abergläubischer Satz besagt: „Wer in Bayern, Franken und Norddeutschland vor dem Sonnenaufgang Hirsebrei und Blutwurst isst, hat das

---

13.8.2019); Barbara Fuhrmann, 150 Jahre Münchner Weißwurst. Geschichten und Empfehlungen, München 2007.

52 Vgl. Richard Bauer, Die Münchner Weißwurst – Legende und Wahrheit. Zum hundertfünfzigsten Jubiläum der Münchner Weißwurst (2007), veröffentlicht: März 2007, in: münchen.de – Das offizielle Stadtportal, URL: https://www.muenchen.de/rathaus/Stadtverwaltung/Direktorium/Stadtarchiv/Stadtgeschichte1/Weisswurst.html (letzter Zugriff: 13.8.2019).

53 Sebastian Brant, Das Narrenschyff. Stultifera navis. Unter Mitarbeit und mit weiteren Zusätzen des Autors aus dem Deutschen übersetzt und ergänzt von Jakob Locher, Basel 1497.

54 Der LII. psalm || inn form eins Liedes/ || widder alle Gottlose Tyrannen vnd verfolger G#[oe]ttliches worts/ Jm || Thon. Ein feste Burgk #[et]c. || Ob sich dieser psalm/ icht ansehen || lasse/ als hette der heylige Geist/ dar= || innen vnsern itzigen Doeg/ Hansen || Wurst abcontrafect/ das w#[oe]lle ein || ieder guthertziger Christ/ vrtheylen. || (Ein ander Lied/ || widder Hans Wurst/ || […] ACh du arger Heintze was hastu || gethan […]) || [v. Martin Luther], Zwickau 1541, Bibliothek der Ludwig-Maximilians-Universität München, Sigel: 19; VD 16, ZV 16124.

ganze Jahr Geld und bleibt fieberfrei."[55] Wie beschrieben, gilt das aber nur für die besagten Gegenden.

Der Ärger und „*der grosse Jammer/ Elend/ Marter und Pein der lieben Wuͤrste/ wie auch die seltzame Namen/ Gebraͤuche und Mißbraͤuche des edlen Getraͤncks/ so ich bißher mit grosser Betruͤbnuͤß und Wehklagen meiner Wurstliebenden Zaͤhne viel Jahr gesehen. Habe derwegen endlich en Herz gefasset/ auff instaͤndiges begehren und sehnliches verlangen vieler der rechten und echten Saͤchsischen Wuͤrste Liebhaber/ denen dieser Schaden auch tieff zu Hertzen geht/ derer Zaͤhne mit girren und knirren den Wurstjammer nebenst nur anschauen und eine geraume Zeit ein hertzliches Verlangen nach der lang gewuͤnschten Wurst= Reformation getragen/ dieses kleine Tractaͤtlein* ediren *wollen/ der gaͤntzlichen Hoffnung/ er werde ja/ wo nicht* in toto/ *doch* ex parte *etwas fruchten.*"[56]

Neben der weltbekannten Lyoner ist ein ganz ähnliches Phänomen zu beobachten, wenn es um Frankfurter Würstchen geht. Bei dieser Wurstsorte verhält es sich allerdings anders. Seit 1860 ist die Bezeichnung als geographische Herkunft geschützt und seit 1929 darf sich nur Frankfurter Würstchen nennen, was im Raum Frankfurt am Main hergestellt wurde.[57] Dementsprechend wäre es zu überdenken, ob Lyoner die richtige Bezeichnung für die Fleischwurst ist, welche andernorts hergestellt wurde. Vielleicht wäre ein Name, wie es beim panierten Kalbsschnitzel der österreichischen Küche[58] üblich ist, nach „Wiener Art" geeigneter. Besteht das Schnitzel

---

55 Hanns Bächtold-Stäubli (Hg.), Handwörterbuch des deutschen Aberglaubens, Bd. 8: Silber–Vulkan, Berlin/New York 1987, S. 308.

56 Marcus Knackwurst, Span-Neue Wurst-Beschreibung/ Darinnen nicht allein von unterschiedlichen Nationen und Geschlechten der wunderbahren Mißgeburten/ der ketzerischen Würste/ sondern auch von verfertigung der echten und rechten Sächsischen Würste gehandelt wird/ Wobey auch angefügt Martyrologia Hordei, Wie das Edle Gersten-Korn so viel Marter ausstehen muß/ und wie sonderlich der liebe Gersten-Safft so schändlich mit dem leidigen Sauffen gemißbraucht wird/ [...], Schweinfuhrt im Lande Wursten 1666, Universitäts- und Landesbibliothek Sachsen-Anhalt Halle, Pon Vf 1513, QK; VD 17, 3:630272B.

57 Vgl. das Urteil des 10. Zivilsenats des Kammergerichts Berlin von 1929, das der Bundesgerichtshof 1955 bestätigte. Demnach ist der Begriff „Frankfurter Würstchen" eine geographische Herkunftsangabe und dementsprechend eine Schutzvoraussetzung für die Würste aus Frankfurt am Main.

58 Maria Anna Neudecker, Allerneuestes allgemeines Kochbuch, oder gründliche Anweisung alle mögliche nahrhafte, geschmackvolle Speisen und Getränke, auf die wohlfeilste Art, ohne Nachtheil der Gesundheit zu bereiten, Prag 1831, S. 48: „Wiener Schnitzel von Kalbfleisch".

nicht aus Kalbsfleisch muss es „Wiener Schnitzel vom Schwein“ genannt werden.[59] So spielt der Markenschutz in Hinsicht auf die regionale Typisierung, die Herkunft und die Produktion des gleichen Produkts aber anderwärts eine unverkennbare Rolle.

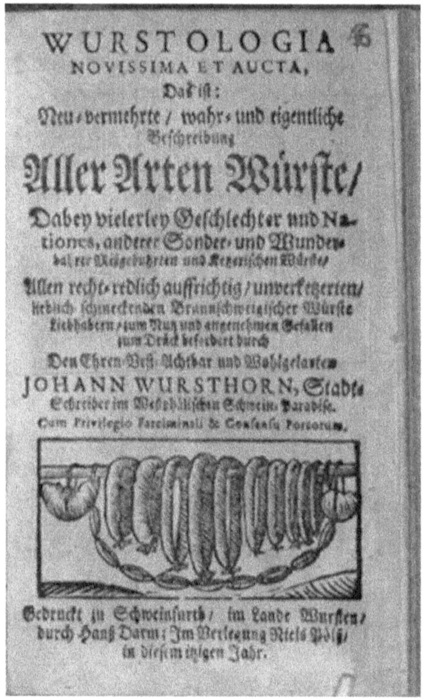

WURSTOLOGIA
NOVISSIMA ET AUCTA,
Das ist:
Neu= vermehrte/ wahr= und eigentliche
Beschreibung
Aller Arten Würste/
Dabey vielerley Geschlechter und Nationes, anderer Sonder= und Wunderbahrer Mißgebuhrten und Ketzerischen Würste/
Allen recht= redlich auffrichtig/ unverketzerten/ lieblich schmeckenden Braunschweigischer Würste Liebhabern/ zum Nutz und angenehmen Gefallen zum Druck befordert durch
Den Ehren=Vest=Achtbar und Wohlgelarten
JOHANN WURSTHORN, Stadt= Schreiber im Westphälischen Schwein= Paradise.
Cum Privilegio Farciminali & Consensu Porcorum.

Gedruckt zu Schweinfurth/ im Lande Wursten/ durch Hanß Darm; Im Verlegung Niels Pölß/ in diesem itzigen Jahr.

Abb. 63: *Marcus Knackwurst, Wurstologia Novissima Et Aucta, Das ist: Neu= vermehrte/ wahr= und eigentliche Beschreibung Aller Arten Wu^e^rste/ Dabey vielerley Geschlechter und Nationes, anderer Sonder= und Wunderbahrer Mißgebuhrten und Ketzerischen Wu^e^rste/ Allen recht= redlich auffrichtig/ unverketzerten/ lieblich= schmeckenden Braunschweigischer Wu^e^rste Liebhabern/ zum Nutz und angenehmen Gefallen zum Dru^e^ck befordert durch Den Ehren= Vest= Achtbar und Wohlgelarten Johann Wursthorn, Stadt= Schreiber im Westpha^e^-lischen Schwein= Paradise. Cum Privilegio Farciminali & Consensu Porcorum, Hamburg 1690, Titelblatt. Bildnachweis: Staats- und Universitätsbibliothek Hamburg, A/31306; VD 17, 18:728878P.*

59 Verwaltungsgericht Arnsberg, Urteil vom 26.10.2009, Aktenzeichen 3 K 3516/08: „Herstellung von panierten Schweineschnitzeln mit dreißigprozentigen Panadeanteil; Verbot des Vertriebs des Produkts ‚Tillman's Wiener Schnitzel vom Schwein‘ wegen Irreführung der Verbraucher; Abweichung von der Verkehrsauffassung hinsichtlich eines Wiener Schnitzels wegen der Herstellung aus Schweinefleisch anstelle der Herstellung aus Kalbsfleisch.“

Dieses Phänomen der Regionalität zeigte schon Marcus Knackwurst um 1700 in seiner *„Tabella Farcminologica.*[60] *oder Wurst= Tabell:/ Worinnen beschrieben wird: 1. Der Schwein= Arth und Grösse./ 2. Die Arth/ wie dieselben zu masten./ 3. Wie dieselbe zu schlachten./ 4. Die Würste und deren 3 fache echte Arth./ 1. Leber= Würste/ 2. Blut= Würste/ 3. Met= Würste (Magen= Würste/ Knack= Würste/ Blasen= Würste). Als nemblich: Lüneburgische Würste/ Märckische Würste/ Meckelburgische Würste/ Holsteinische Würste/ Jüthländische Würste/ Wendische Würste. Welche wieder unterschieden werden in Sack = Würste/ Pot= Würste/ Pluncken= Würste/ Umbgestochen/ oder Kehr= Würste/ Ochsen= Würste. Gänse= Würste/ Schaaf= Würste/ Hechts= Würste/ Deschagen= Würste/ Fuchs= Würste und Wolffs= Würste/ den Jägern und Schützen bekandt/ Klot= Würste/ den Quacksalbern bewust/ und endlich Die Hamburger Stewsisische.*“[61]

---

60 Marcus Knackwurst, Wurstologia Novissima Et Aucta, Das ist: Neu= vermehrte/ wahr= und eigentliche Beschreibung Aller Arten Wuerste/ Dabey vielerley Geschlechter und Nationes, anderer Sonder= und Wunderbahrer Mißgebuhrten und Ketzerischen Wuerste/ Allen recht= redlich auffrichtig/ unverketzerten/ lieblich= schmeckenden Braunschweigischer Wuerste Liebhabern/ zum Nutz und angenehmen Gefallen zum Drueck befordert durch Den Ehren= Vest= Achtbar und Wohlgelarten Johann Wursthorn, Stadt= Schreiber im Westphaelischen Schwein= Paradise. Cum Privilegio Farciminali & Consensu Porcorum, Hamburg 1690, Staats- und Universitätsbibliothek Hamburg, A/31306; VD 17, 18:728878P. Gleiches gilt für Käse: Fritz Käsefäix, Caseologia, oder Außfuehrliche Beschreibung Aller und jeder Auß= und Inlaendischer Kaese/ mit ihren Tugenden und Lastern/ gleichsam wieder einander streitende. Unpartheyischer Weise maenniglichen/ absonderlich aber denen geschwornen Kaese= Rittern zu sonderbarer Nachricht wolmeinend dem Trucke uebergeben, Hamburg 1690, Bayerische Staatsbibliothek München (künftig: BSB), L.eleg.m. 708; VD 17, 12:645624F, in Erläuterung: *„I. Vieler Art seltsamer Wuerme/ so die Menschen plagen und nagen. II. Caffée= und Thée= Logia, dessen Gebrauch und Mißbrauch. III. Biere= Logie, allerhand Biere Krafft und Tugend. IV. Tabaco= Logia, oder deren nuetzlichen Gebrauch und Mißbrauch. V. Caseologia; derer sonderbahre Zubereitung und Geschmack. VI. Wurstologia, derer Composition nach eines jeden Landes Arth. VII. Gaense-Logia, warum man dieselbe*[n] *auff Martini martert un*[d] *bratet/ Alle auff das kurtzweiligste beschrieben/ und jedes in ein besonderes Tractaetlein von einem Liebhaber guter Gesellschafft abgefasset/ und mit Kupffern erleutert.*“ VI. beschreibt hier ebenfalls die unterschiedlichen Zubereitungsweisen nach Regionen.

61 So auch zu lesen in der Rezeptsammlung von: Maria Sophia Schellhammer, Die wol unterwiesene Koechinn/ Das ist: Unterricht/ Wie man alle Speisen/ so nur in Teutschland bekant seyn moegen/ […], Braunschweig 1697, 6. Kapitel: „*Von allerlei Wuerste*“.

CASEOLOGIA,
oder
Außführliche
Beschreibung
Aller und jeder Auß- und Inländischer Käse/ mit ihren Tugenden und Lastern/ gleichsam wieder einander streitende.
Unpartheyischer Weise männiglichen/ absonderlich aber denen geschwornen Käse-Rittern zu sonderbarer Nachricht wolmeinend dem Trucke übergeben
Von
Fritz Käse-Fäixen/ wohlfürgestellten Käse-Visitirern zu Buxtehude.
Gedruckt zu Stanckfort/ bey Hans Käse-Barten/ und eben in diesem Jahr.

Abb. 64: *Fritz Käsefäix, Caseologia, oder Außfuͤhrliche Beschreibung Aller und jeder Auß= und Inlaͤndischer Kaͤse/ mit ihren Tugenden und Lastern/ gleichsam wieder einander streitende. Unpartheyischer Weise maͤnniglichen/ absonderlich aber denen geschwornen Kaͤse= Rittern zu sonderbarer Nachricht wolmeinend dem Trucke uͤbergeben, Hamburg 1690, Titelblatt. Bildnachweis: BSB, L.eleg.m. 708; VD 17, 12:645624F.*

Analog gilt für die Bandbreite der Käsesorten folgendes: „*Der Verworffene Kaͤse-SUMMARIEN. 1. Boͤhmischer Staͤncker Kaͤse./ 2. Limburger Kaͤse./ 3. Meißnische Zwarck und Napff Kaͤse./ 4. Madichter und wuͤrmichter fauler Kaͤse./ 5. Kleine Voigtlaͤndische schmierichte Karn. Kaͤse./ 6. Harter und zaͤher Saͤchsis*[cher] *Ziegen-Kaͤse./ 7. Fauler und harter Schaff-Kaͤse./ 8. Harter alter verfaulter gruͤner Kaͤse./ 9. Alter gebeutzter Kuͤh-Kaͤse./ 10. Buttermilchs-Kaͤse./ 11. Die mancherley Arten mit Bier eingemachte alten Staͤncker-Kaͤse./ 12. Steiffmutzen-Kaͤse./ 13. Hollaͤndische/ Bremische/ Frießlaͤndische und Gruͤnlaͤndische alte Kaͤse./ 14. Alte Nieder-Saͤchsische ausgedorrete harte Kaͤse./ 15. Pollnische/ Preussische/ Westphaͤlische und viel anderer Nationen gantz ungesunde Kaͤse/ u.s.w.*“[62]

Abb. 65: *Titelbild der Caseologia – „Wurst und Kaͤse-LOGIA“, in: Fritz Käsefäix, Fritz Käsefäix, Caseologia, oder Außfuͤhrliche Beschreibung Aller und jeder Auß= und Inlaͤndischer Kaͤse/ mit ihren Tugenden und Lastern/ gleichsam wieder einander streitende. Unpartheyischer Weise maͤnniglichen/ absonderlich aber denen geschwornen Kaͤse= Rittern zu sonderbarer Nachricht wolmeinend dem Trucke uͤbergeben, Hamburg 1690. Bildnachweis: BSB, L.eleg.m. 708; VD 17, 12:645624F.*

Aus diesen Aufzeichnungen des 17. Jahrhunderts geht hervor, dass Wurst- und Käsesorten in einer großen regionalen Abhängigkeit zum Herstellungsgebiet standen. Demzufolge finden sich auch Flurnamen in Anlehnung an unsere hier beschriebenen Produkte. So findet man in der hessischen Gegend um

62 Ebd.

Dotzheim (Gemeinde Wiesbaden) für das 14. Jahrhundert „*an die worste*“[63] und in Moischt (Gemeinde Marburg) für das 18. Jahrhundert „*auf der wurst*“.[64] Die sogenannte „Ahle Wurst“ oder „Rote Wurst“ ist eine nordhessische Spezialität, eine Art luftgetrocknete oder leicht kaltgeräucherte Dauerwurst aus schlachtfrischem Schweinefleisch. Der „Feldkieker“[65] – eine Göttinger Wurst, die seit 2011 geschützt ist – war bereits im Jahr 1488 in Büraberg bei Fritzlar bekannt. Dieser eher grobkörnig schnittfesten Rohwurst ähnlich ist der seit 2013 geschützte „Eichsfelder Feldgieker“ oder „Eichsfelder Feldkieker“[66] – Eichsfeld war ein historisches Territorium und ist eine heutige Kulturlandschaft im südöstlichen Niedersachsen, im nordwestlichen Thüringen und im nordöstlichen Hessen zwischen Harz und Werra. Daran wird die kleine, aber feine Nuance unmissverständlich, die dennoch ihre jeweilige Herkunft geschützt haben will. Die Göttinger Wurst ähnelt zwar dem „Eichsfelder Feldgieker“, obgleich sie nicht in Warmfleischverarbeitung hergestellt wird. Deshalb sind für die Reifung Starterkulturen mit zugesetzten Hefen nötig. Dies führt zu einer anderen Reifung und Geschmacksbildung. Außerdem ist für den „Göttinger Feldkieker“ die Abfüllung in blasenförmige Därme typisch. Ein „Eichsfelder Feldgieker“, der in eine Kalbsblase gefüllt wird, heißt dagegen ausdrücklicher „Eichsfelder Kälberblase“.

Die Ausstellung „9 cm Nürnberg. Eine Kulturgeschichte der Nürnberger Bratwurst“ 2015 im Stadtmuseum im Fembo-Haus zeigte, wie bedeutend die kleine Wurst für die Stadt Nürnberg war und ist. Eine siebenhundertjährige Tradition des kulinarischen Klassikers der ehemaligen Reichsstadt stand damals im Fokus.

Welche Würste nun geschützt sind und welche, für die es unbedingt notwendig wäre, finden im kommenden Abschnitt ihren Platz. Markenschutz bedeutet zugleich auch Herkunftsschutz, zumindest auf geographischer Ebene.

„Das Gütezeichen ‚g.g.A.‘ soll eine Verbindung der landwirtschaftlichen Erzeugnisse und Lebensmittel mit dem Herkunftsgebiet dokumentieren, wobei nur eine der Produktionsstufen – also Erzeugung, Verarbeitung oder Herstellung – im Herkunftsgebiet durchlaufen worden sein muss. Mindestens eine Phase des Produktionsprozesses muss in dem Gebiet erfolgen, während das für ihre Herstellung verwendete Rohmaterial aus einer anderen Region stammen kann. Mit ‚g.g.A.‘

63 Hauptstaatsarchiv Wiesbaden, Abteilung 22, Akten, Nr. 517, pag. 229.
64 Staatsarchiv Marburg, Kataster, 1747.
65 Göttinger Feldkieker, geschützt seit 2011, DE/PGI/0005/0721.
66 DE/PGI/0005/0773.

gekennzeichnete Produkte besitzen somit eine spezifische Eigenschaft oder ein Ansehen, die sie mit einer bestimmten Region verbinden.“[67]

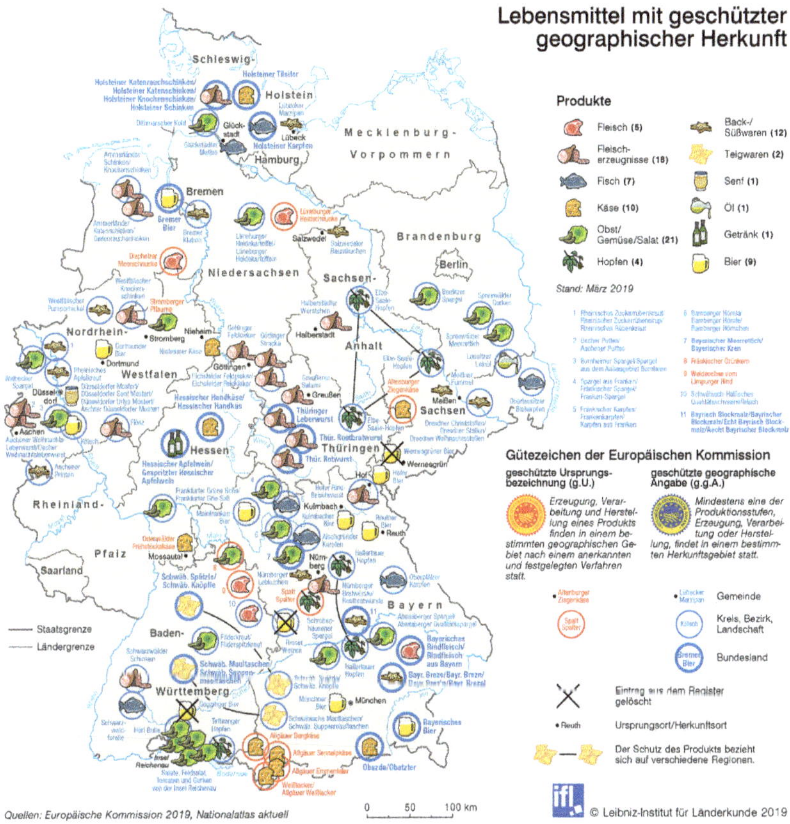

Abb. 66: *Karte der Lebensmittel mit geschützter geografischer Herkunftsangabe in Deutschland, 2019. Bildnachweis: Leibniz-Institut für Landeskunde; Bundesministerium für Ernährung und Landwirtschaft, URL: https://www.bmel.de/DE/themen/landwirtschaft/agrarmaerkte/geschuetzte-bezeichnungen.html.*

67 Schutz von Herkunftsangaben und traditionellen Spezialitäten, veröffentlicht am: 26.6.2019, in: Homepage des Bundesministeriums für Ernährung und Landwirtschaft (BMEL), URL: https://www.bmel.de/DE/Landwirtschaft/Agrarpolitik/1_EU-Marktregelungen/_Texte/GeschuetzteBezeichnungen.html (letzter Zugriff: 13.8.2019).

Die Nürnberger Bratwurst untersteht diesem Schutz:[68]

Dossier-Nr.	Land	Name	Art des Antrags	Datum	Status
DE/PGI/ 0005/0184	Deutschland	Nürnberger Bratwürste/ Nürnberger Rostbratwürste	g.g.A.	16.7.2003	registriert

Dadurch ist die 700-jährige Tradition unantastbar. Doch auch bei diesem „Nationalgericht" geht es im wahrsten Sinne um die Wurst. Der Streit um das älteste Rezept – Stammt es aus Nürnberg oder doch aus Thüringen? – entfacht immer wieder.[69] Die jahrhundertelange Bratwursttradition betrifft die gesamte fränkische Region: unter anderem Nürnberg, Ansbach, Bayreuth, Coburg, Kulmbach, Würzburg und Hof. Diese fränkische Foodscape hat wohl zwischen Weißenburg und Treuchtlingen entlang des Limes in Bayern bereits zur Zeit der Römer ihren Ursprung als sogenannte „*tomacula*".[70] Sachsen, Thüringen und Franken verbindet sogar die Initiative „Bratwurst-Äquator", die unter der Schirmherrschaft des ehemaligen bayerischen Innenministers Günther Beckstein (Amtszeit 1993–2007), der Edeka Handelsgesellschaft Nordbayern-Sachsen-Thüringen mbH und der Gebiets-Winzergenossenschaft Franken eG

68 Verzeichnis für geschützte Lebensmittel der Europäischen Kommission, Landwirtschaft und Ländliche Entwicklung, DOOR, URL: https://ec.europa.eu/agriculture/quality/door/list.html?locale=de&recordStart=0&filter.dossierNumber=&filter.comboName=n%C3%BCrnberger&filterMin.milestone__mask=&filterMin.milestone=&filterMax.milestone__mask=&filterMax.milestone=&filter.country=&filter.category=&filter.type=&filter.status= (letzter Zugriff: 13.8.2019).

69 Bratwurst. Die Franken und ihr Nationalgericht, veröffentlicht am: 5.3.2013, in: Bayerischer Rundfunk (BR.de), URL: https://www.br.de/franken/inhalt/kultur/uf-bratwurst-frankens-nationalgericht100.html (letzter Zugriff: 15.8.2019).

70 Eine Art spätrömische Bratwurst. Vgl. dazu auch: M. Valerii Martialis epigrammaton libri editit F. G. Schneidewin, Bd. 1, Paris u.a. 1842, S. 38, XLI.: „*Quod fumantia qui tomacla raucus* [10] *Circumfert tepidis cocus popinis*". Ein Bratwurstgrill aus römischer Zeit ist heute noch in mehreren Museen zu bestaunen, siehe URL: http://www.antike-tischkultur.de/kochtechnikgitter.html (letzter Zugriff: 19.8.2019); siehe auch: Gaius Petronius Arbiter, Cena Trimalchionis. Das Gastmahl des Trimalchio, hg. und übersetzt von Karl-Wilhelm Weeber, 31,11: „[…] *fuerunt et tomacula ferventia supra craticulam argenteam posita* […]"; Juvenal (X, 355, S. 396): „[…] *candiduli divina tomacula porci* […]", URL: https://www.loebclassics.com/view/juvenal-satires/2004/pb_LCL091.397.xml (letzter Zugriff: 19.8.2019).

(GWF) ins Leben gerufen wurde. Wetteifer um die beste Bratwurst bestand allerdings schon seit eh und je, wie die „Augsburger" zeigt. Zwischen München, Nürnberg und Regensburg wollte nun die ehemalige Römerstadt nicht länger im Schatten stehen. So liest man im Jahr 1725: „[...] *wovor ihm Elias die Dispensation gegeben/ bey jeglicher Tauff= Suppen/ gepreßte Schweins= Koͤpff/ Westphaͤlische Schuncken und gedigene Augspurger Wuͤrste zuessen.*"[71] Diese Wurstsorte ist markenrechtlich nicht geschützt, weshalb die eigentliche Ingredienz des Rindfleisches in Österreich durch Verbreitung der Wurst außerhalb der Stadttore Wiens[72] in ihren Zutaten mutierte.[73]

Nicht nur die „Essenslandschaft", sondern auch die Konfessionalisierung spielt bei der fränkischen Spezialität des fleischlastigen Mittelalters eine große Rolle. Die Reformation ging an Franken und der Fränkischen Bratwurst nicht unbemerkt vorüber. In Franken unterscheidet man in einigen Regionen zwischen „katholischen" und „evangelischen" Bratwürsten. Diese Unterscheidung hat mit der dominierenden Konfession in der jeweiligen Region zu tun. Während in überwiegend katholischen Gebieten, den katholischen Hochstiften Würzburg und Bamberg, die Bratwurst ein sehr feines Brät enthält, ist das Brät in den evangelischen Gebieten Frankens, wie etwa in Coburg, eher grob. Wodurch der Unterschied genau zustande kommt, bleibt ungeklärt; es könnte allerdings der Puritanismus großer Reformatoren etwas Schuld daran tragen. Ist die Bratwurst in der fränkischen Kulinarik wirklich so etwas wie eine *condicio sine qua non*? Im 14. Jahrhundert formulierte der Nürnberger Rat wie folgt: „*Alle sweinen lentpraten sol man in die wurste hacken. Swer si anders verkoufft, der gibt zwên schilling, als dicke und er daz tuͦt.*"[74] Geheime Rezepturen dienten stets

---

71 Johann Daniel Kuttge, Fortsetzung Des Curiosen Fasten-Gesprächs Zwischen Elia Einem Catholischen/ Ezechiel und Habacuc Zwey Lutherischen Studenten [...], Augsburg 1725, S. 74.

72 Codex alimentarius Austriacus, Wien 1912, S. 114; so auch in: Bewährteste Geheimnisse von fünfhundert probatesten Kunst-Stücken aus allen drey Theilen des Natur-Reichs gesammlet mit nöthigen Handgriffen und zum Gebrauch für allerhand Liebhaber an das Licht gegeben, Nürnberg 1771, S. 139.

73 „*Die Augsburger= Wuͤrste, welche aus dem Halsfleische (Zopfen) der Ochsen gemacht werden, sind unter allen die Unschaͤdlichsten.*", in: Der Astrolog, oder: Der wohlerfahrene und zuverlässige Wetterverkündiger. Ein neu eingerichteter Haus=, Familien=, Wirthschafts= und Unterhaltungs= Kalender [...], Wien 1843.

74 Satzungsbuch der Stadt Nürnberg, 1302–ca. 1310, StAN, Reichsstadt Nürnberg, Amts- und Standbücher Nr. 228, Nürnberger Stadtbuch (Satzungsbuch I/A), Zeile 10 ra, 21–26, Digitalisat unter URL: http://www.deutschestextarchiv.de/rem/?d=M338-G1.xml (letzter Zugriff: 30.8.2019).

der Qualitätssicherung und -überwachung. Darüber hinaus waren es auch die Fleisch(er)-Ordnungen, die die Herstellung der Würste dokumentierten oder vorschrieben.[75] Anhand dieser Quellen lässt sich ebenfalls der Bezug zur Städtelandschaft herstellen. Die Räte der Reichsstädte legten dementsprechend ihre Protektion über die kleine oder große Wurst, je nach Region. Diese Satzung bestimmte auch einen Fleischbeschauer, der die Dienstbezeichnung „Würstlein" führte. Bereits damals durften nur spezialisierte Schweinemetzger die Nürnberger Bratwurst herstellen und mussten diese täglich den geschworenen Metzgern und Marktmeistern vorlegen.

75 Vgl. dazu: VD 16, hier in Auswahl: ERnewert vnd ver= || bessert Flaisch Ord= || nung zu Franck= || furt. || (Decret. in Consilio XXIII. || Maij, Anno LIX.), VD 16, ZV 31915; Des Raths zu Leipzig/ || VOrnewerte Ord= || nung vnd Reformation: || I. Wegen der Tracht vnd Kleidung. || II. Wie es mit anstellung der Hochzeiten/ Verl#[oe]bnissen || vnd Kindtauffen sol gehalten werden. || III. Die Vormundschafft sachen belangende. || IIII. Fewr Ordnung. || V. Fleischer Ordnung. || VI. Abdruck der Articul wegen der Dorffschafften. || Auffs New nach Gedruckt. || (Wie es mit der Vni= || uersitet Verwandten zu= || halten.), VD 16, L 1029; Der Stadt || Budissin Fleischer || Ordnung. || Von Newen auff= || gericht, VD 16, ZV 20290; vgl. außerdem: VD 17, hier in Auswahl: Fleischer-Ordnung Der Churfürstlichen Brandenburgischen Residentz- und Haubt-Städte Berlin und Cölln an der Spree: [Geschehen Cölln an der Spree/ den 6. Martii Anno 1656.], VD 17, 7:708595F; Fleischer-Ordnung/ Bey Der Stadt Mersebургk, VD 17, 3:678180L; Wir Rathsmeistere und Rath der Stadt Erffurdt/ fügen allen und jeden unsern Bürgern [...] hiermit zu wissen/ Ob wir uns wol unserer dabevor unter dato den 23. Martii des abgewichenen 1599. Jahres Publicirten Fleischordnung bedächtig zu erinnern wissen [...] Dennoch aber weil in mittelst von dem gantzen Handwerge der Fleischhawer beschwerung einkomen/ wie das sie wegen furfallenden mangel des Viehes unnd thewrem einkauff desselben/ das Fleisch in dem werth wie in obangezogener Ordnung disponirt, hinfürder nicht geben könten [...] Als haben wir [...] hiermit vergönnet und nachgeben/ das daß Rind und Schweinenfleisch ein Pfund fur 12. Pfennige/ das Schöpsen umb 9. Pfennige/ und dan das Kalbfleisch umb 7. Pfennige hinfürder verkaufft werden müge/ Doch mit der [...] bedingung/ das solches lenger nicht/ dan drey Jar wehren/ auch durchaus nichts/ ausserhalb Kopff/ Hals und Hossen beim Rindern/ und der Kopff beim Schöpsenfleisch zugewogen/ die Lämmer aber nach der hand/ und das Kühefleisch besonders/ und auff vorgehenden taxt vorkaufft werden solle/ wie sich dan hierauff umb mehrer/ stetter/ und fester haltung willen/ gegen uns den Rath das gantze Handwerck der Fleischhawer mit hand/ mund und Siegel an eides statt der gestalt vor reversiret und vorplichtet/ Inmassen von wort zu worten folget: [...]. Publicirt unter unserm StadtSecret den 30. Octobris Anno 1602, VD 17, 27:717409W; Deß H. Reichs Freystatt Wormbs Ernewerte Fleisch Ordnung/ Wie es damit in Besichtigung: auch Schätz- und Verkauffung daselbsten gehalten werden soll, VD 17, 27:717208C.

Abb. 67: *Metzger Hans Lengenfelder († 1436), aus: Hausbücher der Mendelschen und Landauerschen Zwölfbrüderhausstiftungen. Bildnachweis: Stadtbibliothek Nürnberg, Amb. 317.2°, fol. 59 v (Mendel I), URL: https://hausbuecher.nuernberg.de/75-Amb-2-317-59-v.*

Der Metzger Hans Lengenfelder († 1436) steht hinter seinem Hacktisch, der sich durch eine besonders dicke Arbeitsplatte auszeichnet, und teilt mit einem schweren Hackbeil ein Bein. Rechts am Tisch, auf dem ein Gewichtsatz steht, hängt an einem mit dem Tisch verbundenen Stab eine Waage mit zwei Waagschalen. Auf dem Ablagetisch, der mit dem Hacktisch verbunden ist, liegen ein Tierkopf, ein Bein und ein flaches Stück Fleisch, während an der Wandhalterung Würste, Rippchen und ein Beil hängen. Ob hier tatsächlich erste Nürnberger Bratwürste abgebildet sind, kann nur vermutet werden.

Auch Fridrich Plecher († 1467) steht breitbeinig mit pelzverbrämtem Wams und erhobener Rechten frei im Raum. In der Hand hält der Metzger den reich

Abb. 68: *Metzger Fridrich Plecher († 1467), aus: Hausbücher der Mendelschen und Landauerschen Zwölfbrüderhausstiftungen. Bildnachweis: Stadtbibliothek Nürnberg, Amb. 317.2°, fol. 85 v (Mendel I), URL: https://hausbuecher.nuernberg.de/75-Amb-2-317-85-v.*

verzierten Handwerkerstab der Metzger, eine Art Szepter mit Blattkrone und liegendem Lamm als Bekrönung. Oben über der Wandstange hängen sechs kleine und eine große Wurst herab. Daran zeigt sich die regelrechte Verehrung des fleischverarbeitenden Handwerks. Im Deutschen Fleischermuseum sind heute noch Obermeisterketten, Siegel und anderes wertvolles Kleinod aus Porzellan oder Kunstgegenstände aus Tierknochen zu bestaunen.[76] Facharbeiter

76 Im Deutschen Fleischermuseum Böblingen, Marktplatz 27, 71032 Böblingen, oder im Internet unter URL: https://fleischermuseum.boeblingen.de/start.html (letzter Zugriff: 29.8.2019) zu finden.

des Metzgerhandwerks waren zum Beispiel die Küter – mittellateinisch *fartores, carnifices.* Sie kümmerten sich um das Schlachten, das Ausweiden und die Verarbeitung der Innereien. Spezieller waren die Wurstmetzger – mittellateinisch *cultellarii.* Hinzu fügen sich diesem Handwerk noch die Garbrater – mittelhochdeutsch *garbraeter, garküchner* / mittellateinisch *assatores.* Sie vertrieben warme Speisen und Würste, jedoch kein Frischfleisch. Weitere Sparten des Metzgerhandwerks waren Speckschneider und Hausschlachter. Der Mittelpunkt einer Stadt war der Marktplatz.[77] Dort befanden sich Wohngebäude der Händler, Speichergebäude und stattliche Gebäude aus Stein der Patrizier. In der Nähe des Marktplatzes befanden sich Wohn- und Arbeitsstätten wie Bäcker und Metzger und eine Kirche neben dem Hauptmarkt. Bis 1759 veranstaltete die Fleischerzunft in Nürnberg Tierhatzen im sogenannten „Fechthaus". Zur Historischen Meile[78] von Nürnberg gehören auch die traditionellen Handwerkerhäuschen „Am Ölberg". Südlich des Burgfelsens wuchs noch im 11. Jahrhundert die salische Siedlung für die Bediensteten sowie für Handwerker und Kaufleute. Sie neigte sich mit ihren engen Gassen an die Höhenlinien des Burgbergs und ist noch heute an den isohypsenparallelen Straßenzügen südlich der Burg offenkundig. Ausgehend von der kleinen Siedlung am Burgberg dehnte sich die Stadt schließlich bis in die sumpfige Pegnitzniederung aus.[79]

Dass die Bratwürste heute nicht mehr allein den Franken „gehören", zeigt der Nürnberger Traditionswirt Werner Behringer[80] mit seiner Schätzung: 1,4 Milliarden Bratwürste verlassen pro Jahr die mehr als 70 Metzgereien und Fabriken Nürnbergs. Ebenso gilt dies für die „Göttinger", welche bereits 1769 in alle Welt verschickt wurden: *„An Fleisch haben wir, nach Verschiedenheit der Jahreszeiten, gute Rinder und Kaᵉlber, trefliche Haᵉmel, wohlgemaᵉstete Schweine; von welchen letzten die Mettwuᵉrste in dem uᵉbrigen Teutschlande, ja in Holland, Engelland, Hollstein,*

77 Max Weber, Wirtschaft und Gesellschaft. Die Wirtschaft und die gesellschaftlichen Ordnungen und Mächte. Nachlaß, Teilbd. 5: Die Stadt (Max Weber Gesamtausgabe I/ 22 – 5), hg. v. Wilfried Nippel, Tübingen 1999.

78 Vgl. Oliver Lewe, Die historische Meile in Nürnberg, veröffentlicht am: 12.12.2019, in: Bayern-online, URL: https://nuernberg.bayern-online.de/die-stadt/historische-meile/ (letzter Zugriff: 20.12.2019).

79 Alfons Baier, Zur Geschichte, Geologie und Hydrologie des Burgberges zu Nürnberg, in: Geologische Blätter für Nordostbayern und angrenzende Gebiete 48/4, 1998, S. 277–300, URL: http://www.angewandte-geologie.geol.uni-erlangen.de/nbg2.htm (letzter Zugriff: 20.8.2019).

80 Unter anderem: Behringer's Bratwursthäusle, Werner Behringer GmbH, Rathausplatz 1, 90403 Nürnberg.

Da^ennemark, Schweden, Rußland, so gar in der Tu^erkey, und Jndien beru^ehmt sind.“[81] Dadurch wird nicht nur die Distanz der Verschickung erkennbar, sondern auch das Vorstellungsvermögen für die Abwandlung in den jeweiligen Regionen geschärft.

Ebenfalls gab es eine Art Reinheitsgebot für die Bratwurst ähnlich wie beim Bayerischen Bier.[82] So ist uns aus der wettinischen Stadt Weimar[83] folgendes überliefert: „*Ouch sullen sie die brotworste lebirwurste unde andir wurste or iglicher bisundern machen von reynem friszchem fleiszche das nicht fynnecht nach wandelbar ist unde sulln daryn nicht hagken milzcen herzce nyren nach keyn ander ungeferte nach frömde fleisz das sich nicht darzcu geburt hinder wen man das queme der ist der stat zcu busse vorfallin vo iglicher wurst bisundern zcwene schlinge denariorum also manche her der gemacht had unde sal darzcu der stat gehorsam halden bisz so lange das dii meistere vor on beten unde dee hantwergke sine busse geben.*“[84]

Derartige Verordnungen oder eben mittelalterliche Marken-Schutzmäntel finden sich in fast allen deutschen Stadt- oder Statutenbüchern.[85] Bedingt durch

---

81 Johann Christoph Gatterer (Hg.), Allgemeine historische Bibliothek von Mitgliedern des königlichen Instituts der historischen Wissenschaften zu Göttingen, Bd. 10, Halle 1769, S. 17.

82 „*Wie das Pier Summer vnnd Wintter auff dem lannd sol geschennckt vnd geprawen werden*“, in: Das buech der gemeinen. Landpot. Landsordnüng. Satzüng. vnd Gebreüch, des Fürstennthumbs. in Obern. vnd Nidern Bairn. Jm Funftzehnhundert vnd Sechtzehendem Jar aufgericht, München 1516, S. 102 f.

83 Zur Stadtgeschichte in Auswahl: Steffen Raßloff, Geschichte der Stadt Weimar, Erfurt 2018; Ders. / Mark Schmidt (Hg.), Orte der Reformation – Weimar, Leipzig 2015.

84 Stadtarchiv Weimar, Statutenbuch HA I-1-51, Weimarer Fleischhauerordnung 1432.

85 So etwa auch für die Reichsstadt Frankfurt am Main: Demnach Wir der Rath dieser desz Heyl. Reichs Stadt Franckfurt am Mayn von Unsern Deputirten zu dem Fleisch-Ampt mißfälligst vernehmen müssen/ daß viele der allhiesigen Bürger […] in grosser Anzahl Kälber in hiesige Stadt bringen liessen/ und davon die Gebühr zu entrichten […] sich weigerten […] befehlen Wir hiemit ernstlich und wollen/ daß hinkünfftig jedermänniglich sothaner Einschleiffung der Kälber sich gäntzlich enthalten […] solle […]. Conclusum in Senatu Dienstags den 25. Jun. 1695, Staatsbibliothek zu Berlin – Preußischer Kulturbesitz, 13 in: 2" Gp 20500; VD 17, 1:016271A; Es wird von wegen eines Wohl-Edlen und Hochweisen Raths hiemit ernstlich befohlen/ daß alle und jede allhiesige Metzgere den angesetzten Accis von dem Wochentlich schlachtendem Viehe alle acht/ oder längstens vierzehen Tage/ dem Löblichen Fleisch-Ampt richtig zahlen und abführen sollen […]. Signatum unter der Stadt Franckfurt Cantzley-Secret den 14. Septembris 1693, Niedersächsische Staats- und Universitätsbibliothek Göttingen, 2 J STAT VI, 6720:1 (47); für die Reichsstadt Worms: Deß H. Reichs Freystatt Wormbs Ernewerte Fleisch Ordnung/ Wie es damit in Besichtigung: auch

die *„gute Policey“*[86] gab es bereits in der Frühen Neuzeit eine Art Lebensmittelüberwachung. In Nürnberg waren die Fleischbänke „auf 92 einheimische und 32 auswärtige begrenzt“ – das seit 1431 bestehende eigene Handwerk der Schweinestecher war (mit besonderer Erlaubnis der Wurstherstellung) auf „40“ verteilt.[87] Eine Gewerbeaufsicht übten die Geschworenen Meister der Geschworenen Handwerke – die Metzger gehörten in Nürnberg auch dazu – aus, die, wie die vereidigten Handwerke, auf eine vom Rat erlassene schriftliche Handwerksordnung vereidigt wurden. Im Fränkischen Kurier liest man für das Jahr 1856: *„Außerdem wurden in Bezug auf Brod, Fleisch und Bier, dann Maas und Gewicht 877 Visitationen vorgenommen, sowie sämmtliche Märkte an jedem Wochenmarkte und die Fleischbänke an jedem Verkaufstage begangen worden sind. Von 335 zum Vermittlungsamte angemeldeten Klagen sind 180 verglichen, 155 aber an das competente Gericht überwiesen worden.“*[88]

Johann Paul Friedrich Richter (1763–1825) – genannt Jean Paul –, der deutsche Schriftsteller, schrieb am 12. März 1813: *„Letzte sind meinem Magen schöne Vergißmeinnicht von Nürnberg; nur die Blutwürste haben den Fehler, den jetzo die Länder nicht haben, zu viel Fett zu enthalten.“*[89] Diese Erinnerung an einen

---

Schätz- und Verkauffung daselbsten gehalten werden soll, Darmbstatt 1619, Thüringer Universitäts- und Landesbibliothek Jena, 4 Bud.Jus germ.155(59); VD 17, 27:717208C; für die Stadt Leipzig: Des Raths zu Leipzig/ || VOrnewerte Ord = || nung vnd Reformation: || I. Wegen der Tracht vnd Kleidung. || II. Wie es mit anstellung der Hochzeiten/ Verl#[oe]bnissen || vnd Kindtauffen sol gehalten werden. || III. Die Vormundschafft sachen belangende. || IIII. Fewr Ordnung. || V. Fleischer Ordnung. || VI. Abdruck der Articul wegen der Dorffschafften. || Auffs New nach Gedruckt. || (Wie es mit der Vni = || uersitet Verwandten zu = || halten.) ||, Leipzig 1596, besonders: 3. enth. Werk: Fleischerordnung. (VD 16, L 1021), Staatsbibliothek zu Berlin – Preußischer Kulturbesitz; VD 16, L 1029. Weitere Nummern: VD 16, L 1013; VD 16, L 1020; VD 16, L 1021; VD 16, L 1026; VD 16, L 1032; VD 16, L 1042; VD 16, L 1050.

86 Wolfgang Wüst (Hg.), Die *„gute“* Policey im Reichskreis. Zur frühmodernen Normensetzung in den Kernregionen des Alten Reiches, Bd. 7: Policeyordnungen in den fränkischen Reichsstädten Nürnberg. Rothenburg o.d. Tauber, Schweinfurt, Weißenburg und (Bad) Windsheim. Ein Quellenwerk, Erlangen 2015.

87 Horst-Dieter Beyerstedt, Art. „Metzgerhandwerke“, in: Michael Diefenbacher / Rudolf Endres (Hg.), Stadtlexikon Nürnberg, Nürnberg ²2000, S. 693.

88 Fränkischer Kurier, Nr. 74, vom 14. März 1856, Tages= Chronik, Nürnberg, 13. März 1856, in: bavarikon, URL: https://www.bavarikon.de/object/bav:BSB-MDZ-00000BSB10505014?cq=%22Reichsstadt+N%C3%BCrnberg%22+fleischb%C3%A4nke&p=1&lang=de (letzter Zugriff: 17.8.2019).

89 An Johann Leonhard Schrag. Bayreuth, 12. März 1813, in: Jean Paul – Sämtliche Briefe digital. In der Fassung der von Eduard Berend herausgegebenen 3. Abteilung der

Besuch oder eine Reise macht deutlich, wie bedeutend die Bratwurst für Nürnberg und ihre Foodscape ist.

In den Quellen finden sich die für die Nahrungszubereitung die am häufigsten gebrauchten Tierarten. So liest man beispielsweise in erster Linie von Rindern und Schweinen, wobei Rinder nicht nur aus dem Umland, sondern bereits im 14. Jahrhundert auch in großen Mengen aus Ungarn, Böhmen und Schlesien bezogen wurden.[90] Der Verkauf von Rindern, Ochsen, Kühen und Schweinen fand außerhalb der Stadt vor dem Frauentor statt. Die Zuchtfrage beschränkte sich mit dem Argument der entstehenden Geruchs- und Abfallbelästigung.[91] Platznot und hygienische Gründe drosselten den Verkauf. Sowohl die Reglements als auch die Implementierung hierzu werden im Satzungsbuch C (ca. 1320–1360) anschaulich. Auch Bäcker und Müller waren in die Schweinezucht involviert, da die Abfallprodukte der Backstube so „weiterverarbeitet" werden konnten.[92] Die Schlachtungen der innerhalb der Stadttore von Privathaushalten gezüchteten Schweine als Hausschlachtung wurde durch die amtlich vereidigten „Schweinestecher" durchgeführt. Ausschließlich Schweinemetzger durften Würste herstellen.[93] Auf den sogenannten „Fleischbänken" – deren Anzahl und Nutzungsrechte waren vom Rat streng kontrolliert – wurden die jeweiligen Metzgereiprodukte angeboten. Verordnungen zu Preisen, Qualität und Frische sowie Strafmaße[94] bei Verstößen waren penibel zu befolgen. Fleischart, Menge und Därme zum Befüllen wurden durch Vorschriften bestimmt.[95] So liest man in der Verordnung von 1607, dass „*Ein gemeine Bratwurst umb 3. Pfenning*" zu verkaufen ist. Des Weiteren „*sollen an den Bratwu^{e}rsten allweg 5. uffs Pfundt gehen/ Auch die beede Pfender befelch haben/ wo^{e}chentlich zum wenigsten/ zweymal*

Historisch-kritischen Ausgabe (1952–1964), im Auftrag der Berlin-Brandenburgischen Akademie der Wissenschaften überarbeitet und herausgegeben von Markus Bernauer, Norbert Miller und Frederike Neuber (2018), URL: http://jeanpaul-edition.de/brief.html?num=VI_737 (letzter Zugriff: 17.8.2019).

90 Carl L. Sachs, Metzgergewerbe und Fleischversorgung der Reichsstadt Nürnberg bis zum Ende des 30jährigen Kriegs, in: MVGN 24, 1922, S. 1–260, hier bes.: S. 163 f.

91 Jegel, Ernährungsfürsorge des Altnürnberger Rates (wie Anm. 26).

92 Sachs, Metzgergewerbe und Fleischversorgung (wie Anm. 90), hier v.a.: S. 41, 44.

93 Ebd., S. 47–49.

94 Mandat vnd Verbott/ Das sich meniglich deß herein schleichens/ vmbtragens/ vnd verhausirens/ deß frembden/ abgestochnen/ vnd vngeschauten Fleisch enthalten soll/ bey straff Zehen Gulden, Decretum in Senatu 22. Junii Anno 1604, Nürnberg 1604, Herzog-August-Bibliothek Wolfenbüttel, Einbl. Xb 2° 26; VD 17, 23:698024C.

95 Ayd/ Der Fleisch= Beschauer: in Sta^{e}tt und Ma^{e}rckten, s.l. [nach 1727], BSB, Kloeckeliana 89,5 a, URN: urn: nbn: de: bvb:12-bsb00097408-7.

herumb zugehen/ die Bratwuͤrst so wol/ als die Roͤsel und Leberwuͤrst auff zu ziehen/ zuschneiden und schauen/ ob sie auch von gutem saubern Schweinen fleisch unnd Speck/ so viel deßen darzu von noͤtten/ gemacht/ und nicht etwan betrieglicher weiß zuerfuͤllung des Gewichts/ zu sehr mit wasser ubergossen und angeschwembt sein/ und do ermelte Pfender einen oder mehr straffbar befinden/ Ist ihnen bey jren pflichten ernstlich eingebunden/ keines zu verschonen/ sondern die jenigen/ so frevendlich wider die ordnung gehandelt/ unverzuͤglich mit Rug fuͤr zunemen.“[96] Diese Gesamtheit von Vorschriften und Bestimmungen für gewisse regionaltypische Produkte ziehen sich durch das späte Mittelalter und die Frühe Neuzeit wie ein roter Faden und finden sich in der Gegenwart innerhalb des Markenschutzes eben wieder. Manche außergewöhnliche Spezialitäten, wie etwa das einstige Matrosengericht „Labskaus“[97] – traditionell mit Spiegelei, Roter Bete, saurer Gurke und Rollmops, meist mit Bratkartoffeln als Beilage – oder der vom ehemaligen Bundeskanzler Helmut Kohl (Amtszeit 1982–1998) verehrte „Pfälzer Saumagen“[98] – ein Fleischgericht aus Schweinefleisch, Brät und Kartoffeln, gegart in einem Schweinemagen – blieben eher regional verwurzelt und sind im süddeutschen Raum kaum auf Speisekarten zu finden. Das mag durchaus an dem Slogan „Das Auge isst mit!“ liegen.

Von den Würsten Europas, aber besonders den fränkischen Bratwürsten oder sogar Nürnbergern verlagern wir uns nun zum dritten Gang des Menüs, dem Dessert.

3. Der dritte Gang – Naschwerk und Fränkische Fastnacht

Ein damit einhergehendes Phänomen ist die verschiedenartige Bezeichnung ein und dergleichen Spezialität, wie zum Beispiel dem Krapfen als Kurzform für unter anderem Berliner Pfannkuchen, Kirchtagskrapfen, Knieküchle (Bauernkrapfen), Krautkrapfen, Punschkrapfen, Schlutzkrapfen, Schmalzgreben, Zigerkrapfen und so weiter.

---

96 Eines Erbarn Raths vernere Ordtnung vnd satzung/ wie alles Fleisch durch das Handwerck der Metzker hie/ zu Werdt/ und Gostenhoff/ dieser zeit vnderschiedlich verkaufft/ vnd bezalt genomen werden soll, Decretum in Senatu den 19. Spetemb. Anno 1607, Nürnberg 1607, BSB, Einbl. V,36 v-30, URN: urn: nbn: de: bvb:12-bsb00099326-9.

97 Johann Christian Fick, Vollständiges Englisch-Deutsches und Deutsch-Englisches Lexicon. Erster Band. Englisch-Deutscher Theil, Erlangen 1802, S. 362.

98 Siehe: David Frogier de Ponlevoy, Ex-Kanzlers Leib- und Magenspeise, veröffentlicht am: 21.12.2006, in: Spiegel Online, URL: https://www.spiegel.de/reise/europa/reisespeisen-ex-kanzlers-leib-und-magenspeise-a-453201.html (letzter Zugriff: 25.8.2019).

In der Etymologie existiert der „Krapfen“ bereits seit dem 9. Jahrhundert als hakenförmiges Gebäck. So findet man im Althochdeutschen den „*krapho*“ und im Mittelhochdeutschen den „*krapfe*“. Bereits Johann Christoph Adelung (1732–1806), der deutsche Bibliothekar, Lexikograph und Germanist, definierte in seinem „Grammatisch-kritischen Wörterbuch der Hochdeutschen Mundart“: „*2. Der Krapf, (1755–1756) des -en, plur. die -en, oder der Krapfen, des -s, plur. ut nom. sing. Diminut. das Kräpfchen, Oberd. das Kräpflein, im gemeinen Leben Kräppel, eine Art runder Kuchen von mancherley Art, welche entweder gefüllt oder ungefüllt sind, entweder in Schmalz oder im Ofen gebacken werden, und auch Krapfkuchen heißen. Bey dem Matthesius Kropf. Entweder mit Kropf von der äußern Dicke, oder auch von ihrer Gestalt, weil man den Rand zackig auszuschneiden, und die Spitzen wechselweise in die Höhe und niederwärts zu biegen pflegt, da sie denn einige Ähnlichkeit mit dem Krapfen oder Haken haben. Im mittlern Lat. Crafus, Craphus, Crato. Indessen scheinen die der Griechen, die Cripisculae der mittlern Lat. und Crouteilles der mittlern Franzosen eine ähnliche Art von Kuchen gewesen zu seyn.*“[99]

Warum gelten jedoch Krapfen als Faschings-Snack? Schon im Mittelalter aßen die Menschen vor Beginn der Fastenzeit[100] – Aschermittwoch bis zum Beginn der Messe vom letzten Abendmahl am Gründonnerstag – noch einmal eine besonders fettige Speise. Das in Schmalz gebackene Hefebällchen hatte also schon damals zur Faschingszeit Hochkonjunktur.

Schon im „*Buͦch von guͦter spîse*“ (ca. 1350) ist der Krapfen Bestandteil mehrerer Rezepte.[101]

Bereits ein Inkunabeldruck des 15. Jahrhunderts aus der Reichsstadt Nürnberg verrät uns einige Details zum Krapfen. Das Rezept lautet wie folgt: „*§ Von krapffen. § Item so du gut krapffen pachen wilt von opffeln vnd von pirn.* […]“. Vermutlich

99 Definition in: Johann Christoph Adelung, Grammatisch-kritisches Wörterbuch der Hochdeutschen Mundart […], Zweyter Theil, von F–L, Leipzig 1796, Sp. 1755, Digitalisat unter URL: https://lexika.digitale-sammlungen.de/adelung/seite/bsb00009132_00880 (letzter Zugriff: 25.8.2019).

100 Grundordnung des Kirchenjahres und des Neuen Römischen Generalkalenders, 1. Kapitel: Das Kirchenjahr, Abschnitt II – Das liturgische Jahr, III. Die Fastenzeit (österliche Bußzeit), Nr. 28, URL: https://www.sbg.ac.at/pth/links-tipps/past_ein/gok/gok.htm (letzter Zugriff: 29.8.2019).

101 Hans Hajek (Hg.), Das buoch von guoter spise. Aus der Würzburg-Münchener Handschrift, Berlin 1958, URL: http://www.staff.uni-giessen.de/gloning/tx/bvgs.htm (letzter Zugriff: 29.08.2018): „*58:1 Einen krapfen.*“, „*59:1 Einen krapfen.*“, „*60:1 Von krapfen.*“, „*61:1 Einen krapfen.*“ Dabei handelt es sich jedoch entweder um mit Fisch gefüllte Krapfen oder Apfelkrapfen, wie man sie heute kennt.

wurden die Äpfel und Birnen hier auch aus der Region bezogen. Ebenfalls erfährt man die Anleitung zur Teigherstellung: „*§ zu machen ein krapffen teick.* [...].“ Damals füllte man die Krapfen nicht ausschließlich mit Marmelade, sondern auch mit Fleisch: „*eß sein walt vogel. ob miltz. lungen. lebern. visch oder* [...].“[102] Bereits im 16. Jahrhundert ist der Krapfen beim Losungsamt der Reichsstadt Nürnberg aufgeführt.

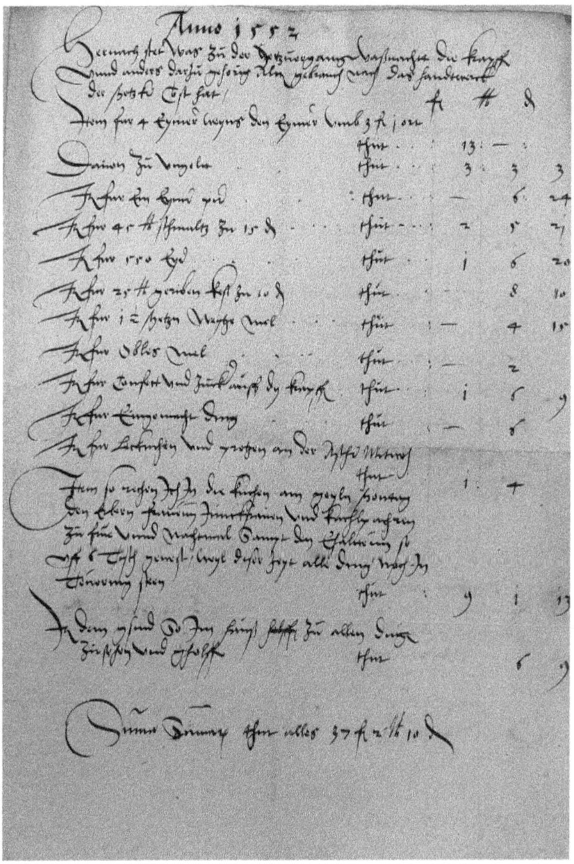
Anno 1552

Abb. 69: *Fastnachtsmahl der Stadtmetzger (1552). Bildnachweis: StAN, Rst. Nürnberg, Losungsamt, Stadtrechnungsbelege, Bündel 182.*

102 Küchenmeisterey, Rezept für die Füllung von Krapfen und der Anfang eines Rezepts für Krapfenteig, BSB, Inkunabeldruck, Nürnberg 1486, S. 41.

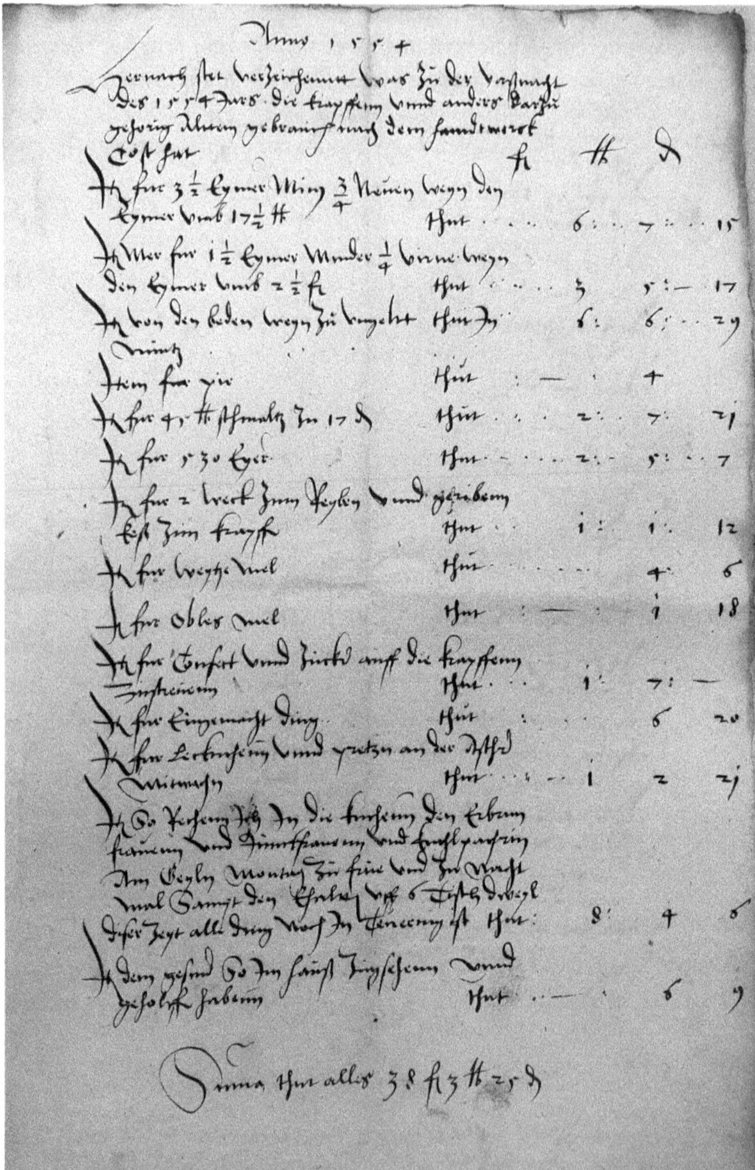

Anno 1554

Summa thut alles 38 fl 3 lb 25 d

Abb. 70: *Fastnachtsmahl der Stadtmetzger (1554). Bildnachweis: StAN, Reichsstadt Nürnberg, Losungsamt, Stadtrechnungsbelege, Bündel 182.*

In den Jahren 1552 und 1554 wird im Fastnachtsmahl der Stadtmetzger auch auf Krapfen nicht verzichtet: „*Item fur Confect vnd Zucker auff dy Krapffen*“ und „*Item fur Confect vnnd Zucker auff die Krapffen zu streuenn tut 1 fl. 7 ort*“.[103]

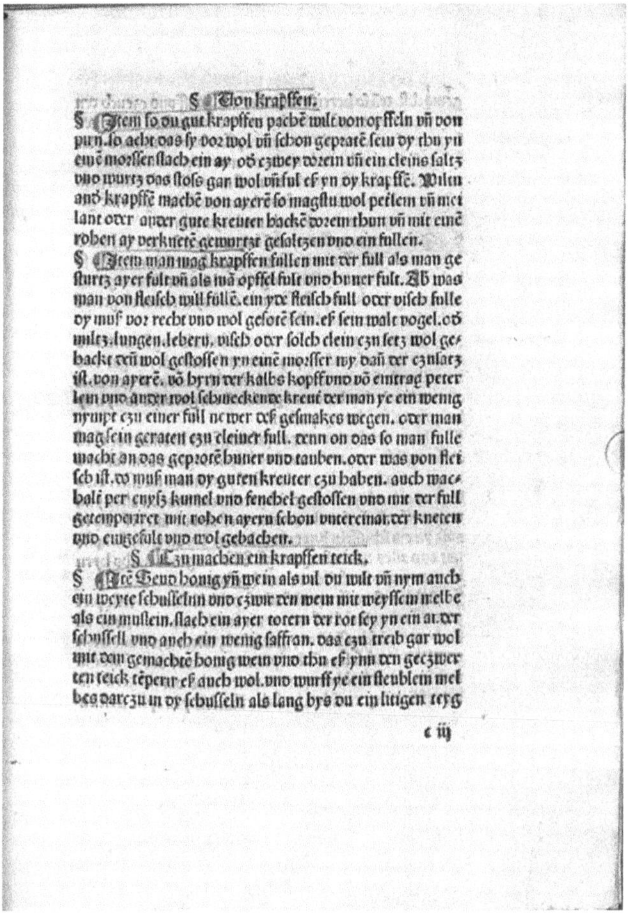

§ Von krapffen.
§ Item so du gut krapffen pachē wilt von opffeln vñ von
pirn. so acht das sy vor wol vñ schon gepratē sein dy thu yn
einē morser slach ein ay od̄ czwey dorein vñ ein cleins saltz
vnd wurtz das stoß gar wol vñ ful es yn dy krapffē. Wiltu
and̄ krapffē machē von ayerē so magstu wol petlein vñ mei
lant oder ander gute kreuter backē dorein thun vñ mit einē
roben ay verknetē gewurtzt gesaltzen vnd ein fullen.
§ Item man mag krapffen fullen mit der full als man ge
sturtz ayer fult vñ als mā opffel fult vnd huner fult. Ab was
man von fleisch will füllē. ein yde fleisch full oder visch fulle
dy muß vor recht vnd wol gesotē sein. es sein walt vogel. od̄
miltz. lungen. lebern. visch oder solch clein czu setz wol ge-
backe deñ wol gestossen yn einē morser wy dañ der czusatz
ist. von ayerē. vō hyrn der kalbs kopff vnd vō eintrag peter
lein vnd ander wol schmeckende kreut der man ye ein wenig
nympt czu einer full ne wer des gesmakes wegen. oder man
mag sein geraten czu cleiner full. denn on das so man fulle
macht an das geprotē huner vnd tauben. oder was von flei
sch ist. do muß man dy guten kreuter czu haben. auch wac-
holē per enyß kumel vnd fenchel gestossen vnd mit der full
getempertert mit roben ayern schon vntereinander kneten
vnd eingefult vnd wol gebachen.
§ Czu machen ein krapffen teick.
§ Itē Seud honig yñ wein als vil du wilt vñ nym auch
ein weyte schusseln vnd czwir den wein mit weyssem melbe
als ein muslein. slach ein ayer toter der rot sey yn ein ander
schussell vnd auch ein wenig saffran. das czu treib gar wol
mit dem gemachtē honig wein vnd thu es ynn den geczwer
ten teick tēperir es auch wol. vnd wurff ye ein steublein mel
bes darczu in dy schusseln als lang bys du ein lutigen teyg

c iij

Abb. 71: *Küchenmeisterey, erstes gedrucktes deutsches Kochbuch – Rezept für die Füllung von Krapfen und ein Rezept für Krapfenteig. Bildnachweis: BSB, Inkunabeldruck, Nürnberg 1486, S. 41.*

---

103 StAN, Reichsstadt Nürnberg, Losungsamt, Stadtrechnungsbelege, Bündel 182, Anno 1552 und 1554.

Diese fetthaltigen Süßgebäcke vor der Fastenzeit wurden abgerundet mit den Lebkuchen und der Fastenbrezel,[104] einer Spezialität, die heute insbesondere in Biberach an der Riß[105] und dem Blauen Ländchen[106] Gang und Gäbe ist und vorwiegend zur Fastenzeit angeboten wird. Dabei handelt es sich um einen leicht gesüßten und mit Zimt aromatisierten Hefeteig, der auf Stroh ausgebacken wird: „*Item fur Lebkuchen vnd pretzen an der Aschen Mittwoch tut 1 fl. 4 ort*".[107]

Aus dem „*Hauß buch*" des Klosters Seligenthal (Landshut) erfährt man „*von der geistlichen regeneration in der faßnacht nach den statuten*" zum Umgang mit Speisen: „*Am schmalzigen sambstag* [Samstag vor Aschermittwoch] *haben wier einer Jeden ain 3 schniten khiehl geben, sambt dem Air Zuntag* [Ostersonntag]. [...] *Am faßnacht sondag, die stunt umb 2 vhr am abent bet, oder sein die Hl. Stiegen gangen, haben den ganzen dag daz Hl. Stilschweigen verhalten, gibt etwas merers zue eßen, alß wirst vund brädl. etc.* [...] *zmorges ain supn* [Rosenmontag und Faschingsdienstag] *ins gmain erlaubt. Khain brandwein.*"[108]

---

104 Anton Hungari (Hg.), Osterglöcklein. Erbauliche Unterhaltungen für den Osterfestkreis im katholischen Kirchenjahre, Frankfurt am Main 1862, S. 49 f.

105 Zur Stadtgeschichte u.a.: Johann Daniel Georg von Memminger, Beschreibung des Oberamts Biberach, Stuttgart/Tübingen 1837; Hans-Hermann Garlepp, Der Bauernkrieg von 1525 um Biberach a.d. Riß. Eine wirtschafts- und sozialgeschichtliche Betrachtung der aufständischen Bauern (Schriften zur europäischen Sozial- und Verfassungsgeschichte 5), Frankfurt am Main u.a. 1987; Dieter Stievermann u.a. (Hg.), Geschichte der Stadt Biberach, Stuttgart 1991.

106 Johann Georg Theodor Grässe, Sagenbuch des Preußischen Staates, Bd. 2, Glogau 1871, S. 715, Nr. 807: „[...] *Theil der hessischen ehemaligen Grafschaft Katzenellenbogen und die Gegend von Nordenstadt bei Wiesbaden, wo die Landmädchen blaue Röckchen und Häubchen und die Burschen blaue Mützen tragen, obwohl nach Andern der blaue schieferartige Boden der dortigen Gebirge die Ursache des Namens gewesen sein soll. Der Sage aber nach wäre ein anderer Grund für die blaue Kleidung der dortigen Bewohner zu finden, nämlich folgender. Es sollen nämlich die Weiber aus Lorch, welche durch die Messen in Frankfurt a.M. mit der neuen Lehre Luthers bekannt geworden waren, dieselbe auch in ihre Heimath zu verpflanzen versucht haben, allein ihre Vaterstadt hing noch zu fest an Rom und so kam es, daß die ganze starke Weberzunft (300 Stühle stark) aus der Stadt vertrieben ward. Sie wandten sich nach der Herrschaft Katzenellenbogen, wo sie der Landgraf Philipp der Großmüthige aufnahm und wo ihr Fleiß das schöne blaue Tuch webte, welches dann die Kleidung der dortigen Bewohner ward.* [...].", URL: http://www.zeno.org/nid/20004953428 (letzter Zugriff: 26.8.2019).

107 StAN, Reichsstadt Nürnberg, Losungsamt, Stadtrechnungsbelege, Bündel 182, Anno 1552.

108 Ain Hauß buch, Bibliothek der Abtei Seligenthal, Hs IV, Seligenthal, 1670–ca. 1720, Digitalisat unter URL: http://daten.digitale-sammlungen.de/~db/0007/bsb00070706/images/index.html?seite=00001&l=de (letzter Zugriff: 29.8.2019), hier: Bildnummer 18.

Ganz frisch in Schmalz goldbraun gebacken und noch lauwarm ist der Krapfen genauso ein Genuss, wie das runde Schmalzgebäck in Altbayern, Franken, Thüringen und Westösterreich: Schmalznudel,[109] Knieküchle,[110] Kiacherl,[111] Kirchweihnudel,[112] Kirtanudel,[113] Knienudel[114] oder Sträubla genannt. Die „Auszogne“ besteht herkömmlich aus Hefeteig, welcher zu einer Kugel gerollt und anschließend so auseinandergezogen wird, dass die Mitte dünn und der Rand dick ist. Einige Leute ziehen den Teig auch über das Knie, um die gewünschte Form zu erhalten. Daher stammt der Name Knieküchle. Anschließend werden sie im heißen Butterschmalz schwimmend ausgebacken und mit Puderzucker bestreut. In Franken gibt es „katholische“ Knieküchla und „evangelische“ Kissen, je nachdem ob sie rund oder viereckig sind. Stets vor dem Hintergrund der kirchlichen Feste standen determinierte Speisen wie Ostereier[115] – Auferstehung und das frische Schlüpfen aus dem Ei als Symbiose des christlichen Glaubens –, Fastenbretzel,[116]

109 Thomas Anlauf, Heiß und fettig, veröffentlicht am: 5.6.2011, in: Süddeutsche Zeitung (SZ.de), URL: https://www.sueddeutsche.de/muenchen/schmalznudel-am-viktualienmarkt-heiss-und-fettig-1.1104910 (letzter Zugriff: 26.8.2019).

110 Art. „Knieküchle (Krapfen, Kiechla)“, in: Tourismusverband Franken e.V. (frankentourismus.de), URL: https://www.frankentourismus.de/rezepte/kniekuechle_krapfen_kiechla-2/ (letzter Zugriff: 26.8.2019).

111 Art. „Kirchweih-Kiachal, Kirchweih-Küchlein (Auszogene)“, in: Ammergauer Alpen GmbH (ammergauer-alpen.de), URL: https://www.ammergauer-alpen.de/Sonderprojekte/Kirchweih-Kiachal-Kirchweih-Kuechlein-Auszogene (letzter Zugriff: 26.8.2019).

112 Karin Trouboukis, In der Ruhe liegt die Kirchweihnudel, veröffentlicht am: 5.10.2017, in: Pfaffenhofener Kurier. Wolnzacher Zeitung. Geisenfelder Zeitung (www.donaukurier.de), URL: https://www.donaukurier.de/lokales/pfaffenhofen/Wolnzach-In-der-Ruhe-liegt-die-Kirchweihnudel;art600,3547996 (letzter Zugriff: 26.8.2019).

113 Art. „Kirchweihnudeln (oberbayerisch: Kirtanudeln)“, URL: https://www.chefkoch.de/rezepte/753511178096088/Kirchweihnudeln.html (letzter Zugriff: 15.10.2020).

114 Art. „‚Knienudeln‘ leicht gemacht“, veröffentlicht am: 17.6.2013, in: Onetz.de, URL: https://www.onetz.de/taennesberg/lokales/knienudeln-leicht-gemacht-d128925.html (letzter Zugriff: 26.8.2019).

115 Damit man nach der langen Fastenzeit die „alten“ übrigen Eier nicht mit den frischen verwechseln konnte, wurden sie zusätzlich rot eingefärbt.

116 Hans Angele, Bauernsprache und Bauerngeräte im Schwäbischen Oberland, Ochsenhausen 2006, S. 166; Hungari, Osterglöcklein (wie Anm. 104), S. 49 f. („*Die Fastenbretzel*“).

Krapfen, Seelenzöpferl[117] oder Gebildebrote wie Christstollen[118] auf der Speisekarte.

Ein weiteres Naschwerk wäre das „Kirschenmännla" oder der „Kirschenmichl", der in ganz Süddeutschland, aber ganz besonders in Franken sehr beliebt ist, denn die Fränkische Schweiz ist das größte Süßkirschenanbaugebiet Deutschlands. In etwaigen Abwandlungen findet sich das einstige „Arme-Leute-Essen" auch in anderen Gebieten wieder.

Wenden wir uns nun dem Osten zu. Hier kann man anstelle der oben geschilderten Süßspeisen sogenannte „Patschkauer Dohlen", ein Hefegebäck, genießen. Einer Überlieferung gemäß mussten die Junggesellen und „alten Jungfern" aus Patschkau[119] am Kirchweihmontag zur Strafe für ihr Ledigsein die Türme scheuern und das dafür benötigte Wasser in einem Sieb hinaufschaffen. Daher schrieb der Bäckermeister und Ehrenbranddirektor Adolf Scholz 1932 nachkommendes Gedicht:

---

117 Die Menschen auf der Erde können die Wartezeit der sogenannten „Armen Seelen" im Fegefeuer nun verkürzen, indem sie ihrer gedenken und für sie beten. Im Laufe der Jahrhunderte sind Allerheiligen und Allerseelen immer mehr zu einem Fest verschmolzen, das am 1. November gefeiert wird. Die Hefe-, Lebkuchen- oder Biskuitzöpfe waren in früheren Zeiten als Gaben für die Armen und sozial Schwachen gedacht. Damit es bei ihrem Kommen an nichts fehlte, stellte man ihnen „Seelenbrote" auf, die regional unterschiedlich auch „Seelenzöpfe", „Seelenwecken" oder „Seelenbrezen" genannt wurden. In einigen Gegenden Bayerns wurde noch bis in die Mitte des vorigen Jahrhunderts den heimkehrenden Seelen in der Nacht ein „Seelenmahl" aufgetischt und „Seelennudeln" zur Stärkung vor die Haustür gelegt. Eine kleine Anleitung dazu findet sich auf der Homepage des Erzbistums München, URL: https://www.erzbistum-muenchen.de/kinder/allerseelen-kinder-backen-seelenbrote (letzter Zugriff: 26.8.2019).

118 Nicht nur der Christstollen – siehe genauer: Art. „Wurde der Weihnachtsstollen in Naumburg erfunden?", in: Museen in Naumburg (mv-naumburg.de), URL: https://www.mv-naumburg.de/baecker (letzter Zugriff: 26.8.2019), Kap. „Das Bäckerprivileg 1329": „*und an desheiligen Crist*[us] *Abende zwey lange weyssene Brothe, die man Stollen nennet, gemacht von eynem halben Scheffel Weysses vns vnd vnsern Nachkommlingen in unsern Hof gelobt haben verbunden zu geben und zu reichen*" –, sondern auch der „Klausenmann" (= heiliger Nikolaus) aus Oberschwaben gehören zu den Gebildbroten. Dazu: IREKS-Arkady-Institut für Bäckereiwissenschaft (Hg.), IREKS-ABC der Bäckerei, Kulmbach [4]1985.

119 Paczków (Patschkau) liegt im südwestlichen Teil der Woiwodschaft Oppeln, etwa 80 Kilometer südwestlich von Oppeln. Hans-Georg Mohr / Leo Schiller, 1254–2004. 750 Jahre Patschkau. Die Geschichte der Stadt Patschkau in Schlesien, Osnabrück 2004.

„Die Potschker Tohla

Ei Potschke hoat's viel Tohla,
's is weit und bräät bekannt.
Drim wärd's wullt ooch vu viela
die Tohlastoadt genannt.
Durt uba uff da Tärma
gefällt's dan Vögeln gutt,
durt hausa sie ei Schärma
mit ihrer junga Brutt.
Doa? Durt die Tohla macha vuul,
's is siehr leicht zu verstiehn,
drim muß ma' oale Joahr' amoal
oan's Ränemacha giehn.
Oam Kermesmontiche gibt's a Fest –
Doa schleicha sich wie Diebe
die aala Jungfern und Junggesell'n
mit Woasser nuff eim Siebe.
Zu woas se durt nuff Woasser troan?
Doas selche derft Euch Wunder hoan:
Ihr werd't zwoar drieber lacha –
nu, doas sull zum Ränemacha!
Und druba, wie im Schwäß geboad't,
die aala Jungfern scheuern,
derweil se nie gedoacht tun hoan,
zur rechta Zeit zu heuern!"[120]

Das Vogelgebäck, die „Patschkauer Dohle",[121] reiht sich ein in die „Ornithologie" zwischen den „Luzerner Weckvogel" und den oberbayrischen „Nikolausvogel". Daran wird wieder sehr stark deutlich, dass ein ähnliches Gebäck in unterschiedlichen Regionen lediglich verschiedene Benennungen trägt, jedoch von der Substanz Gleiches aufweist.

120 Adolf Scholz, Die Potschker Tohla, in: Schlesische Mundart in Gedichten, hg. v. der Heimatgemeinschaft Patschkau und Umgebung im Neisser Kultur- und Heimatbund e.V., URL: https://www.patschkau.de/schlesiche-mundart/ (letzter Zugriff: 26.8.2019).

121 Globus 89, 1906, S. 221 f., URL: https://www.digi-hub.de/viewer/image/DE-11-001832352/239/ und https://www.digi-hub.de/viewer/image/DE-11-001832352/240/#topDocAnchor (letzter Zugriff: 26.8.2019).

z. B. der Vogelkopf. Die früheren Teigschlingen, die den Vogelkörper bilden, wurden vergessen. In der deutschen Heimat stellt man mit ihnen den Vogelkörper noch leidlich gut her, und man vergißt dabei auch niemals das Schwanzende des Vogels, das die jüdische Hausfrau wegläßt. Sichtbar lag das Vorbild für den südrussischen Judenvogel (Abb. 1) in der deutschen Heimat, wie die fünf deutschen Vogelgebäcke in Abb. 2 bis 6 bekunden.

Vogelgebäck.

Abb. 2 ist der Weggenvogel aus Luzern, Abb. 3 die Patschkauer Dohle aus Patschkau bei Neiße in Schlesien, die Karl v. Holtei in seinen schlesischen Gedichten besingt, wobei der Dichter ganz treffend in bezug auf die Bedeutung solcher Gebildbrote sagt:

„Mei lieber Man,
s' kümmt nischte nich uf a Namen ahn;
Ma mag a Ding, wie ma wihl, genennen,
De Hauptsach ihs ebe birsch derkennen."

Wie wir es erkennen, so fällt die Deutung aus.

Die Patschkauer erkannten den Weckvogel der Schweizer (Luzern; Abb. 2 u. 5) oder den Nikolausvogel der Oberbayern (Abb. 4) als die ihnen bekannteren Dohlen. Die südrussischen Juden aber stellten sich die Teigschlingen der deutschen Vorbilder als eine Stufenleiter der Himmelstreppe der Bibel vor; jedes Volk in seiner Art, jeder Bäcker in seiner Weise (vgl. Abb. 6 von der Münchener Bäckerei-Ausstellung 1905).

Hinter den meisten Vogelgebäcken aber steckt eigentlich das Seelenhuhn, das als Opferhuhn namentlich zur Saatzeit (Frühling und Herbst) eine Rolle im Volksbrauche spielt. Je nach der Kultzeit wird der Seelenvogel dann eine Taube, eine Henne, ein Schwan, eine Ente, eine Meise, ein Adler, sogar ein Schmetterling (= Sommervogel). Bei den altgriechischen Adonisfesten im Frühjahre waren die sizilianischen Tempel angefüllt mit Weihbroten in Vogelgestalt, „so viele als Weiber deren auf dem Backbrette zu kneten vermochten", so erzählt der Dichter der sizilianischen Idyllen in den Adoniazusen XV, 117; auch die alten Römer hatten (nach Lobeck, Aglaophamos 1080[h]) ein solches Vogelgebäck, das sie Erneum = *παρὰ τὸν ὄρνεον* nannten.

Bezüglich der übrigen abgebildeten Gebildbrote der südrussischen Juden behalte ich mir spätere Besprechung vor. Höfler.

Abb. 72: *Darstellung von Vogelgebäck: 1. süddeutscher Judenvogel, 2. Weggenvogel aus Luzern, 3. Patschkauer Dohle, 4. oberbayerischer Nikolausvogel, 5. Weckvogel der Schweizer, 6. Teigschlingen als Stufenleiter der Himmelstreppe der Bibel, hier: Münchner Bäckerei-Ausstellung 1905. Bildnachweis: Globus 89, 1906, S. 222, URL: https://www.digi-hub.de/viewer/image/DE-11-001832352/240/#topDocAnchor.*

Im Anschluss an das dreigängige Menü folgt nun ein Resümee zur regionalen Foodscape auf süddeutschem und europäischem Terrain.

4. Ausblick – Essenslandschaft als Teil einer Städtelandschaft

Ratsverlässe, Ratsbücher, Dekrete, Haushaltsbücher, aber auch Aufzeichnungen aus Klöstern und Spitälern sowie Rezeptbücher geben Aufschluss über die Essensgewohnheiten und Nahrungsmittel in Mittelalter und Früher Neuzeit. So reglementierte der Nürnberger Rat Qualität, Verkaufsrecht, Verkaufsort, Menge und Preishöhe sowie die Präsentation der Ware.[122] Daneben spielte stets die Implementierung der Vorschriften eine Rolle. Dies galt auch für den sogenannten „*Fürkauf*". Die Erzeugnisse des Stadtumlandes unterlagen dem Marktzwang

122 Frieser, Zwei spätmittelalterliche Wirtshäuser in Nürnberg (wie Anm. 35), S. 75.

und durften nur auf den Verkaufsplätzen angeboten werden.[123] Der Markt symbolisierte gleichermaßen die Zentralität jeder Stadt.

Gastarbeiter und Einwanderer aus fremden Ländern brachten ihre Tradition, ihr Brauchtum, aber auch Obst- und Gemüsesorten sowie Speisegewohnheiten mit, welche sie in neueröffneten Restaurants offerieren. Sie definieren über das Essen nicht nur Identität oder Identitätserhalt, sondern auch Heimat.[124] Das Nürnberger Knoblauchsland soll auch in Zukunft die „Koexistenz von Natur und Landwirtschaft" bewahren.[125] Das wird weiterhin ein Spagat für Grundstückseigner und Städteplaner sein, diese grüne Kulturlandschaft – Patrizierschlösschen und Wehrkirchen inbegriffen – nicht zu zerstören. Lebensmittel waren nie rein städtisch und kamen vom näheren Umland oder sogar aus dem Ausland in die Region.

Die Foodscape lässt sich in drei Kategorien unterteilen:

1. Produkte, die in der Geschichte lokal verankert waren und es heute noch sind. Darunter fällt etwa im Norden – spezieller in Norddeutschland, Norwegen, Dänemark, Schweden und in der Region um Liverpool – *„Lobscouse, labskaus, als ein Matrossengericht von Poͤckelfleisch, Biskwitt und Zwiebeln"*.[126] Ein weiteres Beispiel wäre das Schmalzgebäck, die schwäbischen beziehungsweise Allgäuer „Nonnenfürzle". Auch die bereits zur Schaffermahlzeit 1545[127] aufgetischten *„Braunkohl und Pinkel"* – geräucherte, grobkörnige Grützwurst mit Grünkohl – haben das Bremer Land nicht verlassen.
2. Produkte, die lokal gebunden waren, aber heute durch die Globalisierung weit verbreitet sind, jedoch unter besonderem politischem Schutz auf Herkunft stehen. Dazu gehören sicherlich die Bratwürste. Die Nürnberger Bratwurst ist zwar geographisch geschützt, jedoch gibt es auch in Hanoi Roster

123 Vgl. Sachs, Metzgergewerbe und Fleischversorgung (wie Anm. 90), S. 106 f.

124 Andreas Otto Weber / Patricia Erkenberg / Brigitte Steinert (Hg.), Kann Spuren von Heimat enthalten. Typische Rezepte der Deutschen aus dem östlichen Europa, München 2018.

125 Alexandra Foghammar / Christine Dierenbach, Von wegen Gurkentruppe: Gemüseerzeuger im Knoblauchsland, in: nuernberg.de – Stadtportal Nürnberg, URL: https://www.nuernberg.de/internet/stadtportal/nh_96_knoblauchsland.html (letzter Zugriff: 29.8.2019).

126 Fick, Vollständiges Englisch-Deutsches und Deutsch-Englisches Lexicon (wie Anm. 97), S. 362.

127 Unter anderem beschrieben bei: Karl H. Schwebel, Haus Seefahrt Bremen. Seine Kaufleute und Kapitäne. 400 Jahre Dienst am deutschen Seemann 1545–1945, Bremen 1947.

namens *xuc xich,*[128] die den Thüringer Bratwürsten ähneln. Zu dieser Gruppe gehören ebenso die geschützten Nürnberger Lebkuchen.

3. Austauschbare Produkte. In diese Kategorie reiht sich die anfangs beschriebene Lyoner ein.

Ein weiteres Phänomen ist die Beschimpfung verschiedener Kulturen, welche nicht zusammenpassen: Die „deutsche Kartoffel", die gar nicht aus Deutschland kommt, sondern aus dem Inkareich, und die Beleidigung der Italiener als „Spaghettifresser" – ein Gericht, dass heute jeder weltweit als Spaghetti Bolognese serviert bekommt.

Durch die Industrialisierung und die Globalisierung öffneten sich mittelalterliche Strukturen. Einerseits gab es zu bestimmten Produkten keine besondere Nachfrage, andererseits boomte die Erfinderschaft, zum Beispiel 1748, als William Cullen die erste künstliche Kühlung an der Universität Glasgow zeigte. Dadurch konnten Würste quer durch ganz Europa transportiert werden, ohne zu verderben. Obendrein „verehelichen" sich auch verschiedene Regionen. So ist das Erfolgsgeheimnis des „Cronuts"[129] anspruchslos: Der Schöpfer, Pâtissier Dominique Ansel, kombinierte zwei beliebte Klassiker der französischen und der US-amerikanischen Backkunst, nämlich das Croissant und den Donut.

Die Frage, die sich zum Schluss stellt: Über welche Distanz kann man eine Foodscape vertreiben, um noch Profit zu machen? Bei der Beantwortung dieser Frage spielen Fracht- und Erzeugerkosten eine übergeordnete Rolle.

Das Essen als Grundbedürfnis, wie es in der Maslowschen Bedürfnispyramide auftaucht, teilt sich in Slow Food[130] – Qualitätskriterien, wie der Verzicht von Geschmacksverstärkern –, ebenso in Junkfood – kalorienreiche Nahrung mit einem ungesund hohen Anteil von Salz, Zucker oder Fett und mit geringem Nährwert (typisches Beispiel: Hamburger) – auf. Dabei spielen regionale Ressourcen eine bestimmte Relevanz, so gibt es das Bodenseefelchen eben nur im Bodensee. Die regionale Diversität prägt die Essenslandschaft ebenso, wie etwa das neumodische *Urban Farming*. In acht ehemaligen Luftschutzbunkern,

---

128 Vgl. dazu: Andrzej Rybak, Michael Campioni, Grillfest in Hanoi, veröffentlicht am: 3.9.2015, in: ZEIT online (zeit.de), URL: https://www.zeit.de/2015/36/michael-campioni-bratwurst-asien (letzter Zugriff: 29.8.2019).

129 Maren Einfeldt, Der Cronut, veröffentlicht am: 9.8.2013, in: essen & trinken (essen-und-trinken.de), URL: https://www.essen-und-trinken.de/news/79947-rtkl-neuer-backtrend-der-cronut (letzter Zugriff: 30.8.2019).

130 Siehe dazu: Slow Food Deutschland e.V. – Slow Food Nürnberg, URL: https://www.slowfood.de/slow_food_vor_ort/nuernberg (letzter Zugriff: 20.8.2019).

33 Meter unter dem Londoner Stadtteil Clapham, liegt die Zukunft des *Urban Farming*. Hier soll *Zero Carbon Food* in Form von Gemüse angebaut werden.[131]

Unsere italienischen Nachbarn setzen in Bezug auf ihre süße Vergangenheit auf sogenannte „Sfogliatelle".[132] Einer Legende zufolge entwickelte 1600 eine Nonne im Kloster Santa Rosa in der Conca dei Marini (Salerno) an der Amalfiküste eher aus Versehen diese ikonische Gebäckspezialität aus Kampanien. So soll sie Reste von feuchtem Grieß mit Ricotta in ihre Rezeptur aufgenommen haben. Dazu kamen getrocknete Früchte, Zitronenlikör, Schmalz und Wein. *Et voilà*: Die Form der Mönchsmütze macht das Produkt der Sfogliatella perfekt. Im Unterschied zum französischen Croissant ist der Blätterteig bei dieser Süßigkeit aufgefächert und nicht parallel verlaufend. Diese neapolitanische regionaltypische Backtradition, die bis in die Moderne reicht – denkt man an den einzigartigen Wohlgeruch in den Gassen Neapels – wurde über die Jahre hinweg lediglich durch unterschiedlichste Cremefüllungen verfeinert.

In der Metropolregion Nürnberg setzt man auf „Biodiversität im Schnapsglas",[133] wenn es um die Kirschblüte in der Fränkischen Schweiz geht. Ansonsten legt man viel Wert auf „Alte Sorten". Die Heimat der Kirschpflaume liegt im Balkan und in Kleinasien. Sie wurde vor mehr als 2.000 Jahren bereits nach Italien eingeführt, seit dem 16. Jahrhundert auch nach Mitteleuropa. Ein Obstbauernhof in der Fränkischen Schweiz hat in einem alten Obstgarten eine wohlschmeckende Sorte entdeckt, die Erzeuger kultivierten diese in einer kleinen Anlage. Daraus machen die Lillinghofer Obstbauern Säfte aus eigener Herstellung.[134] *Upcycling* – aus Alt mach Neu – sollte der neue Slogan der Foodscape lauten, um Dagewesenes nicht aussterben zu lassen.

---

131 Birk Grüling, In London kommt Salat jetzt aus dem Luftschutzbunker, veröffentlicht am: 5.11.2014, in: WirtschaftsWoche (wiwo.de), URL: https://www.wiwo.de/technologie/green/urban-farming-in-london-kommt-salat-jetzt-aus-dem-luftschutzbunker/13550558.html (letzter Zugriff: 29.8.2019).

132 Vgl. Süddeutsche Zeitung Magazin, Nr. 48, vom 29.11.2019, S. 14–21.

133 Andreas Franke, Biodiversität im Schnapsglas, veröffentlicht am: 28.8.2019, in: Nürnberger Nachrichten, Region & Bayern (nordbayern.de), URL: https://www.nordbayern.de/region/biodiversitat-im-schnapsglas-1.9266275/7.4008426 (letzter Zugriff: 29.8.2019).

134 Vgl. Art. „Obst und Säfte aus biologischem Anbau", in: Homepage der Lillinghofer Obstbauern, URL: http://www.lillinghofer-obstbauern.de/index.html (letzter Zugriff: 29.8.2019).

Lina Schröder

„*Das Wasser ist so tief, dass auch große Schiffe anlegen können.*“ Die Topographie als Katalysator für die Stadtentwicklung in den Lage Landen

Abstract: The paper depicts the Lage Landen roughly correlating the present-day Benelux Countries as a macro-region focusing on its medieval and early modern "urban landscape" respectively "landscape of cities". At the current state of research, it is unclear which one of these two terms, both coined by Irsigler (1999), is applicable. It is evident, however, that the topography and the lack of royal sovereignty respectively the power of the church resulted in the development of a federalist governed cultural landscape throughout the region and its provinces in which the cities played a significant role.

Regarding the topography, the following three aspects previously named by contemporary pre-modern protagonists will be discussed: (1) the divergence in soil structure throughout the region, (2) the stream system of the Rhine-Meuse-Scheldt delta, as well as (3) the fact that a quarter of the Lage Landen, particularly the northwestern part, were located below sea level.

Thus, we can conclude that especially the cities as centers of manufacturing, organisation and communication served as crucial hubs in the face of these challenging topographical conditions ensuring nothing short of the survival of the people living in the towns and cities within this complex macro-region.

Ob im vorliegenden Fall der Lage Landen im Sinne Irsiglers von einer Städtelandschaft oder doch besser von einer Stadtlandschaft gesprochen werden muss, lässt sich zum jetzigen Forschungszeitpunkt nicht abschließend klären. Fraglos handelt es sich bei der Begrifflichkeit in beiden Fällen um ein komplexes, in einem bestimmten Raum verankertes Gebilde. Ferner sind es relative Beschreibungsgrößen, da die Zuordnung dieser Bezeichnungen nicht von der jeweiligen Stadtanzahl selbst abhängt, sondern vielmehr von der Form der Abgrenzung bezüglich der sie umgebenden Landschaft.[1] Maßgeblich für die

1 Franz Irsigler, Städtelandschaften und kleine Städte, in: Helmut Flachenecker / Rolf Kießling, Städtelandschaften in Altbayern, Franken und Schwaben. Studien zum Phänomen der Kleinstädte während des Spätmittelalters und der Frühen Neuzeit, München 1999, S. 13–38, hier: S. 32.

Unterscheidung ist hier also der Urbanisierungsgrad, der gleichfalls ein Indikator für die angesprochene begriffliche Relativität darstellt: So ist beispielsweise nicht außer Acht zu lassen, dass ursprüngliche Städtelandschaften im Verlaufe der Jahrhunderte durch zunehmende Urbanisierung zu einer Stadtlandschaft verschmolzen.[2]

Für die gesamtheitliche Betrachtung solcher Stadt-/Städtelandschaften bietet sich grundsätzlich die Vogelperspektive an; nur so kann zugleich auch der jeweilige Bezugsraum erfasst werden. Mit letzterem fallen diverse, die Gestalt und Ausprägung einer solchen Landschaft beeinflussende Rahmenbedingungen und die sich aus diesen herausgebildeten Strukturen ins Auge: herrschaftliche (geistlich, weltlich) sowie kulturelle (zum Beispiel Religion) Strukturen, Topographie und Klima, Bodenschätze und Ressourcen, natürliche Möglichkeiten für die Ausbildung von Verkehrswegen sowie die Verknüpfung der Region mit anderen Herrschaftsräumen (zum Beispiel ökonomisch, politisch). Mit anderen Worten: Eine Stadt- respektive Städtelandschaft muss stets vor dem Hintergrund ihrer jeweiligen Bezugsregion sowie einem spezifischen Betrachtungszeitraum untersucht werden.

Es lassen sich dabei verschiedene Aspekte analysieren: So kann zunächst nach ihrer Entstehung gefragt werden (a): Wieso verdichtet sich in bestimmten Regionen Urbanität und in anderen nicht? Handelt es sich um einen längerfristigen, natürlichen Prozess oder werden die Städte sozusagen im Rahmen herrschaftlicher Maßnahmen auf dem Reißbrett geplant?

Ferner ist ihre weitere Entwicklung in den Blick zu nehmen (b): Was passiert mit den Städten einer Landschaft im Verlaufe der Jahrhunderte? Was und wer sind die diesbezüglich Einfluss nehmenden Faktoren und Akteure? Welche weiteren Umstände wirken sich aus?

Ein anderer Gesichtspunkt ist die Vernetzung der Städte untereinander (c): Sind die Städte einer Stadt- beziehungsweise Städtelandschaft durch ein spezifisches Muster an Vernetzung und Kommunikation gekennzeichnet? Und wenn ja, welche Faktoren sind dafür verantwortlich und wie verändert sich möglicherweise das Beziehungsgeflecht im Verlaufe der Zeit? Im Falle einer Städtelandschaft geht Irsigler beispielsweise von gewissen hierarchischen Beziehungen oder dominierenden Stadttypen aus.[3]

---

2 Ebd.
3 Ebd.

Nicht zuletzt müssen die einzelnen Städte und ihre jeweiligen Strukturen in den Blick genommen werden (d): Welche baulichen, organisatorischen, ökonomischen etc. Gemeinsamkeiten oder Unterschiede weisen sie auf und wie lassen sich diese erklären? Verbindet die Städte einer Stadt- respektive Städtelandschaft mehr als nur die Tatsache, dass sie gemeinsam einen Raum mit verdichteter Urbanität beschreiben – zum Beispiel eine gemeinsame Identität?[4]

Für die Tagung „Die süddeutsche Städtelandschaft im europäischen Vergleich" hatten sich die Organisatoren unter anderem die Diskussion dreier interessanter und zugleich spannender Fragen vorgenommen: (1.) Was meint Städtelandschaft im europäischen Vergleich?, (2.) Was lässt sich im Rahmen eines europäischen Vergleichs über den Typus „Stadt" in Abgrenzung von diversen Siedlungsformen herausfinden? sowie (3.) Welche Rolle kommt dem Menschen als „ewige Konstante" im epochenübergreifenden Konstrukt einer Städtelandschaft zu?

Während die zweite Frage ein altbekanntes Forschungsproblem aufgreift, geht schon die erste über die Erforschung einer einzelnen Städtelandschaft hinaus. Sie lässt sich nur vor dem Hintergrund eines systematischen Vergleichs mehrerer Städtelandschaften beantworten. Die Tagung sollte als Forum für diesbezügliche Diskussionen dienen. Die Corona-Pandemie machte den erforderlichen Austausch unmöglich. Entsprechend bleibt an dieser Stelle lediglich der Blick auf die für die eigene Forschung ausgewählte Stadt- beziehungsweise Städtelandschaft(en), es lassen sich bezüglich der zur Diskussion gestellten Fragen nur Hypothesen formulieren. Die dritte Frage hingegen zielt auf die eingangs erwähnte Entwicklung einer Stadt- respektive Städtelandschaft (b) und lässt möglicherweise über das eigene Betrachtungsbeispiel eine erste Einschätzung zu.

Die eigenen Forschungen führten dazu, den drei Tagungsfragen am Beispiel der Lage Landen exemplarisch und in gebotener Kürze nachzugehen. Bei letzteren handelt es sich um die heutige Benelux-Region sowie das französische Département Nord.[5] Der Raum gilt als Paradebeispiel für eine frühe Urbanisierung und fungiert hier als Bezugsregion für die dortige Stadt-/Städtelandschaft.

---

4 Die Auflistung dient lediglich als Anregung und erhebt keinen Anspruch auf Vollständigkeit.

5 Letzteres soll nur am Rande mitbetrachtet werden, überhaupt liegt der räumliche Schwerpunkt des Beitrags aufgrund der bisher unternommenen eigenen Forschungen auf den heutigen Niederlanden und Belgien.

Die fundierte Beantwortung der Tagungsfragen setzt die Bearbeitung der eingangs erwähnten vier Betrachtungsaspekte (a) bis (d) voraus. Insbesondere (b) und (d) erfordern diesbezüglich intensive Forschungsarbeiten. Einerseits wird ein die gesamte Betrachtungsregion umfassender Längsschnitt benötigt (b), andererseits muss jede einzelne Stadt – zumindest stichprobenartig – in Bezug auf ihre lokalen Strukturen und ihre Entwicklung hin analysiert werden (d).

Erschwerend im Falle der Lage Landen kommt hinzu, dass diese im Gegensatz zu der dortigen Stadt- respektive Städtelandschaft bis dato von der Forschung nur bedingt als gesamte Region wahrgenommen und diskutiert wird. Insbesondere die nationalen Grenzen erschweren den Zugang über die eingangs erwähnte Vogelperspektive.

Eine Aufgabe des hiesigen Beitrags wird also darin bestehen, überzeugende Argumente zusammenzutragen, die perspektivistisch die Lage Landen in ihrer Gesamtheit als Region stützen. Dies ist umso wichtiger, als dass davon ausgegangen wird, dass sich Urbanisierung stets vor dem spezifischen Hintergrund einer Region vollzieht. Die Zerlegung der Region in einzelne, womöglich noch an den nationalen Grenzen orientierte Teilräume wird somit dem Verständnis der dortigen Stadt-/Städtelandschaft nicht gerecht, so der Eindruck. Nationale Forschungsgrenzen überwindend wurde sich daher darum bemüht, soweit wie möglich auch die aktuelle Forschungsliteratur der Nachbarländer in die Betrachtung miteinzubeziehen.[6] Einen Anfang setzend – mehr kann dieser Beitrag hier nicht leisten – wird dabei vor allem die Entstehung der Stadt-/Städtelandschaft fokussiert (a). Die Aspekte (b), (c) und (d) sollen dabei zwar mitgedacht, jedoch nur am Rande betrachtet werden. Wie bereits erwähnt, bedarf es hier noch weiterer, intensiver Forschungen.

Bezüglich der Klärung der drei Tagungsfragen sollten nach dem Willen der Organisatoren „historische Gegebenheiten aus Politik, Geographie, Kunst, Kultur, Musik, Medien, Literatur, Religion und Ökonomie“[7] miteinbezogen werden – im Falle des vorliegenden Beispiels muss vor allem der Topographie dieser Region ein besonderes Augenmerk geschenkt werden. So soll hier in Anlehnung an die Überlegungen van Dams (2016) bezüglich der heutigen Niederlande die

6 Auch hier besteht an keiner Stelle Anspruch auf Vollständigkeit.

7 Beschreibung des Tagungsinhaltes auf H-Soz-Kult, URL: https://www.hsozkult.de/event/id/termine-41958?title=die-sueddeutsche-staedtelandschaft-im-europaeischen-vergleich&recno=2&q=Stadtlandschaft&sort=newestPublished&fq=&total=2 (letzter Zugriff: 27.4.2020). Für Lektorat und anregende Diskussionen bedanke ich mich bei meinem niederländischen Kollegen Tjerk Miedema.

Hypothese gewagt werden,[8] dass die in den Lage Landen entstandene Stadt-/Städtelandschaft insgesamt vor allem ein Ergebnis topographischer Gegebenheiten darstellt. Unter anderem spielt dabei der die regionale Struktur vorgebende Fluss mit seinem ambivalenten Charakter eine wichtige Rolle: „‚Er ist in seiner Längsrichtung ein verbindendes Element, insbesondere war er das in jenen Zeiten, die auf den Wasserweg angewiesen waren. Gleichzeitig ist ein Fluss ein Verkehrshindernis in seiner Querrichtung' – übrigens auch dann, wenn er keine politische Grenze bildet."[9] An dieser Stelle zu ergänzen ist seine einerseits fruchtbare Ausstrahlung, anderseits durch seine Hochwasser Gefahren bergende Eigenschaft.

Das erste Kapitel widmet sich entsprechend neben dem Forschungsstand der Beschreibung der topographischen Rahmenbedingungen dieser Region. Zunächst diskutiert es allerdings, weshalb die aus, streng genommen, fünf Nationen bestehenden Lage Landen nach Meinung der Verfasserin als eine Region anzusehen sind. Zum besseren Nachvollziehen der Argumentation wird hier, das Ziel der Tagung und die eigenen Forschungen aufgreifend, an einigen Stellen vergleichend die Danziger Bucht herangezogen, da sie nicht nur mit den Lage Landen topographische Gemeinsamkeiten, sondern auch forschungstechnisch ähnliche Probleme aufweist.

Im zweiten Kapitel werden die Folgen der regionalen Rahmenbedingungen anhand der sich dort entwickelten Herrschaftsstrukturen aufgezeigt. Im letzten Abschnitt wird schließlich den drei Fragen nachgegangen. Der zeitliche Rahmen der Darstellung liegt dabei schwerpunktmäßig auf dem Mittelalter und der Frühen Neuzeit.

1. Die Lage Landen und ihre regionale Erforschung

1.1. Zum Begriff der „Lage Landen"

Der heute als „Benelux-Region" feststehende Begriff wurde 1958 durch die vertragliche Organisation eines gemeinsamen Wirtschaftsraums (EWG) als

---

8 Petra J. E. M. van Dam, An Amphibious Culture. Coping with floods in the Netherlands, in: Peter Coates / David Moon / Paul Warde (Hg.), Local Places, Global Processes, Oxford 2016, S. 78–93, hier: S. 78.

9 Martin Knoll, „Natürliche" Grenzen? Zur Erfahrung von Region und Territorium in der Frühen Neuzeit, in: Barbara Kuhn / Ursula Winter (Hg.), Grenzen. Annäherungen an einen transdisziplinären Gegenstand, Würzburg 2019, S. 81–99, hier: S. 82.

politisches Sicherheitsunternehmen nach dem Zweiten Weltkrieg zwischen Belgien, Luxemburg und den Niederlanden geprägt.[10] Interessanterweise erfolgte mit der Gründung dieser Wirtschaftsunion die Wiederherstellung des größten Teils der spätmittelalterlichen Geographie – zumindest für ökonomische Zwecke – dieser auch als „Lage Landen" oder „Nederlande" bezeichneten Region:[11] *„Dit deel van Belgica den Catholijcke koninck toebeoorende/ wordt gemeynlijk Nederlande geheeten/ obermits de nedericheydt ende leeghedt nae de wilde Zee steckende."* – *„Dieser Teil von Belgica, welches dem katholischen König gehört/ wird im Gemeinen als Niederlande bezeichnet/ wegen der niedrigen und tiefen Lage zur wilden See."*[12]

Das Zitat des florentinischen Kaufmanns Lodovico Guicciardini (1612) verweist direkt auf die aus dem Niederländischen stammenden Begriffe „Nederlande" oder „Lage Landen" – im Englischen *low lands*, im Spanischen *países bajos*, im Französischen *Les Pais-Bas*. Sie bedeuten soviel wie „niedere Lande" und bezeichnen das an der Nordseeküste liegende Gebiet unter Betonung seiner teilweise unterhalb des Meeresspiegels liegenden Topographie. Die Umschreibung jenes Raumes mit dem Begriff „Lage Landen" erleichtert eine regionale Betrachtung fernab heutiger beziehungsweise sich seit dem 11./12. Jahrhundert verschiebender politisch-herrschaftlicher Grenzen und Zuordnungen innerhalb dieser Region.

Diese entstanden einerseits durch die im Verlaufe der Epochen öfter wechselnde Zugehörigkeit der gesamten Region oder Teile dieser zu unterschiedlichen Herrschaftsdynastien. Entsprechend in der Geschichtswissenschaft Verwendung findende und einen Nichtkenner der regionalen Geschichte eher verwirrende Bezeichnungen wie zum Beispiel „burgundische", „habsburgische", „spanische", „französische" oder „österreichische Niederlande" gehen darauf zurück.

---

10 Wilfried Loth, Europa seit dem Zweiten Weltkrieg. Spaltung und Einigung, in: Dirk Ansorge / Dieter Geuenich / Wilfried Loth, Wegmarken europäischer Zivilisation, S. 372–384, hier: S. 379 f.

11 Dirk Maczkiewitz, Der niederländische Aufstand gegen Spanien (1568–1609). Eine kommunikationswissenschaftliche Analyse (Studien zur Geschichte und Kultur Nordwesteuropas 12), Münster u.a. 2005, S. 31.

12 Zitiert nach: ebd., S. 31 – frei aus dem Niederländischen übersetzt durch die Verfasserin.

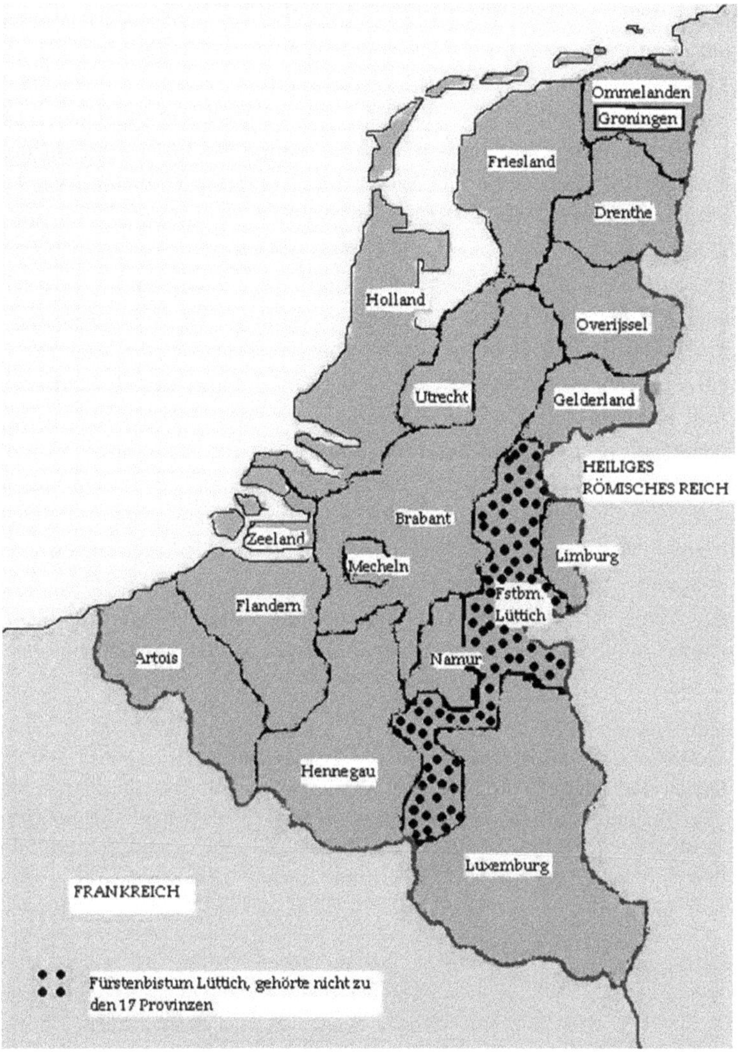

Abb. 73: *Die 17 Provinzen der Lage Landen zu Beginn der Herrschaftsübernahme durch Philipp II. 1555. Diese Abbildung ist weder maßstabsgetreu noch von den Grenzlinien der einzelnen Provinzen her 100 % korrekt. Sie ist eine aus urheberrechtlichen Gründen grobe und einfache Eigenproduktion der Verfasserin, die dazu dienen soll, dem Leser zum besseren Verständnis des Ausgeführten einen ungefähren geographischen Überblick zu ermöglichen. Die Verfasserin rät dazu, zum genauen Nachvollziehen der Grenzlinien, Provinzgrößen etc. einen historischen Atlas zu bemühen. Bildnachweis/Copyright: Lina Schröder.*

Andererseits bestand die zum Burgundischen Reichskreis gehörende Region,[13] ähnlich wie Franken,[14] aus einem Flickenteppich einzelner Herzogtümer, Grafschaften und ein paar wenigen Bistümern – den späteren sogenannten Provinzen, niederländisch *Geweste*. Diese waren seit ihrem offiziellen Zusammenschluss im Burgundischen Vertrag am 26. Juni 1548 unter der Führung des Habsburgers Karl V. die Grafschaft Artois, das Herzogtum Brabant mit der Markgrafschaft Antwerpen, der Grafschaft Löwen, der Grafschaft Brüssel und den Vogteien Nivelles und Gembloux, die Provinz aus Drenthe, Lingen, Wedde und Westerwolde, die Grafschaft Flandern mit den Burggrafschaften Lille, Douai und Orchies, der Herrschaft Tournai und deren Umland, das Herzogtum Geldern mit der Grafschaft Zutphen (spätere Provinz Gelderland), die Herrschaft Groningen, die Grafschaft Hennegau, die Grafschaft Holland, das Herzogtum Limburg mit Dalhem, Valkenburg und Herzogenrath, das Herzogtum Luxemburg, die Herrschaft Mechelen, die Grafschaft Namur, die Ommelande, die Herrschaft Overijssel, das Fürstbistum, später Herrschaft Utrecht, die Herrschaft Westfriesland sowie die Grafschaft Zeeland. Das Fürstbistum Lüttich gehörte nicht dazu.[15]

An dieser Stelle muss die Frage nach der Berechtigung aufgeworfen werden, die gesamten Lage Landen, die ja heutzutage mit den Niederlanden, Belgien, Luxemburg, Teilen Frankreichs und dem Rhein-Maas-Gebiet fünf Nationen umfassen,[16] als eine Region zu begreifen. Es handelt sich hier ursprünglich um

13 Maczkiewitz, Der niederländische Aufstand gegen Spanien (wie Anm. 11), S. 31.

14 Helmut Flachenecker / Stefan Petersen, Kirche vor Ort: Pfarrei, Kloster und Stift im Mittelalter und im Zeitalter der Reformation, in: Werner Freitag / Michael Kißener / Christine Reinle / Sabine Ullmann (Hg.), Handbuch Landesgeschichte, Berlin/Boston 2018, S. 546–577, hier: S. 560; Oliver Auge / Joachim Schneider, Der Niederadel, in: Freitag / Kißener / Reinle / Ullmann, Handbuch Landesgeschichte (wie oben), S. 303–334, hier: S. 311.

15 Maczkiewitz, Der niederländische Aufstand gegen Spanien (wie Anm. 11), S. 39; Johannes Knoll, Geschichtlicher Überblick, in: Ders. (Hg.), Belgien. Geschichte, Politik, Kultur, Wirtschaft, Münster 2007, S. 5–44, hier: S. 6; Lina Schröder, Schnittstelle Niederrhein. Die Gründung der niederländischen Republik. Eine systemtheoretische Betrachtung, Kleve 2013, S. 15 f. Maczkiewitz bemerkte im Jahr 2005, dass die Zuordnung und Beschränkung auf die Zahl 17 in der Vergangenheit zwar hinterfragt, jedoch durch keine andere, unumstrittene, ersetzt worden sei – daran hat sich sowohl diesseits als auch jenseits der Grenzen bis heute nichts geändert. Maczkiewitz, Der niederländische Aufstand gegen Spanien (wie Anm. 11), S. 39.

16 Wenn die Gegend zwischen Rhein und Maas dazugerechnet wird – diesbezügliche Argumente können in diesem Beitrag nur am Rande diskutiert werden –, sind es mit den Deutschen fünf Nationen.

etwa 90.000 km^2 – für eine einzige Region eine ziemlich große Fläche.[17] Zum Vergleich: Das Elsass umfasst etwa 8.280 km^2, Franken (Ober-, Mittel-, Unterfranken) etwa 23.007 km^2, Schwaben ungefähr 9.994 km^2 sowie der Raum rund um die Danziger Bucht herum – in etwa das damalige Ost- und Westpreußen, heute die drei polnischen Woiwodschaften Pommern, Kujawien-Pommern und Emsland-Masuren – circa 62.544 km^2.

Von den aufgezählten Beispielen kommt größentechnisch nur die Danziger Bucht den Lage Landen nahe. Interessanterweise ist auch hier ihre Interpretation als eine Region nicht unumstritten: Als Gründe für das Postulat zweier Einzelregionen Ost- und Westpreußen hielt Bömelburg (1997) beispielsweise fest, dass einerseits alleine schon die ostpreußische und westpreußische Landesgeschichte diese Region aufsplitte. Während die Ostpreußische der Traditionslinie Ordensherrschaft – Herzogtum Preußen – Ostpreußen folge, richte sich die Westpreußische hingegen nach der Linie pommerellische Fürsten – ständischer Widerstand – königliches Preußen – Westpreußen.[18]

Andererseits führt Bömelburg Unterschiede in der Geographie an: Seiner Meinung nach existieren mit Pommerellen, dem Kulmer Land, der Weichsel- und Nogatniederung (Weichseldelta) und dem Ermland alleine schon vier deutlich voneinander abgrenzbare Kleinregionen nebeneinander.[19] Ähnlich ließe sich auch im Falle der Lage Landen argumentieren, denn sicherlich weisen, um nur ein Beispiel herauszugreifen, das Rhein-Maas-Schelde-Delta und die um Lüttich herum sowie Teile Limburgs umfassende Montangegend geographische Unterschiede auf. Auch forschungstechnisch wird die Geschichte der Lage Landen durch die heute existierenden Landesgrenzen zerstückelt.

Unter Zugrundelegung der Aussage Simmels (1903), dass Grenzziehungen nicht als „räumliche Tatsache mit soziologischen Wirkungen, sondern [~~eine~~] [als] soziologische Tatsache, die sich räumlich formt",[20] zu begreifen sind, lässt

---

17 Die heutigen Benelux-Staaten haben in etwa eine Fläche von 75.000 km^2. Die Verluste erklären sich hauptsächlich durch Gebietsabtretungen an Frankreich und Deutschland. Maczkiewitz, Der niederländische Aufstand gegen Spanien (wie Anm. 11), S. 31.

18 Hans-Jürgen Bömelburg, Die königlich preußische bzw. westpreußische Landesgeschichte in der Frühen Neuzeit – Probleme und Tendenzen. Eine Streitschrift, in: Jörg Hackmann, Pommerellen – Preußen – Pomorze Gdańskie. Formen kollektiver Identität in einer deutsch-polnischen Region (Nordost-Archiv, Zeitschrift für Regionalgeschichte VI/2), Lüneburg 1997, S. 607–628, hier: S. 612.

19 Ebd., S. 614.

20 Georg Simmel, Soziologie des Raumes, in: Gustav Schmoller (Hg.), Das Jahrbuch für Gesetzgebung, Verwaltung und Rechtspflege des Deutschen Reiches. Neue Folge, Leipzig 1903, S. 27–71, hier: S. 46.

sich jedoch zumindest die forschungstechnische „Zerschneidung" der hier angesprochenen Räume sowohl im Falle der Lage Landen als auch in Bezug auf die ost- und westpreußische Landesgeschichte guten Gewissens vernachlässigen. Nicht zuletzt verweist auch die Idee vom „Europa der Regionen" auf die Problematik nationaler Staatsgrenzen.[21]

Was aber macht eine Region dann tatsächlich aus? Eine gemeinsame Topographie, ein gemeinsames Klima, eine gemeinsame Sprache, ein gemeinsames Brauchtum, eine gemeinsame Identität, gemeinsame Herrschaftsräume in Mittelalter und Neuzeit, die Tradition eines Sozial- respektive Wirtschaftsraumes, doch eher kirchlich-religiöse Strukturen oder eben, den Bezug zum vorliegenden Thema wieder herstellend, eine Stadt-/Städtelandschaft?

In der Geographie stellt eine Region beispielsweise eine zentrale Bezugsgröße der Wirtschaftsgeographie dar, „um soziale Interaktionen oder Organisationsformen der Produktion zuzuordnen, zu lokalisieren und vergleichen zu können. Es handelt sich hierbei um einen konkreten Ausschnitt der Erdoberfläche, der aufgrund bestimmter Prinzipien oder Strukturen abgrenzbar ist und dadurch von anderen Regionen unterschieden werden kann."[22] Als Abgrenzungsprinzipien wird sich hier ferner dem Verwaltungsprinzip (administrative Einheiten), dem Homogenitäts- (zum Beispiel Klima, Topographie) und dem Funktionalitätskriterium (zum Beispiel Zusammenfassung wirtschaftlicher Faktoren) bedient.[23]

Die Historikerin Martina Steber (2012) definiert Region hingegen als „eine Raumeinheit mittlerer Größe, die durch ihre relative Unbestimmtheit und Fluidität gekennzeichnet ist. Sie bedarf einer Referenzgröße, definiert sich mithin in Abgrenzung von beziehungsweise in Beziehung zu kleineren Raumeinheiten, wie Städten, sowie zu größeren Raumeinheiten, wie etwa dem Nationalstaat."[24] Wie schon die geographische Definition, hebt auch sie Herrschaftsstrukturen, Wirtschaftsbeziehungen, geographische Gegebenheiten, politische, soziale oder religiöse Vernetzungen, aber auch Produktionsregime, Konsumorientierungen, Verkehrsnetze oder Familienbeziehungen hervor.[25] Eine Grundvoraussetzung für die Ausbildung einer Region ist für sie dabei die Kommunikation, außerdem

---

21 Siehe zum Beispiel den Europäischen Ausschuss der Regionen (AdR) in Brüssel.

22 Harald Bathelt / Johannes Glückler, Wirtschaftsgeographie, Stuttgart [4]2018, S. 73.

23 Ebd., S. 73 f.

24 Martina Steber, Region, in: Leibniz-Institut für Europäische Geschichte (Hg.), Europäische Geschichte Online (EGO), Mainz 2012, URN: urn: nbn: de:0159-2012030764; URL: http://www.ieg-ego.eu/steberm-2012-de (letzter Zugriff: 11.5.2020).

25 Ebd.

muss eine Region bewusst als solche von den Zeitgenossen wahrgenommen werden.[26]

Beide Definitionen zeigen, dass es sich hier im Wesentlichen um Kriterienbündel handelt und es auch immer ein Stück weit von der jeweiligen Betrachtungsperspektive abhängt, was als zur Region zugehörig wahrgenommen wird. Es zeigt sich, wie wichtig hier die eingangs erwähnte Vogelperspektive ist. Denn ein Wirtschaftswissenschaftler wird sich wahrscheinlich vornehmlich an den ökonomischen Strukturen orientieren, der Geograph anhand der Topographie, der Klimatologe am Klima und der Historiker vielleicht an den Herrschaftsräumen.

Abb. 74: *Danziger Bucht. Bildnachweis/Copyright: Lina Schröder.*

26 Ebd.

Auf letztere Bezug nehmend, lässt sich sowohl für die Danziger Bucht als auch für die Lage Landen die über viele Jahrhunderte währende Unveränderlichkeit der herrschaftspolitischen „Außengrenzen" hervorheben: Erstere erhielt ihre grundlegende Gestalt zunächst im Rahmen der Deutschordensherrschaft, die in etwa das spätere Ost- und Westpreußen umfasste. Die im Rahmen der über zweihundert Jahre währenden Herrschaft des Ritterordens erfolgte Christianisierung und Siedlungspolitik bildet das Fundament für die gesamte weitere Geschichte und Multikulturalität dieser Region.

Aber auch nach der Vertreibung des Deutschen Ordens wiesen Ost- und Westpreußen zwischen den 1470er Jahren (Eingliederung des Ermlandes) und 1772 vor allem nach Norden *eo ipso*, Osten und Westen hin unveränderte Grenzziehungen auf; sie gehörten zu den Stabilsten in Ostmitteleuropa. Auch nach 1772 blieb Westpreußen nahezu gleich, von der Abtrennung des Ermlandes zu Ostpreußen, der Eingliederung der Kreise Marienwerder und Riesenburg abgesehen.[27] In den Jahren von 1824/29 bis 1878 wurden Ost- und Westpreußen gar wieder zu einer großen Provinz Preußen zusammengelegt. *Summa summarum* hatte sich also seit dem Mittelalter bezüglich der „äußeren Grenzziehung" nichts Wesentliches geändert.

Ähnliches lässt sich auch für die Lage Landen feststellen: Auch wenn eine Zusammenfassung der dortigen Provinzen offiziell erst im Burgundischen Vertrag 1548 erfolgte, hatten sich die unterschiedlichen *Geweste* bereits ab 1100 begonnen zu formieren, verschiedene Grafengeschlechter (Grafen von Holland, von Flandern, von Zeeland etc.) übernahmen die Kontrolle über ihr Lehen.[28]

Die herrschaftspolitische Entwicklung der heutigen Benelux-Region war dabei einerseits eng an die Historie des französischen Königtums geknüpft, wodurch sich der noch heute in Teilen Belgiens (Wallonie) bestehende französische Einfluss in Sprache und Kultur erklären lässt. Andererseits bestimmte dort eine gezielt betriebene Eroberungs- und Heiratspolitik im Vorfeld, welche Ländereien im Erbfall an wen fielen.[29]

Die zunehmende Selbstständigkeit der Region ist laut von Seggern (2018) unter anderem auf die Schwäche des französischen Königtums zurückzuführen: 1413

27 Bömelburg, Preußische bzw. westpreußische Landesgeschichte (wie Anm. 18), S. 609.

28 Roel Zijlmans, Troebele betrekkingen, Grens, scheepvaart- en waterstaatskwesties in de Nederlanden tot 1800, Hilversum 2017, S. 152 f.

29 Harm von Seggern, Geschichte der burgundischen Niederlande, Stuttgart 2018, Kap. 2.2–2.4, S. 31–45; Tanja Michalsky, Projektion und Imagination. Die niederländische Landschaft der Frühen Neuzeit im Diskurs von Geographie und Malerei, München 2011, S. 41.

kam es zu einem Zerwürfnis zwischen den bis dato eng mit der französischen Krone verflochtenen Burgundern,[30] durch welche der burgundische Einfluss in Frankreich ein für alle Mal zurückgedrängt wurde. Stattdessen konzentrierten sie sich nun vollständig auf die Lage Landen.[31] Durch die 1477 erfolgte Heirat der letzten burgundischen Erbin Maria von Burgund mit dem Kaisersohn Maximilian von Österreich fielen nach einem für sie tödlich endenden Jagdunfall schließlich ihre Besitzungen in den Lage Landen an das Haus Habsburg.[32]

Nach dem bereits erwähnten Burgundischen Vertrag 1548 erließ Kaiser Karl V. am 4. November 1549 die Pragmatische Sanktion, in welcher er die einzelnen Provinzen zu einer unteilbaren Herrschaft zusammenfügte und damit die Erbfolge für sein Territorium regelte. Dennoch konnte sein Sohn Philipp II., der auch König von Spanien war, es nicht verhindern, dass sich 1581 sieben der nördlichen Provinzen, nämlich Holland, Zeeland, Utrecht, die Groninger Ommelanden, Friesland, Overijssel und Gelderland von der spanischen Krone lossagten. Aus diesen ging in der Folge die niederländische Republik hervor, was sich jedoch auf die Region in ihrer Gesamtheit nicht weiter auswirkte – außer, dass es jetzt eine weitere Binnengrenze gab.[33]

Spätestens im Rahmen des Wiener Kongresses wurden 1815 jedoch sämtliche Provinzen wieder unter dem Namen „Königreich der Vereinigten Niederlande" geeint, was allerdings nicht lange andauern sollte. Dennoch: Auch die Gründung des Königreichs Belgien 1830 und die vollständige Unabhängigkeit Luxemburgs 1890 veränderten insgesamt die „regionale Grenzziehung" nicht,[34] abgesehen davon, dass zwei weitere nationale „Binnengrenzen" hinzukamen.[35]

---

30 Eine Nebenlinie der französischen Königsfamilie Valois. Von Seggern, Geschichte der burgundischen Niederlande (wie Anm. 29), S. 11; Knoll, Geschichtlicher Überblick (wie Anm. 15), S. 6; Petra Ehm-Schnocks, Burgund und das Reich, München 2002, S. 12; Dieter Scheler, „Die niederen Lande". Der Raum des Niederrheins im späten Mittelalter und in der frühen Neuzeit, in: Ders., Stadt und Kirche, Land und Herrschaft am Niederrhein im Mittelalter und anbrechender Neuzeit, hg. v. Hiram Kümper (Studien und Geschichte zur Kultur Nordwesteuropas 30), Münster 2019, S. 9–27, hier: S. 13.

31 Von Seggern, Geschichte der burgundischen Niederlande (wie Anm. 29), S. 61–70.

32 Ebd., u.a. S. 150 f., 164–170; Maczkiewitz, Der niederländische Aufstand gegen Spanien (wie Anm. 11), S. 31.

33 Die mit dem Burgundischen Vertrag beziehungsweise der Pragmatischen Sanktion erfolgte Abtrennung des Niederrheins hatte bereits die erste Binnengrenze zur Folge. Scheler, „Die niederen Lande" (wie Anm. 30), S. 21.

34 Christoph Driessen, Geschichte der Niederlande. Von der Seemacht zum Trendland, Regensburg ^{2}2016, S. 151.

35 Maczkiewitz, Der niederländische Aufstand gegen Spanien (wie Anm. 11), S. 33 f.

Darüber hinaus zeigen sich die konstanten regionalen Grenzziehungen anhand der bereits erwähnten, schon in Quellen des 16. Jahrhunderts nachweisbaren, historischen und kulturübergreifenden Bezeichnung der Region als *Lage Landen/low lands/países bajos/Les Pais-Bas*. In der überwiegenden Mehrzahl der Publikationen über die Geschichte des belgischen Raums wurde dieses Gebiet so mit der Gesamtheit der siebzehn Provinzen gleichgesetzt.[36] Auch im 17. und 18. Jahrhundert erinnerten zahlreiche Autoren in ihren Schriften an die historische Zusammengehörigkeit und Einheit dieser. Der damit verbundene Mythos blieb somit populär.[37] So bezeichnet auch Cannabich (1820) im *„Handbuch der neuesten Erdbeschreibung"* das Königreich der Niederlande (damit meint er die gesamte heutige Benelux-Region) als *„völlig abgerundeter und geschlossener Staat."*[38]

Hinzukommend wurden zu dieser Zeit die Begriffe „Niederlande" und „Belgien" synonym verwendet.[39] Ein Beispiel von vielen ist Zedlers *„Universal-Lexicon"*, in welchem die Begriffe *„Niederlande"*, *„Niederdeutschland"*, *„Nieder-Germanien"*, *„Belgium"*, *„Germania Inferior"* und *„Les Pais-Bas"* in einem einzigen Artikel zusammengeführt werden.[40]

Auch das heutige Rhein-Maas-Gebiet muss zu dieser Region mitgerechnet werden, da es gleichermaßen wie das benachbarte Limburg und Lüttich durch die großen Ströme geprägt wird und ursprünglich ganz nach Westen hin ausgerichtet war. Erst im Verlaufe der Frühen Neuzeit erfolgte eine allmähliche Orientierung nach Osten hin.[41] Der noch heute existierende deutschsprachige Einfluss im Osten Belgiens respektive auch in Luxemburg unterstreicht diese

---

36 Johannes Knoll, ‚Die belgische Nation'. Patriotismus und Nationalbewußtsein in den Südlichen Niederlanden im späten 18. Jahrhundert (Niederlande-Studien 33), Münster 2003, S. 40.

37 Ebd.

38 Johann G. H. Hassel / Johann G. Fr. Cannabich, Das Königreich Spanien und Portugal und die Niederlande, 2. Abteilung, 3. Bd. (Vollständiges Handbuch der neuesten Erdbeschreibung 9), Weimar 1820, hier: S. 509.

39 Knoll, ‚Die belgische Nation' (wie Anm. 36), S. 45, 47. Eine genaue Betrachtung bezüglich der Entwicklung des Begriffs „Belgien" vollzieht Knoll in seiner genannten Dissertation: Kap. 2, S. 37–55.

40 Ebd., S. 48.

41 Scheler, „Die niederen Lande" (wie Anm. 30), S. 9. Beispielsweise durch die Verlagerung des Rheins: Irmgard Hantsche, Atlas zur Geschichte des Niederrheins, Bd. 1 (Schriftenreihe der Niederrhein-Akademie 4), Bottrop 1999, S. 44 f.

Einschätzung, zudem existieren auffällige Gemeinsamkeiten in Architektur, Kunst und Literatur.[42]

Auf die Lage Landen trifft somit das zu, was Steber als Charakteristikum einer Region bezeichnet hat: Sie wird von den Zeitgenossen als eine solche wahrgenommen, wie gleichfalls eine der frühesten gedruckten Beschreibungen Johannes Cochlaeus, die „*Brevis Germaniae descriptio*" (1512), zeigt. In dieser werden mit unter anderem Holland, Zeeland, Brabant, Flandern, Geldern und den Herzogtümern Kleve und Jülich sowohl heute niederländische, belgische als auch zum Rhein-Maas-Raum zählende deutsche Provinzen beziehungsweise Ortschaften aufgezählt.[43]

---

42 Vgl. hierzu u.a.: Ralf-Peter Fuchs, Konfession und Raum. Die klevisch-geldrische Region im Rhein-Maas-Raum, in: Freitag / Kißener / Reinle / Ullmann, Handbuch Landesgeschichte (wie Anm. 14), S. 578–610, hier: S. 581; Jörg Engelbrecht, Rhein-Maas als grenzüberschreitender Kulturraum, in: Ders. / Stephan Laux (Hg.), Landes- und Reichsgeschichte. Festschrift für Hansgeorg Molitor zum 65. Geburtstag, Bielefeld 2004, S. 11–22, hier: S. 12; Helmut Tervooren, Van der Masen tot op den Rijn. Ein Handbuch zur Geschichte der mittelalterlichen volkssprachlichen Literatur im Raum von Rhein und Maas, Geldern 2005; van Dam, An Amphibious Culture (wie Anm. 8), S. 79; Scheler, „Die niederen Lande" (wie Anm. 30), S. 18. Der sprachliche Einfluss zeigt sich auch ganz deutlich in den mittelalterlichen Quellen, wie z.B. am Rechtsbuch (1430) und ersten Stadtbuch (1441/44) der Stadt Kleve deutlich wird: In den einzelnen Textpassagen befinden sich zahlreiche Wörter aus dem heutigen Niederländisch. Siehe hierzu die folgende Edition: Wolf-Rüdiger Schleidgen (Bearb.), Die ältesten Klever Stadtrechtshandschriften (Klever Archiv. Schriftenreihe des Stadtarchivs Kleve 10), Kleve 1990.

43 Johannes Cochlaeus, Brevis Germaniae descriptio (1512), hg. v. Karl Langosch (Ausgewählte Quellen zur deutschen Geschichte der Neuzeit, Freiherr vom Stein Gedächtnisausgabe 1), Darmstadt 1960, S. 150–157.

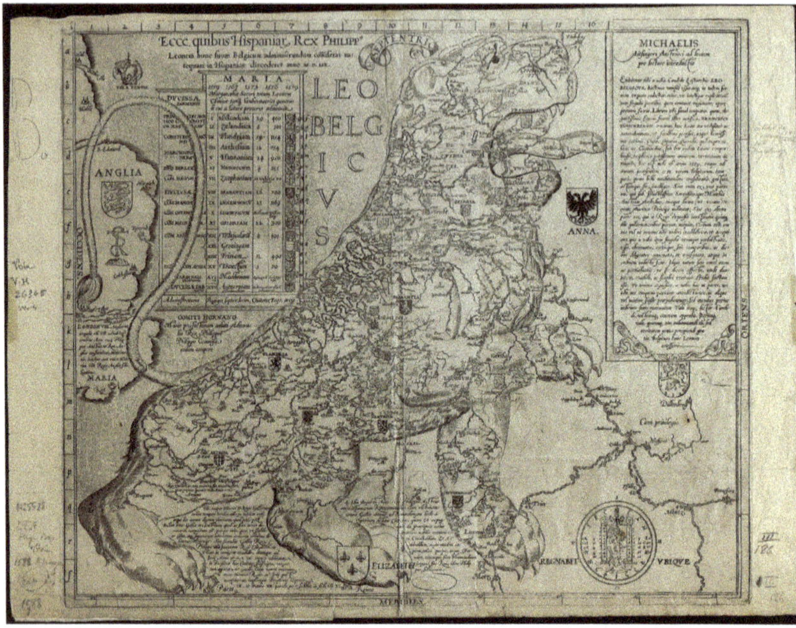

Abb. 75: *„Leo Belgicus" von Franz Hogenberg (1535–1590), 1583. Bildnachweis: Wikimedia Commons (gemeinfrei), URL: https://upload.wikimedia.org/wikipedia/commons/3/36/1583_Leo_Belgicus_Hogenberg.jpg.*

Die anhand der Schriftquellen beschriebene Rezeption spiegelt sich auch in der Kartographie, nämlich im „*Leo Belgicus*". Es handelt sich hierbei um eine Landkarte, welche in Form eines sitzenden Löwen die damaligen Lage Landen mit ihren siebzehn Provinzen erfasst. Das durch diesen abgedeckte Gebiet reicht sogar bis zum Rhein in seiner Nord-Süd-Achse. Die Gegend Rhein-Maas mit ihren Städten Kleve, Moers, Aachen und sogar Köln – um nur einige aufzuzählen –, wurde also durch den Zeichner integriert. Scheler (1998/2019) vermerkt nicht zu Unrecht mit einem Augenzwinkern, dass falls die Bestrebungen Johann I. von Kleve (1448–1481) geglückt wären, die Heirat Marias mit Maximilian von Österreich zu verhindern und sie stattdessen für seinen Sohn Johann II. zu gewinnen, wäre Kleve offiziell Teil Burgunds geworden. Die Grenze Deutschlands zum nordwestlichen Nachbarn verliefe mitunter heute auf der rechten Rheinseite.[44]

44 Scheler, „Die niederen Lande" (wie Anm. 30), S. 14.

Die früheste Zeichnung des belgischen Löwen stammt von dem in Mechelen (Belgien) geborenen Kupferstecher und Radierer Franz Hogenberg aus dem Jahre 1583. Sie wurde durch den österreichischen Kartografen Michael Aitzinger bekannt, der sie in seinem Buch „*De Leone Belgico*" veröffentlichte.[45] Das Motiv des Löwen wirkte schon früh als Symbol für Herrschaft und Macht und nahm bereits zum Zeitpunkt seiner Erschaffung sowohl in den Wappen der Grafschaften von Holland als auch später in denen von Flandern und Limburg seinen Platz ein. Es entwickelte sich zu einem Zeichen der Einigung der verschiedenen Provinzen in Zeiten der großen Unruhen während des „Achtzigjährigen Kriegs" (1568–1648), in denen sich zumindest zwischenzeitlich auf ein gemeinsames Vorgehen gegen die spanische Krone verständigt wurde. Selbst die Generalstaaten nahmen das Tier in ihr Wappen auf.[46] Außerdem stand der Löwe symbolhaft für den Versuch, die ursprünglich siebzehn Provinzen zusammenzuhalten, nachdem 1581 die sieben nördlichen, heute niederländischen Provinzen ihren Abfall von der spanischen Krone verkündet hatten. Er muss entsprechend ein Stück weit als Auftragswerk zum Zweck der Propaganda gesehen werden, zumal Aitzinger als Anhänger der spanischen Partei galt.[47]

Die über die Zeichnung des Löwen demonstrierte Macht kann auch und gerade in direktem Zusammenhang mit der dortigen Städtelandschaft gesehen werden. Denn die zahlreichen entstandenen und zum Teil recht unterschiedlich gearbeiteten und verzierten Ausarbeitungen des „*Leo Belgicus*" haben eine Gemeinsamkeit: Alle enthalten neben den einzelnen Provinzgrenzen, ihren jeweiligen Namen und Wappen, die zahllosen Städte. Einige markieren diese sogar zusätzlich durch einen roten Punkt.

Im Falle der Lage Landen von einer einheitlichen „regionalen Identität, im Sinne einer Nationalen", auszugehen, ist allerdings unmöglich. Die Region zeichnete sich weder vor noch nach dem Burgundischen Vertrag durch eine solche aus. Zwar sahen sowohl Karl V. als auch im Besonderen sein Sohn Philipp II. die Lage Landen als Einheit – dies entsprach jedoch keinesfalls der Realität.[48]

45 Knoll, ‚Die belgische Nation' (wie Anm. 36), S. 41.

46 Schröder, Schnittstelle Niederrhein (wie Anm. 15), S. 17; vgl. hierzu auch: Michalsky, Projektion und Imagination (wie Anm. 29), S. 48 f.

47 Karl H. Salzmann, Aitzing, Michael Freiherr von, in: Neue Deutsche Biographie 1, 1953, S. 119 f., URL: https://www.deutsche-biographie.de/pnd118644173.html#ndbcontent (letzter Zugriff: 3.5.2020).

48 Simon Groenveld, Facetten van der Tachtigjarige Oorlog. Twaalf artikelen over de periode 1559–1652, Hilversum 2018, S. 34.

Im Gegenteil: Auch Groenveld (2018) und Driessen (^{2}2016) betonen, dass es sich hier um siebzehn selbstständige „Staaten" handelte; jeder verfügte über eine eigene Kultur, ein eigenes Besteuerungssystem, eigene Rechte und Regeln.[49] Die Unterschiede fielen selbst aufmerksamen Zeitgenossen wie zum Beispiel Albrecht Dürer auf, wie die Dokumentation seiner Reise in die Lage Landen zeigt: Die Aufzeichnungen über seine Reisestationen Antwerpen, Brüssel und Zeeland führen dem Betrachter die Differenzen der einzelnen Provinzen und Städte sowohl hinsichtlich des Gewerbes (Handelsstadt contra Residenzstadt: Teppichweber, Bildhauer, Goldschmied) als auch der Kultur (niederländisch versus französisch) vor Augen.[50]

In den jeweiligen Provinzen fanden die Einwohner ihre primären sozialen Kontakte in einer Vielheit von Kooperationen, nämlich der Geistlichkeit, des Adels und der städtischen Eliten, deren Vertreter verschiedene Führungspositionen innehatten.[51] Allerdings unterschieden sich die Provinzen bezüglich der Verteilung des Gewichts der jeweiligen Vertreter; zum Beispiel stand zeitweise in Holland eine Stimme der Ritterschaft gegenüber achtzehn Stimmen von Städten, in Zeeland war das Verhältnis eins zu sechs, der Klerus wurde wiederum überhaupt nicht vertreten. In Utrecht sah die Situation wieder anders aus: Hier traten neben der Ritterschaft und fünf Städten auch noch Repräsentanten des Domkapitels als Vertretung der Geistlichkeit auf.[52] Flandern und Brabant waren die Provinzen mit den meisten großen Städten. Holland, Zeeland und Utrecht wiesen eine langsamere Entwicklung auf.[53] Anders war es in den östlichen Provinzen, welche unter deutschem Einfluss standen.[54]

Entsprechend kann auch nach der späteren Nationenbildung, begonnen mit der 1648 erfolgten, offiziellen Anerkennung der niederländischen Republik, nicht von einer „nationalen Identität" ausgegangen werden: Eine solche gab es weder vor der Gründung der niederländischen Republik noch danach, weder vor der Gründung Belgiens noch hinterher. Dies wird nicht nur anhand der sich durch die Geschichte ziehenden Streitigkeiten der verschiedenen, unter ein und

---

49 Ebd., S. 15; Driessen, Geschichte der Niederlande (wie Anm. 34), S. 11.

50 Driessen, Geschichte der Niederlande (wie Anm. 34), S. 11 f.; vgl. auch diverse Zitate in: Gerd Unverfehrt, Da sah ich viel köstliche Dinge. Albrecht Dürers Reise in die Niederlande, Göttingen 2007.

51 Groenveld, Facetten van der Tachtigjarige Oorlog (wie Anm. 48), S. 15.

52 Ebd., S. 22 f.

53 Ebd., S. 16.

54 Ebd.

derselben Nation verweilenden Provinzen untereinander um Freiheiten, Rechte oder Privilegien deutlich.[55] Auch die sich unterhalb Brüssels von Ost nach West mitten durch das heutige Belgien ziehende Sprachgrenze spricht diesbezüglich Bände.[56]

Stattdessen waren die Menschen, Holländer, Friesländer, Brabanter und so weiter, wie schon die zeitgenössischen Darstellungen George Forsters (1790) oder auch die schon viel früher entstandenen Beschreibungen des flandrischen Dichters und Chronisten Georges Chastellain (1405–1475) zeigen. Letzterer unterschied beispielsweise zwischen Holländern, Zeeländern und den Allemands, zu denen er wiederum die Leute aus Geldern, Utrecht und Overijssel zählte.[57]

Überregionale Bekanntheit erlangten die Lage Landen vor allem durch ihre Handelskraft und den sogenannten „Achtzigjährigen Krieg".[58] Bezüglich der in

---

55 Dies zeigte sich unter anderem auch bei den heutigen niederländischen Provinzen: So hielt Groenveld (2018) fest, dass das Jahr 1581 lediglich als Ausgangsjahr des Prozesses des Ausscheidens aus der Habsburger Monarchie gesehen werden muss, denn der Beschluss des Verlasses wurde zum Teil in Abwesenheit einiger Provinzen gefällt. Hinzukommend ließ sich der Beschluss nicht in allen Provinzen sofort umsetzen. Beispielsweise befand sich Zeeland noch 1584 offiziell unter der Herrschaft Philipps II., weil das auf ein Austreten hinweisende Plakat hier noch nicht formell publiziert werden konnte, da Middelburg sich dagegen wehrte. Auch Utrecht ging eigene Wege und stellte im Oktober 1581 eine eigene Akte des Verlasses auf – die Uneinigkeit konnte also nicht größer sein. Groenveld, Facetten van der Tachtigjarige Oorlog (wie Anm. 48), S. 192.

56 Knappe 60 % sind niederländischsprachige Flamen (Norden), knappe 40 % frankophone Wallonen (Süden) und etwa 73.000 deutschsprachig (Osten). An diesem von Vandermotten 2007 publizierten Stand hat sich bis heute nichts Wesentliches geändert. Christian Vandermotten, Die Geographie, in: Knoll, Belgien (wie Anm. 15), S. 73–99, hier: S. 73.

57 Vgl. bezüglich Chastellain die Darstellung in: Scheler, „Die niederen Lande" (wie Anm. 30), S. 18. George Forster beschreibt in seinem Werk „*Ansichten vom Niederrhein*" (1790) die Menschen der verschiedenen Provinzen, zum Beispiel Holland (Bd. 2, S. 447–487), Antwerpen (Bd. 2, S. 219–346), Flandern (Bd. 1, S. 417–457), Brabant (Bd. 2, S. 157–196) oder Lüttich (Bd. 1, S. 326). George Forster, Ansichten vom Niederrhein, von Brabant, Flandern, Holland, England und Frankreich, im April, Mai und Junius 1790, 3 Bde., Berlin 1791–94, hier: Bde. 1 und 2. Vgl. hierzu auch: Jörg Engelbrecht, „Phlegma und überall Phlegma". Die deutsche Sicht auf die Niederlande im 18. und 19. Jahrhundert, in: Nachbarsprache Niederländisch 2, 2007, S. 5–23, hier: S. 8–11.

58 Der „Achtzigjährige Krieg" wird bewusst in Anführungszeichen gesetzt, da seiner Bezeichnung mit einiger Skepsis zu begegnen ist. Vgl. hierzu: Schröder, Schnittstelle Niederrhein (wie Anm. 15), S. 41 f.

der Forschung gängigen, aber nicht unumstrittenen Bezeichnung für die Zeit der großen Unruhen zwischen den Bewohnern der Lage Landen und der spanischen Krone argumentiert Groenveld (2018), dass damit allenfalls „so ungefähr 80“ oder vielleicht auch „sehr lang“ gemeint sein kann.[59] Er fokussiert damit ebenfalls die Region in ihrer Gesamtheit. Während das Jahr 1648 aufgrund des Westfälischen Friedens und der damit einhergehenden Anerkennung der jungen niederländischen Republik als definitiv zu bezeichnen ist, stelle das für den Beginn der Auseinandersetzungen in der Region gebrauchte Jahr 1568 hingegen laut Groenveld das Resultat eines „Klein niederländischen Standpunkts“[60] dar. Dieser gewinne vor allen Dingen immer mehr durch nationalistisch denkende Personen an Kraft und beziehe sich entsprechend nur auf die jetzige niederländische Geschichte. Das Jahr 1568 ist aus einer regional-historischen Perspektive daher nicht unumstritten; es gibt verschiedene Historiker, die andere Jahre als Ausgang benennen, beispielsweise Pieter Borck.[61] Denn militärische Operationen fanden bereits vor 1568 in den südlichen Lage Landen statt. Groenveld (2018) nennt diesbezüglich mehrere Beispiele. Die in diesem Konflikt mit der spanischen Krone anhaltenden Bemühungen Wilhelm von Oraniens ein Auseinanderbrechen der Region zu verhindern, legen übrigens erneut einmal mehr Zeugnis über die zeitgenössische Wahrnehmung der einen großen Region ab.[62]

Wie lässt sich, den Gedankengang abschließend, nun also mit der Bezeichnung „Region“ im Falle der Lage Landen umgehen? Die vorangegangene Argumentation sollte keine Zweifel darüber gelassen haben, dass im Folgenden die Lage Landen als eine Region verstanden wird. Möglicherweise hilft es – sozusagen als ein letzter Überzeugungsversuch –, die Lage Landen mit ihren gewachsenen Strukturen ein Stück weit im Sinne von „Landschaft“ zu denken.[63]

---

59 Groenveld, Facetten van der Tachtigjarige Oorlog (wie Anm. 48), S. 12.

60 Ebd., S. 11.

61 Ebd.

62 Stefan Bildheim, Calvinistische Staatstheorien. Historische Fallstudien zur Präsenz monarchomachischer Denkstrukturen im Mitteleuropa der Frühen Neuzeit, Frankfurt am Main 2001, S. 160; vgl. hierzu auch die 2018 erschienene Sammlung ins moderne Niederländisch übersetzter verschiedener Schriften und Traktate aus dieser Zeit: Luc Panhuysen / René van Stipriaan (Hg.), Ooggetuigen van de Tachtigjarige Oorlog. Van de eerste ruzie tot het laatste kanonschot, Amsterdam/Antwerpen 2018.

63 Hier ist jetzt nicht explizit nur die Stadt- respektive Städtelandschaft gemeint. Zur Auseinandersetzung mit dem Landschaftsbegriff siehe auch: Michalsky, Projektion und Imagination (wie Anm. 29); Steber, Region (wie Anm. 24). An dieser Stelle kann nur eine grobe Skizze erfolgen, da es lediglich um eine Anregung geht, die jedoch an anderer Stelle weitergedacht werden muss.

Für die Definition der Landschaft wird gerne auf Alexander von Humboldt (1769–1859) zurückgegriffen, der die Landschaft als „Totalität aller Aspekte einer Region" umschrieben haben soll.[64] Seine Beschreibung nimmt die Landschaft als Summe aller Aspekte – natürlicher, kultureller, geographischer, geologischer, biologischer ebenso wie künstlerischer Rahmenbedingungen – wahr. Darüber hinaus betont er das menschliche Erfassen und die kollektive und/oder individuelle Erfahrung als wesentliche Elemente der Landschaft.[65]

Diese Beschreibung wurde in etwa dem „Europäischen Landschaftsübereinkommen" (2000/2004) zugrunde gelegt. Hier wird Landschaft als „ein vom Menschen als solches wahrgenommenes Gebiet, dessen Charakter das Ergebnis des Wirkens und Zusammenwirkens natürlicher und/oder anthropogener Faktoren ist",[66] definiert. Entsprechend müssen im Rahmen der Regional- und Landschaftsforschung auch immer die vom Menschen, seinen Ideen und Konzepten hervorgebrachten Antworten auf regionale Bedingungen sowie diese selbst mitberücksichtigt werden. Es handelt sich also um ein über Jahrhunderte entstandenes Konstrukt, gewachsen aus zahllosen spezifischen historischen Situationen. Hochmuth und Rau (2006) beschreiben letztere als „die Vielfalt der stattfindenden Ereignisse, das Zusammentreffen verschiedenster Menschen unterschiedlicher sozialer Herkunft, das dann dort vorhandene Potential an Wechselwirkungen und Austauschprozessen, die anonymisierende Unüberschaubarkeit und Schnelligkeit des Wandels, verbunden mit den Möglichkeiten der Auflösung von Tradition einerseits, Beförderung von Habitusbildungen und der Herausbildung neuer Lebensstile andererseits."[67]

Der Forschungsterminus der *Cultural Landscape* greift diese Aspekte ebenso auf, er umfasst „a geographic area (including both cultural and natural resources and the wildlife or domestic animals therein), associated with a historic event,

64 Olaf Kühne, Landscape Theories. A Brief Introduction, Wiesbaden 2019, S. 10.

65 Michalsky, Projektion und Imagination (wie Anm. 29), S. 23, 32; siehe auch: Spessartprojekt.de. Die europäische Kulturlandschaft Spessart, URL: https://www.spessartprojekt.de/kulturlandschaft/landschaft-als-idee-und-konzept/ (letzter Zugriff: 29.4.2020).

66 Europäisches Landschaftsübereinkommen (Florenz, 20.10.2000), Kapitel I – Allgemeine Bestimmungen, Artikel 1 a) – Begriffsbestimmungen, URL: https://www.coe.int/en/web/conventions/full-list/-/conventions/rms/090000 1680080630 (letzter Zugriff: 29.4.2020). Die Europäische Landschaftskonvention wurde auf Anregung des Ausschusses der Regionen in einem langen Diskussionsprozess entwickelt, sie trat am 1.3.2004 offiziell in Kraft, siehe: Spessartprojekt.de (wie Anm. 65).

67 Christian Hochmuth / Susanne Rau, Stadt – Macht – Räume. Eine Einführung, in: Dies. (Hg.), Machträume der frühneuzeitlichen Stadt (Konflikte und Kultur – Historische Perspektiven 13), Konstanz 2006, S. 13–40, hier: S. 15.

activity, or person or exhibiting other cultural or aesthetic values". Dabei werden vier basale Typen von *Cultural Landscapes* unterschieden: „historic sites, historic designed landscapes, historic vernacular landscapes, and ethnographic landscapes."[68]

Der bereits zitierte Forster unternahm im Frühjahr 1790 gemeinsam mit Alexander von Humboldt eine Reise durch das Rheinland, das Fürstbistum Lüttich, durch das heutige Belgien und die Niederlande sowie durch Nordfrankreich. Die Tatsache, dass er die Begriffe Belgien und Niederlande gleichfalls als Synonyme in seinem Werk „*Ansichten vom Niederrhein*" verwendete, mag die Einschätzung hier rechtfertigen.[69] Die mit dem Begriff der Landschaft assoziierten Gedanken sollten entsprechend auch im Hinblick auf die Beantwortung der Frage nach der generellen Bedeutung von Stadt-/Städtelandschaft zur Grundlage genommen werden.

1.2. Zum Forschungsstand

Der nun folgende Abschnitt liefert ein paar Worte zum Forschungsstand: Die Städteforschung für die hier beschriebene Region ist, wie schon erwähnt, stark durch die nationalen Grenzen geprägt. Einen Versuch, die Region unabhängig von diesen zu erforschen, unternimmt der „Fachinformationsdienst für Niederlandistik, Niederlande-, Belgien- und Luxemburgforschung" (FID Benelux).[70] Ein Blick in den dortigen digitalen Katalog, in welchem diverse Publikationen, die sich mit der Benelux-Region befassen, zusammengetragen sind und der ebenso Hinweise zu den verschiedenen Forschern und deren Teilgebieten enthält, zeigt, dass es sehr viele Arbeiten zu einzelnen Städten gibt. Besonders viele beziehen sich dabei auf die Stadt Luxemburg. Für die heute niederländischen Städte lassen sich vor allem Publikationen von Horst Lademacher und Michael North heranziehen.[71] Im heutigen Belgien ist Stadtgeschichte insbesondere an den Universitäten verankert.[72] Bekannte Werke stammen unter anderem aus der

---

68 National Park Service, URL: https://www.nps.gov/tps/standards/four-treatments/landscape-guidelines/terminology.htm (letzter Zugriff: 2.5.2020).

69 Knoll, ‚Die belgische Nation' (wie Anm. 36), S. 48 f.; Engelbrecht, „Phlegma und überall Phlegma" (wie Anm. 57), S. 5–23, hier: S. 7.

70 Fachinformationsdienst (FID) Benelux, URL: https://www.fid-benelux.de/.

71 Michael North, Geschichte der Niederlande, München ²2013; Horst Lademacher, Geschichte der Niederlande: Politik – Verfassung – Wirtschaft, Darmstadt 1983.

72 Anne-Laure van Bruaene / Bruno Blondé / Marc Boone, Verstedelijking en stadshistoriografie, in: Dies. (Hg.), Gouden Eeuwen. Stad en samenleving in den Lage Landen 1100–1600, Gent 2016, S. 7–29, hier: S. 12.

Feder von van Bruaene, Blondé und Boone (2016), Crouzet-Pavan, Benvenuti und Lecuppre-Desjardin (2008) sowie Stabel (unter anderem 2002/2012). Über den luxemburgischen Raum hat vor allem Pauly (1992/2011) geforscht.[73]

Die von Horst Lademacher herausgegebene Buchreihe „Studien zur Geschichte und Kultur Westeuropas" führt zudem aktuelle Forschungen die Lage Landen betreffend zusammen.

Hinsichtlich der regionalen Besonderheiten hielt Jensen (2018) fest, dass sich der ständige Kampf um die Beherrschung des Wassers in den Niederlanden ein Stück weit identitätsformend auswirkt. Dies zeige sich entsprechend auch in den wissenschaftlichen Diskursen: Auf der einen Seite existieren sogenannte Desasterstudien, auf der anderen Seite vom Nationalismus geprägte Arbeiten.[74] Erstere haben wohl in den letzten Jahren ein bis dato unbekanntes Ausmaß angenommen. Jensen schiebt dies unter anderem auf den Klimawandel, von dem insbesondere die Niederlande durch immer höher ansteigende Meeresspiegel stark bedroht sind. Es handelt sich hier insgesamt um ein interdisziplinäres Untersuchungsgebiet, welches verschiedene Gebiete der Gesellschaftslehre, zum Beispiel Soziologie, Psychologie, Kommunikationswissenschaften oder Politologie, sowie der Naturwissenschaften zusammenführt.[75] Auch diverse Forschungsarbeiten von van Dam (2012/2014/2016) beschäftigen sich mit der „Amphibious Culture" in Bezug auf die heutigen Niederlande.[76] Andere Autoren wie zum Beispiel

---

73 Ebd.; Élisabeth Crouzet-Pavan / Anna Benvenuti / Élodie Lecuppre-Desjardin (Hg.), Villes de Flandre et d'Italie (XIIIe–XVIe siècle). Les enseignements d'une comparaison, Turnhout 2008; Peter Stabel / Ed Taverne, Nederland stedenland: continuïteit en vernieuwing, Rotterdam 2012; Peter Stabel, Holland in de Middeleeuwen. Een regio ontdekt haar geschiedenis, in: Holland: historisch tijdschrift 34/4, 2002, S. 242–249; Ders., Dwarfs among giants: the Flemish urban network in the late Middle Ages (Studies in urban social, economic and political history of the medieval and modern low countries 8), Leuven 1997; Ders., De kleine stad in Vlaanderen: Bevolkingsdynamiek en economische functies van de kleine en secundaire stedelijke centra in het Gentse kwartier (14de–16de eeuw), Brüssel 1995; Michel Pauly, Geschichte Luxemburgs, München 2011; Ders., Die luxemburgischen Städte in zentralörtlicher Perspektive, in: Ders. (Hg.), Les petites villes en Lotharingie (Publications de la Section Historique de l'Institut Grand-Ducal de Luxembourg), Luxemburg 1992, S. 117–162.

74 Lotte Jensen, Wij tegen het water, Een eeuwenoude strijd, Nijmegen 2018, S. 9 f.

75 Ebd.

76 Siehe u.a.: van Dam, An Amphibious Culture (wie Anm. 8); Gerrit R. Boomsma / Petra J. E. M. van Dam, Voor de ongelukkigen van de watersnood. Historisch onderzoek naar de donormotieven bij watersnoden in de negentiende eeuw, in: Tijdschrift voor Waterstaatsgeschiedenis 23/2, 2014, S. 68–81; Petra J. E. M. van Dam, Denken over

von Seggern (2018), Zijlmans (2017), Vandermotten (2007) oder Maczkiewitz (2005) betonen neben dem Wasser immer wieder die unterschiedliche Bodenqualität in der Region.[77]

Die Grundlage für das Verständnis der Lage Landen, das zeigt auch der derzeitige Forschungstrend, liegt also offensichtlich ein Stück weit in den topographischen Rahmenbedingungen der Region. Während die Geologie die gesamte Region betrifft, bezieht sich die Wasserthematik scheinbar „nur" auf etwa ein Viertel der Gesamtregion. Ihre Strahlkraft wirkt jedoch weit in die restliche Region hinein. Alleine die drei ARA-Häfen Amsterdam, Rotterdam und Antwerpen sind auch heute noch auf ihr Hafenhinterland angewiesen.[78] Die Beherrschung und optimale Nutzung des Wassers wurde nicht nur an Stelle eines Nationalgefühls identitätsstiftend, zumindest für einen Teil der dortigen Bewohner, sondern bestimmten über viele Jahrhunderte auch in großem Maße gemeinsam mit der divergierenden Bodentopographie die dortigen Herrschaftsstrukturen mit. Entsprechend soll sich das folgende Kapitel mit ihnen beschäftigen.

1.3. Die topographischen Rahmenbedingungen formen die Landschaft

Über den Zustand der Lage Landen während der Römerzeit ist wenig bekannt, die Beschreibungen von Tacitus, Caesar, Plinius etc. widersprechen sich.[79] Schelde, Maas und Rhein veränderten zudem immer wieder ihr Flussbett, was möglicherweise die Schreiber nicht wussten.[80] Wenn es um eine topographische Beschreibung dieser Kulturlandschaft geht, fallen vorrangig drei Aspekte ins Auge, die auch schon von den Zeitgenossen früherer Epochen beschrieben worden sind:[81] (1.) die divergierende Bodenbeschaffenheit innerhalb der Region, (2.) das Flusssystem mit dem Deltagebiet der drei schon genannten

---

overstromingen, natuurrampen en het idee van de amfibische cultuur, in: Tijdschrift voor Waterstaatsgeschiedenis 22/1–2, 2012, S. 1–10.

77 Zijlmans, Troebele betrekkingen (wie Anm. 28); von Seggern, Geschichte der burgundischen Niederlande (wie Anm. 29); Vandermotten, Die Geographie (wie Anm. 56); Maczkiewitz, Der niederländische Aufstand gegen Spanien (wie Anm. 11).

78 Vgl. hierzu: Lina Schröder, Der Rhein-(Maas-)Schelde-Kanal als geplante Infrastrukturzelle von 1946 bis 1985. Eine Studie zur Infrastruktur- und Netzwerk-Geschichte (Studien zur Geschichte und Kultur Nordwesteuropas 28), Münster 2017.

79 Zijlmans, Troebele betrekkingen (wie Anm. 28), S. 35 f.

80 Ebd., S. 45.

81 Vgl. hierzu u.a.: Forster, Ansichten vom Niederrhein (wie Anm. 57), ab S. 509; Florentii van der Haer, De initiis tumultuum Belgicorum, Duaci 1587; Unverfehrt, Da sah ich viel köstliche Dinge (wie Anm. 50); Johann Jacob Volkmann, Neueste Reisen durch

bedeutenden Ströme Schelde, Maas und Rhein sowie (3.) die Tatsache, dass ein Viertel der Lage Landen, insbesondere der nordwestliche Teil, seit jeher unter dem Meeresspiegel lag.

(1) Verschiedene Autoren haben bereits auf die unterschiedliche Bodenqualität hingewiesen. So gab es durch die Küstengegend sowie den Löß-Gürtel hervorgerufene fruchtbare Böden, welche Teile Brabants, des Hennegaus, der Artois und der Pikardie umfassten. Diese Gegenden trugen nicht umsonst den Namen „Kornkammer der Lage Landen", denn sie eigneten sich zusammen mit dem milden Klima in optimaler Weise für die Landwirtschaft, wenngleich der hier vorwiegend hochgezogene Weizen keinesfalls für die Versorgung der gesamten Region ausreichte.[82]

Weiter nach Osten schreitend ist vor allem das nördliche Brabant und Geldern durch leichte sand- und tonhaltige Böden geprägt,[83] auch Antwerpen und Gent waren von Sandgründen umgeben.[84] Im südlichen Teil Brabants gibt es große Waldgebiete – laut Vandermotten (2007) sind diesbezüglich besonders der Zonienwald südöstlich von Brüssel sowie der Wald von Meerdael südlich von Leuven hervorzuheben. Sie verdanken ihren Erhalt dem historischen Status als herrschaftliche Wälder und Jagdgebiete der Herzöge von Brabant.[85]

Die Gegend von den flämischen Ardennen und dem nördlichen Hennegau bis zum Hageland in der Nähe des flämisch-brabantischen Diest ist durch eine Hügellinie gezeichnet.[86] Das heutige Luxemburg ist wiederum zunächst im

---

die Vereinigten Niederlande, vorzüglich in Absicht auf die Kunstsammlungen, Naturgeschichte, Oekonomie und Manufacturen, 13. Brief, Leipzig 1783.

82 Von Seggern, Geschichte der burgundischen Niederlande (wie Anm. 29), S. 17; Vandermotten, Die Geographie (wie Anm. 56), S. 75; Maczkiewitz, Der niederländische Aufstand gegen Spanien (wie Anm. 11), S. 44; bezüglich der Fruchtbarkeit der Lößböden siehe auch: Gerrit J. Schenk, Das Baby in der Wiege. Hochwasser zwischen Alltag und Katastrophe (ca. 1250–1550). Beobachtungen zur „fluvialen Anthroposphäre" im Kraichgau, in: Kurt Andermann / Gerrit J. Schenk, Wasser. Ressource – Gefahr – Leben (Kraichtaler Kolloquien 12), Ostfildern 2020, S. 79–113, hier: S. 102.

83 Von Seggern, Geschichte der burgundischen Niederlande (wie Anm. 29), S. 17; Vandermotten, Die Geographie (wie Anm. 56), S. 73.

84 Wim Blockmans / Bert de Munck / Peter Stabel, Economische dynamiek: verstedelijking, regionale complementariteit en Europsese roeping, in: van Bruaene / Blondé / Boone, Gouden Eeuwen (wie Anm. 72), S. 31–75, hier: S. 37.

85 Vandermotten, Die Geographie (wie Anm. 56), S. 75.

86 Ebd., S. 74; Maczkiewitz, Der niederländische Aufstand gegen Spanien (wie Anm. 11), S. 43.

Norden durch die Ardennen geprägt, der Teil der Region befindet sich durchschnittlich 400 bis 500 Meter über dem Meeresspiegel und wird durch Flusstäler (zum Beispiel das der Sauer) charakterisiert. Da Landwirtschaft unter solchen Bedingungen nur mit großen Anstrengungen und geringem Ertrag möglich ist, finden sich auf den nährstoffarmen, sauren Braunerden häufig Forstbetriebe sowie weitgehend unbewirtschaftete Wälder.[87] Im Süden dominiert das fruchtbare Gutland; auch dieser Teil der Lage Landen wird von diversen Flüssen, unter anderem Mosel, Sauer, Our und Alzette, geformt.

Das Land östlich von Lüttich gelegen weist ebenso deutliche Höhenunterschiede zwischen den Gipfelregionen und Tälern von Vesdre (Weser) und Maas auf. Seit dem 16. Jahrhundert befand sich dort eine sogenannte Bocagelandschaft – ein Gebiet von Feldern und Wiesen, die mit Wallhecken umzäunt wurden.[88]

Die Gegend nordöstlich von Limburg geht in eine Hochebene mit oft mineralsalzarmen und daher wenig fruchtbaren Böden über. Hier gab es Heidelandschaften, die später durch Kiefernwälder ersetzt wurden.[89] Je weiter der Betrachter seinen Blick nach Norden und Nordwesten wendet, desto mehr wird die Topographie dort durch das Wasser mit seinen moorhaltigen Böden geprägt. Amsterdam lag so inmitten eines Fenns.[90] Der Name Holland leitet sich laut von Seggern direkt von der niederländischen Bezeichnung *Houtland* (= Holzland) ab. In seiner historischen Bedeutung bezog die so bezeichnete Grafschaft das Kennemerland, das Rijnland und das Waterland mit ein,[91] also unter anderem auch Teile des heutigen Noord-Brabant und der Provinz Utrecht.

Da es sich hier, wie auch das folgende Zitat zeigt, überwiegend aber um eine überaus morastige Landschaft handelte und der Holzbedarf dort sehr groß war, blieb von den ursprünglichen Holzbeständen wohl nicht mehr viel übrig. So beschrieb 1783 der deutsche Schriftsteller Johann Jacob Volkmann die beiden Provinzen Nord- und Südholland in seinem Werk „*Neueste Reisen durch die Vereinigten Niederlande*" wie folgt: „*Vor dem Eintritte in Nordholland wird es nicht überflüssig sein, eine kleine Übersicht des ganzen Landes anzustellen. Es ist weit kleiner, als Südholland, und gleichsam eine Halbinsel, die von der Nord= und Südersee und dem IJ umflossen wird, und durch eine schmale Landzunge beim*

87 Maczkiewitz, Der niederländische Aufstand gegen Spanien (wie Anm. 11), S. 44.

88 Vandermotten, Die Geographie (wie Anm. 56), S. 75.

89 Ebd., S. 74.

90 Blockmans / de Munck / Stabel, Economische dynamiek (wie Anm. 84), S. 37; Maczkiewitz, Der niederländische Aufstand gegen Spanien (wie Anm. 11), S. 44 f.

91 Von Seggern, Geschichte der burgundischen Niederlande (wie Anm. 29), S. 18.

Wijkermeere mit Südholland zusammenhängt. An der Nordsee laufen die Dünen hin, das übrige Land ist niedrig, sumpfig, und bestand ehemals aus lauter Morästen und Seen, welche die Einwohner mit Dämmen umgeben, ausgetrocknet, und in herrliche Wiesen verwandelt haben, mittelst deren eine vortreffliche Viehzucht getrieben wird. Weil solche aber doch im Winter unter Wasser stehen, so muß solches im Frühjahre, vermittelst der Schöpf= oder Steertmühlen, weggemahlen werden."[92] Auch fungierten Holland und Zeeland als Zentren des Getreideanbaus schlechthin, der Zeeländer Weizen galt als besonders herausragend.[93]

Der eingangs unternommene Größenvergleich der Lage Landen mit anderen Regionen zeigt, dass sich dort dann doch auf relativ engem Raum sehr viele unterschiedliche Bodenstrukturen nachweisen lassen. Diese Kleinräumigkeit musste automatisch dazu beitragen, dass sich die Menschen aufgrund des Mangels an lokalen landwirtschaftlichen Möglichkeiten schon sehr früh zum Güteraustausch gezwungen sahen, was zum zweiten wichtigen topographischen Aspekt, dem Flusssystem mit dem Flussdelta bestehend aus Rhein, Maas und Schelde, überleitet.

(2) Diese großen Ströme mit ihren zahlreichen Nebenflüssen luden die Menschen geradezu zum Siedlungsbau ein. Das über die Flüsse, aber ebenso über die, durch diese gespeisten, Grachten und Kanäle vorhandene respektive entstandene und stets in Erweiterung begriffene Wasserstraßensystem gestattete insbesondere auch abseits von der Küste gelegenen Ortschaften den erforderlichen Warenaustausch und schuf eine echte Alternative zum beschwerlichen Landtransport. Der Handel wurde im Hoch- und Spätmittelalter vorrangig mit für heutige Verhältnisse kleinen Schiffen oder auch Flößen betrieben. Diese wiesen entsprechend nur einen geringen Tiefgang auf. Auf diese Weise waren selbst ganz kleine Flüsse befahrbar und die Händler fuhren weit die Flüsse hinauf, um möglichst spät umzuladen. In der Konsequenz befanden sich daher nicht nur kleinere, sondern auch die großen Hafen- und Handelsstädte relativ weit im Landesinneren.[94]

---

92 Volkmann, Neueste Reisen durch die Vereinigten Niederlande (wie Anm. 81), S. 348 f.

93 Wilhelm Naudé (Hg.), Getreidehandelspolitik. Bd. 1: Die Getreidehandelspolitik der Europäischen Staaten vom 13. bis zum 18. Jahrhundert, Berlin 1896, S. 423.

94 Von Seggern, Geschichte der burgundischen Niederlande (wie Anm. 29), S. 17; vgl. ferner: Sascha Bütow, Straßen im Fluss. Schifffahrt, Flussnutzung und der lange Wandel der Verkehrsinfrastruktur in der Mark Brandenburg und der Niederlausitz vom 13. bis zum 16. Jahrhundert (Studien zur brandenburgischen und vergleichenden Landesgeschichte 18), Berlin 2015; Peter Ettel / Achim Th. Hack (Hg.), Flusstäler, Flussschifffahrt, Flusshäfen: Befunde aus Antike und Mittelalter. Internationale Tagung im

Entsprechende Beispiele in der Betrachtungsregion sind Brügge, Gent, Antwerpen oder Amsterdam.

Die benötigten und zum Teil durch die Boden- und Ressourcenanordnung an ganz unterschiedlichen Orten gewonnenen Lebensmittel (Gemüse, Fisch, Getreide, Molkereiprodukte, Fleisch etc.) sowie Grundstoffe (Holz, Steine, Torf, Flachs, Wolle, Leder etc.) konnten aufgrund der komfortablen Wasserstraßenführung hin und her verschifft werden. Jeder Stadt wurde auf diese Weise ihr an die Topographie optimal angepasster, wirtschaftlicher Schwerpunkt ermöglicht.[95] Aus den Ardennen wurden im 16. Jahrhundert Naturstein, Schiefer und Holz über die Maas, aus der Umgebung von Lüttich Kalk, Steinkohle und Eisen und aus dem Maas-Rhein-Delta Fisch, Salz, Wein und Milchprodukte verschifft. Dordrecht, Ende des 13. Jahrhundert durch den holländischen Grafen mit dem Stapelrecht ausgestattet, entwickelte sich zur Drehscheibe für den Handel zwischen Maas und Rhein; Venlo wurde zum Umschlag- und Verteilzentrum.[96] Anfang des 16. Jahrhunderts erhielt dann das Gewerbe im Lütticher Land eine immer größere Bedeutung, Lüttich wurde unter anderem zum Zentrum für Eisenproduktion.[97]

Aber auch für die Energiegewinnung waren die Flüsse wichtig. Neben Holz, Wind- und Muskelkraft war das Wasser in der Vormoderne der wichtigste Energielieferant. Auch die größte Arbeitsmaschine des Mittelalters, die Mühle, ließ sich so mithilfe des Wassers betreiben. Sie wurde nicht nur zum Mahlen von Getreide genutzt, sondern seit der sogenannten „industriellen Revolution des Spätmittelalters" auch zum Sägen, Bohren, Hämmern, Walken, Ziehen von Draht und zur Herstellung von Papier.[98] Aber auch anderes Gewerbe war auf das Wasser angewiesen: Weber und Seiler benötigten es zum Einlegen der Pflanzen für die Hanf- und Flachsherstellung (sogenanntes Flachsrösten), Korbflechter griffen auf die Kopfweiden längs der Wasserläufe zurück, die zudem durch ihr

---

Rahmen des DFG-Schwerpunktprogramms 1630 „Häfen von der Römischen Kaiserzeit bis zum Mittelalter" an der Friedrich-Schiller-Universität Jena, 21.–24. Februar 2018, Sonderdruck RGZM – Tagungen 39 (Interdisziplinäre Forschungen zu den Häfen von der Römischen Kaiserzeit bis zum Mittelalter in Europa 7), Mainz 2019.

95 Blockmans / de Munck / Stabel, Economische dynamiek (wie Anm. 84), S. 37.

96 Zijlmans, Troebele betrekkingen (wie Anm. 28), S. 403.

97 Ebd., S. 404.

98 Schenk, Das Baby in der Wiege (wie Anm. 82), S. 106; vgl. hierzu auch: Lina Schröder, Getreidemühlen: tote Infrastrukturzellen im 21. Jahrhundert – vom Herrschaftsspeicher zum Kulturdenkmal, in: Niederrhein-Magazin 23, 2017, S. 11–21.

Wurzelwerk auch vor Erosion schützten.[99] Auch für die Anlage einer Ziegelei war das Wasser laut Ehrhardt (2016) gleich aus zweierlei Gründen notwendig: Zunächst waren für die Herstellung der Ziegel ausreichend Rohstoffe wie Lehm, Sand, Wasser und Brennmaterial erforderlich. Darüber hinaus mussten genügend Abnehmer im näheren Umkreis existieren, im Idealfall eine Wasserstraße zum Abtransport der Ziegel.[100] Auch die Gerber benötigten Wasser zum Reinigen der Häute. Die Wäscher richteten Waschstuben für ihr Gewerbe ein.[101]

Darüber hinaus trugen Fischfang und Fischzucht einen großen Teil zum Überleben bei: Auch Teiche wurden zum Zweck der Fischhaltung angelegt, wie ohnehin jeder geeignete Wasserlauf zum Fischen verwendet wurde.[102] Die Küstennähe ermöglichte den Bewohnern der Lage Landen gleichfalls das Fangen und Versenden von Seefischen.

Selbstverständlich bedeutete so viel Wasser nicht nur ein komfortables Leben, sondern auch Gefahren.[103] Auch wenn die Sturmfluten der Küsten das tiefere Binnenland nicht mehr erreichten, konnten insbesondere langanhaltende, sturzartige Regenfälle schwerwiegende Konsequenzen mit sich bringen: Die Flüsse liefen mit Regenwasser voll und konnten so kein weiteres Grundwasser mehr aufnehmen. Von diesem gab es in den Lage Landen jedoch ohnehin schon mehr als genug. Deichbrüche waren eine regelmäßige Folge. Für die Deiche entlang der Waal sind solche unter anderem für die Jahre 1644, 1651, 1655, 1658, 1663, 1709, 1726, 1728, 1729, 1740, 1741, 1757, 1784, 1795, 1799, 1805, 1809, 1811, 1820, 1827, 1849, 1851 und 1861 schriftlich belegt.[104]

Naturereignisse, wie zum Beispiel das Hochwasser, wurden kulturübergreifend seit frühester Zeit in einen religiösen Zusammenhang gestellt. Schon immer brauchten die Menschen Erklärungen für scheinbar übernatürliche, sich dem menschlichen Einfluss offenkundig entziehende Ereignisse, die noch dazu Leid

99 Schenk, Das Baby in der Wiege (wie Anm. 82), S. 106.

100 Dirk Ehrhardt, Rinnen, Kremper, Biberschwänze. Zur Geschichte der Dachziegel und Ziegelherstellung in Franken, Bad Windsheim 2016, S. 28.

101 Schenk, Das Baby in der Wiege (wie Anm. 82), S. 108.

102 Ebd., S. 106; vgl. hierzu auch: Schröder, Getreidemühlen (wie Anm. 98).

103 Vgl. hierzu u.a.: Lotte Jensen, 'Disaster upon Disaster Inflicted on the Dutch'. Singing about Disasters in the Netherlands, 1600–1900, in: BMGN — Low Countries Historical Review 134/2, 2019, S. 45–70.

104 Alex A. S. van Heezik, Strijd om de rivieren. 200 jaar rivierenbeleid in Nederland of de opkomst en ondergang van het streven naar de normale rivier, Haarlem/Den Haag 2008, FN 45, S. 302; van Dam, An Amphibious Culture (wie Anm. 8), S. 79; Kees Draaisma, Een nieuwe kijk op Caspars dijk, in: It Beaken 79 1/2, 2017, S. 27–105, hier: S. 28. Für den Hinweis auf letztere Publikation bin ich Gertrud Nöth sehr dankbar.

und Not mit sich brachten. Spätestens seit dem Hochmittelalter hatte jedoch laut Schenk (2020) das religiöse Deutungsmuster solcher Ereignisse im Rahmen der Rezeption antiker, aristotelisch geprägter Naturphilosophie Konkurrenz bekommen – nicht jede Überschwemmung wurde so als Vorzeichen einer Sintflut und damit Bestrafung durch Gott interpretiert.[105] Im Gegenteil, es handelte sich, so auch van Heezik (2008), für die Bewohner der Flusslandschaften um ein vertrautes Phänomen.[106] Entlang der Wasserläufe in bestimmten räumlichen Zonen in Abhängigkeit der Jahreszeiten regelmäßig auftretend und dem jährlichen Niederschlagsmuster folgend, waren sie, ebenfalls in Verbindung mit bestimmten Wetterlagen oder im Zusammenhang mit Eisgang und Schneeschmelze, ein Stück weit in ihrer Erscheinung kalkulierbar.[107] Es war normal, dass die Wiesen entlang der Flüsse im Winter unter Wasser standen.[108]

Die Vorhersehbarkeit bot die Möglichkeit, sich vor dieser Gefahr präventiv, zum Beispiel durch Dammbau, zu schützen. Insbesondere die heutigen Niederlande, das Land der Windmühlen, zeugen von diesbezüglicher technischer Meisterleistung vergangener Epochen. Grachten und Mühlengebäude sind auch deshalb noch erhalten, weil sie für die dort lebende Bevölkerung als ständige Symbole des Überlebenskampfes eine Art identitätsstiftendes Denkmal für den gewonnenen Kampf gegen das Wasser darstellen.[109]

(3) Zur Hochwassergefahr als weiterer, schwerwiegender und das ganze dortige Leben bestimmender Faktor kommt die Tatsache hinzu, dass etwa ein gutes Viertel der Betrachtungsregion unter dem Meeresspiegel liegt,[110] hier wieder

---

105 Schenk, Das Baby in der Wiege (wie Anm. 82), S. 90; van Dam, An Amphibious Culture (wie Anm. 8), S. 81.

106 Van Heezik, Strijd om de rivieren (wie Anm. 104), S. 44.

107 Schenk, Das Baby in der Wiege (wie Anm. 82), S. 97.

108 Van Heezik, Strijd om de rivieren (wie Anm. 104), S. 44.

109 Schenk, Das Baby in der Wiege (wie Anm. 82), S. 98. Zum Thema Symbolik und Identität siehe auch die Hinweise Rehbergs auf die im Rahmen der Theorie und Analyse institutioneller Mechanismen (TAIM) erarbeiteten symbolischen Ordnungen. Es werden hier Leib-Symbole (z.B. Rituale als Körperinszenierung), Zeit-Symbole (z.B. Geschichtskonstruktionen), Text-Symbole (Verfassungen, Heilige Schriften) sowie Raum- und Dingsymbole (Türme, Tore etc.) unterschieden. Karl-Siegbert Rehberg, Macht-Räume als Objektivationen sozialer Beziehungen. Institutionenanalytische Perspektiven, in: Christian Hochmuth / Susanne Rau, Machträume der frühneuzeitlichen Stadt (Konflikte und Kultur. Historische Perspektiven 13), Konstanz 2006, S. 41–58, hier: S. 51.

110 Zum Beispiel wird auch noch heute bei den heutigen beiden Provinzen Nord- und Südholland bei der Angabe der Fläche zwischen einer „Land- und Wasserfläche“

mit Dürer exemplarisch für Zeeland historisch belegt: *„Zeeland ist hübsch und wunderlich zu sehen des Wassers halber, denn das steht höher als das Erdreich."*[111]

Erst nach und nach über die Jahrhunderte hinweg konnten also große Teile der Region durch die Trockenlegung von Sümpfen und Mooren vom Menschen erobert werden. Beispielsweise existiert mit Lelystad, der Hauptstadt der aus historischer Perspektive noch jungen niederländischen Provinz Flevoland,[112] eine Stadt, die erst eine 50-jährige Geschichte aufweist, da es vorher das Stück Land dort noch nicht gab. Solche von Wasser umgebenen und mit Grachten durchzogenen Ortschaften mit Häusern auf Holzpfählen sind keine Seltenheit.

Die Lage unterhalb des Meeresspiegels stellte wegen der nur schwer kontrollier- und beherrschbaren Wassermengen die Menschen vor große Probleme und Herausforderungen. Hochwasser, Sturmfluten und Überschwemmungen, aber auch Versandungen bedrohten so in regelmäßigen Abständen die überlebenswichtigen und den Alltag regulierenden Einrichtungen der Ortschaften, die Menschen und das Vieh. So schreibt auch Zijlmans (2017) für das Leben im Rhein-Maas-Schelde-Delta, dass sich dieses durch ständige Landverluste und das eindringende süße oder bracke Wasser insgesamt als schwierig gestaltete.[113]

Sturmfluten wie beispielsweise die der Jahre 1134, 1163, 1248, 1287, 1288, 1334, 1341, 1374, 1393, 1404, 1421, 1477, 1530, 1532, 1570, 1643, 1651, 1682, 1686, 1717, 1825, 1916, nicht zuletzt 1953 sorgten dafür,[114] dass Einrichtungen

---

unterschieden. Insbesondere in Nordholland kommt dabei mit der Wasserfläche fast noch einmal über die Hälfte der Landfläche hinzu.

111 Zitiert nach: Driessen, Geschichte der Niederlande (wie Anm. 34), S. 13; vgl. auch diverse Zitate in: Unverfehrt, Da sah ich viel köstliche Dinge (wie Anm. 50), S. 123.

112 Flevoland wurde am 1. Januar 1986 aus der Öffentlichen Körperschaft Südliche IJsselmeerpolder (Ost- und Südflevoland) und dem Nordwesten der Provinz Overijssel (Noordoostpolder) neu gebildet.

113 Zijlmans, Troebele betrekkingen (wie Anm. 28), S. 76; Maczkiewitz, Der niederländische Aufstand gegen Spanien (wie Anm. 11), S. 45. Bezüglich der Rettung von Vieh wurde nicht selten auf Kirchen als Unterstellmöglichkeiten zurückgegriffen, wie die Ausführungen van Dams und ein Gemälde der Kirche von Oostzaan von James de Rijk aus dem Jahr 1830 zeigen. Van Dam, An Amphibious Culture (wie Anm. 8), S. 87.

114 Dies ist nur ein kleiner Ausschnitt belegter Sturmfluten, die das Gebiet der Lage Landen betrafen, die zeigen soll, dass es sich auch bei diesen Ereignissen um in dieser Region regelmäßig wiederkehrende Naturgeschehnisse handelt. Siehe dazu auch die folgenden Werke: Elisabeth M. K. Gottschalk, Stormvloeden en rivieroverstromingen in Nederland, 3 Bde., Assen 1971–1977; Jan Buisman, Duizend jaar weer, wind en

zerstört wurden, Menschen und Vieh ertranken, Menschen ihres Besitzes verlustig wurden, ganze Dörfer in den Fluten untergingen. Landflächen versanken und mussten mühevoll zurückgeholt werden.[115]

Solche Ereignisse waren jedoch auch Chancen, da auf diese Weise durch Eindeichung und das Anlegen von Kanälen neue Möglichkeiten eröffnet wurden.[116] Beispielsweise entstand wohl in einem solchen Zusammenhang die Westerschelde, sie bestand laut Zijlmans (2017) im Jahr 843 noch nicht. Stattdessen wird angenommen, dass der Durchbruch des Sandrückens bei Rilland wahrscheinlich im 11. Jahrhundert passierte.[117] Weitere Beispiele stellen das IJsselmeer (vormals Zuiderzee) und der Aufstieg Brügges dar: Durch einen Durchbruch bei der Sturmflut 1134 entstand der Zwin. Mithilfe dieses neu entstandenen Zugangs zum Meer entwickelte sich die Stadt ab dem 14. Jahrhundert zum wichtigsten Handelszentrum in Nordwest-Europa.[118] Die herausragende Position verlor sie jedoch ab dem 15. Jahrhundert aufgrund der Versandung des Zwin wieder. Nun rückte Antwerpen an diese Stelle. Es wurde in der Folge zum Zentrum für Textilien, zum größten Umschlagplatz und zum finanziellen Mittelpunkt der Lage Landen. Ebenso kam Gent hinzu. Es ließ einen Kanal anlegen und erhielt entsprechende Begünstigungen durch Karl V.[119]

Die beschriebenen Vorgänge werden immer wieder auch als Grund dafür angegeben, dass im Gegensatz zu anderen König- oder Kaiserreichen sich in den Lage Landen keine dominante Hauptstadt (wie etwa Paris, London oder Berlin) entwickelte. Stattdessen gab es viele verschiedene, bestimmende Produktions- oder Residenzzentren (zum Beispiel Brüssel, Den Haag), verkehrstechnische (zum Beispiel Venlo in Limburg oder Arnhem in Gelderland) und ökonomische Knotenpunkte.[120] Die Städte wechselten sich in Abhängigkeit der Funktionstüchtigkeit ihrer Einrichtungen in ihrer Führungsrolle ab, wie zum Beispiel die

---

water in de Lage Landen. Onder redactie van A. F. V. van Engelen, 6 Bde., Franeker 1995–2015; van Dam, An Amphibious Culture (wie Anm. 8), S. 80.

115 Zijlmans, Troebele betrekkingen (wie Anm. 28), S. 161, 272.

116 Vgl. hierzu auch: Dieter Schott, Stadt in der Geschichtswissenschaft, in: Harald A. Mieg / Christoph Heyl (Hg.), Stadt. Ein interdisziplinäres Handbuch, Stuttgart 2013, S. 120–147, hier: S. 134 f.

117 Zijlmans, Troebele betrekkingen (wie Anm. 28), S. 103.

118 Ebd., S. 162.

119 Ebd., S. 273 f.

120 Bezüglich Gelderlands siehe: Groenveld, Facetten van der Tachtigjarige Oorlog (wie Anm. 48), S. 52.

gate-way-Städte Brügge, Antwerpen und Amsterdam.[121] Auch wenn Karl V. 1531 seine „Hauptstadt" der Lage Landen von Mechelen nach Brüssel verlegte, erhielt Brüssel aufgrund der durch zahlreiche traditionelle Privilegien vorhandenen Macht der weiteren Städte jedoch niemals ein vergleichbares Gewicht wie zum Beispiel London oder Paris – auch Antwerpen kam trotz seines rasanten Aufstiegs zu keiner Zeit eine große politische Rolle zu.[122]

Der Seezugang dieser großen, aber auch vieler anderer kleinerer Häfen schuf zugleich die Möglichkeit der Partizipation an weiter abgelegenen Märkten – der erste Garant einer gewissen wirtschaftlichen Unabhängigkeit der gesamten Region.[123] Im Falle der Lage Landen war sogar noch weitaus mehr möglich, wie das Zitat Naudés (1896) bezüglich der Niederlande zeigt: „*Eine führende Position hinsichtlich der Verschiffung und des Handels nahmen sicherlich die Niederlande mit dem holländischen Hafen Amsterdam und dem Zeeländischen Umschlagplatz Middelburg ein, in beiden hatte auch die Vereinigte Ostindische Kompanie einen Sitz.*"[124] In der Tat war die wichtigste Route im Delta die Hanseroute von Brügge nordostwärts, zwischen Walcheren, längst Noord-Beveland und Schouwen Richtung Holland und weiter zu den Hansestädten.[125]

Aber auch Textil- und Holzhandel spielten eine große Rolle. Ab dem 16. Jahrhundert entwickelten sich ferner die aus der Türkei importierten Tulpen zu einem bedeutenden Handelsgut, ein Austausch mit den, vor allem von der spanischen Krone „verhassten Muselmannen" ließ sich nicht vermeiden.[126]

121 Blockmans / de Munck / Stabel, Economische dynamiek (wie Anm. 84), S. 50.
122 Van Bruaene / Blondé / Boone, Verstedelijking en stadshistoriografie (wie Anm. 72), S. 17.
123 Hans-Werner Niemann, Europäische Wirtschaftsgeschichte. Vom Mittelalter bis heute, Darmstadt 2009, S. 44.
124 Naudé, Getreidehandelspolitik (wie Anm. 93), S. 423.
125 Zijlmans, Troebele betrekkingen (wie Anm. 28), S. 266.
126 Helmut Hetzel (Benelux-Korrespondent mit Standort Den Haag), Was ist typisch niederländisch? Universität Eindhoven bringt Holland-Bild ins Wanken, veröffentlicht am: 16.9.1998, in: Die WELT, URL: https://www.welt.de/print-welt/article630687/Was-ist-typisch-niederlaendisch.html (letzter zugriff: 30.4.2020).

Abb. 76: *Antwerpener Hafen, gezeichnet von Albrecht Dürer (1471–1528), 1520. Bildnachweis: Wikimedia Commons (gemeinfrei), URL: https://upload.wikimedia.org/wikipedia/commons/7/76/Albrecht_Dürer_-_Antwerp_Harbour_-_WGA07090.jpg.*

Mit Antwerpen hatte sich, wie erwähnt, ein großer Umschlagplatz herauskristallisiert, von dem aus das Geschäft florierte – 75 % des gesamten Handels der Lage Landen wickelte sich über diese Stadt ab. Das spanische Königreich gehörte mit 4,7 Millionen Gulden Warenwert zu den größten Partnern.[127]

Die damaligen Zeitgenossen waren sich dieses Umstands durchaus bewusst, wie das folgende Zitat aus dem Jahr 1517 des Sekretärs von Kardinal Luigi d'Aragona, Antonio de Beatis, belegt: „*Antwerpen ist groß und dicht bevölkert, und angesichts der zahlreichen dort niedergelassenen italienischen Kaufleute steht es nicht hinter Bologna zurück. Schöne Straßen und Plätze, zumeist steinerne Häuser, eine prächtige Kirche mit einem Turm, der, wenn er vollendet ist, nicht hinter Straßburg zurücksteht. Die Stadt wird durchkreuzt von Kanälen, was sehr bequem ist.* […] *Es gibt einen schönen Hafen mit zahllosen Schiffen und einem halbkreisförmigen,*

127 Geoffrey Parker, Der Aufstand der Niederlande. Von der Herrschaft der Spanier zur Gründung der Niederländischen Republik 1549–1609, München 1979, S. 19.

gepflasterten Platz, was das Laden und Löschen der Güter vereinfacht. Das Wasser ist so tief, dass auch große Schiffe anlegen können."[128]

Um ein solches Weideland, wie von Volkmann (1896) oben beschrieben, mit derart tiefen Wassern für große Schiffe und eine derartige Handelsaktivität entstehen zu lassen, betreiben und aufrechterhalten zu können, war es jedoch erforderlich, den organisierenden und partizipierenden Menschen zunächst einmal ein sicheres Leben zu gestatten. Es musste also wasserbautechnisch schon sehr früh Großes geleistet werden. Auch hier ermöglichten der Wunsch nach besiedelbarem Land und die Vorhersehbarkeit unliebsamer Naturereignisse präventive Maßnahmen: Grachten, Staudämme, Pumpmühlen, Schleusen oder Deiche prägten entsprechend seit frühster Siedlungszeit das Landschaftsbild.[129] Auch verschiedenen Ämtern wie zum Beispiel dem des Deichgrafen oder gar Deichgenossenschaften, kam in dieser Region eine noch viel größere Bedeutung zu. Die angemessene Umsetzung und Betreuung solcher hochkomplexen Wassersysteme erforderte bezüglich Ausbau und Unterhalt einen beträchtlichen Arbeitsaufwand, eine Unmenge an Kapital und barg üblicherweise auch eine ganze Menge Konfliktpotential.[130]

2. Die Herrschaftsstrukturen – ein Spiegelbild regionaler Rahmenbedingungen

2.1. Städte als Organisationszentren zur Beherrschung der fluvialen und geologischen Topographie

Herrschaft muss sich im Sinne der langsamen Erzeugung eines örtlichen Gleichgewichts stets ein Stück weit den topographischen Bedingungen anpassen. Im Fall der Lage Landen wurden zahlreiche Akteure benötigt, die für das Anlegen etwaiger Schutzeinrichtungen zur Beherrschung der fluvialen Topographie ihre Arbeitskraft zur Verfügung stellten. Hier trifft in vollem Maße zu, was Adam Smith – dabei stets die christlichen Werte vor Augen habend – 1776 festhielt, dass nämlich der Bau und Erhalt derartiger Einrichtungen von keiner einzelnen oder kleinen Anzahl von Personen erwartet werden kann.[131] Die Anlagen mussten gebaut, bedient, überwacht, regelmäßig gewartet und ausgebessert werden. Die diesbezüglich arbeitenden Menschen mussten geschützt und für diese

128 Zitiert nach: Unverfehrt, Da sah ich viel köstliche Dinge (wie Anm. 50), S. 31.
129 Maczkiewitz, Der niederländische Aufstand gegen Spanien (wie Anm. 11), S. 46.
130 Schenk, Das Baby in der Wiege (wie Anm. 82), S. 104.
131 Adam Smith, An Inquiry into the Nature and Causes of the Wealth of Nations, Bd. 2, London 1776, S. 329 f.

Arten von Arbeiten von anderen Tätigkeiten freigestellt, also entsprechend auf anderem Wege versorgt sein, wie die Überlieferung des Deichbaus in Harlingen (Friesland) unter der Leitung des damaligen (1576) Statthalters Caspar de Robles, ein Portugiese in spanischen Diensten, zeigt.[132]

So ließ beispielsweise auf dessen Anweisungen Drost van Espelbach für die Verbesserung des Deichwerkes in Harlingen im September 1572 mehr als 100 Mann aus den Dörfern der ehemaligen, nahegelegenen Gemeinde Barradeel kommen.[133] Zwei Monate später mussten etwa 80 Dorfbewohner das Eis in der am dortigen Deich gelegenen Blockhausgracht brechen[134] – die Anzahl der zur Hilfe gerufenen Arbeiter gibt einen kleinen Einblick bezüglich der Schwere der zu verrichtenden Arbeit. Die Problematik einer angemessenen Versorgung der Arbeiter wird gleichfalls an diesem Exempel deutlich: Im Juni 1574 kam es im Harlinger Deichbauarbeitslager zum Aufruhr, da nicht genügend Essen für die große „Deicharmee" und die Bürger Harlingens vorhanden war.[135]

Es verwundert also nicht, dass sich die Stadt für die langfristige Organisation der wasserbaulichen Anlagen geradezu anbot: „Alle Stadtmauern, hier überhaupt alle massiven Bauwerke zeugen davon, in welchem Maße der Kulturmensch von der Vorsorge für die Zukunft und der Angst vor Risiken bestimmt ist."[136] Die Stadt bringt mit ihren Möglichkeiten der Versorgung, Arbeitsteilung und Spezialisierung einerseits den Menschen Schutz, andererseits ermöglicht sie über ihre Finanzkraft, Vernetzungsaktivitäten und das Arbeitspotential, regionale Bedingungen beherrschbar zu machen. Stadtentwicklung ist laut Braudel (1990) entsprechend kein endogenes, isoliertes Phänomen, sondern stets „Ausdruck der Gesellschaft, die der Stadt von innen, aber auch von außen Zwänge auferlegt"[137] – der geographische Raum gerät zu einem „Element menschlicher Planung."[138]

---

132 Maczkiewitz, Der niederländische Aufstand gegen Spanien (wie Anm. 11), S. 47 f.; Draaisma, Een nieuwe kijk op Caspars dijk (wie Anm. 104), S. 28.

133 Draaisma, Een nieuwe kijk op Caspars dijk (wie Anm. 104), S. 58.

134 Ebd.

135 Ebd., S. 61.

136 Joachim Radkau, Der Wald als Lebenswelt und Konfliktfeld der alten Zeit. Szenen aus hessischen Archivalien in mikro- und makroskopischer Sicht, in: Andreas Hedwig (Hg.), „Weil das Holz eine köstliche Ware ...". Wald und Forst zwischen Mittelalter und Moderne, Marburg 2006, S. 75–103, hier: S. 95.

137 Fernand Braudel, Sozialgeschichte des 15.–18. Jahrhunderts, Bd. 1: Der Alltag, München 1990, S. 570.

138 Helmut Flachenecker, Zwei verschiedene Welten in Franken: Nürnberg und Würzburg zwischen Reichs- und Regionalbildung, in: Michael Gehler (Hg.), Die Macht

Auf über hundertfünfzig Jahre der städtebaulichen Realität des Experiments und der Vision zurückblickend, stellen Noack und Oevermann (2010) fest, dass sich die Diskussionen um den urbanen Raum stets an zwei Hauptlinien entlang entwickelten: „Eine Linie versuchte, anhand des physischen Raums der kohärenten ‚europäischen Stadt' die Kontinuität der Geschichte und die traditionelle Form des urbanen Raums zu stärken. ‚Stadt' wurde hier als Raum für Polyvalenz und funktionale sowie soziale Mannigfaltigkeit verstanden. Die zweite machte sich zum Ziel, ausgehend von einer gesellschaftlichen Idee, die differenzierten und dynamischen sozialen Beziehungen stadträumlich neu zu definieren."[139] Mit Fragen nach der Praxis gesellschaftlicher Konstitution von Raum ist nun in jüngerer Zeit die Untersuchungsperspektive auf die Bewertung und Aneignung von Raum durch die Stadtgesellschaft verlagert worden.

Darunter fällt unter anderem die Erforschung des Zusammenhangs von sozial konstruiertem und materiellem Raum:[140] „De muren die de stedelijke samenleving omsloten, haar poorten die een duidelijke grens trokken tussen de poorters en buitenstaanders, de torens die haar zelfbewustzijn op grote afstand zichtbaar maakten […], al deze bouwwerken materialiseerden de reële verschillen in de leefwijzen tussen burgers en plattelandsbewoners." – „Die Mauern, die das städtische Zusammenleben umschlossen, ihre Tore, die eine deutliche Grenze zwischen städtischen Händlern sowie Grundbesitzern und Außenstehenden zogen, die auf große Distanz ihr Selbstbewusstsein ausstrahlenden Türme […], all diese Bauwerke materialisierten die wirklichen Unterschiede zwischen Bürgern und Landbewohnern."[141]

der Städte. Von der Antike bis zur Gegenwart, Hildesheim/Zürich/New York 2011, S. 493–514, hier: S. 493.

139 Konstanze Noack / Heike Oevermann, Urbaner Raum: Platz – Stadt – Agglomeration, in: Stephan Günzel (Hg.), Raum. Ein interdisziplinäres Handbuch, Berlin/Heidelberg 2010, S. 266–279, hier: S. 271.

140 Michel Pauly / Martina Stercken, Stadtentwicklung im vormodernen Europa. Beobachtungen zu Kontinuitäten und Brüchen, Zürich 2019, S. 7.

141 Van Bruaene / Blondé / Boone, Verstedelijking en stadshistoriografie (wie Anm. 72), S. 8. Hier wird das Stadttor insbesondere in Bezug auf seine die Stadt abriegelnde Funktion angesprochen.

Abb. 77: *Belfried (links) und Kirche (rechts) der heute belgischen Stadt Brügge. Bildnachweis: Wikimedia Commons (gemeinfrei), URL: https://commons.wikimedia.org/wiki/File:Br%C3%BCgge_Blick_vom_Belfried_4.jpg.*

Abb. 78: *Das Rathaus von Middelburg (niederländische Provinz Zeeland). Bildnachweis: Wikimedia Commons (gemeinfrei), URL: https://upload.wikimedia.org/wikipedia/commons/a/a6/Sp%C3%A4tgotisches_Rathaus_in_Middelburg.jpg.*

Der besonders für flämische Städte typische und nicht selten höher als die örtliche Kirche reichende Belfried – ein Glockenturm – ist nur ein Zeugnis städtischen Selbstbewusstseins. Gleiches gilt für die zum Teil überaus prunkvoll ausgestatteten Rathäuser, welche bereits von den Zeitgenossen früherer Epochen hervorgehoben wurden, so zum Beispiel von Dürer: „*In Brüssel ist ein gar köstliches Rathaus, groß mit schön gehauenem Maßwerk und mit einem herrlichen, durchsichtigen Turme.* [...] *Middelburg aber ist eine ansehnliche Stadt, hat ein überaus schönes Rathaus mit einem prächtigen Turm.*“[142]

Insgesamt fungierten die Städte also als physische Orte, in denen das moderne Zusammenleben Gestalt annahm: „Voor velen was en is de band met de eigen stad en haar vele lieux de mémoire concreter dan die met een vortendom, provincie, land of staat.“ – „Für viele war und ist das Band mit der eigenen Stadt und den vielen Erinnerungsorten viel konkreter als ein Fürstentum, eine Provinz, ein Land oder ein Staat.“[143] Bereits zeitgenössische Aufzeichnungen stellen so das Desinteresse der einzelnen Provinzen der Lage Landen an einem Zusammenschluss, die zahlreichen Städte, das Kanalsystem und die Unwegsamkeit des Bodens in einen direkten Zusammenhang.[144] Es scheint so, dass gerade der Föderalismus, häufig in diesem Zusammenhang irrtümlicherweise auch mit dem Poldermodell gleichgesetzt, hier die passende Antwort der Menschen auf die regionale Topographie darstellte – ein landschaftsgestaltender Umstand.

So wiesen die Lage Landen bereits im Spätmittelalter eine starke Urbanisierung auf,[145] wenngleich die in der älteren Forschung kursierende Behauptung, dass in den Lage Landen der Adel kaum vorkam, laut Groenveld (2018) nicht haltbar ist.[146] Die Anfänge dieser Entwicklung gehen auf die erste große Urbanisierungswelle im 11. Jahrhundert zurück.[147] Die städtische Entwicklung erfolgte in einer chronologischen Linie: Zuerst wurden das Maasland, die Artesië und

142 Zitiert nach: Unverfehrt, Da sah ich viel köstliche Dinge (wie Anm. 50), S. 67 bzw. S. 125.

143 Van Bruaene / Blondé / Boone, Verstedelijking en stadshistoriografie (wie Anm. 72), S. 8; Niklas Nitzschke, Baukultur in kleinen Städten und großen Dörfern, in: Doris Schmied (Hg.), Große Dörfer – Kleine Städte, Göttingen 2018, S. 91–103, hier: S. 94.

144 Urte Weeber, Republiken als Blaupause. Venedig, die Niederlande und die Eidgenossenschaft im Reformdiskurs der Frühaufklärung (Ancien Régime, Aufklärung und Revolution 42), Berlin/Boston 2016, S. 117 f.

145 Van Bruaene / Blondé / Boone, Verstedelijking en stadshistoriografie (wie Anm. 72), S. 7; Knoll, Geschichtlicher Überblick (wie Anm. 15), S. 6.

146 Groenveld, Facetten van der Tachtigjarige Oorlog (wie Anm. 48), S. 81.

147 Van Bruaene / Blondé / Boone, Verstedelijking en stadshistoriografie (wie Anm. 72), S. 17.

Flandern besiedelt, dann Brabant, Nordholland, Zeeland und das IJsselgebiet.[148] Mit der Zeit entwickelten sich so beispielsweise im Herzogtum Brabant vier Hauptstädte (Leuven, Brüssel, Antwerpen und 's-Hertogenbosch), in der Grafschaft Flandern, dem politisch-herrschaftlichen Zentrum im 14. Jahrhundert, dominierten sieben große Städte (Atrecht, Sint-Omaars, Dowaai, Rijsel, Gent, Brügge und Ieper) und in der Grafschaft Holland sechs (Dordrecht, Leiden, Haarlem, Amsterdam, Delft, Gouda).[149] Bereits unter den Burgundern (bis 1477) existierten in der Region um die 208 Städte. Ende des 15. Jahrhunderts lebte bereits von etwa 2,6 Millionen Menschen der Lage Landen jeder dritte in einer Stadt. Dies traf insbesondere auf die Grafschaften Flandern und Hennegau, das Herzogtum Brabant und das Fürstbistum Lüttich zu, für die Grafschaft Holland wird diese Zahl sogar auf 44 % geschätzt.[150] Mit Blick auf diesen hohen Urbanisierungsgrad würde es an dieser Stelle nahe liegen, im Sinne der Irsigler'schen Definition von Stadtlandschaft zu sprechen.

Daneben existierten die durch die Mittelgebirge geprägten Fürstentümer Luxemburg, Namur, das zu Brabant gehörende Limburg sowie das eigentliche Herzogtum Burgund und die Franche-Comté. Hier überwog der großgrundbesitzende Adel.[151] In der Provinz Holland entwickelten sich zuerst das Kloster Egmond und dann der Residenzort Den Haag als zentrale Orte von Bedeutung.[152]

Bereits zu dieser Zeit hatten die Städte eine starke wirtschaftliche und dadurch auch politische Position. Die durch Gewerbe und Handel ansässigen, zum Teil in Zünften oder Gilden gut organisierten Handwerker und Händler waren in ihren Aktivitäten nicht zu unterschätzen, wie beispielsweise diverse Aufstände zum Beispiel in Gent, Brügge oder Ypern durch Weber oder Walker zwischen 1336 und 1438 zeigen. Sie brachten den dortigen Herzog zum Teil in arge Bedrängnis.[153]

Ähnlich, allerdings aus ganz anderen Gründen, verhielt es sich übrigens auch mit den Stadtgründungen in der bereits erwähnten Danziger Bucht. Der Deutsche Orden war im Zuge der Christianisierung darauf angewiesen, christliche Bollwerke und Stützpunkte gegen die „heidnischen" Pruzzen zu errichten. Die Betonung liegt hier allerdings vorrangig auf „christliche Bollwerke". Neben den

148 Blockmans / de Munck / Stabel, Economische dynamiek (wie Anm. 84), S. 38.

149 Van Bruaene / Blondé / Boone, Verstedelijking en stadshistoriografie (wie Anm. 72), S. 14, 17.

150 Blockmans / de Munck / Stabel, Economische dynamiek (wie Anm. 84), S. 33.

151 Von Seggern, Geschichte der burgundischen Niederlande (wie Anm. 29), S. 17.

152 Ebd.

153 Ebd., S. 92.

zahlreichen Burgen erfolgte dies ebenfalls über die Gründung von Städten, für deren Besiedelung allerdings gezielt Leute ins Land geholt werden mussten. Über Siedler aus den Lage Landen ist zu diesem Zeitpunkt jedoch nichts bekannt.[154]

An dieser Stelle dem Leser nicht vorenthalten werden soll die Episode, dass 1297 der Deutschordenslandmeister Meinhard von Querfurt via Urkunde das Stadtrecht nach Kulmer Recht an den heute sogenannten, südöstlich von Elbing/Elbląg gelegenen Ort Pazluk verlieh. In dieser wurde vermerkt, dass die Holländer die Stadtgründer gewesen waren. Die Stadt trug entsprechend den Namen Holland, aus dem sich später der Name Preußisch Holland entwickelte.[155] Über die Gründe dieser einmaligen Stadtgründung ist sich die Forschung nicht einig: De Boer (2001) hat in einem Beitrag überzeugend einen Zusammenhang zu einem 1296 in Holland verübten Mordfall an dem Grafen Floris V. von Holland hergestellt. So soll es sich bei den Stadtgründern um wegen Mordes verurteilte holländische Grafen handeln, die 1296 dem Schicksal in ihrer Heimat entgehend die Flucht ergriffen und auf diese Weise nach Preußen gelangt seien.[156] Eine Theorie, auf welche Weise sie es dann bei Meinhard von Querfurt zu einer Stadtgründung bringen konnten, noch dazu mit importiertem Namen und eigenem Stadtsiegel, welches – für eine Stadt ungewöhnlich – einen Ritter auf seinem Pferd zeigt,[157] liefert de Boer allerdings nicht.

Eine andere, vor allem die Literatur des 19. Jahrhunderts beherrschende Meinung geht davon aus, dass Meinhard von Querfurt im Zuge der Eindeichung des Weichseldeltas gezielt Deichbauexperten aus Holland ins Land gerufen hatte, die sich dann in Pazluk niedergelassen hatten.[158] Denn ähnlich wie im Westen der

154 Dick E. H. de Boer, Hatte Vondel doch recht? Ein Versuch zur Erklärung der historischen Rätsel hinsichtlich der Gründung der Stadt „Holland" in Ostpreußen im Jahre 1297, in: Ders. / Rudolf Holbach / Gudrun Gleba (Hg.), „… in guete freuntlichen nachbarlichen verwantnus und hantierung …": Wanderung von Personen, Verbreitung von Ideen, Austausch von Waren in den niederländischen und deutschen Küstenregionen vom 13.–18. Jahrhundert, Oldenburg 2001, S. 205–239, hier: S. 212.

155 Ebd., S. 205 f.

156 Ebd., S. 219 f.

157 Ebd., S. 216.

158 Vgl. den Beitrag von Karl Parey, Meinhard von Querfurt, der Erbauer der Deiche in den Weichsel-Nogat-Niederungen, in: Neue Preußische Provinzial-Blätter, Vierte Folge, Bd. 7, Königsberg 1870, S. 487–501, sowie die Darstellung von Karl Peter Faber, Einiges zur Geschichte und Beschreibung der Stadt Preußisch-Holland, in: Beiträge zur Kunde Preußens, Bd. 2, Königsberg 1819, S. 194–205, hier: S. 198 f.

Lage Landen lagen im 13. Jahrhundert im Norden der Region ganze Landstriche unter Wasser oder waren von Sümpfen überzogen.[159]

Wie dem auch immer sei, nach der Eroberung Pommerellens umfasste, um zum eigentlichen Thema zurückzukehren, das Deutschordensgebiet ein nahezu einheitliches Territorium mit 94 Städten (71 auf Ordensterritoium, 23 in den bischöflichen Territorien).[160]

Auch die kirchliche Organisation war über die vier Bistümer Kulm, Pomesanien, Ermland und Samland (vom Bistum Ermland abgesehen waren alle dem Deutschen Orden inkorporiert) in der Danziger Bucht schon früh ausdifferenziert.[161] Anders als in den Lage Landen – dort gab es bis zur Kirchenreform für die ganze Region sechs Bistümer, nämlich Terwaan, Doornik, Atrecht, Kamerijk, Lüttich und Utrecht (700).[162] Die ersten fünf erstreckten sich jedoch, von Lüttich abgesehen, auf relativ kleine Territorien und erschlossen den Süden mit dem heutigen Belgien und Luxemburg. Für den gesamten Norden und Nordwesten war mit Utrecht lediglich ein einziges Bistum zuständig. Aufgrund der gegebenen Stadtgeographie und des territorialen Flickenteppichs war eine dem damaligen Zeitgeist entsprechende zentrale, flächendeckende kirchliche Organisation kaum möglich. Marnef und van Bruaene (2016) betonen ferner, dass die dortigen Bischöfe, anders als in weiten Teilen des restlichen Europa,[163] kaum Interesse am Landesausbau hatten.[164]

---

159 Jannis Notebaart, Windmühlen. Der Stand der Forschung über das Vorkommen und den Ursprung, Den Haag/Paris 1972, S. 155.

160 Marian Biskup, Der Deutsche Orden und die Freiheiten der großen Städte in Preußen vom 13. bis zur Mitte des 15. Jahrhunderts, in: Udo Arnold (Hg.), Stadt und Orden. Das Verhältnis des Deutschen Ordens zu den Städten in Livland, Preußen und im Deutschen Reich, Marburg 1993, S. 112–127, hier: S. 113; vgl. ferner: Tomasz Jasiński, Die Rolle des Deutschen Ordens bei der Städtegründung in Preußen im 13. Jahrhundert, in: Arnold, Stadt und Orden (wie oben), S. 94–111.

161 Zenon H. Nowak, Neustadtgründungen des Deutschen Ordens in Preußen. Entstehung, Verhältnisse zu den Altstädten, Ende der Eigenständigkeit, in: Arnold, Stadt und Orden (wie Anm. 160), S. 129–142, hier: S. 131, 139.

162 Guido Marnef / Anne-Laure van Bruaene, Civic religion: gemeenschap, identiteit en religieuze vernieuwing, in: van Bruaene / Blondé / Boone, Gouden Eeuwen (wie Anm. 72), S. 165–206, hier: S. 168.

163 Vgl. hierzu u.a.: Frank G. Hirschmann, Stadtplanung, Bauprojekte und Großbaustellen im 10. und 11. Jahrhundert, Teil 3 (Monographien zur Geschichte des Mittelalters 43), Stuttgart 1998, S. 455–531, hier: S. 497.

164 Marnef / van Bruaene, Civic religion (wie Anm. 162), S. 167.

Somit verkehrten insbesondere bis 1200 die Städte in einem „religiösen Vakuum", was die sogenannte Laienfrömmigkeit beförderte. Ab dem 13. Jahrhundert wurden gezielt Klöster oder verwandte religiöse Gemeinschaften (zum Beispiel Franziskaner, Zisterzienser) in den Städten angesiedelt. Auch hier ging die Bewegung wieder vom Süden aus und erreichte nach und nach auch den Rhein-Maas-Raum.[165] Das Zusammenspiel von Laienfrömmigkeit, Klöstern und religiösen Gemeinschaften trug schließlich zur Entwicklung einer eigenen städtisch, religiösen Kultur in den Lage Landen bei (*Devotio moderna*).[166] Insbesondere im Küstenraum waren so über die Jahre viele kleine Kirchen entstanden: Die Menschen lebten dort vor allem von der Fischerei und der Seefahrt.[167] Diese war gefährlich und forderte nicht selten ihre Opfer. Die Kirchen galten als Zufluchtsorte, wo um Schutz und Beistand gebetet wurde.[168]

Während diese religiöse Kultur also im 10. und frühen 11. Jahrhundert in vielen Städten noch von Abteien, Bischöfen und weltlichen Herren geprägt wurde, kennzeichneten sich bereits die Städte des 11. und 12. Jahrhunderts durch ein eigenes soziales, ökonomisches Profil.[169] Von 1050 bis 1270 lagen Reichtum und Macht vor allem in den Händen der Kaufmannsklasse. Im 13. Jahrhundert begannen sich diese städtischen Eliten allmählich sozial und juristisch von den anderen zu unterscheiden. Politik, Macht und Finanzen wurden monopolisiert.[170] Bereits ab 1250 meldete sich vor allem in den Städten Flanderns und Brabants eine neue städtische Schicht, die *poorters* (Gruppe von Händlern, Unternehmern und Grundbesitzern), zu Wort und forderte mehr soziale und politische Rechte.[171]

---

165 Ebd., S. 171; Scheler, „Die niederen Lande" (wie Anm. 30), S. 17.

166 Marnef / van Bruaene, Civic religion (wie Anm. 162), S. 171; Maczkiewitz, Der niederländische Aufstand gegen Spanien (wie Anm. 11), S. 130 f.

167 Im 13. Jahrhundert sollte hier die Fischereitechnik eine bedeutende Entwicklung erfahren. De Boer, Hatte Vondel doch recht? (wie Anm. 154), S. 215.

168 Ähnliches ist beispielsweise auch in den Küstengegenden Dänemarks zu beobachten. In den dortigen Kirchen wird der Bezug zur Seefahrt zum Teil noch deutlicher, denn selbst in jeder kleinen Kirche hängt noch heute von der Decke ein großes, prachtvolles Segelschiff.

169 Bruno Blondé / Frederik Buylaret / Jan Dumolyn / Jord Hanus / Peter Stabel, Samenleven in de stad: sociale relaties tussen ideaal en realiteit, in: van Bruaene / Blondé / Boone, Gouden Eeuwen (wie Anm. 72), S. 77–120, hier: S. 79.

170 Ebd., S. 83.

171 Ebd., S. 89–91; Harm von Seggern, Warum kam es im Spätmittelalter nicht zur Bildung von Stadt-Staaten in den Niederlanden?, in: Gerhard Fouquet / Gabriel Zeilinger (Hg.), Die Urbanisierung Europas von der Antike bis in die Moderne (Kieler

Im 15. Jahrhundert fand erneut eine soziale Bewegung statt. Der Landadel entdeckte die Städte für sich. Ferner bildeten sich, befördert durch die Pestwellen, die Führungsämter besetzende Familienclans einer Mittelschicht heraus. Auch hier ging die Bewegung erneut vom Süden aus.[172] Erst im Jahr 1559 ließ Philipp II. in den Lage Landen eine Diözesanreform durchführen: In den einzelnen Provinzen wurden zahlreiche neue Diözesen eröffnet, welche die katholische Kirche in den Lage Landen stärken sollten.[173] Zu diesem Zeitpunkt hatte jedoch die Reformation bereits Einzug in die Region erhalten. Im Vergleich zum restlichen Europa verfügten die Lage Landen mit ihren Städten über ein dichtes Netzwerk von Druckereien und Buchverkäufern,[174] was die Ausbreitung des protestantischen und humanistischen Gedankenguts beförderte.

Hatten die Städte in den Lage Landen schon sehr früh eine starke Position, wurde diese aber nicht nur durch die fehlende Hand der Kirche, sondern auch durch die ständige Abwesenheit des Königs gestärkt: Nicht zu Unrecht wird in diesem Zusammenhang beispielsweise in der allgemeinen Forschung immer wieder zwischen „königsnahen und -fernen Regionen" unterschieden.[175] Die

---

Werkstücke. Reihe E: Beiträge zur Sozial- und Wirtschaftsgeschichte 7), Frankfurt am Main 2009, S. 207–230, hier: S. 211 f.; vgl. hierzu auch: de Boer, Hatte Vondel doch recht? (wie Anm. 154), FN 12, S. 211.

172 Blondé / Buylaret / Dumolyn / Hanus / Stabel, Samenleven in de stad (wie Anm. 169), S. 91, 95. Welche Rolle im Kulturtransfer dabei Frankreich zukommt, bleibt zu untersuchen. Dass stets der Süden der Lage Landen das Einfallstor bildete, ist auf jeden Fall mit auf die zentrale Rolle des Antwerpener Hafens zurückzuführen.

173 Heinz Schilling, Der Aufstand der Niederlande: Bürgerliche Revolution oder Elitenkonflikt?, in: Hans-Ulrich Wehler (Hg.), 200 Jahre amerikanische Revolution und moderne Revolutionsforschung, Göttingen 1976, S. 177–231, hier: S. 217. Das Bistum Utrecht wurde zum Beispiel zum Erzbistum erhoben, ihm wurden 1559 mit Haarlem, 's-Hertogenbosch, Middelburg, Deventer, Leeuwarden und Groningen gleich mehrere neue Suffraganbistümer unterstellt. Folkert Jan Bakker, De kerk in Drenthe vóór de Reformatie, in: Michiel A. W. Gerding (Hg.), In alle onwetenschap, bijsterije unde wildicheyt. Der Reformatie in Drenthe in de zestiende en zeventiende eeuw, Delft 1998, S. 15–52, hier: S. 18; vgl. hierzu auch: Marnef / van Bruaene, Civic religion (wie Anm. 162), S. 165–206, hier: S. 169.

174 Bert de Munck / Hilde de Ridder-Symoens, Opleiding en kennis: theorie en praktijk in een stedelijke context, in: van Bruaene / Blondé / Boone, Gouden Eeuwen (wie Anm. 72), S. 287–328, hier: S. 302; für die Rhein-Maas-Gegend siehe dazu: Hantsche, Atlas zur Geschichte des Niederrheins (wie Anm. 41), S. 86 f.

175 Vgl. z.B.: Bernd Schneidmüller, Die Welfen, Stuttgart 2000, S. 105–149; Stephan Freund / Christian Warnke (Hg.), Eine königsferne Landschaft? Der Norden des heutigen Sachsen-Anhalt vom 9. bis ins 12. Jahrhundert, Regensburg 2019. Ehm-Schnocks

Abwesenheit des Herrschers, vor allem seit der Regentschaft Philipps II. Mitte des 16. Jahrhunderts, ermöglichte den vor Ort Lebenden ein Stück weit ein unabhängiges Agieren. Ein diesbezügliches Beispiel ist die viel gelobte Toleranz – ein Klischee, dass sich bis heute insbesondere in den Niederlanden gehalten hat.

2.2. Toleranz oder Pragmatismus?

„*Obwohl ich selbst überzeugter Katholik bin, kann ich nicht gutheißen, dass Fürsten über das Gewissen ihrer Untertanen herrschen wollen*",[176] soll Wilhelm von Oranien am 31. Dezember 1564 in einer Ansprache im tagenden Staatsrat in Brüssel der fast schon fanatischen spanischen Proklamation[177] „*un rey, una fe, una ley*" – „*ein Reich, ein Glaube, ein Gesetz*" – entgegengestellt haben. Dieser Satz gilt unter anderem bis heute als Beleg für die frühzeitige Praxis der Toleranz. Schon bald waren die Lage Landen Zufluchtsort für Katholiken, Lutheraner, Täufer, Calvinisten und Spiritualisten aller Art, welche sich vor allen Dingen in den Städten aufhielten.[178]

Bereits zeitgenössische Darstellungen betonen die ausgeübte Freiheit und Toleranz in den einzelnen Provinzen der Lage Landen, beispielsweise auch der anonym publizierte Dialog „*Emanuel-Ernestes*" (1581)[179] oder die aus dem 17. Jahrhundert stammende Schrift „*Les Délices de la Hollande*" von Jean de Parival (1651): Keine Provinz der Welt, so Parival, genieße eine solche Freiheit wie Holland, diese sei noch dazu gepaart mit einer solchen Harmonie, dass keine Bevölkerungsgruppe eine andere dominieren könne.[180] In Holland gäbe es weder Sklaverei noch Leibeigenschaft. Jeder, der einen Fuß auf holländischen Boden setze, sei frei von allen vorherigen Besitzansprüchen seines vorausgegangenen Herrn. Dorfbewohner seien in Holland genauso frei wie Stadtbürger und unterlägen keinerlei lehensrechtlicher Willkür. Ein jeder sei König in seinem

hält dies explizit für diese Region fest: Ehm-Schnocks, Burgund und das Reich (wie Anm. 30), S. 27.

176 Van der Haer, De initiis tumultuum Belgicorum (wie Anm. 81), S. 186 f.

177 Vgl. u.a.: William S. Maltby, El gran Duque de Alba. Un siglo de España y de Europa 1507–1582, Girona 2007, S. 202–204.

178 Vgl. hierzu u.a.: Lademacher, Die Niederlande (wie Anm. 71), S. 22 f.; Parker, Der Aufstand der Niederlande (wie Anm. 127), S. 14–17; North, Geschichte der Niederlande (wie Anm. 71), S. 9–36; Panhuysen / Stipriaan, Ooggetuigen van de Tachtigjarige Oorlog (wie Anm. 62), S. 4 f.

179 Weeber, Republiken als Blaupause (wie Anm. 144), S. 84.

180 Ebd., S. 190.

Hause, genieße Reisefreiheit und das Recht, das Land zu verlassen.[181] Jeder habe Gewissensfreiheit und niemand würde aufgrund seiner Religion verfolgt oder gezwungen, zum reformierten Glauben zu konvertieren. Ein jeder habe, auch in politischen Angelegenheiten, das Recht auf freie Meinungsäußerung und könne nach seiner eigenen Façon leben.[182] Diese Freiheit der Bewohner negiere nicht die Standesunterschiede, Respekt werde aber nur denjenigen gezollt, die die Freiheit der anderen respektierten. Die Freiheit habe einen ständigen Zuzug nach Holland zur Folge, was das Bevölkerungswachstum erkläre, und trage zugleich zum Wohlstand der Provinz bei.[183] Diese spezielle Beschreibung der Provinz Holland gilt allerdings längst nicht für die gesamte Lage Landen. Sie ist lediglich erneut ein Indiz für den angesprochenen, praktizierten Föderalismus.

Was jedoch auf alle Provinzen der Region zutraf, war die Tatsache, dass gerade die gewerblich hoch entwickelten Städte völlig abhängig vom Handel waren – ein Viertel aller in den Lage Landen konsumierten Lebensmittel musste eingeführt werden. Es verwundert also nicht, dass die Region insgesamt stark anfällig für ökonomische Krisen war,[184] wenn mal wieder heftige Fluktuationen, beispielsweise im Kornpreis, das Geschäft durcheinander gebracht hatten. Dieser konnte innerhalb weniger Wochen um das Dreifache steigen, was zu großen Hungersnöten führte. Die einzelnen Provinzen waren demnach zur Vermeidung einer Involvierung in den Handel störender Kriege, die Städte zur Bewahrung des internen Friedens und stetigen Verbesserung ihrer Einrichtungen angehalten.[185]

Alleine schon der ökonomische Austausch sorgte für kulturelle Vielfalt. Seit jeher konnten es sich die Händler nur in seltenen Fällen leisten, sich ihre Geschäftspartner auszusuchen. Große Handelshäfen wie Brügge, Antwerpen und später Amsterdam fungierten entsprechend als Waren- und Kulturumschlagplatz zugleich. Zahlreiche Menschen unterschiedlicher Herkunft und verschiedenen Glaubens liefen via Schiff ein, löschten Ladung, nahmen Neue auf, besuchten vorhandene Kaffeehäuser, empfingen Post, tauschten sich aus und reisten weiter.

An den Börsen wurden Waren gehandelt, wechselten den Besitzer und wurden weiter versandt. Umgekehrt trafen Händler aus den Lage Landen in Häfen und Städten aller Welt mit Menschen unterschiedlicher Kulturen zusammen

181 Ebd.

182 Ebd.

183 Ebd.

184 Lademacher, Geschichte der Niederlande (wie Anm. 71), S. 24 f.; vgl. auch: Parker, Der Aufstand der Niederlande (wie Anm. 127), S. 17–20.

185 Schröder, Schnittstelle Niederrhein (wie Anm. 15), S. 13.

und brachten ihre Erfahrungen in die Heimat mit. Vom Wohlstand angelockt, hielten sich zahlreiche Künstler in den Städten auf oder ließen sich gar nieder. Auch dies wurde durch die damaligen Zeitgenossen wahrgenommen, der Maler und Kunstschriftsteller Karel van Mander hielt so fest (1604): „*Die berühmte und herrliche Stadt Antwerpen, die dem Kaufhandel ihre Blüte verdankt, hat von überallher die bedeutendsten Vertreter unserer Kunst angelockt, die sich in großer Anzahl auch deswegen dorthin begeben haben, weil die Kunst sich gerne in der Nähe des Reichtums aufhält.*“[186]

So bemerkte auch Schenk (2019) im Zusammenhang mit Florenz, welches ähnlich wie die Städte der Lage Landen schon früh gegen Überschwemmungen zu kämpfen hatte: „In diesem Ambiente relativer Sicherheit wuchs die Bereitschaft und für ganz bestimmte, privilegierte Gruppen dank Kapitalakkumulation auch die finanzielle Möglichkeit, etwas aufs Spiel zu setzen, etwas zu riskieren. Dies galt besonders für die Welt der Fernhandelskaufleute, die seit der ‚commercial revolution‘ des 13. Jahrhunderts wegen der fortschreitenden Spezialisierung und Ausdifferenzierung der Gesellschaft zunehmend nicht mehr regelmäßig selbst auf die Reise gingen, sondern von ihren Schreibstuben aus ihre Geschäfte steuern konnten.“[187] Aber nicht nur der Fernhandel dominierte in dieser Region, sondern eben auch der Zwischenhandel. Der Rhein verband die Endpunkte der wichtigsten deutschen Messen des 15. Jahrhunderts, nämlich Frankfurt am Main, Bergen op Zoom und Antwerpen, miteinander. Aufgrund der überaus günstigen Flussverläufe reichte der Handel, wie dargestellt, bis weit in das Hinterland, Deventer und Kampen gerieten so zwischenzeitlich zur Drehscheibe.[188]

Die Ambivalenz der Lage am Fluss war somit den Zeitgenossen weit zurückliegender Epochen grundsätzlich wohl bewusst, wie Schenk (2019) für das 14. Jahrhundert ebenfalls am Beispiel Florenz anhand eines von Robert von Neapel verfassten Trostbriefs zu zeigen vermochte; aus diesem geht die kalkulatorische Kosten-Nutzen-Abwägung, die Forschung spricht in diesem Zusammenhang auch von Risiko-Kulturen, deutlich hervor: „*Warum schwer daran tragen, wenn der Fluss, der seit Anbeginn deiner Stadt zu so vielen Freuden und so*

186 Zitiert nach: Unverfehrt, Da sah ich viel köstliche Dinge (wie Anm. 50), S. 33.

187 Zitiert nach: Gerrit J. Schenk, Die Zukunft zähmen? Die Entstehung eines Risikobegriffs in der Sicherheitskultur spätmittelalterlicher Städte angesichts wiederkehrender „Natur“-Gefahren, in: Benjamin Scheller (Hg.), Kulturen des Risikos im Mittelalter und in der Frühen Neuzeit (Schriften des Historischen Kollegs, Kolloquien 99), Berlin/Boston 2019, S. 195–227, hier: S. 226.

188 Scheler, „Die niederen Lande“ (wie Anm. 30), S. 16.

großer Nützlichkeit beigetragen hat, dir einmal mit ungewöhnlicher Überschwemmung Schaden bringt?"[189]

Dieser ständige Kampf um die Beherrschung des Wassers wirkt laut Jensen (2018) für einen großen Teil der Niederländer noch heute identitätsstiftend, wenngleich dies sicherlich nicht für die Bevölkerung aller Provinzen in gleichem Maße zutrifft. Dennoch genießt er nach wie vor die volle Aufmerksamkeit des Küstenlandes, was sich beispielsweise in zahlreichen populären Publikationen und auch TV-Serien und Filmen widerspiegelt.[190] Er findet auch länderübergreifend Anklang. Der Ausspruch des englischen Journalisten Ben Coates in Bezug auf die heutigen Niederländer verdeutlicht dies zum Beispiel nur zu gut: „The omnipresence of water had […] a profound effect on the Dutch themselves. Earthy and honest, with nothing to hide, the Dutch people I met were as dependable and unexotic as the landscape in which they lived."[191]

Hier tritt also genau das zutage, was Steber (2012) mit Verweis auf Benedict Anderson (1983) in ihrem Beitrag als „imagined communities" bezeichnete: Wie auch Nationen verständigen sich Regionen über „historische Narrative, die Reaktivierung von Traditionsbeständen, Symboliken, Charakterzuschreibungen und Mythen über sich selbst."[192]

Diese kulturelle Konstruktion, wie Jensen (2018) sie bezeichnet, hat selbstverständlich ihren Ursprung in der Historie. Die bereits angesprochene Polder-Metapher lässt sich historisch weit zurückverfolgen: Immer wieder wird in diesem Rahmen die Notwendigkeit, für das Überleben in dieser Polderlandschaft zusammenzuarbeiten, und die Einzigartigkeit des Landes und seiner Bewohner betont.[193]

---

189 Zitiert nach: Schenk, Die Zukunft zähmen? (wie Anm. 187), S. 210; zum Begriff der Risiko-Kulturen siehe den Überblick bei: van Dam, An Amphibious Culture (wie Anm. 8), S. 81.

190 Jensen, Wij tegen het water (wie Anm. 74), S. 7.

191 Zitiert nach: ebd., S. 8.

192 Steber, Region (wie Anm. 24).

193 Jensen, Wij tegen het water (wie Anm. 74), S. 8.

Abb. 79: *„Le Géant de la Côte, kolos van een liggende leeuw" von David Humbert de Superville (1770–1849). Bildnachweis: Universitätsbibliothek Leiden, PK-T-1542, URL: http://hdl.handle.net/1887.1/item:1993281.*

So existieren zahlreiche Schriften und Gemälde, welche über den Kampf gegen das Wasser und die in diesem Zusammenhang erbrachten Heldentaten berichten.[194] Als eine Art Rückbesinnung auf den „*Leo Belgicus*" kann eine nicht fertiggestellte Skizze des niederländischen Künstlers David Humbert de Superville (1770–1849) angesehen werden. Sie wird heute als „Le Géant de la Côte, kolos van een liggende leeuw" – „Der Riese der Küste, Koloss eines liegenden Löwen" – bezeichnet.

194 Vgl. den Aufsatz: Jensen, Wij tegen het water (wie Anm. 74).

Abb. 80: *Ludwig Bonaparte auf dem Deich von Dalem, Reinier Vinkeles (1741–1816), 1809. Bildnachweis: Amsterdamer Rijksmuseum, RP-P-1944–1858, URL: https://www.rijksmuseum.nl/en/collection/RP-P-1944-1858.*

Auch die Obrigkeiten konnten von den nicht selten für die Bevölkerung schlimm ausgehenden Naturereignissen positiven Gebrauch machen, indem sie sich als fürsorgliche und in der Not ihren Untertanen zur Seite stehende Helfer präsentierten. Ein diesbezüglich bekanntes Bild des niederländischen Zeichners und Kupferstechers Reinier Vinkeles stellt so Ludwig Bonaparte auf dem Deich des an der Waal liegenden Dalem (Zuid-Holland) dar. Dort war in der Nacht vom 14. zum 15. Januar 1809 ein Deich gebrochen. Auch der bereits erwähnte Caspar de Robles, dem sogar ein Denkmal gesetzt worden ist, fällt unter diese Rubrik.[195]

Gleichfalls ist die Bedeutung des Wassers für den Handel bis heute fest im Denken der dort lebenden Menschen verankert. Ein diesbezügliches Beispiel ist der

195 Maczkiewitz, Der niederländische Aufstand gegen Spanien (wie Anm. 11), S. 48; siehe auch: Draaisma, Een nieuwe kijk op Caspars dijk (wie Anm. 104).

Kampf zwischen Antwerpen und Rotterdam um den Rhein-Schelde-Kanal (1975 eröffnet). Dieser und der wettlaufartige, fortwährende Ausbau respektive die Vertiefung der Flüsse und Kanäle zur besseren Vernetzung der Häfen mit dem Hinterland beschäftigte über Jahrzehnte nicht nur die beiden genannten Hafenstädte. Gleichfalls involviert waren die belgische Provinz Lüttich, das niederländische *Gewest* Limburg, die deutschen Städte des Rhein-Maas-Raums Aachen, Mönchengladbach und Krefeld und sogar die Europäische Verkehrsministerkonferenz (CEMT). Gerade diese jüngsten Diskussionen haben dabei auch gezeigt, dass sich die einzelnen belgischen und niederländischen Provinzen nach wie vor als Kämpfer in erster Linie für ihre eigenen Belange sehen.[196]

An dieser Stelle soll daher die Arbeitshypothese in den Raum gestellt werden, dass die Topographie, sei es die Bodenbeschaffenheit oder der Kampf um die Beherrschung des Wassers, den Bewohnern der gesamten Lage Landen eine gemeinsame Identität beschert hat, zumindest bis zur offiziellen Gründung Belgiens im Jahr 1830.[197] Ob und inwieweit dann möglicherweise eine bewusste Abgrenzung der Belgier und später ebenso Luxemburger auch und gerade *in puncto* Identität von den Niederlanden stattfand oder ob sich Verschiebungen im Kontext der sogenannten *Shifting-Baselines* ergeben haben, bleibt zu untersuchen.[198]

196 Siehe die Studie der Verfasserin zum Rhein-Maas-Schelde-Kanal: Lina Schröder, Der Rhein-(Maas-)Schelde-Kanal als geplante Infrastrukturzelle (wie Anm. 78).

197 Schon fast in Vergessenheit ist die erste Gründung der „Republik der Vereinigten Belgischen Staaten" 1789/90 geraten, die jedoch nur elf Monate Bestand hatte. Knoll, Geschichtlicher Überblick (wie Anm. 15), S. 7.

198 Lotte Jensen hält beispielsweise Liedgut und Gedichte, welche über die verschiedenen Überflutungen berichten, für besonders wichtige und aussagekräftige Quellen. Möglicherweise lässt sich so über die Entstehung sowie Verbreitung der Lieder und Gedichte auch für den heute belgischen Raum diesbezüglich etwas herausfinden, zumal ja die Dichte der Druckereien und Büchervertriebe in den gesamten Lage Landen, also auch im heutigen Raum Belgiens, sehr hoch war. Jensen, 'Disaster upon Disaster Inflicted on the Dutch' (wie Anm. 103), S. 47. Auch eine Untersuchung bezüglich der Verbreitung der auf die dortigen Wasserverhältnisse zurückgehenden Tradition des Schlittschuhlaufens könnte hier vielleicht weiterhelfen.
Das alles muss jedoch auch ein Stück weit vor dem Hintergrund der von der Umweltforschung angestellten Überlegungen zu den sogenannten *Shifting-Baselines* gedacht werden. Diese bezeichnen kollektiv geprägte und erlernte Referenzpunkte, anhand derer eine Gesellschaft definiert, was von ihr als selbstverständlich angenommen, also auch nicht hinterfragt wird. Eine Annahme in der Umweltforschung geht nun davon aus, dass sich diese Referenzpunkte schleichend verschieben, sodass es erst mehrerer Generationen bedarf, bis die Unterschiede auffallen. Es ist also möglich, dass sich

War es also tatsächlich Toleranz oder reiner Pragmatismus, der hier von allen Seiten so hoch gelobt wurde? Eine Nicht-Duldung oder Zwangsbekehrung wertvolle Arbeit verrichtender, über Kompetenzen verfügender und als Finanziers auftretender Menschen hätte zu deren Abwanderung oder zu Protesten und Revolten geführt, was sich zweifellos auf den zuverlässigen Betrieb der wasserbaulichen Einrichtungen und den Handel auswirken musste. Denn gerade erstere wurden zu großen Teilen von den reichen Kaufleuten finanziert, die ihr Geld gewinnbringend anlegen wollten.[199] Parival (1651) verwies ja explizit auf den durch Zuzug erreichten Wohlstand der Provinz Holland. Zugleich bezeichnet er jedoch die beschriebene Freiheit zutreffend auch als negative Freiheit der Einwohner, als Unabhängigkeit von Zwängen, aber nicht als positive Independenz etwa im Sinne einer politischen Partizipation, zumal Macht und Reichtum in den Händen einer kleinen Gruppe verteilt war.[200]

Ein Blick auf die Revolten Mitte/Ende des 16. und zu Beginn des 17. Jahrhunderts (= „Achtzigjähriger Krieg") zeigt genau die möglichen Auswirkungen solcher Eskapaden. Er bietet Anlass dazu, die hochgelobte Toleranz eher als Pragmatismus zu interpretieren. Als Mitte des 16. Jahrhunderts die Proteste gegen die Politik des spanischen Königs ausbrachen, wurden nämlich nicht nur Seezufahrtswege blockiert, ganze Städte dem Erdboden gleich gemacht, sondern unter anderem auch zahlreiche Deiche durchstochen oder Schleusen geöffnet, um die durch Spanien entsendeten Heere aufzuhalten.[201] Die aufgebrochenen Deiche waren jedoch nicht nur für die Gegner ein Problem. Auch die einheimische Bevölkerung litt unter diesen Bedingungen, weil zum Beispiel unter anderem zahlreiche Wassermühlen während des Kriegs zerstört worden waren, die eigentlich das Wasser vom Land hätten abpumpen sollen.[202] Die Grenzen zwischen Wasser und Land waren nicht mehr erkennbar. Selbst derjenige, der noch ein funktionstüchtiges Schiff besaß, konnte dieses in der Folge nicht benutzen. Außerdem waren die Schleusen verlassen und überflutet und blockierten somit die Wasserwege.[203] Als weitere Kriegsfolgen standen die Kornmühlen in Schutt

---

auch aus dieser Perspektive diesbezügliche Zusammenhänge nachvollziehen lassen. Vgl. hierzu z.B.: Harald Welzer, Klimakriege. Wofür im 21. Jahrhundert getötet wird, Frankfurt am Main ²2008, S. 210.

199 Driessen, Geschichte der Niederlande (wie Anm. 34), S. 168.

200 Weeber, Republiken als Blaupause (wie Anm. 144), S. 190.

201 Vgl. diesbezüglich z.B.: Groenveld, Facetten van der Tachtigjarige Oorlog (wie Anm. 48), S. 111; Michalsky, Projektion und Imagination (wie Anm. 29), S. 44.

202 Vgl. diesbezüglich z.B.: ebd., S. 118.

203 Vgl. diesbezüglich z.B.: ebd., S. 122.

und Asche, sodass die Mehlversorgung zusammenbrach. Groenveld (2018) konnte dies exemplarisch anhand der Stadt Leiden aufzeigen.[204]

Hunger und Not waren die Folgen, wie auch Maximilien de Hénin-Liétard, Graf von Bossu, spanischer Statthalter von Holland (1567–73) und Kommandant der königlichen Flotte, beschrieb: „*Obgleich das Volk hier seinen unmittelbaren Ruin durch den Zehnten Pfennig als alleinigen Grund für den Aufstand angibt,* […] *ist die wirkliche Ursache darin zu suchen, daß in dieser Provinz* [Holland] *zwanzig- oder dreißigtausend Menschen leben, deren einziger Unterhalt in Fischerei und in Seefahrt besteht, und da beides inzwischen völlig zum Erliegen gekommen ist und da die Leute seither alles ausgegeben haben, was sie in der Vergangenheit zu sparen vermochten, werden sie von der Armut schließlich so bedrückt sein, daß es sie nicht länger scheren wird, welchen Preis sie für die Hilfe bezahlen müssen.*“[205] Weiterhin führte er aus: „*Ein allgemeiner Aufstand ist hier zu befürchten, weil die Armen schon sehr an Unterernährung leiden, daß sie nicht mehr ein noch aus wissen, und weil das Getreide aus den Ostseestaaten, das das Hauptnahrungsmittel dieser Provinz ist, nicht mehr bis hierher gelangt.*“[206]

Derartige Konsequenzen bereits früh voraussehend, konnte es also der lokalen, pragmatisch denkenden und vom Handelsgeist getriebenen Elite nicht einfallen,[207] durch unsinnige Eingriffe die gut laufende Wirtschaft und die Stabilisierung der Region im Kampf um die Beherrschung des Wassers zu riskieren. Da halfen selbst Einrichtungen wie der 1430 von Philipp dem Guten gegründete Hoforden zum „Goldenen Vlies“ nichts. Dieser stand auch dem Adel der benachbarten Fürstentümer offen und wurde in der Folge auch als außen- und innenpolitisches Instrument genutzt.[208] Zum Beispiel wurde 1556 Filips van Montmorency, Graf von Hoorn, zusammen mit dem jungen Wilhelm von Oranien in diesen aufgenommen,[209] was beide jedoch nicht daran hindern sollte, sich nachhaltig für den so dringend erforderlichen Föderalismus in ihrer Region gegen den Willen der spanischen Krone zu engagieren. Sie hatten erkannt, dass die dezentralisierte Form des Küsten- und Uferschutzes sowie des so

204 Vgl. diesbezüglich z.B.: ebd., S. 121.

205 Aus dem Spanischen durch die Verfasserin übersetzt und zitiert nach: Santiago Fernández Conti, Los consejos de estado y guerra de la Monarquía hispana en tiempo de Felipe II. 1548–1598, Valladolid 1998, S. 158.

206 Aus dem Spanischen durch die Verfasserin übersetzt und zitiert nach: Fernández Conti, Los consejos de estado y guerra (wie Anm. 205), S. 158.

207 Jensen, Wij tegen het water (wie Anm. 74), S. 43.

208 Von Seggern, Geschichte der burgundischen Niederlande (wie Anm. 29), S. 84.

209 Groenveld, Facetten van der Tachtigjarige Oorlog (wie Anm. 48), S. 49.

überlebenswichtigen Handels eine „Selbstregierung im engen Kreis“[210] erforderlich machte. Denn trotz aller Fortschritte beim Küsten- und Deichschutz blieb das Wasser stets eine der größten Bedrohungen für große Teile der Bevölkerung der Lage Landen.[211]

Über die Regie der Städte *in puncto* Instandhaltung und Ausbau der Wasserbauten stärkten diese damit zugleich ihre politische Position und erschwerten gleichzeitig dem Landesfürsten das Ausbilden lokaler Herrschaftsstrukturen. Die Städte der Lage Landen, möglicherweise einst vorgesehene Bollwerke der lokalen Herrscherelite,[212] verselbstständigten sich in der Folge im Zuge ihrer Konzentration auf eine ökonomische, politische und juristische Organisation. Das Entstehen der Kommunen war also in den Lage Landen insgesamt eine Form „städtischer Emanzipation“ auf Kosten der höheren Obrigkeiten,[213] hervorgerufen durch topographische Rahmenbedingungen.

Der folgende dritte Abschnitt geht abschließend noch einmal explizit auf die eingangs erwähnten Fragen der Tagung ein und stellt zugleich eine Zusammenfassung des Beitrags dar.

3. Der Typus Stadt, Städtelandschaften und die „ewige Konstante Mensch“

3.1. Stadtlandschaft oder Städtelandschaft? Die vier eingangs skizzierten Aspekte (a) bis (d)

In der Einleitung wurde explizit darauf verwiesen, dass eine Stadt- respektive Städtelandschaft stets vor dem Hintergrund einer Bezugsregion analysiert werden muss. In diesem Zusammenhang wurde aufgezeigt, wie wichtig der Blick aus der Vogelperspektive sein kann. Im Falle der Lage Landen wird letztere dringend benötigt, um die nationalen Binnengrenzen innerhalb der Region zu

---

210 Maczkiewitz, Der niederländische Aufstand gegen Spanien (wie Anm. 11), S. 46.

211 Ebd., S. 47.

212 Hochadel: u.a. die Häuser Egmont, Hoorn, Montigny, Hoogstraten (aus dem Westen), Culemborg (aus dem Osten), Bergen (aus dem heutigen belgischen Gebiet) sowie Nassau und Oranien (Königshäuser), Croy, Lalaing, Lannoy und Ligne (aus dem Süden); Niederadel: u.a. die Häuser van Boetzelaar, Brederode (wahrscheinlich aus einer unehelichen Seitenlinie der Grafen von Holland), die Herren von Asperen, die Herren von Bronkenhorst und die Herren Borculo.

213 Marc Boone / Jelle Haemers, Bien commun: bestuur, disciplinering en politieke cultuur, in: van Bruaene / Blondé / Boone, Gouden Eeuwen (wie Anm. 72), S. 121–164, hier: S. 137.

überwinden. Mit der Vogelperspektive war es möglich, mit der Bodenstruktur, dem Flussdelta und der Küstennähe drei zusammenwirkende topographische Aspekte nationenübergreifend mit der dortigen Stadtentwicklung in Verbindung zu bringen. Sie wirkten als eine Art Katalysator, dessen Darstellung Schwerpunkt dieses Beitrags war. So wurde gezeigt, dass sich diese nachhaltig auf die Gestaltung der dortigen Herrschaftsstrukturen, auf die Identität der Bewohner und Kultur auswirkten – in den Lage Landen alle drei vor allem durch Föderalismus gekennzeichnet. Von besonderer Auswirkung zeigten sich in diesem Zusammenhang langfristig unter anderem die Abwesenheit einer zentralen Herrscherfigur (zum Beispiel eines Königs) und die kaum vorhandene Wirkmächtigkeit der Bistümer, die den Akteuren vor Ort einen gewissen Handlungsspielraum boten. Dies fand unter anderem auch in der prunkvollen Gestaltung städtischer Einrichtungen seinen Niederschlag (a).

Ein Streiflicht bezüglich der Entwicklung in den verschiedenen Jahrhunderten stellten die Geschehnisse im Rahmen des „Achtzigjährigen Kriegs“ und die langfristig folgenden nationalen Ausdifferenzierungen dar (b). Auch hier wurde deutlich, dass städtische Entwicklung nur vor dem Hintergrund der jeweiligen Region und dem spezifischen regionalen, aber auch überregionalen Zeitgeschehen verständlich wird.[214]

Gleichfalls kurz angesprochen wurde die besondere Vernetzung der Städte in den Lage Landen über den Handel (c). Dieser verband über das ausgeklügelte System der Wasserwege sämtliche Städte der gesamten Region miteinander. Auch hier sind die Ursprünge mit der divergierenden Bodenbeschaffenheit vor allem in der Topographie zu suchen, die mit den verschiedenen Flüssen jedoch zugleich einen Ausweg zur Verfügung stellte. Wenn nicht schon die topographischen Bedingungen lokale, auf den jeweiligen Ort zugeschnittene Handlungsabläufe benötigen und sinnvoll erscheinen lassen, dann sind föderale und liberale Strukturen spätestens jedoch für einen erfolgreichen Handel erforderlich. Die Organisation im Rahmen der einzelnen *Geweste* band dabei die zu diesem gehörenden Städte über die (regelmäßigen) Ständeversammlungen enger zusammen.

Nur am Rande konnte hier ein Blick auf einzelne Städte und ihre Strukturen geworfen werden (d). Eine Vielzahl, vor allem der küstennahen Städte weisen mit den die Ortschaften untergliedernden Grachten, den prächtigen Rathäusern, der Bauweise der Häuser auf Pfählen, den Deichen oder den zahlreichen Mühlen sicherlich Gemeinsamkeiten auf. Auch die prachtvolle Ausstattung der einzelnen Wohnhäuser und der mit diesen verknüpfte Baustil gehören hierzu und

214 Gleiches gilt übrigens auch für diesen sogenannten „Achtzigjährigen Krieg“ selbst.

verweisen, sofern noch vorhanden, noch heute auf ihre ursprünglichen wohlhabenden Besitzer – die Kaufleute. Für präzisere Aussagen bedürfte es hier jedoch einer separaten Untersuchung.

Ob im vorliegenden Fall der Lage Landen im Sinne Irsiglers nun von einer Städtelandschaft oder doch besser von einer Stadtlandschaft gesprochen werden muss, lässt sich, wie eingangs bereits festgestellt, an dieser Stelle noch nicht abschließend klären.

Irsigler definiert Städtelandschaften als „Raumeinheiten mit einem unterschiedlichen Urbanisierungsgrad; […] definierbar als naturräumliche Großeinheiten oder durch territoriale, kirchliche, sprachlich-dialektale beziehungsweise kulturräumliche Grenzen oder durch gemeinsame Lagemerkmale (Flusslage, küstennahe Lage) oder durch die Verdichtung von Kommunikation. Sie werden entweder durch einen dominanten Stadttyp (zum Beispiel Kathedralstädte, große/kleine Gewerbezentren, kleine Mittelstädte etc.) oder durch eine typische Ausprägung der Städtehierarchie charakterisiert. Der ländliche Raum hat außerhalb des Umlandes der Städte erhebliche Eigenbedeutung in politischer, wirtschaftlicher und kultisch-kultureller Hinsicht."[215] Mit den einzelnen *Gewesten* waren innerhalb der Region territoriale Grenzen gegeben. Die regelmäßigen Ständeversammlungen können so durchaus als Verdichtungsprozess von Kommunikation angesehen werden, zumal festgestellt wurde, dass die Provinzen in einem permanenten Konkurrenzkampf untereinander verwickelt waren. Eine typische Städtehierarchie oder ein dominierender Stadttyp ließ sich jedenfalls auf den ersten Blick nicht ausmachen. Wie das Beispiel Harlingens gezeigt hat, kam aber dem Umland insofern eine tiefere Bedeutung zu, als dass auch die Bewohner der Dörfer für die Instandhaltung der Wasserbauwerke benötigt wurden, ohne die die Städte wiederum nicht hätten agieren können. Als vorläufiges Zwischenergebnis – mehr kann es zum jetzigen Forschungsstand nicht sein – würde also einiges dafür sprechen, die Städte einzelner Provinzen als Städtelandschaft zu begreifen, die mit Blick auf die gesamte Region dann eine große, die Raumeinheit dominierende Stadtlandschaft formen.

3.2. Der Typus Stadt in Abgrenzung von diversen Siedlungsformen am Beispiel der Lage Landen

Schon Braudel (1990) hielt fest, dass in einem gewissen Abstand zu den größeren Zentren zwangsläufig die Kleinstadt auftritt. Dies sei alleine schon mit dem Transportwesen, das den Raum gestaltet und in regelmäßiger Abfolge

215 Irsigler, Städtelandschaften und kleine Städte (wie Anm. 1), S. 32.

„Haltepunkte" entstehen lässt, zu erklären.[216] Siedlungsformen, also insbesondere auch Städte, sind offensichtlich die Antworten der Menschen auf vorgegebene (regionale) Rahmenbedingungen. Unter letzteren werden hier, wie in der Einleitung dargelegt, unter anderem kulturelle, klimatische und topographische Aspekte, natürliche Ressourcen als Basis für Entwicklung und Wirtschaft, die Vernetzungsmöglichkeiten durch Verkehrswege etc. verstanden. Sie führen langfristig zur Ausbildung (regionaler) Herrschafts- beziehungsweise Bistumsstrukturen, steuern die Besiedelungsdichte und lassen entsprechend an die Bedingungen angepasste Siedlungsformen entstehen. Sie formen über die Stadt-Umland-Beziehungen, ihre spezifische Lage, über ihre sozialen, politischen und ökonomischen Strukturen, über die Stadtherrschaft etc. die lokalen Strukturen mit, da die mit ihnen entstandenen Einrichtungen, einmal gebaut, oftmals nicht so ohne weiteres wieder aufgegeben werden können. Der Technikhistoriker bezeichnet dies als Pfadabhängigkeit.[217]

Wenn an dieser Stelle nach einer Unterscheidung der verschiedenen Siedlungsformen gefragt wird, fallen einem sofort die drei für eine Stadt so typischen Aspekte ein: Versorgung, Arbeitsteilung und Spezialisierungsmöglichkeiten. Denn ein reiner Vergleich im Rahmen von Größentypologien als das Unterscheidungsinstrument muss in jedem Fall kritisch hinterfragt werden, da, so auch Schmied und Krämer (2018), unter anderem ein Einwohner mehr oder weniger über die Zugehörigkeit zu den Kategorien Dorf und Stadt entscheiden kann.[218] Wie sieht es in den Lage Landen also mit den scheinbar typischen Aspekten Versorgung, Arbeitsteilung und Spezialisierung aus?

---

216 Braudel, Sozialgeschichte des 15.–18. Jahrhunderts (wie Anm. 137), S. 551.

217 Gerrit J. Schenk, „Knoten" im „Netz"? Überlegungen zur Hafenstadt als „kritischer Infrastruktur", in: Ettel / Hack, Flusstäler, Flussschifffahrt, Flusshäfen (wie Anm. 94), S. 207–236, hier: S. 210; Jens Ivo Engels / Julia Obertreis, Infrastrukturen in der Moderne. Einführung in ein junges Forschungsfeld, in: Rebekka Habermas / Gesine Krüger / Jacob Tanner, Saeculum. Jahrbuch für Universalgeschichte 58/1, 2007, S. 1–12, hier: S. 5.

218 Doris Schmied (Hg.), Große Dörfer – Kleine Städte, Göttingen 2018, S. 2. Auf die verschiedenen und zum Teil wenig gewinnbringenden Definitionsversuche des Stadtbegriffs soll an dieser Stelle erst gar nicht eingegangen werden. Vgl. hierzu: Schenk, „Knoten" im „Netz"? (wie Anm. 217), S. 211, aber auch: Eberhard Isenmann, Die deutsche Stadt im Mittelalter 1150–1550. Stadtgestalt, Recht, Verfassung, Stadtregiment, Kirche, Gesellschaft, Wirtschaft, Köln [2]2012 und Harald A. Mieg / Christoph Heyl (Hg.), Stadt. Ein interdisziplinäres Handbuch, Stuttgart 2013; Franz Irsigler, Städtelandschaften und kleine Städte (wie Anm. 1).

Die Beschreibung der Landschaft hat gezeigt, dass es in den Lage Landen sehr viele kleine Städte gab, die aber keineswegs den Größeren gegenüber zurückstanden, sondern sogar dazu beitrugen, dass sich eine alles überragende, zentrale Hauptstadt gar nicht erst entfalten konnte. Auch unterschiedlich große Dörfer waren in dieser Region seit jeher präsent, denn die topographischen Bedingungen zwangen die Menschen ein Stück weit, in größeren Gemeinschaften zu arbeiten und zu leben. Nur gemeinsam konnten gezielte Maßnahmen gegen das Wasser erfolgreich durchgeführt werden. So hat das Beispiel des Harlinger Deichbaus gezeigt, dass für die durchzuführenden Arbeiten zahlreiche Werktätige aus den umliegenden Dörfern zusammengeholt wurden, deren Versorgung wiederum über Arbeitsteilung sichergestellt werden musste: Wenn die Menschen mit Deichbauen beschäftigt waren, konnten sie nicht zugleich auf dem Feld ihre eigene Ernährung und die ihrer Familie sicherstellen.

Ebenso förderten sich verschiebende Fluss- und Seezugänge und die unterschiedliche Bodenbeschaffenheit im Zusammenhang mit dem Wasserstraßensystem die Existenz von Städten und Dörfern gleichermaßen, denn auch kleinere Siedlungen konnten und mussten im Rahmen ihrer Spezialisierung am Handel und Austausch teilnehmen. Dies zeigt ferner, dass eine einzige Ortschaft in den Lage Landen, sei es nun Dorf oder Stadt, aufgrund der Bodenbeschaffenheit einerseits nicht aus eigener Kraft die Grundversorgung (zum Beispiel mit Getreide) der dort lebenden Bevölkerung immer sicherstellen konnte. Andererseits war es nahezu jeder Siedlung durch das Wasserstraßensystem möglich, eine solche zu organisieren.

Im Falle der Lage Landen lassen sich also Versorgung, Arbeitsteilung und Spezialisierung nicht als eindeutige Unterscheidungskriterien differenter Siedlungsformen heranziehen. Was aber dann macht die Stadt in dieser Region zu etwas Besonderem? Für den untersuchten Zeitraum lässt sich hier eventuell mit der Hypothese arbeiten, dass die Städte durch ihre Finanz- und aktive politische Gestaltungskraft als Motoren regionaler Vernetzung fungierten, wodurch sie sich von anderen Siedlungsformen unterschieden.[219] Die Einschränkung der möglichen Gültigkeit auf einen bestimmten Zeitraum soll zugleich deutlich machen, dass es vielleicht gar nicht möglich oder auch sinnvoll ist, ein immer für alle Epochen gültiges Unterscheidungskriterium zu definieren.

Aber zurück zur Hypothese: An mehreren Beispielen wurde deutlich, dass es vor allem reiche Stadtbürger waren, die das so dringend erforderliche Kapital

---

219 Vgl. hierzu auch: Steber, Region (wie Anm. 24); Schott, Stadt in der Geschichtswissenschaft (wie Anm. 116), S. 140–142.

für die Wasserbauanlagen, auch für Kanäle, zur Verfügung stellten. Van Dam (2016) bringt es auf den Punkt wenn sie schreibt: „Cities, as reserves of labour and capital, could also provide activ aid.“[220] Die Bürger erwarben ihr Kapital wiederum vor allem durch den Im- und Export. Die topographischen Bedingungen ermöglichten dies. Nicht selten fungierten kleinere Ortschaften entsprechend als „verlängerte Werkbank“, wie von Seggern zum Beispiel für viele flämische Städte festhielt.[221]

Auch auf die politische Gestaltungskraft der Städte in ihren unterschiedlichen Provinzen wurde eingegangen. In manchen *Gewesten* waren so zum Teil mehr städtische als geistliche oder adelige Akteure an politischen Entscheidungen beteiligt. Die reichen Bürger stellten durch ihr Geld eine attraktive Gruppe dar, die unbedingt im Kampf um die Beherrschung des Wassers und für den Erhalt des einmal sich entfalteten Wohlstands benötigt wurden. Mit dem Erfolg ihrer Aktivitäten stieg entsprechend auch ihr Selbstbewusstsein, was sich wiederum in den angesprochenen städtischen Bauwerken widerspiegelt. Während des Alten Reichs ist es also insbesondere die soziale Gruppe der Bürger, die eine Stadt von anderen Siedlungsformen unterscheidbar macht. Diese geht in ihrem Wesen über das einer Dorfgemeinschaft hinaus und verbindet die ihr zugehörigen Akteure auf eine ganz eigene Art und Weise miteinander, vielleicht ein wenig vergleichbar mit der sich im 19. Jahrhundert im Zuge der Industrialisierung entwickelnden Arbeiterschaft.

Mit Blick auf die hier zum partiellen Vergleich herangezogene Danziger Bucht fällt außerdem noch ein weiterer Aspekt ins Auge. Wieder ausgehend von den regionalen Rahmenbedingungen und den sich daraus entwickelten Strukturen ist festzustellen, dass sich beide Regionen hinsichtlich der Wassertopographie in gewisser Weise ähneln: In beiden Fällen handelt es sich um Küstenregionen, beide besitzen ein großes Flussdelta, welches Eindeichungen erforderlich machte. In beiden Regionen wurde ein hoher Urbanisierungsgrad erreicht, allerdings auf unterschiedlichem Wege: Während die Lage Landen verschiedene Lehensherren und Bistümer kannte, wurde die Danziger Bucht nur einem einzigen Lehensherrn, nämlich dem Deutschen Orden, übergeben. Noch deutlicher zeigte sich dies anhand der kirchlichen Strukturen, auch hier war der Orden nahezu der alleinige „Herr im Haus“. Diese großzügige Unterstellung des gesamten Gebiets an einen einzigen Befehlshaber vereinfachte die Verwaltung entschieden. Alle

220 Van Dam, An Amphibious Culture (wie Anm. 8), S. 84. Sie führt dies im Folgenden anhand diverser Beispiele für die heutigen niederländischen Städte aus.

221 Von Seggern, Geschichte der burgundischen Niederlande (wie Anm. 29), S. 15.

Städte und Burgen hatten *cum grano salis* den gleichen Stadt- respektive Burgherrn, das Mühlenregal regelte einheitlich für die gesamte Region ihre Handhabung, Betriebsvorschriften und Nutzungsbedingungen.[222]

In beiden Regionen wurden jedoch Städte benötigt, denen zunächst schwerpunktmäßig unterschiedliche Aufgaben zufielen: auf der einen Seite als Finanzierer der Wasserbaumaßnahmen und regionale Versorger, auf der anderen Seite als Kultivierer und Kontrolleure der zunächst heidnischen Bevölkerung. Ein Blick auf die langfristige Entwicklung beider Regionen liefert jedoch am Ende das gleiche Ergebnis, nämlich dominante Städte. In der Danziger Bucht war dies allerdings ein längerfristiger, von dem Gebiet der unteren Weichsel ausgehender Prozess. Sie war eines der wichtigen Tore, über welche „die politischen Muster der städtischen Selbstverwaltung nach Osten vordrangen."[223] Aber auch wenn es von ein paar wenigen Ausnahmen wie zum Beispiel der damaligen Rechtstadt Danzig abgesehen viel länger dauerte als in den Lage Landen, waren die stets stärker werdenden Städte letzten Endes mit für den Niedergang des Deutschen Ordens in dieser Region verantwortlich.[224]

3.3. Die Rolle des Menschen als „ewige Konstante" im Konstrukt der Städtelandschaft

Städte verändern sich und das ist nicht verwunderlich. Wie nun schon mehrfach festgestellt wurde und auch die Überlegungen bezüglich der Toleranz gezeigt haben, sind Städte und die mit ihnen verknüpften lokalen sozialen, kulturellen, politischen und ökonomischen Strukturen die Antwort der Menschen auf vorgegebene (regionale) Rahmenbedingungen. Die diesbezüglich ausgebildeten lokalen Strukturen werden wiederum in Form der städtischen Einrichtungen

222 Vgl. hierzu: Guido Kisch, Das Mühlenrecht im Deutschordensgebiete, in: Zeitschrift für Rechtsgeschichte, Germanische Abteilung 48/1, 1900, S. 176–193; Bernhart Jähnig, Zur Wirtschaftsführung des Deutschen Ordens in Preußen vornehmlich vom 13. bis zum frühen 15. Jahrhundert, in: Udo Arnold, Zur Wirtschaftsentwicklung des Deutschen Ordens im Mittelalter (Quellen und Studien zur Geschichte des Deutschen Ordens 38), Marburg 1989, S. 113–148; Hans Steffen, Das ländliche Mühlenwesen im Deutschordenslande, in: Zeitschrift des Westpreußischen Geschichtsvereins 58, 1918, S. 71–92.

223 Henryk Samsonowicz, Pommerellen im Mittelalter, in: Jörg Hackmann, Pommerellen – Preußen – Pomorze Gdańskie (wie Anm. 18), S. 507–513, hier: S. 509.

224 Vgl. hierzu die bereits zitierten Publikationen: Biskup, Der Deutsche Orden und die Freiheiten der großen Städte (wie Anm. 160); Jasiński, Die Rolle des Deutschen Ordens bei der Städtegründung (wie Anm. 160).

als materielles Resultat intendierter Leitbilder der jeweiligen Stadtgesellschaft augenfällig.

Wenn sich nun die Rahmenbedingungen verändern – und das tun sie mit einigem Abstand regelmäßig –,[225] wirkt sich das wegen der erwähnten Pfadabhängigkeit normalerweise eher langfristig auf die Stadtgestaltung aus. Dabei arbeitet der Mensch als „Transformator", der reagierend sein Leben den neuen Bedingungen anpasst. Bildlich gesprochen, verwandelt er in der Betrachtungsregion zum Beispiel gefährliche, unwägbare, große Wasserströme in handhabbare, gewinnbringende und kontrollierbare Wasserstraßen. Die durch ihn betreuten, weiterentwickelten (städtischen) Einrichtungen haben dabei eine identitätsstiftende Wirkung auf ihre Bewohner. Da die Menschen damals viel unmittelbarer durch das Beisteuern zum Beispiel von Steinen, Wachleistungen oder im Zuge der Ausführung der Reparaturen am Erhalt ihrer Einrichtungen beteiligt waren, gerieten diese ein Stück weit zum Teil ihres Sozialisationsprozesses. So waren auch die meisten Ordnungskonfigurationen, die sich um die Einrichtungen, hier vor allem um die Wasserbauanlagen, bildeten, direkt von der Stadt beschlossen, sodass die Personen hinter den Verordnungen immer persönlich bekannt waren. Die Einrichtungen wurden entsprechend von Generation zu Generation weitergegeben und resultierten dann manchmal sogar zu einem Charakteristikum für eine ganze Region. Im Falle der Lage Landen sind dies zum Beispiel die erwähnten Belfriede, Kanäle oder die zahlreichen Windmühlen. Der Mensch, die „ewige Konstante" im Konstrukt der Stadtlandschaft, sorgte also seit jeher für ihren Erhalt, notfalls Wiederaufbau, aber auch für die Weiterentwicklung und Anpassung der (städtischen) Einrichtungen an die aktuellen Rahmenbedingungen.

Auch in den Lage Landen haben sich Rahmenbedingungen im Verlaufe der Epochen verändert. Geblieben sind nach wie vor der Kampf gegen das Wasser

225 Hierunter fallen unter anderem Katastrophen wie zum Beispiel die Pestwellen, Krisen verursacht durch zum Beispiel klimatische Veränderungen (kleine Eiszeit), Konflikte mit zerstörerischen Ausmaßen wie zum Beispiel Kriege oder Revolutionen, sich verändernde gesellschaftliche Diskurse wie zum Beispiel die Reformation oder auch „technische Revolutionen" wie zum Beispiel die Industrialisierung. Vgl. hierzu mit Perspektive auf die Stadt und ihre Einrichtungen u.a.: Schott, Stadt in der Geschichtswissenschaft (wie Anm. 116), außerdem: Lina Schröder, Herrschaft sichernde Maßnahmen bezüglich Hafen und Siedlung Marktsteft im 18. und 19. Jahrhundert. Eine infrastruktur-historische Betrachtung, in: Biuletyn Polskiej Misji Historycznej = Bulletin der Polnischen Historischen Mission 14, 2019, S. 285–342. Auch in diesem Zusammenhang müssen die von der Umweltforschung angestellten Überlegungen zu den *Shifting-Baselines* berücksichtigt werden. Vgl. hierzu wieder: Welzer, Klimakriege (wie Anm. 198).

einerseits und die ökonomische Bedeutung desselben – insbesondere für den Handel – für die dort lebende Bevölkerung andererseits. Manche der früheren Schifffahrtswege sind inzwischen versandet, andere wurden stattdessen erweitert und ausgebaut. Der Wettlauf um noch größere Schiffe und noch perfektere Abfertigungsmethoden ist nach wie vor im vollen Gange. Für den Antwerpener Umschlagplatz passt so die Beschreibung von 1517 „*Das Wasser ist so tief, dass auch große Schiffe anlegen können*“ noch heute.

Abkürzungen

a.d.	an der
a.M.	am Main
Abb.	Abbildung(en)
Abt./Abtl.	Abteilung
ahd.	althochdeutsch
Anm.	Anmerkung(en)
Art.	Artikel
BayHStA	Bayerisches Hauptstaatsarchiv München
Bd./Bde.	Band/Bände
bearb. v.	bearbeitet von
Bearb.	Bearbeiter/-in
Begr.	Begründer/-in
bes.	besonders
Bl.	Blatt/Blätter
BN	Breves Notitiae
BSB	Bayerische Staatsbibliothek München
bzw.	beziehungsweise
ca.	circa
d.	denarius (Pfennig)
d.Ä.	der Ältere
d.h.	das heißt
Ders./ders.	derselbe
Dies./dies.	dieselbe(n)
Diss.	Dissertation
e.V.	eingetragener Verein
Ebd./ebd.	ebenda
estn.	estnisch
etc.	et cetera
evtl.	eventuell
f.	folgende
fl.	florenus (Gulden)
fol.	Folio (Blatt)
Frhr.	Freiherr
fürstl.	fürstlich
FWFr	Forschungen aus Württembergisch Franken

Gmde.	Gemeinde
Gmkg.	Gemarkung
hg. v.	herausgegeben von
Hg.	Herausgeber/-in
Hl.	Heilige/-r
HONB	Historisches Ortsnamenbuch von Bayern
HRR	Heiliges Römisches Reich (Deutscher Nation)
HStA	Hauptstaatsarchiv
i.d.OPf.	in der Oberpfalz
i.OFr.	in Oberfranken
insbes.	insbesondere
JfL	Jahrbuch für fränkische Landesforschung
Jh.	Jahrhundert
Kap.	Kapitel
königl.	königlich
LA	Landesarchiv
lb h	Pfund Heller
lett.	lettisch
Lit.	Litera
Lkr.	Landkreis
masch.	maschinell
MB	Monumenta Boica
MGH DD	Monumenta Germaniae Historiae Diplomata
MGH SS	Monumenta Germaniae Historiae Scriptores
MGSL	Mitteilungen der Gesellschaft für Salzburger Landeskunde
mhd.	mittelhochdeutsch
MVGN	Mitteilungen des Vereins für Geschichte der Stadt Nürnberg
NA	Notitia Arnonis
ND	Neudruck
NF	Neue Folge
Nr.	Nummer(n)
o.J.	ohne Jahresangabe
o.O.	ohne Ortsangabe
o.S.	ohne Seitenangabe
OÖUB	Oberösterreichisches Urkundenbuch
Opf.	Oberpfalz
OT	Ortsteil
r	recto
Red.	Redakteur/-in

reg.	regierend
Rep.	Repertorium
röm.-dt.	römisch-deutsch(er)
russ.	russisch
S.	Seite(n)
s.	siehe
s.l.	sine loco (ohne Ort)
s.v.	sub voce (unter dem Ausdruck)
Sp.	Spalte(n)
St.	Sankt
StA/StaatsA	Staatsarchiv
StABa	Staatsarchiv Bamberg
StadtA	Stadtarchiv
StAN	Staatsarchiv Nürnberg
StAWü	Staatsarchiv Würzburg
SUB	Salzburger Urkundenbuch
Teilbd.	Teilband
Tom.	Tomus (Band)
u.	und
u.a.	unter anderem/und andere
u.ö.	und öfter
Urk.	Urkunde(n)
URL	Uniform Resource Locator
usw.	und so weiter
v	verso
v. Chr.	vor Christus
v.a.	vor allem
Vgl./vgl.	vergleiche
VSWG	Vierteljahrschrift für Sozial- und Wirtschaftsgeschichte
WFr	Württembergisch Franken
WUB	Wirtembergisches Urkundenbuch
z.B.	zum Beispiel
ZBLG	Zeitschrift für bayerische Landesgeschichte
ZHVS	Zeitschrift des Historischen Vereins für Schwaben

Orts- und Personenregister

B

C

E

H

I

M

O

P

Q

R

S

T

U

X

Y

Z

www.ingramcontent.com/pod-product-compliance
Ingram Content Group UK Ltd.
Pitfield, Milton Keynes, MK11 3LW, UK
UKHW021420091026
15624UKWH00010B/259

9 783631 801178